Moreton Morrell Site

WITHDRAWN

EXPERIENCE + INNOVATION

HELPING TEACHERS
AND STUDENTS
SUCCEED TOGETHER

WILEY

Principles of
ANATOMY &
PHYSIOLOGY

13th Edition

Volume 2—Maintenance and Continuity of the Human Body

International Student Version

Gerard J. Tortora
Bergen Community College

Bryan Derrickson
Valencia Community College

John Wiley & Sons, Inc.

WILEY

Copyright © 2011 John Wiley & Sons (Asia) Pte Ltd

Contributing Subject Matter Expert: Tara D. Foss
Cover image from © Max Delson Martins Santos
Cover Design by Wendy Lai

Founded in 1807, John Wiley & Sons, Inc. has been a valued source of knowledge and understanding for more than 200 years, helping people around the world meet their needs and fulfill their aspirations. Our company is built on a foundation of principles that include responsibility to the communities we serve and where we live and work. In 2008, we launched a Corporate Citizenship Initiative, a global effort to address the environmental, social, economic, and ethical challenges we face in our business. Among the issues we are addressing are carbon impact, paper specifications and procurement, ethical conduct within our business and among our vendors, and community and charitable support. For more information, please visit our website: www.wiley.com/go/citizenship.

ISBN: 978-0-470-92429-7

Printed in Asia

10 9 8 7 6 5 4 3 2 1

HELPING TEACHERS AND STUDENTS SUCCEED TOGETHER

An anatomy and physiology course can be the gateway to a gratifying career in a whole host of health-related professions. It can also be an incredible challenge. Through years of collaboration with students and instructors alike, we have come to intimately understand not only the material but also the evolving dynamics of teaching and learning A&P. So with every new edition, it's our goal to find new ways to help instructors teach more easily and effectively and students to learn in a way that sticks.

We believe we bring together experience and innovation like no one else, offering a unique solution for A&P designed to help instructors and students succeed together. From constantly evolving animations and visualizations to design based on optimal learning to lessons firmly grounded in learning outcomes, everything is designed with the goal of helping instructors like you teach in a way that inspires confidence and resilience in students and better learning outcomes.

The thirteenth edition of **Principles of Anatomy and Physiology,** integrated with **WileyPLUS,** builds students' confidence; it takes the guesswork out of studying by providing students with a clear roadmap (one that tells them what to do, how to do it, and if they did it right). Students will take more initiative, so instructors can have greater impact.

Principles of Anatomy and Physiology 13e continues to offer a balanced presentation of content under the umbrella of our primary and unifying theme of homeostasis, supported by relevant discussions of disruptions to homeostasis. In addition, years of student feedback have convinced us that readers learn anatomy and physiology more readily when they remain mindful of the relationship between structure and function. As a writing team—an anatomist and a physiologist—our very different specializations offer practical advantages in fine-tuning the balance between anatomy and physiology.

On the following pages students will discover the tips and tools needed to make the most of their study time using the integrated text and media. Instructors will gain an overview of the changes to this edition and of the resources available to create dynamic classroom experiences as well as build meaningful assessment opportunities. Both students and instructors will be interested in the outstanding resources available to seamlessly link laboratory activity with lecture presentation and study time.

NOTES TO STUDENTS

The challenges of learning anatomy and physiology can be complex and time-consuming. This textbook and *WileyPLUS for Anatomy and Physiology* have been carefully designed to maximize your study time by simplifying the choices you make in deciding what to study and how to study it, and in assessing your understanding of the content.

Anatomy and Physiology Is a Visual Science

Studying the figures in this book is as important as reading the narrative. The tools described here will help you understand the concepts being presented in any figure and ensure that you get the most out of the visuals.

❶ **LEGEND.** Read this first. It explains what the figure is about.

❷ **KEY CONCEPT STATEMENT.** Indicated by a "key" icon, this reveals a basic idea portrayed in the figure.

❸ **ORIENTATION DIAGRAM.** Added to many figures, this small diagram helps you understand the perspective from which you are viewing a particular piece of anatomical art.

❹ **FIGURE QUESTION.** Found at the bottom of each figure and accompanied by a "question mark" icon, this serves as a self-check to help you understand the material as you go along.

❺ **FUNCTIONS BOX.** Included with selected figures, these provide a brief summary of the functions of the anatomical structure or system depicted.

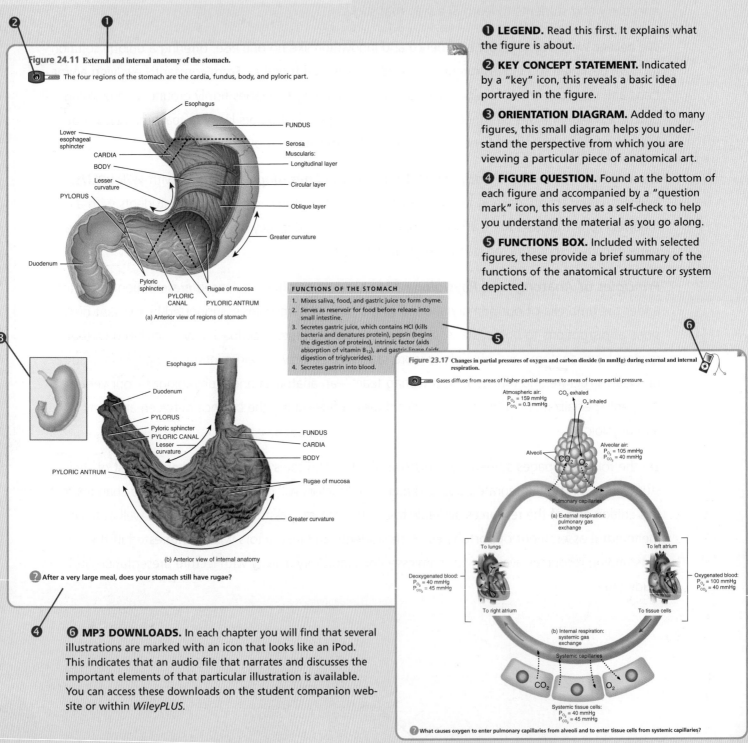

Figure 24.11 External and internal anatomy of the stomach.

The four regions of the stomach are the cardia, fundus, body, and pyloric part.

Esophagus
FUNDUS
Lower esophageal sphincter
Serosa
Muscularis:
CARDIA
Longitudinal layer
BODY
Circular layer
Lesser curvature
Oblique layer
PYLORUS
Greater curvature
Duodenum
Pyloric sphincter
Rugae of mucosa
PYLORIC CANAL
PYLORIC ANTRUM

(a) Anterior view of regions of stomach

FUNCTIONS OF THE STOMACH

1. Mixes saliva, food, and gastric juice to form chyme.
2. Serves as reservoir for food before release into small intestine.
3. Secretes gastric juice, which contains HCl (kills bacteria and denatures protein), pepsin (begins the digestion of proteins), intrinsic factor (aids absorption of vitamin B_{12}), and gastric lipase (aids digestion of triglycerides).
4. Secretes gastrin into blood.

Esophagus
Duodenum
PYLORUS
Pyloric sphincter
PYLORIC CANAL
Lesser curvature
FUNDUS
CARDIA
BODY
PYLORIC ANTRUM
Rugae of mucosa
Greater curvature

(b) Anterior view of internal anatomy

After a very large meal, does your stomach still have rugae?

❻ **MP3 DOWNLOADS.** In each chapter you will find that several illustrations are marked with an icon that looks like an iPod. This indicates that an audio file that narrates and discusses the important elements of that particular illustration is available. You can access these downloads on the student companion website or within *WileyPLUS*.

Figure 23.17 Changes in partial pressures of oxygen and carbon dioxide (in mmHg) during external and internal respiration.

Gases diffuse from areas of higher partial pressure to areas of lower partial pressure.

Atmospheric air:
P_{O_2} = 159 mmHg
P_{CO_2} = 0.3 mmHg
CO_2 exhaled
O_2 inhaled

Alveoli
Alveolar air:
P_{O_2} = 105 mmHg
P_{CO_2} = 40 mmHg

Pulmonary capillaries

(a) External respiration: pulmonary gas exchange

To lungs
To left atrium
Deoxygenated blood:
P_{O_2} = 40 mmHg
P_{CO_2} = 45 mmHg
Oxygenated blood:
P_{O_2} = 100 mmHg
P_{CO_2} = 40 mmHg
To right atrium
To tissue cells

(b) Internal respiration: systemic gas exchange

Systemic capillaries

Systemic tissue cells:
P_{O_2} = 40 mmHg
P_{CO_2} = 45 mmHg

What causes oxygen to enter pulmonary capillaries from alveoli and to enter tissue cells from systemic capillaries?

Studying physiology requires an understanding of the sequence of processes. Correlation of sequential processes in text and art is achieved through the use of special numbered lists in the narrative that correspond to numbered segments in the accompanying figure. This approach is used extensively throughout the book to lend clarity to the flow of complex processes.

Physiology of Hearing

The following events are involved in hearing (Figure 17.22):

1 The auricle directs sound waves into the external auditory canal.

2 When sound waves strike the tympanic membrane, the alternating waves of high and low pressure in the air cause the tympanic membrane to vibrate back and forth. The tympanic membrane vibrates slowly in response to low-frequency (low-pitched) sounds and rapidly in response to high-frequency (high-pitched) sounds.

3 The central area of the tympanic membrane connects to the malleus, which vibrates along with the tympanic membrane. This vibration is transmitted from the malleus to the incus and then to the stapes.

4 As the stapes moves back and forth, its oval-shaped footplate, which is attached via a ligament to the circumference of the oval window, vibrates in the oval window. The vibrations at the oval window are about 20 times more vigorous than the tympanic membrane because the auditory ossicles efficiently transmit small vibrations spread over a large surface area (the tympanic membrane) into larger vibrations at a smaller surface (the oval window).

5 The movement of the stapes at the oval window sets up fluid pressure waves in the perilymph of the cochlea. As the oval window bulges inward, it pushes on the perilymph of the scala vestibuli.

6 Pressure waves are transmitted from the scala vestibuli to the scala tympani and eventually to the round window, causing it to bulge outward into the middle ear. (See **9** in the figure.)

7 The pressure waves travel through the perilymph of the scala vestibuli, then the vestibular membrane, and then move into the endolymph inside the cochlear duct.

8 The pressure waves in the endolymph cause the basilar membrane to vibrate, which moves the hair cells of the spiral organ against the tectorial membrane. This leads to bending of the stereocilia and ultimately to the generation of nerve impulses in first-order neurons in cochlear nerve fibers.

9 Sound waves of various frequencies cause certain regions of the basilar membrane to vibrate more intensely than other regions. Each segment of the basilar membrane is "tuned" for

Figure 17.22 Events in the stimulation of auditory receptors in the right ear. The numbers correspond to the events listed in the text. The cochlea has been uncoiled to more easily visualize the transmission of sound waves and their distortion of the vestibular and basilar membranes of the cochlear duct.

🔑 Hair cells of the spiral organ (organ of Corti) convert a mechanical vibration (stimulus) into an electrical signal (receptor potential).

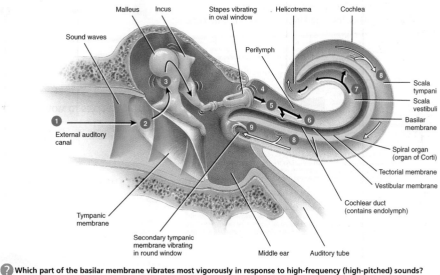

? Which part of the basilar membrane vibrates most vigorously in response to high-frequency (high-pitched) sounds?

WILEY PLUS There are many visual resources within *WileyPLUS*, in addition to the art from your text. These visual aids can help you master the topic you are studying. Examples closely integrated with the reading material include *animations*, *cadaver video clips*, and *Real Anatomy Views*. *Anatomy Drill and Practice* lets you test your knowledge of structures with simple-to-use drag-and-drop labeling exercises or fill-in-the-blank labeling. You can drill and practice on these activities using illustrations from the text, cadaver photographs, histology micrographs, or lab models.

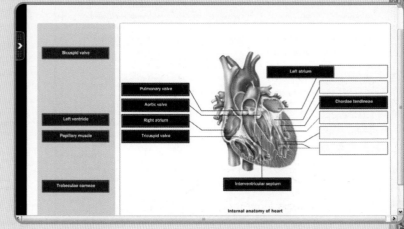

Exhibits Organize Complex Anatomy into Manageable Modules

Many topics in this text have been organized into **Exhibits** that bring together all of the information and elements that you need to learn the complex terminology, anatomy, and the relevance of the anatomy into a simple-to-navigate content module. You will find Exhibits for tissues, bones, joints, skeletal muscles, nerves, and blood vessels. Most exhibits include the following:

❶ Objective to focus your study.

❷ Overview narrative of the structure(s).

❸ Table summarizing key features of the structure(s).

❹ Illustrations and photographs.

❺ Checkpoint Question to assess your understanding.

❻ Clinical Connection to provide relevance for learning the details.

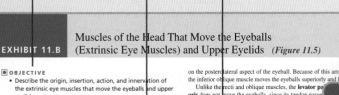

EXHIBIT 11.B — Muscles of the Head That Move the Eyeballs (Extrinsic Eye Muscles) and Upper Eyelids *(Figure 11.5)*

◉ OBJECTIVE

- Describe the origin, insertion, action, and innervation of the extrinsic eye muscles that move the eyeballs and upper eyelids.

Muscles that move the eyeballs are called **extrinsic eye muscles** because they originate outside the eyeballs (in the orbit) and insert on the outer surface of the sclera ("white of the eye") (Figure 11.5). The extrinsic eye muscles are some of the fastest contracting and most precisely controlled skeletal muscles in the body.

Three pairs of extrinsic eye muscles control movements of the eyeballs: (1) superior and inferior recti, (2) lateral and medial recti, and (3) superior and inferior obliques. The four recti muscles (superior, inferior, lateral, and medial) arise from a tendinous ring in the orbit and insert into the sclera of the eye. As their names imply, the **superior** and **inferior recti** move the eyeballs superiorly and inferiorly; the **lateral** and **medial recti** move the eyeballs laterally and medially, respectively.

The actions of the oblique muscles cannot be deduced from their names. The **superior oblique** muscle originates posteriorly near the tendinous ring, then passes anteriorly superior to the medial rectus muscle, and ends in a round tendon. The tendon extends through a pulleylike loop of fibrocartilaginous tissue called the *trochlea* (= pulley) on the anterior and medial part of the roof of the orbit. Finally, the tendon turns and inserts on the posterolateral aspect of the eyeball. Accordingly, the superior oblique muscle moves the eyeballs inferiorly and laterally. The **inferior oblique** muscle originates on the maxilla at the anteromedial aspect of the floor of the orbit. It then passes posteriorly and laterally and inserts

on the posterolateral aspect of the eyeball. Because of this arrangement, the inferior oblique muscle moves the eyeballs superiorly and laterally.

Unlike the recti and oblique muscles, the **levator palpebrae superioris** does not move the eyeballs, since its tendon passes inserts into the upper eyelid. Rather, it raises the upper opens the eyes. It is therefore an antagonist to the which closes the eyes.

⚕ CLINICAL CONNECTION | Strabismus

Strabismus (stra-BIZ-mus; *strabismos* = squint tion in which the two eyeballs are not properly align hereditary or it can be due to birth injuries, poor att muscles, problems with the brain's control center, or le Strabismus can be constant or intermittent. In strab sends an image to a different area of the brain and b usually ignores the messages sent by one of the eyes, becomes weaker; hence "lazy eye," or *amblyopia*, de *strabismus* results when a lesion in the oculomotor (the eyeball to move laterally when at rest, and resul to move the eyeball medially and inferiorly. A lesion (VI) nerve results in *internal strabismus*, a condition i ball moves medially when at rest and cannot move la

Treatment options for strabismus depend on the problem and include surgery, visual therapy (retraining t center), and orthoptics (eye muscle training to straight

MUSCLE	ORIGIN	INSERTION	ACTION	INN
Superior rectus (*rectus* = fascicles parallel to midline)	Common tendinous ring (attached to orbit around optic foramen).	Superior and central part of eyeballs.	Moves eyeballs superiorly (elevation) and medially (adduction), and rotates them medially.	Ocul
Inferior rectus	Same as above.	Inferior and central part of eyeballs.	Moves eyeballs inferiorly (depression) and medially (adduction), and rotates them medially.	Ocul
Lateral rectus	Same as above.	Lateral side of eyeballs.	Moves eyeballs laterally (abduction).	Abdu
Medial rectus	Same as above.	Medial side of eyeballs.	Moves eyeballs medially (adduction).	Ocul
Superior oblique (*oblique* = fascicles diagonal to midline)	Sphenoid bone, superior and medial to common tendinous ring in orbit.	Eyeball between superior and lateral recti. Muscle inserts into superior and lateral surfaces of eyeball via tendon that passes through trochlea.	Moves eyeballs inferiorly (depression) and laterally (abduction), and rotates them medially.	Troc
Inferior oblique	Maxilla in floor of orbit.	Eyeballs between inferior and lateral recti.	Moves eyeballs superiorly (elevation) and laterally (abduction), and rotates them laterally.	Ocul
Levator palpebrae superioris (le-VĀ-tor PAL-pe-brē soo-per'-ē-OR-is; *palpebrae* = eyelids)	Roof of orbit (lesser wing of sphenoid bone).	Skin and tarsal plate of upper eyelids. (opens eyes).	Elevates upper eyelids	Ocul

EXHIBIT 11.B

EXHIBIT 11.B — Muscles of the Head That Move the Eyeballs (Extrinsic Eye Muscles) and Upper Eyelids *(Figure 11.5)* CONTINUED

RELATING MUSCLES TO MOVEMENTS

Arrange the muscles in this exhibit according to their actions on the eyeballs: (1) elevation, (2) depression, (3) abduction, (4) adduction, (5) medial rotation, and (6) lateral rotation. The same muscle may be mentioned more than once.

✔CHECKPOINT

Which muscles that move the eyeballs contract and relax as you look to your left without moving your head?

Figure 11.5 Muscles of the head that move the eyeballs (extrinsic eye muscles) and upper eyelid.

The extrinsic muscles of the eyeball are among the fastest contracting and most precisely controlled skeletal muscles in the body.

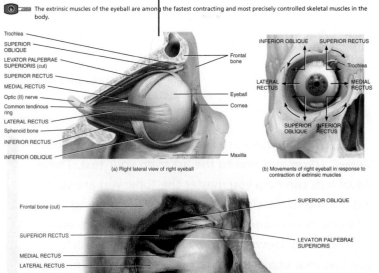

(a) Right lateral view of right eyeball

(b) Movements of right eyeball in response to contraction of extrinsic muscles

(c) Right lateral view of right eyeball

Labels: Trochlea, SUPERIOR OBLIQUE, LEVATOR PALPEBRAE SUPERIORIS (cut), SUPERIOR RECTUS, MEDIAL RECTUS, Optic (II) nerve, Common tendinous ring, LATERAL RECTUS, Sphenoid bone, INFERIOR RECTUS, INFERIOR OBLIQUE, Frontal bone, Eyeball, Cornea, Maxilla, INFERIOR OBLIQUE, SUPERIOR RECTUS, Trochlea, LATERAL RECTUS, MEDIAL RECTUS, SUPERIOR OBLIQUE, INFERIOR RECTUS, Frontal bone (cut), SUPERIOR OBLIQUE, SUPERIOR RECTUS, LEVATOR PALPEBRAE SUPERIORIS, MEDIAL RECTUS, LATERAL RECTUS, INFERIOR RECTUS, INFERIOR OBLIQUE, Zygomatic bone (cut)

❓ How does the inferior oblique muscle move the eyeball superiorly and laterally?

Clinical Discussions Make Your Study Relevant

The relevance of the anatomy and physiology that you are studying is best understood when you make the connection between normal structure and function and what happens when the body doesn't work the way it should. Throughout the chapters of the text you will find **Clinical Connections** that introduce you to interesting clinical perspectives related to the text discussion. In addition, at the end of each body system chapter you will find the **Disorders: Homeostatic Imbalances** section, which includes concise discussions of major diseases and disorders. These provide answers to many of your questions about medical problems. The **Medical Terminology** section that follows includes selected terms dealing with both normal and pathological conditions.

CLINICAL CONNECTION | Arthroplasty

Joints that have been severely damaged by diseases such as arthritis, or by injury, may be replaced surgically with artificial joints in a procedure referred to as **arthroplasty** (AR-thrō-plas'-tē; *arthr-=* joint; *plasty=*plastic repair of). Although most joints in the body can be repaired by arthroplasty, the ones most commonly replaced are the hips, knees, and shoulders. About 400,000 hip replacements and 300,000 knee replacements are performed annually in the United States. During the procedure, the ends of the damaged bones are removed and metal, ceramic, or plastic components are fixed in place. The goals of arthroplasty are to relieve pain and increase range of motion.

Partial hip replacements involve only the femur. **Total hip replacements** involve both the acetabulum and head of the femur (Figures A–C). The damaged portions of the acetabulum and the head of the femur are replaced by prefabricated prostheses (artificial devices). The acetabulum is shaped to accept the new socket, the head of the femur is removed, and the center of the femur is shaped to fit the femoral component. The acetabular component consists of a plas-

tic such as polyethylene, and the femoral component is composed of a metal such as cobalt-chrome, titanium alloys, or stainless steel. These materials are designed to withstand a high degree of stress and to prevent a response by the immune system. Once the appropriate acetabular and femoral components are selected, they are attached to the healthy portion of bone with acrylic cement, which forms an interlocking mechanical bond.

Knee replacements are actually a resurfacing of cartilage and, like hip replacements, may be partial or total. In a **partial knee replacement (PKR)**, also called a **unicompartmental knee replacement**, only one side of the knee joint is replaced. Once the damaged cartilage is removed from the distal end of the femur, the femur is reshaped and a metal femoral component is cemented in place. Then the damaged cartilage from the proximal end of the tibia is removed, along with the meniscus. The tibia is reshaped and fitted with a plastic tibial component that is cemented into place. If the posterior surface of the patella is badly damaged, the patella is replaced with a plastic patellar component.

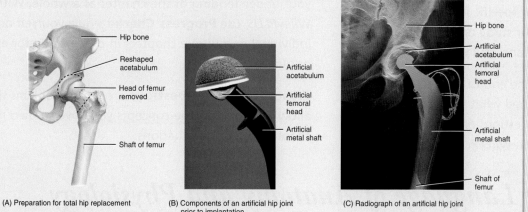

(A) Preparation for total hip replacement

- Hip bone
- Reshaped acetabulum
- Head of femur removed
- Shaft of femur

(B) Components of an artificial hip joint prior to implantation

- Artificial acetabulum
- Artificial femoral head
- Artificial metal shaft

(C) Radiograph of an artificial hip joint

- Hip bone
- Artificial acetabulum
- Artificial femoral head
- Artificial metal shaft
- Shaft of femur

WILEY PLUS *WileyPLUS* offers you opportunities for even further Clinical Connections with animated and interactive case studies that relate specifically to one body system. Look for these under additional chapter resources as an interesting and engaging break from traditional study routines.

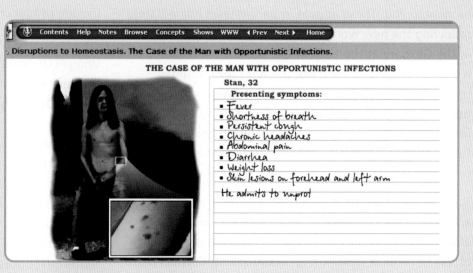

Contents Help Notes Browse Concepts Shows WWW ◀ Prev Next ▶ Home

, Disruptions to Homeostasis. The Case of the Man with Opportunistic Infections.

THE CASE OF THE MAN WITH OPPORTUNISTIC INFECTIONS

Stan, 32

Presenting symptoms:
- Fever
- Shortness of breath
- Persistent cough
- Chronic headaches
- Abdominal pain
- Diarrhea
- Weight loss
- Skin lesions on forehead and left arm

He admits to unprot

Chapter Resources Help You Focus and Review

Your book has a variety of special features that will make your time studying anatomy and physiology a more rewarding experience. These have been developed based on feedback from students—like you—who have used previous editions of the text. Their effectiveness is even further enhanced within *WileyPLUS for Anatomy and Physiology*.

Chapter Introductions set the stage for the content to come. Each chapter starts with a succinct overview of the particular system's role in maintaining homeostasis in your body, followed by an introduction to the chapter content. This opening page concludes with a question that always begins with "Did you ever wonder…?" These questions will capture your interest and encourage you to find the answer in the chapter material to come.

Objectives at the start of each section help you focus on what is important as you read. All of the content within *WileyPLUS* is tagged to these specific learning objectives so that you can organize your study or review what is still not clear in simple, more meaningful ways.

Checkpoint Questions at the end of each section help you assess if you have absorbed what you have read. Take time to review these questions or answer them within the Practice section of each *WileyPLUS* concept module,

where your answers will automatically be graded to let you know where you stand.

Mnemonics are a memory aid that can be particularly helpful when learning specific anatomical features. Mnemonics are included throughout the text—some displayed in figures, tables, or Exhibits, and some included within the text discussion. We encourage you not only to use the mnemonics provided, but also to create your own to help you learn the multitude of terms involved in your study of human anatomy.

Chapter Review and Resource Summary is a helpful table at the end of chapters that offers you a concise summary of the important concepts from the chapter and links each section to the media resources available in *WileyPLUS for Anatomy and Physiology*.

Self-Quiz Questions give you an opportunity to evaluate your understanding of the chapter as a whole. Within *WileyPLUS*, use **Progress Check** to quiz yourself on individual or multiple chapters in preparation for exams or quizzes.

Critical Thinking Questions are word problems that allow you to apply the concepts you have studied in the chapter to specific situations.

Mastering the Language of Anatomy and Physiology

Throughout the text we have included **Pronunciations** and, sometimes, **Word Roots** for many terms that may be new to you. These appear in parentheses immediately following the new words. The pronunciations are repeated in the Glossary at the back of the book. Look at the words carefully and say them out loud several times. Learning to pronounce a new word will help you remember it and make it a useful part of your medical vocabulary. Take a few minutes to read the Pronunciation Key, found at the beginning of the Glossary at the end of this text, so it will be familiar as you encounter new words.

To provide more assistance in learning the language of anatomy, a full **Glossary** of terms with phonetic pronunciations appears at the end of the book. The basic building blocks of medical terminology—**Combining Forms, Word**

Roots, Prefixes, and Suffixes—are listed inside the back cover, accompanied by **Eponyms**, traditional terms that include reference to a person's name, along with the current terminology.

WileyPLUS houses help for you in building your new language skills as well. The **Audio Glossary,** which is always available to you, lets you hear all these new, unfamiliar terms pronounced. Throughout the e-text, these terms can be clicked on and heard pronounced as you read. In addition, you can use the helpful **Mastering Vocabulary** program, which creates electronic flashcards for you of the key terms within each chapter for practice, as well as take a self-quiz specifically on the terms introduced in each chapter.

As active teachers of the course, we recognize both the rewards and challenges in providing a strong foundation for understanding the complexities of the human body. We believe that teaching goes beyond just sharing information. *How* we share information makes all the difference—especially, if as we do, you have an increasingly diverse population of students with varying learning abilities. As we revised this text we focused on those areas that we knew we could enhance to provide greater impact in terms of better learning outcomes. Feedback from many of you, as well as from the students we interact with in our own classrooms, guided us in ensuring that the revisions to the text, along with the powerful new *WileyPLUS for Anatomy and Physiology,* support the needs and challenges you face day to day in your own classrooms.

We focused on several key areas for revision: enhancing the all-important visuals, both drawings and photographs; increasing the use of Exhibits that provide a focused and functional organization of detailed content; adding some new and revising many of the tables to increase their effectiveness; updating and adding clinical material that helps students relate what they are learning to their desired career goals and the world around them; and making narrative changes aimed at increasing student engagement with—and comprehension of—the material.

The Art of Anatomy and Physiology

Illustrations throughout the text have been refined. The color palette for the skulls in Chapter 7, and for the brain and spinal cord throughout the text, has been adjusted for greater impact. Illustrations in each chapter have been revised and updated to provide greater clarity and more saturated colors. Particular emphasis was placed on revised drawings of joints, muscles, and blood vessels.

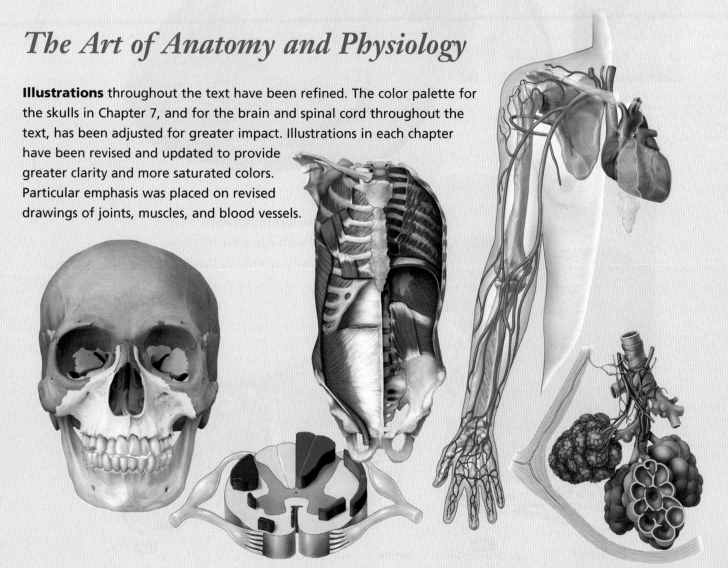

Cadaver photographs are included throughout the text to help students relate the content to real-life images. These are often paired with diagrams to help make the connections. Most of the meticulous dissections and outstanding photography come from Mark Nielsen's lab at the University of Utah.

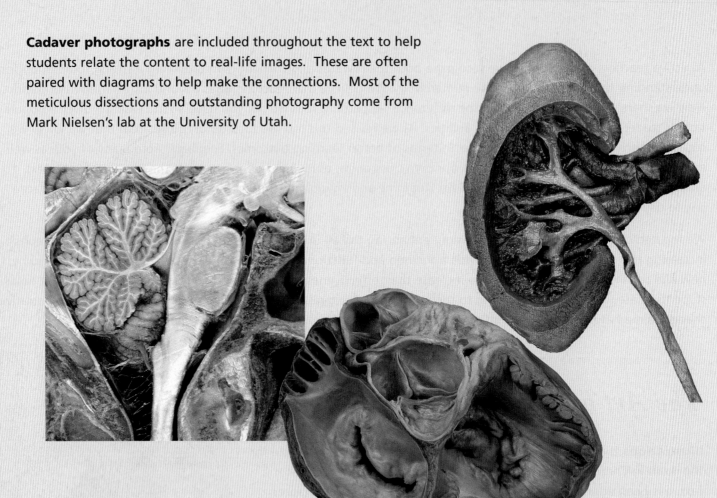

Most tissue **Photomicrographs** have been replaced with exceptionally clear photomicrographs with high-magnification blowups.

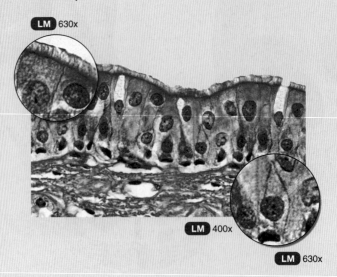

LM 630x

LM 400x

LM 630x

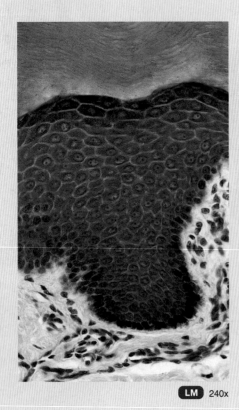

LM 240x

NOTES TO INSTRUCTORS

Exhibits and Tables

The use of the pedagogically designed **Exhibits** has been expanded to include the axial and appendicular skeletons, as well as cranial nerves, providing students with simplified presentations of complex content.

New **Tables,** including Skin Glands, Common Bone Fractures, Summary of the Levels of Organization within a Skeletal Muscle, and Summary of the Respiratory System, have been added, in addition to refinement of many of the existing tables with either new illustrations or rewritten text.

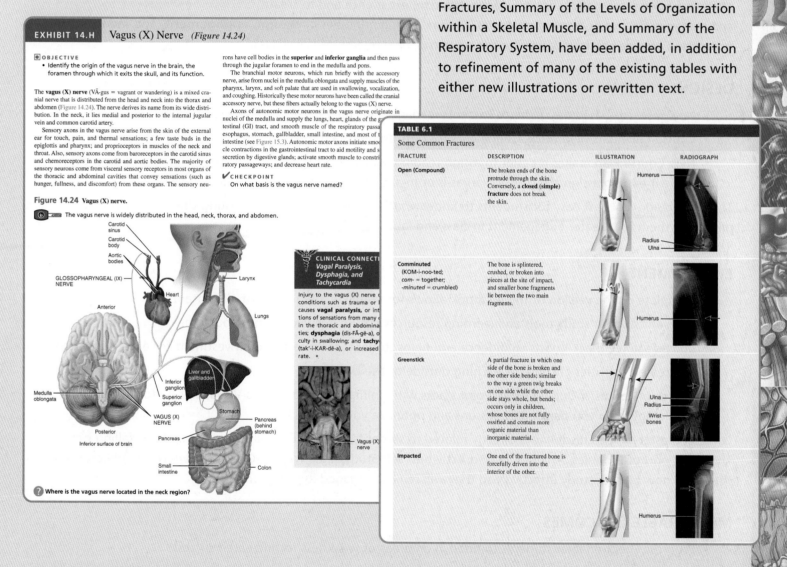

EXHIBIT 14.H Vagus (X) Nerve *(Figure 14.24)*

OBJECTIVE
• Identify the origin of the vagus nerve in the brain, the foramen through which it exits the skull, and its function.

The **vagus (X) nerve** (VĀ-gus = vagrant or wandering) is a mixed cranial nerve that is distributed from the head and neck into the thorax and abdomen (Figure 14.24). The nerve derives its name from its wide distribution. In the neck, it lies medial and posterior to the internal jugular vein and common carotid artery.

Sensory axons in the vagus nerve arise from the skin of the external ear for touch, pain, and thermal sensations; a few taste buds in the epiglottis and pharynx; and proprioceptors in muscles of the neck and throat. Also, sensory axons come from baroreceptors in the carotid sinus and chemoreceptors in the carotid and aortic bodies. The majority of sensory neurons come from visceral sensory receptors in most organs of the thoracic and abdominal cavities that convey sensations (such as hunger, fullness, and discomfort) from these organs. The sensory neu-

rons have cell bodies in the **superior** and **inferior ganglia** and then pass through the jugular foramen to end in the medulla and pons.

The branchial motor neurons, which run briefly with the accessory nerve, arise from nuclei in the medulla oblongata and supply muscles of the pharynx, larynx, and soft palate that are used in swallowing, vocalization, and coughing. Historically these motor neurons have been called the cranial accessory nerve, but these fibers actually belong to the vagus (X) nerve.

Axons of autonomic motor neurons in the vagus nerve originate in nuclei of the medulla and supply the lungs, heart, glands of the gastrointestinal (GI) tract, and smooth muscle of the respiratory passa esophagus, stomach, gallbladder, small intestine, and most of t intestine (see Figure 15.3). Autonomic motor axons initiate smoo cle contractions in the gastrointestinal tract to aid motility and s secretion by digestive glands; activate smooth muscle to constri ratory passageways; and decrease heart rate.

CHECKPOINT
On what basis is the vagus nerve named?

Figure 14.24 Vagus (X) nerve.

The vagus nerve is widely distributed in the head, neck, thorax, and abdomen.

Carotid sinus
Carotid body
Aortic bodies
GLOSSOPHARYNGEAL (IX) NERVE
Larynx
Heart
Anterior
Lungs
Medulla oblongata
Inferior ganglion
Superior ganglion
Liver and gallbladder
Stomach
VAGUS (X) NERVE
Pancreas (behind stomach)
Posterior
Pancreas
Inferior surface of brain
Small intestine
Colon
Vagus (X) nerve

? Where is the vagus nerve located in the neck region?

CLINICAL CONNECTI
Vagal Paralysis, Dysphagia, and Tachycardia

Injury to the vagus (X) nerve c conditions such as trauma or I causes **vagal paralysis,** or int tions of sensations from many in the thoracic and abdomina ties; **dysphagia** (dis-FĀ-gē-a), o culty in swallowing; and **tachy** (tak′-i-KAR-dē-a), or increased rate. •

TABLE 6.1

Some Common Fractures

FRACTURE	DESCRIPTION	ILLUSTRATION	RADIOGRAPH
Open (Compound)	The broken ends of the bone protrude through the skin. Conversely, a **closed (simple) fracture** does not break the skin.	Humerus	Radius Ulna
Comminuted (KOM-i-noo-ted; *com-* = together; *-minuted* = crumbled)	The bone is splintered, crushed, or broken into pieces at the site of impact, and smaller bone fragments lie between the two main fragments.		Humerus
Greenstick	A partial fracture in which one side of the bone is broken and the other side bends; similar to the way a green twig breaks on one side while the other side stays whole, but bends; occurs only in children, whose bones are not fully ossified and contain more organic material than inorganic material.		Ulna Radius Wrist bones
Impacted	One end of the fractured bone is forcefully driven into the interior of the other.		Humerus

Clinical Connections

Your students are fascinated by the **Clinical Connections** to the normal anatomy and physiology that they are learning. You'll find that the text is liberally peppered with engaging discussions of a wide variety of clinical scenarios from disease coverage to tests and procedures. As always, we have updated all of the Clinical Connections and Disorders: Homeostatic Imbalances sections to reflect the most current information. We have added several new Clinical Connections, such as a feature on fibromyalgia, to the text. A complete reference list of the Clinical Connections within each chapter follows the Table of Contents.

WileyPLUS *and You*

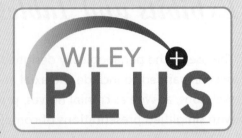

WileyPLUS for Anatomy and Physiology is an innovative, research-based online environment designed for both effective teaching and learning. Utilizing *WileyPLUS* in your course provides your students with an accessible, affordable, and active learning platform and gives you tools and resources to efficiently build presentations for a dynamic classroom experience and to create and manage effective assessment strategies. The underlying principles of **design, engagement,** and **measurable outcomes** provide the foundation for this powerful, new release of *WileyPLUS*.

DESIGN

- New research-based design helps students manage their time better and develop better study skills.
- Course Calendars help track assignments for both students and teachers.
- New Course Plan makes it easier to assign reading, activities, and assessment. Simple drag-and-drop tools make it easy to assign the course plan as-is or in any way that best reflects your course syllabus.

The new design makes it easy for students to know *what* it is they need to do, boosting their confidence and preparing them for greater engagement in class and lab.

ENGAGEMENT

- Complete online version of the textbook allows for seamless integration of all content.
- Relevant student study tools and learning resources ensure positive learning outcomes.
- Immediate feedback boosts confidence and helps students see a return on investment for each study session.
- Precreated activities encourage learning outside of the classroom.
- Course materials, including PowerPoint stacks with animations and Wiley's Visual Library for Anatomy and Physiology, help you personalize lessons and optimize your time.

Concept mastery in this discipline is directly related to students keeping up with the work and not falling behind. The new Concept Modules, Animations and Activities, Self Study, and Progress Checks in *WileyPLUS* will ensure that students know how to study effectively so they will remain engaged and stay on task.

MEASURABLE OUTCOMES

- Progress Check enables students to hone in on areas of weakness for increased success.
- Self-assessment and remediation for all Learning Objectives let students know exactly how their efforts have paid off.
- Instant reports monitor trends in class performance, use of course materials, and student progress toward learning objectives.

With new detailed reporting capabilities, students will know that they are doing it right. With increased confidence, motivation is sustained so students stay on task—success will follow.

RESOURCES FOR INTEGRATING LABORATORY EXPERIENCES

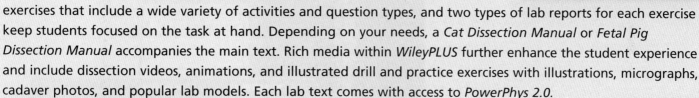

Laboratory Manual for Anatomy and Physiology, 4e

Connie Allen and Valerie Harper

Newly revised, **Laboratory Manual for Anatomy and Physiology** with **WileyPLUS 5.0** engages your students in active learning and focuses on the most important concepts in A&P. Exercises reflect the multiple ways in which students learn and provide guidance for anatomical exploration and application of critical thinking to analyzing physiological processes. A concise narrative, self-contained exercises that include a wide variety of activities and question types, and two types of lab reports for each exercise keep students focused on the task at hand. Depending on your needs, a *Cat Dissection Manual* or *Fetal Pig Dissection Manual* accompanies the main text. Rich media within *WileyPLUS* further enhance the student experience and include dissection videos, animations, and illustrated drill and practice exercises with illustrations, micrographs, cadaver photos, and popular lab models. Each lab text comes with access to *PowerPhys 2.0*.

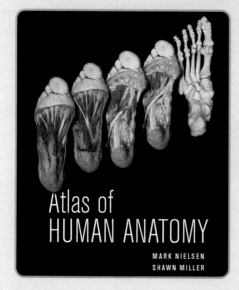

Atlas of Human Anatomy, 1e

Mark Nielsen and Shawn Miller

This new atlas filled with outstanding photographs of meticulously executed dissections of the human body has been developed to be a strong teaching and learning solution, not just a catalog of photographs. Organized around body systems, each chapter includes a narrative overview of the body system and is then followed with detailed photographs that accurately and realistically represent the anatomical structures. Histology is included. *Atlas of Human Anatomy* will work well in your laboratories, as a study companion to your textbook, and as a print companion to the *Real Anatomy* DVD.

Photographic Atlas of the Human Body, 2e

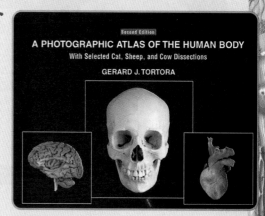

Gerard J. Tortora

Like the new atlas from Nielsen and Miller, this popular atlas is also systemic in its approach to the photographic review of the human body. In addition to the excellent cadaver photographs and micrographs, this atlas also contains selected cat and sheep heart dissections. The high-quality imagery can be used in the classroom, in the laboratory, or for study and review.

RESOURCES FOR INTEGRATING LABORATORY EXPERIENCES

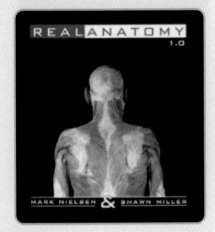

Real Anatomy

Mark Nielsen and Shawn Miller

Real Anatomy is 3-D imaging software that allows you to dissect through multiple layers of a three-dimensional real human body to study and learn the anatomical structures of all body systems.

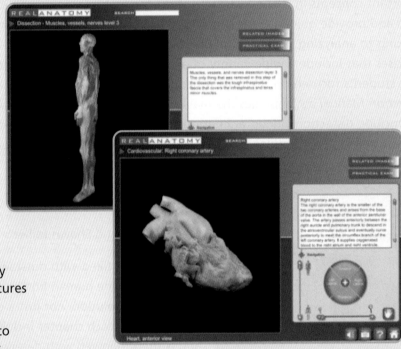

- Dissect through up to 40 layers of the body and discover the relationships of the structures to the whole.

- Rotate the body, as well as major organs, to view the image from multiple perspectives.

- Use a built-in zoom feature to get a closer look at detail.

- A unique approach to highlighting and labeling structures does not obscure the real anatomy in view.

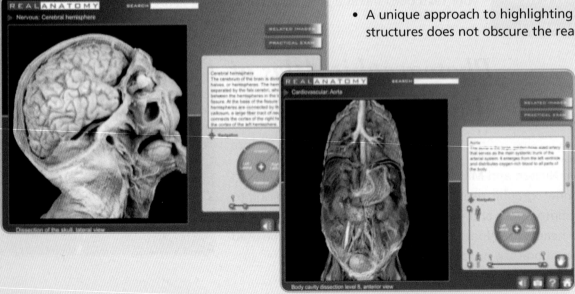

RESOURCES FOR INTEGRATING LABORATORY EXPERIENCES

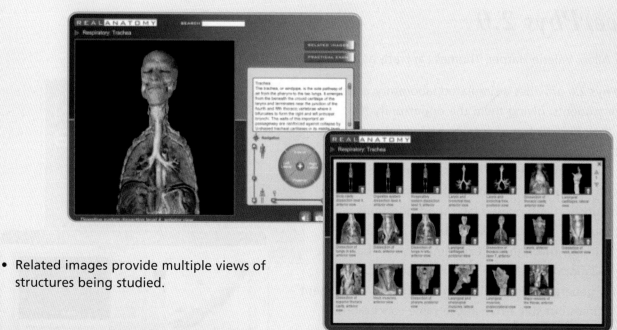

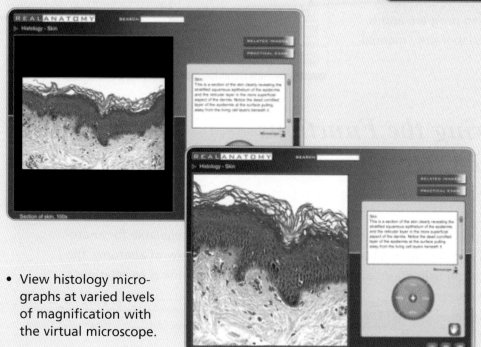

- Related images provide multiple views of structures being studied.

- View histology micrographs at varied levels of magnification with the virtual microscope.

- Snapshots of any image can be saved for use in PowerPoints, quizzes, or handouts.

- Audio pronunciation of all labeled structures is readily available.

Virtual Dissection—100% Real

REALANATOMY

PowerPhys 2.0

Connie Allen, Valerie Harper, Thomas Lancraft, and Yuri Ivlev

PowerPhys 2.0 provides a simulated laboratory experience for students, giving them the opportunity to review their knowledge of core physiological concepts, predict outcomes of an experiment, collect data, analyze it, and report on their findings. This revised edition features a new activity on Homeostatic Imbalance of Thyroid Function and revised lab report questions throughout. An easy-to-use and intuitive interface guides students through the experiments from basic review to laboratory reports. All experiments contain randomly generated data, allowing students to experiment multiple times but still arrive at the same conclusions. A perfect addition to distant learning or hybrid courses, *PowerPhys 2.0* is a stand-alone web-based program and is fully integrated with Allen and Harper's laboratory manual.

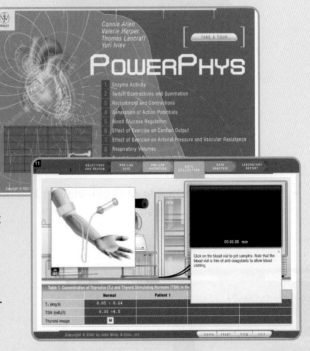

Interactions: Exploring the Functions of the Human Body 3.0

Thomas Lancraft and Frances Frierson

Interactions 3.0 is the most complete program of interactive animations and activities available for anatomy and physiology. A series of modules encompassing all body systems focuses on a review of anatomy, the examination of physiological processes using animations and interactive exercises, and clinical correlations to enhance student understanding. At the heart of **Interactions** is a focus on core principles—*homeostasis, communication, energy flow, fluid flow,* and *boundaries*—that underscore the key relationships between structure and function as well as interrelationships between systems. It is the reinforcement of these fundamental organizing principles that sets this series apart from others. **Interactions** is available on DVD, web-based, or fully integrated within *WileyPLUS*.

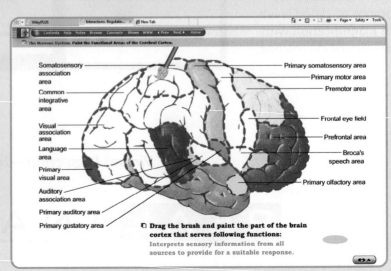

ACKNOWLEDGMENTS

We wish to especially thank several academic colleagues for their helpful contributions to this edition. Creating and implementing the integration of this text with *WileyPLUS for Anatomy and Physiology* was only possible because of the expertise and fine work of the following group of people. We are so very grateful to you:

Sarah Bales
Moraine Valley Community College

Celina Bellanceau
University of South Florida

Curtis DeFriez
Weber State University

Alan Erickson
South Dakota State University

Gibril Fadika
Hampton University

Pamela Fouche
Walters State Community College

Sophia Garcia
Tarrant County College–Trinity River

Clare Hays
Metropolitan State College of Denver

Jason Hunt
Brigham Young University–Idaho

Judy Learn
North Seattle Community College

Jerri K. Lindsey
Tarrant County College

Todd Miller
Hunter College

Erin Morrey
Georgia Perimeter College

Gus Pita
Hunter College

Susan Puglisi
Norwalk Community College

Saeed Rahmanian
Roane State Community College

Lori A. Smith
American River College

Randall Tracy
Worcester State College

Jay Zimmer
South Florida Community College

We are also very grateful to our colleagues who have reviewed the manuscript or participated in focus groups and offered numerous suggestions for improvement:

Charles J. Biggers
University of Memphis

Gladys Bolding
Georgia Perimeter College

Lois Borek
Georgia State University

Betsy Brantley
Lansing Community College

Arthur R. Buckley
University of Cincinnati

Alex Cheroske
Mesa Community College

Robert Comegys
Old Dominion University

Curtis DeFriez
Weber State University

William Dunscombe
Union County College

Heather Dy
Long Beach Community College

Christine Ross Earls
Fairfield University

Angela Edwards
University of South Carolina Allendale

Sharon Ellerton
Queensborough Community College

David Evans
Pennsylvania College of Technology

Gibril Fadika
Hampton University

Sandy Garrett
Texas Woman's University

Michael Harman
Lone Star College

Jane Horlings
Saddleback College

Barbara Hunnicutt
Seminole State College

Jason Hunt
Brigham Young University–Idaho

Alexander T. Imholtz
Prince George's Community College

Amy E. Jetton
Middle Tennessee State University

Becky Keck
Clemson University

Marc LaBella
Ocean County College

Ellen Lathrop-Davis
Community College of Baltimore County

Billy Bob Long
Del Mar College

Wayne M. Mason
Western Kentucky University

Karen McLellan
Indiana Purdue University Fort Wayne

Marie McMahon
Miramar College

Erin Morrey
Georgia Perimeter College

Maria Oehler
Florida State College at Jacksonville

Betsy Ott
Tyler Junior College

Gilbert Pitts
Austin Peay State University

Saeed Rahmanian
Roane State Community College

Terrence J. Ravine
University of South Alabama

Philip D. Reynolds
Troy University

John Roufaiel
SUNY Rockland Community College

Kelly Sexton
North Lake College

Colleen Sinclair
Towson University

Lori A. Smith
American River College

Nora Stevens
Portland Community College

Leo B. Stouder
Broward College

Dennis Strete
McLennan Community College

Peter Susan
Trident Technical College

Jared Taglialatela
Clayton State College

Bonnie J. Tarricone
Ivy Tech Community College

Heather Walker
Clemson University

Janice Webster
Ivy Tech Community College

Delores Wenzel
University of Georgia

Matthew A. Williamson
Georgia Southern University

Finally, our hats are off to everyone at Wiley. We enjoy collaborating with this enthusiastic, dedicated, and talented team of publishing professionals. Our thanks to the entire team: Bonnie Roesch, Executive Editor; Mary Berry and Karen Trost, Developmental Editors; Lorraina Raccuia, Project Editor; Lauren Morris, Program Assistant; Suzanne Ingrao, Outside Production Editor; Hilary Newman, Photo Manager; Claudia Volano, Illustration Coordinator; Anna Melhorn, Senior Illustration Editor; Madelyn Lesure, Senior Designer; Laura Ierardi, LCI Design; Linda Muriello, Senior Media Editor; and Clay Stone, Executive Marketing Manager.

Gerard J. Tortora
Department of Science and Health, S229
Bergen Community College
400 Paramus Road
Paramus, NJ 07652

Bryan Derrickson
Science Department
Valencia Community College
1800 S. Kirkman Rd.
Orlando, FL 32811

ABOUT THE AUTHORS

Courtesy of Heidi Chung.

Gerard J. Tortora is Professor of Biology and former Biology Coordinator at Bergen Community College in Paramus, New Jersey, where he teaches human anatomy and physiology as well as microbiology. He received his bachelor's degree in biology from Fairleigh Dickinson University and his master's degree in science education from Montclair State College. He is a member of many professional organizations, including the Human Anatomy and Physiology Society (HAPS), American Society of Microbiology (ASM), American Association for the Advancement of Science (AAAS), National Education Association (NEA), and Metropolitan Association of College and University Biologists (MACUB).

Above all, Jerry is devoted to his students and their aspirations. In recognition of this commitment, Jerry was the recipient of MACUB's 1992 President's Memorial Award. In 1996, he received a National Institute for Staff and Organizational Development (NISOD) excellence award from the University of Texas and was selected to represent Bergen Community College in a campaign to increase awareness of the contributions of community colleges to higher education.

Jerry is the author of several best-selling science textbooks and laboratory manuals, a calling that often requires an additional 40 hours per week beyond his teaching responsibilities. Nevertheless, he still makes time for four or five weekly aerobic workouts that include biking and running. He also enjoys attending college basketball and professional hockey games and performances at the Metropolitan Opera House.

To my mother, Angelina M. Tortora.
(August 20, 1913–August 14, 2010).
Her love, guidance, faith, support, and example continue to be the cornerstone of my personal and professional life. **G.J.T.**

Bryan Derrickson is Professor of Biology at Valencia Community College in Orlando, Florida, where he teaches human anatomy and physiology as well as general biology and human sexuality. He received his bachelor's degree in biology from Morehouse College and his doctorate in cell biology from Duke University. Bryan's study at Duke was in the Physiology Division within the Department of Cell Biology, so while his degree is in cell biology, his training focused on physiology. At Valencia, he frequently serves on faculty hiring committees. He has served as a member of the Faculty Senate, which is the governing body of the college, and as a member of the Teaching and Learning Academy, which sets the standards for the acquisition of tenure by faculty members. Nationally, he is a member of the Human Anatomy and Physiology Society (HAPS) and the National Association of Biology Teachers (NABT).

Bryan has always wanted to teach. Inspired by several biology professors while in college, he decided to pursue physiology with an eye to teaching at the college level. He is completely dedicated to the success of his students. He particularly enjoys the challenges of his diverse student population, in terms of their age, ethnicity, and academic ability, and he finds being able to reach all of them, despite their differences, a rewarding experience. His students continually recognize Bryan's efforts and care by nominating him for a campus award known as the "Valencia Professor Who Makes Valencia a Better Place to Start." Bryan has received this award three times.

To my family: Rosalind, Hurley, Cherie, and Robb.
Your support and motivation have been invaluable. **B.H.D.**

BRIEF CONTENTS

VOLUME 1

BRIEF CONTENTS

VOLUME 2

VOLUME 2 CONTENTS

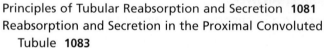

19 THE CARDIOVASCULAR SYSTEM: THE BLOOD

BLOOD AND HOMEOSTASIS *Blood contributes to homeostasis by transporting oxygen, carbon dioxide, nutrients, and hormones to and from your body's cells. It helps regulate body pH and temperature, and provides protection against disease through phagocytosis and the production of antibodies.*

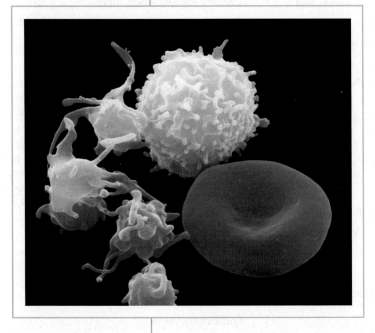

The **cardiovascular system** (*cardio-* = heart; *vascular* = blood or blood vessels) consists of three interrelated components: blood, the heart, and blood vessels. The focus of this chapter is blood; the next two chapters will examine the heart and blood vessels, respectively. Blood transports various substances, helps regulate several life processes, and affords protection against disease. For all of its similarities in origin, composition, and functions, blood is as unique from one person to another as are skin, bone, and hair. Health-care professionals routinely examine and analyze its differences through various blood tests when trying to determine the cause of different diseases. The branch of science concerned with the study of blood, blood-forming tissues, and the disorders associated with them is **hematology** (hēm-a-TOL-ō-jē; *hema-* or *hemato-* = blood; *-logy* = study of).

Did you ever wonder why blood is such a unique substance that can be analyzed to determine if we are healthy, to detect a multitude of infections, and to detect or rule out various diseases and injuries?

19.1 FUNCTIONS AND PROPERTIES OF BLOOD

⊙ OBJECTIVES

• Describe the functions of blood.
• Describe the physical characteristics and principal components of blood.

Most cells of a multicellular organism cannot move around to obtain oxygen and nutrients or eliminate carbon dioxide and other wastes. Instead, these needs are met by two fluids: blood and interstitial fluid. **Blood** is a liquid connective tissue that consists of cells surrounded by a liquid extracellular matrix. The extracellular matrix is called blood plasma, and it suspends various cells and cell fragments. **Interstitial fluid** is the fluid that bathes body cells (see Figure 27.1) and is constantly renewed by the blood. Blood transports oxygen from the lungs and nutrients from the gastrointestinal tract, which diffuse from the blood into the interstitial fluid and then into body cells. Carbon dioxide and other wastes move in the reverse direction, from body cells to interstitial fluid to blood. Blood then transports the wastes to various organs—the lungs, kidneys, and skin—for elimination from the body.

Functions of Blood

Blood has three general functions:

1. *Transportation.* As you just learned, blood transports oxygen from the lungs to the cells of the body and carbon dioxide from the body cells to the lungs for exhalation. It carries nutrients from the gastrointestinal tract to body cells and hormones from endocrine glands to other body cells. Blood also transports heat and waste products to various organs for elimination from the body.

2. *Regulation.* Circulating blood helps maintain homeostasis of all body fluids. Blood helps regulate pH through the use of buffers (chemicals that convert strong acids or bases into weak ones). It also helps adjust body temperature through the heat-absorbing and coolant properties of the water (see Section 2.4) in blood plasma and its variable rate of flow through the skin, where excess heat can be lost from the blood to the environment. In addition, blood osmotic pressure influences the water content of cells, mainly through interactions of dissolved ions and proteins.

3. *Protection.* Blood can clot (become gel-like), which protects against its excessive loss from the cardiovascular system after an injury. In addition, its white blood cells protect against disease by carrying on phagocytosis. Several types of blood proteins, including antibodies, interferons, and complement, help protect against disease in a variety of ways.

Physical Characteristics of Blood

Blood is denser and more viscous (thicker) than water and feels slightly sticky. The temperature of blood is 38°C (100.4°F), about 1°C higher than oral or rectal body temperature, and it has a slightly alkaline pH ranging from 7.35 to 7.45. The color of blood varies with its oxygen content. When saturated with oxygen, it is bright red. When unsaturated with oxygen, it is dark red. Blood constitutes about 20% of extracellular fluid, amounting to 8% of the total body mass. The blood volume is 5 to 6 liters (1.5 gal) in an average-sized adult male and 4 to 5 liters (1.2 gal) in an average-sized adult female. The gender difference in volume is due to differences in body size. Several hormones, regulated by negative feedback, ensure that blood volume and osmotic pressure remain relatively constant. Especially important are the hormones aldosterone, antidiuretic hormone, and atrial natriuretic peptide, which regulate how much water is excreted in the urine (see Section 27.1).

CLINICAL CONNECTION | *Withdrawing Blood*

Blood samples for laboratory testing may be obtained in several ways. The most common procedure is **venipuncture** (vēn'-i-PUNK-chur), withdrawal of blood from a vein using a needle and collecting tube, which contains various additives. A tourniquet is wrapped around the arm above the venipuncture site, which causes blood to accumulate in the vein. This increased blood volume makes the vein stand out. Opening and closing the fist further causes it to stand out, making the venipuncture more successful. A common site for venipuncture is the median cubital vein anterior to the elbow (see Figure 21.25b). Another method of withdrawing blood is through a **finger** or **heel stick.** Diabetic patients who monitor their daily blood sugar typically perform a finger stick, and it is often used for drawing blood from infants and children. In an **arterial stick,** blood is withdrawn from an artery; this test is used to determine the level of oxygen in oxygenated blood. •

Components of Blood

Whole blood has two components: (1) blood plasma, a watery liquid extracellular matrix that contains dissolved substances, and (2) formed elements, which are cells and cell fragments. If a sample of blood is centrifuged (spun) in a small glass tube, the cells (which are more dense) sink to the bottom of the tube while the plasma (which is less dense) forms a layer on top (Figure 19.1a). Blood is about 45% formed elements and 55% blood plasma. Normally, more than 99% of the formed elements are cells named for their red color—red blood cells (RBCs). Pale, colorless white blood cells (WBCs) and platelets occupy less than 1% of the formed elements. Because they are less dense than red blood cells but more dense than blood plasma, they form a very thin **buffy coat** layer between the packed RBCs and plasma in centrifuged blood. Figure 19.1b shows the composition of blood plasma and the numbers of the various types of formed elements in blood.

Blood Plasma

When the formed elements are removed from blood, a straw-colored liquid called **blood plasma** (or simply **plasma**) is left. Blood plasma is about 91.5% water and 8.5% solutes, most of which

Figure 19.1 Components of blood in a normal adult.

Blood is a connective tissue that consists of blood plasma (liquid) plus formed elements (red blood cells, white blood cells, and platelets).

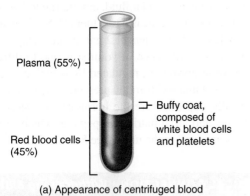

Plasma (55%)

Buffy coat,
composed of
white blood cells
and platelets

Red blood cells
(45%)

(a) Appearance of centrifuged blood

FUNCTIONS OF BLOOD

1. Transports oxygen, carbon dioxide, nutrients, hormones, heat, and wastes.
2. Regulates pH, body temperature, and water content of cells.
3. Protects against blood loss through clotting, and against disease through phagocytic white blood cells and proteins such as antibodies, interferons, and complement.

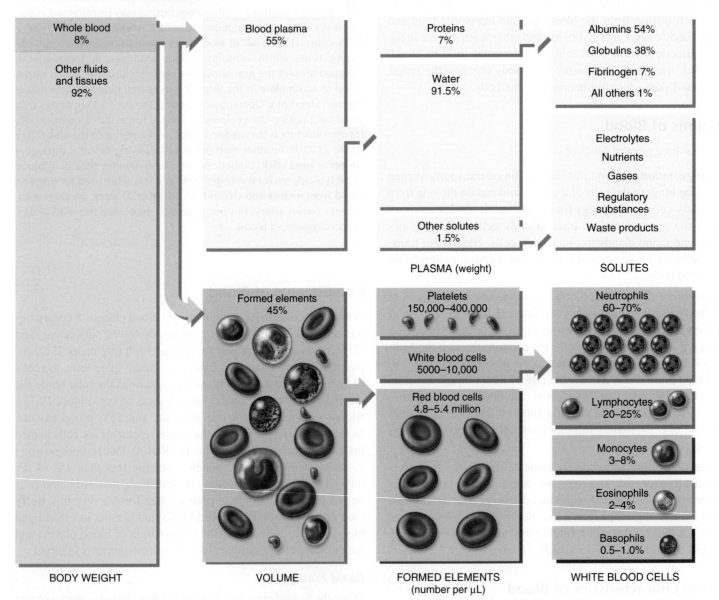

(b) Components of blood

What is the approximate volume of blood in your body?

(7% by weight) are proteins. Some of the proteins in blood plasma are also found elsewhere in the body, but those confined to blood are called **plasma proteins.** Hepatocytes (liver cells) synthesize most of the plasma proteins, which include the **albumins** (al'-BŪ-mins) (54% of plasma proteins), **globulins** (GLOB-ū-lins) (38%), and **fibrinogen** (fī-BRIN-ō-jen) (7%). Certain blood cells develop into cells that produce gamma globulins, an important type of globulin. These plasma proteins are also called **antibodies** or **immunoglobulins** (im'-ū-nō-GLOB-ū-lins) because they are produced during certain immune responses. Foreign substances (antigens) such as bacteria and viruses stimulate production of millions of different antibodies. An antibody binds specifically to the antigen that stimulated its production and thus disables the invading antigen.

Besides proteins, other solutes in plasma include electrolytes, nutrients, regulatory substances such as enzymes and hormones, gases, and waste products such as urea, uric acid, creatinine, ammonia, and bilirubin.

Table 19.1 describes the chemical composition of blood plasma.

Formed Elements

The **formed elements** of the blood include three principal components: **red blood cells (RBCs), white blood cells (WBCs),** and **platelets** (Figure 19.2). RBCs and WBCs are whole cells; platelets are cell fragments. RBCs and platelets have just a few roles, but WBCs have a number of specialized functions. Several distinct types of WBCs—neutrophils, lymphocytes, monocytes, eosinophils, and basophils—each with a unique microscopic appearance, carry out these functions, which are discussed later in this chapter.

Following is the classification of the formed elements in blood:

I. Red blood cells
II. White blood cells
   A. Granular leukocytes (contain conspicuous granules that are visible under a light microscope after staining)
      1. Neutrophils
      2. Eosinophils
      3. Basophils
   B. Agranular leukocytes (no granules are visible under a light microscope after staining)
      1. T and B lymphocytes and natural killer (NK) cells
      2. Monocytes
III. Platelets

The percentage of total blood volume occupied by RBCs is called the **hematocrit** (hē-MAT-ō-krit); a hematocrit of 40 indicates that 40% of the volume of blood is composed of RBCs. The

TABLE 19.1

Substances in Blood Plasma

CONSTITUENT	DESCRIPTION	FUNCTION
Water (91.5%)	Liquid portion of blood.	Solvent and suspending medium. Absorbs, transports, and releases heat.
Plasma proteins (7%)	Most produced by liver.	Responsible for colloid osmotic pressure. Major contributors to blood viscosity. Transport hormones (steroid), fatty acids, and calcium. Help regulate blood pH.
Albumin	Smallest and most numerous of proteins.	
Globulins	Large proteins (plasma cells produce immunoglobulins).	Immunoglobulins help attack viruses and bacteria. Alpha and beta globulins transport iron, lipids, and fat-soluble vitamins.
Fibrinogen	Large protein.	Plays essential role in blood clotting.
Other solutes (1.5%)		
Electrolytes	Inorganic salts; positively charged (cations) Na^+, K^+, Ca^{2+}, Mg^{2+}; negatively charged (anions) Cl^-, HPO_4^{2-}, SO_4^{2-}, HCO_3^-.	Help maintain osmotic pressure and essential roles in cell functions.
Nutrients	Products of digestion, such as amino acids, glucose, fatty acids, glycerol, vitamins, and minerals.	Essential roles in cell functions, growth, and development.
Gases	Oxygen (O_2).	Oxygen is important in many cellular functions.
	Carbon dioxide (CO_2).	Carbon dioxide is involved in the regulation of blood pH.
	Nitrogen (N_2).	Nitrogen has no known function.
Regulatory substances	Enzymes.	Catalyze chemical reactions.
	Hormones.	Regulate metabolism, growth, and development.
	Vitamins.	Cofactors for enzymatic reactions.
Waste products	Urea, uric acid, creatine, creatinine, bilirubin, Ammonia.	Most are breakdown products of protein metabolism that are carried by the blood to organs of excretion.

Figure 19.2 Scanning electron micrograph and photomicrograph of the formed elements of blood.

The formed elements of blood are red blood cells (RBCs), white blood cells (WBCs), and platelets.

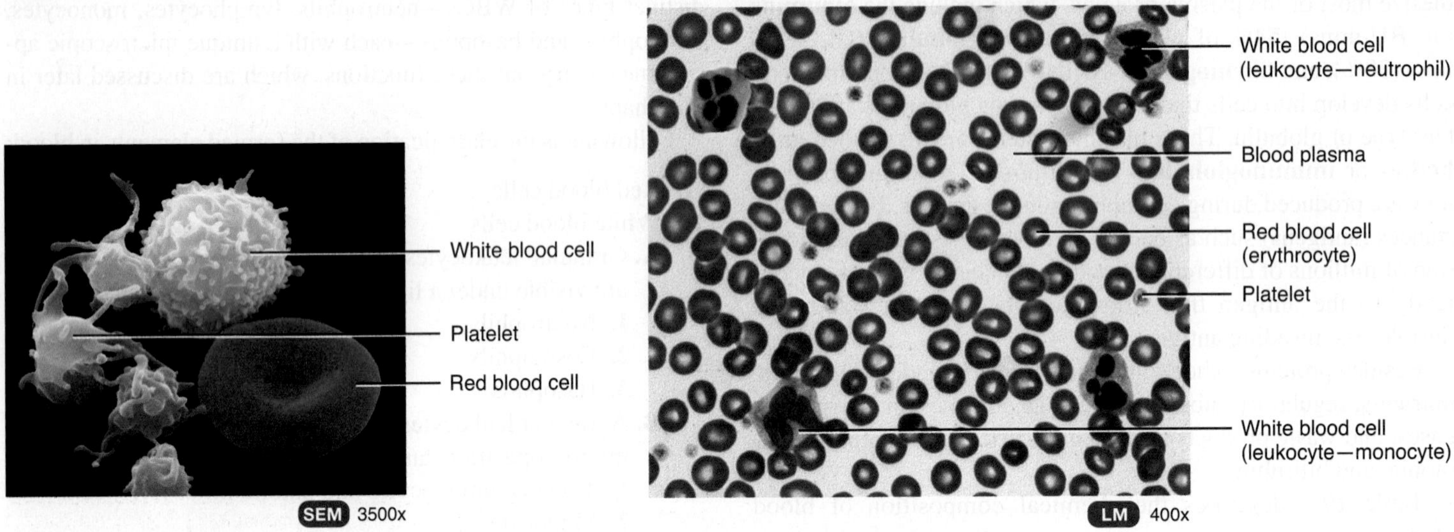

(a) Scanning electron micrograph

SEM 3500x

White blood cell

Platelet

Red blood cell

(b) Blood smear

LM 400x

White blood cell
(leukocyte—neutrophil)

Blood plasma

Red blood cell
(erythrocyte)

Platelet

White blood cell
(leukocyte—monocyte)

? **Which formed elements of the blood are cell fragments?**

normal range of hematocrit for adult females is 38–46% (average = 42); for adult males, it is 40–54% (average = 47). The hormone testosterone, present in much higher concentration in males than in females, stimulates synthesis of erythropoietin (EPO), the hormone that in turn stimulates production of RBCs. Thus, testosterone contributes to higher hematocrits in males. Lower values in women during their reproductive years also may be due to excessive loss of blood during menstruation. A significant drop in hematocrit indicates *anemia,* a lower-than-normal number of RBCs. In *polycythemia* (pol′-ē-sī-THĒ-mē-a) the percentage of RBCs is abnormally high, and the hematocrit may be 65% or higher. This raises the viscosity of blood, which increases the resistance to flow and makes the blood more difficult for the heart to pump. Increased viscosity also contributes to high blood pressure and increased risk of stroke. Causes of polycythemia include abnormal increases in RBC production, tissue hypoxia, dehydration, and blood doping or the use of EPO by athletes.

✔ CHECKPOINT

1. In what ways is blood plasma similar to interstitial fluid? How does it differ?
2. What substances does blood transport?
3. How many kilograms or pounds of blood are there in your body?
4. How does the volume of blood plasma in your body compare to the volume of fluid in a 2-liter bottle of Coke?
5. List the formed elements in blood plasma and describe their functions.
6. What is the significance of lower-than-normal or higher-than-normal hematocrit?

19.2 FORMATION OF BLOOD CELLS

◉ OBJECTIVE
• Explain the origin of blood cells.

Although some lymphocytes have a lifetime measured in years, most formed elements of the blood last only hours, days, or weeks, and must be replaced continually. Negative feedback systems regulate the total number of RBCs and platelets in circulation, and their numbers normally remain steady. The abundance of the different types of WBCs, however, varies in response to challenges by invading pathogens and other foreign antigens.

The process by which the formed elements of blood develop is called **hemopoiesis** (hē-mō-poy-Ē-sis; *-poiesis* = making) or *hematopoiesis.* Before birth, hemopoiesis first occurs in the yolk sac of an embryo and later in the liver, spleen, thymus, and lymph nodes of a fetus. Red bone marrow becomes the primary site of hemopoiesis in the last 3 months before birth, and continues as the source of blood cells after birth and throughout life.

Red bone marrow is a highly vascularized connective tissue located in the microscopic spaces between trabeculae of spongy bone tissue. It is present chiefly in bones of the axial skeleton, pectoral and pelvic girdles, and the proximal epiphyses of the humerus and femur. About 0.05–0.1% of red bone marrow cells are derived from mesenchyme (tissue from which almost all connective tissues develop) and are called **pluripotent stem cells** (ploo-RIP-ō-tent; *pluri-* = several) or *hemocytoblasts.* These cells have the capacity to develop into many different types of cells (Figure 19.3). In newborns, all bone marrow is red and thus

Figure 19.3 Origin, development, and structure of blood cells. A few of the generations of some cell lines have been omitted.

🔑 Blood cell production, called hemopoiesis, occurs mainly in red bone marrow after birth.

Key:

▢ Progenitor cells

▢ Precursor cells or "blasts"

▢ Formed elements of circulating blood

▢ Tissue cells

Key:

CFU–E	Colony-forming unit—erythrocyte
CFU–Meg	Colony-forming unit—megakaryocyte
CFU–GM	Colony-forming unit—granulocyte macrophage

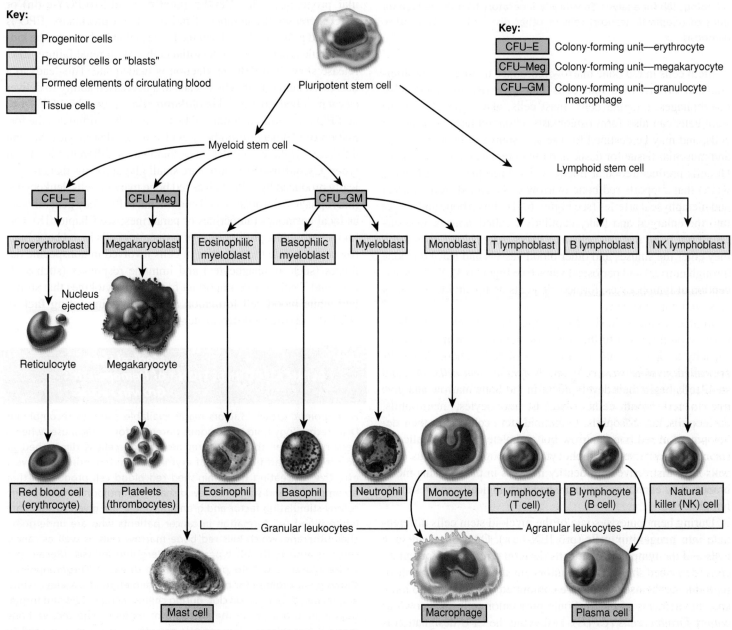

❓ **From which connective tissue cells do pluripotent stem cells develop?**

active in blood cell production. As an individual ages, the rate of blood cell formation decreases; the red bone marrow in the medullary (marrow) cavity of long bones becomes inactive and is replaced by yellow bone marrow, which consists largely of fat cells. Under certain conditions, such as severe bleeding, yellow bone marrow can revert to red bone marrow; this occurs as blood-forming stem cells from the red bone marrow move into the yellow bone marrow, which is then repopulated by pluripotent stem cells.

CLINICAL CONNECTION | *Bone Marrow Examination*

Sometimes a sample of red bone marrow must be obtained in order to diagnose certain blood disorders, such as leukemia and severe anemias. **Bone marrow examination** may involve *bone marrow aspiration* (withdrawal of a small amount of red bone marrow with a fine needle and syringe) or a *bone marrow biopsy* (removal of a core of red bone marrow with a larger needle).

Both types of samples are usually taken from the iliac crest of the hip bone, although samples are sometimes aspirated from the sternum. In young children, bone marrow samples are taken from a vertebra or tibia (shin bone). The tissue or cell sample is then sent to a pathology lab for analysis. Specifically, laboratory technicians look for signs of neoplastic (cancer) cells or other diseased cells to assist in diagnosis. •

Stem cells in red bone marrow reproduce themselves, proliferate, and differentiate into cells that give rise to blood cells, macrophages, reticular cells, mast cells, and adipocytes. Some stem cells can also form osteoblasts, chondroblasts, and muscle cells, and may be destined for use as a source of bone, cartilage, and muscular tissue for tissue and organ replacement. The reticular cells produce reticular fibers, which form the stroma (framework) that supports red bone marrow cells. Blood from nutrient and metaphyseal arteries (see Figure 6.4) enters a bone and passes into the enlarged and leaky capillaries, called *sinuses,* that surround red bone marrow cells and fibers. After blood cells form, they enter the sinuses and other blood vessels and leave the bone through nutrient and periosteal veins (see Figure 6.4). With the exception of lymphocytes, formed elements do not divide once they leave red bone marrow.

In order to form blood cells, pluripotent stem cells in red bone marrow produce two further types of stem cells, which have the capacity to develop into several types of cells. These stem cells are called *myeloid stem cells* and *lymphoid stem cells.* Myeloid stem cells begin their development in red bone marrow and give rise to red blood cells, platelets, monocytes, neutrophils, eosinophils, and basophils. Lymphoid stem cells begin their development in red bone marrow but complete it in lymphatic tissues; they give rise to lymphocytes. Although the various stem cells have distinctive cell identity markers in their plasma membranes, they cannot be distinguished histologically and resemble lymphocytes.

During hemopoiesis, some of the myeloid stem cells differentiate into **progenitor cells** (prō-JEN-i-tor). Other myeloid stem cells and the lymphoid stem cells develop directly into precursor cells (described shortly). Progenitor cells are no longer capable of reproducing themselves and are committed to giving rise to more specific elements of blood. Some progenitor cells are known as *colony-forming units (CFUs).* Following the CFU designation is an abbreviation that indicates the mature elements in blood that they will produce: CFU–E ultimately produces erythrocytes (red blood cells); CFU–Meg produces megakaryocytes, the source of platelets; and CFU–GM ultimately produces granulocytes (specifically, neutrophils) and monocytes (see Figure 19.3). Progenitor cells, like stem cells, resemble lymphocytes and cannot be distinguished by their microscopic appearance alone.

In the next generation, the cells are called **precursor cells,** also known as **blasts.** Over several cell divisions they develop into the actual formed elements of blood. For example, monoblasts develop into monocytes, eosinophilic myeloblasts develop into

eosinophils, and so on. Precursor cells have recognizable microscopic appearances.

Several hormones called **hemopoietic growth factors** (hē-mō-poy-ET-ik) regulate the differentiation and proliferation of particular progenitor cells. **Erythropoietin** (e-rith′-rō-POY-e-tin) or **EPO** increases the number of red blood cell precursors. EPO is produced primarily by cells in the kidneys that lie between the kidney tubules (peritubular interstitial cells). With renal failure, EPO release slows and RBC production is inadequate. This leads to a decreased hematocrit, which leads to a decreased ability to deliver oxygen to body tissues. **Thrombopoietin** (throm′-bō-POY-e-tin) or **TPO** is a hormone produced by the liver that stimulates the formation of platelets (thrombocytes) from megakaryocytes. Several different cytokines regulate development of different blood cell types. **Cytokines** (SĪ-tō-kīns) are small glycoproteins that are typically produced by cells such as red bone marrow cells, leukocytes, macrophages, fibroblasts, and endothelial cells. They generally act as local hormones (autocrines or paracrines; see Chapter 18). Cytokines stimulate proliferation of progenitor cells in red bone marrow and regulate the activities of cells involved in nonspecific defenses (such as phagocytes) and immune responses (such as B cells and T cells). Two important families of cytokines that stimulate white blood cell formation are **colony-stimulating factors (CSFs)** and **interleukins** (in′-ter-LOO-kins).

⚕ CLINICAL CONNECTION | *Medical Uses of Hemopoietic Growth Factors*

Hemopoietic growth factors made available through recombinant DNA technology hold tremendous potential for medical uses when a person's natural ability to form new blood cells is diminished or defective. The artificial form of erythropoietin (epoetin alfa) is very effective in treating the diminished red blood cell production that accompanies end-stage kidney disease. Granulocyte–macrophage colony-stimulating factor and granulocyte CSF are given to stimulate white blood cell formation in cancer patients who are undergoing chemotherapy, which kills red bone marrow cells as well as cancer cells because both cell types are undergoing mitosis. (Recall that white blood cells help protect against disease.) Thrombopoietin shows great promise for preventing the depletion of platelets, which are needed to help blood clot, during chemotherapy. CSFs and thrombopoietin also improve the outcome of patients who receive bone marrow transplants. Hemopoietic growth factors are also used to treat thrombocytopenia in neonates, other clotting disorders, and various types of anemia. Research on these medications is ongoing and shows a great deal of promise. •

✔ CHECKPOINT

7. Which hemopoietic growth factors regulate differentiation and proliferation of CFU–E and formation of platelets from megakaryocytes?
8. Describe the formation of platelets from pluripotent stem cells, including the influence of hormones.

19.3 RED BLOOD CELLS

⦿ OBJECTIVE

• Describe the structure, functions, life cycle, and production of red blood cells.

Red blood cells (RBCs) or **erythrocytes** (e-RITH-rō-sīts; *erythro-* = red; *-cyte* = cell) contain the oxygen-carrying protein **hemoglobin,** which is a pigment that gives whole blood its red color. A healthy adult male has about 5.4 million red blood cells per microliter (μL) of blood,* and a healthy adult female has about 4.8 million. (One drop of blood is about 50 μL.) To maintain normal numbers of RBCs, new mature cells must enter the circulation at the astonishing rate of at least 2 million per second, a pace that balances the equally high rate of RBC destruction.

RBC Anatomy

RBCs are biconcave discs with a diameter of 7–8 μm (Figure 19.4a). Recall that 1 μm = 1/25,000 of an inch or 1/10,000 of a centimeter or 1/1000 of a millimeter. Mature red blood cells have a simple structure. Their plasma membrane is both strong and flexible, which allows them to deform without rupturing as they squeeze through narrow capillaries. As you will see later, certain glycolipids in the plasma membrane of RBCs are antigens that account for the various blood groups such as the ABO and Rh groups. RBCs lack a nucleus and other organelles and can neither reproduce nor carry on extensive metabolic activities. The cytosol of RBCs contains hemoglobin molecules; these important molecules are synthesized before loss of the nucleus during RBC production and constitute about 33% of the cell's weight.

*1 μL = 1 mm^3 = 10^{-6} liter.

RBC Physiology

Red blood cells are highly specialized for their oxygen transport function. Because mature RBCs have no nucleus, all their internal space is available for oxygen transport. Because RBCs lack mitochondria and generate ATP anaerobically (without oxygen), they do not use up any of the oxygen they transport. Even the shape of an RBC facilitates its function. A biconcave disc has a much greater surface area for the diffusion of gas molecules into and out of the RBC than would, say, a sphere or a cube.

Each RBC contains about 280 million hemoglobin molecules. A hemoglobin molecule consists of a protein called **globin,** composed of four polypeptide chains (two alpha and two beta chains); a ring-like nonprotein pigment called a **heme** (Figure 19.4b) is bound to each of the four chains. At the center of each heme ring is an iron ion (Fe^{2+}) that can combine reversibly with one oxygen molecule (Figure 19.4c), allowing each hemoglobin molecule to bind four oxygen molecules. Each oxygen molecule picked up from the lungs is bound to an iron ion. As blood flows through tissue capillaries, the iron–oxygen reaction reverses. Hemoglobin releases oxygen, which diffuses first into the interstitial fluid and then into cells.

Hemoglobin also transports about 23% of the total carbon dioxide, a waste product of metabolism. (The remaining carbon dioxide is dissolved in plasma or carried as bicarbonate ions.) Blood flowing through tissue capillaries picks up carbon dioxide, some of which combines with amino acids in the globin part of hemoglobin. As blood flows through the lungs, the carbon dioxide is released from hemoglobin and then exhaled.

Figure 19.4 **The shapes of a red blood cell (RBC) and a hemoglobin molecule.** In (b), note that each of the four polypeptide chains of a hemoglobin molecule (blue) has one heme group (gold), which contains an iron ion (Fe^{2+}), shown in red.

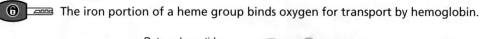

🔑 The iron portion of a heme group binds oxygen for transport by hemoglobin.

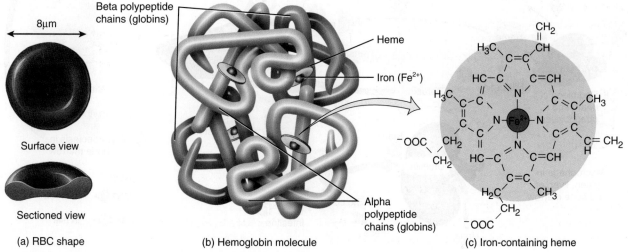

(a) RBC shape   (b) Hemoglobin molecule   (c) Iron-containing heme

❓ **How many molecules of O$_2$ can one hemoglobin molecule transport?**

In addition to its key role in transporting oxygen and carbon dioxide, hemoglobin also plays a role in the regulation of blood flow and blood pressure. The gaseous hormone **nitric oxide (NO),** produced by the endothelial cells that line blood vessels, binds to hemoglobin. Under some circumstances, hemoglobin releases NO. The released NO causes *vasodilation,* an increase in blood vessel diameter that occurs when the smooth muscle in the vessel wall relaxes. Vasodilation improves blood flow and enhances oxygen delivery to cells near the site of NO release.

Red blood cells also contain the enzyme carbonic anhydrase (CA), which catalyzes the conversion of carbon dioxide and water to carbonic acid, which in turn dissociates into H^+ and HCO_3^-. The entire reaction is reversible and is summarized as follows:

$$CO_2 + H_2O \underset{}{\overset{CA}{\rightleftharpoons}} H_2CO_3 \rightleftharpoons H^+ + HCO_3^-$$

| Carbon dioxide | Water | Carbonic acid | Hydrogen ion | Bicarbonate ion |

This reaction is significant for two reasons: (1) It allows about 70% of CO_2 to be transported in blood plasma from tissue cells to the lungs in the form of HCO_3^- (see Chapter 23). (2) It also serves as an important buffer in extracellular fluid (see Chapter 27).

RBC Life Cycle

Red blood cells live only about 120 days because of the wear and tear their plasma membranes undergo as they squeeze through blood capillaries. Without a nucleus and other organelles, RBCs cannot synthesize new components to replace damaged ones. The plasma membrane becomes more fragile with age, and the cells are more likely to burst, especially as they squeeze through narrow channels in the spleen. Ruptured red blood cells are removed from circulation and destroyed by fixed phagocytic macrophages in the spleen and liver, and the breakdown products are recycled and used in numerous metabolic processes, including the formation of new red blood cells. The recycling occurs as follows (Figure 19.5):

1 Macrophages in the spleen, liver, or red bone marrow phagocytize ruptured and worn-out red blood cells.

2 The globin and heme portions of hemoglobin are split apart.

3 Globin is broken down into amino acids, which can be reused to synthesize other proteins.

4 Iron is removed from the heme portion in the form of Fe^{3+}, which associates with the plasma protein **transferrin**

Figure 19.5 Formation and destruction of red blood cells, and the recycling of hemoglobin components. RBCs circulate for about 120 days after leaving red bone marrow before they are phagocytized by macrophages.

The rate of RBC formation by red bone marrow equals the rate of RBC destruction by macrophages.

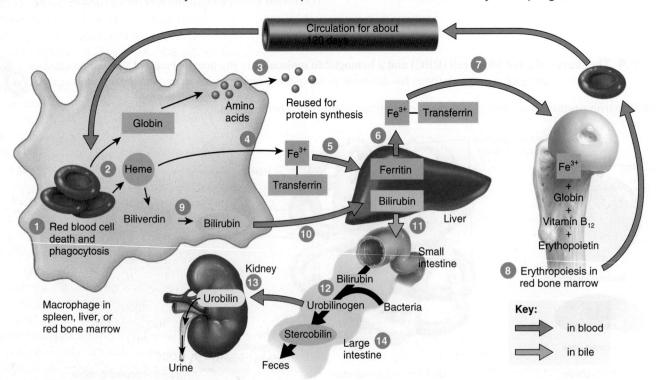

? **What is the function of transferrin?**

(trans-FER-in; *trans-* = across; *-ferr-* = iron), a transporter for Fe^{3+} in the bloodstream.

5 In muscle fibers, liver cells, and macrophages of the spleen and liver, Fe^{3+} detaches from transferrin and attaches to an iron-storage protein called **ferritin** (FER-i-tin).

6 On release from a storage site or absorption from the gastrointestinal tract, Fe^{3+} reattaches to transferrin.

7 The Fe^{3+}–transferrin complex is then carried to red bone marrow, where RBC precursor cells take it up through receptor-mediated endocytosis (see Figure 3.12) for use in hemoglobin synthesis. Iron is needed for the heme portion of the hemoglobin molecule, and amino acids are needed for the globin portion. Vitamin B_{12} is also needed for the synthesis of hemoglobin.

8 Erythropoiesis in red bone marrow results in the production of red blood cells, which enter the circulation.

9 When iron is removed from heme, the non-iron portion of heme is converted to **biliverdin** (bil′-i-VER-din), a green pigment, and then into **bilirubin** (bil′-i-ROO-bin), a yellow-orange pigment.

10 Bilirubin enters the blood and is transported to the liver.

11 Within the liver, bilirubin is released by liver cells into bile, which passes into the small intestine and then into the large intestine.

12 In the large intestine, bacteria convert bilirubin into **urobilinogen** (ūr-ō-bī-LIN-ō-jen).

13 Some urobilinogen is absorbed back into the blood, converted to a yellow pigment called **urobilin** (ūr-ō-BĪ-lin), and excreted in urine.

14 Most urobilinogen is eliminated in feces in the form of a brown pigment called **stercobilin** (ster′-kō-BĪ-lin), which gives feces its characteristic color.

CLINICAL CONNECTION | Iron Overload and Tissue Damage

Because free iron ions (Fe^{2+} and Fe^{3+}) bind to and damage molecules in cells or in the blood, transferrin and ferritin act as protective "protein escorts" during transport and storage of iron ions. As a result, plasma contains virtually no free iron. Furthermore, only small amounts are available inside body cells for use in synthesis of iron-containing molecules such as the cytochrome pigments needed for ATP production in mitochondria (see Figure 25.9). In cases of **iron overload,** the amount of iron present in the body builds up. Because we have no method for eliminating excess iron, any condition that increases dietary iron absorption can cause iron overload. At some point, the proteins transferrin and ferritin become saturated with iron ions, and free iron level rises. Common consequences of iron overload are diseases of the liver, heart, pancreatic islets, and gonads. Iron overload also allows certain iron-dependent microbes to flourish. Such microbes normally are not pathogenic, but they multiply rapidly and can cause lethal effects in a short time when free iron is present. •

Erythropoiesis: Production of RBCs

Erythropoiesis (e-rith′-rō-poy-Ē-sis), the production of RBCs, starts in the red bone marrow with a precursor cell called a **proerythroblast** (pro-e-RITH-ro-blast) (see Figure 19.3). The proerythroblast divides several times, producing cells that begin to synthesize hemoglobin. Ultimately, a cell near the end of the development sequence ejects its nucleus and becomes a **reticulocyte** (re-TIK-ū-lō-sīt). Loss of the nucleus causes the center of the cell to indent, producing the red blood cell's distinctive biconcave shape. Reticulocytes retain some mitochondria, ribosomes, and endoplasmic reticulum. They pass from red bone marrow into the bloodstream by squeezing between the endothelial cells of blood capillaries. Reticulocytes develop into mature red blood cells within 1 to 2 days after their release from red bone marrow.

CLINICAL CONNECTION | Reticulocyte Count

The rate of erythropoiesis is measured by a **reticulocyte count.** Normally, a little less than 1% of the oldest RBCs are replaced by newcomer reticulocytes on any given day. It then takes 1 to 2 days for the reticulocytes to lose the last vestiges of endoplasmic reticulum and become mature RBCs. Thus, reticulocytes account for about 0.5–1.5% of all RBCs in a normal blood sample. A low "retic" count in a person who is anemic might indicate a shortage of erythropoietin or an inability of the red bone marrow to respond to EPO, perhaps because of a nutritional deficiency or leukemia. A high "retic" count might indicate a good red bone marrow response to previous blood loss or to iron therapy in someone who had been iron deficient. It could also point to illegal use of epoetin alfa by an athlete. •

Normally, erythropoiesis and red blood cell destruction proceed at roughly the same pace. If the oxygen-carrying capacity of the blood falls because erythropoiesis is not keeping up with RBC destruction, a negative feedback system steps up RBC production (Figure 19.6). The controlled condition is the amount of oxygen delivered to body tissues. Cellular oxygen deficiency, called **hypoxia** (hī-POKS-ē-a), may occur if too little oxygen enters the blood. For example, the lower oxygen content of air at high altitudes reduces the amount of oxygen in the blood. Oxygen delivery may also fall due to anemia, which has many causes: Lack of iron, lack of certain amino acids, and lack of vitamin B_{12} are but a few (see Disorders: Homeostatic Imbalances at the end of this chapter). Circulatory problems that reduce blood flow to tissues may also reduce oxygen delivery. Whatever the cause, hypoxia stimulates the kidneys to step up the release of erythropoietin, which speeds the development of proerythroblasts into reticulocytes in the red bone marrow. As the number of circulating RBCs increases, more oxygen can be delivered to body tissues.

Premature newborns often exhibit anemia, due in part to inadequate production of erythropoietin. During the first weeks after birth, the liver, not the kidneys, produces most EPO. Because the liver is less sensitive than the kidneys to hypoxia, newborns have a smaller EPO response to anemia than do adults. Because fetal hemoglobin (hemoglobin present at birth) carries up to 30% more oxygen, the loss of fetal hemoglobin, due to insufficient erythropoietin production, makes the anemia worse.

Figure 19.6 Negative feedback regulation of erythropoiesis (red blood cell formation). Lower oxygen content of air at high altitudes, anemia, and circulatory problems may reduce oxygen delivery to body tissues.

🔑 The main stimulus for erythropoiesis is hypoxia, a decrease in the oxygen-carrying capacity of the blood.

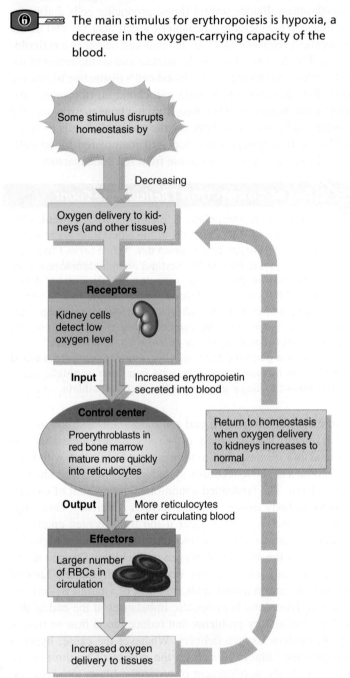

❓ **How might your hematocrit change if you moved from a town at sea level to a high mountain village?**

CLINICAL CONNECTION | *Blood Doping*

Delivery of oxygen to muscles is a limiting factor in muscular feats from weightlifting to running a marathon. As a result, increasing the oxygen-carrying capacity of the blood enhances athletic performance, especially in endurance events. Because RBCs transport oxygen, athletes have tried several means of increasing their RBC count, known as **blood doping** or **artificially induced polycythemia** (an abnormally high number of RBCs), to gain a competitive edge. Athletes have enhanced their RBC production by injecting epoetin alfa (Procrit® or Epogen®), a drug that is used to treat anemia by stimulating the production of RBCs by red bone marrow. Practices that increase the number of RBCs are dangerous because they raise the viscosity of the blood, which increases the resistance to blood flow and makes the blood more difficult for the heart to pump. Increased viscosity also contributes to high blood pressure and increased risk of stroke. During the 1980s, at least 15 competitive cyclists died from heart attacks or strokes linked to suspected use of epoetin alfa. Although the International Olympics Committee bans the use of epoetin alfa, enforcement is difficult because the drug is identical to naturally occurring erythropoietin (EPO).

So-called **natural blood doping** is seemingly the key to the success of marathon runners from Kenya. The average altitude throughout Kenya's highlands is about 2000 meters (6600 feet) above sea level; other areas of Kenya are even higher. Altitude training greatly improves fitness, endurance, and performance. At these higher altitudes, the body increases the production of red blood cells, which means that exercise greatly oxygenates the blood. When these runners compete in Boston, for example, at an altitude just above sea level, the bodies of these runners contain more erythrocytes than do the bodies of competitors who trained in Boston. A number of training camps have been established in Kenya and now attract endurance athletes from all over the world. •

✔ **CHECKPOINT**

9. Describe the size, microscopic appearance, and functions of RBCs.
10. How is hemoglobin recycled?
11. What is erythropoiesis? How does erythropoiesis affect hematocrit? What factors speed up and slow down erythropoiesis?

19.4 WHITE BLOOD CELLS

◉ **OBJECTIVE**

• Describe the structure, functions, and production of white blood cells (WBCs).

Types of WBCs

Unlike red blood cells, white blood cells or **leukocytes** (LOO-kō-sīts; *leuko-* = white) have nuclei and a full complement of other organelles but they do not contain hemoglobin (Figure 19.7). WBCs are classified as either granular or agranular, depending on whether they contain conspicuous chemical-filled cytoplasmic granules (vesicles) that are made visible by staining when viewed through a light microscope. *Granular leukocytes* include neutrophils, eosinophils, and basophils; *agranular leukocytes* include lymphocytes and monocytes. As shown in Figure 19.3, monocytes and granular leukocytes develop from myeloid stem cells. In contrast, lymphocytes develop from lymphoid stem cells.

Granular Leukocytes

After staining, each of the three types of granular leukocytes displays conspicuous granules with distinctive coloration that can be recognized under a light microscope. Granular leukocytes can be distinguished as follows:

- **Neutrophil** (NOO-trō-fil). The granules of a neutrophil are smaller than those of other granular leukocytes, evenly distributed, and pale lilac (Figure 19.7a). Because the granules do not strongly attract either the acidic (red) or basic (blue) stain, these WBCs are neutrophilic (= neutral loving). The nucleus has two to five lobes, connected by very thin strands of nuclear material. As the cells age, the number of nuclear lobes increases. Because older neutrophils thus have several differently shaped nuclear lobes, they are often called *polymorphonuclear leukocytes (PMNs),* polymorphs, or "polys."

- **Eosinophil** (ē-ō-SIN-ō-fil). The large, uniform-sized granules within an eosinophil are *eosinophilic* (= eosin-loving)—they stain red-orange with acidic dyes (Figure 19.7b). The granules usually do not cover or obscure the nucleus, which most often has two lobes connected by either a thin strand or a thick strand of nuclear material.

- **Basophil** (BĀ-sō-fil). The round, variable-sized granules of a basophil are *basophilic* (= basic loving)—they stain blue-purple with basic dyes (Figure 19.7c). The granules commonly obscure the nucleus, which has two lobes.

Agranular Leukocytes

Even though so-called agranular leukocytes possess cytoplasmic granules, the granules are not visible under a light microscope because of their small size and poor staining qualities.

- **Lymphocyte** (LIM-fō-sīt). The nucleus of a lymphocyte stains dark and is round or slightly indented (Figure 19.7d). The cytoplasm stains sky blue and forms a rim around the nucleus. The larger the cell, the more cytoplasm is visible. Lymphocytes are classified by cell diameter as large lymphocytes (10–14 μm) or small lymphocytes (6–9 μm). Although the functional significance of the size difference between small and large lymphocytes is unclear, the distinction is still clinically useful because an increase in the number of large lymphocytes has diagnostic significance in acute viral infections and in some immunodeficiency diseases.

- **Monocyte** (MON-ō-sīt). The nucleus of a monocyte is usually kidney-shaped or horseshoe-shaped, and the cytoplasm is blue-gray and has a foamy appearance (Figure 19.7e). The cytoplasm's color and appearance are due to very fine *azurophilic granules* (az'-ū-rō-FIL-ik; *azur-* = blue; *-philic* = loving), which are lysosomes. Blood is merely a conduit for monocytes, which migrate from the blood into the tissues, where they enlarge and differentiate into **macrophages** (MAK-rō-fā-jez = large eaters). Some become **fixed (tissue) macrophages,** which means they reside in a particular tissue; examples are alveolar macrophages in the lungs or macrophages in the spleen. Others become **wandering macrophages,** which roam the tissues and gather at sites of infection or inflammation.

White blood cells and all other nucleated cells in the body have proteins, called **major histocompatibility (MHC) antigens,** protruding from their plasma membrane into the extracellular fluid. These "cell identity markers" are unique for each person (except identical twins). Although RBCs possess blood group antigens, they lack the MHC antigens.

Functions of WBCs

In a healthy body, some WBCs, especially lymphocytes, can live for several months or years, but most live only a few days. During

Figure 19.7 Types of white blood cells.

The shapes of their nuclei and the staining properties of their cytoplasmic granules distinguish white blood cells from one another.

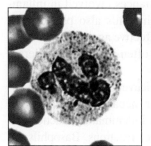 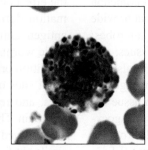 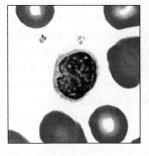 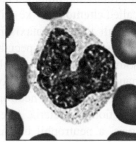

LM all 1600x

   (a) Neutrophil   (b) Eosinophil   (c) Basophil   (d) Lymphocyte   (e) Monocyte

? **Which WBCs are called granular leukocytes? Why?**

a period of infection, phagocytic WBCs may live only a few hours. WBCs are far less numerous than red blood cells; at about 5000–10,000 cells per microliter of blood, they are outnumbered by RBCs by about 700:1. **Leukocytosis** (loo′-kō-sī-TŌ-sis), an increase in the number of WBCs above 10,000/μL, is a normal, protective response to stresses such as invading microbes, strenuous exercise, anesthesia, and surgery. An abnormally low level of white blood cells (below 5000/μL) is termed **leukopenia** (loo′-kō-PĒ-nē-a). It is never beneficial and may be caused by radiation, shock, and certain chemotherapeutic agents.

The skin and mucous membranes of the body are continuously exposed to microbes and their toxins. Some of these microbes can invade deeper tissues to cause disease. Once pathogens enter the body, the general function of white blood cells is to combat them by phagocytosis or immune responses. To accomplish these tasks, many WBCs leave the bloodstream and collect at sites of pathogen invasion or inflammation. Once granular leukocytes and monocytes leave the bloodstream to fight injury or infection, they never return to it. Lymphocytes, on the other hand, continually recirculate—from blood to interstitial spaces of tissues to lymphatic fluid and back to blood. Only 2% of the total lymphocyte population is circulating in the blood at any given time; the rest is in lymphatic fluid and organs such as the skin, lungs, lymph nodes, and spleen.

RBCs are contained within the bloodstream, but WBCs leave the bloodstream by a process termed **emigration** (em′-i-GRĀ-shun; *e-* = out; *-migra-* = wander), also called *diapedesis* (dī-a-pe-DĒ-sis), in which they roll along the endothelium, stick to it, and then squeeze between endothelial cells (Figure 19.8). The precise signals that stimulate emigration through a particular blood vessel vary for the different types of WBCs. Molecules known as **adhesion molecules** help WBCs stick to the endothelium. For example, endothelial cells display adhesion molecules called *selectins* in response to nearby injury and inflammation. Selectins stick to carbohydrates on the surface of neutrophils, causing them to slow down and roll along the endothelial surface. On the neutrophil surface are other adhesion molecules called *integrins,* which tether neutrophils to the endothelium and assist their movement through the blood vessel wall and into the interstitial fluid of the injured tissue.

Neutrophils and macrophages are active in **phagocytosis** (fag′-ō-sī-TŌ-sis); they can ingest bacteria and dispose of dead matter (see Figure 3.13). Several different chemicals released by microbes and inflamed tissues attract phagocytes, a phenomenon called **chemotaxis** (kē-mō-TAK-sis). The substances that provide stimuli for chemotaxis include toxins produced by microbes; kinins, which are specialized products of damaged tissues; and some of the colony-stimulating factors (CSFs). The CSFs also enhance the phagocytic activity of neutrophils and macrophages.

Among WBCs, neutrophils respond most quickly to tissue destruction by bacteria. After engulfing a pathogen during phagocytosis, a neutrophil unleashes several chemicals to destroy the pathogen. These chemicals include the enzyme **lysozyme** (LĪ-sō-zīm′), which destroys certain bacteria, and **strong oxidants,** such as the superoxide anion (O_2^-), hydrogen peroxide (H_2O_2), and the hypochlorite anion (OCl^-), which is similar to household bleach.

Figure 19.8 Emigration of white blood cells.

Adhesion molecules (selectins and integrins) assist the emigration of WBCs from the bloodstream into interstitial fluid.

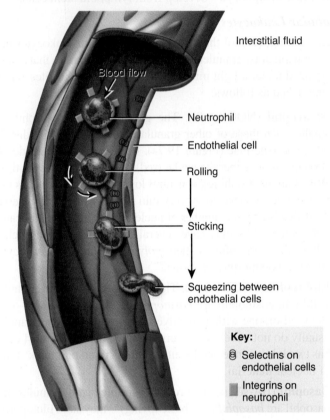

Key:

⑧ Selectins on endothelial cells

▪ Integrins on neutrophil

? In what way is the "traffic pattern" of lymphocytes in the body different from that of other WBCs?

Neutrophils also contain **defensins,** proteins that exhibit a broad range of antibiotic activity against bacteria and fungi. Within a neutrophil, vesicles containing defensins merge with phagosomes containing microbes. Defensins form peptide "spears" that poke holes in microbe membranes; the resulting loss of cellular contents kills the invader.

Eosinophils leave the capillaries and enter tissue fluid. They are believed to release enzymes, such as histaminase, that combat the effects of histamine and other substances involved in inflammation during allergic reactions. Eosinophils also phagocytize antigen–antibody complexes and are effective against certain parasitic worms. A high eosinophil count often indicates an allergic condition or a parasitic infection.

At sites of inflammation, basophils leave capillaries, enter tissues, and release granules that contain heparin, histamine, and serotonin. These substances intensify the inflammatory reaction and are involved in hypersensitivity (allergic) reactions. Basophils are similar in function to mast cells, connective tissue cells that originate from pluripotent stem cells in red bone marrow. Like basophils, mast cells release substances involved in inflammation, including heparin, histamine, and proteases. Mast cells are widely dispersed in the body, particularly in connective tissues of the skin and mucous membranes of the respiratory and gastrointestinal tracts.

Lymphocytes are the major soldiers in immune system battles (described in detail in Chapter 22). Most lymphocytes continually move among lymphoid tissues, lymph, and blood, spending only a few hours at a time in blood. Thus, only a small proportion of the total lymphocytes are present in the blood at any given time. Three main types of lymphocytes are B cells, T cells, and natural killer (NK) cells. B cells are particularly effective in destroying bacteria and inactivating their toxins. T cells attack viruses, fungi, transplanted cells, cancer cells, and some bacteria, and are responsible for transfusion reactions, allergies, and the rejection of transplanted organs. Immune responses carried out by both B cells and T cells help combat infection and provide protection against some diseases. Natural killer cells attack a wide variety of infectious microbes and certain spontaneously arising tumor cells.

Monocytes take longer to reach a site of infection than neutrophils, but they arrive in larger numbers and destroy more microbes. On their arrival, monocytes enlarge and differentiate into wandering macrophages, which clean up cellular debris and microbes by phagocytosis after an infection.

As you have already learned, an increase in the number of circulating WBCs usually indicates inflammation or infection. A physician may order a **differential white blood cell count,** a count of each of the five types of white blood cells, to detect infection or inflammation, determine the effects of possible poisoning by chemicals or drugs, monitor blood disorders (for example, leukemia) and the effects of chemotherapy, or detect allergic reactions and parasitic infections. Because each type of white blood cell plays a different role, determining the *percentage* of each type in the blood assists in diagnosing the condition. Table 19.2 lists the significance of both high and low WBC counts.

✔CHECKPOINT

12. What is the importance of emigration, chemotaxis, and phagocytosis in fighting bacterial invaders?
13. How are leukocytosis and leukopenia different?
14. What is a differential white blood cell count?
15. What functions do granular leukocytes, macrophages, B cells, T cells, and natural killer cells perform?

19.5 PLATELETS

◉ OBJECTIVE

• Describe the structure, function, and origin of platelets.

Besides the immature cell types that develop into erythrocytes and leukocytes, hemopoietic stem cells also differentiate into cells that produce platelets. Under the influence of the hormone **thrombopoietin** (throm'-bo-POY-e-tin), myeloid stem cells develop into megakaryocyte colony-forming cells that in turn develop into precursor cells called *megakaryoblasts* (see Figure 19.3). Megakaryoblasts transform into megakaryocytes, huge cells that splinter into 2000 to 3000 fragments. Each fragment,

TABLE 19.2

Significance of High and Low White Blood Cell Counts

WBC TYPE	HIGH COUNT MAY INDICATE	LOW COUNT MAY INDICATE
Neutrophils	Bacterial infection, burns, stress, inflammation.	Radiation exposure, drug toxicity, vitamin B_{12} deficiency, systemic lupus erythematosus (SLE).
Lymphocytes	Viral infections, some leukemias.	Prolonged illness, immunosuppression, treatment with cortisol.
Monocytes	Viral or fungal infections, tuberculosis, some leukemias, other chronic diseases.	Bone marrow suppression, treatment with cortisol.
Eosinophils	Allergic reactions, parasitic infections, autoimmune diseases.	Drug toxicity, stress.
Basophils	Allergic reactions, leukemias, cancers, hypothyroidism.	Pregnancy, ovulation, stress, hyperthyroidism.

enclosed by a piece of the plasma membrane, is a **platelet (thrombocyte).** Platelets break off from the megakaryocytes in red bone marrow and then enter the blood circulation. Between 150,000 and 400,000 platelets are present in each microliter of blood. Each is irregularly disc-shaped, 2–4 μm in diameter, and has many vesicles but no nucleus.

Their granules contain chemicals that, once released, promote blood clotting. Platelets help stop blood loss from damaged blood vessels by forming a platelet plug. Platelets have a short life span, normally just 5 to 9 days. Aged and dead platelets are removed by fixed macrophages in the spleen and liver.

Table 19.3 summarizes the formed elements in blood.

⚕ **CLINICAL CONNECTION | Complete Blood Count**

A **complete blood count (CBC)** is a very valuable test that screens for anemia and various infections. Usually included are counts of RBCs, WBCs, and platelets per microliter of whole blood; hematocrit; and differential white blood cell count. The amount of hemoglobin in grams per milliliter of blood also is determined. Normal hemoglobin ranges are as follows: infants, 14–20 g/100 mL of blood; adult females, 12–16 g/100 mL of blood; and adult males, 13.5–18 g/100 mL of blood. •

✔CHECKPOINT

16. How do RBCs, WBCs, and platelets compare with respect to size, number per microliter of blood, and life span?

TABLE 19.3

Summary of Formed Elements in Blood

NAME AND APPEARANCE	NUMBER	CHARACTERISTICS*	FUNCTIONS
Red Blood Cells (RBCs) or Erythrocytes	4.8 million/μL in females; 5.4 million/μL in males.	7–8 μm diameter, biconcave discs, without nuclei; live for about 120 days.	Hemoglobin within RBCs transports most oxygen and part of carbon dioxide in blood.
White Blood Cells (WBCs) or Leukocytes	5000–10,000/μL.	Most live for a few hours to a few days.[†]	Combat pathogens and other foreign substances that enter body.
Granular leukocytes			
Neutrophils	60–70% of all WBCs.	10–12 μm diameter; nucleus has 2–5 lobes connected by thin strands of chromatin; cytoplasm has very fine, pale lilac granules.	Phagocytosis. Destruction of bacteria with lysozyme, defensins, and strong oxidants, such as superoxide anion, hydrogen peroxide, and hypochlorite anion.
Eosinophils	2–4% of all WBCs.	10–12 μm diameter; nucleus usually has 2 lobes connected by thick strand of chromatin; large, red-orange granules fill cytoplasm.	Combat effects of histamine in allergic reactions, phagocytize antigen–antibody complexes, and destroy certain parasitic worms.
Basophils	0.5–1% of all WBCs.	8–10 μm diameter; nucleus has 2 lobes; large cytoplasmic granules appear deep blue-purple.	Liberate heparin, histamine, and serotonin in allergic reactions that intensify overall inflammatory response.
Agranular leukocytes			
Lymphocytes (T cells, B cells, and natural killer cells)	20–25% of all WBCs.	Small lymphocytes are 6–9 μm in diameter; large lymphocytes are 10–14 μm in diameter; nucleus is round or slightly indented; cytoplasm forms rim around nucleus that looks sky blue; the larger the cell, the more cytoplasm is visible.	Mediate immune responses, including antigen–antibody reactions. B cells develop into plasma cells, which secrete antibodies. T cells attack invading viruses, cancer cells, and transplanted tissue cells. Natural killer cells attack wide variety of infectious microbes and certain spontaneously arising tumor cells.
Monocytes	3–8% of all WBCs.	12–20 μm diameter; nucleus is kidney or horseshoe shaped; cytoplasm is blue-gray and appears foamy.	Phagocytosis (after transforming into fixed or wandering macrophages).
Platelets (thrombocytes)	150,000–400,000/μL.	2–4 μm diameter cell fragments that live for 5–9 days; contain many vesicles but no nucleus.	Form platelet plug in hemostasis; release chemicals that promote vascular spasm and blood clotting.

*Colors are those seen when using Wright's stain.
[†]Some lymphocytes, called T and B memory cells, can live for many years once they are established.

19.6 STEM CELL TRANSPLANTS FROM BONE MARROW AND CORD BLOOD

◉ **OBJECTIVE**

- Explain the importance of bone marrow transplants and stem cell transplants.

A **bone marrow transplant** is the replacement of cancerous or abnormal red bone marrow with healthy red bone marrow in order to establish normal blood cell counts. In patients with cancer or certain genetic diseases, the defective red bone marrow is destroyed by high doses of chemotherapy and whole body radiation just before the transplant takes place. These treatments kill the cancer cells and destroy the patient's immune system in order to decrease the chance of transplant rejection.

Healthy red bone marrow for transplanting may be supplied by a donor or by the patient when the underlying disease is inactive, as when leukemia is in remission. The red bone marrow from a donor is usually removed from the iliac crest of the hip bone under general anesthesia with a syringe and is then injected into the recipient's vein, much like a blood transfusion. The injected marrow migrates to the recipient's red bone marrow cavities, where the donor's stem cells multiply. If all goes well, the recipient's red bone marrow is replaced entirely by healthy, noncancerous cells.

Bone marrow transplants have been used to treat aplastic anemia, certain types of leukemia, severe combined immunodeficiency disease (SCID), Hodgkin's disease, non-Hodgkin's lymphoma, multiple myeloma, thalassemia, sickle-cell disease, breast cancer, ovarian cancer, testicular cancer, and hemolytic anemia. However, there are some drawbacks. Since the recipient's white blood cells have been completely destroyed by chemotherapy and radiation, the patient is extremely vulnerable to infection. (It takes about 2–3 weeks for transplanted bone marrow to produce enough white blood cells to protect against infection.) In addition, transplanted red bone marrow may produce T cells that attack the recipient's tissues, a reaction called *graft-versus-host disease.* Similarly, any of the recipient's T cells that survived the chemotherapy and radiation can attack donor transplant cells. Another drawback is that patients must take immunosuppressive drugs for life. Because these drugs reduce the level of immune system activity, they increase the risk of infection. Immunosuppressive drugs also have side effects such as fever, muscle aches, headache, nausea, fatigue, depression, high blood pressure, and kidney and liver damage.

A more recent advance for obtaining stem cells involves a **cord-blood transplant.** The connection between the mother and embryo (and later the fetus) is the umbilical cord. Stem cells may be obtained from the umbilical cord shortly after birth. The stem cells are removed from the cord with a syringe and then frozen. Stem cells from the cord have several advantages over those obtained from red bone marrow:

1. They are easily collected following permission of the newborn's parents.

2. They are more abundant than stem cells in red bone marrow.

3. They are less likely to cause graft-versus-host disease, so the match between donor and recipient does not have to be as close as in a bone marrow transplant. This provides a larger number of potential donors.

4. They are less likely to transmit infections.

5. They can be stored indefinitely in cord-blood banks.

✔ **CHECKPOINT**

17. How are cord-blood transplants and bone marrow transplants similar? How do they differ?

19.7 HEMOSTASIS

◉ **OBJECTIVES**

- Describe the three mechanisms that contribute to hemostasis.
- Identify the stages of blood clotting and explain the various factors that promote and inhibit blood clotting.

Hemostasis (hē-mō-STĀ-sis), not to be confused with the very similar term homeostasis, is a sequence of responses that stops bleeding. When blood vessels are damaged or ruptured, the hemostatic response must be quick, localized to the region of damage, and carefully controlled in order to be effective. Three mechanisms reduce blood loss: (1) vascular spasm, (2) platelet plug formation, and (3) blood clotting (coagulation). When successful, hemostasis prevents **hemorrhage** (HEM-o-rij; *-rhage* = burst forth), the loss of a large amount of blood from the vessels. Hemostatic mechanisms can prevent hemorrhage from smaller blood vessels, but extensive hemorrhage from larger vessels usually requires medical intervention.

Vascular Spasm

When arteries or arterioles are damaged, the circularly arranged smooth muscle in their walls contracts immediately, a reaction called **vascular spasm.** This reduces blood loss for several minutes to several hours, during which time the other hemostatic mechanisms go into operation. The spasm is probably caused by damage to the smooth muscle, by substances released from activated platelets, and by reflexes initiated by pain receptors.

Platelet Plug Formation

Considering their small size, platelets store an impressive array of chemicals. Within many vesicles are clotting factors, ADP, ATP, Ca^{2+}, and serotonin. Also present are enzymes that produce thromboxane A2, a prostaglandin; *fibrin-stabilizing factor,* which helps to strengthen a blood clot; lysosomes; some mitochondria; membrane systems that take up and store calcium and provide channels for release of the contents of granules; and glycogen. Also within platelets is **platelet-derived growth factor (PDGF),** a hormone that can cause proliferation of vascular endothelial cells, vascular smooth muscle fibers, and fibroblasts to help repair damaged blood vessel walls.

Platelet plug formation occurs as follows (Figure 19.9):

1 Initially, platelets contact and stick to parts of a damaged blood vessel, such as collagen fibers of the connective tissue

Figure 19.9 Platelet plug formation.

A platelet plug can stop blood loss completely if the hole in a blood vessel is small enough.

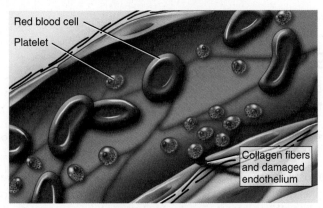

1 Platelet adhesion

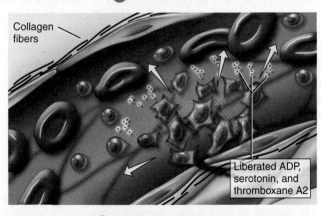

2 Platelet release reaction

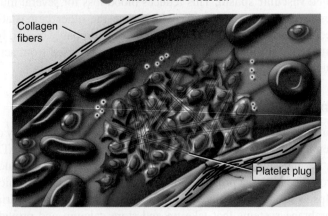

3 Platelet aggregation

? **Along with platelet plug formation, which two mechanisms contribute to hemostasis?**

underlying the damaged endothelial cells. This process is called **platelet adhesion.**

2 Due to adhesion, the platelets become activated, and their characteristics change dramatically. They extend many projections that enable them to contact and interact with one another, and they begin to liberate the contents of their vesicles. This phase is called the **platelet release reaction.** Liberated ADP and thromboxane A2 play a major role by activating nearby platelets. Serotonin and thromboxane A2 function as vasoconstrictors, causing and sustaining contraction of vascular smooth muscle, which decreases blood flow through the injured vessel.

3 The release of ADP makes other platelets in the area sticky, and the stickiness of the newly recruited and activated platelets causes them to adhere to the originally activated platelets. This gathering of platelets is called **platelet aggregation.** Eventually, the accumulation and attachment of large numbers of platelets form a mass called a **platelet plug.**

A platelet plug is very effective in preventing blood loss in a small vessel. Although initially the platelet plug is loose, it becomes quite tight when reinforced by fibrin threads formed during clotting (see Figure 19.10). A platelet plug can stop blood loss completely if the hole in a blood vessel is not too large.

Blood Clotting

Normally, blood remains in its liquid form as long as it stays within its vessels. If it is drawn from the body, however, it thickens and forms a gel. Eventually, the gel separates from the liquid. The straw-colored liquid, called **serum,** is simply blood plasma minus the clotting proteins. The gel is called a **clot.** It consists of a network of insoluble protein fibers called fibrin in which the formed elements of blood are trapped (Figure 19.10).

The process of gel formation, called **clotting** or **coagulation** (kō-ag-u-LĀ-shun), is a series of chemical reactions that culminates in formation of fibrin threads. If blood clots too easily, the result can be **thrombosis**—clotting in an undamaged blood vessel. If the blood takes too long to clot, hemorrhage can occur.

Clotting involves several substances known as **clotting (coagulation) factors.** These factors include calcium ions (Ca^{2+}), several inactive enzymes that are synthesized by hepatocytes (liver cells) and released into the bloodstream, and various molecules associated with platelets or released by damaged tissues. Most clotting factors are identified by Roman numerals that indicate the order of their discovery (not necessarily the order of their participation in the clotting process).

Clotting is a complex cascade of enzymatic reactions in which each clotting factor activates many molecules of the next one in a fixed sequence. Finally, a large quantity of product (the insoluble protein fibrin) is formed. Clotting can be divided into three stages (Figure 19.11):

1 Two pathways, called the extrinsic pathway and the intrinsic pathway (Figures 19.11a, b), which will be described shortly, lead to the formation of prothrombinase. Once prothrombi-

Figure 19.10 **Blood clot formation.** Notice the platelet and red blood cells entrapped in fibrin threads.

A blood clot is a gel that contains formed elements of the blood entangled in fibrin threads.

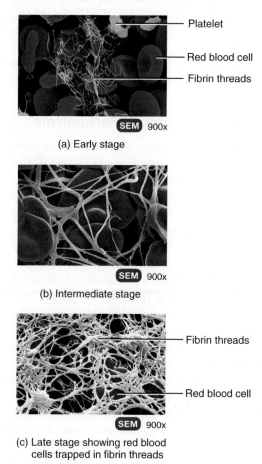

(a) Early stage

SEM 900x

(b) Intermediate stage

SEM 900x

(c) Late stage showing red blood cells trapped in fibrin threads

SEM 900x

Platelet — Red blood cell — Fibrin threads

Fibrin threads — Red blood cell

? **What is serum?**

nase is formed, the steps involved in the next two stages of clotting are the same for both the extrinsic and intrinsic pathways, and together these two stages are referred to as the common pathway.

2 Prothrombinase converts prothrombin (a plasma protein formed by the liver) into the enzyme thrombin.

3 Thrombin converts soluble fibrinogen (another plasma protein formed by the liver) into insoluble fibrin. Fibrin forms the threads of the clot.

The Extrinsic Pathway

The **extrinsic pathway** of blood clotting has fewer steps than the intrinsic pathway and occurs rapidly—within a matter of seconds if trauma is severe. It is so named because a tissue protein called **tissue factor (TF),** also known as **thromboplastin** (throm′-bō-PLAS-tin), leaks into the blood from cells *outside (extrinsic to)* blood vessels and initiates the formation of prothrombinase. TF is a complex mixture of lipoproteins and phospholipids released from the surfaces of damaged cells. In the presence of Ca^{2+}, TF

Figure 19.11 **The blood-clotting cascade.**

In blood clotting, coagulation factors are activated in sequence, resulting in a cascade of reactions that includes positive feedback cycles.

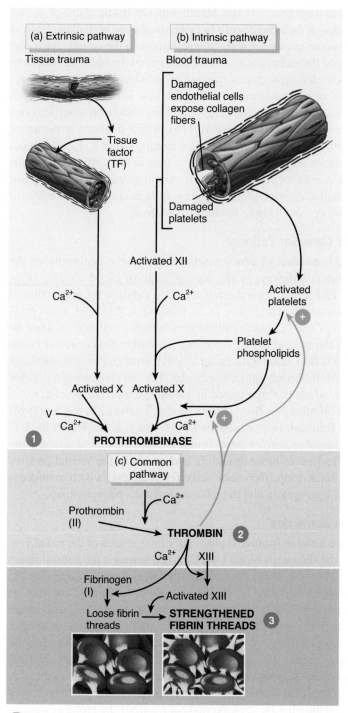

? **What is the outcome of the first stage of blood clotting?**

begins a sequence of reactions that ultimately activates clotting factor X (Figure 19.11a). Once factor X is activated, it combines with factor V in the presence of Ca^{2+} to form the active enzyme prothrombinase, completing the extrinsic pathway.

The Intrinsic Pathway

The **intrinsic pathway** of blood clotting is more complex than the extrinsic pathway, and it occurs more slowly, usually requiring several minutes. The intrinsic pathway is so named because its activators are either in direct contact with blood or contained *within (intrinsic to)* the blood; outside tissue damage is not needed. If endothelial cells become roughened or damaged, blood can come in contact with collagen fibers in the connective tissue around the endothelium of the blood vessel. In addition, trauma to endothelial cells causes damage to platelets, resulting in the release of phospholipids by the platelets. Contact with collagen fibers (or with the glass sides of a blood collection tube) activates clotting factor XII (Figure 19.11b), which begins a sequence of reactions that eventually activates clotting factor X. Platelet phospholipids and Ca^{2+} can also participate in the activation of factor X. Once factor X is activated, it combines with factor V to form the active enzyme prothrombinase (just as occurs in the extrinsic pathway), completing the intrinsic pathway.

The Common Pathway

The formation of prothrombinase marks the beginning of the common pathway. In the second stage of blood clotting (Figure 19.11c), prothrombinase and Ca^{2+} catalyze the conversion of prothrombin to thrombin. In the third stage, thrombin, in the presence of Ca^{2+}, converts fibrinogen, which is soluble, to loose fibrin threads, which are insoluble. Thrombin also activates factor XIII (fibrin stabilizing factor), which strengthens and stabilizes the fibrin threads into a sturdy clot. Plasma contains some factor XIII, which is also released by platelets trapped in the clot.

Thrombin has two positive feedback effects. In the first positive feedback loop, which involves factor V, it accelerates the formation of prothrombinase. Prothrombinase in turn accelerates the production of more thrombin, and so on. In the second positive feedback loop, thrombin activates platelets, which reinforces their aggregation and the release of platelet phospholipids.

Clot Retraction

Once a clot is formed, it plugs the ruptured area of the blood vessel and thus stops blood loss. **Clot retraction** is the consolidation or tightening of the fibrin clot. The fibrin threads attached to the damaged surfaces of the blood vessel gradually contract as platelets pull on them. As the clot retracts, it pulls the edges of the damaged vessel closer together, decreasing the risk of further damage. During retraction, some serum can escape between the fibrin threads, but the formed elements in blood cannot. Normal retraction depends on an adequate number of platelets in the clot, which release factor XIII and other factors, thereby strengthening and stabilizing the clot. Permanent repair of the blood vessel can then take place. In time, fibroblasts form connective tissue in the ruptured area, and new endothelial cells repair the vessel lining.

Role of Vitamin K in Clotting

Normal clotting depends on adequate levels of vitamin K in the body. Although vitamin K is not involved in actual clot formation, it is required for the synthesis of four clotting factors. Normally produced by bacteria that inhabit the large intestine, vitamin K is a fat-soluble vitamin that can be absorbed through the lining of the intestine and into the blood if absorption of lipids is normal. People suffering from disorders that slow absorption of lipids (for example, inadequate release of bile into the small intestine) often experience uncontrolled bleeding as a consequence of vitamin K deficiency.

The various clotting factors, their sources, and the pathways of activation are summarized in Table 19.4.

Hemostatic Control Mechanisms

Many times a day little clots start to form, often at a site of minor roughness or at a developing atherosclerotic plaque inside a blood vessel. Because blood clotting involves amplification and positive feedback cycles, a clot has a tendency to enlarge, creating the potential for impairment of blood flow through undamaged vessels. The **fibrinolytic system** (fī-bri-nō-LIT-ik) dissolves small, inappropriate clots; it also dissolves clots at a site of damage once the damage is repaired. Dissolution of a clot is called **fibrinolysis** (fī-bri-NOL-i-sis). When a clot is formed, an inactive plasma enzyme called **plasminogen** (plaz-MIN-o-jen) is incorporated into the clot. Both body tissues and blood contain substances that can activate plasminogen to **plasmin** or **fibrinolysin** (fī-brin-ō-LĪ-sin), an active plasma enzyme. Among these substances are thrombin, activated factor XII, and tissue plasminogen activator (t-PA), which is synthesized in endothelial cells of most tissues and liberated into the blood. Once plasmin is formed, it can dissolve the clot by digesting fibrin threads and inactivating substances such as fibrinogen, prothrombin, and factors V and XII.

Even though thrombin has a positive feedback effect on blood clotting, clot formation normally remains localized at the site of damage. A clot does not extend beyond a wound site into the general circulation, in part because fibrin absorbs thrombin into the clot. Another reason for localized clot formation is that because of the dispersal of some of the clotting factors by the blood, their concentrations are not high enough to bring about widespread clotting.

Several other mechanisms also control blood clotting. For example, endothelial cells and white blood cells produce a prostaglandin called **prostacyclin** (pros-ta-SĪ-klin) that opposes the actions of thromboxane A2. Prostacyclin is a powerful inhibitor of platelet adhesion and release.

In addition, substances that delay, suppress, or prevent blood clotting, called **anticoagulants** (an′-tī-kō-AG-ū-lants), are present in blood. These include **antithrombin,** which blocks the action of several factors, including XII, X, and II (prothrombin). **Heparin,** an anticoagulant that is produced by mast cells and basophils, combines with antithrombin and increases its effectiveness in blocking thrombin. Another anticoagulant, **activated protein C (APC),** inactivates the two major clotting factors not blocked by antithrombin and enhances activity of plasminogen activators. Babies that lack the ability to produce APC due to a genetic mutation usually die of blood clots in infancy.

TABLE 19.4

Clotting (Coagulation) Factors

NUMBER*	NAME(S)	SOURCE	PATHWAY(S) OF ACTIVATION
I	Fibrinogen.	Liver.	Common.
II	Prothrombin.	Liver.	Common.
III	Tissue factor (thromboplastin).	Damaged tissues and activated platelets.	Extrinsic.
IV	Calcium ions (Ca^{2+}).	Diet, bones, and platelets.	All.
V	Proaccelerin, labile factor, or accelerator globulin (AcG).	Liver and platelets.	Extrinsic and intrinsic.
VII	Serum prothrombin conversion accelerator (SPCA), stable factor, or proconvertin.	Liver.	Extrinsic.
VIII	Antihemophilic factor (AHF), antihemophilic factor A, or antihemophilic globulin (AHG).	Liver.	Intrinsic.
IX	Christmas factor, plasma thromboplastin component (PTC), or antihemophilic factor B.	Liver.	Intrinsic.
X	Stuart factor, Prower factor, or thrombokinase.	Liver.	Extrinsic and intrinsic.
XI	Plasma thromboplastin antecedent (PTA) or antihemophilic factor C.	Liver.	Intrinsic.
XII	Hageman factor, glass factor, contact factor, or antihemophilic factor D.	Liver.	Intrinsic.
XIII	Fibrin-stabilizing factor (FSF).	Liver and platelets.	Common.

*There is no factor VI. Prothrombinase (prothrombin activator) is a combination of activated factors V and X.

CLINICAL CONNECTION | *Anticoagulants*

Patients who are at increased risk of forming blood clots may receive anticoagulants. Examples are heparin or warfarin. Heparin is often administered during hemodialysis and open-heart surgery. **Warfarin (Coumadin®)** acts as an antagonist to vitamin K and thus blocks synthesis of four clotting factors. Warfarin is slower acting than heparin. To prevent clotting in donated blood, blood banks and laboratories often add substances that remove Ca^{2+}; examples are EDTA (ethylenediaminetetraacetic acid) and CPD (citrate phosphate dextrose). •

Intravascular Clotting

Despite the anticoagulating and fibrinolytic mechanisms, blood clots sometimes form within the cardiovascular system. Such clots may be initiated by roughened endothelial surfaces of a blood vessel resulting from atherosclerosis, trauma, or infection. These conditions induce adhesion of platelets. Intravascular clots may also form when blood flows too slowly (stasis), allowing clotting factors to accumulate locally in high enough concentrations to initiate coagulation. Clotting in an unbroken blood vessel (usually a vein) is called **thrombosis** (throm-BŌ-sis; *thromb-* = clot; *-osis* = a condition of). The clot itself, called a **thrombus** (THROM-bus), may dissolve spontaneously. If it remains intact, however, the thrombus may become dislodged and be swept away in the blood. A blood clot, bubble of air, fat from broken bones, or a piece of debris transported by the bloodstream is called an **embolus** (EM-bō-lus; *em-* = in; *-bolus* = a mass). An embolus that breaks away from an arterial wall may lodge in a smaller-diameter artery downstream and block blood flow to a vital organ. When an embolus lodges in the lungs, the condition is called **pulmonary embolism.**

CLINICAL CONNECTION | *Aspirin and Thrombolytic Agents*

In patients with heart and blood vessel disease, the events of hemostasis may occur even without external injury to a blood vessel. At low doses, **aspirin** inhibits vasoconstriction and platelet aggregation by blocking synthesis of thromboxane A2. It also reduces the chance of thrombus formation. Due to these effects, aspirin reduces the risk of transient ischemic attacks (TIA), strokes, myocardial infarction, and blockage of peripheral arteries.

Thrombolytic agents (throm'-bō-LIT-ik) are chemical substances that are injected into the body to dissolve blood clots that have already formed to restore circulation. They either directly or indirectly activate plasminogen. The first thrombolytic agent, approved in 1982 for dissolving clots in the coronary arteries of the heart, was **streptokinase,** which is produced by streptococcal bacteria. A genetically engineered version of human **tissue plasminogen activator (t-PA)** is now used to treat victims of both heart attacks and brain attacks (strokes) that are caused by blood clots. •

✔CHECKPOINT

18. What is hemostasis?
19. How do vascular spasm and platelet plug formation occur?
20. What is fibrinolysis? Why does blood rarely remain clotted inside blood vessels?
21. How do the extrinsic and intrinsic pathways of blood clotting differ?
22. Define each of the following terms: anticoagulant, thrombus, embolus, and thrombolytic agent.

19.8 BLOOD GROUPS AND BLOOD TYPES

◉ OBJECTIVES

• Distinguish between the ABO and Rh blood groups.
• Explain why it is so important to match donor and recipient blood types before administering a transfusion.

The surfaces of erythrocytes contain a genetically determined assortment of **antigens** composed of glycoproteins and glycolipids. These antigens, called **agglutinogens** (a-gloo-TIN-ō-jens), occur in characteristic combinations. Based on the presence or absence of various antigens, blood is categorized into different **blood groups.** Within a given blood group, there may be two or more different **blood types.** There are at least 24 blood groups and more than 100 antigens that can be detected on the surface of red blood cells. Here we discuss two major blood groups—ABO and Rh. Other blood groups include the Lewis, Kell, Kidd, and Duffy

systems. The incidence of ABO and Rh blood types varies among different population groups, as indicated in Table 19.5.

TABLE 19.5

Blood Types in the United States

POPULATION GROUP	BLOOD TYPE (PERCENTAGE)				
	O	A	B	AB	Rh⁺
European-American	45	40	11	4	85
African-American	49	27	20	4	95
Korean-American	32	28	30	10	100
Japanese-American	31	38	21	10	100
Chinese-American	42	27	25	6	100
Native American	79	16	4	1	100

ABO Blood Group

The **ABO blood group** is based on two glycolipid antigens called A and B (Figure 19.12). People whose RBCs display *only antigen A* have **type A** blood. Those who have *only antigen B* are **type B.** Individuals who have *both A and B antigens* are **type AB;** those who have *neither antigen A nor B* are **type O.**

Blood plasma usually contains **antibodies** called **agglutinins** (a-GLOO-ti-nins) that react with the A or B antigens if the two are mixed. These are the **anti-A antibody,** which reacts with antigen A, and the **anti-B antibody,** which reacts with antigen B. The

Figure 19.12 Antigens and antibodies of the ABO blood types.

The antibodies in your plasma do not react with the antigens on your red blood cells.

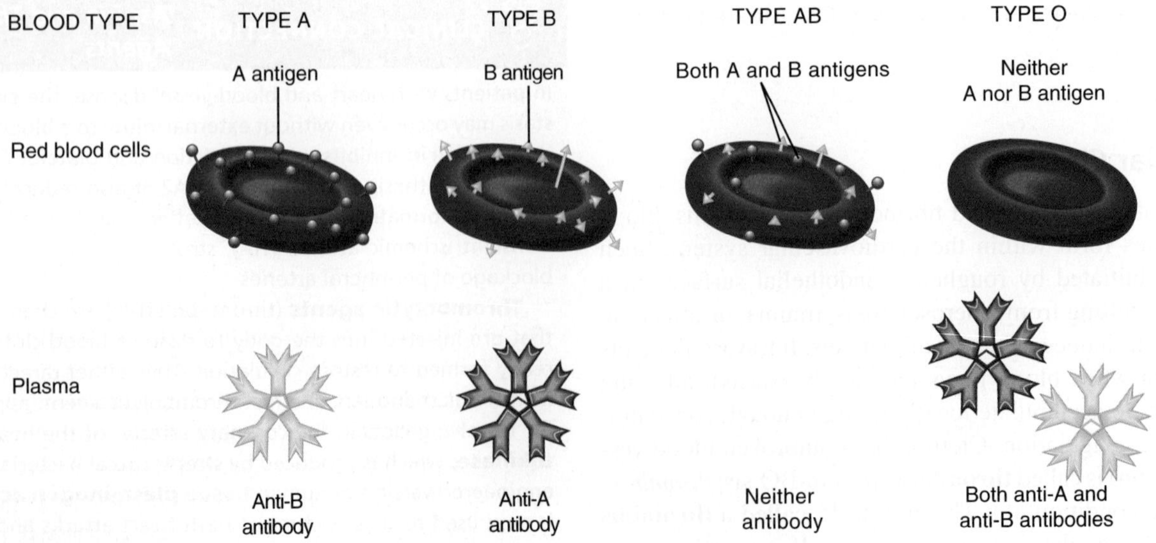

BLOOD TYPE	TYPE A	TYPE B	TYPE AB	TYPE O
	A antigen	B antigen	Both A and B antigens	Neither A nor B antigen
Red blood cells				
Plasma	Anti-B antibody	Anti-A antibody	Neither antibody	Both anti-A and anti-B antibodies

? Which antibodies are usually present in type O blood?

antibodies present in each of the four blood types are shown in Figure 19.12. You do not have antibodies that react with the antigens of your own RBCs, but you do have antibodies for any antigens that your RBCs lack. For example, if your blood type is B, you have B antigens on your red blood cells, and you have anti-A antibodies in your blood plasma. Although agglutinins start to appear in the blood within a few months after birth, the reason for their presence is not clear. Perhaps they are formed in response to bacteria that normally inhabit the gastrointestinal tract. Because the antibodies are large IgM-type antibodies (see Table 22.3) that do not cross the placenta, ABO incompatibility between a mother and her fetus rarely causes problems.

Transfusions

Despite the differences in RBC antigens reflected in the blood group systems, blood is the most easily shared of human tissues, saving many thousands of lives every year through transfusions. A **transfusion** (trans-FŪ-zhun) is the transfer of whole blood or blood components (red blood cells only or blood plasma only) into the bloodstream or directly into the red bone marrow. A transfusion is most often given to alleviate anemia, to increase blood volume (for example, after a severe hemorrhage), or to improve immunity. However, the normal components of one person's RBC plasma membrane can trigger damaging antigen–antibody responses in a transfusion recipient. In an incompatible blood transfusion, antibodies in the recipient's plasma bind to the antigens on the donated RBCs, which causes **agglutination** (a-gloo-ti-NĀ-shun), or clumping, of the RBCs. Agglutination is an antigen–antibody response in which RBCs become cross-linked to one another. (Note that agglutination is not the same as blood clotting.) When these antigen–antibody complexes form, they activate plasma proteins of the complement family (described in Section 22.6). In essence, complement molecules make the plasma membrane of the donated RBCs leaky, causing **hemolysis** (hē-MOL-i-sis) or rupture of the RBCs and the release of hemoglobin into the blood plasma. The liberated hemoglobin may cause kidney damage by clogging the filtration membranes. Although quite rare, it is possible for the viruses that cause AIDS and hepatitis B and C to be transmitted through transfusion of contaminated blood products.

Consider what happens if a person with type A blood receives a transfusion of type B blood. The recipient's blood (type A) contains A antigens on the red blood cells and anti-B antibodies in the plasma. The donor's blood (type B) contains B antigens and anti-A antibodies. In this situation, two things can happen. First, the anti-B antibodies in the recipient's plasma can bind to the B antigens on the donor's erythrocytes, causing agglutination and hemolysis of the red blood cells. Second, the anti-A antibodies in the donor's plasma can bind to the A antigens on the recipient's red blood cells, a less serious reaction because the donor's anti-A antibodies become so diluted in the recipient's plasma that they do not cause significant agglutination and hemolysis of the recipient's RBCs.

Table 19.6 summarizes the interactions of the four blood types of the ABO system.

People with type AB blood do not have anti-A or anti-B antibodies in their blood plasma. They are sometimes called *universal recipients* because theoretically they can receive blood from donors of all four blood types. They have no antibodies to attack antigens on donated RBCs. People with type O blood have neither A nor B antigens on their RBCs and are sometimes called *universal donors* because theoretically they can donate blood to all four ABO blood types. Type O persons requiring blood may receive only type O blood (Table 19.6). In practice, use of the terms universal recipient and universal donor is misleading and dangerous. Blood contains antigens and antibodies other than those associated with the ABO system that can cause transfusion problems. Thus, blood should be carefully cross-matched or screened before transfusion. In about 80% of the population, soluble antigens of the ABO type appear in saliva and other body fluids, in which case blood type can be identified from a sample of saliva.

TABLE 19.6

Summary of ABO Blood Group Interactions

CHARACTERISTIC	BLOOD TYPE			
	A	B	AB	O
Agglutinogen (antigen) on RBCs	A	B	Both A and B	Neither A nor B
Agglutinin (antibody) in plasma	Anti-B	Anti-A	Neither anti-A nor anti-B	Both anti-A and anti-B
Compatible donor blood types (no hemolysis)	A, O	B, O	A, B, AB, O	O
Incompatible donor blood types (hemolysis)	B, AB	A, AB	—	A, B, AB

Rh Blood Group

The **Rh blood group** is so named because the Rh antigen, called **Rh factor,** was first found in the blood of the *Rhesus* monkey. The alleles of three genes may code for the Rh antigen. People whose RBCs have Rh antigens are designated Rh^+ (Rh positive); those who lack Rh antigens are designated Rh^- (Rh negative). Table 19.5 shows the incidence of Rh^+ and Rh^- in various populations. Normally, blood plasma does not contain anti-Rh antibodies. If an Rh^- person receives an Rh^+ blood transfusion, however, the immune system starts to make anti-Rh antibodies that will remain in the blood. If a second transfusion of Rh^+ blood is given later, the previously formed anti-Rh antibodies will cause agglutination and hemolysis of the RBCs in the donated blood, and a severe reaction may occur.

CLINICAL CONNECTION | *Hemolytic Disease of the Newborn*

The most common problem with Rh incompatibility, **hemolytic disease of the newborn (HDN)**, may arise during pregnancy (Figure 19.13). Normally, no direct contact occurs between maternal and fetal blood while a woman is pregnant. However, if a small amount of Rh⁺ blood leaks from the fetus through the placenta into the bloodstream of an Rh⁻ mother, the mother will start to make anti-Rh antibodies. Because the greatest possibility of fetal blood leakage into the maternal circulation occurs at delivery, the firstborn baby usually is not affected. If the mother becomes pregnant again, however, her anti-Rh antibodies can cross the placenta and enter the bloodstream of the fetus. If the fetus is Rh⁻, there is no problem, because Rh⁻ blood does not have the Rh antigen. If the fetus is Rh⁺, however, agglutination and hemolysis brought on by fetal–maternal incompatibility may occur in the fetal blood.

An injection of anti-Rh antibodies called anti-Rh gamma globulin (RhoGAM®) can be given to prevent HDN. Rh⁻ women should receive RhoGAM® before delivery, and soon after every delivery, miscarriage, or abortion. These antibodies bind to and inactivate the fetal Rh antigens before the mother's immune system can respond to the foreign antigens by producing her own anti-Rh antibodies. •

Figure 19.13 Development of hemolytic disease of the newborn (HDN). (a) At birth, a small quantity of fetal blood usually leaks across the placenta into the maternal bloodstream. A problem can arise when the mother is Rh⁻ and the baby is Rh⁺, having inherited an allele for the Rh antigens from the father. (b) On exposure to Rh antigen, the mother's immune system responds by making anti-Rh antibodies. (c) During a subsequent pregnancy, the maternal antibodies cross the placenta into the fetal blood. If the second fetus is Rh⁺, the ensuing antigen–antibody reaction causes agglutination and hemolysis of fetal RBCs. The result is HDN.

HDN occurs when maternal anti-Rh antibodies cross the placenta and cause hemolysis of fetal RBCs.

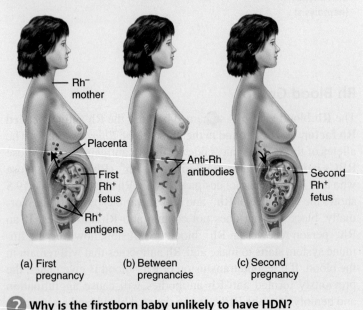

(a) First pregnancy

(b) Between pregnancies

(c) Second pregnancy

? **Why is the firstborn baby unlikely to have HDN?**

Typing and Cross-Matching Blood for Transfusion

To avoid blood-type mismatches, laboratory technicians type the patient's blood and then either cross-match it to potential donor blood or screen it for the presence of antibodies. In the procedure for ABO blood typing, single drops of blood are mixed with different *antisera*, solutions that contain antibodies (Figure 19.14). One drop of blood is mixed with anti-A serum, which contains anti-A antibodies that will agglutinate red blood cells that possess A antigens. Another drop is mixed with anti-B serum, which contains anti-B antibodies that will agglutinate red blood cells that possess B antigens. If the red blood cells agglutinate only when

Figure 19.14 ABO blood typing.

In the procedure for ABO blood typing, blood is mixed with anti-A serum and anti-B serum.

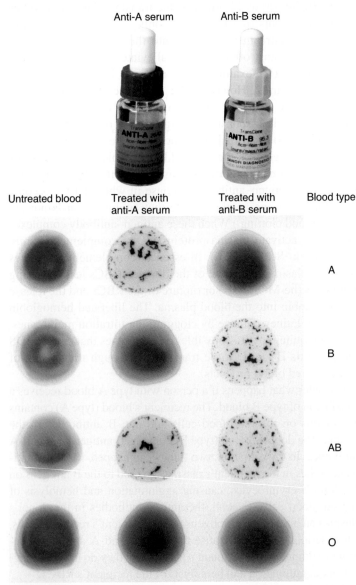

? **What is agglutination?**

mixed with anti-A serum, the blood is type A. If the red blood cells agglutinate only when mixed with anti-B serum, the blood is type B. The blood is type AB if both drops agglutinate; if neither drop agglutinates, the blood is type O.

In the procedure for determining Rh factor, a drop of blood is mixed with antiserum containing antibodies that will agglutinate RBCs displaying Rh antigens. If the blood agglutinates, it is Rh$^+$; no agglutination indicates Rh$^-$.

Once the patient's blood type is known, donor blood of the same ABO and Rh type is selected. In a **cross-match,** the possible donor RBCs are mixed with the recipient's serum. If agglutination does not occur, the recipient does not have antibodies that will attack the donor RBCs. Alternatively, the recipient's serum can be **screened** against a test panel of RBCs having antigens known to cause blood transfusion reactions to detect any antibodies that may be present.

✔ CHECKPOINT

23. What precautions must be taken before giving a blood transfusion?
24. What is hemolysis, and how can it occur after a mismatched blood transfusion?
25. Explain the conditions that may cause hemolytic disease of the newborn.

DISORDERS: HOMEOSTATIC IMBALANCES

Anemia

Anemia (a-NĒ-mē-a) is a condition in which the oxygen-carrying capacity of blood is reduced. All of the many types of anemia are characterized by reduced numbers of RBCs or a decreased amount of hemoglobin in the blood. The person feels fatigued and is intolerant of cold, both of which are related to lack of oxygen needed for ATP and heat production. Also, the skin appears pale, due to the low content of red-colored hemoglobin circulating in skin blood vessels. Among the most important causes and types of anemia are the following:

- *Inadequate absorption of iron, excessive loss of iron, increased iron requirement, or insufficient intake of iron* causes **iron-deficiency anemia,** the most common type of anemia. Women are at greater risk for iron-deficiency anemia due to menstrual blood losses and increased iron demands of the growing fetus during pregnancy. Gastrointestinal losses, such as those that occur with malignancy or ulceration, also contribute to this type of anemia.

- *Inadequate intake of vitamin B$_{12}$ or folic acid* causes **megaloblastic anemia** in which red bone marrow produces large, abnormal red blood cells (megaloblasts). It may also be caused by drugs that alter gastric secretion or are used to treat cancer.

- *Insufficient hemopoiesis* resulting from an inability of the stomach to produce intrinsic factor, which is needed for absorption of vitamin B$_{12}$ in the small intestine, causes **pernicious anemia.**

- *Excessive loss of RBCs* through bleeding resulting from large wounds, stomach ulcers, or especially heavy menstruation leads to **hemorrhagic anemia.**

- *RBC plasma membranes rupture prematurely* in **hemolytic anemia.** The released hemoglobin pours into the plasma and may damage the filtering units (glomeruli) in the kidneys. The condition may result from inherited defects such as abnormal red blood cell enzymes, or from outside agents such as parasites, toxins, or antibodies from incompatible transfused blood.

- *Deficient synthesis of hemoglobin* occurs in **thalassemia** (thal'-a-SĒ-mē-a), a group of hereditary hemolytic anemias. The RBCs are small (microcytic), pale (hypochromic), and short-lived. Thalassemia occurs primarily in populations from countries bordering the Mediterranean Sea.

- *Destruction of red bone marrow* results in **aplastic anemia.** It is caused by toxins, gamma radiation, and certain medications that inhibit enzymes needed for hemopoiesis.

Sickle-Cell Disease

The RBCs of a person with **sickle-cell disease (SCD)** contain Hb-S, an abnormal kind of hemoglobin. When Hb-S gives up oxygen to the interstitial fluid, it forms long, stiff, rodlike structures that bend the erythrocyte into a sickle shape (Figure 19.15). The sickled cells rupture easily. Even though erythropoiesis is stimulated by the loss of the cells, it cannot keep pace with hemolysis. Signs and symptoms of SCD are caused by the sickling of red blood cells. When red blood cells sickle, they break down prematurely (sickled cells die in about 10 to 20 days). This leads to anemia, which can cause shortness of breath, fatigue, paleness, and delayed growth and development in children. The rapid breakdown and loss of blood cells may also cause jaundice, yellowing of the eyes and skin. Sickled cells do not move easily through blood vessels and they tend to stick together and form clumps that cause blockages in blood vessels. This deprives body

Figure 19.15 Red blood cells from a person with sickle-cell disease.

 The red blood cells of a person with sickle-cell disease contain an abnormal type of hemoglobin.

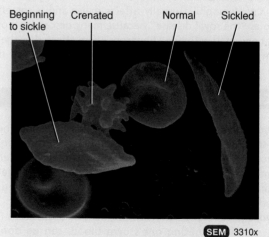

Beginning to sickle   Crenated   Normal   Sickled

SEM 3310x

Red blood cells

 What are some symptoms of sickle-cell disease?

organs of sufficient oxygen and causes pain, for example, in bones and the abdomen; serious infections; and organ damage, especially in the lungs, brain, spleen, and kidneys. Other symptoms of SCD include fever, rapid heart rate, swelling and inflammation of the hands and/or feel, leg ulcers, eye damage, excessive thirst, frequent urination, and painful and prolonged erections in males. Almost all individuals with SCD have painful episodes that can last from hours to days. Some people have one episode every few years; others have several episodes a year. The episodes may range from mild to those that require hospitalization. Any activity that reduces the amount of oxygen in the blood, such as vigorous exercise, may produce a **sickle-cell crisis** (worsening of the anemia, pain in the abdomen and long bones of the limbs, fever, and shortness of breath).

Sickle-cell disease is inherited. People with two sickle-cell genes have severe anemia; those with only one defective gene have the sickle cell trait. Sickle-cell genes are found primarily among populations (or their descendants) that live in the malaria belt around the world, including parts of Mediterranean Europe, sub-Saharan Africa, and tropical Asia. The genes responsible for the tendency of the RBCs to sickle also alter the permeability of the plasma membranes of sickled cells, causing potassium ions to leak out. Low levels of potassium kill the malaria parasites that may infect sickled cells. Because of this effect, a person with one normal gene and one sickle-cell gene has a higher-than-average resistance to malaria. The possession of a single sickle-cell gene thus confers a survival benefit.

Treatment of SCD consists of administration of analgesics to relieve pain, fluid therapy to maintain hydration, oxygen to reduce oxygen deficiency, antibiotics to counter infections, and blood transfusions. People who suffer from SCD have normal fetal hemoglobin (Hb-F), a slightly different form of hemoglobin that predominates at birth and is present in small amounts after birth. In some patients with sickle-cell disease, a drug called hydroxyurea promotes transcription of the normal Hb-F gene, elevates the level of Hb-F, and reduces the chance that the RBCs will sickle. Unfortunately, this drug also has toxic effects on the bone marrow; thus, its safety for long-term use is questionable.

Hemophilia

Hemophilia (hē-mō-FIL-ē-a; *-philia* = loving) is an inherited deficiency of clotting in which bleeding may occur spontaneously or after only minor trauma. It is the oldest known hereditary bleeding disorder; descriptions of the disease are found as early as the second century A.D. Hemophilia usually affects males and is sometimes referred to as "the royal disease" because many descendants of Queen Victoria, beginning with one of her sons, were affected by the disease. Different types of hemophilia are due to deficiencies of different blood clotting factors and exhibit varying degrees of severity, ranging from mild to severe bleeding tendencies. Hemophilia is characterized by spontaneous or traumatic subcutaneous and intramuscular hemorrhaging, nosebleeds, blood in the urine, and hemorrhages in joints that produce pain and tissue damage. Treatment involves transfusions of fresh blood plasma or concentrates of the deficient clotting factor to relieve the tendency to bleed. Another treatment is the drug desmopressin (DDAVP), which can boost the levels of the clotting factors.

Leukemia

The term **leukemia** (loo-KĒ-mē-a; *leuko-* = white) refers to a group of red bone marrow cancers in which abnormal white blood cells multiply uncontrollably. The accumulation of the cancerous white blood cells in red bone marrow interferes with the production of red blood cells, white blood cells, and platelets. As a result the oxygen-carrying capacity of the blood is reduced, an individual is more susceptible to infection, and blood clotting is abnormal. In most leukemias, the cancerous white blood cells spread to the lymph nodes, liver, and spleen, causing them to enlarge. All leukemias produce the usual symptoms of anemia (fatigue, intolerance to cold, and pale skin). In addition, weight loss, fever, night sweats, excessive bleeding, and recurrent infections may occur.

In general, leukemias are classified as **acute** (symptoms develop rapidly) and **chronic** (symptoms may take years to develop). Leukemias are also classified on the basis of the type of white blood cell that becomes malignant. **Lymphoblastic leukemia** (lim-fō-BLAS-tik) involves cells derived from lymphoid stem cells (lymphoblasts) and/or lymphocytes. **Myelogenous leukemia** (mī-e-LOJ-e-nus) involves cells derived from myeloid stem cells (myeloblasts). Combining onset of symptoms and cells involved, there are four types of leukemia:

1. **Acute lymphoblastic leukemia (ALL)** is the most common leukemia in children, but adults can also get it.
2. **Acute myelogenous leukemia (AML)** affects both children and adults.
3. **Chronic lymphoblastic anemia (CLA)** is the most common leukemia in adults, usually those older than 55.
4. **Chronic myelogenous leukemia (CML)** occurs mostly in adults.

The cause of most types of leukemia is unknown. However, certain risk factors have been implicated. These include exposure to radiation or chemotherapy for other cancers, genetics (some genetic disorders such as Down syndrome), environmental factors (smoking and benzene), and microbes such as the human T cell leukemia–lymphoma virus-1 (HTLV-1) and the Epstein-Barr virus.

Treatment options include chemotherapy, radiation, stem cell transplantation, interferon, antibodies, and blood transfusion.

MEDICAL TERMINOLOGY

Acute normovolemic hemodilution (nor-mō-vō-LĒ-mik hē-mō-di-LOO-shun) Removal of blood immediately before surgery and its replacement with a cell-free solution to maintain sufficient blood volume for adequate circulation. At the end of surgery, once bleeding has been controlled, the collected blood is returned to the body.

Autologous preoperative transfusion (aw-TOL-o-gus trans-FŪ-zhun; *auto-* = self) Donating one's own blood; can be done up to 6 weeks before elective surgery. Also called **predonation.** This procedure eliminates the risk of incompatibility and blood-borne disease.

Blood bank A facility that collects and stores a supply of blood for future use by the donor or others. Because blood banks have additional and diverse functions (immunohematology reference work, continuing medical education, bone and tissue storage, and clinical consultation), they are more appropriately referred to as **centers of transfusion medicine.**

Cyanosis (sī-a-NŌ-sis; *cyano-* = blue) Slightly bluish/dark-purple skin discoloration, most easily seen in the nail beds and mucous membranes, due to an increased quantity of *methemoglobin,* hemoglobin not combined with oxygen in systemic blood.

Gamma globulin (GLOB-ū-lin) Solution of immunoglobulins from blood consisting of antibodies that react with specific pathogens, such as viruses. It is prepared by injecting the specific virus into animals, removing blood from the animals after antibodies have accumulated, isolating the antibodies, and injecting them into a human to provide short-term immunity.

Hemochromatosis (hē-mō-krō-ma-TŌ-sis; *chroma* = color) Disorder of iron metabolism characterized by excessive absorption of ingested iron and excess deposits of iron in tissues (especially the liver, heart, pituitary gland, gonads, and pancreas) that result in bronze discoloration of the skin, cirrhosis, diabetes mellitus, and bone and joint abnormalities.

Hemorrhage (HEM-or-ij; *rhegnynai* = bursting forth) Loss of a large amount of blood; can be either internal (from blood vessels into tissues) or external (from blood vessels directly to the surface of the body).

Jaundice (*jaund-* = yellow) An abnormal yellowish discoloration of the sclerae of the eyes, skin, and mucous membranes due to excess bilirubin (yellow-orange pigment) in the blood. The three main categories of jaundice are *prehepatic jaundice,* due to excess pro-

duction of bilirubin; *hepatic jaundice,* abnormal bilirubin processing by the liver caused by congenital liver disease, cirrhosis (scar tissue formation) of the liver, or hepatitis (liver inflammation); and *extrahepatic jaundice,* due to blockage of bile drainage by gallstones or cancer of the bowel or pancreas.

Phlebotomist (fle-BOT-ō-mist; *phlebo-* = vein; *-tom* = cut) A technician who specializes in withdrawing blood.

Septicemia (sep'-ti-SĒ-mē-a; *septic-* = decay; *-emia* = condition of blood) Toxins or disease-causing bacteria in the blood. Also called "blood poisoning."

Thrombocytopenia (throm'-bō-sī-tō-PĒ-nē-a; *-penia* = poverty) Very low platelet count that results in a tendency to bleed from capillaries.

Venesection (vē'-ne-SEK-shun; *ven-* = vein) Opening of a vein for withdrawal of blood. Although **phlebotomy** (fle-BOT-ō-mē) is a synonym for venesection, in clinical practice phlebotomy refers to therapeutic bloodletting, such as the removal of some blood to lower its viscosity in a patient with polycythemia.

Whole blood Blood containing all formed elements, plasma, and plasma solutes in natural concentrations.

CHAPTER REVIEW AND RESOURCE SUMMARY

Review   **Resource**

Introduction

1. The cardiovascular system consists of the blood, heart, and blood vessels.
2. Blood is a liquid connective tissue that consists of cells and cell fragments surrounded by a liquid extracellular matrix (blood plasma).

Anatomy Overview - The Cardiovascular System

19.1 Functions and Properties of Blood

1. Blood transports oxygen, carbon dioxide, nutrients, wastes, and hormones.
2. It helps regulate pH, body temperature, and water content of cells.
3. It provides protection through clotting and by combating toxins and microbes through certain phagocytic white blood cells or specialized blood plasma proteins.
4. Physical characteristics of blood include a viscosity greater than that of water; a temperature of 38°C (100.4°F); and a pH of 7.35–7.45.
5. Blood constitutes about 8% of body weight, and its volume is 4–6 liters in adults.
6. Blood is about 55% blood plasma and 45% formed elements.
7. The hematocrit is the percentage of total blood volume occupied by red blood cells.
8. Blood plasma consists of 91.5% water and 8.5% solutes. Principal solutes include proteins (albumins, globulins, fibrinogen), nutrients, vitamins, hormones, respiratory gases, electrolytes, and waste products.
9. The formed elements in blood include red blood cells (erythrocytes), white blood cells (leukocytes), and platelets.

Anatomy Overview - Blood
Concepts and Connections - Blood

19.2 Formation of Blood Cells

1. Hemopoiesis is the formation of blood cells from hemopoietic stem cells in red bone marrow.
2. Myeloid stem cells form RBCs, platelets, granulocytes, and monocytes. Lymphoid stem cells give rise to lymphocytes.
3. Several hemopoietic growth factors stimulate differentiation and proliferation of the various blood cells.

Anatomy Overview - Erythrocytes
Figure 19.3 - Origin, Development, and Structure of Blood Cells

19.3 Red Blood Cells

1. Mature RBCs are biconcave discs that lack nuclei and contain hemoglobin.
2. The function of the hemoglobin in red blood cells is to transport oxygen and some carbon dioxide.
3. RBCs live about 120 days. A healthy male has about 5.4 million RBCs/μL of blood; a healthy female has about 4.8 million/μL.
4. After phagocytosis of aged RBCs by macrophages, hemoglobin is recycled.

Anatomy Overview - Erythrocytes
Animation - Erythropoietin
Figure 19.4 - The Shapes of a Red Blood Cell and a Hemoglobin Molecule

Review   **Resource**

5. RBC formation, called erythropoiesis, occurs in adult red bone marrow of certain bones. It is stimulated by hypoxia, which stimulates the release of erythropoietin by the kidneys.

6. A reticulocyte count is a diagnostic test that indicates the rate of erythropoiesis.

19.4 White Blood Cells

Anatomy Overview - White
Blood Cells
Figure 19.7 - Types of White
Blood Cells

1. WBCs are nucleated cells. The two principal types are granulocytes (neutrophils, eosinophils, and basophils) and agranulocytes (lymphocytes and monocytes).

2. The general function of WBCs is to combat inflammation and infection. Neutrophils and macrophages (which develop from monocytes) do so through phagocytosis.

3. Eosinophils combat the effects of histamine in allergic reactions, phagocytize antigen–antibody complexes, and combat parasitic worms. Basophils liberate heparin, histamine, and serotonin in allergic reactions that intensify the inflammatory response.

4. B lymphocytes, in response to the presence of foreign substances called antigens, differentiate into plasma cells that produce antibodies. Antibodies attach to the antigens and render them harmless. This antigen–antibody response combats infection and provides immunity. T lymphocytes destroy foreign invaders directly. Natural killer cells attack infectious microbes and tumor cells.

5. Except for lymphocytes, which may live for years, WBCs usually live for only a few hours or a few days. Normal blood contains 5000–10,000 WBCs/μL.

19.5 Platelets

Anatomy Overview - Platelets

1. Platelets (thrombocytes) are disc-shaped cell fragments that splinter from megakaryocytes. Normal blood contains 150,000–400,000 platelets/μL.

2. Platelets help stop blood loss from damaged blood vessels by forming a platelet plug.

19.6 Stem Cell Transplants from Bone Marrow and Cord Blood

1. Bone marrow transplants involve removal of red bone marrow as a source of stem cells from the iliac crest.

2. In a cord-blood transplant, stem cells from the placenta are removed from the umbilical cord.

3. Cord-blood transplants have several advantages over bone marrow transplants.

19.7 Hemostasis

Figure 19.11 - The Blood-
Clotting Cascade

1. Hemostasis refers to the stoppage of bleeding.

2. It involves vascular spasm, platelet plug formation, and blood clotting (coagulation).

3. In vascular spasm, the smooth muscle of a blood vessel wall contracts, which slows blood loss.

4. Platelet plug formation involves the aggregation of platelets to stop bleeding.

5. A clot is a network of insoluble protein fibers (fibrin) in which formed elements of blood are trapped.

6. The chemicals involved in clotting are known as clotting (coagulation) factors.

7. Blood clotting involves a cascade of reactions that may be divided into three stages: formation of prothrombinase, conversion of prothrombin into thrombin, and conversion of soluble fibrinogen into insoluble fibrin.

8. Clotting is initiated by the interplay of the extrinsic and intrinsic pathways of blood clotting.

9. Normal coagulation requires vitamin K and is followed by clot retraction (tightening of the clot) and ultimately fibrinolysis (dissolution of the clot).

10. Clotting in an unbroken blood vessel is called thrombosis. A thrombus that moves from its site of origin is called an embolus.

19.8 Blood Groups and Blood Types

1. ABO and Rh blood groups are genetically determined and based on antigen–antibody responses.

2. In the ABO blood group, the presence or absence of A and B antigens on the surface of RBCs determines blood type.

3. In the Rh system, individuals whose RBCs have Rh antigens are classified as Rh$^+$; those who lack the antigen are Rh$^-$.

4. Hemolytic disease of the newborn (HDN) can occur when an Rh$^-$ mother is pregnant with an Rh$^+$ fetus.

5. Before blood is transfused, a recipient's blood is typed and then either cross-matched to potential donor blood or screened for the presence of antibodies.

SELF-QUIZ QUESTIONS

Fill in the blanks in the following statements.

1. Plasma minus its clotting proteins is termed _____.

2. _____ is the consolidation or tightening of the fibrin clot that helps to bring the edges of a damaged vessel closer together.

Indicate whether the following statements are true or false.

3. Hemoglobin functions in transporting both oxygen and carbon dioxide and in regulating blood pressure.

4. The most numerous white blood cells in a differential white blood cell count of a healthy individual are the neutrophils.

Choose the one best answer to the following questions.

5. Which of the following are *not* required for clot formation? (1) vitamin K, (2) calcium, (3) prostacyclin, (4) plasmin, (5) fibrinogen.
   (a) 1, 2, and 5   (b) 3, 4, and 5   (c) 4 and 5
   (d) 1, 2, and 3   (e) 3 and 4

6. Place the steps involved in hemostasis in the correct order. (1) conversion of fibrinogen into fibrin, (2) conversion of prothrombin into thrombin, (3) adhesion and aggregation of platelets on damaged vessel, (4) prothrombinase formed by extrinsic or intrinsic pathway, (5) reduction of blood loss by initiation of a vascular spasm.
   (a) 5, 3, 4, 2, 1   (b) 5, 4, 3, 1, 2   (c) 3, 5, 4, 2, 1
   (d) 5, 3, 2, 1, 4   (e) 5, 3, 2, 4, 1

7. Which of the following statements explain why red blood cells (RBCs) are highly specialized for oxygen transport? (1) RBCs contain hemoglobin. (2) RBCs lack a nucleus. (3) RBCs have many mitochondria and thus generate ATP aerobically. (4) The biconcave shape of RBCs provides a large surface area for the inward and outward diffusion of gas molecules. (5) RBCs can carry up to four oxygen molecules for each hemoglobin molecule.
   (a) 1, 2, 3, and 5   (b) 1, 2, 4, and 5   (c) 2, 3, 4, and 5
   (d) 1, 3, and 5   (e) 2, 4, and 5

8. Which of the following are *true?* (1) White blood cells leave the bloodstream by emigration. (2) Adhesion molecules help white blood cells stick to the endothelium, which aids emigration. (3) Neutrophils and macrophages are active in phagocytosis. (4) The attraction of phagocytes to microbes and inflamed tissue is termed chemotaxis. (5) Leukopenia is an increase in white blood cell count that occurs during infection.
   (a) 1, 2, 4, and 5   (b) 2, 3, 4, and 5   (c) 1, 2, 3, and 4
   (d) 1, 3, and 5   (e) 1, 2, and 4

9. A person with type A Rh^- blood can receive a blood transfusion from which of the following types? (1) A Rh^+, (2) B Rh^-, (3) AB Rh^-, (4) O Rh^-, (5) A Rh^-.
   (a) 1 only   (b) 3 only   (c) 4 only
   (d) 4 and 5   (e) 1 and 5

10. A person with type B positive blood receives a transfusion of type AB positive blood. What will happen?
    (a) The recipient's antibodies will react with the donor's red blood cells.
    (b) The donor's antigens will destroy the recipient's antibodies.
    (c) The donor's antibodies will react with and destroy all of the recipient's red blood cells.
    (d) The recipient's blood type will change from Rh^+ to Rh^-.
    (e) These blood types are compatible, and the transfusion will be accepted.

11. What happens to the iron (Fe^{3+}) that is released during the breakdown of damaged red blood cells?
    (a) It is used to synthesize proteins.
    (b) It is transported to the liver where it becomes part of bile.
    (c) It is converted into urobilin and excreted in urine.
    (d) It attaches to transferrin and is transported to bone marrow for use in hemoglobin synthesis.
    (e) It is utilized by intestinal bacteria to convert bilirubin into urobilinogen.

12. Which of the following would *not* cause an increase in erythropoietin?
    (a) anemia
    (b) high altitude
    (c) hemorrhage
    (d) donating blood to a blood bank
    (e) polycythemia

13. Match the following:
    _____ (a) the percentage of total blood volume occupied by red blood cells
    _____ (b) the percentage of each type of white blood cell
    _____ (c) measures numbers of RBCs, WBCs, platelets per μ of blood; hematocrit; and differential WBC count
    _____ (d) measures the rate of erythropoiesis
    _____ (e) withdrawal of blood from a vein using a needle and collecting tube
    _____ (f) withdrawal of a small amount of red bone marrow with a fine needle and syringe
    _____ (g) removal of a core of red bone marrow with a large needle

    (1) reticulocyte count
    (2) bone marrow biopsy
    (3) venipuncture
    (4) hematocrit
    (5) bone marrow aspiration
    (6) complete blood count
    (7) differential white blood cell count

14. Match the following:
   ____ (a) contain hemoglobin and function in gas transport
   ____ (b) cell fragments enclosed by a piece of the cell membrane of megakaryocytes; contain clotting factors
   ____ (c) white blood cell showing a kidney-shaped nucleus; capable of phagocytosis
   ____ (d) monocytes that roam the tissues and gather at sites of infection or inflammation
   ____ (e) occur as B cells, T cells, and natural killer cells
   ____ (f) give rise to red blood cells, monocytes, neutrophils, eosinophils, basophils, and platelets
   ____ (g) cells that give rise to all the formed elements of blood; derived from mesenchyme

   (1) lymphocytes
   (2) monocytes
   (3) pluripotent stem cells
   (4) red blood cells
   (5) myeloid stem cells
   (6) platelets
   (7) wandering macrophages

15. Match the following:
   ____ (a) tissue protein that leaks into the blood from cells outside blood vessels and initiates the formation of prothrombinase
   ____ (b) an anticoagulant
   ____ (c) its formation is initiated by either the extrinsic or intrinsic pathway or both; catalyzes the conversion of prothrombin to thrombin
   ____ (d) forms the threads of a clot; produced from fibrinogen
   ____ (e) can dissolve a clot by digesting fibrin threads
   ____ (f) serves as the catalyst to form fibrin; formed from prothrombin

   (1) prothrombinase
   (2) thrombin
   (3) fibrin
   (4) thromboplastin
   (5) plasmin
   (6) heparin

CRITICAL THINKING QUESTIONS

1. Shilpa has recently been on broad-spectrum antibiotics for a recurrent urinary bladder infection. While slicing vegetables, she cut herself and had difficulty stopping the bleeding. How could the antibiotics have played a role in her bleeding?

2. Mrs. Brown is in renal failure. Her recent blood tests indicated a hematocrit of 22. Why is her hematocrit low? What can she be given to raise her hematocrit?

3. Thomas has hepatitis, which is disrupting his liver functions. What kinds of symptoms would he be experiencing based on the role(s) of the liver related to blood?

? ANSWERS TO FIGURE QUESTIONS

19.1 Blood volume is about 8% of your body mass, roughly 5–6 liters in males and 4–5 liters in females. For instance, a 70-kg (150-lb) person has a blood volume of 5.6 liters (70 kg × 8% × 1 liter/kg).

19.2 Platelets are cell fragments.

19.3 Pluripotent stem cells develop from mesenchyme.

19.4 One hemoglobin molecule can transport a maximum of four O_2 molecules, one O_2 bound to each heme group.

19.5 Transferrin is a plasma protein that transports iron in the blood.

19.6 Once you moved to high altitude, your hematocrit would increase due to increased secretion of erythropoietin.

19.7 Neutrophils, eosinophils, and basophils are called granular leukocytes because all have cytoplasmic granules that are visible through a light microscope when stained.

19.8 Lymphocytes recirculate from blood to tissues and back to blood. After leaving the blood, other WBCs remain in the tissues until they die.

19.9 Along with platelet plug formation, vascular spasm and blood clotting contribute to hemostasis.

19.10 Serum is blood plasma minus the clotting proteins.

19.11 The outcome of the first stage of clotting is the formation of prothrombinase.

19.12 Type O blood usually contains both anti-A and anti-B antibodies.

19.13 Because the mother is most likely to start making anti-Rh antibodies after the first baby is already born, that baby suffers no damage.

19.14 Agglutination refers to clumping of red blood cells.

19.15 Some symptoms of sickle-cell disease are anemia, jaundice, bone pain, shortness of breath, rapid heart rate, abdominal pain, fever, and fatigue.

20 THE CARDIOVASCULAR SYSTEM: THE HEART

THE HEART AND HOMEOSTASIS *The heart pumps blood through blood vessels to all body tissues.*

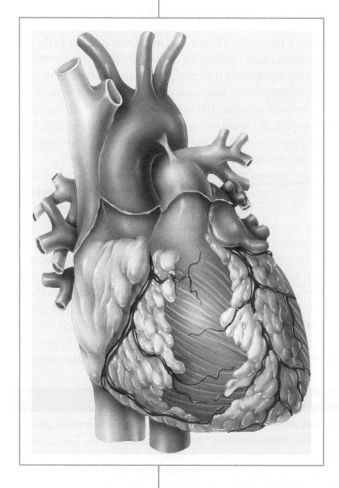

As you learned in the previous chapter, the **cardiovascular system** consists of the blood, the heart, and blood vessels. We also examined the composition and functions of blood, and in this chapter you will learn about the pump that circulates it throughout the body—the heart. For blood to reach body cells and exchange materials with them, it must be pumped continuously by the heart through the body's blood vessels. The heart beats about 100,000 times every day, which adds up to about 35 million beats in a year, and approximately 2.5 billion times in an average lifetime. The left side of the heart pumps blood through an estimated 120,000 km (75,000 mi) of blood vessels, which is equivalent to traveling around the earth's equator about 3 times. The right side of the heart pumps blood through the lungs, enabling blood to pick up oxygen and unload carbon dioxide. Even while you are sleeping, your heart pumps 30 times its own weight each minute, which amounts to about 5 liters (5.3 qt) to the lungs and the same volume to the rest of the body. At this rate, your heart pumps more than about 14,000 liters (3600 gal) of blood in a day, or 5 million liters (1.3 million gal) in a year. You don't spend all your time sleeping, however, and your heart pumps more vigorously when you are active. Thus, the actual blood volume your heart pumps in a single day is much larger.

The scientific study of the normal heart and the diseases associated with it is **cardiology** (kar-dē-OL-ō-jē; *cardio-* = heart; *-logy* = study of). This chapter explores the structure of the heart and the unique properties that permit it to pump for a lifetime without rest.

? *Did you ever wonder about the difference between "good" and "bad" cholesterol?*

20.1 ANATOMY OF THE HEART

◉ **OBJECTIVES**

- Describe the location of the heart.
- Describe the structure of the pericardium and the heart wall.
- Discuss the external and internal anatomy of the chambers of the heart.
- Relate the thickness of the chambers of the heart to their functions.

Location of the Heart

For all its might, the **heart** is relatively small, roughly the same size (but not the same shape) as your closed fist. It is about 12 cm (5 in.) long, 9 cm (3.5 in.) wide at its broadest point, and 6 cm (2.5 in.) thick, with an average mass of 250 g (8 oz) in adult females and 300 g (10 oz) in adult males. The heart rests on the diaphragm, near the midline of the thoracic cavity. Recall that the midline is an imaginary vertical line that divides the body into unequal left and right sides. The heart lies in the **mediastinum** (mē′-dē-as-TĪ-num), an anatomical region that extends from the sternum to the vertebral column, from the first rib to the diaphragm, and between the lungs (Figure 20.1a). About two-thirds of the mass of the heart lies to the left of the body's midline (Figure 20.1b). You can visualize the heart as a cone lying on its side. The pointed **apex** is formed by the tip of the left ventricle (a lower chamber of the heart) and rests on the diaphragm. It is directed anteriorly, inferiorly, and to the left. The **base** of the heart is its posterior surface. It is formed by the atria (upper chambers) of the heart, mostly the left atrium.

In addition to the apex and base, the heart has several distinct surfaces and borders (margins). The **anterior surface** is deep to the sternum and ribs. The **inferior surface** is the part of the heart between the apex and right border and rests mostly on the diaphragm (Figure 20.1b). The **right border** faces the right lung and extends from the inferior surface to the base. The **left border,** also called the *pulmonary border,* faces the left lung and extends from the base to the apex.

⚕ **CLINICAL CONNECTION | *Cardiopulmonary Resuscitation***

Because the heart lies between two rigid structures—the vertebral column and the sternum (Figure 20.1a)—external pressure on the chest (compression) can be used to force blood out of the heart and into the circulation. In cases in which the heart suddenly stops beating, **cardiopulmonary resuscitation** (kar-dē-ō-PUL-mo-nar′-ē re-sus-i-TĀ-shun) or **CPR**—properly applied cardiac compressions, performed with artificial ventilation of the lungs via mouth-to-mouth respiration—saves lives. CPR keeps oxygenated blood circulating until the heart can be restarted.

Researchers have found that chest compressions alone are equally as effective as, if not better than, traditional CPR with lung ventilation. This is good news because it is easier for an emergency dispatcher to give instructions limited to chest compressions to frightened, non-medical bystanders. As public fear of contracting contagious diseases such as hepatitis, HIV, and tuberculosis continues to rise, bystanders are much more likely to perform chest compressions alone than treatment involving mouth-to-mouth rescue breathing. •

Pericardium

The membrane that surrounds and protects the heart is the **pericardium** (per′-i-KAR-dē-um; *peri-* = around). It confines the heart to its position in the mediastinum, while allowing sufficient freedom of movement for vigorous and rapid contraction. The pericardium consists of two main parts: (1) the fibrous pericardium and (2) the serous pericardium (Figure 20.2a). The superficial **fibrous pericardium** is composed of tough, inelastic, dense irregular connective tissue. It resembles a bag that rests on and attaches to the diaphragm; its open end is fused to the connective tissues of the blood vessels entering and leaving the heart. The fibrous pericardium prevents overstretching of the heart, provides protection, and anchors the heart in the mediastinum. The fibrous pericardium near the apex of the heart is partially fused to the central tendon of the diaphragm and therefore movement of the diaphragm, as in deep breathing, facilitates the movement of blood by the heart.

The deeper **serous pericardium** is a thinner, more delicate membrane that forms a double layer around the heart (Figure 20.2a). The outer **parietal layer** of the serous pericardium is fused to the fibrous pericardium. The inner **visceral layer** of the serous pericardium, also called the **epicardium** (ep′-i-KAR-dē-um; *epi-* = on top of), is one of the layers of the heart wall and adheres tightly to the surface of the heart. Between the parietal and visceral layers of the serous pericardium is a thin film of lubricating serous fluid. This slippery secretion of the pericardial cells, known as **pericardial fluid,** reduces friction between the layers of the serous pericardium as the heart moves. The space that contains the few milliliters of pericardial fluid is called the **pericardial cavity.**

⚕ **CLINICAL CONNECTION | *Pericarditis***

Inflammation of the pericardium is called **pericarditis** (per′-i-kar-DĪ-tis). The most common type, *acute pericarditis,* begins suddenly and has no known cause in most cases but is sometimes linked to a viral infection. As a result of irritation to the pericardium, there is chest pain that may extend to the left shoulder and down the left arm (often mistaken for a heart attack) and *pericardial friction rub* (a scratchy or creaking sound heard through a stethoscope as the visceral layer of the serous pericardium rubs against the parietal layer of the serous pericardium). Acute pericarditis usually lasts for about 1 week and is treated with drugs that reduce inflammation and pain, such as ibuprofen or aspirin.

Chronic pericarditis begins gradually and is long-lasting. In one form of this condition, there is a buildup of pericardial fluid. If a great deal of fluid accumulates, this is a life-threatening condition because the fluid compresses the heart, a condition called *cardiac tamponade* (tam′-pon-ĀD). As a result of the compression, ventricular filling is decreased, cardiac output is reduced, venous return to the heart is diminished, blood pressure falls, and breathing is difficult. Most causes of chronic pericarditis involving cardiac tamponade are unknown, but

Figure 20.1 Position of the heart and associated structures in the mediastinum (dashed outline).

The heart is located in the mediastinum, with two-thirds of its mass to the left of the midline.

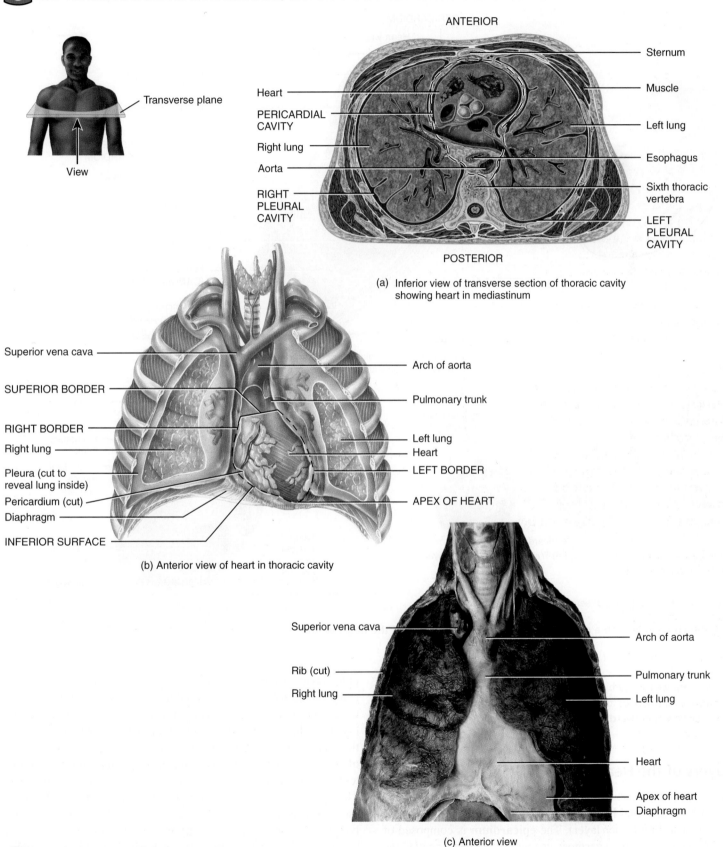

(a) Inferior view of transverse section of thoracic cavity showing heart in mediastinum

(b) Anterior view of heart in thoracic cavity

(c) Anterior view

What is the mediastinum?

Figure 20.2 Pericardium and heart wall.

The pericardium is a triple-layered sac that surrounds and protects the heart.

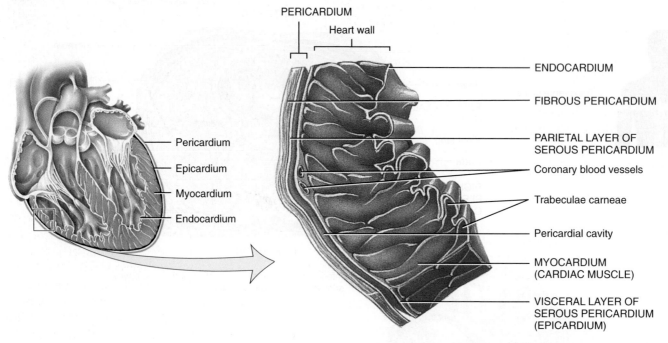

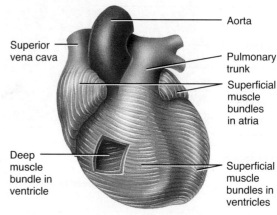

(a) Portion of pericardium and right ventricular heart wall showing divisions of pericardium and layers of heart wall

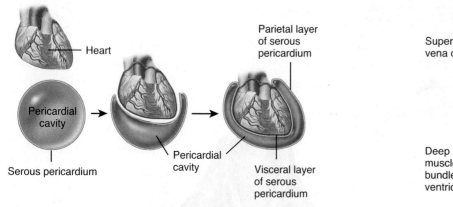

(b) Simplified relationship of serous pericardium to heart

(c) Cardiac muscle bundles of myocardium

? **Which layer is both a part of the pericardium and a part of the heart wall?**

it is sometimes caused by conditions such as cancer and tuberculosis. Treatment consists of draining the excess fluid through a needle passed into the pericardial cavity. •

Layers of the Heart Wall

The wall of the heart consists of three layers (Figure 20.2a): the epicardium (external layer), the myocardium (middle layer), and the endocardium (inner layer). The **epicardium** is composed of two tissue layers. The outermost, as you just learned, is called the *visceral layer of the serous pericardium.* This thin, transparent outer layer of the heart wall is composed of mesothelium. Beneath the mesothelium is a variable layer of delicate fibroelastic tissue and adipose tissue. The adipose tissue predominates and becomes thickest over the ventricular surfaces, where it houses the major coronary and cardiac vessels of the heart. The amount of fat varies from person to person, corresponds to the general extent of body fat in an individual, and typically increases with age. The epicardium imparts a smooth, slippery texture to the outermost surface of the heart. The epicardium contains blood vessels, lymphatics, and vessels that supply the myocardium.

The middle **myocardium** (mī′-ō-KAR-dē-um; *myo-* = muscle) is responsible for the pumping action of the heart and is composed of cardiac muscle tissue. It makes up approximately 95% of the heart wall. The muscle fibers (cells), like those of striated skeletal muscle tissue, are wrapped and bundled with connective tissue sheaths composed of endomysium and perimysium. The cardiac muscle fibers are organized in bundles that swirl diagonally around the heart and generate the strong pumping actions of the heart (Figure 20.2c). Although it is striated like skeletal muscle, recall that cardiac muscle is involuntary like smooth muscle.

The innermost **endocardium** (en′-dō-KAR-dē-um; *endo-* = within) is a thin layer of endothelium overlying a thin layer of connective tissue. It provides a smooth lining for the chambers of the heart and covers the valves of the heart. The smooth endothelial lining minimizes the surface friction as blood passes through the heart. The endocardium is continuous with the endothelial lining of the large blood vessels attached to the heart.

CLINICAL CONNECTION | *Myocarditis and Endocarditis*

Myocarditis (mī-ō-kar-DĪ-tis) is an inflammation of the myocardium that usually occurs as a complication of a viral infection, rheumatic fever, or exposure to radiation or certain chemicals or medications. Myocarditis often has no symptoms. However, if they do occur, they may include fever, fatigue, vague chest pain, irregular or rapid heartbeat, joint pain, and breathlessness. Myocarditis is usually mild and recovery occurs within 2 weeks. Severe cases can lead to cardiac failure and death. Treatment consists of avoiding vigorous exercise, a low-salt diet, electrocardiographic monitoring, and treatment of the cardiac failure. **Endocarditis** (en′-dō-kar-DĪ-tis) refers to an inflammation of the endocardium and typically involves the heart valves. Most cases are caused by bacteria (bacterial endocarditis). Signs and symptoms of endocarditis include fever, heart murmur, irregular or rapid heartbeat, fatigue, loss of appetite, night sweats, and chills. Treatment is with intravenous antibiotics. •

Chambers of the Heart

The heart has four chambers. The two superior receiving chambers are the **atria** (= entry halls or chambers), and the two inferior pumping chambers are the **ventricles** (= little bellies). The paired atria receive blood from blood vessels returning blood to the heart, called veins, while the ventricles eject the blood from the heart into blood vessels called arteries. On the anterior surface of each atrium is a wrinkled pouchlike structure called an **auricle** (OR-i-kul; *auri-* = ear), so named because of its resemblance to a dog's ear (Figure 20.3). Each auricle slightly increases the

Figure 20.3 Structure of the heart: surface features. Throughout this book, blood vessels that carry oxygenated blood (which looks bright red) are colored red, whereas those that carry deoxygenated blood (which looks dark red) are colored blue.

 Sulci are grooves that contain blood vessels and fat and that mark the external boundaries between the various chambers.

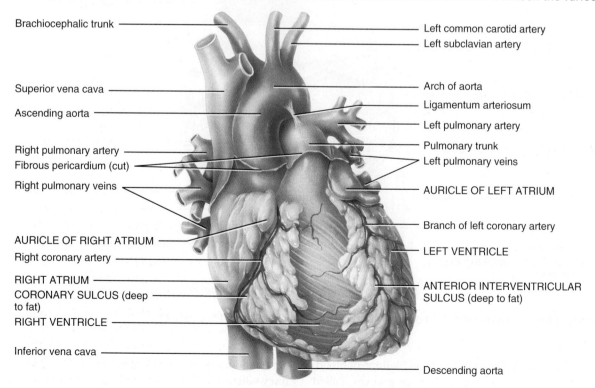

(a) Anterior external view showing surface features

FIGURE 20.3 CONTINUES ▶

■ **FIGURE 20.3 CONTINUED** ▶

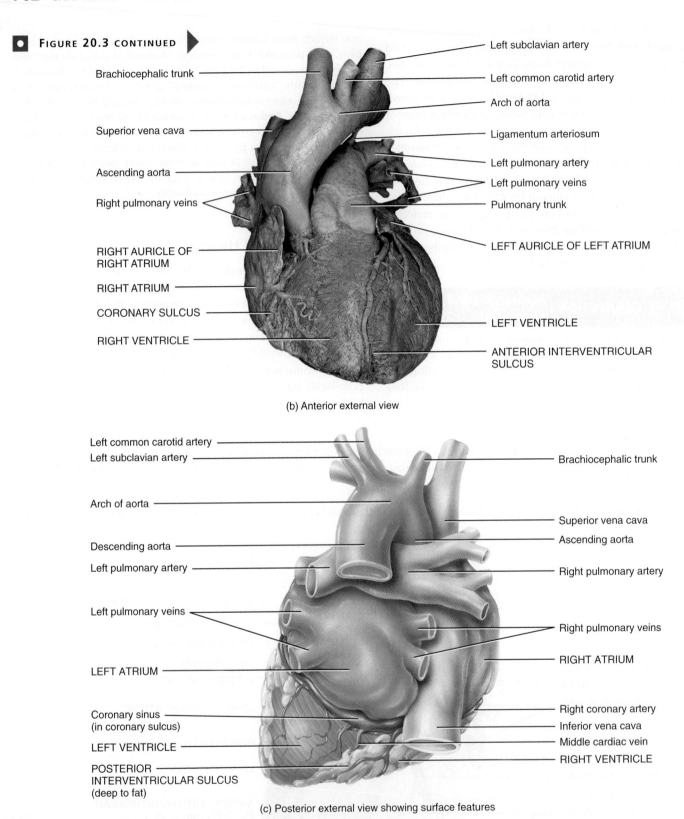

Left subclavian artery

Brachiocephalic trunk

Left common carotid artery

Arch of aorta

Superior vena cava

Ligamentum arteriosum

Ascending aorta

Left pulmonary artery

Left pulmonary veins

Right pulmonary veins

Pulmonary trunk

RIGHT AURICLE OF
RIGHT ATRIUM

LEFT AURICLE OF LEFT ATRIUM

RIGHT ATRIUM

CORONARY SULCUS

LEFT VENTRICLE

RIGHT VENTRICLE

ANTERIOR INTERVENTRICULAR
SULCUS

(b) Anterior external view

Left common carotid artery

Left subclavian artery

Brachiocephalic trunk

Arch of aorta

Superior vena cava

Ascending aorta

Descending aorta

Left pulmonary artery

Right pulmonary artery

Left pulmonary veins

Right pulmonary veins

RIGHT ATRIUM

LEFT ATRIUM

Coronary sinus
(in coronary sulcus)

Right coronary artery

Inferior vena cava

LEFT VENTRICLE

Middle cardiac vein

RIGHT VENTRICLE

POSTERIOR
INTERVENTRICULAR SULCUS
(deep to fat)

(c) Posterior external view showing surface features

? **The coronary sulcus forms an external boundary between which chambers of the heart?**

capacity of an atrium so that it can hold a greater volume of blood. Also on the surface of the heart are a series of grooves, called **sulci** (SUL-sī), that contain coronary blood vessels and a variable amount of fat. Each *sulcus* (SUL-kus; singular) marks the exter-nal boundary between two chambers of the heart. The deep **coronary sulcus** (*coron-* = resembling a crown) encircles most of the heart and marks the external boundary between the superior atria and inferior ventricles. The **anterior interventricular sulcus** (in′-ter-ven-TRIK-ū-lar) is a shallow groove on the anterior surface of the heart that marks the external boundary between the right and left ventricles. This sulcus continues around to the posterior surface of the heart as the **posterior interventricular sulcus,** which marks the external boundary between the ventricles on the posterior aspect of the heart (Figure 20.3c).

Right Atrium

The **right atrium** forms the right border of the heart and receives blood from three veins: the *superior vena cava, inferior vena cava,* and *coronary sinus* (Figure 20.4a). (Veins always carry blood toward the heart.) The right atrium is about 2–3 mm (0.08–0.12 in.) in average thickness. The anterior and posterior walls of the right atrium are very different. The inside of the posterior wall is smooth; the inside of the anterior wall is rough due to the presence of muscular ridges called **pectinate muscles** (PEK-ti-nāt; *pectin* = comb), which also extend into the auricle (Figure 20.4b). Between the right atrium and left atrium is a thin partition called the **interatrial septum** (*inter-* = between; *septum* = a dividing wall or partition). A prominent feature of this septum is

an oval depression called the **fossa ovalis,** the remnant of the *foramen ovale,* an opening in the interatrial septum of the fetal heart that normally closes soon after birth (see Figure 21.30). Blood passes from the right atrium into the right ventricle through a valve that is called the **tricuspid valve** (trī-KUS-pid; *tri-* = three; *-cuspid* = point) because it consists of three leaflets or cusps (Figure 20.4a). It is also called the **right atrioventricular valve** (ā′-trē-ō-ven-TRIK-ū-lar). The valves of the heart are composed of dense connective tissue covered by endocardium.

Right Ventricle

The **right ventricle** is about 4–5 mm (0.16–0.2 in.) in average thickness and forms most of the anterior surface of the heart. The inside of the right ventricle contains a series of ridges formed by raised bundles of cardiac muscle fibers called **trabeculae carneae** (tra-BEK-ū-lē KAR-nē-ē; *trabeculae* = little beams; *carneae* = fleshy; see Figure 20.2a). Some of the trabeculae carneae convey part of the conduction system of the heart, which you will learn about later in this chapter (see Section 20.3). The cusps of the tricuspid valve are connected to tendonlike cords, the

Figure 20.4 **Structure of the heart: internal anatomy.**

Blood flows into the right atrium through the superior vena cava, inferior vena cava, and coronary sinus and into the left atrium through four pulmonary veins.

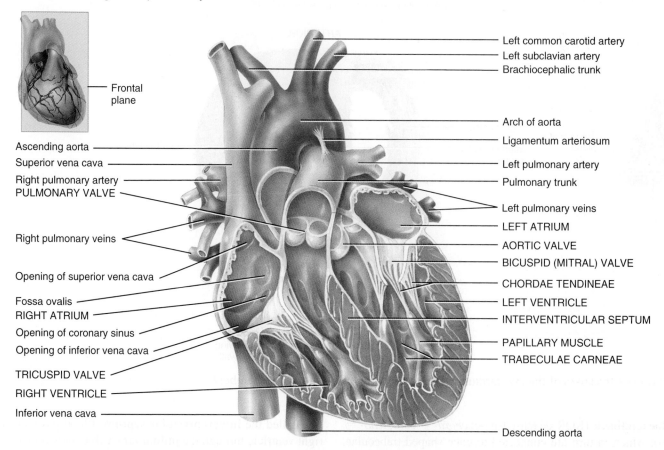

(a) Anterior view of frontal section showing internal anatomy

FIGURE 20.4 CONTINUES ▶

■ **FIGURE 20.4 CONTINUED** ▶

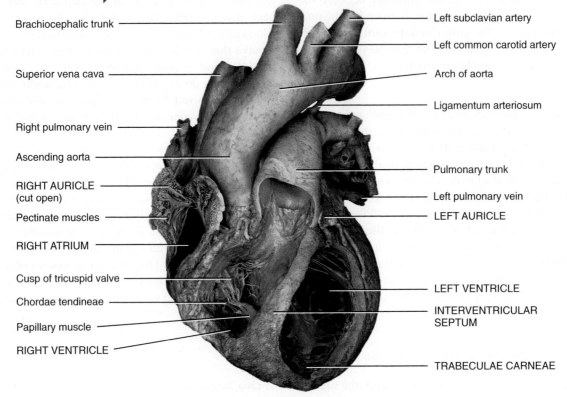

Brachiocephalic trunk

Superior vena cava

Right pulmonary vein

Ascending aorta

RIGHT AURICLE (cut open)

Pectinate muscles

RIGHT ATRIUM

Cusp of tricuspid valve

Chordae tendineae

Papillary muscle

RIGHT VENTRICLE

Left subclavian artery

Left common carotid artery

Arch of aorta

Ligamentum arteriosum

Pulmonary trunk

Left pulmonary vein

LEFT AURICLE

LEFT VENTRICLE

INTERVENTRICULAR SEPTUM

TRABECULAE CARNEAE

(b) Anterior view of partially sectioned heart

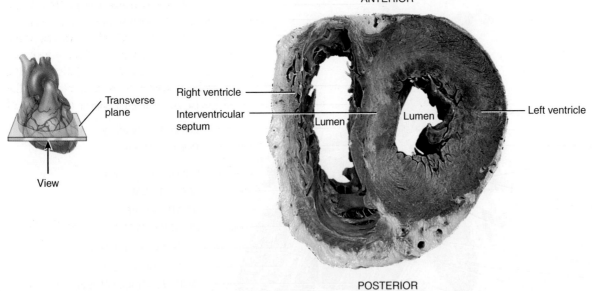

ANTERIOR

Transverse plane

View

Right ventricle

Interventricular septum

Lumen

Lumen

Left ventricle

POSTERIOR

(c) Inferior view of transverse section showing differences in thickness of ventricular walls

❓ **How does thickness of the myocardium relate to the workload of a cardiac chamber?**

chordae tendineae (KOR-dē ten-DI-nē-ē; *chord-* = cord; *tend-* = tendon), which in turn are connected to cone-shaped trabeculae carneae called **papillary muscles** (*papill-* = nipple). Internally, the right ventricle is separated from the left ventricle by a parti-

tion called the **interventricular septum.** Blood passes from the right ventricle through the **pulmonary valve** (*pulmonary semilunar valve*) into a large artery called the *pulmonary trunk,* which divides into right and left *pulmonary arteries* and carries blood to the lungs. Arteries always take blood away from the heart (a mnemonic to help you: artery = away).

Left Atrium

The **left atrium** is about the same thickness as the right atrium and forms most of the base of the heart (Figure 20.4a). It receives blood from the lungs through four *pulmonary veins.* Like the right atrium, the inside of the left atrium has a smooth posterior wall. Because pectinate muscles are confined to the auricle of the left atrium, the anterior wall of the left atrium also is smooth. Blood passes from the left atrium into the left ventricle through the **bicuspid (mitral) valve** (*bi-* = two), which, as its name implies, has two cusps. The term *mitral* refers to the resemblance of the bicuspid valve to a bishop's miter (hat), which is two-sided. It is also called the **left atrioventricular valve.**

Left Ventricle

The **left ventricle** is the thickest chamber of the heart, averaging 10–15 mm (0.4–0.6 in.) and forms the apex of the heart (see Figure 20.1b). Like the right ventricle, the left ventricle contains trabeculae carneae and has chordae tendineae that anchor the cusps of the bicuspid valve to papillary muscles. Blood passes from the left ventricle through the **aortic valve** (*aortic semilunar valve*) into the *ascending aorta* (*aorte* = to suspend, because the aorta once was believed to lift up the heart). Some of the blood in the aorta flows into the *coronary arteries,* which branch from the ascending aorta and carry blood to the heart wall. The remainder of the blood passes into the *arch of the aorta* and *descending aorta* (*thoracic aorta* and *abdominal aorta*). Branches of the arch of the aorta and descending aorta carry blood throughout the body.

During fetal life, a temporary blood vessel, called the *ductus arteriosus,* shunts blood from the pulmonary trunk into the aorta. Hence, only a small amount of blood enters the nonfunctioning fetal lungs (see Figure 21.30). The ductus arteriosus normally closes shortly after birth, leaving a remnant known as the **ligamentum arteriosum** (lig′-a-MEN-tum ar-ter-ē-Ō-sum), which connects the arch of the aorta and pulmonary trunk (Figure 20.4a).

Myocardial Thickness and Function

The thickness of the myocardium of the four chambers varies according to each chamber's function. The thin-walled atria deliver blood under less pressure into the adjacent ventricles. Because the ventricles pump blood under higher pressure over greater distances, their walls are thicker (Figure 20.4a). Although the right and left ventricles act as two separate pumps that simultaneously eject equal volumes of blood, the right side has a much smaller workload. It pumps blood a short distance to the lungs at lower pressure, and the resistance to blood flow is small. The left ventricle pumps blood great distances to all other parts of the body at higher pressure, and the resistance to blood flow is larger. Therefore, the left ventricle works much harder than the right ventricle to maintain the same rate of blood flow. The anatomy of the two ventricles confirms this functional difference—the muscular wall of the left ventricle is considerably thicker than the wall of the right ventricle (Figure 20.4c). Note also that the perimeter of the lumen (space) of the left ventricle is roughly circular, in contrast to that of the right ventricle, which is somewhat crescent-shaped.

Fibrous Skeleton of the Heart

In addition to cardiac muscle tissue, the heart wall also contains dense connective tissue that forms the **fibrous skeleton of the heart** (Figure 20.5). Essentially, the fibrous skeleton consists of four dense connective tissue rings that surround the valves of the heart, fuse with one another, and merge with the interventricular septum. In addition to forming a structural foundation for the heart valves, the fibrous skeleton prevents overstretching of the valves as blood passes through them. It also serves as a point of

Figure 20.5 Fibrous skeleton of the heart. Elements of the fibrous skeleton are shown in capital letters.

 Fibrous rings support the four valves of the heart and are fused to one another.

View

Transverse plane

Pulmonary valve

Left coronary artery

Aortic valve

LEFT FIBROUS TRIGONE

RIGHT FIBROUS TRIGONE

Bicuspid valve

LEFT ATRIOVENTRICULAR FIBROUS RING

PULMONARY FIBROUS RING

CONUS TENDON

AORTIC FIBROUS RING

Right coronary artery

Tricuspid valve

RIGHT ATRIOVENTRICULAR FIBROUS RING

Superior view (the atria have been removed)

? **In what two ways does the fibrous skeleton contribute to the functioning of heart valves?**

insertion for bundles of cardiac muscle fibers and acts as an electrical insulator between the atria and ventricles.

✔**CHECKPOINT**

1. Define each of the following external features of the heart: auricle, coronary sulcus, anterior interventricular sulcus, and posterior interventricular sulcus.
2. Describe the structure of the pericardium and the layers of the wall of the heart.
3. What are the characteristic internal features of each chamber of the heart?
4. Which blood vessels deliver blood to the right and left atria?
5. What is the relationship between wall thickness and function among the various chambers of the heart?
6. What type of tissue composes the fibrous skeleton of the heart, and how is it organized?

20.2 HEART VALVES AND CIRCULATION OF BLOOD

◉ **OBJECTIVES**

• Describe the structure and function of the valves of the heart.
• Outline the flow of blood through the chambers of the heart and through the systemic and pulmonary circulations.
• Discuss the coronary circulation.

As each chamber of the heart contracts, it pushes a volume of blood into a ventricle or out of the heart into an artery. Valves open and close in response to *pressure changes* as the heart contracts and relaxes. Each of the four valves helps ensure the one-way flow of blood by opening to let blood through and then closing to prevent its backflow.

Figure 20.6 Responses of the valves to the pumping of the heart.

🔑 Heart valves prevent the backflow of blood.

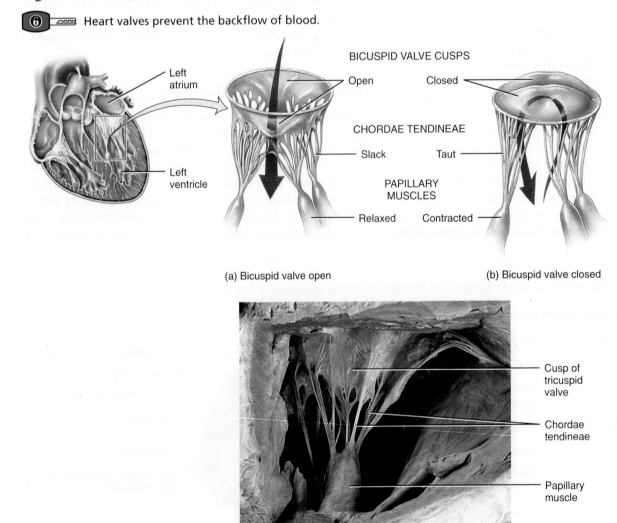

(a) Bicuspid valve open   (b) Bicuspid valve closed

(c) Tricuspid valve open

Operation of the Atrioventricular Valves

Because they are located between an atrium and a ventricle, the tricuspid and bicuspid valves are termed **atrioventricular (AV) valves.** When an AV valve is open, the rounded ends of the cusps project into the ventricle. When the ventricles are relaxed, the papillary muscles are relaxed, the chordae tendineae are slack, and blood moves from a higher pressure in the atria to a lower pressure in the ventricles through open AV valves (Figure 20.6a, d). When the ventricles contract, the pressure of the blood drives the cusps upward until their edges meet and close the opening (Figure 20.6b, e). At the

same time, the papillary muscles contract, which pulls on and tightens the chordae tendineae. This prevents the valve cusps from everting (opening into the atria) in response to the high ventricular pressure. If the AV valves or chordae tendineae are damaged, blood may regurgitate (flow back) into the atria when the ventricles contract.

Operation of the Semilunar Valves

The aortic and pulmonary valves are known as the **semilunar (SL) valves** (sem-ē-LOO-nar; *semi-* = half; *-lunar* = moon-shaped) because they are made up of three crescent moon–shaped

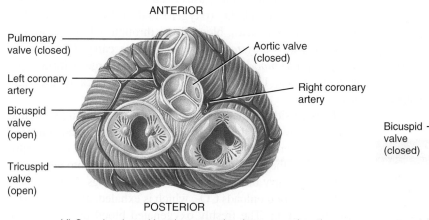

(d) Superior view with atria removed: pulmonary and aortic valves closed, bicuspid and tricuspid valves open

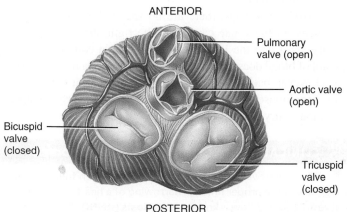

(e) Superior view with atria removed: pulmonary and aortic valves open, bicuspid and tricuspid valves closed

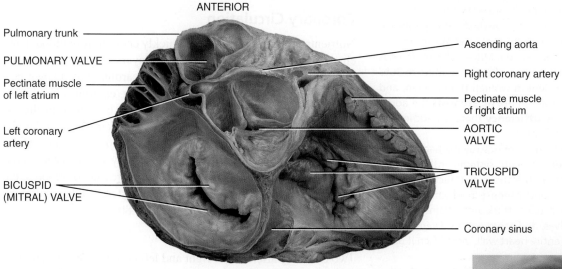

(f) Superior view of atrioventricular and semilunar valves

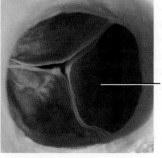

(g) Superior view of aortic valve

How do papillary muscles prevent atrioventricular valve cusps from everting (swinging upward) into the atria?

cusps (Figure 20.6d). Each cusp attaches to the arterial wall by its convex outer margin. The SL valves allow ejection of blood from the heart into arteries but prevent backflow of blood into the ventricles. The free borders of the cusps project into the lumen of the artery. When the ventricles contract, pressure builds up within the chambers. The semilunar valves open when pressure in the ventricles exceeds the pressure in the arteries, permitting ejection of blood from the ventricles into the pulmonary trunk and aorta (Figure 20.6e). As the ventricles relax, blood starts to flow back toward the heart. This backflowing blood fills the valve cusps, which causes the free edges of the semilunar valves to contact each other tightly and close the opening between the ventricle and artery (Figure 20.6d).

Surprisingly perhaps, there are no valves guarding the junctions between the venae cavae and the right atrium or the pulmonary veins and the left atrium. As the atria contract, a small amount of blood does flow backward from the atria into these vessels. However, backflow is minimized by a different mechanism; as the atrial muscle contracts, it compresses and nearly collapses the venous entry points.

☤ CLINICAL CONNECTION | *Heart Valve Disorders*

When heart valves operate normally, they open fully and close completely at the proper times. A narrowing of a heart valve opening that restricts blood flow is known as **stenosis** (ste-NŌ-sis = a narrowing); failure of a valve to close completely is termed **insufficiency** (in'-su-FISH-en-sē) or **incompetence**. In **mitral stenosis,** scar formation or a congenital defect causes narrowing of the mitral valve. One cause of **mitral insufficiency,** in which there is backflow of blood from the left ventricle into the left atrium, is **mitral valve prolapse (MVP).** In MVP one or both cusps of the mitral valve protrude into the left atrium during ventricular contraction. Mitral valve prolapse is one of the most common valvular disorders, affecting as much as 30% of the population. It is more prevalent in women than in men, and does not always pose a serious threat. In **aortic stenosis** the aortic valve is narrowed, and in **aortic insufficiency** there is backflow of blood from the aorta into the left ventricle.

Certain infectious diseases can damage or destroy the heart valves. One example is **rheumatic fever,** an acute systemic inflammatory disease that usually occurs after a streptococcal infection of the throat. The bacteria trigger an immune response in which antibodies produced to destroy the bacteria instead attack and inflame the connective tissues in joints, heart valves, and other organs. Even though rheumatic fever may weaken the entire heart wall, most often it damages the mitral and aortic valves.

If daily activities are affected by symptoms and if a heart valve cannot be repaired surgically, then the valve must be replaced. Tissue valves may be provided by human donors or pigs; sometimes, mechanical replacements are used. In any case, valve replacement involves open heart surgery. The aortic valve is the most commonly replaced heart valve. •

Systemic and Pulmonary Circulations

In postnatal (after birth) circulation, the heart pumps blood into two closed circuits with each beat—**systemic circulation** and **pulmonary circulation** (*pulmon-* = lung). The two circuits are arranged in series: The output of one becomes the input of the other, as would happen if you attached two garden hoses (see Figure 21.17). The left side of the heart is the pump for systemic circulation; it receives bright red *oxygenated* (oxygen-rich) *blood* from the lungs. The left ventricle ejects blood into the *aorta* (Figure 20.7). From the aorta, the blood divides into separate streams, entering progressively smaller *systemic arteries* that carry it to all organs throughout the body—except for the air sacs (alveoli) of the lungs, which are supplied by pulmonary circulation. In systemic tissues, arteries give rise to smaller-diameter *arterioles,* which finally lead into extensive beds of *systemic capillaries.* Exchange of nutrients and gases occurs across the thin capillary walls. Blood unloads O_2 (oxygen) and picks up CO_2 (carbon dioxide). In most cases, blood flows through only one capillary and then enters a *systemic venule.* Venules carry *deoxygenated* (oxygen-poor) *blood* away from tissues and merge to form larger *systemic veins.* Ultimately the blood flows back to the right atrium.

The right side of the heart is the pump for pulmonary circulation; it receives all the dark-red deoxygenated blood returning from the systemic circulation. Blood ejected from the right ventricle flows into the *pulmonary trunk,* which branches into *pulmonary arteries* that carry blood to the right and left lungs. In pulmonary capillaries, blood unloads CO_2, which is exhaled, and picks up O_2 from inhaled air. The freshly oxygenated blood then flows into pulmonary veins and returns to the left atrium.

Coronary Circulation

Nutrients are not able to diffuse quickly enough from blood in the chambers of the heart to supply all the layers of cells that make up the heart wall. For this reason, the myocardium has its own network of blood vessels, the **coronary** or **cardiac circulation** (*coron-* = crown). The **coronary arteries** branch from the ascending aorta and encircle the heart like a crown encircles the head (Figure 20.8a). While the heart is contracting, little blood flows in the coronary arteries because they are squeezed shut. When the heart relaxes, however, the high pressure of blood in the aorta propels blood through the coronary arteries, into capillaries, and then into **coronary veins** (Figure 20.8b).

Coronary Arteries

Two coronary arteries, the right and left coronary arteries, branch from the ascending aorta and supply oxygenated blood to the myocardium (Figure 20.8a). The **left coronary artery** passes inferior to the left auricle and divides into the anterior interventricular and circumflex branches. The **anterior interventricular branch** or *left anterior descending (LAD) artery* is in the anterior interventricular sulcus and supplies oxygenated blood to the walls of both ventricles. The **circumflex branch** (SER-kum-fleks) lies in the coronary sulcus and distributes oxygenated blood to the walls of the left ventricle and left atrium.

The **right coronary artery** supplies small branches (*atrial branches*) to the right atrium. It continues inferior to the right auricle and ultimately divides into the posterior interventricular and marginal branches. The **posterior interventricular branch** follows the posterior interventricular sulcus and supplies the walls of the two ventricles with oxygenated blood. The **marginal branch** beyond the coronary sulcus runs along the right margin of the heart and transports oxygenated blood to the myocardium of the right ventricle.

Most parts of the body receive blood from branches of more than one artery, and where two or more arteries supply the same region, they usually connect. These connections, called **anastomoses** (a-nas′-tō-MŌ-sēs), provide alternate routes, called **collateral circulation,** for blood to reach a particular organ or tissue. The myocardium contains many anastomoses that connect branches of a given coronary artery or extend between branches of different coronary arteries. They provide detours for arterial blood if a main route becomes obstructed. Thus, heart muscle may receive sufficient oxygen even if one of its coronary arteries is partially blocked.

Figure 20.7 Systemic and pulmonary circulations.

The left side of the heart pumps oxygenated blood into the systemic circulation to all tissues of the body except the air sacs (alveoli) of the lungs. The right side of the heart pumps deoxygenated blood into the pulmonary circulation to the air sacs (alveoli) of the lungs.

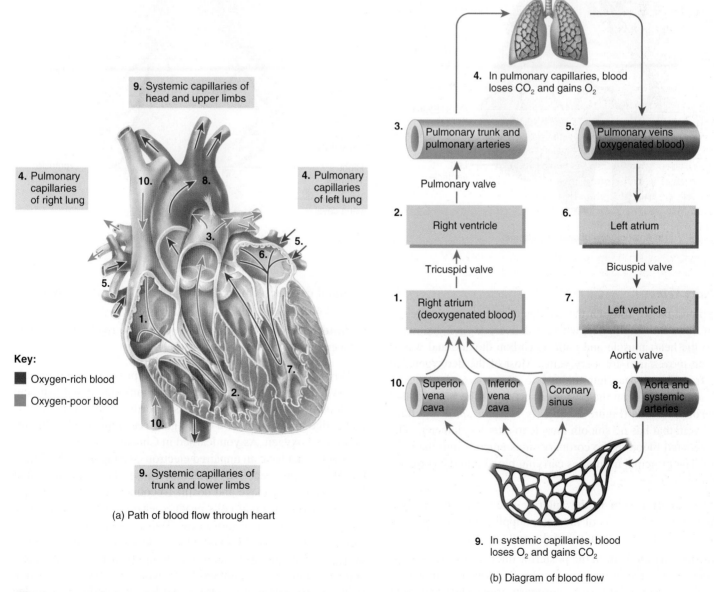

9. Systemic capillaries of head and upper limbs

4. Pulmonary capillaries of right lung

4. Pulmonary capillaries of left lung

Key:
- Oxygen-rich blood
- Oxygen-poor blood

9. Systemic capillaries of trunk and lower limbs

(a) Path of blood flow through heart

4. In pulmonary capillaries, blood loses CO_2 and gains O_2

3. Pulmonary trunk and pulmonary arteries

Pulmonary valve

2. Right ventricle

Tricuspid valve

1. Right atrium (deoxygenated blood)

10. Superior vena cava | Inferior vena cava | Coronary sinus

5. Pulmonary veins (oxygenated blood)

6. Left atrium

Bicuspid valve

7. Left ventricle

Aortic valve

8. Aorta and systemic arteries

9. In systemic capillaries, blood loses O_2 and gains CO_2

(b) Diagram of blood flow

? **Which numbers constitute the pulmonary circulation? Which constitute the systemic circulation?**

Figure 20.8 **The coronary circulation.** The views of the heart from the anterior aspect in (a) and (b) are drawn as if the heart were transparent to reveal blood vessels on the posterior aspect.

The right and left coronary arteries deliver blood to the heart; the coronary veins drain blood from the heart into the coronary sinus.

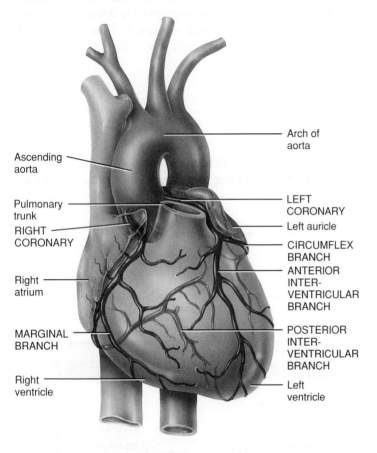

(a) Anterior view of coronary arteries

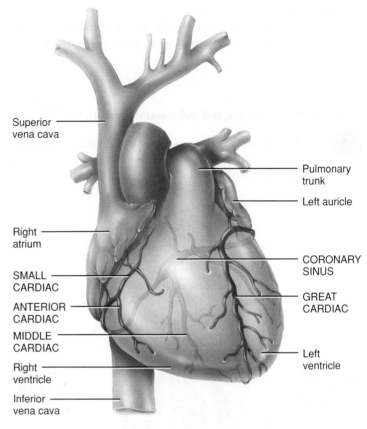

(b) Anterior view of coronary veins

Coronary Veins

After blood passes through the arteries of the coronary circulation, it flows into capillaries, where it delivers oxygen and nutrients to the heart muscle and collects carbon dioxide and waste, and then moves into coronary veins. Most of the deoxygenated blood from the myocardium drains into a large *vascular sinus* in the coronary sulcus on the posterior surface of the heart, called the **coronary sinus** (Figure 20.8b). (A *vascular sinus* is a thin-walled vein that has no smooth muscle to alter its diameter.) The deoxygenated blood in the coronary sinus empties into the right atrium. The principal tributaries carrying blood into the coronary sinus are the following:

• **Great cardiac vein** in the anterior interventricular sulcus, which drains the areas of the heart supplied by the left coronary artery (left and right ventricles and left atrium)

• **Middle cardiac vein** in the posterior interventricular sulcus, which drains the areas supplied by the posterior interventricular branch of the right coronary artery (left and right ventricles)

• **Small cardiac vein** in the coronary sulcus, which drains the right atrium and right ventricle

• **Anterior cardiac veins,** which drain the right ventricle and open directly into the right atrium

When blockage of a coronary artery deprives the heart muscle of oxygen, **reperfusion** (re′-per-FYŪ-zhun), the reestablishment of blood flow, may damage the tissue further. This surprising effect is due to the formation of oxygen **free radicals** from the reintroduced oxygen. As you learned in Chapter 2, free radicals are molecules that have an unpaired electron (see Figure 2.3b). These unstable, highly reactive molecules cause chain reactions that lead to cellular damage and death. To counter the effects of oxygen free radicals, body cells produce enzymes that convert free radicals to less reactive substances. Two such enzymes are *superoxide dismutase* (dis-MŪ-tās) and *catalase* (KAT-a-lās). In addition, nutrients such as vitamin E, vitamin C, beta-carotene, zinc, and selenium serve as antioxidants, which remove oxygen free radicals from circulation. Drugs that lessen reperfusion damage after a heart attack or stroke are currently under development.

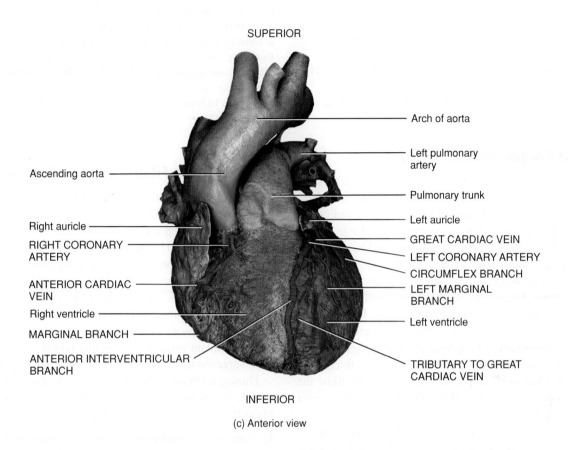

SUPERIOR

Ascending aorta

Right auricle

RIGHT CORONARY
ARTERY

ANTERIOR CARDIAC
VEIN

Right ventricle

MARGINAL BRANCH

ANTERIOR INTERVENTRICULAR
BRANCH

Arch of aorta

Left pulmonary
artery

Pulmonary trunk

Left auricle

GREAT CARDIAC VEIN

LEFT CORONARY ARTERY

CIRCUMFLEX BRANCH

LEFT MARGINAL
BRANCH

Left ventricle

TRIBUTARY TO GREAT
CARDIAC VEIN

INFERIOR

(c) Anterior view

? **Which coronary blood vessel delivers oxygenated blood to the walls of the left atrium and left ventricle?**

CLINICAL CONNECTION | *Myocardial Ischemia and Infarction*

Partial obstruction of blood flow in the coronary arteries may cause **myocardial ischemia** (is-KĒ-mē-a; *ische-* = to obstruct; *-emia* = in the blood), a condition of reduced blood flow to the myocardium. Usually, ischemia causes **hypoxia** (hī-POKS-ē-a = reduced oxygen supply), which may weaken cells without killing them. **Angina pectoris** (an-JĪ-na, or AN-ji-na, PEK-tō-ris), which literally means "strangled chest," is a severe pain that usually accompanies myocardial ischemia. Typically, sufferers describe it as a tightness or squeezing sensation, as though the chest were in a vise. The pain associated with angina pectoris is often referred to the neck, chin, or down the left arm to the elbow. **Silent myocardial ischemia,** ischemic episodes without pain, is particularly dangerous because the person has no forewarning of an impending heart attack.

A complete obstruction to blood flow in a coronary artery may result in a **myocardial infarction (MI)** (in-FARK-shun), commonly called a *heart attack. Infarction* means the death of an area of tissue because of interrupted blood supply. Because the heart tissue distal to the obstruction dies and is replaced by noncontractile scar tissue, the heart muscle loses some of its strength. Depending on the size and location of the infarcted (dead) area, an infarction may disrupt the conduction system of the heart and cause sudden death by triggering ventricular fibrillation. Treatment for a myocardial infarction may involve injection of a thrombolytic (clot-dissolving) agent such as streptokinase or t-PA, plus heparin (an anticoagulant), or performing coronary angioplasty or coronary artery bypass grafting. Fortunately, heart muscle can remain alive in a resting person if it receives as little as 10–15% of its normal blood supply.   •

✔CHECKPOINT

7. What causes the heart valves to open and to close? What supporting structures ensure that the valves operate properly?

8. In correct sequence, which heart chambers, heart valves, and blood vessels would a drop of blood encounter as it flows from the right atrium to the aorta?

9. Which arteries deliver oxygenated blood to the myocardium of the left and right ventricles?

20.3 CARDIAC MUSCLE TISSUE AND THE CARDIAC CONDUCTION SYSTEM

◉ **OBJECTIVES**

• Describe the structural and functional characteristics of cardiac muscle tissue and the cardiac conduction system.

• Explain how an action potential occurs in cardiac contractile fibers.

• Describe the electrical events of a normal electrocardiogram (ECG).

Histology of Cardiac Muscle Tissue

Compared with skeletal muscle fibers, cardiac muscle fibers are shorter in length and less circular in transverse section (Figure 20.9). They also exhibit branching, which gives individual cardiac muscle fibers a "stair-step" appearance (see Table 4.9). A typical cardiac muscle fiber is 50–100 μm long and has a diameter of about 14 μm. Usually one centrally located nucleus is present, although an occasional cell may have two nuclei. The ends of cardiac muscle fibers connect to neighboring fibers by irregular transverse thickenings of the sarcolemma called **intercalated discs** (in-TER-ka-lāt-ed; *intercalat-* = to insert between). The discs contain **desmosomes,** which hold the fibers together, and **gap junctions,** which allow muscle action potentials to conduct from one muscle fiber to its neighbors. Gap junctions allow the entire myocardium of the atria or the ventricles to contract as a single, coordinated unit.

Mitochondria are larger and more numerous in cardiac muscle fibers than in skeletal muscle fibers. In a cardiac muscle fiber, they take up 25% of the cytosolic space; in a skeletal muscle fiber only 2% of the cytosolic space is occupied by mitochondria. Cardiac muscle fibers have the same arrangement of actin and myosin, and the same bands, zones, and Z discs, as skeletal muscle fibers. The transverse tubules of cardiac muscle are wider but less abundant than those of skeletal muscle; the one transverse tubule per sarcomere is located at the Z disc. The sarcoplasmic reticulum of cardiac muscle fibers is somewhat smaller than the SR of skeletal muscle fibers. As a result, cardiac muscle has a smaller intracellular reserve of Ca^{2+}.

⚕ **CLINICAL CONNECTION |** *Regeneration of Heart Cells*

As noted earlier in the chapter, the heart of a heart attack survivor often has regions of infarcted (dead) cardiac muscle tissue that typically are replaced with noncontractile fibrous scar tissue over time. Our inability to repair damage from a heart attack has been attributed to a lack of stem cells in cardiac muscle and to the absence of mitosis in mature cardiac muscle fibers. A recent study of heart transplant recipients by American and Italian scientists, however, provides evidence for significant replacement of heart cells. The researchers studied men who had received a heart from a female, and then looked for the presence of a Y chromosome in heart cells. (All female cells except gametes have two X chromosomes and lack the Y chromosome.) Several years after the transplant surgery, between 7% and 16% of the heart cells in the transplanted tissue, including cardiac muscle fibers and endothelial cells in coronary arterioles and capillaries, had been replaced by the recipient's own cells, as evidenced by the presence of a Y chromosome. The study also revealed cells with some of the characteristics of stem cells in both transplanted hearts and control hearts. Evidently, stem cells can migrate from the blood into the heart and differentiate into functional muscle and endothelial cells. The hope is that researchers can learn how to "turn on" such regeneration of heart cells to treat people with heart failure or cardiomyopathy (diseased heart). •

Autorhythmic Fibers: The Conduction System

An inherent and rhythmical electrical activity is the reason for the heart's lifelong beat. The source of this electrical activity is a network of specialized cardiac muscle fibers called **autorhythmic fibers** (aw'-tō-RITH-mik; *auto-* = self) because they are self-excitable. Autorhythmic fibers repeatedly generate action potentials that trigger heart contractions. They continue to stimulate a heart to beat even after it is removed from the body—for example, to be transplanted into another person—and all of its nerves have been cut. (Note: Surgeons do not attempt to reattach heart nerves during heart transplant operations. For this reason, it has been said that heart surgeons are better "plumbers" than they are "electricians.")

During embryonic development, only about 1% of the cardiac muscle fibers become autorhythmic fibers; these relatively rare fibers have two important functions:

1. They act as a **pacemaker,** setting the rhythm of electrical excitation that causes contraction of the heart.

2. They form the **cardiac conduction system,** a network of specialized cardiac muscle fibers that provide a path for each cycle of cardiac excitation to progress through the heart. The conduction system ensures that cardiac chambers become stimulated to contract in a coordinated manner, which makes the heart an effective pump. As you will see later in the chapter, problems with autorhythmic fibers can result in arrhythmias (abnormal rhythms) in which the heart beats irregularly, too fast, or too slow.

Cardiac action potentials propagate through the conduction system in the following sequence (Figure 20.10a):

❶ Cardiac excitation normally begins in the **sinoatrial (SA) node,** located in the right atrial wall just inferior and lateral to the opening of the superior vena cava. SA node cells do not have a stable resting potential. Rather, they repeatedly depolarize to threshold spontaneously. The spontaneous depolarization is a **pacemaker potential.** When the pacemaker potential reaches threshold, it triggers an action potential (Figure 20.10b). Each action potential from the SA node propagates throughout both atria via gap junctions in the intercalated discs of atrial muscle fibers. Following the action potential, the two atria contract at the same time.

Figure 20.9 **Histology of cardiac muscle tissue.** (See Table 4.9 for a light micrograph of cardiac muscle.)

Cardiac muscle fibers connect to neighboring fibers by intercalated discs, which contain desmosomes and gap junctions.

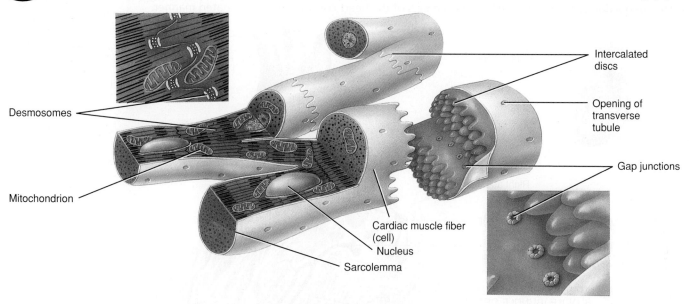

(a) Cardiac muscle fibers

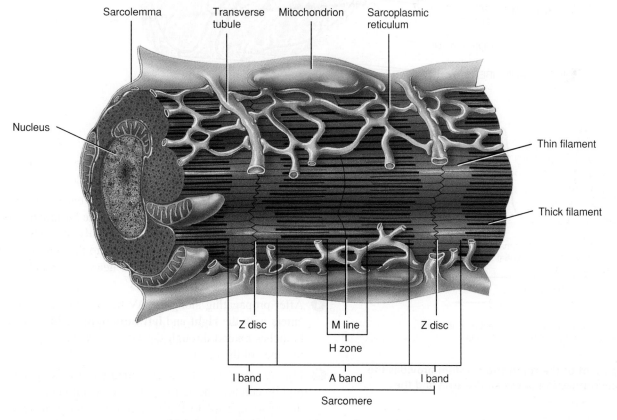

(b) Arrangement of components in a cardiac muscle fiber

What are the functions of intercalated discs in cardiac muscle fibers?

Figure 20.10 The conduction system of the heart. Autorhythmic fibers in the SA node, located in the right atrial wall (a), act as the heart's pacemaker, initiating cardiac action potentials (b) that cause contraction of the heart's chambers.

The conduction system ensures that the chambers of the heart contract in a coordinated manner.

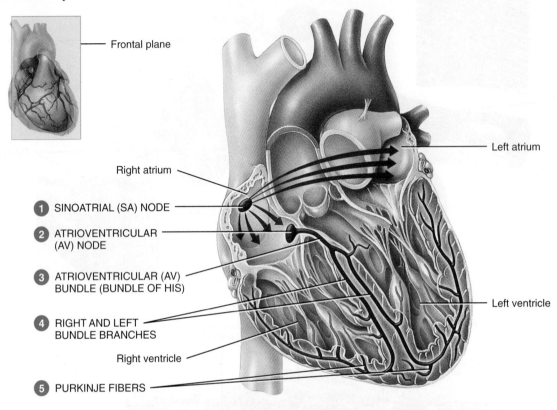

Frontal plane

Right atrium

Left atrium

1 SINOATRIAL (SA) NODE

2 ATRIOVENTRICULAR (AV) NODE

3 ATRIOVENTRICULAR (AV) BUNDLE (BUNDLE OF HIS)

4 RIGHT AND LEFT BUNDLE BRANCHES

Right ventricle

Left ventricle

5 PURKINJE FIBERS

(a) Anterior view of frontal section

+ 10 mV

Action potential

Membrane potential

Threshold

− 60 mV

Pacemaker potential

0   0.8   1.6   2.4

Time (sec) ⟶

(b) Pacemaker potentials and action potentials in autorhythmic fibers of SA node

? **Which component of the conduction system provides the only electrical connection between the atria and the ventricles?**

2 By conducting along atrial muscle fibers, the action potential reaches the **atrioventricular (AV) node,** located in the interatrial septum, just anterior to the opening of the coronary sinus (Figure 20.10a). At the AV node, the action potential slows considerably as a result of various differences in cell structure in the AV node. This delay provides time for the atria to empty their blood into the ventricles.

3 From the AV node, the action potential enters the **atrioventricular (AV) bundle** (also known as the **bundle of His,** pronounced HIZ). This bundle is the only site where action potentials can conduct from the atria to the ventricles. (Elsewhere, the fibrous skeleton of the heart electrically insulates the atria from the ventricles.)

4 After propagating along the AV bundle, the action potential enters both the **right** and **left bundle branches.** The bundle branches extend through the interventricular septum toward the apex of the heart.

5 Finally, the large-diameter **Purkinje fibers** (pur-KIN-jē) rapidly conduct the action potential beginning at the apex of the heart upward to the remainder of the ventricular myocardium. Then the ventricles contract, pushing the blood upward toward the semilunar valves.

On their own, autorhythmic fibers in the SA node would initiate an action potential about every 0.6 second, or 100 times per minute. Thus, the SA node sets the rhythm for contraction of the heart—it is the *natural pacemaker.* This rate is faster than that of any other autorhythmic fibers. Because action potentials from the

SA node spread through the conduction system and stimulate other areas before the other areas are able to generate an action potential at their own, slower rate, the SA node acts as the natural pacemaker of the heart. Nerve impulses from the autonomic nervous system (ANS) and blood-borne hormones (such as epinephrine) *modify the timing and strength* of each heartbeat, but they *do not establish the fundamental rhythm.* In a person at rest, for example, acetylcholine released by the parasympathetic division of the ANS slows SA node pacing to about every 0.8 second or 75 action potentials per minute (Figure 20.10b).

CLINICAL CONNECTION | Artificial Pacemakers

If the SA node becomes damaged or diseased, the slower AV node can pick up the pacemaking task. Its spontaneous pacing rate is 40 to 60 times per minute. If the activity of both nodes is suppressed, the heartbeat may still be maintained by autorhythmic fibers in the ventricles—the AV bundle, a bundle branch, or Purkinje fibers. However, the pacing rate is so slow (20–35 beats per minute) that blood flow to the brain is inadequate. When this condition occurs, normal heart rhythm can be restored and maintained by surgically implanting an **artificial pacemaker,** a device that sends out small electrical currents to stimulate the heart to contract. A pacemaker consists of a battery and impulse generator and is usually implanted beneath the skin just inferior to the clavicle. The pacemaker is connected to one or two flexible leads (wires) that are threaded through the superior vena cava and then passed into the various chambers of the heart. Many of the newer pacemakers, referred to as *activity-adjusted pacemakers,* automatically speed up the heartbeat during exercise. •

Action Potential and Contraction of Contractile Fibers

The action potential initiated by the SA node travels along the conduction system and spreads out to excite the "working" atrial and ventricular muscle fibers, called **contractile fibers.** An action potential occurs in a contractile fiber as follows (Figure 20.11):

1 *Depolarization.* Unlike autorhythmic fibers, contractile fibers have a stable resting membrane potential that is close to -90 mV. When a contractile fiber is brought to threshold by an action potential from neighboring fibers, its **voltage-gated fast Na$^+$ channels** open. These sodium ion channels are referred to as "fast" because they open very rapidly in response to a threshold-level depolarization. Opening of these channels allows Na$^+$ inflow because the cytosol of contractile fibers is electrically more negative than interstitial fluid and Na$^+$ concentration is higher in interstitial fluid. Inflow of Na$^+$ down the electrochemical gradient produces a **rapid depolarization** (dē′-pō-lar-i-ZĀ-shun). Within a few milliseconds, the fast Na$^+$ channels automatically inactivate and Na$^+$ inflow decreases.

2 *Plateau.* The next phase of an action potential in a contractile fiber is the **plateau,** a period of maintained depolarization. It is due in part to opening of **voltage-gated slow Ca^{2+} channels** in the sarcolemma. When these channels open, calcium ions move from the interstitial fluid (which has a higher Ca^{2+} concentration) into the cytosol. This inflow of Ca^{2+} causes even more Ca^{2+} to pour out of the sarcoplasmic reticulum into the cytosol through additional Ca^{2+} channels in the sarcoplasmic reticulum membrane. The increased Ca^{2+}

Figure 20.11 Action potential in a ventricular contractile fiber. The resting membrane potential is about -90 mV.

A long refractory period prevents tetanus in cardiac muscle fibers.

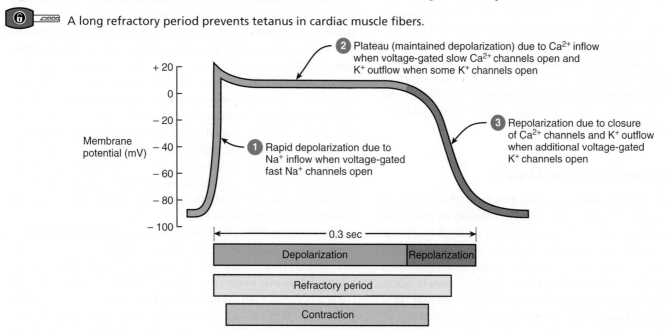

? How does the duration of an action potential in a ventricular contractile fiber compare with that in a skeletal muscle fiber?

concentration in the cytosol ultimately triggers contraction. Several different types of **voltage-gated K$^+$ channels** are also found in the sarcolemma of a contractile fiber. Just before the plateau phase begins, some of these K$^+$ channels open, allowing potassium ions to leave the contractile fiber. Therefore, depolarization is sustained during the plateau phase because Ca^{2+} inflow just balances K$^+$ outflow. The plateau phase lasts for about 0.25 sec, and the membrane potential of the contractile fiber is close to 0 mV. By comparison, depolarization in a neuron or skeletal muscle fiber is much briefer, about 1 msec (0.001 sec), because it lacks a plateau phase.

3 *Repolarization.* The recovery of the resting membrane potential during the **repolarization** (rē′-pō-lar-i-ZĀ-shun) phase of a cardiac action potential resembles that in other excitable cells. After a delay (which is particularly prolonged in cardiac muscle), additional voltage-gated K$^+$ channels open. Outflow of K$^+$ restores the negative resting membrane potential (-90 mV). At the same time, the calcium channels in the sarcolemma and the sarcoplasmic reticulum are closing, which also contributes to repolarization.

The mechanism of contraction is similar in cardiac and skeletal muscle: The electrical activity (action potential) leads to the mechanical response (contraction) after a short delay. As Ca^{2+} concentration rises inside a contractile fiber, Ca^{2+} binds to the regulatory protein troponin, which allows the actin and myosin filaments to begin sliding past one another, and tension starts to develop. Substances that alter the movement of Ca^{2+} through slow Ca^{2+} channels influence the strength of heart contractions. Epinephrine, for example, increases contraction force by enhancing Ca^{2+} flow into the cytosol.

In muscle, the **refractory period** (re-FRAK-to-rē) is the time interval during which a second contraction cannot be triggered. The refractory period of a cardiac muscle fiber lasts longer than the contraction itself (Figure 20.11). As a result, another contraction cannot begin until relaxation is well under way. For this reason, tetanus (maintained contraction) cannot occur in cardiac muscle as it can in skeletal muscle. The advantage is apparent if you consider how the ventricles work. Their pumping function depends on alternating contraction (when they eject blood) and relaxation (when they refill). If heart muscle could undergo tetanus, blood flow would cease.

ATP Production in Cardiac Muscle

In contrast to skeletal muscle, cardiac muscle produces little of the ATP it needs by anaerobic cellular respiration (see Figure 10.11). Instead, it relies almost exclusively on aerobic cellular respiration in its numerous mitochondria. The needed oxygen diffuses from blood in the coronary circulation and is released from myoglobin inside cardiac muscle fibers. Cardiac muscle fibers use several fuels to power mitochondrial ATP production. In a person at rest, the heart's ATP comes mainly from oxidation of fatty acids (60%) and glucose (35%), with smaller contributions from lactic acid, amino acids, and ketone bodies. During exercise, the heart's use of lactic acid, produced by actively contracting skeletal muscles, rises.

Like skeletal muscle, cardiac muscle also produces some ATP from creatine phosphate. One sign that a myocardial infarction (heart attack, see Clinical Connection: Myocardial Ischemia and Infarction) has occurred is the presence in blood of creatine kinase (CK), the enzyme that catalyzes transfer of a phosphate group from creatine phosphate to ADP to make ATP. Normally, CK and other enzymes are confined within cells. Injured or dying cardiac or skeletal muscle fibers release CK into the blood.

Electrocardiogram

As action potentials propagate through the heart, they generate electrical currents that can be detected at the surface of the body. An **electrocardiogram** (e-lek′-trō-KAR-dē-ō-gram), abbreviated either **ECG** or **EKG** (from the German word *Elektrokardiogram*), is a recording of these electrical signals. The ECG is a composite record of action potentials produced by all the heart muscle fibers during each heartbeat. The instrument used to record the changes is an **electrocardiograph.**

In clinical practice, electrodes are positioned on the arms and legs (limb leads) and at six positions on the chest (chest leads) to record the ECG. The electrocardiograph amplifies the heart's electrical signals and produces 12 different tracings from different combinations of limb and chest leads. Each limb and chest electrode records slightly different electrical activity because of the difference in its position relative to the heart. By comparing these records with one another and with normal records, it is possible to determine (1) if the conducting pathway is abnormal, (2) if the heart is enlarged, (3) if certain regions of the heart are damaged, and (4) the cause of chest pain.

In a typical record, three clearly recognizable waves appear with each heartbeat (Figure 20.12). The first, called the **P wave,** is a small upward deflection on the ECG. The P wave represents **atrial depolarization,** which spreads from the SA node through contractile fibers in both atria. The second wave, called the **QRS complex,** begins as a downward deflection, continues as a large, upright, triangular wave, and ends as a downward wave. The QRS complex represents **rapid ventricular depolarization,** as the action potential spreads through ventricular contractile fibers. The third wave is a dome-shaped upward deflection called the **T wave.** It indicates **ventricular repolarization** and occurs just as the ventricles are starting to relax. The T wave is smaller and wider than the QRS complex because repolarization occurs more slowly than depolarization. During the plateau period of steady depolarization, the ECG tracing is flat.

In reading an ECG, the size of the waves can provide clues to abnormalities. Larger P waves indicate enlargement of an atrium; an enlarged Q wave may indicate a myocardial infarction; and an enlarged R wave generally indicates enlarged ventricles. The T wave is flatter than normal when the heart muscle is receiving insufficient oxygen—as, for example, in coronary artery disease. The T wave may be elevated in hyperkalemia (high blood K$^+$ level).

Figure 20.12 **Normal electrocardiogram or ECG (lead II).**
P wave = atrial depolarization; QRS complex = onset of ventricular depolarization; T wave = ventricular repolarization.

An ECG is a recording of the electrical activity that initiates each heartbeat.

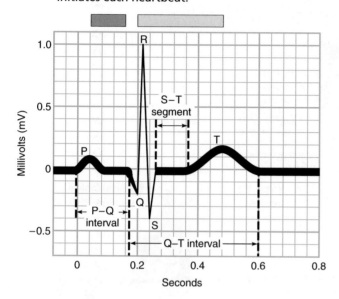

Key:

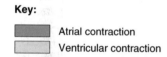

■ Atrial contraction
■ Ventricular contraction

? **What is the significance of an enlarged Q wave?**

Analysis of an ECG also involves measuring the time spans between waves, which are called **intervals** or **segments.** For example, the **P–Q interval** is the time from the beginning of the P wave to the beginning of the QRS complex. It represents the conduction time from the beginning of atrial excitation to the beginning of ventricular excitation. Put another way, the P–Q interval is the time required for the action potential to travel through the atria, atrioventricular node, and the remaining fibers of the conduction system. As the action potential is forced to detour around scar tissue caused by disorders such as coronary artery disease and rheumatic fever, the P–Q interval lengthens.

The **S–T segment,** which begins at the end of the S wave and ends at the beginning of the T wave, represents the time when the ventricular contractile fibers are depolarized during the plateau phase of the action potential. The S–T segment is elevated (above the baseline) in acute myocardial infarction and depressed (below the baseline) when the heart muscle receives insufficient oxygen. The **Q–T interval** extends from the start of the QRS complex to the end of the T wave. It is the time from the beginning of ventricular depolarization to the end of ventricular

repolarization. The Q–T interval may be lengthened by myocardial damage, myocardial ischemia (decreased blood flow), or conduction abnormalities.

Sometimes it is helpful to evaluate the heart's response to the stress of physical exercise (stress testing) (see Disorders: Homeostatic Imbalances at the end of this chapter). Although narrowed coronary arteries may carry adequate oxygenated blood while a person is at rest, they will not be able to meet the heart's increased need for oxygen during strenuous exercise. This situation creates changes that can be seen on an electrocardiogram.

Abnormal heart rhythms and inadequate blood flow to the heart may occur only briefly or unpredictably. To detect these problems, **continuous ambulatory electrocardiographs** are used. With this procedure, a person wears a battery-operated monitor (Holter monitor) that records an ECG continuously for 24 hours. Electrodes attached to the chest are connected to the monitor, and information on the heart's activity is stored in the monitor and retrieved later by medical personnel.

Correlation of ECG Waves with Atrial and Ventricular Systole

As we have seen, the atria and ventricles depolarize and then contract at different times because the conduction system routes cardiac action potentials along a specific pathway. The term **systole** (SIS-tō-lē = contraction) refers to the phase of contraction; the phase of relaxation is **diastole** (dī-AS-tō-lē = dilation or expansion). The ECG waves predict the timing of atrial and ventricular systole and diastole. At a heart rate of 75 beats per minute, the timing is as follows (Figure 20.13):

1 A cardiac action potential arises in the SA node. It propagates throughout the atrial muscle and down to the AV node in about 0.03 sec. As the atrial contractile fibers depolarize, the P wave appears in the ECG.

2 After the P wave begins, the atria contract (atrial systole). Conduction of the action potential slows at the AV node because the fibers there have much smaller diameters and fewer gap junctions. (Traffic slows in a similar way where a four-lane highway narrows to one lane in a construction zone!) The resulting 0.1-sec delay gives the atria time to contract, thus adding to the volume of blood in the ventricles, before ventricular systole begins.

3 The action potential propagates rapidly again after entering the AV bundle. About 0.2 sec after onset of the P wave, it has propagated through the bundle branches, Purkinje fibers, and the entire ventricular myocardium. Depolarization progresses down the septum, upward from the apex, and outward from the endocardial surface, producing the QRS complex. At the same time, atrial repolarization is occurring, but it is not usually evident in an ECG because the larger QRS complex masks it.

4 Contraction of ventricular contractile fibers (ventricular systole) begins shortly after the QRS complex appears and continues during the S–T segment. As contraction proceeds from

Figure 20.13 Timing and route of action potential depolarization and repolarization through the conduction system and myocardium. Green indicates depolarization, and red indicates repolarization.

Depolarization causes contraction and repolarization causes relaxation of cardiac muscle fibers.

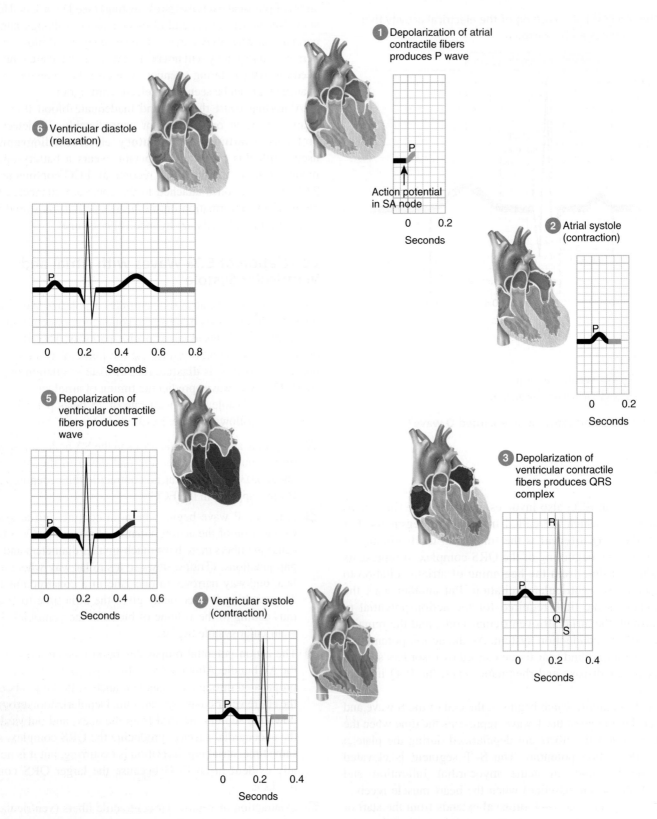

1 Depolarization of atrial contractile fibers produces P wave

Action potential in SA node

6 Ventricular diastole (relaxation)

2 Atrial systole (contraction)

5 Repolarization of ventricular contractile fibers produces T wave

3 Depolarization of ventricular contractile fibers produces QRS complex

4 Ventricular systole (contraction)

? **Where in the conduction system do action potentials propagate most slowly?**

the apex toward the base of the heart, blood is squeezed upward toward the semilunar valves.

5 Repolarization of ventricular contractile fibers begins at the apex and spreads throughout the ventricular myocardium. This produces the T wave in the ECG about 0.4 sec after the onset of the P wave.

6 Shortly after the T wave begins, the ventricles start to relax (ventricular diastole). By 0.6 sec, ventricular repolarization is complete and ventricular contractile fibers are relaxed.

During the next 0.2 sec, contractile fibers in both the atria and ventricles are relaxed. At 0.8 sec, the P wave appears again in the ECG, the atria begin to contract, and the cycle repeats.

As you have just learned, events in the heart occur in cycles that repeat for as long as you live. Next, we will see how the pressure changes associated with relaxation and contraction of the heart chambers allow the heart to alternately fill with blood and then eject blood into the aorta and pulmonary trunk.

✔**CHECKPOINT**

10. How do cardiac muscle fibers differ structurally and functionally from skeletal muscle fibers?
11. In what ways are autorhythmic fibers similar to and different from contractile fibers?
12. What happens during each of the three phases of an action potential in ventricular contractile fibers?
13. In what ways are ECGs helpful in diagnosing cardiac problems?
14. How does each ECG wave, interval, and segment relate to contraction (systole) and relaxation (diastole) of the atria and ventricles?

20.4 THE CARDIAC CYCLE

◉**OBJECTIVES**

• Describe the pressure and volume changes that occur during a cardiac cycle.
• Relate the timing of heart sounds to the ECG waves and pressure changes during systole and diastole.

A single **cardiac cycle** includes all the events associated with one heartbeat. Thus, a cardiac cycle consists of systole and diastole of the atria plus systole and diastole of the ventricles.

Pressure and Volume Changes during the Cardiac Cycle

In each cardiac cycle, the atria and ventricles alternately contract and relax, forcing blood from areas of higher pressure to areas of lower pressure. As a chamber of the heart contracts, blood pressure within it increases. Figure 20.14 shows the relation between the heart's electrical signals (ECG) and changes in atrial pressure, ventricular pressure, aortic pressure, and ventricular volume during the cardiac cycle. The pressures given in the figure apply to the left side of the heart; pressures on the right side are consider-

ably lower. Each ventricle, however, expels the same volume of blood per beat, and the same pattern exists for both pumping chambers. When heart rate is 75 beats/min, a cardiac cycle lasts 0.8 sec. To examine and correlate the events taking place during a cardiac cycle, we will begin with atrial systole.

Atrial Systole

During **atrial systole,** which lasts about 0.1 sec, the atria are contracting. At the same time, the ventricles are relaxed.

1 Depolarization of the SA node causes atrial depolarization, marked by the P wave in the ECG.

2 Atrial depolarization causes atrial systole. As the atria contract, they exert pressure on the blood within, which forces blood through the open AV valves into the ventricles.

3 Atrial systole contributes a final 25 mL of blood to the volume already in each ventricle (about 105 mL). The end of atrial systole is also the end of ventricular diastole (relaxation). Thus, each ventricle contains about 130 mL at the end of its relaxation period (diastole). This blood volume is called the **end-diastolic volume (EDV).**

4 The QRS complex in the ECG marks the onset of ventricular depolarization.

Ventricular Systole

During **ventricular systole,** which lasts about 0.3 sec, the ventricles are contracting. At the same time, the atria are relaxed in **atrial diastole.**

5 Ventricular depolarization causes ventricular systole. As ventricular systole begins, pressure rises inside the ventricles and pushes blood up against the atrioventricular (AV) valves, forcing them shut. For about 0.05 seconds, both the SL (semilunar) and AV valves are closed. This is the period of **isovolumetric contraction** (ī-sō-VOL-u-met′-rik; *iso-* = same). During this interval, cardiac muscle fibers are contracting and exerting force but are not yet shortening. Thus, the muscle contraction is isometric (same length). Moreover, because all four valves are closed, ventricular volume remains the same (isovolumic).

6 Continued contraction of the ventricles causes pressure inside the chambers to rise sharply. When left ventricular pressure surpasses aortic pressure at about 80 millimeters of mercury (mmHg) and right ventricular pressure rises above the pressure in the pulmonary trunk (about 20 mmHg), both SL valves open. At this point, ejection of blood from the heart begins. The period when the SL valves are open is **ventricular ejection** and lasts for about 0.25 sec. The pressure in the left ventricle continues to rise to about 120 mmHg, whereas the pressure in the right ventricle climbs to about 25–30 mmHg.

7 The left ventricle ejects about 70 mL of blood into the aorta and the right ventricle ejects the same volume of blood into the pulmonary trunk. The volume remaining in each ventricle at the end of systole, about 60 mL, is the **end-systolic volume (ESV). Stroke volume,** the volume ejected per beat from

Figure 20.14 Cardiac cycle. (a) ECG. (b) Changes in left atrial pressure (green line), left ventricular pressure (blue line), and aortic pressure (red line) as they relate to the opening and closing of heart valves. (c) Heart sounds. (d) Changes in left ventricular volume. (e) Phases of the cardiac cycle.

A cardiac cycle is composed of all the events associated with one heartbeat.

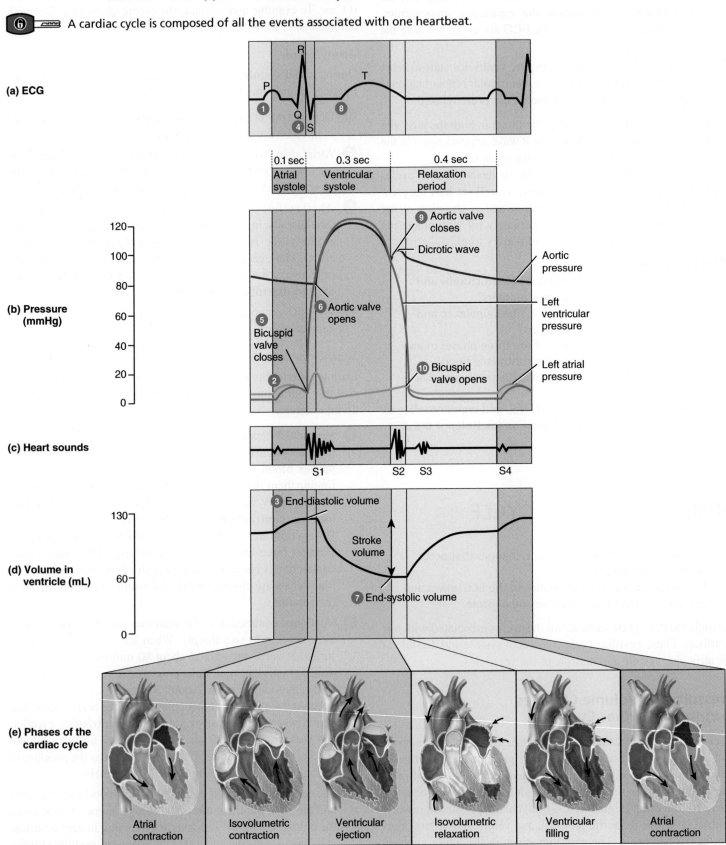

(a) ECG

(b) Pressure (mmHg)

(c) Heart sounds

(d) Volume in ventricle (mL)

(e) Phases of the cardiac cycle

Atrial contraction

Isovolumetric contraction

Ventricular ejection

Isovolumetric relaxation

Ventricular filling

Atrial contraction

? **How much blood remains in each ventricle at the end of ventricular diastole in a resting person? What is this volume called?**

each ventricle, equals end-diastolic volume minus end-systolic volume: SV = EDV − ESV. At rest, the stroke volume is about 130 mL − 60 mL = 70 mL (a little more than 2 oz).

8 The T wave in the ECG marks the onset of ventricular repolarization.

Relaxation Period

During the **relaxation period,** which lasts about 0.4 sec, the atria and the ventricles are both relaxed. As the heart beats faster and faster, the relaxation period becomes shorter and shorter, whereas the durations of atrial systole and ventricular systole shorten only slightly.

9 Ventricular repolarization causes **ventricular diastole.** As the ventricles relax, pressure within the chambers falls, and blood in the aorta and pulmonary trunk begins to flow backward toward the regions of lower pressure in the ventricles. Backflowing blood catches in the valve cusps and closes the SL valves. The aortic valve closes at a pressure of about 100 mmHg. Rebound of blood off the closed cusps of the aortic valve produces the **dicrotic wave** on the aortic pressure curve. After the SL valves close, there is a brief interval when ventricular blood volume does not change because all four valves are closed. This is the period of **isovolumetric relaxation.**

10 As the ventricles continue to relax, the pressure falls quickly. When ventricular pressure drops below atrial pressure, the AV valves open, and **ventricular filling** begins. The major part of ventricular filling occurs just after the AV valves open. Blood that has been flowing into and building up in the atria during ventricular systole then rushes rapidly into the ventricles. At the end of the relaxation period, the ventricles are about three-quarters full. The P wave appears in the ECG, signaling the start of another cardiac cycle.

Heart Sounds

Auscultation (aws-kul-TĀ-shun; *ausculta-* = listening), the act of listening to sounds within the body, is usually done with a stethoscope. The sound of the heartbeat comes primarily from blood turbulence caused by the closing of the heart valves. Smoothly flowing blood is silent. Compare the sounds made by white-water rapids or a waterfall with the silence of a smoothly flowing river. During each cardiac cycle, there are four **heart sounds,** but in a normal heart only the first and second heart sounds (S1 and S2) are loud enough to be heard through a stethoscope. Figure 20.14c shows the timing of heart sounds relative to other events in the cardiac cycle.

The first sound (S1), which can be described as a **lubb** sound, is louder and a bit longer than the second sound. S1 is caused by blood turbulence associated with closure of the AV valves soon after ventricular systole begins. The second sound (S2), which is shorter and not as loud as the first, can be described as a **dupp** sound. S2 is caused by blood turbulence associated with closure of the SL valves at the beginning of ventricular diastole. Although S1 and S2 are due to blood turbulence associated with the closure

of valves, they are best heard at the surface of the chest in locations that are slightly different from the locations of the valves (Figure 20.15). This is because the sound is carried by the blood flow away from the valves. Normally not loud enough to be heard, S3 is due to blood turbulence during rapid ventricular filling, and S4 is due to blood turbulence during atrial systole.

CLINICAL CONNECTION | *Heart Murmurs*

Heart sounds provide valuable information about the mechanical operation of the heart. A **heart murmur** is an abnormal sound consisting of a clicking, rushing, or gurgling noise that either is heard before, between, or after the normal heart sounds, or may mask the normal heart sounds. Heart murmurs in children are extremely common and usually do not represent a health condition. Murmurs are most frequently discovered in children between the ages of 2 and 4. These types of heart murmurs are referred to as *innocent* or *functional heart murmurs;* they often subside or disappear with growth. Although some heart murmurs in adults are innocent, most often an adult murmur indicates a valve disorder. When a heart valve exhibits stenosis, the heart murmur is heard while the valve should be fully open but is not. For example, mitral stenosis (see Clinical Connection: Heart Valve Disorders) produces a murmur during the relaxation period, between S2 and the next S1. An incompetent heart valve, by

Figure 20.15 Heart sounds. Location of valves (blue) and auscultation sites (red) for heart sounds.

 Listening to sounds within the body is called auscultation; it is usually done with a stethoscope.

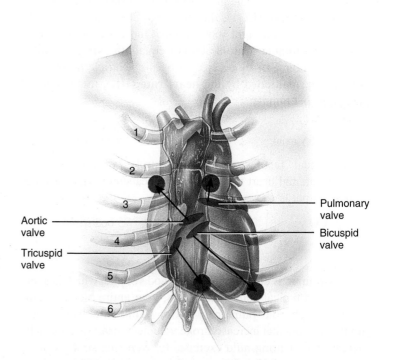

Aortic valve
Tricuspid valve
Pulmonary valve
Bicuspid valve

1
2
3
4
5
6

Anterior view of heart valve locations and auscultation sites

? Which heart sound is related to blood turbulence associated with closure of the atrioventricular valves?

contrast, causes a murmur to appear when the valve should be fully closed but is not. So, a murmur due to mitral incompetence (see Clinical Connection: Heart Valve Disorders) occurs during ventricular systole, between S1 and S2. •

✔ CHECKPOINT

15. Why must left ventricular pressure be greater than aortic pressure during ventricular ejection?
16. Does more blood flow through the coronary arteries during ventricular diastole or ventricular systole? Explain your answer.
17. During which two periods of the cardiac cycle do the heart muscle fibers exhibit isometric contractions?
18. What events produce the four normal heart sounds? Which ones can usually be heard through a stethoscope?

20.5 CARDIAC OUTPUT

◉ OBJECTIVES

• Define cardiac output.
• Describe the factors that affect regulation of stroke volume.
• Outline the factors that affect the regulation of heart rate.

Although the heart has autorhythmic fibers that enable it to beat independently, its operation is governed by events occurring throughout the body. Body cells must receive a certain amount of oxygen from blood each minute to maintain health and life. When cells are metabolically active, as during exercise, they take up even more oxygen from the blood. During rest periods, cellular metabolic need is reduced, and the workload of the heart decreases.

Cardiac output (CO) is the volume of blood ejected from the left ventricle (or the right ventricle) into the aorta (or pulmonary trunk) each minute. Cardiac output equals the **stroke volume (SV),** the volume of blood ejected by the ventricle during each contraction, multiplied by the **heart rate (HR),** the number of heartbeats per minute:

$$\underset{\text{(mL/min)}}{\text{CO}} = \underset{\text{(mL/beat)}}{\text{SV}} \times \underset{\text{(beats/min)}}{\text{HR}}$$

In a typical resting adult male, stroke volume averages 70 mL/beat, and heart rate is about 75 beats/min. Thus, average cardiac output is

$$\begin{aligned}
\text{CO} &= 70 \text{ mL/beat} \times 75 \text{ beats/min}\\
&= 5250 \text{ mL/min}\\
&= 5.25 \text{ L/min}
\end{aligned}$$

This volume is close to the total blood volume, which is about 5 liters in a typical adult male. Thus, your entire blood volume flows through your pulmonary and systemic circulations each minute. Factors that increase stroke volume or heart rate normally increase CO. During mild exercise, for example, stroke volume may increase to 100 mL/beat, and heart rate to 100 beats/min. Cardiac output then would be 10 L/min. During intense (but still not maximal) exercise, the heart rate may accelerate to

150 beats/min, and stroke volume may rise to 130 mL/beat, resulting in a cardiac output of 19.5 L/min.

Cardiac reserve is the difference between a person's maximum cardiac output and cardiac output at rest. The average person has a cardiac reserve of four or five times the resting value. Top endurance athletes may have a cardiac reserve seven or eight times their resting CO. People with severe heart disease may have little or no cardiac reserve, which limits their ability to carry out even the simple tasks of daily living.

Regulation of Stroke Volume

A healthy heart will pump out the blood that entered its chambers during the previous diastole. In other words, if more blood returns to the heart during diastole, then more blood is ejected during the next systole. At rest, the stroke volume is 50–60% of the end-diastolic volume because 40–50% of the blood remains in the ventricles after each contraction (end-systolic volume). Three factors regulate stroke volume and ensure that the left and right ventricles pump equal volumes of blood: (1) **preload,** the degree of stretch on the heart before it contracts; (2) **contractility,** the forcefulness of contraction of individual ventricular muscle fibers; and (3) **afterload,** the pressure that must be exceeded before ejection of blood from the ventricles can occur.

Preload: Effect of Stretching

A greater preload (stretch) on cardiac muscle fibers prior to contraction increases their force of contraction. Preload can be compared to the stretching of a rubber band. The more the rubber band is stretched, the more forcefully it will snap back. Within limits, the more the heart fills with blood during diastole, the greater the force of contraction during systole. This relationship is known as the **Frank–Starling law of the heart.** The preload is proportional to the end-diastolic volume (EDV) (the volume of blood that fills the ventricles at the end of diastole). Normally, the greater the EDV, the more forceful the next contraction.

Two key factors determine EDV: (1) the duration of ventricular diastole and (2) **venous return,** the volume of blood returning to the right ventricle. When heart rate increases, the duration of diastole is shorter. Less filling time means a smaller EDV, and the ventricles may contract before they are adequately filled. By contrast, when venous return increases, a greater volume of blood flows into the ventricles, and the EDV is increased.

When heart rate exceeds about 160 beats/min, stroke volume usually declines due to the short filling time. At such rapid heart rates, EDV is less, and the preload is lower. People who have slow resting heart rates usually have large resting stroke volumes because filling time is prolonged and preload is larger.

The Frank–Starling law of the heart equalizes the output of the right and left ventricles and keeps the same volume of blood flowing to both the systemic and pulmonary circulations. If the left side of the heart pumps a little more blood than the right side, the volume of blood returning to the right ventricle (venous return) increases. The increased EDV causes the right ventricle to contract more forcefully on the next beat, bringing the two sides back into balance.

Contractility

The second factor that influences stroke volume is myocardial **contractility,** the strength of contraction at any given preload. Substances that increase contractility are **positive inotropic agents** (īn′-ō-TRŌ-pik); those that decrease contractility are **negative inotropic agents.** Thus, for a constant preload, the stroke volume increases when a positive inotropic substance is present. Positive inotropic agents often promote Ca^{2+} inflow during cardiac action potentials, which strengthens the force of the next contraction. Stimulation of the sympathetic division of the autonomic nervous system (ANS), hormones such as epinephrine and norepinephrine, increased Ca^{2+} level in the interstitial fluid, and the drug digitalis all have positive inotropic effects. In contrast, inhibition of the sympathetic division of the ANS, anoxia, acidosis, some anesthetics, and increased K^+ level in the interstitial fluid have negative inotropic effects. *Calcium channel blockers* are drugs that can have a negative inotropic effect by reducing Ca^{2+} inflow, thereby decreasing the strength of the heartbeat.

Afterload

Ejection of blood from the heart begins when pressure in the right ventricle exceeds the pressure in the pulmonary trunk (about 20 mmHg), and when the pressure in the left ventricle exceeds the pressure in the aorta (about 80 mmHg). At that point, the higher pressure in the ventricles causes blood to push the semilunar valves open. The pressure that must be overcome before a semilunar valve can open is termed the **afterload.** An increase in afterload causes stroke volume to decrease, so that more blood remains in the ventricles at the end of systole. Conditions that can increase afterload include hypertension (elevated blood pressure) and narrowing of arteries by atherosclerosis (see Disorders: Homoeostatic Imbalances at the end of this chapter).

CLINICAL CONNECTION | *Congestive Heart Failure*

In **congestive heart failure (CHF),** there is a loss of pumping efficiency by the heart. Causes of CHF include coronary artery disease (see Disorders: Homeostatic Imbalances at the end of this chapter), congenital defects, long-term high blood pressure (which increases the afterload), myocardial infarctions (regions of dead heart tissue due to a previous heart attack), and valve disorders. As the pump becomes less effective, more blood remains in the ventricles at the end of each cycle, and gradually the end-diastolic volume (preload) increases. Initially, increased preload may promote increased force of contraction (the Frank–Starling law of the heart), but as the preload increases further, the heart is overstretched and contracts less forcefully. The result is a potentially lethal positive feedback loop: Less-effective pumping leads to even lower pumping capability.

Often, one side of the heart starts to fail before the other. If the left ventricle fails first, it can't pump out all the blood it receives. As a result, blood backs up in the lungs and causes *pulmonary edema,* fluid accumulation in the lungs that can cause suffocation if left untreated. If the right ventricle fails first, blood backs up in the systemic veins and, over time, the kidneys cause an increase in blood volume. In this case, the resulting *peripheral edema* usually is most noticeable in the feet and ankles. •

Regulation of Heart Rate

As you have just learned, cardiac output depends on both heart rate and stroke volume. Adjustments in heart rate are important in the short-term control of cardiac output and blood pressure. The sinoatrial (SA) node initiates contraction and, if left to itself, would set a constant heart rate of about 100 beats/min. However, tissues require different volumes of blood flow under different conditions. During exercise, for example, cardiac output rises to supply working tissues with increased amounts of oxygen and nutrients. Stroke volume may fall if the ventricular myocardium is damaged or if blood volume is reduced by bleeding. In these cases, homeostatic mechanisms maintain adequate cardiac output by increasing the heart rate and contractility. Among the several factors that contribute to regulation of heart rate, the most important are the autonomic nervous system and hormones released by the adrenal medullae (epinephrine and norepinephrine).

Autonomic Regulation of Heart Rate

Nervous system regulation of the heart originates in the **cardiovascular center** in the medulla oblongata. This region of the brain stem receives input from a variety of sensory receptors and from higher brain centers, such as the limbic system and cerebral cortex. The cardiovascular center then directs appropriate output by increasing or decreasing the frequency of nerve impulses in both the sympathetic and parasympathetic branches of the ANS (Figure 20.16).

Even before physical activity begins, especially in competitive situations, heart rate may climb. This anticipatory increase occurs because the limbic system sends nerve impulses to the cardiovascular center in the medulla. As physical activity begins, **proprioceptors** that are monitoring the position of limbs and muscles send nerve impulses at an increased frequency to the cardiovascular center. Proprioceptor input is a major stimulus for the quick rise in heart rate that occurs at the onset of physical activity. Other sensory receptors that provide input to the cardiovascular center include **chemoreceptors,** which monitor chemical changes in the blood, and **baroreceptors,** which monitor the stretching of major arteries and veins caused by the pressure of the blood flowing through them. Important baroreceptors located in the arch of the aorta and in the carotid arteries (see Figure 21.13) detect changes in blood pressure and provide input to the cardiovascular center when it changes. The role of baroreceptors in the regulation of blood pressure is discussed in detail in Chapter 21. Here we focus on the innervation of the heart by the sympathetic and parasympathetic branches of the ANS.

Sympathetic neurons extend from the medulla oblongata into the spinal cord. From the thoracic region of the spinal cord, sympathetic **cardiac accelerator nerves** extend out to the SA node, AV node, and most portions of the myocardium. Impulses in the cardiac accelerator nerves trigger the release of norepinephrine, which binds to beta-1 (β_1) receptors on cardiac muscle fibers. This interaction has two separate effects: (1) In SA (and AV) node fibers, norepinephrine speeds the rate of spontaneous depolarization so that these pacemakers fire impulses more rapidly and heart rate increases; (2) in

Figure 20.16 Nervous system control of the heart.

The cardiovascular center in the medulla oblongata controls both sympathetic and parasympathetic nerves that innervate the heart.

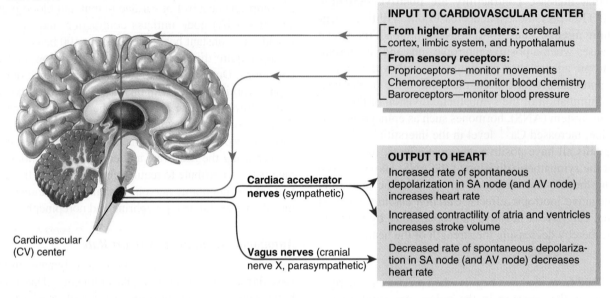

INPUT TO CARDIOVASCULAR CENTER

From higher brain centers: cerebral cortex, limbic system, and hypothalamus

From sensory receptors:
Proprioceptors—monitor movements
Chemoreceptors—monitor blood chemistry
Baroreceptors—monitor blood pressure

Cardiac accelerator nerves (sympathetic)

Vagus nerves (cranial nerve X, parasympathetic)

Cardiovascular (CV) center

OUTPUT TO HEART

Increased rate of spontaneous depolarization in SA node (and AV node) increases heart rate

Increased contractility of atria and ventricles increases stroke volume

Decreased rate of spontaneous depolarization in SA node (and AV node) decreases heart rate

? **What region of the heart is innervated by the sympathetic division but not by the parasympathetic division?**

contractile fibers throughout the atria and ventricles, norepinephrine enhances Ca^{2+} entry through the voltage-gated slow Ca^{2+} channels, thereby increasing contractility. As a result, a greater volume of blood is ejected during systole. With a moderate increase in heart rate, stroke volume does not decline because the increased contractility offsets the decreased preload. With maximal sympathetic stimulation, however, heart rate may reach 200 beats/min in a 20-year-old person. At such a high heart rate, stroke volume is lower than at rest due to the very short filling time. The maximal heart rate declines with age; as a rule, subtracting your age from 220 provides a good estimate of your maximal heart rate in beats per minute.

Parasympathetic nerve impulses reach the heart via the right and left **vagus (X) nerves.** Vagal axons terminate in the SA node, AV node, and atrial myocardium. They release acetylcholine, which decreases heart rate by slowing the rate of spontaneous depolarization in autorhythmic fibers. As only a few vagal fibers innervate ventricular muscle, changes in parasympathetic activity have little effect on contractility of the ventricles.

A continually shifting balance exists between sympathetic and parasympathetic stimulation of the heart. At rest, parasympathetic stimulation predominates. The resting heart rate—about 75 beats/min—is usually lower than the autorhythmic rate of the SA node (about 100 beats/min). With maximal stimulation by the parasympathetic division, the heart can slow to 20 or 30 beats/min, or can even stop momentarily.

Chemical Regulation of Heart Rate

Certain chemicals influence both the basic physiology of cardiac muscle and the heart rate. For example, hypoxia (lowered oxygen level), acidosis (low pH), and alkalosis (high pH) all depress cardiac activity. Several hormones and cations have major effects on the heart:

1. *Hormones.* Epinephrine and norepinephrine (from the adrenal medullae) enhance the heart's pumping effectiveness. These hormones affect cardiac muscle fibers in much the same way as does norepinephrine released by cardiac accelerator nerves—they increase both heart rate and contractility. Exercise, stress, and excitement cause the adrenal medullae to release more hormones. Thyroid hormones also enhance cardiac contractility and increase heart rate. One sign of hyperthyroidism (excessive thyroid hormone) is **tachycardia** (tak′-i-KAR-dē-a), an elevated resting heart rate.

2. *Cations.* Given that differences between intracellular and extracellular concentrations of several cations (for example, Na^+ and K^+) are crucial for the production of action potentials in all nerve and muscle fibers, it is not surprising that ionic imbalances can quickly compromise the pumping effectiveness of the heart. In particular, the relative concentrations of three cations—K^+, Ca^{2+}, and Na^+—have a large effect on cardiac function. Elevated blood levels of K^+ or Na^+ decrease heart rate and contractility. Excess Na^+ blocks Ca^{2+} inflow during cardiac action potentials, thereby decreasing the force of contraction, whereas excess K^+ blocks generation of action potentials. A moderate increase in interstitial (and thus intracellular) Ca^{2+} level speeds heart rate and strengthens the heartbeat.

Other Factors in Heart Rate Regulation

Age, gender, physical fitness, and body temperature also influence resting heart rate. A newborn baby is likely to have a resting heart rate over 120 beats/min; the rate then gradually declines throughout life. Adult females often have slightly higher resting heart rates than adult males, although regular exercise tends to bring resting heart rate down in both sexes. A physically fit person may even exhibit **bradycardia** (brād′-i-KAR-dē-a; *bradys-* = slow), a resting heart rate under 50 beats/min. This is a beneficial effect of endurance-type training because a slowly beating heart is more energy efficient than one that beats more rapidly.

Increased body temperature, as occurs during a fever or strenuous exercise, causes the SA node to discharge impulses more quickly, thereby increasing heart rate. Decreased body temperature decreases heart rate and strength of contraction.

During surgical repair of certain heart abnormalities, it is helpful to slow a patient's heart rate by **hypothermia** (hī-pō-THER-mē-a), in which the person's body is deliberately cooled to a low core temperature. Hypothermia slows metabolism, which reduces the oxygen needs of the tissues, allowing the heart and brain to withstand short periods of interrupted or reduced blood flow during the procedure.

Figure 20.17 summarizes the factors that can increase stroke volume and heart rate to achieve an increase in cardiac output.

Figure 20.17 Factors that increase cardiac output.

🔑 Cardiac output equals stroke volume multiplied by heart rate.

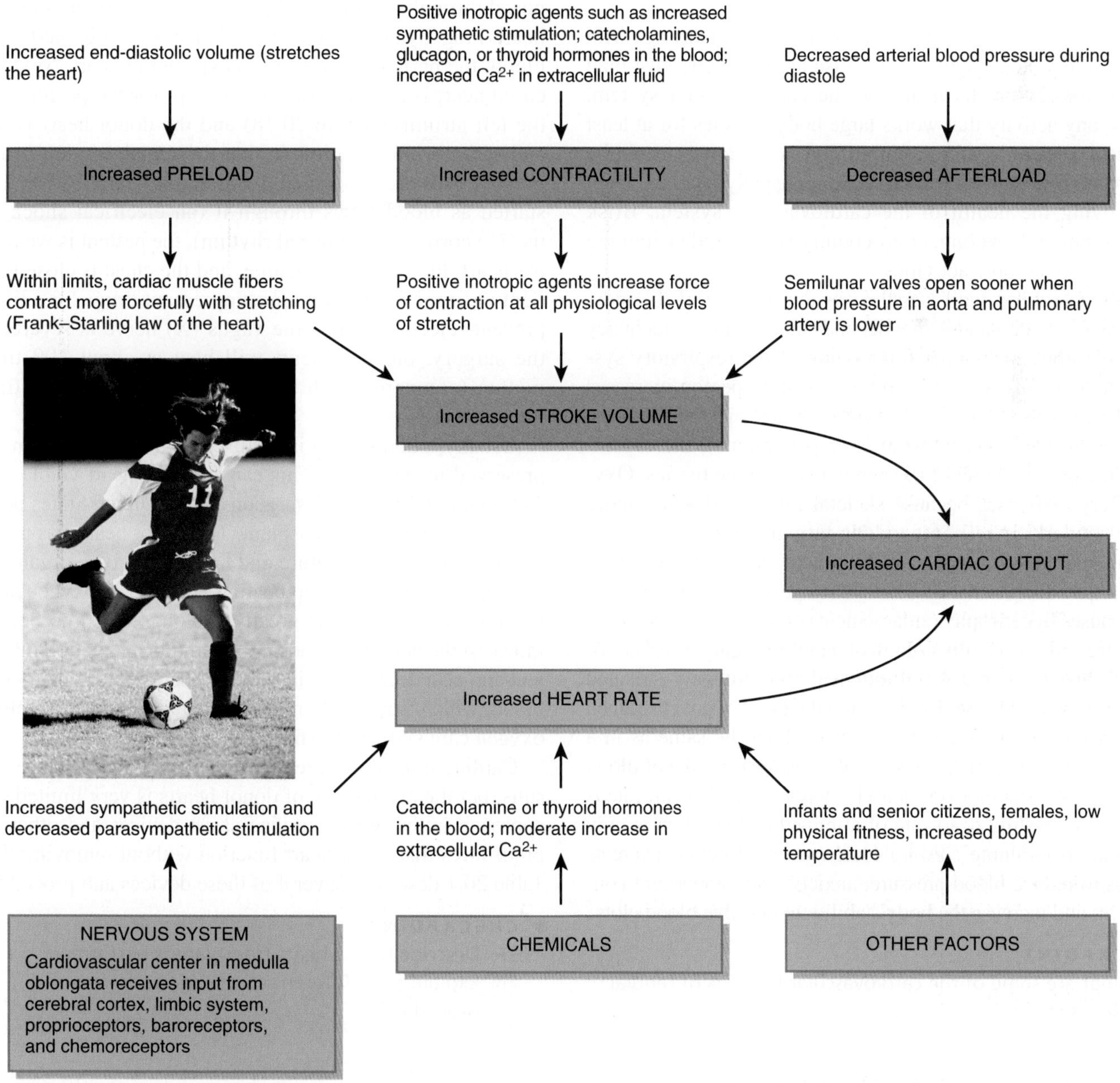

Increased end-diastolic volume (stretches the heart)

Positive inotropic agents such as increased sympathetic stimulation; catecholamines, glucagon, or thyroid hormones in the blood; increased Ca^{2+} in extracellular fluid

Decreased arterial blood pressure during diastole

Increased PRELOAD

Increased CONTRACTILITY

Decreased AFTERLOAD

Within limits, cardiac muscle fibers contract more forcefully with stretching (Frank–Starling law of the heart)

Positive inotropic agents increase force of contraction at all physiological levels of stretch

Semilunar valves open sooner when blood pressure in aorta and pulmonary artery is lower

Increased STROKE VOLUME

Increased CARDIAC OUTPUT

Increased HEART RATE

Increased sympathetic stimulation and decreased parasympathetic stimulation

Catecholamine or thyroid hormones in the blood; moderate increase in extracellular Ca^{2+}

Infants and senior citizens, females, low physical fitness, increased body temperature

NERVOUS SYSTEM
Cardiovascular center in medulla oblongata receives input from cerebral cortex, limbic system, proprioceptors, baroreceptors, and chemoreceptors

CHEMICALS

OTHER FACTORS

❓ When you are exercising, contraction of skeletal muscles helps return blood to the heart more rapidly. Would this tend to increase or decrease stroke volume?

✔**CHECKPOINT**

19. How is cardiac output calculated?
20. Define stroke volume (SV), and explain the factors that regulate it.
21. What is the Frank–Starling law of the heart? What is its significance?
22. Define cardiac reserve. How does it change with training or with heart failure?
23. How do the sympathetic and parasympathetic divisions of the autonomic nervous system adjust heart rate?

20.6 EXERCISE AND THE HEART

◉ **OBJECTIVE**

• Explain how the heart is affected by exercise.

A person's cardiovascular fitness can be improved at any age with regular exercise. Some types of exercise are more effective than others for improving the health of the cardiovascular system. **Aerobics,** any activity that works large body muscles for at least 20 minutes, elevates cardiac output and accelerates metabolic rate. Three to five such sessions a week are usually recommended for improving the health of the cardiovascular system. Brisk walking, running, bicycling, cross-country skiing, and swimming are examples of aerobic activities.

Sustained exercise increases the oxygen demand of the muscles. Whether the demand is met depends mainly on the adequacy of cardiac output and proper functioning of the respiratory system. After several weeks of training, a healthy person increases maximal cardiac output (the amount of blood ejected from the ventricles into their respective arteries per minute), thereby increasing the maximal rate of oxygen delivery to the tissues. Oxygen delivery also rises because skeletal muscles develop more capillary networks in response to long-term training.

During strenuous activity, a well-trained athlete can achieve a cardiac output double that of a sedentary person, in part because training causes hypertrophy (enlargement) of the heart. This condition is referred to as **physiological cardiomegaly** (kar′-dē-ō-MEG-a-lē; *mega* = large). A **pathological cardiomegaly** is related to significant heart disease. Even though the heart of a well-trained athlete is larger, *resting* cardiac output is about the same as in a healthy untrained person, because *stroke volume* (volume of blood pumped by each beat of a ventricle) is increased while heart rate is decreased. The resting heart rate of a trained athlete often is only 40–60 beats per minute (*resting bradycardia*). Regular exercise also helps to reduce blood pressure, anxiety, and depression; control weight; and increase the body's ability to dissolve blood clots.

✔**CHECKPOINT**

24. What are some of the cardiovascular benefits of regular exercise?

20.7 HELP FOR FAILING HEARTS

◉ **OBJECTIVE**

• Describe several techniques used for failing hearts.

As the heart fails, a person has decreasing ability to exercise or even to move around. A variety of surgical techniques and medical devices exist to aid a failing heart. For some patients, even a 10% increase in the volume of blood ejected from the ventricles can mean the difference between being bedridden and having limited mobility.

A **cardiac (heart) transplant** is the replacement of a severely damaged heart with a normal heart from a brain-dead or recently deceased donor. Cardiac transplants are performed on patients with end-stage heart failure or severe coronary artery disease. Once a suitable heart is located, the chest cavity is exposed through a midsternal cut. After the patient is placed on a heart–lung bypass machine, which oxygenates and circulates blood, the pericardium is cut to expose the heart. Next, the diseased heart is removed (usually except for the posterior wall of the left atrium) (Figure 20.18) and the donor heart is trimmed and sutured into position so that the remaining left atrium and great vessels are connected to the donor heart. The new heart is started as blood flows through it (an electrical shock may be used to correct an abnormal rhythm), the patient is weaned from the heart–lung bypass machine, and the chest is closed. The patient must remain on immunosuppressant drugs for a lifetime to prevent rejection. Since the vagus (X) nerve is severed during the surgery, the new heart will beat at about 100 times per minute (compared with a normal rate of about 75 times per minute).

Usually, a donor heart is perfused with a cold solution and then preserved in sterile ice. This can keep the heart viable for about 4–5 hours. In May 2007, surgeons in the United States performed the first beating-heart transplant. The donor heart was maintained at normal body temperature and hooked up to an organ care system that allowed it to keep beating with warm, oxygenated blood flowing through it. This greatly prolongs the time between removal of the heart from the donor and transplantation into a recipient and also decreases injury to the heart while being deprived of blood, which can lead to rejection. The safety and benefits of the oxygen care system are still being evaluated.

Cardiac transplants are common today and produce good results, but the availability of donor hearts is very limited. Another approach is the use of cardiac assist devices and other surgical procedures that assist heart function without removing the heart. Table 20.1 describes several of these devices and procedures.

✔**CHECKPOINT**

25. Describe how a heart transplant is performed.
26. Explain four different cardiac assist devices and procedures.

Figure 20.18 Cardiac transplantation.

Cardiac transplantation is the replacement of a severely damaged heart with a normal heart from a brain-dead or recently deceased donor.

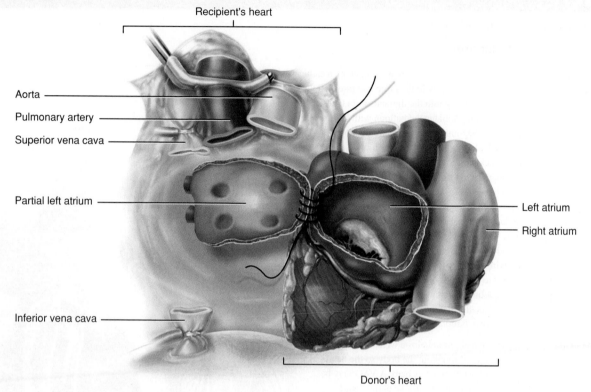

Recipient's heart

Aorta

Pulmonary artery

Superior vena cava

Partial left atrium

Left atrium

Right atrium

Inferior vena cava

Donor's heart

(a) Donor's left atrium is sutured to recipient's left atrium

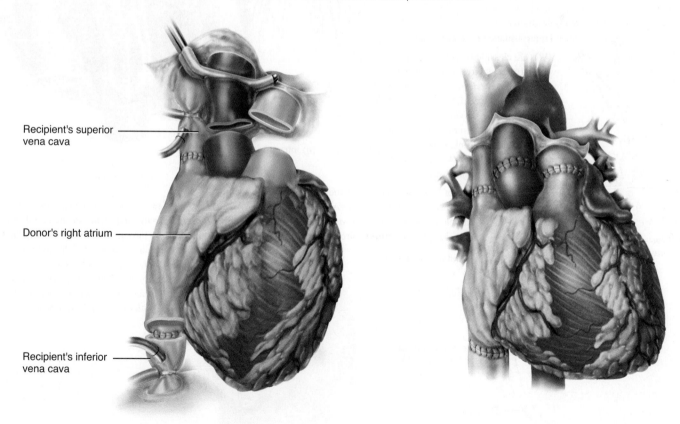

Recipient's superior vena cava

Donor's right atrium

Recipient's inferior vena cava

(b) Donor's right atrium is sutured to recipient's superior and inferior venae cavae

(c) Transplanted heart with sutures

? **Which patients are candidates for cardiac transplantation?**

TABLE 20.1

Cardiac Assist Devices and Procedures

DEVICE	DESCRIPTION
Intra-aortic balloon pump (IABP)	A 40-mL polyurethane balloon mounted on a catheter is inserted into an artery in the groin and threaded through the femoral artery into the thoracic aorta (see a). An external pump inflates the balloon with helium gas at the beginning of ventricular diastole. As the balloon inflates, it pushes blood both backward toward the heart (improves coronary blood flow) and forward toward peripheral tissues. The balloon then is rapidly deflated just before the next ventricular systole, drawing blood out of the left ventricle (making it easier for the left ventricle to eject blood). Because the balloon is inflated between heartbeats, this technique is called *intra-aortic balloon counterpulsation*.
Ventricular assist device (VAD)	A mechanical pump helps a weakened ventricle pump blood throughout the body so the heart does not have to work as hard. A VAD may be used to help a patient survive until a heart transplant can be performed *(bridge to transplant)* or provide an alternative to heart transplantation *(destination therapy)*.

VADs are classified according to the ventricle that requires support. A *left ventricular assist device (LVAD)*, the most common, helps the left ventricle pump blood into the aorta (see b). A *right ventricular assist device (RVAD)* helps the right ventricle pump blood into the pulmonary trunk. A *biventricular assist device (BVAD)* helps both the left and right ventricles perform. The kind of VAD used depends on the patient's specific needs.

To help you understand how a VAD works, see the LVAD (b). An inflow tube attached to the apex of the left ventricle takes blood from the ventricle through a one-way valve into the pump unit. Once the pump fills with blood, an external control system triggers pumping, and blood flows through a one-way valve into an outflow tube that delivers blood into the aorta. The external control system is on a belt around the waist or on a shoulder strap. Some VADs pump at a constant rate; others are coordinated with the person's heartbeat.

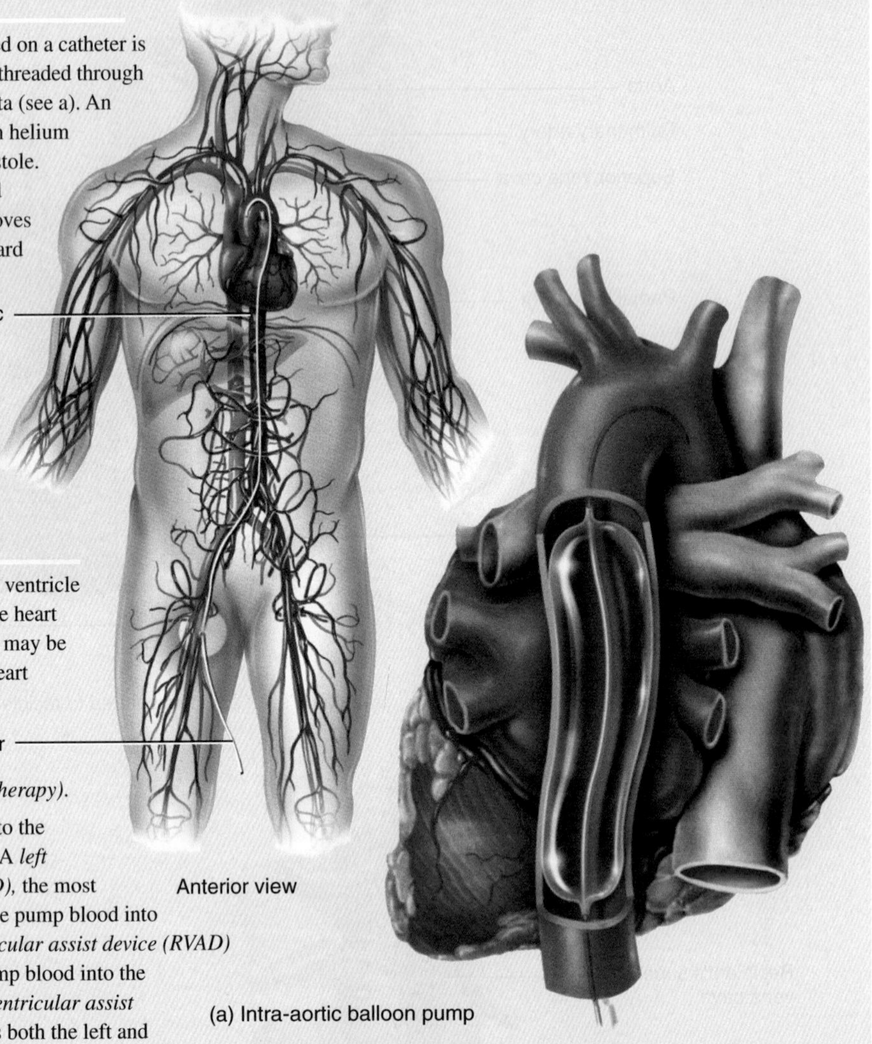

Thoracic aorta

Catheter

Anterior view

(a) Intra-aortic balloon pump

Posterior view

Outflow tube

Inflow tube

Outflow one-way valve

Inflow one-way valve

Pump unit

Parts of left ventricular assist device (LVAD)

Driveline

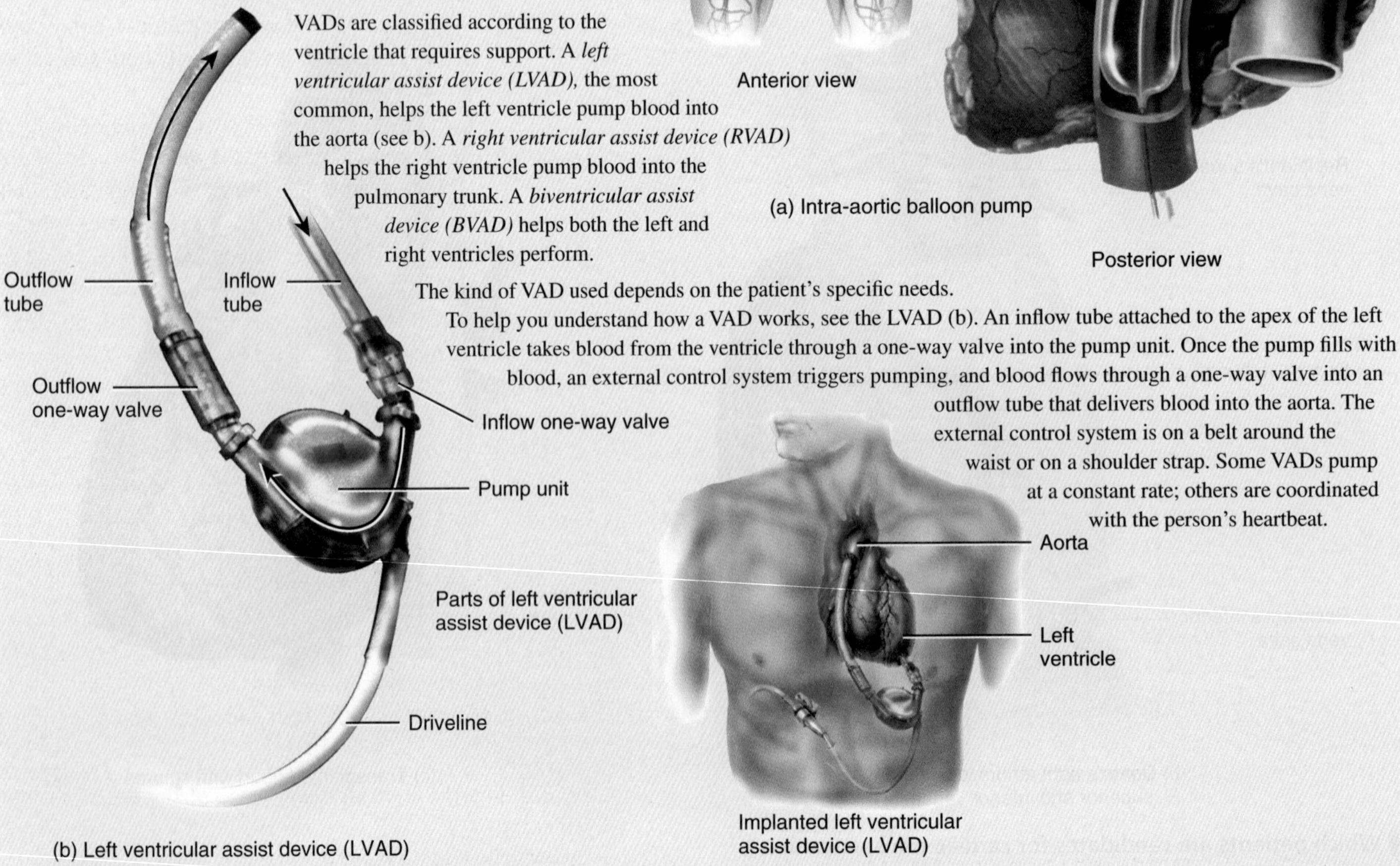

(b) Left ventricular assist device (LVAD)

Aorta

Left ventricle

Implanted left ventricular assist device (LVAD)

TABLE 20.1 CONTINUED

Cardiac Assist Devices and Procedures

DEVICE	DESCRIPTION
Cardiomyoplasty	A large piece of the patient's own skeletal muscle (left latissimus dorsi) is partially freed from connective tissue attachments and wrapped around the heart, leaving the blood and nerve supply intact. An implanted pacemaker stimulates skeletal muscle's motor neurons to cause contraction 10–20 times per minute, in synchrony with some of the heartbeats.
Skeletal muscle assist device	A piece of the patient's own skeletal muscle is used to fashion a pouch inserted between the heart and aorta, functioning as a booster heart. A pacemaker stimulates the muscle's motor neurons to elicit contraction.

20.8 DEVELOPMENT OF THE HEART

◉ OBJECTIVE

- Describe the development of the heart.

Listening to a fetal heartbeat for the first time is an exciting moment for prospective parents, but it is also an important diagnostic tool. The cardiovascular system is one of the first systems to form in an embryo, and the heart is the first functional organ. This order of development is essential because of the need of the rapidly growing embryo to obtain oxygen and nutrients and get rid of wastes. As you will learn shortly, the development of the heart is a complex process, and any disruptions along the way can result in congenital (present at birth) disorders of the heart. Such disorders, described in Disorders: Homeostatic Imbalances at the end of the chapter, are responsible for almost half of all deaths from birth defects.

The *heart* begins its development from **mesoderm** on day 18 or 19 following fertilization. In the head end of the embryo, the heart develops from a group of mesodermal cells called the **cardiogenic area** (kar-dē-ō-JEN-ik; *cardio-* = heart; *-genic* = producing) (Figure 20.19a). In response to signals from the underlying endoderm, the mesoderm in the cardiogenic area forms a pair of elongated strands called **cardiogenic cords.** Shortly after, these cords develop a hollow center and then become known as **endocardial tubes** (Figure 20.19b). With lateral folding of the embryo, the paired endocardial tubes approach each other and fuse into a single tube called the **primitive heart tube** on day 21 following fertilization (Figure 20.19c).

On the twenty-second day, the primitive heart tube develops into five distinct regions and begins to pump blood. From tail end to head end (and the direction of blood flow) they are the (1) **sinus venosus,** (2) **primitive atrium,** (3) **primitive ventricle,** (4) **bulbus cordis,** and (5) **truncus arteriosus.** The sinus venosus initially receives blood from all the veins in the embryo; contractions of the heart begin in this region and follow sequentially in the other regions. Thus, at this stage, the heart consists of a series of unpaired regions. The fates of the regions are as follows:

1. The sinus venosus develops into *part of the right atrium (posterior wall), coronary sinus,* and *sinoatrial (SA) node.*

2. The primitive atrium develops into *part of the right atrium (anterior wall), right auricle, part of the left atrium (anterior wall),* and the *left auricle.*

3. The primitive ventricle gives rise to the *left ventricle.*

4. The bulbus cordis develops into the *right ventricle.*

5. The truncus arteriosus gives rise to the *ascending aorta* and *pulmonary trunk.*

On day 23, the primitive heart tube elongates. Because the bulbus cordis and primitive ventricle grow more rapidly than other parts of the tube and because the atrial and venous ends of the tube are confined by the pericardium, the tube begins to loop and fold. At first, the primitive heart tube assumes a U-shape; later it becomes S-shaped (Figure 20.19e). As a result of these movements, which are completed by day 28, the primitive atria and ventricles of the future heart are reoriented to assume their final adult positions. The remainder of heart development consists of remodeling of the chambers and the formation of septa and valves to form a four-chambered heart.

On about day 28, thickenings of mesoderm of the inner lining of the heart wall, called **endocardial cushions,** appear (Figure 20.20). They grow toward each other, fuse, and divide the single **atrioventricular canal** (region between atria and ventricles) into smaller, separate left and right atrioventricular canals. Also, the *interatrial septum* begins its growth toward the fused endocardial cushions. Ultimately, the interatrial septum and endocardial cushions unite and an opening in the septum, the **foramen ovale** (ō-VAL-ē), develops. The interatrial septum divides the atrial region into a *right atrium* and a *left atrium.* Before birth, the foramen ovale allows most blood entering the right atrium to pass into the left atrium. After birth, it normally closes so that the interatrial septum is a complete partition. The remnant of the foramen ovale is the fossa ovalis (Figure 20.4a). Formation of the *interventricular septum* partitions the ventricular region into a *right ventricle* and a *left ventricle.* Partitioning of the atrioventricular canal, atrial region, and ventricular region is basically complete by the end of the fifth week. The *atrioventricular valves* form between the fifth and eighth weeks. The *semilunar valves* form between the fifth and ninth weeks.

✔CHECKPOINT

27. Why is the cardiovascular system one of the first systems to develop?

28. From which tissue does the heart develop?

Figure 20.19 Development of the heart. Arrows within the structures indicate the direction of blood flow.

The heart begins its development from a group of mesodermal cells called the cardiogenic area during the third week after fertilization.

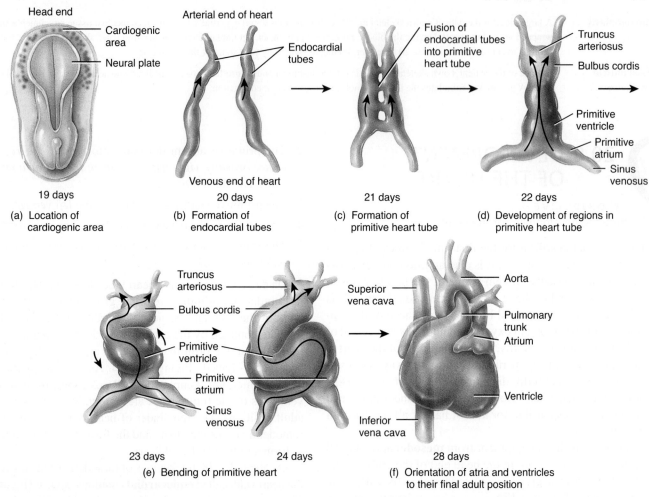

(a) Location of cardiogenic area — 19 days

(b) Formation of endocardial tubes — 20 days

(c) Formation of primitive heart tube — 21 days

(d) Development of regions in primitive heart tube — 22 days

(e) Bending of primitive heart — 23 days, 24 days

(f) Orientation of atria and ventricles to their final adult position — 28 days

When during embryonic development does the primitive heart begin to contract?

Figure 20.20 Partitioning of the heart into four chambers.

Partitioning of the heart begins on about the twenty-eighth day after fertilization.

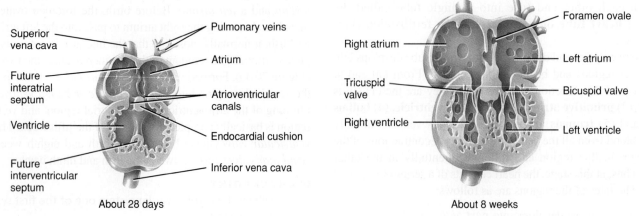

About 28 days

About 8 weeks

When is the partitioning of the heart complete?

DISORDERS: HOMEOSTATIC IMBALANCES

Coronary Artery Disease

Coronary artery disease (CAD) is a serious medical problem that affects about 7 million people annually. Responsible for nearly three-quarters of a million deaths in the United States each year, it is the leading cause of death for both men and women. CAD results from the effects of the accumulation of atherosclerotic plaques (described shortly) in coronary arteries, which leads to a reduction in blood flow to the myocardium. Some individuals have no signs or symptoms; others experience angina pectoris (chest pain), and still others suffer heart attacks.

Risk Factors for CAD

People who possess combinations of certain risk factors are more likely to develop CAD. *Risk factors* (characteristics, symptoms, or signs present in a disease-free person that are statistically associated with a greater chance of developing a disease) include smoking, high blood pressure, diabetes, high cholesterol levels, obesity, "type A" personality, sedentary lifestyle, and a family history of CAD. Most of these can be modified by changing diet and other habits or can be controlled by taking medications. However, other risk factors are unmodifiable (beyond our control), including genetic predisposition (family history of CAD at an early age), age, and gender. For example, adult males are more likely than adult females to develop CAD; after age 70 the risks are roughly equal. Smoking is undoubtedly the number-one risk factor in all CAD-associated diseases, roughly doubling the risk of morbidity and mortality.

Development of Atherosclerotic Plaques

Although the following discussion applies to coronary arteries, the process can also occur in arteries outside the heart. Thickening of the walls of arteries and loss of elasticity are the main characteristics of a group of diseases called **arteriosclerosis** (ar-tē-rē-ō-skle-RŌ-sis; *sclero-* = hardening). One form of arteriosclerosis is **atherosclerosis** (ath-er-ō-skle-RŌ-sis), a progressive disease characterized by the formation in the walls of large and medium-sized arteries of lesions called **atherosclerotic plaques** (ath-er-ō-skle-RO-tik) (Figure 20.21).

To understand how atherosclerotic plaques develop, you will need to learn about the role of molecules produced by the liver and small intestine called **lipoproteins.** These spherical particles consist of an inner core of triglycerides and other lipids and an outer shell of proteins, phospholipids, and cholesterol. Like most lipids, cholesterol does not dissolve in water and must be made water-soluble in order to be transported in the blood. This is accomplished by combining it with lipoproteins. Two major lipoproteins are **low-density lipoproteins (LDLs)** and **high-density lipoproteins (HDLs).** LDLs transport cholesterol from the liver to body cells for use in cell membrane repair and the production of steroid hormones and bile salts. However, excessive amounts of LDLs promote atherosclerosis, so the cholesterol in these particles is commonly known as "bad cholesterol." HDLs, on the other hand, remove excess cholesterol from body cells and transport it to the liver for elimination. Because HDLs decrease blood cholesterol level, the cholesterol in HDLs is commonly referred to as "good cholesterol." Basically, you want your LDL concentration to be low and your HDL concentration to be high.

Inflammation, a defensive response of the body to tissue damage, plays a key role in the development of atherosclerotic plaques. As a result of tissue damage, blood vessels dilate and increase their permeability, and phagocytes, including macrophages, appear in large numbers. The formation of atherosclerotic plaques begins when excess LDLs from the blood accumulate in the inner layer of an artery wall (layer closest to the bloodstream) and the lipids and proteins in the LDLs undergo oxidation (removal of electrons) and the proteins also bind to sugars. In response, endothelial and smooth muscle cells of the artery secrete substances that attract monocytes from the blood and convert them into macrophages. The macrophages then ingest and become so filled with oxidized LDL particles that they have a foamy appearance when viewed microscopically **(foam cells).** T cells (lymphocytes) follow monocytes into the inner lining of an artery, where they release chemicals that intensify the inflammatory response. Together, the foam cells, macrophages, and T cells form a **fatty streak,** the beginning of an atherosclerotic plaque.

Macrophages secrete chemicals that cause smooth muscle cells of the middle layer of an artery to migrate to the top of the atherosclerotic plaque, forming a cap over it and thus walling it off from the blood.

Because most atherosclerotic plaques expand away from the bloodstream rather than into it, blood can still flow through the

Figure 20.21 Photomicrographs of a transverse section of a normal artery and one partially obstructed by an atherosclerotic plaque.

  Inflammation plays a key role in the development of atherosclerotic plaques.

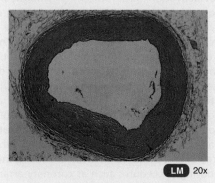

Normal artery

LM 20x

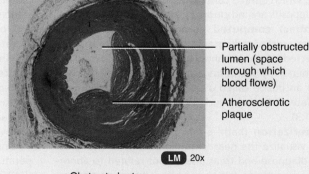

Partially obstructed lumen (space through which blood flows)

Atherosclerotic plaque

LM 20x

Obstructed artery

? What is the role of HDL?

affected artery with relative ease, often for decades. Relatively few heart attacks occur when plaque in a coronary artery expands into the bloodstream and restricts blood flow. Most heart attacks occur when the cap over the plaque breaks open in response to chemicals produced by foam cells. In addition, T cells induce foam cells to produce tissue factor (TF), a chemical that begins the cascade of reactions that result in blood clot formation. If the clot in a coronary artery is large enough, it can significantly decrease or stop the flow of blood and result in a heart attack.

A number of other risk factors (all modifiable) have also been identified as significant predictors of CAD when their levels are elevated. **C-reactive proteins (CRPs)** are proteins produced by the liver or present in blood in an inactive form that are converted to an active form during inflammation. CRPs may play a direct role in the development of atherosclerosis by promoting the uptake of LDLs by macrophages. **Lipoprotein (a)** is an LDL-like particle that binds to endothelial cells, macrophages, and blood platelets; may promote the proliferation of smooth muscle fibers; and inhibits the breakdown of blood clots. **Fibrinogen** is a glycoprotein involved in blood clotting that may help regulate cellular proliferation, vasoconstriction, and platelet aggregation. **Homocysteine** (hō'-mō-SIS-tē-ēn) is an amino acid that may induce blood vessel damage by promoting platelet aggregation and smooth muscle fiber proliferation.

Diagnosis of CAD

Many procedures may be employed to diagnose CAD; the specific procedure used will depend on the signs and symptoms of the individual.

A resting electrocardiogram (see Section 20.3) is the standard test employed to diagnose CAD. **Stress testing** can also be performed. In an *exercise stress test,* the functioning of the heart is monitored when placed under physical stress by exercising using a treadmill, an exercise bicycle, or arm exercises. During the procedure, ECG recordings are monitored continuously and blood pressure is monitored at intervals. A *nonexercise (pharmacologic) stress test* is used for individuals who cannot exercise due to conditions such as arthritis. A medication is injected that stresses the heart to mimic the effects of exercise. During both exercise and nonexercise stress testing, **radionuclide imaging** may be performed to evaluate blood flow through heart muscle (see Table 1.3).

Diagnosis of CAD may also involve **echocardiography** (ek'-ō-kar-dē-OG-ra-fē), a technique that uses ultrasound waves to image the interior of the heart. Echocardiography allows the heart to be seen in motion and can be used to determine the size, shape, and functions of the chambers of the heart; the volume and velocity of blood pumped from the heart; the status of heart valves; the presence of birth defects; and abnormalities of the pericardium. A fairly recent technique for evaluating CAD is **electron beam computerized tomography (EBCT),** which detects calcium deposits in coronary arteries. These calcium deposits are indicators of atherosclerosis.

Coronary (cardiac) computed tomography radiography (CCTA) is a computer-assisted radiography procedure in which a contrast medium is injected into a vein and a beta-blocker is given to decrease heart rate. Then x-ray beams trace an arc around the heart and ultimately produce an image called a *CCTA scan.* This procedure is used primarily to detect blockages such as atherosclerotic plaque or calcium (see Table 1.3).

Cardiac catheterization (kath'-e-ter-i-ZĀ-shun) is an invasive procedure used to visualize the heart's chambers, valves, and great vessels in order to diagnose and treat disease not related to abnormalities of the coronary arteries. It may also be used to measure pressure in the heart and great vessels; to assess cardiac output; to measure the flow of blood through the heart and great vessels; to identify

the location of septal and valvular defects; and to take tissue and blood samples. The basic procedure involves inserting a long, flexible, radiopaque **catheter** (plastic tube) into a peripheral vein (for *right heart catheterization*) or a peripheral artery (for *left heart catheterization*) and guiding it under fluoroscopy (x-ray observation).

Coronary angiography (an'-jē-OG-ra-fē; *angio-* = blood vessel; *-grapho* = to write) is an invasive procedure used to obtain information about the coronary arteries. In the procedure, a catheter is inserted into an artery in the groin or wrist and threaded under fluoroscopy toward the heart and then into the coronary arteries. After the tip of the catheter is in place, a radiopaque contrast medium is injected into the coronary arteries. The radiographs of the arteries, called *angiograms,* appear in motion on a monitor and the information is recorded on a videotape or computer disc. Coronary angiography may be used to visualize coronary arteries (see Table 1.3) and to inject clot-dissolving drugs, such as streptokinase or tissue plasminogen activator (t-PA), into a coronary artery to dissolve an obstructing thrombus.

Treatment of CAD

Treatment options for CAD include **drugs** (antihypertensives, nitroglycerin, beta-blockers, cholesterol-lowering drugs, and clot-dissolving agents) and various surgical and nonsurgical procedures designed to increase the blood supply to the heart.

Coronary artery bypass grafting (CABG) is a surgical procedure in which a blood vessel from another part of the body is attached ("grafted") to a coronary artery to bypass an area of blockage. A piece of the grafted blood vessel is sutured between the aorta and the unblocked portion of the coronary artery (Figure 20.22a). Sometimes multiple blood vessels have to be grafted.

A nonsurgical procedure used to treat CAD is **percutaneous transluminal coronary angioplasty (PTCA)** (*percutaneous* = through the skin; *trans-* = across; *-lumen* = an opening or channel in a tube; *angio-* = blood vessel; *-plasty* = to mold or to shape). In one variation of this procedure, a balloon catheter is inserted into an artery of an arm or leg and gently guided into a coronary artery (Figure 20.22b). While dye is released, angiograms (x-rays of blood vessels) are taken to locate the plaques. Next, the catheter is advanced to the point of obstruction, and a balloonlike device is inflated with air to squash the plaque against the blood vessel wall. Because 30–50% of PTCA-opened arteries fail due to restenosis (renarrowing) within 6 months after the procedure is done, a stent may be inserted via a catheter. A **stent** is a metallic, fine wire tube that is permanently placed in an artery to keep the artery *patent* (open), permitting blood to circulate (Figure 20.22c, d). Restenosis may be due to damage from the procedure itself, for PTCA may damage the arterial wall, leading to platelet activation, proliferation of smooth muscle fibers, and plaque formation. Recently, *drug-coated (drug-eluting) coronary stents* have been used to prevent restenosis. The stents are coated with one of several antiproliferative drugs (drugs that inhibit the proliferation of smooth muscle fibers of the middle layer of an artery) and anti-inflammatory drugs. It has been shown that drug-coated stents reduce the rate of restenosis when compared with bare-metal (noncoated) stents. In addition to balloon and stent angioplasty, laser-emitting catheters are used to vaporize plaques (excimer laser coronary angioplasty or ELCA) and small blades inside catheters are used to remove part of the plaque (directional coronary atherectomy).

One area of current research involves cooling the body's core temperature during procedures such as coronary artery bypass grafting (CABG). There have been some promising results from the application of cold therapy during a cerebral vascular accident (CVA or stroke). This research stemmed from observations of people who had suffered

Figure 20.22 Procedures for reestablishing blood flow in occluded coronary arteries.

 Treatment options for CAD include drugs and various nonsurgical and surgical procedures.

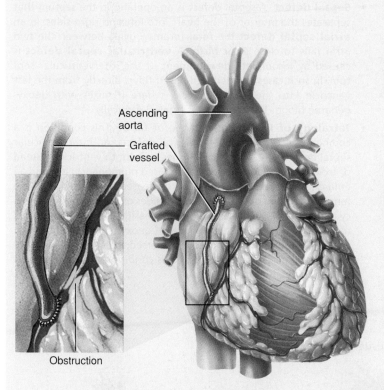

Ascending aorta

Grafted vessel

Obstruction

(a) Coronary artery bypass grafting (CABG)

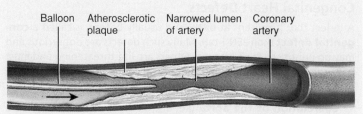

Balloon   Atherosclerotic plaque   Narrowed lumen of artery   Coronary artery

Balloon catheter with uninflated balloon is threaded to obstructed area in artery

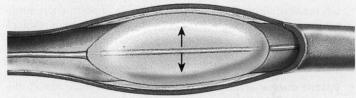

When balloon is inflated, it stretches arterial wall and squashes atherosclerotic plaque

After lumen is widened, balloon is deflated and catheter is withdrawn

(b) Percutaneous transluminal coronary angioplasty (PTCA)

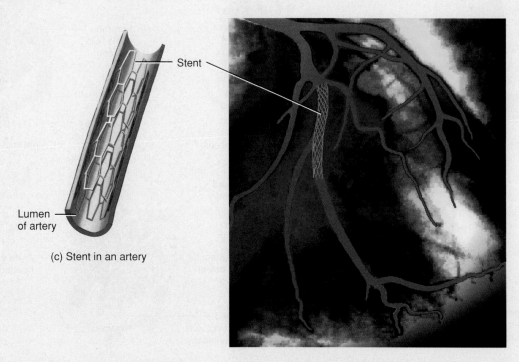

Stent

Lumen of artery

(c) Stent in an artery

(d) Angiogram showing stent in circumflex artery

? **Which diagnostic procedure for CAD is used to visualize coronary blood vessels?**

a hypothermic incident (such as cold-water drowning) and recovered with relatively minimal neurologic deficits.

Congenital Heart Defects

A defect that is present at birth, and usually before, is called a **congenital defect** (kon-JEN-i-tal). Many such defects are not serious and may go unnoticed for a lifetime. Others are life-threatening and must be surgically repaired. Among the several congenital defects that affect the heart are the following (Figure 20.23):

• **Coarctation of the aorta** (kō'-ark-TĀ-shun). In this condition, a segment of the aorta is too narrow, and thus the flow of oxygenated blood to the body is reduced, the left ventricle is forced to pump harder, and high blood pressure develops. Coarctation is usually repaired surgically by removing the area of obstruction. Surgical interventions that are done in childhood may require revisions in adulthood. Another surgical procedure is balloon dilation, insertion and inflation of a device in the aorta to stretch the vessel. A stent can be inserted and left in place to hold the vessel open.

• **Patent ductus arteriosus (PDA)** (PĀ-tent). In some babies, the ductus arteriosus, a temporary blood vessel between the aorta and the pulmonary trunk, remains open rather than closing shortly after birth. As a result, aortic blood flows into the lower-pressure pulmonary trunk, thus increasing the pulmonary trunk blood pressure and overworking both ventricles. In uncomplicated PDA, medication can be used to facilitate the closure of the defect. In more severe cases, surgical intervention may be required.

• **Septal defect.** A septal defect is an opening in the septum that separates the interior of the heart into left and right sides. In an **atrial septal defect** the fetal foramen ovale between the two atria fails to close after birth. A **ventricular septal defect** is caused by incomplete development of the interventricular septum. In such cases, oxygenated blood flows directly from the left ventricle into the right ventricle, where it mixes with deoxygenated blood. The condition is treated surgically.

• **Tetralogy of Fallot** (tet-RAL-ō-jē of fal-Ō). This condition is a combination of four developmental defects: an interventricular septal defect, an aorta that emerges from both ventricles instead of from the left ventricle only, a stenosed pulmonary valve, and an enlarged right ventricle. There is a decreased flow of blood to the lungs and mixing of blood from both sides of the heart. This causes cyanosis, the bluish discoloration most easily seen in nail beds and mucous membranes when the level of deoxygenated hemoglobin is high; in infants, this condition is referred to as "blue baby." Despite the apparent complexity of this condition, surgical repair is usually successful.

Figure 20.23 Congenital heart defects.

🔑 A congenital defect is one that is present at birth, and usually before.

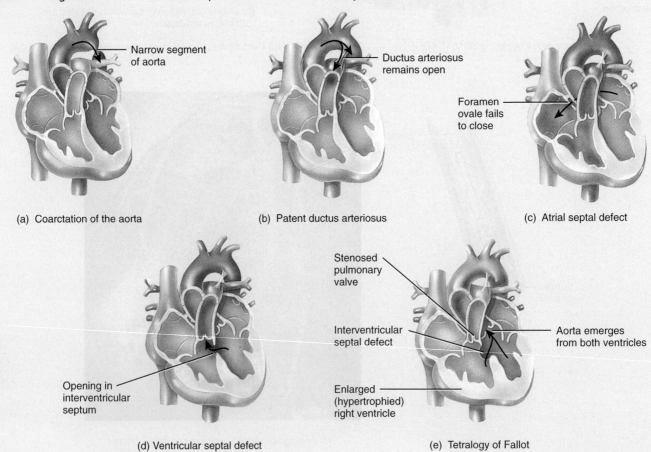

(a) Coarctation of the aorta

(b) Patent ductus arteriosus

(c) Atrial septal defect

(d) Ventricular septal defect

(e) Tetralogy of Fallot

❓ **Which four developmental defects occur in tetralogy of Fallot?**

Arrhythmias

The usual rhythm of heartbeats, established by the SA node, is called **normal sinus rhythm.** The term **arrhythmia** (a-RITH-mē-a) or **dysrhythmia** refers to an abnormal rhythm as a result of a defect in the conduction system of the heart. The heart may beat irregularly, too quickly, or too slowly. Symptoms include chest pain, shortness of breath, lightheadedness, dizziness, and fainting. Arrhythmias may be caused by factors that stimulate the heart such as stress, caffeine, alcohol, nicotine, cocaine, and certain drugs that contain caffeine or other stimulants. Arrhythmias may also be caused by a congenital defect, coronary artery disease, myocardial infarction, hypertension, defective heart valves, rheumatic heart disease, hyperthyroidism, and potassium deficiency.

Arrhythmias are categorized by their speed, rhythm, and origination of the problem. **Bradycardia** (brād´-i-KAR-dē-a; *brady-* = slow) refers to a slow heart rate (below 50 beats per minute); **tachycardia** (tak´-i-KAR-dē-a; *tachy-* = quick) refers to a rapid heart rate (over 100 beats per minute); and **fibrillation** (fi-bri-LĀ-shun) refers to rapid, uncoordinated heartbeats. Arrhythmias that begin in the atria are called **supraventricular** or **atrial arrhythmias;** those that originate in the ventricles are called **ventricular arrhythmias.**

- **Supraventricular tachycardia (SVT)** is a rapid but regular heart rate (160–200 beats per minute) that originates in the atria. The episodes begin and end suddenly and may last from a few minutes to many hours. SVTs can sometimes be stopped by maneuvers that stimulate the vagus (X) nerve and decrease heart rate. These include straining as if having a difficult bowel movement, rubbing the area over the carotid artery in the neck to stimulate the carotid sinus (not recommended for people over 50 since it may cause a stroke), and plunging the face into a bowl of ice-cold water. Treatment may also involve antiarrhythmic drugs and destruction of the abnormal pathway by radiofrequency ablation.

- **Heart block** is an arrhythmia that occurs when the electrical pathways between the atria and ventricles are blocked, slowing the transmission of nerve impulses. The most common site of blockage is the atrioventricular node, a condition called *atrioventricular (AV) block.* In *first-degree AV block,* the P–Q interval is prolonged, usually because conduction through the AV node is slower than normal (Figure 20.24b). In *second-degree AV block,* some of the action potentials from the SA node are not conducted through the AV node. The result is "dropped" beats because excitation doesn't always reach the ventricles. Consequently, there are fewer QRS

Figure 20.24 Representative arrhythmias.

 An arrhythmia is an abnormal rhythm as a result of a defect in the cardiac conduction system.

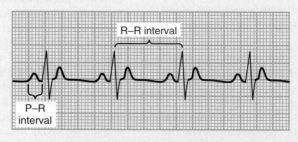

(a) Normal electrocardiogram (ECG)

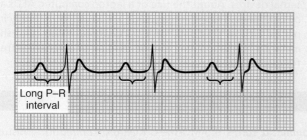

(b) First-degree AV block

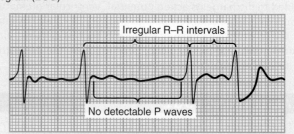

(c) Atrial fibrillation

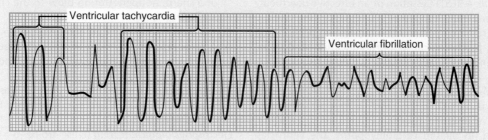

(d) Ventricular tachycardia

(e) Ventricular fibrillation

? **Why is ventricular fibrillation such a serious arrhythmia?**

complexes than P waves on the ECG. In *third-degree (complete) AV block,* no SA node action potentials get through the AV node. Autorhythmic fibers in the atria and ventricles pace the upper and lower chambers separately. With complete AV block, the ventricular contraction rate is less than 40 beats/min.

- **Atrial premature contraction (APC)** is a heartbeat that occurs earlier than expected and briefly interrupts the normal heart rhythm. It often causes a sensation of a skipped heartbeat followed by a more forceful heartbeat. APCs originate in the atrial myocardium and are common in healthy individuals.

- **Atrial flutter** consists of rapid, regular atrial contractions (240–360 beats/min) accompanied by an atrioventricular (AV) block in which some of the nerve impulses from the SA node are not conducted through the AV node.

- **Atrial fibrillation (AF)** is a common arrhythmia, affecting mostly older adults, in which contraction of the atrial fibers is asynchronous (not in unison) so that atrial pumping ceases altogether. The atria may beat 300–600 beats/min. The ventricles may also speed up, resulting in a rapid heartbeat (up to 160 beats/min). The ECG of an individual with atrial fibrillation typically has no clearly defined P waves and irregularly spaced QRS complexes (and P–R intervals) (Figure 20.24c). Since the atria and ventricles do not beat in rhythm, the heartbeat is irregular in timing and strength. In an otherwise strong heart, atrial fibrillation reduces the pumping effectiveness of the heart by 20–30%. The most dangerous complication of atrial fibrillation is stroke since blood may stagnate in the atria and form blood clots. A stroke occurs when part of a blood clot occludes an artery supplying the brain.

- **Ventricular premature contraction,** another form of arrhythmia, arises when an *ectopic focus* (ek-TOP-ik), a region of the heart other than the conduction system, becomes more excitable than normal and causes an occasional abnormal action potential to occur. As a wave of depolarization spreads outward from the ectopic focus, it causes a ventricular premature contraction (beat). The contraction occurs early in diastole before the SA node is normally scheduled to discharge its action potential. Ventricular premature contractions may be relatively benign and may be caused by emotional stress, excessive intake of stimulants such as caffeine, alcohol, or nicotine, and lack of sleep. In other cases, the premature beats may reflect an underlying pathology.

- **Ventricular tachycardia (VT or V-tach)** is an arrhythmia that originates in the ventricles and is characterized by four or more ventricular premature contractions. It causes the ventricles to beat too fast (at least 120 beats/min) (Figure 20.24d). VT is almost always associated with heart disease or a recent myocardial infarction and may develop into a very serious arrhythmia called ventricular fibrillation (described shortly). Sustained VT is dangerous because the ventricles do not fill properly and thus do not pump sufficient blood. The result may be low blood pressure and heart failure.

- **Ventricular fibrillation (VF or V-fib)** is the most deadly arrhythmia, in which contractions of the ventricular fibers are completely asynchronous so that the ventricles quiver rather than contract in a coordinated way. As a result, ventricular pumping stops, blood ejection ceases, and circulatory failure and death occur unless there is immediate medical intervention. During ventricular fibrillation, the ECG has no clearly defined P waves, QRS complexes, or T waves (Figure 20.24e). The most common cause of ventricular fibrillation is inadequate blood flow to the heart due to coronary artery disease, as occurs during a myocardial infarction. Other causes are cardiovascular shock, electrical shock, drowning, and very low potassium levels. Ventricular fibrillation causes unconsciousness in seconds and, if untreated, seizures occur and irreversible brain damage may occur after 5 minutes. Death soon follows. Treatment involves cardiopulmonary resuscitation (CPR) and defibrillation. In **defibrillation** (dē-fib-re-LĀ-shun), also called **cardioversion** (kar′-dē-ō-VER-shun), a strong, brief electrical current is passed to the heart and often can stop the ventricular fibrillation. The electrical shock is generated by a device called a **defibrillator** (de-FIB-ri-lā-tor) and applied via two large paddle-shaped electrodes pressed against the skin of the chest. Patients who face a high risk of dying from heart rhythm disorders now can receive an **automatic implantable cardioverter defibrillator (AICD),** an implanted device that monitors their heart rhythm and delivers a small shock directly to the heart when a life-threatening rhythm disturbance occurs. Thousands of patients around the world have AICDs. Also available are **automated external defibrillators (AEDs)** that function like AICDs, except that they are external devices. About the size of a laptop computer, AEDs are used by emergency response teams and are found increasingly in public places such as stadiums, casinos, airports, hotels, and shopping malls. Defibrillation may also be used as an emergency treatment for cardiac arrest.

MEDICAL TERMINOLOGY

Asystole (ā-SIS-tō-lē; *a-* = without) Failure of the myocardium to contract.

Cardiac arrest (KAR-dē-ak a-REST) Cessation of an effective heartbeat. The heart may be completely stopped or in ventricular fibrillation.

Cardiac rehabilitation (rē-ha-bil-i-TĀ-shun) A supervised program of progressive exercise, psychological support, education, and training to enable a patient to resume normal activities following a myocardial infarction.

Cardiomegaly (kar′-dē-ō-MEG-a-lē; *mega* = large) Heart enlargement.

Cardiomyopathy (kar′-dē-ō-mī-OP-a-thē; *myo-* = muscle; *-pathos* = disease) A progressive disorder in which ventricular structure or function is impaired. In dilated cardiomyopathy, the ventricles enlarge (stretch) and become weaker and reduce the heart's pumping action. In hypertrophic cardiomyopathy, the ventricular walls thicken and the pumping efficiency of the ventricles is reduced.

Commotio cordis (kō-MŌ-shē-ō KOR-dis; *commotio* = disturbance; *cordis* = heart) Damage to the heart, frequently fatal, as a result of a sharp, nonpenetrating blow to the chest while the ventricles are repolarizing.

Cor pulmonale (CP) (KOR pul-mōn-AL-ē; *cor* = heart; *pulmon-* = lung) A term referring to right ventricular hypertrophy from disorders that bring about hypertension (high blood pressure) in the pulmonary circulation.

Ejection fraction The fraction of the end-diastolic volume (EDV) that is ejected during an average heartbeat. Equal to stroke volume (SV) divided by EDV.

Electrophysiological testing (e-lek′-trō-fiz′-ē-ō-LOJ-i-kal) A procedure in which a catheter with an electrode is passed through blood vessels and introduced into the heart. It is used to detect the exact locations of abnormal electrical conduction pathways. Once an abnormal pathway is located, it can be destroyed by sending a current through the electrode, a procedure called *radiofrequency ablation*.

Palpitation (pal'-pi-TĀ-shun) A fluttering of the heart or an abnormal rate or rhythm of the heart about which an individual is aware.

Paroxysmal tachycardia (par'-ok-SIZ-mal tak'-i-KAR-dē-a; *tachy-* = quick) A period of rapid heartbeats that begins and ends suddenly.

Sick sinus syndrome An abnormally functioning SA node that initiates heartbeats too slowly or rapidly, pauses too long between heartbeats, or stops producing heartbeats. Symptoms include lightheadedness, shortness of breath, loss of consciousness, and palpitations. It is caused by degeneration of cells in the SA node and is common in elderly persons. It is also related to coronary artery disease. Treatment consists of drugs to speed up or slow down the heart or implantation of an artificial pacemaker.

Sudden cardiac death The unexpected cessation of circulation and breathing due to an underlying heart disease such as ischemia, myocardial infarction, or a disturbance in cardiac rhythm.

CHAPTER REVIEW AND RESOURCE SUMMARY

Review

Resource

20.1 Anatomy of the Heart

1. The heart is located in the mediastinum; about two-thirds of its mass is to the left of the midline. It is shaped like a cone lying on its side. Its apex is the pointed, inferior part; its base is the broad, superior part.

2. The pericardium is the membrane that surrounds and protects the heart; it consists of an outer fibrous layer and an inner serous pericardium, which is composed of a parietal and a visceral layer. Between the parietal and visceral layers of the serous pericardium is the pericardial cavity, a potential space filled with a few milliliters of lubricating pericardial fluid that reduces friction between the two membranes.

3. Three layers make up the wall of the heart: epicardium (visceral layer of the serous pericardium), myocardium, and endocardium. The epicardium consists of mesothelium and connective tissue, the myocardium is composed of cardiac muscle tissue, and the endocardium consists of endothelium and connective tissue.

4. The heart chambers include two superior chambers, the right and left atria, and two inferior chambers, the right and left ventricles. External features of the heart include the auricles (flaps of each atrium that slightly increase their volume), the coronary sulcus between the atria and ventricles, and the anterior and posterior sulci between the ventricles on the anterior and posterior surfaces of the heart, respectively.

5. The right atrium receives blood from the superior vena cava, inferior vena cava, and coronary sinus. It is separated internally from the left atrium by the interatrial septum, which contains the fossa ovalis. Blood exits the right atrium through the tricuspid valve.

6. The right ventricle receives blood from the right atrium. It is separated internally from the left ventricle by the interventricular septum and pumps blood through the pulmonary valve into the pulmonary trunk.

7. Oxygenated blood enters the left atrium from the pulmonary veins and exits through the bicuspid (mitral) valve.

8. The left ventricle pumps oxygenated blood through the aortic valve into the aorta.

9. The thickness of the myocardium of the four chambers varies according to the chamber's function. The left ventricle, with the highest workload, has the thickest wall.

10. The fibrous skeleton of the heart is dense connective tissue that surrounds and supports the valves of the heart.

Anatomy Overview - The Cardiovascular System
Anatomy Overview - The Heart
Figure 20.3 - Structure of the Heart: Surface Features
Exercise - Paint the Heart

20.2 Heart Valves and Circulation of Blood

1. Heart valves prevent backflow of blood within the heart. The atrioventricular (AV) valves, which lie between atria and ventricles, are the tricuspid valve on the right side of the heart and the bicuspid (mitral) valve on the left. The semilunar (SL) valves are the aortic valve, at the entrance to the aorta, and the pulmonary valve, at the entrance to the pulmonary trunk.

2. The left side of the heart is the pump for systemic circulation, the circulation of blood throughout the body except for the air sacs of the lungs. The left ventricle ejects blood into the aorta, and blood then flows into systemic arteries, arterioles, capillaries, venules, and veins, which carry it back to the right atrium.

3. The right side of the heart is the pump for pulmonary circulation, the circulation of blood through the lungs. The right ventricle ejects blood into the pulmonary trunk, and blood then flows into pulmonary arteries, pulmonary capillaries, and pulmonary veins, which carry it back to the left atrium.

Anatomy Overview - Heart Structures
Anatomy Overview - Pulmonary Circulation
Figure 20.7 - Systemic and Pulmonary Circulations
Exercise - Drag and Drop Blood Flow
Negative Feedback Loop Exercise - Homeostasis of Blood Circulation

Review

4. The coronary circulation provides blood flow to the myocardium. The main arteries of the coronary circulation are the left and right coronary arteries; the main veins are the cardiac veins and the coronary sinus.

20.3 Cardiac Muscle Tissue and the Cardiac Conduction System

1. Cardiac muscle fibers usually contain a single centrally located nucleus. Compared with skeletal muscle fibers, cardiac muscle fibers have more and larger mitochondria, slightly smaller sarcoplasmic reticulum, and wider transverse tubules, which are located at Z discs.
2. Cardiac muscle fibers are connected end-to-end via intercalated discs. Desmosomes in the discs provide strength, and gap junctions allow muscle action potentials to conduct from one muscle fiber to its neighbors.
3. Autorhythmic fibers form the conduction system, cardiac muscle fibers that spontaneously depolarize and generate action potentials.
4. Components of the conduction system are the sinoatrial (SA) node (pacemaker), atrioventricular (AV) node, atrioventricular (AV) bundle (bundle of His), bundle branches, and Purkinje fibers.
5. Phases of an action potential in a ventricular contractile fiber include rapid depolarization, a long plateau, and repolarization.
6. Cardiac muscle tissue has a long refractory period, which prevents tetanus.
7. The record of electrical changes during each cardiac cycle is called an electrocardiogram (ECG). A normal ECG consists of a P wave (atrial depolarization), a QRS complex (onset of ventricular depolarization), and a T wave (ventricular repolarization).
8. The P–Q interval represents the conduction time from the beginning of atrial excitation to the beginning of ventricular excitation. The S–T segment represents the time when ventricular contractile fibers are fully depolarized.

Anatomy Overview - Cardiac Muscle
Animation - Cardiac Conduction
Figure 20.10 - The
  Conduction System
  of the Heart
Exercise - Sequence
  Cardiac Conduction
Exercise - ECG Jigsaw Puzzle

20.4 The Cardiac Cycle

1. A cardiac cycle consists of the systole (contraction) and diastole (relaxation) of both atria, plus the systole and diastole of both ventricles. With an average heartbeat of 75 beats/min, a complete cardiac cycle requires 0.8 sec.
2. The phases of the cardiac cycle are (a) atrial systole, (b) ventricular systole, and (c) relaxation period.
3. S1, the first heart sound (lubb), is caused by blood turbulence associated with the closing of the atrioventricular valves. S2, the second sound (dupp), is caused by blood turbulence associated with the closing of semilunar valves.

Animation - Cardiac Cycle and ECG
Animation - Cardiac Cycle
Exercise - Cardiac Cycle
Concepts and Connections - Cardiac
  Cycle

20.5 Cardiac Output

1. Cardiac output (CO) is the amount of blood ejected per minute by the left ventricle into the aorta (or by the right ventricle into the pulmonary trunk). It is calculated as follows: CO (mL/min) = stroke volume (SV) in mL/beat × heart rate (HR) in beats/min.
2. Stroke volume (SV) is the amount of blood ejected by a ventricle during each systole.
3. Cardiac reserve is the difference between a person's maximum cardiac output and his or her cardiac output at rest.
4. Stroke volume is related to preload (stretch on the heart before it contracts), contractility (forcefulness of contraction), and afterload (pressure that must be exceeded before ventricular ejection can begin).
5. According to the Frank–Starling law of the heart, a greater preload (end-diastolic volume) stretching cardiac muscle fibers just before they contract increases their force of contraction until the stretching becomes excessive.
6. Nervous control of the cardiovascular system originates in the cardiovascular center in the medulla oblongata.
7. Sympathetic impulses increase heart rate and force of contraction; parasympathetic impulses decrease heart rate.
8. Heart rate is affected by hormones (epinephrine, norepinephrine, thyroid hormones), ions (Na^+, K^+, Ca^{2+}), age, gender, physical fitness, and body temperature.

Animation - Cardiac Output
Animation - ANS Effects on Cardiac
  Conduction
Exercise - Cardiac Output Factors
Concepts and Connections - Cardiac
  Output

20.6 Exercise and the Heart

1. Sustained exercise increases oxygen demand on muscles.
2. Among the benefits of aerobic exercise are increased cardiac output, decreased blood pressure, weight control, and increased fibrinolytic activity.

Review

Resource

20.7 Help for Failing Hearts

1. A cardiac (heart) transplant is the replacement of a severely damaged heart with a normal one.
2. Cardiac assist devices and procedures include the intra-aortic balloon pump, the ventricular assist device, cardiomyoplasty, and a skeletal muscle assist device.

20.8 Development of the Heart

Partitioning of the Heart

1. The heart develops from mesoderm.
2. The endocardial tubes develop into the four-chambered heart and great vessels of the heart.

SELF-QUIZ QUESTIONS

Fill in the blanks in the following statements.

1. The chamber of the heart with the thickest myocardium is the _____.

2. The phase of heart contraction is called _____; the phase of relaxation is called _____.

Indicate whether the following statements are true or false.

3. In auscultation, the lubb represents closing of the semilunar valves and the dupp represents closing of the atrioventricular valves.

4. The Frank–Starling law of the heart equalizes the output of the right and left ventricles and keeps the same volume of blood flowing to both the systemic and pulmonary circulations.

Choose the one best answer to the following questions.

5. Which of the following is the correct route of blood through the heart from the systemic circulation to the pulmonary circulation and back to the systemic circulation?
   (a) right atrium, tricuspid valve, right ventricle, pulmonary semilunar valve, left atrium, mitral valve, left ventricle, aortic semilunar valve
   (b) left atrium, tricuspid valve, left ventricle, pulmonary semilunar valve, right atrium, mitral valve, right ventricle, aortic semilunar valve
   (c) left atrium, pulmonary semilunar valve, right atrium, tricuspid valve, left ventricle, aortic semilunar valve, right ventricle, mitral valve
   (d) left ventricle, mitral valve, left atrium, pulmonary semilunar valve, right ventricle, tricuspid valve, right atrium, aortic semilunar valve
   (e) right atrium, mitral valve, right ventricle, pulmonary semilunar valve, left atrium, tricuspid valve, left ventricle, aortic semilunar valve

6. Which of the following represents the correct pathway for conduction of an action potential through the heart?
   (a) AV node, AV bundle, SA node, Purkinje fibers, bundle branches
   (b) AV node, bundle branches, AV bundle, SA node, Purkinje fibers
   (c) SA node, AV node, AV bundle, bundle branches, Purkinje fibers
   (d) SA node, AV bundle, bundle branches, AV node, Purkinje fibers
   (e) SA node, AV node, Purkinje fibers, bundle branches, AV bundle

7. The external boundary between the atria and ventricles is the
   (a) anterior interventricular sulcus.
   (b) interventricular septum.
   (c) interatrial septum.
   (d) coronary sulcus.
   (e) posterior interventricular sulcus.

8. A softball player is found to have a resting cardiac output of 5.0 liters per minute and a heart rate of 50 beats per minute. What is her stroke volume?
   (a) 10 mL   (b) 100 mL
   (c) 1000 mL   (d) 250 mL
   (e) The information given is insufficient to calculate stroke volume.

9. Which of the following are true? (1) ANS regulation of heart rate originates in the cardiovascular center of the medulla oblongata. (2) Proprioceptor input is a major stimulus that accounts for the rapid rise in the heart rate at the onset of physical activity. (3) The vagus nerves release norepinephrine, causing the heart rate to increase. (4) Hormones from the adrenal medulla and the thyroid gland can increase the heart rate. (5) Hypothermia increases the heart rate.
   (a) 1, 2, 3, and 4   (b) 1, 2, and 4
   (c) 2, 3, 4, and 5   (d) 3, 5, and 6
   (e) 1, 2, 4, and 5

10. Which of the following are true concerning action potentials and contraction in the myocardium? (1) The refractory period in a cardiac muscle fiber is very brief. (2) The binding of Ca^{2+} to troponin allows the interaction of actin and myosin filaments, resulting in contraction. (3) Repolarization occurs when the voltage-gated K^+ channels open and calcium channels are closing. (4) Opening of voltage-gated fast Na^+ channels results in depolarization. (5) Opening of voltage-gated slow Ca^{2+} channels results in a period of maintained depolarization, known as the plateau.
   (a) 1, 3, and 5   (b) 2, 3, and 4
   (c) 2 and 5   (d) 3, 4, and 5
   (e) 2, 3, 4, and 5

11. Which of the following would not increase stroke volume?
   (a) increased Ca^{2+} in the interstitial fluid
   (b) epinephrine
   (c) increased K^+ in the interstitial fluid
   (d) increase in venous return
   (e) slow resting heart rate

12. Match the following:
_____ (a) indicates ventricular repolarization
_____ (b) represents the time from the beginning of ventricular depolarization to the end of ventricular repolarization
_____ (c) represents atrial depolarization
_____ (d) represents the time when the ventricular contractile fibers are fully depolarized; occurs during the plateau phase of the action potential
_____ (e) represents the onset of ventricular depolarization
_____ (f) represents the conduction time from the beginning of atrial excitation to the beginning of ventricular excitation

(1) P wave
(2) QRS complex
(3) T wave
(4) P–Q interval
(5) S–T segment
(6) Q–T interval

13. Match the following:
_____ (a) major branch from the ascending aorta; passes inferior to the left auricle
_____ (b) lies in the posterior interventricular sulcus; supplies the walls of the ventricles with oxygenated blood
_____ (c) located in the coronary sulcus on the posterior surface of the heart; receives most of the deoxygenated blood from the myocardium
_____ (d) lies in the coronary sulcus; supplies oxygenated blood to the walls of the right ventricle
_____ (e) lies in the coronary sulcus; drains the right atrium and right ventricle
_____ (f) major branch from the ascending aorta; lies inferior to the right auricle
_____ (g) lies in the posterior interventricular sulcus; drains the right and left ventricles
_____ (h) lies in the anterior interventricular sulcus; supplies oxygenated blood to the walls of both ventricles
_____ (i) lies in the anterior interventricular sulcus; drains the walls of both ventricles and the left atrium
_____ (j) lies in the coronary sulcus; supplies oxygenated blood to the walls of the left ventricle and left atrium
_____ (k) drain the right ventricle and open directly into the right atrium

(1) small cardiac vein
(2) anterior interventricular branch (left anterior descending artery)
(3) anterior cardiac veins
(4) posterior interventricular branch
(5) marginal branch
(6) circumflex branch
(7) middle cardiac vein
(8) left coronary artery
(9) right coronary artery
(10) great cardiac vein
(11) coronary sinus

14. Match the following:
_____ (a) collects oxygenated blood from the pulmonary circulation
_____ (b) pumps deoxygenated blood to the lungs for oxygenation
_____ (c) their contraction pulls on and tightens the chordae tendineae, preventing the valve cusps from everting
_____ (d) cardiac muscle tissue
_____ (e) increase blood-holding capacity of the atria
_____ (f) tendonlike cords connected to the atrioventricular valve cusps which, along with the papillary muscles, prevent valve eversion
_____ (g) the superficial dense irregular connective tissue covering of the heart
_____ (h) outer layer of the serous pericardium; is fused to the fibrous pericardium
_____ (i) endothelial cells lining the interior of the heart; are continuous with the endothelium of the blood vessels
_____ (j) pumps oxygenated blood to all body cells, except the air sacs of the lungs
_____ (k) prevents backflow of blood from the right ventricle into the right atrium
_____ (l) collects deoxygenated blood from the systemic circulation
_____ (m) left atrioventricular valve
_____ (n) the remnant of the foramen ovale, an opening in the interatrial septum of the fetal heart
_____ (o) blood vessels that pierce the heart muscle and supply blood to the cardiac muscle fibers
_____ (p) grooves on the surface of the heart which delineate the external boundaries between the chambers
_____ (q) prevent backflow of blood from the arteries into the ventricles
_____ (r) the gap junction and desmosome connections between individual cardiac muscle fibers
_____ (s) internal wall dividing the chambers of the heart
_____ (t) separate the upper and lower heart chambers, preventing backflow of blood from the ventricles back into the atria
_____ (u) inner visceral layer of the pericardium; adheres tightly to the surface of the heart
_____ (v) ridges formed by raised bundles of cardiac muscle fibers

(1) right atrium
(2) right ventricle
(3) left atrium
(4) left ventricle
(5) tricuspid valve
(6) bicuspid (mitral) valve
(7) chordae tendineae
(8) auricles
(9) papillary muscles
(10) trabeculae carneae
(11) fibrous pericardium
(12) parietal pericardium
(13) epicardium
(14) myocardium
(15) endocardium
(16) atrioventricular valves
(17) semilunar valves
(18) intercalated discs
(19) sulci
(20) septum
(21) fossa ovalis
(22) coronary circulation

15. Match the following:
    _____ (a) amount of blood contained in the ventricles at the end of ventricular relaxation
    _____ (b) period of time when cardiac muscle fibers are contracting and exerting force but not shortening
    _____ (c) amount of blood ejected per beat by each ventricle
    _____ (d) amount of blood remaining in the ventricles following ventricular contraction
    _____ (e) difference between a person's maximum cardiac output and cardiac output at rest
    _____ (f) period of time when semilunar valves are open and blood flows out of the ventricles
    _____ (g) period when all four valves are closed and ventricular blood volume does not change

    (1) cardiac reserve
    (2) stroke volume
    (3) end-diastolic volume (EDV)
    (4) isovolumetric relaxation
    (5) end-systolic volume (ESV)
    (6) ventricular ejection
    (7) isovolumetric contraction

CRITICAL THINKING QUESTIONS

1. Gerald recently visited the dentist to have his teeth cleaned and checked. During the cleaning process, Gerald had some bleeding from his gums. A couple of days later, Gerald developed a fever, rapid heartbeat, sweating, and chills. He visited his family physician, who detected a slight heart murmur. Gerald was given antibiotics and continued to have his heart monitored. How was Gerald's dental visit related to his illness?

2. Unathletic Sylvia makes a resolution to begin an exercise program. She tells you that she wants to make her heart "beat as fast as it can" during exercise. Explain to her why that may not be a good idea.

3. Mr. Perkins is a large, 62-year-old man with a weakness for sweets and fried foods. His idea of exercise is walking to the kitchen for more potato chips to eat while he is watching sports on television. Lately, he's been troubled by chest pains when he walks up stairs. His doctor told him to quit smoking and scheduled cardiac angiography for the next week. What is involved in performing this procedure? Why did the doctor order this test?

? ANSWERS TO FIGURE QUESTIONS

20.1 The mediastinum is the anatomical region that extends from the sternum to the vertebral column, from the first rib to the diaphragm, and between the lungs.

20.2 The visceral layer of the serous pericardium (epicardium) is both a part of the pericardium and a part of the heart wall.

20.3 The coronary sulcus forms a boundary between the atria and ventricles.

20.4 The greater the workload of a heart chamber, the thicker its myocardium.

20.5 The fibrous skeleton attaches to the heart valves and prevents overstretching of the valves as blood passes through them.

20.6 The papillary muscles contract, which pulls on the chordae tendineae and prevents cusps of the atrioventricular valves from everting and letting blood flow back into the atria.

20.7 Numbers 2 (right ventricle) through 6 (left atrium) depict the pulmonary circulation; numbers 7 (left ventricle) through 1 (right atrium) depict the systemic circulation.

20.8 The circumflex artery delivers oxygenated blood to the left atrium and left ventricle.

20.9 The intercalated discs hold the cardiac muscle fibers together and enable action potentials to propagate from one muscle fiber to another.

20.10 The only electrical connection between the atria and the ventricles is the atrioventricular bundle.

20.11 The duration of an action potential is much longer in a ventricular contractile fiber (0.3 sec = 300 msec) than in a skeletal muscle fiber (1–2 msec).

20.12 An enlarged Q wave may indicate a myocardial infarction (heart attack).

20.13 Action potentials propagate most slowly through the AV node.

20.14 The amount of blood in each ventricle at the end of ventricular diastole—called the end-diastolic volume—is about 130 mL in a resting person.

20.15 The first heart sound (S1), or lubb, is associated with closure of the atrioventricular valves.

20.16 The ventricular myocardium receives innervation from the sympathetic division only.

20.17 The skeletal muscle "pump" increases stroke volume by increasing preload (end-diastolic volume).

20.18 Individuals with end-stage heart failure or severe coronary artery disease are candidates for cardiac transplantation.

20.19 The heart begins to contract by the twenty-second day of gestation.

20.20 Partitioning of the heart is complete by the end of the fifth week.

20.21 HDL removes excess cholesterol from body cells and transports it to the liver for elimination.

20.22 Coronary angiography is used to visualize many blood vessels.

20.23 Tetralogy of Fallot involves an interventricular septal defect, an aorta that emerges from both ventricles, a stenosed pulmonary valve, and an enlarged right ventricle.

20.24 Ventricular fibrillation is such a serious arrhythmia because ventricular pumping stops, blood ejection ceases, and circulatory failure and death can occur without immediate medical intervention.

21 THE CARDIOVASCULAR SYSTEM: BLOOD VESSELS AND HEMODYNAMICS

BLOOD VESSELS, HEMODYNAMICS, AND HOMEOSTASIS *Blood vessels contribute to homeostasis by providing the structures for the flow of blood to and from the heart and the exchange of nutrients and wastes in tissues. They also play an important role in adjusting the velocity and volume of blood flow.*

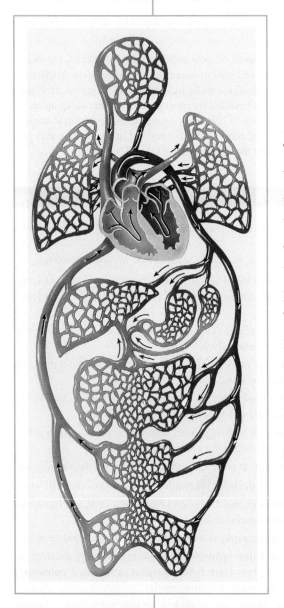

The cardiovascular system contributes to homeostasis of other body systems by transporting and distributing blood throughout the body to deliver materials (such as oxygen, nutrients, and hormones) and carry away wastes. The structures involved in these important tasks are the blood vessels, which form a closed system of tubes that carries blood away from the heart, transports it to the tissues of the body, and then returns it to the heart. The left side of the heart pumps blood through an estimated 100,000 km (60,000 mi) of blood vessels. The right side of the heart pumps blood through the lungs, enabling blood to pick up oxygen and unload carbon dioxide. Chapters 19 and 20 described the composition and functions of blood and the structure and function of the heart. In this chapter, we focus on the structure and functions of the various types of blood vessels; on **hemodynamics** (hē-mō-dī-NAM-iks; *hemo-* = blood; *-dynamics* = power), the forces involved in circulating blood throughout the body; and on the blood vessels that constitute the major circulatory routes.

Did you ever wonder why untreated hypertension has so many damaging effects?

21.1 STRUCTURE AND FUNCTION OF BLOOD VESSELS

● **OBJECTIVES**

- Contrast the structure and function of arteries, arterioles, capillaries, venules, and veins.
- Outline the vessels through which the blood moves in its passage from the heart to the capillaries and back.
- Distinguish between pressure reservoirs and blood reservoirs.

The five main types of blood vessels are arteries, arterioles, capillaries, venules, and veins (see Figure 20.7). **Arteries** (AR-ter-ēz) carry blood *away from the heart* to other organs. Large, elastic arteries leave the heart and divide into medium-sized, muscular arteries that branch out into the various regions of the body. Medium-sized arteries then divide into small arteries, which in turn divide into still smaller arteries called **arterioles** (ar-TĒR-ē-ōls). As the arterioles enter a tissue, they branch into numerous tiny vessels called **capillaries** (KAP-i-lar′-ēz = hairlike). The thin walls of capillaries allow the exchange of substances between the blood and body tissues. Groups of capillaries within a tissue reunite to form small veins called **venules** (VEN-ūls). These in turn merge to form progressively larger blood vessels called veins. **Veins** (VĀNZ) are the blood vessels that convey blood from the tissues *back to the heart.*

CLINICAL CONNECTION | Angiogenesis and Disease

Angiogenesis (an′-jē-ō-JEN-e-sis; *angio-* = blood vessel; *-genesis* = production) refers to the growth of new blood vessels. It is an important process in embryonic and fetal development, and in postnatal life serves important functions such as wound healing, formation of a new uterine lining after menstruation, formation of the corpus luteum after ovulation, and development of blood vessels around obstructed arteries in the coronary circulation. Several proteins (peptides) are known to promote and inhibit angiogenesis.

Clinically angiogenesis is important because cells of a malignant tumor secrete proteins called *tumor angiogenesis factors (TAFs)* that stimulate blood vessel growth to provide nourishment for the tumor cells. Scientists are seeking chemicals that would inhibit angiogenesis and thus stop the growth of tumors. In *diabetic retinopathy* (ret-i-NOP-a-thē), angiogenesis may be important in the development of blood vessels that actually cause blindness, so finding inhibitors of angiogenesis may also prevent the blindness associated with diabetes. •

Basic Structure of a Blood Vessel

The wall of a blood vessel consists of three layers, or tunics, of different tissues: an epithelial inner lining, a middle layer consisting of smooth muscle and elastic connective tissue, and a connective tissue outer covering. The three structural layers of a generalized blood vessel from innermost to outermost are the tunica interna (intima), tunica media, and tunica externa (adventia) (Figure 21.1).

Modifications of this basic design account for the five types of blood vessels and the structural and functional differences among the various vessel types. Always remember that structural variations correlate to the differences in function that occur throughout the cardiovascular system.

Tunica Interna (Intima)

The **tunica interna (intima)** (TOO-ni-ka; *tunic* = garment or coat; *interna or intima* = innermost) forms the inner lining of a blood vessel and is in direct contact with the blood as it flows through the **lumen** (LOO-men), or interior opening, of the vessel (Figure 21.1a, b). Although this layer has multiple parts, these tissue components contribute minimally to the thickness of the vessel wall. Its innermost layer is called *endothelium,* which is continuous with the endocardial lining of the heart. The endothelium is a thin layer of flattened cells that lines the inner surface of the entire cardiovascular system (heart and blood vessels). Until recently, endothelial cells were regarded as little more than a passive barrier between the blood and the remainder of the vessel wall. It is now known that endothelial cells are active participants in a variety of vessel-related activities, including physical influences on blood flow, secretion of locally acting chemical mediators that influence the contractile state of the vessel's overlying smooth muscle, and assistance with capillary permeability. In addition, their smooth luminal surface facilitates efficient blood flow by reducing surface friction.

The second component of the tunica interna is a *basement membrane* deep to the endothelium. It provides a physical support base for the epithelial layer. Its framework of collagen fibers affords the basement membrane significant tensile strength, yet its properties also provide resilience for stretching and recoil. The basement membrane anchors the endothelium to the underlying connective tissue while also regulating molecular movement. It appears to play an important role in guiding cell movements during tissue repair of blood vessel walls. The outermost part of the tunica interna, which forms the boundary between the tunica interna and tunica media, is the *internal elastic lamina (lamina* = thin plate). The internal elastic lamina is a thin sheet of elastic fibers with a variable number of window-like openings that give it the look of Swiss cheese. These openings facilitate diffusion of materials through the tunica interna to the thicker tunica media.

Tunica Media

The **tunica media** (*media* = middle) is a muscular and connective tissue layer that displays the greatest variation among the different vessel types (Figure 21.1a, b). In most vessels, it is a relatively thick layer comprising mainly smooth muscle cells and substantial amounts of elastic fibers. The primary role of the smooth muscle cells, which extend circularly around the lumen like a ring encircles your finger, is to regulate the diameter of the lumen. An increase in sympathetic stimulation typically stimulates the smooth muscle to contract, squeezing the vessel wall and narrowing the lumen. Such a decrease in the diameter of the lumen of a blood vessel is called **vasoconstriction** (vā-sō-kon-STRIK-shun). In contrast, when sympathetic stimulation decreases, or in

Figure 21.1 Comparative structure of blood vessels. The capillary (c) is enlarged relative to the artery (a) and vein (b).

Arteries carry blood from the heart to tissues; veins carry blood from tissues to the heart.

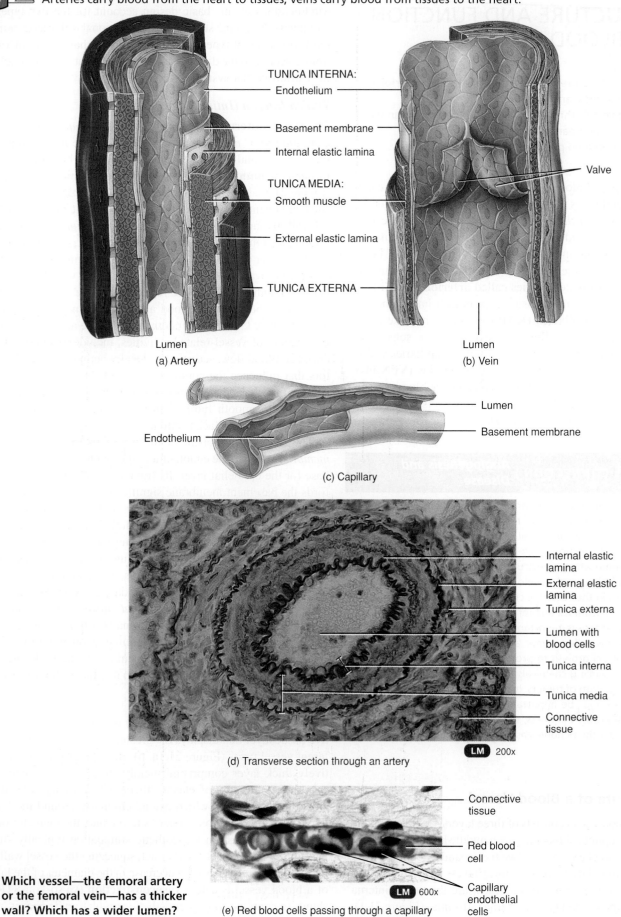

TUNICA INTERNA:
- Endothelium
- Basement membrane
- Internal elastic lamina

TUNICA MEDIA:
- Smooth muscle
- External elastic lamina

TUNICA EXTERNA

Lumen

(a) Artery

Valve

Lumen

(b) Vein

Endothelium

Lumen

Basement membrane

(c) Capillary

Internal elastic lamina
External elastic lamina
Tunica externa
Lumen with blood cells
Tunica interna
Tunica media
Connective tissue

LM 200x

(d) Transverse section through an artery

Connective tissue

Red blood cell

LM 600x
Capillary endothelial cells

(e) Red blood cells passing through a capillary

? **Which vessel—the femoral artery or the femoral vein—has a thicker wall? Which has a wider lumen?**

the presence of certain chemicals (such as nitric oxide, H^+, and lactic acid) or in response to blood pressure, smooth muscle fibers relax. The resulting increase in lumen diameter is called **vasodilation** (vā-sō-dī-LĀ-shun). As you will learn in more detail shortly, the rate of blood flow through different parts of the body is regulated by the extent of smooth muscle contraction in the walls of particular vessels. Furthermore, the extent of smooth muscle contraction in particular vessel types is crucial in the regulation of blood pressure.

In addition to regulating blood flow and blood pressure, smooth muscle contracts when an artery or arteriole is damaged *(vascular spasm)* to help limit loss of blood through the injured vessel if it is small. Smooth muscle cells also help produce the elastic fibers within the tunica media that allow the vessels to stretch and recoil under the applied pressure of the blood.

The tunica media is the most variable of the tunics. As you study the different types of blood vessels in the remainder of this chapter, you will see that the structural differences in this layer account for the many variations in function among the different vessel types. Separating the tunica media from the tunica externa is a network of elastic fibers, the *external elastic lamina,* which is part of the tunica media.

Tunica Externa

The outer covering of a blood vessel, the **tunica externa** (*externa* = outermost), consists of elastic and collagen fibers (Figure 21.1a, b). The tunica externa contains numerous nerves and, especially in larger vessels, tiny blood vessels that supply the tissue of the vessel wall. These small vessels that supply blood to the tissues of the vessel are called **vasa vasorum** (VĀ-sa va-SŌ-rum; *vas* = vessel), or vessels to the vessels. They are easily seen on large vessels such as the aorta. In addition to the important role of supplying the vessel wall with nerves and self-vessels, the tunica externa helps anchor the vessels to surrounding tissues.

Arteries

Because **arteries** (*ar-* = air; *-ter-* = to carry) were found empty at death, in ancient times they were thought to contain only air. The wall of an artery has the three layers of a typical blood vessel, but has a thick muscular-to-elastic tunica media (Figure 21.1a). Due to their plentiful elastic fibers, arteries normally have high *compliance,* which means that their walls stretch easily or expand without tearing in response to a small increase in pressure.

Elastic Arteries

Elastic arteries are the largest arteries in the body, ranging from the garden hose–sized aorta and pulmonary trunk to the finger-sized branches of the aorta. They have the largest diameter among arteries, but their vessel walls (approximately one-tenth of the vessel's total diameter) are relatively thin compared with the overall size of the vessel. These vessels are characterized by well-defined internal and external elastic laminae, along with a thick tunica media that is dominated by elastic fibers, called the **elastic lamellae** (la-MEL-ē = little plates). Elastic arteries

include the two major trunks that exit the heart (the aorta and the pulmonary trunk), along with the aorta's major initial branches, such as the brachiocephalic, subclavian, common carotid, and common iliac arteries (see Figure 21.19a). Elastic arteries perform an important function: They help propel blood onward while the ventricles are relaxing. As blood is ejected from the heart into elastic arteries, their walls stretch, easily accommodating the surge of blood. As they stretch, the elastic fibers momentarily store mechanical energy, functioning as a **pressure reservoir** (REZ-er-vwar) (Figure 21.2a). Then, the elastic fibers recoil and convert stored (potential) energy in the vessel into kinetic energy of the blood. Thus, blood continues to move through the arteries even while the ventricles are relaxed (Figure 21.2b). Because they conduct blood from the heart to medium-sized, more muscular arteries, elastic arteries also are called *conducting arteries.*

Figure 21.2 Pressure reservoir function of elastic arteries.

Recoil of elastic arteries keeps blood flowing during ventricular relaxation (diastole).

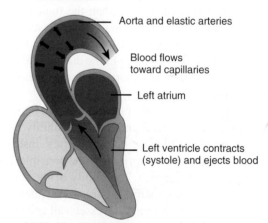

(a) Elastic aorta and arteries stretch during ventricular contraction

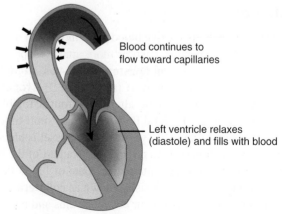

(b) Elastic aorta and arteries recoil during ventricular relaxation

? **In atherosclerosis, the walls of elastic arteries become less compliant (stiffer). What effect does reduced compliance have on the pressure reservoir function of arteries?**

Muscular Arteries

Medium-sized arteries are called **muscular arteries** because their tunica media contains more smooth muscle and fewer elastic fibers than elastic arteries. The large amount of smooth muscle, approximately three-quarters of the total mass, makes the walls of muscular arteries relatively thick. Thus, muscular arteries are capable of greater vasoconstriction and vasodilation to adjust the rate of blood flow. Muscular arteries have a well-defined internal elastic lamina but a thin external elastic lamina. These two elastic laminae form the inner and outer boundaries of the muscular tunica media. In large arteries, the thick tunica media can have as many as 40 layers of circumferentially arranged smooth muscle cells; in smaller arteries there are as few as three layers.

Muscular arteries span a range of sizes from the pencil-sized femoral and axillary arteries to string-sized arteries that enter organs, measuring as little as 0.5 mm in diameter. Compared to elastic arteries, the vessel wall of muscular arteries comprises a larger percentage (25%) of the total vessel diameter. Because the muscular arteries continue to branch and ultimately distribute blood to each of the various organs, they are called **distributing arteries.** Examples include the brachial artery in the arm and radial artery in the forearm (see Figure 21.19a).

The tunica externa is often thicker than the tunica media in muscular arteries. This outer layer contains fibroblasts, collagen fibers, and elastic fibers all oriented longitudinally. The loose structure of this layer permits changes in the diameter of the vessel to take place but also prevents shortening or retraction of the vessel when it is cut.

Because of the reduced amount of elastic tissue in the walls of muscular arteries, these vessels do not have the ability to recoil and help propel the blood like the elastic arteries. Instead, the thick, muscular tunica media is primarily responsible for the functions of the muscular arteries. The ability of the muscle to contract and maintain a state of partial contraction is referred to as *vascular tone*. Vascular tone stiffens the vessel wall and is important in maintaining vessel pressure and efficient blood flow.

Anastomoses

Most tissues of the body receive blood from more than one artery. The union of the branches of two or more arteries supplying the same body region is called an **anastomosis** (a-nas′-tō-MŌ-sis = connecting; plural is *anastomoses*) (see Figure 21.21c). Anastomoses between arteries provide alternative routes for blood to reach a tissue or organ. If blood flow stops for a short time when normal movements compress a vessel, or if a vessel is blocked by disease, injury, or surgery, then circulation to a part of the body is not necessarily stopped. The alternative route of blood flow to a body part through an anastomosis is known as **collateral circulation.** Anastomoses may also occur between veins and between arterioles and venules. Arteries that do not anastomose are known as **end arteries.** Obstruction of an end artery interrupts the blood supply to a whole segment of an organ, producing necrosis (death) of that segment. Alternative blood routes may also be provided by nonanastomosing vessels that supply the same region of the body.

Arterioles

Literally meaning small arteries, **arterioles** are abundant microscopic vessels that regulate the flow of blood into the capillary networks of the body's tissues (see Figure 21.3). The approximately 400 million arterioles have diameters that range in size from 15 μm to 300 μm. The wall thickness of arterioles is one-half of the total vessel diameter.

Arterioles have a thin tunica interna with a thin, fenestrated (with small pores) internal elastic lamina that disappears at the terminal end. The tunica media consists of one to two layers of smooth muscle cells having a circular orientation in the vessel wall. The terminal end of the arteriole, the region called the **metarteriole** (met′-ar-TER-ē-ōl; *meta* = after), tapers toward the capillary junction. At the metarteriole–capillary junction, the distalmost muscle cell forms the **precapillary sphincter** (SFINGK-ter = to bind tight), which monitors the blood flow into the capillary; the other muscle cells in the arteriole regulate the resistance (opposition) to blood flow (see Figure 21.3).

The tunica externa of the arteriole consists of areolar connective tissue containing abundant unmyelinated sympathetic nerves. This sympathetic nerve supply, along with the actions of local chemical mediators, can alter the diameter of arterioles and thus vary the rate of blood flow and resistance through these vessels.

Arterioles play a key role in regulating blood flow from arteries into capillaries by regulating **resistance,** the opposition to blood flow. Because of this they are known as *resistance vessels*. In a blood vessel, resistance is due mainly to friction between blood and the inner walls of blood vessels. When the blood vessel diameter is smaller, the friction is greater, so there is more resistance. Contraction of the smooth muscle of an arteriole causes vasoconstriction, which increases resistance even more and decreases blood flow into capillaries supplied by that arteriole. By contrast, relaxation of the smooth muscle of an arteriole causes vasodilation, which decreases resistance and increases blood flow into capillaries. A change in arteriole diameter can also affect blood pressure: Vasoconstriction of arterioles increases blood pressure, and vasodilation of arterioles decreases blood pressure.

Capillaries

Capillaries (*capillus* = little hair), the smallest of blood vessels, have diameters of 5–10 μm, and form the U-turns that connect the arterial outflow to the venous return (Figure 21.3). Since red blood cells have a diameter of 8 μm, they must often fold on themselves in order to pass single file through the lumens of these vessels. Capillaries form an extensive network, approximately 20 billion in number, of short (hundreds of micrometers in length), branched, interconnecting vessels that course among the individual cells of the body. This network forms an enormous surface area to make contact with the body's cells. The flow of blood from a metarteriole through capillaries and into a **postcapillary venule** (venule that receives blood from a capillary) is called the **microcirculation** (*micro* = small) of the body. The primary function of capillaries is the exchange of substances between the blood and

Figure 21.3 Arteriole, capillaries, and venule. Precapillary sphincters regulate the flow of blood through capillary beds.

In capillaries, nutrients, gases, and wastes are exchanged between the blood and interstitial fluid.

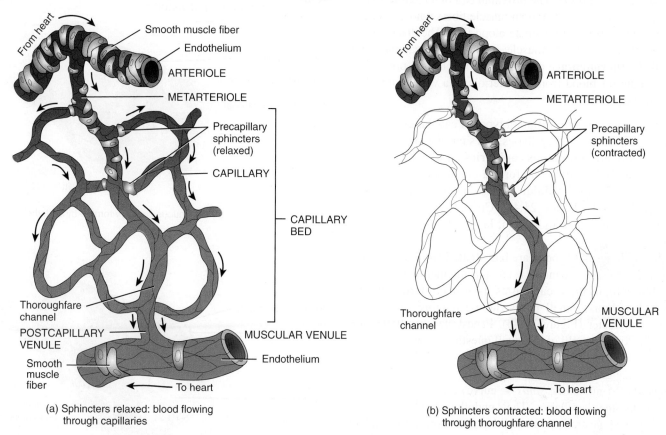

(a) Sphincters relaxed: blood flowing through capillaries

(b) Sphincters contracted: blood flowing through thoroughfare channel

? **Why do metabolically active tissues have extensive capillary networks?**

interstitial fluid. Because of this, these thin-walled vessels are referred to as *exchange vessels.*

Capillaries are found near almost every cell in the body, but their number varies with the metabolic activity of the tissue they serve. Body tissues with high metabolic requirements, such as muscles, the brain, the liver, the kidneys, and the nervous system, use more O_2 and nutrients and thus have extensive capillary networks. Tissues with lower metabolic requirements, such as tendons and ligaments, contain fewer capillaries. Capillaries are absent in a few tissues, such as all covering and lining epithelia, the cornea and lens of the eye, and cartilage.

The structure of capillaries is well suited to their function as exchange vessels because they lack both a tunica media and a tunica externa. Because capillary walls are composed of only a single layer of endothelial cells (see Figure 21.1e) and a basement membrane, a substance in the blood must pass through just one cell layer to reach the interstitial fluid and tissue cells. Exchange of materials occurs only through the walls of capillaries and the beginning of venules; the walls of arteries, arterioles, most venules, and veins present too thick a barrier. Capillaries form extensive branching networks that increase the surface area available for rapid exchange of materials. In most tissues, blood flows through only a small part of the capillary network when metabolic

needs are low. However, when a tissue is active, such as contracting muscle, the entire capillary network fills with blood.

Throughout the body, capillaries function as part of a **capillary bed** (Figure 21.3), a network of 10–100 capillaries that arises from a single metarteriole. In most parts of the body, blood can flow through a capillary network from an arteriole into a venule as follows:

1. *Capillaries.* In this route, blood flows from an arteriole into capillaries and then into venules (postcapillary venules). As noted earlier, at the junctions between the metarteriole and the capillaries are rings of smooth muscle fibers called precapillary sphincters that control the flow of blood through the capillaries. When the precapillary sphincters are relaxed (open), blood flows into the capillaries (Figure 21.3a); when precapillary sphincters contract (close or partially close), blood flow through the capillaries ceases or decreases (Figure 21.3b). Typically, blood flows intermittently through capillaries due to alternating contraction and relaxation of the smooth muscle of metarterioles and the precapillary sphincters. This intermittent contraction and relaxation, which may occur 5 to 10 times per minute, is called **vasomotion** (vā-sō-MŌ-shun). In part, vasomotion is due to chemicals released by the endothelial cells;

nitric oxide is one example. At any given time, blood flows through only about 25% of the capillaries.

2. ***Thoroughfare channel.*** The proximal end of a metarteriole is surrounded by scattered smooth muscle fibers whose contraction and relaxation help regulate blood flow. The distal end of the vessel has no smooth muscle; it resembles a capillary and is called a **thoroughfare channel.** Such a channel provides a direct route for blood from an arteriole to a venule, thus bypassing capillaries.

The body contains three different types of capillaries: continuous capillaries, fenestrated capillaries, and sinusoids (Figure 21.4). Most capillaries are **continuous capillaries,** in which the plasma membranes of endothelial cells form a continuous tube that is interrupted only by **intercellular clefts,** gaps between neighboring endothelial cells (Figure 21.4a). Continuous capillaries are found in the central nervous system, lungs, skin, muscle tissue, and the skin.

Other capillaries of the body are **fenestrated capillaries** (fen'-es-TRĀ-ted; *fenestr-* = window). The plasma membranes of the endothelial cells in these capillaries have many **fenestrations** (fen'-es-TRĀ-shuns), small pores (holes) ranging from 70 to 100 nm in diameter (Figure 21.4b). Fenestrated capillaries are found in the kidneys, villi of the small intestine, choroid plexuses of the ventricles in the brain, ciliary processes of the eyes, and most endocrine glands.

Sinusoids (SĪ-nū-soyds; *sinus* = curve) are wider and more winding than other capillaries. Their endothelial cells may have unusually large fenestrations. In addition to having an incomplete or absent basement membrane (Figure 21.4c), sinusoids have very large intercellular clefts that allow proteins and in some cases even blood cells to pass from a tissue into the bloodstream. For example, newly formed blood cells enter the bloodstream through the sinusoids of red bone marrow. In addition, sinusoids contain specialized lining cells that are adapted to the function of the tissue. Sinusoids in the liver, for example, contain phagocytic cells that remove bacteria and other debris from the blood. The spleen, anterior pituitary, and parathyroid and adrenal glands also have sinusoids.

Usually blood passes from the heart and then in sequence through arteries, arterioles, capillaries, venules, and veins and then back to the heart. In some parts of the body, however, blood passes from one capillary network into another through a vein called a *portal vein.* Such a circulation of blood is called a **portal system.** The name of the portal system gives the name of the second capillary location. For example, there are portal systems associated with the liver (hepatic portal circulation; see Figure 21.28) and the pituitary gland (hypophyseal portal system; see Figure 18.5).

Venules

Unlike their thick-walled arterial counterparts, **venules** (= little vein) and veins have thin walls that do not readily maintain their shape. Venules drain the capillary blood and begin the return flow of blood back toward the heart (see Figure 21.3).

Figure 21.4 **Types of capillaries.**

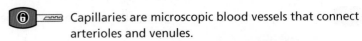 Capillaries are microscopic blood vessels that connect arterioles and venules.

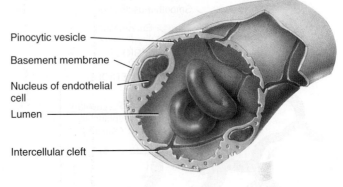

Pinocytic vesicle
Basement membrane
Nucleus of endothelial cell
Lumen
Intercellular cleft

(a) Continuous capillary formed by endothelial cells

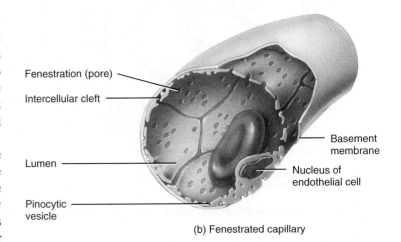

Fenestration (pore)
Intercellular cleft
Lumen
Pinocytic vesicle
Basement membrane
Nucleus of endothelial cell

(b) Fenestrated capillary

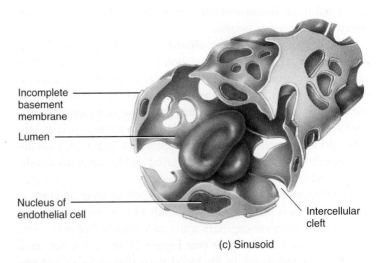

Incomplete basement membrane
Lumen
Nucleus of endothelial cell
Intercellular cleft

(c) Sinusoid

? How do materials move through capillary walls?

As noted earlier, venules that initially receive blood from capillaries are called **postcapillary venules.** They are the smallest venules, measuring 10 μm to 50 μm in diameter, and have loosely organized intercellular junctions (the weakest endothelial contacts encountered along the entire vascular tree) and thus are very porous. They function as significant sites of exchange of nutrients and wastes and white blood cell emigration, and for this reason form part of the microcirculatory exchange unit along with the capillaries.

As the postcapillary venules move away from capillaries, they acquire one or two layers of circularly arranged smooth muscle cells. These **muscular venules** (50 μm to 200 μm) have thicker walls across which exchanges with the interstitial fluid can no longer occur. The thin walls of the postcapillary and muscular venules are the most distensible elements of the vascular system; this allows them to expand and serve as excellent reservoirs for accumulating large volumes of blood. Blood volume increases of 360% have been measured in the postcapillary and muscular venules.

Veins

While **veins** do show structural changes as they increase in size from small to medium to large, the structural changes are not as distinct as they are in arteries. Veins, in general, have very thin walls relative to their total diameter (average thickness is less than one-tenth of the vessel diameter). They range in size from 0.5 mm in diameter for small veins to 3 cm in the large superior and inferior venae cavae entering the heart.

Although veins are composed of essentially the same three layers as arteries, the relative thicknesses of the layers are different. The tunica interna of veins is thinner than that of arteries; the tunica media of veins is much thinner than in arteries, with relatively little smooth muscle and elastic fibers. The tunica externa of veins is the thickest layer and consists of collagen and elastic fibers. Veins lack the internal or external elastic laminae found in arteries (see Figure 22.1b). They are distensible enough to adapt to variations in the volume and pressure of blood passing through them, but are not designed to withstand high pressure. The lumen of a vein is larger than that of a comparable artery, and veins often appear collapsed (flattened) when sectioned.

The pumping action of the heart is a major factor in moving venous blood back to the heart. The contraction of skeletal muscles in the lower limbs also helps boost venous return to the heart (Figure 21.9). The average blood pressure in veins is considerably lower than in arteries. The difference in pressure can be noticed when blood flows from a cut vessel. Blood leaves a cut vein in an even, slow flow but spurts rapidly from a cut artery. Most of the structural differences between arteries and veins reflect this pressure difference. For example, the walls of veins are not as strong as those of arteries.

Many veins, especially those in the limbs, also contain **valves,** thin folds of tunica interna that form flaplike cusps. The valve cusps project into the lumen, pointing toward the heart (see Figure 21.5). The low blood pressure in veins allows blood returning to the heart to slow and even back up; the valves aid in venous return by preventing the backflow of blood.

A **vascular (venous) sinus** is a vein with a thin endothelial wall that has no smooth muscle to alter its diameter. In a vascular sinus, the surrounding dense connective tissue replaces the tunica media

and tunica externa in providing support. For example, dural venous sinuses, which are supported by the dura mater, convey deoxygenated blood from the brain to the heart. Another example of a vascular sinus is the coronary sinus of the heart (see Figure 20.3c).

While veins follow paths similar to those of their arterial counterparts, they differ from arteries in a number of ways, aside from the structures of their walls. First, veins are more numerous than arteries for several reasons. Some veins are paired and accompany medium- to small-sized muscular arteries. These double sets of veins escort the arteries and connect with one another via venous channels called **anastomotic veins** (a-nas′-tō-MOT-ik). The anastomotic veins cross the accompanying artery to form ladderlike rungs between the paired veins (see Figure 21.25c). The greatest number of paired veins occurs within the limbs. The subcutaneous layer deep to the skin is another source of veins. These veins, called **superficial veins,** course through the subcutaneous layer unaccompanied by parallel arteries. Along their course, the superficial veins form small connections (anastomoses) with the **deep veins** that travel between the skeletal muscles. These connections allow communication between the deep and superficial flow of blood. The amount of blood flow through superficial veins varies from location

Figure 21.5 Venous valves.

Valves in veins allow blood to flow in one direction only—toward the heart.

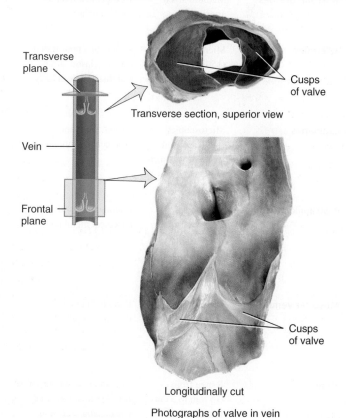

Transverse plane

Vein

Frontal plane

Cusps of valve

Transverse section, superior view

Cusps of valve

Longitudinally cut

Photographs of valve in vein

? **Why are valves more important in arm veins and leg veins than in neck veins?**

to location within the body. In the upper limb, the superficial veins are much larger than the deep veins and serve as the major pathways from the capillaries of the upper limb back to the heart. In the lower limb, the opposite is true; the deep veins serve as the principal return pathways. In fact, one-way valves in small anastomosing vessels allow blood to pass from the superficial to the deep veins, but prevent the blood from passing in the reverse direction. This design has important implications in the development of varicose veins.

In some individuals the superficial veins can be seen as blue-colored tubes passing under the skin. While the venous blood is a deep dark red, the veins appear blue because their thin walls and the tissues of the skin absorb the red-light wavelengths, allowing the blue light to pass through the surface to our eyes where we see them as blue.

A summary of the distinguishing features of blood vessels is presented in Table 21.1.

CLINICAL CONNECTION | *Varicose Veins*

Leaky venous valves can cause veins to become dilated and twisted in appearance, a condition called **varicose veins** (VAR-i-kōs) or **varices** (VAR-i-sēz; *varic-* = a swollen vein). The singular is *varix* (VAR-iks). The condition may occur in the veins of almost any body part, but it is most common in the esophagus, anal canal, and superficial veins of the lower limbs. Those in the lower limbs can range from cosmetic problems to serious medical conditions. The valvular defect may be congenital or may result from mechanical stress (prolonged standing or pregnancy) or aging. The leaking venous valves allow the backflow of blood from the deep veins to the less efficient superficial veins, where the blood pools. This creates pressure that distends the vein and allows fluid to leak into surrounding tissue. As a result, the affected vein and the tissue around it may become inflamed and painfully tender. Veins close to the surface of the legs, especially the

TABLE 21.1

Distinguishing Features of Blood Vessels

BLOOD VESSEL	SIZE	TUNICA INTERNA	TUNICA MEDIA	TUNICA EXTERNA	FUNCTION
Elastic arteries	Largest arteries in the body.	Well-defined internal elastic lamina.	Thick and dominated by elastic fibers; well-defined external elastic lamina.	Thinner than tunica media.	Conduct blood from heart to muscular arteries.
Muscular arteries	Medium-sized arteries.	Well-defined internal elastic lamina.	Thick and dominated by smooth muscle; thin external elastic lamina.	Thicker than tunica media.	Distribute blood to arterioles.
Arterioles	Microscopic (15–300 μm in diameter).	Thin with a fenestrated internal elastic lamina that disappears distally.	One or two layers of circularly oriented smooth muscle; distalmost smooth muscle cell forms a precapillary sphincter.	Loose collagenous connective tissue and sympathetic nerves.	Deliver blood to capillaries and help regulate blood flow from arteries to capillaries.
Capillaries	Microscopic; smallest blood vessels (5–10 μm in diameter).	Endothelium and basement membrane.	None.	None.	Permit exchange of nutrients and wastes between blood and interstitial fluid; distribute blood to postcapillary venules.
Postcapillary venules	Microscopic (10–50 μm in diameter).	Endothelium and basement membrane.	None.	Sparse.	Pass blood into muscular venules; permit exchange of nutrients and wastes between blood and interstitial fluid and function in white blood cell emigration.
Muscular venules	Microscopic (50–200 μm in diameter).	Endothelium and basement membrane.	One or two layers of circularly oriented smooth muscle.	Sparse.	Pass blood into vein; act as reservoirs for accumulating large volumes of blood (along with postcapillary venules).
Veins	Range from 0.5 mm to 3 cm in diameter.	Endothelium and basement membrane; no internal elastic lamina; contain valves; lumen much larger than in accompanying artery.	Much thinner than in arteries; no external elastic lamina.	Thickest of the three layers.	Return blood to heart, facilitated by valves in veins in limbs.

saphenous vein, are highly susceptible to varicosities; deeper veins are not as vulnerable because surrounding skeletal muscles prevent their walls from stretching excessively. Varicose veins in the anal canal are referred to as *hemorrhoids* (HEM-o-royds). Esophageal varices result from dilated veins in the walls of the lower part of the esophagus and sometimes the upper part of the stomach. Bleeding esophageal varices are life-threatening and are usually a result of chronic liver disease.

Several treatment options are available for varicose veins in the lower limbs. *Elastic stockings* (support hose) may be used for individuals with mild symptoms or for whom other options are not recommended. *Sclerotherapy* (skle-rō-THER-a-pē) involves injection of a solution into varicose veins that damages the tunica interna by producing a harmless superficial thrombophlebitis (inflammation involving a blood clot). Healing of the damaged part leads to scar formation that occludes the vein. *Radiofrequency endovenous occlusion* (ō-KLOO-zhun) involves the application of radiofrequency energy to heat up and close off varicose veins. *Laser occlusion* uses laser therapy to shut down veins. In a surgical procedure called *stripping,* veins may be removed. In this procedure, a flexible wire is threaded through the vein and then pulled out to strip (remove) it from the body.

In some individuals the superficial veins can be seen as blue-colored tubes passing under the skin. While the venous blood is a deep dark red, the veins appear blue because their thin walls and the tissue of the skin absorb the red-light wavelengths, allowing the blue light to pass through the surface to our eyes, where we see them as blue. •

Blood Distribution

The largest portion of your blood volume at rest—about 64%—is in systemic veins and venules (Figure 21.6). Systemic arteries and arterioles hold about 13% of the blood volume, systemic capillaries hold about 7%, pulmonary blood vessels hold about 9%, and the heart holds about 7%. Because systemic veins and venules contain a large percentage of the blood volume, they function as **blood reservoirs** from which blood can be diverted quickly if the need arises. For example, during increased muscular activity, the cardiovascular center in the brain stem sends a larger number of sympathetic impulses to veins. The result is *venoconstriction,* constriction of veins, which reduces the volume of blood in reservoirs and allows a greater blood volume to flow to skeletal muscles, where it is needed most. A similar mechanism operates in cases of hemorrhage, when blood volume and pressure decrease; in this case, venoconstriction helps counteract the drop in blood pressure. Among the principal blood reservoirs are the veins of the abdominal organs (especially the liver and spleen) and the veins of the skin.

✔CHECKPOINT
1. What is the function of elastic fibers and smooth muscle in the tunica media of arteries?
2. How are elastic arteries and muscular arteries different?
3. What structural features of capillaries allow the exchange of materials between blood and body cells?
4. What is the difference between pressure reservoirs and blood reservoirs? Why is each important?
5. What is the relationship between anastomoses and collateral circulation?

Figure 21.6 Blood distribution in the cardiovascular system at rest.

 Because systemic veins and venules contain more than half the total blood volume, they are called blood reservoirs.

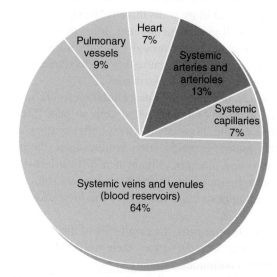

If your total blood volume is 5 liters, what volume is in your venules and veins right now? In your capillaries?

21.2 CAPILLARY EXCHANGE

◉ OBJECTIVE
• Discuss the pressures that cause movement of fluids between capillaries and interstitial spaces.

The mission of the entire cardiovascular system is to keep blood flowing through capillaries to allow **capillary exchange,** the movement of substances between blood and interstitial fluid. The 7% of the blood in systemic capillaries at any given time is continually exchanging materials with interstitial fluid. Substances enter and leave capillaries by three basic mechanisms: diffusion, transcytosis, and bulk flow.

Diffusion

The most important method of capillary exchange is simple diffusion. Many substances, such as oxygen (O_2), carbon dioxide (CO_2), glucose, amino acids, and hormones, enter and leave capillaries by simple diffusion. Because O_2 and nutrients normally are present in higher concentrations in blood, they diffuse down their concentration gradients into interstitial fluid and then into body cells. CO_2 and other wastes released by body cells are present in higher concentrations in interstitial fluid, so they diffuse into blood.

Substances in blood or interstitial fluid can cross the walls of a capillary by diffusing through the intercellular clefts or fenestrations or by diffusing through the endothelial cells (see Figure 21.4). Water-soluble substances such as glucose and amino acids pass across capillary walls through intercellular clefts or fenestrations. Lipid-soluble materials, such as O_2, CO_2, and steroid hormones, may pass across capillary walls directly through the lipid bilayer of endothelial cell plasma membranes. Most plasma proteins and red blood cells cannot pass through capillary walls of continuous and fenestrated capillaries because they are too large to fit through the intercellular clefts and fenestrations.

In sinusoids, however, the intercellular clefts are so large that they allow even proteins and blood cells to pass through their walls. For example, hepatocytes (liver cells) synthesize and release many plasma proteins, such as fibrinogen (the main clotting protein) and albumin, which then diffuse into the bloodstream through sinusoids. In red bone marrow, blood cells are formed (hemopoiesis) and then enter the bloodstream through sinusoids.

In contrast to sinusoids, the capillaries of the brain allow only a few substances to move across their walls. Most areas of the brain contain continuous capillaries; however, these capillaries are very "tight." The endothelial cells of most brain capillaries are sealed together by tight junctions. The resulting blockade to movement of materials into and out of brain capillaries is known as the *blood–brain barrier* (see Section 14.1). In brain areas that lack the blood–brain barrier, for example, the hypothalamus, pineal gland, and pituitary gland, materials undergo capillary exchange more freely.

Transcytosis

A small quantity of material crosses capillary walls by **transcytosis** (tranz′-sī-TŌ-sis; *trans-* = across; *-cyt-* = cell; *-osis* = process). In this process, substances in blood plasma become enclosed within tiny pinocytic vesicles that first enter endothelial cells by endocytosis, then move across the cell and exit on the other side by exocytosis. This method of transport is important mainly for large, lipid-insoluble molecules that cannot cross capillary walls in any other way. For example, the hormone insulin (a small protein) enters the bloodstream by transcytosis, and certain antibodies (also proteins) pass from the maternal circulation into the fetal circulation by transcytosis.

Bulk Flow: Filtration and Reabsorption

Bulk flow is a passive process in which *large* numbers of ions, molecules, or particles in a fluid move together in the same direction. The substances move at rates far greater than can be accounted for by diffusion alone. Bulk flow occurs from an area of higher pressure to an area of lower pressure, and it continues as long as a pressure difference exists. Diffusion is more important for *solute exchange* between blood and interstitial fluid, but bulk flow is more important for regulation of the *relative volumes of blood and interstitial fluid*. Pressure-driven movement of fluid and solutes *from* blood capillaries *into* interstitial fluid is called **filtration.** Pressure-driven movement *from* interstitial fluid *into* blood capillaries is called **reabsorption.**

Two pressures promote filtration: blood hydrostatic pressure (BHP), the pressure generated by the pumping action of the heart, and interstitial fluid osmotic pressure (in′-ter-STISH-al). The main pressure promoting reabsorption of fluid is blood colloid osmotic pressure. The balance of these pressures, called **net filtration pressure (NFP),** determines whether the volumes of blood and interstitial fluid remain steady or change. Overall, the volume of fluid and solutes reabsorbed normally is almost as large as the volume filtered. This near equilibrium is known as **Starling's law of the capillaries.** Let's see how these hydrostatic and osmotic pressures balance.

Within vessels, the hydrostatic pressure is due to the pressure that water in blood plasma exerts against blood vessel walls. The **blood hydrostatic pressure (BHP)** is about 35 millimeters of mercury (mmHg) at the arterial end of a capillary, and about 16 mmHg at the capillary's venous end (Figure 21.7). BHP "pushes" fluid out of capillaries into interstitial fluid. The opposing pressure of the interstitial fluid, called **interstitial fluid hydrostatic pressure (IFHP),** "pushes" fluid from interstitial spaces back into capillaries. However, IFHP is close to zero. (IFHP is difficult to measure, and its reported values vary from small positive values to small negative values.) For our discussion we assume that IFHP equals 0 mmHg all along the capillaries.

The difference in osmotic pressure across a capillary wall is due almost entirely to the presence in blood of plasma proteins, which are too large to pass through either fenestrations or gaps between endothelial cells. **Blood colloid osmotic pressure (BCOP)** is a force caused by the colloidal suspension of these large proteins in plasma that averages 26 mmHg in most capillaries. The effect of BCOP is to "pull" fluid from interstitial spaces into capillaries. Opposing BCOP is **interstitial fluid osmotic pressure (IFOP),** which "pulls" fluid out of capillaries into interstitial fluid. Normally, IFOP is very small—0.1–5 mmHg—because only tiny amounts of protein are present in interstitial fluid. The small amount of protein that leaks from blood plasma into interstitial fluid does not accumulate there because it passes into lymph in lymphatic capillaries and is eventually returned to the blood. For discussion, we can use a value of 1 mmHg for IFOP.

Whether fluids leave or enter capillaries depends on the balance of pressures. If the pressures that push fluid out of capillaries exceed the pressures that pull fluid into capillaries, fluid will move from capillaries into interstitial spaces (filtration). If, however, the pressures that push fluid out of interstitial spaces into capillaries exceed the pressures that pull fluid out of capillaries, then fluid will move from interstitial spaces into capillaries (reabsorption).

The net filtration pressure (NFP), which indicates the direction of fluid movement, is calculated as follows:

$$NFP = (BHP + IFOP) - (BCOP + IFHP)$$

Pressures that promote filtration	Pressures that promote reabsorption

At the arterial end of a capillary,

$$NFP = (35 + 1) \text{ mmHg} - (26 + 0) \text{ mmHg}$$
$$= 36 - 26 \text{ mmHg} = 10 \text{ mmHg}$$

Thus, at the arterial end of a capillary, there is a *net outward pressure* of 10 mmHg, and fluid moves out of the capillary into interstitial spaces (filtration).

At the venous end of a capillary,

$$NFP = (16 + 1) \text{ mmHg} - (26 + 0) \text{ mmHg}$$
$$= 17 - 26 \text{ mmHg} = -9 \text{ mmHg}$$

At the venous end of a capillary, the negative value (−9 mmHg) represents a *net inward pressure,* and fluid moves into the capillary from tissue spaces (reabsorption).

On average, about 85% of the fluid filtered out of capillaries is reabsorbed. The excess filtered fluid and the few plasma proteins that do escape from blood into interstitial fluid enter lymphatic capillaries (see Figure 22.2). As lymph drains into the junction of the jugular and subclavian veins in the upper thorax (see Figure 22.3), these materials return to the blood. Every day about 20 liters of fluid filter out of capillaries in tissues throughout the body. Of this fluid, 17 liters are reabsorbed and 3 liters enter lymphatic capillaries (excluding filtration during urine formation).

Figure 21.7 Dynamics of capillary exchange (Starling's law of the capillaries). Excess filtered fluid drains into lymphatic capillaries.

Blood hydrostatic pressure pushes fluid out of capillaries (filtration), and blood colloid osmotic pressure pulls fluid into capillaries (reabsorption).

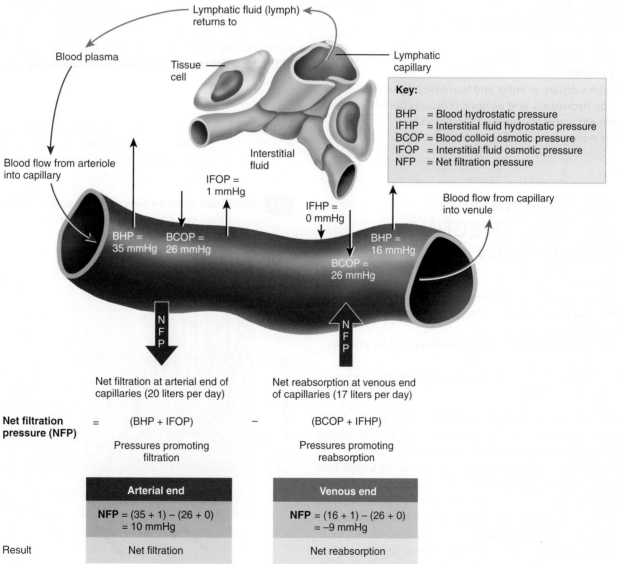

Key:

BHP	= Blood hydrostatic pressure
IFHP	= Interstitial fluid hydrostatic pressure
BCOP	= Blood colloid osmotic pressure
IFOP	= Interstitial fluid osmotic pressure
NFP	= Net filtration pressure

Net filtration at arterial end of capillaries (20 liters per day)

Net reabsorption at venous end of capillaries (17 liters per day)

Net filtration pressure (NFP) = (BHP + IFOP) − (BCOP + IFHP)

Pressures promoting filtration

Pressures promoting reabsorption

Arterial end	**Venous end**
NFP = (35 + 1) − (26 + 0) = 10 mmHg	**NFP** = (16 + 1) − (26 + 0) = −9 mmHg
Net filtration	Net reabsorption

Result

A person who has liver failure cannot synthesize the normal amount of plasma proteins. How does a deficit of plasma proteins affect blood colloid osmotic pressure, and what is the effect on capillary exchange?

If filtration greatly exceeds reabsorption, the result is **edema** (e-DĒ-ma = swelling), an abnormal increase in interstitial fluid volume. Edema is not usually detectable in tissues until interstitial fluid volume has risen to 30% above normal. Edema can result from either excess filtration or inadequate reabsorption.

Two situations may cause excess filtration:

- *Increased capillary blood pressure* causes more fluid to be filtered from capillaries.
- *Increased permeability of capillaries* raises interstitial fluid osmotic pressure by allowing some plasma proteins to escape. Such leakiness may be caused by the destructive effects of chemical, bacterial, thermal, or mechanical agents on capillary walls.

One situation commonly causes inadequate reabsorption:

- *Decreased concentration of plasma proteins* lowers the blood colloid osmotic pressure. Inadequate synthesis or dietary intake or loss of plasma proteins is associated with liver disease, burns, malnutrition (for example, kwashiorkor, see Disorders: Homeostatic Imbalances in Chapter 25), and kidney disease. •

✔ CHECKPOINT

6. How can substances enter and leave blood plasma?
7. How do hydrostatic and osmotic pressures determine fluid movement across the walls of capillaries?
8. Define edema and describe how it develops.

21.3 HEMODYNAMICS: FACTORS AFFECTING BLOOD FLOW

⊙ OBJECTIVES

- Explain the factors that regulate the volume of blood flow.
- Explain how blood pressure changes throughout the cardiovascular system.
- Describe the factors that determine mean arterial pressure and systemic vascular resistance.
- Describe the relationship between cross-sectional area and velocity of blood flow.

Blood flow is the volume of blood that flows through any tissue in a given time period (in mL/min). Total blood flow is cardiac output (CO), the volume of blood that circulates through systemic (or pulmonary) blood vessels each minute. In Chapter 20 we saw that cardiac output depends on heart rate and stroke volume: Cardiac output (CO) = heart rate (HR) × stroke volume (SV). How the cardiac output becomes distributed into circulatory routes that serve various body tissues depends on two more factors: (1) the *pressure difference* that drives the blood flow through a tissue and (2) the *resistance* to blood flow in specific blood vessels. Blood flows from regions of higher pressure to regions of lower pressure; the greater the pressure difference, the greater the blood flow. But the higher the resistance, the smaller the blood flow.

Blood Pressure

As you have just learned, blood flows from regions of higher pressure to regions of lower pressure; the greater the pressure difference, the greater the blood flow. Contraction of the ventricles generates **blood pressure (BP),** the hydrostatic pressure exerted by blood on the walls of a blood vessel. BP is determined by cardiac output (see Section 20.5), blood volume, and vascular resistance (described shortly). BP is highest in the aorta and large systemic arteries; in a resting, young adult, BP rises to about 110 mmHg during systole (ventricular contraction) and drops to about 70 mmHg during diastole (ventricular relaxation). **Systolic blood pressure** (sis-TOL-ik) is the highest pressure attained in arteries during systole, and **diastolic blood pressure** (dī-a-STOL-ik) is the lowest arterial pressure during diastole (Figure 21.8). As blood leaves the aorta and flows through the systemic circulation, its pressure falls progressively as the distance from the left ventricle increases. Blood pressure decreases to about 35 mmHg as blood passes from systemic arteries through systemic arterioles and into capillaries, where the pressure fluctuations disappear. At the

Figure 21.8 Blood pressures in various parts of the cardiovascular system. The dashed line is the mean (average) blood pressure in the aorta, arteries, and arterioles.

Blood pressure rises and falls with each heartbeat in blood vessels leading to capillaries.

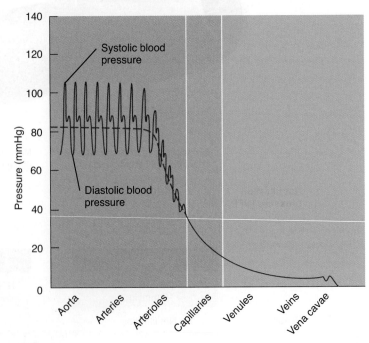

? **Is the mean blood pressure in the aorta closer to systolic or to diastolic pressure?**

venous end of capillaries, blood pressure has dropped to about 16 mmHg. Blood pressure continues to drop as blood enters systemic venules and then veins because these vessels are farthest from the left ventricle. Finally, blood pressure reaches 0 mmHg as blood flows into the right ventricle.

Mean arterial pressure (MAP), the average blood pressure in arteries, is roughly one-third of the way between the diastolic and systolic pressures. It can be estimated as follows:

MAP = diastolic BP + 1/3 (systolic BP − diastolic BP)

Thus, in a person whose BP is 110/70 mmHg, MAP is about 83 mmHg [70 + 1/3(110 − 70)].

We have already seen that cardiac output equals heart rate multiplied by stroke volume. Another way to calculate cardiac output is to divide mean arterial pressure (MAP) by resistance (R): CO = MAP ÷ R. By rearranging the terms of this equation, you can see that MAP = CO × R. If cardiac output rises due to an increase in stroke volume or heart rate, then the mean arterial pressure rises as long as resistance remains steady. Likewise, a decrease in cardiac output causes a decrease in mean arterial pressure if resistance does not change.

Blood pressure also depends on the total volume of blood in the cardiovascular system. The normal volume of blood in an adult is about 5 liters (5.3 qt). Any decrease in this volume, as from hemorrhage, decreases the amount of blood that is circulated through the arteries each minute. A modest decrease can be compensated for by homeostatic mechanisms that help maintain blood pressure (described in Section 21.4), but if the decrease in blood volume is greater than 10% of the total, blood pressure drops. Conversely, anything that increases blood volume, such as water retention in the body, tends to increase blood pressure.

Vascular Resistance

As noted earlier, **vascular resistance** is the opposition to blood flow due to friction between blood and the walls of blood vessels. Vascular resistance depends on (1) size of the blood vessel lumen, (2) blood viscosity, and (3) total blood vessel length.

1. *Size of the lumen.* The smaller the lumen of a blood vessel, the greater its resistance to blood flow. Resistance is inversely proportional to the fourth power of the diameter (d) of the blood vessel's lumen ($R \propto 1/d^4$). The smaller the diameter of the blood vessel, the greater the resistance it offers to blood flow. For example, if the diameter of a blood vessel decreases by one-half, its resistance to blood flow increases 16 times. Vasoconstriction narrows the lumen, and vasodilation widens it. Normally, moment-to-moment fluctuations in blood flow through a given tissue are due to vasoconstriction and vasodilation of the tissue's arterioles. As arterioles dilate, resistance decreases, and blood pressure falls. As arterioles constrict, resistance increases, and blood pressure rises.

2. *Blood viscosity.* The viscosity (vis-KOS-i-tē = thickness) of blood depends mostly on the ratio of red blood cells to plasma (fluid) volume, and to a smaller extent on the concentration of proteins in plasma. The higher the blood's viscosity, the higher the resistance. Any condition that increases the viscosity of blood, such as dehydration or polycythemia (an unusually high number of red blood cells), thus increases blood pressure. A depletion of plasma proteins or red blood cells, due to anemia or hemorrhage, decreases viscosity and thus decreases blood pressure.

3. *Total blood vessel length.* Resistance to blood flow through a vessel is directly proportional to the length of the blood vessel. The longer a blood vessel, the greater the resistance. Obese people often have hypertension (elevated blood pressure) because the additional blood vessels in their adipose tissue increase their total blood vessel length. An estimated 650 km (about 400 miles) of additional blood vessels develop for each extra kilogram (2.2 lb) of fat.

Systemic vascular resistance (SVR), also known as *total peripheral resistance (TPR),* refers to all the vascular resistances offered by systemic blood vessels. The diameters of arteries and veins are large, so their resistance is very small because most of the blood does not come into physical contact with the walls of the blood vessel. The smallest vessels—arterioles, capillaries, and venules—contribute the most resistance. A major function of arterioles is to control SVR—and therefore blood pressure and blood flow to particular tissues—by changing their diameters. Arterioles need to vasodilate or vasoconstrict only slightly to have a large effect on SVR. The main center for regulation of SVR is the vasomotor center in the brain stem (described shortly).

Venous Return

Venous return, the volume of blood flowing back to the heart through the systemic veins, occurs due to the pressure generated by contractions of the heart's left ventricle. The pressure difference from venules (averaging about 16 mmHg) to the right ventricle (0 mmHg), although small, normally is sufficient to cause venous return to the heart. If pressure increases in the right atrium or ventricle, venous return will decrease. One cause of increased pressure in the right atrium is an incompetent (leaky) tricuspid valve, which lets blood regurgitate (flow backward) as the ventricles contract. The result is decreased venous return and buildup of blood on the venous side of the systemic circulation.

When you stand up, for example, at the end of an anatomy and physiology lecture, the pressure pushing blood up the veins in your lower limbs is barely enough to overcome the force of gravity pushing it back down. Besides the heart, two other mechanisms "pump" blood from the lower body back to the heart: (1) the skeletal muscle pump and (2) the respiratory pump. Both pumps depend on the presence of valves in veins.

The **skeletal muscle pump** operates as follows (Figure 21.9):

1. While you are standing at rest, both the venous valve closer to the heart (proximal valve) and the one farther from the heart (distal valve) in this part of the leg are open, and blood flows upward toward the heart.

Figure 21.9   Action of the skeletal muscle pump in returning blood to the heart.

Milking refers to skeletal muscle contractions that drive venous blood toward the heart.

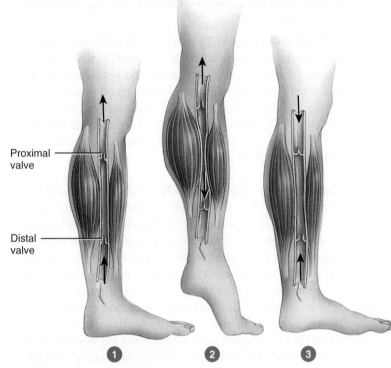

Proximal valve

Distal valve

Aside from cardiac contractions, what mechanisms act as pumps to boost venous return?

② Contraction of leg muscles, such as when you stand on tiptoes or take a step, compresses the vein. The compression pushes blood through the proximal valve, an action called *milking*. At the same time, the distal valve in the uncompressed segment of the vein closes as some blood is pushed against it. People who are immobilized through injury or disease lack these contractions of leg muscles. As a result, their venous return is slower and they may develop circulation problems.

③ Just after muscle relaxation, pressure falls in the previously compressed section of vein, which causes the proximal valve to close. The distal valve now opens because blood pressure in the foot is higher than in the leg, and the vein fills with blood from the foot. The proximal valve then reopens.

The **respiratory pump** is also based on alternating compression and decompression of veins. During inhalation, the diaphragm moves downward, which causes a decrease in pressure in the thoracic cavity and an increase in pressure in the abdominal cavity. As a result, abdominal veins are compressed, and a greater volume of blood moves from the compressed abdominal veins into the decompressed thoracic veins and then into the right atrium. When the pressures reverse during exhalation, the valves in the veins prevent backflow of blood from the thoracic veins to the abdominal veins.

Figure 21.10 summarizes the factors that increase blood pressure through increasing cardiac output or systemic vascular resistance.

Velocity of Blood Flow

Earlier we saw that blood flow is the *volume* of blood that flows through any tissue in a given time period (in mL/min). The speed or *velocity* of blood flow (in cm/sec) is inversely related to the cross-sectional area. Velocity is slowest where the total cross-sectional area is greatest (Figure 21.11). Each time an artery branches, the total cross-sectional area of all its branches is greater than the cross-sectional area of the original vessel, so blood flow becomes slower and slower as blood moves further away from the heart, and is slowest in the capillaries. Conversely, when venules unite to form veins, the total cross-sectional area becomes smaller and flow becomes faster. In an adult, the cross-sectional area of the aorta is only 3–5 cm^2, and the average velocity of the blood there is 40 cm/sec. In capillaries, the total cross-sectional area is 4500–6000 cm^2, and the velocity of blood flow is less than 0.1 cm/sec. In the two venae cavae combined, the cross-sectional area is about 14 cm^2, and the velocity is about 15 cm/sec. Thus, the velocity of blood flow decreases as blood flows from the aorta to arteries to arterioles to capillaries, and increases as it leaves capillaries and returns to the heart. The relatively slow rate of flow through capillaries aids the exchange of materials between blood and interstitial fluid.

Circulation time is the time required for a drop of blood to pass from the right atrium, through the pulmonary circulation, back to the left atrium, through the systemic circulation down to the foot, and back again to the right atrium. In a resting person, circulation time normally is about 1 minute.

CLINICAL CONNECTION | *Syncope*

Syncope (SIN-kō-pē), or fainting, is a sudden, temporary loss of consciousness that is not due to head trauma, followed by spontaneous recovery. It is most commonly due to cerebral ischemia, lack of sufficient blood flow to the brain. Syncope may occur for several reasons:

- *Vasodepressor syncope* is due to sudden emotional stress or real, threatened, or fantasized injury.
- *Situational syncope* is caused by pressure stress associated with urination, defecation, or severe coughing.
- *Drug-induced syncope* may be caused by drugs such as antihypertensives, diuretics, vasodilators, and tranquilizers.
- *Orthostatic hypotension,* an excessive decrease in blood pressure that occurs on standing up, may cause fainting.   •

✔CHECKPOINT

9. Explain how blood pressure and resistance determine volume of blood flow.
10. What is systemic vascular resistance and what factors contribute to it?
11. How is the return of venous blood to the heart accomplished?
12. Why is the velocity of blood flow faster in arteries and veins than in capillaries?

Figure 21.10 **Summary of factors that increase blood pressure.** Changes noted within green boxes increase cardiac output; changes noted within blue boxes increase systemic vascular resistance.

🔐 Increases in cardiac output and increases in systemic vascular resistance will increase mean arterial pressure.

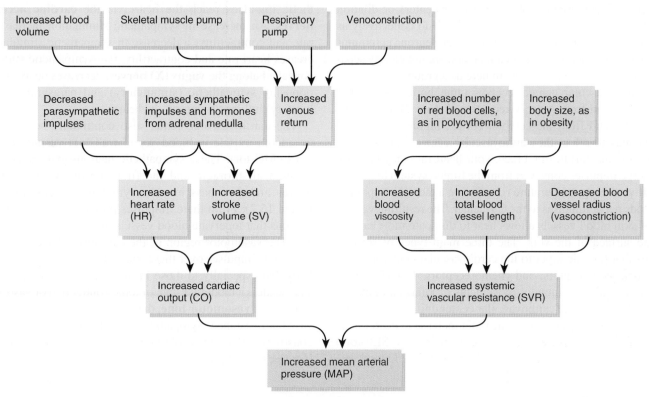

❓ **Which type of blood vessel exerts the major control of systemic vascular resistance, and how does it achieve this?**

Figure 21.11 **Relationship between velocity (speed) of blood flow and total cross-sectional area in different types of blood vessels.**

🔐 Velocity of blood flow is slowest in the capillaries because they have the largest total cross-sectional area.

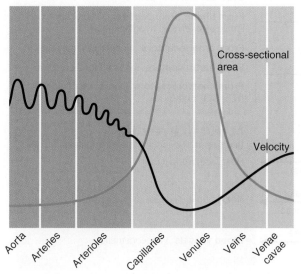

❓ **In which blood vessels is the velocity of flow fastest?**

21.4 CONTROL OF BLOOD PRESSURE AND BLOOD FLOW

◉ **OBJECTIVE**

• Describe how blood pressure is regulated.

Several interconnected negative feedback systems control blood pressure by adjusting heart rate, stroke volume, systemic vascular resistance, and blood volume. Some systems allow rapid adjustments to cope with sudden changes, such as the drop in blood pressure in the brain that occurs when you get out of bed; others act more slowly to provide long-term regulation of blood pressure. The body may also require adjustments to the distribution of blood flow. During exercise, for example, a greater percentage of the total blood flow is diverted to skeletal muscles.

Role of the Cardiovascular Center

In Chapter 20, we noted how the **cardiovascular (CV) center** in the medulla oblongata helps regulate heart rate and stroke volume. The CV center also controls neural, hormonal, and local negative feedback systems that regulate blood pressure and blood flow to specific tissues. Groups of neurons scattered within the

CV center regulate heart rate, contractility (force of contraction) of the ventricles, and blood vessel diameter. Some neurons stimulate the heart (cardiostimulatory center); others inhibit the heart (cardioinhibitory center). Still others control blood vessel diameter by causing constriction (vasoconstrictor center) or dilation (vasodilator center); these neurons are referred to collectively as the vasomotor center. Because the CV center neurons communicate with one another, function together, and are not clearly separated anatomically, we discuss them here as a group.

The cardiovascular center receives input both from higher brain regions and from sensory receptors (Figure 21.12). Nerve impulses descend from the cerebral cortex, limbic system, and hypothalamus to affect the cardiovascular center. For example, even before you start to run a race, your heart rate may increase due to nerve impulses conveyed from the limbic system to the CV center. If your body temperature rises during a race, the hypothalamus sends nerve impulses to the CV center. The resulting vasodilation of skin blood vessels allows heat to dissipate more rapidly from the surface of the skin. The three main types of sensory receptors that provide input to the cardiovascular center are proprioceptors, baroreceptors, and chemoreceptors. *Proprioceptors* (PRŌ-prē-ō-sep′-tors) monitor movements of joints and muscles and provide input to the cardiovascular center during physical activity. Their activity accounts for the rapid increase in heart rate at the beginning of exercise. *Baroreceptors* (bar′-ō-re-SEP-tors) monitor changes in pressure and stretch in the walls of blood ves-

sels, and *chemoreceptors* (kē′-mō-rē-SEP-tors) monitor the concentration of various chemicals in the blood.

Output from the cardiovascular center flows along sympathetic and parasympathetic neurons of the ANS (Figure 21.12). Sympathetic impulses reach the heart via the **cardiac accelerator nerves.** An increase in sympathetic stimulation increases heart rate and contractility; a decrease in sympathetic stimulation decreases heart rate and contractility. Parasympathetic stimulation, conveyed along the **vagus (X) nerves,** decreases heart rate. Thus, opposing sympathetic (stimulatory) and parasympathetic (inhibitory) influences control the heart.

The cardiovascular center also continually sends impulses to smooth muscle in blood vessel walls via **vasomotor nerves** (vā-so-MŌ-tor). These sympathetic neurons exit the spinal cord through all thoracic and the first one or two lumbar spinal nerves and then pass into the sympathetic trunk ganglia (see Figure 15.2). From there, impulses propagate along sympathetic neurons that innervate blood vessels in viscera and peripheral areas. The vasomotor region of the cardiovascular center continually sends impulses over these routes to arterioles throughout the body, but especially to those in the skin and abdominal viscera. The result is a moderate state of tonic contraction or vasoconstriction, called **vasomotor tone,** that sets the resting level of systemic vascular resistance. Sympathetic stimulation of most veins causes constriction that moves blood out of venous blood reservoirs and increases blood pressure.

Figure 21.12 Location and function of the cardiovascular (CV) center in the medulla oblongata. The CV center receives input from higher brain centers, proprioceptors, baroreceptors, and chemoreceptors. Then, it provides output to the sympathetic and parasympathetic divisions of the autonomic nervous system (ANS).

The cardiovascular center is the main region for nervous system regulation of the heart and blood vessels.

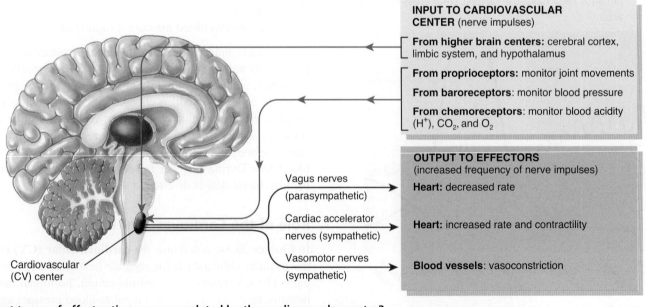

INPUT TO CARDIOVASCULAR CENTER (nerve impulses)

From higher brain centers: cerebral cortex, limbic system, and hypothalamus

From proprioceptors: monitor joint movements

From baroreceptors: monitor blood pressure

From chemoreceptors: monitor blood acidity (H^+), CO_2, and O_2

OUTPUT TO EFFECTORS (increased frequency of nerve impulses)

Vagus nerves (parasympathetic)

Heart: decreased rate

Cardiac accelerator nerves (sympathetic)

Heart: increased rate and contractility

Vasomotor nerves (sympathetic)

Blood vessels: vasoconstriction

Cardiovascular (CV) center

? **What types of effector tissues are regulated by the cardiovascular center?**

Neural Regulation of Blood Pressure

The nervous system regulates blood pressure via negative feedback loops that occur as two types of reflexes: baroreceptor reflexes and chemoreceptor reflexes.

Baroreceptor Reflexes

Baroreceptors, pressure-sensitive sensory receptors, are located in the aorta, internal carotid arteries (arteries in the neck that supply blood to the brain), and other large arteries in the neck and chest. They send impulses to the cardiovascular center to help regulate blood pressure. The two most important **baroreceptor reflexes** are the carotid sinus reflex and the aortic reflex.

Baroreceptors in the wall of the carotid sinuses initiate the **carotid sinus reflex** (ka-ROT-id), which helps regulate blood pressure in the brain. The **carotid sinuses** are small widenings of the right and left internal carotid arteries just above the point where they branch from the common carotid arteries (Figure 21.13). Blood pressure stretches the wall of the carotid sinus, which stimulates the baroreceptors. Nerve impulses propagate from the carotid sinus baroreceptors over sensory axons in the **glossopharyngeal (IX) nerves** (glos′-ō-fa-RIN-jē-al) to the cardiovascular center in the medulla oblongata. Baroreceptors in the wall of the ascending aorta and arch of the aorta initiate the **aortic reflex,** which regulates systemic blood pressure. Nerve impulses from aortic baroreceptors reach the cardiovascular center via sensory axons of the **vagus (X) nerves.**

When blood pressure falls, the baroreceptors are stretched less, and they send nerve impulses at a slower rate to the cardiovascular center (Figure 21.14). In response, the CV center decreases parasympathetic stimulation of the heart by way of motor axons of the vagus nerves and increases sympathetic stimulation of the heart via cardiac accelerator nerves. Another consequence of increased sympathetic stimulation is increased secretion of epinephrine and norepinephrine by the adrenal medulla. As the heart beats faster and more forcefully, and as systemic vascular resistance increases, cardiac output and systemic vascular resistance rise, and blood pressure increases to the normal level.

Conversely, when an increase in pressure is detected, the baroreceptors send impulses at a faster rate. The CV center responds by increasing parasympathetic stimulation and decreasing sympathetic stimulation. The resulting decreases in heart rate and force of contraction reduce the cardiac output. The cardiovascular center also slows the rate at which it sends sympathetic impulses along vasomotor neurons that normally cause vasoconstriction. The resulting vasodilation lowers systemic vascular resistance. Decreased cardiac output and decreased systemic vascular resistance both lower systemic arterial blood pressure to the normal level.

Figure 21.13 ANS innervation of the heart and the baroreceptor reflexes that help regulate blood pressure.

Baroreceptors are pressure-sensitive neurons that monitor stretching.

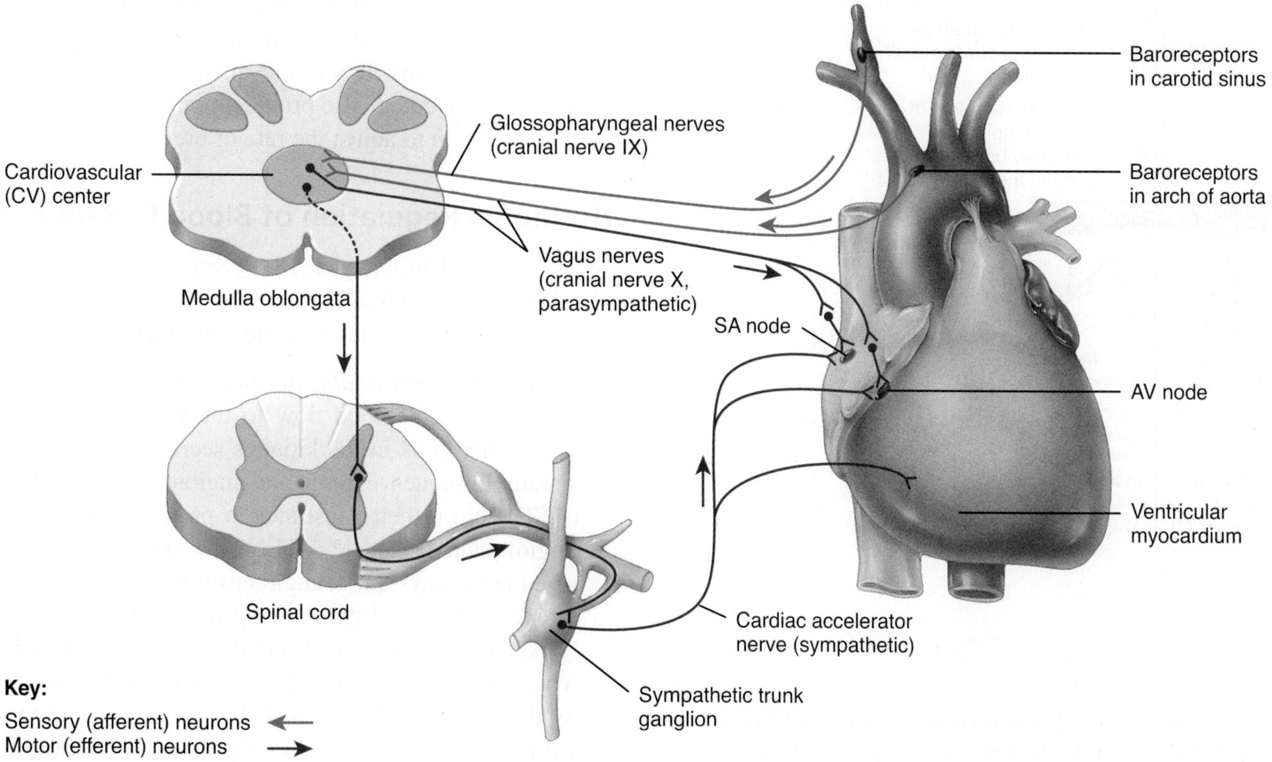

Which cranial nerves conduct impulses to the cardiovascular center from baroreceptors in the carotid sinuses and the arch of the aorta?

Figure 21.14 Negative feedback regulation of blood pressure via baroreceptor reflexes.

⑥ ⊏⊐ When blood pressure decreases, heart rate increases.

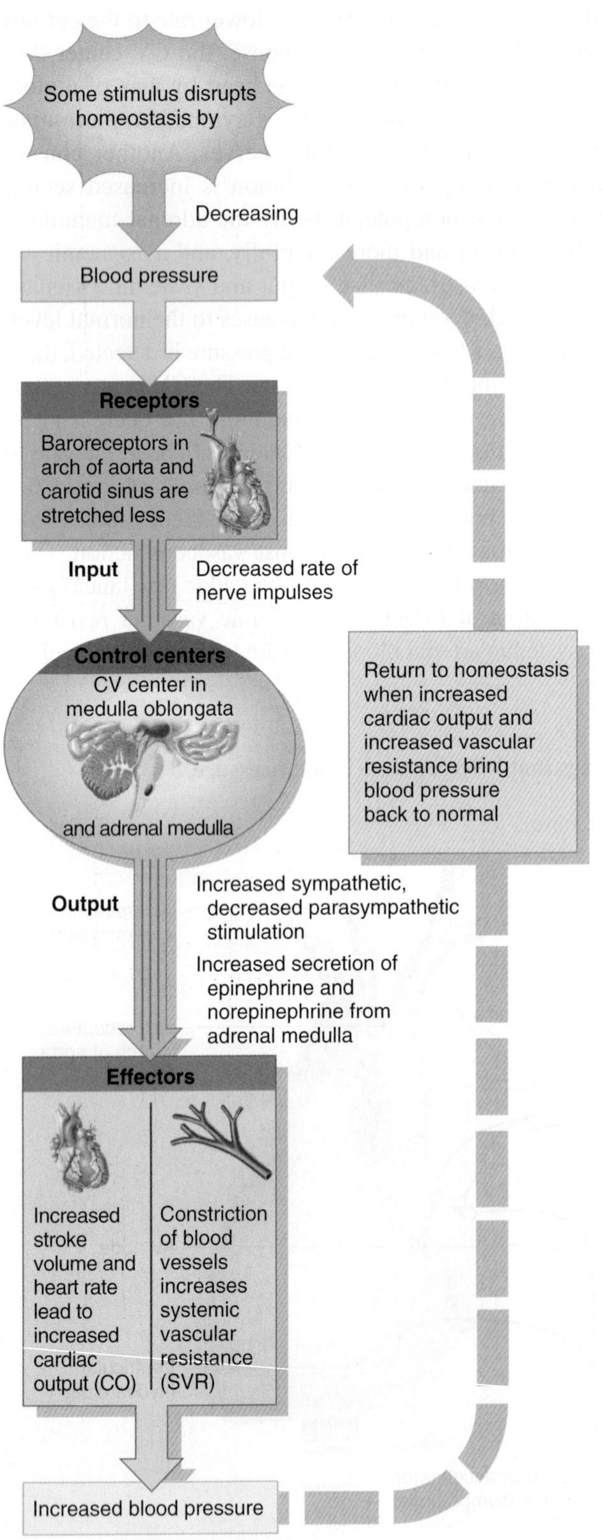

Some stimulus disrupts homeostasis by

Decreasing

Blood pressure

Receptors

Baroreceptors in arch of aorta and carotid sinus are stretched less

Input — Decreased rate of nerve impulses

Control centers

CV center in medulla oblongata

and adrenal medulla

Return to homeostasis when increased cardiac output and increased vascular resistance bring blood pressure back to normal

Output

Increased sympathetic, decreased parasympathetic stimulation

Increased secretion of epinephrine and norepinephrine from adrenal medulla

Effectors

Increased stroke volume and heart rate lead to increased cardiac output (CO)

Constriction of blood vessels increases systemic vascular resistance (SVR)

Increased blood pressure

? **Does this negative feedback cycle represent the changes that occur when you lie down or when you stand up?**

Moving from a prone (lying down) to an erect position decreases blood pressure and blood flow in the head and upper part of the body. The baroreceptor reflexes, however, quickly counteract the drop in pressure. Sometimes these reflexes operate more slowly than normal, especially in the elderly, in which case a person can faint due to reduced brain blood flow after standing up too quickly.

CLINICAL CONNECTION | *Carotid Sinus Massage and Carotid Sinus Syncope*

Because the carotid sinus is close to the anterior surface of the neck, it is possible to stimulate the baroreceptors there by putting pressure on the neck. Physicians sometimes use **carotid sinus massage,** which involves carefully massaging the neck over the carotid sinus, to slow heart rate in a person who has paroxysmal supraventricular tachycardia, a type of tachycardia that originates in the atria. Anything that stretches or puts pressure on the carotid sinus, such as hyperextension of the head, tight collars, or carrying heavy shoulder loads, may also slow heart rate and can cause **carotid sinus syncope,** fainting due to inappropriate stimulation of the carotid sinus baroreceptors. •

Chemoreceptor Reflexes

Chemoreceptors, sensory receptors that monitor the chemical composition of blood, are located close to the baroreceptors of the carotid sinus and arch of the aorta in small structures called **carotid bodies** and **aortic bodies,** respectively. These chemoreceptors detect changes in blood level of O_2, CO_2, and H^+. *Hypoxia* (lowered O_2 availability), *acidosis* (an increase in H^+ concentration), or *hypercapnia* (excess CO_2) stimulates the chemoreceptors to send impulses to the cardiovascular center. In response, the CV center increases sympathetic stimulation to arterioles and veins, producing vasoconstriction and an increase in blood pressure. These chemoreceptors also provide input to the respiratory center in the brain stem to adjust the rate of breathing.

Hormonal Regulation of Blood Pressure

As you learned in Chapter 18, several hormones help regulate blood pressure and blood flow by altering cardiac output, changing systemic vascular resistance, or adjusting the total blood volume:

1. **Renin–angiotensin–aldosterone (RAA) system.** When blood volume falls or blood flow to the kidneys decreases, juxtaglomerular cells in the kidneys secrete **renin** into the bloodstream. In sequence, renin and angiotensin-converting enzyme (ACE) act on their substrates to produce the active hormone **angiotensin II** (an'-jē-ō-TEN-sin), which raises blood pressure in two ways. First, angiotensin II is a potent vasoconstrictor; it raises blood pressure by increasing systemic vascular resistance. Second, it stimulates secretion of **aldosterone,** which increases reabsorption of sodium ions (Na^+) and water by the kidneys. The water reabsorption increases total blood volume, which increases blood pressure. (See Section 21.6.)

2. **Epinephrine and norepinephrine.** In response to sympathetic stimulation, the adrenal medulla releases epinephrine and

norepinephrine. These hormones increase cardiac output by increasing the rate and force of heart contractions. They also cause vasoconstriction of arterioles and veins in the skin and abdominal organs and vasodilation of arterioles in cardiac and skeletal muscle, which helps increase blood flow to muscle during exercise. (See Figure 18.20.)

3. *Antidiuretic hormone (ADH).* ADH is produced by the hypothalamus and released from the posterior pituitary in response to dehydration or decreased blood volume. Among other actions, ADH causes vasoconstriction, which increases blood pressure. For this reason ADH is also called **vasopressin.** (See Figure 18.9.) ADH also promotes movement of water from the lumen of kidney tubules into the bloodstream. This results in an increase in blood volume and a decrease in urine output.

4. *Atrial natriuretic peptide (ANP).* Released by cells in the atria of the heart, ANP lowers blood pressure by causing vasodilation and by promoting the loss of salt and water in the urine, which reduces blood volume.

Table 21.2 summarizes the regulation of blood pressure by hormones.

Autoregulation of Blood Pressure

In each capillary bed, local changes can regulate vasomotion. When vasodilators produce local dilation of arterioles and relaxation of precapillary sphincters, blood flow into capillary networks is increased, which increases O_2 level. Vasoconstrictors have the opposite effect. The ability of a tissue to automatically adjust its blood flow to match its metabolic demands is called **autoregulation** (aw'-tō-reg'-u-LĀ-shun). In tissues such as the heart and skeletal muscle, where the demand for O_2 and nutrients and for the removal of wastes can increase as much as tenfold during physical activity, autoregulation is an important contributor to increased blood flow through the tissue. Autoregulation also controls regional blood flow in the brain; blood distribution to various parts of the brain changes dramatically for different mental and physical activities. During a conversation, for example, blood flow increases to your motor speech areas when you are talking and increases to the auditory areas when you are listening.

Two general types of stimuli cause autoregulatory changes in blood flow:

1. *Physical changes.* Warming promotes vasodilation, and cooling causes vasoconstriction. In addition, smooth muscle in arteriole walls exhibits a **myogenic response** (mī-ō-JEN-ik)—it contracts more forcefully when it is stretched and relaxes when stretching lessens. If, for example, blood flow through an arteriole decreases, stretching of the arteriole walls decreases. As a result, the smooth muscle relaxes and produces vasodilation, which increases blood flow.

2. *Vasodilating and vasoconstricting chemicals.* Several types of cells—including white blood cells, platelets, smooth muscle fibers, macrophages, and endothelial cells—release a wide variety of chemicals that alter blood-vessel diameter. Vasodilating chemicals released by metabolically active tissue cells include K^+, H^+, lactic acid (lactate), and adenosine (from ATP). Another important vasodilator released by endothelial cells is nitric oxide (NO). Tissue trauma or inflammation causes release of vasodilating kinins and histamine. Vasoconstrictors include thromboxane A2, superoxide radicals, serotonin (from platelets), and endothelins (from endothelial cells).

An important difference between the pulmonary and systemic circulations is their autoregulatory response to changes in O_2 level. The walls of blood vessels in the systemic circulation *dilate* in response to low O_2. With vasodilation, O_2 delivery increases, which restores the normal O_2 level. By contrast, the walls of blood vessels in the pulmonary circulation *constrict* in response to low levels of O_2. This response ensures that blood mostly bypasses those alveoli (air sacs) in the lungs that are poorly ventilated by fresh air. Thus, most blood flows to better-ventilated areas of the lung.

✔ CHECKPOINT

13. What are the principal inputs to and outputs from the cardiovascular center?
14. Explain the operation of the carotid sinus reflex and the aortic reflex.
15. What is the role of chemoreceptors in the regulation of blood pressure?
16. How do hormones regulate blood pressure?
17. What is autoregulation, and how does it differ in the systemic and pulmonary circulations?

TABLE 21.2

Blood Pressure Regulation by Hormones

FACTOR INFLUENCING BLOOD PRESSURE	HORMONE	EFFECT ON BLOOD PRESSURE
CARDIAC OUTPUT		
Increased heart rate and contractility	Norepinephrine, epinephrine.	Increase.
SYSTEMIC VASCULAR RESISTANCE		
Vasoconstriction	Angiotensin II, antidiuretic hormone (ADH) or vasopressin, norepinephrine,* epinephrine.†	Increase.
Vasodilation	Atrial natriuretic peptide (ANP), epinephrine,† nitric oxide.	Decrease.
BLOOD VOLUME		
Blood volume increase	Aldosterone, antidiuretic hormone.	Increase.
Blood volume decrease	Atrial natriuretic peptide.	Decrease.

*Acts at α_1 receptors in arterioles of abdomen and skin.
†Acts at β_2 receptors in arterioles of cardiac and skeletal muscle; norepinephrine has a much smaller vasodilating effect.

21.5 CHECKING CIRCULATION

◉ OBJECTIVE

• Define pulse, and define systolic, diastolic, and pulse pressures.

Pulse

The alternate expansion and recoil of elastic arteries after each systole of the left ventricle creates a traveling pressure wave that is called the **pulse.** The pulse is strongest in the arteries closest to the heart, becomes weaker in the arterioles, and disappears altogether in the capillaries. The pulse may be felt in any artery that lies near the surface of the body that can be compressed against a bone or other firm structure. Table 21.3 depicts some common pulse points.

The pulse rate normally is the same as the heart rate, about 70 to 80 beats per minute at rest. **Tachycardia** (tak′-i-KAR-dē-a; *tachy*- = fast) is a rapid resting heart or pulse rate over 100 beats/min. **Bradycardia** (brād′-i-KAR-dē-a; *brady*- = slow) is a slow resting heart or pulse rate under 50 beats/min. Endurance-trained athletes normally exhibit bradycardia.

Measuring Blood Pressure

In clinical use, the term **blood pressure** usually refers to the pressure in arteries generated by the left ventricle during systole and the pressure remaining in the arteries when the ventricle is in diastole. Blood pressure is usually measured in the brachial artery in the left arm (Table 21.3). The device used to measure blood pressure is a **sphygmomanometer** (sfig′-mō-ma-NOM-e-ter; *sphygmo*- = pulse; *-manometer* = instrument used to measure pressure). It consists of a rubber cuff connected to a rubber bulb that is used to inflate the cuff and a meter that registers the pressure in the cuff. With the arm resting on a table so that it is about the same level as the heart, the cuff of the sphygmomanometer is

TABLE 21.3

Pulse Points

STRUCTURE	LOCATION	STRUCTURE	LOCATION
Superficial temporal artery	Medial to ear.	**Femoral artery**	Inferior to inguinal ligament.
Facial artery	Mandible (lower jawbone) on line with corners of mouth.	**Popliteal artery**	Posterior to knee.
		Radial artery	Distal aspect of wrist.
Common carotid artery	Lateral to larynx (voice box).	**Dorsal artery of foot (dorsalis pedis artery)**	Superior to instep of foot.
Brachial artery	Medial side of biceps brachii muscle.		

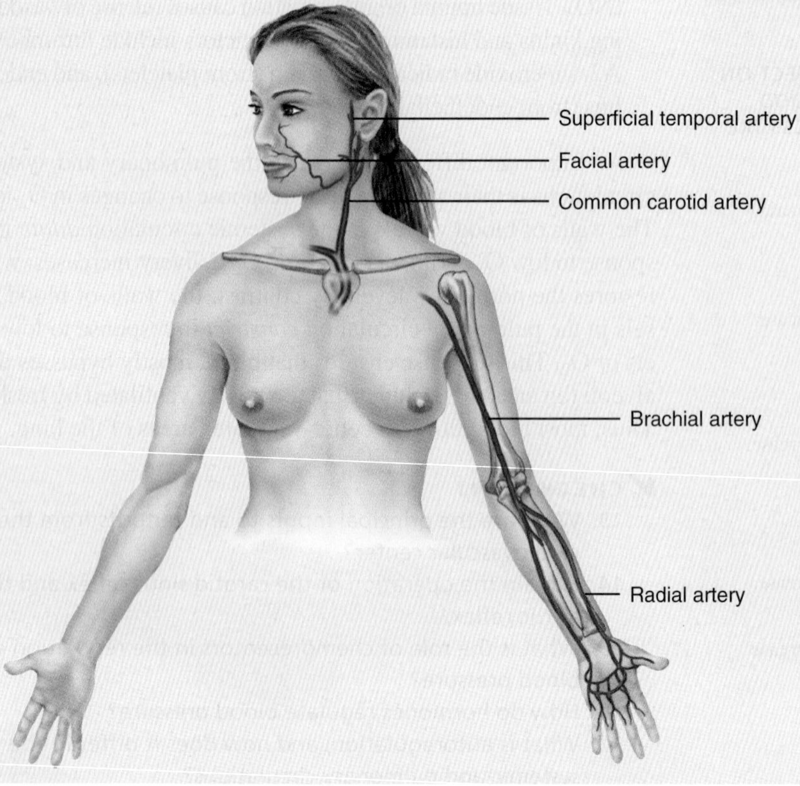

Superficial temporal artery
Facial artery
Common carotid artery
Brachial artery
Radial artery
Femoral artery 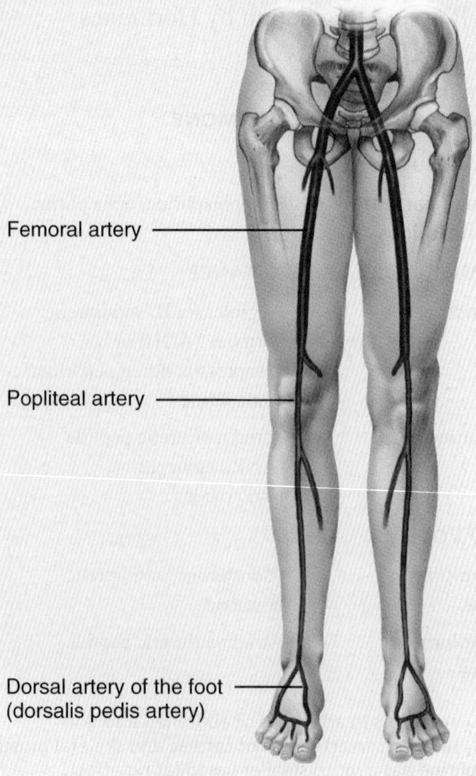
Popliteal artery
Dorsal artery of the foot (dorsalis pedis artery)

wrapped around a bared arm. The cuff is inflated by squeezing the bulb until the brachial artery is compressed and blood flow stops, about 30 mmHg higher than the person's usual systolic pressure. The technician places a stethoscope below the cuff on the brachial artery, and slowly deflates the cuff. When the cuff is deflated enough to allow the artery to open, a spurt of blood passes through, resulting in the first sound heard through the stethoscope. This sound corresponds to **systolic blood pressure (SBP)**, the force of blood pressure on arterial walls just after ventricular contraction (Figure 21.15). As the cuff is deflated further, the sounds suddenly become too faint to be heard through the stethoscope. This level, called the **diastolic blood pressure (DBP)**, represents the force exerted by the blood remaining in arteries during ventricular relaxation. At pressures below diastolic blood pressure, sounds disappear altogether. The various sounds that are heard while taking blood pressure are called **Korotkoff sounds** (kō-ROT-kof).

The normal blood pressure of an adult male is less than 120 mmHg systolic and less than 80 mmHg diastolic. For example, "110 over 70" (written as 110/70) is a normal blood pressure. In young adult females, the pressures are 8 to 10 mmHg less. People who exercise regularly and are in good physical condition may have even lower blood pressures. Thus, blood pressure slightly lower than 120/80 may be a sign of good health and fitness.

The difference between systolic and diastolic pressure is called **pulse pressure.** This pressure, normally about 40 mmHg, provides information about the condition of the cardiovascular system. For example, conditions such as atherosclerosis and patent (open) ductus arteriosus greatly increase pulse pressure. The normal ratio of systolic pressure to diastolic pressure to pulse pressure is about 3:2:1.

Figure 21.15 Relationship of blood pressure changes to cuff pressure.

As the cuff is deflated, sounds first occur at the systolic blood pressure; the sounds suddenly become faint at the diastolic blood pressure.

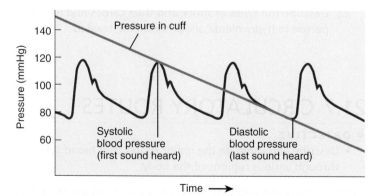

If a blood pressure is reported as "142 over 95," what are the diastolic, systolic, and pulse pressures? Does this person have hypertension as defined in Disorders: Homeostatic Imbalances at the end of the chapter?

✔ **CHECKPOINT**
18. Where may the pulse be felt?
19. What do tachycardia and bradycardia mean?
20. How are systolic and diastolic blood pressures measured with a sphygmomanometer?

21.6 SHOCK AND HOMEOSTASIS

● **OBJECTIVES**

• Define shock, and describe the four types of shock.
• Explain how the body's response to shock is regulated by negative feedback.

Shock is a failure of the cardiovascular system to deliver enough O_2 and nutrients to meet cellular metabolic needs. The causes of shock are many and varied, but all are characterized by inadequate blood flow to body tissues. With inadequate oxygen delivery, cells switch from aerobic to anaerobic production of ATP, and lactic acid accumulates in body fluids. If shock persists, cells and organs become damaged, and cells may die unless proper treatment begins quickly.

Types of Shock

Shock can be of four different types: (1) **hypovolemic shock** (hī-pō-vō-LĒ-mik; *hypo-* = low; *-volemic* = volume) due to decreased blood volume, (2) **cardiogenic shock** (kar′-dē-ō-JEN-ik) due to poor heart function, (3) **vascular shock** due to inappropriate vasodilation, and (4) **obstructive shock** due to obstruction of blood flow.

A common cause of hypovolemic shock is acute (sudden) hemorrhage. The blood loss may be external, as occurs in trauma, or internal, as in rupture of an aortic aneurysm. Loss of body fluids through excessive sweating, diarrhea, or vomiting also can cause hypovolemic shock. Other conditions—for instance, diabetes mellitus—may cause excessive loss of fluid in the urine. Sometimes, hypovolemic shock is due to inadequate intake of fluid. Whatever the cause, when the volume of body fluids falls, venous return to the heart declines, filling of the heart lessens, stroke volume decreases, and cardiac output decreases. Replacing fluid volume as quickly as possible is essential in managing hypovolemic shock.

In cardiogenic shock, the heart fails to pump adequately, most often because of a myocardial infarction (heart attack). Other causes of cardiogenic shock include poor perfusion of the heart (ischemia), heart valve problems, excessive preload or afterload, impaired contractility of heart muscle fibers, and arrhythmias.

Even with normal blood volume and cardiac output, shock may occur if blood pressure drops due to a decrease in systemic vascular resistance. A variety of conditions can cause inappropriate dilation of arterioles or venules. In *anaphylactic shock* (AN-a-fil-lak′-tik), a severe allergic reaction—for example, to a bee sting—releases histamine and other mediators that cause vasodilation. In *neurogenic shock,* vasodilation may occur following trauma to the head that causes malfunction of the cardiovascular center in the medulla. Shock stemming from certain bacterial toxins that produce vasodilation is termed *septic shock*. In the United States, septic shock causes more than 100,000 deaths per year and is the most common cause of death in hospital critical care units.

Obstructive shock occurs when blood flow through a portion of the circulation is blocked. The most common cause is *pulmonary embolism*, a blood clot lodged in a blood vessel of the lungs.

Homeostatic Responses to Shock

The major mechanisms of compensation in shock are *negative feedback systems* that work to return cardiac output and arterial blood pressure to normal. When shock is mild, compensation by homeostatic mechanisms prevents serious damage. In an otherwise healthy person, compensatory mechanisms can maintain adequate blood flow and blood pressure despite an acute blood loss of as much as 10% of total volume. Figure 21.16 shows several of the negative feedback systems that respond to hypovolemic shock.

1. *Activation of the renin–angiotensin–aldosterone system.* Decreased blood flow to the kidneys causes the kidneys to secrete renin and initiates the renin–angiotensin–aldosterone system (see Figure 18.16). Recall that angiotensin II causes vasoconstriction and stimulates the adrenal cortex to secrete aldosterone, a hormone that increases reabsorption of Na^+ and water by the kidneys. The increases in systemic vascular resistance and blood volume help raise blood pressure.

2. *Secretion of antidiuretic hormone.* In response to decreased blood pressure, the posterior pituitary releases more antidiuretic hormone (ADH). ADH enhances water reabsorption by the kidneys, which conserves remaining blood volume. It also causes vasoconstriction, which increases systemic vascular resistance. (See Figure 18.9.)

3. *Activation of the sympathetic division of the ANS.* As blood pressure decreases, the aortic and carotid baroreceptors initiate powerful sympathetic responses throughout the body. One result is marked vasoconstriction of arterioles and veins of the skin, kidneys, and other abdominal viscera. (Vasoconstriction does not occur in the brain or heart.) The constriction of arterioles increases systemic vascular resistance, and the constriction of veins increases venous return. Both effects help maintain an adequate blood pressure. Sympathetic stimulation also increases heart rate and contractility and increases secretion of epinephrine and norepinephrine by the adrenal medulla. These hormones intensify vasoconstriction and increase heart rate and contractility, all of which help raise blood pressure.

4. *Release of local vasodilators.* In response to *hypoxia*, cells liberate vasodilators—including K^+, H^+, lactic acid, adenosine, and nitric oxide—that dilate arterioles and relax precapillary sphincters. Such vasodilation increases local blood flow and may restore O_2 level to normal in part of the body.

However, vasodilation also has the potentially harmful effect of decreasing systemic vascular resistance and thus lowering the blood pressure.

If blood volume drops more than 10–20%, or if the heart cannot bring blood pressure up sufficiently, compensatory mechanisms may fail to maintain adequate blood flow to tissues. At this point, shock becomes life-threatening as damaged cells start to die.

Signs and Symptoms of Shock

Even though the signs and symptoms of shock vary with the severity of the condition, most can be predicted in light of the responses generated by the negative feedback systems that attempt to correct the problem. Among the signs and symptoms of shock are the following:

- Systolic blood pressure is lower than 90 mmHg.
- Resting heart rate is rapid due to sympathetic stimulation and increased blood levels of epinephrine and norepinephrine.
- Pulse is weak and rapid due to reduced cardiac output and fast heart rate.
- Skin is cool, pale, and clammy due to sympathetic constriction of skin blood vessels and sympathetic stimulation of sweating.
- Mental state is altered due to reduced oxygen supply to the brain.
- Urine formation is reduced due to increased levels of aldosterone and antidiuretic hormone (ADH).
- The person is thirsty due to loss of extracellular fluid.
- The pH of blood is low (acidosis) due to buildup of lactic acid.
- The person may have nausea because of impaired blood flow to the digestive organs from sympathetic vasoconstriction.

✔ **CHECKPOINT**

21. Which symptoms of hypovolemic shock relate to actual body fluid loss, and which relate to the negative feedback systems that attempt to maintain blood pressure and blood flow?
22. Describe the types of shock and their causes and how a person in hypovolemic shock should be treated.

21.7 CIRCULATORY ROUTES

◉ **OBJECTIVE**
- Describe and compare the major routes that blood takes through various regions of the body.

Arteries, arterioles, capillaries, venules, and veins are organized into **circulatory routes** that deliver blood throughout the body. Now that you understand the structures of each of these vessel types, we can look at the basic routes the blood takes as it is transported throughout the body.

Figure 21.16 Negative feedback systems that can restore normal blood pressure during hypovolemic shock.

Homeostatic mechanisms can compensate for an acute blood loss of as much as 10% of total blood volume.

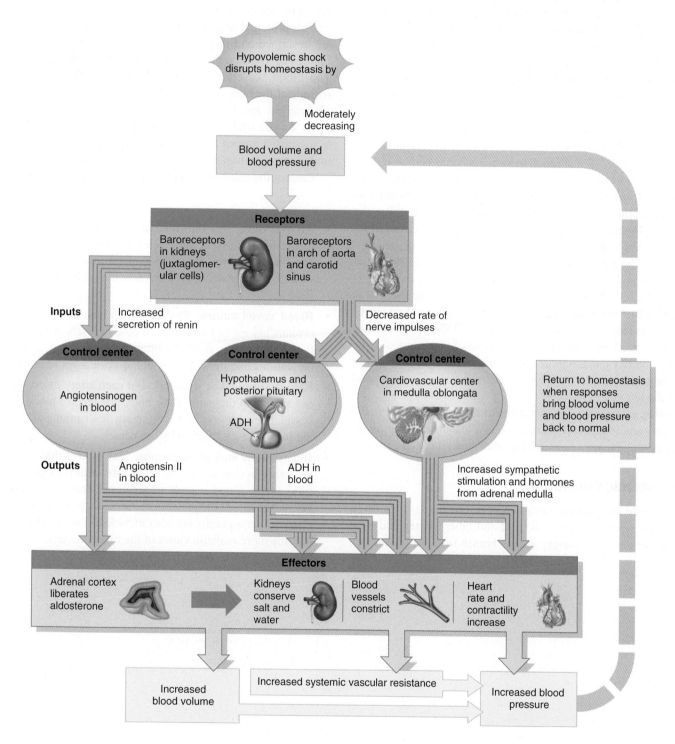

? **Does almost-normal blood pressure in a person who has lost blood indicate that the patient's tissues are receiving adequate perfusion (blood flow)?**

Figure 21.17 shows the circulatory routes for blood flow. The routes are parallel; that is, in most cases a portion of the cardiac output flows separately to each tissue of the body. Thus, each organ receives its own supply of freshly oxygenated blood. The two basic postnatal (after birth) routes for blood flow are the systemic circulation and the pulmonary circulation. The **systemic circulation** includes all arteries and arterioles that carry oxygenated blood from the left ventricle to systemic capillaries, plus the veins and venules that carry deoxygenated blood returning to the right atrium after flowing through body organs. Blood leaving the aorta and flowing through the systemic arteries is a bright red color. As it moves through capillaries, it loses some of its oxygen and picks up carbon dioxide, so that blood in systemic veins is dark red.

Subdivisions of the systemic circulation include the **coronary (cardiac) circulation** (see Figure 20.8), which supplies the myocardium of the heart; **cerebral circulation,** which supplies the brain (see Figure 21.19c); and the **hepatic portal circulation,** which extends from the gastrointestinal tract to the liver (see Figure 21.28). The nutrient arteries to the lungs, such as the bronchial arteries, are also part of the systemic circulation.

When blood returns to the heart from the systemic route, it is pumped out of the right ventricle through the **pulmonary circulation** to the lungs (see Figure 21.29). In capillaries of the air sacs (alveoli) of the lungs, the blood loses some of its carbon dioxide and takes on oxygen. Bright red again, it returns to the left atrium of the heart and reenters the systemic circulation as it is pumped out by the left ventricle.

Another major route—the **fetal circulation**—exists only in the fetus and contains special structures that allow the developing fetus to exchange materials with its mother (see Figure 21.30).

The Systemic Circulation

The **systemic circulation** carries oxygen and nutrients to body tissues and removes carbon dioxide and other wastes and heat from the tissues. All systemic arteries branch from the **aorta.** De-

oxygenated blood returns to the heart through the systemic veins. All the veins of the systemic circulation drain into the **superior vena cava** (KĀ-va), **inferior vena cava,** or **coronary sinus,** which in turn empty into the right atrium.

The principal arteries and veins of the systemic circulation are described and illustrated in Exhibits 21.A through 21.L and Figures 21.18 through 21.27 to assist you in learning their names. The blood vessels are organized in the exhibits according to regions of the body. Figure 21.18a shows an overview of the major arteries, and Figure 21.23 shows an overview of the major veins. As you study the various blood vessels in the exhibits, refer to these two figures to see the relationships of the blood vessels under consideration to other regions of the body.

Each of the exhibits contains the following information:

- **An overview.** This information provides a general orientation to the blood vessels under consideration, with emphasis on how the blood vessels are organized into various regions as well as distinguishing and/or interesting features of the blood vessels.

- **Blood vessel names.** Students often have difficulty with the pronunciations and meanings of blood vessels' names. To learn them more easily, study the phonetic pronunciations and word derivations that indicate how blood vessels get their names.

- **Region supplied or drained.** For each artery listed, there is a description of the parts of the body that receive blood from the vessel. For each vein listed, there is a description of the parts of the body that are drained by the vessel.

- **Illustrations and photographs.** The figures that accompany the exhibits contain several elements. Many include illustrations of the blood vessels under consideration and flow diagrams to indicate the patterns of blood distribution or drainage. Cadaver photographs are also included in selected exhibits to provide more realistic views of the blood vessels.

Figure 21.17 Circulatory routes. Long black arrows indicate the systemic circulation (detailed in Exhibits 21.C–21.L), short black arrows the pulmonary circulation (detailed in Figure 21.29), and red arrows the hepatic portal circulation (detailed in Figure 21.28). Refer to Figure 20.8 for details of the coronary circulation, and to Figure 21.30 for details of the fetal circulation.

 Blood vessels are organized into various routes that deliver blood to tissues of the body.

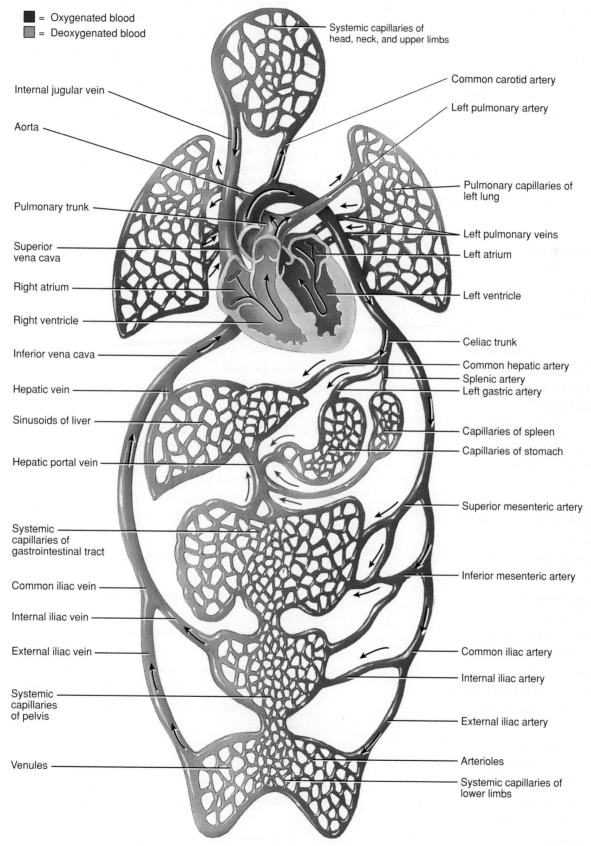

= Oxygenated blood
= Deoxygenated blood

Systemic capillaries of head, neck, and upper limbs
Common carotid artery
Left pulmonary artery
Internal jugular vein
Aorta
Pulmonary capillaries of left lung
Pulmonary trunk
Left pulmonary veins
Superior vena cava
Left atrium
Right atrium
Left ventricle
Right ventricle
Inferior vena cava
Celiac trunk
Common hepatic artery
Hepatic vein
Splenic artery
Left gastric artery
Sinusoids of liver
Capillaries of spleen
Capillaries of stomach
Hepatic portal vein
Systemic capillaries of gastrointestinal tract
Superior mesenteric artery
Common iliac vein
Internal iliac vein
Inferior mesenteric artery
External iliac vein
Common iliac artery
Systemic capillaries of pelvis
Internal iliac artery
External iliac artery
Venules
Arterioles
Systemic capillaries of lower limbs

? **What are the two main circulatory routes?**

EXHIBIT 21.A The Aorta and Its Branches *(Figure 21.18)*

◉ OBJECTIVE

- Identify the four principal divisions of the aorta.
- Locate the major arterial branches arising from each division.

The **aorta** (ā-OR-ta = to lift up) is the largest artery of the body, with a diameter of 2–3 cm (about 1 in.). Its four principal divisions are the ascending aorta, arch of the aorta, thoracic aorta, and abdominal aorta (Figure 21.18). The portion of the aorta that emerges from the left ventricle posterior to the pulmonary trunk is the **ascending aorta.** The beginning of the aorta contains the aortic valve (see Figure 20.4a). The ascending aorta gives off two coronary arteries that supply the myocardium of the heart. Then the ascending aorta arches to the left, forming the **arch of the aorta,** which descends and ends at the level of the intervertebral disc between the fourth and fifth thoracic vertebrae. As

the aorta continues to descend, it lies close to the vertebral bodies and is called the **thoracic aorta.** When the thoracic aorta reaches the bottom of the thorax it passes through the aortic hiatus of the diaphragm to become the **abdominal aorta.** The abdominal aorta descends to the level of the fourth lumbar vertebra where it divides into two **common iliac arteries,** which carry blood to the pelvis and lower limbs. Each division of the aorta gives off arteries that branch into distributing arteries that lead to various organs. Within the organs, the arteries divide into arterioles and then into capillaries that service the systemic tissues (all tissues except the alveoli of the lungs).

✔ CHECKPOINT

What general regions do each of the four principal divisions of the aorta supply?

DIVISION AND BRANCHES	REGION SUPPLIED
ASCENDING AORTA	
Right and left coronary arteries	Heart.
ARCH OF THE AORTA	
Brachiocephalic trunk (brā′-kē-ō-se-FAL-ik)	
Right common carotid artery (ka-ROT-id)	Right side of head and neck.
Right subclavian artery (sub-KLĀ-vē-an)	Right upper limb.
Left common carotid artery	Left side of head and neck.
Left subclavian artery	Left upper limb.
THORACIC AORTA (*THORAC-* = CHEST)	
Pericardial arteries (per-i-KAR-dē-al)	Pericardium.
Bronchial arteries (BRONG-kē-al)	Bronchi of lungs.
Esophageal arteries (e-sof′-a-JĒ-al)	Esophagus.
Mediastinal arteries (mē′-dē-as-TĪ-nal)	Structures in mediastinum.
Posterior intercostal arteries (in′-ter-KOS-tal)	Intercostal and chest muscles.
Subcostal arteries (sub-KOS-tal)	Same as posterior intercostals.
Superior phrenic arteries (FREN-ik)	Superior and posterior surfaces of diaphragm.
ABDOMINAL AORTA	
Inferior phrenic arteries (FREN-ik)	Inferior surface of diaphragm.
Celiac trunk (SĒ-lē-ak)	
Common hepatic artery (he-PAT-ik)	Liver, stomach, duodenum, and pancreas.
Left gastric artery (GAS-trik)	Stomach and esophagus.
Splenic artery (SPLĒN-ik)	Spleen, pancreas, and stomach.
Superior mesenteric artery (MES-en-ter′-ik)	Small intestine, cecum, ascending and transverse colons, and pancreas.
Suprarenal arteries (soo-pra-RĒ-nal)	Adrenal (suprarenal) glands.
Renal arteries (RĒ-nal)	Kidneys.
Gonadal arteries (gō-NAD-al)	
Testicular arteries (tes-TIK-ū-lar)	Testes (male).
Ovarian arteries (ō-VAR-ē-an)	Ovaries (female).
Inferior mesenteric artery	Transverse, descending, and sigmoid colons; rectum.
Common iliac arteries (IL-ē-ak)	
External iliac arteries	Lower limbs.
Internal iliac arteries	Uterus (female), prostate (male), muscles of buttocks, and urinary bladder.

Figure 21.18 Aorta and its principal branches.

All systemic arteries branch from the aorta.

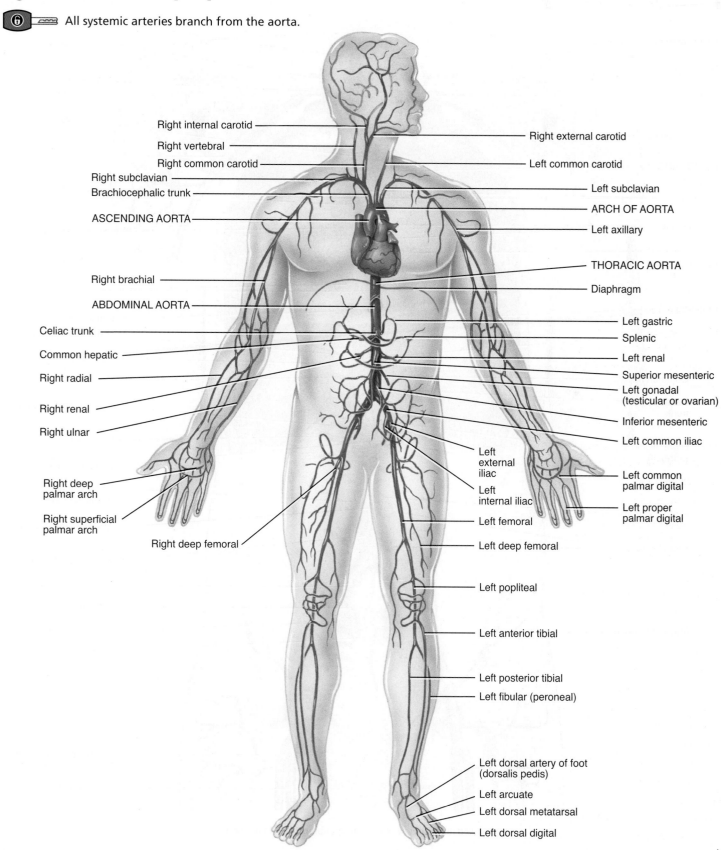

Right internal carotid

Right vertebral

Right common carotid

Right subclavian

Brachiocephalic trunk

ASCENDING AORTA

Right brachial

ABDOMINAL AORTA

Celiac trunk

Common hepatic

Right radial

Right renal

Right ulnar

Right deep palmar arch

Right superficial palmar arch

Right deep femoral

Right external carotid

Left common carotid

Left subclavian

ARCH OF AORTA

Left axillary

THORACIC AORTA

Diaphragm

Left gastric

Splenic

Left renal

Superior mesenteric

Left gonadal (testicular or ovarian)

Inferior mesenteric

Left common iliac

Left external iliac

Left internal iliac

Left common palmar digital

Left proper palmar digital

Left femoral

Left deep femoral

Left popliteal

Left anterior tibial

Left posterior tibial

Left fibular (peroneal)

Left dorsal artery of foot (dorsalis pedis)

Left arcuate

Left dorsal metatarsal

Left dorsal digital

(a) Overall anterior view of the principal branches of aorta

EXHIBIT 21.A CONTINUES ▶

EXHIBIT 21.A 829

EXHIBIT 21.A   The Aorta and Its Branches   *(Figure 21.18)*   CONTINUED

⬛ FIGURE 21.18 CONTINUED ▶

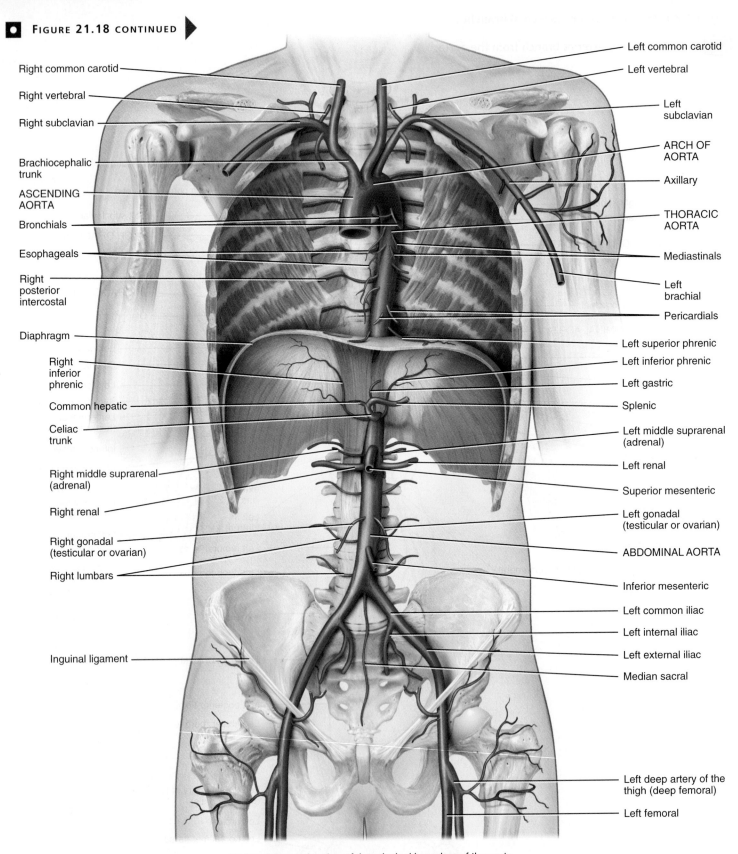

Right common carotid

Right vertebral

Right subclavian

Brachiocephalic trunk

ASCENDING AORTA

Bronchials

Esophageals

Right posterior intercostal

Diaphragm

Right inferior phrenic

Common hepatic

Celiac trunk

Right middle suprarenal (adrenal)

Right renal

Right gonadal (testicular or ovarian)

Right lumbars

Inguinal ligament

Left common carotid

Left vertebral

Left subclavian

ARCH OF AORTA

Axillary

THORACIC AORTA

Mediastinals

Left brachial

Pericardials

Left superior phrenic

Left inferior phrenic

Left gastric

Splenic

Left middle suprarenal (adrenal)

Left renal

Superior mesenteric

Left gonadal (testicular or ovarian)

ABDOMINAL AORTA

Inferior mesenteric

Left common iliac

Left internal iliac

Left external iliac

Median sacral

Left deep artery of the thigh (deep femoral)

Left femoral

(b) Detailed anterior view of the principal branches of the aorta

❓ What are the four subdivisions of the aorta?

830   EXHIBIT 21.A

EXHIBIT 21.B   Ascending Aorta

◉ OBJECTIVE

• Identify the two primary arterial branches of the ascending aorta.

The **ascending aorta** is about 5 cm (2 in.) in length and begins at the aortic valve (see Figure 20.8). It is directed superiorly, slightly anteriorly, and to the right. It ends at the level of the sternal angle, where it becomes the arch of the aorta. The beginning of the ascending aorta is posterior to the pulmonary trunk and right auricle; the right pulmonary artery is posterior to it. At its origin, the ascending aorta contains three dilations called *aortic sinuses.* Two of these, the right and left sinuses, give rise to the right and left coronary arteries, respectively.

The right and left **coronary arteries** (*coron-* = crown) arise from the ascending aorta just superior to the aortic valve (see Figure 20.8). They

form a crownlike ring around the heart, giving off branches to the atrial and ventricular myocardium. The **posterior interventricular branch** (in-ter-ven-TRIK-ū-lar; *inter-* = between) of the right coronary artery supplies both ventricles, and the **marginal branch** supplies the right ventricle. The **anterior interventricular branch,** also known as the **left anterior descending (LAD) branch,** of the left coronary artery supplies both ventricles, and the **circumflex branch** (SER-kum-flex; *circum-* = around; *-flex* = to bend) supplies the left atrium and left ventricle.

✔ CHECKPOINT

Which branches of the coronary arteries supply the left ventricle? Why does the left ventricle have such an extensive arterial blood supply?

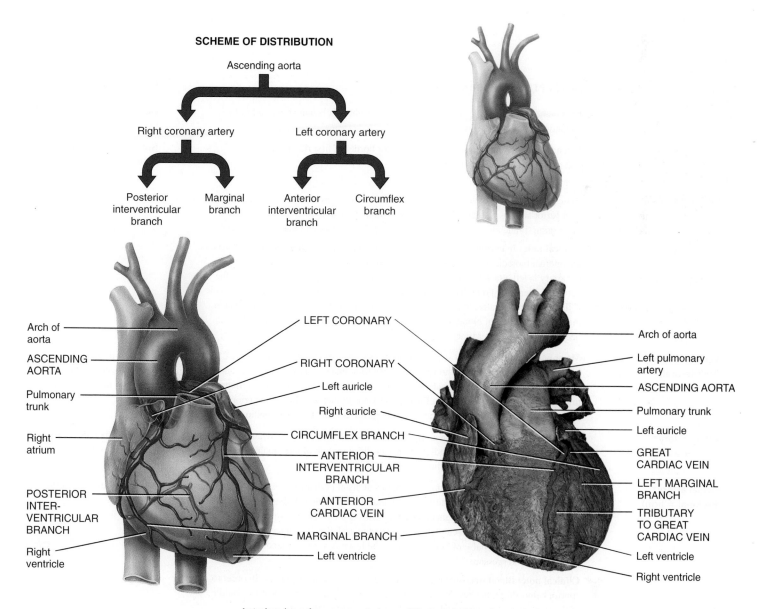

Anterior view of coronary arteries and their major branches

EXHIBIT 21.B   **831**

EXHIBIT 21.C

The Arch of the Aorta *(Figure 21.19)*

◉ **OBJECTIVE**

- Identify the three principal arteries that branch from the arch of the aorta.

The **arch of the aorta** is 4–5 cm (almost 2 in.) in length and is the continuation of the ascending aorta. It emerges from the pericardium posterior to the sternum at the level of the sternal angle (Figure 21.19). The arch of the aorta is directed superiorly and posteriorly to the left and then inferiorly; it ends at the intervertebral disc between the fourth and fifth thoracic vertebrae, where it becomes the thoracic aorta. Three major arteries branch from the superior aspect of the arch of the aorta: the brachiocephalic trunk, the left common carotid, and the left subclavian. The first and largest branch from the arch of the aorta is the **brachiocephalic trunk** (brā′-kē-ō-se-FAL-ik; *brachio-* = arm; *-cephalic* = head). It extends superiorly, bending slightly to the right, and divides at the right sternoclavicular joint to form the right subclavian artery and right common carotid artery. The second branch from the arch of the aorta is the **left common carotid artery** (ka-ROT-id), which divides into the same branches with the same names as the right common carotid artery. The third branch from the arch of the aorta is the **left subclavian artery** (sub-KLĀ-vē-an), which distributes blood to the left vertebral artery and vessels of the left upper limb. Arteries branching from the left subclavian artery are similar in distribution and name to those branching from the right subclavian artery.

✔ **CHECKPOINT**

What general regions do the arteries that arise from the arch of the aorta supply?

BRANCH	DESCRIPTION AND BRANCHES	REGIONS SUPPLIED
Brachiocephalic	First branch of arch of the aorta; divides to form right subclavian artery and right common carotid artery (Figure 21.19a).	Head, neck, upper limb, and thoracic wall.
Right subclavian artery* (sub-KLĀ-ve-an)	Extends from brachiocephalic artery to inferior border of first rib; gives rise to a number of branches at base of neck.	Brain, spinal cord, neck, shoulder, thoracic muscle wall, and scapular muscles.
Internal thoracic (mammary) artery (thor-AS-ik; *thorac-* = chest)	Arises from first part of subclavian artery and descends posterior to costal cartilages of superior six ribs just lateral to sternum; terminates at sixth intercostal space by bifurcating (branching into two arteries) and sends branches into intercostal spaces. ⚕ **Clinical note: In coronary artery bypass grafting,** if only a single vessel is obstructed, the internal thoracic (usually the left) is used to create the bypass. The upper end of the artery is left attached to the subclavian artery and the cut end is connected to the coronary artery at a point distal to the blockage. The lower end of the internal thoracic is tied off. Artery grafts are preferred over vein grafts because arteries can withstand the greater pressure of blood flowing through coronary arteries and are less likely to become obstructed over time.	Anterior thoracic wall.
Vertebral artery (VER-te-bral)	Major branch to brain of right subclavian artery before it passes into axilla (Figure 21.19b); ascends through neck, passes through transverse foramina of cervical vertebrae, and enters skull via foramen magnum to reach inferior surface of brain. Unites with left vertebral artery to form **basilar** (BĀS-i-lar) **artery.** Basilar artery passes along midline of anterior aspect of brain stem and gives off several branches (**posterior cerebral** and **cerebellar arteries**).	Posterior portion of cerebrum, cerebellum, pons, and inner ear.
Axillary artery* (AK-sil-ar-ē = armpit)	Continuation of right subclavian artery into axilla; begins where subclavian artery passes inferior border of first rib and ends as it crosses distal margin of teres major muscle; gives rise to numerous branches in axilla.	Thoracic, shoulder, and scapular muscles and humerus.
Brachial artery* (BRĀ-kē-al = arm)	Continuation of axillary artery into arm; begins at distal border of teres major muscle and terminates by bifurcating into radial and ulnar arteries just distal to bend of elbow; superficial and palpable along medial side of arm. As it descends toward elbow it curves laterally and passes through cubital fossa, a triangular depression anterior to elbow, where you can easily detect pulse of brachial artery and listen to various sounds when taking a person's blood pressure. ⚕ **Clinical note: Blood pressure** is usually measured in the brachial artery. In order to control hemorrhage, the best place to compress the brachial artery is near the middle of the arm where it is superficial and easily pressed against the humerus.	Muscles of arm, humerus, elbow joint.

*This is an example of the practice of giving the same vessel different names as it passes through different regions. See the axillary and brachial arteries.

BRANCH	DESCRIPTION AND BRANCHES	REGIONS SUPPLIED
Radial artery (RĀ-dē-al = radius)	Smaller branch of brachial bifurcation; a direct continuation of brachial artery. Passes along lateral (radial) aspect of forearm and enters wrist where it bifurcates into superficial and deep branches that anastomose with corresponding branches of ulnar artery to form palmar arches of hand. Makes contact with distal end of radius at wrist, where it is covered only by fascia and skin. **Clinical note:** Because of its superficial location at this point, it is a common site for measuring **radial pulse.**	Major blood source to muscles of posterior compartment of forearm.
Ulnar artery (UL-nar = ulna)	Larger branch of brachial artery passes along medial (ulnar) aspect of forearm and then into wrist, where it branches into superficial and deep branches that enter hand. These branches anastomose with corresponding branches of radial artery to form palmar arches of hand.	Major blood source to muscles of anterior compartment of forearm.
Superficial palmar arch (*palma* = palm)	Formed mainly by superficial branch of ulnar artery, with contribution from superficial branch of radial artery; superficial to long flexor tendons of fingers and extends across palm at bases of metacarpals; gives rise to **common palmar digital arteries,** each of which divides into **proper palmar digital arteries.**	Muscles, bones, joints, and skin of palm and fingers.
Deep palmar arch	Arises mainly from deep branch of radial artery, but receives contribution from deep branch of ulnar artery; deep to long flexor tendons of fingers and extends across palm just distal to bases of metacarpals; gives rise to **palmar metacarpal arteries,** which anastomose with common palmar digital arteries from superficial arch.	Muscles, bones, and joints of palm and fingers.
Right common carotid	Begins at bifurcation of brachiocephalic trunk, posterior to right sternoclavicular joint; passes superiorly into neck to supply structures in head (Figure 21.19c); divides into right external and right internal carotid arteries at superior border of larynx (voice box). **Clinical note: Pulse** may be detected in the common carotid artery, just lateral to the larynx. It is convenient to detect a carotid pulse when exercising or when administering cardiopulmonary resuscitation.	Head and neck.
External carotid artery	Begins at superior border of larynx and terminates near temporomandibular joint of parotid gland, where it divides into two branches: superficial temporal and maxillary arteries. **Clinical note:** The **carotid pulse** can be detected in the external carotid artery just anterior to the sternocleidomastoid muscle at the superior border of the larynx.	Major blood source to all structures of head except brain. Supplies skin, connective tissues, muscles, bones, joints, dura and arachnoid mater in head and supplies much of neck anatomy.
Internal carotid artery	Arises from common carotid artery; enters cranial cavity through carotid foramen in temporal bone and emerges in cranial cavity near base of sella turcica of sphenoid bone; gives rise to numerous branches inside cranial cavity and terminates as anterior cerebral arteries. The **anterior cerebral artery** passes forward toward frontal lobe of and cerebrum and **middle cerebral artery** passes laterally between temporal and parietal lobes of cerebrum. Inside cranium (Figure 21.19c), anastomoses of left and right internal carotid arteries via anterior communicating artery between two anterior cerebral arteries, along with internal carotid–basilar artery anastomoses, form an arrangement of blood vessels at base of brain called **cerebral arterial circle (circle of Willis)** (Figure 21.19c). Internal carotid–basilar anastomosis occurs where **posterior communicating arteries** arising from internal carotid artery anastomose with posterior cerebral arteries from basilar artery to link internal carotid blood supply with vertebral blood supply. Cerebral arterial circle equalizes blood pressure to brain and provides alternate routes for blood flow to brain, should arteries become damaged.	Eyeball and other orbital structures, ear, and parts of nose and nasal cavity. Frontal, temporal, parietal lobes of the cerebrum of brain, pituitary gland, and pia mater.
Left common carotid artery	Arises as second branch of arch of the aorta and ascends through mediastinum to enter neck deep to clavicle, then follows similar path to right common carotid artery.	Distribution similar to right common carotid artery.
Left subclavian artery	Arises as third and final branch of arch of the aorta; passes superior and lateral through mediastinum and deep to clavicle at base of neck as it courses toward upper limb; has similar course to right subclavian artery after leaving mediastinum.	Distribution similar to right subclavian artery.

EXHIBIT 21.C CONTINUES

EXHIBIT 21.C **833**

EXHIBIT 21.C   The Arch of the Aorta   *(Figure 21.19)*   CONTINUED

SCHEME OF DISTRIBUTION

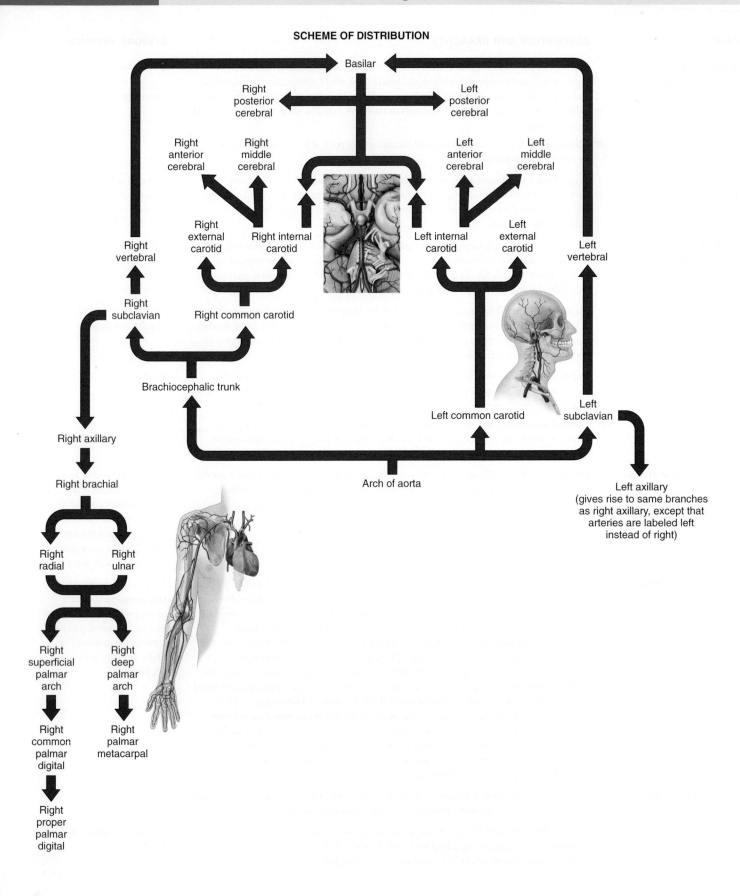

Figure 21.19 Arch of the aorta and its branches. Note in (c) the arteries that constitute the cerebral arterial circle (circle of Willis).

The arch of the aorta ends at the level of the intervertebral disc between the fourth and fifth thoracic vertebrae.

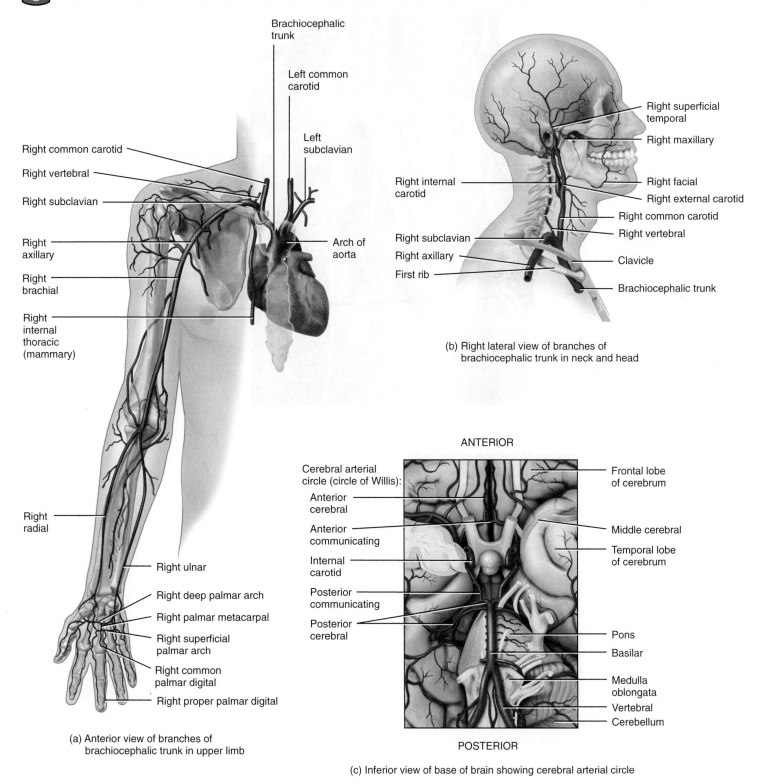

(a) Anterior view of branches of brachiocephalic trunk in upper limb

(b) Right lateral view of branches of brachiocephalic trunk in neck and head

(c) Inferior view of base of brain showing cerebral arterial circle

EXHIBIT 21.C CONTINUES ▶

EXHIBIT 21.C **835**

EXHIBIT 21.C   The Arch of the Aorta   *(Figure 21.19)*   CONTINUED

● **FIGURE 21.19** CONTINUED ▶

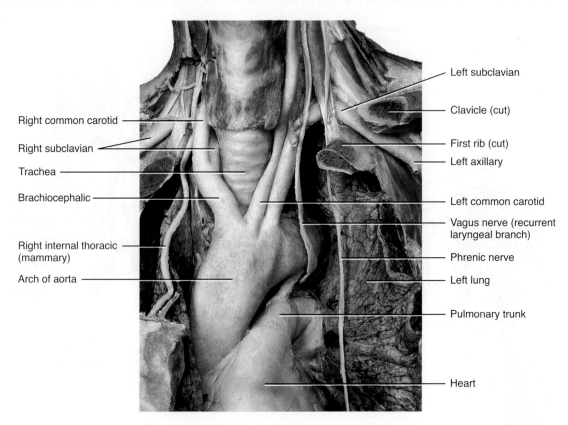

Right common carotid

Right subclavian

Trachea

Brachiocephalic

Right internal thoracic
(mammary)

Arch of aorta

Left subclavian

Clavicle (cut)

First rib (cut)

Left axillary

Left common carotid

Vagus nerve (recurrent
laryngeal branch)

Phrenic nerve

Left lung

Pulmonary trunk

Heart

(d) Anterior view of branches of arch of aorta

? **What are the three major branches of the arch of the aorta, in order of their origination?**

EXHIBIT 21.D | Thoracic Aorta *(Figure 21.20)*

◉ OBJECTIVE

• Identify the visceral and parietal branches of the thoracic aorta.

The **thoracic aorta** is about 20 cm (8 in.) long and is a continuation of the arch of the aorta (Figure 21.20). It begins at the level of the intervertebral disc between the fourth and fifth thoracic vertebrae, where it lies to the left of the vertebral column. As it descends, it moves closer to the midline and extends through an opening in the diaphragm (aortic hiatus), which is located anterior to the vertebral column at the level of the intervertebral disc between the twelfth thoracic and first lumbar vertebrae.

Along its course, the thoracic aorta sends off numerous small arteries, **visceral branches** (VIS-er-al) to viscera, and **parietal branches** (pa-RĪ-e-tal) to body wall structures.

✔ CHECKPOINT

What general regions do the visceral and parietal branches of the thoracic aorta supply?

BRANCH	DESCRIPTION AND BRANCHES	REGIONS SUPPLIED
VISCERAL BRANCHES		
Pericardial arteries (per'-i-KAR-dē-al; *peri-* = around; *-cardia* = heart)	Two to three small arteries that arise from variable levels of thoracic aorta and pass forward to pericardial sac surrounding heart.	Tissues of pericardial sac.
Bronchial arteries (BRONG-kē-al = windpipe)	Arise from thoracic aorta or one of its branches. Right bronchial artery typically arises from third posterior intercostal artery; two left bronchial arteries arise from upper end of thoracic aorta. All follow bronchial tree into lungs.	Supply tissues of bronchial tree and surrounding lung tissue down to level of alveolar ducts.
Esophageal arteries (e-sof'-a-JĒ-al; *eso-* = to carry; *phage* = food)	Four to five arteries that arise from anterior surface of thoracic aorta and pass forward to branch onto esophagus.	All tissues of esophagus.
Mediastinal arteries (mē-dē-as-TĪ-nal)	Arise from various points on thoracic aorta.	Assorted tissues within mediastinum, primarily connective tissue and lymph nodes.
PARIETAL BRANCHES		
Posterior intercostal arteries (in'ter-KOS-tal; *inter-* = between; *-costa* = rib)	Typically, nine pairs of arteries that arise from posterolateral aspect on each side of thoracic aorta. Each passes laterally and then anteriorly through intercostal space, where they will eventually anastomose with anterior branches from internal thoracic arteries.	Skin, muscles, and ribs of thoracic wall. Thoracic vertebrae, meninges, and spinal cord. Mammary glands.
Subcostal arteries (sub-KOS-tal; *sub-* = under)	The lowest segmental branches of thoracic aorta; one on each side passes into thoracic body wall inferior to twelfth rib and courses forward into upper abdominal region of body wall.	Skin, muscles, and ribs. Twelfth thoracic vertebra, meninges, and spinal cord.
Superior phrenic arteries (FREN-ik = pertaining to diaphragm)	Arise from lower end of thoracic aorta and pass onto superior surface of diaphragm.	Diaphragm muscle and pleura covering diaphragm.

SCHEME OF DISTRIBUTION

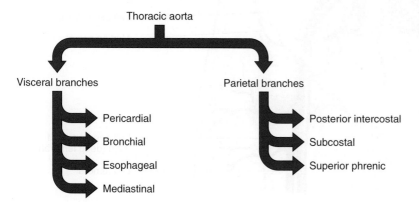

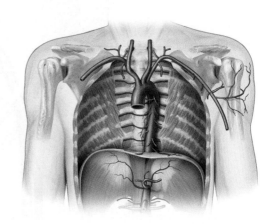

EXHIBIT 21.D CONTINUES ▶

EXHIBIT 21.D **837**

EXHIBIT 21.D Thoracic Aorta *(Figure 21.20)* CONTINUED

Figure 21.20 Thoracic aorta and abdominal aorta and their principal branches.

The thoracic aorta is the continuation of the ascending aorta.

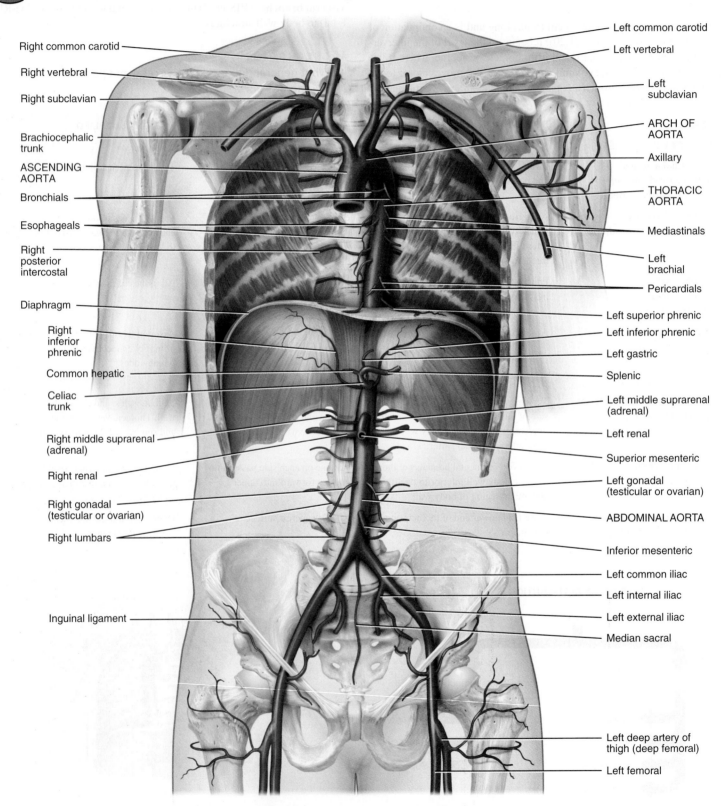

Detailed anterior view of principal branches of aorta

Where does the thoracic aorta begin?

EXHIBIT 21.E

Abdominal Aorta *(Figure 21.21)*

OBJECTIVE

• Identify the visceral and parietal branches of the abdominal aorta.

The **abdominal aorta** is the continuation of the thoracic aorta after it passes through the diaphragm (Figure 21.21). It begins at the aortic hiatus in the diaphragm and ends at about the level of the fourth lumbar

BRANCH	DESCRIPTION AND BRANCHES	REGIONS SUPPLIED
UNPAIRED VISCERAL BRANCHES		
Celiac trunk (artery) (SĒ-lē-ak)	First visceral branch of aorta inferior to diaphragm; arises from abdominal aorta at level of twelfth thoracic vertebra as aorta passes through hiatus in diaphragm; divides into three branches: left gastric, splenic, and common hepatic arteries (Figure 21.21a).	Supplies all organs of gastrointestinal tract that arise from embryonic foregut, that is, from abdominal part of esophagus to duodenum, and also spleen.
	1. Left gastric artery (GAS-trik = stomach). Smallest of three celiac branches arises superiorly to left toward esophagus and then turns to follow lesser curvature of stomach. On lesser curvature of stomach it anastomoses with right gastric artery.	Abdominal part of esophagus and lesser curvature of stomach.
	2. Splenic artery (SPLEN-ik = spleen). Largest branch of celiac trunk arises from left side of celiac trunk distal to left gastric artery, and passes horizontally to left along pancreas. Before reaching spleen, it gives rise to three named arteries:	Spleen, pancreas, fundus and greater curvature of stomach, and greater omentum.
	• **Pancreatic arteries** (pan-krē-AT-ik), a series of small arteries that arise from splenic and descend into tissue of pancreas.	Pancreas.
	• **Left gastroepiploic (gastro-omental) artery** (gas′-trō-ep′-i-PLŌ-ik; *epiplo-* = omentum) arises from terminal end of splenic artery and passes from left to right along greater curvature of stomach.	Greater curvature of stomach and greater omentum.
	• **Short gastric arteries** arise from terminal end of splenic artery and pass onto fundus of stomach.	Fundus of stomach.
	3. Common hepatic artery (he-PAT-ik = liver). Intermediate in size between left gastric and splenic arteries; arises from right side of celiac trunk and gives rise to three arteries:	Liver, gallbladder, lesser omentum, stomach, pancreas, and duodenum.
	• **Proper hepatic artery** branches from common hepatic artery and ascends along bile ducts into liver and gallbladder.	Liver, gallbladder, and lesser omentum.
	• **Right gastric artery** arises from common hepatic artery and curves back to left along lesser curvature of stomach, where it anastomoses with left gastric artery.	Stomach and lesser omentum.
	• **Gastroduodenal artery** (gas′-trō-doo′-ō-DĒ-nal) passes inferiorly toward stomach and duodenum and sends branches along greater curvature of stomach.	Stomach, duodenum, and pancreas.
Superior mesenteric artery (MES-en-ter′-ik; *meso-* = middle; *-enteric* = pertaining to intestines)	Arises from anterior surface of abdominal aorta about 1 cm inferior to celiac trunk at level of first lumbar vertebra (Figure 21.21b); extends inferiorly and anteriorly between layers of mesentery (portion of peritoneum that attaches small intestine to posterior abdominal wall). It anastomoses extensively and has five branches:	Supplies all organs of gastrointestinal tract that arise from embryonic midgut, that is, from duodenum to transverse colon.
	1. Inferior pancreaticoduodenal artery (pan′-krē-at′-i-kō-doo′-ō-DĒ-nal) passes superiorly and to right toward head of pancreas and duodenum.	Pancreas and duodenum.
	2. Jejunal (je-JOO-nal) and **ileal arteries** (IL-ē-al) spread through mesentery to pass to loops of jejunum and ileum (small intestine).	Jejunum and ileum, which is majority of small intestine.
	3. Ileocolic artery (il′-ē-ō-KOL-ik) passes inferiorly and laterally toward right side toward terminal part of ileum, cecum, appendix, and first part of ascending colon.	Terminal part of ileum, cecum, appendix, and first part of ascending colon.
	4. Right colic artery (KOL-ik) passes laterally to right toward ascending colon.	Ascending colon and first part of transverse colon.
	5. Middle colic artery ascends slightly to right toward transverse colon.	Most of transverse colon.
Inferior mesenteric artery	Arises from anterior aspect of abdominal aorta at level of third lumbar vertebra and then passes inferiorly to left of aorta (Figure 21.21c). It anastomoses extensively and has three branches:	Supplies all organs of gastrointestinal tract that arise from embryonic hindgut from transverse colon to rectum.
	1. Left colic artery ascends laterally to left toward distal end of transverse colon and descending colon.	End of transverse colon and descending colon.
	2. Sigmoid arteries (SIG-moyd) descend laterally to left toward sigmoid colon.	Sigmoid colon.
	3. Superior rectal artery (REK-tal) passes inferiorly to superior part of rectum.	Upper part of rectum.

EXHIBIT 21.E CONTINUES ▶

EXHIBIT 21.E **839**

EXHIBIT 21.E Abdominal Aorta *(Figure 21.21)* CONTINUED

vertebra, where it divides into the right and left common iliac arteries. The abdominal aorta lies anterior to the vertebral column.

As with the thoracic aorta, the abdominal aorta gives off **visceral** and **parietal branches.** The unpaired visceral branches arise from the anterior surface of the aorta and include the **celiac trunk** and the **superior mesenteric** and **inferior mesenteric arteries** (see Figure 21.20).

The paired visceral branches arise from the lateral surfaces of the aorta and include the **suprarenal, renal,** and **gonadal arteries.** The lone unpaired parietal branch is the **median sacral artery.** The paired parietal branches arise from the posterolateral surfaces of the aorta and include the **inferior phrenic** and **lumbar arteries.**

✔ **CHECKPOINT**

Name the paired visceral and parietal branches and the unpaired visceral and parietal branches of the abdominal aorta, and indicate the general regions they supply.

BRANCH	DESCRIPTION AND BRANCHES	REGIONS SUPPLIED
PAIRED VISCERAL BRANCHES		
Suprarenal arteries (soo'-pra-RĒ-nal; *supra-* = above; *-ren-* = kidney)	There are typically three pairs (superior, middle, and inferior), but only middle pair originates directly from abdominal aorta (see Figure 21.20). Middle suprarenal arteries arise from abdominal aorta at level of first lumbar vertebra at or superior to renal arteries. Superior suprarenal arteries arise from inferior phrenic arteries, and inferior suprarenal arteries originate from renal arteries.	Suprarenal (adrenal) glands.
Renal arteries (RĒ-nal; *ren* = kidney)	Right and left **renal arteries** usually arise from lateral aspects of abdominal aorta at superior border of second lumbar vertebra, about 1 cm inferior to superior mesenteric artery (see Figure 21.20). Right renal artery, which is longer than left, arises slightly lower than left and passes posterior to right renal vein and inferior vena cava. Left renal artery is posterior to left renal vein and is crossed by inferior mesenteric vein.	All tissues of kidneys.
Gonadal (gō-NAD-al; *gon-* = seed) **arteries** [testicular (tes-TIK-ū-lar) or ovarian (ō-VAR-ē-an)]	Arise from anterior aspect of abdominal aorta at level of second lumbar vertebra just inferior to renal arteries (see Figure 21.20). In males, gonadal arteries are specifically referred to as **testicular arteries.** They descend along posterior abdominal wall to pass through inguinal canal and descend into scrotum. In females, gonadal arteries are called **ovarian arteries.** They are much shorter than testicular arteries and remain within abdominal cavity.	Males: testis, epididymis, ductus deferens, and ureters. Females: ovaries, uterine (fallopian) tubes, and ureters.
UNPAIRED PARIETAL BRANCH		
Median sacral artery (SĀ-kral = pertaining to sacrum)	Arises from posterior surface of abdominal aorta about 1 cm superior to bifurcation (division into two branches) of aorta into right and left common iliac arteries (see Figure 21.20).	Sacrum, coccyx, sacral spinal nerves, and piriformis muscle.
PAIRED PARIETAL BRANCH		
Inferior phrenic arteries (FREN-ik = pertaining to diaphragm)	First paired branches of abdominal aorta; arise immediately superior to origin of celiac trunk (see Figure 21.20). (They may also arise from renal arteries.)	Diaphragm and suprarenal (adrenal) glands.
Lumbar arteries (LUM-bar = pertaining to loin)	Four pairs arise from posterolateral surface of abdominal aorta in a way similar to posterior intercostal arteries of thorax (see Figure 21.20); pass laterally into abdominal muscle wall and curve toward anterior aspect of wall.	Lumbar vertebrae, spinal cord and meninges, skin and muscles of posterior and lateral part of abdominal wall.

SCHEME OF DISTRIBUTION

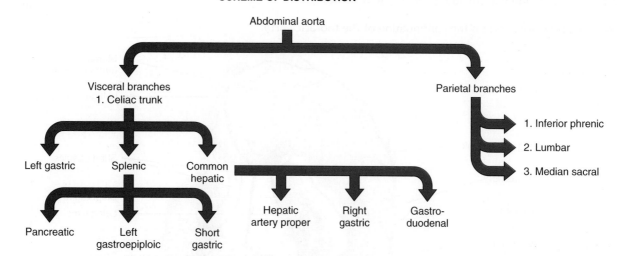

Abdominal aorta

Visceral branches
1. Celiac trunk

Left gastric Splenic Common hepatic

Pancreatic Left gastroepiploic Short gastric

Hepatic artery proper Right gastric Gastro-duodenal

Parietal branches

1. Inferior phrenic
2. Lumbar
3. Median sacral

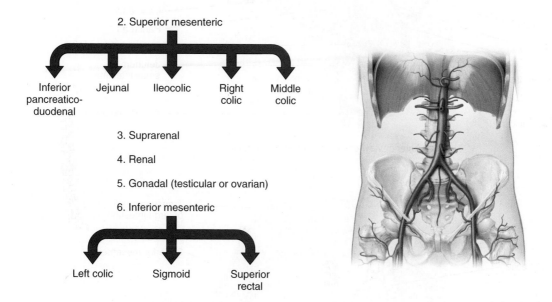

2. Superior mesenteric

Inferior pancreatico-duodenal Jejunal Ileocolic Right colic Middle colic

3. Suprarenal

4. Renal

5. Gonadal (testicular or ovarian)

6. Inferior mesenteric

Left colic Sigmoid Superior rectal

EXHIBIT 21.E CONTINUES ▶

EXHIBIT 21.E **841**

Figure 21.21   **Abdominal aorta and its principal branches.**

The abdominal aorta is the continuation of the thoracic aorta.

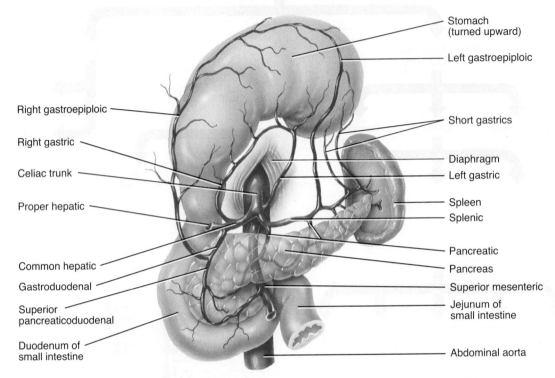

Stomach (turned upward)

Left gastroepiploic

Right gastroepiploic

Short gastrics

Right gastric

Diaphragm

Celiac trunk

Left gastric

Proper hepatic

Spleen

Splenic

Common hepatic

Pancreatic

Gastroduodenal

Pancreas

Superior pancreaticoduodenal

Superior mesenteric

Jejunum of small intestine

Duodenum of small intestine

Abdominal aorta

(a) Anterior view of celiac trunk and its branches

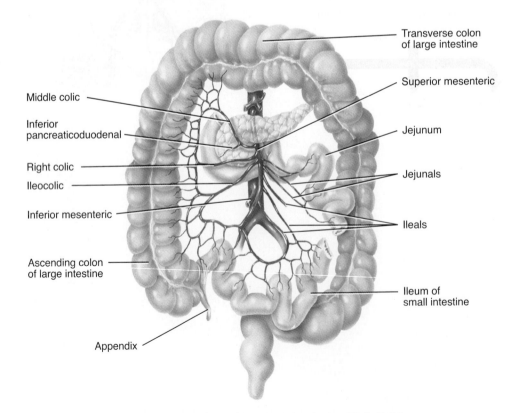

Transverse colon of large intestine

Middle colic

Superior mesenteric

Inferior pancreaticoduodenal

Jejunum

Right colic

Jejunals

Ileocolic

Inferior mesenteric

Ileals

Ascending colon of large intestine

Ileum of small intestine

Appendix

(b) Anterior view of superior mesenteric artery and its branches

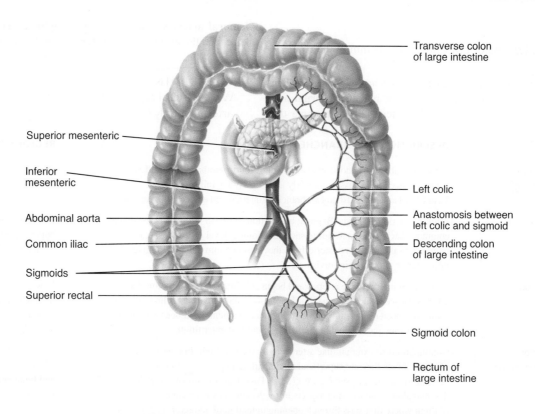

Transverse colon
of large intestine

Superior mesenteric

Inferior
mesenteric

Abdominal aorta

Common iliac

Sigmoids

Superior rectal

Left colic

Anastomosis between
left colic and sigmoid

Descending colon
of large intestine

Sigmoid colon

Rectum of
large intestine

(c) Anterior view of inferior mesenteric artery and its branches

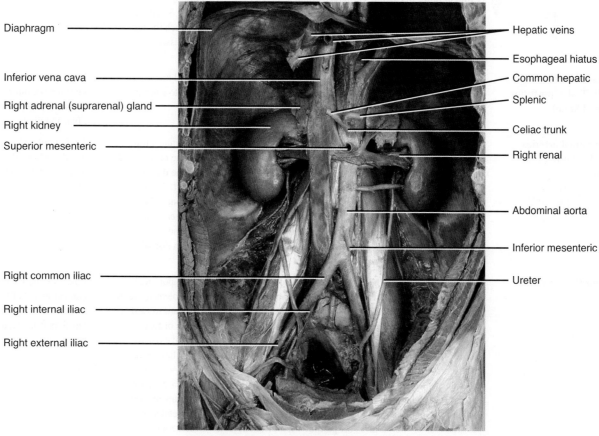

Diaphragm

Inferior vena cava

Right adrenal (suprarenal) gland

Right kidney

Superior mesenteric

Right common iliac

Right internal iliac

Right external iliac

Hepatic veins

Esophageal hiatus

Common hepatic

Splenic

Celiac trunk

Right renal

Abdominal aorta

Inferior mesenteric

Ureter

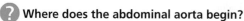

Where does the abdominal aorta begin?

(d) Anterior view of arteries of abdomen and pelvis

EXHIBIT 21.E **843**

EXHIBIT 21.F

Arteries of the Pelvis and Lower Limbs   *(Figure 21.22)*

■ **OBJECTIVE**

- Identify the two major branches of the common iliac arteries.

The abdominal aorta ends by dividing into the right and left **common iliac arteries** (Figure 21.22). These, in turn, divide into the **internal** and **external iliac arteries.** In sequence, the external iliacs become the femoral arteries in the thighs, the **popliteal arteries** posterior to the knee, and the **anterior** and **posterior tibial arteries** in the legs.

✔ **CHECKPOINT**

What general regions do the internal and external iliac arteries supply?

BRANCH	DESCRIPTION AND BRANCHES	REGIONS SUPPLIED
Common iliac arteries (IL-ē-ak = pertaining to ilium)	Arise from abdominal aorta at about level of fourth lumbar vertebra. Each common iliac artery passes inferiorly and slightly laterally for about 5 cm (2 in.) and gives rise to two branches: internal and external iliac arteries.	Pelvic muscle wall, pelvic organs, external genitals, and lower limbs.
Internal iliac arteries	Primary arteries of pelvis. Begin at bifuraction (division into two branches) of common iliac arteries anterior to sacroiliac joint at level of lumbosacral intervertebral disc. Pass posteriorly as they descend into pelvis and divide into anterior and posterior divisions.	Pelvic muscle wall, pelvic organs, buttocks, external genitals, and medial muscles of thigh.
External iliac arteries	Larger than internal iliac arteries and begin at bifurcation of common iliac arteries. Descend along medial border of psoas major muscles following pelvic brim, pass posterior to midportion of inguinal ligaments, and become femoral arteries as they pass beneath inguinal ligament and enter thigh.	Lower abdominal wall, cremaster muscle in males and round ligament of uterus in females, and lower limb.
Femoral arteries (FEM-o-ral=pertaining to thigh)	Continuations of external iliac arteries as they enter thigh. In *femoral triangle* of upper thighs they are superficial along with femoral vein and nerve and deep inguinal lymph nodes (see Figure 11.20a). Pass beneath sartorius muscle as they descend along anteromedial aspects of thighs and follow its course to distal end of thigh where they pass through opening in tendon of adductor magnus muscle to end at posterior aspect of knee, where they become popliteal arteries. **Clinical note:** In **cardiac catheterization,** a catheter is inserted through a blood vessel and advanced into the major vessels to access a heart chamber. A catheter often contains a measuring instrument or other device at its tip. To reach the left side of the heart, the catheter is inserted into the femoral artery and passed into the aorta to the coronary arteries or heart chamber.	Muscles of thigh (quadriceps, adductors, and hamstrings), femur, and ligaments and tendons around knee joint.
Popliteal arteries (pop'-li-TĒ-al = posterior surface of knee)	Continuation of femoral arteries through popliteal fossa (space behind knee). Descend to inferior border of popliteus muscles, where they divide into anterior and posterior tibial arteries.	Muscles of distal thigh, skin of knee region, muscles of proximal leg, knee joint, femur, patella, tibia, and fibula.
Anterior tibial arteries (TIB-ē-al = pertaining to shin)	Descend from bifurcation of popliteal arteries at distal border of popliteus muscles. Smaller than posterior tibial arteries; pass over interosseous membrane of tibia and fibula to descend through anterior muscle compartment of leg; become **dorsal arteries of foot (dorsalis pedis arteries)** at ankles. On dorsum of feet, dorsal arteries of foot give off transverse branch at first medial cuneiform bone called **arcuate arteries** (*arcuat-* = bowed) that run laterally over bases of metatarsals. From arcuate arteries branch **dorsal metatarsal arteries,** which course along metatarsal bones. Dorsal metatarsal arteries terminate by dividing into **dorsal digital arteries,** which pass into toes.	Tibia, fibula, anterior muscles of leg, dorsal muscles of foot, tarsal bones, metatarsal bones, and phalanges.
Posterior tibial arteries	Direct continuations of popliteal arteries, descend from bifurcation of popliteal arteries. Pass down posterior muscular compartment of legs deep to soleus muscles. Pass posterior to medial malleolus at distal end of leg and curve forward toward plantar surface of feet; pass deep to flexor retinaculum on medial side of feet and terminate by branching into medial and lateral plantar arteries. Give rise to **fibular (peroneal) arteries** in middle of leg, which course laterally as they descend into lateral compartment of leg. **Medial plantar arteries** (PLAN-tar = sole) pass along medial side of sole and **lateral plantar arteries** angle toward lateral side of sole and unite with branch of dorsal arteries of foot to form **plantar arch.** Arch begins at base of fifth metatarsal and extends medially across metatarsals. As arch crosses foot, it gives off **plantar metatarsal arteries,** which course along plantar surface of metatarsal bones. These arteries terminate by dividing into **plantar digital arteries** that pass into toes.	Posterior and lateral muscle compartments of leg, plantar muscles of foot, tibia, fibula, tarsal, metatarsal, and phalangeal bones.

SCHEME OF DISTRIBUTION

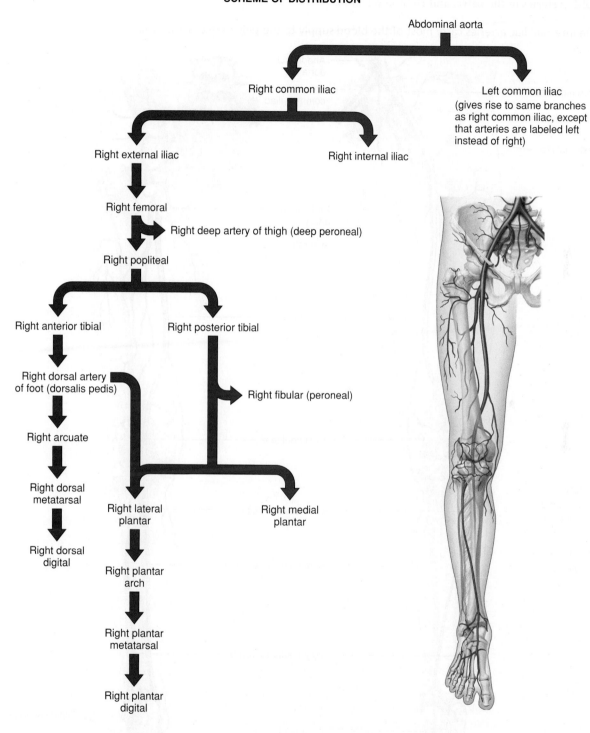

Abdominal aorta

Right common iliac

Left common iliac
(gives rise to same branches
as right common iliac, except
that arteries are labeled left
instead of right)

Right external iliac

Right internal iliac

Right femoral

Right deep artery of thigh (deep peroneal)

Right popliteal

Right anterior tibial

Right posterior tibial

Right dorsal artery
of foot (dorsalis pedis)

Right fibular (peroneal)

Right arcuate

Right dorsal
metatarsal

Right lateral
plantar

Right medial
plantar

Right dorsal
digital

Right plantar
arch

Right plantar
metatarsal

Right plantar
digital

EXHIBIT 21.F CONTINUES ▶

EXHIBIT 21.F **845**

Figure 21.22 **Arteries of the pelvis and right lower limb.**

The internal iliac arteries carry most of the blood supply to the pelvic viscera and wall.

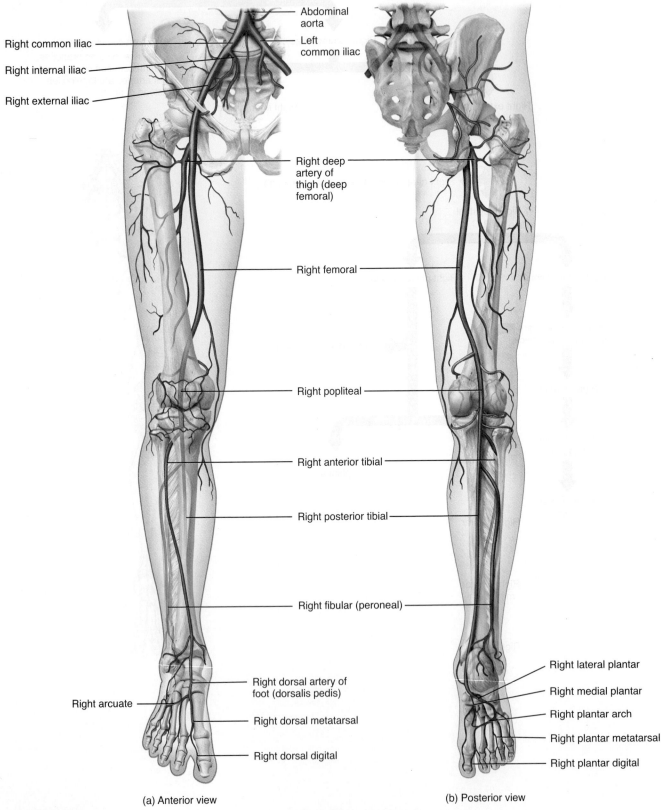

(a) Anterior view

(b) Posterior view

At what point does the abdominal aorta divide into the common iliac arteries?

EXHIBIT 21.G Veins of the Systemic Circulation *(Figure 21.23)*

⦿ **OBJECTIVE**

- Identify the three systemic veins that return deoxygenated blood to the heart.

As you have already learned, arteries distribute blood from the heart to various parts of the body, and veins drain blood away from the various parts and return the blood to the heart. In general, arteries are deep; veins may be superficial or deep. Superficial veins are located just beneath the skin and can be seen easily. Because there are no large superficial arteries, the names of superficial veins do not correspond to those of arteries. Superficial veins are clinically important as sites for withdrawing blood or giving injections. Deep veins generally travel alongside arteries and usually bear the same name. Arteries usually follow definite pathways; veins are more difficult to follow because they connect in irregular networks in which many tributaries merge to form a large vein. Although only one systemic artery, the aorta, takes oxygenated blood away from the heart (left ventricle), three systemic veins, the **coronary sinus, superior vena cava,** and **inferior vena cava,** return deoxygenated blood to the heart (right atrium) (Figure 21.23). The coronary sinus receives blood from the cardiac veins; the superior vena cava receives blood from other veins superior to the diaphragm, except the air sacs (alveoli) of the lungs; the inferior vena cava receives blood from veins inferior to the diaphragm.

✔**CHECKPOINT**

What are the three tributaries of the coronary sinus?

VEINS	DESCRIPTION AND TRIBUTARIES	REGIONS DRAINED
Coronary sinus (KOR-ō-nar-ē; *corona* = crown)	Main vein of heart; receives almost all venous blood from myocardium; located in coronary sulcus (see Figure 20.3c) on posterior aspect of heart and opens into right atrium between orifice of inferior vena cava and tricuspid valve. Wide venous channel into which three veins drain. Receives **great cardiac vein** (from anterior interventricular sulcus) into its left end, and **middle cardiac vein** (from posterior interventricular sulcus) and **small cardiac vein** into its right end. Several **anterior cardiac veins** drain directly into right atrium.	All tissues of heart.
Superior vena cava (SVC) (VĒ-na KĀ-va; *vena* = vein; *cava* = cavelike)	About 7.5 cm (3 in.) long and 2 cm (1 in.) in diameter; empties its blood into superior part of right atrium. Begins posterior to right first costal cartilage by union of right and left brachiocephalic veins and ends at level of right third costal cartilage, where it enters right atrium.	Head, neck, upper limbs, and thorax.
Inferior vena cava (IVC)	Largest vein in body, about 3.5 cm (1.4 in.) in diameter. Begins anterior to fifth lumbar vertebra by union of common iliac veins, ascends behind peritoneum to right of midline, pierces caval opening of diaphragm at level of eighth thoracic vertebra, and enters inferior part of right atrium ⚕ **Clinical note:** The inferior vena cava is commonly **compressed during the later stages of pregnancy** by the enlarging uterus, producing edema of the ankles and feet and temporary varicose veins.	Abdomen, pelvis, and lower limbs.

EXHIBIT 21.G CONTINUES ▶

EXHIBIT 21.G **847**

EXHIBIT 21.G   Veins of the Systemic Circulation   *(Figure 21.23)*   CONTINUED

Figure 21.23 Principal veins.

Deoxygenated blood returns to the heart via the superior vena cava, inferior vena cava, and coronary sinus.

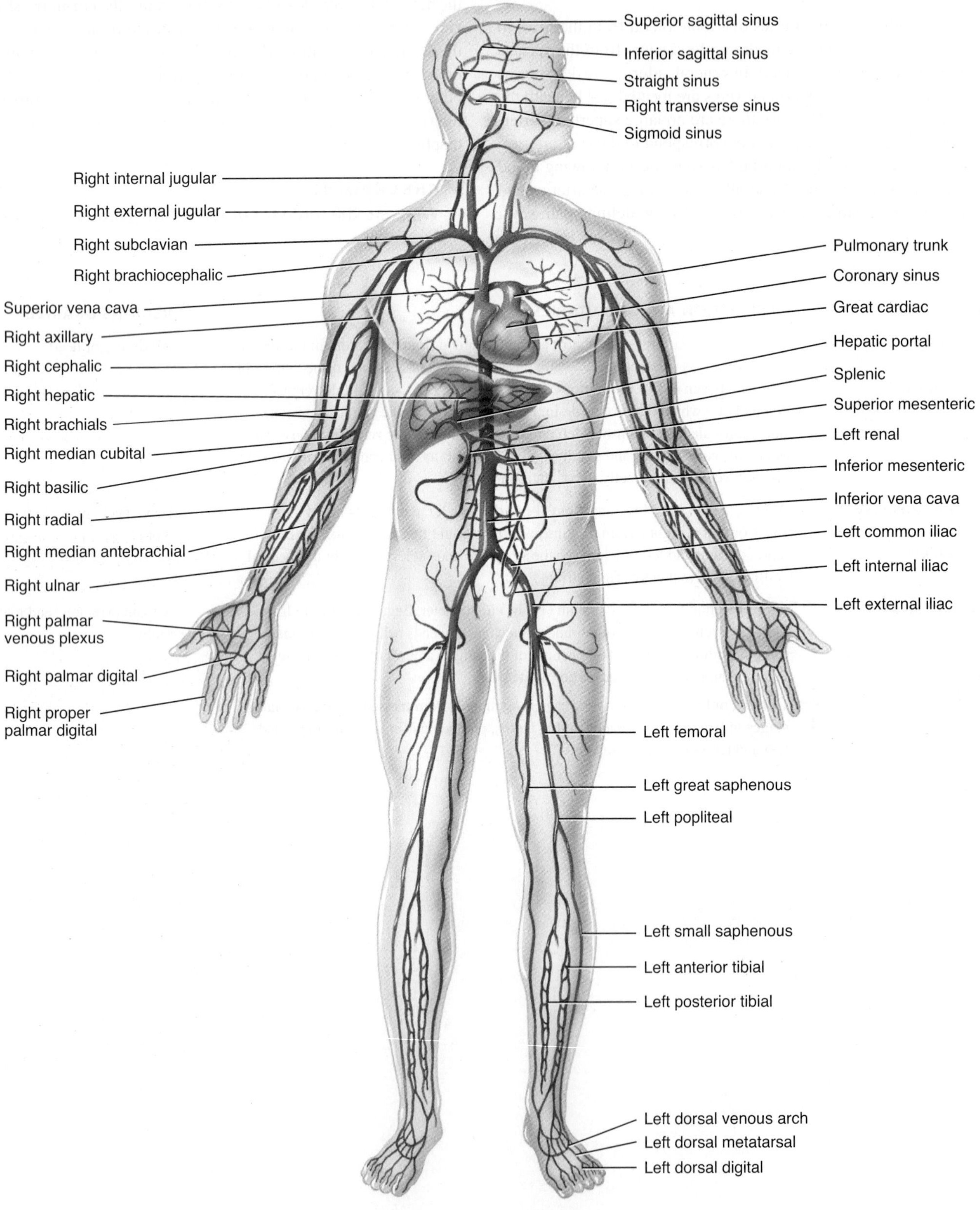

Superior sagittal sinus
Inferior sagittal sinus
Straight sinus
Right transverse sinus
Sigmoid sinus

Right internal jugular
Right external jugular
Right subclavian
Right brachiocephalic
Superior vena cava
Right axillary
Right cephalic
Right hepatic
Right brachials
Right median cubital
Right basilic
Right radial
Right median antebrachial
Right ulnar
Right palmar venous plexus
Right palmar digital
Right proper palmar digital

Pulmonary trunk
Coronary sinus
Great cardiac
Hepatic portal
Splenic
Superior mesenteric
Left renal
Inferior mesenteric
Inferior vena cava
Left common iliac
Left internal iliac
Left external iliac

Left femoral
Left great saphenous
Left popliteal
Left small saphenous
Left anterior tibial
Left posterior tibial
Left dorsal venous arch
Left dorsal metatarsal
Left dorsal digital

Overall anterior view of principal veins

Which general regions of the body are drained by the superior vena cava and the inferior vena cava?

EXHIBIT 21.H   Veins of the Head and Neck   *(Figure 21.24)*

OBJECTIVE

• Identify the three major veins that drain blood from the head.

Most blood draining from the head passes into three pairs of veins: the **internal jugular** (JUG-ū-lar), **external jugular,** and **vertebral veins** (Figure 21.24). Within the cranial cavity, all veins drain into dural ve-nous sinuses and then into the internal jugular veins. **Dural venous sinuses** are endothelial-lined venous channels between layers of the cranial dura mater.

CHECKPOINT

Which general areas are drained by the internal jugular, external jugular, and vertebral veins?

VEINS	DESCRIPTION AND TRIBUTARIES	REGIONS DRAINED
Brachiocephalic veins	(See Exhibit 21.C.)	
Internal jugular veins (JUG-ū-lar = throat)	Begin at base of cranium as sigmoid sinus and inferior petrosal sinus converge at opening of the jugular foramen. Descend within carotid sheath lateral to internal and common carotid arteries, deep to sternocleidomastoid muscles. Receive numerous tributaries from the face and neck. Internal jugular veins anastomose with subclavian veins to form brachiocephalic veins (brā′-kē-ō-se-FAL-ik; *brachi-* = arm; *-cephal-* = head) deep and slightly lateral to sternoclavicular joints. Major dural sinuses that contribute to internal jugular vein are as follows:	Brain, meninges, bones of cranium, muscles and tissues of face and neck.
	1. **Superior sagittal sinus** (SAJ-i-tal = arrow) begins at frontal bone, where it receives vein from nasal cavity, and passes posteriorly to occipital bone along midline of skull deep to sagittal sinus. It usually angles to right and drains into right transverse sinus.	Nasal cavity; superior, lateral, and medial aspects of cerebrum; skull bones; meninges.
	2. **Inferior sagittal sinus** is much smaller than superior sagittal sinus. It begins posterior to attachment of falx cerebri and receives great cerebral vein to become straight sinus.	Medial aspects of cerebrum and diencephalon.
	3. **Straight sinus** runs in tentorium cerebelli and is formed by union of inferior sagittal sinus and great cerebral vein. It typically drains into left transverse sinus.	Medial and inferior aspects of cerebrum and the cerebellum.
	4. **Sigmoid sinuses** (SIG-moyd = S-shaped) are located along posterior aspect of petrous temporal bone. They begin where transverse sinuses and superior petrosal sinuses anastomose and terminate in internal jugular vein at jugular foramen.	Lateral and posterior aspect of cerebrum and the cerebellum.
	5. **Cavernous sinuses** (KAV-er-nus = cavelike) are located on either side of sphenoid bone. Ophthalmic veins from orbits and cerebral veins from cerebral hemispheres, along with other small sinuses, empty into cavernous sinuses. They drain posteriorly to petrosal sinuses to eventually return to internal jugular veins. Cavernous sinuses are unique because they have major blood vessels and nerves passing through them on their way to orbit and face. Oculomotor (III) nerve, trochlear (IV) nerve, ophthalmic and maxillary branches of the trigeminal (V) nerve, abducens (VI) nerve, and internal carotid arteries pass through cavernous sinuses.	Orbits, nasal cavity, frontal regions of cerebrum, and superior aspect of brain stem.
Subclavian veins	(See Exhibit 21.I.)	
External jugular veins	Begin in parotid glands near angle of the mandible. Descend through neck across sternocleidomastoid muscles. Terminate at point opposite middle of clavicles, where they empty into subclavian veins. Become very prominent along side of neck when venous pressure rises, for example, during heavy coughing or straining or in cases of heart failure.	Scalp and skin of head and neck, muscles of face and neck, and oral cavity and pharynx.
Vertebral veins (VER-te-bral = of vertebrae)	Right and left **vertebral veins** originate inferior to occipital condyles. They descend through successive transverse foramina of first six cervical vertebrae and emerge from foramina of sixth cervical vertebra to enter brachiocephalic veins in root of neck.	Cervical vertebrae, cervical spinal cord and meninges, and some deep muscles in neck.

EXHIBIT 21.H CONTINUES

EXHIBIT 21.H   **849**

EXHIBIT 21.H   Veins of the Head and Neck   *(Figure 21.24)*   CONTINUED

SCHEME OF DRAINAGE

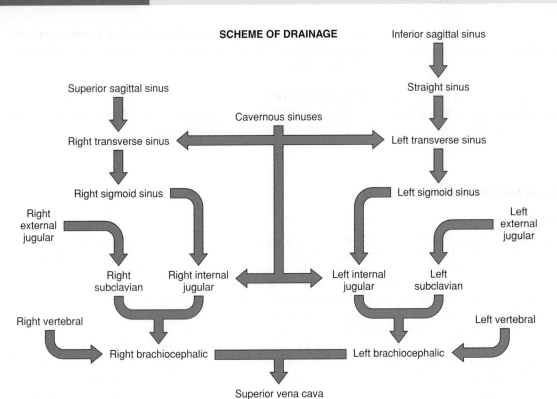

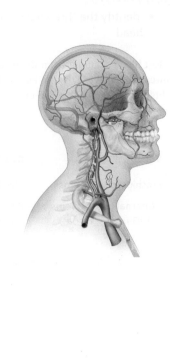

Figure 21.24 **Principal veins of the head and neck.**

🔑 Blood draining from the head passes into the internal jugular, external jugular, and vertebral veins.

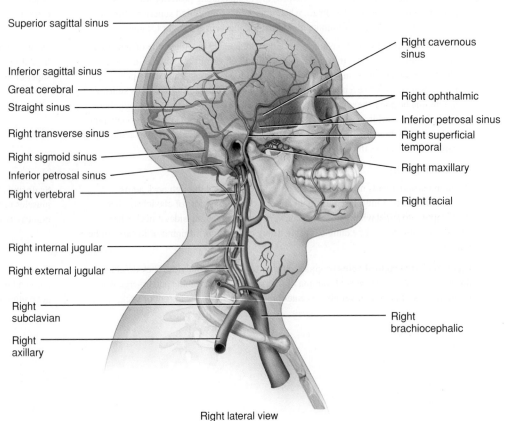

Right lateral view

❓ **Into which veins in the neck does all venous blood in the brain drain?**

EXHIBIT 21.I   Veins of the Upper Limbs *(Figure 21.25)*

◉ OBJECTIVE
- Identify the principal veins that drain the upper limbs.

Both superficial and deep veins return blood from the upper limbs to the heart (Figure 21.25). **Superficial veins** are located just deep to the skin and are often visible. They anastomose extensively with one another and with deep veins, and they do not accompany arteries. Superficial veins are larger than deep veins and return most of the blood from the upper

limbs. **Deep veins** are located deep in the body. They usually accompany arteries and have the same names as the corresponding arteries. Both superficial and deep veins have valves, but valves are more numerous in the deep veins.

✔CHECKPOINT
Where do the cephalic, basilic, median antebrachial, radial, and ulnar veins originate?

VEINS	DESCRIPTION AND TRIBUTARIES	REGIONS DRAINED
DEEP VEINS		
Brachiocephalic veins	(See Exhibit 21.J.)	
Subclavian veins (sub-KLĀ-vē-an; *sub-* = under; *-clavian* = pertaining to clavicle)	Continuations of axillary veins. Pass over first rib deep to clavicle to terminate at sternal end of clavicle, where they unite with internal jugular veins to form brachiocephalic veins. Thoracic duct of lymphatic system delivers lymph into junction between left subclavian and left internal jugular veins. Right lymphatic duct delivers lymph into junction between right subclavian and right internal jugular veins (see Figure 22.3). ⚕ **Clinical note:** In a procedure called **central line placement,** the right subclavian vein is frequently used to administer nutrients and medication and measure venous pressure.	Skin, muscles, bones of arms, shoulders, neck, and superior thoracic wall.
Axillary veins (AK-sil-ār-ē; *axilla* = armpit)	Arise as brachial veins and basilic veins unite near base of axilla (armpit). Ascend to outer borders of first ribs, where they become subclavian veins. Receive numerous tributaries in axilla that correspond to branches of axillary arteries.	Skin, muscles, bones of arm, axilla, shoulder, and superolateral chest wall.
Brachial veins (BRĀ-kē-al; *brachi-* = arm)	Accompany brachial arteries. Begin in anterior aspect of elbow region where radial and ulnar veins join one another. As they ascend through arm, basilic veins join them to form axillary vein near distal border of teres major muscle.	Muscles and bones of elbow and brachial regions.
Ulnar veins (UL-nar = pertaining to ulna)	Begin at **superficial palmar venous arches,** which drain **common palmar digital veins** and **proper palmar digital veins** in fingers. Course along medial aspect of forearms, pass alongside ulnar arteries, and join with radial veins to form brachial veins.	Muscles, bones, and skin of hand, and muscles of medial aspect of forearm.
Radial veins (RĀ-dē-al = pertaining to radius)	Begin at **deep palmar venous arches** (Figure 21.25c), which drain **palmar metacarpal veins** in palms. Drain lateral aspects of forearms and pass alongside radial arteries. Unite with ulnar veins to form brachial veins just inferior to elbow joint.	Muscles and bones of lateral hand and forearm.
SUPERFICIAL VEINS		
Cephalic veins (se-FAL-ik = pertaining to head)	Begin on lateral aspect of **dorsal venous networks of hands (dorsal venous arches),** networks of veins on dorsum of hands formed by dorsal metacarpal veins (Figure 21.25a). These veins in turn drain **dorsal digital veins,** which pass along sides of fingers. Arch around radial side of forearms to anterior surface and ascend through entire limbs along anterolateral surface. End where they join axillary veins, just inferior to clavicles. **Accessory cephalic veins** originate either from venous plexus on dorsum of forearms or from medial aspects of dorsal venous networks of hands, and unite with cephalic veins just inferior to elbow. After receiving median cubital veins, basilic veins continue ascending until they reach middle of arm. There they penetrate tissues deeply and run alongside brachial arteries until they join with brachial veins to form axillary veins.	Integument and superficial muscles of lateral aspect of upper limb.
Basilic veins (ba-SIL-ik = royal, of prime importance)	Begin on medial aspects of dorsal venous networks of hands and ascend along posteromedial surface of forearm and anteromedial surface of arm (Figure 21.25b). Connected to cephalic veins anterior to elbow by **median cubital veins** (*cubital* = pertaining to elbow). ⚕ **Clinical note:** If veins must be **punctured** for an injection, transfusion, or removal of a blood sample, the median cubital veins are preferred.	Integument and superficial muscles of medial aspect of upper limb.
Median antebrachial veins (median veins of forearm) (an′-tē-BRĀ-kē-al; *ante-* = before, in front of; *brachi-* = arm)	Begin in **palmar venous plexuses,** networks of veins on palms. Drain **palmar digital veins** in fingers. Ascend anteriorly in forearms to join basilic or median cubital veins, sometimes both.	Integument and superficial muscles of palm and anterior aspect of upper limb.

EXHIBIT 21.I CONTINUES ▶

EXHIBIT 21.I   **851**

EXHIBIT 21.1   Veins of the Upper Limbs   *(Figure 21.25)*   CONTINUED

SCHEME OF DRAINAGE

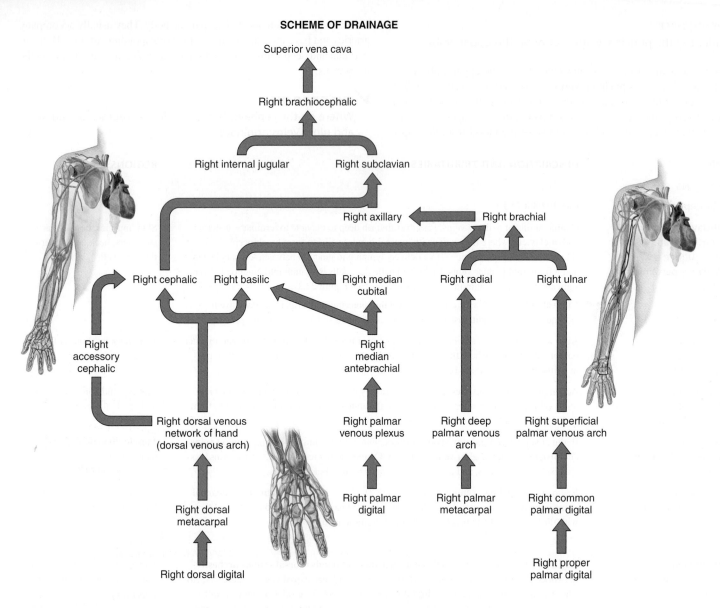

Figure 21.25 **Principal veins of the right upper limb.**

🔑 Deep veins usually accompany arteries that have similar names.

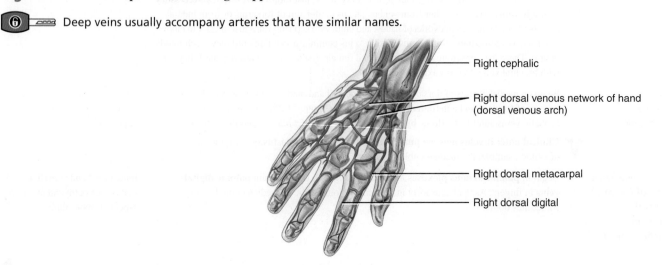

(a) Posterior view of superficial veins of hand

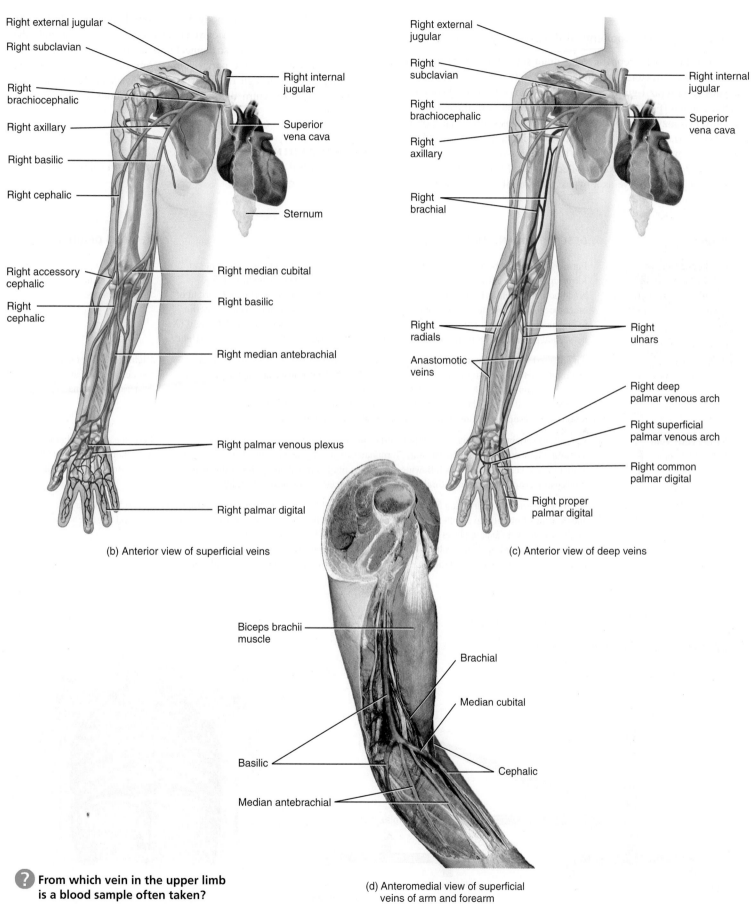

Right external jugular

Right subclavian

Right brachiocephalic

Right axillary

Right basilic

Right cephalic

Right accessory cephalic

Right cephalic

Right internal jugular

Superior vena cava

Sternum

Right median cubital

Right basilic

Right median antebrachial

Right palmar venous plexus

Right palmar digital

(b) Anterior view of superficial veins

Right external jugular

Right subclavian

Right brachiocephalic

Right axillary

Right brachial

Right radials

Anastomotic veins

Right internal jugular

Superior vena cava

Right ulnars

Right deep palmar venous arch

Right superficial palmar venous arch

Right common palmar digital

Right proper palmar digital

(c) Anterior view of deep veins

Biceps brachii muscle

Brachial

Median cubital

Basilic

Cephalic

Median antebrachial

(d) Anteromedial view of superficial veins of arm and forearm

? **From which vein in the upper limb is a blood sample often taken?**

EXHIBIT 21.1   **853**

EXHIBIT 21.J

Veins of the Thorax *(Figure 21.26)*

OBJECTIVE

• Identify the components of the azygos system of veins.

Although the brachiocephalic veins drain some portions of the thorax, most thoracic structures are drained by a network of veins, called the **azygos system** (az-Ī-gus or ā-ZĪ-gus), that runs on either side of the vertebral column (Figure 21.26). The system consists of three veins—the **azygos, hemiazygos,** and **accessory hemiazygos veins**—that show considerable variation in origin, course, tributaries, anastomoses, and termination. Ultimately they empty into the superior vena cava.

The azygos system, besides collecting blood from the thorax and abdominal wall, may serve as a bypass for the inferior vena cava, which drains blood from the lower body. Several small veins directly link the azygos system with the inferior vena cava. Larger veins that drain the lower limbs and abdomen also connect into the azygos system. If the inferior vena cava or hepatic portal vein becomes obstructed, blood that typically passes through the inferior vena cava can detour into the azygos system to return blood from the lower body to the superior vena cava.

✔CHECKPOINT

What is the importance of the azygos system relative to the inferior vena cava?

VEINS	DESCRIPTION AND TRIBUTARIES	REGIONS DRAINED
Brachiocephalic veins (brā′-kē-ō-se-FAL-ik; brachio- = arm; -cephalic = pertaining to head)	Form by union of subclavian and internal jugular veins. Two brachiocephalic veins unite to form superior vena cava. Because superior vena cava is to right of body's midline, left brachiocephalic vein is longer than right. Right brachiocephalic vein is anterior and to right of brachiocephalic trunk and follows more vertical course. Left brachiocephalic vein is anterior to brachiocephalic trunk, left common carotid and left subclavian arteries, trachea, left vagus (X) nerve, and phrenic nerve. It approaches more horizontal position as it passes from left to right.	Head, neck, upper limbs, mammary glands, and superior thorax.
Azygos vein (az-Ī-gus = unpaired)	An unpaired vein that is anterior to vertebral column, slightly to right of midline. Usually begins at junction of right ascending lumbar and right subcostal veins near diaphragm. Arches over root of right lung at level of fourth thoracic vertebra to end in superior vena cava. Receives the following tributaries: **right posterior intercostal, hemiazygos, accessory hemiazygos, esophageal, mediastinal, pericardial,** and **bronchial veins.**	Right side of thoracic wall, thoracic viscera, and posterior abdominal wall.
Hemiazygos vein (HEM-ē-ā-zī-gus; hemi- = half)	Anterior to vertebral column and slightly to left of midline. Often begins at junction of left ascending lumbar and left subcostal veins. Terminates by joining azygos vein at about level of ninth thoracic vertebra. Receives following tributaries: ninth through eleventh **left posterior intercostal, esophageal, mediastinal,** and sometimes **accessory hemiazygos veins.**	Left side of lower thoracic wall, thoracic viscera, and left posterior abdominal wall.
Accessory hemiazygos vein	Anterior to vertebral column and to left of midline. Begins at fourth or fifth intercostal space and descends from fifth to eighth thoracic vertebra or ends in hemiazygos vein. Terminates by joining azygos vein at about level of eighth thoracic vertebra. Receives the following tributaries: fourth through eighth **left posterior intercostal veins** (first through third posterior intercostal veins drain into left brachiocephalic vein), **left bronchial,** and **mediastinal veins.**	Left side of upper thoracic wall and thoracic viscera.

SCHEME OF DRAINAGE

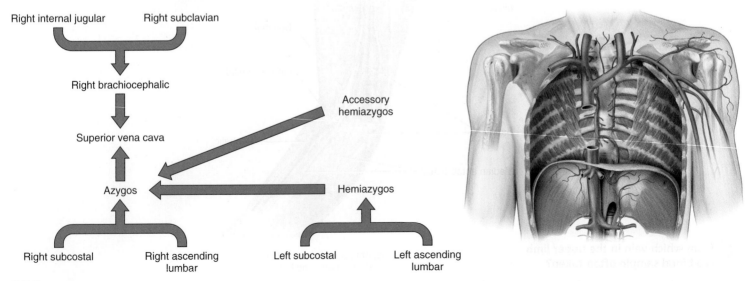

Right internal jugular   Right subclavian

Right brachiocephalic

Superior vena cava

Azygos

Accessory hemiazygos

Hemiazygos

Right subcostal   Right ascending lumbar   Left subcostal   Left ascending lumbar

Figure 21.26 Principal veins of the thorax, abdomen, and pelvis.

Most thoracic structures are drained by the azygos system of veins.

Right internal jugular

Right external jugular

Right brachiocephalic

Right superior intercostal

Superior vena cava

Right posterior intercostal

Azygos

Mediastinals

Bronchial

Pericardial

Diaphragm

Hepatics

Right suprarenal

Right subcostal

Right renal

Right ascending lumbar

Right gonadal (testicular or ovarian)

Right lumbar

Right common iliac

Right internal iliac

Right external iliac

Left internal jugular

Left external jugular

Left subclavian

Left brachiocephalic

Left superior intercostal

Left axillary

Left cephalic

Left posterior intercostal

Left brachial

Accessory hemiazygos

Left basilic

Esophageals

Hemiazygos

Left inferior phrenics

Left suprarenal

Left renal

Left ascending lumbar

Left gonadal (testicular or ovarian)

Inferior vena cava

Left lumbar

Left common iliac

Middle sacral

Left internal iliac

Inguinal ligament

Left external iliac

Left femoral

Anterior view

? **Which vein returns blood from the abdominopelvic viscera to the heart?**

EXHIBIT 21.J **855**

EXHIBIT 21.K   Veins of the Abdomen and Pelvis

● OBJECTIVE

- Identify the principal veins that drain the abdomen and pelvis.

Blood from the abdominal and pelvic viscera and abdominal wall returns to the heart via the inferior vena cava. Many small veins enter the inferior vena cava. Most carry return flow from parietal branches of the abdominal aorta, and their names correspond to the names of the arteries (see also Figure 21.26).

The inferior vena cava does not receive veins directly from the gastrointestinal tract, spleen, pancreas, and gallbladder. These organs pass their blood into a common vein, the **hepatic portal vein,** which delivers the blood to the liver. The superior mesenteric and splenic veins unite to form the hepatic portal vein (see Figure 21.28). This special flow of venous blood, called the **hepatic portal circulation,** is described shortly. After passing through the liver for processing, blood drains into the hepatic veins, which empty into the inferior vena cava.

✔ CHECKPOINT

What structures do the lumbar, gonadal, renal, suprarenal, inferior phrenic, and hepatic veins drain?

VEINS	DESCRIPTION AND TRIBUTARIES	REGIONS DRAINED
Inferior vena cava	(See Exhibit 21.G.)	
Inferior phrenic veins (FREN-ik = pertaining to diaphragm)	Arise on inferior surface of diaphragm. Left inferior phrenic vein usually sends one tributary to left suprarenal vein, which empties into left renal vein, and another tributary into inferior vena cava. Right inferior phrenic vein empties into inferior vena cava.	Inferior surface of diaphragm and adjoining peritoneal tissues.
Hepatic veins (he-PAT-ik = pertaining to liver)	Typically two or three in number. Drain sinusoidal capillaries of liver. Capillaries of liver receive venous blood from capillaries of gastrointestinal organs via hepatic portal vein. **Hepatic portal vein** receives the following tributaries from gastrointestinal organs:	
	1. **Left gastric vein** arises from left side of lesser curvature of stomach and joins left side of hepatic portal vein in lesser omentum.	Terminal esophagus, stomach, liver, gallbladder, spleen, pancreas, small intestine, and large intestine.
	2. **Right gastric vein** arises from right aspect of lesser curvature of stomach and joins hepatic portal vein on its anterior surface within lesser omentum.	Lesser curvature of stomach, abdominal portion of esophagus, stomach, and duodenum.
	3. **Splenic vein** arises in spleen and crosses abdomen transversely posterior to stomach to anastomose with superior mesenteric vein to form hepatic portal vein. It receives **inferior mesenteric vein** near its junction with hepatic portal vein.	Spleen, fundus and greater curvature of stomach, pancreas, greater omentum, descending colon, sigmoid colon, and rectum.
	4. **Superior mesenteric vein** arises from numerous tributaries from most of small intestine and first half of large intestine and ascends to join splenic vein to form hepatic portal vein.	Duodenum, jejunum, ileum, cecum, appendix, ascending colon, and transverse colon.
Lumbar veins (LUM-bar = pertaining to loin)	Usually four on each side; course horizontally through posterior abdominal wall with lumbar arteries. Connect at right angles with right and left **ascending lumbar veins,** which form origin of corresponding azygos or hemiazygos vein. Join ascending lumbar veins and then connect from ascending lumbar veins to inferior vena cava.	Posterior abdominal muscle wall, lumbar vertebrae, spinal cord and spinal nerves (cauda equina) within vertebral canal, and meninges.
Suprarenal veins (soo'-pra-RĒ-nal; supra- = above	Pass medially from adrenal (suprarenal) glands (left suprarenal vein joins left renal vein, and right suprarenal vein joins inferior vena cava).	Adrenal (suprarenal) glands.
Renal veins (RĒ-nal; ren- = kidney)	Pass anterior to renal arteries. Left renal vein is longer than right renal vein and passes anterior to abdominal aorta. It receives left testicular (or ovarian), left inferior phrenic, and usually left suprarenal veins. Right renal vein empties into inferior vena cava posterior to duodenum.	Kidneys.

VEINS	DESCRIPTION AND TRIBUTARIES	REGIONS DRAINED
Gonadal veins (gō-NAD-al; *gon-* = seed) [**testicular** (tes-TIK-ū-lar) or **ovarian** (ō-VAR-ē-an)]	Ascend with gonadal arteries along posterior abdominal wall. Called testicular veins in male. **Testicular veins** drain testes (left testicular vein joins left renal vein, and right testicular vein joins inferior vena cava). Called ovarian veins in female. **Ovarian veins** drain ovaries. Left ovarian vein joins left renal vein, and right ovarian vein joins inferior vena cava.	Testes, epididymis, ductus deferens, ovaries, and ureters.
Common iliac veins (IL-ē-ak = pertaining to ilium)	Formed by union of internal and external iliac veins anterior to sacroiliac joint and anastomose anterior to fifth lumbar vertebra to form inferior vena cava. Right common iliac is much shorter than left and is also more vertical, as inferior vena cava sits to right of midline.	Pelvis, external genitals, and lower limbs.
Internal iliac veins	Begin near superior portion of greater sciatic notch and run medial to their corresponding arteries	Muscles of pelvic wall and gluteal region, pelvic viscera, and external genitals.
External iliac veins	Companions of internal iliac arteries. Begin at inguinal ligaments as continuations of femoral veins. End anterior to sacroiliac joints where they join with internal iliac veins to form common iliac veins.	Lower abdominal wall anteriorly, cremaster muscle in males, and external genitals and lower limb.

SCHEME OF DRAINAGE

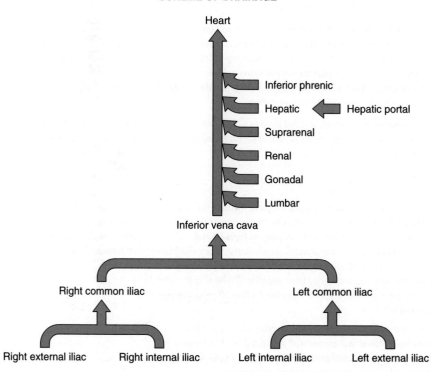

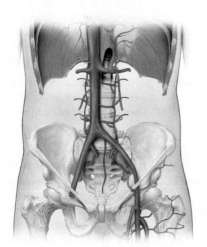

EXHIBIT 21.K **857**

EXHIBIT 21.L

Veins of the Lower Limbs *(Figure 21.27)*

◉ OBJECTIVE

- Identify the principal superficial and deep veins that drain the lower limbs.

As with the upper limbs, blood from the lower limbs is drained by both **superficial** and **deep veins.** The superficial veins often anastomose with one another and with deep veins along their length. Deep veins, for the

VEINS	DESCRIPTION AND TRIBUTARIES	REGIONS DRAINED
DEEP VEINS		
Common iliac veins	(See Exhibit 21.K.)	
External iliac veins	(See Exhibit 21.K.)	
Femoral veins (FEM-o-ral)	Accompany femoral arteries and are continuations of popliteal veins just superior to knee where veins pass through opening in adductor magnus muscle. Ascend deep to sartorius muscle and emerge from beneath muscle in femoral triangle at proximal end of thigh. Receive **deep veins of thigh (deep femoral veins)** and great saphenous veins just before penetrating abdominal wall. Pass below inguinal ligament and enter abdominopelvic region to become external iliac veins. ☤ **Clinical note:** In order to take **blood samples** or **pressure recordings** from the right side of the heart, a catheter is inserted into the femoral vein as it passes through the femoral triangle. The catheter passes through the external and common iliac veins, then into the inferior vena cava, and finally into the right atrium.	Skin, lymph nodes, muscles, and bones of thigh, and external genitals.
Popliteal veins (pop'-li-TĒ-al = pertaining to hollow behind knee)	Formed by union of anterior and posterior tibial veins at proximal end of leg; ascend through popliteal fossa with popliteal arteries and tibial nerve. Terminate where they pass through window in adductor magnus muscle and pass to front of knee to become femoral veins. Also receive blood from small saphenous veins and tributaries that correspond to branches of popliteal artery.	Knee joint and skin, muscles, and bones around knee joint.
Posterior tibial veins (TIB-ē-al)	Begin posterior to medial malleolus at union of **medial** and **lateral plantar veins** from plantar surface of foot. Ascend through leg with posterior tibial artery and tibial nerve deep to soleus muscle. Join posterior tibial veins about two-thirds of way up leg. Join anterior tibial veins near top of interosseous membrane to form popliteal veins. On plantar surface of foot **plantar digital veins** unite to form **plantar metatarsal veins,** which parallel metatarsals. They in turn unite to form **deep plantar venous arches.** Medial and lateral plantar veins emerge from deep plantar venous arches.	Skin, muscles, and bones on plantar surface of foot, and skin, muscles, and bones from posterior and lateral aspects of leg.
Anterior tibial veins	Arise in dorsal venous arch and accompany anterior tibial artery. Ascend deep to tibialis anterior muscle on anterior surface of interosseous membrane. Pass through opening at superior end of interosseous membrane to join posterior tibial veins to form popliteal veins.	Dorsal surface of foot, ankle joint, anterior aspect of leg, knee joint, and tibiofibular joint.
SUPERFICIAL VEINS		
Great (long) saphenous veins (sa-FĒ-nus = clearly visible)	Longest veins in body; ascend from foot to groin in subcutaneous layer. Begin at medial end of dorsal venous arches of foot. **Dorsal venous arches** (VĒ-nus) are networks of veins on dorsum of foot formed by **dorsal digital veins,** which collect blood from toes, and then unite in pairs to form **dorsal metatarsal veins,** which parallel metatarsals. As dorsal metatarsal veins approach foot, they combine to form dorsal venous arches. Pass anterior to medial malleolus of tibia and then superiorly along medial aspect of leg and thigh just deep to skin. Receive tributaries from superficial tissues and connect with deep veins as well. Empty into femoral veins at groin. Have from 10 to 20 valves along their length, with more located in leg than thigh. ☤ **Clinical note:** These veins are more likely to be subject to **varicosities** than other veins in the lower limbs because they must support a long column of blood and are not well supported by skeletal muscles. The great saphenous veins are often used for prolonged administration of intravenous fluids. This is particularly important in very young children and in patients of any age who are in shock and whose veins are collapsed. In **coronary artery bypass grafting,** if multiple blood vessels need to be grafted, sections of the great saphenous vein are used along with at least one artery as a graft (see first **Clinical Note** in Exhibit 21.C). After the great saphenous vein is removed and divided into sections, the sections are used to bypass the blockages. The vein grafts are reversed so that the valves do not obstruct the flow of blood.	Integumentary tissues and superficial muscles of lower limbs, groin, and lower abdominal wall.
Small saphenous veins	Begin at lateral aspect of dorsal venous arches of foot. Pass posterior to lateral malleolus of fibula and ascend deep to skin along posterior aspect of leg. Empty into popliteal veins in popliteal fossa, posterior to knee. Have from 9 to 12 valves. May communicate with great saphenous veins in proximal thigh.	Integumentary tissues and superficial muscles of foot and posterior aspect of leg.

most part, have the same names as corresponding arteries (Figure 21.27). All veins of the lower limbs have valves, which are more numerous than in veins of the upper limbs.

✔CHECKPOINT
What is the clinical importance of the great saphenous veins?

SCHEME OF DRAINAGE

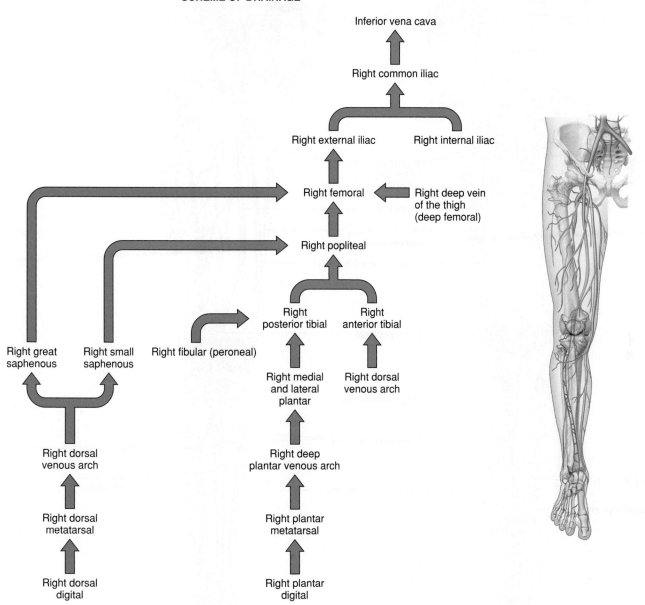

EXHIBIT 21.L CONTINUES ▶

EXHIBIT 21.L **859**

EXHIBIT 21.L | **Veins of the Lower Limbs** *(Figure 21.27)* CONTINUED

Figure 21.27 Principal veins of the pelvis and lower limbs.

Deep veins usually bear the names of their companion arteries.

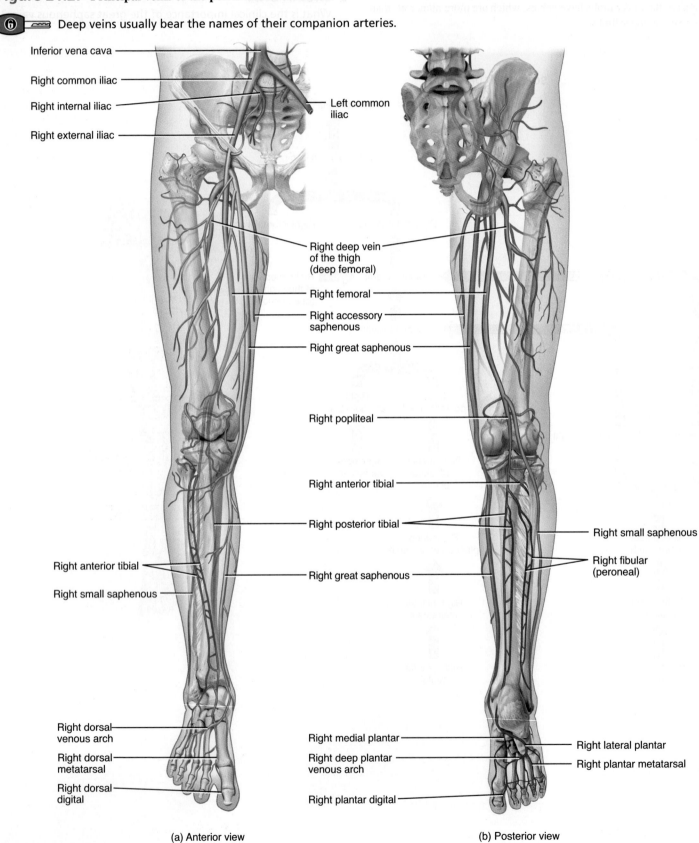

Inferior vena cava

Right common iliac

Right internal iliac

Left common iliac

Right external iliac

Right deep vein of the thigh (deep femoral)

Right femoral

Right accessory saphenous

Right great saphenous

Right popliteal

Right anterior tibial

Right posterior tibial

Right small saphenous

Right anterior tibial

Right fibular (peroneal)

Right great saphenous

Right small saphenous

Right dorsal venous arch

Right dorsal metatarsal

Right dorsal digital

Right medial plantar

Right lateral plantar

Right deep plantar venous arch

Right plantar metatarsal

Right plantar digital

(a) Anterior view

(b) Posterior view

Which veins of the lower limb are superficial?

The Hepatic Portal Circulation

The **hepatic portal circulation** (he-PAT-ik) carries venous blood from the gastrointestinal organs and spleen to the liver. A vein that carries blood from one capillary network to another is called a **portal vein.** The **hepatic portal vein** (*hepat-* = liver) receives blood from capillaries of gastrointestinal organs and the spleen and delivers it to the sinusoids of the liver (Figure 21.28). After a meal, hepatic portal blood is rich in nutrients absorbed from the gastrointestinal tract. The liver stores some of them and modifies others before they pass into the general circulation. For example, the liver converts glucose into glycogen for storage, reducing

blood glucose level shortly after a meal. The liver also detoxifies harmful substances, such as alcohol, that have been absorbed from the gastrointestinal tract and destroys bacteria by phagocytosis.

The superior mesenteric and splenic veins unite to form the hepatic portal vein. The **superior mesenteric vein** (mez-en-TER-ik) drains blood from the small intestine and portions of the large intestine, stomach, and pancreas through the *jejunal, ileal, ileocolic* (il′-ē-ō-KOL-ik)*, right colic, middle colic, pancreaticoduodenal* (pan-krē-at′-i-kō-doo′-ō-DĒ-nal)*,* and *right gastroepiploic veins* (gas′-trō-ep′-i-PLŌ-ik). The **splenic vein** drains blood from the stomach, pancreas, and portions of the large intestine through the *short gastric, left gastroepiploic, pancreatic,* and *inferior*

Figure 21.28 Hepatic portal circulation. A schematic diagram of blood flow through the liver, including arterial circulation, is shown in (b). As usual, deoxygenated blood is indicated in blue, and oxygenated blood in red.

🔑 The hepatic portal circulation delivers venous blood from the organs of the gastrointestinal tract and spleen to the liver.

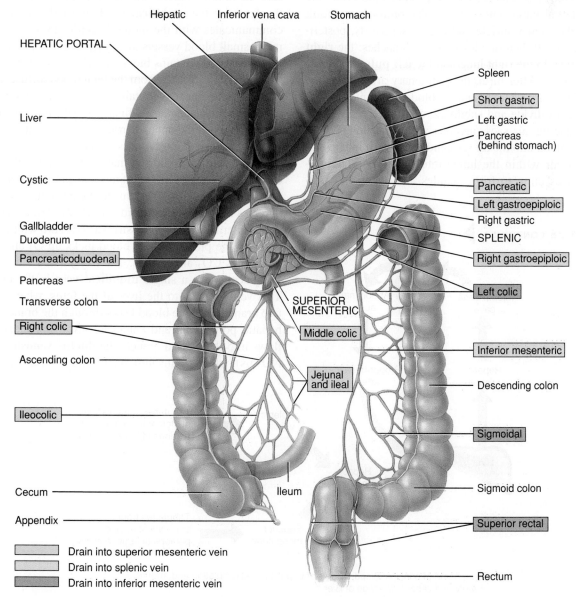

Drain into superior mesenteric vein
Drain into splenic vein
Drain into inferior mesenteric vein

(a) Anterior view of veins draining into hepatic portal vein

FIGURE 21.28 CONTINUES ▶

mesenteric veins. The inferior mesenteric vein, which passes into the splenic vein, drains portions of the large intestine through the *superior rectal, sigmoidal,* and *left colic veins.* The *right* and *left gastric veins,* which open directly into the hepatic portal vein, drain the stomach. The *cystic vein,* which also opens into the hepatic portal vein, drains the gallbladder.

At the same time the liver is receiving nutrient-rich but deoxygenated blood via the hepatic portal vein, it is also receiving oxygenated blood via the hepatic artery, a branch of the celiac trunk. The oxygenated blood mixes with the deoxygenated blood in sinusoids. Eventually, blood leaves the sinusoids of the liver through the **hepatic veins,** which drain into the inferior vena cava.

The Pulmonary Circulation

The **pulmonary circulation** (PUL-mō-ner′-ē; *pulmo-* = lung) carries deoxygenated blood from the right ventricle to the air sacs (alveoli) within the lungs and returns oxygenated blood from the air sacs to the left atrium (Figure 21.29). The **pulmonary trunk** emerges from the right ventricle and passes superiorly, posteriorly, and to the left. It then divides into two branches: the **right pulmonary artery** to the right lung and the **left pulmonary artery** to the left lung. After birth, the pulmonary arteries are the only arteries that carry deoxygenated blood. On entering the lungs, the branches divide and subdivide until finally they form capillaries around the air sacs (alveoli) within the lungs. CO_2 passes from the blood into the air sacs and is exhaled. Inhaled O_2 passes from the air within the lungs into the blood. The pulmonary capillaries unite to form venules and eventually **pul-**monary veins, which exit the lungs and carry the oxygenated blood to the left atrium. Two left and two right pulmonary veins enter the left atrium. After birth, the pulmonary veins are the only veins that carry oxygenated blood. Contractions of the left ventricle then eject the oxygenated blood into the systemic circulation.

The Fetal Circulation

The circulatory system of a fetus, called the **fetal circulation,** exists only in the fetus and contains special structures that allow the developing fetus to exchange materials with its mother (Figure 21.30). It differs from the postnatal (after birth) circulation because the lungs, kidneys, and gastrointestinal organs do not begin to function until birth. The fetus obtains O_2 and nutrients from the maternal blood and eliminates CO_2 and other wastes into it.

The exchange of materials between fetal and maternal circulations occurs through the **placenta** (pla-SEN-ta), which forms inside the mother's uterus and attaches to the umbilicus (navel) of the fetus by the **umbilical cord** (um-BIL-i-kal). The placenta communicates with the mother's cardiovascular system through many small blood vessels that emerge from the uterine wall. The umbilical cord contains blood vessels that branch into capillaries in the placenta. Wastes from the fetal blood diffuse out of the capillaries, into spaces containing maternal blood (intervillous spaces) in the placenta, and finally into the mother's uterine veins. Nutrients travel the opposite route—from the maternal blood vessels to the intervillous spaces to the fetal capillaries. Normally, there is no direct mixing of maternal and fetal blood because all exchanges occur by diffusion through capillary walls.

Blood passes from the fetus to the placenta via two **umbilical arteries** in the umbilical cord (Figure 21.30a, c). These branches of the internal iliac (hypogastric) arteries are within the umbilical cord. At the placenta, fetal blood picks up O_2 and nutrients and eliminates CO_2 and wastes. The oxygenated blood returns from the placenta via a single **umbilical vein** in the umbilical cord. This vein ascends to the liver of the fetus, where it divides into two branches. Some blood flows through the branch that joins the hepatic portal vein and enters the liver, but most of the blood flows into the second branch, the **ductus venosus** (DUK-tus ve-NŌ-sus), which drains into the inferior vena cava.

☐ FIGURE 21.28 CONTINUED ▶

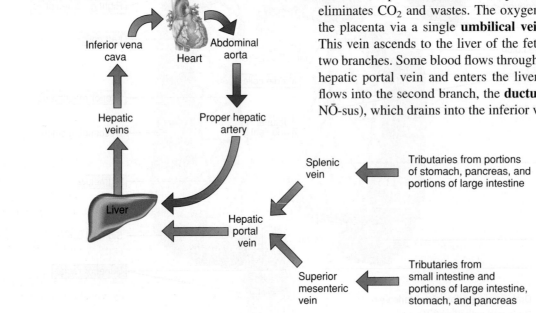

(b) Scheme of principal blood vessels of hepatic portal circulation and arterial supply and venous drainage of liver

❓ Which veins carry blood away from the liver?

Figure 21.29 Pulmonary circulation.

The pulmonary circulation brings deoxygenated blood from the right ventricle to the lungs and returns oxygenated blood from the lungs to the left atrium.

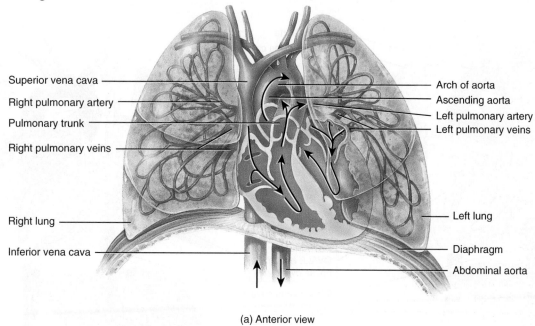

Superior vena cava
Right pulmonary artery
Pulmonary trunk
Right pulmonary veins
Right lung
Inferior vena cava

Arch of aorta
Ascending aorta
Left pulmonary artery
Left pulmonary veins
Left lung
Diaphragm
Abdominal aorta

(a) Anterior view

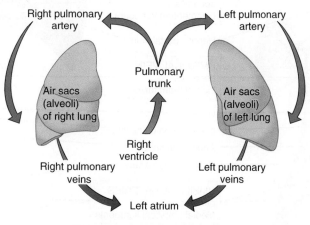

Right pulmonary artery
Left pulmonary artery
Pulmonary trunk
Air sacs (alveoli) of right lung
Air sacs (alveoli) of left lung
Right ventricle
Right pulmonary veins
Left pulmonary veins
Left atrium

(b) Scheme of pulmonary circulation

? After birth, which are the only arteries that carry deoxygenated blood?

Deoxygenated blood returning from lower body regions of the fetus mingles with oxygenated blood from the ductus venosus in the inferior vena cava. This mixed blood then enters the right atrium. Deoxygenated blood returning from upper body regions of the fetus enters the superior vena cava and also passes into the right atrium.

Most of the fetal blood does not pass from the right ventricle to the lungs, as it does in postnatal circulation, because an opening called the **foramen ovale** (fō-RĀ-men ō-VAL-ē) exists in the septum between the right and left atria. Most of the blood that enters the right atrium passes through the foramen ovale into the left atrium and joins the systemic circulation. The blood that does pass into the right ventricle is pumped into the pulmonary trunk, but little of this blood reaches the nonfunctioning fetal lungs. In-

stead, most is sent through the **ductus arteriosus** (ar-tē-rē-Ō-sus), a vessel that connects the pulmonary trunk with the aorta. The blood in the aorta is carried to all fetal tissues through the systemic circulation. When the common iliac arteries branch into the external and internal iliacs, part of the blood flows into the internal iliacs, into the umbilical arteries, and back to the placenta for another exchange of materials.

After birth, when pulmonary (lung), renal (kidney), and digestive functions begin, the following vascular changes occur (Figure 21.30b):

1. When the umbilical cord is tied off, blood no longer flows through the umbilical arteries, they fill with connective tissue,

Figure 21.30 Fetal circulation and changes at birth. The gold boxes between parts (a) and (b) describe the fate of certain fetal structures once postnatal circulation is established.

The lungs and gastrointestinal organs do not begin to function until birth.

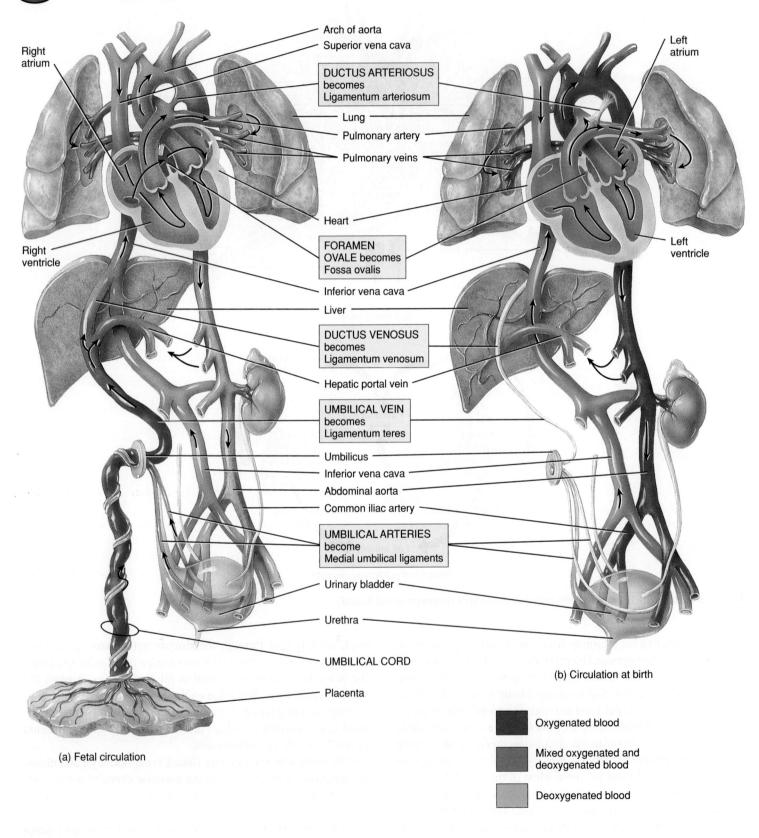

Arch of aorta
Superior vena cava

Right atrium

DUCTUS ARTERIOSUS becomes **Ligamentum arteriosum**

Left atrium

Lung
Pulmonary artery
Pulmonary veins

Heart

Right ventricle

FORAMEN OVALE becomes Fossa ovalis

Left ventricle

Inferior vena cava
Liver

DUCTUS VENOSUS becomes **Ligamentum venosum**

Hepatic portal vein

UMBILICAL VEIN becomes **Ligamentum teres**

Umbilicus
Inferior vena cava
Abdominal aorta
Common iliac artery

UMBILICAL ARTERIES become **Medial umbilical ligaments**

Urinary bladder

Urethra

UMBILICAL CORD

Placenta

(a) Fetal circulation

(b) Circulation at birth

■ Oxygenated blood

■ Mixed oxygenated and deoxygenated blood

■ Deoxygenated blood

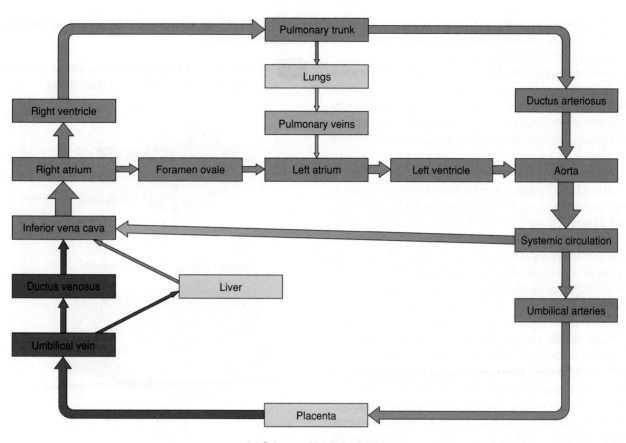

(c) Scheme of fetal circulation

 Which structure provides for exchange of materials between mother and fetus?

and the distal portions of the umbilical arteries become fibrous cords called the **medial umbilical ligaments.** Although the arteries are closed functionally only a few minutes after birth, complete obliteration of the lumens may take 2 to 3 months.

2. The umbilical vein collapses but remains as the **ligamentum teres** (TE-rez) **(round ligament),** a structure that attaches the umbilicus to the liver.

3. The ductus venosus collapses but remains as the **ligamentum venosum** (ve-NŌ-sum), a fibrous cord on the inferior surface of the liver.

4. The placenta is expelled as the **"afterbirth."**

5. The foramen ovale normally closes shortly after birth to become the **fossa ovalis,** a depression in the interatrial septum. When an infant takes its first breath, the lungs expand and blood flow to the lungs increases. Blood returning from the lungs to the heart increases pressure in the left atrium. This closes the foramen ovale by pushing the valve that guards it against the interatrial septum. Permanent closure occurs in about a year.

6. The ductus arteriosus closes by vasoconstriction almost immediately after birth and becomes the **ligamentum arteriosum** (ar-tēr′-ē-Ō-sum). Complete anatomical obliteration of the lumen takes 1 to 3 months.

✔ **CHECKPOINT**

23. Diagram the hepatic portal circulation. Why is this route important?

24. Diagram the route of the pulmonary circulation.

25. Discuss the anatomy and physiology of the fetal circulation. Indicate the function of the umbilical arteries, umbilical vein, ductus venosus, foramen ovale, and ductus arteriosus.

21.8 DEVELOPMENT OF BLOOD VESSELS AND BLOOD

◉ **OBJECTIVE**

• Describe the development of blood vessels and blood.

The development of blood cells and the formation of blood vessels begins outside the embryo as early as 15 to 16 days in the **mesoderm** of the wall of the yolk sac, chorion, and connecting stalk. About 2 days later, blood vessels form within the embryo. The early formation of the cardiovascular system is linked to the small amount of yolk in the ovum and yolk sac. As the embryo develops rapidly during the third week, there is a greater need to develop a cardiovascular system to supply sufficient nutrients to the embryo and remove wastes from it.

Blood vessels and blood cells develop from the same precursor cell, called a **hemangioblast** (hē-MAN-jē-ō-blast; *hema-* = blood; *-blast* = immature stage). Once mesenchyme develops into hemangioblasts, they can give rise to cells that produce blood vessels (angioblasts) or cells that produce blood cells (pluripotent stem cells).

Blood vessels develop from **angioblasts** (AN-jē-ō-blasts), which are derived from hemangioblasts. Angioblasts aggregate to form isolated masses and cords throughout the embryonic discs called **blood islands** (Figure 21.31). Spaces soon appear in the islands and become the lumens of the blood vessels. Some of the angioblasts immediately around the spaces give rise to the *endothelial lining of the blood vessels.* Angioblasts around the endothelium form the *tunics* (interna, media, and externa) of the larger blood vessels. Growth and fusion of blood islands form an extensive network of blood vessels throughout the embryo. By continuous branching, blood vessels outside the embryo connect with those inside the embryo, linking the embryo with the placenta.

Blood cells develop from **pluripotent stem cells** (ploo-RIP-ō-tent) derived from hemangioblasts. This development occurs in the walls of blood vessels in the yolk sac, chorion, and allantois at about 3 weeks after fertilization. Blood formation in the embryo itself begins at about the fifth week in the liver and the twelfth week in the spleen, red bone marrow, and thymus.

✔CHECKPOINT

26. What are the sites of blood cell production outside the embryo and within the embryo?

21.9 AGING AND THE CARDIOVASCULAR SYSTEM

◉OBJECTIVE

• Explain the effects of aging on the cardiovascular system.

General changes in the cardiovascular system associated with aging include decreased compliance (distensibility) of the aorta, reduction in cardiac muscle fiber size, progressive loss of cardiac muscular strength, reduced cardiac output, a decline in maximum heart rate, and an increase in systolic blood pressure. Total blood cholesterol tends to increase with age, as does low-density lipoprotein (LDL); high-density lipoprotein (HDL) tends to decrease. There is an increase in the incidence of coronary artery disease (CAD), the major cause of heart disease and death in older Americans. Congestive heart failure (CHF), a set of symptoms associated with impaired pumping of the heart, is also prevalent in older individuals. Changes in blood vessels that serve brain tissue—for example, atherosclerosis—reduce nourishment to the brain and result in malfunction or death of brain cells. By

Figure 21.31 Development of blood vessels and blood cells from blood islands.

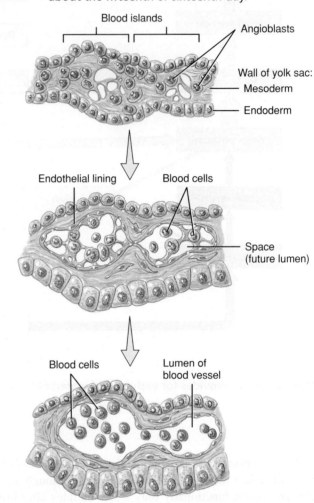

🔑 Blood vessel development begins in the embryo on about the fifteenth or sixteenth day.

❓ From which germ cell layer are blood vessels and blood derived?

age 80, cerebral blood flow is 20% less and renal blood flow is 50% less than in the same person at age 30 because of the effects of aging on blood vessels.

✔CHECKPOINT

27. How does aging affect the heart?

• • •

To appreciate the many ways the blood, heart, and blood vessels contribute to homeostasis of other body systems, examine *Focus on Homeostasis: The Cardiovascular System.*

BODY SYSTEM	CONTRIBUTION OF THE CARDIOVASCULAR SYSTEM

For all body systems
The heart pumps blood through blood vessels to body tissues, delivering oxygen and nutrients and removing wastes by means of capillary exchange. Circulating blood keeps body tissues at a proper temperature.

Integumentary system
Blood delivers clotting factors and white blood cells that aid in hemostasis when skin is damaged and contribute to repair of injured skin. Changes in skin blood flow contribute to body temperature regulation by adjusting the amount of heat loss via the skin. Blood flowing in skin may give skin a pink hue.

Skeletal system
Blood delivers calcium and phosphate ions that are needed for building bone extracellular matrix, hormones that govern building and breakdown of bone extracellular matrix, and erythropoietin that stimulates production of red blood cells by red bone marrow.

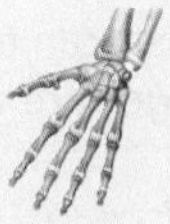

Muscular system
Blood circulating through exercising muscle removes heat and lactic acid.

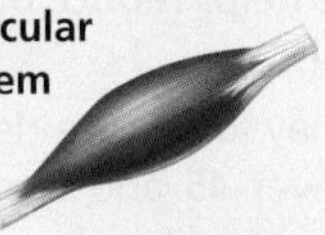

Nervous system
Endothelial cells lining choroid plexuses in brain ventricles help produce cerebrospinal fluid (CSF) and contribute to the blood–brain barrier.

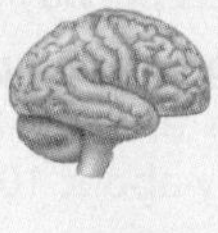

Endocrine system
Circulating blood delivers most hormones to their target tissues. Atrial cells secrete atrial natriuretic peptide.

Lymphatic system and immunity
Circulating blood distributes lymphocytes, antibodies, and macrophages that carry out immune functions. Lymph forms from excess interstitial fluid, which filters from blood plasma due to blood pressure generated by the heart.

Respiratory system
Circulating blood transports oxygen from the lungs to body tissues and carbon dioxide to the lungs for exhalation.

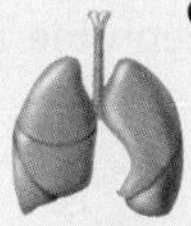

Digestive system
Blood carries newly absorbed nutrients and water to liver. Blood distributes hormones that aid digestion.

Urinary system
Heart and blood vessels deliver 20% of the resting cardiac output to the kidneys, where blood is filtered, needed substances are reabsorbed, and unneeded substances remain as part of urine, which is excreted.

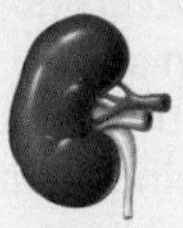

Reproductive systems
Vasodilation of arterioles in penis and clitoris cause erection during sexual intercourse. Blood distributes hormones that regulate reproductive functions.

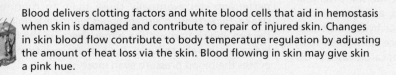

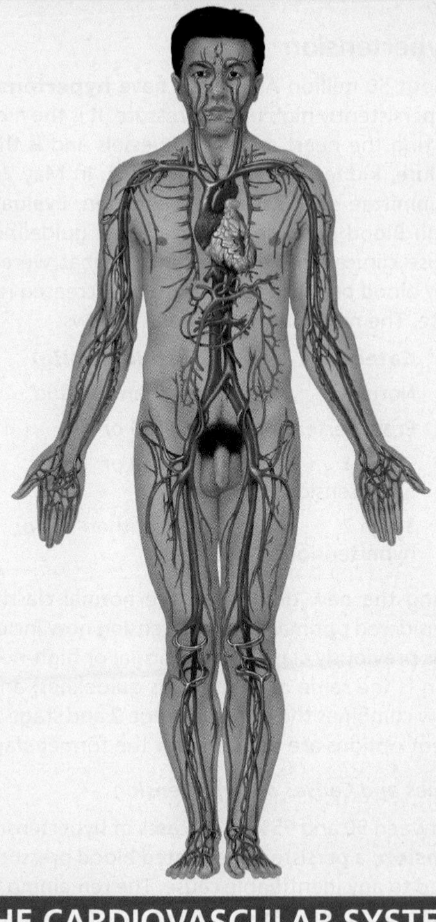

THE CARDIOVASCULAR SYSTEM

DISORDERS: HOMEOSTATIC IMBALANCES

Hypertension

About 50 million Americans have **hypertension** (hī'-per-TEN-shun), or persistently high blood pressure. It is the most common disorder affecting the heart and blood vessels and is the major cause of heart failure, kidney disease, and stroke. In May 2003, the Joint National Committee on Prevention, Detection, Evaluation, and Treatment of High Blood Pressure published new guidelines for hypertension because clinical studies have linked what were once considered fairly low blood pressure readings to an increased risk of cardiovascular disease. The new guidelines are as follows:

Category	Systolic (mmHg)	Diastolic (mmHg)
Normal	Less than 120 *and*	Less than 80
Prehypertension	120–139 *or*	80–89
Stage 1 hypertension	140–159 *or*	90–99
Stage 2 hypertension	Greater than 160 *or*	Greater than 100

Using the new guidelines, the normal classification was previously considered optimal; prehypertension now includes many more individuals previously classified as normal or high-normal; stage 1 hypertension is the same as in previous guidelines; and stage 2 hypertension now combines the previous stage 2 and stage 3 categories since treatment options are the same for the former stages 2 and 3.

Types and Causes of Hypertension

Between 90 and 95% of all cases of hypertension are **primary hypertension**, a persistently elevated blood pressure that cannot be attributed to any identifiable cause. The remaining 5–10% of cases are **secondary hypertension**, which has an identifiable underlying cause. Several disorders cause secondary hypertension:

* *Obstruction of renal blood flow* or disorders that damage renal tissue may cause the kidneys to release excessive amounts of renin into the blood. The resulting high level of angiotensin II causes vasoconstriction, thus increasing systemic vascular resistance.

* *Hypersecretion of aldosterone*—resulting, for instance, from a tumor of the adrenal cortex—stimulates excess reabsorption of salt and water by the kidneys, which increases the volume of body fluids.

* *Hypersecretion of epinephrine and norepinephrine* may occur by a **pheochromocytoma** (fē-ō-krō'-mō-sī-TŌ-ma), a tumor of the adrenal medulla. Epinephrine and norepinephrine increase heart rate and contractility and increase systemic vascular resistance.

Damaging Effects of Untreated Hypertension

High blood pressure is known as the "silent killer" because it can cause considerable damage to the blood vessels, heart, brain, and kidneys before it causes pain or other noticeable symptoms. It is a major risk factor for the number-one (heart disease) and number-three (stroke) causes of death in the United States. In blood vessels, hypertension causes thickening of the tunica media, accelerates development of atherosclerosis and coronary artery disease, and increases systemic vascular resistance. In the heart, hypertension increases the afterload, which forces the ventricles to work harder to eject blood.

The normal response to an increased workload due to vigorous and regular exercise is hypertrophy of the myocardium, especially in the wall of the left ventricle. This is a positive effect that makes the heart a more efficient pump. An increased afterload, however, leads to myocardial hypertrophy that is accompanied by muscle damage and fibrosis (a buildup of collagen fibers between the muscle fibers). As a result, the left ventricle enlarges, weakens, and dilates. Because arteries in the brain are usually less protected by surrounding tissues than are the major arteries in other parts of the body, prolonged hypertension can eventually cause them to rupture, resulting in a stroke. Hypertension also damages kidney arterioles, causing them to thicken, which narrows the lumen; because the blood supply to the kidneys is thereby reduced, the kidneys secrete more renin, which elevates the blood pressure even more.

Lifestyle Changes to Reduce Hypertension

Although several categories of drugs (described next) can reduce elevated blood pressure, the following lifestyle changes are also effective in managing hypertension:

* *Lose weight.* This is the best treatment for high blood pressure short of using drugs. Loss of even a few pounds helps reduce blood pressure in overweight hypertensive individuals.

* *Limit alcohol intake.* Drinking in moderation may lower the risk of coronary heart disease, mainly among males over 45 and females over 55. Moderation is defined as no more than one 12-oz beer per day for females and no more than two 12-oz beers per day for males.

* *Exercise.* Becoming more physically fit by engaging in moderate activity (such as brisk walking) several times a week for 30 to 45 minutes can lower systolic blood pressure by about 10 mmHg.

* *Reduce intake of sodium (salt).* Roughly half the people with hypertension are "salt sensitive." For them, a high-salt diet appears to promote hypertension, and a low-salt diet can lower their blood pressure.

* *Maintain recommended dietary intake of potassium, calcium, and magnesium.* Higher levels of potassium, calcium, and magnesium in the diet are associated with a lower risk of hypertension.

* *Don't smoke or quit smoking.* Smoking has devastating effects on the heart and can augment the damaging effects of high blood pressure by promoting vasoconstriction.

* *Manage stress.* Various meditation and biofeedback techniques help some people reduce high blood pressure. These methods may work by decreasing the daily release of epinephrine and norepinephrine by the adrenal medulla.

Drug Treatment of Hypertension

Drugs having several different mechanisms of action are effective in lowering blood pressure. Many people are successfully treated with *diuretics* (dī-ū-RET-iks), agents that decrease blood pressure by decreasing blood volume, because they increase elimination of water and salt in the urine. *ACE (angiotensin-converting enzyme) inhibitors* block formation of angiotensin II and thereby promote vasodilation and decrease the secretion of aldosterone. *Beta blockers* (BĀ-ta) reduce blood pressure by inhibiting the secretion of renin and by decreasing heart rate and contractility. *Vasodilators* relax the smooth muscle in arterial walls, causing vasodilation and lowering blood pressure by lowering systemic vascular resistance. An important category of vasodilators are the *calcium channel blockers,* which slow the inflow of Ca^{2+} into vascular smooth muscle cells. They reduce the heart's workload by slowing Ca^{2+} entry into pacemaker cells and regular myocardial fibers, thereby decreasing heart rate and the force of myocardial contraction.

MEDICAL TERMINOLOGY

Aneurysm (AN-ū-rizm) A thin, weakened section of the wall of an artery or a vein that bulges outward, forming a balloonlike sac. Common causes are atherosclerosis, syphilis, congenital blood vessel defects, and trauma. If untreated, the aneurysm enlarges and the blood vessel wall becomes so thin that it bursts. The result is massive hemorrhage with shock, severe pain, stroke, or death. Treatment may involve surgery in which the weakened area of the blood vessel is removed and replaced with a graft of synthetic material.

Aortography (ā'-or-TOG-ra-fē) X-ray examination of the aorta and its main branches after injection of a radiopaque dye.

Carotid endarterectomy (ka-ROT-id end'-ar-ter-EK-tō-mē) The removal of atherosclerotic plaque from the carotid artery to restore greater blood flow to the brain.

Claudication (klaw'-di-KĀ-shun) Pain and lameness or limping caused by defective circulation of the blood in the vessels of the limbs.

Deep vein thrombosis **(DVT)** The presence of a thrombus (blood clot) in a deep vein of the lower limbs. It may lead to (1) pulmonary embolism, if the thrombus dislodges and then lodges within the pulmonary arterial blood flow, and (2) postphlebitic syndrome, which consists of edema, pain, and skin changes due to destruction of venous valves.

Doppler ultrasound scanning Imaging technique commonly used to measure blood flow. A transducer is placed on the skin and an image is displayed on a monitor that provides the exact position and severity of a blockage.

Femoral angiography An imaging technique in which a contrast medium is injected into the femoral artery and spreads to other arteries in the lower limb, and then a series of radiographs are taken of one or more sites. It is used to diagnose narrowing or blockage of arteries in the lower limbs.

Hypotension (hī-pō-TEN-shun) Low blood pressure; most commonly used to describe an acute drop in blood pressure, as occurs during excessive blood loss.

Normotensive (nor'-mō-TEN-siv) Characterized by normal blood pressure.

Occlusion (ō-KLOO-zhun) The closure or obstruction of the lumen of a structure such as a blood vessel. An example is an atherosclerotic plaque in an artery.

Orthostatic hypotension (or'-thō-STAT-ik; *ortho-* = straight; *-static* = causing to stand) An excessive lowering of systemic blood pressure when a person assumes an erect or semierect posture; it is usually a sign of a disease. May be caused by excessive fluid loss, certain drugs, and cardiovascular or neurogenic factors. Also called **postural hypotension.**

Phlebitis (fle-BĪ-tis; *phleb-* = vein) Inflammation of a vein, often in a leg.

Thrombectomy (throm-BEK-tō-mē; *thrombo-* = clot) An operation to remove a blood clot from a blood vessel.

Thrombophlebitis (throm'-bō-fle-BĪ-tis) Inflammation of a vein involving clot formation. Superficial thrombophlebitis occurs in veins under the skin, especially in the calf.

Venipuncture (VEN-i-punk-chur; *vena-* = vein) The puncture of a vein, usually to withdraw blood for analysis or to introduce a solution, for example, an antibiotic. The median cubital vein is frequently used.

White coat (office) hypertension A stress-induced syndrome found in patients who have elevated blood pressure when being examined by health-care personnel, but otherwise have normal blood pressure.

CHAPTER REVIEW AND RESOURCE SUMMARY

Review

Resource

21.1 Structure and Function of Blood Vessels

1. Arteries carry blood away from the heart. The wall of an artery consists of a tunica interna, a tunica media (which maintains elasticity and contractility), and a tunica externa. Large arteries are termed elastic (conducting) arteries, and medium-sized arteries are called muscular (distributing) arteries.

2. Many arteries anastomose: The distal ends of two or more vessels unite. An alternative blood route from an anastomosis is called collateral circulation. Arteries that do not anastomose are called end arteries.

3. Arterioles are small arteries that deliver blood to capillaries. Through constriction and dilation, arterioles assume a key role in regulating blood flow from arteries into capillaries and in altering arterial blood pressure.

4. Capillaries are microscopic blood vessels through which materials are exchanged between blood and tissue cells; some capillaries are continuous, and others are fenestrated. Capillaries branch to form an extensive network throughout a tissue. This network increases surface area, allowing a rapid exchange of large quantities of materials.

5. Precapillary sphincters regulate blood flow through capillaries.

6. Microscopic blood vessels in the liver are called sinusoids.

7. Venules are small vessels that continue from capillaries and merge to form veins.

8. Veins consist of the same three tunics as arteries but have a thinner tunica interna and a thinner tunica media. The lumen of a vein is also larger than that of a comparable artery. Veins contain valves to prevent backflow of blood. Weak valves can lead to varicose veins.

9. Vascular (venous) sinuses are veins with very thin walls.

10. Systemic veins are collectively called blood reservoirs because they hold a large volume of blood. If the need arises, this blood can be shifted into other blood vessels through vasoconstriction of veins. The principal blood reservoirs are the veins of the abdominal organs (liver and spleen) and skin.

Anatomy Overview - The Cardiovascular System
Anatomy Overview - Arteries and Arterioles
Anatomy Overview - Capillaries
Figure 21.1 - Comparative Structure of Blood Vessels 
Exercise - Vein Archery
Exercise - Concentrate on Blood and Vessels

Review

21.2 Capillary Exchange

1. Substances enter and leave capillaries by diffusion, transcytosis, or bulk flow.
2. The movement of water and solutes (except proteins) through capillary walls depends on hydrostatic and osmotic pressures.
3. The near equilibrium between filtration and reabsorption in capillaries is called Starling's law of the capillaries.
4. Edema is an abnormal increase in interstitial fluid.

Animation - Capillary Exchange
Animation - Lymph Flow
Figure 21.7 - Dynamics of Capillary Exchange
Exercise - Capillary Exchange Pick 'em
Concepts and Connections - Capillary Exchange

21.3 Hemodynamics: Factors Affecting Blood Flow

1. The velocity of blood flow is inversely related to the cross-sectional area of blood vessels; blood flows slowest where cross-sectional area is greatest. The velocity of blood flow decreases from the aorta to arteries to capillaries and increases in venules and veins.
2. Blood pressure and resistance determine blood flow.
3. Blood flows from regions of higher to lower pressure. The higher the resistance, however, the lower the blood flow.
4. Cardiac output equals the mean arterial pressure divided by total resistance (CO = MAP ÷ R).
5. Blood pressure is the pressure exerted on the walls of a blood vessel.
6. Factors that affect blood pressure are cardiac output, blood volume, viscosity, resistance, and the elasticity of arteries.
7. As blood leaves the aorta and flows through the systemic circulation, its pressure progressively falls to 0 mmHg by the time it reaches the right ventricle.
8. Resistance depends on blood vessel diameter, blood viscosity, and total blood vessel length.
9. Venous return depends on pressure differences between the venules and the right ventricle.
10. Blood return to the heart is maintained by several factors, including skeletal muscular contractions, valves in veins (especially in the limbs), and pressure changes associated with breathing.

Animation - Structures Affecting Circulation
Figure 21.10 - Summary of Factors That Increase Blood Pressure
Concepts and Connections - Blood Flow

21.4 Control of Blood Pressure and Blood Flow

1. The cardiovascular (CV) center is a group of neurons in the medulla oblongata that regulates heart rate, contractility, and blood vessel diameter.
2. The cardiovascular center receives input from higher brain regions and sensory receptors (baroreceptors and chemoreceptors).
3. Output from the cardiovascular center flows along sympathetic and parasympathetic axons. Sympathetic impulses propagated along cardioaccelerator nerves increase heart rate and contractility; parasympathetic impulses propagated along vagus nerves decrease heart rate.
4. Baroreceptors monitor blood pressure, and chemoreceptors monitor blood levels of O_2, CO_2, and hydrogen ions. The carotid sinus reflex helps regulate blood pressure in the brain. The aortic reflex regulates general systemic blood pressure.
5. Hormones that help regulate blood pressure are epinephrine, norepinephrine, ADH (antidiuretic hormone), angiotensin II, and ANP (atrial natriuretic peptide).
6. Autoregulation refers to local, automatic adjustments of blood flow in a given region to meet a particular tissue's need.
7. O_2 level is the principal stimulus for autoregulation.

Animation - Vascular Regulation
Animation - Regulating Blood Pressure
Exercise - Regulate BP with Hormones
Exercise - Regulate BP with Nervous Impulses
Concepts and Connections - Blood Pressure Regulation

21.5 Checking Circulation

1. Pulse is the alternate expansion and elastic recoil of an artery wall with each heartbeat. It may be felt in any artery that lies near the surface or over a hard tissue.
2. A normal resting pulse (heart) rate is 70–80 beats/min.
3. Blood pressure is the pressure exerted by blood on the wall of an artery when the left ventricle undergoes systole and then diastole. It is measured by the use of a sphygmomanometer.
4. Systolic blood pressure (SBP) is the arterial blood pressure during ventricular contraction. Diastolic blood pressure (DBP) is the arterial blood pressure during ventricular relaxation. Normal blood pressure is less than 120/80.
5. Pulse pressure is the difference between systolic and diastolic blood pressure. It normally is about 40 mmHg.

Animation - MABP

Review	Resource

21.6 Shock and Homeostasis

1. Shock is a failure of the cardiovascular system to deliver enough O_2 and nutrients to meet the metabolic needs of cells.
2. Types of shock include hypovolemic, cardiogenic, vascular, and obstructive.
3. Signs and symptoms of shock include systolic blood pressure less than 90 mmHg; rapid resting heart rate; weak, rapid pulse; clammy, cool, pale skin; sweating; hypotension; altered mental state; decreased urinary output; thirst; and acidosis.

21.7 Circulatory Routes

Anatomy Overview - Comparison of Circulatory Routes
Anatomy Overview - Pulmonary Circulation
Figure 21.28 - Hepatic Portal Circulation
Figure 21.30 - Fetal Circulation and Changes at Birth

1. The two main circulatory routes are the systemic and pulmonary circulations.
2. Among the subdivisions of the systemic circulation are the coronary (cardiac) circulation and the hepatic portal circulation.
3. The systemic circulation carries oxygenated blood from the left ventricle through the aorta to all parts of the body, including some lung tissue, but *not* the air sacs (alveoli) of the lungs, and returns the deoxygenated blood to the right atrium.
4. The aorta is divided into the ascending aorta, the arch of the aorta, and the descending aorta. Each section gives off arteries that branch to supply the whole body.
5. Blood returns to the heart through the systemic veins. All veins of the systemic circulation drain into the superior or inferior venae cavae or the coronary sinus, which in turn empty into the right atrium.
6. The major blood vessels of the systemic circulation may be reviewed in Exhibits 21.A–21.L.
7. The hepatic portal circulation directs venous blood from the gastrointestinal organs and spleen into the hepatic portal vein of the liver before it returns to the heart. It enables the liver to utilize nutrients and detoxify harmful substances in the blood.
8. The pulmonary circulation takes deoxygenated blood from the right ventricle to the alveoli within the lungs and returns oxygenated blood from the alveoli to the left atrium.
9. Fetal circulation exists only in the fetus. It involves the exchange of materials between fetus and mother via the placenta.
10. The fetus derives O_2 and nutrients from and eliminates CO_2 and wastes into maternal blood. At birth, when pulmonary (lung), digestive, and liver functions begin to operate, the special structures of fetal circulation are no longer needed.

21.8 Development of Blood Vessels and Blood

1. Blood vessels develop from mesenchyme (hemangioblasts → angioblasts → blood islands) in mesoderm called blood islands.
2. Blood cells also develop from mesenchyme (hemangioblasts → pluripotent stem cells).
3. The development of blood cells from pluripotent stem cells derived from angioblasts occurs in the walls of blood vessels in the yolk sac, chorion, and allantois at about 3 weeks after fertilization. Within the embryo, blood is produced by the liver at about the fifth week and in the spleen, red bone marrow, and thymus at about the twelfth week.

21.9 Aging and the Cardiovascular System

Animation - Regulating Blood Pressure

1. General changes associated with aging include reduced compliance (distensibility) of blood vessels, reduction in cardiac muscle size, reduced cardiac output, and increased systolic blood pressure.
2. The incidence of coronary artery disease (CAD), congestive heart failure (CHF), and atherosclerosis increases with age.

SELF-QUIZ QUESTIONS

Fill in the blanks in the following statements.

1. The _____ reflex helps maintain normal blood pressure in the brain; the _____ reflex governs general systemic blood pressure.

2. In addition to the pressure created by contraction of the left ventricle, venous return is aided by the _____ and the _____, both of which depend on the presence of valves in the veins.

Indicate whether the following statements are true or false.

3. Baroreceptors and chemoreceptors are located in the aorta and carotid arteries.

4. The most important method of capillary exchange is simple diffusion.

Choose the one best answer to the following questions.

5. Which of the following are *not* true? (1) Muscular arteries are also known as conducting arteries. (2) Capillaries play a key role in regulating resistance. (3) The flow of blood through true capillaries is controlled by precapillary sphincters. (4) The lumen of an artery is larger than in a comparable vein. (5) Elastic arteries help propel blood. (6) The tunica media of arteries is thicker than the tunica media of veins.
   (a) 2, 3, and 6   (b) 1, 2, and 4   (c) 1, 2, 4, and 6
   (d) 3, 4, and 5   (e) 1, 2, 3, and 4

6. Which of the following are *true* concerning capillary exchange? (1) Large, lipid-insoluble molecules cross capillary walls by transcytosis. (2) The blood hydrostatic pressure promotes reabsorption of fluid into the capillaries. (3) If the pressures that promote filtration are greater than the pressures that promote reabsorption, fluid will move out of a capillary and into interstitial spaces. (4) A negative net filtration pressure results in reabsorption of fluid from interstitial spaces into a capillary. (5) The difference in osmotic pressure across a capillary wall is due primarily to red blood cells.
   (a) 1, 3, and 4   (b) 1, 2, 3, 4, and 5   (c) 1, 2, 3, and 4
   (d) 3 and 4   (e) 2, 4, and 5

7. Which of the following would *not* increase vascular resistance? (1) vasodilation, (2) polycythemia, (3) obesity, (4) dehydration, (5) anemia.
   (a) 1 and 2   (b) 1, 3, and 4   (c) 1 and 5
   (d) 1, 4, and 5   (e) 1 only

8. Capillary exchange is enhanced by (1) the slow rate of flow through the capillaries, (2) a small cross-sectional area, (3) the thinness of capillary walls, (4) the respiratory pump, (5) extensive branching, which increases the surface area.
   (a) 1, 2, 3, 4, and 5   (b) 1, 2, 3, and 5   (c) 1 and 3
   (d) 3 and 5   (e) 1, 3, and 5

9. Systemic vascular resistance depends on which of the following factors? (1) blood viscosity, (2) total blood vessel length, (3) size of the lumen, (4) type of blood vessel, (5) oxygen concentration of the blood.
   (a) 1, 2, and 3   (b) 2, 3, and 4   (c) 3, 4, and 5
   (d) 1, 3, and 5   (e) 2, 4, and 5

10. Which of the following help regulate blood pressure and help control regional blood flow? (1) baroreceptor and chemoreceptor reflexes, (2) hormones, (3) autoregulation, (4) H^+ concentration of blood, (5) oxygen concentration of the blood.
   (a) 1, 2, and 4   (b) 2, 4, and 5   (c) 1, 4, and 5
   (d) 1, 2, 3, 4, and 5   (e) 3, 4, and 5

11. For each of the following, indicate if it causes vasoconstriction or vasodilation. Use D for vasodilation and C for vasoconstriction.
   (a) atrial natriuretic peptide   (b) ADH
   (c) decrease in body temperature   (d) lactic acid
   (e) histamine   (f) hypoxia
   (g) hypercapnia   (h) angiotensin II
   (i) nitric oxide   (j) decreased sympathetic
   (k) acidosis   impulses

12. Match the following:
   ____ (a) pressure generated by the pumping of the heart; pushes fluids out of capillaries
   ____ (b) pressure created by proteins present in the interstitial fluid; pulls fluid out of capillaries
   ____ (c) balance of pressure; determines whether blood volume and interstitial fluid remain steady or change
   ____ (d) force due to presence of plasma proteins; pulls fluid into capillaries from interstitial spaces
   ____ (e) pressure due to fluid in interstitial spaces; pushes fluid back into capillaries

   (1) net filtration pressure
   (2) blood hydrostatic pressure
   (3) interstitial fluid hydrostatic pressure
   (4) blood colloid osmotic pressure
   (5) interstitial fluid osmotic pressure

13. Match the following:
   ____ (a) supplies blood to the kidney
   ____ (b) drains blood from the small intestine, portions of the large intestine, stomach, and pancreas
   ____ (c) main blood supply to arm; commonly used to measure blood pressure
   ____ (d) supply blood to the free lower limbs
   ____ (e) drain oxygenated blood from the lungs and send it to the left atrium
   ____ (f) supplies blood to the stomach, liver, and pancreas
   ____ (g) supply blood to the brain
   ____ (h) supplies blood to the large intestine
   ____ (i) drain blood from the head
   ____ (j) detours venous blood from the gastrointestinal organs and spleen through the liver before it returns to the heart
   ____ (k) drains most of the thorax and abdominal wall; can serve as a bypass for the inferior vena cava
   ____ (l) a part of the venous circulation of the leg; a vessel used in heart bypass surgery
   ____ (m) carry deoxygenated blood from the right ventricle to the lungs

   (1) superior mesenteric vein
   (2) inferior mesenteric artery
   (3) pulmonary veins
   (4) brachial artery
   (5) hepatic portal circulation
   (6) carotid arteries
   (7) jugular veins
   (8) celiac trunk
   (9) common iliac arteries
   (10) azygos veins
   (11) renal artery
   (12) great saphenous vein
   (13) pulmonary arteries

14. Match the following:
    ____ (a) a traveling pressure wave created by the alternate expansion and recoil of elastic arteries after each systole of the left ventricle
    ____ (b) the lowest blood pressure in arteries during ventricular relaxation
    ____ (c) a slow resting heart rate or pulse rate
    ____ (d) an inadequate cardiac output that results in a failure of the cardiovascular system to deliver enough oxygen and nutrients to meet the metabolic needs of body cells
    ____ (e) a rapid resting heart rate or pulse rate
    ____ (f) the highest force with which blood pushes against arterial walls as a result of ventricular contraction

    (1) shock
    (2) pulse
    (3) tachycardia
    (4) bradycardia
    (5) systolic blood pressure
    (6) diastolic blood pressure

15. Match the following (some answers will be used more than once):
    ____ (a) returns oxygenated blood from the placenta to the fetal liver
    ____ (b) an opening in the septum between the right and left atria
    ____ (c) becomes the ligamentum venosum after birth
    ____ (d) pass blood from the fetus to the placenta
    ____ (e) bypasses the nonfunctioning lungs; becomes the ligamentum arteriosum at birth
    ____ (f) become the medial umbilical ligaments at birth
    ____ (g) transports oxygenated blood into the inferior vena cava
    ____ (h) becomes the ligamentum teres at birth
    ____ (i) becomes the fossa ovalis after birth

    (1) ductus venosus
    (2) ductus arteriosus
    (3) foramen ovale
    (4) umbilical arteries
    (5) umbilical vein

CRITICAL THINKING QUESTIONS

1. Kim Sung was told that her baby was born with a hole in the upper chambers of his heart. Is this something Kim Sung should worry about?

2. Michael was brought into the emergency room suffering from a gunshot wound. He is bleeding profusely and exhibits the following: systolic blood pressure is 40 mmHg; weak pulse of 200 beats per minute; cool, pale, and clammy skin. Michael is not producing urine but is asking for water. He is confused and disoriented. What is his diagnosis and what, specifically, is causing these symptoms?

3. Maureen's job entails standing on a concrete floor for 10-hour days on an assembly line. Lately she has noticed swelling in her ankles at the end of the day and some tenderness in her calves. What do you suspect is Maureen's problem and how could she help counteract the problem?

? ANSWERS TO FIGURE QUESTIONS

21.1 The femoral artery has the thicker wall; the femoral vein has the wider lumen.

21.2 Due to atherosclerosis, less energy is stored in the less-compliant elastic arteries during systole; thus, the heart must pump harder to maintain the same rate of blood flow.

21.3 Metabolically active tissues use O_2 and produce wastes more rapidly than inactive tissues, so they require more extensive capillary networks.

21.4 Materials cross capillary walls through intercellular clefts and fenestrations, via transcytosis in pinocytic vesicles, and through the plasma membranes of endothelial cells.

21.5 Valves are more important in arm veins and leg veins than in neck veins because, when you are standing, gravity causes pooling of blood in the veins of the free limbs but aids the flow of blood in neck veins back toward the heart.

21.6 Blood volume in venules and veins is about 64% of 5 liters, or 3.2 liters; blood volume in capillaries is about 7% of 5 liters, or 350 mL.

21.7 Blood colloid osmotic pressure is lower than normal in a person with a low level of plasma proteins, and therefore capillary reabsorption is low. The result is edema.

21.8 Mean blood pressure in the aorta is closer to diastolic than to systolic pressure.

21.9 The skeletal muscle pump and respiratory pump aid venous return.

21.10 Vasodilation and vasoconstriction of arterioles are the main regulators of systemic vascular resistance.

21.11 Velocity of blood flow is fastest in the aorta and arteries.

21.12 The effector tissues regulated by the cardiovascular center are cardiac muscle in the heart and smooth muscle in blood vessel walls.

21.13 Impulses to the cardiovascular center pass from baroreceptors in the carotid sinuses via the glossopharyngeal (IX) nerves and from baroreceptors in the arch of the aorta via the vagus (X) nerves.

21.14 It represents a change that occurs when you stand up because gravity causes pooling of blood in leg veins once you are upright, decreasing the blood pressure in your upper body.

21.15 Diastolic blood pressure = 95 mmHg; systolic blood pressure = 142 mmHg; pulse pressure = 47 mmHg. This person has stage I hypertension because the systolic blood pressure is greater than 140 mmHg and the diastolic blood pressure is greater than 90 mmHg.

21.16 Almost-normal blood pressure in a person who has lost blood does not necessarily indicate that the patient's tissues are receiving adequate blood flow; if systemic vascular resistance has increased greatly, tissue perfusion may be inadequate.

21.17 The two main circulatory routes are the systemic circulation and the pulmonary circulation.

21.18 The subdivisions of the aorta are the ascending aorta, arch of the aorta, thoracic aorta, and abdominal aorta.

21.19 Branches of the arch of the aorta (in order of origination) are the brachiocephalic trunk, left common carotid artery, and left subclavian artery.

21.20 The thoracic aorta begins at the level of the intervertebral disc between T4 and T5.

21.21 The abdominal aorta begins at the aortic hiatus in the diaphragm.

21.22 The abdominal aorta divides into the common iliac arteries at about the level of L4.

21.23 The superior vena cava drains regions above the diaphragm, and the inferior vena cava drains regions below the diaphragm.

21.24 All venous blood in the brain drains into the internal jugular veins.

21.25 The median cubital vein of the upper limb is often used for withdrawing blood.

21.26 The inferior vena cava returns blood from abdominopelvic viscera to the heart.

21.27 Superficial veins of the lower limbs are the dorsal venous arches and the great saphenous and small saphenous veins.

21.28 The hepatic veins carry blood away from the liver.

21.29 After birth, the pulmonary arteries are the only arteries that carry deoxygenated blood.

21.30 Exchange of materials between mother and fetus occurs across the placenta.

21.31 Blood vessels and blood are derived from mesoderm.

22

THE LYMPHATIC SYSTEM AND IMMUNITY

THE LYMPHATIC SYSTEM, DISEASE RESISTANCE, AND HOMEOSTASIS *The lymphatic system contributes to homeostasis by draining interstitial fluid as well as providing the mechanisms for defense against disease.*

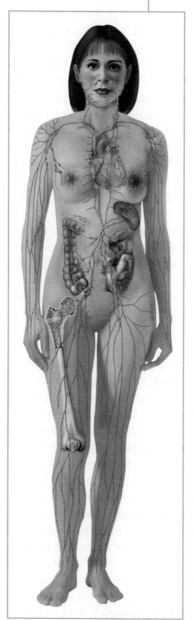

Maintaining homeostasis in the body requires continual combat against harmful agents in our internal and external environment. Despite constant exposure to a variety of **pathogens** (PATH-ō-jens)—disease-producing microbes such as bacteria and viruses—most people remain healthy. The body surface also endures cuts and bumps, exposure to ultraviolet rays, chemical toxins, and minor burns with an array of defensive ploys.

Immunity (i-MŪ-ni-tē) or **resistance** is the ability to ward off damage or disease through our defenses. Vulnerability or lack of resistance is termed **susceptibility.** The two general types of immunity are (1) innate and (2) adaptive. **Innate (nonspecific) immunity** refers to defenses that are present at birth. Innate immunity does not involve specific recognition of a microbe and acts against all microbes in the same way. Among the components of innate immunity are the first line of defense (the physical and chemical barriers of the skin and mucous membranes) and the second line of defense (antimicrobial substances, natural killer cells, phagocytes, inflammation, and fever). Innate immune responses represent immunity's early warning system and are designed to prevent microbes from gaining access into the body and to help eliminate those that do gain access.

Adaptive (specific) immunity refers to defenses that involve specific recognition of a microbe once it has breached the innate immunity defenses. Adaptive immunity is based on a specific response to a specific microbe; that is, it adapts or adjusts to handle a specific microbe. Adaptive immunity involves lymphocytes (a type of white blood cell) called T lymphocytes (T cells) and B lymphocytes (B cells).

The body system responsible for adaptive immunity (and some aspects of innate immunity) is the lymphatic system. This system is closely allied with the cardiovascular system, and it also functions with the digestive system in the absorption of fatty foods. In this chapter, we explore the mechanisms that provide defenses against intruders and promote the repair of damaged body tissues.

?

Did you ever wonder how cancer can spread from one part of the body to another?

22.1 LYMPHATIC SYSTEM STRUCTURE AND FUNCTION

⦿ **OBJECTIVES**

- List the components and major functions of the lymphatic system.
- Describe the organization of lymphatic vessels.
- Explain the formation and flow of lymph.
- Compare the structure and functions of the primary and secondary lymphatic organs and tissues.

The **lymphatic system** (lim-FAT-ik) consists of a fluid called lymph, vessels called lymphatic vessels that transport the lymph, a number of structures and organs containing lymphatic tissue (lymphocytes within a filtering tissue), and red bone marrow (Figure 22.1). The lymphatic system assists in circulating body fluids and helps defend the body against disease-causing agents. As you will see shortly, most components of blood plasma filter through blood capillary walls to form interstitial fluid. After interstitial fluid passes into lymphatic vessels, it is called **lymph** (LIMF = clear fluid). The major difference between interstitial fluid and lymph is location: Interstitial fluid is found between cells, and lymph is located within lymphatic vessels and lymphatic tissue.

Lymphatic tissue is a specialized form of reticular connective tissue (see Table 4.4) that contains large numbers of lymphocytes. Recall from Chapter 19 that lymphocytes are agranular white blood cells (see Section 19.4). Two types of lymphocytes participate in adaptive immune responses: B cells and T cells.

Functions of the Lymphatic System

The lymphatic system has three primary functions:

1. *Drains excess interstitial fluid.* Lymphatic vessels drain excess interstitial fluid from tissue spaces and return it to the blood.

2. *Transports dietary lipids.* Lymphatic vessels transport lipids and lipid-soluble vitamins (A, D, E, and K) absorbed by the gastrointestinal tract.

3. *Carries out immune responses.* Lymphatic tissue initiates highly specific responses directed against particular microbes or abnormal cells.

Lymphatic Vessels and Lymph Circulation

Lymphatic vessels begin as **lymphatic capillaries.** These capillaries, which are located in the spaces between cells, are closed at one end (Figure 22.2). Just as blood capillaries converge to form venules and then veins, lymphatic capillaries unite to form larger **lymphatic vessels** (see Figure 22.1), which resemble small veins in structure but have thinner walls and more valves. At intervals along the lymphatic vessels, lymph flows through lymph nodes, encapsulated bean-shaped organs consisting of masses of B cells and T cells. In the skin, lymphatic vessels lie in the subcutaneous tissue and generally follow the same route as veins; lymphatic vessels of the viscera generally follow arteries, forming plexuses (networks) around them. Tissues that lack lymphatic capillaries include avascular tissues (such as cartilage, the epidermis, and the cornea of the eye), the central nervous system, portions of the spleen, and red bone marrow.

Lymphatic Capillaries

Lymphatic capillaries have greater permeability than blood capillaries and thus can absorb large molecules such as proteins and lipids. Lymphatic capillaries are also slightly larger in diameter than blood capillaries and have a unique one-way structure that permits interstitial fluid to flow into them but not out. The ends of endothelial cells that make up the wall of a lymphatic capillary overlap (Figure 22.2b). When pressure is greater in the interstitial fluid than in lymph, the cells separate slightly, like the opening of a one-way swinging door, and interstitial fluid enters the lymphatic capillary. When pressure is greater inside the lymphatic capillary, the cells adhere more closely, and lymph cannot escape back into interstitial fluid. The pressure is relieved as lymph moves further down the lymphatic capillary. Attached to the lymphatic capillaries are *anchoring filaments,* which contain elastic fibers. They extend out from the lymphatic capillary, attaching lymphatic endothelial cells to surrounding tissues. When excess interstitial fluid accumulates and causes tissue swelling, the anchoring filaments are pulled, making the openings between cells even larger so that more fluid can flow into the lymphatic capillary.

In the small intestine, specialized lymphatic capillaries called **lacteals** (LAK-tē-als; *lact-* = milky) carry dietary lipids into lymphatic vessels and ultimately into the blood (see Figure 24.20). The presence of these lipids causes the lymph draining from the small intestine to appear creamy white; such lymph is referred to as **chyle** (KĪL = juice). Elsewhere, lymph is a clear, pale-yellow fluid.

Lymph Trunks and Ducts

As you have already learned, lymph passes from lymphatic capillaries into lymphatic vessels and then through lymph nodes. As lymphatic vessels exit lymph nodes in a particular region of the body, they unite to form **lymph trunks.** The principal trunks are the lumbar, intestinal, bronchomediastinal, subclavian, and jugular trunks (Figure 22.3). The **lumbar trunks** drain lymph from the lower limbs, the wall and viscera of the pelvis, the kidneys, the adrenal glands, and the abdominal wall. The **intestinal trunk** drains lymph from the stomach, intestines, pancreas, spleen, and part of the liver. The **bronchomediastinal trunks** (brong-kō-mē′-dē-as-TĪ-nal) drain lymph from the thoracic wall, lung, and heart. The **subclavian trunks** drain the upper limbs. The **jugular trunks** drain the head and neck.

Lymph passes from lymph trunks into two main channels, the thoracic duct and the right lymphatic duct, and then drains into venous blood. The **thoracic (left lymphatic) duct** is about 38–45 cm (15–18 in.) long and begins as a dilation called the **cisterna chyli** (sis-TER-na KĪ-lē; *cisterna* = cavity or reservoir) anterior to the second lumbar vertebra. The thoracic duct is the main duct for the return of lymph to blood. The cisterna chyli receives

Figure 22.1 Components of the lymphatic system.

The lymphatic system consists of lymph, lymphatic vessels, lymphatic tissues, and red bone marrow.

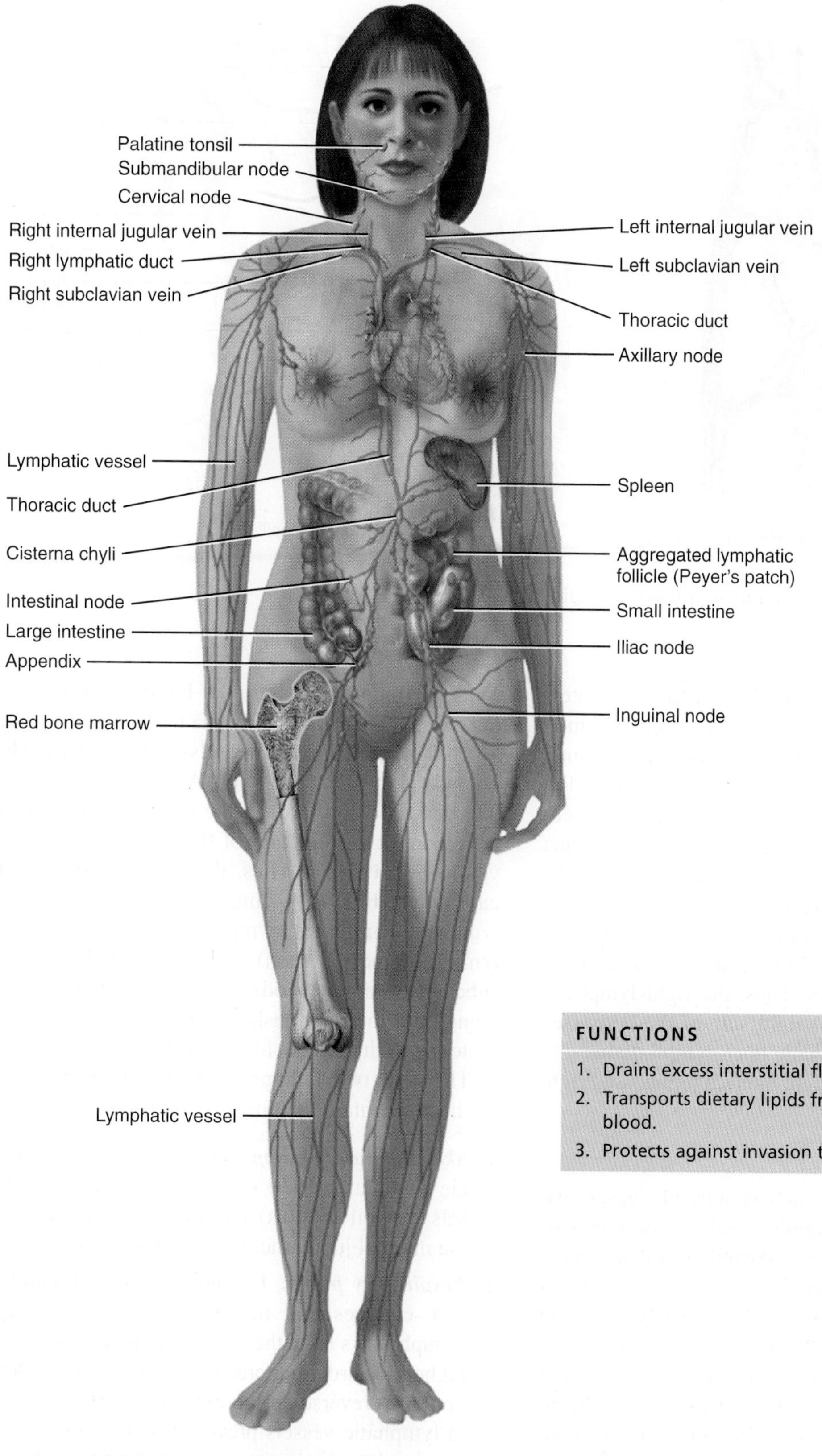

Palatine tonsil
Submandibular node
Cervical node
Right internal jugular vein
Right lymphatic duct
Right subclavian vein

Left internal jugular vein
Left subclavian vein
Thoracic duct
Axillary node

Lymphatic vessel
Thoracic duct
Cisterna chyli
Intestinal node
Large intestine
Appendix

Spleen

Aggregated lymphatic follicle (Peyer's patch)
Small intestine
Iliac node

Red bone marrow

Inguinal node

Lymphatic vessel

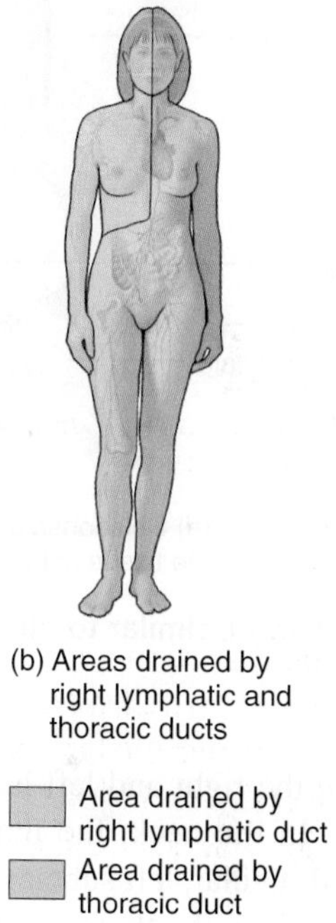

(b) Areas drained by right lymphatic and thoracic ducts

Area drained by right lymphatic duct

Area drained by thoracic duct

(a) Anterior view of principal components of lymphatic system

FUNCTIONS

1. Drains excess interstitial fluid.
2. Transports dietary lipids from the gastrointestinal tract to the blood.
3. Protects against invasion through immune responses.

? What tissue contains stem cells that develop into lymphocytes?

Figure 22.2 Lymphatic capillaries.

🔑 Lymphatic capillaries are found throughout the body except in avascular tissues, the central nervous system, portions of the spleen, and bone marrow.

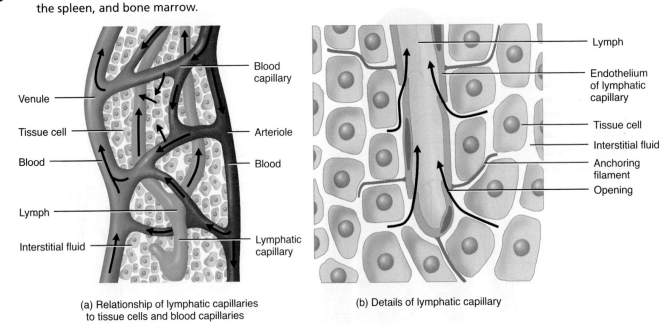

(a) Relationship of lymphatic capillaries to tissue cells and blood capillaries

(b) Details of lymphatic capillary

❓ **Is lymph more similar to blood plasma or to interstitial fluid? Why?**

lymph from the right and left lumbar trunks and from the intestinal trunk. In the neck, the thoracic duct also receives lymph from the left jugular, left subclavian, and left bronchomediastinal trunks. Therefore, the thoracic duct receives lymph from the left side of the head, neck, and chest, the left upper limb, and the entire body inferior to the ribs (see Figure 22.1b). The thoracic duct in turn drains lymph into venous blood at the junction of the left internal jugular and left subclavian veins.

The **right lymphatic duct** (Figure 22.3) is about 1.2 cm (0.5 in.) long and receives lymph from the right jugular, right subclavian, and right bronchomediastinal trunks. Thus, the right lymphatic duct receives lymph from the upper right side of the body (see Figure 22.1b). From the right lymphatic duct, lymph drains into venous blood at the junction of the right internal jugular and right subclavian veins.

Formation and Flow of Lymph

Most components of blood plasma, such as nutrients, gases, and hormones, filter freely through the capillary walls to form interstitial fluid, but more fluid filters out of blood capillaries than returns to them by reabsorption (see Figure 21.7). The excess filtered fluid—about 3 liters per day—drains into lymphatic vessels and becomes lymph. Because most plasma proteins are too large to leave blood vessels, interstitial fluid contains only a small amount of protein. Proteins that do leave blood plasma cannot return to the blood by diffusion because the concentration gradient (high level of proteins inside blood capillaries, low level outside) opposes such movement. The proteins can, however, move readily through the more permeable lymphatic capillaries into lymph.

Thus, an important function of lymphatic vessels is to return the lost plasma proteins and plasma to the bloodstream.

Like veins, lymphatic vessels contain valves, which ensure the one-way movement of lymph. As noted previously, lymph drains into venous blood through the right lymphatic duct and the thoracic duct at the junction of the internal jugular and subclavian veins (Figure 22.3). Thus, the sequence of fluid flow is blood capillaries (blood) → interstitial spaces (interstitial fluid) → lymphatic capillaries (lymph) → lymphatic vessels (lymph) → lymphatic ducts (lymph) → junction of the internal jugular and subclavian veins (blood). Figure 22.4 illustrates this sequence, along with the relationship of the lymphatic and cardiovascular systems. Both systems form a very efficient circulatory system.

The same two "pumps" that aid the return of venous blood to the heart maintain the flow of lymph.

1. **Skeletal muscle pump.** The "milking action" of skeletal muscle contractions (see Figure 21.9) compresses lymphatic vessels (as well as veins) and forces lymph toward the junction of the internal jugular and subclavian veins.

2. **Respiratory pump.** Lymph flow is also maintained by pressure changes that occur during inhalation (breathing in). Lymph flows from the abdominal region, where the pressure is higher, toward the thoracic region, where it is lower. When the pressures reverse during exhalation (breathing out), the valves in lymphatic vessels prevent backflow of lymph. In addition, when a lymphatic vessel distends, the smooth muscle in its wall contracts, which helps move lymph from one segment of the vessel to the next.

Figure 22.3 Routes for drainage of lymph from lymph trunks into the thoracic and right lymphatic ducts.

All lymph returns to the bloodstream through the thoracic (left) lymphatic duct and right lymphatic duct.

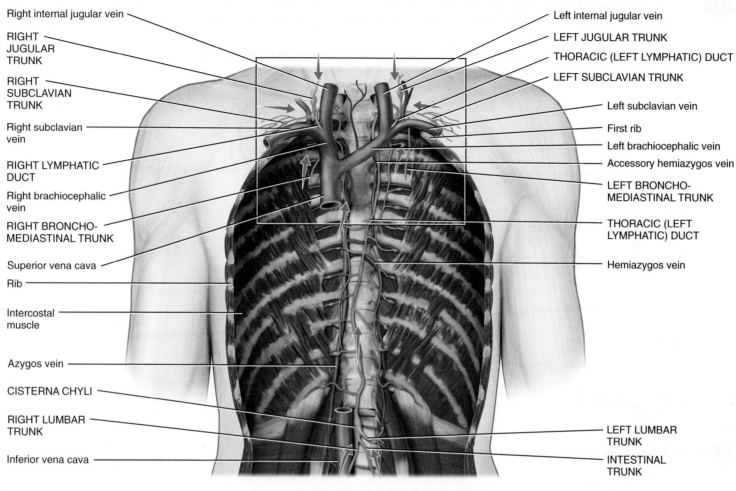

Right internal jugular vein

RIGHT JUGULAR TRUNK

RIGHT SUBCLAVIAN TRUNK

Right subclavian vein

RIGHT LYMPHATIC DUCT

Right brachiocephalic vein

RIGHT BRONCHO-MEDIASTINAL TRUNK

Superior vena cava

Rib

Intercostal muscle

Azygos vein

CISTERNA CHYLI

RIGHT LUMBAR TRUNK

Inferior vena cava

Left internal jugular vein

LEFT JUGULAR TRUNK

THORACIC (LEFT LYMPHATIC) DUCT

LEFT SUBCLAVIAN TRUNK

Left subclavian vein

First rib

Left brachiocephalic vein

Accessory hemiazygos vein

LEFT BRONCHO-MEDIASTINAL TRUNK

THORACIC (LEFT LYMPHATIC) DUCT

Hemiazygos vein

LEFT LUMBAR TRUNK

INTESTINAL TRUNK

(a) Overall anterior view

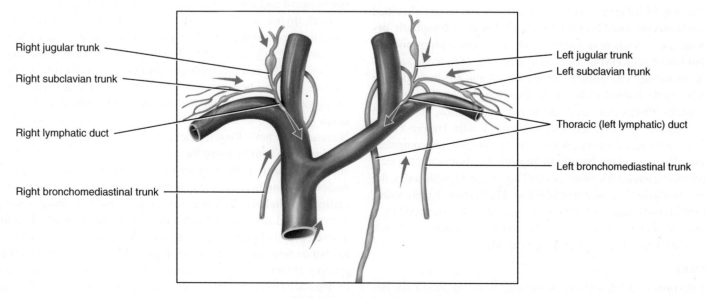

Right jugular trunk

Right subclavian trunk

Right lymphatic duct

Right bronchomediastinal trunk

Left jugular trunk

Left subclavian trunk

Thoracic (left lymphatic) duct

Left bronchomediastinal trunk

(b) Detailed anterior view

Which lymphatic vessels empty into the cisterna chyli, and which duct receives lymph from the cisterna chyli?

Figure 22.4 Schematic diagram showing the relationship of the lymphatic system to the cardiovascular system. Arrows indicate the direction of flow of lymph and blood.

 The sequence of fluid flow is blood capillaries (blood) → interstitial spaces (interstitial fluid) → lymphatic capillaries (lymph) → lymphatic vessels (lymph) → lymphatic ducts (lymph) → junction of the internal jugular and subclavian veins (blood).

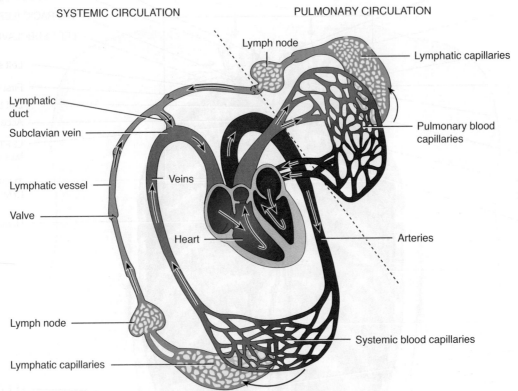

? Does inhalation promote or hinder the flow of lymph?

Lymphatic Organs and Tissues

The widely distributed lymphatic organs and tissues are classified into two groups based on their functions. **Primary lymphatic organs** are the sites where stem cells divide and become **immunocompetent** (im′-ū-nō-KOM-pe-tent), that is, capable of mounting an immune response. The primary lymphatic organs are the red bone marrow (in flat bones and the epiphyses of long bones of adults) and the thymus. Pluripotent stem cells in red bone marrow give rise to mature, immunocompetent B cells and to pre-T cells. The pre-T cells in turn migrate to the thymus, where they become immunocompetent T cells. The **secondary lymphatic organs** and **tissues** are the sites where most immune responses occur. They include lymph nodes, the spleen, and lymphatic nodules (follicles). The thymus, lymph nodes, and spleen are considered organs because each is surrounded by a connective tissue capsule; lymphatic nodules, in contrast, are not considered organs because they lack a capsule.

Thymus

The **thymus** is a bilobed organ located in the mediastinum between the sternum and the aorta (Figure 22.5a). An enveloping layer of connective tissue holds the two lobes closely together, but a connective tissue **capsule** separates the two. Extensions of the

capsule, called **trabeculae** (tra-BEK-ū-lē = little beams), penetrate inward and divide each lobe into **lobules** (Figure 22.5b).

Each thymic lobule consists of a deeply staining outer cortex and a lighter-staining central medulla (Figure 22.5b). The **cortex** is composed of large numbers of T cells and scattered dendritic cells, epithelial cells, and macrophages. Immature T cells (pre-T cells) migrate from red bone marrow to the cortex of the thymus, where they proliferate and begin to mature. **Dendritic cells** (den-DRIT-ik; *dendr-* = a tree), which are derived from monocytes, and so named because they have long, branched projections that resemble the dendrites of a neuron, assist the maturation process. As you will see shortly, dendritic cells in other parts of the body, such as lymph nodes, play another key role in immune responses. Each of the specialized **epithelial cells** in the cortex has several long processes that surround and serve as a framework for as many as 50 T cells. These epithelial cells help "educate" the pre-T cells in a process known as positive selection (see Figure 22.22). Additionally, they produce thymic hormones that are thought to aid in the maturation of T cells. Only about 2% of developing T cells survive in the cortex. The remaining cells die via apoptosis (programmed cell death). Thymic **macrophages** help clear out the debris of dead and dying cells. The surviving T cells enter the medulla.

Figure 22.5 Thymus.

The bilobed thymus is largest at puberty and then the functional portion atrophies with age.

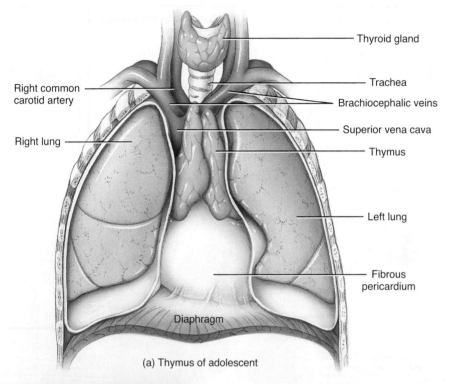

Right common carotid artery

Right lung

Thyroid gland

Trachea

Brachiocephalic veins

Superior vena cava

Thymus

Left lung

Fibrous pericardium

Diaphragm

(a) Thymus of adolescent

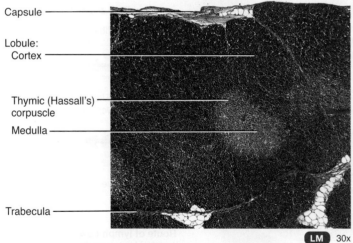

Capsule

Lobule:
Cortex

Thymic (Hassall's) corpuscle

Medulla

Trabecula

LM 30x

(b) Thymic lobules

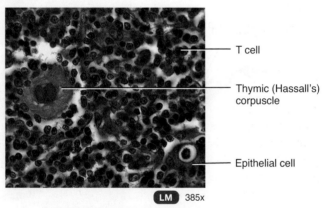

T cell

Thymic (Hassall's) corpuscle

Epithelial cell

LM 385x

(c) Details of thymic medulla

? Which type of lymphocytes mature in the thymus?

The **medulla** consists of widely scattered, more mature T cells, epithelial cells, dendritic cells, and macrophages (Figure 22.5c). Some of the epithelial cells become arranged into concentric layers of flat cells that degenerate and become filled with keratohyalin granules and keratin. These clusters are called **thymic (Hassall's) corpuscles.** Although their role is uncertain, they may serve as sites of T cell death in the medulla. T cells that leave the thymus via the blood migrate to lymph nodes, the spleen, and other lymphatic tissues, where they colonize parts of these organs and tissues.

Because of its high content of lymphoid tissue and a rich blood supply, the thymus has a reddish appearance in a living body. With age, however, fatty infiltrations replace the lymphoid tissue and the thymus takes on more of the yellowish color of the invading fat, giving the false impression of reduced size. However, the actual size of the thymus, defined by its connective tissue capsule, does not change. In infants, the thymus has a mass of about 70 g (2.3 oz). It is after puberty that adipose and areolar connective tissue begin to replace the thymic tissue. By the time a person reaches maturity, the functional portion of the gland is reduced

Figure 22.6 Structure of a lymph node. Arrows indicate the direction of lymph flow through a lymph node.

Lymph nodes are present throughout the body, usually clustered in groups.

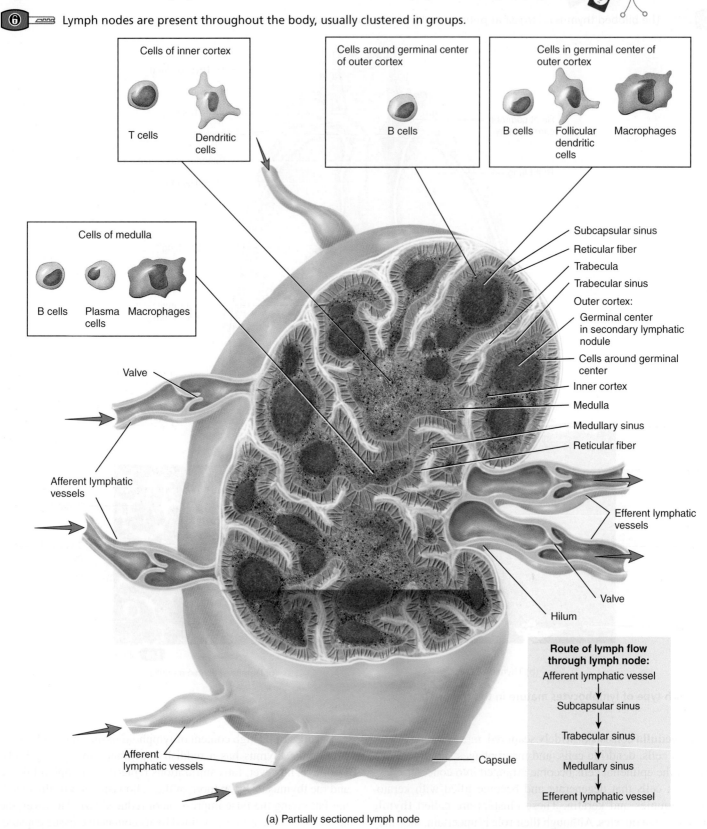

Cells of inner cortex

T cells   Dendritic cells

Cells around germinal center of outer cortex

B cells

Cells in germinal center of outer cortex

B cells   Follicular dendritic cells   Macrophages

Cells of medulla

B cells   Plasma cells   Macrophages

Subcapsular sinus

Reticular fiber

Trabecula

Trabecular sinus

Outer cortex:

Germinal center in secondary lymphatic nodule

Cells around germinal center

Inner cortex

Medulla

Medullary sinus

Reticular fiber

Valve

Afferent lymphatic vessels

Efferent lymphatic vessels

Valve

Hilum

Route of lymph flow through lymph node:

Afferent lymphatic vessel
↓
Subcapsular sinus
↓
Trabecular sinus
↓
Medullary sinus
↓
Efferent lymphatic vessel

Afferent lymphatic vessels

Capsule

(a) Partially sectioned lymph node

considerably, and in old age the functional portion may weigh only 3 g (0.1 oz). Before the thymus atrophies, it populates the secondary lymphatic organs and tissues with T cells. However, some T cells continue to proliferate in the thymus throughout an individual's lifetime, but this number decreases with age.

Lymph Nodes

Located along lymphatic vessels are about 600 bean-shaped **lymph nodes.** They are scattered throughout the body, both superficially and deep, and usually occur in groups (see Figure 22.1). Large groups of lymph nodes are present near the mammary glands and in the axillae and groin.

Lymph nodes are 1–25 mm (0.04–1 in.) long and, like the thymus, are covered by a **capsule** of dense connective tissue that extends into the node (Figure 22.6). The capsular extensions, called **trabeculae** (tra-BEK-ū-lē), divide the node into compartments, provide support, and provide a route for blood vessels into the interior of a node. Internal to the capsule is a supporting network of reticular fibers and fibroblasts. The capsule, trabeculae, reticular fibers, and fibroblasts constitute the *stroma* (supporting framework of connective tissue) of a lymph node.

The *parenchyma* (functioning part) of a lymph node is divided into a superficial cortex and a deep medulla. The cortex consists of an outer cortex and an inner cortex. Within the **outer cortex** are

egg-shaped aggregates of B cells called **lymphatic nodules (follicles).** A lymphatic nodule consisting chiefly of B cells is called a *primary lymphatic nodule.* Most lymphatic nodules in the outer cortex are *secondary lymphatic nodules* (Figure 22.6), which form in response to an antigen (a foreign substance) and are sites of plasma cell and memory B cell formation. After B cells in a primary lymphatic nodule recognize an antigen, the primary lymphatic nodule develops into a secondary lymphatic nodule. The center of a secondary lymphatic nodule contains a region of light-staining cells called a *germinal center.* In the germinal center are B cells, follicular dendritic cells (a special type of dendritic cell), and macrophages. When follicular dendritic cells "present" an antigen (described later in the chapter), B cells proliferate and develop into antibody-producing plasma cells or develop into memory B cells. Memory B cells persist after an initial immune response and "remember" having encountered a specific antigen. B cells that do not develop properly undergo apoptosis (programmed cell death) and are destroyed by macrophages. The region of a secondary lymphatic nodule surrounding the germinal center is composed of dense accumulations of B cells that have migrated away from their site of origin within the nodule.

The **inner cortex** does not contain lymphatic nodules. It consists mainly of T cells and dendritic cells that enter a lymph node from other tissues. The dendritic cells present antigens to T cells, causing their proliferation. The newly formed T cells then migrate from the lymph node to areas of the body where there is antigenic activity.

The **medulla** of a lymph node contains B cells, antibody-producing plasma cells that have migrated out of the cortex into the medulla, and macrophages. The various cells are embedded in a network of reticular fibers and reticular cells.

As you have already learned, lymph flows through a node in one direction only (Figure 22.6a). It enters through several

Capsule
Subcapsular sinus
Outer cortex
Trabecular sinus
Germinal center in secondary lymphatic nodule
Trabecula
Inner cortex
Medullary sinus
Medulla

LM 40x

(b) Portion of a lymph node

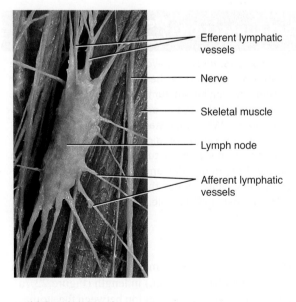

Efferent lymphatic vessels
Nerve
Skeletal muscle
Lymph node
Afferent lymphatic vessels

(c) Anterior view of inguinal lymph node

 What happens to foreign substances in lymph that enter a lymph node?

afferent lymphatic vessels (AF-er-ent; *afferent* = to carry toward), which penetrate the convex surface of the node at several points. The afferent vessels contain valves that open toward the center of the node, directing the lymph *inward*. Within the node, lymph enters **sinuses,** a series of irregular channels that contain branching reticular fibers, lymphocytes, and macrophages. From the afferent lymphatic vessels, lymph flows into the **subcapsular sinus** (sub-KAP-soo-lar), immediately beneath the capsule. From here the lymph flows through **trabecular sinuses** (tra-BEK-ū-lar), which extend through the cortex parallel to the trabeculae, and into **medullary sinuses,** which extend through the medulla. The medullary sinuses drain into one or two **efferent lymphatic vessels** (EF-er-ent; *efferent* = to carry away), which are wider and fewer in number than afferent vessels. They contain valves that open away from the center of the lymph node to convey lymph, antibodies secreted by plasma cells, and activated T cells *out* of the node. Efferent lymphatic vessels emerge from one side of the lymph node at a slight depression called a **hilum** (HĪ-lum) or **hilus.** Blood vessels also enter and leave the node at the hilum.

Lymph nodes function as a type of filter. As lymph enters one end of a lymph node, foreign substances are trapped by the reticular fibers within the sinuses of the node. Then macrophages destroy some foreign substances by phagocytosis, while lymphocytes destroy others by immune responses. The filtered lymph then leaves the other end of the lymph node. Since there are many afferent lymphatic vessels that bring lymph into a lymph node and only one or two efferent lymphatic vessels that transport lymph out of a lymph node, the slow flow of lymph within the lymph nodes allows additional time for lymph to be filtered. Additionally, all lymph flows through multiple lymph nodes on its path through the lymph vessels. This exposes the lymph to multiple filtering events before returning to the blood.

CLINICAL CONNECTION | *Metastasis through Lymphatic Vessels*

Metastasis (me-TAS-ta-sis; *meta-* = beyond; *-stasis* = to stand), the spread of a disease from one part of the body to another, can occur via lymphatic vessels. All malignant tumors eventually metastasize. Cancer cells may travel in the blood or lymph and establish new tumors where they lodge. When metastasis occurs via lymphatic vessels, secondary tumor sites can be predicted according to the direction of lymph flow from the primary tumor site. Cancerous lymph nodes feel enlarged, firm, nontender, and fixed to underlying structures. By contrast, most lymph nodes that are enlarged due to an infection are softer, tender, and movable. •

Spleen

The oval **spleen** is the largest single mass of lymphatic tissue in the body, measuring about 12 cm (5 in.) in length (Figure 22.7a). It is located in the left hypochondriac region between the stomach and diaphragm. The superior surface of the spleen is smooth and convex and conforms to the concave surface of the diaphragm. Neighboring organs make indentations in the visceral surface of

the spleen—the *gastric impression* (stomach), the *renal impression* (left kidney), and the *colic impression* (left colic flexure of large intestine). Like lymph nodes, the spleen has a hilum. Through it pass the splenic artery, splenic vein, and efferent lymphatic vessels.

A capsule of dense connective tissue surrounds the spleen and is covered in turn by a serous membrane, the visceral peritoneum. Trabeculae extend inward from the capsule. The capsule plus trabeculae, reticular fibers, and fibroblasts constitute the stroma of the spleen; the parenchyma of the spleen consists of two different kinds of tissue called white pulp and red pulp (Figure 22.7b, c). **White pulp** is lymphatic tissue, consisting mostly of lymphocytes and macrophages arranged around branches of the splenic artery called **central arteries.** The **red pulp** consists of bloodfilled **venous sinuses** and cords of splenic tissue called **splenic (Billroth's) cords.** Splenic cords consist of red blood cells, macrophages, lymphocytes, plasma cells, and granulocytes. Veins are closely associated with the red pulp.

Blood flowing into the spleen through the splenic artery enters the central arteries of the white pulp. Within the white pulp, B cells and T cells carry out immune functions, similar to lymph nodes, while spleen macrophages destroy blood-borne pathogens by phagocytosis. Within the red pulp, the spleen performs three functions related to blood cells: (1) removal by macrophages of ruptured, worn out, or defective blood cells and platelets; (2) storage of platelets, up to one-third of the body's supply; and (3) production of blood cells (hemopoiesis) during fetal life.

CLINICAL CONNECTION | *Ruptured Spleen*

The spleen is the organ most often damaged in cases of abdominal trauma. Severe blows over the inferior left chest or superior abdomen can fracture the protecting ribs. Such crushing injury may result in a **ruptured spleen,** which causes significant hemorrhage and shock. Prompt removal of the spleen, called a **splenectomy** (splē-NEK-tō-mē), is needed to prevent death due to bleeding. Other structures, particularly red bone marrow and the liver, can take over some functions normally carried out by the spleen. Immune functions, however, decrease in the absence of a spleen. The spleen's absence also places the patient at higher risk for **sepsis** (a blood infection) due to loss of the filtering and phagocytic functions of the spleen. To reduce the risk of sepsis, patients who have undergone a splenectomy take prophylactic (preventive) antibiotics before any invasive procedures. •

Lymphatic Nodules

Lymphatic nodules (follicles) are egg-shaped masses of lymphatic tissue that are not surrounded by a capsule. Because they are scattered throughout the lamina propria (connective tissue) of mucous membranes lining the gastrointestinal, urinary, and reproductive tracts and the respiratory airways, lymphatic nodules in these areas are also referred to as **mucosa-associated lymphatic tissue (MALT).**

Although many lymphatic nodules are small and solitary, some occur in multiple large aggregations in specific parts of the body. Among these are the tonsils in the pharyngeal region

Figure 22.7 Structure of the spleen.

The spleen is the largest single mass of lymphatic tissue in the body.

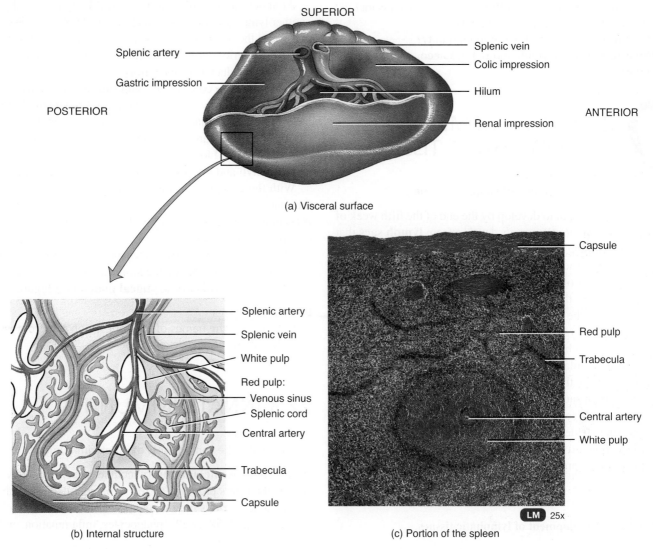

(a) Visceral surface

(b) Internal structure

(c) Portion of the spleen

? After birth, what are the main functions of the spleen?

and the aggregated lymphatic follicles (Peyer's patches) in the ileum of the small intestine. Aggregations of lymphatic nodules also occur in the appendix. Usually there are five **tonsils,** which form a ring at the junction of the oral cavity and oropharynx and at the junction of the nasal cavity and nasopharynx (see Figure 23.2b). The tonsils are strategically positioned to participate in immune responses against inhaled or ingested foreign substances. The single **pharyngeal tonsil** (fa-RIN-jē-al) or **adenoid** is embedded in the posterior wall of the nasopharynx. The two **palatine tonsils** (PAL-a-tīn) lie at the posterior region of the oral cavity, one on either side; these are the tonsils commonly removed in a tonsillectomy. The paired **lingual tonsils** (LIN-gwal), located at the base of the tongue, may also require removal during a tonsillectomy.

CLINICAL CONNECTION | Tonsillitis

Tonsillitis is an infection or inflammation of the tonsils. Most often, it is caused by a virus, but it may also be caused by the same bacteria that cause strep throat. The principal symptom of tonsillitis is a sore throat. Additionally, fever, swollen lymph nodes, nasal congestion, difficulty in swallowing, and headache may also occur. Tonsillitis of viral origin usually resolves on its own. Bacterial tonsillitis is typically treated with antibiotics. **Tonsillectomy** (ton-si-LEK-tō-mē; *ectomy* = incision), the removal of a tonsil, may be indicated for individuals who do not respond to other treatments. Such individuals usually have tonsillitis lasting for more than 3 months (despite medication), obstructed air pathways, and difficulty in swallowing and talking. It appears that tonsillectomy does not interfere with a person's response to subsequent infections. •

✔**CHECKPOINT**

1. How are interstitial fluid and lymph similar, and how do they differ?
2. How do lymphatic vessels differ in structure from veins?
3. Diagram the route of lymph circulation.
4. What is the role of the thymus in immunity?
5. What functions do lymph nodes, the spleen, and the tonsils serve?

22.2 DEVELOPMENT OF LYMPHATIC TISSUES

◉ **OBJECTIVE**

• Describe the development of lymphatic tissues.

Lymphatic tissues begin to develop by the end of the fifth week of embryonic life. *Lymphatic vessels* develop from **lymph sacs** that arise from developing veins, which are derived from **mesoderm.**

The first lymph sacs to appear are the paired **jugular lymph sacs** at the junction of the internal jugular and subclavian veins (Figure 22.8). From the jugular lymph sacs, lymphatic capillary plexuses spread to the thorax, upper limbs, neck, and head. Some of the plexuses enlarge and form lymphatic vessels in their respective regions. Each jugular lymph sac retains at least one connection with its jugular vein, the left one developing into the superior portion of the thoracic duct (left lymphatic duct).

The next lymph sac to appear is the unpaired **retroperitoneal lymph sac** (ret′-rō-per′-i-tō-NĒ-al) at the root of the mesentery of the intestine. It develops from the primitive vena cava and mesonephric (primitive kidney) veins. Capillary plexuses and lymphatic vessels spread from the retroperitoneal lymph sac to the abdominal viscera and diaphragm. The sac establishes connections with the cisterna chyli but loses its connections with neighboring veins.

At about the time the retroperitoneal lymph sac is developing, another lymph sac, the **cisterna chyli** (sis-TER-na KĪ-lē), develops inferior to the diaphragm on the posterior abdominal wall. It gives rise to the inferior portion of the *thoracic duct* and the *cisterna chyli* of the thoracic duct. Like the retroperitoneal lymph sac, the cisterna chyli also loses its connections with surrounding veins.

The last of the lymph sacs, the paired **posterior lymph sacs,** develop from the iliac veins. The posterior lymph sacs produce capillary plexuses and lymphatic vessels of the abdominal wall, pelvic region, and lower limbs. The posterior lymph sacs join the cisterna chyli and lose their connections with adjacent veins.

With the exception of the anterior part of the sac from which the cisterna chyli develops, all lymph sacs become invaded by **mesenchymal cells** (me-SENG-kī-mal) and are converted into groups of *lymph nodes.*

The *spleen* develops from mesenchymal cells between layers of the dorsal mesentery of the stomach. The *thymus* arises as an outgrowth of the **third pharyngeal pouch** (see Figure 18.21a).

✔**CHECKPOINT**

6. What are the names of the four lymph sacs from which lymphatic vessels develop?

22.3 INNATE IMMUNITY

◉ **OBJECTIVE**

• Describe the components of innate immunity.

Innate (nonspecific) immunity includes the external physical and chemical barriers provided by the skin and mucous membranes. It also includes various internal defenses, such as antimicrobial substances, natural killer cells, phagocytes, inflammation, and fever.

First Line of Defense: Skin and Mucous Membranes

The skin and mucous membranes of the body are the first line of defense against pathogens. These structures provide both physical and chemical barriers that discourage pathogens and foreign substances from penetrating the body and causing disease.

With its many layers of closely packed, keratinized cells, the outer epithelial layer of the skin—the **epidermis**—provides a formidable physical barrier to the entrance of microbes (see Figure 5.1). In addition, periodic shedding of epidermal cells helps remove microbes at the skin surface. Bacteria rarely penetrate the intact surface of healthy epidermis. If this surface is broken by cuts, burns, or punctures, however, pathogens can penetrate the epidermis and invade adjacent tissues or circulate in the blood to other parts of the body.

The epithelial layer of **mucous membranes,** which line body cavities, secretes a fluid called **mucus** that lubricates and moistens the cavity surface. Because mucus is slightly viscous, it traps

Figure 22.8 Development of lymphatic tissues.

🔟 ⚏ Lymphatic tissues are derived from mesoderm.

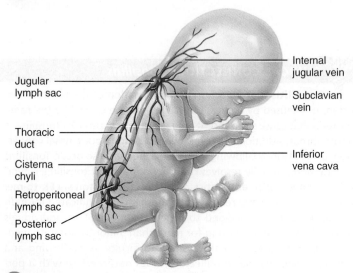

Jugular lymph sac
Thoracic duct
Cisterna chyli
Retroperitoneal lymph sac
Posterior lymph sac
Internal jugular vein
Subclavian vein
Inferior vena cava

❓ **When do lymphatic tissues begin to develop?**

many microbes and foreign substances. The mucous membrane of the nose has mucus-coated **hairs** that trap and filter microbes, dust, and pollutants from inhaled air. The mucous membrane of the upper respiratory tract contains **cilia,** microscopic hairlike projections on the surface of the epithelial cells. The waving action of cilia propels inhaled dust and microbes that have become trapped in mucus toward the throat. Coughing and sneezing accelerate movement of mucus and its entrapped pathogens out of the body. Swallowing mucus sends pathogens to the stomach, where gastric juice destroys them.

Other fluids produced by various organs also help protect epithelial surfaces of the skin and mucous membranes. The **lacrimal apparatus** (LAK-ri-mal) of the eyes (see Figure 17.6) manufactures and drains away tears in response to irritants. Blinking spreads tears over the surface of the eyeball, and the continual washing action of tears helps to dilute microbes and keep them from settling on the surface of the eye. Tears also contain **lysozyme** (LĪ-so-zīm), an enzyme capable of breaking down the cell walls of certain bacteria. Besides tears, lysozyme is present in saliva, perspiration, nasal secretions, and tissue fluids. **Saliva,** produced by the salivary glands, washes microbes from the surfaces of the teeth and from the mucous membrane of the mouth, much as tears wash the eyes. The flow of saliva reduces colonization of the mouth by microbes.

The cleansing of the urethra by the **flow of urine** retards microbial colonization of the urinary system. **Vaginal secretions** likewise move microbes out of the body in females. **Defecation** and **vomiting** also expel microbes. For example, in response to some microbial toxins, the smooth muscle of the lower gastrointestinal tract contracts vigorously; the resulting diarrhea rapidly expels many of the microbes.

Certain chemicals also contribute to the high degree of resistance of the skin and mucous membranes to microbial invasion. Sebaceous (oil) glands of the skin secrete an oily substance called **sebum** that forms a protective film over the surface of the skin. The unsaturated fatty acids in sebum inhibit the growth of certain pathogenic bacteria and fungi. The acidity of the skin (pH 3–5) is caused in part by the secretion of fatty acids and lactic acid. **Perspiration** helps flush microbes from the surface of the skin. **Gastric juice,** produced by the glands of the stomach, is a mixture of hydrochloric acid, enzymes, and mucus. The strong acidity of gastric juice (pH 1.2–3.0) destroys many bacteria and most bacterial toxins. Vaginal secretions also are slightly acidic, which discourages bacterial growth.

Second Line of Defense: Internal Defenses

When pathogens penetrate the physical and chemical barriers of the skin and mucous membranes, they encounter a second line of defense: internal antimicrobial substances, phagocytes, natural killer cells, inflammation, and fever.

Antimicrobial Substances

There are four main types of **antimicrobial substances** that discourage microbial growth: interferons, complement, iron-binding proteins, and antimicrobial proteins.

1. Lymphocytes, macrophages, and fibroblasts infected with viruses produce proteins called **interferons** (in′-ter-FER-ons), or **IFNs.** Once released by virus-infected cells, IFNs diffuse to uninfected neighboring cells, where they induce synthesis of antiviral proteins that interfere with viral replication. Although IFNs do not prevent viruses from attaching to and penetrating host cells, they do stop replication. Viruses can cause disease only if they can replicate within body cells. IFNs are an important defense against infection by many different viruses. The three types of interferons are alpha-, beta-, and gamma-IFN.

2. A group of normally inactive proteins in blood plasma and on the plasma membranes makes up the **complement system.** When activated, these proteins "complement" or enhance certain immune reactions (see Section 22.6). The complement system causes cytolysis (bursting) of microbes, promotes phagocytosis, and contributes to inflammation.

3. **Iron-binding proteins** inhibit the growth of certain bacteria by reducing the amount of available iron. Examples include *transferrin* (found in blood and tissue fluids), *lactoferrin* (found in milk, saliva, and mucus), *ferritin* (found in the liver, spleen, and red bone marrow), and *hemoglobin* (found in red blood cells).

4. **Antimicrobial proteins (AMPs)** are short peptides that have a broad spectrum of antimicrobial activity. Examples of AMPs are *dermicidin* (der-ma-SĪ-din) (produced by sweat glands), *defensins* and *cathelicidins* (cath-el-i-SĪ-dins) (produced by neutrophils, macrophages, and epithelia), and *thrombocidin* (throm′-bō-SĪ-din) (produced by platelets). Besides killing a wide range of microbes, AMPs can attract dendritic cells and mast cells, which participate in immune responses. Interestingly enough, microbes exposed to AMPs do not appear to develop resistance, as often happens with antibiotics.

Natural Killer Cells and Phagocytes

When microbes penetrate the skin and mucous membranes or bypass the antimicrobial substances in blood, the next nonspecific defense consists of natural killer cells and phagocytes. About 5–10% of lymphocytes in the blood are **natural killer (NK) cells.** They are also present in the spleen, lymph nodes, and red bone marrow. NK cells lack the membrane molecules that identify B and T cells, but they have the ability to kill a wide variety of infected body cells and certain tumor cells. NK cells attack any body cells that display abnormal or unusual plasma membrane proteins.

The binding of NK cells to a target cell, such as an infected human cell, causes the release of granules containing toxic substances from NK cells. Some granules contain a protein called **perforin** (PER-for-in) that inserts into the plasma membrane of the target cell and creates channels (perforations) in the membrane. As a result, extracellular fluid flows into the target cell and the cell bursts, a process called **cytolysis** (sī-TOL-i-sis; *cyto-* = cell; *-lysis* = loosening). Other granules of NK cells release **granzymes** (GRAN-zīms), which are protein-digesting enzymes that induce the target cell to undergo apoptosis, or self-destruction. This type of attack kills infected cells, but not the microbes inside the cells; the released microbes, which may or may not be intact, can be destroyed by phagocytes.

Phagocytes (FAG-ō-sīts; *phago-* = eat; *-cytes* = cells) are specialized cells that perform **phagocytosis** (fag-ō-sī-TŌ-sis; *-osis* = process), the ingestion of microbes or other particles such as cellular debris (see Figure 3.13). The two major types of phagocytes are **neutrophils** and **macrophages** (MAK-rō-fā-jez). When an infection occurs, neutrophils and monocytes migrate to the infected area. During this migration, the monocytes enlarge and develop into actively phagocytic macrophages called **wandering macrophages.** Other macrophages, called **fixed macrophages,** stand guard in specific tissues. Among the fixed macrophages are *histiocytes* (HIS-tē-ō-sīts) (connective tissue macrophages), *stellate reticuloendothelial cells* (STEL-āt re-tik′-ū-lō-en-dō-THĒ-lē-al) or *Kupffer cells* (KOOP-fer) in the liver, *alveolar macrophages* in the lungs, *microglia* in the nervous system, and *tissue macrophages* in the spleen, lymph nodes, and red bone marrow. In addition to being an innate defense mechanism, phagocytosis plays a vital role in adaptive immunity, as discussed later in the chapter.

CLINICAL CONNECTION | *Microbial Evasion of Phagocytosis*

Some microbes, such as the bacteria that cause pneumonia, have extracellular structures called capsules that prevent adherence. This makes it physically difficult for phagocytes to engulf the microbes. Other microbes, such as the toxin-producing bacteria that cause one kind of food poisoning, may be ingested but not killed; instead, the toxins they produce (leukocidins) may kill the phagocytes by causing the release of the phagocyte's own lysosomal enzymes into its cytoplasm. Still other microbes—such as the bacteria that cause tuberculosis—inhibit fusion of phagosomes and lysosomes and thus prevent exposure of the microbes to lysosomal enzymes. These bacteria apparently can also use chemicals in their cell walls to counter the effects of lethal oxidants produced by phagocytes. Subsequent multiplication of the microbes within phagosomes may eventually destroy the phagocyte. •

Phagocytosis occurs in five phases: chemotaxis, adherence, ingestion, digestion, and killing (Figure 22.9):

1 *Chemotaxis.* Phagocytosis begins with **chemotaxis** (kē-mō-TAK-sis), a chemically stimulated movement of phagocytes to a site of damage. Chemicals that attract phagocytes might come from invading microbes, white blood cells, damaged tissue cells, or activated complement proteins.

2 *Adherence* (ad-HER-ents). Attachment of the phagocyte to the microbe or other foreign material is termed **adherence.** The binding of complement proteins to the invading pathogen enhances adherence.

3 *Ingestion.* The plasma membrane of the phagocyte extends projections, called **pseudopods** (SOO-dō-pods), that engulf the microbe in a process called **ingestion.** When the pseudopods meet, they fuse, surrounding the microorganism with a sac called a **phagosome** (FAG-ō-sōm).

4 *Digestion.* The phagosome enters the cytoplasm and merges with lysosomes to form a single, larger structure called a **phagolysosome** (fag-ō-LĪ-sō-sōm). The lysosome contributes lysozyme, which breaks down microbial cell walls, and other digestive enzymes that degrade carbohydrates, proteins, lipids, and nucleic acids. The phagocyte also forms lethal oxidants, such as superoxide anion (O_2^-), hypochlorite anion (OCl^-), and hydrogen peroxide (H_2O_2), in a process called an **oxidative burst.**

5 *Killing.* The chemical onslaught provided by lysozyme, digestive enzymes, and oxidants within a phagolysosome quickly kills many types of microbes. Any materials that cannot be degraded further remain in structures called **residual bodies.**

Inflammation

Inflammation is a nonspecific, defensive response of the body to tissue damage. Among the conditions that may produce inflammation are pathogens, abrasions, chemical irritations, distortion or disturbances of cells, and extreme temperatures. The four characteristic signs and symptoms of inflammation are **redness, pain, heat,** and **swelling.** Inflammation can also cause a **loss of function** in the injured area (for example, the inability to detect sensations), depending on the site and extent of the injury. Inflammation is an attempt to dispose of microbes, toxins, or foreign material at the site of injury, to prevent their spread to other tissues, and to prepare the site for tissue repair in an attempt to restore tissue homeostasis.

Because inflammation is one of the body's nonspecific defense mechanisms, the response of a tissue to a cut is similar to the response to damage caused by burns, radiation, or bacterial or viral invasion. In each case, the inflammatory response has three basic stages: (1) vasodilation and increased permeability of blood vessels, (2) emigration (movement) of phagocytes from the blood into interstitial fluid, and, ultimately, (3) tissue repair.

VASODILATION AND INCREASED BLOOD VESSEL PERMEABILITY Two immediate changes occur in the blood vessels in a region of tissue injury: **vasodilation** (increase in the diameter) of arterioles and **increased permeability** of capillaries (Figure 22.10). Increased permeability means that substances normally retained in blood are permitted to pass from the blood vessels. Vasodilation allows more blood to flow through the damaged area, and increased permeability permits defensive proteins such as antibodies and clotting factors to enter the injured area from the blood. The increased blood flow also helps remove microbial toxins and dead cells.

Among the substances that contribute to vasodilation, increased permeability, and other aspects of the inflammatory response are the following:

• *Histamine.* In response to injury, mast cells in connective tissue and basophils and platelets in blood release histamine. Neutrophils and macrophages attracted to the site of injury also stimulate the release of histamine, which causes vasodilation and increased permeability of blood vessels.

• *Kinins.* These polypeptides, formed in blood from inactive precursors called kininogens, induce vasodilation and increased

Figure 22.9 **Phagocytosis of a microbe.**

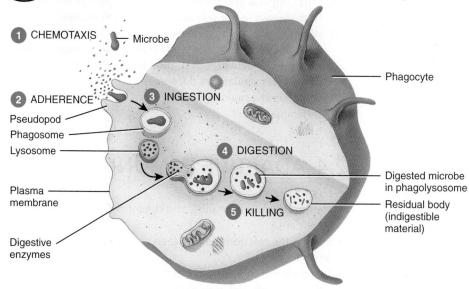 The major types of phagocytes are neutrophils and macrophages.

1 CHEMOTAXIS — Microbe

2 ADHERENCE
Pseudopod
Phagosome
Lysosome

3 INGESTION

Plasma membrane

Digestive enzymes

4 DIGESTION

5 KILLING

Phagocyte

Digested microbe in phagolysosome

Residual body (indigestible material)

(a) Phases of phagocytosis

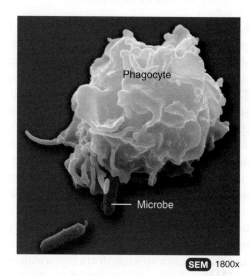

Phagocyte

Microbe

SEM 1800x

(b) Phagocyte (white blood cell) engulfing microbe.

? **What chemicals are responsible for killing ingested microbes?**

permeability and serve as chemotactic agents for phagocytes. An example of a kinin is bradykinin.

- *Prostaglandins (PGs)* (pros′-ta-GLAN-dins). These lipids, especially those of the E series, are released by damaged cells and intensify the effects of histamine and kinins. PGs also may stimulate the emigration of phagocytes through capillary walls.

- *Leukotrienes (LTs)* (loo′-kō-TRĪ-ēns). Produced by basophils and mast cells, LTs cause increased permeability; they also function in adherence of phagocytes to pathogens and as chemotactic agents that attract phagocytes.

- *Complement.* Different components of the complement system stimulate histamine release, attract neutrophils by chemotaxis, and promote phagocytosis; some components can also destroy bacteria.

Dilation of arterioles and increased permeability of capillaries produce three of the signs and symptoms of inflammation: heat, redness (erythema), and swelling (edema). Heat and redness result from the large amount of blood that accumulates in the damaged area. As the local temperature rises slightly, metabolic reactions proceed more rapidly and release additional heat. Edema results from increased permeability of blood vessels, which permits more fluid to move from blood plasma into tissue spaces.

Pain is a prime symptom of inflammation. It results from injury to neurons and from toxic chemicals released by microbes. Kinins affect some nerve endings, causing much of the pain associated with inflammation. Prostaglandins intensify and prolong the pain associated with inflammation. Pain may also be due to increased pressure from edema.

Figure 22.10 **Inflammation.**

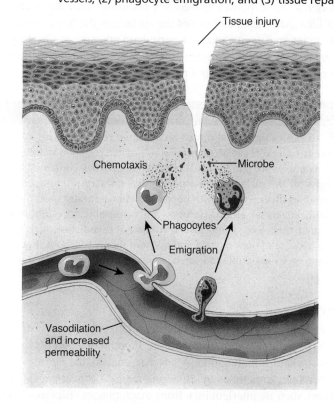 The three stages of inflammation are as follows: (1) vasodilation and increased permeability of blood vessels, (2) phagocyte emigration, and (3) tissue repair.

Tissue injury

Chemotaxis

Microbe

Phagocytes

Emigration

Vasodilation and increased permeability

Phagocytes migrate from blood to site of tissue injury

? **What causes each of the following signs and symptoms of inflammation: redness, pain, heat, and swelling?**

The increased permeability of capillaries allows leakage of blood-clotting factors into tissues. The clotting sequence is set into motion, and fibrinogen is ultimately converted to an insoluble, thick mesh of fibrin threads that localizes and traps invading microbes and blocks their spread.

EMIGRATION OF PHAGOCYTES Within an hour after the inflammatory process starts, phagocytes appear on the scene. As large amounts of blood accumulate, neutrophils begin to stick to the inner surface of the endothelium (lining) of blood vessels (Figure 22.10). Then the neutrophils begin to squeeze through the wall of the blood vessel to reach the damaged area. This process, called **emigration** (em′-i-GRĀ-shun), depends on chemotaxis. Neutrophils attempt to destroy the invading microbes by phagocytosis. A steady stream of neutrophils is ensured by the production and release of additional cells from red bone marrow. Such an increase in white blood cells in the blood is termed **leukocytosis** (loo-kō-sī-TŌ-sis).

Although neutrophils predominate in the early stages of infection, they die off rapidly. As the inflammatory response continues, monocytes follow the neutrophils into the infected area. Once in the tissue, monocytes transform into wandering macrophages that add to the phagocytic activity of the fixed macrophages already present. True to their name, macrophages are much more potent phagocytes than neutrophils. They are large enough to engulf damaged tissue, worn-out neutrophils, and invading microbes.

Eventually, macrophages also die. Within a few days, a pocket of dead phagocytes and damaged tissue forms; this collection of dead cells and fluid is called **pus.** Pus formation occurs in most inflammatory responses and usually continues until the infection subsides. At times, pus reaches the surface of the body or drains into an internal cavity and is dispersed; on other occasions the pus remains even after the infection is terminated. In this case, the pus is gradually destroyed over a period of days and is absorbed.

CLINICAL CONNECTION | *Abscesses and Ulcers*

If pus cannot drain out of an inflamed region, the result is an **abscess**—an excessive accumulation of pus in a confined space. Common examples are pimples and boils. When superficial inflamed tissue sloughs off the surface of an organ or tissue, the resulting open sore is called an **ulcer.** People with poor circulation—for instance, diabetics with advanced atherosclerosis—are susceptible to ulcers in the tissues of their legs. These ulcers, which are called stasis ulcers, develop because of poor oxygen and nutrient supply to tissues that then become very susceptible to even a very mild injury or an infection. •

Fever

Fever is an abnormally high body temperature that occurs because the hypothalamic thermostat is reset. It commonly occurs during infection and inflammation. Many bacterial toxins elevate body temperature, sometimes by triggering release of fever-causing cytokines such as interleukin-1 from macrophages. Elevated body temperature intensifies the effects of interferons, inhibits the growth of some microbes, and speeds up body reactions that aid repair.

Table 22.1 summarizes the components of innate immunity.

✔**CHECKPOINT**

7. What physical and chemical factors provide protection from disease in the skin and mucous membranes?
8. What internal defenses provide protection against microbes that penetrate the skin and mucous membranes?
9. How are the activities of natural killer cells and phagocytes similar and different?
10. What are the main signs, symptoms, and stages of inflammation?

22.4 ADAPTIVE IMMUNITY

◉ **OBJECTIVES**
• Define adaptive immunity, and describe how T cells and B cells arise.
• Explain the relationship between an antigen and an antibody.
• Compare the functions of cell-mediated immunity and antibody-mediated immunity.

The ability of the body to defend itself against specific invading agents such as bacteria, toxins, viruses, and foreign tissues is called **adaptive (specific) immunity.** Substances that are recognized as foreign and provoke immune responses are called **antigens (Ags)** (AN-ti-jens), meaning **anti**body **gen**erators. Two properties distinguish adaptive immunity from innate immunity: (1) *specificity* for particular foreign molecules (antigens), which also involves distinguishing self from nonself molecules, and (2) *memory* for most previously encountered antigens so that a second encounter prompts an even more rapid and vigorous response. The branch of science that deals with the responses of the body when challenged by antigens is called **immunology** (im′-ū-NOL-ō-jē; *immuno-* = free from service or exempt; *-logy* = study of). The **immune system** includes the cells and tissues that carry out immune responses.

Maturation of T Cells and B Cells

Adaptive immunity involves lymphocytes called **B cells** and **T cells.** Both develop in primary lymphatic organs (red bone marrow and the thymus) from pluripotent stem cells that originate in red bone marrow (see Figure 19.3). B cells complete their development in red bone marrow, a process that continues throughout life. T cells develop from pre-T cells that migrate from red bone marrow into the thymus, where they mature (Figure 22.11). Most T cells arise before puberty, but they continue to mature and leave the thymus throughout life. B cells and T cells are named based on where they mature. In birds, B cells mature in an organ called the *bursa of Fabricius.* Although this organ is not present in humans, the term *B cell* is still used, but the letter *B* stands for *bursa equivalent,* which is the red bone marrow since that is the location in humans where B cells mature. T cells are so named because they mature in the *thymus* gland.

Before T cells leave the thymus or B cells leave red bone marrow, they develop **immunocompetence** (im′-ū-nō-KOM-petens), the ability to carry out adaptive immune responses. This

TABLE 22.1

Summary of Innate Defenses

COMPONENT	FUNCTIONS
FIRST LINE OF DEFENSE: SKIN AND MUCOUS MEMBRANES	
Physical Factors	
Epidermis of skin	Forms physical barrier to entrance of microbes.
Mucous membranes	Inhibit entrance of many microbes, but not as effective as intact skin.
Mucus	Traps microbes in respiratory and gastrointestinal tracts.
Hairs	Filter out microbes and dust in nose.
Cilia	Together with mucus, trap and remove microbes and dust from upper respiratory tract.
Lacrimal apparatus	Tears dilute and wash away irritating substances and microbes.
Saliva	Washes microbes from surfaces of teeth and mucous membranes of mouth.
Urine	Washes microbes from urethra.
Defecation and vomiting	Expel microbes from body.
Chemical Factors	
Sebum	Forms protective acidic film over skin surface that inhibits growth of many microbes.
Lysozyme	Antimicrobial substance in perspiration, tears, saliva, nasal secretions, and tissue fluids.
Gastric juice	Destroys bacteria and most toxins in stomach.
Vaginal secretions	Slight acidity discourages bacterial growth; flush microbes out of vagina.
SECOND LINE OF DEFENSE: INTERNAL DEFENSES	
Antimicrobial Substances	
Interferons (IFNs)	Protect uninfected host cells from viral infection.
Complement system	Causes cytolysis of microbes; promotes phagocytosis; contributes to inflammation.
Iron-binding proteins	Inhibit growth of certain bacteria by reducing amount of available iron.
Antimicrobial proteins (AMPs)	Have broad-spectrum antimicrobial activities and attract dendritic cells and mast cells.
Natural killer (NK) cells	Kill infected target cells by releasing granules that contain perforin and granzymes; phagocytes then kill released microbes.
Phagocytes	Ingest foreign particulate matter.
Inflammation	Confines and destroys microbes; initiates tissue repair.
Fever	Intensifies effects of interferons; inhibits growth of some microbes; speeds up body reactions that aid repair.

means that B cells and T cells begin to make several distinctive proteins that are inserted into their plasma membranes. Some of these proteins function as **antigen receptors**—molecules capable of recognizing specific antigens (Figure 22.11).

There are two major types of mature T cells that exit the thymus: **helper T cells** and **cytotoxic T cells** (Figure 22.11). Helper T cells are also known as **CD4 T cells,** which means that, in addition to antigen receptors, their plasma membranes include a protein called CD4. Cytotoxic T cells are also referred to as **CD8 T cells** because their plasma membranes not only contain antigen receptors but also a protein known as CD8. As we will see later in this chapter, these two types of T cells have very different functions.

Types of Adaptive Immunity

There are two types of adaptive immunity: cell-mediated immunity and antibody-mediated immunity. Both types of adaptive immunity are triggered by antigens. In **cell-mediated immunity,** cytotoxic T cells directly attack invading antigens. In **antibody-mediated immunity,** B cells transform into plasma cells, which synthesize and secrete specific proteins called **antibodies (Abs)** or **immunoglobulins (Igs)** (im'-ū-nō-GLOB-ū-lins). A given antibody can bind to and inactivate a specific antigen. Helper T cells aid the immune responses of both cell-mediated and antibody-mediated immunity.

Cell-mediated immunity is particularly effective against (1) intracellular pathogens, which include any viruses, bacteria, or fungi that are inside cells; (2) some cancer cells; and (3) foreign tissue transplants. Thus, cell-mediated immunity always involves cells attacking cells. Antibody-mediated immunity works mainly against extracellular pathogens, which include any viruses, bacteria, or fungi that are in body fluids outside cells. Since antibody-mediated immunity involves antibodies that bind to antigens in body *humors* or fluids (such as blood and lymph), it is also referred to as *humoral immunity*.

In most cases, when a particular antigen initially enters the body, there is only a small group of lymphocytes with the correct antigen receptors to respond to that antigen; this small group of cells includes a few helper T cells, cytotoxic T cells, and B cells. Depending on its location, a given antigen can provoke both types of adaptive immune responses. This is due to the fact that when a specific antigen invades the body, there are usually many copies of that antigen spread throughout the body's tissues and fluids. Some copies of the antigen may be present inside body cells (which provokes a cell-mediated immune response by cytotoxic T cells), while other copies of the antigen may be present in extracellular fluid (which provokes an antibody-mediated immune response by B cells). Thus, cell-mediated and antibody-mediated immune responses often work together to get rid of the large number of copies of a particular antigen from the body.

Clonal Selection: The Principle

As you just learned, when a specific antigen is present in the body, there are usually many copies of that antigen located throughout the body's tissues and fluids. The numerous copies of the antigen

Figure 22.11 **B cells and pre-T cells arise from pluripotent stem cells in red bone marrow.** B cells and T cells develop in primary lymphatic tissues (red bone marrow and the thymus) and are activated in secondary lymphatic organs and tissues (lymph nodes, spleen, and lymphatic nodules). Once activated, each type of lymphocyte forms a clone of cells that can recognize a specific antigen. For simplicity, antigen receptors, CD4 proteins, and CD8 proteins are not shown in the plasma membranes of the cells of the lymphocyte clones.

The two types of adaptive immunity are cell-mediated immunity and antibody-mediated immunity.

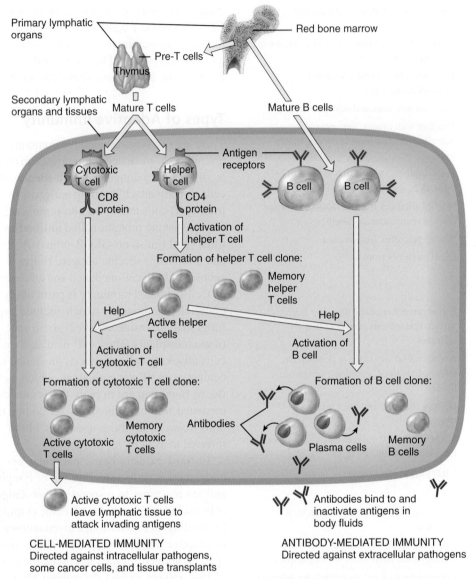

CELL-MEDIATED IMMUNITY
Directed against intracellular pathogens, some cancer cells, and tissue transplants

ANTIBODY-MEDIATED IMMUNITY
Directed against extracellular pathogens

? **Which type of T cell participates in both cell-mediated and antibody-mediated immune responses?**

initially outnumber the small group of helper T cells, cytotoxic T cells, and B cells with the correct antigen receptors to respond to that antigen. Therefore, once each of these lymphocytes encounters a copy of the antigen and receives stimulatory cues, it subsequently undergoes clonal selection. **Clonal selection** is the process by which a lymphocyte **proliferates** (divides) and **differentiates** (forms more highly specialized cells) in response to a specific antigen. The result of clonal selection is the formation of a population of identical cells, called a **clone,** that can recognize the same specific antigen as the original lymphocyte (Figure 22.11). Before the first exposure to a given antigen, only a few lymphocytes are able to recognize it, but once clonal selection

occurs, there are thousands of lymphocytes that can respond to that antigen. Clonal selection of lymphocytes occurs in the secondary lymphatic organs and tissues. The swollen tonsils or lymph nodes in your neck you experienced the last time you were sick were probably caused by clonal selection of lymphocytes participating in an immune response.

A lymphocyte that undergoes clonal selection gives rise to two major types of cells in the clone: effector cells and memory cells. The thousands of **effector cells** of a lymphocyte clone carry out immune responses that ultimately result in the destruction or inactivation of the antigen. Effector cells include **active helper T cells,** which are part of a helper T cell clone; **active cytotoxic T** cells, which are part of a cytotoxic T cell clone; and **plasma cells,** which are part of a B cell clone. Most effector cells eventually die after the immune response has been completed.

Memory cells do not actively participate in the initial immune response to the antigen. However, if the same antigen enters the body again in the future, the thousands of memory cells of a lymphocyte clone are available to initiate a far swifter reaction than occurred during the first invasion. The memory cells respond to the antigen by proliferating and differentiating into more effector cells and more memory cells. Consequently, the second response to the antigen is usually so fast and so vigorous that the antigen is destroyed before any signs or symptoms of disease can occur. Memory cells include **memory helper T cells,** which are part of a helper T cell clone; **memory cytotoxic T cells,** which are part of a cytotoxic T cell clone; and **memory B cells,** which are part of a B cell clone. Most memory cells do not die at the end of an immune response. Instead, they have long life spans (often lasting for decades). The functions of effector cells and memory cells are described in more detail later in this chapter.

Antigens and Antigen Receptors

Antigens have two important characteristics: immunogenicity and reactivity. **Immunogenicity** (im-ū-nō-je-NIS-i-tē; -*genic* = producing) is the ability to provoke an immune response by stimulating the production of specific antibodies, the proliferation of specific T cells, or both. The term **antigen** derives from its function as an *anti*body *gen*erator. **Reactivity** is the ability of the antigen to react specifically with the antibodies or cells it provoked. Strictly speaking, immunologists define antigens as substances that have reactivity; substances with both immunogenicity and reactivity are considered **complete antigens.** Commonly, however, the term *antigen* implies both immunogenicity and reactivity, and we use the word in this way.

Entire microbes or parts of microbes may act as antigens. Chemical components of bacterial structures such as flagella, capsules, and cell walls are antigenic, as are bacterial toxins. Nonmicrobial examples of antigens include chemical components of pollen, egg white, incompatible blood cells, and transplanted tissues and organs. The huge variety of antigens in the environment provides myriad opportunities for provoking immune responses. Typically, just certain small parts of a large antigen molecule act as the triggers for immune responses. These small parts are called **epitopes** (EP-i-tōps), or *antigenic determinants* (Figure 22.12). Most antigens have many epitopes, each of which induces production of a specific antibody or activates a specific T cell.

Antigens that get past the innate defenses generally follow one of three routes into lymphatic tissue: (1) Most antigens that enter the bloodstream (for example, through an injured blood vessel) are trapped as they flow through the spleen. (2) Antigens that penetrate the skin enter lymphatic vessels and lodge in lymph nodes. (3) Antigens that penetrate mucous membranes are entrapped by mucosa-associated lymphatic tissue (MALT).

Chemical Nature of Antigens

Antigens are large, complex molecules. Most often, they are proteins. However, nucleic acids, lipoproteins, glycoproteins, and cer-

Figure 22.12 Epitopes (antigenic determinants).

 Most antigens have several epitopes that induce the production of different antibodies or activate different T cells.

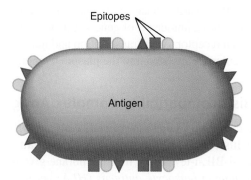

Epitopes

Antigen

? **What is the difference between an epitope and a hapten?**

tain large polysaccharides may also act as antigens. Complete antigens usually have large molecular weights of 10,000 daltons or more, but large molecules that have simple, repeating subunits—for example, cellulose and most plastics—are not usually antigenic. This is why plastic materials can be used in artificial heart valves or joints.

A smaller substance that has reactivity but lacks immunogenicity is called a **hapten** (HAP-ten = to grasp). A hapten can stimulate an immune response only if it is attached to a larger carrier molecule. An example is the small lipid toxin in poison ivy, which triggers an immune response after combining with a body protein. Likewise, some drugs, such as penicillin, may combine with proteins in the body to form immunogenic complexes. Such hapten-stimulated immune responses are responsible for some allergic reactions to drugs and other substances in the environment (see Disorders: Homeostatic Imbalances at the end of the chapter).

As a rule, antigens are foreign substances; they are not usually part of body tissues. However, sometimes the immune system fails to distinguish "friend" (self) from "foe" (nonself). The result is an autoimmune disorder (see Disorders: Homeostatic Imbalances at the end of the chapter) in which self-molecules or cells are attacked as though they were foreign.

Diversity of Antigen Receptors

An amazing feature of the human immune system is its ability to recognize and bind to at least a billion (10^9) different epitopes. Before a particular antigen ever enters the body, T cells and B cells that can recognize and respond to that intruder are ready and waiting. Cells of the immune system can even recognize artificially made molecules that do not exist in nature. The basis for the ability to recognize so many epitopes is an equally large diversity of antigen receptors. Given that human cells contain only about 35,000 genes, how could a billion or more different antigen receptors possibly be generated?

The answer to this puzzle turned out to be simple in concept. The diversity of antigen receptors in both B cells and T cells is the result of shuffling and rearranging a few hundred versions of several small gene segments. This process is called **genetic**

recombination. The gene segments are put together in different combinations as the lymphocytes are developing from stem cells in red bone marrow and the thymus. The situation is similar to shuffling a deck of 52 cards and then dealing out three cards. If you did this over and over, you could generate many more than 52 different sets of three cards. Because of genetic recombination, each B cell or T cell has a unique set of gene segments that codes for its unique antigen receptor. After transcription and translation, the receptor molecules are inserted into the plasma membrane.

Major Histocompatibility Complex Antigens

Located in the plasma membrane of body cells are "self-antigens," the **major histocompatibility complex (MHC) antigens** (his′-tō-kom-pat′-i-BIL-i-tē). These transmembrane glycoproteins are also called *human leukocyte antigens (HLA)* because they were first identified on white blood cells. Unless you have an identical twin, your MHC antigens are unique. Thousands to several hundred thousand MHC molecules mark the surface of each of your body cells except red blood cells. Although MHC antigens are the reason that tissues may be rejected when they are transplanted from one person to another, their normal function is to help T cells recognize that an antigen is foreign, not self. Such recognition is an important first step in any adaptive immune response.

The two types of major histocompatibility complex antigens are class I and class II. Class I MHC (MHC-I) molecules are built into the plasma membranes of all body cells except red blood cells. Class II MHC (MHC-II) molecules appear on the surface of antigen-presenting cells (described in the next section).

Pathways of Antigen Processing

For an immune response to occur, B cells and T cells must recognize that a foreign antigen is present. B cells can recognize and bind to antigens in lymph, interstitial fluid, or blood plasma. T cells only recognize fragments of antigenic proteins that are processed and presented in a certain way. In **antigen processing,** antigenic proteins are broken down into peptide fragments that then associate with MHC molecules. Next the antigen–MHC complex is inserted into the plasma membrane of a body cell. The insertion of the complex into the plasma membrane is called **antigen presentation.** When a peptide fragment comes from a *self-protein,* T cells ignore the antigen–MHC complex. However, if the peptide fragment comes from a *foreign protein,* T cells recognize the antigen–MHC complex as an intruder, and an immune response takes place. Antigen processing and presentation occurs in two ways, depending on whether the antigen is located outside or inside body cells.

Processing of Exogenous Antigens

Foreign antigens that are present in fluids *outside* body cells are termed **exogenous antigens** (ex-OG-e-nus). They include intruders such as bacteria and bacterial toxins, parasitic worms, inhaled pollen and dust, and viruses that have not yet infected a body cell. A special class of cells called **antigen-presenting cells (APCs)** process and present exogenous antigens. APCs include dendritic cells, macrophages, and B cells. They are strategically located in places where antigens are likely to penetrate the innate defenses and enter the body, such as the epidermis and dermis of the skin (Langerhans cells are a type of dendritic cell); mucous membranes that line the respiratory, gastrointestinal, urinary, and reproductive tracts; and lymph nodes. After processing and presenting an antigen, APCs migrate from tissues via lymphatic vessels to lymph nodes.

The steps in the processing and presenting of an exogenous antigen by an antigen-presenting cell occur as follows (Figure 22.13):

1 *Ingestion of the antigen.* Antigen-presenting cells ingest exogenous antigens by phagocytosis or endocytosis. Ingestion could occur almost anywhere in the body that invaders, such as microbes, have penetrated the innate defenses.

2 *Digestion of antigen into peptide fragments.* Within the phagosome or endosome, protein-digesting enzymes split large antigens into short peptide fragments.

3 *Synthesis of MHC-II molecules.* At the same time, the APC synthesizes MHC-II molecules at the endoplasmic reticulum (ER).

4 *Packaging of MHC-II molecules.* Once synthesized, the MHC-II molecules are packaged into vesicles.

5 *Fusion of vesicles.* The vesicles containing antigen peptide fragments and MHC-II molecules merge and fuse.

6 *Binding of peptide fragments to MHC-II molecules.* After fusion of the two types of vesicles, antigen peptide fragments bind to MHC-II molecules.

7 *Insertion of antigen–MHC-II complexes into the plasma membrane.* The combined vesicle that contains antigen–MHC-II complexes undergoes exocytosis. As a result, the antigen–MHC-II complexes are inserted into the plasma membrane.

After processing an antigen, the antigen-presenting cell migrates to lymphatic tissue to present the antigen to T cells. Within lymphatic tissue, a small number of T cells that have compatibly shaped receptors recognize and bind to the antigen fragment–MHC-II complex, triggering an adaptive immune response. The presentation of exogenous antigen together with MHC-II molecules by antigen-presenting cells informs T cells that intruders are present in the body and that combative action should begin.

Processing of Endogenous Antigens

Foreign antigens that are present *inside* body cells are termed **endogenous antigens** (en-DOJ-e-nus). Such antigens may be viral proteins produced after a virus infects the cell and takes over the cell's metabolic machinery, toxins produced from intracellular bacteria, or abnormal proteins synthesized by a cancerous cell.

The steps in the processing and presenting of an endogenous antigen by an infected body cell occur as follows (Figure 22.14):

1 *Digestion of antigen into peptide fragments.* Within the infected cell, protein-digesting enzymes split the endogenous antigen into short peptide fragments.

2 *Synthesis of MHC-I molecules.* At the same time, the infected cell synthesizes MHC-I molecules at the endoplasmic reticulum (ER).

Figure 22.13 Processing and presenting of exogenous antigen by an antigen-presenting cell (APC).

Fragments of exogenous antigens are processed and then presented with MHC-II molecules on the surface of an antigen-presenting cell (APC).

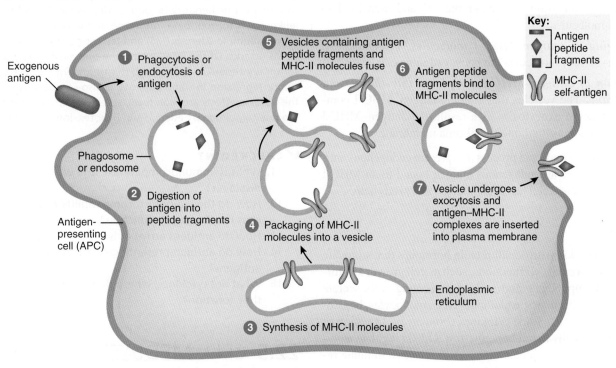

Key:
- Antigen peptide fragments
- MHC-II self-antigen

1 Phagocytosis or endocytosis of antigen

Exogenous antigen

Phagosome or endosome

Antigen-presenting cell (APC)

2 Digestion of antigen into peptide fragments

5 Vesicles containing antigen peptide fragments and MHC-II molecules fuse

6 Antigen peptide fragments bind to MHC-II molecules

7 Vesicle undergoes exocytosis and antigen–MHC-II complexes are inserted into plasma membrane

4 Packaging of MHC-II molecules into a vesicle

Endoplasmic reticulum

3 Synthesis of MHC-II molecules

APCs present exogenous antigens in association with MHC-II molecules

What types of cells are APCs, and where in the body are they found?

Figure 22.14 Processing and presenting of endogenous antigen by an infected body cell.

Fragments of endogenous antigens are processed and then presented with MHC-I proteins on the surface of an infected body cell.

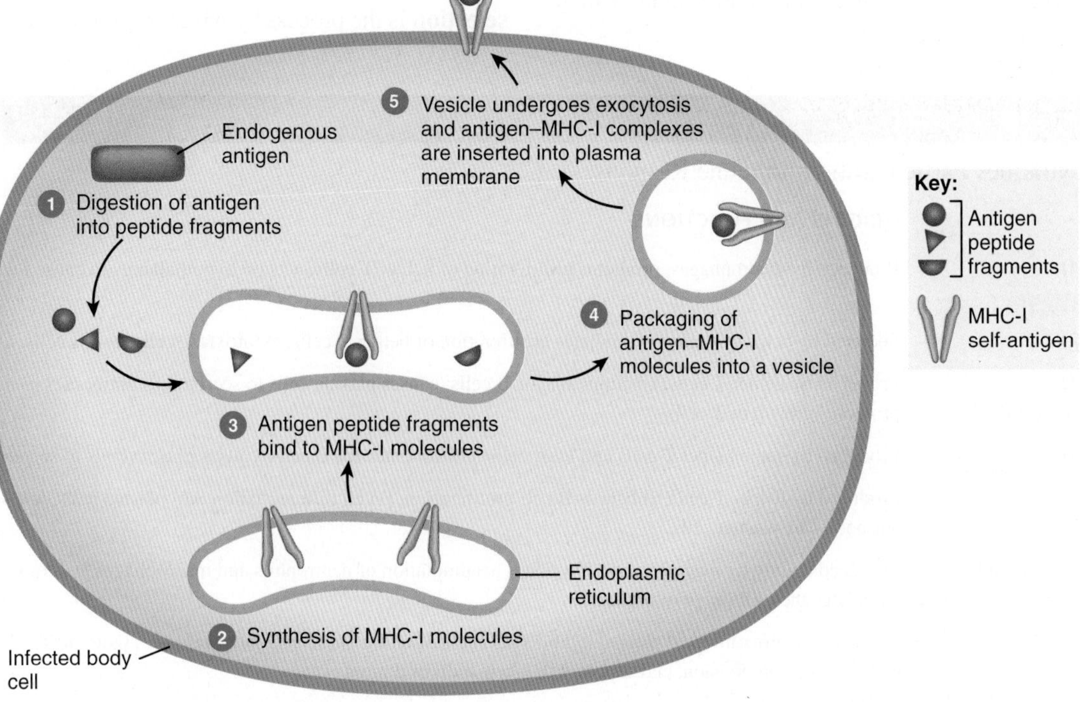

5 Vesicle undergoes exocytosis and antigen–MHC-I complexes are inserted into plasma membrane

Endogenous antigen

1 Digestion of antigen into peptide fragments

4 Packaging of antigen–MHC-I molecules into a vesicle

3 Antigen peptide fragments bind to MHC-I molecules

Endoplasmic reticulum

2 Synthesis of MHC-I molecules

Infected body cell

Key:
- Antigen peptide fragments
- MHC-I self-antigen

Infected body cells present endogenous antigens in association with MHC-I molecules

What are some examples of endogenous antigens?

895

③ *Binding of peptide fragments to MHC-I molecules.* The antigen peptide fragments enter the ER and then bind to MHC-I molecules.

④ *Packaging of antigen–MHC-I molecules.* From the ER, antigen–MHC-I molecules are packaged into vesicles.

⑤ *Insertion of antigen–MHC-I complexes into the plasma membrane.* The vesicles that contain antigen–MHC-I complexes undergo exocytosis. As a result, the antigen–MHC-I complexes are inserted into the plasma membrane.

Most cells of the body can process and present endogenous antigens. The display of an endogenous antigen bound to an MHC-I molecule signals that a cell has been infected and needs help.

Cytokines

Cytokines (SĪ-tō-kīns) are small protein hormones that stimulate or inhibit many normal cell functions, such as cell growth and differentiation. Lymphocytes and antigen-presenting cells secrete cytokines, as do fibroblasts, endothelial cells, monocytes, hepatocytes, and kidney cells. Some cytokines stimulate proliferation of progenitor blood cells in red bone marrow. Others regulate activities of cells involved in innate defenses or adaptive immune responses, as described in Table 22.2.

CLINICAL CONNECTION | *Cytokine Therapy*

Cytokine therapy is the use of cytokines to treat medical conditions. Interferons were the first cytokines shown to have limited effects against some human cancers. Alpha-interferon (Intron A®) is approved in the United States for treating Kaposi (KAP-ō-sē) sarcoma, a cancer that often occurs in patients infected with HIV, the virus that causes AIDS. Other approved uses for alpha-interferon include treating

---

genital herpes caused by the herpes virus; treating hepatitis B and C, caused by the hepatitis B and C viruses; and treating hairy cell leukemia. A form of beta-interferon (Betaseron®) slows the progression of multiple sclerosis (MS) and lessens the frequency and severity of MS attacks. Of the interleukins, the one most widely used to fight cancer is interleukin-2. Although this treatment is effective in causing tumor regression in some patients, it also can be very toxic. Among the adverse effects are high fever, severe weakness, difficulty breathing due to pulmonary edema, and hypotension leading to shock. •

✔CHECKPOINT

11. What is immunocompetence, and which body cells display it?
12. How do the major histocompatibility complex class I and class II self-antigens function?
13. How do antigens arrive at lymphatic tissues?
14. How do antigen-presenting cells process exogenous antigens?
15. What are cytokines, where do they arise, and how do they function?

22.5 CELL-MEDIATED IMMUNITY

◉ OBJECTIVES

• Outline the steps in a cell-mediated immune response.
• Distinguish between the action of natural killer cells and cytotoxic T cells.
• Define immunological surveillance.

A cell-mediated immune response begins with *activation* of a small number of T cells by a specific antigen. Once a T cell has been activated, it undergoes clonal selection. Recall that **clonal selection** is the process by which a lymphocyte proliferates (divides

---

TABLE 22.2

Summary of Cytokines Participating in Immune Responses

CYTOKINE	ORIGINS AND FUNCTIONS
Interleukin-1 (IL-1) (in'-ter-LOO-kin)	Produced by macrophages; promotes proliferation of helper T cells; acts on hypothalamus to cause fever.
Interleukin-2 (IL-2)	Secreted by helper T cells; costimulates proliferation of helper T cells, cytotoxic T cells, and B cells; activates NK cells.
Interleukin-4 (IL-4) (B cell stimulating factor)	Produced by helper T cells; costimulator for B cells; causes plasma cells to secrete IgE antibodies (see Table 22.3); promotes growth of T cells.
Interleukin-5 (IL-5)	Produced by some helper T cells and mast cells; costimulator for B cells; causes plasma cells to secrete IgA antibodies.
Interleukin-6 (IL-6)	Produced by helper T cells; enhances B cell proliferation, B cell differentiation into plasma cells, and secretion of antibodies by plasma cells.
Tumor necrosis factor (TNF) (ne-KRŌ-sis)	Produced mainly by macrophages; stimulates accumulation of neutrophils and macrophages at sites of inflammation and stimulates their killing of microbes.
Interferons (IFNs) (in'-ter-FĒR-ons)	Produced by virus-infected cells to inhibit viral replication in uninfected cells; activate cytotoxic T cells and natural killer cells, inhibit cell division, and suppress the formation of tumors.
Macrophage migration inhibiting factor	Produced by cytotoxic T cells; prevents macrophages from leaving site of infection.

several times) and differentiates (forms more highly specialized cells) in response to a specific antigen. The result of clonal selection is the formation of a **clone** of cells that can recognize the same antigen as the original lymphocyte (see Figure 22.11). Some of the cells of a T cell clone become effector cells, while other cells of the clone become memory cells. The effector cells of a T cell clone carry out immune responses that ultimately result in *elimination* of the intruder.

Activation of T Cells

At any given time, most T cells are inactive. As you learned in the last section, antigen receptors on the surface of T cells, called **T-cell receptors (TCRs),** recognize and bind to specific foreign antigen fragments that are presented in antigen–MHC complexes. There are millions of different T cells; each has its own unique TCRs that can recognize a specific antigen–MHC complex. When an antigen enters the body, only a few T cells have TCRs that can recognize and bind to the antigen. Antigen recognition also involves other surface proteins on T cells, the CD4 or CD8 proteins. These proteins interact with the MHC antigens and help maintain the TCR–MHC coupling. For this reason, they are referred to as *coreceptors.* Antigen recognition by a TCR with CD4 or CD8 proteins is the *first signal* in activation of a T cell.

A T cell becomes activated only if it binds to the foreign antigen and at the same time receives a *second signal,* a process known as **costimulation.** Of the more than 20 known costimulators, some are cytokines, such as **interleukin-2.** Other costimulators include pairs of plasma membrane molecules, one on the surface of the T cell and a second on the surface of an antigen-presenting cell, that enable the two cells to adhere to one another for a period of time.

The need for two signals to activate a T cell is a little like starting and driving a car: When you insert the correct key (antigen) in the ignition (TCR) and turn it, the car starts (recognition of specific antigen), but it cannot move forward until you move the gear shift into drive (costimulation). The need for costimulation may prevent immune responses from occurring accidentally. Different costimulators affect the activated T cell in different ways, just as shifting a car into reverse has a different effect than shifting it into drive. Moreover, recognition (antigen binding to a receptor) without costimulation leads to a prolonged *state of inactivity* called **anergy** (AN-er-jē) in both T cells and B cells. Anergy is rather like leaving a car in neutral gear with its engine running until it's out of gas!

Once a T cell has received these two signals (antigen recognition and costimulation), it is **activated.** An activated T cell subsequently undergoes clonal selection.

Activation and Clonal Selection of Helper T Cells

Most T cells that display CD4 develop into **helper T cells,** also known as **CD4 T cells.** Inactive (resting) helper T cells recognize exogenous antigen fragments associated with major histocompatibility complex class II (MHC-II) molecules at the surface of an

APC (Figure 22.15). With the aid of the CD4 protein, the helper T cell and APC interact with each other (antigenic recognition), costimulation occurs, and the helper T cell becomes activated.

Once activated, the helper T cell undergoes clonal selection (Figure 22.15). The result is the formation of a clone of helper T cells that consists of active helper T cells and memory helper T cells. Within hours after costimulation, **active helper T cells** start secreting a variety of cytokines (see Table 22.2). One very important cytokine produced by helper T cells is interleukin-2 (IL-2), which is needed for virtually all immune responses and is the prime

Figure 22.15 Activation and clonal selection of a helper T cell.

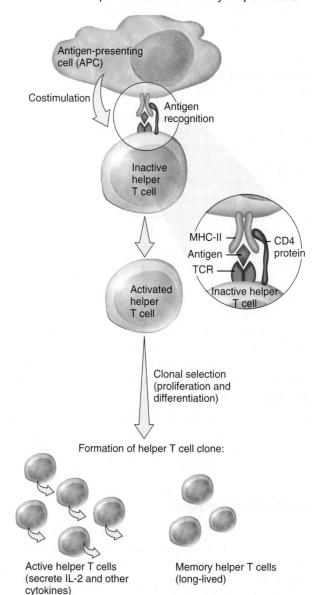

Once a helper T cell is activated, it forms a clone of active helper T cells and memory helper T cells.

Antigen-presenting cell (APC)

Costimulation

Antigen recognition

Inactive helper T cell

MHC-II
Antigen
TCR
CD4 protein
Inactive helper T cell

Activated helper T cell

Clonal selection (proliferation and differentiation)

Formation of helper T cell clone:

Active helper T cells (secrete IL-2 and other cytokines)

Memory helper T cells (long-lived)

? **What are the first and second signals in activation of a T cell?**

trigger of T cell proliferation. IL-2 can act as a costimulator for resting helper T cells or cytotoxic T cells, and it enhances activation and proliferation of T cells, B cells, and natural killer cells. Some actions of interleukin-2 provide a good example of a beneficial positive feedback system. As noted earlier, activation of a helper T cell stimulates it to start secreting IL-2, which then acts in an autocrine manner by binding to IL-2 receptors on the plasma membrane of the cell that secreted it. One effect is stimulation of cell division. As the helper T cells proliferate, a positive feedback effect occurs because they secrete more IL-2, which causes further cell division. IL-2 may also act in a paracrine manner by binding to IL-2 receptors on neighboring helper T cells, cytotoxic T cells, or B cells. If any of these neighboring cells have already become bound to a copy of the same antigen, IL-2 serves as a costimulator.

The **memory helper T cells** of a helper T cell clone are not active cells. However, if the same antigen enters the body again in the future, memory helper T cells can quickly proliferate and differentiate into more active helper T cells and more memory helper T cells.

Activation and Clonal Selection of Cytotoxic T Cells

Most T cells that display CD8 develop into **cytotoxic T cells** (sī'-tō-TOK-sik), also termed **CD8 T cells.** Cytotoxic T cells recognize foreign antigens combined with major histocompatibility complex class I (MHC-I) molecules on the surface of (1) body cells infected by microbes, (2) some tumor cells, and (3) cells of a tissue transplant (Figure 22.16). Recognition requires the TCR and CD8 protein to maintain the coupling with MHC-I. Following antigenic recognition, costimulation occurs. In order to become activated, cytotoxic T cells require costimulation by interleukin-2 or other cytokines produced by active helper T cells that have already become bound to copies of the same antigen. (Recall that helper T cells are activated by antigen associated with MHC-II molecules.) Thus, *maximal activation* of cytotoxic T cells requires presentation of antigen associated with both MHC-I and MHC-II molecules.

Once activated, the cytotoxic T cell undergoes clonal selection. The result is the formation of a clone of cytotoxic T cells that consists of active cytotoxic T cells and memory cytotoxic T cells. **Active cytotoxic T cells** attack other body cells that have been infected with the antigen. **Memory cytotoxic T cells** do not attack infected body cells. Instead, they can quickly proliferate and differentiate into more active cytotoxic T cells and more memory cytotoxic T cells if the same antigen enters the body at a future time.

Elimination of Invaders

Cytotoxic T cells are the soldiers that march forth to do battle with foreign invaders in cell-mediated immune responses. They leave secondary lymphatic organs and tissues and migrate to seek out and destroy infected target cells, cancer cells, and transplanted cells (Figure 22.17). Cytotoxic T cells recognize and attach to target cells. Then, the cytotoxic T cells deliver a "lethal hit" that kills the target cells.

Figure 22.16 Activation and clonal selection of a cytotoxic T cell.

 Once a cytotoxic T cell is activated, it forms a clone of active cytotoxic T cells and memory cytotoxic T cells.

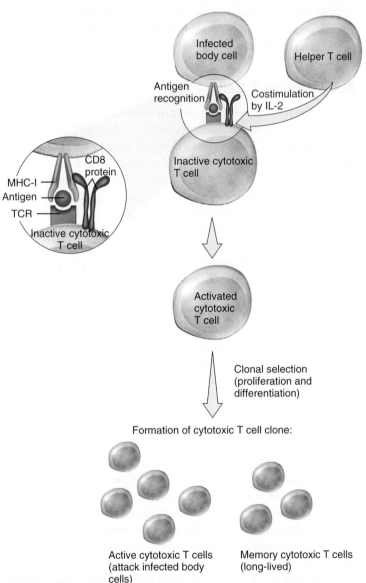

? **What is the function of the CD8 protein of a cytotoxic T cell?**

Cytotoxic T cells kill infected target body cells much like natural killer cells do. The major difference is that cytotoxic T cells have receptors specific for a particular microbe and thus kill only target body cells infected with *one* particular type of microbe; natural killer cells can destroy a wide variety of microbe-infected body cells. Cytotoxic T cells have two principal mechanisms for killing infected target cells.

1. Cytotoxic T cells, using receptors on their surfaces, recognize and bind to infected target cells that have microbial antigens displayed on their surface. The cytotoxic T cell then releases

Figure 22.17 Activity of cytotoxic T cells. After delivering a "lethal hit," a cytotoxic T cell can detach and attack another infected target cell displaying the same antigen.

Cytotoxic T cells release granzymes that trigger apoptosis and perforin that triggers cytolysis of infected target cells.

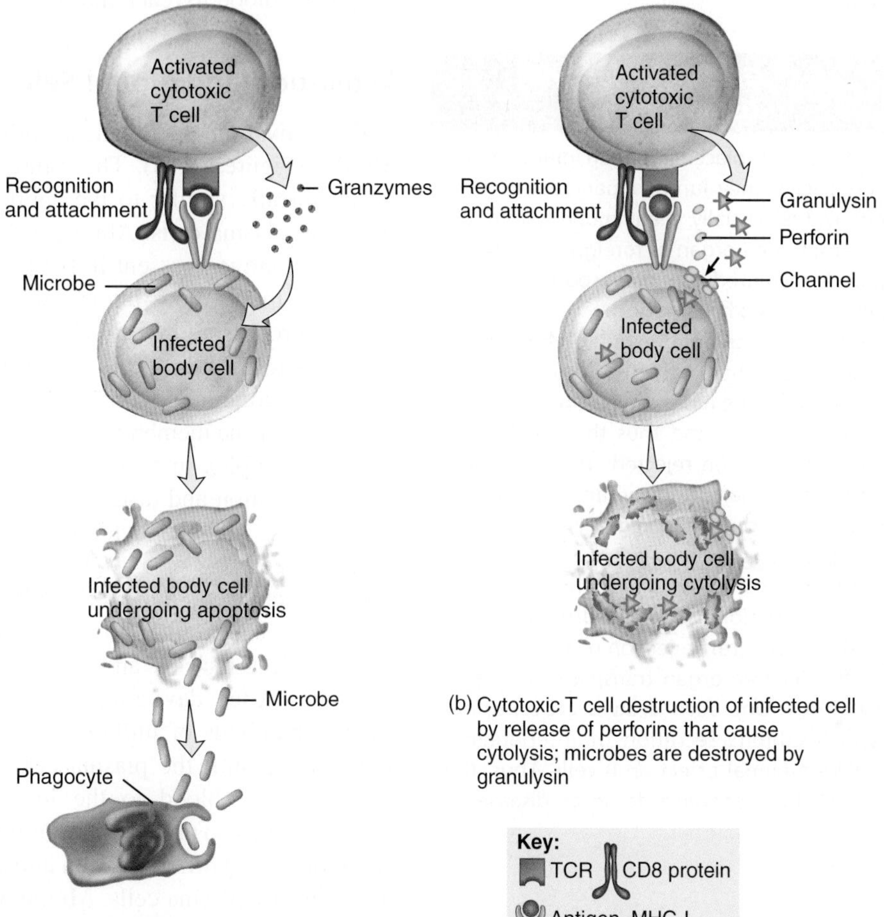

(a) Cytotoxic T cell destruction of infected cell by release of granzymes that cause apoptosis; released microbes are destroyed by phagocyte

(b) Cytotoxic T cell destruction of infected cell by release of perforins that cause cytolysis; microbes are destroyed by granulysin

Key:
- TCR
- CD8 protein
- Antigen–MHC-I complex

In addition to cells infected by microbes, what other types of target cells are attacked by cytotoxic T cells?

granzymes, protein-digesting enzymes that trigger apoptosis (Figure 22.17a). Once the infected cell is destroyed, the released microbes are killed by phagocytes.

2. Alternatively, cytotoxic T cells bind to infected body cells and release two proteins from their granules: perforin and granulysin. **Perforin** inserts into the plasma membrane of the target cell and creates channels in the membrane (Figure 22.17b). As a result, extracellular fluid flows into the target cell and cytolysis (cell bursting) occurs. Other granules in cytotoxic T cells release **granulysin** (gran′-ū-LĪ-sin), which enters through the channels and destroys the microbes by creating holes in their plasma membranes. Cytotoxic T cells may also destroy target cells by releasing a toxic molecule called **lymphotoxin** (lim′-fō-TOK-sin), which activates enzymes in the target cell. These enzymes cause the target cell's DNA to fragment and the cell dies. In addition, cytotoxic T cells se-

crete gamma-interferon, which attracts and activates phagocytic cells, and macrophage migration inhibition factor, which prevents migration of phagocytes from the infection site. After detaching from a target cell, a cytotoxic T cell can seek out and destroy another target cell.

Immunological Surveillance

When a normal cell transforms into a cancerous cell, it often displays novel cell surface components called **tumor antigens.** These molecules are rarely, if ever, displayed on the surface of normal cells. If the immune system recognizes a tumor antigen as nonself, it can destroy any cancer cells carrying that antigen. Such immune responses, called **immunological surveillance** (im′-ū-nō-LOJ-i-kul sur-VĀ-lants), are carried out by cytotoxic T cells, macrophages, and natural killer cells. Immunological surveillance is most effective in eliminating tumor cells due to cancer-causing viruses. For this reason, transplant recipients who are taking immunosuppressive drugs to prevent transplant rejection have an increased incidence of virus-associated cancers. Their risk for other types of cancer is not increased.

CLINICAL CONNECTION | *Graft Rejection and Tissue Typing*

Organ transplantation involves the replacement of an injured or diseased organ, such as the heart, liver, kidney, lungs, or pancreas, with an organ donated by another individual. Usually, the immune system recognizes the proteins in the transplanted organ as foreign and mounts both cell-mediated and antibody-mediated immune responses against them. This phenomenon is known as **graft rejection.**

The success of an organ or tissue transplant depends on **histocompatibility** (his'-tō-kom-pat-i-BIL-i-tē)—that is, the tissue compatibility between the donor and the recipient. The more similar the MHC antigens, the greater the histocompatibility, and thus the greater the probability that the transplant will not be rejected. **Tissue typing (histocompatibility testing)** is done before any organ transplant. In the United States, a nationwide computerized registry helps physicians select the most histocompatible and needy organ transplant recipients whenever donor organs become available. The closer the match between the major histocompatibility complex proteins of the donor and recipient, the weaker is the graft rejection response.

To reduce the risk of graft rejection, organ transplant recipients receive immunosuppressive drugs. One such drug is *cyclosporine,* derived from a fungus, which inhibits secretion of interleukin-2 by helper T cells but has only a minimal effect on B cells. Thus, the risk of rejection is diminished while resistance to some diseases is maintained. •

✔**CHECKPOINT**

16. What are the functions of helper, cytotoxic, and memory T cells?
17. How do cytotoxic T cells kill infected target cells?
18. How is immunological surveillance useful?

22.6 ANTIBODY-MEDIATED IMMUNITY

◉ **OBJECTIVES**

- Describe the steps in an antibody-mediated immune response.
- List the chemical characteristics and actions of antibodies.
- Explain how the complement system operates.
- Distinguish between a primary response and a secondary response to infection.

The body contains not only millions of different T cells but also millions of different B cells, each capable of responding to a specific antigen. Cytotoxic T cells leave lymphatic tissues to seek out and destroy a foreign antigen, but B cells stay put. In the presence of a foreign antigen, a specific B cell in a lymph node, the spleen,

or mucosa-associated lymphatic tissue becomes activated. Then it undergoes clonal selection, forming a clone of plasma cells and memory cells. Plasma cells are the effector cells of a B cell clone; they secrete specific antibodies, which in turn circulate in the lymph and blood to reach the site of invasion.

Activation and Clonal Selection of B Cells

During activation of a B cell, an antigen binds to **B-cell receptors (BCRs)** (Figure 22.18). These integral transmembrane proteins are chemically similar to the antibodies that eventually are secreted by plasma cells. Although B cells can respond to an unprocessed antigen present in lymph or interstitial fluid, their response is much more intense when they process the antigen. Antigen processing in a B cell occurs in the following way: The antigen is taken into the B cell, broken down into peptide fragments and combined with MHC-II self-antigens, and moved to the B cell plasma membrane. Helper T cells recognize the antigen–MHC-II complex and deliver the costimulation needed for B cell proliferation and differentiation. The helper T cell produces interleukin-2 and other cytokines that function as costimulators to activate B cells.

Once activated, a B cell undergoes clonal selection (Figure 22.18). The result is the formation of a clone of B cells that consists of plasma cells and memory B cells. **Plasma cells** secrete antibodies. A few days after exposure to an antigen, a plasma cell secretes hundreds of millions of antibodies each day for about 4 or 5 days, until the plasma cell dies. Most antibodies travel in lymph and blood to the invasion site. Interleukin-4 and interleukin-6, also produced by helper T cells, enhance B cell proliferation, B cell differentiation into plasma cells, and secretion of antibodies by plasma cells. **Memory B cells** do not secrete antibodies. Instead, they can quickly proliferate and differentiate into more plasma cells and more memory B cells should the same antigen reappear at a future time.

Different antigens stimulate different B cells to develop into plasma cells and their accompanying memory B cells. All of the B cells of a particular clone are capable of secreting only one type of antibody, which is identical to the antigen receptor displayed by the B cell that first responded. Each specific antigen activates only those B cells that are predestined (by the combination of gene segments they carry) to secrete antibody specific to that antigen. Antibodies produced by a clone of plasma cells enter the circulation and form antigen–antibody complexes with the antigen that initiated their production.

Antibodies

An **antibody (Ab)** can combine specifically with the epitope on the antigen that triggered its production. The antibody's structure matches its antigen much as a lock accepts a specific key. In theory, plasma cells could secrete as many different antibodies as there are different B-cell receptors because the same recombined gene segments code for both the BCR and the antibodies eventually secreted by plasma cells.

Figure 22.18 Activation and clonal selection of B cells. Plasma cells are actually much larger than B cells.

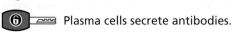 Plasma cells secrete antibodies.

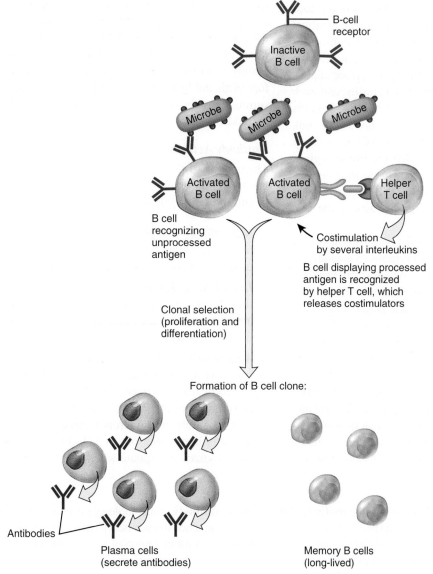

? How many different kinds of antibodies will be secreted by the plasma cells in the clone shown here?

CLINICAL CONNECTION | Severe Combined Immunodeficiency Disease

Severe combined immunodeficiency disease (SCID) (im′-ū-nō-de-FISH-en-sē) is a rare inherited disorder in which both B cells and T cells are missing or inactive. Scientists have now identified mutations in several genes that are responsible for some types of SCID. In some cases, an infusion of red bone marrow cells from a sibling having very similar MHC (HLA) antigens can provide normal stem cells that give rise to normal B and T cells. The result can be a complete cure. Less than 30% of afflicted patients, however, have a compatible sibling who could serve as a donor. The disorder, which occurs more frequently in males, is also known as *bubble boy disease*, named for David Vetter, who was born with the condition and lived behind plastic barriers to protect him from microbes. He died at age 12 in 1984. The chances of a child born with SCID are about 1 in 500,000 and until recent years, it was always fatal. Children with SCID have virtually no defenses against microbes. Treatment consists of bringing any current infections under control, bolstering nutrition, bone marrow transplant (provides stem cells to make new B and T cells), enzymatic replacement therapy (injections of polyethylene glycol–linked adenosine deaminase, or PE-ADA), and gene therapy. In this technique, the most common approach is to insert a normal gene into a genome to replace a nonfunctional gene. The normal gene is usually delivered by a virus. Then, the normal gene would produce B and T cells to provide sufficient immunity. •

Antibody Structure

Antibodies belong to a group of glycoproteins called globulins, and for this reason they are also known as **immunoglobulins (Igs)** (im′-ū-nō-GLOB-ū-lins). Most antibodies contain four polypeptide chains (Figure 22.19). Two of the chains are identical to each other and are called **heavy (H) chains;** each consists of about 450 amino acids. Short carbohydrate chains are attached to each heavy polypeptide chain. The two other polypeptide chains, also identical to each other, are called **light (L) chains,** and each consists of about 220 amino acids. A disulfide bond (S—S) holds each light chain to a heavy chain. Two disulfide bonds also link the midregion of the two heavy chains; this part of the antibody displays considerable flexibility and is called the **hinge region.** Because the antibody "arms" can move somewhat as the hinge region bends, an antibody can assume either a T shape or a Y shape (Figure 22.19a, b). Beyond the hinge region, parts of the two heavy chains form the **stem region.**

Within each H and L chain are two distinct regions. The tips of the H and L chains, called the **variable (V) regions,** constitute the **antigen-binding site.** The variable region, which is different for each kind of antibody, is the part of the antibody that recognizes and attaches specifically to a particular antigen. Because most antibodies have two antigen-binding sites, they are said to be *bivalent*. Flexibility at the hinge allows the antibody to simultaneously bind to two epitopes that are some distance apart—for example, on the surface of a microbe.

The remainder of each H and L chain, called the **constant (C) region,** is nearly the same in all antibodies of the same class and is responsible for the type of antigen–antibody reaction that occurs. However, the constant region of the H chain differs from one class of antibody to another, and its structure serves as a basis for distinguishing five different classes, designated IgG, IgA, IgM, IgD, and IgE. Each class has a distinct chemical structure and a specific biological role. Because they appear first and are relatively short-lived, IgM antibodies indicate a recent invasion. In a sick patient, the responsible pathogen may be suggested by the presence of high levels of IgM specific to a particular organism. Resistance of the fetus and newborn baby to infection stems mainly from maternal IgG antibodies that cross the placenta before birth and IgA antibodies in breast milk after birth. Table 22.3 summarizes the structures and functions of the five classes of antibodies.

Antibody Actions

The actions of the five classes of immunoglobulins differ somewhat, but all of them act to disable antigens in some way. Actions of antibodies include the following:

- *Neutralizing antigen.* The reaction of antibody with antigen blocks or neutralizes some bacterial toxins and prevents attachment of some viruses to body cells.

- *Immobilizing bacteria.* If antibodies form against antigens on the cilia or flagella of motile bacteria, the antigen–antibody reaction may cause the bacteria to lose their motility, which limits their spread into nearby tissues.

- *Agglutinating and precipitating antigen.* Because antibodies have two or more sites for binding to antigen, the antigen–antibody reaction may cross-link pathogens to one another, causing agglutination (clumping together). Phagocytic cells ingest agglutinated microbes more readily. Likewise, soluble antigens may come out of solution and form a more easily phagocytized precipitate when cross-linked by antibodies.

Figure 22.19 Chemical structure of the immunoglobulin G (IgG) class of antibody. Each molecule is composed of four polypeptide chains (two heavy and two light) plus a short carbohydrate chain attached to each heavy chain. In (a), each circle represents one amino acid. In (b), V_L = variable regions of light chain, C_L = constant region of light chain, V_H = variable region of heavy chain, and C_H = constant region of heavy chain.

An antibody combines only with the epitope on the antigen that triggered its production.

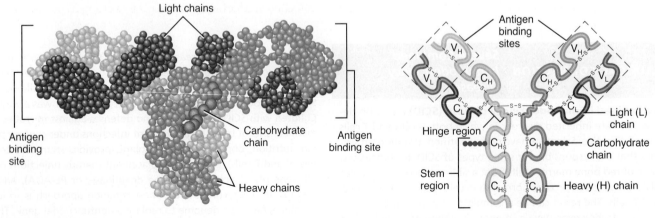

(a) Model of IgG molecule

(b) Diagram of IgG heavy and light chains

 What is the function of the variable regions in an antibody molecule?

TABLE 22.3

Classes of Immunoglobulins (Igs)

NAME AND STRUCTURE	CHARACTERISTICS AND FUNCTIONS
IgG	Most abundant, about 80% of all antibodies in blood; found in blood, lymph, and intestines; monomer (one-unit) structure. Protects against bacteria and viruses by enhancing phagocytosis, neutralizing toxins, and triggering complement system. Is the only class of antibody to cross placenta from mother to fetus, conferring considerable immune protection in newborns.
IgA	Found mainly in sweat, tears, saliva, mucus, breast milk, and gastrointestinal secretions. Smaller quantities are present in blood and lymph. Makes up 10–15% of all antibodies in blood; occurs as monomers and dimers (two units). Levels decrease during stress, lowering resistance to infection. Provides localized protection of mucous membranes against bacteria and viruses.
IgM	About 5–10% of all antibodies in blood; also found in lymph. Occurs as pentamers (five units); first antibody class to be secreted by plasma cells after initial exposure to any antigen. Activates complement and causes agglutination and lysis of microbes. Also present as monomers on surfaces of B cells, where they serve as antigen receptors. In blood plasma, anti-A and anti-B antibodies of ABO blood group, which bind to A and B antigens during incompatible blood transfusions, are also IgM antibodies (see Figure 19.12).
IgD	Mainly found on surfaces of B cells as antigen receptors, where it occurs as monomers; involved in activation of B cells. About 0.2% of all antibodies in blood.
IgE	Less than 0.1% of all antibodies in blood; occurs as monomers; located on mast cells and basophils. Involved in allergic and hypersensitivity reactions; provides protection against parasitic worms.

- *Activating complement.* Antigen–antibody complexes initiate the classical pathway of the complement system (discussed shortly).

- *Enhancing phagocytosis.* The stem region of an antibody acts as a flag that attracts phagocytes once antigens have bound to the antibody's variable region. Antibodies enhance the activity of phagocytes by causing agglutination and precipitation, by activating complement, and by coating microbes so that they are more susceptible to phagocytosis.

CLINICAL CONNECTION | *Monoclonal Antibodies*

The antibodies produced against a given antigen by plasma cells can be harvested from an individual's blood. However, because an antigen typically has many epitopes, several different clones of plasma cells produce different antibodies against the antigen. If a single plasma cell could be isolated and induced to proliferate into a clone of identical plasma cells, then a large quantity of identical antibodies could be produced. Unfortunately, lymphocytes and plasma cells are difficult to grow in culture, so scientists sidestepped this difficulty by fusing B cells with tumor cells that grow easily and proliferate endlessly. The resulting hybrid cell is called a **hybridoma** (hī-bri-DŌ-ma). Hybridomas are long-term sources of large quantities of pure, identical antibodies, called

monoclonal antibodies (MAbs) (mon′-ō-KLŌ-nal) because they come from a single clone of identical cells. One clinical use of monoclonal antibodies is for measuring levels of a drug in a patient's blood. Other uses include the diagnosis of strep throat, pregnancy, allergies, and diseases such as hepatitis, rabies, and some sexually transmitted diseases. MAbs have also been used to detect cancer at an early stage and to ascertain the extent of metastasis. They may also be useful in preparing vaccines to counteract the rejection associated with transplants, to treat autoimmune diseases, and perhaps to treat AIDS. •

Role of the Complement System in Immunity

The **complement system** (KOM-ple-ment) is a defensive system made up of over 30 proteins produced by the liver and found circulating in blood plasma and within tissues throughout the body. Collectively, the complement proteins destroy microbes by causing phagocytosis, cytolysis, and inflammation; they also prevent excessive damage to body tissues.

Most complement proteins are designated by an uppercase letter C, numbered C1 through C9, named for the order in which they were discovered. The C1–C9 complement proteins are inactive and become activated only when split by enzymes into active fragments, which are indicated by lowercase letters *a* and *b*. For example, inactive complement protein C3 is split into the activated fragments, C3a and C3b. The active fragments carry out the destructive actions of the C1–C9 complement proteins. Other complement proteins are referred to as factors B, D, and P (properdin).

Complement proteins act in a *cascade*—one reaction triggers another reaction, which in turn triggers another reaction, and so on. With each succeeding reaction, more and more product is formed so that the net effect is amplified many times.

Complement activation may begin by three different pathways (described shortly), all of which activate C3. Once activated, C3 begins a cascade of reactions that brings about phagocytosis, cytolysis, and inflammation as follows (Figure 22.20):

1 Inactivated C3 splits into activated C3a and C3b.

2 C3b binds to the surface of a microbe and receptors on phagocytes attach to the C3b. Thus C3b enhances **phagocytosis** by coating a microbe, a process called **opsonization** (op-sō-ni-ZĀ-shun). Opsonization promotes attachment of a phagocyte to a microbe.

3 C3b also initiates a series of reactions that bring about cytolysis. First, C3b splits C5. The C5b fragment then binds to C6 and C7, which attach to the plasma membrane of an invading microbe. Then C8 and several C9 molecules join the other complement proteins and together form a cylinder-shaped **membrane attack complex,** which inserts into the plasma membrane.

4 The membrane attack complex creates channels in the plasma membrane that result in **cytolysis,** the bursting of the microbial cells due to the inflow of extracellular fluid through the channels.

5 C3a and C5a bind to mast cells and cause them to release histamine that increases blood vessel permeability during **inflammation.** C5a also attracts phagocytes to the site of inflammation (chemotaxis).

Figure 22.20 Complement activation and results of activation. (Adapted from Tortora, Funke, and Case, *Microbiology: An Introduction, Eighth Edition,* Figure 16.10, Pearson Benjamin-Cummings, 2004.)

When activated, complement proteins enhance phagocytosis, cytolysis, and inflammation.

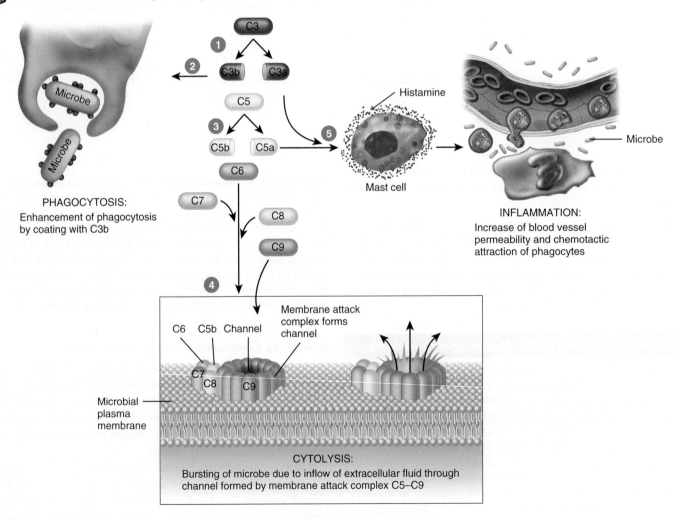

Which pathway for activation of complement involves antibodies? Explain why.

C3 can be activated in three ways: (1) The **classical pathway** starts when antibodies bind to antigens (microbes). The antigen–antibody complex binds and activates C1. Eventually, C3 is activated and the C3 fragments initiate phagocytosis, cytolysis, and inflammation. (2) The **alternative pathway** does not involve antibodies. It is initiated by an interaction between lipid–carbohydrate complexes on the surface of microbes and complement protein factors B, D, and P. This interaction activates C3. (3) In the **lectin pathway,** macrophages that digest microbes release chemicals that cause the liver to produce proteins called **lectins.** Lectins bind to the carbohydrates on the surface of microbes, ultimately causing the activation of C3.

Once complement is activated, proteins in blood and on body cells such as blood cells break down activated C3. In this way, its destructive capabilities cease very quickly so that damage to body cells is minimized.

Immunological Memory

A hallmark of immune responses is memory for specific antigens that have triggered immune responses in the past. Immunological memory is due to the presence of long-lasting antibodies and very long-lived lymphocytes that arise during clonal selection of antigen-stimulated B cells and T cells.

Immune responses, whether cell-mediated or antibody-mediated, are much quicker and more intense after a second or subsequent exposure to an antigen than after the first exposure. Initially, only a few cells have the correct specificity to respond, and the immune response may take several days to build to maximum intensity. Because thousands of memory cells exist after an initial encounter with an antigen, the next time the same antigen appears they can proliferate and differentiate into helper T cells, cytotoxic T cells, or plasma cells within hours.

One measure of immunological memory is *antibody titer* (TĪ-ter), the amount of antibody in serum. After an initial contact with an antigen, no antibodies are present for a period of several days. Then, a slow rise in the antibody titer occurs, first IgM and then IgG, followed by a gradual decline in antibody titer (Figure 22.21). This is the **primary response.**

Memory cells may remain for decades. Every new encounter with the same antigen results in a rapid proliferation of memory cells. After subsequent encounters, the antibody titer is far greater than during a primary response and consists mainly of IgG antibodies. This accelerated, more intense response is called the **secondary response.** Antibodies produced during a secondary response have an even higher affinity for the antigen than those produced during a primary response, and thus they are more successful in disposing of it.

Primary and secondary responses occur during microbial infection. When you recover from an infection without taking antimicrobial drugs, it is usually because of the primary response. If the same microbe infects you later, the secondary response could be so swift that the microbes are destroyed before you exhibit any signs or symptoms of infection.

Figure 22.21 Production of antibodies in the primary (after first exposure) and secondary (after second exposure) responses to a given antigen.

Immunological memory is the basis for successful immunization by vaccination.

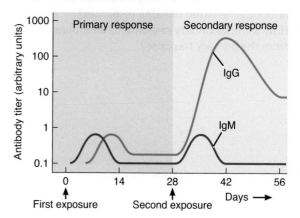

According to this graph, how much more IgG is circulating in the blood in the secondary response than in the primary response? (*Hint:* Notice that each mark on the antibody titer axis represents a 10-fold increase.)

Immunological memory provides the basis for immunization by vaccination against certain diseases (for example, polio). When you receive the vaccine, which may contain **attenuated** (weakened) or killed whole microbes or portions of microbes, your B cells and T cells are activated. Should you subsequently encounter the living pathogen as an infecting microbe, your body initiates a secondary response.

Table 22.4 summarizes the various ways to acquire adaptive immunity.

TABLE 22.4

Ways to Acquire Adaptive Immunity

METHOD	DESCRIPTION
Naturally acquired active immunity	Following exposure to a microbe, antigen recognition by B cells and T cells and costimulation lead to formation of antibody-secreting plasma cells, cytotoxic T cells, and B and T memory cells.
Naturally acquired passive immunity	IgG antibodies are transferred from mother to fetus across placenta, or IgA antibodies are transferred from mother to baby in milk during breast-feeding.
Artificially acquired active immunity	Antigens introduced during vaccination stimulate cell-mediated and antibody-mediated immune responses, leading to production of memory cells. Antigens are pretreated to be immunogenic but not pathogenic (they will trigger an immune response but not cause significant illness).
Artificially acquired passive immunity	Intravenous injection of immunoglobulins (antibodies).

✔ CHECKPOINT

19. How do the five classes of antibodies differ in structure and function?
20. How are cell-mediated and antibody-mediated immune responses similar and different?
21. In what ways does the complement system augment antibody-mediated immune responses?
22. How is the secondary response to an antigen different from the primary response?

22.7 SELF-RECOGNITION AND SELF-TOLERANCE

◉ **OBJECTIVE**

• Describe how self-recognition and self-tolerance develop.

To function properly, your T cells must have two traits: (1) They must be able to *recognize* your own major histocompatibility complex (MHC) proteins, a process known as **self-recognition,** and (2) they must *lack reactivity* to peptide fragments from your own proteins, a condition known as **self-tolerance** (Figure 22.22). B cells also display self-tolerance. Loss of self-tolerance leads to the development of autoimmune diseases (see Disorders: Homeostatic Imbalances at the end of the chapter).

Pre-T cells in the thymus develop the capability for self-recognition via **positive selection** (Figure 22.22a). In this process, some pre-T cells express T-cell receptors (TCRs) that

Figure 22.22 Development of self-recognition and self-tolerance. MHC = major histocompatibility complex, TCR = T-cell receptor.

🔑 Positive selection allows recognition of self-MHC proteins; negative selection provides self-tolerance of your own peptides and other self-antigens.

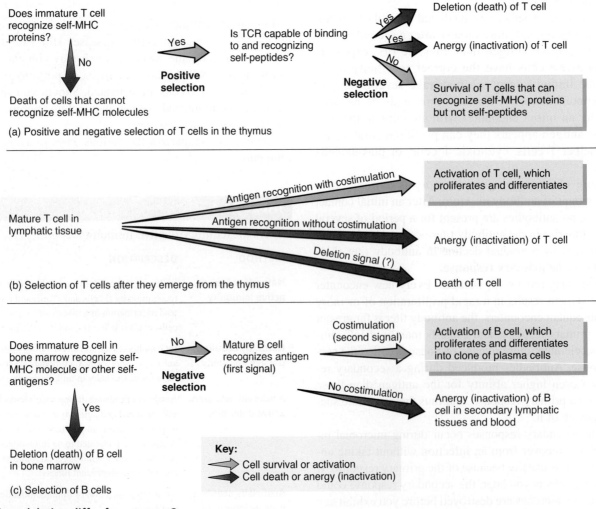

(a) Positive and negative selection of T cells in the thymus

(b) Selection of T cells after they emerge from the thymus

(c) Selection of B cells

Key:
▷ Cell survival or activation
▶ Cell death or anergy (inactivation)

❓ **How does deletion differ from anergy?**

interact with self-MHC proteins on epithelial cells in the thymic cortex. Because of this interaction, the T cells can recognize the MHC part of an antigen–MHC complex. These T cells survive. Other immature T cells that fail to interact with thymic epithelial cells are not able to recognize self-MHC proteins. These cells undergo apoptosis.

The development of self-tolerance occurs by a weeding-out process called **negative selection** in which the T cells interact with dendritic cells located at the junction of the cortex and medulla in the thymus. In this process, T cells with receptors that recognize self-peptide fragments or other self-antigens are eliminated or inactivated (Figure 22.22a). The T cells selected to survive do not respond to self-antigens, the fragments of molecules that are normally present in the body. Negative selection occurs via both deletion and anergy. In **deletion,** self-reactive T cells undergo apoptosis and die; in **anergy** they remain alive but are unresponsive to antigenic stimulation. Only 1–5% of the immature T cells in the thymus receive the proper signals to survive apoptosis during both positive and negative selection and emerge as mature, immunocompetent T cells.

Once T cells have emerged from the thymus, they may still encounter an unfamiliar self-protein; in such cases they may also become anergic if there is no costimulator (Figure 22.22b). Deletion of self-reactive T cells may also occur after they leave the thymus.

B cells also develop tolerance through deletion and anergy (Figure 22.22c). While B cells are developing in bone marrow, those cells exhibiting antigen receptors that recognize common self-antigens (such as MHC proteins or blood group antigens) are deleted. Once B cells are released into the blood, however, anergy appears to be the main mechanism for preventing responses to self-proteins. When B cells encounter an antigen not associated with an antigen-presenting cell, the necessary costimulation signal often is missing. In this case, the B cell is likely to become anergic (inactivated) rather than activated.

Table 22.5 summarizes the activities of cells involved in adaptive immune responses.

CLINICAL CONNECTION | Cancer Immunology

Although the immune system responds to cancerous cells, often immunity provides inadequate protection, as evidenced by the number of people dying each year from cancer. Over the past 25 years, considerable research has focused on *cancer immunology,* the study of ways to use immune responses for detecting, monitoring, and treating cancer. For example, some tumors of the colon release *carcinoembryonic antigen (CEA)* into the blood, and prostate cancer cells release *prostate-specific antigen (PSA).* Detecting these antigens in blood does not provide definitive diagnosis of cancer, because both antigens are also released in certain noncancerous conditions. However, high levels of cancer-related antigens in the blood often do indicate the presence of a malignant tumor.

Finding ways to induce our immune system to mount vigorous attacks against cancerous cells has been an elusive goal. Many different techniques have been tried, with only modest success. In one method, inactive lymphocytes are removed in a blood sample and cultured with interleukin-2. The resulting *lymphokine-activated killer (LAK) cells* are then transfused back into the patient's blood. Although LAK cells have produced dramatic improvement in a few cases, severe complications affect most patients. In another method, lymphocytes procured from a small biopsy sample of a tumor are cultured with interleukin-2. After

TABLE 22.5

Summary of Functions of Cells Participating in Adaptive Immune Responses

CELL	FUNCTIONS
ANTIGEN-PRESENTING CELLS (APCs)	
Macrophage	Processing and presentation of foreign antigens to T cells; secretion of interleukin-1, which stimulates secretion of interleukin-2 by helper T cells and induces proliferation of B cells; secretion of interferons that stimulate T cell growth.
Dendritic cell	Processes and presents antigen to T cells and B cells; found in mucous membranes, skin, lymph nodes.
B cell	Processes and presents antigen to helper T cells.
LYMPHOCYTES	
Cytotoxic T cell	Kills host target cells by releasing granzymes that induce apoptosis, perforin that forms channels to cause cytolysis, granulysin that destroys microbes, lymphotoxin that destroys target cell DNA, gamma-interferon that attracts macrophages and increases their phagocytic activity, and macrophage migration inhibition factor that prevents macrophage migration from site of infection.
Helper T cell	Cooperates with B cells to amplify antibody production by plasma cells and secretes interleukin-2, which stimulates proliferation of T cells and B cells. May secrete gamma-IFN and tumor necrosis factor (TNF), which stimulate inflammatory response.
Memory T cell	Remains in lymphatic tissue and recognizes original invading antigens, even years after first encounter.
B cell	Differentiates into antibody-producing plasma cell.
Plasma cell	Descendant of B cell that produces and secretes antibodies.
Memory B cell	Descendant of B cell that remains after immune response and is ready to respond rapidly and forcefully should the same antigen enter body in future.

their proliferation in culture, such *tumor-infiltrating lymphocytes (TILs)* are reinjected. About a quarter of patients with malignant melanoma and renal-cell carcinoma who received TIL therapy showed significant improvement. The many studies currently under way provide reason to hope that immune-based methods will eventually lead to cures for cancer. •

✔**CHECKPOINT**

23. What do positive selection, negative selection, and anergy accomplish?

22.8 STRESS AND IMMUNITY

◉ **OBJECTIVE**

• Describe the effects of stress on immunity.

The field of **psychoneuroimmunology (PNI)** deals with communication pathways that link the nervous, endocrine, and immune systems. PNI research appears to justify what people have long observed: Your thoughts, feelings, moods, and beliefs influence your level of health and the course of disease. For example, cortisol, a hormone secreted by the adrenal cortex in association with the stress response, inhibits immune system activity.

If you want to observe the relationship between lifestyle and immune function, visit a college campus. As the semester progresses and the workload accumulates, an increasing number of students can be found in the waiting rooms of student health services. When work and stress pile up, health habits can change. Many people smoke or consume more alcohol when stressed, two habits detrimental to optimal immune function. Under stress, people are less likely to eat well or exercise regularly, two habits that enhance immunity.

People resistant to the negative health effects of stress are more likely to experience a sense of control over the future, a commitment to their work, expectations of generally positive outcomes for themselves, and feelings of social support. To increase your stress resistance, cultivate an optimistic outlook, get involved in your work, and build good relationships with others.

Adequate sleep and relaxation are especially important for a healthy immune system. But when there aren't enough hours in the day, you may be tempted to steal some from the night. While skipping sleep may give you a few more hours of productive time in the short run, in the long run you end up even farther behind, especially if getting sick keeps you out of commission for several days, blurs your concentration, and blocks your creativity.

Even if you make time to get 8 hours of sleep, stress can cause insomnia. If you find yourself tossing and turning at night, it's time to improve your stress management and relaxation skills! Be sure to unwind from the day before going to bed.

✔**CHECKPOINT**

24. Have you ever observed a connection between stress and illness in your own life?

22.9 AGING AND THE IMMUNE SYSTEM

◉ **OBJECTIVE**

• Describe the effects of aging on the immune system.

With advancing age, most people become more susceptible to all types of infections and malignancies. Their response to vaccines is decreased, and they tend to produce more autoantibodies (antibodies against their body's own molecules). In addition, the immune system exhibits lowered levels of function. For example, T cells become less responsive to antigens, and fewer T cells respond to infections. This may result from age-related atrophy of the thymus or decreased production of thymic hormones. Because the T cell population decreases with age, B cells are also less responsive. Consequently, antibody levels do not increase as rapidly in response to a challenge by an antigen, resulting in increased susceptibility to various infections. It is for this key reason that elderly individuals are encouraged to get influenza (flu) vaccinations each year.

✔**CHECKPOINT**

25. How are T cells affected by aging?

• • •

To appreciate the many ways that the lymphatic system contributes to homeostasis of other body systems, examine *Focus on Homeostasis: The Lymphatic System and Immunity*.

Next, in Chapter 23, we will explore the structure and function of the respiratory system and see how its operation is regulated by the nervous system. Most importantly, the respiratory system provides for gas exchange—taking in oxygen and blowing off carbon dioxide. The cardiovascular system aids gas exchange by transporting blood containing these gases between the lungs and tissue cells.

BODY SYSTEM	CONTRIBUTION OF THE LYMPHATIC SYSTEM AND IMMUNITY

For all body systems
B cells, T cells, and antibodies protect all body systems from attack by harmful foreign invaders (pathogens), foreign cells, and cancer cells.

Integumentary system
Lymphatic vessels drain excess interstitial fluid and leaked plasma proteins from dermis of skin. Immune system cells (Langerhans cells) in skin help protect skin. Lymphatic tissue provides IgA antibodies in sweat.

Skeletal system
Lymphatic vessels drain excess interstitial fluid and leaked plasma proteins from connective tissue around bones.

Muscular system
Lymphatic vessels drain excess interstitial fluid and leaked plasma proteins from muscles.

Endocrine system
Flow of lymph helps distribute some hormones and cytokines. Lymphatic vessels drain excess interstitial fluid and leaked plasma proteins from endocrine glands.

Cardiovascular system
Lymph returns excess fluid filtered from blood capillaries and leaked plasma proteins to venous blood. Macrophages in spleen destroy aged red blood cells and remove debris in blood.

Respiratory system
Tonsils, alveolar macrophages, and MALT (mucosa-associated lymphatic tissue) help protect lungs from pathogens. Lymphatic vessels drain excess interstitial fluid from lungs.

Digestive system
Tonsils and MALT help defend against toxins and pathogens that penetrate the body from the gastrointestinal tract. Digestive system provides IgA antibodies in saliva and gastrointestinal secretions. Lymphatic vessels pick up absorbed dietary lipids and fat-soluble vitamins from the small intestine and transport them to the blood. Lymphatic vessels drain excess interstitial fluid and leaked plasma proteins from organs of the digestive system.

Urinary system
Lymphatic vessels drain excess interstitial fluid and leaked plasma proteins from organs of the urinary system. MALT helps defend against toxins and pathogens that penetrate the body via the urethra.

Reproductive systems
Lymphatic vessels drain excess interstitial fluid and leaked plasma proteins from organs of the reproductive system. MALT helps defend against toxins and pathogens that penetrate the body via the vagina and penis. In females, sperm deposited in the vagina are not attacked as foreign invaders due to inhibition of immune responses. IgG antibodies can cross the placenta to provide protection to a developing fetus. Lymphatic tissue provides IgA antibodies in the milk of a nursing mother.

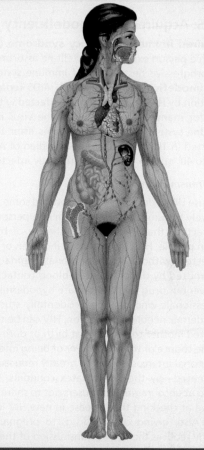

THE LYMPHATIC SYSTEM AND IMMUNITY

DISORDERS: HOMEOSTATIC IMBALANCES

AIDS: Acquired Immunodeficiency Syndrome

Acquired immunodeficiency syndrome (AIDS) is a condition in which a person experiences a telltale assortment of infections due to the progressive destruction of immune system cells by the **human immunodeficiency virus (HIV).** AIDS represents the end stage of infection by HIV. A person who is infected with HIV may be symptom-free for many years, even while the virus is actively attacking the immune system. In the two decades after the first five cases were reported in 1981, 22 million people died of AIDS. Worldwide, 35 million to 40 million people are currently infected with HIV.

HIV Transmission

Because HIV is present in the blood and some body fluids, it is most effectively transmitted (spread from one person to another) by actions or practices that involve the exchange of blood or body fluids between people. HIV is transmitted in semen or vaginal fluid during unprotected (without a condom) anal, vaginal, or oral sex. HIV also is transmitted by direct blood-to-blood contact, such as occurs among intravenous drug users who share hypodermic needles or health-care professionals who may be accidentally stuck by HIV-contaminated hypodermic needles. In addition, HIV can be transmitted from an HIV-infected mother to her baby at birth or during breast-feeding.

The chance of transmitting or of being infected by HIV during vaginal or anal intercourse can be greatly reduced—although not entirely eliminated—by the use of latex condoms. Public health programs aimed at encouraging drug users not to share needles have proved effective at checking the increase in new HIV infections in this population. Also, giving certain drugs to pregnant HIV-infected women greatly reduces the risk of transmission of the virus to their babies.

HIV is a very fragile virus; it cannot survive for long outside the human body. The virus is not transmitted by insect bites. One cannot become infected by casual physical contact with an HIV-infected person, such as by hugging or sharing household items. The virus can be eliminated from personal care items and medical equipment by exposing them to heat (135°F for 10 minutes) or by cleaning them with common disinfectants such as hydrogen peroxide, rubbing alcohol, household bleach, or germicidal cleansers such as Betadine or Hibiclens. Standard dishwashing and clothes washing also kill HIV.

HIV: Structure and Infection

HIV consists of an inner core of ribonucleic acid (RNA) covered by a protein coat (capsid). HIV is classified as a **retrovirus** (RET-rō-vī-rus) since its genetic information is carried in RNA instead of DNA. Surrounding the HIV capsid is an envelope composed of a lipid bilayer that is penetrated by glycoproteins (Figure 22.23).

Outside a living host cell, a virus is unable to replicate. However, when a virus infects and enters a host cell, it uses the host cell's enzymes and ribosomes to make thousands of copies of the virus. New viruses eventually leave and then infect other cells. HIV infection of a host cell begins with the binding of HIV glycoproteins to receptors in the host cell's plasma membrane. This causes the cell to transport the virus into its cytoplasm via receptor-mediated endocytosis. Once inside the host cell, HIV sheds its protein coat, and a viral enzyme called **reverse transcriptase** (tran-SKRIP-tās') reads the viral RNA strand and makes a DNA copy. The viral DNA copy then becomes integrated into the host cell's DNA. Thus, the viral DNA is duplicated along with the host cell's DNA during normal cell division. In addition, the viral DNA can cause the infected cell to begin producing millions of copies of viral RNA and to assemble new protein coats for each copy. The

Figure 22.23 Human immunodeficiency virus (HIV), the causative agent of AIDS.

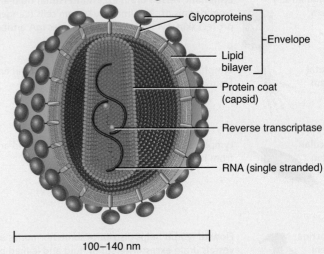

HIV is most effectively transmitted by practices that involve the exchange of body fluids.

- Glycoproteins
- Lipid bilayer
} Envelope
- Protein coat (capsid)
- Reverse transcriptase
- RNA (single stranded)

100–140 nm

Human immunodeficiency virus (HIV)

? Which cells of the immune system are attacked by HIV?

new HIV copies bud off from the cell's plasma membrane and circulate in the blood to infect other cells.

HIV mainly damages helper T cells, and it does so in various ways. Over 10 billion viral copies may be produced each day. The viruses bud so rapidly from an infected cell's plasma membrane that cell lysis eventually occurs. In addition, the body's defenses attack the infected cells, killing them but not all the viruses they harbor. In most HIV-infected individuals, helper T cells are initially replaced as fast as they are destroyed. After several years, however, the body's ability to replace helper T cells is slowly exhausted, and the number of helper T cells in circulation progressively declines.

Signs, Symptoms, and Diagnosis of HIV Infection

Soon after being infected with HIV, most people experience a brief flulike illness. Common signs and symptoms are fever, fatigue, rash, headache, joint pain, sore throat, and swollen lymph nodes. About 50% of infected people also experience night sweats. As early as 3 to 4 weeks after HIV infection, plasma cells begin secreting antibodies against HIV. These antibodies are detectable in blood plasma and form the basis for some of the screening tests for HIV. When people test "HIV-positive," it usually means they have antibodies to HIV antigens in their bloodstream.

Progression to AIDS

After a period of 2 to 10 years, HIV destroys enough helper T cells that most infected people begin to experience symptoms of immunodeficiency. HIV-infected people commonly have enlarged lymph nodes and experience persistent fatigue, involuntary weight loss, night sweats, skin rashes, diarrhea, and various lesions of the mouth and gums. In addition, the virus may begin to infect neurons in the brain, affecting the person's memory and producing visual disturbances.

As the immune system slowly collapses, an HIV-infected person becomes susceptible to a host of *opportunistic infections.* These are diseases caused by microorganisms that are normally held in check but now proliferate because of the defective immune system. AIDS is diagnosed when the helper T cell count drops below 200 cells per microliter (= cubic millimeter) of blood or when opportunistic infections arise, whichever occurs first. In time, opportunistic infections usually are the cause of death.

Treatment of HIV Infection

At present, infection with HIV cannot be cured. Vaccines designed to block new HIV infections and to reduce the viral load (the number of copies of HIV RNA in a microliter of blood plasma) in those who are already infected are in clinical trials. Meanwhile, two categories of drugs have proved successful in extending the life of many HIV-infected people:

1. **Reverse transcriptase inhibitors** interfere with the action of the reverse transcriptase enzyme, the enzyme that the virus uses to convert its RNA into a DNA copy. Among the drugs in this category are zidovudine (ZDV, previously called AZT), didanosine (ddI), and stavudine (trade name d4T®). Trizivir, approved in 2000 for treatment of HIV infection, combines three reverse transcriptase inhibitors in one pill.

2. **Protease inhibitors** interfere with the action of protease, a viral enzyme that cuts proteins into pieces to assemble the protein coat of newly produced HIV particles. Drugs in this category include nelfinavir, saquinavir, ritonavir, and indinavir.

In 1996, physicians treating HIV-infected patients widely adopted *highly active antiretroviral therapy (HAART)*—a combination of two differently acting reverse transcriptase inhibitors and one protease inhibitor. Most HIV-infected individuals receiving HAART experience a drastic reduction in viral load and an increase in the number of helper T cells in their blood. Not only does HAART delay the progression of HIV infection to AIDS, but many individuals with AIDS have seen the remission or disappearance of opportunistic infections and an apparent return to health. Unfortunately, HAART is very costly (exceeding $10,000 per year), the dosing schedule is grueling, and not all people can tolerate the toxic side effects of these drugs. Although HIV may virtually disappear from the blood with drug treatment (and thus a blood test may be "negative" for HIV), the virus typically still lurks in various lymphatic tissues. In such cases, the infected person can still transmit the virus to another person.

Allergic Reactions

A person who is overly reactive to a substance that is tolerated by most other people is said to be **allergic** or **hypersensitive.** Whenever an allergic reaction takes place, some tissue injury occurs. The antigens that induce an allergic reaction are called **allergens** (AL-er-jens). Common allergens include certain foods (milk, peanuts, shellfish, eggs), antibiotics (penicillin, tetracycline), vaccines (pertussis, typhoid), venoms (honeybee, wasp, snake), cosmetics, chemicals in plants such as poison ivy, pollens, dust, molds, iodine-containing dyes used in certain x-ray procedures, and even microbes.

There are four basic types of hypersensitivity reactions: type I (anaphylactic), type II (cytotoxic), type III (immune-complex), and type IV (cell-mediated). The first three are antibody-mediated immune responses; the last is a cell-mediated immune response.

Type I (anaphylactic) reactions (AN-a-fil-lak'-tik) are the most common and occur within a few minutes after a person sensitized to an allergen is re-exposed to it. In response to the first exposure to certain allergens, some people produce IgE antibodies that bind to the surface of mast cells and basophils. The next time the same allergen enters the body, it attaches to the IgE antibodies already present. In response, the mast cells and basophils release histamine, prostaglandins, leukotrienes, and kinins. Collectively, these mediators cause vasodilation, increased blood capillary permeability, increased smooth muscle contraction in the airways of the lungs, and increased mucus secretion. As a result, a person may experience inflammatory responses, difficulty in breathing through the constricted airways, and a runny nose from excess mucus secretion. In **anaphylactic shock,** which may occur in a susceptible individual who has just received a triggering drug or been stung by a wasp, wheezing and shortness of breath as airways constrict are usually accompanied by shock due to vasodilation and fluid loss from blood. This life-threatening emergency is usually treated by injecting epinephrine to dilate the airways and strengthen the heartbeat.

Type II (cytotoxic) reactions are caused by antibodies (IgG or IgM) directed against antigens on a person's blood cells (red blood cells, lymphocytes, or platelets) or tissue cells. The reaction of antibodies and antigens usually leads to activation of complement. Type II reactions, which may occur in incompatible blood transfusion reactions, damage cells by causing lysis.

Type III (immune-complex) reactions involve antigens, antibodies (IgA or IgM), and complement. When certain ratios of antigen to antibody occur, the immune complexes are small enough to escape phagocytosis, but they become trapped in the basement membrane under the endothelium of blood vessels, where they activate complement and cause inflammation. Glomerulonephritis and rheumatoid arthritis (RA) arise in this way.

Type IV (cell-mediated) reactions or **delayed hypersensitivity reactions** usually appear 12–72 hours after exposure to an allergen. Type IV reactions occur when allergens are taken up by antigen-presenting cells (such as Langerhans cells in the skin) that migrate to lymph nodes and present the allergen to T cells, which then proliferate. Some of the new T cells return to the site of allergen entry into the body, where they produce gamma-interferon, which activates macrophages, and tumor necrosis factor, which stimulates an inflammatory response. Intracellular bacteria such as *Mycobacterium tuberculosis* (mī-kō-bak-TE-rē-um too-ber'-ku-LŌ-sis) trigger this type of cell-mediated immune response, as do certain haptens, such as poison ivy toxin. The skin test for tuberculosis also is a delayed hypersensitivity reaction.

Autoimmune Diseases

In an **autoimmune disease** (aw-tō-i-MŪN) or **autoimmunity,** the immune system fails to display self-tolerance and attacks the person's own tissues. Autoimmune diseases usually arise in early adulthood and are common, afflicting an estimated 5% of adults in North America and Europe. Females suffer autoimmune diseases twice as often as males. Recall that self-reactive B cells and T cells normally are deleted or undergo anergy during negative selection (see Figure 22.22). Apparently, this process is not 100% effective. Under the influence of unknown environmental triggers and certain genes that make some people more susceptible, self-tolerance breaks down, leading to activation of self-reactive clones of T cells and B cells. These cells then generate cell-mediated or antibody-mediated immune responses against self-antigens.

A variety of mechanisms produce different autoimmune diseases. Some involve production of **autoantibodies,** antibodies that bind to and stimulate or block self-antigens. For example, autoantibodies that mimic TSH (thyroid-stimulating hormone) are present in Graves disease and stimulate secretion of thyroid hormones (thus producing hyperthyroidism); autoantibodies that bind to and block acetylcholine receptors cause the muscle weakness characteristic of myasthenia gravis. Other autoimmune diseases involve activation of cytotoxic T cells that destroy certain body cells. Examples include type 1 diabetes mellitus, in which T cells attack the insulin-producing pancreatic beta cells, and multiple sclerosis (MS), in which T cells attack myelin sheaths around axons of neurons. Inappropriate activation of helper T cells or excessive production of gamma-interferon also occur in certain autoimmune diseases. Other autoimmune disorders include rheumatoid arthritis (RA), systemic lupus erythematosus (SLE), rheumatic fever, hemolytic and pernicious anemias, Addison's disease, Hashimoto's thyroiditis, and ulcerative colitis.

Therapies for various autoimmune diseases include removal of the thymus gland (thymectomy), injections of beta-interferon, immunosuppressive drugs, and plasmapheresis, in which the person's blood plasma is filtered to remove antibodies and antigen–antibody complexes.

Infectious Mononucleosis

Infectious mononucleosis (mon'-ō-noo-klē-Ō-sis) or "mono" is a contagious disease caused by the *Epstein–Barr virus (EBV)*. It occurs mainly in children and young adults, and more often in females than in males. The virus most commonly enters the body through intimate oral contact such as kissing, which accounts for its common name, the "kissing disease." EBV then multiplies in lymphatic tissues and filters into the blood, where it infects and multiplies in B cells, the primary host cells. Because of this infection, the B cells become so enlarged and abnormal in appearance that they resemble monocytes, the primary reason for the term **mononucleosis.** In addition to an elevated white blood cell count with an abnormally high percentage of lymphocytes, signs and symptoms include fatigue, headache, dizziness, sore throat, enlarged and tender lymph nodes, and fever. There is no cure for infectious mononucleosis, but the disease usually runs its course in a few weeks.

Lymphomas

Lymphomas (lim-FŌ-mas; *lymph-* = clear water; *-oma* = tumor) are cancers of the lymphatic organs, especially the lymph nodes. Most have no known cause. The two main types of lymphomas are Hodgkin disease and non-Hodgkin lymphoma.

Hodgkin disease (HD) (HOJ-kin) is characterized by a painless, nontender enlargement of one or more lymph nodes, most commonly in the neck, chest, and axilla. If the disease has metastasized from these sites, fever, night sweats, weight loss, and bone pain also occur. HD primarily affects individuals between ages 15 and 35 and those over 60, and it is more common in males. If diagnosed early, HD has a 90–95% cure rate.

Non-Hodgkin lymphoma (NHL), which is more common than HD, occurs in all age groups, the incidence increasing with age to a maximum between ages 45 and 70. NHL may start the same way as HD but may also include an enlarged spleen, anemia, and general malaise. Up to half of all individuals with NHL are cured or survive for a lengthy period. Treatment options for both HD and NHL include radiation therapy, chemotherapy, and bone marrow transplantation.

Systemic Lupus Erythematosus

Systemic lupus erythematosus (er'-e-thē'-ma-TŌ-sus), **SLE,** or simply **lupus** (*lupus* = wolf) is a chronic autoimmune, inflammatory disease that affects multiple body systems. Lupus is characterized by periods of active disease and remission; symptoms range from mild to life-threatening. Lupus most often develops between ages 15 and 44 and is 10–15 times more common in females than in males. It is also 2–3 times more common in African Americans, Hispanics, Asian Americans, and Native Americans than in European Americans. Although the cause of SLE is not known, both a genetic predisposition to the disease and environmental factors (infections, antibiotics, ultraviolet light, stress, and hormones) may trigger it. Sex hormones appear to influence the development of SLE. The disorder often occurs in females who exhibit extremely low levels of androgens.

Signs and symptoms of SLE include joint pain, muscle pain, chest pain with deep breaths, headaches, pale or purple fingers or toes, kidney dysfunction, low blood cell count, nerve or brain dysfunction, slight fever, fatigue, oral ulcers, weight loss, swelling in the legs or around the eyes, enlarged lymph nodes and spleen, photosensitivity, rapid loss of large amounts of scalp hair, and sometimes an eruption across the bridge of the nose and cheeks called a "butterfly rash." The erosive nature of some of the SLE skin lesions was thought to resemble the damage inflicted by the bite of a wolf—thus, the term *lupus.*

Two immunological features of SLE are excessive activation of B cells and inappropriate production of autoantibodies against DNA (anti-DNA antibodies) and other components of cellular nuclei such as histone proteins. Triggers of B cell activation are thought to include various chemicals and drugs, viral and bacterial antigens, and exposure to sunlight. Circulating complexes of abnormal autoantibodies and their "antigens" cause damage in tissues throughout the body. Kidney damage occurs as the complexes become trapped in the basement membrane of kidney capillaries, obstructing blood filtering. Renal failure is the most common cause of death.

There is no cure for lupus, but drug therapy can minimize symptoms, reduce inflammation, and forestall flare-ups. The most commonly used lupus medications are pain relievers (nonsteroidal anti-inflammatory drugs such as aspirin and ibuprofen), antimalarial drugs (hydroxychloroquine), and corticosteroids (prednisone and hydrocortisone).

MEDICAL TERMINOLOGY

Adenitis (ad'-e-NĪ-tis; *aden-* = gland; *-itis* = inflammation of) Enlarged, tender, and inflamed lymph nodes resulting from an infection.

Allograft (AL-ō-graft; *allo-* = other) A transplant between genetically distinct individuals of the same species. Skin transplants from other people and blood transfusions are allografts.

Autograft (AW-tō-graft; *auto-* = self) A transplant in which one's own tissue is grafted to another part of the body (such as skin grafts for burn treatment or plastic surgery).

Chronic fatigue syndrome (CFS) A disorder, usually occurring in young adults and primarily in females, characterized by (1) extreme fatigue that impairs normal activities for at least 6 months and (2) the absence of other known diseases (cancer, infections, drug abuse, toxicity, or psychiatric disorders) that might produce similar symptoms.

Gamma globulin (GLOB-ū-lin) Suspension of immunoglobulins from blood consisting of antibodies that react with a specific pathogen. It is prepared by injecting the pathogen into animals, removing blood from the animals after antibodies have been produced, isolating the antibodies, and injecting them into a human to provide short-term immunity.

Hypersplenism (hī-per-SPLEN-izm; *hyper-* = over) Abnormal splenic activity due to splenic enlargement and associated with an increased rate of destruction of normal blood cells.

Lymphadenopathy (lim-fad'-e-NOP-a-thē; *lymph-* = clear fluid; *-pathy* = disease) Enlarged, sometimes tender lymph glands as a response to infection; also called swollen glands.

Lymphangitis (lim-fan-JĪ-tis; *-itis* = inflammation of) Inflammation of lymphatic vessels.

Lymphedema (lim'-fe-DĒ-ma; *edema* = swelling) Accumulation of lymph in lymphatic vessels, causing painless swelling of a limb.
Splenomegaly (splē'-nō-MEG-a-lē; *mega-* = large) Enlarged spleen.
Xenograft (ZEN-ō-graft; *xeno-* = strange or foreign) A transplant between animals of different species. Xenografts from porcine (pig)

or bovine (cow) tissue may be used in humans as a physiological dressing for severe burns. Other xenografts include pig heart valves and baboon hearts.

CHAPTER REVIEW AND RESOURCE SUMMARY

Review

Resource

Introduction

1. The ability to ward off disease is called immunity (resistance). Lack of resistance is called susceptibility.
2. The two general types of immunity are (a) innate and (b) adaptive.
3. Innate immunity refers to a wide variety of body responses to a wide range of pathogens.
4. Adaptive immunity involves activation of specific lymphocytes to combat a particular foreign substance.

Anatomy Overview - The Lymphatic System and Disease Resistance

22.1 Lymphatic System Structure and Function

1. The lymphatic system carries out immune responses and consists of lymph, lymphatic vessels, and structures and organs that contain lymphatic tissue (specialized reticular tissue containing many lymphocytes). The lymphatic system drains interstitial fluid, transports dietary lipids, and protects against invasion through immune responses.
2. Lymphatic vessels begin as closed-ended lymphatic capillaries in tissue spaces between cells. Interstitial fluid drains into lymphatic capillaries, thus forming lymph. Lymphatic capillaries merge to form larger vessels, called lymphatic vessels, which convey lymph into and out of lymph nodes.
3. The route of lymph flow is from lymphatic capillaries to lymphatic vessels to lymph trunks to the thoracic duct (left lymphatic duct) and right lymphatic duct to the subclavian veins.
4. Lymph flows because of skeletal muscle contractions and respiratory movements. Valves in lymphatic vessels also aid flow of lymph.
5. The primary lymphatic organs are red bone marrow and the thymus. Secondary lymphatic organs are lymph nodes, the spleen, and lymphatic nodules.
6. The thymus lies between the sternum and the large blood vessels above the heart. It is the site of T cell maturation.
7. Lymph nodes are encapsulated, egg-shaped structures located along lymphatic vessels. Lymph enters lymph nodes through afferent lymphatic vessels, is filtered, and exits through efferent lymphatic vessels. Lymph nodes are the site of proliferation of B cells and T cells.
8. The spleen is the largest single mass of lymphatic tissue in the body. Within the spleen, B cells and T cells carry out immune functions and macrophages destroy blood-borne pathogens and worn-out red blood cells by phagocytosis.
9. Lymphatic nodules are scattered throughout the mucosa of the gastrointestinal, respiratory, urinary, and reproductive tracts. This lymphatic tissue is termed mucosa-associated lymphatic tissue (MALT).

Anatomy Overview - Lymphatic Vessels
Anatomy Overview - The Thymus
Anatomy Overview - The Spleen and Lymph Nodes
Animation - Lymph Formation and Flow
Animation - Lymphatic System Functions
Figure 22.5 - Thymus
Figure 22.6 - Structure of a Lymph Node
Exercise - Lymphatic Highway
Concepts and Connections - Lymphatic System

22.2 Development of Lymphatic Tissues

1. Lymphatic vessels develop from lymph sacs, which arise from developing veins. Thus, they are derived from mesoderm.
2. Lymph nodes develop from lymph sacs that become invaded by mesenchymal cells.

22.3 Innate Immunity

1. Innate immunity includes physical factors, chemical factors, antimicrobial proteins, natural killer cells, phagocytes, inflammation, and fever.
2. The skin and mucous membranes are the first line of defense against entry of pathogens.
3. Antimicrobial substances include interferons, the complement system, iron-binding proteins, and antimicrobial proteins.
4. Natural killer cells and phagocytes attack and kill pathogens and defective cells in the body.
5. Inflammation aids disposal of microbes, toxins, or foreign material at the site of an injury, and prepares the site for tissue repair.

Anatomy Overview - The Integument and Disease Resistance
Animation - Introduction to Disease Resistance
Animation - Complement Proteins
Concepts and Connections: Complement Proteins
Exercise - Integument vs. Disease
Exercise - Microbe Massacre
Concepts and Connections - Nonspecific Disease Resistance

Review	Resource

6. Fever intensifies the antiviral effects of interferons, inhibits growth of some microbes, and speeds up body reactions that aid repair.

7. Table 22.1 summarizes the innate defenses.

22.4 Adaptive Immunity

1. Adaptive immunity involves lymphocytes called B cells and T cells. B cells and T cells arise from stem cells in red bone marrow. B cells mature in red bone marrow, whereas T cells mature in the thymus gland.

2. Before B cells leave the red bone marrow or T cells leave the thymus, they develop immunocompetence, the ability to carry out adaptive immune responses. This process involves the insertion of antigen receptors into their plasma membranes. Antigen receptors are molecules that are capable of recognizing specific antigens.

3. Two major types of mature T cells exit the thymus: helper T cells (also known as CD4 T cells) and cytotoxic T cells (also referred to as CD8 T cells).

4. There are two types of adaptive immunity: cell-mediated immunity and antibody-mediated immunity. In cell-mediated immune responses, cytotoxic T cells directly attack invading antigens; in antibody-mediated immune responses, B cells transform into plasma cells that secrete antibodies.

5. Clonal selection is the process by which a lymphocyte proliferates and differentiates in response to a specific antigen. The result of clonal selection is the formation of a clone of cells that can recognize the same specific antigen as the original lymphocyte.

6. A lymphocyte that undergoes clonal selection gives rise to two major types of cells in the clone: effector cells and memory cells. The effector cells of a lymphocyte clone carry out immune responses that ultimately result in the destruction or inactivation of the antigen. Effector cells include active helper T cells, which are part of a helper T cell clone; active cytotoxic T cells, which are part of a cytotoxic T cell clone; and plasma cells, which are part of a B cell clone. The memory cells of a lymphocyte clone do not actively participate in the initial immune response. However, if the antigen reappears in the body in the future, the memory cells can quickly respond to the antigen by proliferating and differentiating into more effector cells and more memory cells. Memory cells include memory helper T cells, which are part of a helper T cell clone; memory cytotoxic T cells, which are part of a cytotoxic T cell clone; and memory B cells, which are part of a B cell clone.

7. Antigens (Ags) are chemical substances that are recognized as foreign by the immune system. Antigen receptors exhibit great diversity due to genetic recombination.

8. "Self-antigens" called major histocompatibility complex (MHC) antigens are unique to each person's body cells. All cells except red blood cells display MHC-I molecules. Antigen-presenting cells (APCs) display MHC-II molecules. APCs include macrophages, B cells, and dendritic cells.

9. Exogenous antigens (formed outside body cells) are presented with MHC-II molecules; endogenous antigens (formed inside body cells) are presented with MHC-I molecules.

10. Cytokines are small protein hormones that may stimulate or inhibit many normal cell functions such as growth and differentiation. Other cytokines regulate immune responses (see Table 22.2).

22.5 Cell-Mediated Immunity

1. A cell-mediated immune response begins with activation of a small number of T cells by a specific antigen.

2. During the activation process, T-cell receptors (TCRs) recognize antigen fragments associated with MHC molecules on the surface of a body cell.

3. Activation of T cells also requires costimulation, either by cytokines such as interleukin-2 or by pairs of plasma membrane molecules.

4. Once a T cell has been activated, it undergoes clonal selection. The result of clonal selection is the formation of a clone of effector cells and memory cells. The effector cells of a T cell clone carry out immune responses that ultimately result in elimination of the antigen.

5. Helper T cells display CD4 protein, recognize antigen fragments associated with MHC-II molecules, and secrete several cytokines, most importantly interleukin-2, which acts as a costimulator for other helper T cells, cytotoxic T cells, and B cells.

6. Cytotoxic T cells display CD8 protein and recognize antigen fragments associated with MHC-I molecules.

7. Active cytotoxic T cells eliminate invaders by (1) releasing granzymes that cause target cell apoptosis (phagocytes then kill the microbes) and (2) releasing perforin, which causes cytolysis, and granulysin that destroys the microbes.

8. Cytotoxic T cells, macrophages, and natural killer cells carry out immunological surveillance, recognizing and destroying cancerous cells that display tumor antigens.

Review	Resource

22.6 Antibody-Mediated Immunity

1. An antibody-mediated immune response begins with activation of a B cell by a specific antigen.
2. B cells can respond to unprocessed antigens, but their response is more intense when they process the antigen. Interleukin-2 and other cytokines secreted by helper T cells provide costimulation for activation of B cells.
3. Once activated, a B cell undergoes clonal selection, forming a clone of plasma cells and memory cells. Plasma cells are the effector cells of a B cell clone; they secrete antibodies.
4. An antibody (Ab) is a protein that combines specifically with the antigen that triggered its production.
5. Antibodies consist of heavy and light chains and variable and constant regions.
6. Based on chemistry and structure, antibodies are grouped into five principal classes (IgG, IgA, IgM, IgD, and IgE), each with specific biological roles.
7. Actions of antibodies include neutralization of antigen, immobilization of bacteria, agglutination and precipitation of antigen, activation of complement, and enhancement of phagocytosis.
8. Complement is a group of proteins that complement immune responses and help clear antigens from the body.
9. Immunization against certain microbes is possible because memory B cells and memory T cells remain after a primary response to an antigen. The secondary response provides protection should the same microbe enter the body again.

Resource:
Animation - Antibody-Mediated Immunity
Animation - Primary and Secondary Infections
Animation - Complement Proteins
Figure 22.18 - Activation and Clonal Selection of B Cells
Exercise - Antibody Ambush
Exercise - Disease Resistance Fighters
Exercise - Concentrate on Disease Resistance

22.7 Self-Recognition and Self-Tolerance

1. T cells undergo positive selection to ensure that they can recognize self-MHC proteins (self-recognition), and negative selection to ensure that they do not react to other self-proteins (self-tolerance). Negative selection involves both deletion and anergy.
2. B cells develop tolerance through deletion and anergy.

Resource:
Anatomy Overview - Antigens and Antibodies

22.8 Stress and Immunity

1. Psychoneuroimmunology (PNI) deals with communication pathways that link the nervous, endocrine, and immune systems. Thoughts, feelings, moods, and beliefs influence health and the course of disease.
2. Under stress, people are less likely to eat well or exercise regularly, two habits that enhance immunity.

22.9 Aging and the Immune System

1. With advancing age, individuals become more susceptible to infections and malignancies, respond less well to vaccines, and produce more autoantibodies.
2. Immune responses also diminish with age.

SELF-QUIZ QUESTIONS

Fill in the blanks in the following statements.

1. The first line of defense of innate immunity against pathogens consists of the _____ and _____; the second line of defense consists of _____, _____, and _____.

2. Substances that are recognized as foreign and provoke immune responses are known as _____.

Indicate whether the following statements are true or false.

3. Your body's ability to ward off damage or disease through your defenses is known as resistance; vulnerability to disease is susceptibility.

4. Your T cells must be able to recognize your own MHC molecules, a process known as self-recognition, and lack reactivity to peptide fragments from your own proteins, a condition known as self-tolerance.

Choose the one best answer to the following questions.

5. Trace the sequence of fluid from blood vessel to blood vessel by way of the lymphatic system. (1) lymphatic vessels, (2) blood capillaries, (3) subclavian veins, (4) lymphatic capillaries, (5) interstitial spaces, (6) arteries, (7) lymphatic ducts.
   (a) 2, 5, 4, 1, 7, 6, 3 (b) 3, 6, 2, 4, 5, 1, 7 (c) 6, 2, 5, 4, 1, 7, 3
   (d) 6, 2, 5, 4, 7, 1, 3 (e) 2, 5, 4, 7, 1, 3, 6

6. Which of the following describe lymph nodes? (1) Lymph enters the nodes through efferent lymphatic vessels and leaves through afferent lymphatic vessels. (2) The outer cortex consists of lymphatic nodules that contain B cells and are the sites of plasma cell and memory B cell formation. (3) The inner cortex contains lymphatic nodules with mature T cells. (4) The reticular fibers within the sinuses of the lymph nodes trap foreign substances in the lymph. (5) The sinuses of lymph nodes are known as red pulp.
   (a) 1, 2, 3, and 4 (b) 2, 4, and 5 (c) 1, 2, 3, 4, and 5
   (d) 2 and 4 (e) 1, 2, and 4

7. Which of the following statements are *correct*? (1) Lymphatic vessels are found throughout the body, except in avascular tissues, the CNS, portions of the spleen, and red bone marrow. (2) Lymphatic capillaries allow interstitial fluid to flow into them but not out of them. (3) Anchoring filaments attach lymphatic endothelial cells to surrounding tissues. (4) Lymphatic vessels freely receive all the components of blood, including the formed elements. (5) Lymph ducts directly connect to blood vessels by way of the subclavian veins.
   (a) 1, 3, 4, and 5   (b) 2, 3, 4, and 5   (c) 1, 2, 3, and 4
   (d) 1, 2, 4, and 5   (e) 1, 2, 3, and 5

8. Which of the following are physical factors that help fight pathogens and disease? (1) numerous layers of the epidermis, (2) mucus of mucous membranes, (3) saliva, (4) interferons, (5) complement.
   (a) 1, 3, and 4   (b) 2, 4, and 5   (c) 1, 4, and 5
   (d) 1, 2, and 3   (e) 1, 2, and 4

9. Which of the following are functions of antibodies? (1) neutralization of antigens, (2) immobilization of bacteria, (3) agglutination and precipitation of antigens, (4) activation of complement, (5) enhancement of phagocytosis.
   (a) 1, 3, and 4   (b) 2, 4, and 5   (c) 1, 2, 3, and 4
   (d) 1, 2, 3, and 5   (e) 1, 2, 3, 4, and 5

10. Which of the following are *true*? (1) Lymphatic vessels resemble arteries. (2) Lymph is very similar to interstitial fluid. (3) Lacteals are specialized lymphatic capillaries responsible for transporting dietary lipids. (4) Lymph is normally a cloudy, pale yellow fluid. (5) The thoracic duct drains lymph from the upper right side of the body. (6) Lymph flow is maintained by skeletal muscle contractions, one-way valves, and breathing movements.
    (a) 1, 2, 5, and 6   (b) 2, 3, and 6   (c) 2, 3, 4, and 6
    (d) 2, 4, and 6   (e) 3, 5, and 6

11. Place the stages of phagocytosis in the correct order of occurrence. (1) formation of phagolysosome, (2) adherence to microbe, (3) destruction of microbe, (4) ingestion to form a phagosome, (5) chemotactic attraction of phagocyte.
    (a) 2, 5, 4, 1, 3   (b) 4, 5, 2, 1, 3   (c) 5, 2, 4, 1, 3
    (d) 5, 4, 2, 3, 1   (e) 2, 5, 1, 4, 3

12. Place in order the steps involved in cell-mediated immune response to an exogenous antigen.
    (a) costimulation and activation of helper T cells
    (b) presentation of antigen to helper T cells
    (c) elimination of invaders through the release of granzymes, perforin, granulysin, or lymphotoxin or by attraction and activation of phagocytes
    (d) proliferation and differentiation of helper T cells to produce a helper T cell clone
    (e) antigen processing by dendritic cells, macrophages, or B cells
    (f) recognition of antigen fragments associated with MHC-II molecules by T-cell receptors
    (g) secretion of cytokines such as interleukin-2 by activated helper T cells
    (h) migration of antigen-presenting cells to lymphatic tissue
    (i) activation of cytotoxic T cells

13. Match the following:
    ____ (a) encapsulated bean-shaped structures located along the length of lymphatic vessels; contain T and B cells, macrophages, and follicular dendritic cells; filter lymph
    ____ (b) produces pre-T cells and B cells; found in flat bones and epiphyses of long bones
    ____ (c) clusters of lymphatic nodules involved in immune responses against inhaled or ingested foreign substances
    ____ (d) the single largest mass of lymphatic tissue in the body; consists of red and white pulp
    ____ (e) responsible for the maturation of T cells
    ____ (f) lymphatic nodules associated with mucous membranes of the digestive, urinary, reproductive, and respiratory systems
    ____ (g) nonencapsulated clusters of lymphocytes

    (1) red bone marrow
    (2) thymus
    (3) lymph nodes
    (4) spleen
    (5) mucosa-associated lymphatic tissue
    (6) lymphatic nodules
    (7) tonsils

14. Match the following:
    ____ (a) recognize foreign antigens combined with MHC-I molecules on the surface of body cells infected by microbes, some tumor cells, and cells of a tissue transplant; display CD8 proteins
    ____ (b) are programmed to recognize the reappearance of a previously encountered antigen
    ____ (c) differentiate into plasma cells that secrete specific antibodies
    ____ (d) process and present exogenous antigens; include macrophages, B cells, and dendritic cells
    ____ (e) secrete cytokines as costimulators; display CD4 proteins
    ____ (f) ingest microbes or any foreign particulate matter; include neutrophils and macrophages
    ____ (g) lymphocytes that have the ability to kill a wide variety of infectious microbes plus certain spontaneously arising tumor cells; lack antigen receptors

    (1) active helper T cells
    (2) cytotoxic T cells
    (3) memory T cells
    (4) B cells
    (5) NK cells
    (6) phagocytes
    (7) antigen-presenting cells

CRITICAL THINKING QUESTIONS

1. Esperanza watched as her mother got her "flu shot." "Why do you need a shot if you're not sick?" she asked. "So I won't get sick," answered her mom. Explain how the influenza vaccination prevents illness.

2. Due to the presence of breast cancer, Mrs. Franco had a right radical mastectomy in which her right breast, underlying muscle, and right axillary lymph nodes and vessels were removed. Now she is experiencing severe swelling in her right arm. Why did the surgeon remove lymph tissue as well as the breast? Why is Mrs. Franco's right arm swollen?

3. Tariq's little sister has the mumps. Tariq can't remember if he has had mumps or not, but he is feeling slightly feverish. How could Tariq's doctor determine if he is getting sick with mumps or if he has previously had mumps?

? ANSWERS TO FIGURE QUESTIONS

22.1 Red bone marrow contains stem cells that develop into lymphocytes.

22.2 Lymph is more similar to interstitial fluid than to blood plasma because the protein content of lymph is low.

22.3 The left and right lumbar trunks and the intestinal trunk empty into the cisterna chyli, which then drains into the thoracic duct.

22.4 Inhalation promotes the movement of lymph from abdominal lymphatic vessels toward the thoracic region because the pressure in the vessels of the thoracic region is lower than the pressure in the abdominal region when a person inhales.

22.5 T cells mature in the thymus.

22.6 Foreign substances in lymph that enter a lymph node may be phagocytized by macrophages or attacked by lymphocytes that mount immune responses.

22.7 White pulp of the spleen functions in immunity; red pulp of the spleen performs functions related to blood cells.

22.8 Lymphatic tissues begin to develop by the end of the fifth week of gestation.

22.9 Lysozyme, digestive enzymes, and oxidants can kill microbes ingested during phagocytosis.

22.10 Redness results from increased blood flow due to vasodilation; pain, from injury of nerve fibers, irritation by microbial toxins, kinins, and prostaglandins, and pressure due to edema; heat, from increased blood flow and heat released by locally increased metabolic reactions; swelling, from leakage of fluid from capillaries due to increased permeability.

22.11 Helper T cells participate in both cell-mediated and antibody-mediated immune responses.

22.12 Epitopes are small immunogenic parts of a larger antigen; haptens are small molecules that become immunogenic only when they attach to a body protein.

22.13 APCs include macrophages in tissues throughout the body, B cells in blood and lymphatic tissue, and dendritic cells in mucous membranes and the skin.

22.14 Endogenous antigens include viral proteins, toxins from intracellular bacteria, and abnormal proteins synthesized by a cancerous cell.

22.15 The first signal in T cell activation is antigen binding to a TCR; the second signal is a costimulator, such as a cytokine or another pair of plasma membrane molecules.

22.16 The CD8 protein of a cytotoxic T cell binds to the MHC-I molecule of an infected body cell to help anchor the T cell receptor (TCR)–antigen interaction so that antigen recognition can occur.

22.17 Cytotoxic T cells attack some tumor cells and transplanted tissue cells, as well as cells infected by microbes.

22.18 Since all of the plasma cells in this figure are part of the same clone, they secrete just one kind of antibody.

22.19 The variable regions recognize and bind to a specific antigen.

22.20 The classical pathway for the activation of complement is linked to antibody-mediated immunity because Ag–Ab complexes activate C1.

22.21 At peak secretion, approximately 1000 times more IgG is produced in the secondary response than in the primary response.

22.22 In deletion, self-reactive T cells or B cells die; in anergy, T cells or B cells are alive but are unresponsive to antigenic stimulation.

22.23 HIV attacks helper T cells.

23 THE RESPIRATORY SYSTEM

THE RESPIRATORY SYSTEM AND HOMEOSTASIS *The respiratory system contributes to homeostasis by providing for the exchange of gases—oxygen and carbon dioxide—between the atmospheric air, blood, and tissue cells. It also helps adjust the pH of body fluids.*

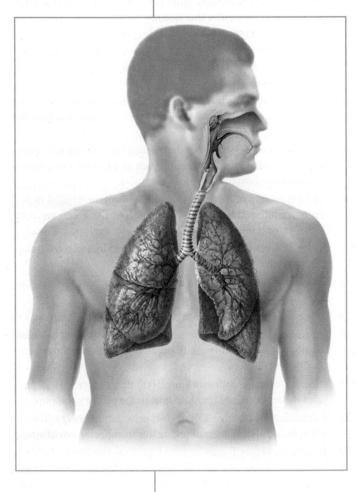

Your body's cells continually use oxygen (O_2) for the metabolic reactions that release energy from nutrient molecules and produce ATP. At the same time, these reactions release carbon dioxide (CO_2). Because an excessive amount of CO_2 produces acidity that can be toxic to cells, excess CO_2 must be eliminated quickly and efficiently. The cardiovascular and respiratory systems cooperate to supply O_2 and eliminate CO_2. The respiratory system provides for gas exchange—intake of O_2 and elimination of CO_2—and the cardiovascular system transports blood containing the gases between the lungs and body cells. Failure of either system disrupts homeostasis by causing rapid death of cells from oxygen starvation and buildup of waste products. In addition to functioning in gas exchange, the respiratory system also participates in regulating blood pH, contains receptors for the sense of smell, filters inspired air, produces sounds, and rids the body of some water and heat in exhaled air. As in the digestive and urinary systems, which will be covered in subsequent chapters, in the respiratory system there is an extensive area of contact between the external environment and capillary blood vessels.

?

Did you ever wonder how smoking affects the respiratory system?

23.1 RESPIRATORY SYSTEM ANATOMY

OBJECTIVES

- Describe the anatomy and histology of the nose, pharynx, larynx, trachea, bronchi, and lungs.
- Identify the functions of each respiratory system structure.

The **respiratory system** (RES-pi-ra-tōr-ē) consists of the nose, pharynx (throat), larynx (voice box), trachea (windpipe), bronchi, and lungs (Figure 23.1). Its parts can be classified according to either structure or function. *Structurally,* the respiratory system consists of two parts: (1) The **upper respiratory system** includes the nose, nasal cavity, pharynx, and associated structures. (2) The **lower respiratory system** includes the larynx, trachea, bronchi, and lungs. *Functionally,* the respiratory system also consists of two parts: (1) The **conducting zone** consists of a series of interconnecting cavities and tubes both outside and within the lungs. These include the nose, nasal cavity, pharynx, larynx, trachea, bronchi, bronchioles, and terminal bronchioles; their function is to filter, warm, and moisten air and conduct it into the lungs. (2) The **respiratory zone** consists of tubes and tissues within the lungs where gas exchange occurs. These include the respiratory bronchioles, alveolar ducts, alveolar sacs, and alveoli and are the main sites of gas exchange between air and blood.

The branch of medicine that deals with the diagnosis and treatment of diseases of the ears, nose, and throat (ENT) is called **otorhinolaryngology** (ō′-tō-rī′-nō-lar′-in-GOL-ō-jē; *oto-* = ear; *-rhino-* = nose; *-laryngo-* = voice box; *-logy* = study of). A **pulmonologist** (pul′-mo-NOL-ō-jist) is a specialist in the diagnosis and treatment of diseases of the lungs.

Nose

The **nose** is a specialized organ at the entrance of the respiratory system that is divided into an external portion and an internal portion called the nasal cavity. The **external nose** is the portion of the nose visible on the face and consists of a supporting framework of bone and hyaline cartilage covered with muscle and skin and lined by a mucous membrane. The frontal bone, nasal bones, and maxillae form the *bony framework* of the external nose (Figure 23.2a). The *cartilaginous framework* of the external nose consists of the **septal nasal cartilage,** which forms the anterior portion of the nasal septum; the **lateral nasal cartilages** inferior to the nasal bones; and the **alar cartilages** (Ā-lar), which form a portion of the walls of the nostrils. Because it consists of pliable hyaline cartilage, the cartilaginous framework of the external nose is somewhat flexible. On the undersurface of the external nose are two openings called the **external nares** (NĀ-rez; singular is **naris**) or **nostrils.** Figure 23.3 shows the surface anatomy of the nose.

The interior structures of the external nose have three functions: (1) warming, moistening, and filtering incoming air; (2) detecting olfactory stimuli; and (3) modifying speech vibrations as they pass through the large, hollow resonating chambers. *Resonance* refers to prolonging, amplifying, or modifying a sound by vibration.

CLINICAL CONNECTION | *Rhinoplasty*

Rhinoplasty (RĪ-nō-plas′-tē; *-plasty* = to mold or to shape), commonly called a "nose job," is a surgical procedure in which the shape of the external nose is altered. Although rhinoplasty is often done for cosmetic reasons, it is sometimes performed to repair a fractured nose or a deviated nasal septum. In the procedure, both local and general anesthetics are given. Instruments are then inserted through the nostrils, the nasal cartilage is reshaped, and the nasal bones are fractured and repositioned to achieve the desired shape. An internal packing and splint are inserted to keep the nose in the desired position as it heals. •

The **nasal cavity** is a large space in the anterior aspect of the skull that lies inferior to the nasal bone and superior to the oral cavity; it is lined with muscle and mucous membrane. Anteriorly, the nasal cavity merges with the external nose, and posteriorly it communicates with the pharynx through two openings called the **internal nares** or **choanae** (kō-Ā-nē) (see Figure 23.2b). Ducts from the *paranasal sinuses* (which drain mucus) and the *nasolacrimal ducts* (which drain tears) also open into the nasal cavity. Recall from Chapter 7 that the paranasal sinuses are cavities in certain cranial and facial bones lined with mucous membranes that are continuous with the lining of the nasal cavity. Skull bones containing the paranasal sinuses are the frontal, sphenoid, ethmoid, and maxillae. Besides producing mucus, the paranasal sinuses serve as resonating chambers for sound as we speak or sing. The lateral walls of the internal nose are formed by the ethmoid, maxillae, lacrimal, palatine, and inferior nasal conchae bones (see Figure 7.9); the ethmoid bone also forms the roof. The palatine bones and palatine processes of the maxillae, which together constitute the hard palate, form the floor of the internal nose.

The bony and cartilaginous framework of the nose help to keep the vestibule and nasal cavity **patent,** that is, open or unobstructed. The nasal cavity is divided into a larger, inferior *respiratory region* and a smaller, superior *olfactory region*. The respiratory region is lined with pseudostratified ciliated columnar epithelium with numerous goblet cells, which is frequently called the **respiratory epithelium** (see Table 4.1). The anterior portion of the nasal cavity just inside the nostrils, called the **nasal vestibule,** is surrounded by cartilage; the superior part of the nasal cavity is surrounded by bone. A vertical partition, the **nasal septum,** divides the nasal cavity into right and left sides. The anterior portion of the nasal septum consists primarily of hyaline cartilage; the remainder is formed by the vomer, perpendicular plate of the ethmoid, maxillae, and palatine bones (see Figure 7.11).

When air enters the nostrils, it passes first through the vestibule, which is lined by skin containing coarse hairs that filter out large dust particles. Three shelves formed by projections of the superior, middle, and inferior **nasal conchae** extend out of each lateral wall of the nasal cavity. The conchae, almost reaching the nasal septum, subdivide each side of the nasal cavity into a series of groovelike passageways—the **superior, middle,** and **inferior meatuses** (mē-Ā-tus-ēz = openings or passages; singular is **meatus**). Mucous membrane lines the nasal cavity and its

Figure 23.1 Structures of the respiratory system.

🔑 The upper respiratory system includes the nose, nasal cavity, pharynx, and associated structures; the lower respiratory system includes the larynx, trachea, bronchi, and lungs.

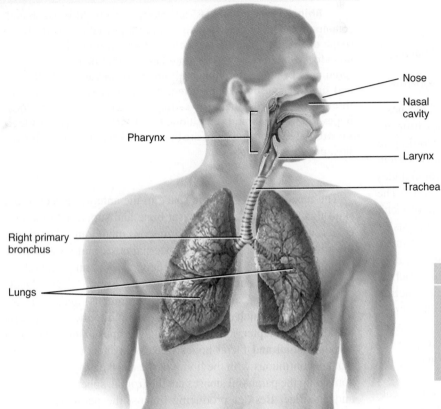

Nose

Nasal cavity

Pharynx

Larynx

Trachea

Right primary bronchus

Lungs

FUNCTIONS OF THE RESPIRATORY SYSTEM

1. Provides for gas exchange: intake of O_2 for delivery to body cells and removal of CO_2 produced by body cells.
2. Helps regulate blood pH.
3. Contains receptors for sense of smell, filters inspired air, produces vocal sounds (phonation), and excretes small amounts of water and heat.

(a) Anterior view showing organs of respiration

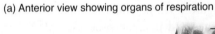

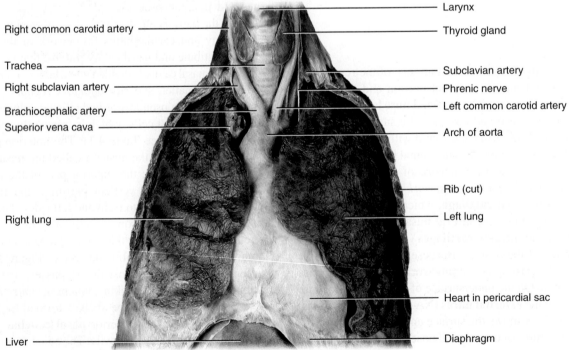

Right common carotid artery

Trachea

Right subclavian artery

Brachiocephalic artery

Superior vena cava

Right lung

Liver

Larynx

Thyroid gland

Subclavian artery

Phrenic nerve

Left common carotid artery

Arch of aorta

Rib (cut)

Left lung

Heart in pericardial sac

Diaphragm

(b) Anterior view of lungs and heart after removal of anterolateral thoracic wall and pleura

❓ **Which structures are part of the conducting zone of the respiratory system?**

Figure 23.2 Respiratory structures in the head and neck.

As air passes through the nose; it is warmed, filtered, and moistened; and olfaction occurs.

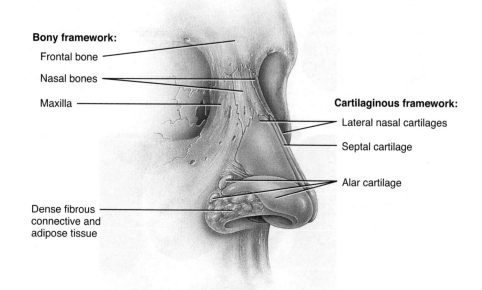

Bony framework:

Frontal bone

Nasal bones

Maxilla

Cartilaginous framework:

Lateral nasal cartilages

Septal cartilage

Alar cartilage

Dense fibrous connective and adipose tissue

(a) Anterolateral view of external portion of nose showing cartilaginous and bony framework

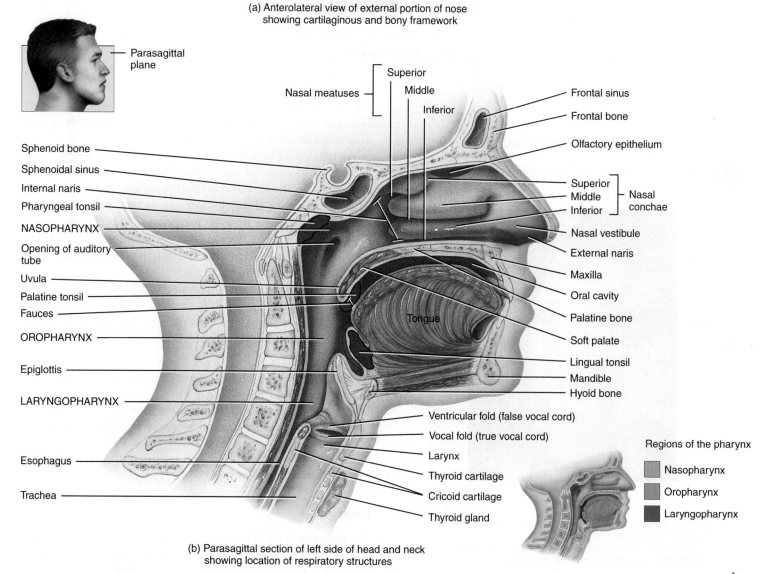

Parasagittal plane

Nasal meatuses
- Superior
- Middle
- Inferior

Frontal sinus

Frontal bone

Olfactory epithelium

Sphenoid bone

Sphenoidal sinus

Internal naris

Pharyngeal tonsil

NASOPHARYNX

Opening of auditory tube

Uvula

Palatine tonsil

Fauces

OROPHARYNX

Epiglottis

LARYNGOPHARYNX

Esophagus

Trachea

Superior
Middle
Inferior
Nasal conchae

Nasal vestibule

External naris

Maxilla

Oral cavity

Palatine bone

Soft palate

Lingual tonsil

Mandible

Hyoid bone

Tongue

Ventricular fold (false vocal cord)

Vocal fold (true vocal cord)

Larynx

Thyroid cartilage

Cricoid cartilage

Thyroid gland

Regions of the pharynx

Nasopharynx

Oropharynx

Laryngopharynx

(b) Parasagittal section of left side of head and neck showing location of respiratory structures

FIGURE 23.2 CONTINUES ▶

■ **FIGURE 23.2 CONTINUED** ▶

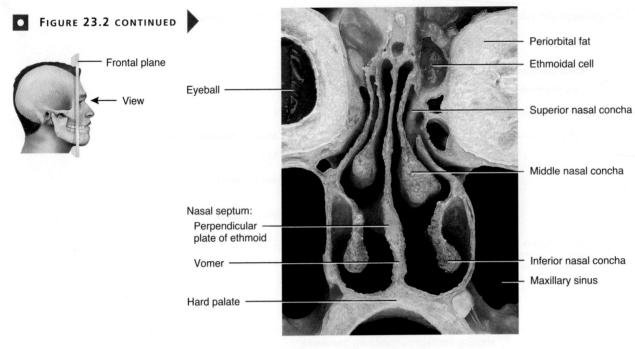

Frontal plane

View

Eyeball

Periorbital fat

Ethmoidal cell

Superior nasal concha

Middle nasal concha

Nasal septum:
Perpendicular plate of ethmoid

Vomer

Inferior nasal concha

Maxillary sinus

Hard palate

(c) Frontal section showing conchae

? **What is the path taken by air molecules into and through the nose?**

Figure 23.3 Surface anatomy of the nose.

The external nose has a cartilaginous framework and a bony framework.

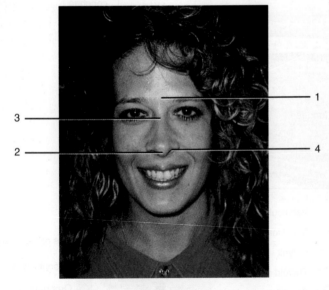

Anterior view

1. **Root**: Superior attachment of the nose to the frontal bone
2. **Apex**: Tip of nose
3. **Bridge**: Bony framework of nose formed by nasal bones
4. **External naris**: Nostril; external opening into nasal cavity

? **Which part of the nose is attached to the frontal bone?**

shelves. The arrangement of conchae and meatuses increases surface area in the internal nose and prevents dehydration by trapping water droplets during exhalation.

As inhaled air whirls around the conchae and meatuses, it is warmed by blood in the capillaries. Mucus secreted by the goblet cells moistens the air and traps dust particles. Drainage from the nasolacrimal ducts also helps moisten the air, and is sometimes assisted by secretions from the paranasal sinuses. The cilia move the mucus and trapped dust particles toward the pharynx, at which point they can be swallowed or spit out, thus removing the particles from the respiratory tract.

The olfactory receptors, supporting cells, and basal cells lie in the respiratory region, which is near the superior nasal conchae and adjacent septum. These cells make up the **olfactory epithelium.** It contains cilia but no goblet cells.

✔ CHECKPOINT

1. What functions do the respiratory and cardiovascular systems have in common?
2. What structural and functional features are different in the upper and lower respiratory systems? Which are the same?
3. Compare the structure and functions of the external nose and the internal nose.

Pharynx

The **pharynx** (FAR-inks), or throat, is a funnel-shaped tube about 13 cm (5 in.) long that starts at the internal nares and extends to the level of the cricoid cartilage, the most inferior cartilage of the larynx (voice box) (see Figure 23.2b). The pharynx lies just posterior to the nasal and oral cavities, superior to the larynx, and just anterior to the cervical vertebrae. Its wall is composed of skeletal muscles and is lined with a mucous membrane. Relaxed skeletal muscles help keep the pharynx patent. Contraction of the skeletal muscles assists in deglutition (swallowing). The pharynx functions as a passageway for air and food, provides a resonating chamber for speech sounds, and houses the tonsils, which participate in immunological reactions against foreign invaders.

The pharynx can be divided into three anatomical regions: (1) nasopharynx, (2) oropharynx, and (3) laryngopharynx. (See the lower orientation diagram in Figure 23.2b.) The muscles of the entire pharynx are arranged in two layers, an outer circular layer and an inner longitudinal layer.

The superior portion of the pharynx, called the **nasopharynx,** lies posterior to the nasal cavity and extends to the soft palate. The **soft palate,** which forms the posterior portion of the roof of the mouth, is an arch-shaped muscular partition between the nasopharynx and oropharynx that is lined by mucous membrane. There are five openings in its wall: two internal nares, two openings that lead into the auditory (pharyngotympanic) tubes (commonly known as the eustachian tubes), and the opening into the oropharynx. The posterior wall also contains the **pharyngeal tonsil** (fa-RIN-je-al), or **adenoid.** Through the internal nares, the nasopharynx receives air from the nasal cavity along with packages of dust-laden mucus. The nasopharynx is lined with pseudostratified ciliated columnar epithelium, and the cilia move the mucus down toward the most inferior part of the pharynx. The nasopharynx also exchanges small amounts of air with the auditory tubes to equalize air pressure between the pharynx and the middle ear.

The intermediate portion of the pharynx, the **oropharynx,** lies posterior to the oral cavity and extends from the soft palate inferiorly to the level of the hyoid bone. It has only one opening into it, the **fauces** (FAW-sēz = throat), the opening from the mouth. This portion of the pharynx has both respiratory and digestive functions, serving as a common passageway for air, food, and drink. Because the oropharynx is subject to abrasion by food particles, it is lined with nonkeratinized stratified squamous epithelium. Two pairs of tonsils, the **palatine** and **lingual tonsils,** are found in the oropharynx.

The inferior portion of the pharynx, the **laryngopharynx** (la-ring'-gō-FAR-ingks), or **hypopharynx,** begins at the level of the hyoid bone. At its inferior end it opens into the esophagus (food tube) posteriorly and the larynx (voice box) anteriorly. Like the oropharynx, the laryngopharynx is both a respiratory and a digestive pathway and is lined by nonkeratinized stratified squamous epithelium.

Larynx

The **larynx** (LAR-inks), or voice box, is a short passageway that connects the laryngopharynx with the trachea. It lies in the midline of the neck anterior to the esophagus and the fourth through sixth cervical vertebrae (C4–C6).

The wall of the larynx is composed of nine pieces of cartilage (Figure 23.4). Three occur singly (thyroid cartilage, epiglottis, and cricoid cartilage), and three occur in pairs (arytenoid, cuneiform, and corniculate cartilages). Of the paired cartilages, the arytenoid cartilages are the most important because they influence changes in position and tension of the vocal folds (true vocal cords for speech). The extrinsic muscles of the larynx connect the cartilages to other structures in the throat; the intrinsic muscles connect the cartilages to one another. The **cavity of the larynx** is the space that extends from the entrance into the larynx down to the inferior border of the cricoid cartilage (described shortly). The portion of the cavity of the larynx above the vocal folds (true vocal cords) is called the **vestibule of the larynx** (Figure 23.4d).

The **thyroid cartilage (Adam's apple)** consists of two fused plates of hyaline cartilage that form the anterior wall of the larynx and give it a triangular shape. It is present in both males and females but is usually larger in males due to the influence of male sex hormones on its growth during puberty. The ligament that connects the thyroid cartilage to the hyoid bone is called the **thyrohyoid membrane.**

The **epiglottis** (epi- = over; -glottis = tongue) is a large, leaf-shaped piece of elastic cartilage that is covered with epithelium (see also Figure 23.2b). The "stem" of the epiglottis is the tapered inferior portion that is attached to the anterior rim of the thyroid cartilage and hyoid bone. The broad superior "leaf" portion of the epiglottis is unattached and is free to move up and down like a trap door. During swallowing, the pharynx and larynx rise. Elevation of the pharynx widens it to receive food or drink; elevation of the larynx causes the epiglottis to move down and form a lid over the glottis, closing it off. The **glottis** consists of a pair of folds of mucous membrane, the vocal folds (true vocal cords) in the larynx, and the space between them called the **rima glottidis** (RĪ-ma GLOT-ti-dis). The closing of the larynx in this way during swallowing routes liquids and foods into the esophagus and keeps them out of the larynx and airways. When small particles of dust, smoke, food, or liquids pass into the larynx, a cough reflex occurs, usually expelling the material.

The **cricoid cartilage** (KRĪ-koyd = ringlike) is a ring of hyaline cartilage that forms the inferior wall of the larynx. It is attached to the first ring of cartilage of the trachea by the **cricotracheal ligament** (krī'-kō-TRĀ-kē-al). The thyroid cartilage is connected to the cricoid cartilage by the **cricothyroid ligament.** The cricoid cartilage is the landmark for making an emergency airway called a tracheotomy (see Clinical Connection: Tracheotomy and Intubation).

The paired **arytenoid cartilages** (ar'-i-TĒ-noyd = ladlelike) are triangular pieces of mostly hyaline cartilage located at the posterior, superior border of the cricoid cartilage. They form synovial joints with the cricoid cartilage and have a wide range of mobility.

The paired **corniculate cartilages** (kor-NIK-ū-lāt = shaped like a small horn), horn-shaped pieces of elastic cartilage, are located at the apex of each arytenoid cartilage. The paired **cuneiform cartilages** (KŪ-nē-i-form = wedge-shaped), club-shaped elastic cartilages anterior to the corniculate cartilages, support the vocal folds and lateral aspects of the epiglottis.

The lining of the larynx superior to the vocal folds is nonkeratinized stratified squamous epithelium. The lining of the larynx inferior to the vocal folds is pseudostratified ciliated columnar epithelium consisting of ciliated columnar cells, goblet cells, and basal cells. The mucus produced by the goblet cells helps trap dust not removed in the upper passages. The cilia in the upper respiratory tract

Figure 23.4 Larynx.

The larynx is composed of nine pieces of cartilage.

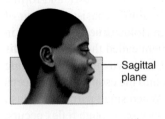

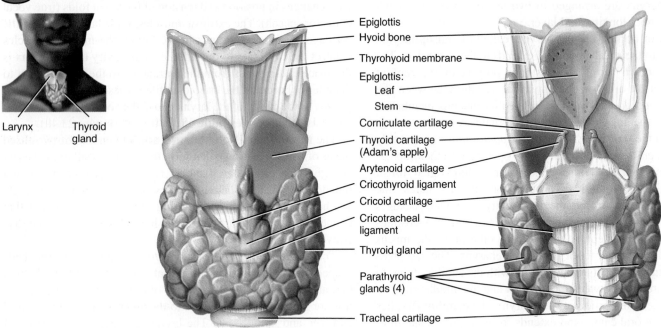

Larynx   Thyroid gland

Epiglottis
Hyoid bone
Thyrohyoid membrane
Epiglottis:
    Leaf
    Stem
Corniculate cartilage
Thyroid cartilage (Adam's apple)
Arytenoid cartilage
Cricothyroid ligament
Cricoid cartilage
Cricotracheal ligament
Thyroid gland
Parathyroid glands (4)
Tracheal cartilage

(a) Anterior view

(b) Posterior view

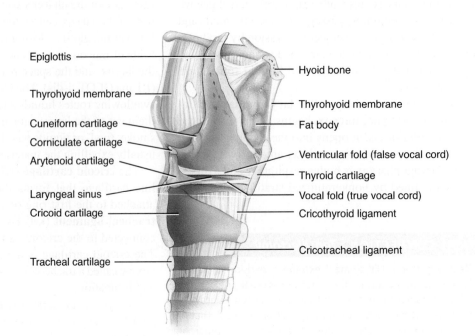

Sagittal plane

Epiglottis
Thyrohyoid membrane
Cuneiform cartilage
Corniculate cartilage
Arytenoid cartilage
Laryngeal sinus
Cricoid cartilage
Tracheal cartilage

Hyoid bone
Thyrohyoid membrane
Fat body
Ventricular fold (false vocal cord)
Thyroid cartilage
Vocal fold (true vocal cord)
Cricothyroid ligament
Cricotracheal ligament

(c) Sagittal section

move mucus and trapped particles *down* toward the pharynx; the cilia in the lower respiratory tract move them *up* toward the pharynx.

The Structures of Voice Production

The mucous membrane of the larynx forms two pairs of folds (Figure 23.4c): a superior pair called the **ventricular folds (false vocal cords)** and an inferior pair called the **vocal folds (true vocal cords).** The space between the ventricular folds is known as the **rima vestibuli.** The **laryngeal sinus (ventricle)** is a lateral expansion of the middle portion of the laryngeal cavity inferior to the ventricular folds and superior to the vocal folds (see Figure 23.2b). While the ventricular folds do not function in voice production, they do have other important functional roles. When the ventricular folds are brought together, they function in holding the breath against pressure in the thoracic cavity, such as might occur when a person strains to lift a heavy object.

The vocal folds are the principal structures of voice production. Deep to the mucous membrane of the vocal folds, which is nonkeratinized stratified squamous epithelium, are bands of elastic ligaments stretched between the rigid cartilages of the larynx like the strings on a guitar. Intrinsic laryngeal muscles attach to both the rigid cartilages and the vocal folds. When the muscles contract they move the cartilages, which pulls the elastic ligaments tight, and this stretches the vocal folds out into the airways so that the rima glottidis is narrowed. Contracting and relaxing the muscles varies the tension in the vocal folds, much like loosening or tightening a guitar string. Air passing through the larynx vibrates the folds and produces sound (phonation) by setting up sound waves in the column of air in the pharynx, nose, and mouth. The variation in the pitch of the sound is related to the tension in the vocal folds. The greater the pressure of air, the louder the sound produced by the vibrating vocal folds.

When the intrinsic muscles of the larynx contract, they pull on the arytenoid cartilages, which causes the cartilages to pivot and slide. Contraction of the posterior cricoarytenoid muscles, for example, moves the vocal folds apart (abduction), thereby opening the rima glottidis (Figure 23.5a). By contrast, contraction of the lateral cricoarytenoid muscles moves the vocal folds together (adduction), thereby closing the rima glottidis (Figure 23.5b). Other intrinsic muscles can elongate (and place tension on) or shorten (and relax) the vocal folds.

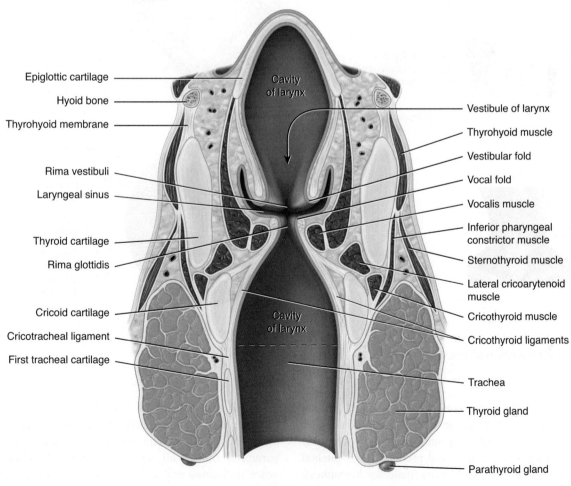

(d) Frontal section

❓ **How does the epiglottis prevent aspiration of foods and liquids?**

Figure 23.5 Movement of the vocal folds.

🔒 ▭ The glottis consists of a pair of folds of mucous membrane in the larynx (the vocal folds) and the space between them (the rima glottidis).

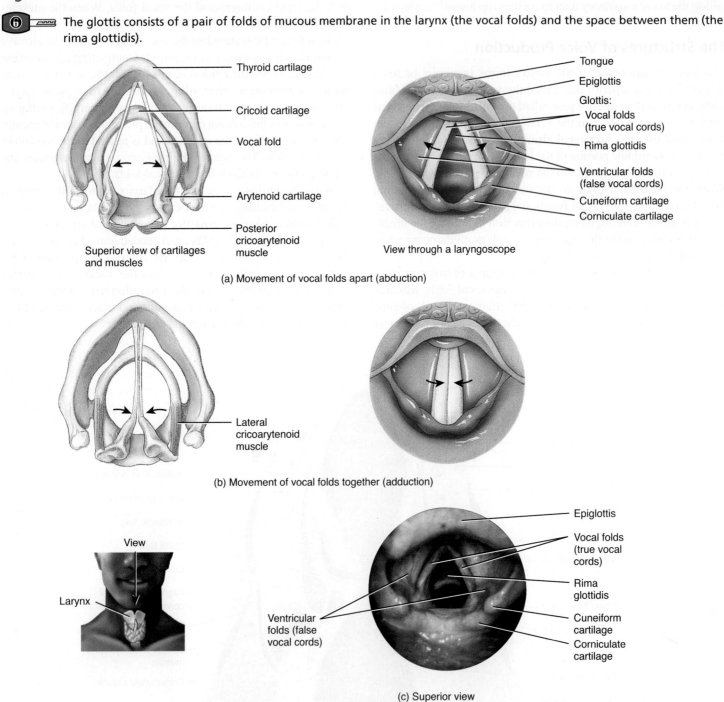

Thyroid cartilage

Cricoid cartilage

Vocal fold

Arytenoid cartilage

Posterior cricoarytenoid muscle

Superior view of cartilages and muscles

Tongue

Epiglottis

Glottis:
Vocal folds (true vocal cords)
Rima glottidis
Ventricular folds (false vocal cords)
Cuneiform cartilage
Corniculate cartilage

View through a laryngoscope

(a) Movement of vocal folds apart (abduction)

Lateral cricoarytenoid muscle

(b) Movement of vocal folds together (adduction)

View

Larynx

Epiglottis

Vocal folds (true vocal cords)

Rima glottidis

Cuneiform cartilage

Corniculate cartilage

Ventricular folds (false vocal cords)

(c) Superior view

❓ **What is the main function of the vocal folds?**

Pitch is controlled by the tension on the vocal folds. If they are pulled taut by the muscles, they vibrate more rapidly, and a higher pitch results. Decreasing the muscular tension on the vocal folds causes them to vibrate more slowly and produce lower-pitched sounds. Due to the influence of androgens (male sex hormones), vocal folds are usually thicker and longer in males than in females, and therefore they vibrate more slowly. This is why a man's voice generally has a lower range of pitch than that of a woman.

Sound originates from the vibration of the vocal folds, but other structures are necessary for converting the sound into recognizable speech. The pharynx, mouth, nasal cavity, and paranasal sinuses all act as resonating chambers that give the voice its human and individual quality. We produce the vowel sounds by constricting and relaxing the muscles in the wall of the pharynx. Muscles of the face, tongue, and lips help us enunciate words.

Whispering is accomplished by closing all but the posterior portion of the rima glottidis. Because the vocal folds do not vibrate during whispering, there is no pitch to this form of speech. However, we can still produce intelligible speech while whispering by changing the shape of the oral cavity as we enunciate. As the size of the oral cavity changes, its resonance qualities change, which imparts a vowel-like pitch to the air as it rushes toward the lips.

Trachea

The **trachea** (TRĀ-kē-a = sturdy), or **windpipe,** is a tubular passageway for air that is about 12 cm (5 in.) long and 2.5 cm (1 in.) in diameter. It is located anterior to the esophagus (Figure 23.6)

and extends from the larynx to the superior border of the fifth thoracic vertebra (T5), where it divides into right and left primary bronchi (see Figure 23.7).

The layers of the tracheal wall, from deep to superficial, are the (1) mucosa, (2) submucosa, (3) hyaline cartilage, and (4) adventitia (composed of areolar connective tissue). The mucosa of the trachea consists of an epithelial layer of pseudostratified ciliated columnar epithelium and an underlying layer of lamina propria that contains elastic and reticular fibers. It provides the same protection against dust as the membrane lining the nasal cavity and larynx. The submucosa consists of areolar connective tissue that contains seromucous glands and their ducts.

The 16–20 incomplete, horizontal rings of hyaline cartilage resemble the letter **C,** are stacked one above another, and are connected together by dense connective tissue. They may be felt through the skin inferior to the larynx. The open part of each **C**-shaped cartilage ring faces posteriorly toward the esophagus (Figure 23.6) and is spanned by a *fibromuscular membrane.* Within this membrane are transverse smooth muscle fibers, called the **trachealis muscle** (trā-kē-Ā-lis), and elastic connective tissue that allow the diameter of the trachea to change subtly during inhalation and exhalation, which is important in maintaining efficient airflow. The solid **C**-shaped cartilage rings provide a semirigid support to maintain patency so that the tracheal wall does not collapse inward (especially during inhalation) and obstruct the air passageway. The adventitia of the trachea consists of areolar connective tissue that joins the trachea to surrounding tissues.

Figure 23.6 Location of the trachea in relation to the esophagus.

The trachea is anterior to the esophagus and extends from the larynx to the superior border of the fifth thoracic vertebra.

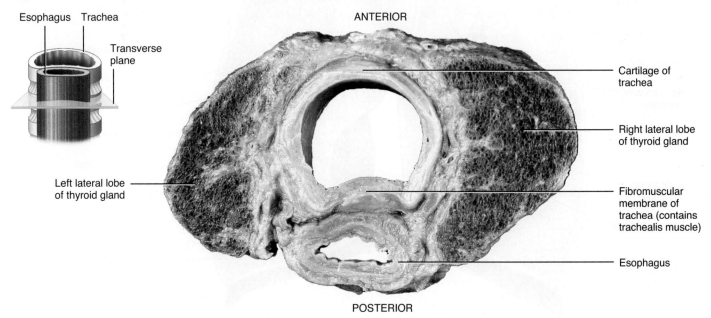

Esophagus   Trachea
Transverse plane

ANTERIOR

Cartilage of trachea

Right lateral lobe of thyroid gland

Left lateral lobe of thyroid gland

Fibromuscular membrane of trachea (contains trachealis muscle)

Esophagus

POSTERIOR

Superior view of transverse section of thyroid gland, trachea, and esophagus

? **What is the benefit of not having complete rings of tracheal cartilage between the trachea and the esophagus?**

CLINICAL CONNECTION | *Tracheotomy and Intubation*

Several conditions may block airflow by obstructing the trachea. The rings of cartilage that support the trachea may be accidentally crushed, the mucous membrane may become inflamed and swell so much that it closes off the passageway, excess mucus secreted by inflamed membranes may clog the lower respiratory passages, a large object may be aspirated (breathed in), or a cancerous tumor may protrude into the airway. Two methods are used to reestablish airflow past a tracheal obstruction. If the obstruction is above the level of the larynx, a **tracheotomy** (trā-kē-O-tō-mē) may be performed. In this procedure, also called a *tracheostomy,* a skin incision is followed by a short longitudinal incision into the trachea below the cricoid cartilage. A tracheal tube is then inserted to create an emergency air passageway. The second method is **intubation** (in'-too-BĀ-shun), in which a tube is inserted into the mouth or nose and passed inferiorly through the larynx and trachea. The firm wall of the tube pushes aside any flexible obstruction, and the lumen of the tube provides a passageway for air; any mucus clogging the trachea can be suctioned out through the tube. •

Bronchi

At the superior border of the fifth thoracic vertebra, the trachea divides into a **right primary bronchus** (BRONG-kus = windpipe), which goes into the right lung, and a **left primary bronchus,** which goes into the left lung (Figure 23.7). The right primary bronchus is more vertical, shorter, and wider than the left. As a result, an aspirated object is more likely to enter and lodge in the right primary bronchus than the left. Like the trachea, the primary

Figure 23.7 Branching of airways from the trachea: the bronchial tree.

🔑 The bronchial tree begins at the trachea and ends at the terminal bronchioles.

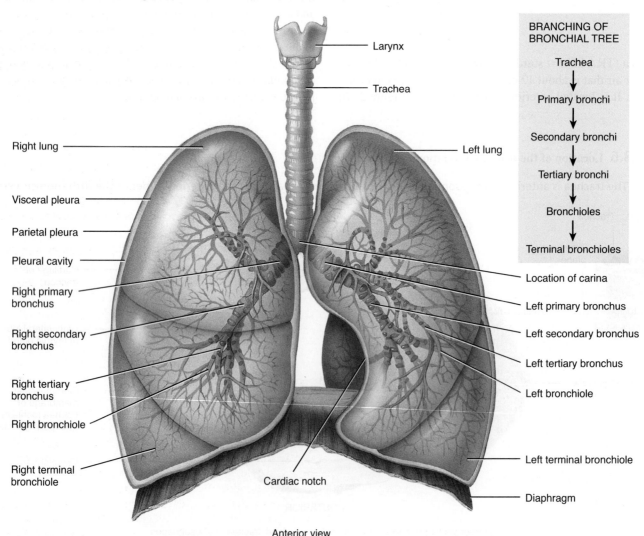

Anterior view

❓ **How many lobes and secondary bronchi are present in each lung?**

bronchi (BRONG-kī) contain incomplete rings of cartilage and are lined by pseudostratified ciliated columnar epithelium.

At the point where the trachea divides into right and left primary bronchi an internal ridge called the **carina** (ka-RĪ-na = keel of a boat) is formed by a posterior and somewhat inferior projection of the last tracheal cartilage. The mucous membrane of the carina is one of the most sensitive areas of the entire larynx and trachea for triggering a cough reflex. Widening and distortion of the carina is a serious sign because it usually indicates a carcinoma of the lymph nodes around the region where the trachea divides.

On entering the lungs, the primary bronchi divide to form smaller bronchi—the **secondary (lobar) bronchi,** one for each lobe of the lung. (The right lung has three lobes; the left lung has two.) The secondary bronchi continue to branch, forming still smaller bronchi, called **tertiary (segmental) bronchi** (TER-shē-e-rē), that divide into **bronchioles.** Bronchioles in turn branch repeatedly, and the smallest ones branch into even smaller tubes called **terminal bronchioles.** These bronchioles contain *Clara cells,* columnar, nonciliated cells interspersed among the epithelial cells. Clara cells may protect against harmful effects of inhaled toxins and carcinogens, produce surfactant (discussed shortly), and function as stem cells (reserve cells), which give rise to various cells of the epithelium. The terminal bronchioles represent the end of the conducting zone of the respiratory system. This extensive branching from the trachea through the terminal bronchioles resembles an inverted tree and is commonly referred to as the **bronchial tree.**

As the branching becomes more extensive in the bronchial tree, several structural changes may be noted.

1. The mucous membrane in the bronchial tree changes from pseudostratified ciliated columnar epithelium in the primary bronchi, secondary bronchi, and tertiary bronchi to ciliated simple columnar epithelium with some goblet cells in larger bronchioles, to mostly ciliated simple cuboidal epithelium with no goblet cells in smaller bronchioles, to mostly nonciliated simple cuboidal epithelium in terminal bronchioles. Recall that ciliated epithelium of the respiratory membrane removes inhaled particles in two ways. Mucus produced by goblet cells traps the particles, and the cilia move the mucus and trapped particles toward the pharynx for removal. In regions where nonciliated simple cuboidal epithelium is present, inhaled particles are removed by macrophages.

2. Plates of cartilage gradually replace the incomplete rings of cartilage in primary bronchi and finally disappear in the distal bronchioles.

3. As the amount of cartilage decreases, the amount of smooth muscle increases. Smooth muscle encircles the lumen in spiral bands and helps maintain patency. However, because there is no supporting cartilage, muscle spasms can close off the airways. This is what happens during an asthma attack, which can be a life-threatening situation.

During exercise, activity in the sympathetic division of the autonomic nervous system (ANS) increases and the adrenal medulla releases the hormones epinephrine and norepinephrine; both of these events cause relaxation of smooth muscle in the bronchioles, which dilates the airways. Because air reaches the alveoli more quickly, lung ventilation improves. The parasympathetic division of the ANS and mediators of allergic reactions such as histamine have the opposite effect, causing contraction of bronchiolar smooth muscle, which results in constriction of distal bronchioles.

✔CHECKPOINT

4. List the roles of each of the three anatomical regions of the pharynx in respiration.
5. How does the larynx function in respiration and voice production?
6. Describe the location, structure, and function of the trachea.
7. Describe the structure of the bronchial tree.

Lungs

The **lungs** (= lightweights, because they float) are paired cone-shaped organs in the thoracic cavity. They are separated from each other by the heart and other structures of the mediastinum, which divides the thoracic cavity into two anatomically distinct chambers. As a result, if trauma causes one lung to collapse, the other may remain expanded. Each lung is enclosed and protected by a double-layered serous membrane called the **pleural membrane** (PLOOR-al; *pleur-* = side). The superficial layer, called the **parietal pleura,** lines the wall of the thoracic cavity; the deep layer, the **visceral pleura,** covers the lungs themselves (Figure 23.8). Between the visceral and parietal pleurae is a small space, the **pleural cavity,** which contains a small amount of lubricating fluid secreted by the membranes. This pleural fluid reduces friction between the membranes, allowing them to slide easily over one another during breathing. Pleural fluid also causes the two membranes to adhere to one another just as a film of water causes two glass microscope slides to stick together, a phenomenon called surface tension. Separate pleural cavities surround the left and right lungs. Inflammation of the pleural membrane, called **pleurisy** or **pleuritis,** may in its early stages cause pain due to friction between the parietal and visceral layers of the pleura. If the inflammation persists, excess fluid accumulates in the pleural space, a condition known as **pleural effusion.**

CLINICAL CONNECTION | *Pneumothorax and Hemothorax*

In certain conditions, the pleural cavities may fill with air (**pneumothorax;** noo′-mō-THOR-aks; *pneumo-* = air or breath), blood (**hemothorax**), or pus. Air in the pleural cavities, most commonly introduced in a surgical opening of the chest or as a result of a stab or gunshot wound, may cause the lungs to collapse. This collapse of a part of a lung, or rarely an entire lung, is called **atelectasis** (at′-e-LEK-ta-sis; *ateles-* = incomplete; *-ectasis* = expansion). The goal of treatment is the evacuation of air (or blood) from the pleural space, which allows the lung to reinflate. A small pneumothorax may resolve on its own, but it is often necessary to insert a chest tube to assist in evacuation. •

Figure 23.8 Relationship of the pleural membranes to the lungs.

The parietal pleura lines the thoracic cavity, and the visceral pleura covers the lungs.

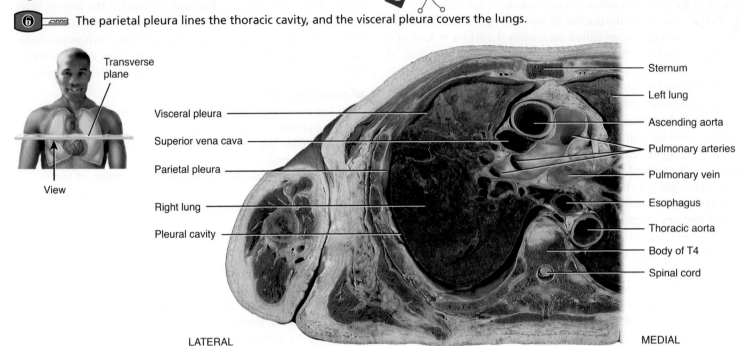

Inferior view of transverse section through thoracic cavity
showing pleural cavity and pleural membranes

? **What type of membrane is the pleural membrane?**

The lungs extend from the diaphragm to just slightly superior to the clavicles and lie against the ribs anteriorly and posteriorly (Figure 23.9a). The broad inferior portion of the lung, the **base,** is concave and fits over the convex area of the diaphragm. The narrow superior portion of the lung is the **apex.** The surface of the lung lying against the ribs, the **costal surface,** matches the rounded curvature of the ribs. The **mediastinal (medial) surface** of each lung contains a region, the **hilum** (or **hilus**), through which bronchi, pulmonary blood vessels, lymphatic vessels, and nerves enter and exit (Figure 23.9e). These structures are held together by the pleura and connective tissue and constitute the **root** of the lung. Medially, the left lung also contains a concavity, the **cardiac notch,** in which the apex of the heart lies. Due to the space occupied by the heart, the left lung is about 10% smaller than the right lung. Although the right lung is thicker and broader, it is also somewhat shorter than the left lung because the diaphragm is higher on the right side, accommodating the liver that lies inferior to it.

The lungs almost fill the thorax (Figure 23.9a). The apex of the lungs lies superior to the medial third of the clavicles, and this is the only area that can be palpated. The anterior, lateral, and posterior surfaces of the lungs lie against the ribs. The base of the lungs extends from the sixth costal cartilage anteriorly to the spinous process of the tenth thoracic vertebra posteriorly. The pleura extends about 5 cm (2 in.) below the base from the sixth costal cartilage anteriorly to the twelfth rib posteriorly. Thus, the lungs

do not completely fill the pleural cavity in this area. Removal of excessive fluid in the pleural cavity can be accomplished without injuring lung tissue by inserting a needle anteriorly through the seventh intercostal space, a procedure called **thoracentesis** (thor′-a-sen-TĒ-sis; *-centesis* = puncture). The needle is passed along the superior border of the lower rib to avoid damage to the intercostal nerves and blood vessels. Inferior to the seventh intercostal space there is danger of penetrating the diaphragm.

Lobes, Fissures, and Lobules

One or two fissures divide each lung into lobes (Figure 23.9b–e). Both lungs have an **oblique fissure,** which extends inferiorly and anteriorly; the right lung also has a **horizontal fissure.** The oblique fissure in the left lung separates the **superior lobe** from the **inferior lobe.** In the right lung, the superior part of the oblique fissure separates the superior lobe from the inferior lobe; the inferior part of the oblique fissure separates the inferior lobe from the **middle lobe,** which is bordered superiorly by the horizontal fissure.

Each lobe receives its own secondary (lobar) bronchus. Thus, the right primary bronchus gives rise to three secondary (lobar) bronchi called the **superior, middle,** and **inferior secondary (lobar) bronchi,** and the left primary bronchus gives rise to **superior** and **inferior secondary (lobar) bronchi.** Within the lung, the secondary bronchi give rise to the **tertiary (segmental) bronchi,** which are constant in both origin and distribution—there

Figure 23.9 Surface anatomy of the lungs.

The oblique fissure divides the left lung into two lobes. The oblique and horizontal fissures divide the right lung into three lobes.

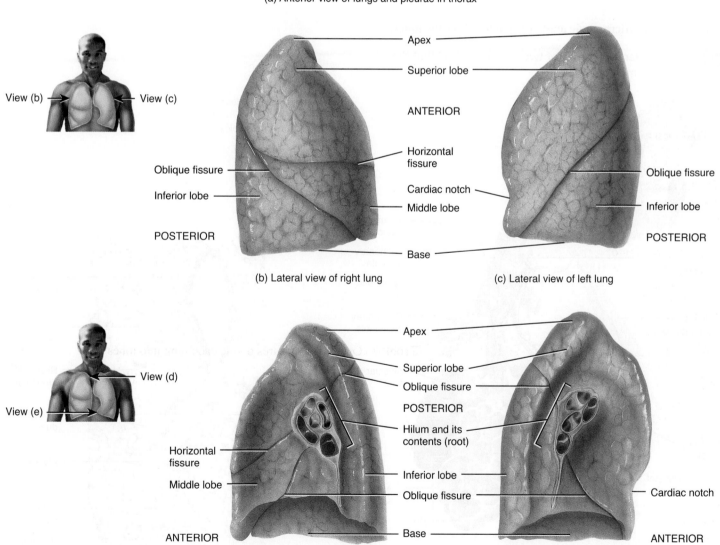

(a) Anterior view of lungs and pleurae in thorax

(b) Lateral view of right lung

(c) Lateral view of left lung

(d) Medial view of right lung

(e) Medial view of left lung

Why are the right and left lungs slightly different in size and shape?

are 10 tertiary bronchi in each lung. The segment of lung tissue that each tertiary bronchus supplies is called a **bronchopulmonary segment** (brong'-kō-PUL-mō-nar'-ē). Bronchial and pulmonary disorders (such as tumors or abscesses) that are localized in a bronchopulmonary segment may be surgically removed without seriously disrupting the surrounding lung tissue.

Each bronchopulmonary segment of the lungs has many small compartments called **lobules;** each lobule is wrapped in elastic connective tissue and contains a lymphatic vessel, an arteriole, a venule, and a branch from a terminal bronchiole (Figure 23.10a). Terminal bronchioles subdivide into microscopic branches called **respiratory bronchioles** (Figure 23.10b). They also have alveoli (described shortly) budding from their walls. Alveoli participate in gas exchange, and thus respiratory bronchioles begin the respiratory zone of the respiratory system. As the respiratory bronchioles penetrate more deeply into the lungs, the epithelial lining changes from simple cuboidal to simple squamous. Respiratory bronchioles in turn subdivide into several (2–11) **alveolar ducts** (al-VĒ-ō-lar), which consist of simple squamous epithelium. The respiratory passages from the trachea to the alveolar ducts contain about 25 orders of branching; branching that from the trachea into primary bronchi is called first-order branching, that from primary bronchi into secondary bronchi is called second-order branching, and so on down to the alveolar ducts.

Alveoli

Around the circumference of the alveolar ducts are numerous alveoli and alveolar sacs. An **alveolus** (al-VĒ-ō-lus) is a cup-shaped outpouching lined by simple squamous epithelium and supported by a thin elastic basement membrane; an **alveolar sac** consists of two or more alveoli that share a common opening (Figure 23.10a, b). The walls of alveoli consist of two types of alveolar epithelial cells (Figure 23.11). The more numerous **type I alveolar cells** are simple squamous epithelial cells that form a nearly continuous lining of the alveolar wall. **Type II alveolar cells,** also called **septal cells,** are fewer in number and are found

Figure 23.10 Microscopic anatomy of a lobule of the lungs.

Alveolar sacs consist of two or more alveoli that share a common opening.

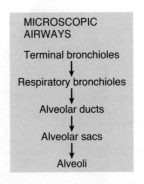

MICROSCOPIC AIRWAYS

Terminal bronchioles
↓
Respiratory bronchioles
↓
Alveolar ducts
↓
Alveolar sacs
↓
Alveoli

Labels (diagram a): Pulmonary venule; Elastic connective tissue; Pulmonary capillary; Visceral pleura; Alveoli; Terminal bronchiole; Pulmonary arteriole; Lymphatic vessel; Respiratory bronchiole; Alveoli; Alveolar ducts; Alveolar sac

(a) Diagram of portion of lobule of lung

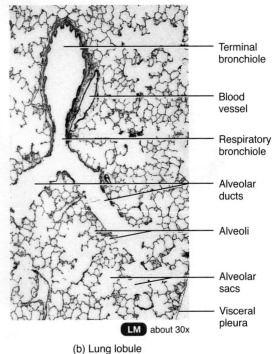

Labels (b): Terminal bronchiole; Blood vessel; Respiratory bronchiole; Alveolar ducts; Alveoli; Alveolar sacs; Visceral pleura

LM about 30x

(b) Lung lobule

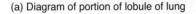

? **What types of cells make up the wall of an alveolus?**

Figure 23.11 Structural components of an alveolus. The respiratory membrane consists of a layer of type I and type II alveolar cells, an epithelial basement membrane, a capillary basement membrane, and the capillary endothelium.

The exchange of respiratory gases occurs by diffusion across the respiratory membrane.

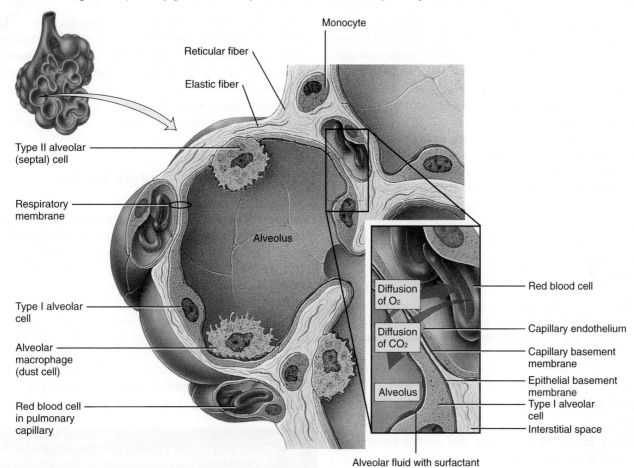

Monocyte

Reticular fiber

Elastic fiber

Type II alveolar (septal) cell

Respiratory membrane

Alveolus

Type I alveolar cell

Alveolar macrophage (dust cell)

Red blood cell in pulmonary capillary

Diffusion of O_2

Diffusion of CO_2

Alveolus

Red blood cell

Capillary endothelium

Capillary basement membrane

Epithelial basement membrane

Type I alveolar cell

Interstitial space

Alveolar fluid with surfactant

(a) Section through alveolus showing cellular components

(b) Details of respiratory membrane

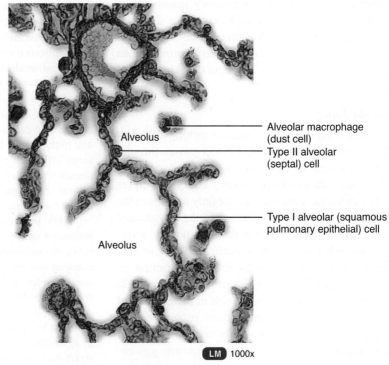

Alveolus

Alveolar macrophage (dust cell)

Type II alveolar (septal) cell

Type I alveolar (squamous pulmonary epithelial) cell

Alveolus

LM 1000x

(c) Details of several alveoli

How thick is the respiratory membrane?

between type I alveolar cells. The thin type I alveolar cells are the main sites of gas exchange. Type II alveolar cells, rounded or cuboidal epithelial cells with free surfaces containing microvilli, secrete alveolar fluid, which keeps the surface between the cells and the air moist. Included in the alveolar fluid is **surfactant** (sur-FAK-tant), a complex mixture of phospholipids and lipoproteins. Surfactant lowers the surface tension of alveolar fluid, which reduces the tendency of alveoli to collapse and thus maintains their patency (described later).

Associated with the alveolar wall are **alveolar macrophages (dust cells),** phagocytes that remove fine dust particles and other debris from the alveolar spaces. Also present are fibroblasts that produce reticular and elastic fibers. Underlying the layer of type I alveolar cells is an elastic basement membrane. On the outer surface of the alveoli, the lobule's arteriole and venule disperse into a network of blood capillaries (see Figure 23.10a) that consist of a single layer of endothelial cells and basement membrane.

The exchange of O_2 and CO_2 between the air spaces in the lungs and the blood takes place by diffusion across the alveolar and capillary walls, which together form the **respiratory membrane.** Extending from the alveolar air space to blood plasma, the respiratory membrane consists of four layers (Figure 23.11b):

1. A layer of type I and type II alveolar cells and associated alveolar macrophages that constitutes the **alveolar wall**

2. An **epithelial basement membrane** underlying the alveolar wall

3. A **capillary basement membrane** that is often fused to the epithelial basement membrane

4. The **capillary endothelium**

Despite having several layers, the respiratory membrane is very thin—only 0.5 μm thick, about one-sixteenth the diameter of a red blood cell—to allow rapid diffusion of gases. It has been estimated that the lungs contain 300 million alveoli, providing an immense surface area of 70 m^2 (750 ft^2)—about the size of a racquetball court—for gas exchange.

Blood Supply to the Lungs

The lungs receive blood via two sets of arteries: pulmonary arteries and bronchial arteries. Deoxygenated blood passes through the pulmonary trunk, which divides into a left pulmonary artery that enters the left lung and a right pulmonary artery that enters the right lung. (The pulmonary arteries are the only arteries in the body that carry deoxygenated blood.) Return of the oxygenated blood to the heart occurs by way of the four pulmonary veins, which drain into the left atrium (see Figure 21.29). A unique feature of pulmonary blood vessels is their constriction in response to localized hypoxia (low O_2 level). In all other body tissues, hypoxia causes dilation of blood vessels to increase blood flow. In the lungs, however, vasoconstriction in response to hypoxia diverts pulmonary blood from poorly ventilated areas of the lungs to well-ventilated regions for more efficient gas exchange. This phenomenon is known as **ventilation–perfusion coupling** (per-FYU-zhun) because the perfusion (blood flow) to

each area of the lungs matches the extent of ventilation (airflow) to alveoli in that area.

Bronchial arteries, which branch from the aorta, deliver oxygenated blood to the lungs. This blood mainly perfuses the muscular walls of the bronchi and bronchioles. Connections exist between branches of the bronchial arteries and branches of the pulmonary arteries, however; most blood returns to the heart via pulmonary veins. Some blood, however, drains into bronchial veins, branches of the azygos system, and returns to the heart via the superior vena cava.

Patency of the Respiratory System

Throughout the discussion of the respiratory organs, several examples were given of structures or secretions that help to maintain patency of the system so that air passageways are kept free of obstruction. These included the bony and cartilaginous frameworks of the nose, skeletal muscles of the pharynx, cartilages of the larynx, C-shaped rings of cartilage in the trachea and bronchi, smooth muscle in the bronchioles, and surfactant in the alveoli.

Unfortunately, there are also factors that can compromise patency. These include crushing injuries to bone and cartilage, a deviated nasal septum, nasal polyps, inflammation of mucous membranes, spasms of smooth muscle, and a deficiency of surfactant.

A summary of the epithelial linings and special features of the organs of the respiratory system is presented in Table 23.1.

CLINICAL CONNECTION | Coryza, Seasonal Influenza, and H1N1 Influenza

Hundreds of viruses can cause coryza (ko-RĪ-za), or the **common cold,** but a group of viruses called *rhinoviruses* (RĪ-nō-vī-rus-es) is responsible for about 40% of all colds in adults. Typical symptoms include sneezing, excessive nasal secretion, dry cough, and congestion. The uncomplicated common cold is not usually accompanied by a fever. Complications include sinusitis, asthma, bronchitis, ear infections, and laryngitis. Recent investigations suggest an association between emotional stress and the common cold. The higher the stress level, the greater the frequency and duration of colds. **Seasonal influenza (flu)** is also caused by a virus. Its symptoms include chills, fever (usually higher than 101°F = 39°C), headache, and muscular aches. Seasonal influenza can become life-threatening and may develop into pneumonia. It is important to recognize that influenza is a respiratory disease, not a gastrointestinal (GI) disease. Many people mistakenly report having seasonal flu when they are suffering from a GI illness.

H1N1 influenza (flu), also known as *swine flu,* is a type of influenza caused by a new virus called *influenza H1N1.* The term swine flu originated because early laboratory testing indicated that many of the genes in the new virus were similar to ones found in pigs (swine) in North America. However, subsequent testing revealed that the new virus is very different from the one that circulates in North American pigs.

H1N1 flu is a respiratory disorder first detected in the United States in April 2009. In June 2009, the World Health Organization declared H1N1 flu to be a *global pandemic disease* (a disease that affects large numbers of individuals in a short period of time and that occurs world-

TABLE 23.1

Summary of the Respiratory System

STRUCTURE	EPITHELIUM	CILIA	GOBLET CELLS	SPECIAL FEATURES
NOSE				
Vestibule	Nonkeratinized stratified squamous.	No.	No.	Contains numerous hairs.
Respiratory region	Pseudostratified ciliated columnar.	Yes.	Yes.	Contains conchae and meatuses.
Olfactory region	Olfactory epithelium (olfactory receptors).	Yes.	No.	Functions in olfaction.
PHARYNX				
Nasopharynx	Pseudostratified ciliated columnar.	Yes.	Yes.	Passageway for air; contains internal nares, openings for auditory tubes, and pharyngeal tonsil.
Oropharynx	Nonkeratinized stratified squamous.	No.	No.	Passageway for both air and food and drink; contains opening from mouth (fauces).
Laryngopharynx	Nonkeratinized stratified squamous.	No.	No.	Passageway for both air and food and drink.
LARYNX	Nonkeratinized stratified squamous above the vocal folds; pseudostratified ciliated columnar below the vocal folds.	No above folds; yes below folds.	No above folds; yes below folds.	Passageway for air; contains vocal folds for voice production.
TRACHEA	Pseudostratified ciliated columnar.	Yes.	Yes.	Passageway for air; contains C-shaped rings of cartilage to keep trachea open.
BRONCHI				
Primary bronchi	Pseudostratified ciliated columnar.	Yes.	Yes.	Passageway for air; contain C-shaped rings of cartilage to maintain patency.
Secondary bronchi	Pseudostratified ciliated columnar.	Yes.	Yes.	Passageway for air; contain plates of cartilage to maintain patency.
Tertiary bronchi	Pseudostratified ciliated columnar.	Yes.	Yes.	Passageway for air; contain plates of cartilage to maintain patency.
Larger bronchioles	Ciliated simple columnar.	Yes.	Yes.	Passageway for air; contain more smooth muscle than in the bronchi.
Smaller bronchioles	Ciliated simple columnar.	Yes.	No.	Passageway for air; contain more smooth muscle than in the larger bronchioles.
Terminal bronchioles	Nonciliated simple columnar.	No.	No.	Passageway for air; contain more smooth muscle than in the smaller bronchioles.
LUNGS				
Respiratory bronchioles	Simple cuboidal to simple squamous.	No.	No.	Passageway for air; gas exchange.
Alveolar ducts	Simple squamous.	No.	No.	Passageway for air; gas exchange; produce surfactant.
Alveoli	Simple squamous.	No.	No.	Passageway for air; gas exchange; produce surfactant to maintain patency.

☐ Conducting structures

☐ Respiratory structures

wide). The virus is spread in the same way that seasonal flu spreads: from person to person through coughing or sneezing or by touching infected objects and then touching one's mouth or nose. Most individuals infected with the virus have mild disease and recover without medical treatment, but some people have severe disease and have even died. The symptoms of H1N1 flu include fever, cough, runny or stuffy nose, headache, body aches, chills, and fatigue. Some people also have vomiting and diarrhea. Most people who have been hospitalized for H1N1 flu have had one or more preexisting medical conditions such as diabetes, heart disease, asthma, kidney disease, or pregnancy.

People infected with the virus can infect others from 1 day before symptoms occur to 5–7 days or more after they occur. Treatment of H1N1 flu involves taking antiviral drugs, such as Tamiflu® and Relenza®. A vaccine is also available, but the H1N1 flu vaccine is not a substitute for seasonal flu vaccines. In order to prevent infection, the Centers for Disease Control and Prevention (CDC) recommends washing your hands often with soap and water or with an alcohol-based hand cleaner; covering your mouth and nose with a tissue when coughing or sneezing and disposing of the tissue; avoiding touching your mouth, nose, or eyes; avoiding close contact (within 6 feet) with people who have flu-like symptoms; and staying home for 7 days after symptoms begin or for 24 hours after being symptom-free, whichever is longer. •

✔**CHECKPOINT**

8. Where are the lungs located? Distinguish the parietal pleura from the visceral pleura.
9. Define each of the following parts of a lung: base, apex, costal surface, medial surface, hilum, root, cardiac notch, lobe, and lobule.
10. What is a bronchopulmonary segment?
11. Describe the histology and function of the respiratory membrane.

23.2 PULMONARY VENTILATION

◉ **OBJECTIVE**

• Describe the events that cause inhalation and exhalation.

The process of gas exchange in the body, called **respiration**, has three basic steps:

1. **Pulmonary ventilation** (*pulmon-* = lung), or **breathing,** is the inhalation (inflow) and exhalation (outflow) of air and involves the exchange of air between the atmosphere and the alveoli of the lungs.

2. **External (pulmonary) respiration** is the exchange of gases between the alveoli of the lungs and the blood in pulmonary capillaries across the respiratory membrane. In this process, pulmonary capillary blood gains O_2 and loses CO_2.

3. **Internal (tissue) respiration** is the exchange of gases between blood in systemic capillaries and tissue cells. In this step the blood loses O_2 and gains CO_2. Within cells, the metabolic reactions that consume O_2 and give off CO_2 during the production of ATP are termed *cellular respiration* (discussed in Chapter 25).

In pulmonary ventilation, air flows between the atmosphere and the alveoli of the lungs because of alternating pressure differences created by contraction and relaxation of respiratory muscles. The rate of airflow and the amount of effort needed for breathing are also influenced by alveolar surface tension, compliance of the lungs, and airway resistance.

Pressure Changes during Pulmonary Ventilation

Air moves into the lungs when the air pressure inside the lungs is less than the air pressure in the atmosphere. Air moves out of the lungs when the air pressure inside the lungs is greater than the air pressure in the atmosphere.

Inhalation

Breathing in is called **inhalation (inspiration).** Just before each inhalation, the air pressure inside the lungs is equal to the air pressure of the atmosphere, which at sea level is about 760 millimeters of mercury (mmHg), or 1 atmosphere (atm). For air to flow into the lungs, the pressure inside the alveoli must become lower than the atmospheric pressure. This condition is achieved by increasing the size of the lungs.

The pressure of a gas in a closed container is inversely proportional to the volume of the container. This means that if the size of a closed container is increased, the pressure of the gas inside the container decreases, and that if the size of the container is decreased, then the pressure inside it increases. This inverse relationship between volume and pressure, called **Boyle's law,** may be demonstrated as follows (Figure 23.12): Suppose we place a gas in a cylinder that has a movable piston and a pressure gauge, and that the initial pressure created by the gas molecules striking the wall of the container is 1 atm. If the piston is pushed down, the gas is compressed into a smaller volume, so that the same number of gas molecules strike less wall area. The gauge shows that the pressure doubles as the gas is compressed to half its original volume. In other words, the same number of molecules in half the volume produces twice the pressure. Conversely, if the piston is raised to increase the volume, the pressure decreases. Thus, the pressure of a gas varies inversely with volume.

Differences in pressure caused by changes in lung volume force air into our lungs when we inhale and out when we exhale. For inhalation to occur, the lungs must expand, which increases lung volume and thus decreases the pressure in the lungs to below atmospheric pressure. The first step in expanding the lungs during normal quiet inhalation involves contraction of the main muscles of inhalation, the diaphragm and external intercostals (Figure 23.13).

Figure 23.12 Boyle's law.

🔑 ▭ The volume of a gas varies inversely with its pressure.

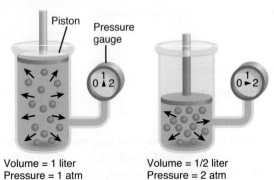

Volume = 1 liter
Pressure = 1 atm

Volume = 1/2 liter
Pressure = 2 atm

❓ **If the volume is decreased from 1 liter to ¼ liter, how would the pressure change?**

Figure 23.13 Muscles of inhalation and exhalation and their actions. The pectoralis minor muscle (not shown here) is illustrated in Figure 11.14a.

During deep, labored breathing, accessory muscles of inhalation (sternocleidomastoids, scalenes, and pectoralis minors) participate.

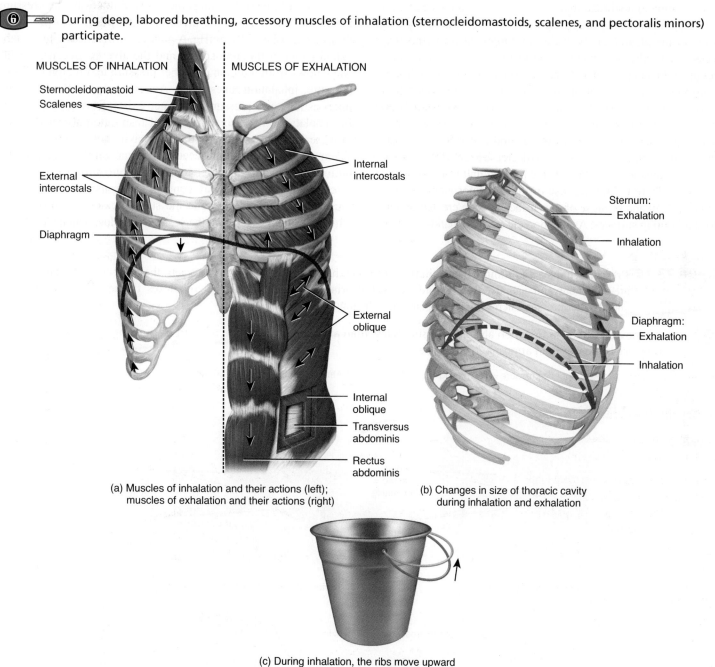

(a) Muscles of inhalation and their actions (left); muscles of exhalation and their actions (right)

(b) Changes in size of thoracic cavity during inhalation and exhalation

(c) During inhalation, the ribs move upward and outward like the handle on a bucket

Right now, what is the main muscle that is powering your breathing?

The most important muscle of inhalation is the diaphragm, the dome-shaped skeletal muscle that forms the floor of the thoracic cavity. It is innervated by fibers of the phrenic nerves, which emerge from the spinal cord at cervical levels 3, 4, and 5. Contraction of the diaphragm causes it to flatten, lowering its dome. This increases the vertical diameter of the thoracic cavity. During normal quiet inhalation, the diaphragm descends about 1 cm (0.4 in.), producing a pressure difference of 1–3 mmHg and the inhalation of about 500 mL of air. In strenuous breathing, the diaphragm may descend 10 cm (4 in.), which produces a pressure difference of 100 mmHg and the inhalation of 2–3 liters of air. Contraction of the diaphragm is responsible for about 75% of the air that enters the lungs during quiet breathing. Advanced pregnancy, excessive obesity, or confining abdominal clothing can prevent complete descent of the diaphragm.

The next most important muscles of inhalation are the external intercostals. When these muscles contract, they elevate the ribs. As a result, there is an increase in the anteroposterior and lateral diameters of the chest cavity. Contraction of the external inter-costals is responsible for about 25% of the air that enters the lungs during normal quiet breathing.

During quiet inhalations, the pressure between the two pleural layers in the pleural cavity, called **intrapleural (intrathoracic) pressure,** is always subatmospheric (lower than atmospheric pressure). Just before inhalation, it is about 4 mmHg less than the atmospheric pressure, or about 756 mmHg at an atmospheric pressure of 760 mmHg (Figure 23.14). As the diaphragm and external intercostals contract and the overall size of the thoracic cavity increases, the volume of the pleural cavity also increases, which causes intrapleural pressure to decrease to about 754 mmHg. During expansion of the thorax, the parietal and visceral pleurae normally adhere tightly because of the subatmospheric pressure between them and because of the surface tension created by their moist adjoining surfaces. As the thoracic cavity expands, the pari-

etal pleura lining the cavity is pulled outward in all directions, and the visceral pleura and lungs are pulled along with it.

As the volume of the lungs increases in this way, the pressure inside the lungs, called the **alveolar (intrapulmonic) pressure,** drops from 760 to 758 mmHg. A pressure difference is thus established between the atmosphere and the alveoli. Because air always flows from a region of higher pressure to a region of lower pressure, inhalation takes place. Air continues to flow into the lungs as long as a pressure difference exists. During deep, forceful inhalations, accessory muscles of inspiration also participate in increasing the size of the thoracic cavity (see Figure 23.13a). The muscles are so named because they make little, if any, contribution during normal quiet inhalation, but during exercise or forced ventilation they may contract vigorously. The accessory muscles of inhalation include the sternocleidomastoid muscles, which elevate the sternum; the scalene muscles, which elevate the

Figure 23.14 Pressure changes in pulmonary ventilation. During inhalation, the diaphragm contracts, the chest expands, the lungs are pulled outward, and alveolar pressure decreases. During exhalation, the diaphragm relaxes, the lungs recoil inward, and alveolar pressure increases, forcing air out of the lungs.

Air moves into the lungs when alveolar pressure is less than atmospheric pressure, and out of the lungs when alveolar pressure is greater than atmospheric pressure.

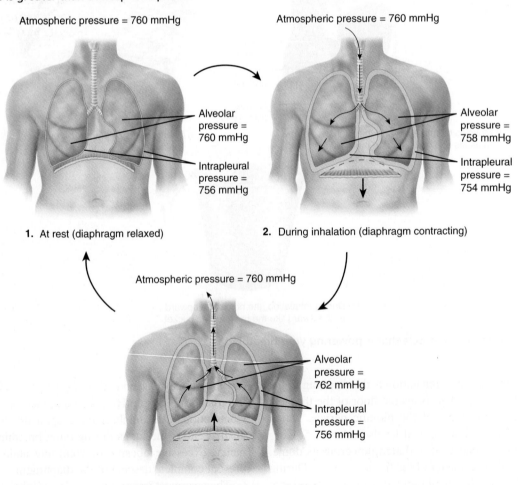

1. At rest (diaphragm relaxed)

Atmospheric pressure = 760 mmHg

Alveolar pressure = 760 mmHg

Intrapleural pressure = 756 mmHg

2. During inhalation (diaphragm contracting)

Atmospheric pressure = 760 mmHg

Alveolar pressure = 758 mmHg

Intrapleural pressure = 754 mmHg

3. During exhalation (diaphragm relaxing)

Atmospheric pressure = 760 mmHg

Alveolar pressure = 762 mmHg

Intrapleural pressure = 756 mmHg

? **How does the intrapleural pressure change during a normal, quiet breath?**

first two ribs; and the pectoralis minor muscles, which elevate the third through fifth ribs. Because both normal quiet inhalation and inhalation during exercise or forced ventilation involve muscular contraction, the process of inhalation is said to be *active*.

Figure 23.15a summarizes the events of inhalation.

Exhalation

Breathing out, called **exhalation (expiration),** is also due to a pressure gradient, but in this case the gradient is in the opposite direction: The pressure in the lungs is greater than the pressure of the atmosphere. Normal exhalation during quiet breathing, unlike inhalation, is a *passive process* because no muscular contractions are involved. Instead, exhalation results from **elastic recoil** of the chest wall and lungs, both of which have a natural tendency to spring back after they have been stretched. Two inwardly directed forces contribute to elastic recoil: (1) the recoil of elastic fibers that were stretched during inhalation and (2) the inward pull of surface tension due to the film of alveolar fluid.

Exhalation starts when the inspiratory muscles relax. As the diaphragm relaxes, its dome moves superiorly owing to its elasticity. As the external intercostals relax, the ribs are depressed. These movements decrease the vertical, lateral, and anteroposterior diameters of the thoracic cavity, which decreases lung volume. In turn, the alveolar pressure increases to about 762 mmHg. Air then flows from the area of higher pressure in the alveoli to the area of lower pressure in the atmosphere (see Figure 23.14).

Exhalation becomes active only during forceful breathing, as occurs while playing a wind instrument or during exercise. During these times, muscles of exhalation—the abdominals and internal intercostals (see Figure 23.13a)—contract, which increases pressure in the abdominal region and thorax. Contraction of the abdominal muscles moves the inferior ribs downward and compresses the abdominal viscera, thereby forcing the diaphragm superiorly. Contraction of the internal intercostals, which extend inferiorly and posteriorly between adjacent ribs, pulls the ribs inferiorly. Although intrapleural pressure is always less than alveolar pressure, it may briefly exceed atmospheric pressure during a forceful exhalation, such as during a cough.

Figure 23.15b summarizes the events of exhalation.

Other Factors Affecting Pulmonary Ventilation

As you have just learned, air pressure differences drive airflow during inhalation and exhalation. However, three other factors affect the rate of airflow and the ease of pulmonary ventilation: surface tension of the alveolar fluid, compliance of the lungs, and airway resistance.

Surface Tension of Alveolar Fluid

As noted earlier, a thin layer of alveolar fluid coats the luminal surface of alveoli and exerts a force known as **surface tension.** Surface tension arises at all air–water interfaces because the polar

Figure 23.15 Summary of events of inhalation and exhalation.

Inhalation and exhalation are caused by changes in alveolar pressure.

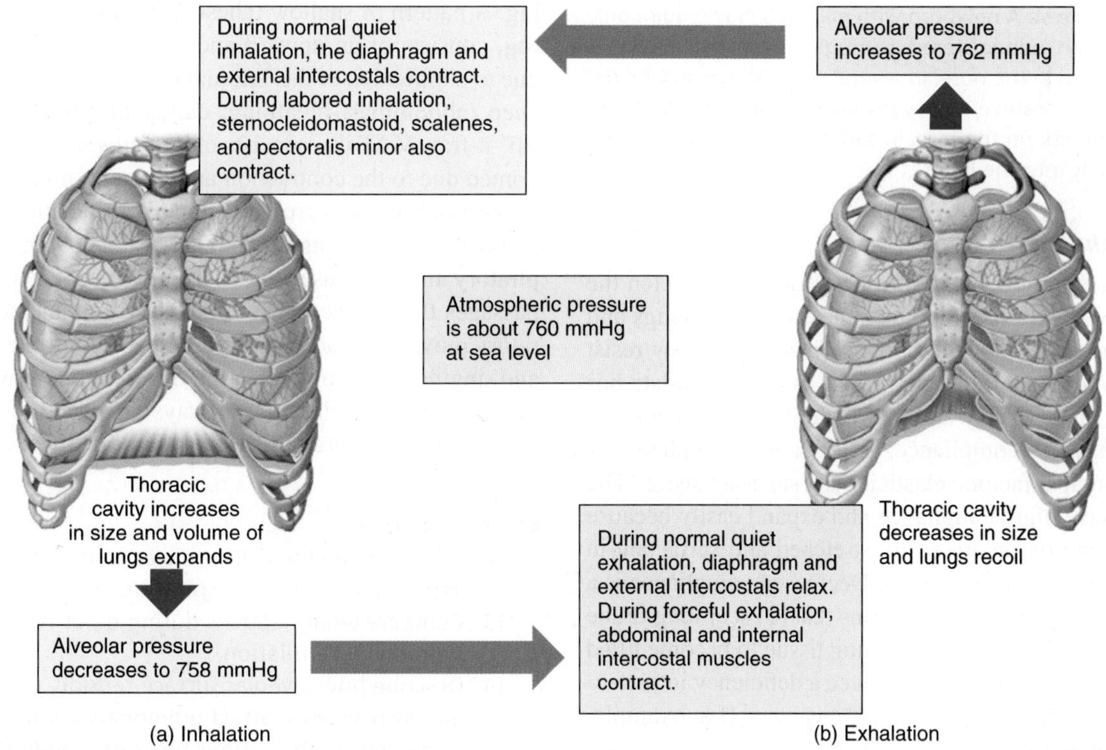

During normal quiet inhalation, the diaphragm and external intercostals contract. During labored inhalation, sternocleidomastoid, scalenes, and pectoralis minor also contract.

Alveolar pressure increases to 762 mmHg

Atmospheric pressure is about 760 mmHg at sea level

Thoracic cavity increases in size and volume of lungs expands

Alveolar pressure decreases to 758 mmHg

During normal quiet exhalation, diaphragm and external intercostals relax. During forceful exhalation, abdominal and internal intercostal muscles contract.

Thoracic cavity decreases in size and lungs recoil

(a) Inhalation

(b) Exhalation

? What is normal atmospheric pressure at sea level?

water molecules are more strongly attracted to each other than they are to gas molecules in the air. When liquid surrounds a sphere of air, as in an alveolus or a soap bubble, surface tension produces an inwardly directed force. Soap bubbles "burst" because they collapse inward due to surface tension. In the lungs, surface tension causes the alveoli to assume the smallest possible diameter. During breathing, surface tension must be overcome to expand the lungs during each inhalation. Surface tension also accounts for two-thirds of lung elastic recoil, which decreases the size of alveoli during exhalation.

The surfactant (a mixture of phospholipids and lipoproteins) present in alveolar fluid reduces its surface tension below the surface tension of pure water. A deficiency of surfactant in premature infants causes *respiratory distress syndrome,* in which the surface tension of alveolar fluid is greatly increased, so that many alveoli collapse at the end of each exhalation. Great effort is then needed at the next inhalation to reopen the collapsed alveoli.

CLINICAL CONNECTION | *Respiratory Distress Syndrome*

Respiratory distress syndrome (RDS) is a breathing disorder of premature newborns in which the alveoli do not remain open due to a lack of surfactant. Recall that surfactant reduces surface tension and is necessary to prevent the collapse of alveoli during exhalation. The more premature the newborn, the greater the chance that RDS will develop. The condition is also more common in infants whose mothers have diabetes and in males; it also occurs more often in European Americans than African Americans. Symptoms of RDS include labored and irregular breathing, flaring of the nostrils during inhalation, grunting during exhalation, and perhaps a blue skin color. Besides the symptoms, RDS is diagnosed on the basis of chest radiographs and a blood test. A newborn with mild RDS may require only supplemental oxygen administered through an oxygen hood or through a tube placed in the nose. In severe cases oxygen may be delivered by continuous positive airway pressure (CPAP) through tubes in the nostrils or a mask on the face. In such cases surfactant may be administered directly into the lungs. •

Compliance of the Lungs

Compliance refers to how much effort is required to stretch the lungs and chest wall. High compliance means that the lungs and chest wall expand easily; low compliance means that they resist expansion. By analogy, a thin balloon that is easy to inflate has high compliance, and a heavy and stiff balloon that takes a lot of effort to inflate has low compliance. In the lungs, compliance is related to two principal factors: elasticity and surface tension. The lungs normally have high compliance and expand easily because elastic fibers in lung tissue are easily stretched and surfactant in alveolar fluid reduces surface tension. Decreased compliance is a common feature in pulmonary conditions that (1) scar lung tissue (for example, tuberculosis), (2) cause lung tissue to become filled with fluid (pulmonary edema), (3) produce a deficiency in surfactant, or (4) impede lung expansion in any way (for example, paralysis of the intercostal muscles). Decreased lung compliance occurs in emphysema (see Disorders: Homeostatic Imbalances

at the end of the chapter) due to destruction of elastic fibers in alveolar walls.

Airway Resistance

Like the flow of blood through blood vessels, the rate of airflow through the airways depends on both the pressure difference and the resistance: Airflow equals the pressure difference between the alveoli and the atmosphere divided by the resistance. The walls of the airways, especially the bronchioles, offer some resistance to the normal flow of air into and out of the lungs. As the lungs expand during inhalation, the bronchioles enlarge because their walls are pulled outward in all directions. Larger-diameter airways have decreased resistance. Airway resistance then increases during exhalation as the diameter of bronchioles decreases. Airway diameter is also regulated by the degree of contraction or relaxation of smooth muscle in the walls of the airways. Signals from the sympathetic division of the autonomic nervous system cause relaxation of this smooth muscle, which results in bronchodilation and decreased resistance.

Any condition that narrows or obstructs the airways increases resistance, so that more pressure is required to maintain the same airflow. The hallmark of asthma or chronic obstructive pulmonary disease (COPD)—emphysema or chronic bronchitis—is increased airway resistance due to obstruction or collapse of airways.

Breathing Patterns and Modified Respiratory Movements

The term for the normal pattern of quiet breathing is **eupnea** (ūp-NĒ-a; *eu-* = good, easy, or normal; *-pnea* = breath). Eupnea can consist of shallow, deep, or combined shallow and deep breathing. A pattern of shallow (chest) breathing, called **costal breathing,** consists of an upward and outward movement of the chest due to contraction of the external intercostal muscles. A pattern of deep (abdominal) breathing, called **diaphragmatic breathing** (dī′-a-frag-MAT-ik), consists of the outward movement of the abdomen due to the contraction and descent of the diaphragm.

Respirations also provide humans with methods for expressing emotions such as laughing, sighing, and sobbing. Moreover, respiratory air can be used to expel foreign matter from the lower air passages through actions such as sneezing and coughing. Respiratory movements are also modified and controlled during talking and singing. Some of the modified respiratory movements that express emotion or clear the airways are listed in Table 23.2. All these movements are reflexes, but some of them also can be initiated voluntarily.

✔CHECKPOINT

12. What are the basic differences among pulmonary ventilation, external respiration, and internal respiration?
13. Compare what happens during quiet versus forceful pulmonary ventilation.
14. Describe how alveolar surface tension, compliance, and airway resistance affect pulmonary ventilation.
15. Demonstrate the various types of modified respiratory movements.

TABLE 23.2

Modified Respiratory Movements

MOVEMENT	DESCRIPTION
Coughing	A long-drawn and deep inhalation followed by a complete closure of the rima glottidis, which results in a strong exhalation that suddenly pushes the rima glottidis open and sends a blast of air through the upper respiratory passages. Stimulus for this reflex act may be a foreign body lodged in the larynx, trachea, or epiglottis.
Sneezing	Spasmodic contraction of muscles of exhalation that forcefully expels air through the nose and mouth. Stimulus may be an irritation of the nasal mucosa.
Sighing	A long-drawn and deep inhalation immediately followed by a shorter but forceful exhalation.
Yawning	A deep inhalation through the widely opened mouth producing an exaggerated depression of the mandible. It may be stimulated by drowsiness, or someone else's yawning, but the precise cause is unknown.
Sobbing	A series of convulsive inhalations followed by a single prolonged exhalation. The rima glottidis closes earlier than normal after each inhalation so only a little air enters the lungs with each inhalation.
Crying	An inhalation followed by many short convulsive exhalations, during which the rima glottidis remains open and the vocal folds vibrate; accompanied by characteristic facial expressions and tears.
Laughing	The same basic movements as crying, but the rhythm of the movements and the facial expressions usually differ from those of crying. Laughing and crying are sometimes indistinguishable.
Hiccupping	Spasmodic contraction of the diaphragm followed by a spasmodic closure of the rima glottidis, which produces a sharp sound on inhalation. Stimulus is usually irritation of the sensory nerve endings of the gastrointestinal tract.
Valsalva (val-SAL-va) maneuver	Forced exhalation against a closed rima glottidis as may occur during periods of straining while defecating.
Pressurizing the middle ear	The nose and mouth are held closed and air from the lungs is forced through the pharyngotympanic tube into the middle ear. Employed by those snorkeling or scuba diving during descent to equalize the pressure of the middle ear with that of the external environment.

23.3 LUNG VOLUMES AND CAPACITIES

● OBJECTIVES

- Explain the difference between tidal volume, inspiratory reserve volume, expiratory reserve volume, and residual volume.
- Differentiate between inspiratory capacity, functional residual capacity, vital capacity, and total lung capacity.

While at rest, a healthy adult averages 12 breaths a minute, with each inhalation and exhalation moving about 500 mL of air into and out of the lungs. The volume of one breath is called the **tidal volume (V_T).** The **minute ventilation (MV)**—the total volume of air inhaled and exhaled each minute—is respiratory rate multiplied by tidal volume:

$$MV = 12 \text{ breaths/min} \times 500 \text{ mL/breath}$$
$$= 6 \text{ liters/min}$$

A lower-than-normal minute ventilation usually is a sign of pulmonary malfunction. The apparatus commonly used to measure the volume of air exchanged during breathing and the respiratory rate is a **spirometer** (spī-ROM-e-ter; *spiro-* = breathe; *-meter* = measuring device) or **respirometer** (res′-pi-ROM-e-ter). The record is called a **spirogram.** Inhalation is recorded as an upward deflection, and exhalation is recorded as a downward deflection (Figure 23.16).

Tidal volume varies considerably from one person to another and in the same person at different times. In a typical adult, about 70% of the tidal volume (350 mL) actually reaches the respiratory zone of the respiratory system—the respiratory bronchioles, alveolar ducts, alveolar sacs, and alveoli—and participates in external respiration. The other 30% (150 mL) remains in the conducting airways of the nose, pharynx, larynx, trachea, bronchi, bronchioles, and terminal bronchioles. Collectively, the conducting airways with air that does not undergo respiratory exchange are known as the **anatomic (respiratory) dead space.** (An easy rule of thumb for determining the volume of your anatomic dead space is that it is about the same in milliliters as your ideal weight in pounds.) Not all of the minute ventilation can be used in gas exchange because some of it remains in the anatomic dead space. The **alveolar ventilation rate** is the volume of air per minute that actually reaches the respiratory zone. In the example just given, alveolar ventilation rate would be 350 mL/breath $\times$ 12 breaths/min = 4200 mL/min.

Several other lung volumes are defined relative to forceful breathing. In general, these volumes are larger in males, taller individuals, and younger adults, and smaller in females, shorter individuals, and the elderly. Various disorders also may be diagnosed by comparison of actual and predicted normal values for a

Figure 23.16 Spirogram of lung volumes and capacities. The average values for a healthy adult male and female are indicated, with the values for a female in parentheses. Note that the spirogram is read from right (start of record) to left (end of record).

🔑 ⚙️ Lung capacities are combinations of various lung volumes.

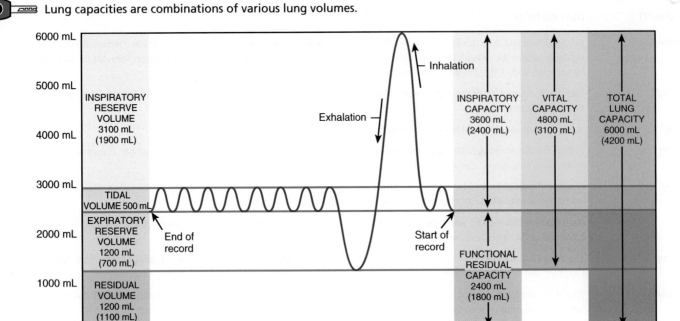

LUNG VOLUMES   LUNG CAPACITIES

❓ **If you breathe in as deeply as possible and then exhale as much air as you can, which lung capacity have you demonstrated?**

patient's gender, height, and age. The values given here are averages for young adults.

By taking a very deep breath, you can inhale a good deal more than 500 mL. This additional inhaled air, called the **inspiratory reserve volume,** is about 3100 mL in an average adult male and 1900 mL in an average adult female (Figure 23.16). Even more air can be inhaled if inhalation follows forced exhalation. If you inhale normally and then exhale as forcibly as possible, you should be able to push out considerably more air in addition to the 500 mL of tidal volume. The extra 1200 mL in males and 700 mL in females is called the **expiratory reserve volume.** The $FEV_{1.0}$ is the **forced expiratory volume in 1 second,** the volume of air that can be exhaled from the lungs in 1 second with maximal effort following a maximal inhalation. Typically, chronic obstructive pulmonary disease (COPD) greatly reduces $FEV_{1.0}$ because COPD increases airway resistance.

Even after the expiratory reserve volume is exhaled, considerable air remains in the lungs because the subatmospheric intrapleural pressure keeps the alveoli slightly inflated, and some air also remains in the noncollapsible airways. This volume, which cannot be measured by spirometry, is called the **residual volume** (re-ZID-u-al) and amounts to about 1200 mL in males and 1100 mL in females.

If the thoracic cavity is opened, the intrapleural pressure rises to equal the atmospheric pressure and forces out some of the

residual volume. The air remaining is called the **minimal volume.** Minimal volume provides a medical and legal tool for determining whether a baby is born dead (stillborn) or died after birth. The presence of minimal volume can be demonstrated by placing a piece of lung in water and observing if it floats. Fetal lungs contain no air, so the lung of a stillborn baby will not float in water.

Lung capacities are combinations of specific lung volumes (Figure 23.16). **Inspiratory capacity** is the sum of tidal volume and inspiratory reserve volume (500 mL + 3100 mL = 3600 mL in males and 500 mL + 1900 mL = 2400 mL in females). **Functional residual capacity** is the sum of residual volume and expiratory reserve volume (1200 mL + 1200 mL = 2400 mL in males and 1100 mL + 700 mL = 1800 mL in females). **Vital capacity** is the sum of inspiratory reserve volume, tidal volume, and expiratory reserve volume (4800 mL in males and 3100 mL in females). Finally, **total lung capacity** is the sum of vital capacity and residual volume (4800 mL + 1200 mL = 6000 mL in males and 3100 mL + 1100 mL = 4200 mL in females).

✔ CHECKPOINT

16. What is a spirometer?
17. What is the difference between a lung volume and a lung capacity?
18. How is minute ventilation calculated?
19. Define alveolar ventilation rate and $FEV_{1.0}$.

23.4 EXCHANGE OF OXYGEN AND CARBON DIOXIDE

◉ **OBJECTIVES**

- Explain Dalton's law and Henry's law.
- Describe the exchange of oxygen and carbon dioxide in external and internal respiration.

The exchange of oxygen and carbon dioxide between alveolar air and pulmonary blood occurs via passive diffusion, which is governed by the behavior of gases as described by two gas laws, Dalton's law and Henry's law. Dalton's law is important for understanding how gases move down their pressure differences by diffusion, and Henry's law helps explain how the solubility of a gas relates to its diffusion.

Gas Laws: Dalton's Law and Henry's Law

According to **Dalton's law,** each gas in a mixture of gases exerts its own pressure as if no other gases were present. The pressure of a specific gas in a mixture is called its *partial pressure* (P_x); the subscript is the chemical formula of the gas. The total pressure of the mixture is calculated simply by adding all the partial pressures. Atmospheric air is a mixture of gases—nitrogen (N_2), oxygen (O_2), argon (Ar), carbon dioxide (CO_2), variable amounts of water vapor (H_2O), plus other gases present in small quantities. Atmospheric pressure is the sum of the pressures of all these gases:

$$\text{Atmospheric pressure (760 mmHg)}$$
$$= P_{N_2} + P_{O_2} + P_{Ar} + P_{H_2O} + P_{CO_2} + P_{\text{other gases}}$$

We can determine the partial pressure exerted by each component in the mixture by multiplying the percentage of the gas in the mixture by the total pressure of the mixture. Atmospheric air is 78.6% nitrogen, 20.9% oxygen, 0.93% argon, 0.04% carbon dioxide, and 0.06% other gases; and a variable amount of water vapor is also present. The amount of water varies from practically 0% over a desert to 4% over the ocean, about 0.4% on a cool, dry day. Thus, the partial pressures of the gases in inhaled air are as follows:

$$P_{N_2} = 0.786 \times 760 \text{ mmHg} = 597.4 \text{ mmHg}$$
$$P_{O_2} = 0.209 \times 760 \text{ mmHg} = 158.8 \text{ mmHg}$$
$$P_{Ar} = 0.0009 \times 760 \text{ mmHg} = 0.7 \text{ mmHg}$$
$$P_{H_2O} = 0.003 \times 760 \text{ mmHg} = 2.3 \text{ mmHg}$$
$$P_{CO_2} = 0.0004 \times 760 \text{ mmHg} = 0.3 \text{ mmHg}$$
$$P_{\text{other gases}} = 0.0006 \times 760 \text{ mmHg} = 0.5 \text{ mmHg}$$
$$\text{Total} = 760.0 \text{ mmHg}$$

These partial pressures determine the movement of O_2 and CO_2 between the atmosphere and lungs, between the lungs and blood, and between the blood and body cells. Each gas diffuses across a permeable membrane from the area where its partial pressure is greater to the area where its partial pressure is less. The greater the difference in partial pressure, the faster the rate of diffusion.

Compared with inhaled air, alveolar air has less O_2 (13.6% versus 20.9%) and more CO_2 (5.2% versus 0.04%) for two reasons. First, gas exchange in the alveoli increases the CO_2 content and decreases the O_2 content of alveolar air. Second, when air is inhaled it becomes humidified as it passes along the moist mucosal linings. As water vapor content of the air increases, the relative percentage that is O_2 decreases. In contrast, exhaled air contains more O_2 than alveolar air (16% versus 13.6%) and less CO_2 (4.5% versus 5.2%) because some of the exhaled air was in the anatomic dead space and did not participate in gas exchange. Exhaled air is a mixture of alveolar air and inhaled air that was in the anatomic dead space.

Henry's law states that the quantity of a gas that will dissolve in a liquid is proportional to the partial pressure of the gas and its solubility. In body fluids, the ability of a gas to stay in solution is greater when its partial pressure is higher and when it has a high solubility in water. The higher the partial pressure of a gas over a liquid and the higher the solubility, the more gas will stay in solution. In comparison to oxygen, much more CO_2 is dissolved in blood plasma because the solubility of CO_2 is 24 times greater than that of O_2. Even though the air we breathe contains mostly N_2, this gas has no known effect on bodily functions, and at sea level pressure very little of it dissolves in blood plasma because its solubility is very low.

An everyday experience gives a demonstration of Henry's law. You have probably noticed that a soft drink makes a hissing sound when the top of the container is removed, and bubbles rise to the surface for some time afterward. The gas dissolved in carbonated beverages is CO_2. Because the soft drink is bottled or canned under high pressure and capped, the CO_2 remains dissolved as long as the container is unopened. Once you remove the cap, the pressure decreases and the gas begins to bubble out of solution.

Henry's law explains two conditions resulting from changes in the solubility of nitrogen in body fluids. Even though the air we breathe contains about 79% nitrogen, this gas has no known effect on bodily functions, and very little of it dissolves in blood plasma because of its low solubility at sea level pressure. As the total air pressure increases, the partial pressures of all its gases increase. When a scuba diver breathes air under high pressure, the nitrogen in the mixture can have serious negative effects. Because the partial pressure of nitrogen is higher in a mixture of compressed air than in air at sea level pressure, a considerable amount of nitrogen dissolves in plasma and interstitial fluid. Excessive amounts of dissolved nitrogen may produce giddiness and other symptoms similar to alcohol intoxication. The condition is called **nitrogen narcosis** or "rapture of the deep."

If a diver comes to the surface slowly, the dissolved nitrogen can be eliminated by exhaling it. However, if the ascent is too rapid, nitrogen comes out of solution too quickly and forms gas bubbles in the tissues, resulting in **decompression sickness** (the **bends**). The effects of decompression sickness typically result from bubbles in nervous tissue and can be mild or severe, depending on the number of bubbles formed. Symptoms include joint pain, especially in the arms and legs, dizziness, shortness of breath, extreme fatigue, paralysis, and unconsciousness.

External and Internal Respiration

External respiration or **pulmonary gas exchange** is the diffusion of O_2 from air in the alveoli of the lungs to blood in pulmonary capillaries and the diffusion of CO_2 in the opposite direction (Figure 23.17a). External respiration in the lungs converts **deoxygenated blood** (depleted of some O_2) coming from the right side of the heart into **oxygenated blood** (saturated with O_2) that returns to the left side of the heart (see Figure 21.29). As blood flows through the pulmonary capillaries, it picks up O_2 from alveolar air and unloads CO_2 into alveolar air. Although this process is commonly called an "exchange" of gases, each gas diffuses independently from the area where its partial pressure is higher to the area where its partial pressure is lower.

As Figure 23.17a shows, O_2 diffuses from alveolar air, where its partial pressure is 105 mmHg, into the blood in pulmonary capillaries, where P_{O_2} is only 40 mmHg in a resting person. If you have been exercising, the P_{O_2} will be even lower because contracting muscle fibers are using more O_2. Diffusion continues until the P_{O_2} of pulmonary capillary blood increases to match the P_{O_2} of alveolar air, 105 mmHg. Because blood leaving pulmonary capillaries near alveolar air spaces mixes with a small volume of blood that has flowed through conducting portions of the respiratory system, where gas exchange does not occur, the P_{O_2} of blood in the pulmonary veins is slightly less than the P_{O_2} in pulmonary capillaries, about 100 mmHg.

While O_2 is diffusing from alveolar air into deoxygenated blood, CO_2 is diffusing in the opposite direction. The P_{CO_2} of deoxygenated blood is 45 mmHg in a resting person, and the P_{CO_2} of alveolar air is 40 mmHg. Because of this difference in P_{CO_2}, carbon dioxide diffuses from deoxygenated blood into the alveoli until the P_{CO_2} of the blood decreases to 40 mmHg. Exhalation keeps alveolar P_{CO_2} at 40 mmHg. Oxygenated blood returning to the left side of the heart in the pulmonary veins thus has a P_{CO_2} of 40 mmHg.

The number of capillaries near alveoli in the lungs is very large, and blood flows slowly enough through these capillaries that it picks up a maximal amount of O_2. During vigorous exercise, when cardiac output is increased, blood flows more rapidly through both the systemic and pulmonary circulations. As a result, blood's transit time in the pulmonary capillaries is shorter. Still, the P_{O_2} of blood in the pulmonary veins normally reaches 100 mmHg. In diseases that decrease the rate of gas diffusion, however, the blood may not come into full equilibrium with alveolar air, especially during exercise. When this happens, the P_{O_2} declines and P_{CO_2} rises in systemic arterial blood.

The left ventricle pumps oxygenated blood into the aorta and through the systemic arteries to systemic capillaries. The exchange of O_2 and CO_2 between systemic capillaries and tissue cells is called **internal respiration** or **systemic gas exchange** (Figure 23.17b). As O_2 leaves the bloodstream, oxygenated blood is converted into deoxygenated blood. Unlike external respiration, which occurs only in the lungs, internal respiration occurs in tissues throughout the body.

The P_{O_2} of blood pumped into systemic capillaries is higher (100 mmHg) than the P_{O_2} in tissue cells (40 mmHg at rest) because the cells constantly use O_2 to produce ATP. Due to this pressure difference, oxygen diffuses out of the capillaries into tissue cells and blood P_{O_2} drops to 40 mmHg by the time the blood exits systemic capillaries.

While O_2 diffuses from the systemic capillaries into tissue cells, CO_2 diffuses in the opposite direction. Because tissue cells are constantly producing CO_2, the P_{CO_2} of cells (45 mmHg at rest) is higher than that of systemic capillary blood (40 mmHg). As a result, CO_2 diffuses from tissue cells through interstitial fluid into systemic capillaries until the P_{CO_2} in the blood increases to 45 mmHg. The deoxygenated blood then returns to the heart and is pumped to the lungs for another cycle of external respiration.

In a person at rest, tissue cells, on average, need only 25% of the available O_2 in oxygenated blood; despite its name, deoxygenated blood retains 75% of its O_2 content. During exercise, more O_2 diffuses from the blood into metabolically active cells, such as contracting skeletal muscle fibers. Active cells use more O_2 for ATP production, causing the O_2 content of deoxygenated blood to drop below 75%.

The *rate* of pulmonary and systemic gas exchange depends on several factors.

• ***Partial pressure difference of the gases.*** Alveolar P_{O_2} must be higher than blood P_{O_2} for oxygen to diffuse from alveolar air into the blood. The rate of diffusion is faster when the difference between P_{O_2} in alveolar air and pulmonary capillary blood is larger; diffusion is slower when the difference is smaller. The differences between P_{O_2} and P_{CO_2} in alveolar air versus pulmonary blood increase during exercise. The larger partial pressure differences accelerate the rates of gas diffusion. The partial pressures of O_2 and CO_2 in alveolar air also depend on the rate of airflow into and out of the lungs. Certain drugs (such as morphine) slow ventilation, thereby decreasing the amount of O_2 and CO_2 that can be exchanged between alveolar air and blood. With increasing altitude, the total atmospheric pressure decreases, as does the partial pressure of O_2—from 159 mmHg at sea level, to 110 mmHg at 10,000 ft, to 73 mmHg at 20,000 ft. Although O_2 still is 20.9% of the

Figure 23.17 **Changes in partial pressures of oxygen and carbon dioxide (in mmHg) during external and internal respiration.**

Gases diffuse from areas of higher partial pressure to areas of lower partial pressure.

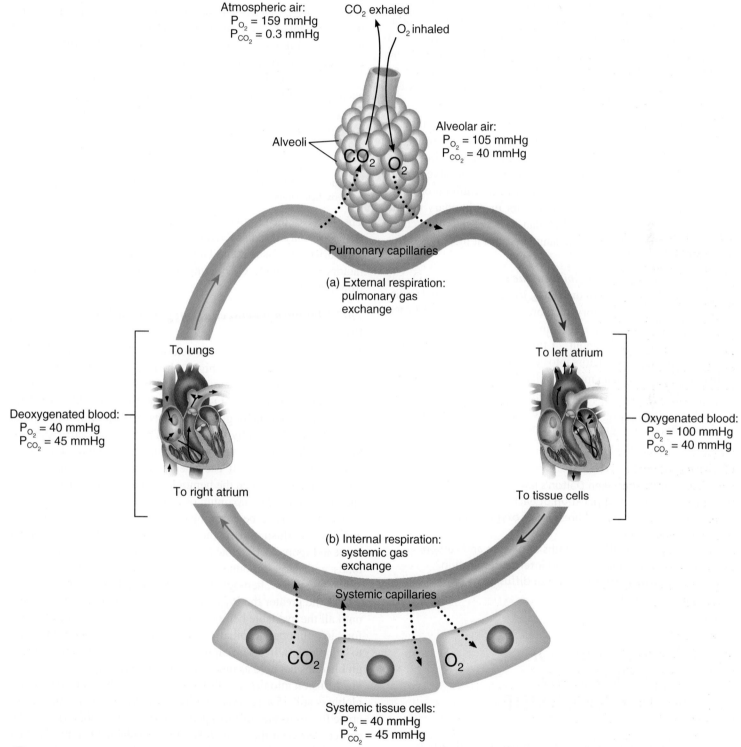

Atmospheric air:
P_{O_2} = 159 mmHg
P_{CO_2} = 0.3 mmHg

CO_2 exhaled
O_2 inhaled

Alveolar air:
P_{O_2} = 105 mmHg
P_{CO_2} = 40 mmHg

Alveoli

CO_2 O_2

Pulmonary capillaries

(a) External respiration: pulmonary gas exchange

To lungs

To left atrium

Deoxygenated blood:
P_{O_2} = 40 mmHg
P_{CO_2} = 45 mmHg

Oxygenated blood:
P_{O_2} = 100 mmHg
P_{CO_2} = 40 mmHg

To right atrium

To tissue cells

(b) Internal respiration: systemic gas exchange

Systemic capillaries

CO_2 O_2

Systemic tissue cells:
P_{O_2} = 40 mmHg
P_{CO_2} = 45 mmHg

? **What causes oxygen to enter pulmonary capillaries from alveoli and to enter tissue cells from systemic capillaries?**

total, the P_{O_2} of inhaled air decreases with increasing altitude. Alveolar P_{O_2} decreases correspondingly, and O_2 diffuses into the blood more slowly. The common signs and symptoms of **high altitude sickness**—shortness of breath, headache, fatigue, insomnia, nausea, and dizziness—are due to a lower level of oxygen in the blood.

- *Surface area available for gas exchange.* As you learned earlier in the chapter, the surface area of the alveoli is huge (about $70 m^2$ or $750 ft^2$). In addition, many capillaries surround each alveolus, so many that as much as 900 mL of blood is able to participate in gas exchange at any instant. Any pulmonary disorder that decreases the functional surface area of the respiratory membranes decreases the rate of external respiration. In emphysema (see Disorders: Homeostatic Imbalances at the end of the chapter), for example, alveolar walls disintegrate, so surface area is smaller than normal and pulmonary gas exchange is slowed.

- *Diffusion distance.* The respiratory membrane is very thin, so diffusion occurs quickly. Also, the capillaries are so narrow that the red blood cells must pass through them in single file, which minimizes the diffusion distance from an alveolar air space to hemoglobin inside red blood cells. Buildup of interstitial fluid between alveoli, as occurs in pulmonary edema (see Disorders: Homeostatic Imbalances at the end of the chapter), slows the rate of gas exchange because it increases diffusion distance.

- *Molecular weight and solubility of the gases.* Because O_2 has a lower molecular weight than CO_2, it could be expected to diffuse across the respiratory membrane about 1.2 times faster. However, the solubility of CO_2 in the fluid portions of the respiratory membrane is about 24 times greater than that of O_2. Taking both of these factors into account, net outward CO_2 diffusion occurs 20 times more rapidly than net inward O_2 diffusion. Consequently, when diffusion is slower than normal—for example, in emphysema or pulmonary edema—O_2 insufficiency (hypoxia) typically occurs before there is significant retention of CO_2 (hypercapnia).

✔CHECKPOINT

20. Distinguish between Dalton's law and Henry's law and give a practical application of each.
21. How does the partial pressure of oxygen change as altitude changes?
22. What are the diffusion paths of oxygen and carbon dioxide during external and internal respiration?
23. What factors affect the rate of diffusion of oxygen and carbon dioxide?

23.5 TRANSPORT OF OXYGEN AND CARBON DIOXIDE

◉ OBJECTIVE

- Describe how the blood transports oxygen and carbon dioxide.

As you have already learned, the blood transports gases between the lungs and body tissues. When O_2 and CO_2 enter the blood, certain chemical reactions occur that aid in gas transport and gas exchange.

Oxygen Transport

Oxygen does not dissolve easily in water, so only about 1.5% of inhaled O_2 is dissolved in blood plasma, which is mostly water. About 98.5% of blood O_2 is bound to hemoglobin in red blood cells (Figure 23.18). Each 100 mL of oxygenated blood contains the equivalent of 20 mL of gaseous O_2. Using the percentages just given, the amount dissolved in the plasma is 0.3 mL and the amount bound to hemoglobin is 19.7 mL.

The heme portion of hemoglobin contains four atoms of iron, each capable of binding to a molecule of O_2 (see Figure 19.4b, c). Oxygen and hemoglobin bind in an easily reversible reaction to form **oxyhemoglobin:**

$$\underset{\substack{\text{Reduced hemoglobin} \\ \text{(deoxyhemoglobin)}}}{Hb} + \underset{\text{Oxygen}}{O_2} \underset{\substack{\text{Dissociation} \\ \text{of } O_2}}{\overset{\text{Binding of } O_2}{\rightleftharpoons}} \underset{\text{Oxyhemoglobin}}{Hb\text{—}O_2}$$

The 98.5% of the O_2 that is bound to hemoglobin is trapped inside RBCs, so only the dissolved O_2 (1.5%) can diffuse out of tissue capillaries into tissue cells. Thus, it is important to understand the factors that promote O_2 binding to and dissociation (separation) from hemoglobin.

The Relationship between Hemoglobin and Oxygen Partial Pressure

The most important factor that determines how much O_2 binds to hemoglobin is the P_{O_2}; the higher the P_{O_2}, the more O_2 combines with Hb. When reduced hemoglobin (Hb) is completely converted to oxyhemoglobin (Hb—O_2), the hemoglobin is said to be **fully saturated;** when hemoglobin consists of a mixture of Hb and Hb—O_2, it is **partially saturated.** The **percent saturation of hemoglobin** expresses the average saturation of hemoglobin with oxygen. For instance, if each hemoglobin molecule has bound two O_2 molecules, then the hemoglobin is 50% saturated because each Hb can bind a maximum of four O_2.

The relationship between the percent saturation of hemoglobin and P_{O_2} is illustrated in the oxygen–hemoglobin dissociation curve in Figure 23.19. Note that when the P_{O_2} is high, hemoglobin binds with large amounts of O_2 and is almost 100% saturated. When P_{O_2} is low, hemoglobin is only partially saturated. In other words, the greater the P_{O_2}, the more O_2 will bind to hemoglobin, until all the available hemoglobin molecules are saturated. Therefore, in pulmonary capillaries, where P_{O_2} is high, a lot of O_2 binds to hemoglobin. In tissue capillaries, where the P_{O_2} is lower, hemoglobin does not hold as much O_2, and the dissolved O_2 is unloaded via diffusion into tissue cells (see Figure 23.18b). Note that hemoglobin is still 75% saturated with O_2 at a P_{O_2} of 40 mmHg, the average P_{O_2} of tissue cells in a person at rest. This is the basis for the earlier statement that only 25% of the available O_2 unloads from hemoglobin and is used by tissue cells under resting conditions.

When the P_{O_2} is between 60 and 100 mmHg, hemoglobin is 90% or more saturated with O_2 (Figure 23.19). Thus, blood picks up a nearly full load of O_2 from the lungs even when the P_{O_2} of alveolar air is as low as 60 mmHg. The Hb–P_{O_2} curve explains why people can still perform well at high altitudes or when they

Figure 23.18 **Transport of oxygen (O$_2$) and carbon dioxide (CO$_2$) in the blood.**

Most O$_2$ is transported by hemoglobin as oxyhemoglobin (Hb—O$_2$) within red blood cells; most CO$_2$ is transported in blood plasma as bicarbonate ions (HCO$_3^-$).

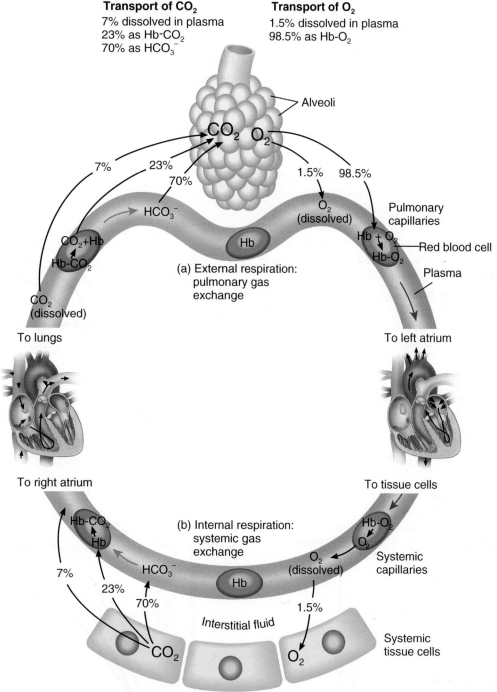

Transport of CO$_2$
7% dissolved in plasma
23% as Hb-CO$_2$
70% as HCO$_3^-$

Transport of O$_2$
1.5% dissolved in plasma
98.5% as Hb-O$_2$

Alveoli

CO$_2$ O$_2$

7% 23% 1.5% 98.5%

70%

HCO$_3^-$ O$_2$ (dissolved) Pulmonary capillaries

CO$_2$+Hb Hb Hb + O$_2$ Red blood cell

Hb-CO$_2$ Hb-O$_2$

(a) External respiration: pulmonary gas exchange Plasma

CO$_2$ (dissolved)

To lungs To left atrium

To right atrium To tissue cells

Hb-CO$_2$ Hb-O$_2$

Hb **(b) Internal respiration: systemic gas exchange** O$_2$

7% HCO$_3^-$ Hb O$_2$ (dissolved) Systemic capillaries

23% 70% 1.5%

Interstitial fluid

CO$_2$ O$_2$ Systemic tissue cells

? **What is the most important factor that determines how much O$_2$ binds to hemoglobin?**

have certain cardiac and pulmonary diseases, even though P_{O_2} may drop as low as 60 mmHg. Note also in the curve that at a considerably lower P_{O_2} of 40 mmHg, hemoglobin is still 75% saturated with O$_2$. However, oxygen saturation of Hb drops to 35% at 20 mmHg. Between 40 and 20 mmHg, large amounts of O$_2$ are released from hemoglobin in response to only small decreases in P_{O_2}. In active tissues such as contracting muscles, P_{O_2} may drop well below 40 mmHg. Then, a large percentage of the O$_2$ is released from hemoglobin, providing more O$_2$ to metabolically active tissues.

Other Factors Affecting the Affinity of Hemoglobin for Oxygen

Although P_{O_2} is the most important factor that determines the percent O_2 saturation of hemoglobin, several other factors influence the tightness or **affinity** with which hemoglobin binds O_2. In effect, these factors shift the entire curve either to the left (higher affinity) or to the right (lower affinity). The changing affinity of hemoglobin for O_2 is another example of how homeostatic mechanisms adjust body activities to cellular needs. Each one makes sense if you keep in mind that metabolically active tissue cells need O_2 and produce acids, CO_2, and heat as wastes. The following four factors affect the affinity of hemoglobin for O_2:

1. **Acidity (pH).** As acidity increases (pH decreases), the affinity of hemoglobin for O_2 decreases, and O_2 dissociates more readily from hemoglobin (Figure 23.20a). In other words, increasing acidity enhances the unloading of oxygen from hemoglobin. The main acids produced by metabolically active tissues are lactic acid and carbonic acid. When pH decreases, the entire oxygen–hemoglobin dissociation curve shifts to the right; at any given P_{O_2}, Hb is less saturated with O_2, a change termed the **Bohr effect** (BŌR). The Bohr effect works both ways: An increase in H^+ in blood causes O_2 to unload from hemoglobin, and the binding of O_2 to hemoglobin causes unloading of H^+ from hemoglobin. The explanation for the Bohr effect is that hemoglobin can act as a buffer for hydrogen ions

(H^+). But when H^+ ions bind to amino acids in hemoglobin, they alter its structure slightly, decreasing its oxygen-carrying capacity. Thus, lowered pH drives O_2 off hemoglobin, making more O_2 available for tissue cells. By contrast, elevated pH increases the affinity of hemoglobin for O_2 and shifts the oxygen–hemoglobin dissociation curve to the left.

Figure 23.20 Oxygen–hemoglobin dissociation curves showing the relationship of (a) pH and (b) P_{CO_2} to hemoglobin saturation at normal body temperature. As pH increases or P_{CO_2} decreases, O_2 combines more tightly with hemoglobin, so that less is available to tissues. The broken lines emphasize these relationships.

🔑 As pH decreases or P_{CO_2} increases, the affinity of hemoglobin for O_2 declines, so less O_2 combines with hemoglobin and more is available to tissues.

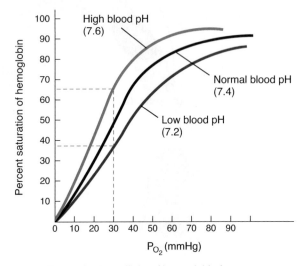

(a) Effect of pH on affinity of hemoglobin for oxygen

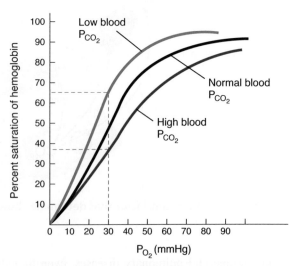

(b) Effect of P_{CO_2} on affinity of hemoglobin for oxygen

Figure 23.19 Oxygen–hemoglobin dissociation curve showing the relationship between hemoglobin saturation and P_{O_2} at normal body temperature.

🔑 As P_{O_2} increases, more O_2 combines with hemoglobin.

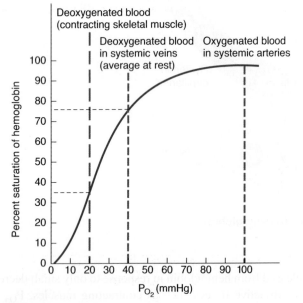

❓ What point on the curve represents blood in your pulmonary veins right now? In your pulmonary veins if you were jogging?

❓ In comparison to the value when you are sitting, is the affinity of your hemoglobin for O_2 higher or lower when you are exercising? How does this benefit you?

2. ***Partial pressure of carbon dioxide.*** CO_2 also can bind to hemoglobin, and the effect is similar to that of H^+ (shifting the curve to the right). As P_{CO_2} rises, hemoglobin releases O_2 more readily (Figure 23.20b). P_{CO_2} and pH are related factors because low blood pH (acidity) results from high P_{CO_2}. As CO_2 enters the blood, much of it is temporarily converted to carbonic acid (H_2CO_3), a reaction catalyzed by an enzyme in red blood cells called *carbonic anhydrase (CA):*

$$CO_2 + H_2O \overset{CA}{\rightleftharpoons} H_2CO_3 \rightleftharpoons H^+ + HCO_3^-$$

Carbon dioxide   Water   Carbonic acid   Hydrogen ion   Bicarbonate ion

The carbonic acid thus formed in red blood cells dissociates into hydrogen ions and bicarbonate ions. As the H^+ concentration increases, pH decreases. Thus, an increased P_{CO_2} produces a more acidic environment, which helps release O_2 from hemoglobin. During exercise, lactic acid—a by-product of anaerobic metabolism within muscles—also decreases blood pH. Decreased P_{CO_2} (and elevated pH) shifts the saturation curve to the left.

3. ***Temperature.*** Within limits, as temperature increases, so does the amount of O_2 released from hemoglobin (Figure 23.21). Heat is a by-product of the metabolic reactions of all cells, and the heat released by contracting muscle fibers tends to raise body temperature. Metabolically active cells require more O_2 and liberate more acids and heat. The acids and heat in turn promote release of O_2 from oxyhemoglobin. Fever produces a similar result. In contrast, during hypothermia (lowered body temperature) cellular metabolism slows, the need for O_2 is re-

duced, and more O_2 remains bound to hemoglobin (a shift to the left in the saturation curve).

4. ***BPG.*** A substance found in red blood cells called **2,3-bis-phosphoglycerate (BPG)** (bis'-fos-fō-GLIS-e-rāt), previously called diphosphoglycerate (DPG), decreases the affinity of hemoglobin for O_2 and thus helps unload O_2 from hemoglobin. BPG is formed in red blood cells when they break down glucose to produce ATP in a process called glycolysis (described in Section 25.3). When BPG combines with hemoglobin by binding to the terminal amino groups of the two beta globin chains, the hemoglobin binds O_2 less tightly at the heme group sites. The greater the level of BPG, the more O_2 is unloaded from hemoglobin. Certain hormones, such as thyroxine, human growth hormone, epinephrine, norepinephrine, and testosterone, increase the formation of BPG. The level of BPG also is higher in people living at higher altitudes.

Oxygen Affinity of Fetal and Adult Hemoglobin

Fetal hemoglobin (Hb-F) differs from **adult hemoglobin (Hb-A)** in structure and in its affinity for O_2. Hb-F has a higher affinity for O_2 because it binds BPG less strongly. Thus, when P_{O_2} is low, Hb-F can carry up to 30% more O_2 than maternal Hb-A (Figure 23.22). As the maternal blood enters the placenta, O_2 is readily transferred to fetal blood. This is very important because the O_2 saturation in maternal blood in the placenta is quite low, and the fetus might suffer hypoxia were it not for the greater affinity of fetal hemoglobin for O_2.

Figure 23.21 Oxygen–hemoglobin dissociation curves showing the effect of temperature changes.

As temperature increases, the affinity of hemoglobin for O_2 decreases.

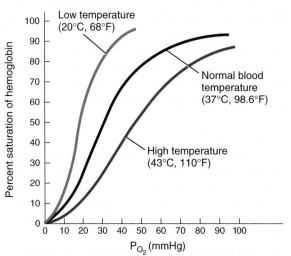

? Is O_2 more available or less available to tissue cells when you have a fever? Why?

Figure 23.22 Oxygen–hemoglobin dissociation curves comparing fetal and maternal hemoglobin.

Fetal hemoglobin has a higher affinity for O_2 than does adult hemoglobin.

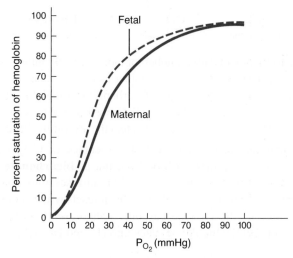

? The P_{O_2} of placental blood is about 40 mmHg. What are the O_2 saturations of maternal and fetal hemoglobin at this P_{O_2}?

Carbon monoxide (CO) is a colorless and odorless gas found in exhaust fumes from automobiles, gas furnaces, and space heaters and in tobacco smoke. It is a by-product of the combustion of carbon-containing materials such as coal, gas, and wood. CO binds to the heme group of hemoglobin, just as O_2 does, except that the binding of carbon monoxide to hemoglobin is over 200 times as strong as the binding of O_2 to hemoglobin. Thus, at a concentration as small as 0.1% (P_{CO} = 0.5 mmHg), CO will combine with half the available hemoglobin molecules and reduce the oxygen-carrying capacity of the blood by 50%. Elevated blood levels of CO cause **carbon monoxide poisoning,** which can cause the lips and oral mucosa to appear bright, cherry-red (the color of hemoglobin with carbon monoxide bound to it). Without prompt treatment, carbon monoxide poisoning is fatal. It is possible to rescue a victim of CO poisoning by administering pure oxygen, which speeds up the separation of carbon monoxide from hemoglobin. •

Carbon Dioxide Transport

Under normal resting conditions, each 100 mL of deoxygenated blood contains the equivalent of 53 mL of gaseous CO_2, which is transported in the blood in three main forms (see Figure 23.18):

1. *Dissolved CO_2.* The smallest percentage—about 7%—is dissolved in blood plasma. On reaching the lungs, it diffuses into alveolar air and is exhaled.

2. *Carbamino compounds.* A somewhat higher percentage, about 23%, combines with the amino groups of amino acids and proteins in blood to form **carbamino compounds** (kar-BAM-i-nō). Because the most prevalent protein in blood is hemoglobin (inside red blood cells), most of the CO_2 transported in this manner is bound to hemoglobin. The main CO_2 binding sites are the terminal amino acids in the two alpha and two beta globin chains. Hemoglobin that has bound CO_2 is termed **carbaminohemoglobin (Hb—CO_2):**

$$Hb + CO_2 \rightleftharpoons Hb—CO_2$$

Hemoglobin   Carbon dioxide   Carbaminohemoglobin

The formation of carbaminohemoglobin is greatly influenced by P_{CO_2}. For example, in tissue capillaries P_{CO_2} is relatively high, which promotes formation of carbaminohemoglobin. But in pulmonary capillaries, P_{CO_2} is relatively low, and the CO_2 readily splits apart from globin and enters the alveoli by diffusion.

3. *Bicarbonate ions.* The greatest percentage of CO_2—about 70%—is transported in blood plasma as **bicarbonate ions** (HCO_3^-) (bī'-KAR-bo-nāt). As CO_2 diffuses into systemic capillaries and enters red blood cells, it reacts with water in the presence of the enzyme carbonic anhydrase (CA) to form carbonic acid, which dissociates into H^+ and HCO_3^-:

$$CO_2 + H_2O \overset{CA}{\rightleftharpoons} H_2CO_3 \rightleftharpoons H^+ + HCO_3^-$$

Carbon dioxide   Water   Carbonic acid   Hydrogen ion   Bicarbonate ion

Thus, as blood picks up CO_2, HCO_3^- accumulates inside RBCs. Some HCO_3^- moves out into the blood plasma, down its concentration gradient. In exchange, chloride ions (Cl^-) move from plasma into the RBCs. This exchange of negative ions, which maintains the electrical balance between blood plasma and RBC cytosol, is known as the **chloride shift** (Figure 23.23b). The net effect of these reactions is that CO_2 is removed from tissue cells and transported in blood plasma as HCO_3^-. As blood passes through pulmonary capillaries in the lungs, all these reactions reverse and CO_2 is exhaled.

The amount of CO_2 that can be transported in the blood is influenced by the percent saturation of hemoglobin with oxygen. The lower the amount of oxyhemoglobin (Hb—O_2), the higher the CO_2-carrying capacity of the blood, a relationship known as the **Haldane effect.** Two characteristics of deoxyhemoglobin give rise to the Haldane effect: (1) Deoxyhemoglobin binds to and thus transports more CO_2 than does Hb—O_2. (2) Deoxyhemoglobin also buffers more H^+ than does Hb—O_2, thereby removing H^+ from solution and promoting conversion of CO_2 to HCO_3^- via the reaction catalyzed by carbonic anhydrase.

Summary of Gas Exchange and Transport in Lungs and Tissues

Deoxygenated blood returning to the pulmonary capillaries in the lungs (Figure 23.23a) contains CO_2 dissolved in blood plasma, CO_2 combined with globin as carbaminohemoglobin (Hb—CO_2), and CO_2 incorporated into HCO_3^- within RBCs. The RBCs have also picked up H^+, some of which binds to and therefore is buffered by hemoglobin (Hb—H). As blood passes through the pulmonary capillaries, molecules of CO_2 dissolved in blood plasma and CO_2 that dissociates from the globin portion of hemoglobin diffuse into alveolar air and are exhaled. At the same time, inhaled O_2 is diffusing from alveolar air into RBCs and is binding to hemoglobin to form oxyhemoglobin (Hb—O_2). Carbon dioxide also is released from HCO_3^- when H^+ combines with HCO_3^- inside RBCs. The H_2CO_3 formed from this reaction then splits into CO_2, which is exhaled, and H_2O. As the concentration of HCO_3^- declines inside RBCs in pulmonary capillaries, HCO_3^- diffuses in from the blood plasma, in exchange for Cl^-. In sum, oxygenated blood leaving the lungs has increased O_2 content and decreased amounts of CO_2 and H^+. In systemic capillaries, as cells use O_2 and produce CO_2, the chemical reactions reverse (Figure 23.23b).

✔**CHECKPOINT**

24. In a resting person, how many O_2 molecules are attached to each hemoglobin molecule, on average, in blood in the pulmonary arteries? In blood in the pulmonary veins?

25. What is the relationship between hemoglobin and P_{O_2}? How do temperature, H^+, P_{CO_2}, and BPG influence the affinity of Hb for O_2?

26. Why can hemoglobin unload more oxygen as blood flows through capillaries of metabolically active tissues, such as skeletal muscle during exercise, than is unloaded at rest?

Figure 23.23 Summary of chemical reactions that occur during gas exchange. (a) As carbon dioxide (CO_2) is exhaled, hemoglobin (Hb) inside red blood cells in pulmonary capillaries unloads CO_2 and picks up O_2 from alveolar air. Binding of O_2 to Hb—H releases hydrogen ions (H^+). Bicarbonate ions (HCO_3^-) pass into the RBC and bind to released H^+, forming carbonic acid (H_2CO_3). The H_2CO_3 dissociates into water (H_2O) and CO_2, and the CO_2 diffuses from blood into alveolar air. To maintain electrical balance, a chloride ion (Cl^-) exits the RBC for each HCO_3^- that enters (reverse chloride shift). (b) CO_2 diffuses out of tissue cells that produce it and enters red blood cells, where some of it binds to hemoglobin, forming carbaminohemoglobin (Hb—CO_2). This reaction causes O_2 to dissociate from oxyhemoglobin (Hb—O_2). Other molecules of CO_2 combine with water to produce bicarbonate ions (HCO_3^-) and hydrogen ions (H^+). As Hb buffers H^+, the Hb releases O_2 (Bohr effect). To maintain electrical balance, a chloride ion (Cl^-) enters the RBC for each HCO_3^- that exits (chloride shift).

Hemoglobin inside red blood cells transports O_2, CO_2, and H^+.

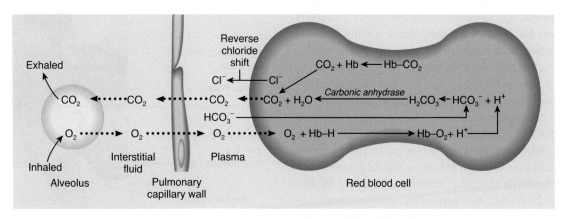

(a) Exchange of O_2 and CO_2 in pulmonary capillaries (external respiration)

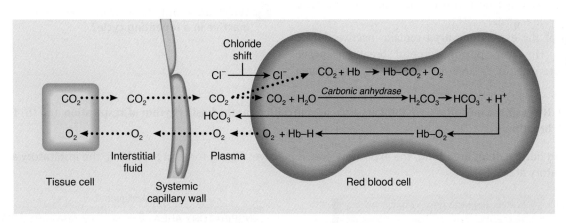

(b) Exchange of O_2 and CO_2 in systemic capillaries (internal respiration)

? **Would you expect the concentration of HCO_3^- to be higher in blood plasma taken from a systemic artery or a systemic vein?**

23.6 CONTROL OF RESPIRATION

● OBJECTIVES
- Explain how the nervous system controls breathing.
- List the factors that can alter the rate and depth of breathing.

At rest, about 200 mL of O_2 is used each minute by body cells. During strenuous exercise, however, O_2 use typically increases 15- to 20-fold in normal healthy adults, and as much as 30-fold in elite endurance-trained athletes. Several mechanisms help match respiratory effort to metabolic demand.

Respiratory Center

The size of the thorax is altered by the action of the respiratory muscles, which contract as a result of nerve impulses transmitted to them from centers in the brain and relax in the absence of nerve impulses. These nerve impulses are sent from clusters of neurons located bilaterally in the medulla oblongata and pons of the brain stem. This widely dispersed group of neurons, collectively called the **respiratory center,** can be divided into three areas on the basis of their functions: (1) the medullary rhythmicity area in the medulla oblongata; (2) the pneumotaxic area in the pons; and (3) the apneustic area, also in the pons (Figure 23.24).

Medullary Rhythmicity Area

The function of the **medullary rhythmicity area** (rith-MIS-i-tē) is to control the basic rhythm of respiration. There are inspiratory and expiratory areas within the medullary rhythmicity area. Figure 23.25 shows the relationships of the inspiratory and expiratory areas during normal quiet breathing and forceful breathing.

During quiet breathing, inhalation lasts for about 2 seconds and exhalation lasts for about 3 seconds. Nerve impulses generated in the **inspiratory area** establish the basic rhythm of breathing. While the inspiratory area is active, it generates nerve impulses for about 2 seconds (Figure 23.25a). The impulses propagate to the external intercostal muscles via intercostal nerves and to the diaphragm via the phrenic nerves. When the nerve impulses reach the diaphragm and external intercostal muscles, the muscles contract and inhalation occurs. Even when all incoming nerve connections to the inspiratory area are cut or blocked, neurons in this area still rhythmically discharge impulses that cause inhalation. At the end of 2 seconds, the inspiratory area becomes inactive and nerve impulses cease. With no impulses arriving, the diaphragm and external intercostal muscles relax for about 3 seconds, allowing passive elastic recoil of the lungs and thoracic wall. Then, the cycle repeats.

The neurons of the **expiratory area** remain inactive during quiet breathing. However, during forceful breathing nerve impulses from the inspiratory area activate the expiratory area

Figure 23.24 Locations of areas of the respiratory center.

 The respiratory center is composed of neurons in the medullary rhythmicity area in the medulla oblongata plus the pneumotaxic and apneustic areas in the pons.

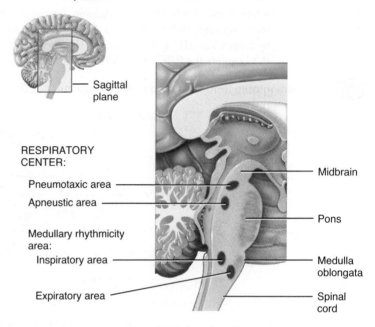

RESPIRATORY CENTER:

Pneumotaxic area

Apneustic area

Medullary rhythmicity area:
  Inspiratory area

  Expiratory area

Sagittal plane

Midbrain

Pons

Medulla oblongata

Spinal cord

Sagittal section of brain stem

? **Which area contains neurons that are active and then inactive in a repeating cycle?**

Figure 23.25 Roles of the medullary rhythmicity area in controlling (a) the basic rhythm of respiration and (b) forceful breathing.

 During normal quiet breathing, the expiratory area is inactive; during forceful breathing, the inspiratory area activates the expiratory area.

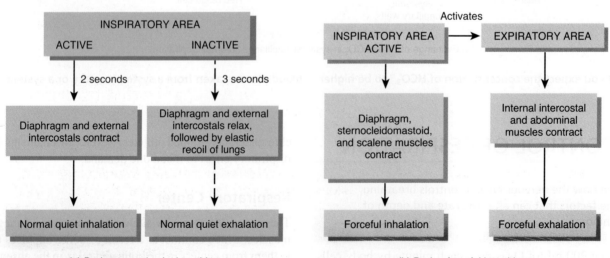

INSPIRATORY AREA	
ACTIVE	INACTIVE
2 seconds	3 seconds
Diaphragm and external intercostals contract	Diaphragm and external intercostals relax, followed by elastic recoil of lungs
Normal quiet inhalation	Normal quiet exhalation

(a) During normal quiet breathing

Activates

INSPIRATORY AREA ACTIVE	→	EXPIRATORY AREA
Diaphragm, sternocleidomastoid, and scalene muscles contract		Internal intercostal and abdominal muscles contract
Forceful inhalation		Forceful exhalation

(b) During forceful breathing

? **Which nerves convey impulses from the respiratory center to the diaphragm?**

(Figure 23.25b). Impulses from the expiratory area cause contraction of the internal intercostal and abdominal muscles, which decreases the size of the thoracic cavity and causes forceful exhalation.

Pneumotaxic Area

Although the medullary rhythmicity area controls the basic rhythm of respiration, other sites in the brain stem help coordinate the transition between inhalation and exhalation. One of these sites is the **pneumotaxic area** (noo-mō-TAK-sik; *pneumo-* = air or breath; *-taxic* = arrangement) in the upper pons (see Figure 23.24), which transmits inhibitory impulses to the inspiratory area. The major effect of these nerve impulses is to help turn off the inspiratory area before the lungs become too full of air. In other words, the impulses shorten the duration of inhalation. When the pneumotaxic area is more active, breathing rate is more rapid.

Apneustic Area

Another part of the brain stem that coordinates the transition between inhalation and exhalation is the **apneustic area** (ap-NOO-stik) in the lower pons (see Figure 23.24). This area sends stimulatory impulses to the inspiratory area that activate it and prolong inhalation. The result is a long, deep inhalation. When the pneumotaxic area is active, it overrides signals from the apneustic area.

Regulation of the Respiratory Center

The basic rhythm of respiration set and coordinated by the inspiratory area can be modified in response to inputs from other brain regions, receptors in the peripheral nervous system, and other factors.

Cortical Influences on Respiration

Because the cerebral cortex has connections with the respiratory center, we can voluntarily alter our pattern of breathing. We can even refuse to breathe at all for a short time. Voluntary control is protective because it enables us to prevent water or irritating gases from entering the lungs. The ability to not breathe, however, is limited by the buildup of CO_2 and H^+ in the body. When P_{CO_2} and H^+ concentrations increase to a certain level, the inspiratory area is strongly stimulated, nerve impulses are sent along the phrenic and intercostal nerves to inspiratory muscles, and breathing resumes, whether the person wants it to or not. It is impossible for small children to kill themselves by voluntarily holding their breath, even though many have tried in order to get their way. If breath is held long enough to cause fainting, breathing resumes when consciousness is lost. Nerve impulses from the hypothalamus and limbic system also stimulate the respiratory center, allowing emotional stimuli to alter respirations as, for example, in laughing and crying.

Chemoreceptor Regulation of Respiration

Certain chemical stimuli modulate how quickly and how deeply we breathe. The respiratory system functions to maintain proper levels of CO_2 and O_2 and is very responsive to changes in the levels of these gases in body fluids. We introduced sensory neurons that are responsive to chemicals, called **chemoreceptors** (kē'-mō-rē-SEP-tors), in Chapter 21. Chemoreceptors in two locations of the respiratory system monitor levels of CO_2, H^+, and O_2 and provide input to the respiratory center (Figure 23.26). **Central chemoreceptors** are located in or near the medulla oblongata in

Figure 23.26 Locations of peripheral chemoreceptors.

Chemoreceptors are sensory neurons that respond to changes in the levels of certain chemicals in the body.

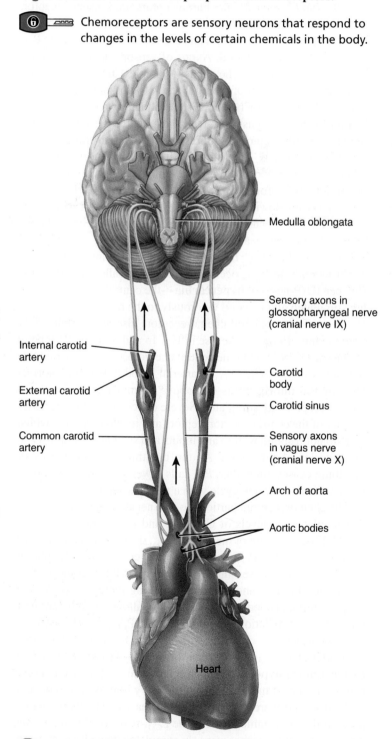

Medulla oblongata

Sensory axons in glossopharyngeal nerve (cranial nerve IX)

Internal carotid artery

External carotid artery

Common carotid artery

Carotid body

Carotid sinus

Sensory axons in vagus nerve (cranial nerve X)

Arch of aorta

Aortic bodies

Heart

? **Which chemicals stimulate peripheral chemoreceptors?**

the *central* nervous system. They respond to changes in H$^+$ concentration or P$_{CO_2}$, or both, in cerebrospinal fluid. **Peripheral chemoreceptors** are located in the **aortic bodies,** clusters of chemoreceptors located in the wall of the arch of the aorta, and in the **carotid bodies,** which are oval nodules in the wall of the left and right common carotid arteries where they divide into the internal and external carotid arteries. (The chemoreceptors of the aortic bodies are located close to the aortic baroreceptors, and the carotid bodies are located close to the carotid sinus baroreceptors. Recall from Chapter 21 that baroreceptors are sensory receptors that monitor blood pressure.) These chemoreceptors are part of the *peripheral* nervous system and are sensitive to changes in P$_{O_2}$, H$^+$, and P$_{CO_2}$ in the blood. Axons of sensory neurons from the aortic bodies are part of the vagus (X) nerves, and those from the carotid bodies are part of the right and left glossopharyngeal (IX) nerves. Recall from Chapter 17 that olfactory receptors for the sense of smell and gustatory receptor cells for the sense of taste are also chemoreceptors. Both respond to external stimuli.

Because CO$_2$ is lipid-soluble, it easily diffuses into cells where, in the presence of carbonic anhydrase, it combines with water (H$_2$O) to form carbonic acid (H$_2$CO$_3$). Carbonic acid quickly breaks down into H$^+$ and HCO$_3^-$. Thus, an increase in CO$_2$ in the blood causes an increase in H$^+$ inside cells, and a decrease in CO$_2$ causes a decrease in H$^+$.

Normally, the P$_{CO_2}$ in arterial blood is 40 mmHg. If even a slight increase in P$_{CO_2}$ occurs—a condition called **hypercapnia** (hī′-per-KAP-nē-a) or **hypercarbia**—the central chemoreceptors are stimulated and respond vigorously to the resulting increase in H$^+$ level. The peripheral chemoreceptors also are stimulated by both the high P$_{CO_2}$ and the rise in H$^+$. In addition, the peripheral chemoreceptors (but not the central chemoreceptors) respond to a deficiency of O$_2$. When P$_{O_2}$ in arterial blood falls from a normal level of 100 mmHg but is still above 50 mmHg, the peripheral chemoreceptors are stimulated. Severe deficiency of O$_2$ depresses activity of the central chemoreceptors and inspiratory area, which then do not respond well to any inputs and send fewer impulses to the muscles of inhalation. As the breathing rate decreases or breathing ceases altogether, P$_{O_2}$ falls lower and lower, establishing a positive feedback cycle with a possibly fatal result.

The chemoreceptors participate in a negative feedback system that regulates the levels of CO$_2$, O$_2$, and H$^+$ in the blood (Figure 23.27). As a result of increased P$_{CO_2}$, decreased pH (increased H$^+$), or decreased P$_{O_2}$, input from the central and peripheral chemoreceptors causes the inspiratory area to become highly active, and the rate and depth of breathing increase. Rapid and deep breathing, called **hyperventilation,** allows the inhalation of more O$_2$ and exhalation of more CO$_2$ until P$_{CO_2}$ and H$^+$ are lowered to normal.

If arterial P$_{CO_2}$ is lower than 40 mmHg—a condition called **hypocapnia** or **hypocarbia**—the central and peripheral chemoreceptors are not stimulated, and stimulatory impulses are not sent to the inspiratory area. As a result, the area sets its own moderate pace until CO$_2$ accumulates and the P$_{CO_2}$ rises to 40 mmHg. The inspiratory center is more strongly stimulated when P$_{CO_2}$ is rising above normal than when P$_{O_2}$ is falling below normal. As a result,

Figure 23.27 Regulation of breathing in response to changes in blood P$_{CO_2}$, P$_{O_2}$, and pH (H$^+$) via negative feedback control.

An increase in arterial blood P$_{CO_2}$ stimulates the inspiratory center.

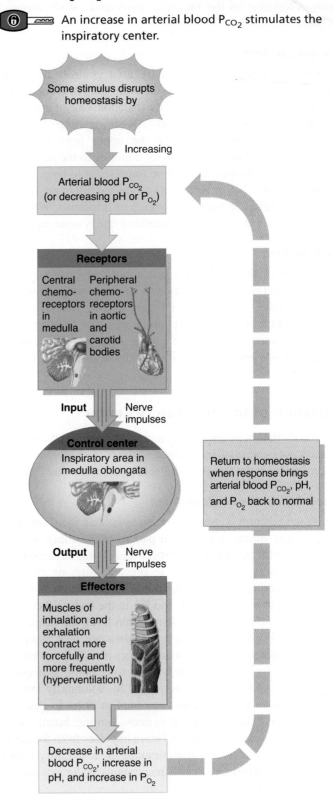

What is the normal arterial blood P$_{CO_2}$?

people who hyperventilate voluntarily and cause hypocapnia can hold their breath for an unusually long period. Swimmers were once encouraged to hyperventilate just before diving in to compete. However, this practice is risky because the O_2 level may fall dangerously low and cause fainting before the P_{CO_2} rises high enough to stimulate inhalation. If you faint on land you may suffer bumps and bruises, but if you faint in the water you could drown.

CLINICAL CONNECTION | Hypoxia

Hypoxia (hī-POK-sē-a; *hypo-* = under) is a deficiency of O_2 at the tissue level. Based on the cause, we can classify hypoxia into four types, as follows:

1. **Hypoxic hypoxia** is caused by a low P_{O_2} in arterial blood as a result of high altitude, airway obstruction, or fluid in the lungs.

2. In **anemic hypoxia,** too little functioning hemoglobin is present in the blood, which reduces O_2 transport to tissue cells. Among the causes are hemorrhage, anemia, and failure of hemoglobin to carry its normal complement of O_2, as in carbon monoxide poisoning.

3. In **ischemic hypoxia** (is-KĒ-mik), blood flow to a tissue is so reduced that too little O_2 is delivered to it, even though P_{O_2} and oxyhemoglobin levels are normal.

4. In **histotoxic hypoxia** (his-tō-TOK-sik), the blood delivers adequate O_2 to tissues, but the tissues are unable to use it properly because of the action of some toxic agent. One cause is cyanide poisoning, in which cyanide blocks an enzyme required for the use of O_2 during ATP synthesis. •

Proprioceptor Stimulation of Respiration

As soon as you start exercising, your rate and depth of breathing increase, even before changes in P_{O_2}, P_{CO_2}, or H^+ level occur. The main stimulus for these quick changes in respiratory effort is input from proprioceptors, which monitor movement of joints and muscles. Nerve impulses from the proprioceptors stimulate the inspiratory area of the medulla oblongata. At the same time, axon collaterals (branches) of upper motor neurons that originate in the primary motor cortex (precentral gyrus) also feed excitatory impulses into the inspiratory area.

The Inflation Reflex

Similar to those in the blood vessels, stretch-sensitive receptors called **baroreceptors** (bar'-ō-re-SEP-tors) or **stretch receptors** are located in the walls of bronchi and bronchioles. When these receptors become stretched during overinflation of the lungs, nerve impulses are sent along the vagus (X) nerves to the inspiratory and apneustic areas. In response, the inspiratory area is inhibited directly, and the apneustic area is inhibited from activating the inspiratory area. As a result, exhalation begins. As air leaves the lungs during exhalation, the lungs deflate and the stretch receptors are no longer stimulated. Thus, the inspiratory and apneustic areas are no longer inhibited, and a new inhalation begins. This reflex, referred to as the **inflation (Hering–Breuer) reflex** (HER-ing BROY-er), is mainly a protective mechanism for preventing excessive inflation of the lungs rather than a key component in the normal regulation of respiration.

Other Influences on Respiration

Other factors that contribute to regulation of respiration include the following:

• *Limbic system stimulation.* Anticipation of activity or emotional anxiety may stimulate the limbic system, which then sends excitatory input to the inspiratory area, increasing the rate and depth of ventilation.

• *Temperature.* An increase in body temperature, as occurs during a fever or vigorous muscular exercise, increases the rate of respiration. A decrease in body temperature decreases respiratory rate. A sudden cold stimulus (such as plunging into cold water) causes temporary **apnea** (AP-nē-a; *a-* = without; *-pnea* = breath), an absence of breathing.

• *Pain.* A sudden, severe pain brings about brief apnea, but a prolonged somatic pain increases respiratory rate. Visceral pain may slow the rate of respiration.

• *Stretching the anal sphincter muscle.* This action increases the respiratory rate and is sometimes used to stimulate respiration in a newborn baby or a person who has stopped breathing.

• *Irritation of airways.* Physical or chemical irritation of the pharynx or larynx brings about an immediate cessation of breathing followed by coughing or sneezing.

• *Blood pressure.* The carotid and aortic baroreceptors that detect changes in blood pressure have a small effect on respiration. A sudden rise in blood pressure decreases the rate of respiration, and a drop in blood pressure increases the respiratory rate.

Table 23.3 summarizes the stimuli that affect the rate and depth of ventilation.

✔ CHECKPOINT

27. How does the medullary rhythmicity area regulate respiration?
28. How are the apneustic and pneumotaxic areas related to the control of respiration?
29. How do the cerebral cortex, levels of CO_2 and O_2, proprioceptors, inflation reflex, temperature changes, pain, and irritation of the airways modify respiration?

23.7 EXERCISE AND THE RESPIRATORY SYSTEM

◉ OBJECTIVE
• Describe the effects of exercise on the respiratory system.

The respiratory and cardiovascular systems make adjustments in response to both the intensity and duration of exercise. The effects of exercise on the heart are discussed in Chapter 20. Here we focus on how exercise affects the respiratory system.

Recall that the heart pumps the same amount of blood to the lungs as to all the rest of the body. Thus, as cardiac output rises, the blood flow to the lungs, termed **pulmonary perfusion,** increases as well. In addition, the O_2 **diffusing capacity,** a

TABLE 23.3

Summary of Stimuli That Affect Ventilation Rate and Depth

STIMULI THAT INCREASE VENTILATION RATE AND DEPTH	STIMULI THAT DECREASE VENTILATION RATE AND DEPTH
Voluntary hyperventilation controlled by cerebral cortex and anticipation of activity by stimulation of limbic system.	Voluntary hypoventilation controlled by cerebral cortex.
Increase in arterial blood P_{CO_2} above 40 mmHg (causes an increase in H^+) detected by peripheral and central chemoreceptors.	Decrease in arterial blood P_{CO_2} below 40 mmHg (causes a decrease in H^+) detected by peripheral and central chemoreceptors.
Decrease in arterial blood P_{O_2} from 105 mmHg to 50 mmHg.	Decrease in arterial blood P_{O_2} below 50 mmHg.
Increased activity of proprioceptors.	Decreased activity of proprioceptors.
Increase in body temperature.	Decrease in body temperature (decreases respiration rate), sudden cold stimulus (causes apnea).
Prolonged pain.	Severe pain (causes apnea).
Decrease in blood pressure.	Increase in blood pressure.
Stretching of anal sphincter.	Irritation of pharynx or larynx by touch or chemicals (causes brief apnea followed by coughing or sneezing).

measure of the rate at which O_2 can diffuse from alveolar air into the blood, may increase threefold during maximal exercise because more pulmonary capillaries become maximally perfused. As a result, there is a greater surface area available for diffusion of O_2 into pulmonary blood capillaries.

When muscles contract during exercise, they consume large amounts of O_2 and produce large amounts of CO_2. During vigorous exercise, O_2 consumption and pulmonary ventilation both increase dramatically. At the onset of exercise, an abrupt increase in pulmonary ventilation is followed by a more gradual increase. With moderate exercise, the increase is due mostly to an increase in the depth of ventilation rather than to increased breathing rate. When exercise is more strenuous, the frequency of breathing also increases.

The abrupt increase in ventilation at the start of exercise is due to *neural* changes that send excitatory impulses to the inspiratory area in the medulla oblongata. These changes include (1) anticipation of the activity, which stimulates the limbic system; (2) sensory impulses from proprioceptors in muscles, tendons, and joints; and (3) motor impulses from the primary motor cortex (precentral gyrus). The more gradual increase in ventilation during moderate exercise is due to *chemical* and *physical* changes in the bloodstream, including (1) slightly decreased P_{O_2}, due to increased O_2 consumption; (2) slightly increased P_{CO_2}, due to increased CO_2 production by contracting muscle fibers; and (3) increased temperature, due to liberation of more heat as more O_2 is utilized. During strenuous exercise, HCO_3^- buffers H^+ released by lactic acid in a reaction that liberates CO_2, which further increases P_{CO_2}.

At the end of an exercise session, an abrupt decrease in pulmonary ventilation is followed by a more gradual decline to the resting level. The initial decrease is due mainly to changes in neural factors when movement stops or slows; the more gradual phase reflects the slower return of blood chemistry levels and temperature to the resting state.

CLINICAL CONNECTION | Effects of Smoking on the Respiratory System

Smoking may cause a person to become easily "winded" during even moderate exercise because several factors decrease respiratory efficiency in smokers: (1) Nicotine constricts terminal bronchioles, which decreases airflow into and out of the lungs. (2) Carbon monoxide in smoke binds to hemoglobin and reduces its oxygen-carrying capability. (3) Irritants in smoke cause increased mucus secretion by the mucosa of the bronchial tree and swelling of the mucosal lining, both of which impede airflow into and out of the lungs. (4) Irritants in smoke also inhibit the movement of cilia and destroy cilia in the lining of the respiratory system. Thus, excess mucus and foreign debris are not easily removed, which further adds to the difficulty in breathing. The irritants can also convert the normal respiratory epithelium into stratified squamous epithelium, which lacks cilia and goblet cells. (5) With time, smoking leads to destruction of elastic fibers in the lungs and is the prime cause of emphysema (described in Disorders: Homeostatic Imbalances at the end of the chapter). These changes cause collapse of small bronchioles and trapping of air in alveoli at the end of exhalation. The result is less efficient gas exchange. •

✔**CHECKPOINT**

30. How does exercise affect the inspiratory area?

23.8 DEVELOPMENT OF THE RESPIRATORY SYSTEM

◉ **OBJECTIVE**

• Describe the development of the respiratory system.

The development of the mouth and pharynx are discussed in Chapter 24. Here we consider the development of the other structures of the respiratory system that you learned about in this chapter.

At about 4 weeks of development, the respiratory system begins as an outgrowth of the foregut (precursor of some digestive organs) just anterior to the pharynx. This outgrowth is called the **respiratory diverticulum** (dī-ver-TIK-ū-lum) or **lung bud** (Figure 23.28). The **endoderm** lining the respiratory diverticulum gives rise to the epithelium and glands of the trachea, bronchi, and alveoli. **Mesoderm** surrounding the respiratory diverticulum gives rise to the connective tissue, cartilage, and smooth muscle of these structures.

The epithelial lining of the *larynx* develops from the **endoderm** of the respiratory diverticulum; the cartilages and muscles originate from the **fourth** and **sixth pharyngeal arches,** swellings on the surface of the embryo.

As the respiratory diverticulum elongates, its distal end enlarges to form a globular **tracheal bud,** which gives rise to the *trachea.* Soon after, the tracheal bud divides into **bronchial buds,** which branch repeatedly and develop with the *bronchi.* By 24 weeks, 17 orders of branches have formed and *respiratory bronchioles* have developed.

During weeks 6 to 16, all major elements of the *lungs* have formed, except for those involved in gaseous exchange (respiratory bronchioles, alveolar ducts, and alveoli). Since respiration is not possible at this stage, fetuses born during this time cannot survive.

During weeks 16 to 26, lung tissue becomes highly vascular and respiratory bronchioles, alveolar ducts, and some primitive alveoli develop. Although it is possible for a fetus born near the end of this period to survive if given intensive care, death frequently occurs due to the immaturity of the respiratory and other systems.

From 26 weeks to birth, many more primitive alveoli develop; they consist of type I alveolar cells (main sites of gaseous exchange) and type II surfactant-producing cells. Blood capillaries also establish close contact with the primitive alveoli. Recall that surfactant is necessary to lower surface tension of alveolar fluid and thus reduce the tendency of alveoli to collapse on exhalation. Although surfactant production begins by 20 weeks, it is present in only small quantities. Amounts sufficient to permit survival of a premature (preterm) infant are not produced until 26 to 28 weeks' gestation. Infants born before 26 to 28 weeks are at high risk of respiratory distress syndrome (RDS), in which the alveoli collapse during exhalation and must be reinflated during inhalation (see Clinical Connection: Respiratory Distress Syndrome in Section 23.2).

At about 30 weeks, mature alveoli develop. However, it is estimated that only about one-sixth of the full complement of alveoli develop before birth; the remainder develop after birth during the first 8 years.

As the lungs develop, they acquire their *pleural sacs.* The *visceral pleura* and the *parietal pleura* develop from **mesoderm.** The space between the pleural layers is the *pleural cavity.*

During development, breathing movements of the fetus cause the aspiration of fluid into the lungs. This fluid is a mixture of amniotic fluid, mucus from bronchial glands, and surfactant. At birth, the lungs are about half-filled with fluid. When breathing begins at birth, most of the fluid is rapidly reabsorbed by blood and lymph capillaries and a small amount is expelled through the nose and mouth during delivery.

Figure 23.28 Development of the bronchial tubes and lungs.

🔒 The respiratory system develops from endoderm and mesoderm.

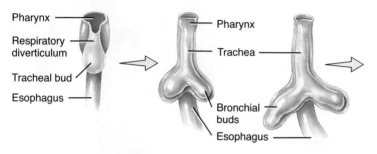

Pharynx
Respiratory diverticulum
Tracheal bud
Esophagus

Pharynx
Trachea
Bronchial buds
Esophagus

Fourth week

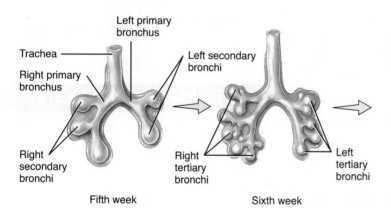

Left primary bronchus
Trachea
Right primary bronchus
Left secondary bronchi
Right secondary bronchi

Right tertiary bronchi
Left tertiary bronchi

Fifth week

Sixth week

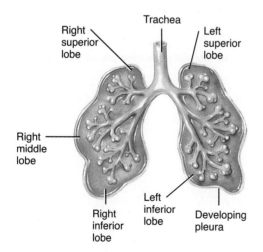

Trachea
Right superior lobe
Left superior lobe
Right middle lobe
Right inferior lobe
Left inferior lobe
Developing pleura

Eighth week

❓ **When does the respiratory system begin to develop in an embryo?**

✔ **CHECKPOINT**

31. What structures develop from the laryngotracheal bud?

BODY SYSTEM	CONTRIBUTION OF THE RESPIRATORY SYSTEM

For all body systems
Provides oxygen and removes carbon dioxide. Helps adjust pH of body fluids through exhalation of carbon dioxide.

Muscular system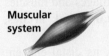
Increased rate and depth of breathing support increased activity of skeletal muscles during exercise.

Nervous system
Nose contains receptors for sense of smell (olfaction). Vibrations of air flowing across vocal folds produce sounds for speech.

Endocrine system
Angiotensin-converting enzyme (ACE) in lungs catalyzes formation of the hormone angiotensin II from angiotensin I.

Cardiovascular system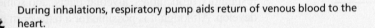
During inhalations, respiratory pump aids return of venous blood to the heart.

Lymphatic system and immunity
Hairs in nose, cilia and mucus in trachea, bronchi, and smaller airways, and alveolar macrophages contribute to nonspecific resistance to disease. Pharynx (throat) contains lymphatic tissue (tonsils). Respiratory pump (during inhalation) promotes flow of lymph.

Digestive system
Forceful contraction of respiratory muscles can assist in defecation.

Urinary system
Together, respiratory and urinary systems regulate pH of body fluids.

Reproductive systems
Increased rate and depth of breathing support activity during sexual intercourse. Internal respiration provides oxygen to developing fetus.

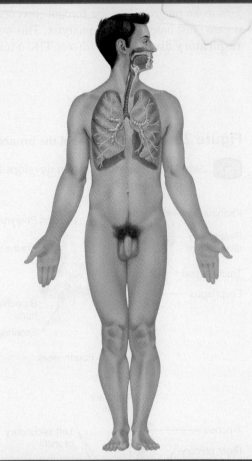

THE RESPIRATORY SYSTEM

23.9 AGING AND THE RESPIRATORY SYSTEM

⊙ **OBJECTIVE**

• Describe the effects of aging on the respiratory system.

With advancing age, the airways and tissues of the respiratory tract, including the alveoli, become less elastic and more rigid; the chest wall becomes more rigid as well. The result is a decrease in lung capacity. In fact, vital capacity (the maximum amount of air that can be expired after maximal inhalation) can decrease as much as 35% by age 70. A decrease in blood level of O_2, decreased activity of alveolar macrophages, and diminished ciliary action of the epithelium lining the respiratory tract occur. Because of these age-related factors, elderly people are more susceptible to pneumonia, bronchitis, emphysema, and other pulmonary disorders. Age-related changes in the structure and functions of the lung can also contribute to an older person's reduced ability to perform vigorous exercises, such as running.

✔ **CHECKPOINT**

32. What accounts for the decrease in lung capacity with aging?

• • •

To appreciate the many ways that the respiratory system contributes to homeostasis of other body systems, examine *Focus on Homeostasis: The Respiratory System.* Next, in Chapter 24, we will see how the digestive system makes nutrients available to body cells so that oxygen provided by the respiratory system can be used for ATP production.

DISORDERS: HOMEOSTATIC IMBALANCES

Asthma

Asthma (AZ-ma = panting) is a disorder characterized by chronic airway inflammation, airway hypersensitivity to a variety of stimuli, and airway obstruction. It is at least partially reversible, either spontaneously or with treatment. Asthma affects 3–5% of the U.S. population and is more common in children than in adults. Airway obstruction may be due to smooth muscle spasms in the walls of smaller bronchi and bronchioles, edema of the mucosa of the airways, increased mucus secretion, and/or damage to the epithelium of the airway.

Individuals with asthma typically react to concentrations of agents too low to cause symptoms in people without asthma. Sometimes the trigger is an allergen such as pollen, house dust mites, molds, or a particular food. Other common triggers of asthma attacks are emotional upset, aspirin, sulfiting agents (used in wine and beer and to keep greens fresh in salad bars), exercise, and breathing cold air or cigarette smoke. In the early phase (acute) response, smooth muscle spasm is accompanied by excessive secretion of mucus that may clog the bronchi and bronchioles and worsen the attack. The late phase (chronic) response is characterized by inflammation, fibrosis, edema, and necrosis (death) of bronchial epithelial cells. A host of mediator chemicals, including leukotrienes, prostaglandins, thromboxane, platelet-activating factor, and histamine, take part.

Symptoms include difficult breathing, coughing, wheezing, chest tightness, tachycardia, fatigue, moist skin, and anxiety. An acute attack is treated by giving an inhaled beta₂-adrenergic agonist (albuterol) to help relax smooth muscle in the bronchioles and open up the airways. This drug mimics the effect of sympathetic stimulation, that is, it causes bronchodilation. However, long-term therapy of asthma strives to suppress the underlying inflammation. The anti-inflammatory drugs that are used most often are inhaled corticosteroids (glucocorticoids), cromolyn sodium (Intal®), and leukotriene blockers (Accolate®).

Chronic Obstructive Pulmonary Disease

Chronic obstructive pulmonary disease (COPD) is a type of respiratory disorder characterized by chronic and recurrent obstruction of airflow, which increases airway resistance. COPD affects about 30 million Americans and is the fourth leading cause of death behind heart disease, cancer, and cerebrovascular disease. The principal types of COPD are emphysema and chronic bronchitis. In most cases, COPD is preventable because its most common cause is cigarette smoking or breathing secondhand smoke. Other causes include air pollution, pulmonary infection, occupational exposure to dusts and gases, and genetic factors. Because men, on average, have more years of exposure to cigarette smoke than women, they are twice as likely as women to suffer from COPD; still, the incidence of COPD in women has risen sixfold in the past 50 years, a reflection of increased smoking among women.

Emphysema

Emphysema (em-fi-SĒ-ma = blown up or full of air) is a disorder characterized by destruction of the walls of the alveoli, producing abnormally large air spaces that remain filled with air during exhalation. With less surface area for gas exchange, O_2 diffusion across the damaged respiratory membrane is reduced. Blood O_2 level is somewhat lowered, and any mild exercise that raises the O_2 requirements of the cells leaves the patient breathless. As increasing numbers of alveolar walls are damaged, lung elastic recoil decreases due to loss of elastic fibers, and an increasing amount of air becomes trapped in the lungs at the end of exhalation. Over several years, added exertion during inhalation increases the size of the chest cage, resulting in a "barrel chest."

Emphysema is generally caused by a long-term irritation; cigarette smoke, air pollution, and occupational exposure to industrial dust are the most common irritants. Some destruction of alveolar sacs may be caused by an enzyme imbalance. Treatment consists of cessation of smoking, removal of other environmental irritants, exercise training under careful medical supervision, breathing exercises, use of bronchodilators, and oxygen therapy.

Chronic Bronchitis

Chronic bronchitis is a disorder characterized by excessive secretion of bronchial mucus accompanied by a productive cough (sputum is raised) that lasts for at least 3 months of the year for two successive years. Cigarette smoking is the leading cause of chronic bronchitis. Inhaled irritants lead to chronic inflammation with an increase in the size and number of mucous glands and goblet cells in the airway epithelium. The thickened and excessive mucus produced narrows the airway and impairs ciliary function. Thus, inhaled pathogens become embedded in airway secretions and multiply rapidly. Besides a productive cough, symptoms of chronic bronchitis are shortness of breath, wheezing, cyanosis, and pulmonary hypertension. Treatment for chronic bronchitis is similar to that for emphysema.

Lung Cancer

In the United States, **lung cancer** is the leading cause of cancer death in both males and females, accounting for 160,000 deaths annually. At the time of diagnosis, lung cancer is usually well advanced, with distant metastases present in about 55% of patients, and regional lymph node involvement in an additional 25%. Most people with lung cancer die within a year of the initial diagnosis; the overall survival rate is only 10–15%. Cigarette smoke is the most common cause of lung cancer. Roughly 85% of lung cancer cases are related to smoking, and the disease is 10 to 30 times more common in smokers than nonsmokers. Exposure to secondhand smoke is also associated with lung cancer and heart disease. In the United States, secondhand smoke causes an estimated 4000 deaths a year from lung cancer, and nearly 40,000 deaths a year from heart disease. Other causes of lung cancer are ionizing radiation and inhaled irritants, such as asbestos and radon gas. Emphysema is a common precursor to the development of lung cancer.

The most common type of lung cancer, **bronchogenic carcinoma** (brong'-kō-JEN-ik), starts in the epithelium of the bronchial tubes. Bronchogenic tumors are named based on where they arise. For example, *adenocarcinomas* (ad-en-ō-kar-si-NŌ-mas; *adeno-* = gland) develop in peripheral areas of the lungs from bronchial glands and alveolar cells, *squamous cell carcinomas* develop from the squamous cells in the epithelium of larger bronchial tubes, and *small (oat) cell carcinomas* develop from epithelial cells in primary bronchi near the hilus of the lungs that get their name due to their flat cell shape with little cytoplasm. They tend to involve the mediastinum early on. Depending on the type, bronchogenic tumors may be aggressive, locally invasive, and undergo widespread metastasis. The tumors begin as epithelial lesions that grow to form masses that obstruct the bronchial tubes or invade adjacent lung tissue. Bronchogenic carcinomas metastasize to lymph nodes, the brain, bones, liver, and other organs.

Symptoms of lung cancer are related to the location of the tumor. These may include a chronic cough, spitting blood from the respiratory tract, wheezing, shortness of breath, chest pain, hoarseness, difficulty swallowing, weight loss, anorexia, fatigue, bone pain, confusion, problems with balance, headache, anemia, thrombocytopenia, and jaundice.

Treatment consists of partial or complete surgical removal of a diseased lung (pulmonectomy), radiation therapy, and chemotherapy.

Pneumonia

Pneumonia (noo-MŌ-ne-a) is an acute infection or inflammation of the alveoli. It is the most common infectious cause of death in the United States, where an estimated 4 million cases occur annually. When certain microbes enter the lungs of susceptible individuals, they release damaging toxins, stimulating inflammation and immune responses that have damaging side effects. The toxins and immune response damage alveoli and bronchial mucous membranes; inflammation and edema cause the alveoli to fill with fluid, interfering with ventilation and gas exchange.

The most common cause of pneumonia is the pneumococcal bacterium *Streptococcus pneumoniae* (strep'-tō-KOK-us noo-MŌ-ne-i), but other microbes may also cause pneumonia. Those who are most susceptible to pneumonia are the elderly, infants, immunocompromised individuals (AIDS or cancer patients, or those taking immunosuppressive drugs), cigarette smokers, and individuals with an obstructive lung disease. Most cases of pneumonia are preceded by an upper respiratory infection that often is viral. Individuals then develop fever, chills, productive or dry cough, malaise, chest pain, and sometimes dyspnea (difficult breathing) and hemoptysis (spitting blood).

Treatment may involve antibiotics, bronchodilators, oxygen therapy, increased fluid intake, and chest physiotherapy (percussion, vibration, and postural drainage).

Tuberculosis

The bacterium *Mycobacterium tuberculosis* (mī'-kō-bak'-TI-rē-um) produces an infectious, communicable disease called **tuberculosis (TB)** that most often affects the lungs and the pleurae but may involve other parts of the body. Once the bacteria are inside the lungs, they multiply and cause inflammation, which stimulates neutrophils and macrophages to migrate to the area and engulf the bacteria to prevent their spread. If the immune system is not impaired, the bacteria remain dormant for life, but impaired immunity may enable the bacteria to escape into blood and lymph to infect other organs. In many people, symptoms—fatigue, weight loss, lethargy, anorexia, a low-grade fever, night sweats, cough, dyspnea, chest pain, and hemoptysis—do not develop until the disease is advanced.

During the past several years, the incidence of TB in the United States has risen dramatically. Perhaps the single most important factor related to this increase is the spread of the human immunodeficiency virus (HIV). People infected with HIV are much more likely to develop tuberculosis because their immune systems are impaired. Among the other factors that have contributed to the increased number of cases are homelessness, increased drug abuse, increased immigration from countries with a high prevalence of tuberculosis, increased crowding in housing among the poor, and airborne transmission of tuberculosis in prisons and shelters. In addition, recent outbreaks of tuberculosis involving multi-drug-resistant strains of *Mycobacterium tuberculosis* have occurred because patients fail to complete their antibiotic and other treatment regimens. TB is treated with the medication isoniazid.

Pulmonary Edema

Pulmonary edema is an abnormal accumulation of fluid in the interstitial spaces and alveoli of the lungs. The edema may arise from increased permeability of the pulmonary capillaries (pulmonary origin) or increased pressure in the pulmonary capillaries (cardiac origin); the latter cause may coincide with congestive heart failure. The most common symptom is dyspnea. Others include wheezing, tachypnea (rapid breathing rate), restlessness, a feeling of suffocation, cyanosis, pallor (paleness), diaphoresis (excessive perspiration), and pulmonary hypertension. Treatment consists of administering oxygen, drugs that dilate the bronchioles and lower blood pressure, diuretics to rid the body of excess fluid, and drugs that correct acid–base imbalance; suctioning of airways; and mechanical ventilation. One of the recent culprits in the development of pulmonary edema was found to be "phen-fen" diet pills.

Asbestos-Related Diseases

Asbestos-related diseases are serious lung disorders that develop as a result of inhaling asbestos particles decades earlier. When asbestos particles are inhaled, they penetrate lung tissue. In response, white blood cells attempt to destroy them by phagocytosis. However, the fibers usually destroy the white blood cells and scarring of lung tissue may follow. Asbestos-related diseases include **asbestosis** (as'-bes-TŌ-sis) (widespread scarring of lung tissue), **diffuse pleural thickening** (thickening of the pleurae), and **mesothelioma** (mez'-o-thē-lē-Ō-ma) (cancer of the pleurae or, less commonly, the peritoneum).

Sudden Infant Death Syndrome

Sudden infant death syndrome (SIDS) is the sudden, unexpected death of an apparently healthy infant during sleep. It rarely occurs before 2 weeks or after 6 months of age, with the peak incidence between the second and fourth months. SIDS is more common in premature infants, male babies, low-birth-weight babies, babies of drug users or smokers, babies who have stopped breathing and have had to be resuscitated, babies with upper respiratory tract infections, and babies who have had a sibling die of SIDS. African-American and Native American babies are at higher risk. The exact cause of SIDS is unknown. However, it may be due to an abnormality in the mechanisms that control respiration or low levels of oxygen in the blood. SIDS may also be linked to hypoxia while sleeping in a prone position (on the stomach) and the rebreathing of exhaled air trapped in a depression of a mattress. It is recommended that for the first 6 months infants be placed on their backs for sleeping ("back to sleep").

Severe Acute Respiratory Syndrome

Severe acute respiratory syndrome (SARS) is an example of an *emerging infectious disease,* that is, a disease that is new or changing. Other examples of emerging infectious diseases are West Nile encephalitis, mad cow disease, and AIDS. SARS first appeared in southern China in late 2002 and has subsequently spread worldwide. It is a respiratory illness caused by a new variety of coronavirus. Symptoms of SARS include fever, malaise, muscle aches, nonproductive (dry) cough, difficulty in breathing, chills, headache, and diarrhea. About 10–20% of patients require mechanical ventilation and in some cases death may result. The disease is primarily spread through person-to-person contact. There is no effective treatment for SARS and the death rate is 5–10%, usually among the elderly and in persons with other medical problems.

MEDICAL TERMINOLOGY

Abdominal thrust maneuver First-aid procedure designed to clear the airways of obstructing objects. It is performed by applying a quick upward thrust between the navel and costal margin that causes sudden elevation of the diaphragm and forceful, rapid expulsion of air in the lungs; this action forces air out the trachea to eject the obstructing object. The abdominal thrust maneuver is also used to expel water from the lungs of near-drowning victims before resuscitation is begun.

Asphyxia (as-FIK-sē-a; *sphyxia* = pulse) Oxygen starvation due to low atmospheric oxygen or interference with ventilation, external respiration, or internal respiration.

Aspiration (as'-pi-RĀ-shun) Inhalation of a foreign substance such as water, food, or a foreign body into the bronchial tree; also, the drawing of a substance in or out by suction.

Avian influenza A respiratory disorder that has resulted in the deaths of hundreds of millions of birds worldwide. It is usually transmitted from one bird to another bird through their droppings, saliva, and nasal secretions. Currently, avian influenza is difficult to transmit from birds to humans; the few humans who have died from avian influenza have had close contact with infected birds. Also called **bird flu.**

Black lung disease A condition in which the lungs appear black instead of pink due to inhalation of coal dust over a period of many years. Most often it affects people who work in the coal industry.

Bronchiectasis (bron'-kē-EK-ta-sis; *-ektasis* = stretching) A chronic dilation of the bronchi or bronchioles resulting from damage to the bronchial wall, for example, from respiratory infections.

Bronchography (bron-KOG-ra-fē) An imaging technique used to visualize the bronchial tree using x-rays. After an opaque contrast medium is inhaled through an intratracheal catheter, radiographs of the chest in various positions are taken, and the developed film, a **bronchogram** (BRON-kō-gram), provides a picture of the bronchial tree.

Bronchoscopy (bron-KOS-kō-pē) Visual examination of the bronchi through a **bronchoscope,** an illuminated, flexible tubular instrument that is passed through the mouth (or nose), larynx, and trachea into the bronchi. The examiner can view the interior of the trachea and bronchi to biopsy a tumor, clear an obstructing object or secretions from an airway, take cultures or smears for microscopic examination, stop bleeding, or deliver drugs.

Cheyne–Stokes respiration (CHĀN STŌKS res'-pi-RĀ-shun) A repeated cycle of irregular breathing that begins with shallow breaths that increase in depth and rapidity and then decrease and cease altogether for 15 to 20 seconds. Cheyne–Stokes is normal in infants; it is also often seen just before death from pulmonary, cerebral, cardiac, and kidney disease.

Dyspnea (DISP-nē-a; *dys-* = painful, difficult) Painful or labored breathing.

Epistaxis (ep'-i-STAK-sis) Loss of blood from the nose due to trauma, infection, allergy, malignant growths, or bleeding disorders. It can be arrested by cautery with silver nitrate, electrocautery, or firm packing. Also called **nosebleed.**

Hypoventilation (*hypo-* = below) Slow and shallow breathing.

Mechanical ventilation The use of an automatically cycling device (ventilator or respirator) to assist breathing. A plastic tube is inserted into the nose or mouth and the tube is attached to a device that forces air into the lungs. Exhalation occurs passively due to the elastic recoil of the lungs.

Rales (RĀLS) Sounds sometimes heard in the lungs that resemble bubbling or rattling. Rales are to the lungs what murmurs are to the heart. Different types are due to the presence of an abnormal type or amount of fluid or mucus within the bronchi or alveoli, or to bronchoconstriction that causes turbulent airflow.

Respirator (RES-pi-rā'-tor) An apparatus fitted to a mask over the nose and mouth, or hooked directly to an endotracheal or tracheotomy tube, that is used to assist or support ventilation or to provide nebulized medication to the air passages.

Respiratory failure A condition in which the respiratory system either cannot supply sufficient O_2 to maintain metabolism or cannot eliminate enough CO_2 to prevent respiratory acidosis (a lower-than-normal pH in interstitial fluid).

Rhinitis (rī-NĪ-tis; *rhin-* = nose) Chronic or acute inflammation of the mucous membrane of the nose due to viruses, bacteria, or irritants. Excessive mucus production produces a runny nose, nasal congestion, and postnasal drip.

Sleep apnea (AP-nē-a; *a-* = without; *-pnea* = breath) A disorder in which a person repeatedly stops breathing for 10 or more seconds while sleeping. Most often, it occurs because loss of muscle tone in pharyngeal muscles allows the airway to collapse.

Sputum (SPŪ-tum = to spit) Mucus and other fluids from the air passages that is expectorated (expelled by coughing).

Strep throat Inflammation of the pharynx caused by the bacterium *Streptococcus pyogenes*. It may also involve the tonsils and middle ear.

Tachypnea (tak′-ip-NĒ-a; *tachy-* = rapid; *-pnea* = breath) Rapid breathing rate.

Wheeze (HWĒZ) A whistling, squeaking, or musical high-pitched sound during breathing resulting from a partially obstructed airway.

CHAPTER REVIEW AND RESOURCE SUMMARY

Review

23.1 Respiratory System Anatomy

1. The respiratory system consists of the nose, pharynx, larynx, trachea, bronchi, and lungs. They act with the cardiovascular system to supply oxygen (O_2) and remove carbon dioxide (CO_2) from the blood.
2. The external portion of the nose is made of cartilage and skin and is lined with a mucous membrane. Openings to the exterior are the external nares. The internal portion of the nose communicates with the paranasal sinuses and nasopharynx through the internal nares. The nasal cavity is divided by a septum. The anterior portion of the cavity is called the vestibule. The nose warms, moistens, and filters air and functions in olfaction and speech.
3. The pharynx (throat) is a muscular tube lined by a mucous membrane. The anatomic regions are the nasopharynx, oropharynx, and laryngopharynx. The nasopharynx functions in respiration. The oropharynx and laryngopharynx function both in digestion and in respiration.
4. The larynx (voice box) is a passageway that connects the pharynx with the trachea. It contains the thyroid cartilage (Adam's apple); the epiglottis, which prevents food from entering the larynx; the cricoid cartilage, which connects the larynx and trachea; and the paired arytenoid, corniculate, and cuneiform cartilages. The larynx contains vocal folds, which produce sound as they vibrate. Taut folds produce high pitches, and relaxed ones produce low pitches.
5. The trachea (windpipe) extends from the larynx to the primary bronchi. It is composed of C-shaped rings of cartilage and smooth muscle and is lined with pseudostratified ciliated columnar epithelium.
6. The bronchial tree consists of the trachea, primary bronchi, secondary bronchi, tertiary bronchi, bronchioles, and terminal bronchioles. Walls of bronchi contain rings of cartilage; walls of bronchioles contain increasingly smaller plates of cartilage and increasing amounts of smooth muscle.
7. Lungs are paired organs in the thoracic cavity enclosed by the pleural membrane. The parietal pleura is the superficial layer that lines the thoracic cavity; the visceral pleura is the deep layer that covers the lungs. The right lung has three lobes separated by two fissures; the left lung has two lobes separated by one fissure and a depression, the cardiac notch.
8. Secondary bronchi give rise to branches called segmental bronchi, which supply segments of lung tissue called bronchopulmonary segments. Each bronchopulmonary segment consists of lobules, which contain lymphatics, arterioles, venules, terminal bronchioles, respiratory bronchioles, alveolar ducts, alveolar sacs, and alveoli.
9. Alveolar walls consist of type I alveolar cells, type II alveolar cells, and associated alveolar macrophages.
10. Gas exchange occurs across the respiratory membranes.

23.2 Pulmonary Ventilation

1. Pulmonary ventilation, or breathing, consists of inhalation and exhalation.
2. The movement of air into and out of the lungs depends on pressure changes governed in part by Boyle's law, which states that the volume of a gas varies inversely with pressure, assuming that temperature remains constant.
3. Inhalation occurs when alveolar pressure falls below atmospheric pressure. Contraction of the diaphragm and external intercostals increases the size of the thorax, thereby decreasing the intrapleural pressure so that the lungs expand. Expansion of the lungs decreases alveolar pressure so that air moves down a pressure gradient from the atmosphere into the lungs.
4. During forceful inhalation, accessory muscles of inhalation (sternocleidomastoids, scalenes, and pectoralis minors) are also used.
5. Exhalation occurs when alveolar pressure is higher than atmospheric pressure. Relaxation of the diaphragm and external intercostals results in elastic recoil of the chest wall and lungs, which increases intrapleural pressure; lung volume decreases and alveolar pressure increases, so air moves from the lungs to the atmosphere.

Resource

WILEY PLUS

Anatomy Overview - Overview of Respiratory Organs
Anatomy Overview - Respiratory Tissues
Figure 23.4 - The Larynx
Figure 23.8 - Relationship of the Pleural Membranes to the Lungs
Figure 23.11 - Structural Components of an Alveolus
Exercise - Concentrate on Respiratory Structures
Exercise - Paint the Lung
Concepts and Connections - Functional Anatomy of the Respiratory System

Animation - Pulmonary Ventilation
Exercise - The Airway
Exercise - Build an Airway
Concepts and Connections - Ventilation

Review	Resource

6. Forceful exhalation involves contraction of the internal intercostal and abdominal muscles.
7. The surface tension exerted by alveolar fluid is decreased by the presence of surfactant.
8. Compliance is the ease with which the lungs and thoracic wall can expand.
9. The walls of the airways offer some resistance to breathing.
10. Normal quiet breathing is termed eupnea; other patterns are costal breathing and diaphragmatic breathing. Modified respiratory movements, such as coughing, sneezing, sighing, yawning, sobbing, crying, laughing, and hiccupping, are used to express emotions and to clear the airways. (See Table 23.2.)

23.3 Lung Volumes and Capacities

1. Lung volumes exchanged during breathing and the rate of respiration are measured with a spirometer.
2. Lung volumes measured by spirometry include tidal volume, minute ventilation, alveolar ventilation rate, inspiratory reserve volume, expiratory reserve volume, and $FEV_{1.0}$. Other lung volumes include anatomic dead space, residual volume, and minimal volume.
3. Lung capacities, the sum of two or more lung volumes, include inspiratory, functional, residual, vital, and total lung capacities.

Animation - Regulation of Ventilation: Basic Rhythms

23.4 Exchange of Oxygen and Carbon Dioxide

1. The partial pressure of a gas is the pressure exerted by that gas in a mixture of gases. It is symbolized by P_x, where the subscript is the chemical formula of the gas.
2. According to Dalton's law, each gas in a mixture of gases exerts its own pressure as if all the other gases were not present.
3. Henry's law states that the quantity of a gas that will dissolve in a liquid is proportional to the partial pressure of the gas and its solubility (given that the temperature remains constant).
4. In internal and external respiration, O_2 and CO_2 diffuse from areas of higher partial pressures to areas of lower partial pressures.
5. External respiration or pulmonary gas exchange is the exchange of gases between alveoli and pulmonary blood capillaries. It depends on partial pressure differences, a large surface area for gas exchange, a small diffusion distance across the respiratory membrane, and the rate of airflow into and out of the lungs.
6. Internal respiration or systemic gas exchange is the exchange of gases between systemic blood capillaries and tissue cells.

Animation - Gas Exchange Introduction
Animation - Gas Exchange: Internal and External Respiration
Figure 23.17 - Changes in Partial Pressures of Oxygen and Carbon Dioxide
Exercise - Gas Exchange Match-Up

23.5 Transport of Oxygen and Carbon Dioxide

1. In each 100 mL of oxygenated blood, 1.5% of the O_2 is dissolved in blood plasma and 98.5% is bound to hemoglobin as oxyhemoglobin (Hb—O_2).
2. The binding of O_2 to hemoglobin is affected by P_{O_2}, acidity (pH), P_{CO_2}, temperature, and 2,3-bisphosphoglycerate (BPG).
3. Fetal hemoglobin differs from adult hemoglobin in structure and has a higher affinity for O_2.
4. In each 100 mL of deoxygenated blood, 7% of CO_2 is dissolved in blood plasma, 23% combines with hemoglobin as carbaminohemoglobin (Hb—CO_2), and 70% is converted to bicarbonate ions (HCO_3^-).
5. In an acidic environment, hemoglobin's affinity for O_2 is lower, and O_2 dissociates more readily from it (Bohr effect).
6. In the presence of O_2, less CO_2 binds to hemoglobin (Haldane effect).

Animation - Gas Transport
Figure 23.23 - Summary of Chemical Reactions That Occur during Gas Exchange
Exercise - Carbon Dioxide Transport Try-Out
Exercise - Oxygen Transport Try-Out
Exercise - Concentrate on Respiration
Concepts and Connections - Carbon Dioxide Transport
Concepts and Connections - Oxygen Transport

23.6 Control of Respiration

1. The respiratory center consists of a medullary rhythmicity area in the medulla oblongata and a pneumotaxic area and an apneustic area in the pons.
2. The inspiratory area sets the basic rhythm of respiration.
3. The pneumotaxic and apneustic areas coordinate the transition between inhalation and exhalation.
4. Respirations may be modified by a number of factors, including cortical influences; the inflation reflex; chemical stimuli, such as O_2 and CO_2 and H^+ levels; proprioceptor input; blood pressure changes; limbic system stimulation; temperature; pain; and irritation to the airways. (See Table 23.3.)

Anatomy Overview - Respiratory Control Center
Animation - Regulation of Ventilation
Animation - Control of Ventilation Rate and Blood Chemistry
Animation - Role of Respiratory System in pH Regulation
Exercise - Respiration and pH Reflex
Concepts and Connections - Ventilation

23.7 Exercise and the Respiratory System

1. The rate and depth of ventilation change in response to both the intensity and duration of exercise.
2. An increase in pulmonary perfusion and O_2-diffusing capacity occurs during exercise.
3. The abrupt increase in ventilation at the start of exercise is due to neural changes that send excitatory

Review   **Resource**

impulses to the inspiratory area in the medulla oblongata. The more gradual increase in ventilation during moderate exercise is due to chemical and physical changes in the bloodstream.

23.8 Development of the Respiratory System

1. The respiratory system begins as an outgrowth of endoderm called the respiratory diverticulum.
2. Smooth muscle, cartilage, and connective tissue of the bronchial tubes and pleural sacs develop from mesoderm.

23.9 Aging and the Respiratory System

1. Aging results in decreased vital capacity, decreased blood level of O_2, and diminished alveolar macrophage activity.
2. Elderly people are more susceptible to pneumonia, emphysema, bronchitis, and other pulmonary disorders.

SELF-QUIZ QUESTIONS

Fill in the blanks in the following statements.

1. Oxygen in blood is carried primarily in the form of _____; carbon dioxide is carried as _____, _____, and _____.

2. Write the equation for the chemical reaction that occurs for the transport of carbon dioxide as bicarbonate ions in blood: _____.

Indicate whether the following statements are true or false.

3. The three basic steps of respiration are pulmonary ventilation, external respiration, and cellular respiration.

4. For inhalation to occur, air pressure in the alveoli must be less than atmospheric pressure; for exhalation to occur, air pressure in the alveoli must be greater than atmospheric pressure.

Choose the one best answer to the following questions.

5. What structural changes occur from primary bronchi to terminal bronchioles? (1) The mucous membrane changes from pseudostratified ciliated columnar epithelium to nonciliated simple cuboidal epithelium. (2) The number of goblet cells increases. (3) The amount of smooth muscle increases. (4) Incomplete rings of cartilage disappear. (5) The amount of branching decreases.
   (a) 1, 2, 3, 4, and 5   (b) 2, 3, and 4   (c) 1, 3, and 4
   (d) 1, 3, 4, and 5   (e) 1, 2, 3, and 4

6. Which of the following would cause oxygen to dissociate more readily from hemoglobin? (1) low P_{O_2}, (2) an increase in H^+ in blood, (3) hypercapnia, (4) hypothermia, (5) low levels of BPG (2,3-bisphosphoglycerate).
   (a) 1 and 2   (b) 2, 3, and 4   (c) 1, 2, 3, and 5
   (d) 1, 3, and 5   (e) 1, 2, and 3

7. Which of the following statements are *correct?* (1) Normal exhalation during quiet breathing is an active process involving intensive muscle contraction. (2) Passive exhalation results from elastic recoil of the chest wall and lungs. (3) Air flow during breathing is due to a pressure gradient between the lungs and the atmospheric air. (4) During normal breathing, the pressure between the two pleural layers (intrapleural pressure) is always subatmospheric. (5) Surface tension of alveolar fluid facilitates inhalation.
   (a) 1, 2, and 3   (b) 2, 3, and 4   (c) 3, 4, and 5
   (d) 1, 3, and 5   (e) 2, 3, and 5

8. Which of the following factors affect the rate of external respiration? (1) partial pressure differences of the gases, (2) surface area for gas exchange, (3) diffusion distance, (4) solubility and molecular weight of the gases, (5) presence of bisphosphoglycerate (BPG).
   (a) 1, 2, and 3   (b) 2, 4, and 5   (c) 1, 2, 4, and 5
   (d) 1, 2, 3, and 4   (e) 2, 3, 4, and 5

9. The most important factor in determining the percent oxygen saturation of hemoglobin is
   (a) the partial pressure of oxygen.
   (b) acidity.
   (c) the partial pressure of carbon dioxide.
   (d) temperature.
   (e) BPG.

10. Which of the following statements are *true?* (1) Central chemoreceptors are stimulated by changes in P_{CO_2}, H^+, and P_{O_2}. (2) Respiratory rate increases during the initial onset of exercise due to input to the inspiratory area from proprioceptors. (3) When baroreceptors in the lungs are stimulated, the expiratory area is activated. (4) Stimulation of the limbic system can result in excitation of the inspiratory area. (5) Sudden severe pain causes brief apnea, while prolonged somatic pain causes an increase in respiratory rate. (6) The respiratory rate increases during fever.
    (a) 1, 2, 3, and 6   (b) 1, 4, and 5   (c) 1, 2, 4, 5, and 6
    (d) 2, 3, 4, 5, and 6   (e) 2, 4, 5, and 6

11. Place the steps for normal inhalation in order.
    (a) decrease in intrapleural pressure to 754 mmHg
    (b) increase in the size of the thoracic cavity
    (c) flow of air from higher to lower pressure
    (d) outward pull of pleurae, resulting in lung expansion
    (e) stimulation of primary breathing muscles by phrenic and intercostal nerves
    (f) decrease in alveolar pressure to 758 mmHg
    (g) contraction of the diaphragm and external intercostals
    (h) increase in the volume of the pleural cavity

12. Match the following:

____ (a) total volume of air inhaled and exhaled each minute

____ (b) tidal volume + inspiratory reserve volume + expiratory reserve volume

____ (c) additional amount of air inhaled beyond tidal volume when taking a very deep breath

____ (d) residual volume + expiratory reserve volume

____ (e) amount of air remaining in lungs after expiratory reserve volume is expelled

____ (f) tidal volume + inspiratory reserve volume

____ (g) vital capacity + residual volume

____ (h) volume of air in one breath

____ (i) amount of air exhaled in forced exhalation following a normal exhalation

____ (j) provides a medical and legal tool for determining if a baby was born dead or died after birth

(1) tidal volume
(2) residual volume
(3) minute ventilation
(4) expiratory reserve volume
(5) inspiratory reserve volume
(6) minimal volume
(7) inspiratory capacity
(8) vital capacity
(9) functional residual capacity
(10) total lung capacity

13. Match the following:

____ (a) functions as a passageway for air and food, provides a resonating chamber for speech sounds, and houses the tonsils

____ (b) site of external respiration

____ (c) connects the laryngopharynx with the trachea; houses the vocal cords

____ (d) serous membrane that surrounds the lungs

____ (e) functions in warming, moistening, and filtering air; receives olfactory stimuli; is a resonating chamber for sound

____ (f) simple squamous epithelial cells that form a continuous lining of the alveolar wall; sites of gas exchange

____ (g) forms anterior wall of the larynx

____ (h) a tubular passageway for air connecting the larynx to the bronchi

____ (i) secrete alveolar fluid and surfactant

____ (j) forms inferior wall of larynx; landmark for tracheotomy

____ (k) prevents food or fluid from entering the airways

____ (l) air passageways entering the lungs

____ (m) ridge covered by a sensitive mucous membrane; irritation triggers cough reflex

(1) nose
(2) pharynx
(3) larynx
(4) epiglottis
(5) trachea
(6) bronchi
(7) carina
(8) cricoid cartilage
(9) pleura
(10) thyroid cartilage
(11) alveoli
(12) type I alveolar cells
(13) type II alveolar cells

14. Match the following:

____ (a) a deficiency of oxygen at the tissue level

____ (b) above-normal partial pressure of carbon dioxide

____ (c) normal quiet breathing

____ (d) deep, abdominal breathing

____ (e) the ease with which the lungs and thoracic wall can be expanded

____ (f) hypoxia-induced vasoconstriction to divert pulmonary blood from poorly ventilated to well-ventilated regions of the lungs

____ (g) absence of breathing

____ (h) rapid and deep breathing

____ (i) shallow, chest breathing

(1) eupnea
(2) apnea
(3) hyperventilation
(4) costal breathing
(5) diaphragmatic breathing
(6) compliance
(7) hypoxia
(8) hypercapnia
(9) ventilation--- perfusion coupling

15. Match the following:

____ (a) prevents excessive inflation of the lungs

____ (b) the lower the amount of oxyhemoglobin, the higher the carbon dioxide–carrying capacity of the blood

____ (c) controls the basic rhythm of respiration

____ (d) active during normal inhalation; sends nerve impulses to external intercostals and diaphragm

____ (e) sends stimulatory impulses to the inspiratory area that activate it and prolong inhalation

____ (f) as acidity increases, the affinity of hemoglobin for oxygen decreases and oxygen dissociates more readily from hemoglobin; shifts oxygen-dissociation curve to the right

____ (g) active during forceful exhalation

____ (h) pressure of a gas in a closed container is inversely proportional to the volume of the container

____ (i) transmits inhibitory impulses to turn off the inspiratory area before the lungs become too full of air

____ (j) the quantity of a gas that dissolves in a liquid is proportional to the partial pressure of the gas and its solubility

____ (k) relates to the partial pressure of a gas in a mixture of gases whereby each gas in a mixture exerts its own pressure as if all the other gases were not present

(1) Bohr effect
(2) Dalton's law
(3) medullary rhythmicity area
(4) inspiratory area
(5) expiratory area
(6) apneustic area
(7) pneumotaxic area
(8) Henry's law
(9) inflation (Hering–Breuer) reflex
(10) Boyle's law
(11) Haldane effect

CRITICAL THINKING QUESTIONS

1. Aretha loves to sing. Right now she has a cold, a severely runny nose, and a "sore throat" that is affecting her ability to sing and talk. What structures are involved and how are they affected by her cold?

2. Ms. Brown has smoked cigarettes for years and is having breathing difficulties. She has been diagnosed with emphysema. Describe specific kinds of structural changes you would expect to observe in Mrs. Brown's respiratory system. How are air flow and gas exchange affected by these structural changes?

3. The Robinson family went to bed one frigid winter night and were found deceased the next day. A squirrel's nest was found in their chimney. What happened to the Robinsons?

? ANSWERS TO FIGURE QUESTIONS

23.1 The conducting zone of the respiratory system includes the nose, pharynx, larynx, trachea, bronchi, and bronchioles (except the respiratory bronchioles).

23.2 The path of air is external nares → vestibule → nasal cavity → internal nares.

23.3 The root of the nose attaches it to the frontal bone.

23.4 During swallowing, the epiglottis closes over the rima glottidis, the entrance to the trachea, to prevent aspiration of food and liquids into the lungs.

23.5 The main function of the vocal folds is voice production.

23.6 Because the tissues between the esophagus and trachea are soft, the esophagus can bulge and press against the trachea during swallowing.

23.7 The left lung has two lobes and two secondary bronchi; the right lung has three of each.

23.8 The pleural membrane is a serous membrane.

23.9 Because two-thirds of the heart lies to the left of the midline, the left lung contains a cardiac notch to accommodate the presence of the heart. The right lung is shorter than the left because the diaphragm is higher on the right side to accommodate the liver.

23.10 The wall of an alveolus is made up of type I alveolar cells, type II alveolar cells, and associated alveolar macrophages.

23.11 The respiratory membrane averages 0.5 μm in thickness.

23.12 The pressure would increase fourfold, to 4 atm.

23.13 If you are at rest while reading, your diaphragm is responsible for about 75% of each inhalation.

23.14 At the start of inhalation, intrapleural pressure is about 756 mmHg. With contraction of the diaphragm, it decreases to about 754 mmHg as the volume of the space between the two pleural layers expands. With relaxation of the diaphragm, it increases back to 756 mmHg.

23.15 Normal atmospheric pressure at sea level is 760 mmHg.

23.16 Breathing in and then exhaling as much air as possible demonstrates vital capacity.

23.17 A difference in P_{O_2} promotes oxygen diffusion into pulmonary capillaries from alveoli and into tissue cells from systemic capillaries.

23.18 The most important factor that determines how much O_2 binds to hemoglobin is the P_{O_2}.

23.19 Both during exercise and at rest, hemoglobin in your pulmonary veins would be fully saturated with O_2, a point that is at the upper right of the curve.

23.20 Because lactic acid (lactate) and CO_2 are produced by active skeletal muscles, blood pH decreases slightly and P_{CO_2} increases when you are actively exercising. The result is lowered affinity of hemoglobin for O_2, so more O_2 is available to the working muscles.

23.21 O_2 is more available to your tissue cells when you have a fever because the affinity of hemoglobin for O_2 decreases with increasing temperature.

23.22 At a P_{O_2} of 40 mmHg, fetal Hb is 80% saturated with O_2 and maternal Hb is about 75% saturated.

23.23 Blood in a systemic vein would have a higher concentration of HCO_3^-.

23.24 The medullary inspiratory area contains autorhythmic neurons that are active and then inactive in a repeating cycle.

23.25 The phrenic nerves innervate the diaphragm.

23.26 Peripheral chemoreceptors are responsive to changes in blood levels of oxygen, carbon dioxide, and H^+.

23.27 Normal arterial P_{CO_2} is 40 mmHg.

23.28 The respiratory system begins to develop about 4 weeks after fertilization.

24 THE DIGESTIVE SYSTEM

THE DIGESTIVE SYSTEM AND HOMEOSTASIS *The digestive system contributes to homeostasis by breaking down food into forms that can be absorbed and used by body cells. It also absorbs water, vitamins, and minerals, and it eliminates wastes from the body.*

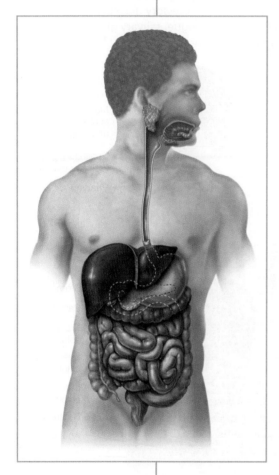

The food we eat contains a variety of nutrients, which are used for building new body tissues and repairing damaged tissues. Food is also vital to life because it is our only source of chemical energy. However, most of the food we eat consists of molecules that are too large to be used by body cells. Therefore, foods must be broken down into molecules that are small enough to enter body cells, a process known as **digestion** (*dis-* = apart; *-gerere* = to carry). The organs involved in the breakdown of food—collectively called the **digestive system**— are the focus of this chapter. Like the respiratory system, the digestive system is a tubular system. It extends from the mouth to the anus, forms an extensive surface area in contact with the external environment, and is closely associated with the cardiovascular system. The combination of extensive environmental exposure and close association with blood vessels is essential for processing the food that we eat.

The medical specialty that deals with the structure, function, diagnosis, and treatment of diseases of the stomach and intestines is called **gastroenterology** (gas′-trō-en′-ter-OL-ō-jē; *gastro-* = stomach; *-entero-* = intestines; *-logy* = study of). The medical specialty that deals with the diagnosis and treatment of disorders of the rectum and anus is called **proctology** (prok-TOL-ō-jē; *proct-* = rectum).

? *Did you ever wonder why some people are sensitive to dairy products?*

24.1 OVERVIEW OF THE DIGESTIVE SYSTEM

⦿ **OBJECTIVES**

- Identify the organs of the digestive system.
- Describe the basic processes performed by the digestive system.

Two groups of organs compose the digestive system (Figure 24.1): the gastrointestinal (GI) tract and the accessory digestive organs. The **gastrointestinal (GI) tract,** or **alimentary canal** (*alimentary* = nourishment), is a continuous tube that extends from the mouth to the anus through the thoracic and abdominopelvic cavities. Organs of the gastrointestinal tract include the mouth, most of the pharynx, esophagus, stomach, small intestine, and large intestine. The length of the GI tract is about 5–7 meters (16.5–23 ft) in a living person when the muscles along the wall of the GI tract organs are in a state of *tonus* (sustained contraction). It is longer in a cadaver (about 7–9 meters or 23–29.5 ft) because of the loss of muscle tone after death. The **accessory digestive organs** include the teeth, tongue, salivary glands, liver, gallbladder, and pancreas. Teeth aid in the physical breakdown of food, and the tongue assists in chewing and swallowing. The other accessory digestive organs, however, never come into direct contact with food. They produce or store secretions that flow into the GI tract through ducts; the secretions aid in the chemical breakdown of food.

The GI tract contains food from the time it is eaten until it is digested and absorbed or eliminated. Muscular contractions in the wall of the GI tract physically break down the food by churning it and propel the food along the tract, from the esophagus to the anus. The contractions also help to dissolve foods by mixing them with fluids secreted into the tract. Enzymes secreted by accessory digestive organs and cells that line the tract break down the food chemically.

Overall, the digestive system performs six basic processes:

1. *Ingestion.* This process involves taking foods and liquids into the mouth (eating).
2. *Secretion.* Each day, cells within the walls of the GI tract and accessory digestive organs secrete a total of about 7 liters of water, acid, buffers, and enzymes into the lumen (interior space) of the tract.
3. *Mixing and propulsion.* Alternating contractions and relaxations of smooth muscle in the walls of the GI tract mix food and secretions and propel them toward the anus. This capability of the GI tract to mix and move material along its length is called **motility** (mō-TIL-i-tē).
4. *Digestion.* Mechanical and chemical processes break down ingested food into small molecules. In **mechanical digestion** the teeth cut and grind food before it is swallowed, and then smooth muscles of the stomach and small intestine churn the food. As a result, food molecules become dissolved and thoroughly mixed with digestive enzymes. In **chemical digestion** the large carbohydrate, lipid, protein, and nucleic acid molecules in food are split into smaller molecules by hydrolysis

(see Figure 2.15). Digestive enzymes produced by the salivary glands, tongue, stomach, pancreas, and small intestine catalyze these catabolic reactions. A few substances in food can be absorbed without chemical digestion. These include vitamins, ions, cholesterol, and water.

5. *Absorption.* The entrance of ingested and secreted fluids, ions, and the products of digestion into the epithelial cells lining the lumen of the GI tract is called **absorption** (ab-SŌRP-shun).

Figure 24.1 Organs of the digestive system.

 Organs of the gastrointestinal (GI) tract are the mouth, pharynx, esophagus, stomach, small intestine, and large intestine. Accessory digestive organs include the teeth, tongue, salivary glands, liver, gallbladder, and pancreas.

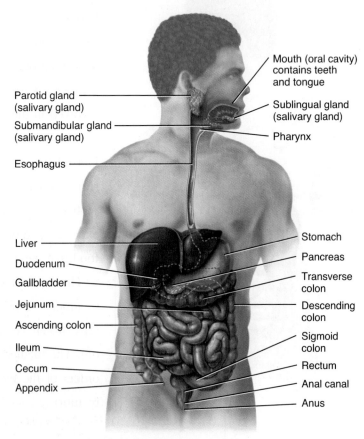

(a) Right lateral view of head and neck and anterior view of trunk

FUNCTIONS OF THE DIGESTIVE SYSTEM

1. Ingestion: taking food into mouth.
2. Secretion: release of water, acid, buffers, and enzymes into lumen of GI tract.
3. Mixing and propulsion: churning and propulsion of food through GI tract.
4. Digestion: mechanical and chemical breakdown of food.
5. Absorption: passage of digested products from GI tract into blood and lymph.
6. Defecation: elimination of feces from GI tract.

The absorbed substances pass into blood or lymph and circulate to cells throughout the body.

6. *Defecation.* Wastes, indigestible substances, bacteria, cells sloughed from the lining of the GI tract, and digested materials that were not absorbed in their journey through the digestive tract leave the body through the anus in a process called **defecation** (def′-e-KĀ-shun). The eliminated material is termed **feces** (FĒ-sēz) or **stool.**

✔**CHECKPOINT**

1. Which components of the digestive system are GI tract organs, and which are accessory digestive organs?
2. Which organs of the digestive system come in contact with food, and what are some of their digestive functions?
3. Which kinds of food molecules undergo chemical digestion, and which do not?

24.2 LAYERS OF THE GI TRACT

◉ **OBJECTIVE**

• Describe the structure and function of the layers that form the wall of the gastrointestinal tract.

The wall of the GI tract from the lower esophagus to the anal canal has the same basic, four-layered arrangement of tissues.

The four layers of the tract, from deep to superficial, are the mucosa, submucosa, muscularis, and serosa/adventitia (Figure 24.2).

Mucosa

The **mucosa,** or inner lining of the GI tract, is a mucous membrane. It is composed of (1) a layer of epithelium in direct contact with the contents of the GI tract, (2) a layer of connective tissue called the lamina propria, and (3) a thin layer of smooth muscle (muscularis mucosae).

1. The **epithelium** in the mouth, pharynx, esophagus, and anal canal is mainly nonkeratinized stratified squamous epithelium that serves a protective function. Simple columnar epithelium, which functions in secretion and absorption, lines the stomach and intestines. The tight junctions that firmly seal neighboring simple columnar epithelial cells to one another restrict leakage between the cells. The rate of renewal of GI tract epithelial cells is rapid: Every 5 to 7 days they slough off and are replaced by new cells. Located among the epithelial cells are exocrine cells that secrete mucus and fluid into the lumen of the tract, and several types of endocrine cells, collectively called **enteroendocrine cells** (en′-ter-ō-EN-dō-krin), which secrete hormones.

2. The **lamina propria** (*lamina* = thin, flat plate; *propria* = one's own) is areolar connective tissue containing many

SUPERIOR

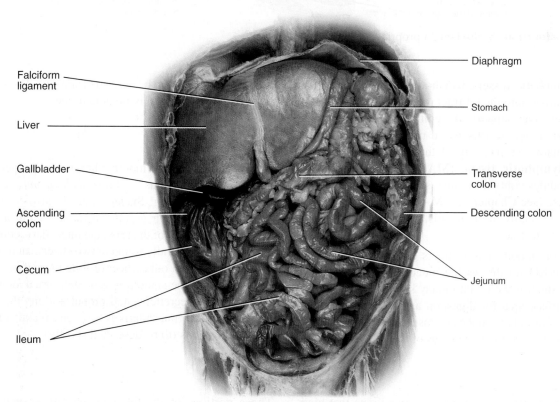

Falciform ligament

Liver

Gallbladder

Ascending colon

Cecum

Ileum

Diaphragm

Stomach

Transverse colon

Descending colon

Jejunum

(b) Anterior view

❓ **Which structures of the digestive system secrete digestive enzymes?**

Figure 24.2 Layers of the gastrointestinal tract. Variations in this basic plan may be seen in the esophagus (Figure 24.9), stomach (Figure 24.12), small intestine (Figure 24.19), and large intestine (Figure 24.24).

The four layers of the GI tract, from deep to superficial, are the mucosa, submucosa, muscularis, and serosa.

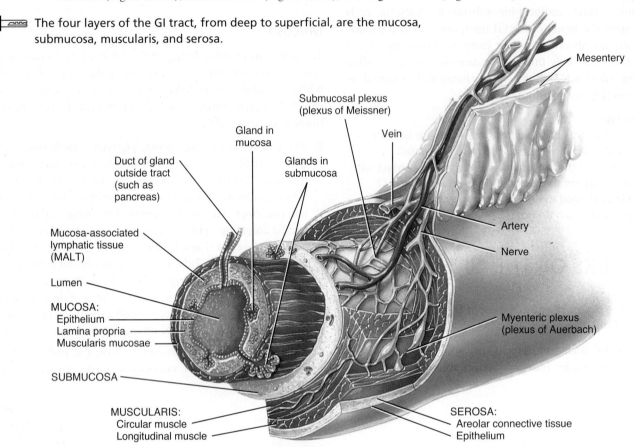

What are the functions of the lamina propria?

blood and lymphatic vessels, which are the routes by which nutrients absorbed into the GI tract reach the other tissues of the body. This layer supports the epithelium and binds it to the muscularis mucosae (discussed next). The lamina propria also contains the majority of the cells of the **mucosa-associated lymphatic tissue (MALT).** These prominent lymphatic nodules contain immune system cells that protect against disease (see Chapter 22). MALT is present all along the GI tract, especially in the tonsils, small intestine, appendix, and large intestine.

3. A thin layer of smooth muscle fibers called the **muscularis mucosae** (mū-KŌ-sē) throws the mucous membrane of the stomach and small intestine into many small folds, which increase the surface area for digestion and absorption. Movements of the muscularis mucosae ensure that all absorptive cells are fully exposed to the contents of the GI tract.

Submucosa

The **submucosa** consists of areolar connective tissue that binds the mucosa to the muscularis. It contains many blood and lymphatic vessels that receive absorbed food molecules. Also located in the submucosa is an extensive network of neurons known as the

submucosal plexus (to be described shortly). The submucosa may also contain glands and lymphatic tissue.

Muscularis

The **muscularis** of the mouth, pharynx, and superior and middle parts of the esophagus contains *skeletal muscle* that produces voluntary swallowing. Skeletal muscle also forms the external anal sphincter, which permits voluntary control of defecation. Throughout the rest of the tract, the muscularis consists of *smooth muscle* that is generally found in two sheets: an inner sheet of circular fibers and an outer sheet of longitudinal fibers. Involuntary contractions of the smooth muscle help break down food, mix it with digestive secretions, and propel it along the tract. Between the layers of the muscularis is a second plexus of neurons—the myenteric plexus (to be described shortly).

Serosa

Those portions of the GI tract that are suspended in the abdominopelvic cavity have a superficial layer called the **serosa.** As its name implies, the serosa is a serous membrane composed of areolar connective tissue and simple squamous epithelium (mesothelium). The serosa is also called the *visceral peritoneum* because it forms a portion of the peritoneum, which we examine in detail shortly. The esophagus lacks a serosa; instead only a single layer of areolar connective tissue called the *adventitia* forms the superficial layer of this organ.

✔CHECKPOINT

4. Where along the GI tract is the muscularis composed of skeletal muscle? Is control of this skeletal muscle voluntary or involuntary?

5. Name the four layers of the gastrointestinal tract, and describe their functions.

24.3 NEURAL INNERVATION OF THE GI TRACT

◉ OBJECTIVE

• Describe the nerve supply of the GI tract.

The gastrointestinal tract is regulated by an intrinsic set of nerves known as the enteric nervous system and by an extrinsic set of nerves that are part of the autonomic nervous system.

Enteric Nervous System

We first introduced you to the **enteric nervous system (ENS),** the "brain of the gut," in Chapter 12. It consists of about 100 million neurons that extend from the esophagus to the anus. The neurons of the ENS are arranged into two plexuses: the myenteric plexus and submucosal plexus (see Figure 24.2). The **myenteric plexus** (*myo-* = muscle), or *plexus of Auerbach* (OW-er-bak), is located between the longitudinal and circular smooth muscle layers of the muscularis. The **submucosal plexus,** or *plexus of Meissner* (MĪS-ner), is found within the submucosa. The plexuses of the ENS consist of motor neurons, interneurons, and sensory neurons (Figure 24.3). Because the motor neurons of the myenteric plexus supply the longitudinal and circular smooth muscle layers of the muscularis, this plexus mostly controls GI tract motility (movement), particularly the frequency and strength of contraction of the muscularis. The motor neurons of the submucosal plexus supply the secretory cells of the mucosal epithelium, controlling the secretions of the organs of the GI tract. The interneurons of the ENS interconnect the neurons of the myenteric and submucosal plexuses. The sensory neurons of the ENS supply the mucosal epithelium. Some of these sensory neurons function as *chemoreceptors,* receptors that are activated by the presence of certain chemicals in food located in the lumen of a GI organ. Other sensory neurons function as *stretch receptors,* receptors that are activated when food distends (stretches) the wall of a GI organ.

Autonomic Nervous System

Although the neurons of the ENS can function independently, they are subject to regulation by the neurons of the autonomic nervous system. The vagus (X) nerves supply parasympathetic fibers to

Figure 24.3 Organization of the enteric nervous system.

 The enteric nervous system consists of neurons arranged into the myenteric and submucosal plexuses.

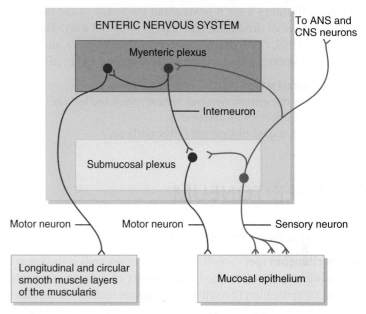

? What are the functions of the myenteric and submucosal plexuses of the enteric nervous system?

most parts of the GI tract, with the exception of the last half of the large intestine, which is supplied with parasympathetic fibers from the sacral spinal cord. The parasympathetic nerves that supply the GI tract form neural connections with the ENS. Parasympathetic preganglionic neurons of the vagus or pelvic splanchnic nerves synapse with parasympathetic postganglionic neurons located in the myenteric and submucosal plexuses. Some of the parasympathetic postganglionic neurons in turn synapse with neurons in the ENS; others directly innervate smooth muscle and glands within the wall of the GI tract. In general, stimulation of the parasympathetic nerves that innervate the GI tract causes an increase in GI secretion and motility by increasing the activity of ENS neurons.

Sympathetic nerves that supply the GI tract arise from the thoracic and upper lumbar regions of the spinal cord. Like the parasympathetic nerves, these sympathetic nerves form neural connections with the ENS. Sympathetic postganglionic neurons synapse with neurons located in the myenteric plexus and the submucosal plexus. In general, the sympathetic nerves that supply the GI tract cause a decrease in GI secretion and motility by inhibiting the neurons of the ENS. Emotions such as anger, fear, and anxiety may slow digestion because they stimulate the sympathetic nerves that supply the GI tract.

Gastrointestinal Reflex Pathways

Many neurons of the ENS are components of *GI (gastrointestinal) reflex pathways* that regulate GI secretion and motility in response to stimuli present in the lumen of the GI tract. The initial components of a typical GI reflex pathway are sensory receptors (such as chemoreceptors and stretch receptors) that are associated with the sensory neurons of the ENS. The axons of these sensory neurons can synapse with other neurons located in the ENS, CNS, or ANS, informing these regions about the nature of the contents and the degree of distension (stretching) of the GI tract. The neurons of the ENS, CNS, or ANS subsequently activate or inhibit GI glands and smooth muscle, altering GI secretion and motility.

✔ **CHECKPOINT**

6. How is the enteric nervous system regulated by the autonomic nervous system?
7. What is a gastrointestinal reflex pathway?

24.4 PERITONEUM

◉ **OBJECTIVE**

• Describe the peritoneum and its folds.

The **peritoneum** (per'-i-tō-NĒ-um; *peri-* = around) is the largest serous membrane of the body; it consists of a layer of simple squamous epithelium (mesothelium) with an underlying support-

ing layer of areolar connective tissue. The peritoneum is divided into the **parietal peritoneum,** which lines the wall of the abdominopelvic cavity, and the **visceral peritoneum,** which covers some of the organs in the cavity and is their serosa (Figure 24.4a). The slim space containing lubricating serous fluid that is between the parietal and visceral portions of the peritoneum is called the **peritoneal cavity.** In certain diseases, the peritoneal cavity may become distended by the accumulation of several liters of fluid, a condition called **ascites** (a-SĪ-tēz).

As you will see shortly, some organs lie on the posterior abdominal wall and are covered by peritoneum only on their anterior surfaces; they are not in the peritoneal cavity. Such organs, including the kidneys, ascending and descending colons of the large intestine, duodenum of the small intestine, and pancreas, are said to be **retroperitoneal** (*retro-* = behind).

Unlike the pericardium and pleurae, which smoothly cover the heart and lungs, the peritoneum contains large folds that weave between the viscera. The folds bind the organs to one another and to the walls of the abdominal cavity. They also contain blood vessels, lymphatic vessels, and nerves that supply the abdominal organs. There are five major peritoneal folds: the greater omentum, falciform ligament, lesser omentum, mesentery, and mesocolon:

Figure 24.4 Relationship of the peritoneal folds to one another and to organs of the digestive system. The size of the peritoneal cavity in (a) is exaggerated for emphasis.

🔑 The peritoneum is the largest serous membrane in the body.

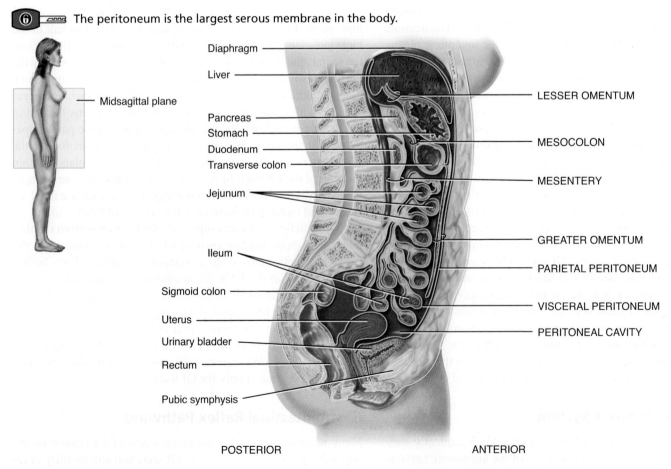

Diaphragm
Liver
Midsagittal plane
LESSER OMENTUM
Pancreas
Stomach
Duodenum
MESOCOLON
Transverse colon
Jejunum
MESENTERY
Ileum
GREATER OMENTUM
PARIETAL PERITONEUM
Sigmoid colon
Uterus
VISCERAL PERITONEUM
Urinary bladder
PERITONEAL CAVITY
Rectum
Pubic symphysis

POSTERIOR   ANTERIOR

(a) Midsagittal section showing the peritoneal folds

1. The **greater omentum** (ō-MEN-tum = fat skin), the largest peritoneal fold, drapes over the transverse colon and coils of the small intestine like a "fatty apron" (Figure 24.4a, d). The greater omentum is a double sheet that folds back on itself, giving it a total of four layers. From attachments along the stomach and duodenum, the greater omentum extends downward anterior to the small intestine, then turns and extends upward and attaches to the transverse colon. The greater omentum normally contains a considerable amount of adipose tissue. Its adipose tissue content can greatly expand with weight gain, contributing to the characteristic "beer belly" seen in some overweight individuals.

The many lymph nodes of the greater omentum contribute macrophages and antibody-producing plasma cells that help combat and contain infections of the GI tract.

2. The **falciform ligament** (FAL-si-form; *falc-* = sickle-shaped) attaches the liver to the anterior abdominal wall and diaphragm (Figure 24.4b). The liver is the only digestive organ that is attached to the anterior abdominal wall.

3. The **lesser omentum** arises as an anterior fold in the serosa of the stomach and duodenum, and it suspends the stomach and duodenum from the liver (Figure 24.4a, c). It is the pathway

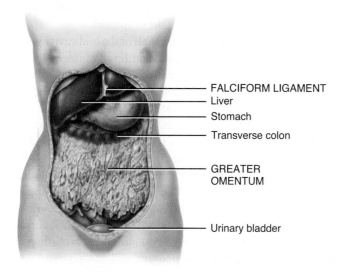

- FALCIFORM LIGAMENT
- Liver
- Stomach
- Transverse colon
- GREATER OMENTUM
- Urinary bladder

(b) Anterior view

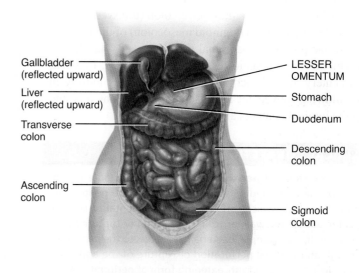

- Gallbladder (reflected upward)
- Liver (reflected upward)
- Transverse colon
- Ascending colon
- LESSER OMENTUM
- Stomach
- Duodenum
- Descending colon
- Sigmoid colon

(c) Lesser omentum, anterior view (liver and gallbladder lifted)

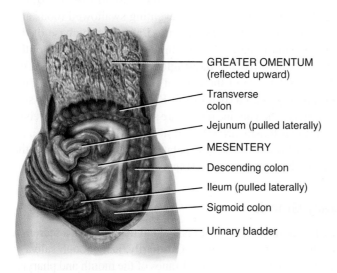

- GREATER OMENTUM (reflected upward)
- Transverse colon
- Jejunum (pulled laterally)
- MESENTERY
- Descending colon
- Ileum (pulled laterally)
- Sigmoid colon
- Urinary bladder

(d) Anterior view (greater omentum lifted and small intestine reflected to right side)

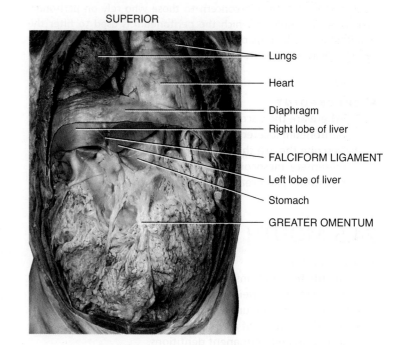

SUPERIOR

- Lungs
- Heart
- Diaphragm
- Right lobe of liver
- FALCIFORM LIGAMENT
- Left lobe of liver
- Stomach
- GREATER OMENTUM

(e) Anterior view

? **Which peritoneal fold binds the small intestine to the posterior abdominal wall?**

for blood vessels entering the liver and contains the hepatic portal vein, common hepatic artery, and common bile duct, along with some lymph nodes.

4. A fan-shaped fold of the peritoneum, called the **mesentery** (MEZ-en-ter′-ē; *mes-* = middle), binds the jejunum and ileum of the small intestine to the posterior abdominal wall (Figure 24.4a, d). This is the largest peritoneal fold and is typically laden with fat and contributes extensively to the large abdomen in obese individuals. It extends from the posterior abdominal wall to wrap around the small intestine and then returns to its origin, forming a double-layered structure. Between the two layers are blood and lymphatic vessels and lymph nodes.

5. Two separate folds of peritoneum, called the **mesocolon** (mez′-ō-KŌ-lon), bind the transverse colon (transverse mesocolon) and sigmoid colon (sigmoid mesocolon) of the large intestine to the posterior abdominal wall (Figure 24.4a). It also carries blood and lymphatic vessels to the intestines. Together, the mesentery and mesocolon hold the intestines loosely in place, allowing movement as muscular contractions mix and move the luminal contents along the GI tract.

CLINICAL CONNECTION | *Peritonitis*

A common cause of **peritonitis** (per′-i-tō-NĪ-tis), an acute inflammation of the peritoneum, is contamination of the peritoneum by infectious microbes, which can result from accidental or surgical wounds in the abdominal wall, or from perforation or rupture of abdominal organs. If, for example, bacteria gain access to the peritoneal cavity through an intestinal perforation or rupture of the appendix, they can produce an acute, life-threatening form of peritonitis. A less serious (but still painful) form of peritonitis can result from the rubbing together of inflamed peritoneal surfaces. The increased risk of peritonitis is of particular concern to those who rely on peritoneal dialysis, a procedure in which the peritoneum is used to filter the blood when the kidneys do not function properly (see Clinical Connection: Dialysis in Chapter 26). •

✔ CHECKPOINT

8. Where are the visceral peritoneum and parietal peritoneum located?
9. Describe the attachment sites and functions of the mesentery, mesocolon, falciform ligament, lesser omentum, and greater omentum.

24.5 MOUTH

☉ OBJECTIVES

• Identify the locations of the salivary glands, and describe the functions of their secretions.
• Describe the structure and functions of the tongue.
• Identify the parts of a typical tooth, and compare deciduous and permanent dentitions.

The **mouth,** also referred to as the **oral** or **buccal cavity** (BUK-al; *bucca* = cheeks), is formed by the cheeks, hard and soft palates, and tongue (Figure 24.5). The **cheeks** form the lateral walls of the oral cavity. They are covered externally by skin and internally by a mucous membrane, which consists of nonkeratinized stratified squamous epithelium. Buccinator muscles and connective tissue lie between the skin and mucous membranes of the cheeks. The anterior portions of the cheeks end at the lips.

The **lips** or **labia** (= fleshy borders) are fleshy folds surrounding the opening of the mouth. They contain the orbicularis oris muscle and are covered externally by skin and internally by a mucous membrane. The inner surface of each lip is attached to its corresponding gum by a midline fold of mucous membrane called the **labial frenulum** (LĀ-bē-al FREN-ū-lum; *frenulum* = small bridle). During chewing, contraction of the buccinator muscles in the cheeks and orbicularis oris muscle in the lips helps keep food between the upper and lower teeth. These muscles also assist in speech.

The **oral vestibule** (= entrance to a canal) of the oral cavity is the space bounded externally by the cheeks and lips and internally by the gums and teeth. The **oral cavity proper** is the space that extends from the gums and teeth to the **fauces** (FAW-sēs = passages), the opening between the oral cavity and the oropharynx (throat).

The **palate** is a wall or septum that separates the oral cavity from the nasal cavity, and forms the roof of the mouth. This important structure makes it possible to chew and breathe at the same time. The **hard palate**—the anterior portion of the roof of the mouth—is formed by the maxillae and palatine bones and is covered by a mucous membrane; it forms a bony partition between the oral and nasal cavities. The **soft palate,** which forms the posterior portion of the roof of the mouth, is an arch-shaped muscular partition between the oropharynx and nasopharynx that is lined with mucous membrane.

Hanging from the free border of the soft palate is a conical muscular process called the **uvula** (Ū-vū-la = little grape). During swallowing, the soft palate and uvula are drawn superiorly, closing off the nasopharynx and preventing swallowed foods and liquids from entering the nasal cavity. Lateral to the base of the uvula are two muscular folds that run down the lateral sides of the soft palate: Anteriorly, the **palatoglossal arch** (pal-a-tō-GLOS-al) extends to the side of the base of the tongue; posteriorly, the **palatopharyngeal arch** (pal-a-tō-fa-RIN-jē-al) extends to the side of the pharynx. The palatine tonsils are situated between the arches, and the lingual tonsils are situated at the base of the tongue. At the posterior border of the soft palate, the mouth opens into the oropharynx through the fauces (Figure 24.5).

Salivary Glands

A **salivary gland** (SAL-i-vār-ē) is a gland that releases a secretion called saliva into the oral cavity. Ordinarily, just enough saliva is secreted to keep the mucous membranes of the mouth and pharynx moist and to cleanse the mouth and teeth. When food enters the mouth, however, secretion of saliva increases, and it lubricates, dissolves, and begins the chemical breakdown of the food.

The mucous membrane of the mouth and tongue contains many small salivary glands that open directly, or indirectly via short ducts, to the oral cavity. These glands include *labial, buccal,*

Figure 24.5 Structures of the mouth (oral cavity).

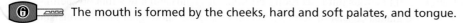 The mouth is formed by the cheeks, hard and soft palates, and tongue.

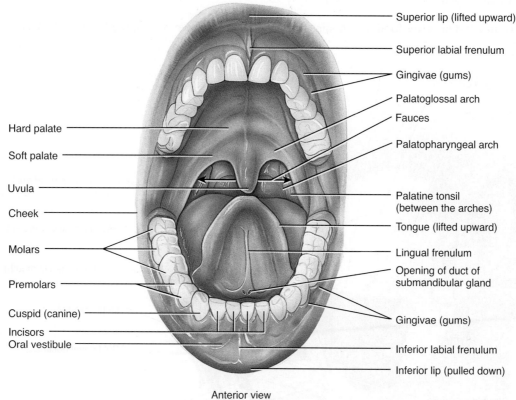

- Superior lip (lifted upward)
- Superior labial frenulum
- Gingivae (gums)
- Palatoglossal arch
- Fauces
- Palatopharyngeal arch
- Hard palate
- Soft palate
- Uvula
- Cheek
- Palatine tonsil (between the arches)
- Tongue (lifted upward)
- Molars
- Lingual frenulum
- Opening of duct of submandibular gland
- Premolars
- Cuspid (canine)
- Gingivae (gums)
- Incisors
- Oral vestibule
- Inferior labial frenulum
- Inferior lip (pulled down)

Anterior view

? **What is the function of the uvula?**

and *palatal glands* in the lips, cheeks, and palate, respectively, and *lingual glands* in the tongue, all of which make a small contribution to saliva.

However, most saliva is secreted by the **major salivary glands,** which lie beyond the oral mucosa, into ducts that lead to the oral cavity. There are three pairs of major salivary glands: the parotid, submandibular, and sublingual glands (Figure 24.6a). The **parotid glands** (pa-ROT-id; *par-* = near; *to-* = ear) are located inferior and anterior to the ears, between the skin and the masseter muscle. Each secretes saliva into the oral cavity via a **parotid duct** that pierces the buccinator muscle to open into the vestibule opposite the second maxillary (upper) molar tooth. The **submandibular glands** (sub′-man-DIB-ū-lar) are found in the floor of the mouth; they are medial and partly inferior to the body of the mandible. Their ducts, the **submandibular ducts,** run under the mucosa on either side of the midline of the floor of the mouth and enter the oral cavity proper lateral to the lingual frenulum. The **sublingual glands** (sub-LING-gwal) are beneath the tongue and superior to the submandibular glands. Their ducts, the **lesser sublingual ducts,** open into the floor of the mouth in the oral cavity proper.

Composition and Functions of Saliva

Chemically, **saliva** is 99.5% water and 0.5% solutes. Among the solutes are ions, including sodium, potassium, chloride, bicarbon-

ate, and phosphate. Also present are some dissolved gases and various organic substances, including urea and uric acid, mucus, immunoglobulin A, the bacteriolytic enzyme lysozyme, and salivary amylase, a digestive enzyme that acts on starch.

Not all salivary glands supply the same ingredients. The parotid glands secrete a watery (serous) liquid containing salivary amylase. Because the submandibular glands contain cells similar to those found in the parotid glands, plus some mucous cells, they secrete a fluid that contains amylase but is thickened with mucus. The sublingual glands contain mostly mucous cells, so they secrete a much thicker fluid that contributes only a small amount of salivary amylase.

The water in saliva provides a medium for dissolving foods so that they can be tasted by gustatory receptors and so that digestive reactions can begin. Chloride ions in the saliva activate **salivary amylase** (AM-i-lās), an enzyme that starts the breakdown of starch in the mouth into maltose, maltotriose, and α-dextrin. Bicarbonate and phosphate ions buffer acidic foods that enter the mouth, so saliva is only slightly acidic (pH 6.35–6.85). Salivary glands (like the sweat glands of the skin) help remove waste molecules from the body, which accounts for the presence of urea and uric acid in saliva. Mucus lubricates food so it can be moved around easily in the mouth, formed into a ball, and swallowed. Immunoglobulin A (IgA) prevents attachment of microbes so

Figure 24.6 **The three major salivary glands—parotid, sublingual, and submandibular.** The submandibular glands, shown in the light micrograph (b), consist mostly of serous acini (serous fluid–secreting portions of gland) and a few mucous acini (mucus-secreting portions of gland); the parotid glands consist of serous acini only; and the sublingual glands consist of mostly mucous acini and a few serous acini.

Saliva lubricates and dissolves foods and begins the chemical breakdown of carbohydrates and lipids.

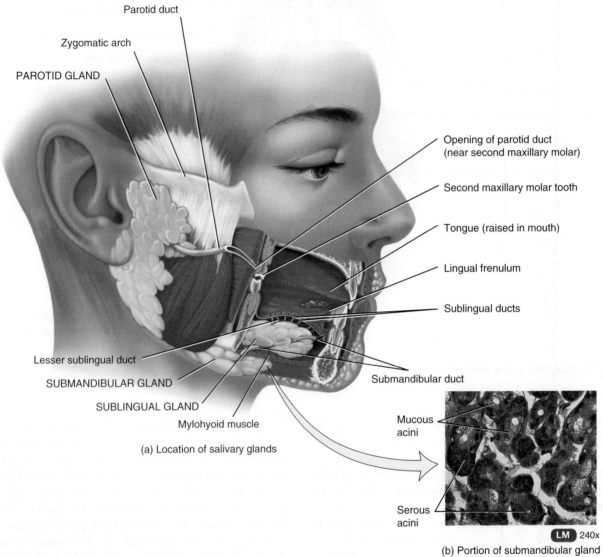

Parotid duct
Zygomatic arch
PAROTID GLAND
Opening of parotid duct (near second maxillary molar)
Second maxillary molar tooth
Tongue (raised in mouth)
Lingual frenulum
Sublingual ducts
Lesser sublingual duct
SUBMANDIBULAR GLAND
SUBLINGUAL GLAND
Mylohyoid muscle
Submandibular duct

(a) Location of salivary glands

Mucous acini
Serous acini

LM 240x

(b) Portion of submandibular gland

? **What is the function of the chloride ions in saliva?**

they cannot penetrate the epithelium, and the enzyme lysozyme kills bacteria; however, these substances are not present in large enough quantities to eliminate all oral bacteria.

Salivation

The secretion of saliva, called **salivation** (sal-i-VĀ-shun), is controlled by the autonomic nervous system. Amounts of saliva secreted daily vary considerably but average 1000–1500 mL (1–1.6 qt). Normally, parasympathetic stimulation promotes continuous secretion of a moderate amount of saliva, which keeps the mucous membranes moist and lubricates the movements of the tongue and lips during speech. The saliva is then swallowed and helps moisten the esophagus. Eventually, most components of

saliva are reabsorbed, which prevents fluid loss. Sympathetic stimulation dominates during stress, resulting in dryness of the mouth. If the body becomes dehydrated, the salivary glands stop secreting saliva to conserve water; the resulting dryness in the mouth contributes to the sensation of thirst. Drinking not only restores the homeostasis of body water but also moistens the mouth.

The feel and taste of food also are potent stimulators of salivary gland secretions. Chemicals in the food stimulate receptors in taste buds on the tongue, and impulses are conveyed from the taste buds to two salivary nuclei in the brain stem (**superior** and **inferior salivatory nuclei**). Returning parasympathetic impulses in fibers of the facial (VII) and glossopharyngeal (IX) nerves stimulate the secretion of saliva. Saliva continues to be secreted heavily for some time after food is swallowed; this flow of saliva washes out the mouth and dilutes and buffers the remnants of irritating chemicals such as that tasty (but hot!) salsa. The smell, sight, sound, or thought of food may also stimulate secretion of saliva.

CLINICAL CONNECTION | *Mumps*

Although any of the salivary glands may be the target of a nasopharyngeal infection, the mumps virus *(paramyxovirus)* typically attacks the parotid glands. **Mumps** is an inflammation and enlargement of the parotid glands accompanied by moderate fever, malaise (general discomfort), and extreme pain in the throat, especially when swallowing sour foods or acidic juices. Swelling occurs on one or both sides of the face, just anterior to the ramus of the mandible. In about 30% of males past puberty, the testes may also become inflamed; sterility rarely occurs because testicular involvement is usually unilateral (one testis only). Since a vaccine became available for mumps in 1967, the incidence of the disease has declined dramatically. •

Tongue

The **tongue** is an accessory digestive organ composed of skeletal muscle covered with mucous membrane. Together with its associated muscles, it forms the floor of the oral cavity. The tongue is divided into symmetrical lateral halves by a median septum that extends its entire length, and it is attached inferiorly to the hyoid bone, styloid process of the temporal bone, and mandible. Each half of the tongue consists of an identical complement of extrinsic and intrinsic muscles.

The **extrinsic muscles** of the tongue, which originate outside the tongue (attach to bones in the area) and insert into connective tissues in the tongue, include the *hyoglossus, genioglossus,* and *styloglossus muscles* (see Figure 11.7). The extrinsic muscles move the tongue from side to side and in and out to maneuver food for chewing, shape the food into a rounded mass, and force the food to the back of the mouth for swallowing. They also form the floor of the mouth and hold the tongue in position. The **intrinsic muscles** originate in and insert into connective tissue within the tongue. They alter the shape and size of the tongue for speech and swallowing. The intrinsic muscles include the *longitudinalis superior, longitudinalis inferior, transversus linguae,* and *verticalis linguae muscles.* The **lingual frenulum** (*lingua* = the tongue), a fold of mucous membrane in the midline of the undersurface of the tongue, is attached to the floor of the mouth and aids in limiting the movement of the tongue posteriorly (see Figures 24.5 and 24.6). If a person's lingual frenulum is abnormally short or rigid—a condition called **ankyloglossia** (ang′-kē-lō-GLOSS-ē-a)—the person is said to be "tongue-tied" because of the resulting impairment to speech.

The dorsum (upper surface) and lateral surfaces of the tongue are covered with **papillae** (pa-PIL-ē = nipple-shaped projections), projections of the lamina propria covered with stratified squamous epithelium (see Figure 17.3). Many papillae contain taste buds, the receptors for gustation (taste). Some papillae lack taste buds, but they contain receptors for touch and increase

friction between the tongue and food, making it easier for the tongue to move food in the oral cavity. The different types of taste buds are described in detail in Chapter 17. **Lingual glands** in the lamina propria of the tongue secrete both mucus and a watery serous fluid that contains the enzyme **lingual lipase,** which acts on as much as 30% of dietary triglycerides (fats and oils) and converts them to simpler fatty acids and diglycerides.

Teeth

The **teeth,** or **dentes** (Figure 24.7), are accessory digestive organs located in sockets of the alveolar processes of the mandible and maxillae. The alveolar processes are covered by the **gingivae** (JIN-ji-vē), or gums, which extend slightly into each socket. The sockets are lined by the **periodontal ligament** (per′-ē-ō-DON-tal; *odont-* = tooth) or **membrane,** which consists of dense fibrous connective tissue that anchors the teeth to the socket walls and acts as a shock absorber during chewing.

A typical tooth has three major external regions: the crown, root, and neck. The **crown** is the visible portion above the level of

---

Figure 24.7 A typical tooth and surrounding structures.

Teeth are anchored in sockets of the alveolar processes of the mandible and maxillae.

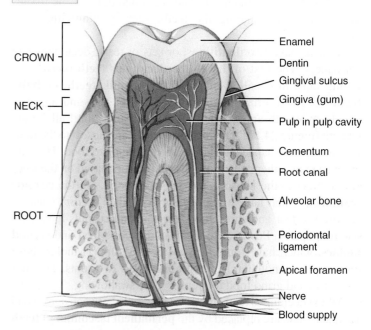

Sagittal section of a mandibular (lower) molar

? **What type of tissue is the main component of teeth?**

the gums. Embedded in the socket are one to three **roots.** The **neck** is the constricted junction of the crown and root near the gum line.

Internally, **dentin** forms the majority of the tooth. Dentin consists of a calcified connective tissue that gives the tooth its basic shape and rigidity. It is harder than bone because of its higher content of calcium salts (70% of dry weight).

The dentin of the crown is covered by **enamel,** which consists primarily of calcium phosphate and calcium carbonate. Enamel is also harder than bone because of its even higher content of calcium salts (about 95% of dry weight). In fact, enamel is the hardest substance in the body. It serves to protect the tooth from the wear and tear of chewing. It also protects against acids that can easily dissolve dentin. The dentin of the root is covered by **cementum,** another bonelike substance, which attaches the root to the periodontal ligament.

The dentin of a tooth encloses a space. The enlarged part of the space, the **pulp cavity,** lies within the crown and is filled with **pulp,** a connective tissue containing blood vessels, nerves, and lymphatic vessels. Narrow extensions of the pulp cavity, called **root canals,** run through the root of the tooth. Each root canal has an opening at its base, the **apical foramen,** through which blood vessels, lymphatic vessels, and nerves extend. The blood vessels bring nourishment, the lymphatic vessels offer protection, and the nerves provide sensation.

The branch of dentistry that is concerned with the prevention, diagnosis, and treatment of diseases that affect the pulp, root, periodontal ligament, and alveolar bone is known as **endodontics** (en′-dō-DON-tiks; *endo-* = within). **Orthodontics** (or′-thō-DON-tiks; *ortho-* = straight) is a branch of dentistry that is concerned with the prevention and correction of abnormally aligned teeth; **periodontics** (per′-ē-ō-DON-tiks) is a branch of dentistry concerned with the treatment of abnormal conditions of the tissues immediately surrounding the teeth, such as gingivitis (gum disease).

Humans have two **dentitions,** or sets of teeth: deciduous and permanent. The first of these—the **deciduous teeth** (*decidu-* = falling out), also called **primary teeth, milk teeth,** or **baby teeth**—begin to erupt at about 6 months of age, and approximately two teeth appear each month thereafter, until all 20 are present (Figure 24.8a). The **incisors,** which are closest to the midline, are chisel-shaped and adapted for cutting into food. They are referred to as either **central** or **lateral incisors** based on their position. Next to the incisors, moving posteriorly, are the **cuspids (canines),** which have a pointed surface called a *cusp.* Cuspids are used to tear and shred food. Incisors and cuspids have only one root apiece. Posterior to the cuspids lie the **first** and **second molars,** which have four cusps. Maxillary (upper) molars have three roots; mandibular (lower) molars have two roots. The molars crush and grind food to prepare it for swallowing.

All the deciduous teeth are lost—generally between ages 6 and 12 years—and are replaced by the **permanent (secondary) teeth** (Figure 24.8b). The permanent dentition contains 32 teeth that erupt between age 6 and adulthood. The pattern resembles the deciduous dentition, with the following exceptions. The deciduous

molars are replaced by the **first** and **second premolars (bicuspids),** which have two cusps and one root (upper first premolars have two roots) and are used for crushing and grinding. The permanent molars, which erupt into the mouth posterior to the premolars, do not replace any deciduous teeth and erupt as the jaw grows to accommodate them—the **first molars** at age 6 (six-year molars), the **second molars** at age 12 (twelve-year molars), and the **third molars (wisdom teeth)** after age 17 or not at all.

Often the human jaw does not have enough room posterior to the second molars to accommodate the eruption of the third molars. In this case, the third molars remain embedded in the alveolar bone and are said to be *impacted.* They often cause pressure and pain and must be removed surgically. In some people, third molars may be dwarfed in size or may not develop at all.

CLINICAL CONNECTION | *Root Canal Therapy*

Root canal therapy is a multistep procedure in which all traces of pulp tissue are removed from the pulp cavity and root canals of a badly diseased tooth. After a hole is made in the tooth, the root canals are filed out and irrigated to remove bacteria. Then, the canals are treated with medication and sealed tightly. The damaged crown is then repaired. •

Mechanical and Chemical Digestion in the Mouth

Mechanical digestion in the mouth results from chewing, or **mastication** (mas′-ti-KĀ-shun = to chew), in which food is manipulated by the tongue, ground by the teeth, and mixed with saliva. As a result, the food is reduced to a soft, flexible, easily swallowed mass called a **bolus** (= lump). Food molecules begin to dissolve in the water in saliva, an important activity because enzymes can react with food molecules in a liquid medium only.

Two enzymes, salivary amylase and lingual lipase, contribute to chemical digestion in the mouth. Salivary amylase, which is secreted by the salivary glands, initiates the breakdown of starch. Dietary carbohydrates are either monosaccharide and disaccharide sugars or complex polysaccharides such as starches. Most of the carbohydrates we eat are starches, but only monosaccharides can be absorbed into the bloodstream. Thus, ingested disaccharides and starches must be broken down into monosaccharides. The function of salivary amylase is to begin starch digestion by breaking down starch into smaller molecules such as the disaccharide maltose, the trisaccharide maltotriose, and short-chain glucose polymers called α-dextrins. Even though food is usually swallowed too quickly for all the starches to be broken down in the mouth, salivary amylase in the swallowed food continues to act on the starches for about another hour, at which time stomach acids inactivate it. Saliva also contains **lingual lipase** (LĪ-pās), which is secreted by lingual glands in the tongue. This enzyme becomes activated in the acidic environment of the stomach and thus starts to work after food is swallowed. It breaks down dietary triglycerides (fats and oils) into fatty acids and diglycerides. A diglyceride consists of a glycerol molecule that is attached to two fatty acids.

Table 24.1 summarizes the digestive activities in the mouth.

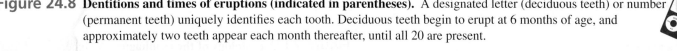

Figure 24.8 Dentitions and times of eruptions (indicated in parentheses). A designated letter (deciduous teeth) or number (permanent teeth) uniquely identifies each tooth. Deciduous teeth begin to erupt at 6 months of age, and approximately two teeth appear each month thereafter, until all 20 are present.

🔑 There are 20 teeth in a complete deciduous set and 32 teeth in a complete permanent set.

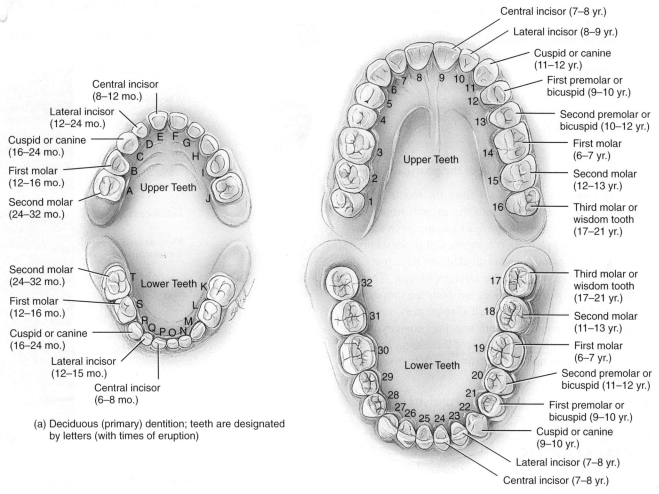

(a) Deciduous (primary) dentition; teeth are designated by letters (with times of eruption)

(b) Permanent (secondary) dentition; teeth are designated by numbers (with times of eruption)

❓ **Which permanent teeth do not replace any deciduous teeth?**

TABLE 24.1

Summary of Digestive Activities in the Mouth

STRUCTURE	ACTIVITY	RESULT
Cheeks and lips	Keep food between teeth.	Foods uniformly chewed during mastication.
Salivary glands	Secrete saliva.	Lining of mouth and pharynx moistened and lubricated. Saliva softens, moistens, and dissolves food and cleanses mouth and teeth. Salivary amylase splits starch into smaller fragments (maltose, maltotriose, and α-dextrins).
Tongue		
Extrinsic tongue muscles	Move tongue from side to side and in and out.	Food maneuvered for mastication, shaped into bolus, and maneuvered for swallowing.
Intrinsic tongue muscles	Alter shape of tongue.	Swallowing and speech.
Taste buds	Serve as receptors for gustation (taste) and presence of food in mouth.	Secretion of saliva stimulated by nerve impulses from taste buds to salivatory nuclei in brain stem to salivary glands.
Lingual glands	Secrete lingual lipase.	Triglycerides broken down into fatty acids and diglycerides.
Teeth	Cut, tear, and pulverize food.	Solid foods reduced to smaller particles for swallowing.

979

✔ CHECKPOINT

10. What structures form the mouth?
11. How are the major salivary glands distinguished on the basis of location?
12. How is the secretion of saliva regulated?
13. What functions do incisors, cuspids, premolars, and molars perform?

24.6 PHARYNX

◉ OBJECTIVE

• Describe the location and function of the pharynx.

When food is first swallowed, it passes from the mouth into the **pharynx** (= throat), a funnel-shaped tube that extends from the internal nares to the esophagus posteriorly and to the larynx anteriorly (see Figure 23.2). The pharynx is composed of skeletal muscle and lined by mucous membrane, and is divided into three parts: the nasopharynx, the oropharynx, and the laryngopharynx. The nasopharynx functions only in respiration, but both the oropharynx and laryngopharynx have digestive as well as respiratory functions. Swallowed food passes from the mouth into the oropharynx and laryngopharynx; the muscular contractions of these areas help propel food into the esophagus and then into the stomach.

✔ CHECKPOINT

14. To which two organ systems does the pharynx belong?

24.7 ESOPHAGUS

◉ OBJECTIVE

• Describe the location, anatomy, histology, and functions of the esophagus.

The **esophagus** (e-SOF-a-gus = eating gullet) is a collapsible muscular tube, about 25 cm (10 in.) long, that lies posterior to the trachea. The esophagus begins at the inferior end of the laryngopharynx, passes through the inferior aspect of the neck, and enters the mediastinum anterior to the vertebral column. Then it pierces the diaphragm through an opening called the **esophageal hiatus** (e-sof-a-JĒ-al hī-Ā-tus), and ends in the superior portion of the stomach (see Figure 24.1). Sometimes, part of the stomach protrudes above the diaphragm through the esophageal hiatus. This condition, termed a **hiatus hernia** (HER-nē-a), is described in the Medical Terminology section at the end of the chapter.

Histology of the Esophagus

The **mucosa** of the esophagus consists of nonkeratinized stratified squamous epithelium, lamina propria (areolar connective tissue), and a muscularis mucosae (smooth muscle) (Figure 24.9). Near the stomach, the mucosa of the esophagus also contains mucous glands. The stratified squamous epithelium associated with the lips, mouth, tongue, oropharynx, laryngopharynx, and esophagus affords considerable protection against abrasion and wear and tear

Figure 24.9 Histology of the esophagus. A higher-magnification view of nonkeratinized stratified squamous epithelium is shown in Table 4.1.

🔑 The esophagus secretes mucus and transports food to the stomach.

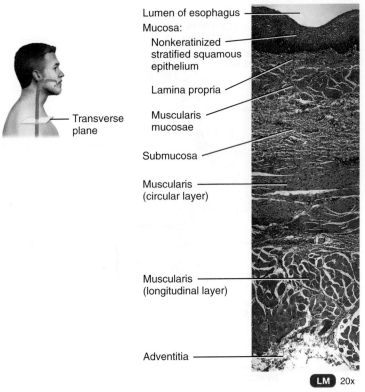

- Lumen of esophagus
- Mucosa:
  - Nonkeratinized stratified squamous epithelium
  - Lamina propria
  - Muscularis mucosae
- Submucosa
- Muscularis (circular layer)
- Muscularis (longitudinal layer)
- Adventitia

— Transverse plane

LM 20x

Wall of the esophagus

❓ **In which layers of the esophagus are the glands that secrete lubricating mucus located?**

from food particles that are chewed, mixed with secretions, and swallowed. The **submucosa** contains areolar connective tissue, blood vessels, and mucous glands. The **muscularis** of the superior third of the esophagus is skeletal muscle, the intermediate third is skeletal and smooth muscle, and the inferior third is smooth muscle. At each end of the esophagus, the muscularis becomes slightly more prominent and forms two sphincters—the **upper esophageal sphincter (UES)** (e-sof'-a-JĒ-al) or **valve,** which consists of skeletal muscle, and the **lower esophageal sphincter (LES)** or **valve,** which consists of smooth muscle. The upper esophageal sphincter regulates the movement of food from the pharynx into the esophagus; the lower esophageal sphincter regulates the movement of food from the esophagus into the stomach. The superficial layer of the esophagus is known as the **adventitia** (ad-ven-TISH-a), rather than the serosa as in the stomach and intestines, because the areolar connective tissue of this layer is not covered by mesothelium and because the connective tissue merges with the connective tissue of surrounding structures of the mediastinum through which it passes. The adventitia attaches the esophagus to surrounding structures.

Physiology of the Esophagus

The esophagus secretes mucus and transports food into the stomach. It does not produce digestive enzymes, and it does not carry on absorption.

✔**CHECKPOINT**

15. Describe the location and histology of the esophagus. What is its role in digestion?
16. What are the functions of the upper and lower esophageal sphincters?

24.8 DEGLUTITION

◉ **OBJECTIVE**

• Describe the three phases of deglutition.

The movement of food from the mouth into the stomach is achieved by the act of swallowing, or **deglutition** (dē-gloo-TISH-un) (Figure 24.10). Deglutition is facilitated by the secretion of saliva and mucus and involves the mouth, pharynx, and esophagus. Swallowing occurs in three stages: (1) the voluntary stage, in which the bolus is passed into the oropharynx; (2) the pharyngeal stage, the involuntary passage of the bolus through the pharynx into the esophagus; and (3) the esophageal stage, the involuntary passage of the bolus through the esophagus into the stomach.

Swallowing starts when the bolus is forced to the back of the oral cavity and into the oropharynx by the movement of the tongue upward and backward against the palate; these actions constitute the **voluntary stage** of swallowing. With the passage of the bolus into the oropharynx, the involuntary **pharyngeal stage** of swallowing begins (Figure 24.10b). The bolus stimulates receptors in the oropharynx, which send impulses to the **deglutition center** in the medulla oblongata and lower pons of the brain stem. The returning impulses cause the soft palate and uvula to move upward to close off the nasopharynx, which prevents swallowed foods and liquids from entering the nasal cavity. In addition, the epiglottis closes off the opening to the larynx, which prevents the bolus from entering the rest of the respiratory tract. The bolus moves through the oropharynx and the laryngopharynx. Once the upper esophageal sphincter relaxes, the bolus moves into the esophagus.

The **esophageal stage** of swallowing begins once the bolus enters the esophagus. During this phase, **peristalsis** (per'-i-STAL-sis; *stalsis* = constriction), a progression of coordinated contractions and relaxations of the circular and longitudinal layers of the muscularis, pushes the bolus onward (Figure 24.10c). (Peristalsis occurs in other tubular structures, including other parts of the GI tract and the ureters, bile ducts, and uterine tubes; in the esophagus it is controlled by the medulla oblongata.) In the section of the esophagus just superior to the bolus, the circular muscle fibers contract, constricting the esophageal wall and squeezing the bolus

Figure 24.10 Deglutition (swallowing). During the pharyngeal stage (b) the tongue rises against the palate, the nasopharynx is closed off, the larynx rises, the epiglottis seals off the larynx, and the bolus is passed into the esophagus. During the esophageal stage (c), food moves through the esophagus into the stomach via peristalsis.

🔑 Deglutition is a mechanism that moves food from the mouth into the stomach.

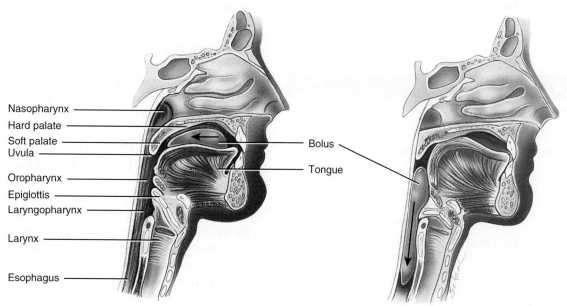

(a) Position of structures before swallowing   (b) During pharyngeal stage of swallowing

FIGURE 24.10 CONTINUES ▶

■ FIGURE 24.10 CONTINUED ▶

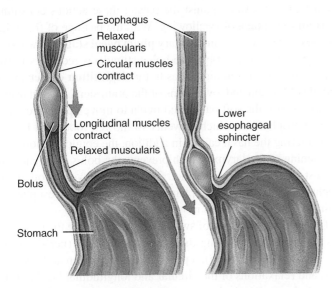

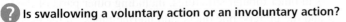

(c) Anterior view of frontal sections of peristalsis in esophagus

? Is swallowing a voluntary action or an involuntary action?

toward the stomach. Meanwhile, longitudinal fibers inferior to the bolus also contract, which shortens this inferior section and pushes its walls outward so it can receive the bolus. The contractions are repeated in waves that push the food toward the stomach. As the bolus approaches the end of the esophagus, the lower esophageal sphincter relaxes and the bolus moves into the stomach. Mucus secreted by esophageal glands lubricates the bolus and reduces friction. The passage of solid or semisolid food from the mouth to the stomach takes 4 to 8 seconds; very soft foods and liquids pass through in about 1 second.

Table 24.2 summarizes the digestive activities of the pharynx and esophagus.

TABLE 24.2

Summary of Digestive Activities in the Pharynx and Esophagus

STRUCTURE	ACTIVITY	RESULT
Pharynx	Pharyngeal stage of deglutition.	Moves bolus from oropharynx to laryngopharynx and into esophagus; closes air passageways.
Esophagus	Relaxation of upper esophageal sphincter.	Permits entry of bolus from laryngopharynx into esophagus.
	Esophageal stage of deglutition (peristalsis).	Pushes bolus down esophagus.
	Relaxation of lower esophageal sphincter.	Permits entry of bolus into stomach.
	Secretion of mucus.	Lubricates esophagus for smooth passage of bolus.

If the lower esophageal sphincter fails to close adequately after food has entered the stomach, the stomach contents can reflux (back up) into the inferior portion of the esophagus. This condition is known as **gastroesophageal reflux disease (GERD)** (gas′-trō-e-sof-a-JĒ-al). Hydrochloric acid (HCl) from the stomach contents can irritate the esophageal wall, resulting in a burning sensation that is called **heartburn** because it is experienced in a region very near the heart; it is unrelated to any cardiac problem. Drinking alcohol and smoking can cause the sphincter to relax, worsening the problem. The symptoms of GERD often can be controlled by avoiding foods that strongly stimulate stomach acid secretion (coffee, chocolate, tomatoes, fatty foods, orange juice, peppermint, spearmint, and onions). Other acid-reducing strategies include taking over-the-counter histamine-2 (H_2) blockers such as Tagamet HB® or Pepcid AC® 30 to 60 minutes before eating to block acid secretion, and neutralizing acid that has already been secreted with antacids such as Tums® or Maalox®. Symptoms are less likely to occur if food is eaten in smaller amounts and if the person does not lie down immediately after a meal. GERD may be associated with cancer of the esophagus. •

✔ CHECKPOINT

17. What does deglutition mean?
18. What occurs during the voluntary and pharyngeal phases of swallowing?
19. Does peristalsis "push" or "pull" food along the gastrointestinal tract?

24.9 STOMACH

◉ OBJECTIVE

• Describe the location, anatomy, histology, and functions of the stomach.

The **stomach** is a J-shaped enlargement of the GI tract directly inferior to the diaphragm in the abdomen. The stomach connects the esophagus to the duodenum, the first part of the small intestine (Figure 24.11). Because a meal can be eaten much more quickly than the intestines can digest and absorb it, one of the functions of the stomach is to serve as a mixing chamber and holding reservoir. At appropriate intervals after food is ingested, the stomach forces a small quantity of material into the first portion of the small intestine. The position and size of the stomach vary continually; the diaphragm pushes it inferiorly with each inhalation and pulls it superiorly with each exhalation. Empty, it is about the size of a large sausage, but it is the most distensible part of the GI tract and can accommodate a large quantity of food. In the stomach, digestion of starch and triglycerides continues, digestion of proteins begins, the semisolid bolus is converted to a liquid, and certain substances are absorbed.

Anatomy of the Stomach

The stomach has four main regions: the cardia, fundus, body, and pyloric part (Figure 24.11). The **cardia** (KAR-dē-a) surrounds the

Figure 24.11 External and internal anatomy of the stomach.

The four regions of the stomach are the cardia, fundus, body, and pyloric part.

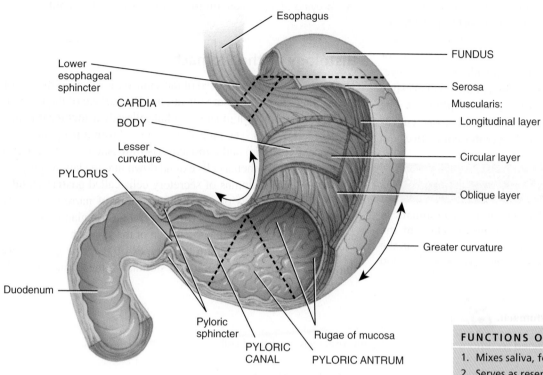

(a) Anterior view of regions of stomach

FUNCTIONS OF THE STOMACH

1. Mixes saliva, food, and gastric juice to form chyme.
2. Serves as reservoir for food before release into small intestine.
3. Secretes gastric juice, which contains HCl (kills bacteria and denatures protein), pepsin (begins the digestion of proteins), intrinsic factor (aids absorption of vitamin B_{12}), and gastric lipase (aids digestion of triglycerides).
4. Secretes gastrin into blood.

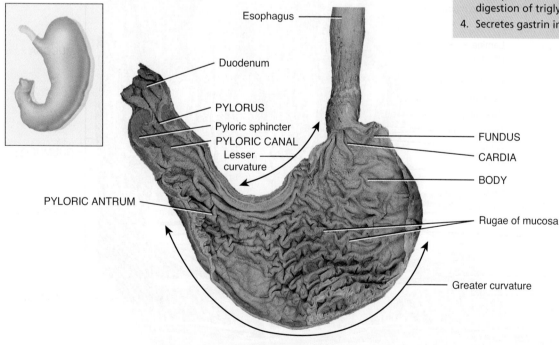

(b) Anterior view of internal anatomy

? **After a very large meal, does your stomach still have rugae?**

superior opening of the stomach. The rounded portion superior to and to the left of the cardia is the **fundus** (FUN-dus). Inferior to the fundus is the large central portion of the stomach, the **body.** The **pyloric part** is divisible into three regions. The first region, the **pyloric antrum,** connects to the body of the stomach. The second region, the **pyloric canal,** leads to the third region, the **pylorus** (pī-LOR-us; *pyl-* = gate; *-orus* = guard), which in turn connects to the duodenun. When the stomach is empty, the mucosa lies in large folds, or **rugae** (ROO-gē = wrinkles), that can be seen with the unaided eye. The pylorus communicates with the duodenum of the small intestine via a smooth muscle sphincter called the **pyloric sphincter.** The concave medial border of the stomach is called the **lesser curvature;** the convex lateral border is called the **greater curvature.**

⚕ CLINICAL CONNECTION | *Pylorospasm and Pyloric Stenosis*

Two abnormalities of the pyloric sphincter can occur in infants. In **pylorospasm** (pī-LOR-ō-spazm), the smooth muscle fibers of the sphincter fail to relax normally, so food does not pass easily from the stomach to the small intestine, the stomach becomes overly full, and the infant vomits often to relieve the pressure. Pylorospasm is treated by drugs that relax the muscle fibers of the pyloric sphincter. **Pyloric stenosis** (ste-NŌ-sis) is a narrowing of the pyloric sphincter that must be corrected surgically. The hallmark symptom is *projectile vomiting*—the spraying of liquid vomitus some distance from the infant. •

Histology of the Stomach

The stomach wall is composed of the same basic layers as the rest of the GI tract, with certain modifications. The surface of the **mucosa** is a layer of simple columnar epithelial cells called **surface mucous cells** (Figure 24.12). The mucosa contains a **lamina propria** (areolar connective tissue) and a **muscularis mucosae** (smooth muscle) (Figure 24.12). Epithelial cells extend down into the lamina propria, where they form columns of secretory cells called **gastric glands.** Several gastric glands open into the bottom of narrow channels called **gastric pits.** Secretions from several gastric glands flow into each gastric pit and then into the lumen of the stomach.

The gastric glands contain three types of *exocrine gland cells* that secrete their products into the stomach lumen: mucous neck

Figure 24.12 Histology of the stomach.

🔒 Gastric juice is the combined secretions of mucous cells, parietal cells, and chief cells.

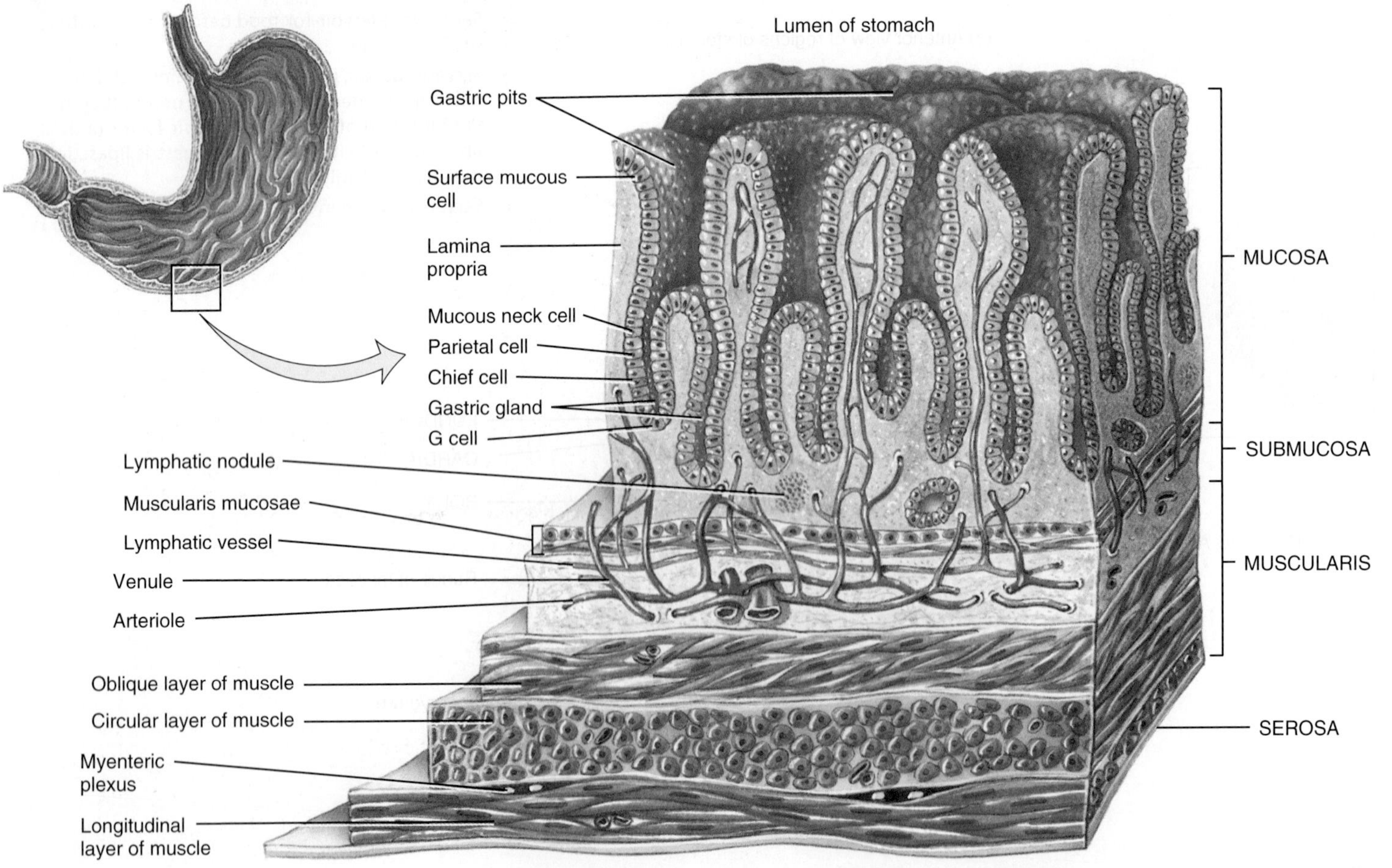

Lumen of stomach

Gastric pits

Surface mucous cell

Lamina propria

Mucous neck cell

Parietal cell

Chief cell

Gastric gland

G cell

Lymphatic nodule

Muscularis mucosae

Lymphatic vessel

Venule

Arteriole

Oblique layer of muscle

Circular layer of muscle

Myenteric plexus

Longitudinal layer of muscle

MUCOSA

SUBMUCOSA

MUSCULARIS

SEROSA

(a) Three-dimensional view of layers of stomach

cells, chief cells, and parietal cells. Both surface mucous cells and **mucous neck cells** secrete mucus (Figure 24.12b). **Parietal cells** produce intrinsic factor (needed for absorption of vitamin B_{12}) and hydrochloric acid. The **chief cells** secrete pepsinogen and gastric lipase. The secretions of the mucous, parietal, and chief cells form **gastric juice,** which totals 2000–3000 mL (roughly 2–3 qt) per day. In addition, gastric glands include a type of enteroendocrine cell, the **G cell,** which is located mainly in the pyloric antrum and secretes the hormone gastrin into the bloodstream. As we will see shortly, this hormone stimulates several aspects of gastric activity.

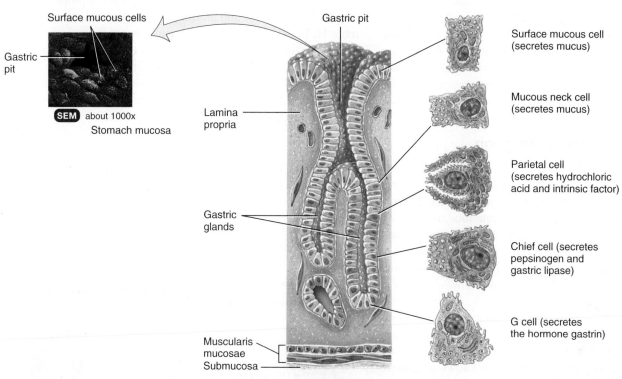

(b) Sectional view of stomach mucosa showing gastric glands and cell types

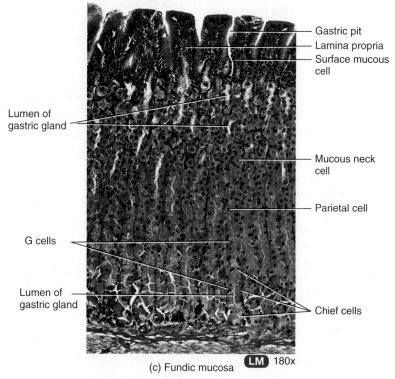

(c) Fundic mucosa

? Where is HCl secreted, and what are its functions?

Three additional layers lie deep to the mucosa. The **submucosa** of the stomach is composed of areolar connective tissue. The **muscularis** has three layers of smooth muscle (rather than the two found in the esophagus and small and large intestines): an outer longitudinal layer, a middle circular layer, and an inner oblique layer. The oblique layer is limited primarily to the body of the stomach. The **serosa** is composed of simple squamous epithelium (mesothelium) and areolar connective tissue; the portion of the serosa covering the stomach is part of the visceral peritoneum. At the lesser curvature of the stomach, the visceral peritoneum extends upward to the liver as the lesser omentum. At the greater curvature of the stomach, the visceral peritoneum continues downward as the greater omentum and drapes over the intestines.

Mechanical and Chemical Digestion in the Stomach

Several minutes after food enters the stomach, gentle, rippling, peristaltic movements called **mixing waves** pass over the stomach every 15 to 25 seconds. These waves macerate food, mix it with secretions of the gastric glands, and reduce it to a soupy liquid called **chyme** (KĪM = juice). Few mixing waves are observed in the fundus, which primarily has a storage function. As digestion proceeds in the stomach, more vigorous mixing waves begin at the body of the stomach and intensify as they reach the pylorus. The pyloric sphincter normally remains almost, but not completely, closed. As food reaches the pylorus, each mixing wave periodically forces about 3 mL of chyme into the duodenum through the pyloric sphincter, a phenomenon known as **gastric emptying.** Most of the chyme is forced back into the body of the stomach, where mixing continues. The next wave pushes the chyme forward again and forces a little more into the duodenum. These forward and backward movements of the gastric contents are responsible for most mixing in the stomach.

Foods may remain in the fundus for about an hour without becoming mixed with gastric juice. During this time, digestion by salivary amylase continues. Soon, however, the churning action mixes chyme with acidic gastric juice, inactivating salivary amylase and activating lingual lipase, which starts to digest triglycerides into fatty acids and diglycerides.

Although parietal cells secrete hydrogen ions (H^+) and chloride ions (Cl^-) separately into the stomach lumen, the net effect is secretion of hydrochloric acid (HCl). **Proton pumps** powered by H^+/K^+ ATPases actively transport H^+ into the lumen while bringing potassium ions (K^+) into the cell (Figure 24.13). At the same time, Cl^- and K^+ diffuse out into the lumen through Cl^- and K^+ channels in the apical membrane. The enzyme *carbonic anhydrase,* which is especially plentiful in parietal cells, catalyzes the formation of carbonic acid (H_2CO_3) from water (H_2O) and carbon dioxide (CO_2). As carbonic acid dissociates, it provides a ready source of H^+ for the proton pumps but also generates bicarbonate ions (HCO_3^-). As HCO_3^- builds up in the cytosol, it exits the parietal cell in exchange for Cl^- via Cl^-/HCO_3^- antiporters in the basolateral membrane (next to the lamina propria). HCO_3^- diffuses into nearby blood capillaries.

This "alkaline tide" of bicarbonate ions entering the bloodstream after a meal may be large enough to elevate blood pH slightly and make urine more alkaline.

HCl secretion by parietal cells can be stimulated by several sources: acetylcholine (ACh) released by parasympathetic neurons, gastrin secreted by G cells, and histamine, which is a paracrine substance released by mast cells in the nearby lamina propria (Figure 24.14). Acetylcholine and gastrin stimulate parietal cells to secrete more HCl in the presence of histamine. In other words, histamine acts synergistically, enhancing the effects of acetylcholine and gastrin. Receptors for all three substances are present in the plasma membrane of parietal cells. The histamine receptors on parietal cells are called H_2 receptors; they mediate different responses than do the H_1 receptors involved in allergic responses.

Figure 24.13 Secretion of HCl (hydrochloric acid) by parietal cells in the stomach.

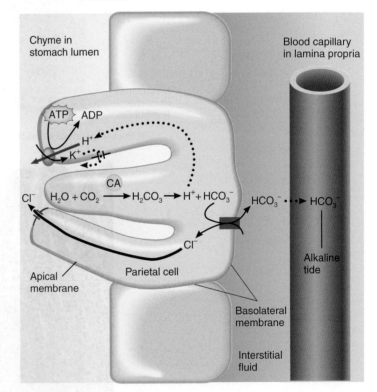

Proton pumps, powered by ATP, secrete the H^+; Cl^- diffuses into the stomach lumen through Cl^- channels.

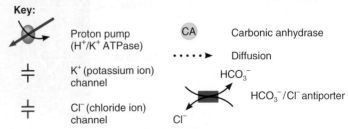

? **What molecule is the source of the hydrogen ions that are secreted into gastric juice?**

The strongly acidic fluid of the stomach kills many microbes in food. HCl partially denatures (unfolds) proteins in food and stimulates the secretion of hormones that promote the flow of bile and pancreatic juice. Enzymatic digestion of proteins also begins in the stomach. The only proteolytic (protein-digesting) enzyme in the stomach is **pepsin,** which is secreted by chief cells. Pepsin severs certain peptide bonds between amino acids, breaking down a protein chain of many amino acids into smaller peptide fragments. Pepsin is most effective in the very acidic environment of the stomach (pH 2); it becomes inactive at a higher pH.

What keeps pepsin from digesting the protein in stomach cells along with the food? First, pepsin is secreted in an inactive form called *pepsinogen;* in this form, it cannot digest the proteins in the chief cells that produce it. Pepsinogen is not converted into active pepsin until it comes in contact with hydrochloric acid secreted by parietal cells or active pepsin molecules. Second, the stomach epithelial cells are protected from gastric juices by a layer 1–3 mm thick of alkaline mucus secreted by surface mucous cells and mucous neck cells.

Another enzyme of the stomach is **gastric lipase,** which splits the short-chain triglycerides (fats and oils) in fat molecules (such as those found in milk) into fatty acids and monoglycerides. A monoglyceride consists of a glycerol molecule that is attached to one fatty acid molecule. This enzyme, which has a limited role in the adult stomach, operates best at a pH of 5–6. More important than either lingual lipase or gastric lipase is pancreatic lipase, an enzyme secreted by the pancreas into the small intestine.

Only a small amount of nutrients are absorbed in the stomach because its epithelial cells are impermeable to most materials. However, mucous cells of the stomach absorb some water, ions, and short-chain fatty acids, as well as certain drugs (especially aspirin) and alcohol.

Within 2 to 4 hours after eating a meal, the stomach has emptied its contents into the duodenum. Foods rich in carbohydrate spend the least time in the stomach; high-protein foods remain somewhat longer, and emptying is slowest after a fat-laden meal containing large amounts of triglycerides.

Table 24.3 summarizes the digestive activities of the stomach.

Figure 24.14 Regulation of HCl secretion.

HCl secretion by parietal cells can be stimulated by several sources: acetylcholine (ACh), gastrin, and histamine.

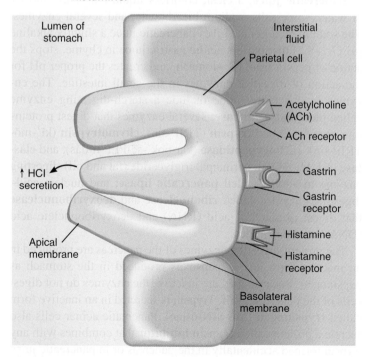

Among the sources that stimulate HCl secretion, which one is a paracrine agent that is released by mast cells in the lamina propria?

TABLE 24.3

Summary of Digestive Activities in the Stomach

STRUCTURE	ACTIVITY	RESULT
Mucosa		
Surface mucous cells and mucous neck cells	Secrete mucus.	Forms protective barrier that prevents digestion of stomach wall.
	Absorption.	Small quantity of water, ions, short-chain fatty acids, and some drugs enter bloodstream.
Parietal cells	Secrete intrinsic factor.	Needed for absorption of vitamin B_{12} (used in red blood cell formation, or erythropoiesis).
	Secrete hydrochloric acid.	Kills microbes in food; denatures proteins; converts pepsinogen into pepsin.
Chief cells	Secrete pepsinogen.	Pepsin (activated form) breaks down proteins into peptides.
	Secrete gastric lipase.	Splits triglycerides into fatty acids and monoglycerides.
G cells	Secrete gastrin.	Stimulates parietal cells to secrete HCl and chief cells to secrete pepsinogen; contracts lower esophageal sphincter, increases motility of stomach, and relaxes pyloric sphincter.
Muscularis	Mixing waves (gentle peristaltic movements).	Churns and physically breaks down food and mixes it with gastric juice, forming chyme. Forces chyme through pyloric sphincter.
Pyloric sphincter	Opens to permit passage of chyme into duodenum.	Regulates passage of chyme from stomach to duodenum; prevents backflow of chyme from duodenum to stomach.

Vomiting or *emesis* is the forcible expulsion of the contents of the upper GI tract (stomach and sometimes duodenum) through the mouth. The strongest stimuli for vomiting are irritation and distension of the stomach; other stimuli include unpleasant sights, general anesthesia, dizziness, and certain drugs such as morphine and derivatives of digitalis. Nerve impulses are transmitted to the vomiting center in the medulla oblongata, and returning impulses propagate to the upper GI tract organs, diaphragm, and abdominal muscles. Vomiting involves squeezing the stomach between the diaphragm and abdominal muscles and expelling the contents through open esophageal sphincters. Prolonged vomiting, especially in infants and elderly people, can be serious because the loss of acidic gastric juice can lead to alkalosis (higher than normal blood pH), dehydration, and damage to the esophagus and teeth. •

✔ CHECKPOINT

20. Compare the epithelium of the esophagus with that of the stomach. How is each adapted to the function of the organ?
21. What is the importance of rugae, surface mucous cells, mucous neck cells, chief cells, parietal cells, and G cells in the stomach?
22. What is the role of pepsin? Why is it secreted in an inactive form?
23. What are the functions of gastric lipase and lingual lipase in the stomach?

24.10 PANCREAS

◉ **OBJECTIVE**

• Describe the location, anatomy, histology, and function of the pancreas.

From the stomach, chyme passes into the small intestine. Because chemical digestion in the small intestine depends on activities of the pancreas, liver, and gallbladder, we first consider the activities of these accessory digestive organs and their contributions to digestion in the small intestine.

Anatomy of the Pancreas

The **pancreas** (*pan-* = all; *-creas* = flesh), a retroperitoneal gland that is about 12–15 cm (5–6 in.) long and 2.5 cm (1 in.) thick, lies posterior to the greater curvature of the stomach. The pancreas consists of a head, a body, and a tail and is usually connected to the duodenum by two ducts (Figure 24.15a). The **head** is the expanded portion of the organ near the curve of the duodenum; superior to and to the left of the head are the central **body** and the tapering **tail.**

Pancreatic juices are secreted by exocrine cells into small ducts that ultimately unite to form two larger ducts, the pancreatic duct and the accessory duct. These in turn convey the secretions into the small intestine. The **pancreatic duct,** or **duct of Wirsung** (VĒR-sung), is the larger of the two ducts. In most people, the pancreatic duct joins the common bile duct from the liver and gallbladder and enters the duodenum as a dilated common duct called the **hepatopancreatic ampulla** (hep′-a-tō-pan-krē-A-tik), or **ampulla of Vater** (FAH-ter). The ampulla opens on an elevation of the duodenal mucosa known as the **major duodenal papilla,** which lies about 10 cm (4 in.) inferior to the pyloric sphincter of the stomach. The passage of pancreatic juice and bile through the hepatopancreatic ampulla into the small intestine is regulated by a mass of smooth muscle surrounding the ampulla known as the **sphincter of the hepatopancreatic ampulla,** or **sphincter of Oddi** (OD-ē). The other major duct of the pancreas, the **accessory duct (duct of Santorini),** leads from the pancreas and empties into the duodenum about 2.5 cm (1 in.) superior to the hepatopancreatic ampulla.

Histology of the Pancreas

The pancreas is made up of small clusters of glandular epithelial cells. About 99% of the clusters, called **acini** (AS-i-nī), constitute the *exocrine* portion of the organ (see Figure 18.18b, c). The cells within acini secrete a mixture of fluid and digestive enzymes called pancreatic juice. The remaining 1% of the clusters, called **pancreatic islets** (Ī-lets) **(islets of Langerhans),** form the *endocrine* portion of the pancreas. These cells secrete the hormones glucagon, insulin, somatostatin, and pancreatic polypeptide. The functions of these hormones are discussed in Chapter 18.

Composition and Functions of Pancreatic Juice

Each day the pancreas produces 1200–1500 mL (about 1.2–1.5 qt) of **pancreatic juice,** a clear, colorless liquid consisting mostly of water, some salts, sodium bicarbonate, and several enzymes. The sodium bicarbonate gives pancreatic juice a slightly alkaline pH (7.1–8.2) that buffers acidic gastric juice in chyme, stops the action of pepsin from the stomach, and creates the proper pH for the action of digestive enzymes in the small intestine. The enzymes in pancreatic juice include a starch-digesting enzyme called **pancreatic amylase;** several enzymes that digest proteins into peptides called **trypsin** (TRIP-sin), **chymotrypsin** (kī′-mō-TRIP-sin), **carboxypeptidase** (kar-bok′-sē-PEP-ti-dās), and **elastase** (ē-LAS-tās); the principal triglyceride (fat and oil)-digesting enzyme in adults, called **pancreatic lipase;** and nucleic acid–digesting enzymes called **ribonuclease** and **deoxyribonuclease** that digest ribonucleic acid (RNA) and deoxyribonucleic acid (DNA) into nucleotides.

The protein-digesting enzymes of the pancreas are produced in an inactive form just as pepsin is produced in the stomach as pepsinogen. Because they are inactive, the enzymes do not digest cells of the pancreas itself. Trypsin is secreted in an inactive form called **trypsinogen** (trip-SIN-ō-jen). Pancreatic acinar cells also secrete a protein called **trypsin inhibitor** that combines with any trypsin formed accidentally in the pancreas or in pancreatic juice and blocks its enzymatic activity. When trypsinogen reaches the lumen of the small intestine, it encounters an activating brush-border enzyme called **enterokinase** (en′-ter-ō-KĪ-nās), which

Figure 24.15 **Relation of the pancreas to the liver, gallbladder, and duodenum.** The inset (b) shows details of the common bile duct and pancreatic duct forming the hepatopancreatic ampulla (ampulla of Vater) and emptying into the duodenum.

Pancreatic enzymes digest starches (polysaccharides), proteins, triglycerides, and nucleic acids.

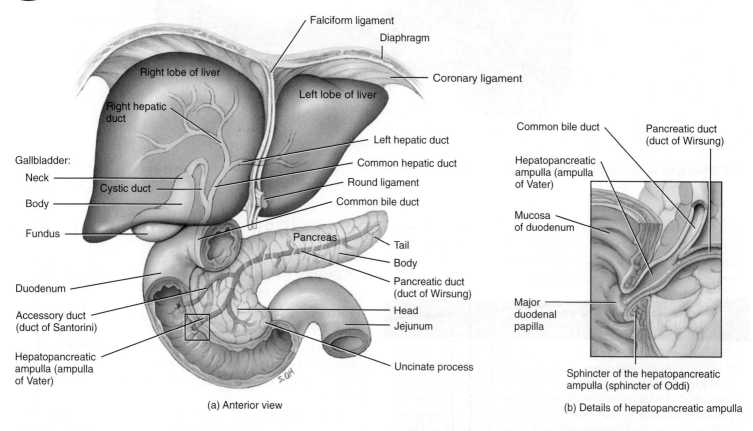

(a) Anterior view

(b) Details of hepatopancreatic ampulla

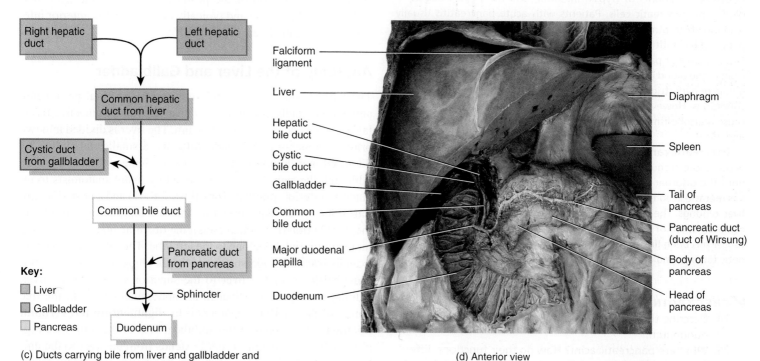

(c) Ducts carrying bile from liver and gallbladder and pancreatic juice from pancreas to duodenum

(d) Anterior view

FIGURE 24.15 CONTINUES ▶

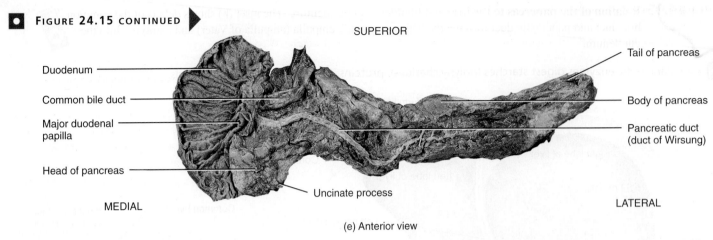

SUPERIOR

Duodenum

Common bile duct

Major duodenal papilla

Head of pancreas

Tail of pancreas

Body of pancreas

Pancreatic duct (duct of Wirsung)

Uncinate process

MEDIAL

LATERAL

(e) Anterior view

? **What type of fluid is found in the pancreatic duct? The common bile duct? The hepatopancreatic ampulla?**

splits off part of the trypsinogen molecule to form trypsin. In turn, trypsin acts on the inactive precursors (called **chymotrypsinogen, procarboxypeptidase,** and **proelastase**) to produce chymotrypsin, carboxypeptidase, and elastase, respectively.

CLINICAL CONNECTION | *Pancreatitis and Pancreatic Cancer*

Inflammation of the pancreas, as may occur in association with alcohol abuse or chronic gallstones, is called **pancreatitis** (pan'-krē-a-TĪ-tis). In a more severe condition known as **acute pancreatitis,** which is associated with heavy alcohol intake or biliary tract obstruction, the pancreatic cells may release either trypsin instead of trypsinogen or insufficient amounts of trypsin inhibitor, and the trypsin begins to digest the pancreatic cells. Patients with acute pancreatitis usually respond to treatment, but recurrent attacks are the rule. In some people pancreatitis is idiopathic, meaning that the cause is unknown. Other causes of pancreatitis include cystic fibrosis, high levels of calcium in the blood (hypercalcemia), high levels of blood fats (hyperlipidemia or hypertriglyceridemia), some drugs, and certain autoimmune conditions. However, in roughly 70% of adults with pancreatitis, the cause is alcoholism. Often the first episode happens between ages 30 and 40.

Pancreatic cancer usually affects people over 50 years of age and occurs more frequently in males. Typically, there are few symptoms until the disorder reaches an advanced stage and often not until it has metastasized to other parts of the body such as the lymph nodes, liver, or lungs. The disease is nearly always fatal and is the fourth most common cause of death from cancer in the United States. Pancreatic cancer has been linked to fatty foods, high alcohol consumption, genetic factors, smoking, and chronic pancreatitis. •

✔CHECKPOINT

24. Describe the duct system connecting the pancreas to the duodenum.

25. What are pancreatic acini? How do their functions differ from those of the pancreatic islets (islets of Langerhans)?

26. What are the digestive functions of the components of pancreatic juice?

24.11 LIVER AND GALLBLADDER

◉ **OBJECTIVE**

• Describe the location, anatomy, histology, and functions of the liver and gallbladder.

The **liver** is the heaviest gland of the body, weighing about 1.4 kg (about 3 lb) in an average adult. Of all of the organs of the body, it is second only to the skin in size. The liver is inferior to the diaphragm and occupies most of the right hypochondriac and part of the epigastric regions of the abdominopelvic cavity (see Figure 1.12b).

The **gallbladder** (*gall-* = bile) is a pear-shaped sac that is located in a depression of the posterior surface of the liver. It is 7–10 cm (3–4 in.) long and typically hangs from the anterior inferior margin of the liver (Figure 24.15a).

Anatomy of the Liver and Gallbladder

The liver is almost completely covered by visceral peritoneum and is completely covered by a dense irregular connective tissue layer that lies deep to the peritoneum. The liver is divided into two principal lobes—a large **right lobe** and a smaller **left lobe**—by the **falciform ligament,** a fold of the mesentery (Figure 24.15a). Although the right lobe is considered by many anatomists to include an inferior **quadrate lobe** (kwa-DRĀT) and a posterior **caudate lobe** (KAW-dāt), based on internal morphology (primarily the distribution of blood vessels), the quadrate and caudate lobes more appropriately belong to the left lobe. The falciform ligament extends from the undersurface of the diaphragm between the two principal lobes of the liver to the superior surface of the liver, helping to suspend the liver in the abdominal cavity. In the free border of the falciform ligament is the **ligamentum teres (round ligament),** a remnant of the umbilical vein of the fetus (see Figure 21.30a, b); this fibrous cord extends from the liver to the umbilicus. The right and left **coronary ligaments** are narrow extensions of the parietal peritoneum that suspend the liver from the diaphragm.

The parts of the gallbladder include the broad **fundus,** which projects inferiorly beyond the inferior border of the liver; the **body,** the central portion; and the **neck,** the tapered portion. The body and neck project superiorly.

Histology of the Liver and Gallbladder

Histologically, the liver is composed of several components (Figure 24.16a–c):

1. **Hepatocytes** (*hepat-* = liver; *-cytes* = cells). Hepatocytes are the major functional cells of the liver and perform a wide array of metabolic, secretory, and endocrine functions. These are specialized epithelial cells with 5 to 12 sides that make up about 80% of the volume of the liver. Hepatocytes form complex three-dimensional arrangements called **hepatic laminae** (LAM-i-nē). The hepatic laminae are plates of hepatocytes one cell thick bordered on either side by the endothelial-lined vascular spaces called hepatic sinusoids. The hepatic laminae are highly branched, irregular structures. Grooves in the cell membranes between neighboring hepatocytes provide spaces for canaliculi (described next) into which the hepatocytes secrete bile. Bile, a yellow, brownish, or olive-green liquid secreted by hepatocytes, serves as both an excretory product and a digestive secretion.

2. **Bile canaliculi** (kan-a-LIK-ū-li = small canals). These are small ducts between hepatocytes that collect bile produced by the hepatocytes. From bile canaliculi, bile passes into **bile ductules** and then **bile ducts.** The bile ducts merge and eventually form the larger **right** and **left hepatic ducts,** which unite and exit the liver as the **common hepatic duct** (see Figure 24.15). The common hepatic duct joins the **cystic duct** (*cystic* = bladder) from the gallbladder to form the **common bile duct.** From here, bile enters the small intestine to participate in digestion.

3. **Hepatic sinusoids.** These are highly permeable blood capillaries between rows of hepatocytes that receive oxygenated blood from branches of the hepatic artery and nutrient-rich deoxygenated blood from branches of the hepatic portal vein. Recall that the hepatic portal vein brings venous blood from the gastrointestinal organs and spleen into the liver. Hepatic sinusoids converge and deliver blood into a **central vein.** From central veins the blood flows into the **hepatic veins,** which

Figure 24.16 **Histology of the liver.**

Histologically, the liver is composed of hepatocytes, bile canaliculi, and hepatic sinusoids.

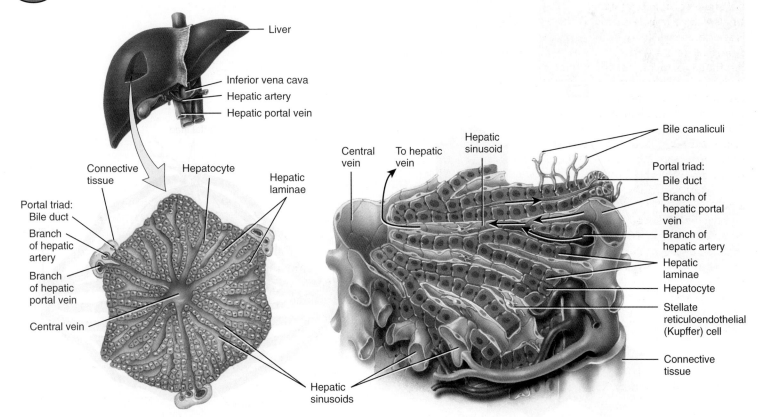

(a) Overview of histological components of liver

(b) Details of histological components of liver

FIGURE 24.16 CONTINUES ▶

⬛ FIGURE 24.16 CONTINUED ▶

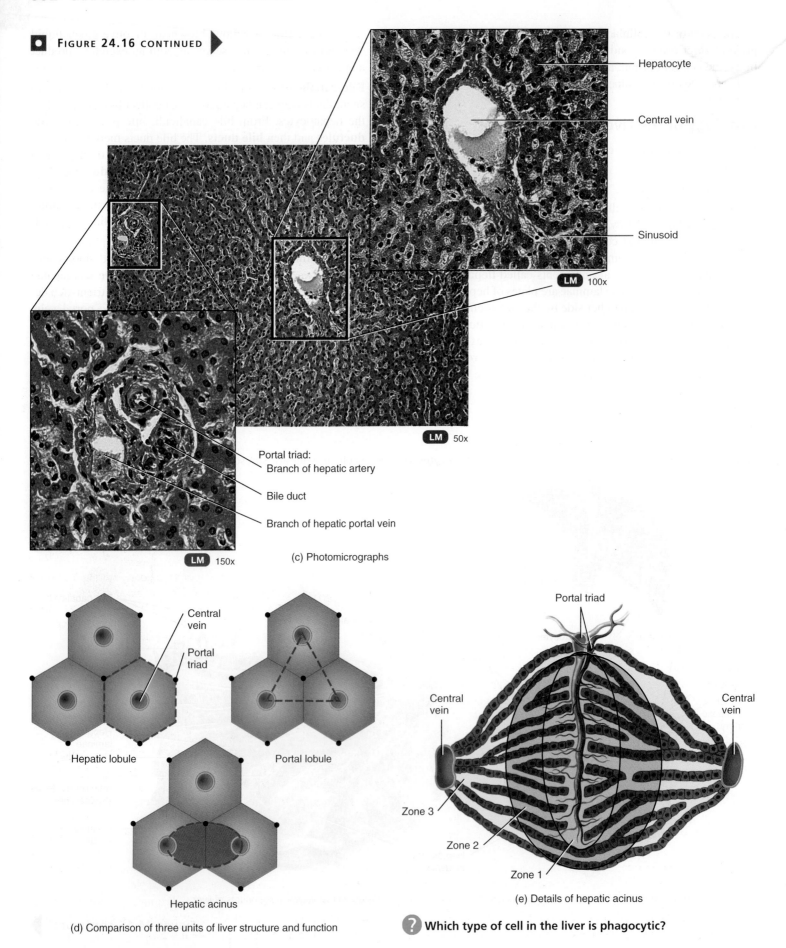

Hepatocyte

Central vein

Sinusoid

LM 100x

LM 50x

Portal triad:
Branch of hepatic artery

Bile duct

Branch of hepatic portal vein

LM 150x

(c) Photomicrographs

Central vein

Portal triad

Hepatic lobule

Portal lobule

Hepatic acinus

(d) Comparison of three units of liver structure and function

Portal triad

Central vein

Central vein

Zone 3

Zone 2

Zone 1

(e) Details of hepatic acinus

❓ **Which type of cell in the liver is phagocytic?**

drain into the inferior vena cava (see Figure 21.28). In contrast to blood, which flows toward a central vein, bile flows in the opposite direction. Also present in the hepatic sinusoids are fixed phagocytes called **stellate reticuloendothelial (Kupffer) cells** (STEL-āt re-tik′-ū-lō-en′-dō-THĒ-lē-al [KŪP-fer]), which destroy worn-out white and red blood cells, bacteria, and other foreign matter in the venous blood draining from the gastrointestinal tract.

Together, a bile duct, branch of the hepatic artery, and branch of the hepatic vein are referred to as a **portal triad** (*tri* = three).

The hepatocytes, bile duct system, and hepatic sinusoids can be organized into anatomical and functional units in three different ways:

1. **Hepatic lobule.** For years, anatomists described the hepatic lobule as the functional unit of the liver. According to this model, each hepatic lobule is shaped like a hexagon (six-sided structure) (Figure 24.16d, left). At its center is the central vein, and radiating out from it are rows of hepatocytes and hepatic sinusoids. Located at three corners of the hexagon is a portal triad. This model is based on a description of the liver of adult pigs. In the human liver it is difficult to find such well-defined hepatic lobules surrounded by thick layers of connective tissue.

2. **Portal lobule.** This model emphasized the exocrine function of the liver, that is, bile secretion. Accordingly, the bile duct of a portal triad is taken as the center of the portal lobule. The portal lobule is triangular in shape and is defined by three imaginary straight lines that connect three central veins that are closest to the portal triad (Figure 24.16d, center). This model has not gained widespread acceptance.

3. **Hepatic acinus.** In recent years, the preferred structural and functional unit of the liver is the hepatic **acinus** (AS-i-nus). Each hepatic acinus is an approximately oval mass that includes portions of two neighboring hepatic lobules. The short axis of the hepatic acinus is defined by branches of the portal triad—branches of the hepatic artery, vein, and bile ducts—that run along the border of the hepatic lobules. The long axis of the acinus is defined by two imaginary curved lines, which connect the two central veins closest to the short axis (Figure 24.16d, right). Hepatocytes in the hepatic acinus are arranged in three zones around the short axis, with no sharp boundaries between them (Figure 24.16e). Cells in zone 1 are closest to the branches of the portal triad and the first to receive incoming oxygen, nutrients, and toxins from incoming blood. These cells are the first ones to take up glucose and store it as glycogen after a meal and break down glycogen to glucose during fasting. They are also the first to show morphological changes following bile duct obstruction or exposure to toxic substances. Zone 1 cells are the last ones to die if circulation is impaired and the first ones to regenerate. Cells in zone 3 are farthest from branches of the portal triad and are the last to show the effects of bile obstruction or exposure to toxins, the first ones to show the effects of impaired circulation, and the

last ones to regenerate. Zone 3 cells also are the first to show evidence of fat accumulation. Cells in zone 2 have structural and functional characteristics intermediate between the cells in zones 1 and 3.

The hepatic acinus is the smallest structural and functional unit of the liver. Its popularity and appeal are based on the fact that it provides a logical description and interpretation of (1) patterns of glycogen storage and release and (2) toxic effects, degeneration, and regeneration relative to the proximity of the acinar zones to branches of the portal triad.

The mucosa of the gallbladder consists of simple columnar epithelium arranged in rugae resembling those of the stomach. The wall of the gallbladder lacks a submucosa. The middle, muscular coat of the wall consists of smooth muscle fibers. Contraction of the smooth muscle fibers ejects the contents of the gallbladder into the **cystic duct.** The gallbladder's outer coat is the visceral peritoneum. The functions of the gallbladder are to store and concentrate the bile produced by the liver (up to tenfold) until it is needed in the small intestine. In the concentration process, water and ions are absorbed by the gallbladder mucosa.

CLINICAL CONNECTION | *Jaundice*

Jaundice (JAWN-dis = yellowed) is a yellowish coloration of the sclerae (whites of the eyes), skin, and mucous membranes due to a buildup of a yellow compound called bilirubin. After bilirubin is formed from the breakdown of the heme pigment in aged red blood cells, it is transported to the liver, where it is processed and eventually excreted into bile. The three main categories of jaundice are (1) *prehepatic jaundice,* due to excess production of bilirubin; (2) *hepatic jaundice,* due to congenital liver disease, cirrhosis of the liver, or hepatitis; and (3) *extrahepatic jaundice,* due to blockage of bile drainage by gallstones or cancer of the bowel or the pancreas.

Because the liver of a newborn functions poorly for the first week or so, many babies experience a mild form of jaundice called *neonatal (physiological) jaundice* that disappears as the liver matures. Usually, it is treated by exposing the infant to blue light, which converts bilirubin into substances the kidneys can excrete. •

Blood Supply of the Liver

The liver receives blood from two sources (Figure 24.17). From the hepatic artery it obtains oxygenated blood, and from the hepatic portal vein it receives deoxygenated blood containing newly absorbed nutrients, drugs, and possibly microbes and toxins from the gastrointestinal tract (see Figure 21.28). Branches of both the hepatic artery and the hepatic portal vein carry blood into liver sinusoids, where oxygen, most of the nutrients, and certain toxic substances are taken up by the hepatocytes. Products manufactured by the hepatocytes and nutrients needed by other cells are secreted back into the blood, which then drains into the central vein and eventually passes into a hepatic vein. Because blood from the gastrointestinal tract passes through the liver as part of the hepatic portal circulation, the liver is often a site for metastasis of cancer that originates in the GI tract.

Figure 24.17 Hepatic blood flow: sources, path through the liver, and return to the heart.

The liver receives oxygenated blood via the hepatic artery and nutrient-rich deoxygenated blood via the hepatic portal vein.

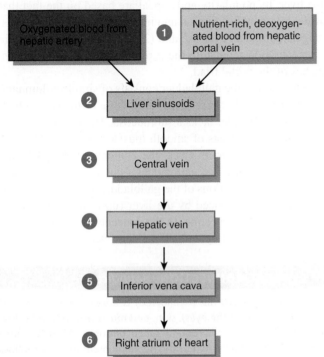

? During the first few hours after a meal, how does the chemical composition of blood change as it flows through the liver sinusoids?

CLINICAL CONNECTION | Liver Function Tests

Liver function tests are blood tests designed to determine the presence of certain chemicals released by liver cells. These include albumin globulinase, alanine aminotransferase (ALT), aspartate aminotransferase (AST), alkaline phosphatase (ALP), gamma-glutamyl-transpeptidase (GGT), and bilirubin. The tests are used to evaluate and monitor liver disease or damage. Common causes of elevated liver enzymes include nonsteroidal anti-inflammatory drugs, cholesterol-lowering medications, some antibiotics, alcohol, diabetes, infections (viral hepatitis and mononucleosis), gallstones, tumors of the liver, and excessive use of herbal supplements such as kava, comfrey, pennyroyal, dandelion root, skullcap, and ephedra. •

Functions of the Liver and Gallbladder

Each day, hepatocytes secrete 800–1000 mL (about 1 qt) of **bile,** a yellow, brownish, or olive-green liquid. It has a pH of 7.6–8.6 and consists mostly of water, bile salts, cholesterol, a phospholipid called lecithin, bile pigments, and several ions.

The principal bile pigment is **bilirubin** (bil-i-ROO-bin). The phagocytosis of aged red blood cells liberates iron, globin, and bilirubin (derived from heme) (see Figure 19.5). The iron and globin are recycled; the bilirubin is secreted into the bile and is eventually broken down in the intestine. One of its breakdown products—**stercobilin** (ster-kō-BĪ-lin)—gives feces their normal brown color.

Bile is partially an excretory product and partially a digestive secretion. Bile salts, which are sodium salts and potassium salts of bile acids (mostly chenodeoxycholic acid and cholic acid), play a role in **emulsification** (ē-mul′-si-fi-KĀ-shun), the breakdown of large lipid globules into a suspension of small lipid globules. The small lipid globules present a very large surface area that allows pancreatic lipase to more rapidly accomplish digestion of triglycerides. Bile salts also aid in the absorption of lipids following their digestion.

Although hepatocytes continually release bile, they increase production and secretion when the portal blood contains more bile acids; thus, as digestion and absorption continue in the small intestine, bile release increases. Between meals, after most absorption has occurred, bile flows into the gallbladder for storage because the sphincter of the hepatopancreatic ampulla (sphincter of Oddi; see Figure 24.15) closes off the entrance to the duodenum. The sphincter surrounds the hepatopancreatic ampulla.

CLINICAL CONNECTION | Gallstones

If bile contains either insufficient bile salts or lecithin or excessive cholesterol, the cholesterol may crystallize to form **gallstones.** As they grow in size and number, gallstones may cause minimal, intermittent, or complete obstruction to the flow of bile from the gallbladder into the duodenum. Treatment consists of using gallstone-dissolving drugs, lithotripsy (shock-wave therapy), or surgery. For people with recurrent gallstones or for whom drugs or lithotripsy is not indicated, **cholecystectomy** (kō′-lē-sis-TEK-tō-mē)—the removal of the gallbladder and its contents—is necessary. More than half a million cholecystectomies are performed each year in the United States. To prevent side effects resulting from a loss of the gallbladder, patients should make lifestyle and dietary changes, including the following: (1) limiting the intake of saturated fat; (2) avoiding the consumption of alcoholic beverages; (3) eating smaller amounts of food during a meal and eating five to six smaller meals per day instead of two to three larger meals; and (4) taking vitamin and mineral supplements. •

In addition to secreting bile, which is needed for absorption of dietary fats, the liver performs many other vital functions:

• **Carbohydrate metabolism.** The liver is especially important in maintaining a normal blood glucose level. When blood glucose is low, the liver can break down glycogen to glucose and release the glucose into the bloodstream. The liver can also convert certain amino acids and lactic acid to glucose, and it can convert other sugars, such as fructose and galactose, into glucose. When blood glucose is high, as occurs just after eating a meal, the liver converts glucose to glycogen and triglycerides for storage.

- *Lipid metabolism.* Hepatocytes store some triglycerides; break down fatty acids to generate ATP; synthesize lipoproteins, which transport fatty acids, triglycerides, and cholesterol to and from body cells; synthesize cholesterol; and use cholesterol to make bile salts.

- *Protein metabolism.* Hepatocytes *deaminate* (remove the amino group, NH_2, from) amino acids so that the amino acids can be used for ATP production or converted to carbohydrates or fats. The resulting toxic ammonia (NH_3) is then converted into the much less toxic urea, which is excreted in urine. Hepatocytes also synthesize most plasma proteins, such as alpha and beta globulins, albumin, prothrombin, and fibrinogen.

- *Processing of drugs and hormones.* The liver can detoxify substances such as alcohol and excrete drugs such as penicillin, erythromycin, and sulfonamides into bile. It can also chemically alter or excrete thyroid hormones and steroid hormones such as estrogens and aldosterone.

- *Excretion of bilirubin.* As previously noted, bilirubin, derived from the heme of aged red blood cells, is absorbed by the liver from the blood and secreted into bile. Most of the bilirubin in bile is metabolized in the small intestine by bacteria and eliminated in feces.

- *Synthesis of bile salts.* Bile salts are used in the small intestine for the emulsification and absorption of lipids.

- *Storage.* In addition to glycogen, the liver is a prime storage site for certain vitamins (A, B_{12}, D, E, and K) and minerals (iron and copper), which are released from the liver when needed elsewhere in the body.

- *Phagocytosis.* The stellate reticuloendothelial (Kupffer) cells of the liver phagocytize aged red blood cells, white blood cells, and some bacteria.

- *Activation of vitamin D.* The skin, liver, and kidneys participate in synthesizing the active form of vitamin D.

The liver functions related to metabolism are discussed more fully in Chapter 25.

✔CHECKPOINT

27. Draw and label a diagram of the cell zones of a hepatic acinus.
28. Describe the pathways of blood flow into, through, and out of the liver.
29. How are the liver and gallbladder connected to the duodenum?
30. Once bile has been formed by the liver, how is it collected and transported to the gallbladder for storage?
31. Describe the major functions of the liver and gallbladder.

24.12 SMALL INTESTINE

◉ OBJECTIVES

- Describe the location and structure of the small intestine.
- Identify the functions of the small intestine.

Most digestion and absorption of nutrients occur in a long tube called the **small intestine.** Because of this, its structure is specially adapted for these functions. Its length alone provides a large surface area for digestion and absorption, and that area is further increased by circular folds, villi, and microvilli. The small intestine begins at the pyloric sphincter of the stomach, coils through the central and inferior part of the abdominal cavity, and eventually opens into the large intestine. It averages 2.5 cm (1 in.) in diameter; its length is about 3 m (10 ft) in a living person and about 6.5 m (21 ft) in a cadaver due to the loss of smooth muscle tone after death.

Anatomy of the Small Intestine

The small intestine is divided into three regions (Figure 24.18). The **duodenum** (doo′-ō-DĒ-num or doo-OD-e-num), the shortest region, is retroperitoneal. It starts at the pyloric sphincter of the stomach and is in the form of a C-shaped tube that extends about 25 cm (10 in.) until it merges with the jejunum. *Duodenum* means "12"; it is so named because it is about as long as the width of 12 fingers. The **jejunum** (je-JOO-num) is about 1 m (3 ft) long and extends to the ileum. *Jejunum* means "empty," which is how it is found at death. The final and longest region of the small intestine, the **ileum** (IL-ē-um = twisted), measures about 2 m (6 ft) and joins the large intestine at a smooth muscle sphincter called the **ileocecal sphincter (valve)** (il′-ē-ō-SĒ-kal).

Histology of the Small Intestine

The wall of the small intestine is composed of the same four layers that make up most of the GI tract: mucosa, submucosa, muscularis, and serosa (Figure 24.19b). The **mucosa** is composed of a layer of epithelium, lamina propria, and muscularis mucosae. The epithelial layer of the small intestinal mucosa consists of simple columnar epithelium that contains many types of cells (Figure 24.19c). **Absorptive cells** of the epithelium digest and absorb nutrients in small intestinal chyme. Also present in the epithelium are **goblet cells,** which secrete mucus. The small intestinal mucosa contains many deep crevices lined with glandular epithelium. Cells lining the crevices form the **intestinal glands,** or **crypts of Lieberkühn** (LĒ-ber-kūn), and secrete intestinal juice (to be discussed shortly). Besides absorptive cells and goblet cells, the intestinal glands also contain paneth cells and enteroendocrine cells. **Paneth cells** secrete lysozyme, a bactericidal enzyme, and are capable of phagocytosis. Paneth cells may have a role in regulating the microbial population in the small intestine. Three types of enteroendocrine cells are found in the intestinal glands of the small intestine: **S cells, CCK cells,** and **K cells,** which secrete the hormones **secretin** (se-KRĒ-tin), **cholecystokinin** (kō-lē-sis′-tō-KĪN-in)

Figure 24.18 Anatomy of the small intestine. (a) Regions of the small intestine are the duodenum, jejunum, and ileum. (b) Circular folds increase the surface area for digestion and absorption in the small intestine.

🔑 ▭ Most digestion and absorption occur in the small intestine.

FUNCTIONS OF THE SMALL INTESTINE

1. Segmentations mix chyme with digestive juices and bring food into contact with mucosa for absorption; peristalsis propels chyme through small intestine.
2. Completes digestion of carbohydrates, proteins, and lipids; begins and completes digestion of nucleic acids.
3. Absorbs about 90% of nutrients and water that pass through digestive system.

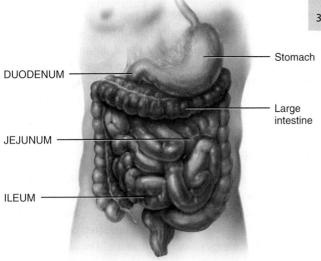

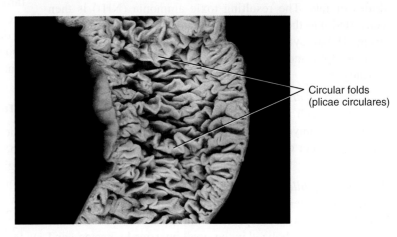

(a) Anterior view of external anatomy

(b) Internal anatomy of jejunum

❓ **Which portion of the small intestine is the longest?**

or **CCK,** and **glucose-dependent insulinotropic peptide** (in-soo-lin′-ō-TRŌ-pik) or **GIP,** respectively.

The lamina propria of the small intestinal mucosa contains areolar connective tissue and has an abundance of mucosa-associated lymphoid tissue (MALT). **Solitary lymphatic nodules** are most numerous in the distal part of the ileum (see Figure 24.20c). Groups of lymphatic nodules referred to as **aggregated lymphatic follicles,** or **Peyer's patches** (PĪ-erz), are also present in the ileum. The muscularis mucosae of the small intestinal mucosa consists of smooth muscle.

The **submucosa** of the duodenum contains **duodenal glands,** also called **Brunner's glands** (BRUN-erz) (Figure 24.20a), which secrete an alkaline mucus that helps neutralize gastric acid in the chyme. Sometimes the lymphatic tissue of the lamina propria extends through the muscularis mucosae into the submucosa. The **muscularis** of the small intestine consists of two layers of smooth muscle. The outer, thinner layer contains longitudinal fibers; the inner, thicker layer contains circular fibers. Except for a major portion of the duodenum, the **serosa** (or visceral peritoneum) completely surrounds the small intestine.

Even though the wall of the small intestine is composed of the same four basic layers as the rest of the GI tract, special structural features of the small intestine facilitate the process of digestion and absorption. These structural features include circular folds, villi, and microvilli. **Circular folds** or **plicae circulares** are folds of the mucosa and submucosa (see Figures 24.18b and 24.19a). These permanent ridges, which are about 10 mm (0.4 in.) long, begin near the proximal portion of the duodenum and end at about the midportion of the ileum. Some extend all the way around the circumference of the intestine; others extend only part of the way around. Circular folds enhance absorption by increasing surface area and causing the chyme to spiral, rather than move in a straight line, as it passes through the small intestine.

Also present in the small intestine are **villi** (= tufts of hair), which are fingerlike projections of the mucosa that are 0.5–1 mm long (see Figure 24.19b, c). The large number of villi (20–40 per square millimeter) vastly increases the surface area of the epithelium available for absorption and digestion and gives the intestinal mucosa a velvety appearance. Each villus (singular form) is covered by epithelium and has a core of lamina propria; embedded in the connective tissue of the lamina propria are an arteriole, a venule, a blood capillary network, and a **lacteal** (LAK-tē-al = milky), which is a lymphatic capillary (see Figure 24.19c). Nutrients absorbed by the epithelial cells covering the villus pass through the wall of a capillary or a lacteal to enter blood or lymph, respectively.

Figure 24.19 **Histology of the small intestine.**

Circular folds, villi, and microvilli increase the surface area of the small intestine for digestion and absorption.

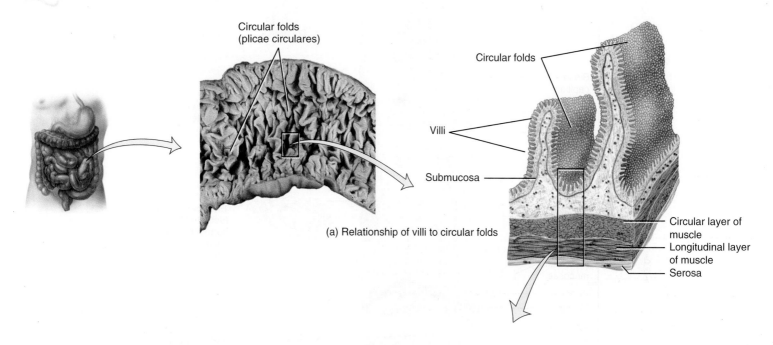

Circular folds
(plicae circulares)

Circular folds

Villi

Submucosa

Circular layer of
muscle

Longitudinal layer
of muscle

Serosa

(a) Relationship of villi to circular folds

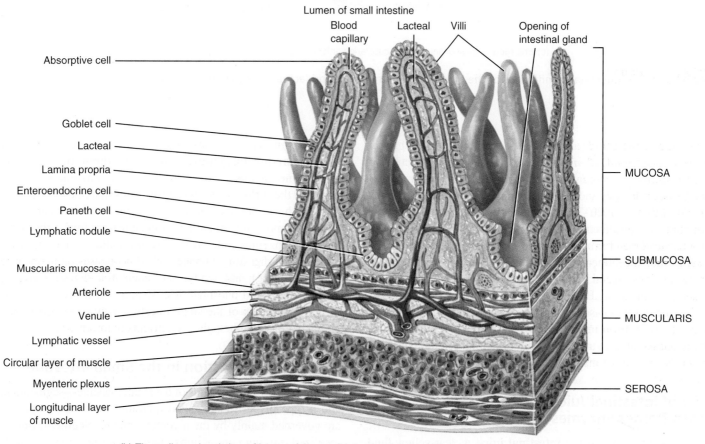

Lumen of small intestine

Blood
capillary

Lacteal

Villi

Opening of
intestinal gland

Absorptive cell

Goblet cell

Lacteal

Lamina propria

Enteroendocrine cell

Paneth cell

Lymphatic nodule

Muscularis mucosae

Arteriole

Venule

Lymphatic vessel

Circular layer of muscle

Myenteric plexus

Longitudinal layer
of muscle

MUCOSA

SUBMUCOSA

MUSCULARIS

SEROSA

(b) Three-dimensional view of layers of the small intestine showing villi

FIGURE 24.19 CONTINUES ▶

■ FIGURE 24.19 CONTINUED ▶

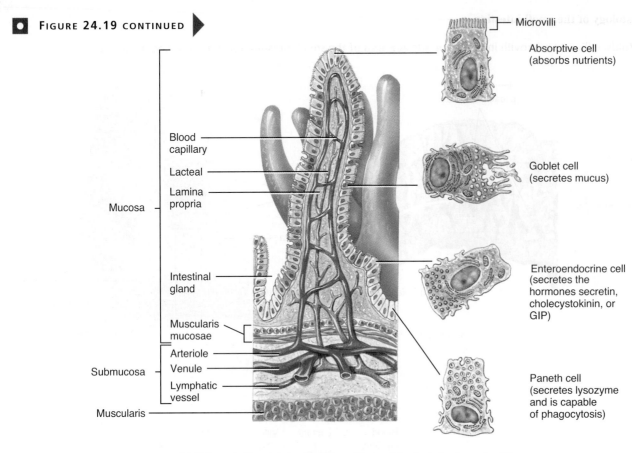

- Microvilli
- Absorptive cell (absorbs nutrients)
- Mucosa
  - Blood capillary
  - Lacteal
  - Lamina propria
  - Intestinal gland
  - Muscularis mucosae
- Goblet cell (secretes mucus)
- Enteroendocrine cell (secretes the hormones secretin, cholecystokinin, or GIP)
- Submucosa
  - Arteriole
  - Venule
  - Lymphatic vessel
- Muscularis
- Paneth cell (secretes lysozyme and is capable of phagocytosis)

(c) Enlarged villus showing lacteal, capillaries, intestinal glands, and cell types

? **What is the functional significance of the blood capillary network and lacteal in the center of each villus?**

Besides circular folds and villi, the small intestine also has **microvilli** (mī-krō-VIL-ī; *micro-* = small), which are projections of the apical (free) membrane of the absorptive cells. Each microvillus is a 1-μm-long cylindrical, membrane-covered projection that contains a bundle of 20–30 actin filaments. When viewed through a light microscope, the microvilli are too small to be seen individually; instead they form a fuzzy line, called the **brush border,** extending into the lumen of the small intestine (Figure 24.20d). There are an estimated 200 million microvilli per square millimeter of small intestine. Because the microvilli greatly increase the surface area of the plasma membrane, larger amounts of digested nutrients can diffuse into absorptive cells in a given period. The brush border also contains several brush-border enzymes that have digestive functions (discussed shortly).

Role of Intestinal Juice and Brush-Border Enzymes

About 1–2 liters (1–2 qt) of **intestinal juice,** a clear yellow fluid, is secreted each day. Intestinal juice contains water and mucus and is slightly alkaline (pH 7.6). Together, pancreatic and intestinal juices provide a liquid medium that aids the absorption of substances from chyme in the small intestine. The absorptive cells of the small intestine synthesize several digestive enzymes, called **brush-border enzymes,** and insert them in the plasma membrane of the microvilli. Thus, some enzymatic digestion occurs at the surface of the absorptive cells that line the villi, rather than in the lumen exclusively, as occurs in other parts of the GI tract. Among the brush-border enzymes are four carbohydrate-digesting enzymes called α-dextrinase, maltase, sucrase, and lactase; protein-digesting enzymes called peptidases (aminopeptidase and dipeptidase); and two types of nucleotide-digesting enzymes, nucleosidases and phosphatases. Also, as absorptive cells slough off into the lumen of the small intestine, they break apart and release enzymes that help digest nutrients in the chyme.

Mechanical Digestion in the Small Intestine

The two types of movements of the small intestine—segmentations and a type of peristalsis called migrating motility complexes—are governed mainly by the myenteric plexus. **Segmentations** are localized, mixing contractions that occur in portions of intestine distended by a large volume of chyme. Segmentations mix chyme with the digestive juices and bring the particles of food into contact with the mucosa for absorption; they do not push the intestinal contents along the tract. A segmentation starts with the contractions

Figure 24.20 Histology of the duodenum and ileum.

Microvilli in the small intestine contain several brush-border enzymes that help digest nutrients.

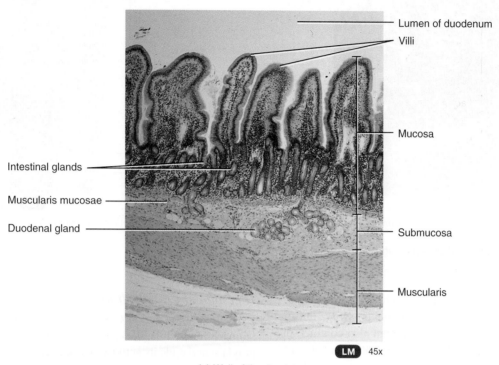

Lumen of duodenum

Villi

Mucosa

Intestinal glands

Muscularis mucosae

Duodenal gland

Submucosa

Muscularis

LM 45x

(a) Wall of the duodenum

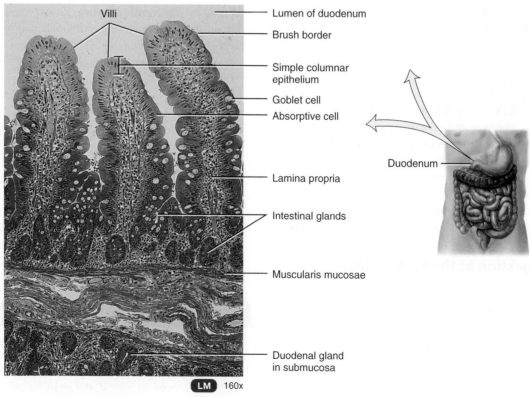

Villi

Lumen of duodenum

Brush border

Simple columnar epithelium

Goblet cell

Absorptive cell

Duodenum

Lamina propria

Intestinal glands

Muscularis mucosae

Duodenal gland in submucosa

LM 160x

(b) Three villi from duodenum

FIGURE 24.20 CONTINUES ▶

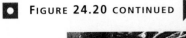

FIGURE 24.20 CONTINUED

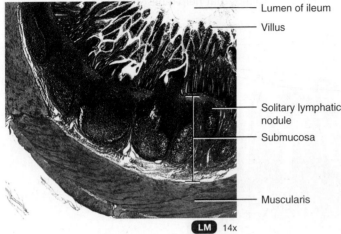

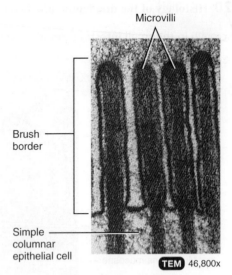

— Lumen of ileum
— Villus
— Solitary lymphatic nodule
— Submucosa
— Muscularis

LM 14x

(c) Lymphatic nodules in ileum

Microvilli
Brush border
Simple columnar epithelial cell

TEM 46,800x

(d) Several microvilli from duodenum

? **What is the function of the fluid secreted by duodenal (Brunner's) glands?**

of circular muscle fibers in a portion of the small intestine, an action that constricts the intestine into segments. Next, muscle fibers that encircle the middle of each segment also contract, dividing each segment again. Finally, the fibers that first contracted relax, and each small segment unites with an adjoining small segment so that large segments are formed again. As this sequence of events repeats, the chyme sloshes back and forth. Segmentations occur most rapidly in the duodenum, about 12 times per minute, and progressively slow to about 8 times per minute in the ileum. This movement is similar to alternately squeezing the middle and then the ends of a capped tube of toothpaste.

After most of a meal has been absorbed, which lessens distension of the wall of the small intestine, segmentation stops and peristalsis begins. The type of peristalsis that occurs in the small intestine, termed a **migrating motility complex (MMC),** begins in the lower portion of the stomach and pushes chyme forward along a short stretch of small intestine before dying out. The MMC slowly migrates down the small intestine, reaching the end of the ileum in 90–120 minutes. Then another MMC begins in the stomach. Altogether, chyme remains in the small intestine for 3–5 hours.

Chemical Digestion in the Small Intestine

In the mouth, salivary amylase converts starch (a polysaccharide) to maltose (a disaccharide), maltotriose (a trisaccharide), and α-dextrins (short-chain, branched fragments of starch with 5–10 glucose units). In the stomach, pepsin converts proteins to peptides (small fragments of proteins), and lingual and gastric lipases convert some triglycerides into fatty acids, diglycerides, and monoglycerides. Thus, chyme entering the small intestine contains partially digested carbohydrates, proteins, and lipids. The completion of the digestion of carbohydrates, proteins, and lipids is a collective effort of pancreatic juice, bile, and intestinal juice in the small intestine.

Digestion of Carbohydrates

Even though the action of **salivary amylase** may continue in the stomach for a while, the acidic pH of the stomach destroys salivary amylase and ends its activity. Thus, only a few starches are broken down by the time chyme leaves the stomach. Those starches not already broken down into maltose, maltotriose, and α-dextrins are cleaved by **pancreatic amylase,** an enzyme in pancreatic juice that acts in the small intestine. Although pancreatic amylase acts on both glycogen and starches, it has no effect on another polysaccharide called cellulose, an indigestible plant fiber that is commonly referred to as "roughage" as it moves through the digestive system. After amylase (either salivary or pancreatic) has split starch into smaller fragments, a brush-border enzyme called **α-dextrinase** acts on the resulting α-dextrins, clipping off one glucose unit at a time.

Ingested molecules of sucrose, lactose, and maltose—three disaccharides—are not acted on until they reach the small intestine. Three brush-border enzymes digest the disaccharides into monosaccharides. **Sucrase** breaks sucrose into a molecule of glucose and a molecule of fructose; **lactase** digests lactose into a molecule of glucose and a molecule of galactose; and **maltase** splits maltose and maltotriose into two or three molecules of glucose, respectively. Digestion of carbohydrates ends with the production of monosaccharides, which the digestive system is able to absorb.

⚕ **CLINICAL CONNECTION** | *Lactose Intolerance*

In some people the absorptive cells of the small intestine fail to produce enough lactase, which, as you just learned, is essential for the digestion of lactose. This results in a condition called **lactose intolerance,** in which undigested lactose in chyme causes fluid to be retained in the feces; bacterial fermentation of the undigested lactose results in the production of gases. Symptoms of lactose intolerance include diarrhea, gas, bloating, and abdominal cramps after consumption of milk and other dairy products. The symptoms can be relatively minor or serious enough to require medical attention. The *hydrogen breath test* is often used to aid in diagnosis of lactose intolerance. Very little hydrogen can be detected in the breath of a normal person, but hydrogen is among the gases produced when undigested lactose in the colon is fermented by bacteria. The hydrogen is absorbed from the intestines and carried through the bloodstream to the lungs, where it is exhaled. Persons with lactose intolerance should select a diet that restricts lactose (but not calcium) and take dietary supplements to aid in the digestion of lactose. •

Digestion of Proteins

Recall that protein digestion starts in the stomach, where proteins are fragmented into peptides by the action of **pepsin.** Enzymes in pancreatic juice—**trypsin, chymotrypsin, carboxypeptidase,** and **elastase**—continue to break down proteins into peptides. Although all these enzymes convert whole proteins into peptides, their actions differ somewhat because each splits peptide bonds between different amino acids. Trypsin, chymotrypsin, and elastase all cleave the peptide bond between a specific amino acid and its neighbor; carboxypeptidase splits off the amino acid at the carboxyl end of a peptide. Protein digestion is completed by two **peptidases** in the brush border: aminopeptidase and dipeptidase. **Aminopeptidase** cleaves off the amino acid at the amino end of a peptide. **Dipeptidase** splits dipeptides (two amino acids joined by a peptide bond) into single amino acids.

Digestion of Lipids

The most abundant lipids in the diet are triglycerides, which consist of a molecule of glycerol bonded to three fatty acid molecules (see Figure 2.17). Enzymes that split triglycerides and phospholipids are called **lipases.** Recall that there are three types of lipases that can participate in lipid digestion: **lingual lipase, gastric lipase,** and **pancreatic lipase.** Although some lipid digestion occurs in the stomach through the action of lingual and gastric lipases, most occurs in the small intestine through the action of pancreatic lipase. Triglycerides are broken down by pancreatic lipase into fatty acids and monoglycerides. The liberated fatty acids can be either short-chain fatty acids (with fewer than 10–12 carbons) or long-chain fatty acids.

Before a large lipid globule containing triglycerides can be digested in the small intestine, it must first undergo **emulsification**—a process in which the large lipid globule is broken down into several small lipid globules. Recall that bile contains bile salts, the sodium salts and potassium salts of bile acids (mainly chenodeoxycholic acid and cholic acid). Bile salts are **amphipathic** (am′-fē-PATH-ik), which means that each bile salt has a hydrophobic (nonpolar) region and a hydrophilic (polar) region. The amphipathic nature of bile salts allows them to emulsify a large lipid globule: The hydrophobic regions of bile salts interact with the large lipid globule, while the hydrophilic regions of bile salts interact with the watery intestinal chyme. Consequently, the large lipid globule is broken apart into several small lipid globules, each about 1 μm in diameter. The small lipid globules formed from emulsification provide a large surface area that allows pancreatic lipase to function more effectively.

Digestion of Nucleic Acids

Pancreatic juice contains two nucleases: **ribonuclease,** which digests RNA, and **deoxyribonuclease,** which digests DNA. The nucleotides that result from the action of the two nucleases are further digested by brush-border enzymes called **nucleosidases** (noo′-klē-ō-SĪ-dās-ez) and **phosphatases** (FOS-fa-tās′-ez) into pentoses, phosphates, and nitrogenous bases. These products are absorbed via active transport.

Absorption in the Small Intestine

All the chemical and mechanical phases of digestion from the mouth through the small intestine are directed toward changing food into forms that can pass through the absorptive epithelial cells lining the mucosa and into the underlying blood and lymphatic vessels. These forms are monosaccharides (glucose, fructose, and galactose) from carbohydrates; single amino acids, dipeptides, and tripeptides from proteins; and fatty acids, glycerol, and monoglycerides from triglycerides. Passage of these digested nutrients from the gastrointestinal tract into the blood or lymph is called **absorption.**

Absorption of materials occurs via diffusion, facilitated diffusion, osmosis, and active transport. About 90% of all absorption of nutrients occurs in the small intestine; the other 10% occurs in the stomach and large intestine. Any undigested or unabsorbed material left in the small intestine passes on to the large intestine.

Absorption of Monosaccharides

All carbohydrates are absorbed as monosaccharides. The capacity of the small intestine to absorb monosaccharides is huge—an estimated 120 grams per hour. As a result, all dietary carbohydrates that are digested normally are absorbed, leaving only indigestible cellulose and fibers in the feces. Monosaccharides pass from the lumen through the apical membrane via *facilitated diffusion* or *active transport.* Fructose, a monosaccharide found in fruits, is transported via *facilitated diffusion;* glucose and galactose are transported into absorptive cells of the villi via *secondary active transport* that is coupled to the active transport of Na^+ (Figure 24.21a). The transporter has binding sites for one glucose molecule and two sodium ions; unless all three sites are filled, neither substance is transported. Galactose competes with glucose to ride the same transporter. (Because both Na^+ and glucose or galactose move in the same direction, this is a *symporter.*) Monosaccharides then move out of the absorptive cells through their basolateral surfaces via *facilitated diffusion* and enter the capillaries of the villi (Figure 24.21b).

Absorption of Amino Acids, Dipeptides, and Tripeptides

Most proteins are absorbed as amino acids via *active transport* processes that occur mainly in the duodenum and jejunum. About half of the absorbed amino acids are present in food; the other half come from the body itself as proteins in digestive juices and dead cells that slough off the mucosal surface! Normally, 95–98% of the protein present in the small intestine is digested and absorbed. Different transporters carry different types of amino acids. Some

Figure 24.21 Absorption of digested nutrients in the small intestine. For simplicity, all digested foods are shown in the lumen of the small intestine, even though some nutrients are digested by brush-border enzymes.

Long-chain fatty acids and monoglycerides are absorbed into lacteals; other products of digestion enter blood capillaries.

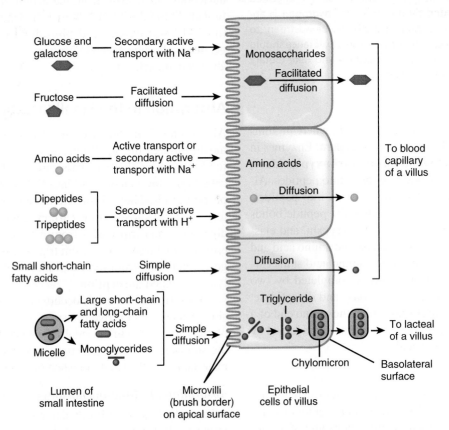

(a) Mechanisms for movement of nutrients through absorptive epithelial cells of villi

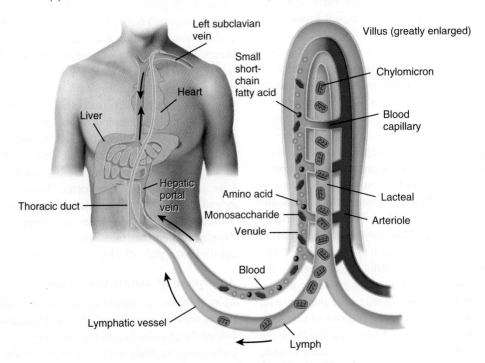

(b) Movement of absorbed nutrients into blood and lymph

? **A monoglyceride may be larger than an amino acid. Why can monoglycerides be absorbed by simple diffusion, but amino acids cannot?**

amino acids enter absorptive cells of the villi via Na^+-dependent secondary active transport processes that are similar to the glucose transporter; other amino acids are actively transported by themselves. At least one symporter brings in dipeptides and tripeptides together with H^+; the peptides then are hydrolyzed to single amino acids inside the absorptive cells. Amino acids move out of the absorptive cells via diffusion and enter capillaries of the villus (Figure 24.21). Both monosaccharides and amino acids are transported in the blood to the liver by way of the hepatic portal system. If not removed by hepatocytes, they enter the general circulation.

Absorption of Lipids

All dietary lipids are absorbed via *simple diffusion*. Adults absorb about 95% of the lipids present in the small intestine; due to their lower production of bile, newborn infants absorb only about 85% of lipids. As a result of their emulsification and digestion, triglycerides are mainly broken down into monoglycerides and fatty acids, which can be either short-chain fatty acids or long-chain fatty acids. Although short-chain fatty acids are hydrophobic, they are very small but do vary in size. Small short-chain fatty acids contain less than 10–12 carbon atoms and are more water-soluble. Thus, they can dissolve in the watery intestinal chyme, pass through the absorptive cells via simple diffusion, and follow the same route taken by monosaccharides and amino acids into a blood capillary of a villus (Figure 24.21a).

Large short-chain fatty acids (with more than 10–12 carbon atoms), long-chain fatty acids, and monoglycerides are larger and hydrophobic, and since they are not water-soluble, they have difficulty being suspended in the watery environment of the intestinal chyme. Besides their role in emulsification, bile salts also help to make these large short-chain fatty acids, long-chain fatty acids, and monoglycerides more soluble. The bile salts in intestinal chyme surround them, forming tiny spheres called **micelles** (mī-SELZ = small morsels), each of which is 2–10 nm in diameter and includes 20–50 bile salt molecules (Figure 24.21a). Micelles are formed due to the amphipathic nature of bile salts: The hydrophobic regions of bile salts interact with the large short-chain fatty acids, long-chain fatty acids, and monoglycerides, and the hydrophilic regions of bile salts interact with the watery intestinal chyme. Once formed, the micelles move from the interior of the small intestinal lumen to the brush border of the absorptive cells. At that point, the large short-chain fatty acids, long-chain fatty acids, and monoglycerides diffuse out of the micelles into the absorptive cells, leaving the micelles behind in the chyme. The micelles continually repeat this ferrying function as they move from the brush border back through the chyme to the interior of the small intestinal lumen to pick up more of the large short-chain fatty acids, long-chain fatty acids, and monoglycerides. Micelles also solubilize other large hydrophobic molecules such as fat-soluble vitamins (A, D, E, and K) and cholesterol that may be present in intestinal chyme, and aid in their absorption. These fat-soluble vitamins and cholesterol molecules are packed in the micelles along with the long-chain fatty acids and monoglycerides.

Once inside the absorptive cells, long-chain fatty acids and monoglycerides are recombined to form triglycerides, which aggregate into globules along with phospholipids and cholesterol and become coated with proteins. These large spherical masses, about 80 nm in diameter, are called **chylomicrons** (kī-lō-MĪ-krons). Chylomicrons leave the absorptive cell via exocytosis. Because they are so large and bulky, chylomicrons cannot enter blood capillaries—the pores in the walls of blood capillaries are too small. Instead, chylomicrons enter lacteals, which have much larger pores than blood capillaries. From lacteals, chylomicrons are transported by way of lymphatic vessels to the thoracic duct and enter the blood at the left subclavian vein (Figure 24.21b). The hydrophilic protein coat that surrounds each chylomicron keeps the chylomicrons suspended in blood and prevents them from sticking to each other.

Within 10 minutes after absorption, about half of the chylomicrons have already been removed from the blood as they pass through blood capillaries in the liver and adipose tissue. This removal is accomplished by an enzyme attached to the apical surface of capillary endothelial cells, called **lipoprotein lipase,** that breaks down triglycerides in chylomicrons and other lipoproteins into fatty acids and glycerol. The fatty acids diffuse into hepatocytes and adipose cells and combine with glycerol during resynthesis of triglycerides. Two or three hours after a meal, few chylomicrons remain in the blood.

After participating in the emulsification and absorption of lipids, 90–95% of the bile salts are reabsorbed by active transport in the final segment of the small intestine (ileum) and returned by the blood to the liver through the hepatic portal system for recycling. This cycle of bile salt secretion by hepatocytes into bile, reabsorption by the ileum, and resecretion into bile is called the **enterohepatic circulation** (en′-ter-ō-he-PAT-ik). Insufficient bile salts, due either to obstruction of the bile ducts or removal of the gallbladder, can result in the loss of up to 40% of dietary lipids in feces due to diminished lipid absorption. When lipids are not absorbed properly, the fat-soluble vitamins are not adequately absorbed.

Absorption of Electrolytes

Many of the electrolytes absorbed by the small intestine come from gastrointestinal secretions, and some are part of ingested foods and liquids. Recall that electrolytes are compounds that separate into ions in water and conduct electricity. Sodium ions are actively transported out of absorptive cells by basolateral sodium–potassium pumps (Na^+/K^+ ATPases) after they have moved into absorptive cells via diffusion and secondary active transport. Thus, most of the sodium ions (Na^+) in gastrointestinal secretions are reclaimed and not lost in the feces. Negatively charged bicarbonate, chloride, iodide, and nitrate ions can passively follow Na^+ or be actively transported. Calcium ions also are absorbed actively in a process stimulated by calcitriol. Other electrolytes such as iron, potassium, magnesium, and phosphate ions also are absorbed via active transport mechanisms.

Absorption of Vitamins

As you have just learned, the fat-soluble vitamins A, D, E, and K are included with ingested dietary lipids in micelles and are absorbed via simple diffusion. Most water-soluble vitamins, such as most B vitamins and vitamin C, also are absorbed via simple diffusion. Vitamin B_{12}, however, combines with intrinsic factor produced by the stomach, and the combination is absorbed in the ileum via an active transport mechanism.

Absorption of Water

The total volume of fluid that enters the small intestine each day—about 9.3 liters (9.8 qt)—comes from ingestion of liquids (about 2.3 liters) and from various gastrointestinal secretions (about 7.0 liters). Figure 24.22 depicts the amounts of fluid ingested, secreted, absorbed, and excreted by the GI tract. The small intestine absorbs about 8.3 liters of the fluid; the remainder passes into the large intestine, where most of the rest of it—about 0.9 liter—is also absorbed. Only 0.1 liter (100 mL) of water is excreted in the feces each day.

All water absorption in the GI tract occurs via *osmosis* from the lumen of the intestines through absorptive cells and into blood

Figure 24.22 Daily volumes of fluid ingested, secreted, absorbed, and excreted from the GI tract.

🔑 All water absorption in the GI tract occurs via osmosis.

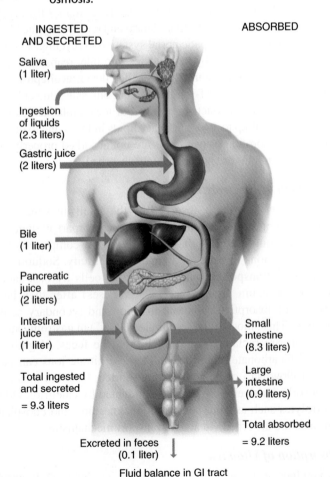

Fluid balance in GI tract

❓ **Which two organs of the digestive system secrete the most fluid?**

capillaries. Because water can move across the intestinal mucosa in both directions, the absorption of water from the small intestine depends on the absorption of electrolytes and nutrients to maintain

TABLE 24.4	
Summary of Digestive Activities in the Pancreas, Liver, Gallbladder, and Small Intestine	
STRUCTURE	**ACTIVITY**
Pancreas	Delivers pancreatic juice into duodenum via pancreatic duct to assist absorption (see Table 24.5 for pancreatic enzymes and their functions).
Liver	Produces bile (bile salts) necessary for emulsification and absorption of lipids.
Gallbladder	Stores, concentrates, and delivers bile into duodenum via common bile duct.
Small intestine	Major site of digestion and absorption of nutrients and water in gastrointestinal tract.
Mucosa/submucosa	
Intestinal glands	Secrete intestinal juice to assist absorption.
Absorptive cells	Digest and absorb nutrients.
Goblet cells	Secrete mucus.
Enteroendocrine cells (S, CCK, K)	Secrete secretin, cholecystokinin, and glucose-dependent insulinotropic peptide.
Paneth cells	Secrete lysozyme (bactericidal enzyme) and phagocytosis.
Duodenal (Brunner's) glands	Secrete alkaline fluid to buffer stomach acids, and mucus for protection and lubrication.
Circular folds	Folds of mucosa and submucosa that increase surface area for digestion and absorption.
Villi	Fingerlike projections of mucosa that are sites of absorption of digested food and increase surface area for digestion and absorption.
Microvilli	Microscopic, membrane-covered projections of absorptive epithelial cells that contain brush-border enzymes (listed in Table 24.5) and that increase surface area for digestion and absorption.
Muscularis	
Segmentation	Type of peristalsis: alternating contractions of circular smooth muscle fibers that produce segmentation and resegmentation of sections of small intestine; mixes chyme with digestive juices and brings food into contact with mucosa for absorption.
Migrating motility complex (MMC)	Type of peristalsis: waves of contraction and relaxation of circular and longitudinal smooth muscle fibers passing the length of the small intestine; moves chyme toward ileocecal sphincter.

an osmotic balance with the blood. The absorbed electrolytes, monosaccharides, and amino acids establish a concentration gradient for water that promotes water absorption via osmosis.

Table 24.4 summarizes the digestive activities of the pancreas, liver, gallbladder, and small intestine and Table 24.5 summarizes the digestive enzymes and their functions in the digestive system.

CLINICAL CONNECTION | *Absorption of Alcohol*

The intoxicating and incapacitating effects of alcohol depend on the blood alcohol level. Because it is lipid-soluble, alcohol begins to be absorbed in the stomach. However, the surface area available for absorption is much greater in the small intestine than in the

TABLE 24.5

Summary of Digestive Enzymes

ENZYME	SOURCE	SUBSTRATES	PRODUCTS
SALIVA			
Salivary amylase	Salivary glands.	Starches (polysaccharides).	Maltose (disaccharide), maltotriose (trisaccharide), and α-dextrins.
Lingual lipase	Lingual glands in tongue.	Triglycerides (fats and oils) and other lipids.	Fatty acids and diglycerides.
GASTRIC JUICE			
Pepsin (activated from pepsinogen by pepsin and hydrochloric acid)	Stomach chief cells.	Proteins.	Peptides.
Gastric lipase	Stomach chief cells.	Triglycerides (fats and oils).	Fatty acids and monoglycerides.
PANCREATIC JUICE			
Pancreatic amylase	Pancreatic acinar cells.	Starches (polysaccharides).	Maltose (disaccharide), maltotriose (trisaccharide), and α-dextrins.
Trypsin (activated from trypsinogen by enterokinase)	Pancreatic acinar cells.	Proteins.	Peptides.
Chymotrypsin (activated from chymotrypsinogen by trypsin)	Pancreatic acinar cells.	Proteins.	Peptides.
Elastase (activated from proelastase by trypsin)	Pancreatic acinar cells.	Proteins.	Peptides.
Carboxypeptidase (activated from procarboxypeptidase by trypsin)	Pancreatic acinar cells.	Amino acid at carboxyl end of peptides.	Amino acids and peptides.
Pancreatic lipase	Pancreatic acinar cells.	Triglycerides (fats and oils) that have been emulsified by bile salts.	Fatty acids and monoglycerides.
Nucleases			
Ribonuclease	Pancreatic acinar cells.	Ribonucleic acid.	Nucleotides.
Deoxyribonuclease	Pancreatic acinar cells.	Deoxyribonucleic acid.	Nucleotides.
BRUSH-BORDER ENZYMES IN MICROVILLI PLASMA MEMBRANE			
α-**Dextrinase**	Small intestine.	α-Dextrins.	Glucose.
Maltase	Small intestine.	Maltose.	Glucose.
Sucrase	Small intestine.	Sucrose.	Glucose and fructose.
Lactase	Small intestine.	Lactose.	Glucose and galactose.
Enterokinase	Small intestine.	Trypsinogen.	Trypsin.
Peptidases			
Aminopeptidase	Small intestine.	Amino acid at amino end of peptides.	Amino acids and peptides.
Dipeptidase	Small intestine.	Dipeptides.	Amino acids.
Nucleosidases and phosphatases	Small intestine.	Nucleotides.	Nitrogenous bases, pentoses, and phosphates.

stomach, so when alcohol passes into the duodenum, it is absorbed more rapidly. Thus, the longer the alcohol remains in the stomach, the more slowly blood alcohol level rises. Because fatty acids in chyme slow gastric emptying, blood alcohol level will rise more slowly when fat-rich foods, such as pizza, hamburgers, or nachos, are consumed with alcoholic beverages. Also, the enzyme alcohol dehydrogenase, which is present in gastric mucosa cells, breaks down some of the alcohol to acetaldehyde, which is not intoxicating. When the rate of gastric emptying is slower, proportionally more alcohol will be absorbed and converted to acetaldehyde in the stomach, and thus less alcohol will reach the bloodstream. Given identical consumption of alcohol, females often develop higher blood alcohol levels (and therefore experience greater intoxication) than males of comparable size because the activity of gastric alcohol dehydrogenase is up to 60% lower in females than in males. Asian males may also have lower levels of this gastric enzyme. •

✔ CHECKPOINT

32. List the regions of the small intestine and describe their functions.
33. In what ways are the mucosa and submucosa of the small intestine adapted for digestion and absorption?
34. Describe the types of movement that occur in the small intestine.
35. Explain the functions of pancreatic amylase, aminopeptidase, gastric lipase, and deoxyribonuclease.
36. What is the difference between digestion and absorption? How are the end products of carbohydrate, protein, and lipid digestion absorbed?
37. By what routes do absorbed nutrients reach the liver?
38. Describe the absorption of electrolytes, vitamins, and water by the small intestine.

24.13 LARGE INTESTINE

◉ OBJECTIVE

• Describe the anatomy, histology, and functions of the large intestine.

The large intestine is the terminal portion of the GI tract. The overall functions of the large intestine are the completion of absorption, the production of certain vitamins, the formation of feces, and the expulsion of feces from the body.

Anatomy of the Large Intestine

The **large intestine,** which is about 1.5 m (5 ft) long and 6.5 cm (2.5 in.) in diameter, extends from the ileum to the anus. It is attached to the posterior abdominal wall by its **mesocolon** (mez'-ō-KŌ-lon), which is a double layer of peritoneum (see Figure 24.4a). Structurally, the four major regions of the large intestine are the cecum, colon, rectum, and anal canal (Figure 24.23a).

The opening from the ileum into the large intestine is guarded by a fold of mucous membrane called the ileocecal sphincter (valve), which allows materials from the small intestine to pass into the large intestine. Hanging inferior to the ileocecal valve is

the **cecum,** a small pouch about 6 cm (2.4 in.) long. Attached to the cecum is a twisted, coiled tube, measuring about 8 cm (3 in.) in length, called the **appendix** or **vermiform appendix** (VER-mi-form; *vermiform* = worm-shaped; *appendix* = appendage). The mesentery of the appendix, called the **mesoappendix** (mez-ō-a-PEN-diks), attaches the appendix to the inferior part of the mesentery of the ileum.

Inflammation of the appendix, termed **appendicitis,** is preceded by obstruction of the lumen of the appendix by chyme, inflammation, a foreign body, a carcinoma of the cecum, stenosis, or kinking of the organ. It is characterized by high fever, elevated white blood cell count, and a neutrophil count higher than 75%. The infection that follows may result in edema and ischemia and may progress to gangrene and perforation within 24 hours. Typically, appendicitis begins with referred pain in the umbilical region of the abdomen, followed by anorexia (loss of appetite for food), nausea, and vomiting. After several hours the pain localizes in the right lower quadrant (RLQ) and is continuous, dull or severe, and intensified by coughing, sneezing, or body movements. Early appendectomy (removal of the appendix) is recommended because it is safer to operate than to risk rupture, peritonitis, and gangrene. Although it required major abdominal surgery in the past, today appendectomies are usually performed laparoscopically. •

The open end of the cecum merges with a long tube called the **colon** (= food passage), which is divided into ascending, transverse, descending, and sigmoid portions. Both the ascending and descending colon are retroperitoneal; the transverse and sigmoid colon are not. True to its name, the **ascending colon** ascends on the right side of the abdomen, reaches the inferior surface of the liver, and turns abruptly to the left to form the **right colic (hepatic) flexure.** The colon continues across the abdomen to the left side as the **transverse colon.** It curves beneath the inferior end of the spleen on the left side as the **left colic (splenic) flexure** and passes inferiorly to the level of the iliac crest as the **descending colon.** The **sigmoid colon** (*sigm-* = S-shaped) begins near the left iliac crest, projects medially to the midline, and terminates as the rectum at about the level of the third sacral vertebra.

The **rectum,** the last 20 cm (8 in.) of the GI tract, lies anterior to the sacrum and coccyx. The terminal 2–3 cm (1 in.) of the rectum is called the **anal canal** (Figure 24.23b). The mucous membrane of the anal canal is arranged in longitudinal folds called **anal columns** that contain a network of arteries and veins. The opening of the anal canal to the exterior, called the **anus,** is guarded by an **internal anal sphincter** of smooth muscle (involuntary) and an **external anal sphincter** of skeletal muscle (voluntary). Normally these sphincters keep the anus closed except during the elimination of feces.

Histology of the Large Intestine

The wall of the large intestine contains the typical four layers found in the rest of the GI tract: mucosa, submucosa, muscularis,

Figure 24.23 Anatomy of the large intestine.

The regions of the large intestine are the cecum, colon, rectum, and anal canal.

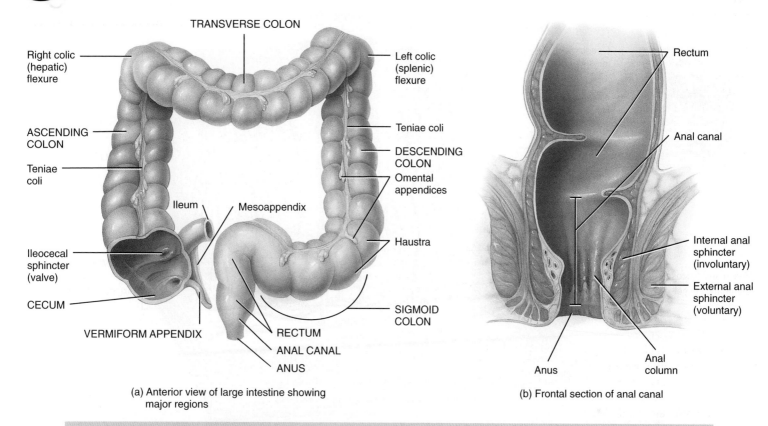

(a) Anterior view of large intestine showing major regions

(b) Frontal section of anal canal

FUNCTIONS OF THE LARGE INTESTINE
1. Haustral churning, peristalsis, and mass peristalsis drive contents of colon into rectum.
2. Bacteria in large intestine convert proteins to amino acids, break down amino acids, and produce some B vitamins and vitamin K.
3. Absorbing some water, ions, and vitamins.
4. Forming feces.
5. Defecating (emptying rectum).

? **Which portions of the colon are retroperitoneal?**

and serosa. The **mucosa** consists of simple columnar epithelium, lamina propria (areolar connective tissue), and muscularis mucosae (smooth muscle) (Figure 24.24a). The epithelium contains mostly absorptive and goblet cells (Figure 24.24b, c). The absorptive cells function primarily in water absorption; the goblet cells secrete mucus that lubricates the passage of the colonic contents. Both absorptive and goblet cells are located in long, straight, tubular intestinal glands (crypts of Lieberkühn) that extend the full thickness of the mucosa. Solitary lymphatic nodules are also found in the lamina propria of the mucosa and may extend through the muscularis mucosae into the submucosa. Compared to the small intestine, the mucosa of the large intestine does not have as many structural adaptations that increase surface area. There are no circular folds or villi; however, microvilli are present on the absorptive cells. Consequently, much more absorption occurs in the small intestine than in the large intestine.

The **submucosa** of the large intestine consists of areolar connective tissue. The **muscularis** consists of an external layer of longitudinal smooth muscle and an internal layer of circular smooth muscle. Unlike other parts of the GI tract, portions of the longitudinal muscles are thickened, forming three conspicuous bands called the **teniae coli** (TĒ-nē-ē KŌ-lī; *teniae* = flat bands) that run most of the length of the large intestine (see Figure 24.23a). The teniae coli are separated by portions of the wall with less or no longitudinal muscle. Tonic contractions of the bands gather the colon into a series of pouches called **haustra** (HAWS-tra = shaped like pouches; singular is *haustrum*), which give the colon a puckered appearance. A single layer of circular smooth muscle lies between teniae coli. The **serosa** of the large intestine is part of the visceral peritoneum. Small pouches of visceral peritoneum filled with fat are attached to teniae coli and are called **omental (fatty) appendices.**

Figure 24.24 Histology of the large intestine.

🔑 Intestinal glands formed by simple columnar epithelial cells and goblet cells extend the full thickness of the mucosa.

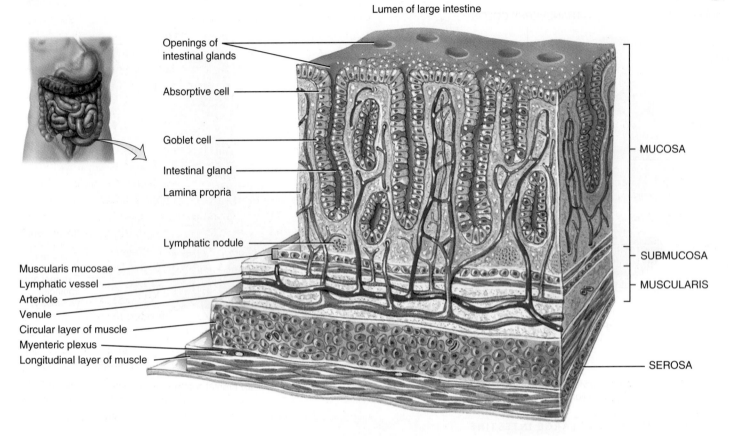

Lumen of large intestine

Openings of
intestinal glands

Absorptive cell

Goblet cell

Intestinal gland

Lamina propria

Lymphatic nodule

Muscularis mucosae
Lymphatic vessel
Arteriole
Venule
Circular layer of muscle
Myenteric plexus
Longitudinal layer of muscle

MUCOSA

SUBMUCOSA

MUSCULARIS

SEROSA

(a) Three-dimensional view of layers of large intestine

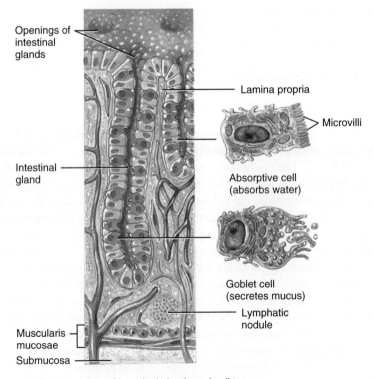

Openings of
intestinal
glands

Lamina propria

Microvilli

Intestinal
gland

Absorptive cell
(absorbs water)

Goblet cell
(secretes mucus)

Lymphatic
nodule

Muscularis
mucosae

Submucosa

(b) Sectional view of intestinal glands and cell types

point, the walls contract and squeeze the contents into the next haustrum. **Peristalsis** also occurs, although at a slower rate (3–12 contractions per minute) than in more proximal portions of the tract. A final type of movement is **mass peristalsis,** a strong peristaltic wave that begins at about the middle of the transverse colon and quickly drives the contents of the colon into the rectum. Because food in the stomach initiates this **gastrocolic reflex** in the colon, mass peristalsis usually takes place three or four times a day, during or immediately after a meal.

Mechanical Digestion in the Large Intestine

The passage of chyme from the ileum into the cecum is regulated by the action of the ileocecal sphincter. Normally, the valve remains partially closed so that the passage of chyme into the cecum usually occurs slowly. Immediately after a meal, a **gastroileal reflex** (gas′-trō-IL-ē-al) intensifies peristalsis in the ileum and forces any chyme into the cecum. The hormone gastrin also relaxes the sphincter. Whenever the cecum is distended, the degree of contraction of the ileocecal sphincter intensifies.

Movements of the colon begin when substances pass the ileocecal sphincter. Because chyme moves through the small intestine at a fairly constant rate, the time required for a meal to pass into the colon is determined by gastric emptying time. As food passes through the ileocecal sphincter, it fills the cecum and accumulates in the ascending colon.

One movement characteristic of the large intestine is **haustral churning.** In this process, the haustra remain relaxed and become distended while they fill up. When the distension reaches a certain

Chemical Digestion in the Large Intestine

The final stage of digestion occurs in the colon through the activity of bacteria that inhabit the lumen. Mucus is secreted by the glands of the large intestine, but no enzymes are secreted. Chyme is prepared for elimination by the action of bacteria, which ferment any remaining carbohydrates and release hydrogen, carbon dioxide, and methane gases. These gases contribute to flatus (gas) in the colon, termed *flatulence* when it is excessive. Bacteria also convert any remaining proteins to amino acids and break down the amino acids into simpler substances: indole, skatole, hydrogen sulfide, and fatty acids. Some of the indole and skatole is eliminated in the feces and contributes to their odor; the rest is absorbed and transported to the liver, where these compounds are converted to less toxic compounds and excreted in the urine. Bacteria also decompose bilirubin to simpler pigments, including stercobilin, which gives feces their brown color. Bacterial products that are absorbed in the colon include several vitamins needed for normal metabolism, among them some B vitamins and vitamin K.

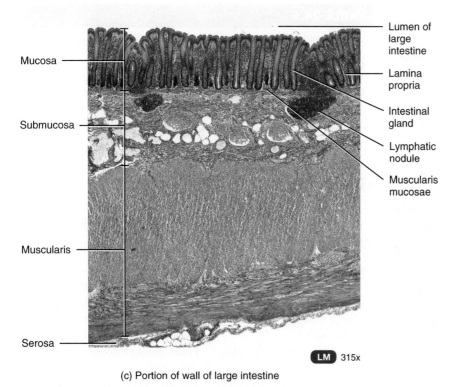

Mucosa

Submucosa

Muscularis

Serosa

Lumen of large intestine

Lamina propria

Intestinal gland

Lymphatic nodule

Muscularis mucosae

LM 315x

(c) Portion of wall of large intestine

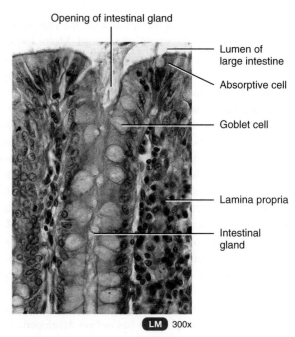

Opening of intestinal gland

Lumen of large intestine

Absorptive cell

Goblet cell

Lamina propria

Intestinal gland

LM 300x

(d) Details of mucosa of large intestine

? **What is the function of the goblet cells in the large intestine?**

Absorption and Feces Formation in the Large Intestine

By the time chyme has remained in the large intestine 3–10 hours, it has become solid or semisolid because of water absorption and is now called **feces.** Chemically, feces consist of water, inorganic salts, sloughed-off epithelial cells from the mucosa of the gastrointestinal tract, bacteria, products of bacterial decomposition, unabsorbed digested materials, and indigestible parts of food.

Although 90% of all water absorption occurs in the small intestine, the large intestine absorbs enough to make it an important organ in maintaining the body's water balance. Of the 0.5–1.0 liter of water that enters the large intestine, all but about 100–200 mL is normally absorbed via osmosis. The large intestine also absorbs ions, including sodium and chloride, and some vitamins.

CLINICAL CONNECTION | Occult Blood

The term **occult blood** refers to blood that is hidden; it is not detectable by the human eye. The main diagnostic value of occult blood testing is to screen for colorectal cancer. Two substances often examined for occult blood are feces and urine. Several types of products are available for at-home testing for hidden blood in feces. The tests are based on color changes when reagents are added to feces. The presence of occult blood in urine may be detected at home by using dip-and-read reagent strips. •

The Defecation Reflex

Mass peristaltic movements push fecal material from the sigmoid colon into the rectum. The resulting distension of the rectal wall stimulates stretch receptors, which initiates a **defecation reflex** that empties the rectum. The defecation reflex occurs as follows: In response to distension of the rectal wall, the receptors send sensory nerve impulses to the sacral spinal cord. Motor impulses from the cord travel along parasympathetic nerves back to the descending colon, sigmoid colon, rectum, and anus. The resulting contraction of the longitudinal rectal muscles shortens the rectum, thereby increasing the pressure within it. This pressure, along with voluntary contractions of the diaphragm and abdominal muscles, plus parasympathetic stimulation, opens the internal anal sphincter.

The external anal sphincter is voluntarily controlled. If it is voluntarily relaxed, defecation occurs and the feces are expelled through the anus; if it is voluntarily constricted, defecation can be postponed. Voluntary contractions of the diaphragm and abdominal muscles aid defecation by increasing the pressure within the abdomen, which pushes the walls of the sigmoid colon and rectum inward. If defecation does not occur, the feces back up into the sigmoid colon until the next wave of mass peristalsis stimulates the stretch receptors, again creating the urge to defecate. In infants, the defecation reflex causes automatic emptying of the rectum because voluntary control of the external anal sphincter has not yet developed.

The amount of bowel movements that a person has over a given period of time depends on various factors such as diet, health, and stress. The normal range of bowel activity varies from two or three bowel movements per day to three or four bowel movements per week.

Diarrhea (dī-a-RĒ-a; *dia-* = through; *-rrhea* = flow) is an increase in the frequency, volume, and fluid content of the feces caused by increased motility of and decreased absorption by the intestines. When chyme passes too quickly through the small intestine and feces pass too quickly through the large intestine, there is not enough time for absorption. Frequent diarrhea can result in dehydration and electrolyte imbalances. Excessive motility may be caused by lactose intolerance, stress, and microbes that irritate the gastrointestinal mucosa.

Constipation (kon-sti-PĀ-shun; *con-* = together; *-stip-* = to press) refers to infrequent or difficult defecation caused by decreased motility of the intestines. Because the feces remain in the colon for prolonged periods, excessive water absorption occurs, and the feces become dry and hard. Constipation may be caused by poor habits (delaying defecation), spasms of the colon, insufficient fiber in the diet, inadequate fluid intake, lack of exercise, emotional stress, and certain drugs. A common treatment is a mild laxative, such as milk of magnesia, which induces defecation. However, many physicians maintain that laxatives are habit-forming, and that adding fiber to the diet, increasing the amount of exercise, and increasing fluid intake are safer ways of controlling this common problem.

Table 24.6 summarizes the digestive activities in the large intestine, and Table 24.7 summarizes the functions of all digestive system organs.

TABLE 24.6

Summary of Digestive Activities in the Large Intestine

STRUCTURE	ACTIVITY	FUNCTION(S)
Lumen	Bacterial activity.	Breaks down undigested carbohydrates, proteins, and amino acids into products that can be expelled in feces or absorbed and detoxified by liver; synthesizes certain B vitamins and vitamin K.
Mucosa	Secretes mucus.	Lubricates colon; protects mucosa.
	Absorption.	Water absorption solidifies feces and contributes to body's water balance; solutes absorbed include ions and some vitamins.
Muscularis	Haustral churning.	Moves contents from haustrum to haustrum by muscular contractions.
	Peristalsis.	Moves contents along length of colon by contractions of circular and longitudinal muscles.
	Mass peristalsis.	Forces contents into sigmoid colon and rectum.
	Defecation reflex.	Eliminates feces by contractions in sigmoid colon and rectum.

TABLE 24.7

Summary of Organs of the Digestive System and Their Functions

ORGAN	FUNCTION(S)
Tongue	Maneuvers food for mastication, shapes food into a bolus, maneuvers food for deglutition, detects sensations for taste, and initiates digestion of triglycerides.
Salivary glands	Saliva produced by these glands softens, moistens, and dissolves foods; cleanses mouth and teeth; initiates the digestion of starch.
Teeth	Cut, tear, and pulverize food to reduce solids to smaller particles for swallowing.
Pancreas	Pancreatic juice buffers acidic gastric juice in chyme, stops the action of pepsin from the stomach, creates the proper pH for digestion in the small intestine, and participates in the digestion of carbohydrates, proteins, triglycerides, and nucleic acids.
Liver	Produces bile, which is required for the emulsification and absorption of lipids in the small intestine.
Gallbladder	Stores and concentrates bile and releases it into the small intestine.
Mouth	See the functions of the tongue, salivary glands, and teeth, all of which are in the mouth. Additionally, the lips and cheeks keep food between the teeth during mastication, and buccal glands lining the mouth produce saliva.
Pharynx	Receives a bolus from the oral cavity and passes it into the esophagus.
Esophagus	Receives a bolus from the pharynx and moves it into the stomach; this requires relaxation of the upper esophageal sphincter and secretion of mucus.
Stomach	Mixing waves combine saliva, food, and gastric juice, which activates pepsin, initiates protein digestion, kills microbes in food, helps absorb vitamin B_{12}, contracts the lower esophageal sphincter, increases stomach motility, relaxes the pyloric sphincter, and moves chyme into the small intestine.
Small intestine	Segmentation mixes chyme with digestive juices; peristalsis propels chyme toward the ileocecal sphincter; digestive secretions from the small intestine, pancreas, and liver complete the digestion of carbohydrates, proteins, lipids, and nucleic acids; circular folds, villi, and microvilli help absorb about 90 percent of digested nutrients.
Large intestine	Haustral churning, peristalsis, and mass peristalsis drive the colonic contents into the rectum; bacteria produce some B vitamins and vitamin K; absorption of some water, ions, and vitamins occurs; defecation.

CLINICAL CONNECTION | *Dietary Fiber*

Dietary fiber consists of indigestible plant carbohydrates—such as cellulose, lignin, and pectin—found in fruits, vegetables, grains, and beans. **Insoluble fiber,** which does not dissolve in water, includes the woody or structural parts of plants such as the skins of fruits and vegetables and the bran coating around wheat and corn kernels. Insoluble fiber passes through the GI tract largely unchanged but speeds up the passage of material through the tract. **Soluble fiber,** which does dissolve in water, forms a gel that slows the passage of material through the tract. It is found in abundance in beans, oats, barley, broccoli, prunes, apples, and citrus fruits.

People who choose a fiber-rich diet may reduce their risk of developing obesity, diabetes, atherosclerosis, gallstones, hemorrhoids, diverticulitis, appendicitis, and colorectal cancer. Soluble fiber also may help lower blood cholesterol. The liver normally converts cholesterol to bile salts, which are released into the small intestine to help fat digestion. Having accomplished their task, the bile salts are reabsorbed by the small intestine and recycled back to the liver. Since soluble fiber binds to bile salts to prevent their reabsorption, the liver makes more bile salts to replace those lost in feces. Thus, the liver uses more cholesterol to make more bile salts and blood cholesterol level is lowered. •

✔ CHECKPOINT

39. What are the major regions of the large intestine?
40. How does the muscularis of the large intestine differ from that of the rest of the gastrointestinal tract? What are haustra?
41. Describe the mechanical movements that occur in the large intestine.
42. What is defecation and how does it occur?
43. What activities occur in the large intestine to change its contents into feces?

24.14 PHASES OF DIGESTION

◉ OBJECTIVES

• Describe the three phases of digestion.
• Describe the major hormones regulating digestive activities.

Digestive activities occur in three overlapping phases: the cephalic phase, the gastric phase, and the intestinal phase.

Cephalic Phase

During the **cephalic phase** of digestion, the smell, sight, thought, or initial taste of food activates neural centers in the cerebral cortex, hypothalamus, and brain stem. The brain stem then activates the facial (VII), glossopharyngeal (IX), and vagus (X) nerves. The facial and glossopharyngeal nerves stimulate the salivary glands to secrete saliva, while the vagus nerves stimulate the gastric glands to secrete gastric juice. The purpose of the cephalic phase of digestion is to prepare the mouth and stomach for food that is about to be eaten.

Gastric Phase

Once food reaches the stomach, the **gastric phase** of digestion begins. Neural and hormonal mechanisms regulate the gastric phase of digestion to promote gastric secretion and gastric motility.

- *Neural regulation.* Food of any kind distends the stomach and stimulates stretch receptors in its walls. Chemoreceptors in the stomach monitor the pH of the stomach chyme. When the stomach walls are distended or pH increases because proteins have entered the stomach and buffered some of the stomach acid, the stretch receptors and chemoreceptors are activated, and a neural negative feedback loop is set in motion (Figure 24.25). From the stretch receptors and chemoreceptors, nerve impulses propagate to the submucosal plexus, where they activate parasympathetic and enteric neurons. The resulting nerve impulses cause waves of peristalsis and continue to stimulate the flow of gastric juice from gastric glands. The peristaltic waves mix the food with gastric juice; when the waves become strong enough, a small quantity of chyme undergoes gastric emptying into the duodenum. The pH of the stomach chyme decreases (becomes more acidic) and the distension of the stomach walls lessens because chyme has passed into the small intestine, suppressing secretion of gastric juice.

- *Hormonal regulation.* Gastric secretion during the gastric phase is also regulated by the hormone **gastrin.** Gastrin is released from the **G cells** of the gastric glands in response to several stimuli: distension of the stomach by chyme, partially digested proteins in chyme, the high pH of chyme due to the presence of food in the stomach, caffeine in gastric chyme, and acetylcholine released from parasympathetic neurons. Once it is released, gastrin enters the bloodstream, makes a round-trip through the body, and finally reaches its target organs in the digestive system. Gastrin stimulates gastric glands to secrete large amounts of gastric juice. It also strengthens the contraction of the lower esophageal sphincter to prevent reflux of acid chyme into the esophagus, increases motility of the stomach, and relaxes the pyloric sphincter, which promotes gastric emptying. Gastrin secretion is inhibited when the pH of gastric juice drops below 2.0 and is stimulated when the pH rises. This negative feedback mechanism helps provide an optimal low pH for the functioning of pepsin, the killing of microbes, and the denaturing of proteins in the stomach.

Intestinal Phase

The **intestinal phase** of digestion begins once food enters the small intestine. In contrast to reflexes initiated during the cephalic

Figure 24.25 Neural negative feedback regulation of the pH of gastric juice and gastric motility during the gastric phase of digestion.

 Food entering the stomach stimulates secretion of gastric juice and causes vigorous waves of peristalsis.

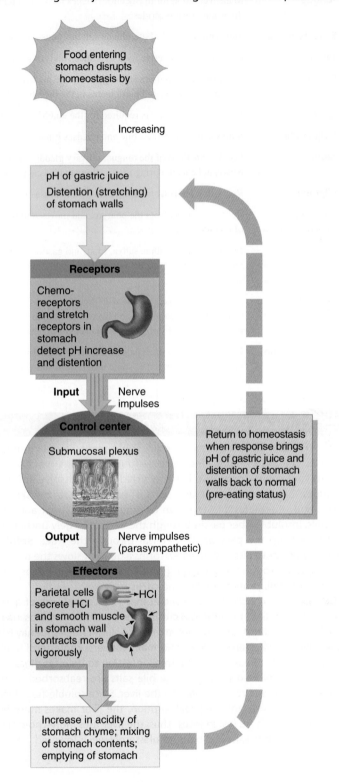

? Why does food initially cause the pH of the gastric juice to rise?

and gastric phases, which stimulate stomach secretory activity and motility, those occurring during the intestinal phase have inhibitory effects that slow the exit of chyme from the stomach. This prevents the duodenum from being overloaded with more chyme than it can handle. In addition, responses occurring during the intestinal phase promote the continued digestion of foods that have reached the small intestine. These activities of the intestinal phase of digestion are regulated by neural and hormonal mechanisms.

- *Neural regulation.* Distension of the duodenum by the presence of chyme causes the **enterogastric reflex** (en′-ter-ō-GAS-trik). Stretch receptors in the duodenal wall send nerve impulses to the medulla oblongata, where they inhibit parasympathetic stimulation and stimulate the sympathetic nerves to the stomach. As a result, gastric motility is inhibited and there is an increase in the contraction of the pyloric sphincter, which decreases gastric emptying.

- *Hormonal regulation.* The intestinal phase of digestion is mediated by two major hormones secreted by the small intestine: cholecystokinin and secretin. **Cholecystokinin (CCK)** is secreted by the **CCK cells** of the small intestinal crypts of Lieberkühn in response to chyme containing amino acids from partially digested proteins and fatty acids from partially digested triglycerides. CCK stimulates secretion of pancreatic juice that is rich in digestive enzymes. It also causes contraction of the wall of the gallbladder, which squeezes stored bile out of the gallbladder into the cystic duct and through the common bile duct. In addition, CCK causes relaxation of the sphincter of the hepatopancreatic ampulla (sphincter of Oddi), which allows pancreatic juice and bile to flow into the duodenum. CCK also slows gastric emptying by promoting contraction of the pyloric sphincter, produces satiety (a feeling of fullness) by acting on the hypothalamus in the brain, promotes normal growth and maintenance of the pancreas, and enhances the effects of secretin. Acidic chyme entering the duodenum stimulates the release of **secretin** from the **S cells** of the small intestinal crypts of Lieberkühn. In turn, secretin stimulates the flow of pancreatic juice that is rich in bicarbonate (HCO_3^-) ions to buffer the acidic chyme that enters the duodenum from the small intestine. Besides this major effect, secretin inhibits secretion of gastric juice, promotes normal growth and maintenance of the pancreas, and enhances the effects of CCK. Overall, secretin causes buffering of acid in chyme that reaches the duodenum and slows production of acid in the stomach.

Other Hormones of the Digestive System

Besides gastrin, CCK, and secretin, at least 10 other so-called gut hormones are secreted by and have effects on the GI tract. They include *motilin, substance P,* and *bombesin,* which stimulate motility of the intestines; *vasoactive intestinal polypeptide (VIP),* which stimulates secretion of ions and water by the intestines and inhibits gastric acid secretion; *gastrin-releasing peptide,* which stimulates release of gastrin; and *somatostatin,* which inhibits gastrin release. Some of these hormones are thought to act as local hormones (paracrines), whereas others are secreted into the blood or even into the lumen of the GI tract. The physiological roles of these and other gut hormones are still under investigation.

Table 24.8 summarizes the major hormones that control digestion.

✔**CHECKPOINT**

44. What is the purpose of the cephalic phase of digestion?
45. Describe the role of gastrin in the gastric phase of digestion.
46. Outline the steps of the enterogastric reflex.
47. Explain the roles of CCK and secretin in the intestinal phase of digestion.

TABLE 24.8

Major Hormones That Control Digestion

HORMONE	STIMULUS AND SITE OF SECRETION	ACTIONS
Gastrin	Distension of stomach, partially digested proteins and caffeine in stomach, and high pH of stomach chyme stimulate gastrin secretion by enteroendocrine G cells, located mainly in mucosa of pyloric antrum of stomach.	*Major effects:* Promotes secretion of gastric juice, increases gastric motility, promotes growth of gastric mucosa. *Minor effects:* Constricts lower esophageal sphincter, relaxes pyloric sphincter.
Secretin	Acidic (high H^+ level) chyme that enters small intestine stimulates secretion of secretin by enteroendocrine S cells in the mucosa of duodenum.	*Major effects:* Stimulates secretion of pancreatic juice and bile that are rich in HCO_3^- (bicarbonate ions). *Minor effects:* Inhibits secretion of gastric juice, promotes normal growth and maintenance of pancreas, enhances effects of CCK.
Cholecystokinin (CCK)	Partially digested proteins (amino acids), triglycerides, and fatty acids that enter small intestine stimulate secretion of CCK by enteroendocrine CCK cells in mucosa of small intestine; CCK is also released in brain.	*Major effects:* Stimulates secretion of pancreatic juice rich in digestive enzymes, causes ejection of bile from gallbladder and opening of sphincter of the hepatopancreatic ampulla (sphincter of Oddi), induces satiety (feeling full to satisfaction). *Minor effects:* Inhibits gastric emptying, promotes normal growth and maintenance of pancreas, enhances effects of secretin.

24.15 DEVELOPMENT OF THE DIGESTIVE SYSTEM

◉ OBJECTIVE

• Describe the development of the digestive system.

During the fourth week of development, the cells of the **endo-derm** form a cavity called the **primitive gut,** the forerunner of the gastrointestinal tract (see Figure 29.12b). Soon afterward the mesoderm forms and splits into two layers (somatic and splanchnic), as shown in Figure 29.9d. The splanchnic mesoderm associates with the endoderm of the primitive gut; as a result, the primitive gut has a double-layered wall. The **endodermal layer** gives rise to the *epithelial lining* and *glands* of most of the gastrointestinal tract; the **mesodermal layer** produces the *smooth muscle* and *connective tissue* of the tract.

The primitive gut elongates and differentiates into an anterior **foregut,** an intermediate **midgut,** and a posterior **hindgut** (see Figure 29.12c). Until the fifth week of development, the midgut opens into the yolk sac; after that time, the yolk sac constricts and detaches from the midgut, and the midgut seals. In the region of the foregut, a depression consisting of ectoderm, the **stomodeum** (stō-mō-DĒ-um), appears (see Figure 29.12d), which develops into the *oral cavity*. The **oropharyngeal membrane** (or'-ō-fa-RIN-jē-al) is a depression of fused ectoderm and endoderm on the surface of the embryo that separates the foregut from the stomodeum. The membrane ruptures during the fourth week of development, so that the foregut is continuous with the outside of the embryo through the oral cavity. Another depression consisting of ectoderm, the **proctodeum** (prok-tō-DĒ-um), forms in the hindgut and goes on to develop into the *anus* (see Figure 29.12d). The **cloacal membrane** (klō-Ā-kul) is a fused membrane of ectoderm and endoderm that separates the hindgut from the proctodeum. After it ruptures during the seventh week, the hindgut is continuous with the outside of the embryo through the anus. Thus, the gastrointestinal tract forms a continuous tube from mouth to anus.

The foregut develops into the *pharynx, esophagus, stomach,* and *part of the duodenum*. The midgut is transformed into the *remainder of the duodenum,* the *jejunum,* the *ileum,* and *portions of the large intestine* (cecum, appendix, ascending colon, and most of the transverse colon). The hindgut develops into the *remainder of the large intestine,* except for a portion of the anal canal that is derived from the proctodeum.

As development progresses, the endoderm at various places along the foregut develops into hollow buds that grow into the mesoderm. These buds will develop into the *salivary glands, liver, gallbladder,* and *pancreas*. Each of these organs retains a connection with the gastrointestinal tract through ducts.

✔CHECKPOINT

48. What structures develop from the foregut, midgut, and hindgut?

24.16 AGING AND THE DIGESTIVE SYSTEM

◉ OBJECTIVE

• Describe the effects of aging on the digestive system.

Overall changes of the digestive system associated with aging include decreased secretory mechanisms, decreased motility of the digestive organs, loss of strength and tone of the muscular tissue and its supporting structures, changes in neurosensory feedback regarding enzyme and hormone release, and diminished response to pain and internal sensations. In the upper portion of the GI tract, common changes include reduced sensitivity to mouth irritations and sores, loss of taste, periodontal disease, difficulty in swallowing, hiatal hernia, gastritis, and peptic ulcer disease. Changes that may appear in the small intestine include duodenal ulcers, malabsorption, and maldigestion. Other pathologies that increase in incidence with age are appendicitis, gallbladder problems, jaundice, cirrhosis, and acute pancreatitis. Large intestinal changes such as constipation, hemorrhoids, and diverticular disease may also occur. Cancer of the colon or rectum is quite common, as are bowel obstructions and impactions.

✔CHECKPOINT

49. What are the general effects of aging on the digestive system?

• • •

Now that our exploration of the digestive system is complete, you can appreciate the many ways that this system contributes to homeostasis of other body systems by examining *Focus on Homeostasis: The Digestive System*. Next, in Chapter 25, you will discover how the nutrients absorbed by the GI tract enter into metabolic reactions in the body tissues.

BODY SYSTEM	CONTRIBUTION OF THE DIGESTIVE SYSTEM

For all body systems
The digestive system breaks down dietary nutrients into forms that can be absorbed and used by body cells for producing ATP and building body tissues. Absorbs water, minerals, and vitamins needed for growth and function of body tissues; and eliminates wastes from body tissues in feces.

Integumentary system
Small intestine absorbs vitamin D, which skin and kidneys modify to produce the hormone calcitriol. Excess dietary calories are stored as triglycerides in adipose cells in dermis and subcutaneous layer.

Skeletal system
Small intestine absorbs dietary calcium and phosphorus salts needed to build bone extracellular matrix.

Muscular system
Liver can convert lactic acid (produced by muscles during exercise) to glucose.

Nervous system
Gluconeogenesis (synthesis of new glucose molecules) in liver plus digestion and absorption of dietary carbohydrates provide glucose, needed for ATP production by neurons.

Endocrine system
Liver inactivates some hormones, ending their activity. Pancreatic islets release insulin and glucagon. Cells in mucosa of stomach and small intestine release hormones that regulate digestive activities. Liver produces angiotensinogen.

Cardiovascular system
GI tract absorbs water that helps maintain blood volume and iron that is needed for synthesis of hemoglobin in red blood cells. Bilirubin from hemoglobin breakdown is partially excreted in feces. Liver synthesizes most plasma proteins.

Lymphatic system and immunity
Acidity of gastric juice destroys bacteria and most toxins in stomach. Lymphatic nodules in lamina propria of mucosa of gastrointestinal tract (MALT) destroy microbes.

Respiratory system
Pressure of abdominal organs against diaphragm helps expel air quickly during forced exhalation.

Urinary system
Absorption of water by GI tract provides water needed to excrete waste products in urine.

Reproductive systems
Digestion and absorption provide adequate nutrients, including fats, for normal development of reproductive structures, for production of gametes (oocytes and sperm), and for fetal growth and development during pregnancy.

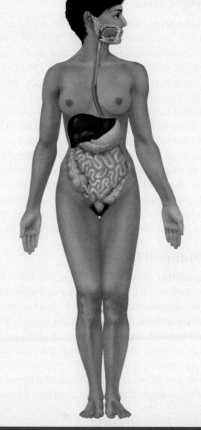

THE DIGESTIVE SYSTEM

DISORDERS: HOMEOSTATIC IMBALANCES

Dental Caries

Dental caries (KAR-ēz), or tooth decay, involves a gradual demineralization (softening) of the enamel and dentin. If untreated, microorganisms may invade the pulp, causing inflammation and infection, with subsequent death of the pulp and abscess of the alveolar bone surrounding the root's apex, requiring root canal therapy (see Section 24.5).

Dental caries begin when bacteria, acting on sugars, produce acids that demineralize the enamel. **Dextran,** a sticky polysaccharide produced from sucrose, causes the bacteria to stick to the teeth. Masses of bacterial cells, dextran, and other debris adhering to teeth constitute **dental plaque** (PLAK). Saliva cannot reach the tooth surface to buffer the acid because the plaque covers the teeth. Brushing the teeth after eating removes the plaque from flat surfaces before the bacteria can produce acids. Dentists also recommend that the plaque between the teeth be removed every 24 hours with dental floss.

Periodontal Disease

Periodontal disease is a collective term for a variety of conditions characterized by inflammation and degeneration of the gingivae, alveolar bone, periodontal ligament, and cementum. In one such condition, called **pyorrhea,** initial symptoms include enlargement and inflammation of the soft tissue and bleeding of the gums. Without treatment, the soft tissue may deteriorate and the alveolar bone may be resorbed, causing loosening of the teeth and recession of the gums. Periodontal diseases are often caused by poor oral hygiene; by local irritants, such as bacteria, impacted food, and cigarette smoke; or by a poor "bite."

Peptic Ulcer Disease

In the United States, 5–10% of the population develops **peptic ulcer disease (PUD).** An **ulcer** is a craterlike lesion in a membrane; ulcers that develop in areas of the GI tract exposed to acidic gastric juice are called **peptic ulcers.** The most common complication of peptic ulcers is bleeding, which can lead to anemia if enough blood is lost. In acute cases, peptic ulcers can lead to shock and death. Three distinct causes of PUD are recognized: (1) the bacterium *Helicobacter pylori* (hel-i-kō-BAK-ter pī-Lō-rē); (2) nonsteroidal anti-inflammatory drugs (NSAIDs) such as aspirin; and (3) hypersecretion of HCl, as occurs in Zollinger–Ellison syndrome (ZOL-in-jer EL-i-son), a gastrin-producing tumor, usually of the pancreas.

Helicobacter pylori (previously named *Campylobacter pylori*) is the most frequent cause of PUD. The bacterium produces an enzyme called urease, which splits urea into ammonia and carbon dioxide. While shielding the bacterium from the acidity of the stomach, the ammonia also damages the protective mucous layer of the stomach and the underlying gastric cells. The microbe also produces catalase, an enzyme that may protect *H. pylori* from phagocytosis by neutrophils, plus several adhesion proteins that allow the bacterium to attach itself to gastric cells.

Several therapeutic approaches are helpful in the treatment of PUD. Cigarette smoke, alcohol, caffeine, and NSAIDs should be avoided because they can impair mucosal defensive mechanisms, which increases mucosal susceptibility to the damaging effects of HCl. In cases associated with *H. pylori,* treatment with an antibiotic drug often resolves the problem. Oral antacids such as Tums® or Maalox® can help temporarily by buffering gastric acid. When hypersecretion of HCl is the cause of PUD, H_2 blockers (such as Tagamet®) or proton pump inhibitors such as omeprazole (Prilosec®), which block secretion of H^+ from parietal cells, may be used.

Diverticular Disease

In **diverticular disease** (dī'-ver-TIK-ū-lar), saclike outpouchings of the wall of the colon, termed **diverticula,** occur in places where the muscularis has weakened and may become inflamed. Development of diverticula is known as **diverticulosis.** Many people who develop diverticulosis have no symptoms and experience no complications. Of those people known to have diverticulosis, 10–25% eventually develop an inflammation known as **diverticulitis** (dī'-ver-tik-ū-Lī-tis). This condition may be characterized by pain, either constipation or increased frequency of defecation, nausea, vomiting, and low-grade fever. Because diets low in fiber contribute to development of diverticulitis, patients who change to high-fiber diets show marked relief of symptoms. In severe cases, affected portions of the colon may require surgical removal. If diverticula rupture, the release of bacteria into the abdominal cavity can cause peritonitis.

Colorectal Cancer

Colorectal cancer is among the deadliest of malignancies, ranking second to lung cancer in males and third after lung cancer and breast cancer in females. Genetics plays a very important role; an inherited predisposition contributes to more than half of all cases of colorectal cancer. Intake of alcohol and diets high in animal fat and protein are associated with increased risk of colorectal cancer; dietary fiber, retinoids, calcium, and selenium may be protective. Signs and symptoms of colorectal cancer include diarrhea, constipation, cramping, abdominal pain, and rectal bleeding, either visible or occult (hidden in feces). Precancerous growths on the mucosal surface, called **polyps,** also increase the risk of developing colorectal cancer. Screening for colorectal cancer includes testing for blood in the feces, digital rectal examination, sigmoidoscopy, colonoscopy, and barium enema. Tumors may be removed endoscopically or surgically.

Hepatitis

Hepatitis is an inflammation of the liver that can be caused by viruses, drugs, and chemicals, including alcohol. Clinically, several types of viral hepatitis are recognized.

Hepatitis A (infectious hepatitis) is caused by the hepatitis A virus and is spread via fecal contamination of objects such as food, clothing, toys, and eating utensils (fecal–oral route). It is generally a mild disease of children and young adults characterized by loss of appetite, malaise, nausea, diarrhea, fever, and chills. Eventually, jaundice appears. This type of hepatitis does not cause lasting liver damage. Most people recover in 4 to 6 weeks.

Hepatitis B is caused by the hepatitis B virus and is spread primarily by sexual contact and contaminated syringes and transfusion equipment. It can also be spread via saliva and tears. Hepatitis B virus can be present for years or even a lifetime, and it can produce cirrhosis and possibly cancer of the liver. Individuals who harbor the active hepatitis B virus also become carriers. Vaccines produced through recombinant DNA technology (for example, Recombivax HB®) are available to prevent hepatitis B infection.

Hepatitis C, caused by the hepatitis C virus, is clinically similar to hepatitis B. Hepatitis C can cause cirrhosis and possibly liver cancer. In developed nations, donated blood is screened for the presence of hepatitis B and C.

Hepatitis D is caused by the hepatitis D virus. It is transmitted like hepatitis B, and in fact a person must have been co-infected with hepatitis B before contracting hepatitis D. Hepatitis D results in severe liver damage and has a higher fatality rate than infection with hepatitis B virus alone.

Hepatitis E is caused by the hepatitis E virus and is spread like hepatitis A. Although it does not cause chronic liver disease, hepatitis E virus has a very high mortality rate among pregnant women.

MEDICAL TERMINOLOGY

Achalasia (ak'-a-LĀ-zē-a; *a-* = without; *-chalasis* = relaxation) A condition caused by malfunction of the myenteric plexus in which the lower esophageal sphincter fails to relax normally as food approaches. A whole meal may become lodged in the esophagus and enter the stomach very slowly. Distension of the esophagus results in chest pain that is often confused with pain originating from the heart.

Bariatric surgery (bar'-ē-AT-rik; *baros-* = weight; *-iatreia* = medical treatment) A surgical procedure that limits the amount of food that can be ingested and absorbed in order to bring about a significant weight loss in obese individuals. The most commonly performed type of bariatric surgery is called *gastric bypass surgery*. In one variation of this procedure, the stomach is reduced in size by making a small pouch at the top of the stomach about the size of a walnut. The pouch, which is only 5–10% of the stomach, is sealed off from the rest of the stomach using surgical staples or a plastic band. The pouch is connected to the jejunum of the small intestine, thus bypassing the rest of the stomach and the duodenum. The result is that smaller amounts of food are ingested and fewer nutrients are absorbed in the small intestine. This leads to weight loss.

Borborygmus (bor'-bō-RIG-mus) A rumbling noise caused by the propulsion of gas through the intestines.

Canker sore (KANG-ker) Painful ulcer on the mucous membrane of the mouth that affects females more often than males, usually between ages 10 and 40; may be an autoimmune reaction or a food allergy.

Cirrhosis Distorted or scarred liver as a result of chronic inflammation due to hepatitis, chemicals that destroy hepatocytes, parasites that infect the liver, or alcoholism; the hepatocytes are replaced by fibrous or adipose connective tissue. Symptoms include jaundice, edema in the legs, uncontrolled bleeding, and increased sensitivity to drugs.

Colitis (kō-LĪ-tis) Inflammation of the mucosa of the colon and rectum in which absorption of water and salts is reduced, producing watery, bloody feces and, in severe cases, dehydration and salt depletion. Spasms of the irritated muscularis produce cramps. It is thought to be an autoimmune condition.

Colonoscopy (kō-lon-OS-kō-pē; *-skopes* = to view) The visual examination of the lining of the colon using an elongated, flexible, fiber-optic endoscope called a *colonoscope*. It is used to detect disorders such as polyps, cancer, and diverticulosis; to take tissue samples; and to remove small polyps. Most tumors of the large intestine occur in the rectum.

Colostomy (kō-LOS-tō-mē; *-stomy* = provide an opening) The diversion of feces through an opening in the colon, creating a surgical "stoma" (artificial opening) that is made in the exterior of the abdominal wall. This opening serves as a substitute anus through which feces are eliminated into a bag worn on the abdomen.

Dysphagia (dis-FĀ-jē-a; *dys-* = abnormal; *-phagia* = to eat) Difficulty in swallowing that may be caused by inflammation, paralysis, obstruction, or trauma.

Flatus (FLĀ-tus) Air (gas) in the stomach or intestine, usually expelled through the anus. If the gas is expelled through the mouth, it is called **eructation** or **belching** (burping). Flatus may result from gas released during the breakdown of foods in the stomach or from swallowing air or gas-containing substances such as carbonated drinks.

Food poisoning A sudden illness caused by ingesting food or drink contaminated by an infectious microbe (bacterium, virus, or protozoan) or a toxin (poison). The most common cause of food poisoning is the toxin produced by the bacterium *Staphylococcus aureus.* Most types of food poisoning cause diarrhea and/or vomiting, often associated with abdominal pain.

Gastroenteritis (gas'-trō-en-ter-Ī-tis; *gastro-* = stomach; *-enteron-* = intestine; *-itis* = inflammation) Inflammation of the lining of the stomach and intestine (especially the small intestine). It is usually caused by a viral or bacterial infection that may be acquired by contaminated food or water or by people in close contact. Symptoms include diarrhea, vomiting, fever, loss of appetite, cramps, and abdominal discomfort.

Gastroscopy (gas-TROS-kō-pē; *-scopy* = to view with a lighted instrument) Endoscopic examination of the stomach in which the examiner can view the interior of the stomach directly to evaluate an ulcer, tumor, inflammation, or source of bleeding.

Halitosis (hal'-i-TŌ-sis; *halitus-* = breath; *-osis* = condition) A foul odor from the mouth; also called **bad breath.**

Heartburn A burning sensation in a region near the heart due to irritation of the mucosa of the esophagus from hydrochloric acid in stomach contents. It is caused by failure of the lower esophageal sphincter to close properly, so that the stomach contents enter the inferior esophagus. It is not related to any cardiac problem.

Hemorrhoids (HEM-ō-royds; *hemo-* = blood; *-rhoia* = flow) Varicosed (enlarged and inflamed) superior rectal veins. Hemorrhoids develop when the veins are put under pressure and become engorged with blood. If the pressure continues, the wall of the vein stretches. Such a distended vessel oozes blood; bleeding or itching is usually the first sign that a hemorrhoid has developed. Stretching of a vein also favors clot formation, further aggravating swelling and pain. Hemorrhoids may be caused by constipation, which may be brought on by low-fiber diets. Also, repeated straining during defecation forces blood down into the rectal veins, increasing pressure in those veins and possibly causing hemorrhoids. Also called **piles.**

Hernia (HER-nē-a) Protrusion of all or part of an organ through a membrane or cavity wall, usually the abdominal cavity. *Hiatus (diaphragmatic) hernia* is the protrusion of a part of the stomach into the thoracic cavity through the esophageal hiatus of the diaphragm. *Inguinal hernia* is the protrusion of the hernial sac into the inguinal opening; it may contain a portion of the bowel in an advanced stage and may extend into the scrotal compartment in males, causing strangulation of the herniated part.

Inflammatory bowel disease (in-FLAM-a-tō'-rē BOW-el) Inflammation of the gastrointestinal tract that exists in two forms. (1) ***Crohn's disease*** is an inflammation of any part of the gastrointestinal tract in which the inflammation extends from the mucosa through the submucosa, muscularis, and serosa. (2) ***Ulcerative colitis*** is an inflammation of the mucosa of the colon and rectum, usually accompanied by rectal bleeding. Curiously, cigarette smoking increases the risk of Crohn's disease but decreases the risk of ulcerative colitis.

Irritable bowel syndrome (IBS) Disease of the entire gastrointestinal tract in which a person reacts to stress by developing symptoms (such as cramping and abdominal pain) associated with alternating patterns of diarrhea and constipation. Excessive amounts of mucus may appear in feces; other symptoms include flatulence, nausea, and loss of appetite. The condition is also known as **irritable colon** or **spastic colitis.**

Malabsorption (mal-ab-SORP-shun; *mal-* = bad) A number of disorders in which nutrients from food are not absorbed properly. It may be due to disorders that result in the inadequate breakdown of food during digestion (due to inadequate digestive enzymes or juices), damage to the lining of the small intestine (from surgery, infections, and drugs like neomycin and alcohol), and impairment of motility. Symptoms may include diarrhea, weight loss, weakness, vitamin deficiencies, and bone demineralization.

Malocclusion (mal'-ō-KLOO-zhun; *mal-* = bad; *-occlusion* = to fit together) Condition in which the surfaces of the maxillary (upper) and mandibular (lower) teeth fit together poorly.

Nausea (NAW-sē-a; *nausia* = seasickness) Discomfort characterized by a loss of appetite and the sensation of impending vomiting. Its causes include local irritation of the gastrointestinal tract, a systemic disease, brain disease or injury, overexertion, or the effects of medication or drug overdosage.

Traveler's diarrhea Infectious disease of the gastrointestinal tract that results in loose, urgent bowel movements, cramping, abdominal pain, malaise, nausea, and occasionally fever and dehydration. It is acquired through ingestion of food or water contaminated with fecal material typically containing bacteria (especially *Escherichia coli*); viruses or protozoan parasites are less common causes.

CHAPTER REVIEW AND RESOURCE SUMMARY

Review

Resource

Introduction

1. The breaking down of larger food molecules into smaller molecules is called digestion.
2. The organs involved in the breakdown of food are collectively known as the digestive system.

24.1 Overview of the Digestive System

1. The digestive system is composed of two main groups of organs: the gastrointestinal (GI) tract and accessory digestive organs.
2. The GI tract is a continuous tube extending from the mouth to the anus.
3. The accessory digestive organs include the teeth, tongue, salivary glands, liver, gallbladder, and pancreas.
4. Digestion includes six basic processes: ingestion, secretion, mixing and propulsion, mechanical and chemical digestion, absorption, and defecation.
5. Mechanical digestion consists of mastication and movements of the gastrointestinal tract that aid chemical digestion.
6. Chemical digestion is a series of hydrolysis reactions that break down large carbohydrates, lipids, proteins, and nucleic acids in foods into smaller molecules that are usable by body cells.

Anatomy Overview - The Digestive System
Animation - Chemical Digestion: Enzymes
Exercise - Concentrate on Digestion

24.2 Layers of the GI Tract

1. The basic arrangement of layers in most of the gastrointestinal tract, from deep to superficial, is the mucosa, submucosa, muscularis, and serosa.
2. Associated with the lamina propria of the mucosa are extensive patches of lymphatic tissue called mucosa-associated lymphoid tissue (MALT).

Anatomy Overview - GI Tract Histology
Figure 24.2 - Layers of the Gastrointestinal Tract
Exercise - Paint the Gastrointestinal Tract

24.3 Neural Innervation of the GI Tract

1. The gastrointestinal tract is regulated by an intrinsic set of nerves known as the enteric nervous system (ENS) and by an extrinsic set of nerves that are part of the autonomic nervous system (ANS).
2. The ENS consists of neurons arranged into two plexuses: the myenteric plexus and the submucosal plexus.
3. The myenteric plexus, which is located between the longitudinal and circular smooth muscle layers of the muscularis, regulates GI tract motility.
4. The submucosal plexus, which is located in the submucosa, regulates GI secretion.
5. Although the neurons of the ENS can function independently, they are subject to regulation by the neurons of the ANS.
6. Parasympathetic fibers of the vagus (X) nerves and pelvic splanchnic nerves increase GI tract secretion and motility by increasing the activity of ENS neurons.
7. Sympathetic fibers from the thoracic and upper lumbar regions of the spinal cord decrease GI tract secretion and motility by inhibiting ENS neurons.

.Animation - Neural Regulation of Mechanical Digestion
Animation - Enterogastric Reflex

Review   **Resource**

24.4 Peritoneum

1. The peritoneum is the largest serous membrane of the body; it lines the wall of the abdominal cavity and covers some abdominal organs.
2. Folds of the peritoneum include the mesentery, mesocolon, falciform ligament, lesser omentum, and greater omentum.

24.5 Mouth

1. The mouth is formed by the cheeks, hard and soft palates, lips, and tongue.
2. The vestibule is the space bounded externally by the cheeks and lips and internally by the teeth and gums.
3. The oral cavity proper extends from the vestibule to the fauces.
4. The tongue, together with its associated muscles, forms the floor of the oral cavity. It is composed of skeletal muscle covered with mucous membrane. The upper surface and sides of the tongue are covered with papillae, some of which contain taste buds.
5. The major portion of saliva is secreted by the major salivary glands, which lie outside the mouth and pour their contents into ducts that empty into the oral cavity. There are three pairs of major salivary glands: parotid, submandibular, and sublingual glands.
6. Saliva lubricates food and starts the chemical digestion of carbohydrates. Salivation is controlled by the nervous system.
7. The teeth (dentes) project into the mouth and are adapted for mechanical digestion.
8. A typical tooth consists of three principal regions: crown, root, and neck. Teeth are composed primarily of dentin and are covered by enamel, the hardest substance in the body. There are two dentitions: deciduous and permanent.
9. Through mastication, food is mixed with saliva and shaped into a soft, flexible mass called a bolus. Salivary amylase then begins the digestion of starches, and lingual lipase acts on triglycerides.

Anatomy Overview - Oral Cavity
Anatomy Overview - Salivary Glands
Animation - Mastication
Figure 24.8 - Dentitions and Times of Eruption

24.6 Pharynx

1. The pharynx is a funnel-shaped tube that extends from the internal nares to the esophagus posteriorly and to the larynx anteriorly.
2. The pharynx has both respiratory and digestive functions.

Anatomy Overview - Pharynx and Esophagus

24.7 Esophagus

1. The esophagus is a collapsible, muscular tube that connects the pharynx to the stomach.
2. It contains an upper and a lower esophageal sphincter.

Anatomy Overview - Esophagus Histology

24.8 Deglutition

1. Deglutition, or swallowing, moves a bolus from the mouth to the stomach.
2. Swallowing consists of a voluntary stage, a pharyngeal stage (involuntary), and an esophageal stage (involuntary).

Animation - Deglutition
Animation - Neural Regulation of Mechanical Digestion

24.9 Stomach

1. The stomach connects the esophagus to the duodenum.
2. The principal anatomical regions of the stomach are the cardia, fundus, body, and pylorus.
3. Adaptations of the stomach for digestion include rugae; glands that produce mucus, hydrochloric acid, pepsin, gastric lipase, and intrinsic factor; and a three-layered muscularis.
4. Mechanical digestion consists of mixing waves.
5. Chemical digestion consists mostly of the conversion of proteins into peptides by pepsin.
6. The stomach wall is impermeable to most substances.
7. Among the substances the stomach can absorb are water, certain ions, drugs, and alcohol.

Anatomy Overviews - The Stomach; Stomach Histology
Animations - Stomach Peristalsis; Protein Digestion in the Stomach; Lipid Digestion in the Stomach; Chemical Digestion: Gastric Acid
Figure 24.12 - Histology of the Stomach
Figure 24.13 - Secretion of HCl by Parietal Cells in the Stomach
Exercise - Role of Digestive Chemicals

24.10 Pancreas

1. The pancreas consists of a head, a body, and a tail and is connected to the duodenum via the pancreatic duct and accessory duct.
2. Endocrine pancreatic islets (islets of Langerhans) secrete hormones, and exocrine acini secrete pancreatic juice.
3. Pancreatic juice contains enzymes that digest starch (pancreatic amylase), proteins (trypsin, chymotrypsin, carboxypeptidase, and elastase), triglycerides (pancreatic lipase), and nucleic acids (ribonuclease and deoxyribonuclease).

Anatomy Overviews - Pancreas; Pancreas Histology
Animations - Carbohydrate Digestion: Pancreas; Protein Digestion: Pancreatic Juice; Lipid Digestion: Bile Salts and Pancreatic Lipase
Figure 24.15 - Relationship of the Pancreas to the Liver, Gallbladder, and Duodenum

Review	**Resource**

24.11 Liver and Gallbladder

1. The liver has left and right lobes; the left lobe includes a quadrate lobe and a caudate lobe. The gallbladder is a sac located in a depression on the posterior surface of the liver that stores and concentrates bile.
2. The lobes of the liver are made up of lobules that contain hepatocytes (liver cells), sinusoids, stellate reticuloendothelial (Kupffer) cells, and a central vein.
3. Hepatocytes produce bile that is carried by a duct system to the gallbladder for concentration and temporary storage.
4. Bile's contribution to digestion is the emulsification of dietary lipids.
5. The liver also functions in carbohydrate, lipid, and protein metabolism; processing of drugs and hormones; excretion of bilirubin; synthesis of bile salts; storage of vitamins and minerals; phagocytosis; and activation of vitamin D.

24.12 Small Intestine

1. The small intestine extends from the pyloric sphincter to the ileocecal sphincter. It is divided into duodenum, jejunum, and ileum.
2. Its glands secrete fluid and mucus, and the circular folds, villi, and microvilli of its wall provide a large surface area for digestion and absorption.
3. Brush-border enzymes digest α-dextrins, maltose, sucrose, lactose, peptides, and nucleotides at the surface of mucosal epithelial cells.
4. Pancreatic and intestinal brush-border enzymes break down starches into maltose, maltotriose, and α-dextrins (pancreatic amylase), α-dextrins into glucose (α-dextrinase), maltose to glucose (maltase), sucrose to glucose and fructose (sucrase), lactose to glucose and galactose (lactase), and proteins into peptides (trypsin, chymotrypsin, and elastase). Also, enzymes break off amino acids at the carboxyl ends of peptides (carboxypeptidases) and break off amino acids at the amino ends of peptides (aminopeptidases). Finally, enzymes split dipeptides into amino acids (dipeptidases), triglycerides to fatty acids and monoglycerides (lipases), and nucleotides to pentoses and nitrogenous bases (nucleosidases and phosphatases).
5. Mechanical digestion in the small intestine involves segmentation and migrating motility complexes.
6. Absorption occurs via diffusion, facilitated diffusion, osmosis, and active transport; most absorption occurs in the small intestine.
7. Monosaccharides, amino acids, and short-chain fatty acids pass into the blood capillaries.
8. Long-chain fatty acids and monoglycerides are absorbed from micelles, resynthesized to triglycerides, and formed into chylomicrons.
9. Chylomicrons move into lymph in the lacteal of a villus.
10. The small intestine also absorbs electrolytes, vitamins, and water.

24.13 Large Intestine

1. The large intestine extends from the ileocecal sphincter to the anus.
2. Its regions include the cecum, colon, rectum, and anal canal.
3. The mucosa contains many goblet cells, and the muscularis consists of teniae coli and haustra.
4. Mechanical movements of the large intestine include haustral churning, peristalsis, and mass peristalsis.
5. The last stages of chemical digestion occur in the large intestine through bacterial action. Substances are further broken down, and some vitamins are synthesized.
6. The large intestine absorbs water, ions, and vitamins.
7. Feces consist of water, inorganic salts, epithelial cells, bacteria, and undigested foods.
8. The elimination of feces from the rectum is called defecation.
9. Defecation is a reflex action aided by voluntary contractions of the diaphragm and abdominal muscles and relaxation of the external anal sphincter.

24.14 Phases of Digestion

1. Digestive activities occur in three overlapping phases: cephalic phase, gastric phase, and intestinal phase.
2. During the cephalic phase of digestion, salivary glands secrete saliva and gastric glands secrete gastric juice in order to prepare the mouth and stomach for food that is about to be eaten.
3. The presence of food in the stomach causes the gastric phase of digestion, which promotes gastric juice secretion and gastric motility.

Review

4. During the intestinal phase of digestion, food is digested in the small intestine. In addition, gastric motility and gastric secretion decrease in order to slow the exit of chyme from the stomach, which prevents the small intestine from being overloaded with more chyme than it can handle.
5. The activities that occur during the various phases of digestion are coordinated by neural pathways and by hormones. Table 24.8 summarizes the major hormones that control digestion.

24.15 Development of the Digestive System

1. The endoderm of the primitive gut forms the epithelium and glands of most of the gastrointestinal tract.
2. The mesoderm of the primitive gut forms the smooth muscle and connective tissue of the gastrointestinal tract.

24.16 Aging and the Digestive System

1. General changes include decreased secretory mechanisms, decreased motility, and loss of tone.
2. Specific changes may include loss of taste, pyorrhea, hernias, peptic ulcer disease, constipation, hemorrhoids, and diverticular diseases.

SELF-QUIZ QUESTIONS

Fill in the blanks in the following statements.

1. The end products of chemical digestion of carbohydrates are _____, of proteins are _____, of lipids are _____ and _____, and of nucleic acids are _____, _____, and _____.

2. List the mechanisms of absorption of materials in the small intestine: _____, _____, _____, and _____.

Indicate whether the following statements are true or false.

3. The soft palate, uvula, and epiglottis prevent swallowed foods and liquids from entering the respiratory passages.

4. The coordinated contractions and relaxations of the muscularis, which propels materials through the GI tract, is known as peristalsis.

Choose the one best answer to the following questions.

5. Which of the following are mismatched?
   (a) chemical digestion: splitting food molecules into simple substances by hydrolysis with the assistance of digestive enzymes
   (b) motility: mechanical processes that break apart ingested food into small molecules
   (c) ingestion: taking foods and liquids into the mouth
   (d) propulsion: movement of food through GI tract due to smooth muscle contraction
   (e) absorption: passage into blood or lymph of ions, fluids and small molecules via the epithelial lining of the GI tract lumen

6. Which of the following are *true* concerning the peritoneum? (1) The kidneys and pancreas are retroperitoneal. (2) The greater omentum is the largest of the peritoneal folds. (3) The lesser omentum binds the large intestine to the posterior abdominal wall. (4) The falciform ligament attaches the liver to the anterior abdominal wall and diaphragm. (5) The mesentery is associated with the jejunum and ileum.
   (a) 1, 2, 3, and 5   (b) 1, 2, and 5   (c) 2 and 5
   (d) 1, 2, 4, and 5   (e) 3, 4, and 5

7. When a surgeon makes an incision in the small intestine, in what order would the physician encounter these structures? (1) epithelium, (2) submucosa, (3) serosa, (4) muscularis, (5) lamina propria, (6) muscularis mucosae.
   (a) 3, 4, 5, 6, 2, 1   (b) 1, 2, 3, 4, 6, 5   (c) 1, 5, 6, 2, 4, 3
   (d) 5, 1, 2, 6, 4, 3   (e) 3, 4, 2, 6, 5, 1

8. Which of the following are functions of the liver? (1) carbohydrate, lipid, and protein metabolism, (2) nucleic acid metabolism, (3) excretion of bilirubin, (4) synthesis of bile salts, (5) activation of vitamin D.
   (a) 1, 2, 3, and 5   (b) 1, 2, 3, and 4   (c) 1, 3, 4, and 5
   (d) 2, 3, 4, and 5   (e) 1, 2, 4, and 5

9. Which of the following statements regarding the regulation of gastric secretion and motility are *true*? (1) The sight, smell, taste, or thought of food can initiate the cephalic phase of gastric activity. (2) The gastric phase begins when food enters the small intestine. (3) Once activated, stretch receptors and chemoreceptors in the stomach trigger the flow of gastric juice and peristalsis. (4) The intestinal phase reflexes inhibit gastric activity. (5) The enterogastric reflex stimulates gastric emptying.
   (a) 1, 3, and 4   (b) 2, 4, and 5   (c) 1, 3, 4, and 5
   (d) 1, 2, and 5   (e) 1, 2, 3, and 4

10. Which of the following are *true*? (1) Segmentations in the small intestine help propel chyme through the intestinal tract. (2) The migrating motility complex is a type of peristalsis in the small intestine. (3) The large surface area for absorption in the small intestine is due to the presence of circular folds, villi, and microvilli. (4) The mucus-producing cells of the small intestine are paneth cells. (5) Most long-chain fatty acid and monoglyceride absorption in the small intestine requires the presence of bile salts.
   (a) 1, 2, and 3   (b) 2, 3, and 5   (c) 1, 2, 3, 4, and 5
   (d) 1, 3, and 5   (e) 1, 2, 3, and 5

11. The release of feces from the large intestine is dependent on (1) stretching of the rectal walls, (2) voluntary relaxation of the

external anal sphincter, (3) involuntary contraction of the diaphragm and abdominal muscles, (4) activity of the intestinal bacteria, (5) sympathetic stimulation of the internal sphincter.

(a) 2, 4, and 5   (b) 1, 2, and 5   (c) 1, 2, 3, and 5

(d) 1 and 2   (e) 3, 4, and 5

12. Which of the following is *not* true concerning the liver?
    (a) The left hepatic duct joins the cystic duct from the gallbladder.
    (b) As blood passes through the sinusoids, it is processed by hepatocytes and phagocytes.
    (c) Processed blood returns from the liver to systemic circulation through the hepatic veins.
    (d) The liver receives oxygenated blood through the hepatic artery.
    (e) The hepatic portal vein delivers deoxygenated blood from the GI tract to the liver.

13. Match the following:
    ____ (a) collapsed, muscular tube involved in deglutition and peristalsis
    ____ (b) coiled tube attached to the cecum
    ____ (c) contains duodenal glands in the submucosa
    ____ (d) produces and secretes bile
    ____ (e) contains aggregated lymphatic follicles in the mucosa
    ____ (f) responsible for ingestion, mastication, and deglutition
    ____ (g) responsible for churning, peristalsis, storage, and chemical digestion with the enzyme pepsin
    ____ (h) storage area for bile
    ____ (i) contain acini that release juices containing several digestive enzymes for protein, carbohydrate, lipid, and nucleic acid digestion and sodium bicarbonate to buffer stomach acid
    ____ (j) composed of enamel, dentin, and pulp cavity; used in mastication
    ____ (k) passageway for food, fluid, and air; involved in deglutition
    ____ (l) forms a semisolid waste material through haustral churning and peristalsis
    ____ (m) forces the food to the back of the mouth for swallowing; places food in contact with the teeth
    ____ (n) produce a fluid in the mouth that helps cleanse the mouth and teeth and that lubricates, dissolves, and begins the chemical breakdown of food

    (1) mouth
    (2) teeth
    (3) salivary glands
    (4) pharynx
    (5) esophagus
    (6) tongue
    (7) stomach
    (8) duodenum
    (9) ileum
    (10) colon
    (11) liver
    (12) gallbladder
    (13) appendix
    (14) pancreas

14. Match the following:
    ____ (a) an activating brush-border enzyme that splits off part of the trypsinogen molecule to form trypsin, a protease
    ____ (b) an enzyme that initiates carbohydrate digestion in the mouth
    ____ (c) the principal triglyceride-digesting enzyme in adults
    ____ (d) stimulates secretion of gastric juices and promotes gastric emptying
    ____ (e) secreted by chief cells in the stomach; a proteolytic enzyme
    ____ (f) stimulates the flow of pancreatic juice rich in bicarbonates; decreases gastric secretions
    ____ (g) a nonenzymatic fat-emulsifying agent
    ____ (h) causes contraction of the gallbladder and stimulates the production of pancreatic juice rich in digestive enzymes
    ____ (i) inhibits gastrin release
    ____ (j) stimulates secretion of ions and water by the intestines and inhibits gastric acid secretion
    ____ (k) secreted by glands in the tongue; begins breakdown of triglycerides in the stomach

    (1) gastrin
    (2) cholecystokinin
    (3) secretin
    (4) enterokinase
    (5) pepsin
    (6) salivary amylase
    (7) pancreatic lipase
    (8) lingual lipase
    (9) bile
    (10) vasoactive intestinal polypeptide
    (11) somatostatin

CRITICAL THINKING QUESTIONS

1. Why would you *not* want to completely suppress HCl secretion in the stomach?

2. Trey has cystic fibrosis, a genetic disorder that is characterized by the production of excessive mucus, affecting several body systems (e.g., respiratory, digestive, reproductive). In the digestive system, the excess mucus blocks bile ducts in the liver and pancreatic ducts. How would this affect Trey's digestive processes?

3. Antonio had dinner at his favorite Italian restaurant. His menu consisted of a salad, a large plate of spaghetti, garlic bread, and wine. For dessert, he consumed "death by chocolate" cake and a cup of coffee. He topped off his evening with a cigarette and brandy. He returned home and, while lying on his couch watching television, he experienced a pain in his chest. He called 911 because he was certain he was having a heart attack. Antonio was told his heart was fine, but he needed to watch his diet. What happened to Antonio?

? ANSWERS TO FIGURE QUESTIONS

24.1 Digestive enzymes are produced by the salivary glands, tongue, stomach, pancreas, and small intestine.

24.2 The lamina propria has the following functions: (1) It contains blood vessels and lymphatic vessels, which are the routes by which nutrients are absorbed from the GI tract; (2) it supports the mucosal epithelium and binds it to the muscularis mucosae; and (3) it contains mucosa-associated lymphatic tissue (MALT), which helps protect against disease.

24.3 The neurons of the myenteric plexus regulate GI tract motility, and the neurons of the submucosal plexus regulate GI secretion.

24.4 Mesentery binds the small intestine to the posterior abdominal wall.

24.5 The uvula helps prevent foods and liquids from entering the nasal cavity during swallowing.

24.6 Chloride ions in saliva activate salivary amylase.

24.7 The main component of teeth is connective tissue, specifically dentin.

24.8 The first, second, and third molars do not replace any deciduous teeth.

24.9 The esophageal mucosa and submucosa contain mucus-secreting glands.

24.10 Both. Initiation of swallowing is voluntary and the action is carried out by skeletal muscles. Completion of swallowing—moving a bolus along the esophagus and into the stomach—is involuntary and involves peristalsis by smooth muscle.

24.11 After a large meal, the rugae stretch and disappear as the stomach fills.

24.12 Parietal cells in gastric glands secrete HCl, which is a component of gastric juice. HCl kills microbes in food, denatures proteins, and converts pepsinogen into pepsin.

24.13 Hydrogen ions secreted into gastric juice are derived from carbonic acid (H_2CO_3).

24.14 Histamine is a paracrine agent released by mast cells in the lamina propria.

24.15 The pancreatic duct contains pancreatic juice (fluid and digestive enzymes); the common bile duct contains bile; the hepatopancreatic ampulla contains pancreatic juice and bile.

24.16 The phagocytic cell in the liver is the stellate reticuloendothelial (Kupffer) cell.

24.17 While a meal is being absorbed, nutrients, O_2, and certain toxic substances are removed by hepatocytes from blood flowing through liver sinusoids.

24.18 The ileum is the longest part of the small intestine.

24.19 Nutrients being absorbed in the small intestine enter the blood via capillaries or the lymph via lacteals.

24.20 The fluid secreted by duodenal (Brunner's) glands—alkaline mucus—neutralizes gastric acid and protects the mucosal lining of the duodenum.

24.21 Because monoglycerides are hydrophobic (nonpolar) molecules, they can dissolve in and diffuse through the lipid bilayer of the plasma membrane.

24.22 The stomach and pancreas are the two digestive system organs that secrete the largest volumes of fluid.

24.23 The ascending and descending portions of the colon are retroperitoneal.

24.24 Goblet cells in the large intestine secrete mucus to lubricate colonic contents.

24.25 The pH of gastric juice rises due to the buffering action of some amino acids in food proteins.

25 METABOLISM AND NUTRITION

METABOLISM, NUTRITION, AND HOMEOSTASIS *Metabolic reactions contribute to homeostasis by harvesting chemical energy from consumed nutrients for use in the body's growth, repair, and normal functioning.*

The food we eat is our only source of energy for running, walking, and even breathing. Many molecules needed to maintain cells and tissues can be made from simpler precursors by the body's metabolic reactions; others—the essential amino acids, essential fatty acids, vitamins, and minerals—must be obtained from our food. As you learned in Chapter 24, carbohydrates, lipids, and proteins in food are digested by enzymes and absorbed in the gastrointestinal tract. The products of digestion that reach body cells are monosaccharides, fatty acids, glycerol, monoglycerides, and amino acids. Some minerals and many vitamins are part of enzyme systems that catalyze the breakdown and synthesis of carbohydrates, lipids, and proteins. Food molecules absorbed by the gastrointestinal (GI) tract have three main fates:

1. Most food molecules are used to *supply energy* for sustaining life processes, such as active transport, DNA replication, protein synthesis, muscle contraction, maintenance of body temperature, and mitosis.

2. Some food molecules *serve as building blocks* for the synthesis of more complex structural or functional molecules, such as muscle proteins, hormones, and enzymes.

3. Other food molecules are *stored for future use.* For example, glycogen is stored in liver cells, and triglycerides are stored in adipose cells.

In this chapter we discuss how metabolic reactions harvest the chemical energy stored in foods; how each group of food molecules contributes to the body's growth, repair, and energy needs; and how heat and energy balance is maintained in the body. Finally, we explore some aspects of nutrition to discover why you should opt for fish instead of a burger the next time you eat out.

? *Did you ever wonder how fasting and starvation affect the body?*

25.1 METABOLIC REACTIONS

◉ OBJECTIVES

- Define metabolism.
- Explain the role of ATP in anabolism and catabolism.

Metabolism (me-TAB-ō-lizm; *metabol-* = change) refers to all of the chemical reactions that occur in the body. There are two types of metabolism: catabolism and anabolism. Those chemical reactions that break down complex organic molecules into simpler ones are collectively known as **catabolism** (ka-TAB-ō-lizm; *cata-* = downward). Overall, catabolic (decomposition) reactions are *exergonic;* they produce more energy than they consume, releasing the chemical energy stored in organic molecules. Important sets of catabolic reactions occur in glycolysis, the Krebs cycle, and the electron transport chain, each of which will be discussed later in the chapter.

Chemical reactions that combine simple molecules and monomers to form the body's complex structural and functional components are collectively known as **anabolism** (a-NAB-ō-lizm; *ana-* = upward). Examples of anabolic reactions are the formation of peptide bonds between amino acids during protein synthesis, the building of fatty acids into phospholipids that form the plasma membrane bilayer, and the linkage of glucose monomers to form glycogen. Anabolic reactions are *endergonic;* they consume more energy than they produce.

Metabolism is an energy-balancing act between catabolic (decomposition) reactions and anabolic (synthesis) reactions. The molecule that participates most often in energy exchanges in living cells is **ATP (adenosine triphosphate),** which couples energy-releasing catabolic reactions to energy-requiring anabolic reactions.

The metabolic reactions that occur depend on which enzymes are active in a particular cell at a particular time, or even in a particular part of the cell. Catabolic reactions can be occurring in the mitochondria of a cell at the same time as anabolic reactions are taking place in the endoplasmic reticulum.

A molecule synthesized in an anabolic reaction has a limited lifetime. With few exceptions, it will eventually be broken down and its component atoms recycled into other molecules or excreted from the body. Recycling of biological molecules occurs continuously in living tissues, more rapidly in some than in others. Individual cells may be refurbished molecule by molecule, or a whole tissue may be rebuilt cell by cell.

Coupling of Catabolism and Anabolism by ATP

The chemical reactions of living systems depend on the efficient transfer of manageable amounts of energy from one molecule to another. The molecule that most often performs this task is ATP, the "energy currency" of a living cell. Like money, it is readily available to "buy" cellular activities; it is spent and earned over and over. A typical cell has about a billion molecules of ATP, each of which typically lasts for less than a minute before being used. Thus, ATP is not a long-term storage form of currency, like gold

in a vault, but rather convenient cash for moment-to-moment transactions.

Recall from Chapter 2 that a molecule of ATP consists of an adenine molecule, a ribose molecule, and three phosphate groups bonded to one another (see Figure 2.25). Figure 25.1 shows how ATP links anabolic and catabolic reactions. When the terminal phosphate group is split off ATP, adenosine diphosphate (ADP) and a phosphate group (symbolized as Ⓟ) are formed. Some of the energy released is used to drive anabolic reactions such as the formation of glycogen from glucose. In addition, energy from complex molecules is used in catabolic reactions to combine ADP and a phosphate group to resynthesize ATP:

$$ADP + Ⓟ + energy \longrightarrow ATP$$

About 40% of the energy released in catabolism is used for cellular functions; the rest is converted to heat, some of which helps maintain normal body temperature. Excess heat is lost to the environment. Compared with machines, which typically convert only 10–20% of energy into work, the 40% efficiency of the body's metabolism is impressive. Still, the body has a continuous need to take in and process external sources of energy so that cells can synthesize enough ATP to sustain life.

✔CHECKPOINT

1. What is metabolism? Distinguish between anabolism and catabolism, and give examples of each.
2. How does ATP link anabolism and catabolism?

Figure 25.1 Role of ATP in linking anabolic and catabolic reactions. When complex molecules and polymers are split apart (catabolism, at left), some of the energy is transferred to form ATP and the rest is given off as heat. When simple molecules and monomers are combined to form complex molecules (anabolism, at right), ATP provides the energy for synthesis, and again some energy is given off as heat.

🔑 The coupling of energy-releasing and energy-requiring reactions is achieved through ATP.

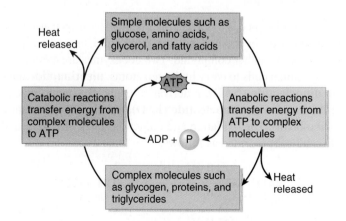

❓ In a pancreatic cell that produces digestive enzymes, does anabolism or catabolism predominate?

25.2 ENERGY TRANSFER

● OBJECTIVES

- Describe oxidation–reduction reactions.
- Explain the role of ATP in metabolism.

Various catabolic reactions transfer energy into the "high-energy" phosphate bonds of ATP. Although the amount of energy in these bonds is not exceptionally large, it can be released quickly and easily. Before discussing metabolic pathways, it is important to understand how this transfer of energy occurs. Two important aspects of energy transfer are oxidation–reduction reactions and mechanisms of ATP generation.

Oxidation–Reduction Reactions

Oxidation (ok′-si-DĀ-shun) is the *removal of electrons* from an atom or molecule; the result is a *decrease* in the potential energy of the atom or molecule. Because most biological oxidation reactions involve the loss of hydrogen atoms, they are called *dehydrogenation reactions*. An example of an oxidation reaction is the conversion of lactic acid into pyruvic acid:

$$
\begin{array}{ccc}
\text{COOH} & & \text{COOH} \\
| & \xrightarrow[\text{Remove 2 H (H}^+ + \text{H}^-)]{\text{Oxidation}} & | \\
\text{H—C—OH} & & \text{C=O} \\
| & & | \\
\text{CH}_3 & & \text{CH}_3 \\
\text{Lactic acid} & & \text{Pyruvic acid}
\end{array}
$$

In the preceding reaction, 2 H ($H^+ + H^-$) means that two neutral hydrogen atoms (2 H) are removed as one hydrogen ion (H^+) plus one hydride ion (H^-).

Reduction (rē-DUK-shun) is the opposite of oxidation; it is the *addition of electrons* to a molecule. Reduction results in an *increase* in the potential energy of the molecule. An example of a reduction reaction is the conversion of pyruvic acid into lactic acid:

$$
\begin{array}{ccc}
\text{COOH} & & \text{COOH} \\
| & \xrightarrow[\text{Add 2 H (H}^+ + \text{H}^-)]{\text{Reduction}} & | \\
\text{C=O} & & \text{H—C—OH} \\
| & & | \\
\text{CH}_3 & & \text{CH}_3 \\
\text{Pyruvic acid} & & \text{Lactic acid}
\end{array}
$$

When a substance is oxidized, the liberated hydrogen atoms do not remain free in the cell but are transferred immediately by coenzymes to another compound. Two coenzymes are commonly used by animal cells to carry hydrogen atoms: **nicotinamide adenine dinucleotide (NAD)**, a derivative of the B vitamin niacin, and **flavin adenine dinucleotide (FAD)**, a derivative of vitamin B_2 (riboflavin). The oxidation and reduction states of NAD^+ and FAD can be represented as follows:

$$
\underset{\text{Oxidized}}{NAD^+} \underset{-2\,H\,(H^+ + H^-)}{\overset{+2\,H\,(H^+ + H^-)}{\rightleftharpoons}} \underset{\text{Reduced}}{NADH + H^+}
$$

$$
\underset{\text{Oxidized}}{FAD} \underset{-2\,H\,(H^+ + H^-)}{\overset{+2\,H\,(H^+ + H^-)}{\rightleftharpoons}} \underset{\text{Reduced}}{FADH_2}
$$

When NAD^+ is reduced to $NADH + H^+$, the NAD^+ gains a hydride ion (H^-), neutralizing its charge, and the H^+ is released into the surrounding solution. When NADH is oxidized to NAD^+, the loss of the hydride ion results in one less hydrogen atom and an additional positive charge. FAD is reduced to $FADH_2$ when it gains a hydrogen ion and a hydride ion, and $FADH_2$ is oxidized to FAD when it loses the same two ions.

Oxidation and reduction reactions are always coupled; each time one substance is oxidized, another is simultaneously reduced. Such paired reactions are called **oxidation–reduction** or **redox reactions.** For example, when lactic acid is *oxidized* to form pyruvic acid, the two hydrogen atoms removed in the reaction are used to *reduce* NAD^+. This coupled redox reaction may be written as follows:

$$
\begin{array}{cc}
\underset{\text{Reduced}}{\text{Lactic acid}} & \underset{\text{Oxidized}}{NAD^+} \\
& \\
\underset{\text{Oxidized}}{\text{Pyruvic acid}} & \underset{\text{Reduced}}{NADH + H^+}
\end{array}
$$

An important point to remember about oxidation–reduction reactions is that oxidation is usually an exergonic (energy-releasing) reaction. Cells use multistep biochemical reactions to release energy from energy-rich, highly reduced compounds (with many hydrogen atoms) to lower-energy, highly oxidized compounds (with many oxygen atoms or multiple bonds). For example, when a cell oxidizes a molecule of glucose ($C_6H_{12}O_6$), the energy in the glucose molecule is removed in a stepwise manner. Ultimately, some of the energy is captured by transferring it to ATP, which then serves as an energy source for energy-requiring reactions within the cell. Compounds with many hydrogen atoms such as glucose contain more chemical potential energy than oxidized compounds. For this reason, glucose is a valuable nutrient.

Mechanisms of ATP Generation

Some of the energy released during oxidation reactions is captured within a cell when ATP is formed. Briefly, a phosphate group (Ⓟ) is added to ADP, with an input of energy, to form ATP. The two high-energy phosphate bonds that can be used to transfer energy are indicated by "squiggles" ($\sim$):

$$
\underset{\text{ADP}}{\text{Adenosine — Ⓟ} \sim \text{Ⓟ}} + \text{Ⓟ} + \text{energy} \longrightarrow
$$

$$
\underset{\text{ATP}}{\text{Adenosine — Ⓟ} \sim \text{Ⓟ} \sim \text{Ⓟ}}
$$

The high-energy phosphate bond that attaches the third phosphate group contains the energy stored in this reaction. The addition of a phosphate group to a molecule, called **phosphorylation** (fos′-for-i-LĀ-shun), increases its potential energy. Organisms use three mechanisms of phosphorylation to generate ATP:

1. **Substrate-level phosphorylation** generates ATP by transferring a high-energy phosphate group from an intermediate phosphorylated metabolic compound—a substrate—directly to ADP. In human cells, this process occurs in the cytosol.

2. **Oxidative phosphorylation** removes electrons from organic compounds and passes them through a series of electron acceptors, called the **electron transport chain,** to molecules of oxygen (O_2). This process occurs in the inner mitochondrial membrane of cells.

3. **Photophosphorylation** occurs only in chlorophyll-containing plant cells or in certain bacteria that contain other light-absorbing pigments.

✔**CHECKPOINT**

3. How is a hydride ion different from a hydrogen ion? What is the involvement of both ions in redox reactions?
4. What are three ways that ATP can be generated?

25.3 CARBOHYDRATE METABOLISM

◉ **OBJECTIVE**

• Describe the fate, metabolism, and functions of carbohydrates.

As you learned in Chapter 24, both polysaccharides and disaccharides are hydrolyzed into the monosaccharides glucose (about 80%), fructose, and galactose during the digestion of carbohydrates. (Some fructose is converted into glucose as it is absorbed through the intestinal epithelial cells.) Hepatocytes (liver cells) convert most of the remaining fructose and practically all the galactose to glucose. So the story of carbohydrate metabolism is really the story of glucose metabolism. Because negative feedback systems maintain blood glucose at about 90 mg/100 mL of plasma (5 mmol/liter), a total of 2–3 g of glucose normally circulates in the blood.

The Fate of Glucose

Beçause glucose is the body's preferred source for synthesizing ATP, its use depends on the needs of body cells, which include the following:

• *ATP production.* In body cells that require immediate energy, glucose is oxidized to produce ATP. Glucose not needed for immediate ATP production can enter one of several other metabolic pathways.

• *Amino acid synthesis.* Cells throughout the body can use glucose to form several amino acids, which then can be incorporated into proteins.

• *Glycogen synthesis.* Hepatocytes and muscle fibers can perform **glycogenesis** (glī′-kō-JEN-e-sis; *glyco-* = sugar or sweet; *-genesis* = to generate), in which hundreds of glucose monomers are combined to form the polysaccharide glycogen. Total storage capacity of glycogen is about 125 g in the liver and 375 g in skeletal muscles.

• *Triglyceride synthesis.* When the glycogen storage areas are filled up, hepatocytes can transform the glucose to glycerol and

fatty acids that can be used for **lipogenesis** (lip-ō-JEN-e-sis), the synthesis of triglycerides. Triglycerides then are deposited in adipose tissue, which has virtually unlimited storage capacity.

Glucose Movement into Cells

Before glucose can be used by body cells, it must first pass through the plasma membrane and enter the cytosol. Glucose absorption in the gastrointestinal tract (and kidney tubules) is accomplished via secondary active transport (Na^+–glucose symporters). Glucose entry into most other body cells occurs via GluT molecules, a family of transporters that bring glucose into cells via facilitated diffusion (see Section 3.3). A high level of insulin increases the insertion of one type of GluT, called GluT4, into the plasma membranes of most body cells, thereby increasing the rate of facilitated diffusion of glucose into cells. In neurons and hepatocytes, however, another type of GluT is always present in the plasma membrane, so glucose entry is always "turned on." On entering a cell, glucose becomes phosphorylated. Because GluT cannot transport phosphorylated glucose, this reaction traps glucose within the cell.

Glucose Catabolism

The oxidation of glucose to produce ATP is also known as **cellular respiration,** and it involves four sets of reactions: glycolysis, the formation of acetyl coenzyme A, the Krebs cycle, and the electron transport chain (Figure 25.2).

1 *Glycolysis* is a set of reactions in which one glucose molecule is oxidized and two molecules of pyruvic acid are produced. The reactions also produce two molecules of ATP and two energy-containing NADH + H$^+$. Because glycolysis does not require oxygen, it is a way to produce ATP anaerobically (without oxygen) and is known as **anaerobic cellular respiration** (an-ar-Ō-bik; *an-* = not; *-aer-* = air; *-bios* = life).

2 *Formation of acetyl coenzyme A* is a transition step that prepares pyruvic acid for entrance into the Krebs cycle. This step also produces energy-containing NADH + H$^+$ plus carbon dioxide (CO_2).

3 *Krebs cycle reactions* oxidize acetyl coenzyme A and produce CO_2, ATP, NADH + H$^+$, and $FADH_2$.

4 *Electron transport chain reactions* oxidize NADH + H$^+$ and $FADH_2$ and transfer their electrons through a series of electron carriers. The Krebs cycle and the electron transport chain both require oxygen to produce ATP and are collectively known as **aerobic cellular respiration.**

Glycolysis

During **glycolysis** (glī-KOL-i-sis; *-lysis* = breakdown), chemical reactions split a 6-carbon molecule of glucose into two 3-carbon molecules of pyruvic acid (Figure 25.3). Even though glycolysis consumes two ATP molecules, it produces four ATP molecules, for a net gain of two ATP molecules for each glucose molecule that is oxidized.

Figure 25.2 **Overview of cellular respiration (oxidation of glucose).** A modified version of this figure appears in several places in this chapter to indicate the relationships of particular reactions to the overall process of cellular respiration.

The oxidation of glucose involves glycolysis, the formation of acetyl coenzyme A, the Krebs cycle, and the electron transport chain.

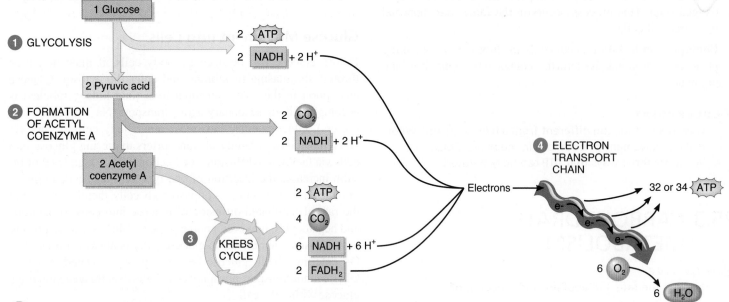

? **Which of the four processes shown here is also called anaerobic cellular respiration?**

Figure 25.3 **Cellular respiration begins with glycolysis.**

During glycolysis, each molecule of glucose is converted to two molecules of pyruvic acid.

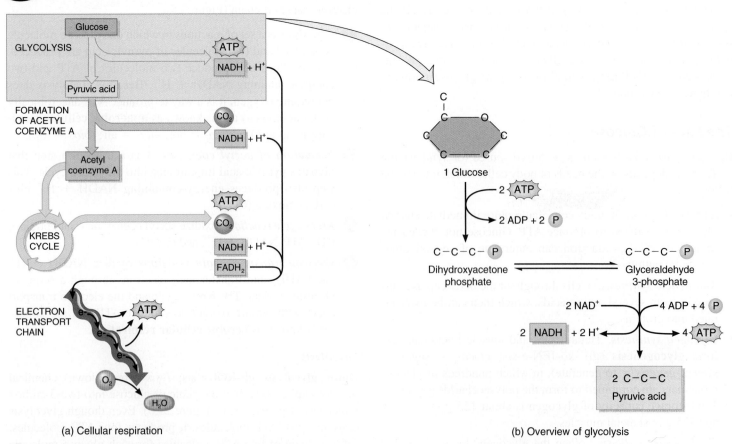

(a) Cellular respiration

(b) Overview of glycolysis

? **For each glucose molecule that undergoes glycolysis, how many ATP molecules are generated?**

1028

Figure 25.4 shows the 10 reactions that glycolysis comprises. In the first half of the sequence (reactions ① through ⑤), energy in the form of ATP is "invested" and the 6-carbon glucose is split into two 3-carbon molecules of glyceraldehyde 3-phosphate.

Phosphofructokinase (fos′-fō-fruk′-tō-KĪ-nās), the enzyme that catalyzes step ③, is the key regulator of the rate of glycolysis. The activity of this enzyme is high when ADP concentration is high, in which case ATP is produced rapidly. When the activity of

Figure 25.4 The 10 reactions of glycolysis. ① Glucose is phosphorylated, using a phosphate group from an ATP molecule to form glucose 6-phosphate. ② Glucose 6-phosphate is converted to fructose 6-phosphate. ③ A second ATP is used to add a second phosphate group to fructose 6-phosphate to form fructose 1,6-bisphosphate. ④ and ⑤ Fructose splits into two 3-carbon molecules, glyceraldehyde 3-phosphate (G 3-P) and dihydroxyacetone phosphate, each having one phosphate group. ⑥ Oxidation occurs as two molecules of NAD⁺ accept two pairs of electrons and hydrogen ions from two molecules of G 3-P to form two molecules of NADH. Many body cells use the two NADH produced in this step to generate four ATPs in the electron transport chain. Hepatocytes, kidney cells, and cardiac muscle fibers can generate six ATPs from the two NADH. A second phosphate group attaches to G 3-P, forming 1,3-bisphosphoglyceric acid (BPG). ⑦ through ⑩ These reactions generate four molecules of ATP and produce two molecules of pyruvic acid (pyruvate*).

🔑 Glycolysis results in a net gain of two ATP, two NADH, and two H⁺.

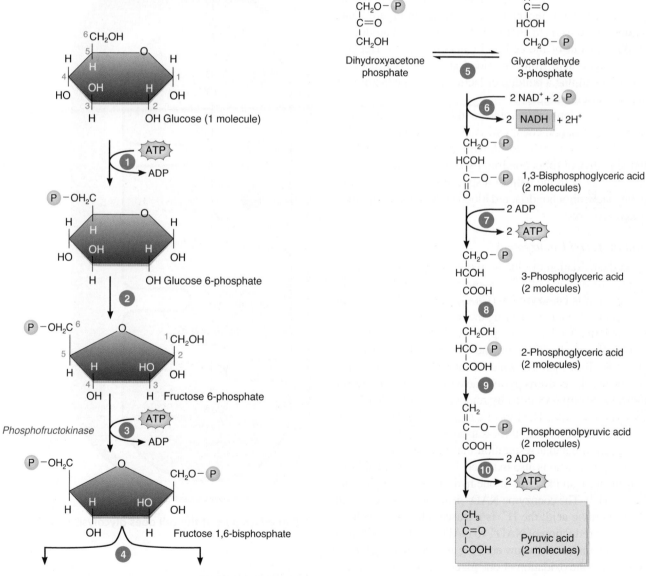

❓ **Why is the enzyme that catalyzes step ③ called a kinase?**

*The carboxyl groups (—COOH) of intermediates in glycolysis and in the citric acid cycle are mostly ionized at the pH of body fluids to —COO⁻. The suffix "-ic acid" indicates the non-ionized form, whereas the ending "-ate" indicates the ionized form. Although the "-ate" names are more correct, we will use the "acid" names because these terms are more familiar.

phosphofructokinase is low, most glucose does not enter the reactions of glycolysis but instead undergoes conversion to glycogen for storage. In the second half of the sequence (reactions ⑥ through ⑩), the two glyceraldehyde 3-phosphate molecules are converted to two pyruvic acid molecules and ATP is generated.

The Fate of Pyruvic Acid

The fate of pyruvic acid produced during glycolysis depends on the availability of oxygen (Figure 25.5). If oxygen is scarce (anaerobic conditions)—for example, in skeletal muscle fibers during strenuous exercise—then pyruvic acid is reduced via an anaerobic pathway by the addition of two hydrogen atoms to form lactic acid (lactate):

$$2 \text{ Pyruvic acid} + 2 \text{ NADH} + 2 \text{ H}^+ \longrightarrow 2 \text{ Lactic acid} + 2 \text{ NAD}^+$$
Oxidized   Reduced

This reaction regenerates the NAD^+ that was used in the oxidation of glyceraldehyde 3-phosphate (see step ⑥ in Figure 25.4) and thus allows glycolysis to continue. As lactic acid is produced, it rapidly diffuses out of the cell and enters the blood. Hepatocytes remove lactic acid from the blood and convert it back to pyruvic acid. Recall that a buildup of lactic acid is one factor that contributes to muscle fatigue.

When oxygen is plentiful (aerobic conditions), most cells convert pyruvic acid to acetyl coenzyme A. This molecule links glycolysis, which occurs in the cytosol, with the Krebs cycle, which occurs in the matrix of mitochondria. Pyruvic acid enters the mitochondrial matrix with the help of a special transporter protein. Because they lack mitochondria, red blood cells can only produce ATP through glycolysis.

Formation of Acetyl Coenzyme A

Each step in the oxidation of glucose requires a different enzyme, and often a coenzyme as well. The coenzyme used at this point in cellular respiration is **coenzyme A (CoA),** which is derived from pantothenic acid, a B vitamin. During the transitional step between glycolysis and the Krebs cycle, pyruvic acid is prepared for entrance into the cycle. The enzyme *pyruvate dehydrogenase* (pī-ROO-vāt dē-HĪ-drō-jen-ās), which is located exclusively in the mitochondrial matrix, converts pyruvic acid to a 2-carbon fragment called an **acetyl group** (AS-e-til) by removing a molecule of carbon dioxide (Figure 25.5). The loss of a molecule of CO_2 by a substance is called **decarboxylation** (dē-kar-bok′-si-LĀ-shun). This is the first reaction in cellular respiration that releases CO_2. During this reaction, pyruvic acid is also oxidized. Each pyruvic acid loses two hydrogen atoms in the form of one hydride ion (H^-) plus one hydrogen ion (H^+). The coenzyme NAD^+ is reduced as it picks up the H^- from pyruvic acid; the H^+ is released into the mitochondrial matrix. The reduction of NAD^+ to $NADH + H^+$ is indicated in Figure 25.5 by the curved arrow entering and then leaving the reaction. Recall that the oxidation of one glucose molecule produces two molecules of pyruvic acid, so for each molecule of glucose, two molecules of carbon dioxide are lost and two $NADH + H^+$ are produced. The acetyl group attaches to coenzyme A, producing a molecule called **acetyl coenzyme A (acetyl CoA).**

Figure 25.5 Fate of pyruvic acid.

 When oxygen is plentiful, pyruvic acid enters mitochondria, is converted to acetyl coenzyme A, and enters the Krebs cycle (aerobic pathway). When oxygen is scarce, most pyruvic acid is converted to lactic acid via an anaerobic pathway.

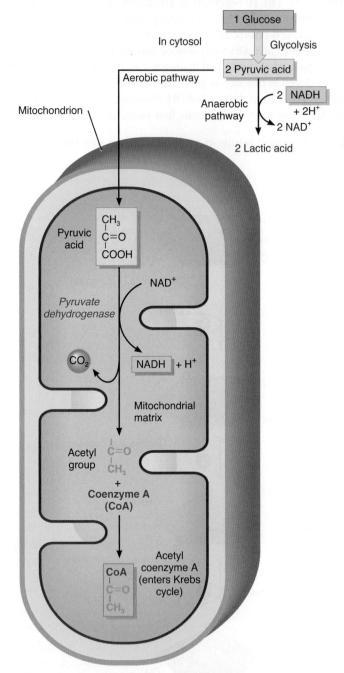

 In which part of the cell does glycolysis occur?

The Krebs Cycle

Once the pyruvic acid has undergone decarboxylation and the remaining acetyl group has attached to CoA, the resulting compound (acetyl CoA) is ready to enter the Krebs cycle (Figure 25.6). The

Figure 25.6 **After formation of acetyl coenzyme A, the next stage of cellular respiration is the Krebs cycle.**

Reactions of the Krebs cycle occur in the matrix of mitochondria.

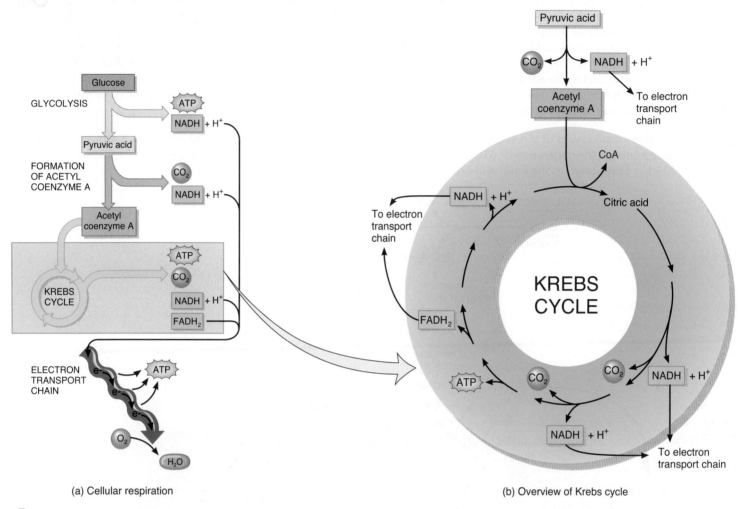

(a) Cellular respiration

(b) Overview of Krebs cycle

When in cellular respiration is carbon dioxide given off? What happens to this gas?

Krebs cycle—named for the biochemist Hans Krebs, who described these reactions in the 1930s—is also known as the **citric acid cycle,** for the first molecule formed when an acetyl group joins the cycle. The reactions occur in the matrix of mitochondria and consist of a series of oxidation–reduction reactions and decarboxylation reactions that release CO_2. In the Krebs cycle, the oxidation–reduction reactions transfer chemical energy, in the form of electrons, to two coenzymes—NAD^+ and FAD. The pyruvic acid derivatives are oxidized, and the coenzymes are reduced. In addition, one step generates ATP. Figure 25.7 shows the reactions of the Krebs cycle in more detail.

The reduced coenzymes (NADH and $FADH_2$) are the most important outcome of the Krebs cycle because they contain the energy originally stored in glucose and then in pyruvic acid. Overall, for every acetyl CoA that enters the Krebs cycle, three NADH, three H^+, and one $FADH_2$ are produced by oxidation–reduction reactions, and one molecule of ATP is generated by

substrate-level phosphorylation (see Figure 25.6). In the electron transport chain, the three NADH + three H^+ will later yield nine ATP molecules, and the $FADH_2$ will later yield two ATP molecules. Thus, each "turn" of the Krebs cycle eventually generates 12 molecules of ATP. Because each glucose molecule provides two acetyl CoA molecules, glucose catabolism via the Krebs cycle and the electron transport chain yields 24 molecules of ATP per glucose molecule.

Liberation of CO_2 occurs as pyruvic acid is converted to acetyl CoA and during the two decarboxylation reactions of the Krebs cycle (see Figure 25.6). Because each molecule of glucose generates two molecules of pyruvic acid, six molecules of CO_2 are liberated from each original glucose molecule catabolized along this pathway. The molecules of CO_2 diffuse out of the mitochondria, through the cytosol and plasma membrane, and then into the blood. Blood transports the CO_2 to the lungs, where it eventually is exhaled.

Figure 25.7 The eight reactions of the Krebs cycle. ❶ *Entry of the acetyl group.* The chemical bond that attaches the acetyl group to coenzyme A (CoA) breaks, and the 2-carbon acetyl group attaches to a 4-carbon molecule of oxaloacetic acid to form a 6-carbon molecule called citric acid. CoA is free to combine with another acetyl group from pyruvic acid and repeat the process. ❷ ***Isomerization.*** Citric acid undergoes isomerization to isocitric acid, which has the same molecular formula as citrate. Notice, however, that the hydroxyl group (—OH) is attached to a different carbon. ❸ ***Oxidative decarboxylation.*** Isocitric acid is oxidized and loses a molecule of CO_2, forming alpha-ketoglutaric acid. The H^+ from the oxidation is passed on to NAD^+, which is reduced to $NADH + H^+$. ❹ ***Oxidative decarboxylation.*** Alpha-ketoglutaric acid is oxidized, loses a molecule of CO_2, and picks up CoA to form succinyl-CoA. ❺ ***Substrate-level phosphorylation.*** CoA is displaced by a phosphate group, which is then transferred to guanosine diphosphate (GDP) to form guanosine triphosphate (GTP). GTP can donate a phosphate group to ADP to form ATP. ❻ ***Dehydrogenation.*** Succinic acid is oxidized to fumaric acid as two of its hydrogen atoms are transferred to the coenzyme flavin adenine dinucleotide (FAD), which is reduced to $FADH_2$. ❼ ***Hydration.*** Fumaric acid is converted to malic acid by the addition of a molecule of water. ❽ ***Dehydrogenation.*** In the final step in the cycle, malic acid is oxidized to re-form oxaloacetic acid. Two hydrogen atoms are removed and one is transferred to NAD^+, which is reduced to $NADH + H^+$. The regenerated oxaloacetic acid can combine with another molecule of acetyl CoA, beginning a new cycle.

🔑 The three main results of the Krebs cycle are the production of reduced coenzymes (NADH and $FADH_2$), which contain stored energy; the generation of GTP, a high-energy compound that is used to produce ATP; and the formation of CO_2, which is transported to the lungs and exhaled.

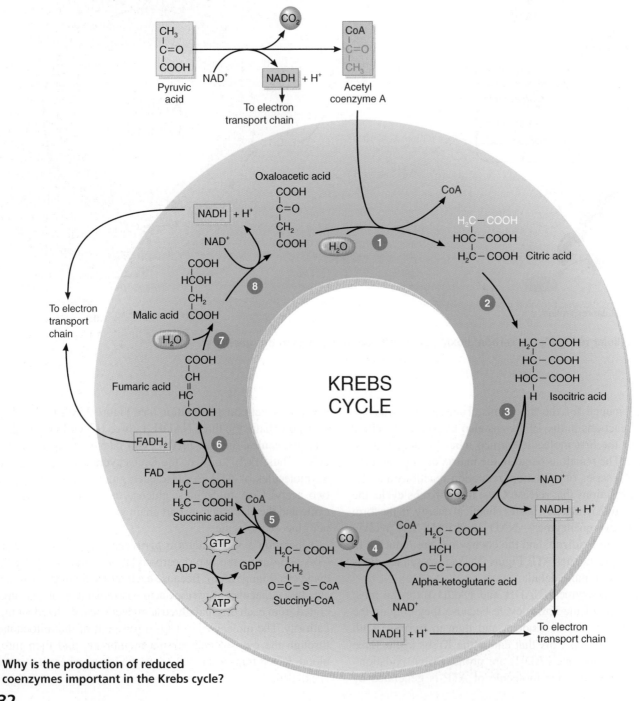

❓ **Why is the production of reduced coenzymes important in the Krebs cycle?**

The Electron Transport Chain

The **electron transport chain** is a series of **electron carriers,** integral membrane proteins in the inner mitochondrial membrane. This membrane is folded into cristae that increase its surface area, accommodating thousands of copies of the transport chain in each mitochondrion. Each carrier in the chain is reduced as it picks up electrons and oxidized as it gives up electrons. As electrons pass through the chain, a series of exergonic reactions release small amounts of energy; this energy is used to form ATP. In aerobic cellular respiration, the final electron acceptor of the chain is oxygen. Because this mechanism of ATP generation links chemical reactions (the passage of electrons along the transport chain) with the pumping of hydrogen ions, it is called **chemiosmosis** (kem′-ē-oz-MŌ-sis; *chemi-* = chemical; *-osmosis* = pushing). Briefly, chemiosmosis works as follows (Figure 25.8):

① Energy from NADH + H$^+$ passes along the electron transport chain and is used to pump H$^+$ from the matrix of the mitochondrion into the space between the inner and outer mitochondrial membranes. This mechanism is called a **proton pump** because H$^+$ ions consist of a single proton.

② A high concentration of H$^+$ accumulates between the inner and outer mitochondrial membranes.

③ ATP synthesis then occurs as hydrogen ions flow back into the mitochondrial matrix through a special type of H$^+$ channel in the inner membrane.

ELECTRON CARRIERS Several types of molecules and atoms serve as electron carriers:

- **Flavin mononucleotide (FMN)** (FLĀ-vin mon′-ō-NOO-klē-ō-tīd) is a flavoprotein derived from riboflavin (vitamin B$_2$).
- **Cytochromes** (SĪ-tō-krōmz) are proteins with an iron-containing group (heme) capable of existing alternately in a reduced form (Fe^{2+}) and an oxidized form (Fe^{3+}). The cytochromes involved in the electron transport chain include cytochrome *b* (cyt *b*), cytochrome c_1 (cyt c_1), cytochrome *c* (cyt *c*), cytochrome *a* (cyt *a*), and cytochrome a_3 (cyt a_3).
- **Iron–sulfur (Fe-S) centers** contain either two or four iron atoms bound to sulfur atoms that form an electron transfer center within a protein.
- **Copper (Cu) atoms** bound to two proteins in the chain also participate in electron transfer.
- **Coenzyme Q,** symbolized **Q,** is a nonprotein, low-molecular-weight carrier that is mobile in the lipid bilayer of the inner membrane.

STEPS IN ELECTRON TRANSPORT AND CHEMIOSMOTIC ATP GENERATION Within the inner mitochondrial membrane, the carriers of the electron transport chain are clustered into three complexes, each of which acts as a proton pump that expels H$^+$ from the mitochondrial matrix and helps create an electrochemical gradient of H$^+$. Each of the three proton pumps transports electrons and pumps H$^+$, as shown in Figure 25.9. Notice that oxygen is used to help form water in step **③**. This is the only point in aerobic cellular respiration where O$_2$ is consumed. **Cyanide** is a deadly poison because it binds to the cytochrome oxidase complex and blocks this last step in electron transport.

The pumping of H$^+$ produces both a concentration gradient of protons and an electrical gradient. The buildup of H$^+$ makes one side of the inner mitochondrial membrane positively charged compared with the other side. The resulting electrochemical gradient has potential energy, called the *proton motive force.* Proton channels in the inner mitochondrial membrane allow H$^+$ to flow back across the membrane, driven by the proton motive force. As H$^+$ flow back, they generate ATP because the H$^+$ channels also include an enzyme called **ATP synthase** (SIN-thās). The enzyme uses the proton motive force to synthesize ATP from ADP and Ⓟ. The process of chemiosmosis is responsible for most of the ATP produced during cellular respiration.

Figure 25.8 Chemiosmosis.

In chemiosmosis, ATP is produced when hydrogen ions diffuse back into the mitochondrial matrix.

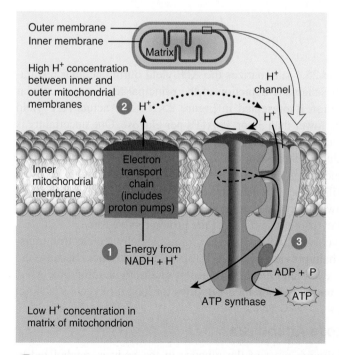

Summary of Cellular Respiration

The various electron transfers in the electron transport chain generate either 32 or 34 ATP molecules from each molecule of

? **What is the energy source that powers the proton pumps?**

Figure 25.9   The actions of the three proton pumps and ATP synthase in the inner membrane of mitochondria.   Each pump is a complex of three or more electron carriers. ❶ The first proton pump is the *NADH dehydrogenase complex,* which contains flavin mononucleotide (FMN) and five or more Fe-S centers. NADH + H$^+$ is oxidized to NAD$^+$, and FMN is reduced to FMNH$_2$, which in turn is oxidized as it passes electrons to the iron–sulfur centers. Q, which is mobile in the membrane, shuttles electrons to the second pump complex. ❷ The second proton pump is the *cytochrome b–c$_1$ complex,* which contains cytochromes and an iron–sulfur center. Electrons are passed successively from Q to cyt *b*, to Fe-S, to cyt *c$_1$*. The mobile shuttle that passes electrons from the second pump complex to the third is cytochrome *c* (cyt *c*). ❸ The third proton pump is the *cytochrome oxidase complex,* which contains cytochromes *a* and *a$_3$* and two copper atoms. Electrons pass from cyt *c*, to Cu, to cyt *a*, and finally to cyt *a$_3$*. Cyt *a$_3$* passes its electrons to one-half of a molecule of oxygen (O$_2$), which becomes negatively charged and then picks up two H$^+$ from the surrounding medium to form H$_2$O.

🔑 As the three proton pumps pass electrons from one carrier to the next, they also move protons (H$^+$) from the matrix into the space between the inner and outer mitochondrial membranes. As protons flow back into the mitochondrial matrix through the H$^+$ channel in ATP synthase, ATP is synthesized.

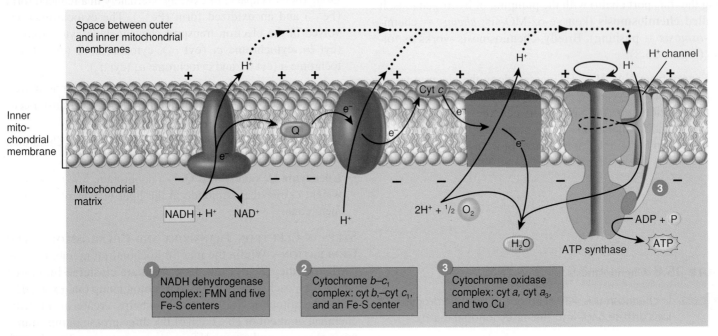

❓ **Where is the concentration of H$^+$ highest?**

glucose that is oxidized: either 28 or 30* from the 10 molecules of NADH + H$^+$ and two from each of the two molecules of FADH$_2$ (four total). Thus, during cellular respiration, 36 or 38 ATPs can be generated from one molecule of glucose. Note that two of those ATPs come from substrate-level phosphorylation in glycolysis, and two come from substrate-level phosphorylation in the Krebs cycle. The overall reaction is

$$C_6H_{12}O_6 + 6\,O_2 + 36 \text{ or } 38 \text{ ADPs} + 36 \text{ or } 38\;ⓟ \longrightarrow$$
Glucose   Oxygen

$$6\,CO_2 + 6\,H_2O + 36 \text{ or } 38 \text{ ATPs}$$
Carbon dioxide Water

*The two NADH produced in the cytosol during glycolysis cannot enter mitochondria. Instead, they donate their electrons to one of two transfer molecules, known as the malate shuttle and the glycerol phosphate shuttle. In cells of the liver, kidneys, and heart, use of the malate shuttle results in three ATPs synthesized for each NADH. In other body cells, such as skeletal muscle fibers and neurons, use of the glycerol phosphate shuttle results in two ATPs synthesized for each NADH.

Table 25.1 summarizes the ATP yield during cellular respiration. A schematic depiction of the principal reactions of cellular respiration is presented in Figure 25.10. The actual ATP yield may be lower than 36 or 38 ATPs per glucose. One uncertainty is the exact number of H$^+$ that must be pumped out to generate one ATP during chemiosmosis. In addition, the ATP generated in mitochondria must be transported out of these organelles into the cytosol for use elsewhere in a cell. Exporting ATP in exchange for the inward movement of ADP formed from metabolic reactions in the cytosol uses up part of the proton motive force.

Glycolysis, the Krebs cycle, and especially the electron transport chain provide all the ATP for cellular activities. Because the Krebs cycle and electron transport chain are aerobic processes, cells cannot carry on their activities for long if oxygen is lacking.

Glucose Anabolism

Even though most of the glucose in the body is catabolized to generate ATP, glucose may take part in or be formed via several

Figure 25.10 Summary of the principal reactions of cellular respiration. ETC = electron transport chain and chemiosmosis.

Except for glycolysis, which occurs in the cytosol, all other reactions of cellular respiration occur within mitochondria.

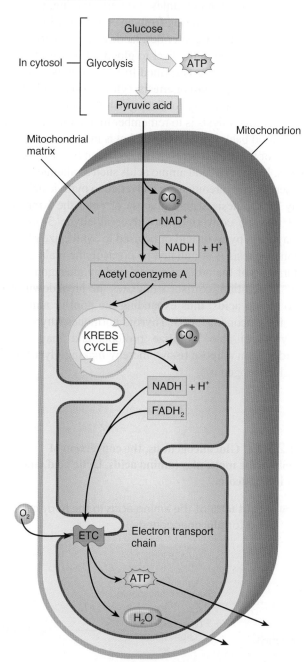

How many molecules of O_2 are used, and how many molecules of CO_2 are produced during the complete oxidation of one glucose molecule?

TABLE 25.1

Summary of ATP Produced in Cellular Respiration

SOURCE	ATP YIELD PER GLUCOSE MOLECULE (PROCESS)
GLYCOLYSIS	
Oxidation of one glucose molecule to two pyruvic acid molecules	2 ATPs (substrate-level phosphorylation).
Production of 2 NADH + H$^+$	4 or 6 ATPs (oxidative phosphorylation in electron transport chain).
FORMATION OF TWO MOLECULES OF ACETYL COENZYME A	
2 NADH + 2 H$^+$	6 ATPs (oxidative phosphorylation in electron transport chain).
KREBS CYCLE AND ELECTRON TRANSPORT CHAIN	
Oxidation of succinyl-CoA to succinic acid	2 GTPs that are converted to 2 ATPs (substrate-level phosphorylation).
Production of 6 NADH + 6 H$^+$	18 ATPs (oxidative phosphorylation in electron transport chain).
Production of 2 FADH$_2$	4 ATPs (oxidative phosphorylation in electron transport chain).
Total	36 or 38 ATPs per glucose molecule (theoretical maximum).

Glucose Storage: Glycogenesis

If glucose is not needed immediately for ATP production, it combines with many other molecules of glucose to form **glycogen,** a polysaccharide that is the only stored form of carbohydrate in our bodies. The hormone insulin, from pancreatic beta cells, stimulates hepatocytes and skeletal muscle cells to carry out **glycogenesis** (glī′-kō-JEN-e-sis), the synthesis of glycogen (Figure 25.11). The body can store about 500 g (about 1.1 lb) of glycogen, roughly 75% in skeletal muscle fibers and the rest in liver cells. During glycogenesis, glucose is first phosphorylated to glucose 6-phosphate by hexokinase. Glucose 6-phosphate is converted to glucose 1-phosphate, then to uridine diphosphate glucose, and finally to glycogen.

Glucose Release: Glycogenolysis

When body activities require ATP, glycogen stored in hepatocytes is broken down into glucose and released into the blood to be transported to cells, where it will be catabolized by the processes of cellular respiration already described. The process of splitting glycogen into its glucose subunits is called **glycogenolysis** (glī′-kō-je-NOL-e-sis). (Note: Do not confuse *glycogenolysis,* which is the breakdown of glycogen to glucose, with *glycolysis,* the 10 reactions that convert glucose to pyruvic acid.)

Glycogenolysis is not a simple reversal of the steps of glycogenesis (Figure 25.11). It begins by splitting glucose molecules off the branched glycogen molecule via phosphorylation to form glucose 1-phosphate. Phosphorylase, the enzyme that catalyzes this reaction, is activated by glucagon from pancreatic alpha cells and

anabolic reactions. One is the synthesis of glycogen; another is the synthesis of new glucose molecules from some of the products of protein and lipid breakdown.

Figure 25.11 Glycogenesis and glycogenolysis.

The glycogenesis pathway converts glucose into glycogen; the glycogenolysis pathway breaks down glycogen into glucose.

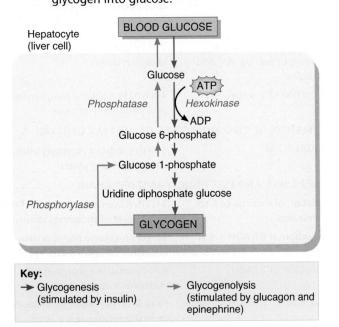

Key:
→ Glycogenesis (stimulated by insulin)
→ Glycogenolysis (stimulated by glucagon and epinephrine)

? Besides hepatocytes, which body cells can synthesize glycogen? Why can't they release glucose into the blood?

epinephrine from the adrenal medulla. Glucose 1-phosphate is then converted to glucose 6-phosphate and finally to glucose, which leaves hepatocytes via glucose transporters (GluT) in the plasma membrane. Phosphorylated glucose molecules cannot ride aboard the GluT transporters, however, and *phosphatase,* the enzyme that converts glucose 6-phosphate into glucose, is absent in skeletal muscle cells. Thus, hepatocytes, which have phosphatase, can release glucose derived from glycogen to the bloodstream, but skeletal muscle cells cannot. In skeletal muscle cells, glycogen is broken down into glucose 1-phosphate, which is then catabolized for ATP production via glycolysis and the Krebs cycle. However, the lactic acid produced by glycolysis in muscle cells can be converted to glucose in the liver. In this way, muscle glycogen can be an indirect source of blood glucose.

CLINICAL CONNECTION | *Carbohydrate Loading*

The amount of glycogen stored in the liver and skeletal muscles varies and can be completely exhausted during long-term athletic endeavors. Thus, many marathon runners and other endurance athletes follow a precise exercise and dietary regimen that includes eating large amounts of complex carbohydrates, such as pasta and potatoes, in the 3 days before an event. This practice, called **carbohydrate loading,** helps maximize the amount of glycogen available for ATP production in muscles. For athletic events lasting more than an hour, carbohydrate loading has been shown to increase an athlete's endurance. The increased endurance is due to increased glycogenolysis, which results in more glucose that can be catabolized for energy. •

Formation of Glucose from Proteins and Fats: Gluconeogenesis

When your liver runs low on glycogen, it is time to eat. If you don't, your body starts catabolizing triglycerides (fats) and proteins. Actually, the body normally catabolizes some of its triglycerides and proteins, but large-scale triglyceride and protein catabolism does not happen unless you are starving, eating very few carbohydrates, or suffering from an endocrine disorder.

The glycerol part of triglycerides, lactic acid, and certain amino acids can be converted in the liver to glucose (Figure 25.12). The process by which glucose is formed from these noncarbohydrate sources is called **gluconeogenesis** (gloo′-kō-nē′-ō-JEN-e-sis; *neo-* = new). An easy way to distinguish this term from glycogenesis or glycogenolysis is to remember that in this case glucose is not converted back from glycogen, but is instead *newly formed*. About 60% of the amino acids in the body can be used for gluconeogenesis. Lactic acid and amino acids such as alanine, cysteine, glycine, serine, and threonine are converted to pyruvic acid, which then may be synthesized into glucose or enter the Krebs cycle. Glycerol may be converted into glyceraldehyde 3-phosphate, which may form pyruvic acid or be used to synthesize glucose.

Gluconeogenesis is stimulated by cortisol, the main glucocorticoid hormone of the adrenal cortex, and by glucagon from the pancreas. In addition, cortisol stimulates the breakdown of proteins into amino acids, thus expanding the pool of amino acids available for gluconeogenesis. Thyroid hormones (thyroxine and triiodothyronine) also mobilize proteins and may mobilize triglycerides from adipose tissue, thereby making glycerol available for gluconeogenesis.

Figure 25.12 Gluconeogenesis, the conversion of noncarbohydrate molecules (amino acids, lactic acid, and glycerol) into glucose.

About 60% of the amino acids in the body can be used for gluconeogenesis.

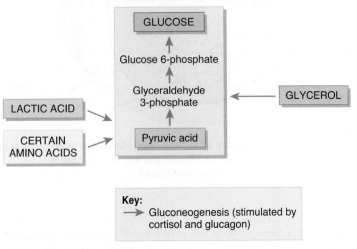

Key:
→ Gluconeogenesis (stimulated by cortisol and glucagon)

? What cells can carry out gluconeogenesis and glycogenesis?

✔ CHECKPOINT

5. How does glucose move into or out of body cells?
6. What happens during glycolysis?
7. How is acetyl coenzyme A formed?
8. Outline the principal events and outcomes of the Krebs cycle.
9. What happens in the electron transport chain and why is this process called chemiosmosis?
10. Which reactions produce ATP during the complete oxidation of a molecule of glucose?
11. Under what circumstances do glycogenesis and glycogenolysis occur?
12. What is gluconeogenesis, and why is it important?

25.4 LIPID METABOLISM

◉ **OBJECTIVES**

- Describe the lipoproteins that transport lipids in the blood.
- Describe the fate, metabolism, and functions of lipids.

Transport of Lipids by Lipoproteins

Most lipids, such as triglycerides, are nonpolar and therefore very hydrophobic molecules. They do not dissolve in water. To be transported in watery blood, such molecules first must be made more water-soluble by combining them with proteins produced by the liver and intestine. The lipid and protein combinations thus formed are **lipoproteins** (lip′-ō-PRŌ-tēns), spherical particles with an outer shell of proteins, phospholipids, and cholesterol molecules surrounding an inner core of triglycerides and other lipids (Figure 25.13). The proteins in the outer shell are called **apoproteins (apo)** (ap-ō-PRŌ-tēns) and are designated by the letters A, B, C, D, and E plus a number. In addition to helping solubilize the lipoprotein in body fluids, each apoprotein has specific functions.

Each of the several types of lipoproteins has different functions, but all essentially are transport vehicles. They provide delivery and pickup services so that lipids can be available when cells need them or removed from circulation when they are not needed. Lipoproteins are categorized and named mainly according to their density, which varies with the ratio of lipids (which have a low density) to proteins (which have a high density). From largest and lightest to smallest and heaviest, the four major classes of lipoproteins are chylomicrons, very-low-density lipoproteins (VLDLs), low-density lipoproteins (LDLs), and high-density lipoproteins (HDLs).

Chylomicrons (kī′-lō-MĪ-krons), which form in mucosal epithelial cells of the small intestine, transport *dietary* (ingested) lipids to adipose tissue for storage. They contain about 1–2% proteins, 85% triglycerides, 7% phospholipids, and 6–7% cholesterol, plus a small amount of fat-soluble vitamins. Chylomicrons enter lacteals of intestinal villi and are carried by lymph into venous blood and then into the systemic circulation. Their presence gives blood plasma a milky appearance, but they remain in the blood for only a few minutes. As chylomicrons circulate through the capillaries of adipose tissue, one of their apoproteins, **apo C-2,** activates *endothelial lipoprotein lipase,* an enzyme that removes fatty acids from chylomicron triglycerides. The free fatty acids are then taken up by adipocytes for synthesis and storage as triglycerides and by muscle cells for ATP production. Hepatocytes remove chylomicron remnants from the blood via receptor-mediated endocytosis, in which another chylomicron apoprotein, **apo E,** is the docking protein.

Very-low-density lipoproteins (VLDLs), which form in hepatocytes, contain mainly *endogenous* (made in the body) lipids. VLDLs contain about 10% proteins, 50% triglycerides, 20% phospholipids, and 20% cholesterol. VLDLs transport triglycerides synthesized in hepatocytes to adipocytes for storage. Like chylomicrons, they lose triglycerides as their apo C-2 activates endothelial lipoprotein lipase, and the resulting fatty acids are taken up by adipocytes for storage and by muscle cells for ATP production. As they deposit some of their triglycerides in adipose cells, VLDLs are converted to LDLs.

Low-density lipoproteins (LDLs) contain 25% proteins, 5% triglycerides, 20% phospholipids, and 50% cholesterol. They carry about 75% of the total cholesterol in blood and deliver it to cells throughout the body for use in repair of cell membranes and synthesis of steroid hormones and bile salts. LDLs contain a single apoprotein, **apo B100,** which is the docking protein that binds to LDL receptors on the plasma membrane of body cells so that LDL can enter the cell via receptor-mediated endocytosis. Within

Figure 25.13 A lipoprotein. Shown here is a VLDL.

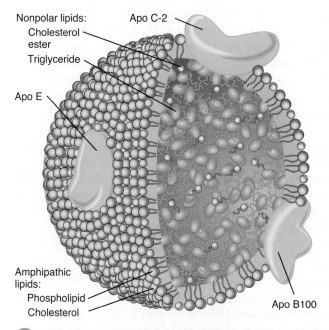

🔑 A single layer of amphipathic phospholipids, cholesterol, and proteins surrounds a core of nonpolar lipids.

Nonpolar lipids:
Cholesterol ester
Triglyceride

Apo C-2

Apo E

Amphipathic lipids:
Phospholipid
Cholesterol

Apo B100

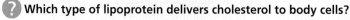

❓ **Which type of lipoprotein delivers cholesterol to body cells?**

the cell, the LDL is broken down, and the cholesterol is released to serve the cell's needs. Once a cell has sufficient cholesterol for its activities, a negative feedback system inhibits the cell's synthesis of new LDL receptors.

When present in excessive numbers, LDLs also deposit cholesterol in and around smooth muscle fibers in arteries, forming fatty plaques that increase the risk of coronary artery disease (see Disorders: Homeostatic Imbalances at the end of Chapter 20). For this reason, the cholesterol in LDLs, called LDL-cholesterol, is known as "bad" cholesterol. Because some people have too few LDL receptors, their body cells remove LDL from the blood less efficiently; as a result, their plasma LDL level is abnormally high, and they are more likely to develop fatty plaques. Eating a high-fat diet increases the production of VLDLs, which elevates the LDL level and increases the formation of fatty plaques.

High-density lipoproteins (HDLs), which contain 40–45% proteins, 5–10% triglycerides, 30% phospholipids, and 20% cholesterol, remove excess cholesterol from body cells and the blood and transport it to the liver for elimination. Because HDLs prevent accumulation of cholesterol in the blood, a high HDL level is associated with decreased risk of coronary artery disease. For this reason, HDL-cholesterol is known as "good" cholesterol.

Sources and Significance of Blood Cholesterol

There are two sources of cholesterol in the body. Some is present in foods (eggs, dairy products, organ meats, beef, pork, and processed luncheon meats), but most is synthesized by hepatocytes. Fatty foods that don't contain any cholesterol at all can still dramatically increase blood cholesterol level in two ways. First, a high intake of dietary fats stimulates reabsorption of cholesterol-containing bile back into the blood, so less cholesterol is lost in the feces. Second, when saturated fats are broken down in the body, hepatocytes use some of the breakdown products to make cholesterol.

A lipid profile test usually measures total cholesterol (TC), HDL-cholesterol, and triglycerides (VLDLs). LDL-cholesterol then is calculated by using the following formula: LDL-cholesterol = TC − HDL-cholesterol − (triglycerides/5). In the United States, blood cholesterol is usually measured in milligrams per deciliter (mg/dL); a deciliter is 0.1 liter or 100 mL. For adults, desirable levels of blood cholesterol are total cholesterol under 200 mg/dL, LDL-cholesterol under 130 mg/dL, and HDL-cholesterol over 40 mg/dL. Normally, triglycerides are in the range of 10–190 mg/dL.

As total cholesterol level increases, the risk of coronary artery disease begins to rise. When total cholesterol is above 200 mg/dL (5.2 mmol/liter), the risk of a heart attack doubles with every 50 mg/dL (1.3 mmol/liter) increase in total cholesterol. Total cholesterol of 200–239 mg/dL and LDL of 130–159 mg/dL are borderline-high; total cholesterol above 239 mg/dL and LDL above 159 mg/dL are classified as high blood cholesterol. The ratio of total cholesterol to HDL-cholesterol predicts the risk of developing coronary artery disease. For example, a person with a total cholesterol of 180 mg/dL and HDL of 60 mg/dL has a risk ratio of 3. Ratios above 4 are considered undesirable; the higher the ratio, the greater the risk of developing coronary artery disease.

Among the therapies used to reduce blood cholesterol level are exercise, diet, and drugs. Regular physical activity at aerobic and nearly aerobic levels raises HDL level. Dietary changes are aimed at reducing the intake of total fat, saturated fats, and cholesterol. Drugs used to treat high blood cholesterol levels include cholestyramine (Questran) and colestipol (Colestid), which promote excretion of bile in the feces; nicotinic acid (Liponicin); and the "statin" drugs—atorvastatin (Lipitor), lovastatin (Mevacor), and simvastatin (Zocor), which block the key enzyme (HMG-CoA reductase) needed for cholesterol synthesis.

The Fate of Lipids

Lipids, like carbohydrates, may be oxidized to produce ATP. If the body has no immediate need to use lipids in this way, they are stored in adipose tissue (fat depots) throughout the body and in the liver. A few lipids are used as structural molecules or to synthesize other essential substances. Some examples include phospholipids, which are constituents of plasma membranes; lipoproteins, which are used to transport cholesterol throughout the body; thromboplastin, which is needed for blood clotting; and myelin sheaths, which speed up nerve impulse conduction. Two **essential fatty acids** that the body cannot synthesize are linoleic acid and linolenic acid. Dietary sources include vegetable oils and leafy vegetables. The various functions of lipids in the body may be reviewed in Table 2.7.

Triglyceride Storage

A major function of adipose tissue is to remove triglycerides from chylomicrons and VLDLs and store them until they are needed for ATP production in other parts of the body. Triglycerides stored in adipose tissue constitute 98% of all body energy reserves. They are stored more readily than glycogen, in part because triglycerides are hydrophobic and do not exert osmotic pressure on cell membranes. Adipose tissue also insulates and protects various parts of the body. Adipocytes in the subcutaneous layer contain about 50% of the stored triglycerides. Other adipose tissues account for the other half: about 12% around the kidneys, 10–15% in the omenta, 15% in genital areas, 5–8% between muscles, and 5% behind the eyes, in the sulci of the heart, and attached to the outside of the large intestine. Triglycerides in adipose tissue are continually broken down and resynthesized. Thus, the triglycerides stored in adipose tissue today are not the same molecules that were present last month because they are continually released from storage, transported in the blood, and redeposited in other adipose tissue cells.

Lipid Catabolism: Lipolysis

In order for muscle, liver, and adipose tissue to oxidize the fatty acids derived from triglycerides to produce ATP, the triglycerides must first be split into glycerol and fatty acids, a process called **lipolysis** (li-POL-i-sis). Lipolysis is catalyzed by enzymes called **lipases.** Epinephrine and norepinephrine enhance triglyceride breakdown into fatty acids and glycerol. These hormones are released when sympathetic tone increases, as occurs, for example, during exercise. Other lipolytic hormones include cortisol, thyroid hormones, and insulinlike growth factors. By contrast, insulin inhibits lipolysis.

The glycerol and fatty acids that result from lipolysis are catabolized via different pathways (Figure 25.14). Glycerol is converted by many cells of the body to glyceraldehyde 3-phosphate, one of the compounds also formed during the catabolism of glucose. If ATP supply in a cell is high, glyceraldehyde 3-phosphate is converted into glucose, an example of gluconeogenesis. If ATP supply in a cell is low, glyceraldehyde 3-phosphate enters the catabolic pathway to pyruvic acid.

Fatty acids are catabolized differently than glycerol and yield more ATP. The first stage in fatty acid catabolism is a series of reactions, collectively called **beta oxidation** (BĀ-ta), that occurs in the matrix of mitochondria. Enzymes remove two carbon atoms at a time from the long chain of carbon atoms composing a fatty acid and attach the resulting two-carbon fragment to coenzyme A, forming acetyl CoA. Then, acetyl CoA enters the Krebs cycle (Figure 25.14). A 16-carbon fatty acid such as palmitic acid can yield as many as 129 ATPs on its complete oxidation via beta oxidation, the Krebs cycle, and the electron transport chain.

As part of normal fatty acid catabolism, hepatocytes can take two acetyl CoA molecules at a time and condense them to form **acetoacetic acid** (as′-ē-tō-a-SĒ-tik). This reaction liberates the bulky CoA portion, which cannot diffuse out of cells. Some acetoacetic acid is converted into **beta-hydroxybutyric acid** (hī-drok-sē-bū-TIR-ik) and **acetone** (AS-e-tōn). The formation of these three substances, collectively known as **ketone bodies** (KĒ-tōn), is called **ketogenesis** (kē-tō-JEN-e-sis) (Figure 25.14). Because ketone bodies freely diffuse through plasma membranes, they leave hepatocytes and enter the bloodstream.

Other cells take up acetoacetic acid and attach its four carbons to two coenzyme A molecules to form two acetyl CoA molecules, which can then enter the Krebs cycle for oxidation. Heart muscle and the cortex (outer part) of the kidneys use acetoacetic acid in preference to glucose for generating ATP. Hepatocytes, which make acetoacetic acid, cannot use it for ATP production because they lack the enzyme that transfers acetoacetic acid back to coenzyme A.

Lipid Anabolism: Lipogenesis

Liver cells and adipose cells can synthesize lipids from glucose or amino acids through **lipogenesis** (lip-ō-JEN-e-sis) (Figure 25.14), which is stimulated by insulin. Lipogenesis occurs when individuals consume more calories than are needed to satisfy their ATP needs. Excess dietary carbohydrates, proteins, and fats all have the same fate—they are converted into triglycerides.

Figure 25.14 Pathways of lipid metabolism. Glycerol may be converted to glyceraldehyde 3-phosphate, which can then be converted to glucose or enter the Krebs cycle for oxidation. Fatty acids undergo beta oxidation and enter the Krebs cycle via acetyl coenzyme A. The synthesis of lipids from glucose or amino acids is called lipogenesis.

Glycerol and fatty acids are catabolized in separate pathways.

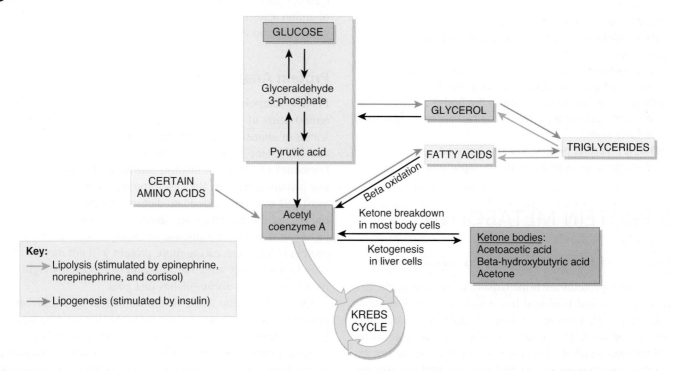

What types of cells can carry out lipogenesis, beta oxidation, and lipolysis? What type of cell can carry out ketogenesis?

Certain amino acids can undergo the following reactions: amino acids → acetyl CoA → fatty acids → triglycerides. The use of glucose to form lipids takes place via two pathways: (1) glucose → glyceraldehyde 3-phosphate → glycerol and (2) glucose → glyceraldehyde 3-phosphate → acetyl CoA → fatty acids. The resulting glycerol and fatty acids can undergo anabolic reactions to become stored triglycerides, or they can go through a series of anabolic reactions to produce other lipids such as lipoproteins, phospholipids, and cholesterol.

CLINICAL CONNECTION | *Ketosis*

The level of ketone bodies in the blood normally is very low because other tissues use them for ATP production as fast as they are generated from the breakdown of fatty acids in the liver. During periods of excessive beta oxidation, however, the production of ketone bodies exceeds their uptake and use by body cells. This might occur after a meal rich in triglycerides, or during fasting or starvation, because few carbohydrates are available for catabolism. Excessive beta oxidation may also occur in poorly controlled or untreated diabetes mellitus for two reasons: (1) Because adequate glucose cannot get into cells, triglycerides are used for ATP production, and (2) because insulin normally inhibits lipolysis, a lack of insulin accelerates the pace of lipolysis. When the concentration of ketone bodies in the blood rises above normal—a condition called ketosis—the ketone bodies, most of which are acids, must be buffered. If too many accumulate, they decrease the concentration of buffers, such as bicarbonate ions, and blood pH falls. Extreme or prolonged ketosis can lead to **acidosis (ketoacidosis),** an abnormally low blood pH. The decreased blood pH in turn causes depression of the central nervous system, which can result in disorientation, coma, and even death if the condition is not treated. When a diabetic becomes seriously insulin-deficient, one of the telltale signs is the sweet smell on the breath from the ketone body acetone. •

✔ CHECKPOINT

13. What are the functions of the apoproteins in lipoproteins?
14. Which lipoprotein particles contain "good" and "bad" cholesterol, and why are these terms used?
15. Where are triglycerides stored in the body?
16. Explain the principal events of the catabolism of glycerol and fatty acids.
17. What are ketone bodies? What is ketosis?
18. Define lipogenesis and explain its importance.

25.5 PROTEIN METABOLISM

◉ OBJECTIVE

• Describe the fate, metabolism, and functions of proteins.

During digestion, proteins are broken down into amino acids. Unlike carbohydrates and triglycerides, which are stored, proteins are not warehoused for future use. Instead, amino acids are either oxidized to produce ATP or used to synthesize new proteins for body growth and repair. Excess dietary amino acids are not excreted in the urine or feces but instead are converted into glucose (gluconeogenesis) or triglycerides (lipogenesis).

The Fate of Proteins

The active transport of amino acids into body cells is stimulated by insulinlike growth factors (IGFs) and insulin. Almost immediately after digestion, amino acids are reassembled into proteins. Many proteins function as enzymes; others are involved in transportation (hemoglobin) or serve as antibodies, clotting chemicals (fibrinogen), hormones (insulin), or contractile elements in muscle fibers (actin and myosin). Several proteins serve as structural components of the body (collagen, elastin, and keratin). The various functions of proteins in the body may be reviewed in Table 2.8.

Protein Catabolism

A certain amount of protein catabolism occurs in the body each day, stimulated mainly by cortisol from the adrenal cortex. Proteins from worn-out cells (such as red blood cells) are broken down into amino acids. Some amino acids are converted into other amino acids, peptide bonds are re-formed, and new proteins are synthesized as part of the recycling process. Hepatocytes convert some amino acids to fatty acids, ketone bodies, or glucose. Cells throughout the body oxidize a small amount of amino acids to generate ATP via the Krebs cycle and the electron transport chain. However, before amino acids can be oxidized, they must first be converted to molecules that are part of the Krebs cycle or can enter the Krebs cycle, such as acetyl CoA (Figure 25.15). Before amino acids can enter the Krebs cycle, their amino group (NH_2) must first be removed—a process called **deamination** (dē-am′-i-NĀ-shun). Deamination occurs in hepatocytes and produces ammonia (NH_3). The liver cells then convert the highly toxic ammonia to urea, a relatively harmless substance that is excreted in the urine. The conversion of amino acids into glucose (gluconeogenesis) may be reviewed in Figure 25.12; the conversion of amino acids into fatty acids (lipogenesis) or ketone bodies (ketogenesis) is shown in Figure 25.14.

Protein Anabolism

Protein anabolism, the formation of peptide bonds between amino acids to produce new proteins, is carried out on the ribosomes of almost every cell in the body, directed by the cells' DNA and RNA (see Figure 3.29). Insulinlike growth factors, thyroid hormones (T_3 and T_4), insulin, estrogen, and testosterone stimulate protein synthesis. Because proteins are a main component of most cell structures, adequate dietary protein is especially essential during the growth years, during pregnancy, and when tissue has been damaged by disease or injury. Once dietary intake of protein is adequate, eating more protein will not increase bone or muscle mass; only a regular program of forceful, weight-bearing muscular activity accomplishes that goal.

Of the 20 amino acids in the human body, 10 are **essential amino acids:** They must be present in the diet because they cannot be synthesized in the body in adequate amounts. It is *essential* to include them in your diet. Humans are unable to synthesize eight amino acids (isoleucine, leucine, lysine, methionine, phenylalanine, threonine, tryptophan, and valine) and synthesize two others

Figure 25.15 Points at which amino acids (yellow boxes) enter the Krebs cycle for oxidation.

Before amino acids can be catabolized, they must first be converted to various substances that can enter the Krebs cycle.

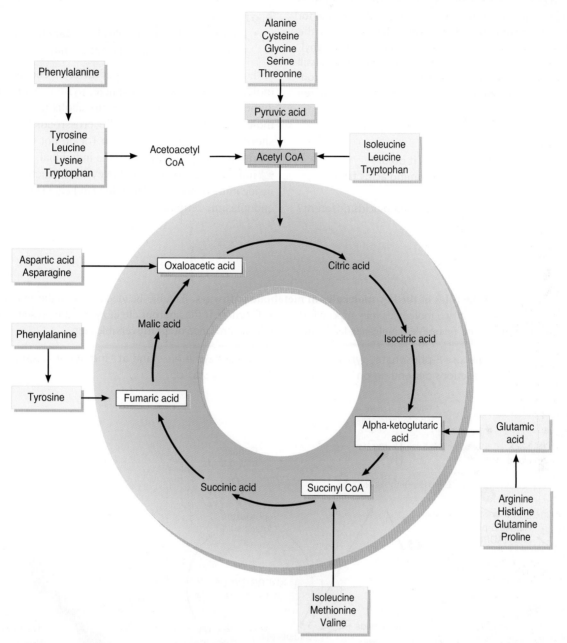

What group is removed from an amino acid before it can enter the Krebs cycle, and what is this process called?

(arginine and histidine) in inadequate amounts, especially in childhood. A **complete protein** contains sufficient amounts of all essential amino acids. Beef, fish, poultry, eggs, and milk are examples of foods that contain complete proteins. An **incomplete protein** does not contain all essential amino acids. Examples of incomplete proteins are leafy green vegetables, legumes (beans and peas), and grains. **Nonessential amino acids** can be synthesized by body cells. They are formed by **transamination** (trans′-am-i-NĀ-shun), the transfer of an amino group from an amino acid to pyruvic acid or to an acid in the Krebs cycle. Once the ap-

propriate essential and nonessential amino acids are present in cells, protein synthesis occurs rapidly.

CLINICAL CONNECTION | Phenylketonuria

Phenylketonuria (fen′-il-kē′-tō-NOO-rē-a) or **PKU** is a genetic error of protein metabolism characterized by elevated blood levels of the amino acid phenylalanine. Most children with phenylketonuria have a mutation in the gene that codes for the enzyme phenylalanine hydroxylase, the enzyme needed to convert phenylalanine into the amino acid tyrosine, which can enter the Krebs cycle (Figure 25.15).

Because the enzyme is deficient, phenylalanine cannot be metabolized, and what is not used in protein synthesis builds up in the blood. If untreated, the disorder causes vomiting, rashes, seizures, growth deficiency, and severe mental retardation. Newborns are screened for PKU, and mental retardation can be prevented by restricting the affected child to a diet that supplies only the amount of phenylalanine needed for growth, although learning disabilities may still ensue. Because the artificial sweetener aspartame (NutraSweet) contains phenylalanine, its consumption must be restricted in children with PKU. •

✔ **CHECKPOINT**

19. What is deamination and why does it occur?
20. What are the possible fates of the amino acids from protein catabolism?
21. How are essential and nonessential amino acids different?

25.6 KEY MOLECULES AT METABOLIC CROSSROADS

◉ **OBJECTIVE**

• Identify the key molecules in metabolism, and describe the reactions and the products they may form.

Although there are thousands of different chemicals in cells, three molecules—glucose 6-phosphate, pyruvic acid, and acetyl coenzyme A—play pivotal roles in metabolism (Figure 25.16). These molecules stand at "metabolic crossroads"; as you will learn shortly, the reactions that occur (or do not occur) depend on the nutritional or activity status of the individual. Reactions ❶ through ❼ in Figure 25.16 occur in the cytosol, reactions ❽ and ❾ occur inside mitochondria, and reactions indicated by ❿ occur on smooth endoplasmic reticulum.

Figure 25.16 Summary of the roles of the key molecules in metabolic pathways. Double-headed arrows indicate that reactions between two molecules may proceed in either direction, if the appropriate enzymes are present and the conditions are favorable; single-headed arrows signify the presence of an irreversible step.

 Three molecules—glucose 6-phosphate, pyruvic acid, and acetyl coenzyme A—stand at "metabolic crossroads." They can undergo different reactions depending on your nutritional or activity status.

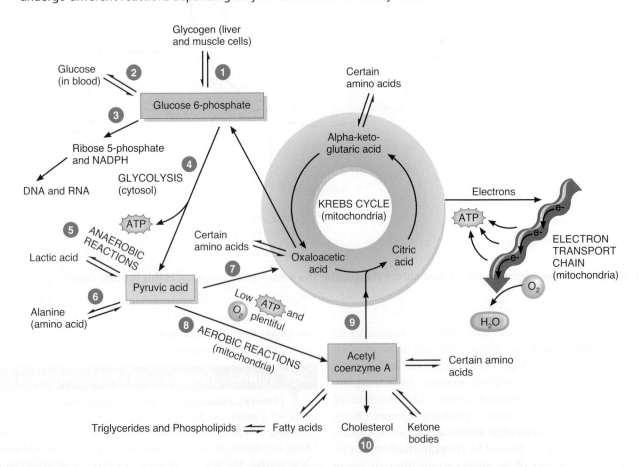

❓ Which substance is the gateway into the Krebs cycle for molecules that are being oxidized to generate ATP?

The Role of Glucose 6-Phosphate

Shortly after glucose enters a body cell, a kinase converts it to **glucose 6-phosphate.** Four possible fates await glucose 6-phosphate (see Figure 25.16):

1 *Synthesis of glycogen.* When glucose is abundant in the bloodstream, as it is just after a meal, a large amount of glucose 6-phosphate is used to synthesize glycogen, the storage form of carbohydrate in animals. Subsequent breakdown of glycogen into glucose 6-phosphate occurs through a slightly different series of reactions. Synthesis and breakdown of glycogen occur mainly in skeletal muscle fibers and hepatocytes.

2 *Release of glucose into the bloodstream.* If the enzyme glucose 6-phosphatase is present and active, glucose 6-phosphate can be dephosphorylated to glucose. Once glucose is released from the phosphate group, it can leave the cell and enter the bloodstream. Hepatocytes are the main cells that can provide glucose to the bloodstream in this way.

3 *Synthesis of nucleic acids.* Glucose 6-phosphate is the precursor used by cells throughout the body to make ribose 5-phosphate, a 5-carbon sugar that is needed for synthesis of RNA (ribonucleic acid) and DNA (deoxyribonucleic acid). The same sequence of reactions also produces NADPH. This molecule is a hydrogen and electron donor in certain reduction reactions, such as synthesis of fatty acids and steroid hormones.

4 *Glycolysis.* Some ATP is produced anaerobically via glycolysis, in which glucose 6-phosphate is converted to pyruvic acid, another key molecule in metabolism. Most body cells carry out glycolysis.

The Role of Pyruvic Acid

Each 6-carbon molecule of glucose that undergoes glycolysis yields two 3-carbon molecules of **pyruvic acid** (pī-ROO-vik). This molecule, like glucose 6-phosphate, stands at a metabolic crossroads: Given enough oxygen, the aerobic (oxygen-consuming) reactions of cellular respiration can proceed; if oxygen is in short supply, anaerobic reactions can occur (Figure 25.16):

5 *Production of lactic acid.* When oxygen is in short supply in a tissue, as in actively contracting skeletal or cardiac muscle, some pyruvic acid is changed to lactic acid. The lactic acid then diffuses into the bloodstream and is taken up by hepatocytes, which eventually convert it back to pyruvic acid.

6 *Production of alanine.* Carbohydrate and protein metabolism are linked by pyruvic acid. Through transamination, an amino group (—NH_2) can either be added to pyruvic acid (a carbohydrate) to produce the amino acid alanine, or be removed from alanine to generate pyruvic acid.

7 *Gluconeogenesis.* Pyruvic acid and certain amino acids also can be converted to oxaloacetic acid, one of the Krebs cycle intermediates, which in turn can be used to form glucose 6-phosphate. This sequence of gluconeogenesis reactions bypasses certain one-way reactions of glycolysis.

The Role of Acetyl Coenzyme A

8 When the ATP level in a cell is low but oxygen is plentiful, most pyruvic acid streams toward ATP-producing reactions—the Krebs cycle and electron transport chain—via conversion to **acetyl coenzyme A.**

9 *Entry into the Krebs cycle.* Acetyl CoA is the vehicle for 2-carbon acetyl groups to enter the Krebs cycle. Oxidative Krebs cycle reactions convert acetyl CoA to CO_2 and produce reduced coenzymes (NADH and $FADH_2$) that transfer electrons into the electron transport chain. Oxidative reactions in the electron transport chain in turn generate ATP. Most fuel molecules that will be oxidized to generate ATP—glucose, fatty acids, and ketone bodies—are first converted to acetyl CoA.

10 *Synthesis of lipids.* Acetyl CoA also can be used for synthesis of certain lipids, including fatty acids, ketone bodies, and cholesterol. Because pyruvic acid can be converted to acetyl CoA, carbohydrates can be turned into triglycerides; this metabolic pathway stores some excess carbohydrate calories as fat. Mammals, including humans, cannot reconvert acetyl CoA to pyruvic acid, however, so fatty acids cannot be used to generate glucose or other carbohydrate molecules.

Table 25.2 is a summary of carbohydrate, lipid, and protein metabolism.

✔**CHECKPOINT**

22. What are the possible fates of glucose 6-phosphate, pyruvic acid, and acetyl coenzyme A in a cell?

25.7 METABOLIC ADAPTATIONS

◉ **OBJECTIVE**

• Compare metabolism during the absorptive and postabsorptive states.

Regulation of metabolic reactions depends both on the chemical environment within body cells, such as the levels of ATP and oxygen, and on signals from the nervous and endocrine systems. Some aspects of metabolism depend on how much time has passed since the last meal. During the **absorptive state,** ingested nutrients are entering the bloodstream, and glucose is readily available for ATP production. During the **postabsorptive state,** absorption of nutrients from the GI tract is complete, and energy needs must be met by fuels already in the body. A typical meal requires about 4 hours for complete absorption; given three meals a day, the absorptive state exists for about 12 hours each day. Assuming no between-meal snacks, the other 12 hours—typically late morning, late afternoon, and most of the night—are spent in the postabsorptive state.

Because the nervous system and red blood cells continue to depend on glucose for ATP production during the postabsorptive state, maintaining a steady blood glucose level is critical during this period. Hormones are the major regulators of metabolism in

1044 CHAPTER 25 • METABOLISM AND NUTRITION

TABLE 25.2

Summary of Metabolism

PROCESS	COMMENTS
CARBOHYDRATES	
Glucose catabolism	Complete oxidation of glucose (cellular respiration) is chief source of ATP in cells; consists of glycolysis, Krebs cycle, and electron transport chain. Complete oxidation of 1 molecule of glucose yields maximum of 36 or 38 molecules of ATP.
Glycolysis	Conversion of glucose into pyruvic acid results in production of some ATP. Reactions do not require oxygen (anaerobic cellular respiration).
Krebs cycle	Cycle includes series of oxidation–reduction reactions in which coenzymes (NAD^+ and FAD) pick up hydrogen ions and hydride ions from oxidized organic acids; some ATP produced. CO_2 and H_2O are by-products. Reactions are aerobic.
Electron transport chain	Third set of reactions in glucose catabolism: another series of oxidation–reduction reactions, in which electrons are passed from one carrier to next; most ATP produced. Reactions require oxygen (aerobic cellular respiration).
Glucose anabolism	Some glucose is converted into glycogen (glycogenesis) for storage if not needed immediately for ATP production. Glycogen can be reconverted to glucose (glycogenolysis). Conversion of amino acids, glycerol, and lactic acid into glucose is called gluconeogenesis.
LIPIDS	
Triglyceride catabolism	Triglycerides are broken down into glycerol and fatty acids. Glycerol may be converted into glucose (gluconeogenesis) or catabolized via glycolysis. Fatty acids are catabolized via beta oxidation into acetyl coenzyme A that can enter Krebs cycle for ATP production or be converted into ketone bodies (ketogenesis).
Triglyceride anabolism	Synthesis of triglycerides from glucose and fatty acids is called lipogenesis. Triglycerides are stored in adipose tissue.
PROTEINS	
Protein catabolism	Amino acids are oxidized via Krebs cycle after deamination. Ammonia resulting from deamination is converted into urea in liver, passed into blood, and excreted in urine. Amino acids may be converted into glucose (gluconeogenesis), fatty acids, or ketone bodies.
Protein anabolism	Protein synthesis is directed by DNA and utilizes cells' RNA and ribosomes.

each state. The effects of insulin dominate in the absorptive state; several other hormones regulate metabolism in the postabsorptive state. During fasting and starvation, many body cells turn to ketone bodies for ATP production, as noted in the Clinical Connection on Ketosis.

Metabolism during the Absorptive State

Soon after a meal, nutrients start to enter the blood. Recall that ingested food reaches the bloodstream mainly as glucose, amino acids, and triglycerides (in chylomicrons). Two metabolic hallmarks of the absorptive state are the oxidation of glucose for ATP production, which occurs in most body cells, and the storage of excess fuel molecules for future between-meal use, which occurs mainly in hepatocytes, adipocytes, and skeletal muscle fibers.

Absorptive State Reactions

The following reactions dominate during the absorptive state (Figure 25.17):

1 About 50% of the glucose absorbed from a typical meal is oxidized by cells throughout the body to produce ATP via glycolysis, the Krebs cycle, and the electron transport chain.

2 Most glucose that enters hepatocytes is converted to glycogen. Small amounts may be used for synthesis of fatty acids and glyceraldehyde 3-phosphate.

3 Some fatty acids and triglycerides synthesized in the liver remain there, but hepatocytes package most into VLDLs, which carry lipids to adipose tissue for storage.

4 Adipocytes also take up glucose not picked up by the liver and convert it into triglycerides for storage. Overall, about 40% of the glucose absorbed from a meal is converted to triglycerides, and about 10% is stored as glycogen in skeletal muscles and hepatocytes.

5 Most dietary lipids (mainly triglycerides and fatty acids) are stored in adipose tissue; only a small portion is used for synthesis reactions. Adipocytes obtain the lipids from chylomicrons, from VLDLs, and from their own synthesis reactions.

6 Many absorbed amino acids that enter hepatocytes are deaminated to keto acids, which can either enter the Krebs cycle for ATP production or be used to synthesize glucose or fatty acids.

7 Some amino acids that enter hepatocytes are used to synthesize proteins (for example, plasma proteins).

Figure 25.17 Principal metabolic pathways during the absorptive state.

During the absorptive state, most body cells produce ATP by oxidizing glucose to CO_2 and H_2O.

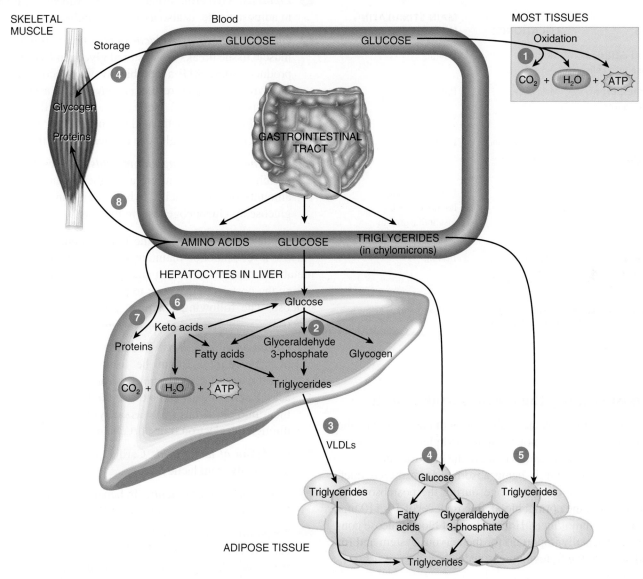

Are the reactions shown in this figure mainly anabolic or catabolic?

8. Amino acids not taken up by hepatocytes are used in other body cells (such as muscle cells) for synthesis of proteins or regulatory chemicals such as hormones or enzymes.

Regulation of Metabolism during the Absorptive State

Soon after a meal, glucose-dependent insulinotropic peptide (GIP), plus the rising blood levels of glucose and certain amino acids, stimulates pancreatic beta cells to release insulin. In general, insulin increases the activity of enzymes needed for anabolism and the synthesis of storage molecules; at the same time it decreases the activity of enzymes needed for catabolic or breakdown reactions. Insulin promotes the entry of glucose and amino acids into cells of many tissues, and it stimulates the phosphorylation of glucose in hepatocytes and the conversion of glucose 6-phosphate to glycogen in both liver and muscle cells. In liver and adipose tissue, insulin enhances the synthesis of triglycerides, and in cells throughout the body insulin stimulates protein synthesis. (See Section 18.10 to review the effects of insulin.) Insulinlike growth factors and the thyroid hormones (T_3 and T_4) also stimulate protein synthesis. Table 25.3 summarizes the hormonal regulation of metabolism in the absorptive state.

TABLE 25.3

Hormonal Regulation of Metabolism in the Absorptive State

PROCESS	LOCATION(S)	MAIN STIMULATING HORMONE(S)
Facilitated diffusion of glucose into cells	Most cells.	Insulin.*
Active transport of amino acids into cells	Most cells.	Insulin.
Glycogenesis (glycogen synthesis)	Hepatocytes and muscle fibers.	Insulin.
Protein synthesis	All body cells.	Insulin, thyroid hormones, and insulinlike growth factors.
Lipogenesis (triglyceride synthesis)	Adipose cells and hepatocytes.	Insulin.

*Facilitated diffusion of glucose into hepatocytes (liver cells) and neurons is always "turned on" and does not require insulin.

Metabolism during the Postabsorptive State

About 4 hours after the last meal, absorption of nutrients from the small intestine is nearly complete, and blood glucose level starts to fall because glucose continues to leave the bloodstream and enter body cells while none is being absorbed from the GI tract. Thus, the main metabolic challenge during the postabsorptive state is to maintain the normal blood glucose level of 70–110 mg/100 mL (3.9–6.1 mmol/liter). Homeostasis of blood glucose concentration is especially important for the nervous system and for red blood cells for the following reasons:

- The dominant fuel molecule for ATP production in the nervous system is glucose, because fatty acids are unable to pass the blood–brain barrier.

- Red blood cells derive all their ATP from glycolysis of glucose because they have no mitochondria, so the Krebs cycle and the electron transport chain are not available to them.

Postabsorptive State Reactions

During the postabsorptive state, both *glucose production* and *glucose conservation* help maintain blood glucose level: Hepatocytes produce glucose molecules and export them into the blood, and other body cells switch from glucose to alternative fuels for ATP production to conserve scarce glucose. The major reactions of the postabsorptive state that produce glucose are the following (Figure 25.18):

1 **Breakdown of liver glycogen.** During fasting, a major source of blood glucose is liver glycogen, which can provide about a

4-hour supply of glucose. Liver glycogen is continually being formed and broken down as needed.

2 **Lipolysis.** Glycerol, produced by breakdown of triglycerides in adipose tissue, is also used to form glucose.

3 **Gluconeogenesis using lactic acid.** During exercise, skeletal muscle tissue breaks down stored glycogen (see step 9) and produces some ATP anaerobically via glycolysis. Some of the pyruvic acid that results is converted to acetyl CoA, and some is converted to lactic acid, which diffuses into the blood. In the liver, lactic acid can be used for gluconeogenesis, and the resulting glucose is released into the blood.

4 **Gluconeogenesis using amino acids.** Modest breakdown of proteins in skeletal muscle and other tissues releases large amounts of amino acids, which then can be converted to glucose by gluconeogenesis in the liver.

Despite all of these ways the body produces glucose, blood glucose level cannot be maintained for very long without further metabolic changes. Thus, a major adjustment must be made during the postabsorptive state to produce ATP while conserving glucose. The following reactions produce ATP without using glucose:

5 **Oxidation of fatty acids.** The fatty acids released by lipolysis of triglycerides cannot be used for glucose production because acetyl CoA cannot be readily converted to pyruvic acid. But most cells can oxidize the fatty acids directly, feed them into the Krebs cycle as acetyl CoA, and produce ATP through the electron transport chain.

6 **Oxidation of lactic acid.** Cardiac muscle can produce ATP aerobically from lactic acid.

7 **Oxidation of amino acids.** In hepatocytes, amino acids may be oxidized directly to produce ATP.

8 **Oxidation of ketone bodies.** Hepatocytes also convert fatty acids to ketone bodies, which can be used by the heart, kidneys, and other tissues for ATP production.

9 **Breakdown of muscle glycogen.** Skeletal muscle cells break down glycogen to glucose 6-phosphate, which undergoes glycolysis and provides ATP for muscle contraction.

Regulation of Metabolism during the Postabsorptive State

Both hormones and the sympathetic division of the autonomic nervous system (ANS) regulate metabolism during the postabsorptive state. The hormones that regulate postabsorptive state metabolism sometimes are called anti-insulin hormones because they counter the effects of insulin during the absorptive state. As blood glucose level declines, the secretion of insulin falls and the release of anti-insulin hormones rises.

When blood glucose concentration starts to drop, the pancreatic alpha cells release glucagon at a faster rate, and the beta cells secrete insulin more slowly. The primary target tissue of glucagon is the liver; the major effect is increased release of glucose into the bloodstream due to gluconeogenesis and glycogenolysis.

Figure 25.18 Principal metabolic pathways during the postabsorptive state.

The principal function of postabsorptive state reactions is to maintain a normal blood glucose level.

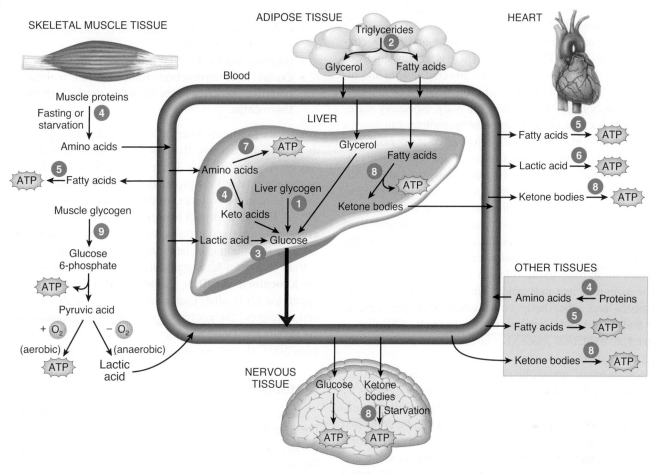

What processes directly elevate blood glucose during the postabsorptive state, and where does each occur?

Low blood glucose also activates the sympathetic branch of the ANS. Glucose-sensitive neurons in the hypothalamus detect low blood glucose and increase sympathetic output. As a result, sympathetic nerve endings release the neurotransmitter norepinephrine, and the adrenal medulla releases two catecholamine hormones—epinephrine and norepinephrine—into the bloodstream. Like glucagon, epinephrine stimulates glycogen breakdown. Epinephrine and norepinephrine are both potent stimulators of lipolysis. These actions of the catecholamines help to increase glucose and free fatty acid levels in the blood. As a result, muscle uses more fatty acids for ATP production, and more glucose is available to the nervous system. Table 25.4 summarizes the hormonal regulation of metabolism in the postabsorptive state.

Metabolism during Fasting and Starvation

The term **fasting** means going without food for many hours or a few days; **starvation** implies weeks or months of food deprivation or inadequate food intake. People can survive without food for 2 months or more if they drink enough water to prevent

TABLE 25.4

Hormonal Regulation of Metabolism in the Postabsorptive State

PROCESS	LOCATION(S)	MAIN STIMULATING HORMONE(S)
Glycogenolysis (glycogen breakdown)	Hepatocytes and skeletal muscle fibers.	Glucagon and epinephrine.
Lipolysis (triglyceride breakdown)	Adipocytes.	Epinephrine, norepinephrine, cortisol, insulinlike growth factors, thyroid hormones, and others.
Protein breakdown	Most body cells, but especially skeletal muscle fibers.	Cortisol.
Gluconeogenesis (synthesis of glucose from noncarbohydrates)	Hepatocytes and kidney cortex cells.	Glucagon and cortisol.

dehydration. Although glycogen stores are depleted within a few hours of beginning a fast, catabolism of stored triglycerides and structural proteins can provide energy for several weeks. The amount of adipose tissue the body contains determines the lifespan possible without food.

During fasting and starvation, nervous tissue and RBCs continue to use glucose for ATP production. There is a ready supply of amino acids for gluconeogenesis because lowered insulin and increased cortisol levels slow the pace of protein synthesis and promote protein catabolism. Most cells in the body, especially skeletal muscle cells because of their high protein content, can spare a fair amount of protein before their performance is adversely affected. During the first few days of fasting, protein catabolism outpaces protein synthesis by about 75 grams daily as some of the "old" amino acids are being deaminated and used for gluconeogenesis and "new" (that is, dietary) amino acids are lacking.

By the second day of a fast, blood glucose level has stabilized at about 65 mg/100 mL (3.6 mmol/liter); at the same time the level of fatty acids in plasma has risen fourfold. Lipolysis of triglycerides in adipose tissue releases glycerol, which is used for gluconeogenesis, and fatty acids. The fatty acids diffuse into muscle fibers and other body cells, where they are used to produce acetyl CoA, which enters the Krebs cycle. ATP then is synthesized as oxidation proceeds via the Krebs cycle and the electron transport chain.

The most dramatic metabolic change that occurs with fasting and starvation is the increase in the formation of ketone bodies by hepatocytes. During fasting, only small amounts of glucose undergo glycolysis to pyruvic acid, which in turn can be converted to oxaloacetic acid. Acetyl CoA enters the Krebs cycle by combining with oxaloacetic acid (see Figure 25.16); when oxaloacetic acid is scarce due to fasting, only some of the available acetyl CoA can enter the Krebs cycle. Surplus acetyl CoA is used for ketogenesis, mainly in hepatocytes. Ketone body production thus increases as catabolism of fatty acids rises. Lipid-soluble ketone bodies can diffuse through plasma membranes and across the blood–brain barrier and be used as an alternative fuel for ATP production, especially by cardiac and skeletal muscle fibers and neurons. Normally, only a trace of ketone bodies (0.01 mmol/liter) are present in the blood, so they are a negligible fuel source. After 2 days of fasting, however, the level of ketones is 100–300 times higher and supplies roughly a third of the brain's fuel for ATP production. By 40 days of starvation, ketones provide up to two-thirds of the brain's energy needs. The presence of ketones actually reduces the use of glucose for ATP production, which in turn decreases the demand for gluconeogenesis and slows the catabolism of muscle proteins later in starvation to about 20 grams daily.

✔ CHECKPOINT

23. What are the roles of insulin, glucagon, epinephrine, insulinlike growth factors, thyroxine, cortisol, estrogen, and testosterone in regulation of metabolism?

24. Why is ketogenesis more significant during fasting or starvation than during normal absorptive and postabsorptive states?

25.8 HEAT AND ENERGY BALANCE

● OBJECTIVES

• Define basal metabolic rate (BMR), and explain several factors that affect it.
• Describe the factors that influence body heat production.
• Explain how normal body temperature is maintained by negative feedback loops involving the hypothalamic thermostat.

Your body produces more or less heat depending on the rates of its metabolic reactions. Because homeostasis of body temperature can be maintained only if the rate of heat loss from the body equals the rate of heat production by metabolism, it is important to understand the ways in which heat can be lost, gained, or conserved. **Heat** is a form of energy that can be measured as **temperature** and expressed in units called calories. A **calorie (cal)** is defined as the amount of heat required to raise the temperature of 1 gram of water 1°C. Because the calorie is a relatively small unit, the **kilocalorie (kcal)** or **Calorie (Cal)** (always spelled with an uppercase C) is often used to measure the body's metabolic rate and to express the energy content of foods. A kilocalorie equals 1000 calories. Thus, when we say that a particular food item contains 500 Calories, we are actually referring to kilocalories.

Metabolic Rate

The overall rate at which metabolic reactions use energy is termed the **metabolic rate.** As you have already learned, some of the energy is used to produce ATP, and some is released as heat. Because many factors affect metabolic rate, it is measured under standard conditions, with the body in a quiet, resting, and fasting condition called the **basal state.** The measurement obtained under these conditions is the **basal metabolic rate (BMR).** The most common way to determine BMR is by measuring the amount of oxygen used per kilocalorie of food metabolized. When the body uses 1 liter of oxygen to oxidize a typical dietary mixture of triglycerides, carbohydrates, and proteins, about 4.8 Cal of energy is released. BMR is 1200–1800 Cal/day in adults, or about 24 Cal/kg of body mass in adult males and 22 Cal/kg in adult females. The added calories needed to support daily activities, such as digestion and walking, range from 500 Cal for a small, relatively sedentary person to over 3000 Cal for a person in training for Olympic-level competitions or mountain climbing.

Body Temperature Homeostasis

Despite wide fluctuations in environmental temperature, homeostatic mechanisms can maintain a normal range for internal body temperature. If the rate of body heat production equals the rate of heat loss, the body maintains a constant core temperature near 37°C (98.6°F). **Core temperature** is the temperature in body structures deep to the skin and subcutaneous layer. **Shell temperature** is the temperature near the body surface—in the skin and subcutaneous layer. Depending on the environmental temperature, shell temperature is 1–6°C lower than core temperature. A core temperature that is too high kills by denaturing body proteins; a core temperature that is too low causes cardiac arrhythmias that result in death.

Heat Production

The production of body heat is proportional to metabolic rate. Several factors affect the metabolic rate and thus the rate of heat production:

- **Exercise.** During strenuous exercise, the metabolic rate may increase to as much as 15 times the basal rate. In well-trained athletes, the rate may increase up to 20 times.

- **Hormones.** Thyroid hormones (thyroxine and triiodothyronine) are the main regulators of BMR; BMR increases as the blood levels of thyroid hormones rise. The response to changing levels of thyroid hormones is slow, however, taking several days to appear. Thyroid hormones increase BMR in part by stimulating aerobic cellular respiration. As cells use more oxygen to produce ATP, more heat is given off, and body temperature rises. Other hormones have minor effects on BMR. Testosterone, insulin, and human growth hormone can increase the metabolic rate by 5–15%.

- **Nervous system.** During exercise or in a stressful situation, the sympathetic division of the autonomic nervous system is stimulated. Its postganglionic neurons release norepinephrine (NE), and it also stimulates release of the hormones epinephrine and norepinephrine by the adrenal medulla. Both epinephrine and norepinephrine increase the metabolic rate of body cells.

- **Body temperature.** The higher the body temperature, the higher the metabolic rate. Each 1°C rise in core temperature increases the rate of biochemical reactions by about 10%. As a result, metabolic rate may be increased substantially during a fever.

- **Ingestion of food.** The ingestion of food raises the metabolic rate 10–20% due to the energy "costs" of digesting, absorbing, and storing nutrients. This effect, *food-induced thermogenesis,* is greatest after eating a high-protein meal and is less after eating carbohydrates and lipids.

- **Age.** The metabolic rate of a child, in relation to its size, is about double that of an elderly person due to the high rates of reactions related to growth.

- **Other factors.** Other factors that affect metabolic rate are gender (lower in females, except during pregnancy and lactation), climate (lower in tropical regions), sleeping (lower), and malnutrition (lower).

Mechanisms of Heat Transfer

Maintaining normal body temperature depends on the ability to lose heat to the environment at the same rate as it is produced by metabolic reactions. Heat can be transferred from the body to its surroundings in four ways: via conduction, convection, radiation, and evaporation.

1. **Conduction** is the heat exchange that occurs between molecules of two materials that are in direct contact with each other. At rest, about 3% of body heat is lost via conduction to solid materials in contact with the body, such as a chair, clothing, and jewelry. Heat can also be gained via conduction—for example, while soaking in a hot tub. Because water conducts heat 20 times more effectively than air, heat loss or heat gain via conduction is much greater when the body is submerged in cold or hot water.

2. **Convection** (kon-VEK-shun) is the transfer of heat by the movement of a fluid (a gas or a liquid) between areas of different temperature. The contact of air or water with your body results in heat transfer by both conduction and convection. When cool air makes contact with the body, it becomes warmed and therefore less dense and is carried away by convection currents created as the less dense air rises. The faster the air moves—for example, by a breeze or a fan—the faster the rate of convection. At rest, about 15% of body heat is lost to the air via conduction and convection.

3. **Radiation** is the transfer of heat in the form of infrared rays between a warmer object and a cooler one without physical contact. Your body loses heat by radiating more infrared waves than it absorbs from cooler objects. If surrounding objects are warmer than you are, you absorb more heat than you lose by radiation. In a room at 21°C (70°F), about 60% of heat loss occurs via radiation in a resting person.

4. **Evaporation** is the conversion of a liquid to a vapor. Every milliliter of evaporating water takes with it a great deal of heat—about 0.58 Cal/mL. Under typical resting conditions, about 22% of heat loss occurs through evaporation of about 700 mL of water per day—300 mL in exhaled air and 400 mL from the skin surface. Because we are not normally aware of this water loss through the skin and mucous membranes of the mouth and respiratory system, it is termed **insensible water loss** (in-SEN-si-bel). The rate of evaporation is inversely related to relative humidity, the ratio of the actual amount of moisture in the air to the maximum amount it can hold at a given temperature. The higher the relative humidity, the lower the rate of evaporation. At 100% humidity, heat is gained via condensation of water on the skin surface as fast as heat is lost via evaporation. Evaporation provides the main defense against overheating during exercise. Under extreme conditions, a maximum of about 3 liters of sweat can be produced each hour, removing more than 1700 Calories of heat if all of it evaporates. (Note: Sweat that drips off the body rather than evaporating removes very little heat.)

Hypothalamic Thermostat

The control center that functions as the body's thermostat is a group of neurons in the anterior part of the hypothalamus, the **preoptic area.** This area receives impulses from thermoreceptors in the skin and mucous membranes and in the hypothalamus. Neurons of the preoptic area generate nerve impulses at a higher frequency when blood temperature increases, and at a lower frequency when blood temperature decreases.

Nerve impulses from the preoptic area propagate to two other parts of the hypothalamus known as the **heat-losing center** and the **heat-promoting center,** which, when stimulated by the preoptic area, set into operation a series of responses that lower body temperature and raise body temperature, respectively.

Thermoregulation

If core temperature declines, mechanisms that help conserve heat and increase heat production act via several negative feedback loops to raise the body temperature to normal (Figure 25.19). Thermoreceptors in the skin and hypothalamus send nerve impulses to the preoptic area and the heat-promoting center in the hypothalamus, and to hypothalamic neurosecretory cells that produce thyrotropin-releasing hormone (TRH). In response, the hypothalamus discharges nerve impulses and secretes TRH, which in turn stimulates thyrotrophs in the anterior pituitary gland to

Figure 25.19 Negative feedback mechanisms that conserve heat and increase heat production.

Core temperature is the temperature in body structures deep to the skin and subcutaneous layer; shell temperature is the temperature near the body surface.

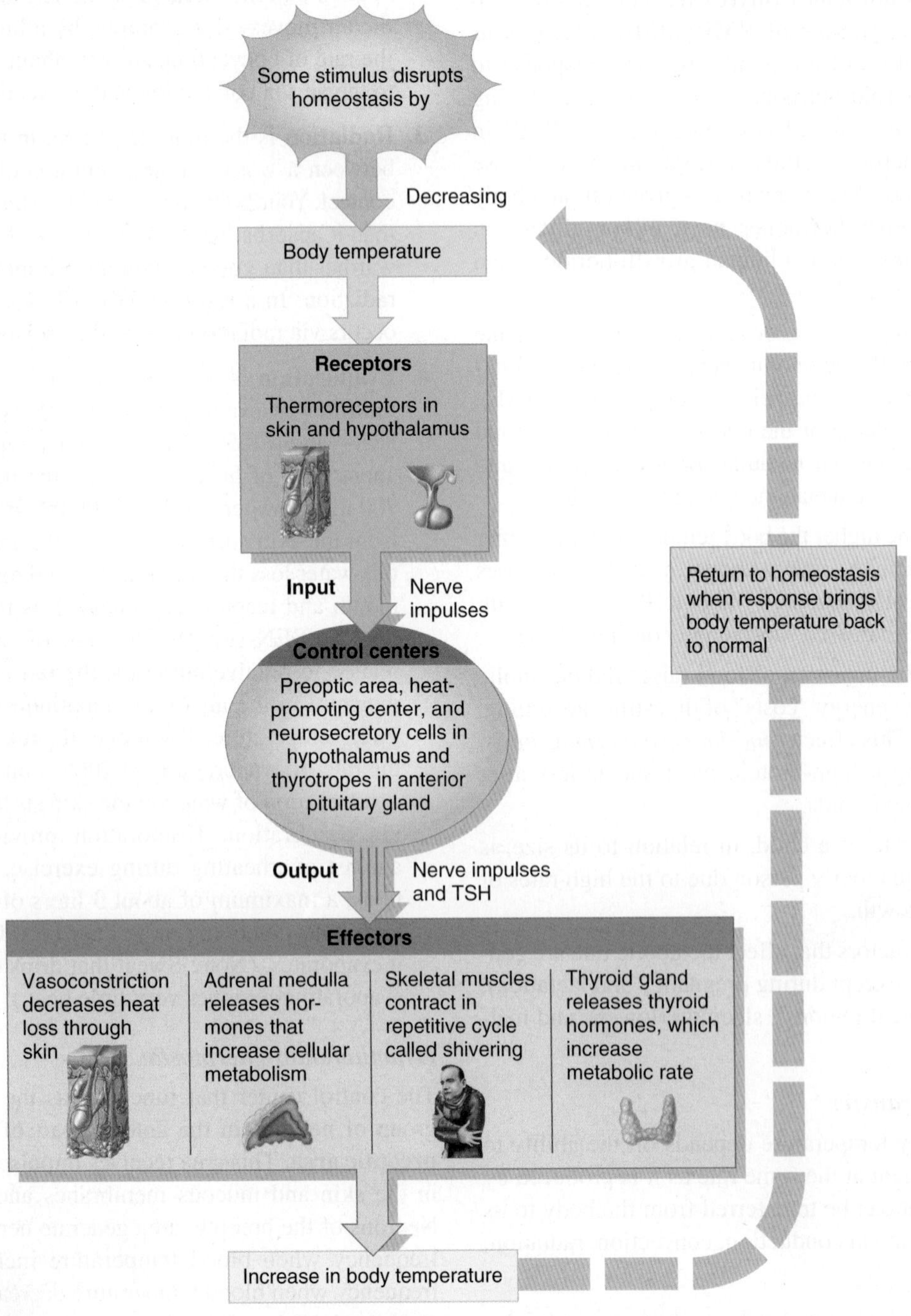

What factors can increase metabolic rate and thus increase the rate of heat production?

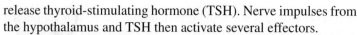

release thyroid-stimulating hormone (TSH). Nerve impulses from the hypothalamus and TSH then activate several effectors.

Each effector responds in a way that helps increase core temperature to the normal value:

- Nerve impulses from the heat-promoting center stimulate sympathetic nerves that cause blood vessels of the skin to constrict. Vasoconstriction decreases the flow of warm blood, and thus the transfer of heat, from the internal organs to the skin. Slowing the rate of heat loss allows the internal body temperature to increase as metabolic reactions continue to produce heat.

- Nerve impulses in sympathetic nerves leading to the adrenal medulla stimulate the release of epinephrine and norepinephrine into the blood. The hormones in turn bring about an increase in cellular metabolism, which increases heat production.

- The heat-promoting center stimulates parts of the brain that increase muscle tone and hence heat production. As muscle tone increases in one muscle (the agonist), the small contractions stretch muscle spindles in its antagonist, initiating a stretch reflex. The resulting contraction in the antagonist stretches muscle spindles in the agonist, and it too develops a stretch reflex. This repetitive cycle—called **shivering**—greatly increases the rate of heat production. During maximal shivering, body heat production can rise to about four times the basal rate in just a few minutes.

- The thyroid gland responds to TSH by releasing more thyroid hormones into the blood. As increased levels of thyroid hormones slowly increase the metabolic rate, body temperature rises.

If core body temperature rises above normal, a negative feedback loop opposite to the one depicted in Figure 25.19 goes into action. The higher temperature of the blood stimulates thermoreceptors that send nerve impulses to the preoptic area, which in turn stimulate the heat-losing center and inhibit the heat-promoting center. Nerve impulses from the heat-losing center cause dilation of blood vessels in the skin. The skin becomes warm, and the excess heat is lost to the environment via radiation and conduction as an increased volume of blood flows from the warmer core of the body into the cooler skin. At the same time, metabolic rate decreases, and shivering does not occur. The high temperature of the blood stimulates sweat glands of the skin via hypothalamic activation of sympathetic nerves. As the water in perspiration evaporates from the surface of the skin, the skin is cooled. All these responses counteract heat-promoting effects and help return body temperature to normal.

core body temperature falls: sensation of cold, shivering, confusion, vasoconstriction, muscle rigidity, bradycardia, acidosis, hypoventilation, hypotension, loss of spontaneous movement, coma, and death (usually caused by cardiac arrhythmias). Because the elderly have reduced metabolic protection against a cold environment coupled with a reduced perception of cold, they are at greater risk for developing hypothermia. •

Energy Homeostasis and Regulation of Food Intake

Most mature animals and many men and women maintain **energy homeostasis,** the precise matching of energy intake (in food) to energy expenditure over time. When the energy content of food balances the energy used by all the cells of the body, body weight remains constant (unless there is a gain or loss of water). In many people, weight stability persists despite large day-to-day variations in activity and food intake. In the more affluent nations, however, a large fraction of the population is overweight. Easy access to tasty, high-calorie foods and a "couch potato" lifestyle promote weight gain. Being overweight increases the risk of dying from a variety of cardiovascular and metabolic disorders, including hypertension, varicose veins, diabetes mellitus, arthritis, certain cancers, and gallbladder disease.

Energy intake depends only on the amount of food consumed (and absorbed), but three components contribute to total energy expenditure:

1. The basal metabolic rate accounts for about 60% of energy expenditure.

2. Physical activity typically adds 30–35% but can be lower in sedentary people. The energy expenditure is partly from voluntary exercise, such as walking, and partly from **nonexercise activity thermogenesis (NEAT)** (ther-mō-JEN-e-sis), the energy costs for maintaining muscle tone, posture while sitting or standing, and involuntary fidgeting movements.

3. **Food-induced thermogenesis,** the heat produced while food is being digested, absorbed, and stored, represents 5–10% of total energy expenditure.

The major site of stored chemical energy in the body is adipose tissue. When energy use exceeds energy input, triglycerides in adipose tissue are catabolized to provide the extra energy, and when energy input exceeds energy expenditure, triglycerides are stored. Over time, the amount of stored triglycerides indicates the excess of energy intake over energy expenditure. Even small differences add up over time. A gain of 20 lb (9 kg) between ages 25 and 55 represents only a tiny imbalance, about 0.3% more energy intake in food than energy expenditure.

Clearly, negative feedback mechanisms are regulating both our energy intake and our energy expenditure. But no sensory receptors exist to monitor our weight or size. How, then, is food intake regulated? The answer to this question is incomplete, but important advances in understanding regulation of food intake have occurred recently. It depends on many factors, including neural and endocrine signals, levels of certain nutrients in the blood, psychological elements such as stress or depression, signals from the GI tract and the special senses, and neural connections between the hypothalamus and other parts of the brain.

Within the hypothalamus are clusters of neurons that play key roles in regulating food intake. **Satiety** (sa-TĪ-i-tē) is a feeling of fullness accompanied by lack of desire to eat. Two hypothalamic areas involved in regulation of food intake are the *arcuate nucleus* and the *paraventricular nucleus* (see Figure 14.10). There is a mouse gene, named *obese*, that causes overeating and severe obesity in its mutated form. The product of this gene is the hormone **leptin.** In both mice and humans, leptin helps decrease **adiposity** (ad-i-POS-i-tē), total body-fat mass. Leptin is synthesized and secreted by adipocytes in proportion to adiposity; as more triglycerides are stored, more leptin is secreted into the bloodstream. Leptin acts on the hypothalamus to inhibit circuits that stimulate eating while also activating circuits that increase energy expenditure. The hormone insulin has a similar, but smaller, effect. Both leptin and insulin are able to pass through the blood–brain barrier.

When leptin and insulin levels are *low,* neurons that extend from the arcuate nucleus to the paraventricular nucleus release a neurotransmitter called **neuropeptide Y** that stimulates food intake. Other neurons that extend between the arcuate and paraventricular nuclei release a neurotransmitter called **melanocortin** (mel-an-ō-KOR-tin), which is similar to melanocyte-stimulating hormone (MSH). Leptin stimulates release of melanocortin, which acts to inhibit food intake. Although leptin, neuropeptide Y, and melanocortin are key signaling molecules for maintaining energy homeostasis, several other hormones and neurotransmitters also contribute. An understanding of the brain circuits involved is still far from complete. Other areas of the hypothalamus plus nuclei in the brain stem, limbic system, and cerebral cortex take part.

Achieving energy homeostasis requires regulation of energy intake. Most increases and decreases in food intake are due to changes in meal size rather than changes in number of meals. Many experiments have demonstrated the presence of satiety signals, chemical or neural changes that help terminate eating when "fullness" is attained. For example, an increase in blood glucose level, as occurs after a meal, decreases appetite. Several hormones, such as glucagon, cholecystokinin, estrogens, and epinephrine (acting via beta receptors), act to signal satiety and to increase energy expenditure. Distension of the GI tract, particularly the stomach and duodenum, also contributes to termination of food intake. Other hormones increase appetite and decrease energy expenditure. These include growth hormone–releasing hormone (GHRH), androgens, glucocorticoids, epinephrine (acting via alpha receptors), and progesterone.

CLINICAL CONNECTION | *Emotional Eating*

In addition to keeping us alive, eating serves countless psychological, social, and cultural purposes. We eat to celebrate, punish, comfort, defy, and deny. Eating in response to emotional drives, such as feeling stressed, bored, or tired, is called **emotional eating.**

Emotional eating is so common that, within limits, it is considered well within the range of normal behavior. Who hasn't at one time or another headed for the refrigerator after a bad day? Problems arise when emotional eating becomes so excessive that it interferes with health. Physical health problems include obesity and associated disorders such as hypertension and heart disease. Psychological health problems include poor self-esteem, an inability to cope effectively with feelings of stress, and in extreme cases, eating disorders such as anorexia nervosa, bulimia, and obesity.

Eating provides comfort and solace, numbing pain and "feeding the hungry heart." Eating may provide a biochemical "fix" as well. Emotional eaters typically overeat carbohydrate foods (sweets and starches), which may raise brain serotonin levels and lead to feelings of relaxation. Food becomes a way to self-medicate when negative emotions arise. •

✔ CHECKPOINT

25. Define a kilocalorie (kcal). How is the unit used? How does it relate to a calorie?
26. Distinguish between core temperature and shell temperature.
27. In what ways can a person lose heat to or gain heat from the surroundings? How is it possible for a person to lose heat on a sunny beach when the temperature is 40°C (104°F) and the humidity is 85%?
28. What does the term energy homeostasis mean?
29. How is food intake regulated?

25.9 NUTRITION

● OBJECTIVES

• Describe how to select foods to maintain a healthy diet.
• Compare the sources, functions, and importance of minerals and vitamins in metabolism.

Nutrients are chemical substances in food that body cells use for growth, maintenance, and repair. The six main types of nutrients are water, carbohydrates, lipids, proteins, minerals, and vitamins. Water is the nutrient needed in the largest amount—about 2–3 liters per day. As the most abundant compound in the body, water provides the medium in which most metabolic reactions occur, and it also participates in some reactions (for example, hydrolysis reactions). The important roles of water in the body can be reviewed in Section 2.4. Three organic nutrients—carbohydrates, lipids, and proteins—provide the energy needed for metabolic reactions and serve as building blocks to make body structures. Some minerals and many vitamins are components of the enzyme systems that catalyze metabolic reactions. *Essential nutrients* are specific nutrient molecules that the body cannot make in sufficient quantity to meet its needs and thus must be obtained from the diet. Some amino acids, fatty acids, vitamins, and minerals are essential nutrients.

Next, we discuss some guidelines for healthy eating and the roles of minerals and vitamins in metabolism.

Guidelines for Healthy Eating

Each gram of protein or carbohydrate in food provides about 4 Calories; 1 gram of fat (lipids) provides about 9 Calories. We do not know with certainty what levels and types of carbohydrate, fat, and protein are optimal in the diet. Different populations around the world eat radically different diets that are adapted to their particular lifestyles. However, many experts recommend the following distribution of calories: 50–60% from carbohydrates, with less than 15% from simple sugars; less than 30% from fats (triglycerides are the main type of dietary fat), with no more than 10% as saturated fats; and about 12–15% from proteins.

The guidelines for healthy eating are to:

- Eat a variety of foods.
- Maintain a healthy weight.
- Choose foods low in fat, saturated fat, and cholesterol.
- Eat plenty of vegetables, fruits, and grain products.
- Use sugars in moderation only.

In 2005, the United States Department of Agriculture (USDA) introduced a food pyramid called **MyPyramid,** which represents a *personalized* approach to making healthy food choices and maintaining regular physical activity. By consulting a chart, it is possible to determine your calorie level based on your gender, age, and activity level. Once this is determined, you can choose the type and amount of food to be consumed.

If you carefully examine the MyPyramid in Figure 25.20 you will note that the six color bands represent the five basic food groups plus oils. Foods from all bands are needed each day. Also note that the overall size of the bands suggests the proportion of food a person should choose on a daily basis. The wider base of each band represents foods with little or no solid fats or added sugars; these foods should be selected more often. The narrower top of each band represents foods with more added sugars and solid fats, which should be selected less frequently. The person climbing the steps is a reminder of the need for daily physical activity.

As an example of how MyPyramid works, let's assume based on consulting a chart that the calorie level of an 18-year-old moderately active female is 2000 Calories and that of an 18-year-old moderately active male is 2800 Calories. Accordingly, it is suggested that the following foods should be chosen in the following amounts:

Calorie Level	2000	2800
Fruits (includes all fresh, frozen, canned, and dried fruits and fruit juices)	2 cups	2.5 cups
Vegetables (includes all fresh, frozen, canned, and dried vegetables and vegetable juices)	2.5 cups	3.5 cups
Grains (includes all foods made from wheat, rice, oats, cornmeal, and barley such as bread, cereals, oatmeal, rice, pasta, crackers, tortillas, and grits)	6 oz	10 oz
Meats and beans (includes lean meat, poultry, fish, eggs, peanut butter, beans, nuts, and seeds)	5.5 oz	7 oz
Milk group (includes milk products and foods made from milk that retain their calcium content such as cheeses and yogurt)	3 cups	3 cups
Oils (choose mostly fats that contain monounsaturated and polyunsaturated fatty acids such as fish, nuts, seeds, and vegetable oils)	6 tsp	8 tsp

Figure 25.20 MyPyramid.

MyPyramid is a personalized approach to making healthy food choices and maintaining regular physical activity.

GRAINS VEGETABLES FRUITS OILS MILK MEAT & BEANS

What does the wider base of each band mean?

In addition, you should choose and prepare foods with little salt. In fact, sodium intake should be less than 2300 mg per day. If you choose to drink alcohol, it should be consumed in moderation (no more than 1 drink per day for women and 2 drinks per day for men). A drink is defined as 12 oz of regular beer, 5 oz of wine, or $1^1/_2$ oz of 80 proof distilled spirits.

Minerals

Minerals are inorganic elements that occur naturally in Earth's crust. In the body they appear in combination with one another, in combination with organic compounds, or as ions in solution. Minerals constitute about 4% of total body mass and are concentrated most heavily in the skeleton. Minerals with known functions in the body include calcium, phosphorus, potassium, sulfur, sodium, chloride, magnesium, iron, iodide, manganese, copper, cobalt, zinc, fluoride, selenium, and chromium. Table 25.5 describes the vital functions of these minerals. Note that the body generally uses the ions of the minerals rather than the non-ionized form. Some minerals, such as chlorine, are toxic or even fatal if ingested in the non-ionized form. Other minerals—aluminum, boron, silicon, and molybdenum—are present but their functions are unclear. Typical diets supply adequate amounts of potassium, sodium, chloride, and magnesium. Some attention must be paid to eating foods that provide enough calcium, phosphorus, iron, and iodide. Excess amounts of most minerals are excreted in the urine and feces.

Calcium and phosphorus form part of the matrix of bone. Because minerals do not form long-chain compounds, they are otherwise poor building materials. A major role of minerals is to help regulate enzymatic reactions. Calcium, iron, magnesium, and manganese are constituents of some coenzymes. Magnesium also serves as a catalyst for the conversion of ADP to ATP. Minerals such as sodium and phosphorus work in buffer systems, which help control the pH of body fluids. Sodium also helps regulate the osmosis of water and, along with other ions, is involved in the generation of nerve impulses.

Vitamins

Organic nutrients required in small amounts to maintain growth and normal metabolism are called **vitamins.** Unlike carbohydrates, lipids, or proteins, vitamins do not provide energy or serve as the body's building materials. Most vitamins with known functions are coenzymes.

Most vitamins cannot be synthesized by the body and must be ingested in food. Other vitamins, such as vitamin K, are produced by bacteria in the GI tract and then absorbed. The body can assemble some vitamins if the raw materials, called **provitamins,** are provided. For example, vitamin A is produced by the body from the provitamin beta-carotene, a chemical present in yellow vegetables such as carrots and in dark green vegetables such as spinach. No single food contains all the required vitamins—one of the best reasons to eat a varied diet.

Vitamins are divided into two main groups: fat-soluble and water-soluble. The **fat-soluble vitamins,** vitamins A, D, E, and K, are absorbed along with other dietary lipids in the small intestine and packaged into chylomicrons. They cannot be absorbed in adequate quantity unless they are ingested with other lipids. Fat-soluble vitamins may be stored in cells, particularly hepatocytes. The **water-soluble vitamins,** including several B vitamins and vitamin C, are dissolved in body fluids. Excess quantities of these vitamins are not stored but instead are excreted in the urine.

Besides their other functions, three vitamins—C, E, and beta-carotene (a provitamin)—are termed **antioxidant vitamins** because they inactivate oxygen free radicals. Recall that free radicals are highly reactive ions or molecules that carry an unpaired electron in their outermost electron shell (see Figure 2.3). Free radicals damage cell membranes, DNA, and other cellular structures and contribute to the formation of artery-narrowing atherosclerotic plaques. Some free radicals arise naturally in the body, and others come from environmental hazards such as tobacco smoke and radiation. Antioxidant vitamins are thought to play a role in protecting against some kinds of cancer, reducing the buildup of atherosclerotic plaque, delaying some effects of aging, and decreasing the chance of cataract formation in the lens of the eyes. Table 25.6 lists the major vitamins, their sources, their functions, and related deficiency disorders.

CLINICAL CONNECTION | *Vitamin and Mineral Supplements*

Most nutritionists recommend eating a balanced diet that includes a variety of foods rather than taking vitamin or mineral supplements, except in special circumstances. Common examples of necessary supplementations include iron for women who have excessive menstrual bleeding; iron and calcium for women who are pregnant or breast-feeding; folic acid (folate) for all women who may become pregnant, to reduce the risk of fetal neural tube defects; calcium for most adults, because they do not receive the recommended amount in their diets; and vitamin B$_{12}$ for strict vegetarians, who eat no meat. Because most North Americans do not ingest in their food the high levels of antioxidant vitamins thought to have beneficial effects, some experts recommend supplementing vitamins C and E. More is not always better, however; larger doses of vitamins or minerals can be very harmful.

Hypervitaminosis (hī-per-vī-ta-mi-NŌ-sis; *hyper-* = too much or above) refers to dietary intake of a vitamin that exceeds the ability of the body to utilize, store, or excrete the vitamin. Since water-soluble vitamins are not stored in the body, few of them cause any problems related to hypervitaminosis. However, because lipid-soluble vitamins are stored in the body, excessive consumption may cause problems. For example, excess intake of vitamin A can cause drowsiness, general weakness, irritability, headache, vomiting, dry and peeling skin, partial hair loss, joint pain, liver and spleen enlargement, coma, and even death. Excessive intake of vitamin D may result in loss of appetite, nausea, vomiting, excessive thirst, general weakness, irritability, hypertension, and kidney damage and malfunction. **Hypovitaminosis** (*hypo-* = too little or below), or vitamin deficiency, is discussed in Table 25.6 for the various vitamins. •

TABLE 25.5

Minerals Vital to the Body

MINERAL	COMMENTS	IMPORTANCE
Calcium	Most abundant mineral in body. Appears in combination with phosphates. About 99% stored in bone and teeth. Blood Ca^{2+} level controlled by parathyroid hormone (PTH). Calcitriol promotes absorption of dietary calcium. Excess excreted in feces and urine. Sources: milk, egg yolk, shellfish, leafy green vegetables.	Formation of bones and teeth, blood clotting, normal muscle and nerve activity, endocytosis and exocytosis, cellular motility, chromosome movement during cell division, glycogen metabolism, release of neurotransmitters and hormones.
Phosphorus	About 80% found in bones and teeth as phosphate salts. Blood phosphate level controlled by parathyroid hormone (PTH). Excess excreted in urine; small amount eliminated in feces. Sources: dairy products, meat, fish, poultry, nuts.	Formation of bones and teeth. Phosphates ($H_2PO_4^-$, HPO_4^{2-}, and PO_4^{3-}) constitute a major buffer system of blood. Role in muscle contraction and nerve activity. Component of many enzymes. Involved in energy transfer (ATP). Component of DNA and RNA.
Potassium	Major cation (K^+) in intracellular fluid. Excess excreted in urine. Present in most foods (meats, fish, poultry, fruits, nuts).	Needed for generation and conduction of action potentials in neurons and muscle fibers.
Sulfur	Component of many proteins (such as insulin and chondroitin sulfate), electron carriers in electron transport chain, and some vitamins (thiamine and biotin). Excreted in urine. Sources: beef, liver, lamb, fish, poultry, eggs, cheese, beans.	As component of hormones and vitamins, regulates various body activities. Needed for ATP production by electron transport chain.
Sodium	Most abundant cation (Na^+) in extracellular fluids; some found in bones. Excreted in urine and perspiration. Normal intake of NaCl (table salt) supplies more than required amounts.	Strongly affects distribution of water through osmosis. Part of bicarbonate buffer system. Functions in nerve and muscle action potential conduction.
Chloride	Major anion (Cl^-) in extracellular fluid. Excess excreted in urine. Sources: table salt (NaCl), soy sauce, processed foods.	Role in acid–base balance of blood, water balance, and formation of HCl in stomach.
Magnesium	Important cation (Mg^{2+}) in intracellular fluid. Excreted in urine and feces. Widespread in various foods, such as green leafy vegetables, seafood, and whole-grain cereals.	Required for normal functioning of muscle and nervous tissue. Participates in bone formation. Constituent of many coenzymes.
Iron	About 66% found in hemoglobin of blood. Normal losses of iron occur by shedding of hair, epithelial cells, and mucosal cells, and in sweat, urine, feces, bile, and blood lost during menstruation. Sources: meat, liver, shellfish, egg yolk, beans, legumes, dried fruits, nuts, cereals.	As component of hemoglobin, reversibly binds O_2. Component of cytochromes involved in electron transport chain.
Iodide	Essential component of thyroid hormones. Excreted in urine. Sources: seafood, iodized salt, vegetables grown in iodine-rich soils.	Required by thyroid gland to synthesize thyroid hormones, which regulate metabolic rate.
Manganese	Some stored in liver and spleen. Most excreted in feces.	Activates several enzymes. Needed for hemoglobin synthesis, urea formation, growth, reproduction, lactation, bone formation, and possibly production and release of insulin, and inhibition of cell damage.
Copper	Some stored in liver and spleen. Most excreted in feces. Sources: eggs, whole-wheat flour, beans, beets, liver, fish, spinach, asparagus.	Required with iron for synthesis of hemoglobin. Component of coenzymes in electron transport chain and enzyme necessary for melanin formation.
Cobalt	Constituent of vitamin B_{12}.	As part of vitamin B_{12}, required for erythropoiesis.
Zinc	Important component of certain enzymes. Widespread in many foods, especially meats.	As component of carbonic anhydrase, important in carbon dioxide metabolism. Necessary for normal growth and wound healing, normal taste sensations and appetite, and normal sperm counts in males. As component of peptidases, involved in protein digestion.
Fluoride	Components of bones, teeth, other tissues.	Appears to improve tooth structure and inhibit tooth decay.
Selenium	Important component of certain enzymes. Sources: seafood, meat, chicken, tomatoes, egg yolk, milk, mushrooms, garlic, cereal grains grown in selenium-rich soil.	Needed for synthesis of thyroid hormones, sperm motility, and proper functioning of immune system. Also functions as antioxidant. Prevents chromosome breakage and may play role in preventing certain birth defects, miscarriage, prostate cancer, and coronary artery disease.
Chromium	Found in high concentrations in brewer's yeast. Also found in wine and some brands of beer.	Needed for normal activity of insulin in carbohydrate and lipid metabolism.

TABLE 25.6

The Principal Vitamins

VITAMIN	COMMENT AND SOURCE	FUNCTIONS	DEFICIENCY SYMPTOMS AND DISORDERS
Fat-soluble	All require bile salts and some dietary lipids for adequate absorption.		
A	Formed from provitamin beta-carotene (and other provitamins) in GI tract. Stored in liver. Sources of carotene and other provitamins: orange, yellow, and green vegetables. Sources of vitamin A: liver, milk.	Maintains general health and vigor of epithelial cells. Beta-carotene acts as antioxidant to inactivate free radicals. Essential for formation of light-sensitive pigments in photoreceptors of retina. Aids in growth of bones and teeth by helping to regulate activity of osteoblasts and osteoclasts.	Deficiency results in atrophy and keratinization of epithelium, leading to dry skin and hair; increased incidence of ear, sinus, respiratory, urinary, and digestive system infections; inability to gain weight; drying of cornea; and skin sores. **Night blindness** (decreased ability for dark adaptation). Slow and faulty development of bones and teeth.
D	Sunlight converts 7-dehydrocholesterol in skin to cholecalciferol (vitamin D_3). A liver enzyme then converts cholecalciferol to 25-hydroxycholecalciferol. A second enzyme in kidneys converts 25-hydroxycholecalciferol to calcitriol (1,25-dihydroxycalciferol), the active form of vitamin D. Most excreted in bile. Dietary sources: fish-liver oils, egg yolk, fortified milk.	Essential for absorption of calcium and phosphorus from GI tract. Works with parathyroid hormone (PTH) to maintain Ca^{2+} homeostasis.	Defective utilization of calcium by bones leads to **rickets** in children and **osteomalacia** in adults. Possible loss of muscle tone.
E (tocopherols)	Stored in liver, adipose tissue, and muscles. Sources: fresh nuts and wheat germ, seed oils, green leafy vegetables.	Inhibits catabolism of certain fatty acids that help form cell structures, especially membranes. Involved in formation of DNA, RNA, and red blood cells. May promote wound healing, contribute to normal structure and functioning of nervous system, and prevent scarring. May help protect liver from toxic chemicals such as carbon tetrachloride. Acts as antioxidant to inactivate free radicals.	May cause oxidation of monounsaturated fats, resulting in abnormal structure and function of mitochondria, lysosomes, and plasma membranes. Possible consequence is hemolytic anemia.
K	Produced by intestinal bacteria. Stored in liver and spleen. Dietary sources: spinach, cauliflower, cabbage, liver.	Coenzyme essential for synthesis of several clotting factors by liver, including prothrombin.	Delayed clotting time results in excessive bleeding.
Water-soluble	Dissolved in body fluids. Most not stored in body. Excess intake eliminated in urine.		
B_1 (thiamine)	Rapidly destroyed by heat. Sources: whole-grain products, eggs, pork, nuts, liver, yeast.	Acts as coenzyme for many different enzymes that break carbon-to-carbon bonds and are involved in carbohydrate metabolism of pyruvic acid to CO_2 and H_2O. Essential for synthesis of neurotransmitter acetylcholine.	Improper carbohydrate metabolism leads to buildup of pyruvic and lactic acids and insufficient production of ATP for muscle and nerve cells. Deficiency leads to (1) **beriberi,** partial paralysis of smooth muscle of GI tract, causing digestive disturbances; skeletal muscle paralysis; and atrophy of limbs; (2) **polyneuritis,** due to degeneration of myelin sheaths; impaired reflexes, impaired sense of touch, stunted growth in children, and poor appetite.
B_2 (riboflavin)	Small amounts supplied by bacteria of GI tract. Dietary sources: yeast, liver, beef, veal, lamb, eggs, whole-grain products, asparagus, peas, beets, peanuts.	Component of certain coenzymes (for example, FAD and FMN) in carbohydrate and protein metabolism, especially in cells of eye, integument, mucosa of intestine, and blood.	Deficiency may lead to improper utilization of oxygen, resulting in blurred vision, cataracts, and corneal ulcerations. Also dermatitis and cracking of skin, lesions of intestinal mucosa, and one type of anemia.

TABLE 25.6 CONTINUED

The Principal Vitamins

VITAMIN	COMMENT AND SOURCE	FUNCTIONS	DEFICIENCY SYMPTOMS AND DISORDERS
Water-soluble (continued)			
Niacin (nicotinamide)	Derived from amino acid tryptophan. Sources: yeast, meats, liver, fish, whole-grain products, peas, beans, nuts.	Essential component of NAD and NADP, coenzymes in oxidation–reduction reactions. In lipid metabolism, inhibits production of cholesterol and assists in triglyceride breakdown.	Principal deficiency is **pellagra,** characterized by dermatitis, diarrhea, and psychological disturbances.
B$_6$ (pyridoxine)	Synthesized by bacteria of GI tract. Stored in liver, muscle, and brain. Other sources: salmon, yeast, tomatoes, yellow corn, spinach, whole grain products, liver, yogurt.	Essential coenzyme for normal amino acid metabolism. Assists production of circulating antibodies. May function as coenzyme in triglyceride metabolism.	Most common deficiency symptom is dermatitis of eyes, nose, and mouth. Other symptoms are retarded growth and nausea.
B$_{12}$ (cyanocobalamin)	Only B vitamin not found in vegetables; only vitamin containing cobalt. Absorption from GI tract depends on intrinsic factor secreted by gastric mucosa. Sources: liver, kidney, milk, eggs, cheese, meat.	Coenzyme necessary for red blood cell formation, formation of amino acid methionine, entrance of some amino acids into Krebs cycle, and manufacture of choline (used to synthesize acetylcholine).	Pernicious anemia, neuropsychiatric abnormalities (ataxia, memory loss, weakness, personality and mood changes, and abnormal sensations), and impaired activity of osteoblasts.
Pantothenic acid	Some produced by bacteria of GI tract. Stored primarily in liver and kidneys. Other sources: kidneys, liver, yeast, green vegetables, cereal.	Constituent of coenzyme A, which is essential for transfer of acetyl group from pyruvic acid into Krebs cycle, conversion of lipids and amino acids into glucose, and synthesis of cholesterol and steroid hormones.	Fatigue, muscle spasms, insufficient production of adrenal steroid hormones, vomiting, and insomnia.
Folic acid (folate, folacin)	Synthesized by bacteria of GI tract. Dietary sources: green leafy vegetables, broccoli, asparagus, breads, dried beans, citrus fruits.	Component of enzyme systems synthesizing nitrogenous bases of DNA and RNA. Essential for normal production of red and white blood cells.	Production of abnormally large red blood cells (macrocytic anemia). Higher risk of neural tube defects in babies born to folate-deficient mothers.
Biotin	Synthesized by bacteria of GI tract. Dietary sources include yeast, liver, egg yolk, kidneys.	Essential coenzyme for conversion of pyruvic acid to oxaloacetic acid and synthesis of fatty acids and purines.	Mental depression, muscular pain, dermatitis, fatigue, and nausea.
C (ascorbic acid)	Rapidly destroyed by heat. Some stored in glandular tissue and plasma. Sources: citrus fruits, tomatoes, green vegetables.	Promotes protein synthesis, including laying down of collagen in formation of connective tissue. As coenzyme, may combine with poisons, rendering them harmless until excreted. Works with antibodies, promotes wound healing, and functions as an antioxidant.	Scurvy; anemia; many symptoms related to poor collagen formation, including tender swollen gums, loosening of teeth (alveolar processes also deteriorate), poor wound healing, bleeding (vessel walls are fragile because of connective tissue degeneration), and retardation of growth.

✔ CHECKPOINT

30. What is a nutrient?
31. Describe the food guide pyramid and give examples of foods from each food group.
32. What is a mineral? Briefly describe the functions of the following minerals: calcium, phosphorus, potassium, sulfur, sodium, chloride, magnesium, iron, iodine, copper, zinc, fluoride, manganese, cobalt, chromium, and selenium.
33. Define vitamin. Explain how we obtain vitamins. Distinguish between a fat-soluble vitamin and a water-soluble vitamin.
34. For each of the following vitamins, indicate its principal function and the effect(s) of deficiency: A, D, E, K, B$_1$, B$_2$, niacin, B$_6$, B$_{12}$, pantothenic acid, folic acid, biotin, and C.

DISORDERS: HOMEOSTATIC IMBALANCES

Anorexia Nervosa

Anorexia nervosa is a chronic disorder characterized by self-induced weight loss, negative perception of body image, and physiological changes that result from nutritional depletion. Patients with anorexia nervosa have a fixation on weight control and often insist on having a bowel movement every day despite inadequate food intake. They often abuse laxatives, which worsens the fluid and electrolyte imbalances and nutrient deficiencies. The disorder is found predominantly in young, single females, and it may be inherited. Abnormal patterns of menstruation, amenorrhea (absence of menstruation), and a lowered basal metabolic rate reflect the depressant effects of starvation. Individuals may become emaciated and may ultimately die of starvation or one of its complications. Also associated with the disorder are osteoporosis, depression, and brain abnormalities coupled with impaired mental performance. Treatment consists of psychotherapy and dietary regulation.

Fever

A **fever** is an elevation of core temperature caused by a resetting of the hypothalamic thermostat. The most common causes of fever are viral or bacterial infections and bacterial toxins; other causes are ovulation, excessive secretion of thyroid hormones, tumors, and reactions to vaccines. When phagocytes ingest certain bacteria, they are stimulated to secrete a **pyrogen** (PĪ-rō-gen; *pyro-* = fire; *-gen* = produce), a fever-producing substance. One pyrogen is interleukin-1. It circulates to the hypothalamus and induces neurons of the preoptic area to secrete prostaglandins. Some prostaglandins can reset the hypothalamic thermostat at a higher temperature, and temperature-regulating reflex mechanisms then act to bring the core body temperature up to this new setting. *Antipyretics* are agents that relieve or reduce fever. Examples include aspirin, acetaminophen (Tylenol), and ibuprofen (Advil), all of which reduce fever by inhibiting synthesis of certain prostaglandins.

Suppose that due to production of pyrogens the thermostat is reset at 39°C (103°F). Now the heat-promoting mechanisms (vasoconstriction, increased metabolism, shivering) are operating at full force. Thus, even though core temperature is climbing higher than normal—say, 38°C (101°F)—the skin remains cold, and shivering occurs. This condition, called a **chill,** is a definite sign that core temperature is rising. After several hours, core temperature reaches the setting of the thermostat, and the chills disappear. But now the body will continue to regulate temperature at 39°C (103°F). When the pyrogens disappear, the thermostat is reset at normal—37.0°C (98.6°F). Because core temperature is high in the beginning, the heat-losing mechanisms (vasodilation and sweating) go into operation to decrease core temperature. The skin becomes warm, and the person begins to sweat. This phase of the fever is called the **crisis,** and it indicates that core temperature is falling.

Although death results if core temperature rises above 44–46°C (112–114°F), up to a point, fever is beneficial. For example, a higher temperature intensifies the effects of interferons and the phagocytic activities of macrophages while hindering replication of some pathogens. Because fever increases heart rate, infection-fighting white blood cells are delivered to sites of infection more rapidly. In addition, antibody production and T cell proliferation increase. Moreover, heat speeds up the rate of chemical reactions, which may help body cells repair themselves more quickly.

Obesity

Obesity is body weight more than 20% above a desirable standard due to an excessive accumulation of adipose tissue. About one-third of the adult population in the United States is obese. (An athlete may be *overweight* due to higher-than-normal amounts of muscle tissue without being obese.) Even moderate obesity is hazardous to health; it is a risk factor in cardiovascular disease, hypertension, pulmonary disease, non-insulin-dependent diabetes mellitus, arthritis, certain cancers (breast, uterus, and colon), varicose veins, and gallbladder disease.

In a few cases, obesity may result from trauma of or tumors in the food-regulating centers in the hypothalamus. In most cases of obesity, no specific cause can be identified. Contributing factors include genetic factors, eating habits taught early in life, overeating to relieve tension, and social customs. Studies indicate that some obese people burn fewer calories during digestion and absorption of a meal, a smaller food-induced thermogenesis effect. Additionally, obese people who lose weight require about 15% fewer calories to maintain normal body weight than do people who have never been obese. Interestingly, people who gain weight easily when deliberately fed excess calories exhibit less NEAT (nonexercise activity thermogenesis, such as occurs with fidgeting) than people who resist weight gains in the face of excess calories. Although leptin suppresses appetite and produces satiety in experimental animals, it is not deficient in most obese people.

Most surplus calories in the diet are converted to triglycerides and stored in adipose cells. Initially, the adipocytes increase in size, but at a maximal size, they divide. As a result, proliferation of adipocytes occurs in extreme obesity. The enzyme endothelial lipoprotein lipase regulates triglyceride storage. The enzyme is very active in abdominal fat but less active in hip fat. Accumulation of fat in the abdomen is associated with higher blood cholesterol level and other cardiac risk factors because adipose cells in this area appear to be more metabolically active.

Treatment of obesity is difficult because most people who are successful at losing weight gain it back within 2 years. Yet, even modest weight loss is associated with health benefits. Treatments for obesity include behavior modification programs, very-low-calorie diets, drugs, and surgery. Behavior modification programs, offered at many hospitals, strive to alter eating behaviors and increase exercise activity. The nutrition program includes a "heart-healthy" diet that includes abundant vegetables but is low in fats, especially saturated fats. A typical exercise program suggests walking for 30 minutes a day, five to seven times a week. Regular exercise enhances both weight loss and weight-loss maintenance. Very-low-calorie (VLC) diets include 400 to 800 kcal/day in a commercially made liquid mixture. The VLC diet is usually prescribed for 12 weeks, under close medical supervision. Two drugs are available to treat obesity. Sibutramine is an appetite suppressant that works by inhibiting reuptake of serotonin and norepinephrine in brain areas that govern eating behavior. Orlistat works by inhibiting the lipases released into the lumen of the GI tract. With less lipase activity, fewer dietary triglycerides are absorbed. For those with extreme obesity who have not responded to other treatments, a surgical procedure may be considered. The two operations most commonly performed—gastric bypass and gastroplasty—both greatly reduce the stomach size so that it can hold just a tiny quantity of food.

MEDICAL TERMINOLOGY

Bulimia (*bu-* = ox; *-limia* = hunger) or *binge–purge syndrome* A disorder that typically affects young, single, middle-class, white females, characterized by overeating at least twice a week followed by purging by self-induced vomiting, strict dieting or fasting, vigorous exercise, or use of laxatives or diuretics; it occurs in response to fears of being overweight or to stress, depression, and physiological disorders such as hypothalamic tumors.

Heat cramps Cramps that result from profuse sweating. The salt lost in sweat causes painful contractions of muscles; such cramps tend to occur in muscles used while working but do not appear until the person relaxes once the work is done. Drinking salted liquids usually leads to rapid improvement.

Heat exhaustion (heat prostration) A condition in which the core temperature is generally normal, or a little below, and the skin is cool and moist due to profuse perspiration. Heat exhaustion is usually characterized by loss of fluid and electrolytes, especially salt (NaCl). The salt loss results in muscle cramps, dizziness, vomiting, and fainting; fluid loss may cause low blood pressure. Complete rest, rehydration, and electrolyte replacement are recommended.

Heatstroke (sunstroke) A severe and often fatal disorder caused by exposure to high temperatures, especially when the relative humidity is high, which makes it difficult for the body to lose heat. Blood flow to the skin is decreased, perspiration is greatly reduced, and body temperature rises sharply because of failure of the hypothalamic thermostat. Body temperature may reach 43°C (110°F). Treatment, which must be undertaken immediately, consists of cooling the body by immersing the victim in cool water and by administering fluids and electrolytes.

Kwashiorkor (kwash-ē-OR-kor) A disorder in which protein intake is deficient despite normal or nearly normal caloric intake, characterized by edema of the abdomen, enlarged liver, decreased blood pressure, low pulse rate, lower-than-normal body temperature, and sometimes mental retardation. Because the main protein in corn (zein) lacks two essential amino acids, which are needed for growth and tissue repair, many African children whose diet consists largely of cornmeal develop kwashiorkor.

Malnutrition (*mal-* = bad) An imbalance of total caloric intake or intake of specific nutrients, which can be either inadequate or excessive.

Marasmus (mar-AZ-mus) A type of protein–calorie undernutrition that results from inadequate intake of both protein and calories. Its characteristics include retarded growth, low weight, muscle wasting, emaciation, dry skin, and thin, dry, dull hair.

CHAPTER REVIEW AND RESOURCE SUMMARY

Review	Resource
Introduction	Anatomy Overview - Role of Nutrients
1. Our only source of energy for performing biological work is the food we eat. Food also provides essential substances that we cannot synthesize.	
2. Most food molecules absorbed by the gastrointestinal tract are used to supply energy for life processes, serve as building blocks during synthesis of complex molecules, or are stored for future use.	
25.1 Metabolic Reactions	Animation - Introduction to Metabolism
1. Metabolism refers to all chemical reactions of the body and is of two types: catabolism and anabolism.	Animation - Regulation of Metabolism
2. Catabolism is the term for reactions that break down complex organic compounds into simple ones. Overall, catabolic reactions are exergonic; they produce more energy than they consume.	Animation - Carbohydrate Metabolism
3. Chemical reactions that combine simple molecules into more complex ones that form the body's structural and functional components are collectively known as anabolism. Overall, anabolic reactions are endergonic; they consume more energy than they produce.	Animation - Types of Reactions and Equilibrium
4. The coupling of anabolism and catabolism occurs via ATP.	
25.2 Energy Transfer	Animation - Types of Reactions and Equilibrium
1. Oxidation is the removal of electrons from a substance; reduction is the addition of electrons to a substance.	Animation - Enzyme Functions and ATP
2. Two coenzymes that carry hydrogen atoms during coupled oxidation–reduction reactions are nicotinamide adenine dinucleotide (NAD) and flavin adenine dinucleotide (FAD).	Exercise - Predict ATP Production
3. ATP can be generated via substrate-level phosphorylation, oxidative phosphorylation, and photophosphorylation.	
25.3 Carbohydrate Metabolism	Anatomy Overview - Role of Nutrients
1. During digestion, polysaccharides and disaccharides are hydrolyzed into the monosaccharides glucose (about 80%), fructose, and galactose; the latter two are then converted to glucose. Some glucose is oxidized by cells to provide ATP. Glucose also can be used to synthesize amino acids, glycogen, and triglycerides.	Animation - Cell Respiration Animation - Carbohydrate Metabolism Figure 25.2 - Overview of Cellular Respiration

| **Review** | **Resource** |

2. Glucose moves into most body cells via facilitated diffusion through glucose transporters (GluT) and becomes phosphorylated to glucose 6-phosphate. In muscle cells, this process is stimulated by insulin. Glucose entry into neurons and hepatocytes is always "turned on."

3. Cellular respiration, the complete oxidation of glucose to CO_2 and H_2O, involves glycolysis, the Krebs cycle, and the electron transport chain.

4. Glycolysis is the breakdown of glucose into two molecules of pyruvic acid; there is a net production of two molecules of ATP.

5. When oxygen is in short supply, pyruvic acid is reduced to lactic acid; under aerobic conditions, pyruvic acid enters the Krebs cycle. Pyruvic acid is prepared for entrance into the Krebs cycle by conversion to a 2-carbon acetyl group followed by the addition of coenzyme A to form acetyl coenzyme A. The Krebs cycle involves decarboxylations, oxidations, and reductions of various organic acids. Each molecule of pyruvic acid that is converted to acetyl coenzyme A and then enters the Krebs cycle produces three molecules of CO_2, four molecules of NADH and four H^+, one molecule of $FADH_2$, and one molecule of ATP. The energy originally stored in glucose and then in pyruvic acid is transferred primarily to the reduced coenzymes NADH and $FADH_2$.

6. The electron transport chain involves a series of oxidation–reduction reactions in which the energy in NADH and $FADH_2$ is liberated and transferred to ATP. The electron carriers include FMN, cytochromes, iron–sulfur centers, copper atoms, and coenzyme Q. The electron transport chain yields a maximum of 32 or 34 molecules of ATP and six molecules of H_2O.

7. Table 25.1 summarizes the ATP yield during cellular respiration. The complete oxidation of glucose can be represented as follows:

$$C_6H_{12}O_6 + 6\ O_2 + 36\ \text{or}\ 38\ \text{ADPs} + 36\ \text{or}\ 38\ \circled{P} \longrightarrow 6\ CO_2 + 6\ H_2O + 36\ \text{or}\ 38\ \text{ATPs}$$

8. The conversion of glucose to glycogen for storage in the liver and skeletal muscle is called glycogenesis. It is stimulated by insulin.

9. The conversion of glycogen to glucose is called glycogenolysis. It occurs between meals and is stimulated by glucagon and epinephrine.

10. Gluconeogenesis is the conversion of noncarbohydrate molecules into glucose. It is stimulated by cortisol and glucagon.

Resource column:
Figure 25.3 - Cellular Respiration Begins with Glycolysis
Figure 25.10 - Principal Reactions of Cellular Respiration
Exercise - Glucose Catabolism Sequence
Exercise - Glucose Catabolism Substrates and Products
Concepts and Connections - Glucose Catabolism

25.4 Lipid Metabolism

1. Lipoproteins transport lipids in the bloodstream. Types of lipoproteins include chylomicrons, which carry dietary lipids to adipose tissue; very-low-density lipoproteins (VLDLs), which carry triglycerides from the liver to adipose tissue; low-density lipoproteins (LDLs), which deliver cholesterol to body cells; and high-density lipoproteins (HDLs), which remove excess cholesterol from body cells and transport it to the liver for elimination.

2. Cholesterol in the blood comes from two sources: from food and from synthesis by the liver.

3. Lipids may be oxidized to produce ATP or stored as triglycerides in adipose tissue, mostly in the subcutaneous layer.

4. A few lipids are used as structural molecules or to synthesize essential molecules.

5. Adipose tissue contains lipases that catalyze the deposition of triglycerides from chylomicrons and hydrolyze triglycerides into fatty acids and glycerol.

6. In lipolysis, triglycerides are split into fatty acids and glycerol and released from adipose tissue under the influence of epinephrine, norepinephrine, cortisol, thyroid hormones, and insulinlike growth factors.

7. Glycerol can be converted into glucose by conversion into glyceraldehyde 3-phosphate.

8. In beta oxidation of fatty acids, carbon atoms are removed in pairs from fatty acid chains; the resulting molecules of acetyl coenzyme A enter the Krebs cycle.

9. The conversion of glucose or amino acids into lipids is called lipogenesis; it is stimulated by insulin.

Resource column:
Anatomy Overview - Role of Nutrients
Animation - Lipid Metabolism

25.5 Protein Metabolism

1. During digestion, proteins are hydrolyzed into amino acids, which enter the liver via the hepatic portal vein.

2. Amino acids, under the influence of insulinlike growth factors and insulin, enter body cells via active transport.

3. Inside cells, amino acids are synthesized into proteins that function as enzymes, hormones, structural elements, and so forth; stored as fat or glycogen; or used for energy.

Resource column:
Anatomy Overview - Role of Nutrients
Animation - Protein Metabolism

Review	Resource

4. Before amino acids can be catabolized, they must be deaminated and converted to substances that can enter the Krebs cycle.

5. Amino acids may also be converted into glucose, fatty acids, and ketone bodies.

6. Protein synthesis is stimulated by insulinlike growth factors, thyroid hormones, insulin, estrogen, and testosterone.

7. Table 25.2 summarizes carbohydrate, lipid, and protein metabolism.

25.6 Key Molecules at Metabolic Crossroads

1. Three molecules play a key role in metabolism: glucose 6-phosphate, pyruvic acid, and acetyl coenzyme A.

2. Glucose 6-phosphate may be converted to glucose, glycogen, ribose 5-phosphate, and pyruvic acid.

3. When ATP is low and oxygen is plentiful, pyruvic acid is converted to acetyl coenzyme A; when oxygen supply is low, pyruvic acid is converted to lactic acid. Carbohydrate and protein metabolism are linked by pyruvic acid.

4. Acetyl coenzyme A is the molecule that enters the Krebs cycle; it is also used to synthesize fatty acids, ketone bodies, and cholesterol.

Animation - Cell Respiration
Figure 25.16 - Roles of Key Molecules in Metabolic Pathways

25.7 Metabolic Adaptations

1. During the absorptive state, ingested nutrients enter the blood and lymph from the GI tract.

2. During the absorptive state, blood glucose is oxidized to form ATP, and glucose transported to the liver is converted to glycogen or triglycerides. Most triglycerides are stored in adipose tissue. Amino acids in hepatocytes are converted to carbohydrates, fats, and proteins. Table 25.3 summarizes the hormonal regulation of metabolism during the absorptive state.

3. During the postabsorptive state, absorption is complete and the ATP needs of the body are satisfied by nutrients already present in the body. The major task is to maintain normal blood glucose level by converting glycogen in the liver and skeletal muscle into glucose, converting glycerol into glucose, and converting amino acids into glucose. Fatty acids, ketone bodies, and amino acids are oxidized to supply ATP. Table 25.4 summarizes the hormonal regulation of metabolism during the postabsorptive state.

4. Fasting is going without food for a few days; starvation implies weeks or months of inadequate food intake. During fasting and starvation, fatty acids and ketone bodies are increasingly utilized for ATP production.

Animation - Maintaining Normal Blood Glucose Levels
Exercise - Alternative Metabolic Pathways

25.8 Heat and Energy Balance

1. Measurement of the metabolic rate under basal conditions is called the basal metabolic rate (BMR).

2. A calorie (cal) is the amount of energy required to raise the temperature of 1 g of water 1°C.

3. Because the calorie is a relatively small unit, the kilocalorie (kcal) or Calorie (Cal) is often used to measure the body's metabolic rate and to express the energy content of foods; a kilocalorie equals 1000 Calories.

4. Normal core temperature is maintained by a delicate balance between heat-producing and heat-losing mechanisms.

5. Exercise, hormones, the nervous system, body temperature, ingestion of food, age, gender, climate, sleep, and malnutrition affect metabolic rate.

6. Mechanisms of heat transfer include conduction, convection, radiation, and evaporation. Conduction is the transfer of heat between two substances or objects in contact with each other. Convection is the transfer of heat by a liquid or gas between areas of different temperatures. Radiation is the transfer of heat from a warmer object to a cooler object without physical contact. Evaporation is the conversion of a liquid to a vapor; in the process, heat is lost.

7. The hypothalamic thermostat is in the preoptic area.

8. Responses that produce, conserve, or retain heat when core temperature falls are vasoconstriction; release of epinephrine, norepinephrine, and thyroid hormones; and shivering.

9. Responses that increase heat loss when core temperature increases include vasodilation, decreased metabolic rate, and evaporation of perspiration.

10. Two nuclei in the hypothalamus that help regulate food intake are the arcuate and paraventricular nuclei. The hormone leptin, released by adipocytes, inhibits release of neuropeptide Y from the arcuate nucleus and thereby decreases food intake. Melanocortin also decreases food intake.

Animation - Metabolic Rate, Heat, and Temperature Regulation
Animation - Regulation of Metabolism
Concepts and Connections - Hormonal Control of Metabolic Activities
Concepts and Connections - Nutrient Utilization Pathways

Review	Resource
25.9 Nutrition 1. Nutrients include water, carbohydrates, lipids, proteins, minerals, and vitamins. 2. Nutrition experts suggest dietary calories be 50–60% from carbohydrates, 30% or less from fats, and 12–15% from proteins. 3. The MyPyramid guide represents a personalized approach to making healthy food choices and maintaining regular physical activity. 4. Minerals known to perform essential functions include calcium, phosphorus, potassium, sulfur, sodium, chloride, magnesium, iron, iodide, manganese, copper, cobalt, zinc, fluoride, selenium, and chromium. Their functions are summarized in Table 25.5. 5. Vitamins are organic nutrients that maintain growth and normal metabolism. Many function in enzyme systems. 6. Fat-soluble vitamins are absorbed with fats and include vitamins A, D, E, and K; water-soluble vitamins include the B vitamins and vitamin C. 7. The functions and deficiency disorders of the principal vitamins are summarized in Table 25.6.	BBC Video - Study on Vitamin Supplements Animation - Enzyme Functions and ATP Part 1

SELF-QUIZ QUESTIONS

Fill in the blanks in the following statements.

1. The thermostat and food intake regulating center of the body is in the _____ of the brain.

2. The three key molecules of metabolism are _____, _____, and _____.

Indicate whether the following statements are true or false.

3. Foods that we eat are used to supply energy for life processes, serve as building blocks for synthesis reactions, or are stored for future use.

4. Vitamins A, B, D, and K are fat-soluble vitamins.

Choose the one best answer to the following questions.

5. NAD^+ and FAD (1) are both derivatives of B vitamins, (2) are used to carry hydrogen atoms released during oxidation reactions, (3) become NADH and $FADH_2$ in their reduced forms, (4) act as coenzymes in the Krebs cycle, (5) are the final electron acceptors in the electron transport chain.
   (a) 1, 2, 3, 4, and 5   (b) 2, 3, and 4   (c) 2 and 4
   (d) 1, 2, and 3   (e) 1, 2, 3, and 4

6. During glycolysis, (1) a 6-carbon glucose is split into two 3-carbon pyruvic acids, (2) there is a net gain of two ATP molecules, (3) two NADH molecules are oxidized, (4) moderately high levels of oxygen are needed, (5) the activity of phosphofructokinase determines the rate of the chemical reactions.
   (a) 1, 2, and 3   (b) 1 and 2   (c) 1, 2, and 5
   (d) 2, 3, 4, and 5   (e) 1, 2, 3, 4, and 5

7. If glucose is not needed for immediate ATP production, it can be used for (1) vitamin synthesis, (2) amino acid synthesis, (3) gluconeogenesis, (4) glycogenesis, (5) lipogenesis.
   (a) 1, 3, and 5   (b) 2, 4, and 5   (c) 2, 3, 4, and 5
   (d) 1, 2, and 3   (e) 2 and 5

8. Which of the following is the *correct* sequence for the oxidation of glucose to produce ATP?
   (a) electron transport chain, Krebs cycle, glycolysis, formation of acetyl CoA

   (b) Krebs cycle, formation of acetyl CoA, electron transport chain, glycolysis
   (c) glycolysis, electron transport chain, Krebs cycle, formation of acetyl CoA
   (d) glycolysis, formation of acetyl CoA, Krebs cycle, electron transport chain
   (e) formation of acetyl CoA, Krebs cycle, glycolysis, electron transport chain

9. Which of the following would you *not* expect to experience during fasting or starvation?
   (a) decrease in plasma fatty acid levels
   (b) increase in ketone body formation
   (c) lipolysis
   (d) increased use of ketones for ATP production in the brain
   (e) depletion of glycogen

10. If core body temperature rises above normal, which of the following would occur to cool the body? (1) dilation of vessels in the skin, (2) increased radiation and conduction of heat to the environment, (3) increased metabolic rate, (4) evaporation of perspiration, (5) increased secretion of thyroid hormones.
   (a) 3, 4, and 5   (b) 1, 2, and 4   (c) 1, 2, and 5
   (d) 1, 2, 3, 4 and 5   (e) 1, 2, 4, and 5

11. In which of the following situations would the metabolic rate increase? (1) sleep, (2) after ingesting food, (3) increased secretion of thyroid hormones, (4) parasympathetic nervous system stimulation, (5) fever.
   (a) 3 and 4   (b) 1, 3, and 5   (c) 2 and 3
   (d) 2, 3, and 4   (e) 2, 3, and 5

12. Which of the following are absorptive state reactions? (1) aerobic cellular respiration, (2) glycogenesis, (3) glycogenolysis, (4) gluconeogenesis using lactic acid, (5) lipolysis.
   (a) 1 and 2   (b) 2 and 3   (c) 3 and 4
   (d) 4 and 5   (e) 1 and 5

13. Match the hormones with the reactions they regulate (answers may be used more than once; some reactions have more than one answer):

_____ (a) gluconeogenesis
_____ (b) glycogenesis
_____ (c) glycogenolysis
_____ (d) lipolysis
_____ (e) lipogenesis
_____ (f) protein catabolism
_____ (g) protein anabolism

(1) insulin
(2) cortisol
(3) glucagon
(4) thyroid hormones
(5) epinephrine
(6) insulinlike growth factors

14. Match the following:

_____ (a) deliver cholesterol to body cells for use in repair of membranes and synthesis of steroid hormones and bile salts
_____ (b) remove excess cholesterol from body cells and transport it to the liver for elimination
_____ (c) organic nutrients required in small amounts for growth and normal metabolism
_____ (d) the energy-transferring molecule of the body
_____ (e) nutrient molecules that can be oxidized to produce ATP or stored in adipose tissue
_____ (f) transport endogenous lipids to adipocytes for storage
_____ (g) the body's preferred source for synthesizing ATP
_____ (h) composed of amino acids and are the primary regulatory molecules in the body
_____ (i) acetoacetic acid, beta-hydroxybutyric acid, and acetone
_____ (j) hormone secreted by adipocytes that acts to decrease total body-fat mass
_____ (k) neurotransmitter that stimulates food intake
_____ (l) inorganic substances that perform many vital functions in the body
_____ (m) carriers of electrons in the electron transport chain

(1) leptin
(2) minerals
(3) glucose
(4) lipids
(5) proteins
(6) neuropeptide Y
(7) cytochromes
(8) ketone bodies
(9) low-density lipoproteins
(10) ATP
(11) vitamins
(12) high-density lipoproteins
(13) very-low-density lipoproteins

15. Match the following:

_____ (a) the mechanism of ATP generation that links chemical reactions with pumping of hydrogen ions
_____ (b) the removal of electrons from an atom or molecule resulting in a decrease in potential energy
_____ (c) the transfer of an amino group from an amino acid to a substance such as pyruvic acid
_____ (d) the formation of glucose from noncarbohydrate sources
_____ (e) refers to all the chemical reactions in the body
_____ (f) the oxidation of glucose to produce ATP
_____ (g) the splitting of a triglyceride into glycerol and fatty acids
_____ (h) the synthesis of lipids
_____ (i) the addition of electrons to a molecule resulting in an increase in potential energy content of the molecule
_____ (j) the formation of ketone bodies
_____ (k) the breakdown of glycogen back to glucose
_____ (l) exergonic chemical reactions that break down complex organic molecules into simpler ones
_____ (m) overall rate at which metabolic reactions use energy
_____ (n) the breakdown of glucose into two molecules of pyruvic acid
_____ (o) removal of CO_2 from a molecule
_____ (p) endergonic chemical reactions that combine simple molecules and monomers to make more complex ones
_____ (q) the addition of a phosphate group to a molecule
_____ (r) the removal of the amino group from an amino acid
_____ (s) the cleavage of one pair of carbon atoms at a time from a fatty acid
_____ (t) the conversion of glucose into glycogen

(1) metabolism
(2) catabolism
(3) beta oxidation
(4) lipolysis
(5) phosphorylation
(6) glycolysis
(7) cellular respiration
(8) transamination
(9) anabolism
(10) lipogenesis
(11) glycogenolysis
(12) glycogenesis
(13) metabolic rate
(14) ketogenesis
(15) oxidation
(16) reduction
(17) chemiosmosis
(18) deamination
(19) gluconeogenesis
(20) decarboxylation

CRITICAL THINKING QUESTIONS

1. Jane Doe's deceased body was found at her dining room table. Her death was considered suspicious. Lab results from the medical investigation revealed cyanide in her blood. How did the cyanide cause her death?

2. During a recent physical, 55-year-old Glenn's blood serum lab results showed the following: total cholesterol = 300 mg/dL; LDL = 175 mg/dL; HDL = 20 mg/dL. Interpret these results for Glenn and indicate to him what changes, if any, he needs to make in his lifestyle. Why are these changes important?

3. Marissa has joined a weight loss program. As part of the program, she regularly submits a urine sample which is tested for ketones. She went to the clinic today, had her urine checked, and was confronted by the nurse who accused Marissa of "cheating" on her diet. How did the nurse know Marissa was not following her diet?

? ANSWERS TO FIGURE QUESTIONS

25.1 In pancreatic acinar cells, anabolism predominates because the primary activity is synthesis of complex molecules (digestive enzymes).

25.2 Glycolysis is also called anaerobic cellular respiration.

25.3 The reactions of glycolysis consume two molecules of ATP but generate four molecules of ATP, for a net gain of two.

25.4 Kinases are enzymes that phosphorylate (add phosphate to) their substrate.

25.5 Glycolysis occurs in the cytosol.

25.6 CO_2 is given off during the production of acetyl coenzyme A and during the Krebs cycle. It diffuses into the blood, is transported by the blood to the lungs, and is exhaled.

25.7 The production of reduced coenzymes is important in the Krebs cycle because they will subsequently yield ATP in the electron transport chain.

25.8 The energy source that powers the proton pumps is electrons provided by NADH + H^+.

25.9 The concentration of H^+ is highest in the space between the inner and outer mitochondrial membranes.

25.10 During the complete oxidation of one glucose molecule, six molecules of O_2 are used and six molecules of CO_2 are produced.

25.11 Skeletal muscle fibers can synthesize glycogen, but they cannot release glucose into the blood because they lack the enzyme phosphatase required to remove the phosphate group from glucose.

25.12 Hepatocytes can carry out gluconeogenesis and glycogenesis.

25.13 LDLs deliver cholesterol to body cells.

25.14 Hepatocytes and adipose cells carry out lipogenesis, beta oxidation, and lipolysis; hepatocytes carry out ketogenesis.

25.15 Before an amino acid can enter the Krebs cycle, an amino group must be removed via deamination.

25.16 Acetyl coenzyme A is the gateway into the Krebs cycle for molecules being oxidized to generate ATP.

25.17 Reactions of the absorptive state are mainly anabolic.

25.18 Processes that directly elevate blood glucose during the postabsorptive state include lipolysis (in adipocytes and hepatocytes), gluconeogenesis (in hepatocytes), and glycogenolysis (in hepatocytes).

25.19 Exercise, the sympathetic nervous system, hormones (epinephrine, norepinephrine, thyroxine, testosterone, human growth hormone), elevated body temperature, and ingestion of food increase metabolic rate, which results in an increase in body temperature.

25.20 The wider base of each band represents foods with little or no solid fats or added sugars.

26 THE URINARY SYSTEM

THE URINARY SYSTEM AND HOMEOSTASIS *The urinary system contributes to homeostasis by altering blood composition, pH, volume, and pressure; maintaining blood osmolarity; excreting wastes and foreign substances; and producing hormones.*

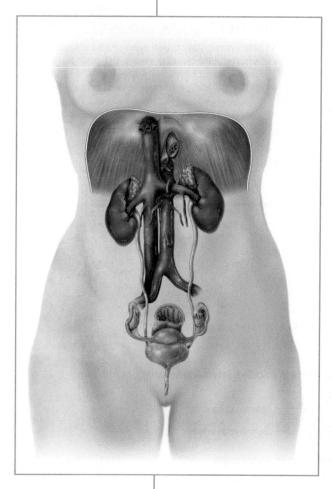

The **urinary system** consists of two kidneys, two ureters, one urinary bladder, and one urethra (Figure 26.1). After the kidneys filter blood plasma, they return most of the water and solutes to the bloodstream. The remaining water and solutes constitute **urine,** which passes through the ureters and is stored in the urinary bladder until it is excreted from the body through the urethra. **Nephrology** (nef-ROL-ō-jē; *nephr-* = kidney; *-ology* = study of) is the scientific study of the anatomy, physiology, and pathology of the kidneys. The branch of medicine that deals with the male and female urinary systems and the male reproductive system is called **urology** (ū-ROL-ō-jē; *uro-* = urine). A physician who specializes in this branch of medicine is called a **urologist** (ū-ROL-ō-jist).

? *Did you ever wonder how diuretics work and why they are used?*

Figure 26.1 Organs of the urinary system in a female.

Urine formed by the kidneys passes first into the ureters, then to the urinary bladder for storage, and finally through the urethra for elimination from the body.

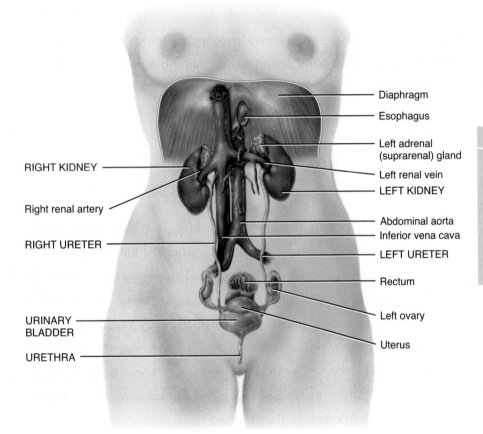

- Diaphragm
- Esophagus
- Left adrenal (suprarenal) gland
- Left renal vein
- LEFT KIDNEY
- Abdominal aorta
- Inferior vena cava
- LEFT URETER
- Rectum
- Left ovary
- Uterus

RIGHT KIDNEY

Right renal artery

RIGHT URETER

URINARY BLADDER

URETHRA

FUNCTIONS OF THE URINARY SYSTEM

1. Kidneys regulate blood volume and composition; help regulate blood pressure, pH, and glucose levels; produce two hormones (calcitriol and erythropoietin); and excrete wastes in urine.
2. Ureters transport urine from kidneys to urinary bladder.
3. Urinary bladder stores urine and expels it into urethra.
4. Urethra discharges urine from body.

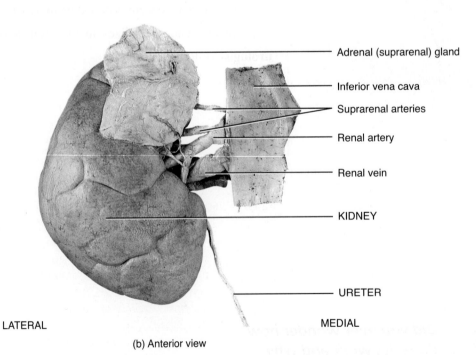

- Adrenal (suprarenal) gland
- Inferior vena cava
- Suprarenal arteries
- Renal artery
- Renal vein
- KIDNEY
- URETER

LATERAL   MEDIAL

(b) Anterior view

? **Which organs constitute the urinary system?**

26.1 OVERVIEW OF KIDNEY FUNCTIONS

◉ **OBJECTIVE**

• List the functions of the kidneys.

The kidneys do the major work of the urinary system. The other parts of the system are mainly passageways and storage areas. Functions of the kidneys include the following:

• **Regulation of blood ionic composition.** The kidneys help regulate the blood levels of several ions, most importantly sodium ions (Na^+), potassium ions (K^+), calcium ions (Ca^{2+}), chloride ions (Cl^-), and phosphate ions (HPO_4^{2-}).

• **Regulation of blood pH.** The kidneys excrete a variable amount of hydrogen ions (H^+) into the urine and conserve bicarbonate ions (HCO_3^-), which are an important buffer of H^+ in the blood. Both of these activities help regulate blood pH.

• **Regulation of blood volume.** The kidneys adjust blood volume by conserving or eliminating water in the urine. An increase in blood volume increases blood pressure; a decrease in blood volume decreases blood pressure.

• **Regulation of blood pressure.** The kidneys also help regulate blood pressure by secreting the enzyme renin, which activates the renin–angiotensin–aldosterone pathway (see Figure 18.16). Increased renin causes an increase in blood pressure.

• **Maintenance of blood osmolarity.** By separately regulating loss of water and loss of solutes in the urine, the kidneys maintain a relatively constant blood osmolarity close to 300 milliosmoles per liter (mOsm/liter).*

• **Production of hormones.** The kidneys produce two hormones. *Calcitriol,* the active form of vitamin D, helps regulate calcium homeostasis (see Figure 18.14), and *erythropoietin* stimulates the production of red blood cells (see Figure 19.5).

• **Regulation of blood glucose level.** Like the liver, the kidneys can use the amino acid glutamine in *gluconeogenesis,* the synthesis of new glucose molecules. They can then release glucose into the blood to help maintain a normal blood glucose level.

• **Excretion of wastes and foreign substances.** By forming urine, the kidneys help excrete **wastes**—substances that have no useful function in the body. Some wastes excreted in urine result from metabolic reactions in the body. These include ammonia and urea from the deamination of amino acids; bilirubin

from the catabolism of hemoglobin; creatinine from the breakdown of creatine phosphate in muscle fibers; and uric acid from the catabolism of nucleic acids. Other wastes excreted in urine are foreign substances from the diet, such as drugs and environmental toxins.

✔ **CHECKPOINT**

1. What are wastes, and how do the kidneys participate in their removal from the body?

26.2 ANATOMY AND HISTOLOGY OF THE KIDNEYS

◉ **OBJECTIVES**

• Describe the external and internal gross anatomical features of the kidneys.
• Trace the path of blood flow through the kidneys.
• Describe the structure of renal corpuscles and renal tubules.

The paired **kidneys** are reddish, kidney bean–shaped organs located just above the waist between the peritoneum and the posterior wall of the abdomen. Because their position is posterior to the peritoneum of the abdominal cavity, they are said to be **retroperitoneal** (re′-trō-per-i-tō-NĒ-al; *retro-* = behind) organs (Figure 26.2). The kidneys are located between the levels of the last thoracic and third lumbar vertebrae, a position where they are partially protected by the eleventh and twelfth pairs of ribs. Unfortunately, if these lower ribs are fractured, they can puncture the kidneys and cause significant, and even life-threatening, damage. The right kidney is slightly lower than the left (see Figure 26.1) because the liver occupies considerable space on the right side superior to the kidney.

External Anatomy of the Kidneys

A typical adult kidney is 10–12 cm (4–5 in.) long, 5–7 cm (2–3 in.) wide, and 3 cm (1 in.) thick—about the size of a bar of bath soap—and has a mass of 135–150 g (4.5–5 oz). The concave medial border of each kidney faces the vertebral column (see Figure 26.1). Near the center of the concave border is an indentation called the **renal hilum** (RĒ-nal HĪ-lum; *renal* = kidney) or **hilus** (see Figure 26.3), through which the ureter emerges from the kidney along with blood vessels, lymphatic vessels, and nerves.

Three layers of tissue surround each kidney (Figure 26.2). The deep layer, the **renal capsule,** is a smooth, transparent sheet of dense irregular connective tissue that is continuous with the outer coat of the ureter. It serves as a barrier against trauma and helps maintain the shape of the kidney. The middle layer, the **adipose capsule,** is a mass of fatty tissue surrounding the renal capsule. It also protects the kidney from trauma and holds it firmly in place within the abdominal cavity. The superficial layer, the **renal fascia** (FASH-ē-a), is another thin layer of dense irregular connective tissue that anchors the kidney to the surrounding structures and to the abdominal wall. On the anterior surface of the kidneys, the renal fascia is deep to the peritoneum.

---

*The **osmolarity** of a solution is a measure of the total number of dissolved particles per liter of solution. The particles may be molecules, ions, or a mixture of both. To calculate osmolarity, multiply molarity (see Section 2.4) by the number of particles per molecule, once the molecule dissolves. A similar term, *osmolality,* is the number of particles of solute per *kilogram* of water. Because it is easier to measure volumes of solutions than to determine the mass of water they contain, osmolarity is used more commonly than osmolality. Most body fluids and solutions used clinically are dilute, in which case there is less than a 1% difference between the two measures.

Figure 26.2 Position and coverings of the kidneys.

🔑 The kidneys are surrounded by a renal capsule, adipose capsule, and renal fascia.

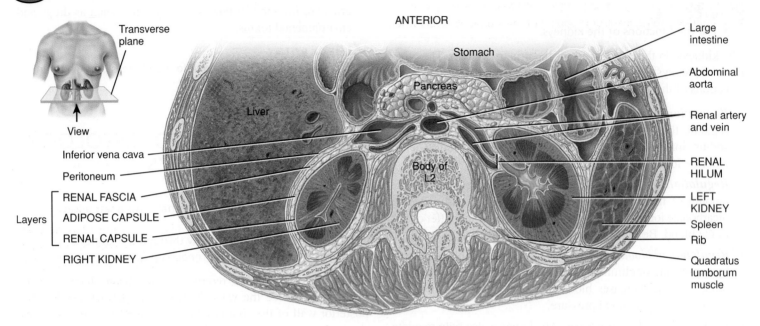

ANTERIOR

Transverse plane

View

Stomach

Pancreas

Liver

Large intestine

Abdominal aorta

Renal artery and vein

Inferior vena cava

Peritoneum

Body of L2

RENAL HILUM

Layers
- RENAL FASCIA
- ADIPOSE CAPSULE
- RENAL CAPSULE

RIGHT KIDNEY

LEFT KIDNEY

Spleen

Rib

Quadratus lumborum muscle

POSTERIOR

(a) Inferior view of transverse section of abdomen (L2)

SUPERIOR

Parasagittal plane

Diaphragm

Twelfth rib

Right kidney

Quadratus lumborum muscle

Hip bone

Lung

Liver

Adrenal (suprarenal) gland

Peritoneum

RENAL FASCIA

ADIPOSE CAPSULE } Layers

RENAL CAPSULE

Large intestine

POSTERIOR

ANTERIOR

(b) Parasagittal section through right kidney

❓ **Why are the kidneys said to be retroperitoneal?**

Nephroptosis (nef'-rōp-TŌ-sis; *-ptosis* = falling), or **floating kidney,** is an inferior displacement or dropping of the kidney. It occurs when the kidney slips from its normal position because it is not securely held in place by adjacent organs or its covering of fat. Nephroptosis develops most often in very thin people whose adipose capsule or renal fascia is deficient. It is dangerous because the ureter may kink and block urine flow. The resulting backup of urine puts pressure on the kidney, which damages the tissue. Twisting of the ureter also causes pain. Nephroptosis is very common; about one in four people has some degree of weakening of the fibrous bands that hold the kidney in place. It is 10 times more common in females than males. Because it happens during life it is very easy to distinguish from congenital anomalies. •

Internal Anatomy of the Kidneys

A frontal section through the kidney reveals two distinct regions: a superficial, light red area called the **renal cortex** (*cortex* = rind or bark) and a deep, darker reddish-brown inner region called the **renal medulla** (*medulla* = inner portion) (Figure 26.3). The renal medulla consists of several cone-shaped **renal pyramids.** The base (wider end) of each pyramid faces the renal cortex, and its apex (narrower end), called a **renal papilla,** points toward the renal hilum. The renal cortex is the smooth-textured area extending from the renal capsule to the bases of the renal pyramids and into the spaces between them. It is divided into an outer *cortical zone* and an inner *juxtamedullary zone* (juks'-ta-MED-ū-la-rē). Those portions of the renal cortex that extend between renal pyramids are called **renal columns.** A **renal lobe** consists of a renal pyramid, its overlying area of renal cortex, and one-half of each adjacent renal column.

Together, the renal cortex and renal pyramids of the renal medulla constitute the **parenchyma** (pa-RENG-ki-ma) or functional portion of the kidney. Within the parenchyma are the functional units of the kidney—about 1 million microscopic structures called **nephrons** (NEF-rons). Filtrate (filtered fluid) formed by the nephrons drains into large **papillary ducts,** which extend through the renal papillae of the pyramids. The papillary ducts drain into cuplike structures called **minor** and **major calyces** (KĀ-li-sēz = cups; singular is *calyx,* pronounced KĀ-liks). Each kidney has 8 to 18 minor calyces and 2 or 3 major calyces. A minor calyx receives urine from the papillary ducts of one renal papilla and delivers it to a major calyx. Once the filtrate enters the calyces it becomes urine because no further reabsorption can occur. The reason for this is that the simple epithelium of the nephron and ducts becomes transitional epithelium in the calyces. From the major calyces, urine drains into a single large cavity called the **renal pelvis** (*pelv-* = basin) and then out through the ureter to the urinary bladder.

The hilum expands into a cavity within the kidney called the **renal sinus,** which contains part of the renal pelvis, the calyces, and branches of the renal blood vessels and nerves. Adipose tissue helps stabilize the position of these structures in the renal sinus.

Blood and Nerve Supply of the Kidneys

Because the kidneys remove wastes from the blood and regulate its volume and ionic composition, it is not surprising that they are abundantly supplied with blood vessels. Although the kidneys constitute less than 0.5% of total body mass, they receive 20–25% of the resting cardiac output via the right and left **renal arteries** (Figure 26.4). In adults, **renal blood flow,** the blood flow through both kidneys, is about 1200 mL per minute.

Within the kidney, the renal artery divides into several **segmental arteries** (seg-MEN-tal), which supply different segments (areas) of the kidney. Each segmental artery gives off several branches that enter the parenchyma and pass through the renal columns between the renal pyramids as the **interlobar arteries** (in'-ter-LŌ-bar). At the bases of the renal pyramids, the interlobar arteries arch between the renal medulla and cortex; here they are known as the **arcuate arteries** (AR-kū-āt = shaped like a bow). Divisions of the arcuate arteries produce a series of **interlobular arteries** (in'-ter-LOB-ū-lar). These arteries are so named because they pass between renal lobules. Interlobular arteries enter the renal cortex and give off branches called **afferent arterioles** (AF-er-ent; *af-* = toward; *-ferrent* = to carry).

Each nephron receives one afferent arteriole, which divides into a tangled, ball-shaped capillary network called the **glomerulus** (glō-MER-ū-lus = little ball; plural is *glomeruli*). The glomerular capillaries then reunite to form an **efferent arteriole** (EF-er-ent; *ef-* = out) that carries blood out of the glomerulus. Glomerular capillaries are unique among capillaries in the body because they are positioned between two arterioles, rather than between an arteriole and a venule. Because they are capillary networks and they also play an important role in urine formation, the glomeruli are considered part of both the cardiovascular and the urinary systems.

The efferent arterioles divide to form the **peritubular capillaries** (per-i-TOOB-ū-lar; *peri-* = around), which surround tubular parts of the nephron in the renal cortex. Extending from some efferent arterioles are long loop-shaped capillaries called **vasa recta** (VĀ-sa REK-ta; *vasa* = vessels; *recta* = straight) that supply tubular portions of the nephron in the renal medulla (see Figure 26.5b).

The peritubular capillaries eventually reunite to form **peritubular venules** and then **interlobular veins,** which also receive blood from the vasa recta. Then the blood drains through the **arcuate veins** to the **interlobar veins** running between the renal pyramids. Blood leaves the kidney through a single **renal vein** that exits at the renal hilum and carries venous blood to the inferior vena cava.

Many renal nerves originate in the *renal ganglion* and pass through the *renal plexus* into the kidneys along with the renal arteries. Renal nerves are part of the sympathetic division of the autonomic nervous system. Most are vasomotor nerves that regulate the flow of blood through the kidney by causing vasodilation or vasoconstriction of renal arterioles.

Figure 26.3 Internal anatomy of the kidneys.

The two main regions of the kidney parenchyma are the renal cortex and the renal pyramids in the renal medulla.

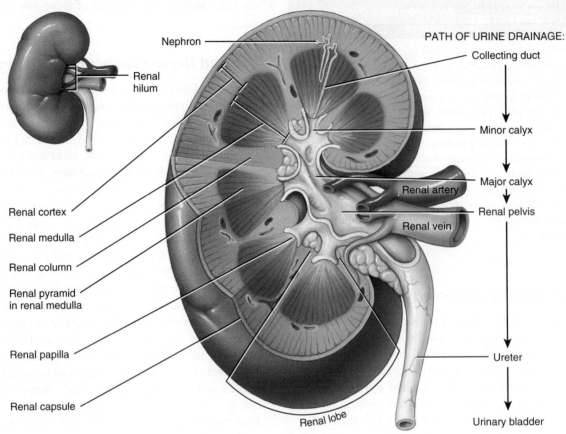

Nephron

Renal hilum

PATH OF URINE DRAINAGE:

Collecting duct

Minor calyx

Major calyx

Renal artery

Renal pelvis

Renal vein

Renal cortex

Renal medulla

Renal column

Renal pyramid in renal medulla

Renal papilla

Renal capsule

Renal lobe

Ureter

Urinary bladder

(a) Anterior view of dissection of right kidney

SUPERIOR

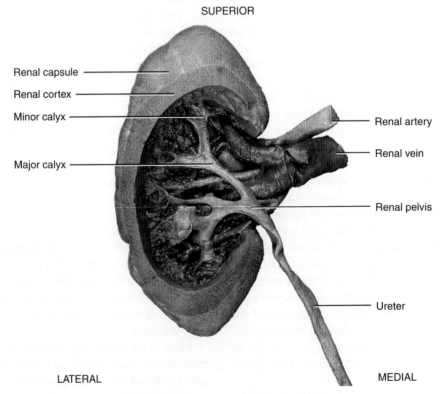

Renal capsule

Renal cortex

Minor calyx

Major calyx

Renal artery

Renal vein

Renal pelvis

Ureter

LATERAL

MEDIAL

(b) Posterior view of dissection of left kidney

? **What structures pass through the renal hilum?**

Figure 26.4 Blood supply of the kidneys.

The renal arteries deliver 20–25% of the resting cardiac output to the kidneys.

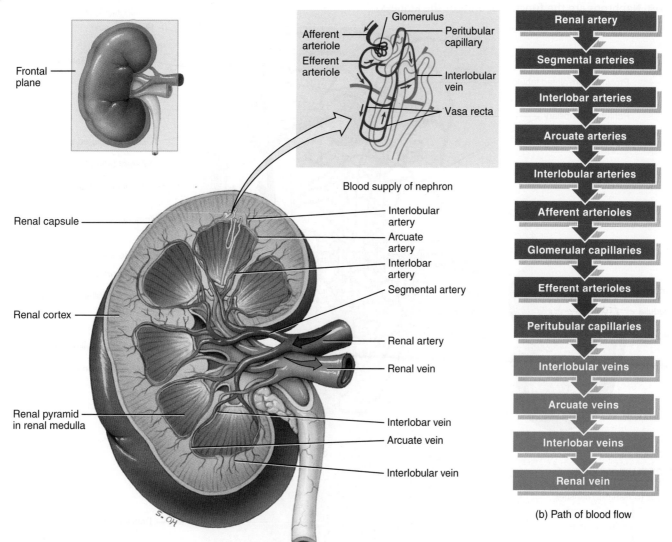

Blood supply of nephron

(a) Frontal section of right kidney

(b) Path of blood flow

What volume of blood enters the renal arteries per minute?

CLINICAL CONNECTION | Kidney Transplant

A **kidney transplant** is the transfer of a kidney from a donor to a recipient whose kidneys no longer function. In the procedure, the donor kidney is placed in the pelvis of the recipient through an abdominal incision. The renal artery and vein of the transplanted kidney are attached to a nearby artery or vein in the pelvis of the recipient and the ureter of the transplanted kidney is then attached to the urinary bladder. During a kidney transplant, the patient receives only one donor kidney, since only one kidney is needed to maintain sufficient renal function. The nonfunctioning diseased kidneys are usually left in place. As with all organ transplants, kidney transplant recipients must be ever vigilant for signs of infection or organ rejection. The transplant recipient will take immunosuppressive drugs for the rest of his or her life to avoid rejection of the "foreign" organ. •

The Nephron

Parts of a Nephron

Nephrons (NEF-rons) are the functional units of the kidneys. Each nephron (Figure 26.5) consists of two parts: a **renal corpuscle** (KOR-pus-sul = tiny body), where blood plasma is filtered, and a **renal tubule** into which the filtered fluid passes. The two components of a renal corpuscle are the **glomerulus** (glō-MER-ū-lus) (capillary network) and the **glomerular (Bowman's) capsule,** a double-walled epithelial cup that surrounds the glomerular capillaries. Blood plasma is filtered in the glomerular capsule, and then the filtered fluid passes into the renal tubule, which has three main sections. In the order that fluid passes through them, the renal tubule consists of a (1) **proximal convoluted**

Figure 26.5 **The structure of nephrons and associated blood vessels.** Note that the collecting duct and papillary duct are not part of a nephron.

Nephrons are the functional units of the kidneys.

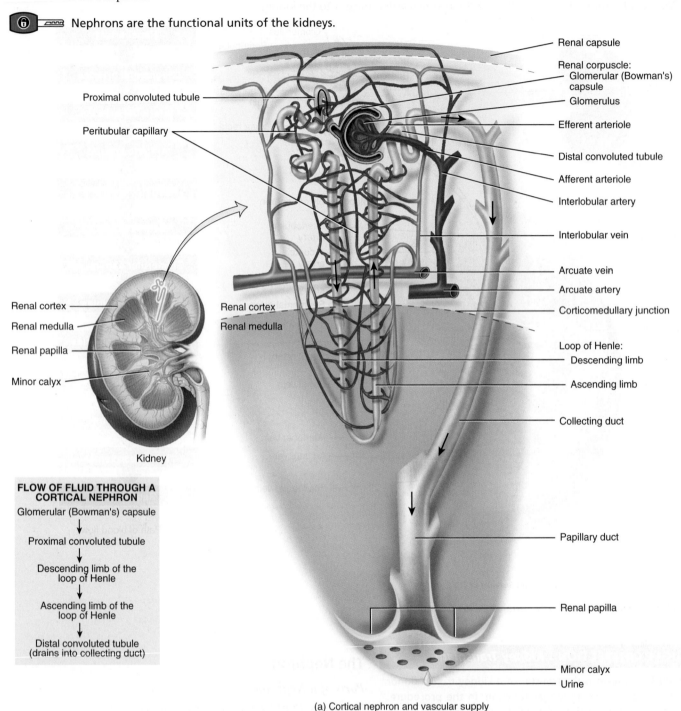

Proximal convoluted tubule

Peritubular capillary

Renal capsule

Renal corpuscle:
Glomerular (Bowman's) capsule

Glomerulus

Efferent arteriole

Distal convoluted tubule

Afferent arteriole

Interlobular artery

Interlobular vein

Arcuate vein

Arcuate artery

Corticomedullary junction

Loop of Henle:
Descending limb

Ascending limb

Collecting duct

Papillary duct

Renal papilla

Minor calyx

Urine

Renal cortex
Renal medulla
Renal papilla
Minor calyx

Kidney

Renal cortex
Renal medulla

FLOW OF FLUID THROUGH A CORTICAL NEPHRON

Glomerular (Bowman's) capsule
↓
Proximal convoluted tubule
↓
Descending limb of the loop of Henle
↓
Ascending limb of the loop of Henle
↓
Distal convoluted tubule
(drains into collecting duct)

(a) Cortical nephron and vascular supply

tubule or **PCT** (kon′-vō-LOOT-ed), (2) **loop of Henle (nephron loop),** and (3) **distal convoluted tubule** or **DCT.** *Proximal* denotes the part of the tubule attached to the glomerular capsule, and *distal* denotes the part that is further away. *Convoluted* means the tubule is tightly coiled rather than straight. The renal corpuscle and both convoluted tubules lie within the renal cortex; the loop of Henle extends into the renal medulla, makes a hairpin turn, and then returns to the renal cortex.

The distal convoluted tubules of several nephrons empty into a single **collecting duct.** Collecting ducts then unite and converge into several hundred large **papillary ducts** (PAP-i-lar′-ē), which drain into the minor calyces. The collecting ducts and papillary ducts extend from the renal cortex through the renal medulla to the renal pelvis. So one kidney has about 1 million nephrons, but a much smaller number of collecting ducts and even fewer papillary ducts.

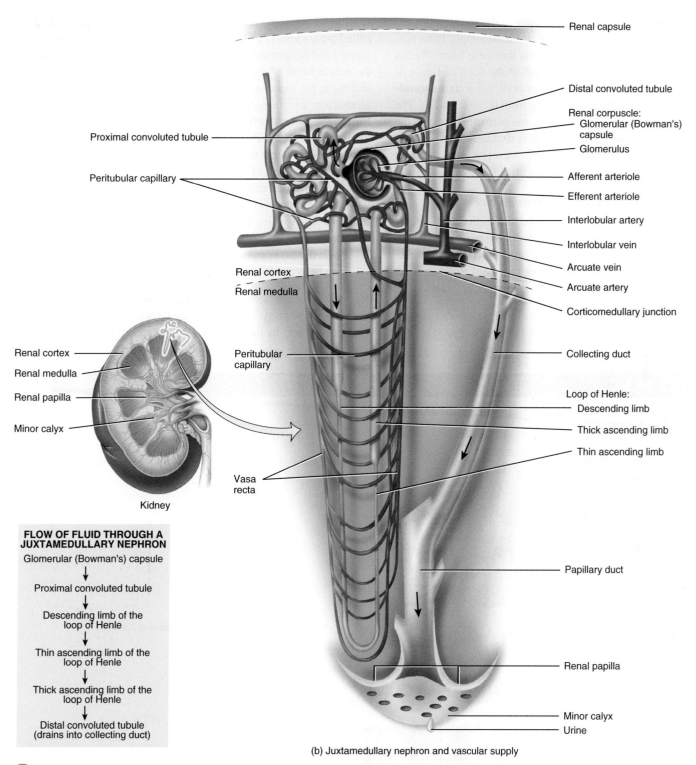

(b) Juxtamedullary nephron and vascular supply

FLOW OF FLUID THROUGH A JUXTAMEDULLARY NEPHRON

Glomerular (Bowman's) capsule

↓

Proximal convoluted tubule

↓

Descending limb of the loop of Henle

↓

Thin ascending limb of the loop of Henle

↓

Thick ascending limb of the loop of Henle

↓

Distal convoluted tubule (drains into collecting duct)

? **What are the basic differences between cortical and juxtamedullary nephrons?**

In a nephron, the loop of Henle connects the proximal and distal convoluted tubules. The first part of the loop of Henle dips into the renal medulla, where it is called the **descending limb of the loop of Henle** (Figure 26.5). It then makes that hairpin turn and returns to the renal cortex as the **ascending limb of the loop of Henle.** About 80–85% of the nephrons are **cortical nephrons** (KOR-ti-kul). Their renal corpuscles lie in the outer portion of the renal cor-

tex, and they have *short* loops of Henle that lie mainly in the cortex and penetrate only into the outer region of the renal medulla (Figure 26.5a). The short loops of Henle receive their blood supply from peritubular capillaries that arise from efferent arterioles. The other 15–20% of the nephrons are **juxtamedullary nephrons** (juks′-ta-MED-ū-lar′-e; *juxta-* = near to). Their renal corpuscles lie deep in the cortex, close to the medulla, and they have a *long* loop of Henle that extends into the deepest region of the medulla (Figure 26.5b). Long loops of Henle receive their blood supply from peritubular capillaries and from the vasa recta that arise from efferent arterioles. In addition, the ascending limb of the loop of Henle of juxtamedullary nephrons consists of two portions: a **thin ascending limb** followed by a **thick ascending limb** (Figure 26.5b). The lumen of the thin ascending limb is the same as in other areas of the renal tubule; it is only the epithelium that is thinner. Nephrons with long loops of Henle enable the kidneys to excrete very dilute or very concentrated urine (described in Section 26.6).

Histology of the Nephron and Collecting Duct

A single layer of epithelial cells forms the entire wall of the glomerular capsule, renal tubule, and ducts. However, each part has distinctive histological features that reflect its particular functions. We will discuss them in the order that fluid flows through them: glomerular capsule, renal tubule, and collecting duct.

GLOMERULAR CAPSULE The glomerular (Bowman's) capsule consists of visceral and parietal layers (Figure 26.6a). The visceral layer consists of modified simple squamous epithelial cells called **podocytes** (PŌD-ō-sīts; *podo-* = foot; *-cytes* = cells). The many footlike projections of these cells (pedicels) wrap around the single layer of endothelial cells of the glomerular capillaries and form the inner wall of the capsule. The parietal layer of the glomerular capsule consists of simple squamous epithelium and forms the outer wall of the capsule. Fluid filtered from the glomerular capillaries enters the **capsular (Bowman's) space,** the space between the two layers of the glomerular capsule, which is the lumen of the urinary tube. Think of the relationship between the glomerulus and glomerular capsule in the following way. The glomerulus is a fist punched into a limp balloon (the glomerular capsule) until the fist is covered by two layers of balloon (the layer of the balloon touching the fist is the visceral layer and the layer not against the fist is the parietal layer) with a space in between (the inside of the balloon), the capsular space.

RENAL TUBULE AND COLLECTING DUCT Table 26.1 illustrates the histology of the cells that form the renal tubule and collecting duct. In the proximal convoluted tubule, the cells are

---

TABLE 26.1

Histological Features of the Renal Tubule and Collecting Duct

REGION AND HISTOLOGY		DESCRIPTION
Proximal convoluted tubule (PCT)	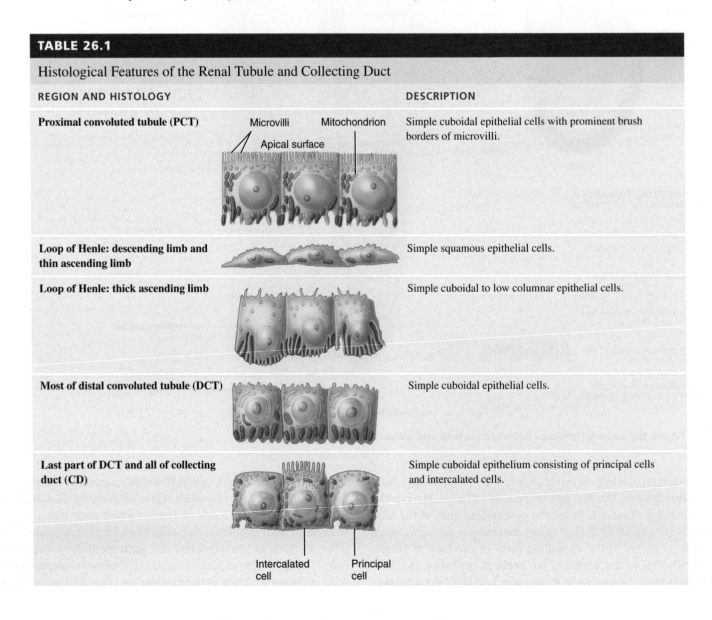 Microvilli, Mitochondrion, Apical surface	Simple cuboidal epithelial cells with prominent brush borders of microvilli.
Loop of Henle: descending limb and thin ascending limb		Simple squamous epithelial cells.
Loop of Henle: thick ascending limb		Simple cuboidal to low columnar epithelial cells.
Most of distal convoluted tubule (DCT)		Simple cuboidal epithelial cells.
Last part of DCT and all of collecting duct (CD)	Intercalated cell, Principal cell	Simple cuboidal epithelium consisting of principal cells and intercalated cells.

Figure 26.6 Histology of a renal corpuscle.

A renal corpuscle consists of a glomerular (Bowman's) capsule and a glomerulus.

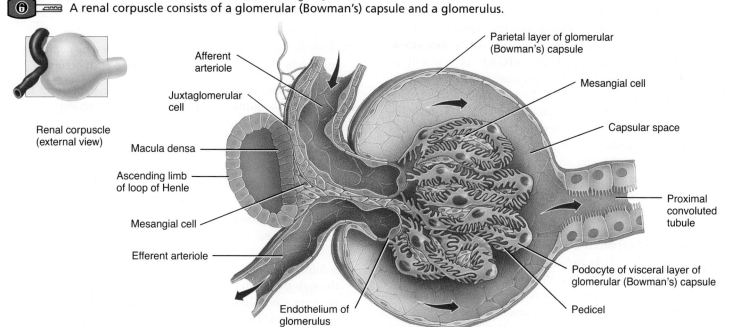

Renal corpuscle (external view)

Afferent arteriole

Juxtaglomerular cell

Macula densa

Ascending limb of loop of Henle

Mesangial cell

Efferent arteriole

Endothelium of glomerulus

Parietal layer of glomerular (Bowman's) capsule

Mesangial cell

Capsular space

Proximal convoluted tubule

Podocyte of visceral layer of glomerular (Bowman's) capsule

Pedicel

(a) Renal corpuscle (internal view)

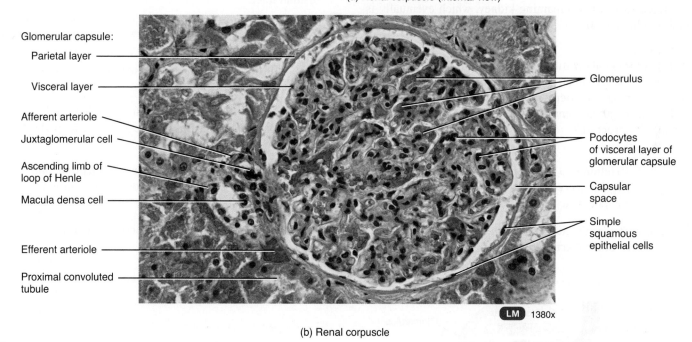

Glomerular capsule:

Parietal layer

Visceral layer

Afferent arteriole

Juxtaglomerular cell

Ascending limb of loop of Henle

Macula densa cell

Efferent arteriole

Proximal convoluted tubule

Glomerulus

Podocytes of visceral layer of glomerular capsule

Capsular space

Simple squamous epithelial cells

LM 1380x

(b) Renal corpuscle

Is the photomicrograph in (b) from a section through the renal cortex or renal medulla? How can you tell?

simple cuboidal epithelial cells with a prominent brush border of microvilli on their apical surface (surface facing the lumen). These microvilli, like those of the small intestine, increase the surface area for reabsorption and secretion. The descending limb of the loop of Henle and the first part of the ascending limb of the loop of Henle (the thin ascending limb) are composed of simple squamous epithelium. (Recall that cortical or short-loop nephrons lack the thin ascending limb.) The thick ascending limb of the loop of Henle is composed of simple cuboidal to low columnar epithelium.

In each nephron, the final part of the ascending limb of the loop of Henle makes contact with the afferent arteriole serving that renal corpuscle (Figure 26.6a). Because the columnar tubule cells in this region are crowded together, they are known as the macula densa (MAK-ū-la DEN-sa; *macula* = spot; *densa* = dense). Alongside the macula densa, the wall of the afferent arteriole (and sometimes the efferent arteriole) contains modified smooth muscle fibers called **juxtaglomerular (JG) cells** (juks'-ta-glō-MER-ū-lar). Together with the macula densa, they constitute the **juxtaglomerular apparatus (JGA).** As you will see later, the JGA helps regulate blood pressure within the kidneys. The distal convoluted tubule (DCT) begins a short distance past the macula densa. In the last part of the DCT and continuing into the collecting ducts, two different types of cells are present. Most are **principal cells,** which have receptors for both antidiuretic hormone (ADH) and aldosterone, two hormones that regulate their functions. A smaller number are **intercalated cells** (in-TER-ka-lā-ted), which play a role in the homeostasis of blood pH. The collecting ducts drain into large papillary ducts, which are lined by simple columnar epithelium.

The number of nephrons is constant from birth. Any increase in kidney size is due solely to the growth of individual nephrons. If nephrons are injured or become diseased, new ones do not form. Signs of kidney dysfunction usually do not become apparent until function declines to less than 25% of normal because the remaining functional nephrons adapt to handle a larger-than-normal load. Surgical removal of one kidney, for example, stimulates hypertrophy (enlargement) of the remaining kidney, which eventually is able to filter blood at 80% of the rate of two normal kidneys.

✔ CHECKPOINT

2. What is the renal capsule and why is it important?
3. What are the two main parts of a nephron?
4. What are the components of the renal tubule?
5. Where is the juxtaglomerular apparatus (JGA) located, and what is its structure?

26.3 OVERVIEW OF RENAL PHYSIOLOGY

◉ OBJECTIVE
• Identify the three basic functions performed by nephrons and collecting ducts, and indicate where each occurs.

To produce urine, nephrons and collecting ducts perform three basic processes—glomerular filtration, tubular reabsorption, and tubular secretion (Figure 26.7):

❶ *Glomerular filtration.* In the first step of urine production, water and most solutes in blood plasma move across the wall of glomerular capillaries, where they are filtered and move into the glomerular capsule and then into the renal tubule.

❷ *Tubular reabsorption.* As filtered fluid flows through the renal tubules and through the collecting ducts, tubule cells reabsorb about 99% of the filtered water and many useful solutes. The water and solutes return to the blood as it flows through the peritubular capillaries and vasa recta. Note that the term *reabsorption* refers to the return of substances to the bloodstream. The term *absorption,* by contrast, means entry of new substances into the body, as occurs in the gastrointestinal tract.

❸ *Tubular secretion.* As filtered fluid flows through the renal tubules and collecting ducts, the renal tubule and duct cells secrete other materials, such as wastes, drugs, and excess ions, into the fluid. Notice that tubular secretion *removes a substance from the blood.*

Solutes and the fluid that drain into the minor and major calyces and renal pelvis constitute urine and are excreted. The rate

Figure 26.7 **Relation of a nephron's structure to its three basic functions: glomerular filtration, tubular reabsorption, and tubular secretion.** Excreted substances remain in the urine and subsequently leave the body. For any substance S, excretion rate of S = filtration rate of S − reabsorption rate of S + secretion rate of S.

🔑 Glomerular filtration occurs in the renal corpuscle. Tubular reabsorption and tubular secretion occur all along the renal tubule and collecting duct.

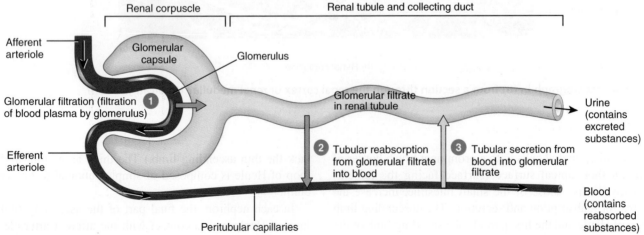

❓ **When cells of the renal tubules secrete the drug penicillin, is the drug being added to or removed from the bloodstream?**

of urinary excretion of any solute is equal to its rate of glomerular filtration, plus its rate of secretion, minus its rate of reabsorption.

By filtering, reabsorbing, and secreting, nephrons help maintain homeostasis of the blood's volume and composition. The situation is somewhat analogous to a recycling center: Garbage trucks dump refuse into an input hopper, where the smaller refuse passes onto a conveyor belt (glomerular filtration of plasma). As the conveyor belt carries the garbage along, workers remove useful items, such as aluminum cans, plastics, and glass containers (reabsorption). Other workers place additional garbage left at the center and larger items onto the conveyor belt (secretion). At the end of the belt, all remaining garbage falls into a truck for transport to the landfill (excretion of wastes in urine).

✔ CHECKPOINT

6. What is the difference between tubular reabsorption and tubular secretion?

26.4 GLOMERULAR FILTRATION

◉ OBJECTIVES
- Describe the filtration membrane.
- Discuss the pressures that promote and oppose glomerular filtration.

The fluid that enters the capsular space is called the **glomerular filtrate.** The fraction of blood plasma in the afferent arterioles

of the kidneys that becomes glomerular filtrate is the **filtration fraction.** Although a filtration fraction of 0.16–0.20 (16–20%) is typical, the value varies considerably in both health and disease. On average, the daily volume of glomerular filtrate in adults is 150 liters in females and 180 liters in males. More than 99% of the glomerular filtrate returns to the bloodstream via tubular reabsorption, so only 1–2 liters (about 1–2 qt) is excreted as urine.

The Filtration Membrane

Together, the glomerular capillaries and the podocytes, which completely encircle the capillaries, form a leaky barrier known as the **filtration membrane.** This sandwichlike assembly permits filtration of water and small solutes but prevents filtration of most plasma proteins, blood cells, and platelets. Substances filtered from the blood cross three filtration barriers—a glomerular endothelial cell, the basal lamina, and a filtration slit formed by a podocyte (Figure 26.8):

1 Glomerular endothelial cells are quite leaky because they have large **fenestrations** (fen′-es-TRĀ-shuns) (pores) that measure 0.07–0.1 μm in diameter. This size permits all solutes in blood plasma to exit glomerular capillaries but prevents filtration of blood cells and platelets. Located among the glomerular capillaries and in the cleft between afferent and efferent arterioles are **mesangial cells** (mes-AN-jē-al; *mes-* =

Figure 26.8 The filtration membrane. The size of the endothelial fenestrations and filtration slits in (a) have been exaggerated for emphasis.

🔒 ▭ During glomerular filtration, water and solutes pass from blood plasma into the capsular space.

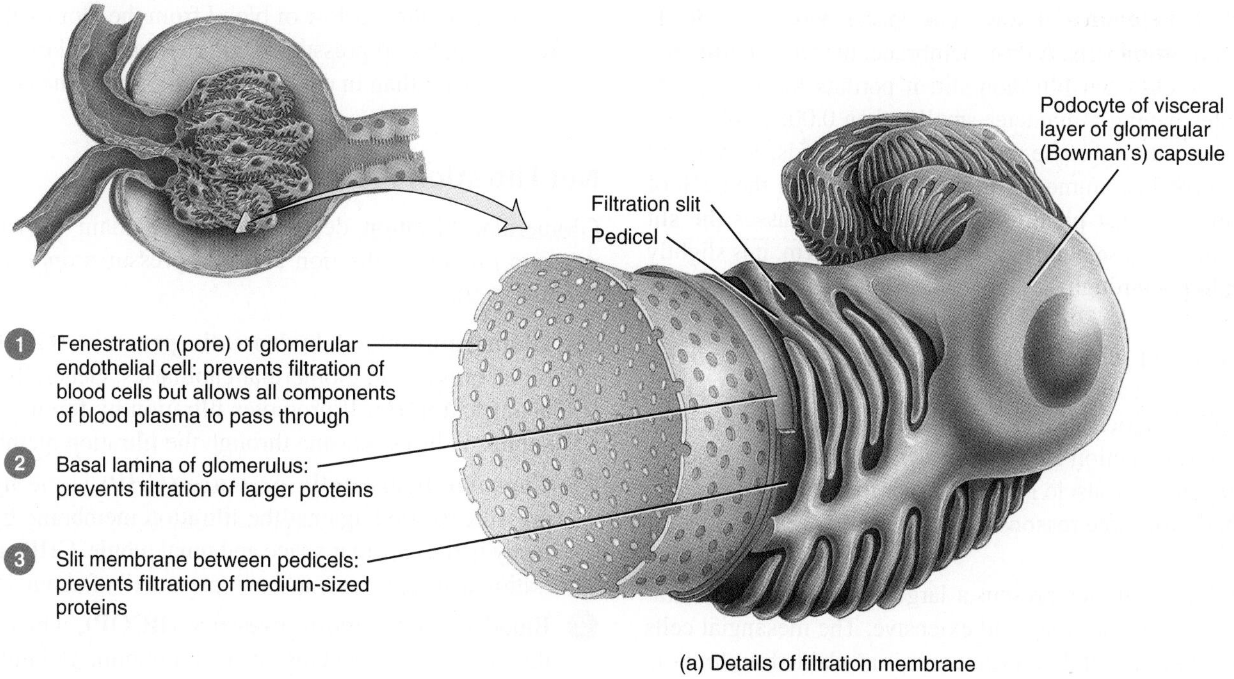

1 Fenestration (pore) of glomerular endothelial cell: prevents filtration of blood cells but allows all components of blood plasma to pass through

2 Basal lamina of glomerulus: prevents filtration of larger proteins

3 Slit membrane between pedicels: prevents filtration of medium-sized proteins

Filtration slit
Pedicel

Podocyte of visceral layer of glomerular (Bowman's) capsule

(a) Details of filtration membrane

FIGURE 26.8 CONTINUES ▶

FIGURE 26.8 CONTINUED

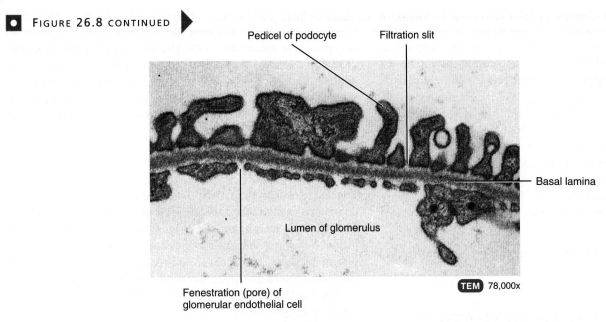

Pedicel of podocyte

Filtration slit

Basal lamina

Lumen of glomerulus

Fenestration (pore) of glomerular endothelial cell

TEM 78,000x

(b) Filtration membrane

? Which part of the filtration membrane prevents red blood cells from entering the capsular space?

in the middle; -*angi*- = blood vessel) (see Figure 26.6a). These contractile cells help regulate glomerular filtration.

❷ The **basal lamina,** a layer of acellular material between the endothelium and the podocytes, consists of minute collagen fibers and proteoglycans in a glycoprotein matrix; negative charges in the matrix prevent filtration of larger negatively charged plasma proteins.

❸ Extending from each podocyte are thousands of footlike processes termed **pedicels** (PED-i-sels = little feet) that wrap around glomerular capillaries. The spaces between pedicels are the **filtration slits.** A thin membrane, the **slit membrane,** extends across each filtration slit; it permits the passage of molecules having a diameter smaller than 0.006–0.007 μm, including water, glucose, vitamins, amino acids, very small plasma proteins, ammonia, urea, and ions. Less than 1% of albumin, the most plentiful plasma protein, passes the slit membrane because, with a diameter of 0.007 μm, it is slightly too big to get through.

The principle of *filtration*—the use of pressure to force fluids and solutes through a membrane—is the same in glomerular capillaries as in capillaries elsewhere in the body (see Starling's law of the capillaries, Section 21.2). However, the volume of fluid filtered by the renal corpuscle is much larger than in other capillaries of the body for three reasons:

1. Glomerular capillaries present a large surface area for filtration because they are long and extensive. The mesangial cells regulate how much of this surface area is available for filtration. When mesangial cells are relaxed, surface area is maximal, and glomerular filtration is very high. Contraction of mesangial

cells reduces the available surface area, and glomerular filtration decreases.

2. The filtration membrane is thin and porous. Despite having several layers, the thickness of the filtration membrane is only 0.1 mm. Glomerular capillaries also are about 50 times leakier than capillaries in most other tissues, mainly because of their large fenestrations.

3. Glomerular capillary blood pressure is high. Because the efferent arteriole is smaller in diameter than the afferent arteriole, resistance to the outflow of blood from the glomerulus is high. As a result, blood pressure in glomerular capillaries is considerably higher than in capillaries elsewhere in the body.

Net Filtration Pressure

Glomerular filtration depends on three main pressures. One pressure *promotes* filtration and two pressures *oppose* filtration (Figure 26.9).

❶ **Glomerular blood hydrostatic pressure (GBHP)** is the blood pressure in glomerular capillaries. Generally, GBHP is about 55 mmHg. It promotes filtration by forcing water and solutes in blood plasma through the filtration membrane.

❷ **Capsular hydrostatic pressure (CHP)** is the hydrostatic pressure exerted against the filtration membrane by fluid already in the capsular space and renal tubule. CHP opposes filtration and represents a "back pressure" of about 15 mmHg.

❸ **Blood colloid osmotic pressure (BCOP),** which is due to the presence of proteins such as albumin, globulins, and fibrinogen in blood plasma, also opposes filtration. The average BCOP in glomerular capillaries is 30 mmHg.

Figure 26.9 **The pressures that drive glomerular filtration.** Taken together, these pressures determine net filtration pressure (NFP).

🔑 Glomerular blood hydrostatic pressure promotes filtration, whereas capsular hydrostatic pressure and blood colloid osmotic pressure oppose filtration.

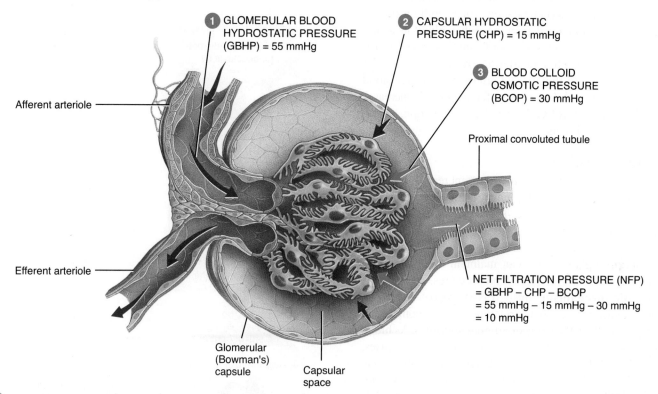

1 GLOMERULAR BLOOD HYDROSTATIC PRESSURE (GBHP) = 55 mmHg

2 CAPSULAR HYDROSTATIC PRESSURE (CHP) = 15 mmHg

3 BLOOD COLLOID OSMOTIC PRESSURE (BCOP) = 30 mmHg

Proximal convoluted tubule

Afferent arteriole

Efferent arteriole

NET FILTRATION PRESSURE (NFP)
= GBHP – CHP – BCOP
= 55 mmHg – 15 mmHg – 30 mmHg
= 10 mmHg

Glomerular (Bowman's) capsule

Capsular space

❓ **Suppose a tumor is pressing on and obstructing the right ureter. What effect might this have on CHP and thus on NFP in the right kidney? Would the left kidney also be affected?**

Net filtration pressure (NFP), the total pressure that promotes filtration, is determined as follows:

Net filtration pressure (NFP) = GBHP − CHP − BCOP

By substituting the values just given, normal NFP may be calculated:

$$NFP = 55 \text{ mmHg} - 15 \text{ mmHg} - 30 \text{ mmHg}$$
$$= 10 \text{ mmHg}$$

Thus, a pressure of only 10 mmHg causes a normal amount of blood plasma (minus plasma proteins) to filter from the glomerulus into the capsular space.

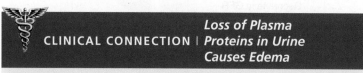

CLINICAL CONNECTION | *Loss of Plasma Proteins in Urine Causes Edema*

In some kidney diseases, glomerular capillaries are damaged and become so permeable that plasma proteins enter glomerular filtrate. As a result, the filtrate exerts a colloid osmotic pressure that draws water out of the blood. In this situation, the NFP increases, which means more fluid is filtered. At the same time, blood colloid osmotic pressure decreases because plasma proteins are being lost in the urine. Because more fluid filters out of blood capillaries into tissues throughout the body than returns via reabsorption, blood volume decreases and interstitial fluid volume increases. Thus, loss of plasma proteins in urine causes *edema*, an abnormally high volume of interstitial fluid. •

Glomerular Filtration Rate

The amount of filtrate formed in all the renal corpuscles of both kidneys each minute is the **glomerular filtration rate (GFR).** In adults, the GFR averages 125 mL/min in males and 105 mL/min in females. Homeostasis of body fluids requires that the kidneys maintain a relatively constant GFR. If the GFR is too high, needed substances may pass so quickly through the renal tubules that some are not reabsorbed and are lost in the urine. If the GFR is too low, nearly all the filtrate may be reabsorbed and certain waste products may not be adequately excreted.

GFR is directly related to the pressures that determine net filtration pressure; any change in net filtration pressure will affect GFR. Severe blood loss, for example, reduces mean arterial blood pressure and decreases the glomerular blood hydrostatic pressure. Filtration ceases if glomerular blood hydrostatic pressure drops to 45 mmHg because the opposing pressures add up to 45 mmHg. Amazingly, when systemic blood pressure rises above normal, net filtration pressure and GFR increase very little. GFR is nearly constant when the mean arterial blood pressure is anywhere between 80 and 180 mmHg.

The mechanisms that regulate glomerular filtration rate operate in two main ways: (1) by adjusting blood flow into and out of the glomerulus and (2) by altering the glomerular capillary surface area available for filtration. GFR increases when blood flow into the glomerular capillaries increases. Coordinated control of the diameter of both afferent and efferent arterioles regulates glomerular blood flow. Constriction of the afferent arteriole decreases blood flow into the glomerulus; dilation of the afferent arteriole increases it. Three mechanisms control GFR: renal autoregulation, neural regulation, and hormonal regulation.

Renal Autoregulation of GFR

The kidneys themselves help maintain a constant renal blood flow and GFR despite normal, everyday changes in blood pressure, like those that occur during exercise. This capability is called **renal autoregulation** (aw′-tō-reg′-ū-LĀ-shun) and consists of two mechanisms—the myogenic mechanism and tubuloglomerular feedback. Working together, they can maintain nearly constant GFR over a wide range of systemic blood pressures.

The **myogenic mechanism** (mī-ō-JEN-ik; *myo-* = muscle; *-genic* = producing) occurs when stretching triggers contraction of smooth muscle cells in the walls of afferent arterioles. As blood pressure rises, GFR also rises because renal blood flow increases. However, the elevated blood pressure stretches the walls of the afferent arterioles. In response, smooth muscle fibers in the wall of the afferent arteriole contract, which narrows the arteriole's lumen. As a result, renal blood flow decreases, thus reducing GFR to its previous level. Conversely, when arterial blood pressure drops, the smooth muscle cells are stretched less and thus relax. The afferent arterioles dilate, renal blood flow increases, and GFR increases. The myogenic mechanism normalizes renal blood flow and GFR within seconds after a change in blood pressure.

The second contributor to renal autoregulation, **tubuloglomerular feedback** (too′-bū-lō-glō-MER-ū-lar), is so named because part of the renal tubules—the macula densa—provides feedback to the glomerulus (Figure 26.10). When GFR is above normal due to elevated systemic blood pressure, filtered fluid flows more rapidly along the renal tubules. As a result, the proximal convoluted tubule and loop of Henle have less time to reabsorb Na^+, Cl^-, and water. Macula densa cells are thought to detect the increased delivery of Na^+, Cl^-, and water and to inhibit release of nitric oxide (NO) from cells in the juxtaglomerular apparatus (JGA). Because NO causes vasodilation, afferent arterioles constrict when the level of NO declines. As a result, less blood flows into the glomerular capillaries, and GFR decreases. When blood pressure falls, causing GFR to be lower than normal, the opposite sequence of events occurs, although to a lesser degree. Tubuloglomerular feedback operates more slowly than the myogenic mechanism.

Neural Regulation of GFR

Like most blood vessels of the body, those of the kidneys are supplied by sympathetic ANS fibers that release norepinephrine. Nor-

Figure 26.10 Tubuloglomerular feedback.

Macula densa cells of the juxtaglomerular apparatus (JGA) provide negative feedback regulation of the glomerular filtration rate.

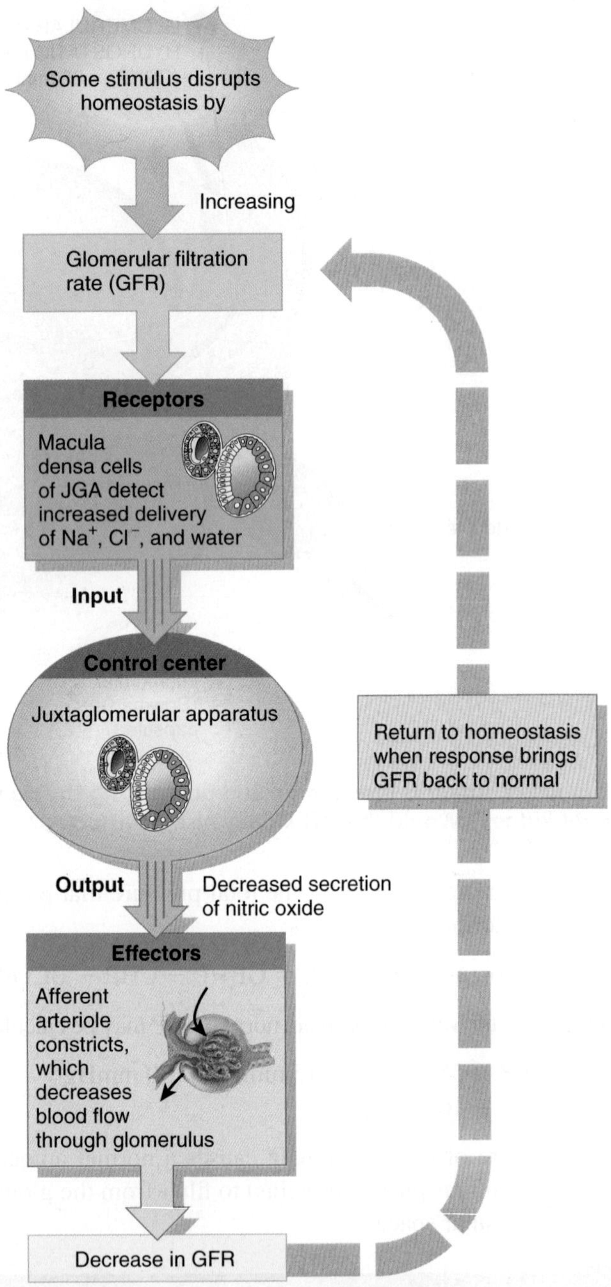

Why is this process termed autoregulation?

epinephrine causes vasoconstriction through the activation of α_1 receptors, which are particularly plentiful in the smooth muscle fibers of afferent arterioles. At rest, sympathetic stimulation is moderately low, the afferent and efferent arterioles are dilated, and renal autoregulation of GFR prevails. With moderate sympathetic stimulation, both afferent and efferent arterioles constrict to the same degree. Blood flow into and out of the glomerulus is restricted to the same extent, which decreases GFR only slightly. With greater sympathetic stimulation, however, as occurs during exercise or hemorrhage, vasoconstriction of the afferent arterioles predominates. As a result, blood flow into glomerular capillaries is greatly decreased, and GFR drops. This lowering of renal blood flow has two consequences: (1) It reduces urine output, which helps conserve blood volume. (2) It permits greater blood flow to other body tissues.

Hormonal Regulation of GFR

Two hormones contribute to regulation of GFR. Angiotensin II reduces GFR; atrial natriuretic peptide (ANP) increases GFR. **Angiotensin II** (an′-jē-ō-TEN-sin) is a very potent vasoconstrictor that narrows both afferent and efferent arterioles and reduces renal blood flow, thereby decreasing GFR. Cells in the atria of the heart secrete **atrial natriuretic peptide (ANP)** (na′-trē-ū-RET-ik). Stretching of the atria, as occurs when blood volume increases, stimulates secretion of ANP. By causing relaxation of the glomerular mesangial cells, ANP increases the capillary surface area available for filtration. Glomerular filtration rate rises as the surface area increases.

Table 26.2 summarizes the regulation of glomerular filtration rate.

✔CHECKPOINT

7. If the urinary excretion rate of a drug such as penicillin is greater than the rate at which it is filtered at the glomerulus, how else is it getting into the urine?
8. What is the major chemical difference between blood plasma and glomerular filtrate?
9. Why is there much greater filtration through glomerular capillaries than through capillaries elsewhere in the body?
10. Write the equation for the calculation of net filtration pressure (NFP) and explain the meaning of each term.
11. How is glomerular filtration rate regulated?

26.5 TUBULAR REABSORPTION AND TUBULAR SECRETION

◉ OBJECTIVES

• Describe the routes and mechanisms of tubular reabsorption and secretion.
• Describe how specific segments of the renal tubule and collecting duct reabsorb water and solutes.
• Describe how specific segments of the renal tubule and collecting duct secrete solutes into the urine.

Principles of Tubular Reabsorption and Secretion

The volume of fluid entering the proximal convoluted tubules in just half an hour is greater than the total blood plasma volume because the normal rate of glomerular filtration is so high. Obviously some of this fluid must be returned somehow to the bloodstream. Reabsorption—the return of most of the filtered water and many of the filtered solutes to the bloodstream—is the second basic function of the nephron and collecting duct. Normally, about 99% of the filtered water is reabsorbed. Epithelial cells all along the renal tubule and duct carry out reabsorption, but proximal convoluted tubule cells make the largest contribution. Solutes that are reabsorbed by both active and passive processes include glucose, amino acids, urea, and ions such as Na^+ (sodium), K^+ (potassium), Ca^{2+} (calcium), Cl^- (chloride), HCO_3^- (bicarbonate), and HPO_4^{2-} (phosphate). Once fluid passes through the proximal convoluted tubule, cells located more distally fine-tune the reabsorption processes to maintain homeostatic balances of water and selected ions. Most small proteins and peptides that pass through the filter also are reabsorbed, usually via pinocytosis.

TABLE 26.2

Regulation of Glomerular Filtration Rate (GFR)

TYPE OF REGULATION	MAJOR STIMULUS	MECHANISM AND SITE OF ACTION	EFFECT ON GFR
Renal autoregulation			
Myogenic mechanism	Increased stretching of smooth muscle fibers in afferent arteriole walls due to increased blood pressure.	Stretched smooth muscle fibers contract, thereby narrowing lumen of afferent arterioles.	Decrease.
Tubuloglomerular feedback	Rapid delivery of Na^+ and Cl^- to the macula densa due to high systemic blood pressure.	Decreased release of nitric oxide (NO) by juxtaglomerular apparatus causes constriction of afferent arterioles.	Decrease.
Neural regulation	Increase in activity level of renal sympathetic nerves releases norepinephrine.	Constriction of afferent arterioles through activation of α_1 receptors and increased release of renin.	Decrease.
Hormone regulation			
Angiotensin II	Decreased blood volume or blood pressure stimulates production of angiotensin II.	Constriction of afferent and efferent arterioles.	Decrease.
Atrial natriuretic peptide (ANP)	Stretching of atria of heart stimulates secretion of ANP.	Relaxation of mesangial cells in glomerulus increases capillary surface area available for filtration.	Increase.

TABLE 26.3

Substances Filtered, Reabsorbed, and Excreted in Urine

SUBSTANCE	FILTERED* (ENTERS GLOMERULAR CAPSULE PER DAY)	REABSORBED (RETURNED TO BLOOD PER DAY)	URINE (EXCRETED PER DAY)
Water	180 liters	178–179 liters	1–2 liters
Proteins	2.0 g	1.9 g	0.1 g
Sodium ions (Na$^+$)	579 g	575 g	4 g
Chloride ions (Cl$^-$)	640 g	633.7 g	6.3 g
Bicarbonate ions (HCO$_3$$^-$)	275 g	274.97 g	0.03 g
Glucose	162 g	162 g	0 g
Urea	54 g	24 g	30 g[†]
Potassium ions (K$^+$)	29.6 g	29.6 g	2.0 g[‡]
Uric acid	8.5 g	7.7 g	0.8 g
Creatinine	1.6 g	0 g	1.6 g

*Assuming GFR is 180 liters per day.   [†]In addition to being filtered and reabsorbed, urea is secreted.
[‡]After virtually all filtered K$^+$ is reabsorbed in the convoluted tubules and loop of Henle, a variable amount of K$^+$ is secreted by principal cells in the collecting duct.

To appreciate the magnitude of tubular reabsorption, look at Table 26.3 and compare the amounts of substances that are filtered, reabsorbed, and excreted in urine.

The third function of nephrons and collecting ducts is tubular secretion, the transfer of materials from the blood and tubule cells into glomerular filtrate. Secreted substances include hydrogen ions (H$^+$), K$^+$, ammonium ions (NH$_4$$^+$), creatinine, and certain drugs such as penicillin. Tubular secretion has two important outcomes: (1) The secretion of H$^+$ helps control blood pH. (2) The secretion of other substances helps eliminate them from the body.

As a result of tubular secretion, certain substances pass from blood into urine and may be detected by a urinalysis (see Section 26.7). It is especially important to test athletes for the presence of performance-enhancing drugs such as anabolic steroids, plasma expanders, erythropoietin, hCG, hGH, and amphetamines. Urine tests can also be used to detect the presence of alcohol or illegal drugs such as marijuana, cocaine, and heroin.

Reabsorption Routes

A substance being reabsorbed from the fluid in the tubule lumen can take one of two routes before entering a peritubular capillary: It can move *between* adjacent tubule cells or *through* an individual tubule cell (Figure 26.11). Along the renal tubule, tight junctions surround and join neighboring cells to one another, much like the plastic rings that hold a six-pack of soda cans together. The **apical membrane** (the tops of the soda cans) contacts the tubular fluid, and the **basolateral membrane** (the bottoms and sides of the soda cans) contacts interstitial fluid at the base and sides of the cell.

Fluid can leak *between* the cells in a passive process known as **paracellular reabsorption** (par′-a-SEL-ū-lar; *para-* = beside). Even though the epithelial cells are connected by tight junctions, the tight junctions between cells in the proximal convoluted

Figure 26.11 Reabsorption routes: paracellular reabsorption and transcellular reabsorption.

In paracellular reabsorption, water and solutes in tubular fluid return to the bloodstream by moving between tubule cells; in transcellular reabsorption, solutes and water in tubular fluid return to the bloodstream by passing through a tubule cell.

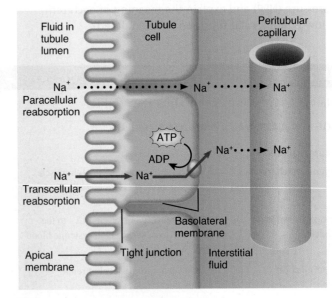

Key:

••••▶ Diffusion

——▶ Active transport

⊘ Sodium–potassium pump (Na$^+$/K$^+$ ATPase)

? **What is the main function of the tight junctions between tubule cells?**

tubules are "leaky" and permit some reabsorbed substances to pass between cells into peritubular capillaries. In some parts of the renal tubule, the paracellular route is thought to account for up to 50% of the reabsorption of certain ions and the water that accompanies them via osmosis. In **transcellular reabsorption** (trans′-SEL-ū-lar; *trans-* = across), a substance passes from the fluid in the tubular lumen *through* the apical membrane of a tubule cell, across the cytosol, and out into interstitial fluid through the basolateral membrane.

Transport Mechanisms

When renal cells transport solutes out of or into tubular fluid, they move specific substances in one direction only. Not surprisingly, different types of transport proteins are present in the apical and basolateral membranes. The tight junctions form a barrier that prevents mixing of proteins in the apical and basolateral membrane compartments. Reabsorption of Na^+ by the renal tubules is especially important because of the large number of sodium ions that pass through the glomerular filters.

Cells lining the renal tubules, like other cells throughout the body, have a low concentration of Na^+ in their cytosol due to the activity of sodium–potassium pumps (Na^+/K^+ ATPases). These pumps are located in the basolateral membranes and eject Na^+ from the renal tubule cells (Figure 26.11). The absence of sodium–potassium pumps in the apical membrane ensures that reabsorption of Na^+ is a one-way process. Most sodium ions that cross the apical membrane will be pumped into interstitial fluid at the base and sides of the cell. The amount of ATP used by sodium–potassium pumps in the renal tubules is about 6% of the total ATP consumption of the body at rest. This may not sound like much, but it is about the same amount of energy used by the diaphragm as it contracts during quiet breathing.

As we noted in Chapter 3, transport of materials across membranes may be either active or passive. Recall that in **primary active transport** the energy derived from hydrolysis of ATP is used to "pump" a substance across a membrane; the sodium–potassium pump is one such pump. In **secondary active transport** the energy stored in an ion's electrochemical gradient, rather than hydrolysis of ATP, drives another substance across a membrane. Secondary active transport couples the movement of an ion down its electrochemical gradient to the "uphill" movement of a second substance against its electrochemical gradient. *Symporters* are membrane proteins that move two or more substances in the same direction across a membrane. *Antiporters* move two or more substances in opposite directions across a membrane. Each type of transporter has an upper limit on how fast it can work, just as an escalator has a limit on how many people it can carry from one level to another in a given period. This limit, called the **transport maximum (T_m),** is measured in mg/min.

Solute reabsorption drives water reabsorption because all water reabsorption occurs via osmosis. About 90% of the reabsorption of water filtered by the kidneys occurs along with the reabsorption of solutes such as Na^+, Cl^-, and glucose. Water reabsorbed with solutes in tubular fluid is termed **obligatory water reabsorption** (ob-LIG-a-tor′-ē) because the water is "obliged" to fol-

low the solutes when they are reabsorbed. This type of water reabsorption occurs in the proximal convoluted tubule and the descending limb of the loop of Henle because these segments of the nephron are always permeable to water. Reabsorption of the final 10% of the water, a total of 10–20 liters per day, is termed **facultative water reabsorption** (FAK-ul-tā′-tiv). The word *facultative* means "capable of adapting to a need." Facultative water reabsorption is regulated by antidiuretic hormone and occurs mainly in the collecting ducts.

Now that we have discussed the principles of renal transport, we will follow the filtered fluid from the proximal convoluted tubule, into the loop of Henle, on to the distal convoluted tubule, and through the collecting ducts. In each segment, we will examine where and how specific substances are reabsorbed and secreted. The filtered fluid becomes *tubular fluid* once it enters the proximal convoluted tubule. The composition of tubular fluid changes as it flows along the nephron tubule and through the collecting duct due to reabsorption and secretion. The fluid that drains from papillary ducts into the renal pelvis is *urine*.

Reabsorption and Secretion in the Proximal Convoluted Tubule

The largest amount of solute and water reabsorption from filtered fluid occurs in the proximal convoluted tubules, which reabsorb 65% of the filtered water, Na^+, and K^+; 100% of most filtered organic solutes such as glucose and amino acids; 50% of the filtered Cl^-; 80–90% of the filtered HCO_3^-; 50% of the filtered urea; and a variable amount of the filtered Ca^{2+}, Mg^{2+}, and HPO_4^{2-} (phosphate). In addition, proximal convoluted tubules secrete a variable amount of H^+ ions, ammonium ions (NH_4^+), and urea.

Most solute reabsorption in the proximal convoluted tubule (PCT) involves Na^+. Na^+ transport occurs via symport and antiport mechanisms in the proximal convoluted tubule. Normally, filtered glucose, amino acids, lactic acid, water-soluble vitamins, and other nutrients are not lost in the urine. Rather, they are completely reabsorbed in the first half of the proximal convoluted tubule by several types of **Na^+ symporters** located in the apical membrane. Figure 26.12 depicts the operation of one such symporter, the **Na^+–glucose symporter** in the apical membrane of a

Figure 26.12 Reabsorption of glucose by Na^+–glucose symporters in cells of the proximal convoluted tubule (PCT).

🔑🔲 Normally, all filtered glucose is reabsorbed in the PCT.

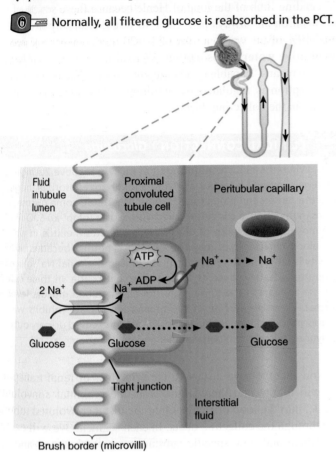

Key:

▭ Na^+–glucose symporter

⊏⊐ Glucose facilitated diffusion transporter

•••• ▸ Diffusion

⊘ Sodium–potassium pump

❓ **How does filtered glucose enter and leave a PCT cell?**

cell in the PCT. Two Na^+ and a molecule of glucose attach to the symporter protein, which carries them from the tubular fluid into the tubule cell. The glucose molecules then exit the basolateral membrane via facilitated diffusion and they diffuse into peritubular capillaries. Other Na^+ symporters in the PCT reclaim filtered HPO_4^{2-} (phosphate) and SO_4^{2-} (sulfate) ions, all amino acids, and lactic acid in a similar way.

In another secondary active transport process, the **Na^+/H^+ antiporters** carry filtered Na^+ down its concentration gradient into a PCT cell as H^+ is moved from the cytosol into the lumen (Figure 26.13a), causing Na^+ to be reabsorbed into blood and H^+ to be secreted into tubular fluid. PCT cells produce the H^+ needed to keep the antiporters running in the following way. Carbon dioxide (CO_2) diffuses from peritubular blood or tubular fluid or is produced by metabolic reactions within the cells. As also occurs in red blood cells (see Figure 23.23), the enzyme *carbonic anhydrase (CA)* (an-HĪ-drās) catalyzes the reaction of CO_2 with

Figure 26.13 Actions of Na^+/H^+ antiporters in proximal convoluted tubule cells. (a) Reabsorption of sodium ions (Na^+) and secretion of hydrogen ions (H^+) via secondary active transport through the apical membrane; (b) reabsorption of bicarbonate ions (HCO_3^-) via facilitated diffusion through the basolateral membrane. CO_2 = carbon dioxide; H_2CO_3 = carbonic acid; CA = carbonic anhydrase.

🔑🔲 Na^+/H^+ antiporters promote transcellular reabsorption of Na^+ and secretion of H^+.

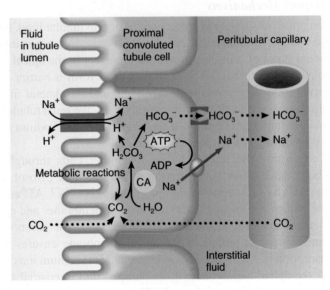

(a) Na^+ reabsorption and H^+ secretion

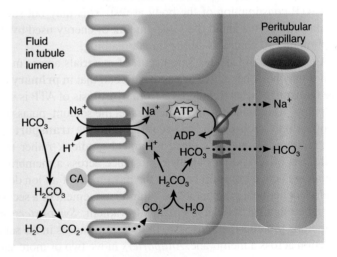

(b) HCO_3^- reabsorption

Key:

▬ Na^+/H^+ antiporter

⊏⊐ HCO_3^- facilitated diffusion transporter

•••• ▸ Diffusion

⊘ Sodium–potassium pump

❓ **Which step in Na^+ movement in part (a) is promoted by the electrochemical gradient?**

water (H_2O) to form carbonic acid (H_2CO_3), which then dissociates into H^+ and HCO_3^-:

$$CO_2 + H_2O \xrightarrow{\text{Carbonic anhydrase}} H_2CO_3 \longrightarrow H^+ + HCO_3^-$$

Most of the HCO_3^- in filtered fluid is reabsorbed in proximal convoluted tubules, thereby safeguarding the body's supply of an important buffer (Figure 26.13b). After H^+ is secreted into the fluid within the lumen of the proximal convoluted tubule, it reacts with filtered HCO_3^- to form H_2CO_3, which readily dissociates into CO_2 and H_2O. Carbon dioxide then diffuses into the tubule cells and joins with H_2O to form H_2CO_3, which dissociates into H^+ and HCO_3^-. As the level of HCO_3^- rises in the cytosol, it exits via facilitated diffusion transporters in the basolateral membrane and diffuses into the blood with Na^+. Thus, for every H^+ secreted into the tubular fluid of the proximal convoluted tubule, one HCO_3^- and one Na^+ are reabsorbed.

Solute reabsorption in proximal convoluted tubules promotes osmosis of water. Each reabsorbed solute increases the osmolarity, first inside the tubule cell, then in interstitial fluid, and finally in the blood. Water thus moves rapidly from the tubular fluid, via both the paracellular and transcellular routes, into the peritubular capillaries and restores osmotic balance (Figure 26.14). In other words, reabsorption of the solutes creates an osmotic gradient that promotes the reabsorption of water via osmosis. Cells lining the proximal convoluted tubule and the descending limb of the loop of Henle are especially permeable to water because they have many molecules of **aquaporin-1** (ak-kwa-PŌR-in). This integral protein in the plasma membrane is a water channel that greatly increases the rate of water movement across the apical and basolateral membranes.

As water leaves the tubular fluid, the concentrations of the remaining filtered solutes increase. In the second half of the PCT, electrochemical gradients for Cl^-, K^+, Ca^{2+}, Mg^{2+}, and urea promote their passive diffusion into peritubular capillaries via both paracellular and transcellular routes. Among these ions, Cl^- is present in the highest concentration. Diffusion of negatively charged Cl^- into interstitial fluid via the paracellular route makes the interstitial fluid electrically more negative than the tubular fluid. This negativity promotes passive paracellular reabsorption of cations, such as K^+, Ca^{2+}, and Mg^{2+}.

Ammonia (NH_3) is a poisonous waste product derived from the deamination (removal of an amino group) of various amino acids, a reaction that occurs mainly in hepatocytes (liver cells). Hepatocytes convert most of this ammonia to urea, a less-toxic compound. Although tiny amounts of urea and ammonia are present in sweat, most excretion of these nitrogen-containing waste products occurs via the urine. Urea and ammonia in blood are both filtered at the glomerulus and secreted by proximal convoluted tubule cells into the tubular fluid.

Proximal convoluted tubule cells can produce additional NH_3 by deaminating the amino acid glutamine in a reaction that also generates HCO_3^-. The NH_3 quickly binds H^+ to become an ammonium ion (NH_4^+), which can substitute for H^+ aboard Na^+/H^+ antiporters in the apical membrane and be secreted into the tubular fluid. The HCO_3^- generated in this reaction moves through the basolateral membrane and then diffuses into the bloodstream, providing additional buffers in blood plasma.

Reabsorption in the Loop of Henle

Because all of the proximal convoluted tubules reabsorb about 65% of the filtered water (about 80 mL/min), fluid enters the next part of the nephron, the loop of Henle, at a rate of 40–45 mL/min. The chemical composition of the tubular fluid now is quite different from that of glomerular filtrate because glucose, amino acids, and other nutrients are no longer present. The osmolarity of the tubular fluid is still close to the osmolarity of blood, however, because reabsorption of water by osmosis keeps pace with reabsorption of solutes all along the proximal convoluted tubule.

The loop of Henle reabsorbs about 15% of the filtered water, 20–30% of the filtered Na^+ and K^+, 35% of the filtered Cl^-, 10–20% of the filtered HCO_3^-, and a variable amount of the filtered Ca^{2+} and Mg^{2+}. Here, for the first time, reabsorption of water via osmosis is *not* automatically coupled to reabsorption of filtered solutes because part of the loop of Henle is relatively impermeable to water. The loop of Henle thus sets the stage for *independent* regulation of both the *volume* and *osmolarity* of body fluids.

The apical membranes of cells in the thick ascending limb of the loop of Henle have **Na^+–K^+–$2Cl^-$ symporters** that simultaneously reclaim one Na^+, one K^+, and two Cl^- from the fluid in the tubular lumen (Figure 26.15). Na^+ that is actively transported into interstitial fluid at the base and sides of the cell diffuses into the vasa recta. Cl^- moves through leakage channels in the basolateral membrane into interstitial fluid and then into the vasa

Figure 26.14 Passive reabsorption of Cl^-, K^+, Ca^{2+}, Mg^{2+}, urea, and water in the second half of the proximal convoluted tubule.

Electrochemical gradients promote passive reabsorption of solutes via both paracellular and transcellular routes.

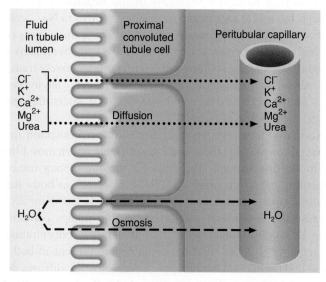

By what mechanism is water reabsorbed from tubular fluid?

Figure 26.15 Na⁺–K⁺–2Cl⁻ symporter in the thick ascending limb of the loop of Henle.

Cells in the thick ascending limb have symporters that simultaneously reabsorb one Na^+, one K^+, and two Cl^-.

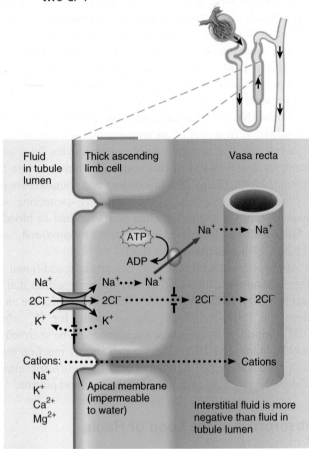

Fluid in tubule lumen

Thick ascending limb cell

Vasa recta

ATP

ADP

Na^+ ···· ▶ Na^+

Na^+
$2Cl^-$
K^+

Na^+ ···▶ Na^+
$2Cl^-$ ··········· $2Cl^-$ ···· ▶ $2Cl^-$
K^+

Cations: ·················· ▶ Cations
Na^+
K^+
Ca^{2+}
Mg^{2+}

Apical membrane (impermeable to water)

Interstitial fluid is more negative than fluid in tubule lumen

Key:

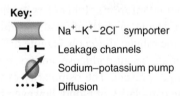 Na⁺–K⁺–2Cl⁻ symporter

Leakage channels

Sodium–potassium pump

····▶ Diffusion

Why is this process considered secondary active transport? Does water reabsorption accompany ion reabsorption in this region of the nephron?

recta. Because many K^+ leakage channels are present in the apical membrane, most K^+ brought in by the symporters moves down its concentration gradient back into the tubular fluid. Thus, the main effect of the Na⁺–K⁺–2Cl⁻ symporters is reabsorption of Na^+ and Cl^-.

The movement of positively charged K^+ into the tubular fluid through the apical membrane channels leaves the interstitial fluid and blood with more negative charges relative to fluid in the ascending limb of the loop of Henle. This relative negativity promotes reabsorption of cations—Na^+, K^+, Ca^{2+}, and Mg^{2+}—via the paracellular route.

Although about 15% of the filtered water is reabsorbed in the *descending* limb of the loop of Henle, little or no water is reabsorbed in the *ascending* limb. In this segment of the tubule, the apical membranes are virtually impermeable to water. Because ions but not water molecules are reabsorbed, the osmolarity of the tubular fluid decreases progressively as fluid flows toward the end of the ascending limb.

Reabsorption in the Early Distal Convoluted Tubule

Fluid enters the distal convoluted tubules at a rate of about 25 mL/min because 80% of the filtered water has now been reabsorbed. The early or initial part of the distal convoluted tubule (DCT) reabsorbs about 10–15% of the filtered water, 5% of the filtered Na^+, and 5% of the filtered Cl^-. Reabsorption of Na^+ and Cl^- occurs by means of **Na⁺–Cl⁻ symporters** in the apical membranes. Sodium–potassium pumps and Cl^- leakage channels in the basolateral membranes then permit reabsorption of Na^+ and Cl^- into the peritubular capillaries. The early DCT also is a major site where parathyroid hormone (PTH) stimulates reabsorption of Ca^{2+}. The amount of Ca^{2+} reabsorption in the early DCT varies depending on the body's needs.

Reabsorption and Secretion in the Late Distal Convoluted Tubule and Collecting Duct

By the time fluid reaches the end of the distal convoluted tubule, 90–95% of the filtered solutes and water have returned to the bloodstream. Recall that two different types of cells—principal cells and intercalated cells—are present at the late or terminal part of the distal convoluted tubule and throughout the collecting duct. The principal cells reabsorb Na^+ and secrete K^+; the intercalated cells reabsorb K^+ and HCO_3^- and secrete H^+. In the late distal convoluted tubules and collecting ducts, the amount of water and solute reabsorption and the amount of solute secretion vary depending on the body's needs.

In contrast to earlier segments of the nephron, Na^+ passes through the apical membrane of principal cells via Na^+ leakage channels rather than by means of symporters or antiporters (Figure 26.16). The concentration of Na^+ in the cytosol remains low, as usual, because the sodium–potassium pumps actively transport Na^+ across the basolateral membranes. Then Na^+ passively diffuses into the peritubular capillaries from the interstitial spaces around the tubule cells.

Normally, transcellular and paracellular reabsorption in the proximal convoluted tubule and loop of Henle return most filtered K^+ to the bloodstream. To adjust for varying dietary intake of potassium and to maintain a stable level of K^+ in body fluids, principal cells secrete a variable amount of K^+ (Figure 26.16). Because the basolateral sodium–potassium pumps continually bring K^+ into principal cells, the intracellular concentration of K^+ remains high. K^+ leakage channels are present in both the apical and basolateral membranes. Thus, some K^+ diffuses down its concentration gradient into the tubular fluid, where the K^+

Figure 26.16 Reabsorption of Na⁺ and secretion of K⁺ by principal cells in the last part of the distal convoluted tubule and in the collecting duct.

In the apical membrane of principal cells, Na⁺ leakage channels allow entry of Na⁺ while K⁺ leakage channels allow exit of K⁺ into the tubular fluid.

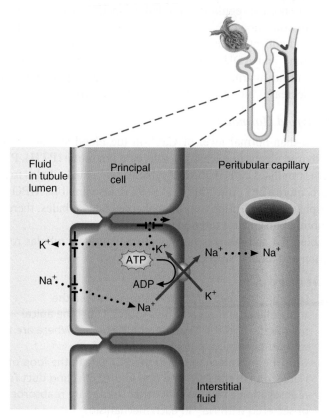

Key:

∙∙∙∙∙► Diffusion

⊣ ⊢ Leakage channels

✖ Sodium–potassium pump

? **Which hormone stimulates reabsorption and secretion by principal cells, and how does this hormone exert its effect?**

concentration is very low. This secretion mechanism is the main source of K⁺ excreted in the urine.

Hormonal Regulation of Tubular Reabsorption and Tubular Secretion

Five hormones affect the extent of Na⁺, Cl⁻, Ca²⁺, and water reabsorption as well as K⁺ secretion by the renal tubules. These hormones include angiotensin II, aldosterone, antidiuretic hormone, atrial natriuretic peptide, and parathyroid hormone.

Renin–Angiotensin–Aldosterone System

When blood volume and blood pressure decrease, the walls of the afferent arterioles are stretched less, and the juxtaglomerular cells

secrete the enzyme **renin** (RĒ-nin) into the blood. Sympathetic stimulation also directly stimulates release of renin from juxtaglomerular cells. Renin clips off a 10–amino acid peptide called angiotensin I from angiotensinogen, which is synthesized by hepatocytes (see Figure 18.16). By clipping off two more amino acids, *angiotensin-converting enzyme (ACE)* (an′-jē-ō-TEN-sin) converts angiotensin I to **angiotensin II,** which is the active form of the hormone.

Angiotensin II affects renal physiology in three main ways:

1. It decreases the glomerular filtration rate by causing vasoconstriction of the afferent arterioles.

2. It enhances reabsorption of Na⁺, Cl⁻, and water in the proximal convoluted tubule by stimulating the activity of Na⁺/H⁺ antiporters.

3. It stimulates the adrenal cortex to release **aldosterone** (al-DOS-ter-ōn), a hormone that in turn stimulates the principal cells in the collecting ducts to reabsorb more Na⁺ and Cl⁻ and secrete more K⁺. The osmotic consequence of reabsorbing more Na⁺ and Cl⁻ is that more water is reabsorbed, which causes an increase in blood volume and blood pressure.

Antidiuretic Hormone

Antidiuretic hormone (ADH or **vasopressin)** is released by the posterior pituitary. It regulates facultative water reabsorption by increasing the water permeability of principal cells in the last part of the distal convoluted tubule and throughout the collecting duct. In the absence of ADH, the apical membranes of principal cells have a very low permeability to water. Within principal cells are tiny vesicles containing many copies of a water channel protein known as **aquaporin-2.*** ADH stimulates insertion of the aquaporin-2–containing vesicles into the apical membranes via exocytosis. As a result, the water permeability of the principal cell's apical membrane increases, and water molecules move more rapidly from the tubular fluid into the cells. Because the basolateral membranes are always relatively permeable to water, water molecules then move rapidly into the blood. The kidneys can produce as little as 400–500 mL of very concentrated urine each day when ADH concentration is maximal, for instance, during severe dehydration. When ADH level declines, the aquaporin-2 channels are removed from the apical membrane via endocytosis. The kidneys produce a large volume of dilute urine when ADH level is low.

A negative feedback system involving ADH regulates facultative water reabsorption (Figure 26.17). When the osmolarity or osmotic pressure of plasma and interstitial fluid increases—that is, when water concentration decreases—by as little as 1%, osmoreceptors in the hypothalamus detect the change. Their nerve impulses stimulate secretion of more ADH into the blood, and the principal cells become more permeable to water. As facultative water reabsorption increases, plasma osmolarity decreases to normal. A second powerful stimulus for ADH secretion is a decrease in blood volume, as occurs in hemorrhaging or severe

*ADH does not govern the previously mentioned water channel (aquaporin-1).

Figure 26.17 Negative feedback regulation of facultative water reabsorption by ADH.

Most water reabsorption (90%) is obligatory; 10% is facultative.

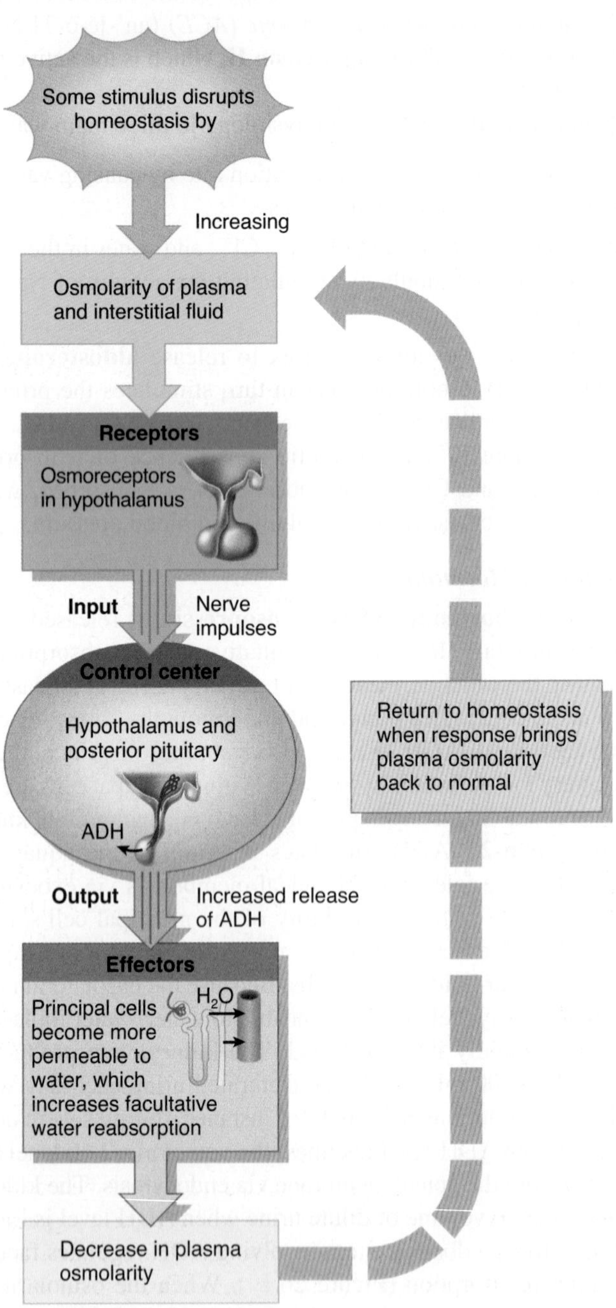

Some stimulus disrupts homeostasis by

Increasing

Osmolarity of plasma and interstitial fluid

Receptors

Osmoreceptors in hypothalamus

Input — Nerve impulses

Control center

Hypothalamus and posterior pituitary

ADH

Return to homeostasis when response brings plasma osmolarity back to normal

Output — Increased release of ADH

Effectors

Principal cells become more permeable to water, which increases facultative water reabsorption

H_2O

Decrease in plasma osmolarity

? In addition to ADH, which other hormones contribute to the regulation of water reabsorption?

dehydration. In the pathological absence of ADH activity, a condition known as *diabetes insipidus,* a person may excrete up to 20 liters of very dilute urine daily.

Atrial Natriuretic Peptide

A large increase in blood volume promotes release of **atrial natriuretic peptide (ANP)** from the heart. Although the importance of ANP in normal regulation of tubular function is unclear, it can inhibit reabsorption of Na^+ and water in the proximal convoluted tubule and collecting duct. ANP also suppresses the secretion of aldosterone and ADH. These effects increase the excretion of Na^+ in urine (natriuresis) and increase urine output (diuresis), which decreases blood volume and blood pressure.

Parathyroid Hormone

A lower-than-normal level of Ca^{2+} in the blood stimulates the parathyroid glands to release **parathyroid hormone (PTH).** PTH in turn stimulates cells in the early distal convoluted tubules to reabsorb more Ca^{2+} into the blood. PTH also inhibits HPO_4^{2-} (phosphate) reabsorption in proximal convoluted tubules, thereby promoting phosphate excretion.

Table 26.4 summarizes hormonal regulation of tubular reabsorption and tubular secretion.

✔ CHECKPOINT

12. Diagram the reabsorption of substances via the transcellular and paracellular routes. Label the apical membrane and the basolateral membrane. Where are the sodium–potassium pumps located?
13. Describe two mechanisms in the PCT, one in the loop of Henle, one in the DCT, and one in the collecting duct for reabsorption of Na^+. What other solutes are reabsorbed or secreted with Na^+ in each mechanism?
14. How do intercalated cells secrete hydrogen ions?
15. Graph the percentages of filtered water and filtered Na^+ that are reabsorbed in the PCT, loop of Henle, DCT, and collecting duct. Indicate which hormones, if any, regulate reabsorption in each segment.

26.6 PRODUCTION OF DILUTE AND CONCENTRATED URINE

◉ OBJECTIVE

• Describe how the renal tubule and collecting ducts produce dilute and concentrated urine.

Even though your fluid intake can be highly variable, the total volume of fluid in your body normally remains stable. Homeostasis of body fluid volume depends in large part on the ability of the kidneys to regulate the rate of water loss in urine. Normally functioning kidneys produce a large volume of dilute urine when fluid intake is high, and a small volume of concentrated urine when fluid intake is low or fluid loss is large. ADH controls whether dilute urine or concentrated urine is formed. In the absence of ADH, urine is very dilute. However, a high level of ADH stimulates

TABLE 26.4

Hormonal Regulation of Tubular Reabsorption and Tubular Secretion

HORMONE	MAJOR STIMULI THAT TRIGGER RELEASE	MECHANISM AND SITE OF ACTION	EFFECTS
Angiotensin II	Low blood volume or low blood pressure stimulates renin-induced production of angiotensin II.	Stimulates activity of Na^+/H^+ antiporters in proximal tubule cells.	Increases reabsorption of Na^+, other solutes, and water, which increases blood volume and blood pressure.
Aldosterone	Increased angiotensin II level and increased level of plasma K^+ promote release of aldosterone by adrenal cortex.	Enhances activity of sodium–potassium pumps in basolateral membrane and Na^+ channels in apical membrane of principal cells in collecting duct.	Increases secretion of K^+ and reabsorption of Na^+, Cl^-; increases reabsorption of water, which increases blood volume and blood pressure.
Antidiuretic hormone (ADH) or vasopressin	Increased osmolarity of extracellular fluid or decreased blood volume promotes release of ADH from posterior pituitary gland.	Stimulates insertion of water channel proteins (aquaporin-2) into apical membranes of principal cells.	Increases facultative reabsorption of water, which decreases osmolarity of body fluids.
Atrial natriuretic peptide (ANP)	Stretching of atria of heart stimulates ANP secretion.	Suppresses reabsorption of Na^+ and water in proximal tubule and collecting duct; inhibits secretion of aldosterone and ADH.	Increases excretion of Na^+ in urine (natriuresis); increases urine output (diuresis) and thus decreases blood volume and blood pressure.
Parathyroid hormone (PTH)	Decreased level of plasma Ca^{2+} promotes release of PTH from parathyroid glands.	Stimulates opening of Ca^{2+} channels in apical membranes of early distal tubule cells.	Increases reabsorption of Ca^{2+}.

reabsorption of more water into blood, producing a concentrated urine.

Formation of Dilute Urine

Glomerular filtrate has the same ratio of water and solute particles as blood; its osmolarity is about 300 mOsm/liter. As previously noted, fluid leaving the proximal convoluted tubule is still isotonic to plasma. When *dilute* urine is being formed (Figure 26.18), the osmolarity of the fluid in the tubular lumen *increases* as it flows down the descending limb of the loop of Henle, *decreases* as it flows up the ascending limb, and *decreases* still more as it flows through the rest of the nephron and collecting duct. These changes in osmolarity result from the following conditions along the path of tubular fluid:

1. Because the osmolarity of the interstitial fluid of the renal medulla becomes progressively greater, more and more water is reabsorbed by osmosis as tubular fluid flows along the descending limb toward the tip of the loop. (The source of this medullary osmotic gradient is explained shortly.) As a result, the fluid remaining in the lumen becomes progressively more concentrated.

2. Cells lining the thick ascending limb of the loop have symporters that actively reabsorb Na^+, K^+, and Cl^- from the tubular fluid (see Figure 26.15). The ions pass from the tubular fluid into thick ascending limb cells, then into interstitial fluid, and finally some diffuse into the blood inside the vasa recta.

3. Although solutes are being reabsorbed in the thick ascending limb, the water permeability of this portion of the nephron is always quite low, so water cannot follow by osmosis. As solutes—but not water molecules—are leaving the tubular fluid, its osmolarity drops to about 150 mOsm/liter. The fluid entering the distal convoluted tubule is thus more dilute than plasma.

4. While the fluid continues flowing along the distal convoluted tubule, additional solutes but only a few water molecules are reabsorbed. The early distal convoluted tubule cells are not very permeable to water and are not regulated by ADH.

5. Finally, the principal cells of the late distal convoluted tubules and collecting ducts are impermeable to water when the ADH level is very low. Thus, tubular fluid becomes progressively more dilute as it flows onward. By the time the tubular fluid drains into the renal pelvis, its concentration can be as low as 65–70 mOsm/liter. This is four times more dilute than blood plasma or glomerular filtrate.

Formation of Concentrated Urine

When water intake is low or water loss is high (such as during heavy sweating), the kidneys must conserve water while still eliminating wastes and excess ions. Under the influence of ADH, the kidneys produce a small volume of highly concentrated urine. Urine can be four times more concentrated (up to 1200 mOsm/liter) than blood plasma or glomerular filtrate (300 mOsm/liter).

Figure 26.18 Formation of dilute urine. Numbers indicate osmolarity in milliosmoles per liter (mOsm/liter). Heavy brown lines in the ascending limb of the loop of Henle and in the distal convoluted tubule indicate impermeability to water; heavy blue lines indicate the last part of the distal convoluted tubule and the collecting duct, which are impermeable to water in the absence of ADH; light blue areas around the nephron represent interstitial fluid.

🔑 When the ADH level is low, urine is dilute and has an osmolarity less than the osmolarity of blood.

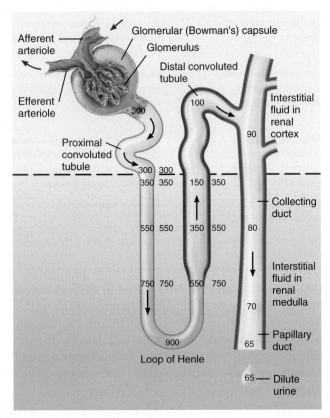

❓ Which portions of the renal tubule and collecting duct reabsorb more solutes than water to produce dilute urine?

The ability of ADH to cause excretion of concentrated urine depends on the presence of an **osmotic gradient** of solutes in the interstitial fluid of the renal medulla. Notice in Figure 26.19 that the solute concentration of the interstitial fluid in the kidney increases from about 300 mOsm/liter in the renal cortex to about 1200 mOsm/liter deep in the renal medulla. The three major solutes that contribute to this high osmolarity are Na^+, Cl^-, and urea. Two main factors contribute to building and maintaining this osmotic gradient: (1) differences in solute and water permeability and reabsorption in different sections of the long loops of Henle and the collecting ducts, and (2) the countercurrent flow of fluid through tube-shaped structures in the renal medulla. *Countercurrent flow* refers to the flow of fluid in opposite directions. This occurs when fluid flowing in one tube runs counter (opposite) to fluid flowing in a nearby parallel tube. Examples of countercur-

rent flow include the flow of tubular fluid through the descending and ascending limbs of the loop of Henle and the flow of blood through the ascending and descending parts of the vasa recta. Two types of **countercurrent mechanisms** exist in the kidneys: countercurrent multiplication and countercurrent exchange.

Countercurrent Multiplication

Countercurrent multiplication is the process by which a progressively increasing osmotic gradient is formed in the interstitial fluid of the renal medulla as a result of countercurrent flow. Countercurrent multiplication involves the long loops of Henle of juxtamedullary nephrons. Note in Figure 26.19a that the descending limb of the loop of Henle carries tubular fluid from the renal cortex deep into the medulla, and the ascending limb carries it in the opposite direction. Since countercurrent flow through the descending and ascending limbs of the long loop of Henle establishes the osmotic gradient in the renal medulla, the long loop of Henle is said to function as a **countercurrent multiplier.** The kidneys use this osmotic gradient to excrete concentrated urine.

Production of concentrated urine by the kidneys occurs in the following way (Figure 26.19):

1 *Symporters in thick ascending limb cells of the loop of Henle cause a buildup of Na^+ and Cl^- in the renal medulla.* In the thick ascending limb of the loop of Henle, the Na^+–K^+–$2Cl^-$ symporters reabsorb Na^+ and Cl^- from the tubular fluid (Figure 26.19a). Water is not reabsorbed in this segment, however, because the cells are impermeable to water. As a result, there is a buildup of Na^+ and Cl^- ions in the interstitial fluid of the medulla.

2 *Countercurrent flow through the descending and ascending limbs of the loop of Henle establishes an osmotic gradient in the renal medulla.* Since tubular fluid constantly moves from the descending limb to the thick ascending limb of the loop of Henle, the thick ascending limb is constantly reabsorbing Na^+ and Cl^-. Consequently, the reabsorbed Na^+ and Cl^- become increasingly concentrated in the interstitial fluid of the medulla, which results in the formation of an osmotic gradient that ranges from 300 mOsm/liter in the outer medulla to 1200 mOsm/liter deep in the inner medulla. The descending limb of the loop of Henle is very permeable to water but impermeable to solutes except urea. Because the osmolarity of the interstitial fluid outside the descending limb is higher than the tubular fluid within it, water moves out of the descending limb via osmosis. This causes the osmolarity of the tubular fluid to increase. As the fluid continues along the descending limb, its osmolarity increases even more: At the hairpin turn of the loop, the osmolarity can be as high as 1200 mOsm/liter in juxtamedullary nephrons. As you have already learned, the ascending limb of the loop is impermeable to water, but its symporters reabsorb Na^+ and Cl^- from the tubular fluid into the interstitial fluid of the renal medulla, so the osmolarity of the tubular fluid progressively decreases as it flows through the ascending limb. At the junction of the medulla and cortex, the osmolarity of the tubular fluid has

Figure 26.19 **Mechanism of urine concentration in long-loop juxtamedullary nephrons.** The green line indicates the presence of Na^+–K^+–$2Cl^-$ symporters that simultaneously reabsorb these ions into the interstitial fluid of the renal medulla; this portion of the nephron is also relatively impermeable to water and urea. All concentrations are in milliosmoles per liter (mOsm/liter).

🔑 The formation of concentrated urine depends on high concentrations of solutes in interstitial fluid in the renal medulla.

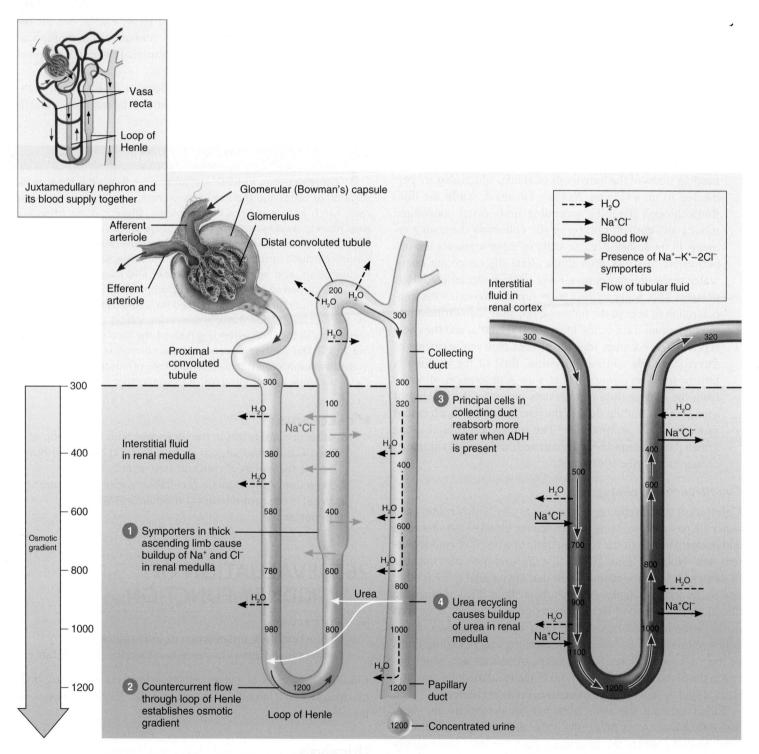

(a) Reabsorption of Na^+, Cl^-, and water in long-loop juxtamedullary nephron

(b) Recycling of salts and urea in vasa recta

❓ **Which solutes are the main contributors to the high osmolarity of interstitial fluid in the renal medulla?**

fallen to about 100 mOsm/liter. Overall, tubular fluid becomes progressively more concentrated as it flows along the descending limb and progressively more dilute as it moves along the ascending limb.

3 *Cells in the collecting ducts reabsorb more water and urea.* When ADH increases the water permeability of the principal cells, water quickly moves via osmosis out of the collecting duct tubular fluid, into the interstitial fluid of the inner medulla, and then into the vasa recta. With loss of water, the urea left behind in the tubular fluid of the collecting duct becomes increasingly concentrated. Because duct cells deep in the medulla are permeable to it, urea diffuses from the fluid in the duct into the interstitial fluid of the medulla.

4 *Urea recycling causes a buildup of urea in the renal medulla.* As urea accumulates in the interstitial fluid, some of it diffuses into the tubular fluid in the descending and thin ascending limbs of the long loops of Henle, which also are permeable to urea (Figure 26.19a). However, while the fluid flows through the thick ascending limb, distal convoluted tubule, and cortical portion of the collecting duct, urea remains in the lumen because cells in these segments are impermeable to it. As fluid flows along the collecting ducts, water reabsorption continues via osmosis because ADH is present. This water reabsorption *further increases* the concentration of urea in the tubular fluid, more urea diffuses into the interstitial fluid of the inner renal medulla, and the cycle repeats. The constant transfer of urea between segments of the renal tubule and the interstitial fluid of the medulla is termed *urea recycling*. In this way, reabsorption of water from the tubular fluid of the ducts promotes the buildup of urea in the interstitial fluid of the renal medulla, which in turn promotes water reabsorption. The solutes left behind in the lumen thus become very concentrated, and a small volume of concentrated urine is excreted.

Countercurrent Exchange

Countercurrent exchange is the process by which solutes and water are passively exchanged between the blood of the vasa recta and interstitial fluid of the renal medulla as a result of countercurrent flow. Note in Figure 26.19b that the vasa recta also consists of descending and ascending limbs that are parallel to each other and to the loop of Henle. Just as tubular fluid flows in opposite directions in the loop of Henle, blood flows in opposite directions in the ascending and descending parts of the vasa recta. Since countercurrent flow between the descending and ascending limbs of the vasa recta allows for exchange of solutes and water between the blood and interstitial fluid of the renal medulla, the vasa recta is said to function as a **countercurrent exchanger.**

Blood entering the vasa recta has an osmolarity of about 300 mOsm/liter. As it flows along the descending part into the renal medulla, where the interstitial fluid becomes increasingly concentrated, Na^+, Cl^-, and urea diffuse from interstitial fluid into the blood and water diffuses from the blood into the interstitial fluid. But after its osmolarity increases, the blood flows into the ascending part of the vasa recta. Here blood flows through a region where the interstitial fluid becomes increasingly less concentrated. As a result Na^+, Cl^-, and urea diffuse from the blood back into interstitial fluid, and water diffuses from interstitial fluid back into the vasa recta. The osmolarity of blood leaving the vasa recta is only slightly higher than the osmolarity of blood entering the vasa recta. Thus, the vasa recta provides oxygen and nutrients to the renal medulla without washing out or diminishing the osmotic gradient. The long loop of Henle *establishes* the osmotic gradient in the renal medulla by countercurrent multiplication, but the vasa recta *maintains* the osmotic gradient in the renal medulla by countercurrent exchange.

Figure 26.20 summarizes the processes of filtration, reabsorption, and secretion in each segment of the nephron and collecting duct.

CLINICAL CONNECTION | *Diuretics*

Diuretics (dī-ū-RET-iks) are substances that slow renal reabsorption of water and thereby cause *diuresis,* an elevated urine flow rate, which in turn reduces blood volume. Diuretic drugs often are prescribed to treat *hypertension* (high blood pressure) because lowering blood volume usually reduces blood pressure. Naturally occurring diuretics include *caffeine* in coffee, tea, and sodas, which inhibits Na^+ reabsorption, and *alcohol* in beer, wine, and mixed drinks, which inhibits secretion of ADH. Most diuretic drugs act by interfering with a mechanism for reabsorption of filtered Na^+. For example, loop diuretics, such as furosemide (Lasix®), selectively inhibit the Na^+–K^+–$2Cl^-$ symporters in the thick ascending limb of the loop of Henle (see Figure 26.15). The thiazide diuretics, such as chlorothiazide (Diuril®), act in the distal convoluted tubule, where they promote loss of Na^+ and Cl^- in the urine by inhibiting Na^+–Cl^- symporters. •

✔ CHECKPOINT

16. How do symporters in the ascending limb of the loop of Henle and principal cells in the collecting duct contribute to the formation of concentrated urine?
17. How does ADH regulate facultative water reabsorption?
18. What is the countercurrent mechanism? Why is it important?

26.7 EVALUATION OF KIDNEY FUNCTION

◉ OBJECTIVES
• Define urinalysis and describe its importance.
• Define renal plasma clearance and describe its importance.

Routine assessment of kidney function involves evaluating both the quantity and quality of urine and the levels of wastes in the blood.

Urinalysis

An analysis of the volume and physical, chemical, and microscopic properties of urine, called a **urinalysis** (ū-ri-NAL-i-sis), reveals much about the state of the body. Table 26.5 summarizes

Figure 26.20 Summary of filtration, reabsorption, and secretion in the nephron and collecting duct.

Filtration occurs in the renal corpuscle; reabsorption occurs all along the renal tubule and collecting ducts.

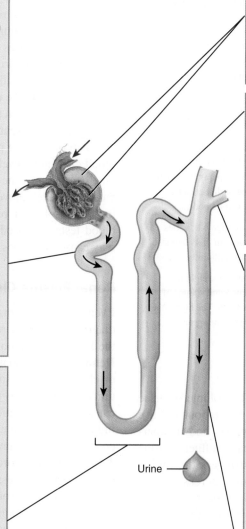

PROXIMAL CONVOLUTED TUBULE

Reabsorption (into blood) of filtered:

Water	65% (osmosis)
Na^+	65% (sodium–potassium pumps, symporters, antiporters)
K^+	65% (diffusion)
Glucose	100% (symporters and facilitated diffusion)
Amino acids	100% (symporters and facilitated diffusion)
Cl^-	50% (diffusion)
HCO_3^-	80–90% (facilitated diffusion)
Urea	50% (diffusion)
Ca^{2+}, Mg^{2+}	variable (diffusion)

Secretion (into urine) of:

H^+	variable (antiporters)
NH_4^+	variable, increases in acidosis (antiporters)
Urea	variable (diffusion)
Creatinine	small amount

At end of PCT, tubular fluid is still isotonic to blood (300 mOsm/liter).

LOOP OF HENLE

Reabsorption (into blood) of:

Water	15% (osmosis in descending limb)
Na^+	20–30% (symporters in ascending limb)
K^+	20–30% (symporters in ascending limb)
Cl^-	35% (symporters in ascending limb)
HCO_3^-	10–20% (facilitated diffusion)
Ca^{2+}, Mg^{2+}	variable (diffusion)

Secretion (into urine) of:

Urea	variable (recycling from collecting duct)

At end of loop of Henle, tubular fluid is hypotonic (100–150 mOsm/liter).

RENAL CORPUSCLE

Glomerular filtration rate:
105–125 mL/min of fluid that is isotonic to blood

Filtered substances: water and all solutes present in blood (except proteins) including ions, glucose, amino acids, creatinine, uric acid

EARLY DISTAL CONVOLUTED TUBULE

Reabsorption (into blood) of:

Water	10–15% (osmosis)
Na^+	5% (symporters)
Cl^-	5% (symporters)
Ca^{2+}	variable (stimulated by parathyroid hormone)

LATE DISTAL CONVOLUTED TUBULE AND COLLECTING DUCT

Reabsorption (into blood) of:

Water	5–9% (insertion of water channels stimulated by ADH)
Na^+	1–4% (sodium–potassium pumps and sodium channels stimulated by aldosterone)
HCO_3^-	variable amount, depends on H^+ secretion (antiporters)
Urea	variable (recycling to loop of Henle)

Secretion (into urine) of:

K^+	variable amount to adjust for dietary intake (leakage channels)
H^+	variable amounts to maintain acid–base homeostasis (H^+ pumps)

Tubular fluid leaving the collecting duct is dilute when ADH level is low and concentrated when ADH level is high.

Urine

In which segments of the nephron and collecting duct does secretion occur?

TABLE 26.5

Characteristics of Normal Urine

CHARACTERISTIC	DESCRIPTION
Volume	One to two liters in 24 hours; varies considerably.
Color	Yellow or amber; varies with urine concentration and diet. Color due to urochrome (pigment produced from breakdown of bile) and urobilin (from breakdown of hemoglobin). Concentrated urine is darker in color. Color affected by diet (reddish from beets), medications, and certain diseases. Kidney stones may produce blood in urine.
Turbidity	Transparent when freshly voided; becomes turbid (cloudy) on standing.
Odor	Mildly aromatic; becomes ammonia-like on standing. Some people inherit ability to form methylmercaptan from digested asparagus, which gives characteristic odor. Urine of diabetics has fruity odor due to presence of ketone bodies.
pH	Ranges between 4.6 and 8.0; average 6.0; varies considerably with diet. High-protein diets increase acidity; vegetarian diets increase alkalinity.
Specific gravity (density)	Specific gravity (density) is ratio of weight of volume of substance to weight of equal volume of distilled water. In urine, 1.001–1.035. The higher the concentration of solutes, the higher the specific gravity.

the major characteristics of normal urine. The volume of urine eliminated per day in a normal adult is 1–2 liters (about 1–2 qt). Fluid intake, blood pressure, blood osmolarity, diet, body temperature, diuretics, mental state, and general health influence urine volume. For example, low blood pressure triggers the renin–angiotensin–aldosterone pathway. Aldosterone increases reabsorption of water and salts in the renal tubules and decreases urine volume. By contrast, when blood osmolarity decreases—for example, after drinking a large volume of water—secretion of ADH is inhibited and a larger volume of urine is excreted.

Water accounts for about 95% of the total volume of urine. The remaining 5% consists of electrolytes, solutes derived from cellular metabolism, and exogenous substances such as drugs. Normal urine is virtually protein-free. Typical solutes normally present in urine include filtered and secreted electrolytes that are not reabsorbed, urea (from breakdown of proteins), creatinine (from breakdown of creatine phosphate in muscle fibers), uric acid (from breakdown of nucleic acids), urobilinogen (from breakdown of hemoglobin), and small quantities of other substances, such as fatty acids, pigments, enzymes, and hormones.

If disease alters body metabolism or kidney function, traces of substances not normally present may appear in the urine, or normal constituents may appear in abnormal amounts. Table 26.6 lists several abnormal constituents in urine that may be detected as part of a urinalysis. Normal values of urine components and the

clinical implications of deviations from normal are listed in Appendix D.

Blood Tests

Two blood-screening tests can provide information about kidney function. One is the **blood urea nitrogen (BUN)** test, which measures the blood nitrogen that is part of the urea resulting from catabolism and deamination of amino acids. When glomerular filtration rate decreases severely, as may occur with renal disease or obstruction of the urinary tract, BUN rises steeply. One strategy in treating such patients is to minimize their protein intake, thereby reducing the rate of urea production.

Another test often used to evaluate kidney function is measurement of **plasma creatinine** (krē-AT-i-nin), which results from catabolism of creatine phosphate in skeletal muscle. Normally, the blood creatinine level remains steady because the rate of creatinine excretion in the urine equals its discharge from muscle. A creatinine level above 1.5 mg/dL (135 mmol/liter) usually is an indication of poor renal function. Normal values for selected blood tests are listed in Appendix C along with situations that may cause the values to increase or decrease.

Renal Plasma Clearance

Even more useful than BUN and blood creatinine values in the diagnosis of kidney problems is an evaluation of how effectively the kidneys are removing a given substance from blood plasma. **Renal plasma clearance** is the volume of blood that is "cleaned" or cleared of a substance per unit of time, usually expressed in units of *milliliters per minute*. High renal plasma clearance indicates efficient excretion of a substance in the urine; low clearance indicates inefficient excretion. For example, the clearance of glucose normally is zero because it is completely reabsorbed (see Table 26.3); therefore, glucose is not excreted at all. Knowing a drug's clearance is essential for determining the correct dosage. If clearance is high (one example is penicillin), then the dosage must also be high, and the drug must be given several times a day to maintain an adequate therapeutic level in the blood.

The following equation is used to calculate clearance:

$$\text{Renal plasma clearance of substance S} = \left(\frac{U \times V}{P} \right)$$

where U and P are the concentrations of the substance in urine and plasma, respectively (both expressed in the same units, such as mg/mL), and V is the urine flow rate in mL/min.

The clearance of a solute depends on the three basic processes of a nephron: glomerular filtration, tubular reabsorption, and tubular secretion. Consider a substance that is filtered but neither reabsorbed nor secreted. Its clearance equals the glomerular filtration rate because all the molecules that pass the filtration membrane appear in the urine. This is the situation for the plant polysaccharide **inulin** (IN-ū-lin); it easily passes the filter, it is not reabsorbed, and it is not secreted. (Do not confuse inulin with the hormone insulin, which is produced by the pancreas.) Typically,

TABLE 26.6

Summary of Abnormal Constituents in Urine

ABNORMAL CONSTITUENT	COMMENTS
Albumin	Normal constituent of plasma; usually appears in only very small amounts in urine because it is too large to pass through capillary fenestrations. Presence of excessive albumin in urine—**albuminuria** (al'-bū-mi-NOO-rē-a)—indicates increase in permeability of filtration membranes due to injury or disease, increased blood pressure, or irritation of kidney cells by substances such as bacterial toxins, ether, or heavy metals.
Glucose	Presence of glucose in urine—**glucosuria** (gloo-kō-SOO-rē-a)—usually indicates diabetes mellitus. Occasionally caused by stress, which can cause excessive epinephrine secretion. Epinephrine stimulates breakdown of glycogen and liberation of glucose from liver.
Red blood cells (erythrocytes)	Presence of red blood cells in urine—**hematuria** (hēm-a-TOO-rē-a)—generally indicates pathological condition. One cause is acute inflammation of urinary organs due to disease or irritation from kidney stones. Other causes: tumors, trauma, kidney disease, contamination of sample by menstrual blood.
Ketone bodies	High levels of ketone bodies in urine—**ketonuria** (kē-tō-NOO-rē-a)—may indicate diabetes mellitus, anorexia, starvation, or too little carbohydrate in diet.
Bilirubin	When red blood cells are destroyed by macrophages, the globin portion of hemoglobin is split off and heme is converted to biliverdin. Most biliverdin is converted to bilirubin, which gives bile its major pigmentation. Above-normal level of bilirubin in urine is called **bilirubinuria** (bil'-ē-roo-bi-NOO-rē-a).
Urobilinogen	Presence of urobilinogen (breakdown product of hemoglobin) in urine is called **urobilinogenuria** (ū'-rō-bi-lin'-ō-je-NOO-rē-a). Trace amounts are normal, but elevated urobilinogen may be due to hemolytic or pernicious anemia, infectious hepatitis, biliary obstruction, jaundice, cirrhosis, congestive heart failure, or infectious mononucleosis.
Casts	**Casts** are tiny masses of material that have hardened and assumed shape of lumen of tubule in which they formed, from which they are flushed when filtrate builds up behind them. Casts are named after cells or substances that compose them or based on appearance (for example, white blood cell casts, red blood cell casts, and epithelial cell casts that contain cells from walls of tubules).
Microbes	Number and type of bacteria vary with specific urinary tract infections. One of the most common is *E. coli.* Most common fungus is yeast *Candida albicans,* cause of vaginitis. Most frequent protozoan is *Trichomonas vaginalis,* cause of vaginitis in females and urethritis in males.

the clearance of inulin is about 125 mL/min, which equals the GFR. Clinically, the clearance of inulin can be used to determine the GFR. The clearance of inulin is obtained in the following way: Inulin is administered intravenously and then the concentrations of inulin in plasma and urine are measured along with the urine flow rate. Although using the clearance of inulin is an accurate method for determining the GFR, it has a few drawbacks: Inulin is not produced by the body and it must be infused continuously while clearance measurements are being determined. Measuring the creatinine clearance, however, is an easier way to assess the GFR because creatinine is a substance that is naturally produced by the body as an end product of muscle metabolism. Once creatinine is filtered, it is not reabsorbed, and is secreted only to a very small extent. Because there is a small amount of creatinine secretion, the creatinine clearance is only a close estimate of the GFR and is not as accurate as using the inulin clearance to determine the GFR. The creatinine clearance is normally about 120–140 mL/min.

The clearance of the organic anion *para*-**aminohippuric acid** (**PAH**) (par'-a-a-mē'-nō-hi-PYOOR-ik) is also of clinical importance. After PAH is administered intravenously, it is filtered and secreted in a single pass through the kidneys. Thus, the clearance of PAH is used to measure **renal plasma flow,** the amount of plasma that passes through the kidneys in one minute. Typically, the renal plasma flow is 650 mL per minute, which is about 55% of the renal blood flow (1200 mL per minute).

CLINICAL CONNECTION | Dialysis

If a person's kidneys are so impaired by disease or injury that he or she is unable to function adequately, then blood must be cleansed artificially by **dialysis** (dī-AL-i-sis; *dialyo* = to separate), the separation of large solutes from smaller ones by diffusion through a selectively permeable membrane. One method of dialysis is **hemodialysis** (hē-mō-dī-AL-i-sis; *hemo-* = blood), which directly filters the patient's blood by removing wastes and excess electrolytes and fluid and then returning the cleansed blood to the patient. Blood removed from the body is delivered to a *hemodialyzer* (artificial kidney). Inside the hemodialyzer, blood flows through a *dialysis membrane,* which contains pores large enough to permit the diffusion of small solutes. A special solution, called the *dialysate* (dī-AL-i-sāt), is pumped into the hemodialyzer so that it surrounds the dialysis membrane. The dialysate is specially formulated to maintain diffusion gradients that remove wastes from the blood (for example, urea, creatinine, uric acid, excess phosphate, potassium, and sulfate ions) and add needed substances (for example, glucose and bicarbonate ions) to it. The cleansed blood is passed through an air embolus detector to remove air and then returned to the body. An anticoagulant (heparin) is added to prevent blood from clotting in the hemodialyzer. As a rule, most people on hemodialysis require about 6–12 hours a week, typically divided into three sessions.

Another method of dialysis, called **peritoneal dialysis** (per′-i-tō-NĒ-al), uses the peritoneum of the abdominal cavity as the dialysis membrane to filter the blood. The peritoneum has a large surface area and numerous blood vessels, and is a very effective filter. A catheter is inserted into the peritoneal cavity and connected to a bag of dialysate. The fluid flows into the peritoneal cavity by gravity and is left there for sufficient time to permit wastes and excess electrolytes and fluids to diffuse into the dialysate. Then the dialysate is drained out into a bag, discarded, and replaced with fresh dialysate.

Each cycle is called an *exchange*. One variation of peritoneal dialysis, called **continuous ambulatory peritoneal dialysis (CAPD),** can be performed at home. Usually, the dialysate is drained and replenished four times a day and once at night during sleep. Between exchanges the person can move about freely with the dialysate in the peritoneal cavity. •

✔**CHECKPOINT**

19. What are the characteristics of normal urine?
20. What chemical substances normally are present in urine?
21. How may kidney function be evaluated?
22. Why are the renal plasma clearances of glucose, urea, and creatinine different? How does each clearance compare to glomerular filtration rate?

26.8 URINE TRANSPORTATION, STORAGE, AND ELIMINATION

◉ **OBJECTIVE**

• Describe the anatomy, histology, and physiology of the ureters, urinary bladder, and urethra.

From collecting ducts, urine drains into the minor calyces, which join to become major calyces that unite to form the renal pelvis (see Figure 26.3). From the renal pelvis, urine first drains into the ureters and then into the urinary bladder. Urine is then discharged from the body through the single urethra (see Figure 26.1).

Ureters

Each of the two **ureters** (Ū-rē-ters) transports urine from the renal pelvis of one kidney to the urinary bladder. Peristaltic contractions of the muscular walls of the ureters push urine toward the urinary bladder, but hydrostatic pressure and gravity also contribute. Peristaltic waves that pass from the renal pelvis to the urinary bladder vary in frequency from one to five per minute, depending on how fast urine is being formed.

The ureters are 25–30 cm (10–12 in.) long and are thick-walled, narrow tubes that vary in diameter from 1 mm to 10 mm along their course between the renal pelvis and the urinary bladder. Like the kidneys, the ureters are retroperitoneal. At the base of the urinary bladder, the ureters curve medially and pass obliquely through the wall of the posterior aspect of the urinary bladder (Figure 26.21).

Figure 26.21 Ureters, urinary bladder, and urethra in a female.

🔒 Urine is stored in the urinary bladder before being expelled by micturition.

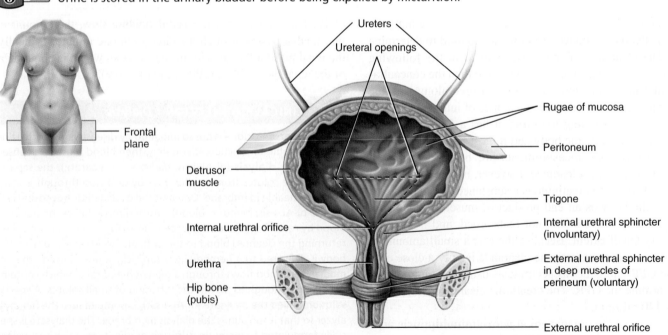

Anterior view of frontal section

❓ **What is a lack of voluntary control over micturition called?**

Even though there is no anatomical valve at the opening of each ureter into the urinary bladder, a physiological one is quite effective. As the urinary bladder fills with urine, pressure within it compresses the oblique openings into the ureters and prevents the backflow of urine. When this physiological valve is not operating properly, it is possible for microbes to travel up the ureters from the urinary bladder to infect one or both kidneys.

Three layers of tissue form the wall of the ureters. The deepest coat, the **mucosa,** is a mucous membrane with **transitional epithelium** (see Table 4.1) and an underlying **lamina propria** of areolar connective tissue with considerable collagen, elastic fibers, and lymphatic tissue. Transitional epithelium is able to stretch—a marked advantage for any organ that must accommodate a variable volume of fluid. Mucus secreted by the goblet cells of the mucosa prevents the cells from coming in contact with urine, the solute concentration and pH of which may differ drastically from the cytosol of cells that form the wall of the ureters.

Throughout most of the length of the ureters, the intermediate coat, the **muscularis,** is composed of inner longitudinal and outer circular layers of smooth muscle fibers. This arrangement is opposite to that of the gastrointestinal tract, which contains inner circular and outer longitudinal layers. The muscularis of the distal third of the ureters also contains an outer layer of longitudinal muscle fibers. Thus, the muscularis in the distal third of the ureter is inner longitudinal, middle circular, and outer longitudinal. Peristalsis is the major function of the muscularis.

The superficial coat of the ureters is the **adventitia,** a layer of areolar connective tissue containing blood vessels, lymphatic vessels, and nerves that serve the muscularis and mucosa. The adventitia blends in with surrounding connective tissue and anchors the ureters in place.

Urinary Bladder

The **urinary bladder** is a hollow, distensible muscular organ situated in the pelvic cavity posterior to the pubic symphysis. In males, it is directly anterior to the rectum; in females, it is anterior to the vagina and inferior to the uterus (see Figure 26.22). Folds of the peritoneum hold the urinary bladder in position. When slightly distended due to the accumulation of urine, the urinary bladder is spherical. When it is empty, it collapses. As urine volume increases, it becomes pear-shaped and rises into the abdominal cavity. Urinary bladder capacity averages 700–800 mL. It is smaller in females because the uterus occupies the space just superior to the urinary bladder.

Anatomy and Histology of the Urinary Bladder

In the floor of the urinary bladder is a small triangular area called the **trigone** (TRĪ-gōn = triangle). The two posterior corners of the trigone contain the two ureteral openings; the opening into the urethra, the **internal urethral orifice** (OR-i-fis), lies in the anterior corner (see Figure 26.21). Because its mucosa is firmly bound to the muscularis, the trigone has a smooth appearance.

Three coats make up the wall of the urinary bladder. The deepest is the **mucosa,** a mucous membrane composed of **transitional epithelium** and an underlying **lamina propria** similar to that of the ureters. Rugae (the folds in the mucosa) are also present to permit expansion of the urinary bladder. Surrounding the mucosa is the intermediate **muscularis,** also called the **detrusor muscle** (de-TROO-ser = to push down), which consists of three layers of smooth muscle fibers: the inner longitudinal, middle circular, and outer longitudinal layers. Around the opening to the urethra the circular fibers form an **internal urethral sphincter;** inferior to it is the **external urethral sphincter,** which is composed of skeletal muscle and is a modification of the deep muscles of the perineum (see Figure 11.12). The most superficial coat of the urinary bladder on the posterior and inferior surfaces is the **adventitia,** a layer of areolar connective tissue that is continuous with that of the ureters. Over the superior surface of the urinary bladder is the **serosa,** a layer of visceral peritoneum.

The Micturition Reflex

Discharge of urine from the urinary bladder, called **micturition** (mik-choo-RISH-un; *mictur-* = urinate), is also known as *urination* or *voiding*. Micturition occurs via a combination of involuntary and voluntary muscle contractions. When the volume of urine in the urinary bladder exceeds 200–400 mL, pressure within the bladder increases considerably, and stretch receptors in its wall transmit nerve impulses into the spinal cord. These impulses propagate to the **micturition center** in sacral spinal cord segments S2 and S3 and trigger a spinal reflex called the **micturition reflex.** In this reflex arc, parasympathetic impulses from the micturition center propagate to the urinary bladder wall and internal urethral sphincter. The nerve impulses cause *contraction* of the detrusor muscle and *relaxation* of the internal urethral sphincter muscle. Simultaneously, the micturition center inhibits somatic motor neurons that innervate skeletal muscle in the external urethral sphincter. On contraction of the urinary bladder wall and relaxation of the sphincters, urination takes place. Urinary bladder filling causes a sensation of fullness that initiates a conscious desire to urinate before the micturition reflex actually occurs. Although emptying of the urinary bladder is a reflex, in early childhood we learn to initiate it and stop it voluntarily. Through learned control of the external urethral sphincter muscle and certain muscles of the pelvic floor, the cerebral cortex can initiate micturition or delay its occurrence for a limited period.

CLINICAL CONNECTION | Cystoscopy

Cystoscopy (sis-TOS-kō-pē; *cysto-* = bladder; *-skopy* = to examine) is a very important procedure for direct examination of the mucosa of the urethra and urinary bladder and prostate in males. In the procedure, a *cystoscope* (a flexible narrow tube with a light) is inserted into the urethra to examine the structures through which it passes. With special attachments, tissue samples can be removed for examination (biopsy) and small stones can be removed. Cystoscopy is useful for evaluating urinary bladder problems such as cancer and infections. It can also evaluate the degree of obstruction resulting from an enlarged prostate. •

Figure 26.22 Comparison between female and male urethras.

The female urethra is about 4 cm (1.5 in.) in length, while the male urethra is about 20 cm (8 in.) in length.

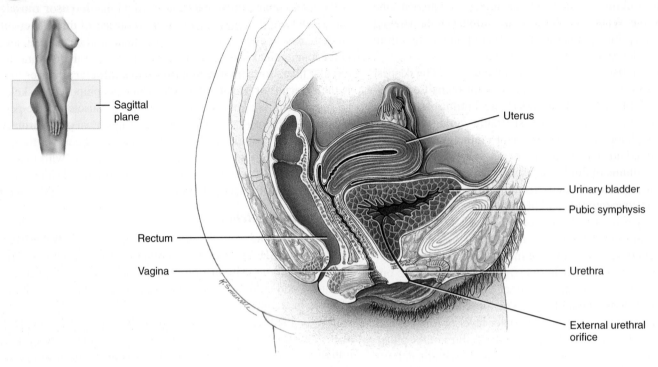

Sagittal plane

Uterus

Urinary bladder

Pubic symphysis

Rectum

Vagina

Urethra

External urethral orifice

(a) Sagittal section, female

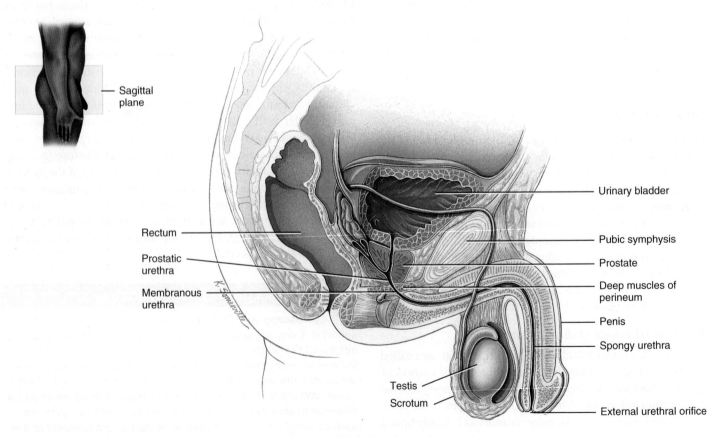

Sagittal plane

Urinary bladder

Rectum

Pubic symphysis

Prostatic urethra

Prostate

Membranous urethra

Deep muscles of perineum

Penis

Spongy urethra

Testis

Scrotum

External urethral orifice

(b) Sagittal section, male

? **What are the three subdivisions of the male urethra?**

Urethra

The **urethra** (ū-RĒ-thra) is a small tube leading from the internal urethral orifice in the floor of the urinary bladder to the exterior of the body. In both males and females, the urethra is the terminal portion of the urinary system and the passageway for discharging urine from the body. In males, it discharges semen (fluid that contains sperm) as well.

In females, the urethra lies directly posterior to the pubic symphysis; is directed obliquely, inferiorly, and anteriorly; and has a length of 4 cm (1.5 in.) (Figure 26.22a). The opening of the urethra to the exterior, the **external urethral orifice,** is located between the clitoris and the vaginal opening (see Figure 28.11a). The wall of the female urethra consists of a deep **mucosa** and a superficial **muscularis.** The mucosa is a mucous membrane composed of **epithelium** and **lamina propria** (areolar connective tissue with elastic fibers and a plexus of veins). The muscularis consists of circularly arranged smooth muscle fibers and is continuous with that of the urinary bladder. Near the urinary bladder, the mucosa contains transitional epithelium that is continuous with that of the urinary bladder; near the external urethral orifice, the epithelium is nonkeratinized stratified squamous epithelium. Between these areas, the mucosa contains stratified columnar or pseudostratified columnar epithelium.

In males, the urethra also extends from the internal urethral orifice to the exterior, but its length and passage through the body are considerably different than in females (Figure 26.22b). The male urethra first passes through the prostate, then through the deep muscles of the perineum, and finally through the penis, a distance of about 20 cm (8 in.).

The male urethra, which also consists of a deep **mucosa** and a superficial **muscularis,** is subdivided into three anatomical regions: (1) The **prostatic urethra** passes through the prostate. (2) The **membranous (intermediate) urethra,** the shortest portion, passes through the deep muscles of the perineum. (3) The **spongy urethra,** the longest portion, passes through the penis. The epithelium of the prostatic urethra is continuous with that of the urinary bladder and consists of transitional epithelium that becomes stratified columnar or pseudostratified columnar epithelium more distally. The mucosa of the membranous urethra contains stratified columnar or pseudostratified columnar epithelium. The epithelium of the spongy urethra is stratified columnar or pseudostratified columnar epithelium, except near the external urethral orifice. There it is nonkeratinized stratified squamous epithelium. The **lamina propria** of the male urethra is areolar connective tissue with elastic fibers and a plexus of veins.

The muscularis of the prostatic urethra is composed of mostly circular smooth muscle fibers superficial to the lamina propria; these circular fibers help form the internal urethral sphincter of the urinary bladder. The muscularis of the membranous urethra consists of circularly arranged skeletal muscle fibers of the deep muscles of the perineum that help form the external urethral sphincter of the urinary bladder.

Several glands and other structures associated with reproduction deliver their contents into the male urethra (see Figure 28.9).

The prostatic urethra contains the openings of (1) ducts that transport secretions from the **prostate** and (2) the **seminal vesicles** and **ductus (vas) deferens,** which deliver sperm into the urethra and provide secretions that both neutralize the acidity of the female reproductive tract and contribute to sperm motility and viability. The openings of the ducts of the **bulbourethral (Cowper's) glands** (bul′-bō-ū-RĒ-thral) empty into the spongy urethra. They deliver an alkaline substance prior to ejaculation that neutralizes the acidity of the urethra. The glands also secrete mucus, which lubricates the end of the penis during sexual arousal. Throughout the urethra, but especially in the spongy urethra, the openings of the ducts of **urethral (Littré) glands** (LĒ-trē) discharge mucus during sexual arousal and ejaculation.

CLINICAL CONNECTION | *Urinary Incontinence*

A lack of voluntary control over micturition is called **urinary incontinence** (in-KON-ti-nens). In infants and children under 2–3 years old, incontinence is normal because neurons to the external urethral sphincter muscle are not completely developed; voiding occurs whenever the urinary bladder is sufficiently distended to stimulate the micturition reflex. Urinary incontinence also occurs in adults. There are four types of urinary incontinence—stress, urge, overflow, and functional. **Stress incontinence** is the most common type of incontinence in young and middle-aged females, and results from weakness of the deep muscles of the pelvic floor. As a result, any physical stress that increases abdominal pressure, such as coughing, sneezing, laughing, exercising, straining, lifting heavy objects, and pregnancy, causes leakage of urine from the urinary bladder. **Urge incontinence** is most common in older people and is characterized by an abrupt and intense urge to urinate followed by an involuntary loss of urine. It may be caused by irritation of the urinary bladder wall by infection or kidney stones, stroke, multiple sclerosis, spinal cord injury, or anxiety. **Overflow incontinence** refers to the involuntary leakage of small amounts of urine caused by some type of blockage or weak contractions of the musculature of the urinary bladder. When urine flow is blocked (for example, from an enlarged prostate or kidney stones) or the urinary bladder muscles can no longer contract, the urinary bladder becomes overfilled and the pressure inside increases until small amounts of urine dribble out. **Functional incontinence** is urine loss resulting from the inability to get to a toilet facility in time as a result of conditions such as stroke, severe arthritis, and Alzheimer disease. Choosing the right treatment option depends on correct diagnosis of the type of incontinence. Treatments include Kegel exercises (see Clinical Connection: Injury of Levator Ani and Urinary Stress Incontinence in Chapter 11), urinary bladder training, medication, and possibly even surgery. •

A summary of the organs of the urinary system is presented in Table 26.7.

✔CHECKPOINT

23. What forces help propel urine from the renal pelvis to the urinary bladder?
24. What is micturition? How does the micturition reflex occur?
25. How do the location, length, and histology of the urethra compare in males and females?

TABLE 26.7

Summary of Urinary System Organs

STRUCTURE	LOCATION	DESCRIPTION	FUNCTION
Kidneys	Posterior abdomen between last thoracic and third lumbar vertebrae posterior to peritoneum (retroperitoneal). Lie against eleventh and twelfth ribs.	Solid, reddish, bean-shaped organs. Internal structure: three tubular systems (arteries, veins, urinary tubes).	Regulate blood volume and composition, help regulate blood pressure, synthesize glucose, release erythropoietin, participate in vitamin D synthesis, excrete wastes in urine.
Ureters	Posterior to peritoneum (retroperitoneal); descend from kidney to urinary bladder along anterior surface of psoas major muscle and cross back of pelvis to reach inferoposterior surface of urinary bladder anterior to sacrum.	Thick, muscular walled tubes with three structural layers: mucosa of transitional epithelium, muscularis with circular and longitudinal layers of smooth muscle, adventitia of areolar connective tissue.	Transport tubes that move urine from kidneys to urinary bladder.
Urinary bladder	In pelvic cavity anterior to sacrum and rectum in males and sacrum, rectum, and vagina in females and posterior to pubis in both sexes. In males, superior surface covered with parietal peritoneum; in females, uterus covers superior aspect.	Hollow, distensible, muscular organ with variable shape depending on how much urine it contains. Three basic layers: inner mucosa of transitional epithelium, middle smooth muscle coat (detrusor muscle), outer adventitia or serosa over superior aspect in males.	Storage organ that temporarily stores urine until convenient to discharge from body.
Urethra	Exits urinary bladder in both sexes. In females, runs through perineal floor of pelvis to exit between labia minora. In males, passes through prostate, then perineal floor of pelvis, and then penis to exit at its tip.	Thin-walled tubes with three structural layers: inner mucosa that consists of transitional, stratified columnar, and stratified squamous epithelium; thin middle layer of circular smooth muscle; thin connective tissue exterior.	Drainage tube that transports stored urine from body.

26.9 WASTE MANAGEMENT IN OTHER BODY SYSTEMS

☉ **OBJECTIVE**

• Describe the ways that body wastes are handled.

As we have seen, just one of the many functions of the urinary system is to help rid the body of some kinds of waste materials. Besides the kidneys, several other tissues, organs, and processes contribute to the temporary confinement of wastes, the transport of waste materials for disposal, the recycling of materials, and the excretion of excess or toxic substances in the body. These waste management systems include the following:

• **Body buffers.** Buffers in body fluids bind excess hydrogen ions (H^+), thereby preventing an increase in the acidity of body fluids. Buffers, like wastebaskets, have a limited capacity; eventually the H^+, like the paper in a wastebasket, must be eliminated from the body by excretion.

• **Blood.** The bloodstream provides pickup and delivery services for the transport of wastes, in much the same way that garbage trucks and sewer lines serve a community.

• **Liver.** The liver is the primary site for metabolic recycling, as occurs, for example, in the conversion of amino acids into glucose or of glucose into fatty acids. The liver also converts toxic substances into less toxic ones, such as ammonia into urea. These functions of the liver are described in Chapters 24 and 25.

• **Lungs.** With each exhalation, the lungs excrete CO_2, and expel heat and a little water vapor.

• **Sweat (sudoriferous) glands.** Especially during exercise, sweat glands in the skin help eliminate excess heat, water, and CO_2, plus small quantities of salts and urea as well.

• **Gastrointestinal tract.** Through defecation, the gastrointestinal tract excretes solid, undigested foods; wastes; some CO_2; water; salts; and heat.

✔ **CHECKPOINT**

26. What roles do the liver and lungs play in the elimination of wastes?

 # 26.10 DEVELOPMENT OF THE URINARY SYSTEM

☉ **OBJECTIVE**

• Describe the development of the urinary system.

Starting in the third week of fetal development, a portion of the mesoderm along the posterior aspect of the embryo, the **intermediate mesoderm,** differentiates into the kidneys. The intermedi-

Figure 26.23 Development of the urinary system.

🔑 Three pairs of kidneys form within intermediate mesoderm in succession: pronephros, mesonephros, and metanephros.

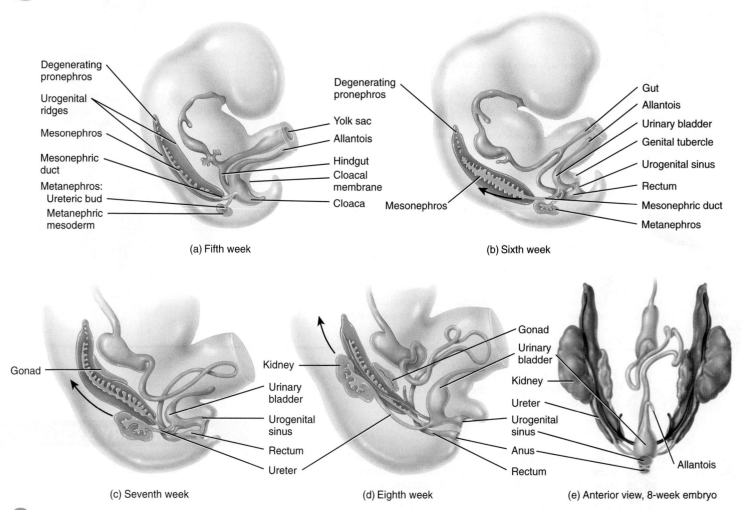

Degenerating pronephros

Urogenital ridges

Mesonephros

Mesonephric duct

Metanephros:
Ureteric bud
Metanephric mesoderm

Yolk sac
Allantois
Hindgut
Cloacal membrane
Cloaca

(a) Fifth week

Degenerating pronephros

Mesonephros

Gut
Allantois
Urinary bladder
Genital tubercle
Urogenital sinus
Rectum
Mesonephric duct
Metanephros

(b) Sixth week

Gonad

Kidney
Urinary bladder
Urogenital sinus
Rectum
Ureter

(c) Seventh week

Gonad
Urinary bladder
Kidney
Ureter
Urogenital sinus
Anus
Rectum

(d) Eighth week

Gonad
Urinary bladder
Kidney
Ureter
Urogenital sinus
Allantois

(e) Anterior view, 8-week embryo

❓ **When do the kidneys begin to develop?**

ate mesoderm is located in paired elevations called **urogenital ridges** (ū-rō-JEN-i-tal). Three pairs of kidneys form within the intermediate mesoderm in succession: the pronephros, the mesonephros, and the metanephros (Figure 26.23). Only the last pair remains as the functional kidneys of the newborn.

The first kidney to form, the **pronephros** (prō-NEF-rōs; *pro-* = before; *-nephros* = kidney), is the most superior of the three and has an associated **pronephric duct.** This duct empties into the **cloaca** (klō-Ā-ka), the expanded terminal part of the hindgut, which functions as a common outlet for the urinary, digestive, and reproductive ducts. The pronephros begins to degenerate during the fourth week and is completely gone by the sixth week.

The second kidney, the **mesonephros** (mez'-ō-NEF-rōs; *meso-* = middle), replaces the pronephros. The retained portion of the pronephric duct, which connects to the mesonephros, develops into the **mesonephric duct.** The mesonephros begins to degenerate by the sixth week and is almost gone by the eighth week.

At about the fifth week, a mesodermal outgrowth, called a **ureteric bud** (ū-rē-TER-ik), develops from the distal portion of

the mesonephric duct near the cloaca. The **metanephros** (met-a-NEF-rōs; *meta-* = after), or ultimate kidney, develops from the ureteric bud and metanephric mesoderm. The ureteric bud forms the *collecting ducts, calyces, renal pelvis,* and *ureter.* The **metanephric mesoderm** (met'-a-NEF-rik) forms the *nephrons* of the kidneys. By the third month, the fetal kidneys begin excreting urine into the surrounding amniotic fluid; indeed, fetal urine makes up most of the amniotic fluid.

During development, the cloaca divides into a **urogenital sinus,** into which urinary and genital ducts empty, and a *rectum* that discharges into the anal canal. The *urinary bladder* develops from the urogenital sinus. In females, the *urethra* develops as a result of lengthening of the short duct that extends from the urinary bladder to the urogenital sinus. In males, the urethra is considerably longer and more complicated, but it is also derived from the urogenital sinus.

Although the metanephric kidneys form in the pelvis, they ascend to their ultimate destination in the abdomen. As they do so, they receive renal blood vessels. Although the inferior blood

Focus on Homeostasis

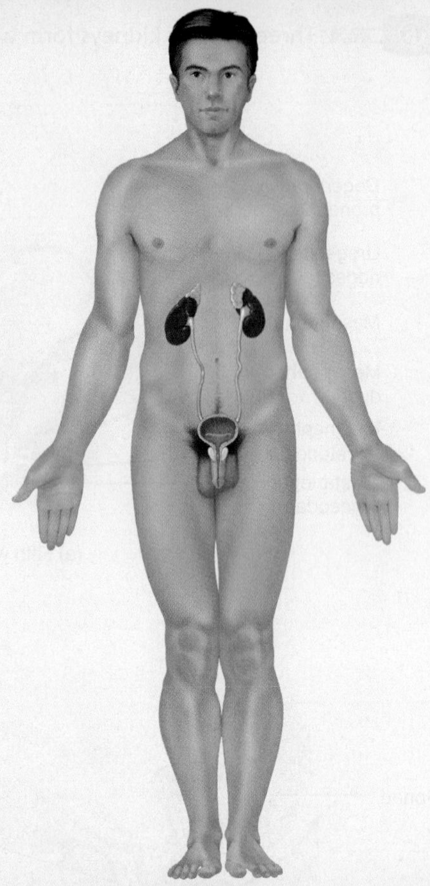

BODY SYSTEM	CONTRIBUTION OF THE URINARY SYSTEM

 For all body systems
Kidneys regulate the volume, composition, and pH of body fluids by removing wastes and excess substances from blood and excreting them in urine; the ureters transport urine from the kidneys to the urinary bladder, which stores urine until it is eliminated through the urethra.

 Integumentary system
Kidneys and skin both contribute to synthesis of calcitriol, the active form of vitamin D.

 Skeletal system
Kidneys help adjust levels of blood calcium and phosphates, needed for building extracellular bone matrix.

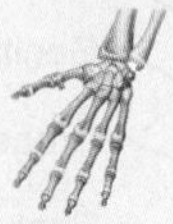

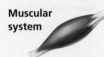 **Muscular system**
Kidneys help adjust level of blood calcium, needed for contraction of muscle.

 Nervous system
Kidneys perform gluconeogenesis, which provides glucose for ATP production in neurons, especially during fasting or starvation.

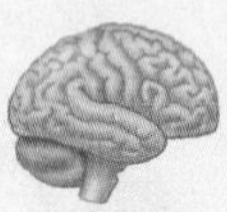

 Endocrine system
Kidneys participate in synthesis of calcitriol, the active form of vitamin D, and release erythropoietin, the hormone that stimulates production of red blood cells.

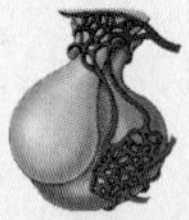

 Cardiovascular system
By increasing or decreasing their reabsorption of water filtered from blood, the kidneys help adjust blood volume and blood pressure; renin released by juxtaglomerular cells in the kidneys raises blood pressure; some bilirubin from hemoglobin breakdown is converted to a yellow pigment (urobilin), which is excreted in urine.

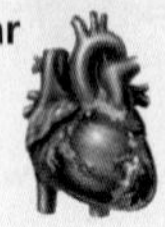

 Lymphatic system and immunity
By increasing or decreasing their reabsorption of water filtered from blood, the kidneys help adjust the volume of interstitial fluid and lymph; urine flushes microbes out of the urethra.

 Respiratory system
Kidneys and lungs cooperate in adjusting pH of body fluids.

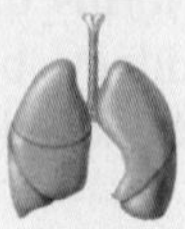

 Digestive system
Kidneys help synthesize calcitriol, the active form of vitamin D, which is needed for absorption of dietary calcium.

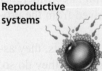 **Reproductive systems**
In males, the portion of the urethra that extends through the prostate and penis is a passageway for semen as well as urine.

THE URINARY SYSTEM

1102

vessels usually degenerate as superior ones appear, sometimes the inferior vessels do not degenerate. Consequently, some individuals (about 30%) develop multiple renal vessels.

In a condition called **unilateral renal agenesis** (a-JEN-e-sis; *a-* = without; *-genesis* = production; *unilateral* = one side) only one kidney develops (usually the right) due to the absence of a ureteric bud. The condition occurs once in every 1000 newborn infants and usually affects males more than females. Other kidney abnormalities that occur during development are **malrotated kidneys** (the hilum faces anteriorly, posteriorly, or laterally instead of medially); **ectopic kidney** (one or both kidneys may be in an abnormal position, usually inferior); and **horseshoe kidney** (the fusion of the two kidneys, usually inferiorly, into a single U-shaped kidney).

✔ **CHECKPOINT**

27. Which type of embryonic tissue develops into nephrons?
28. Which tissue gives rise to collecting ducts, calyces, renal pelves, and ureters?

26.11 AGING AND THE URINARY SYSTEM

◉ **OBJECTIVE**

• Describe the effects of aging on the urinary system.

With aging, the kidneys shrink in size, have a decreased blood flow, and filter less blood. These age-related changes in kidney size and function seem to be linked to a progressive reduction in blood supply to the kidneys as an individual gets older; for example, blood vessels such as the glomeruli become damaged or decrease in number. The mass of the two kidneys decreases from an average of nearly 300 g in 20-year-olds to less than 200 g by age 80, a decrease of about one-third. Likewise, renal blood flow and filtration rate decline by 50% between ages 40 and 70. By age 80, about 40% of glomeruli are not functioning and thus filtration, reabsorption, and secretion decrease. Kidney diseases that become more common with age include acute and chronic kidney inflammations and renal calculi (kidney stones). Because the sensation of thirst diminishes with age, older individuals also are susceptible to dehydration. Urinary bladder changes that occur with aging include a reduction in size and capacity and weakening of the muscles. Urinary tract infections are more common among the elderly, as are polyuria (excessive urine production), nocturia (excessive urination at night), increased frequency of urination, dysuria (painful urination), urinary retention or incontinence, and hematuria (blood in the urine).

✔ **CHECKPOINT**

29. To what extent do kidney mass and filtration rate decrease with age?

• • •

To appreciate the many ways that the urinary system contributes to homeostasis of other body systems, examine *Focus on Homeostasis: The Urinary System.* Next, in Chapter 27, we will see how the kidneys and lungs contribute to maintenance of homeostasis of body fluid volume, electrolyte levels in body fluids, and acid–base balance.

 DISORDERS: HOMEOSTATIC IMBALANCES

Renal Calculi

The crystals of salts present in urine occasionally precipitate and solidify into insoluble stones called **renal calculi** (KAL-kū-lī = pebbles) or **kidney stones.** They commonly contain crystals of calcium oxalate, uric acid, or calcium phosphate. Conditions leading to calculus formation include the ingestion of excessive calcium, low water intake, abnormally alkaline or acidic urine, and overactivity of the parathyroid glands. When a stone lodges in a narrow passage, such as a ureter, the pain can be intense. **Shock-wave lithotripsy** (LITH-ō-trip′-sē; *litho-* = stone) is a procedure that uses high-energy shock waves to disintegrate kidney stones and offers an alternative to surgical removal. Once the kidney stone is located using x-rays, a device called a *lithotripter* delivers brief, high-intensity sound waves through a water- or gel-filled cushion placed under the back. Over a period of 30 to 60 minutes, 1000 or more shock waves pulverize the stone, creating fragments that are small enough to wash out in the urine.

Urinary Tract Infections

The term **urinary tract infection (UTI)** is used to describe either an infection of a part of the urinary system or the presence of large numbers of microbes in urine. UTIs are more common in females due to the shorter length of the urethra. Symptoms include painful or burning urination, urgent and frequent urination, low back pain, and bedwetting. UTIs include *urethritis* (ū-rē-THRĪ-tis), inflammation of the urethra; *cystitis* (sis-TĪ-tis), inflammation of the urinary bladder; and *pyelonephritis* (pī-e-lō-ne-FRĪ-tis), inflammation of the kidneys. If pyelonephritis becomes chronic, scar tissue can form in the kidneys and severely impair their function. Drinking cranberry juice can prevent the attachment of *E. coli* bacteria to the lining of the urinary bladder so that they are more readily flushed away during urination.

Glomerular Diseases

A variety of conditions may damage the kidney glomeruli, either directly or indirectly because of disease elsewhere in the body. Typically, the filtration membrane sustains damage, and its permeability increases.

Glomerulonephritis (glō-mer′-ū-lō-ne-FRĪ-tis) is an inflammation of the kidney that involves the glomeruli. One of the most common causes is an allergic reaction to the toxins produced by streptococcal bacteria that have recently infected another part of the body, especially the throat. The glomeruli become so inflamed, swollen, and engorged with blood that the filtration membranes allow blood cells and plasma proteins to enter the filtrate. As a result, the urine contains many erythrocytes (hematuria) and a lot of protein. The glomeruli may be permanently damaged, leading to chronic renal failure.

Nephrotic syndrome (nef-ROT-ik) is a condition characterized by *proteinuria* (prō-tēn-OO-rē-a), protein in the urine, and *hyperlipidemia* (hī′-per-lip-i-DĒ-mē-a), high blood levels of cholesterol, phospholipids, and triglycerides. The proteinuria is due to an increased permeability of the filtration membrane, which permits proteins, especially albumin, to escape from blood into urine. Loss of albumin results in *hypoalbuminemia* (hī'-pō-al-bū-mi-NĒ-mē-a), low blood albumin level, once liver production of albumin fails to meet increased urinary losses. Edema, usually seen around the eyes, ankles, feet, and abdomen, occurs in nephrotic syndrome because loss of albumin from the blood decreases blood colloid osmotic pressure. Nephrotic syndrome is associated with several glomerular diseases of unknown cause, as well as with systemic disorders such as diabetes mellitus, systemic lupus erythematosus (SLE), a variety of cancers, and AIDS.

Renal Failure

Renal failure is a decrease or cessation of glomerular filtration. In **acute renal failure (ARF),** the kidneys abruptly stop working entirely (or almost entirely). The main feature of ARF is the suppression of urine flow, usually characterized either by *oliguria* (ol-i-GŪ-rē-a), daily urine output between 50 mL and 250 mL, or by *anuria* (an-Ū-rē-a), daily urine output less than 50 mL. Causes include low blood volume (for example, due to hemorrhage), decreased cardiac output, damaged renal tubules, kidney stones, the dyes used to visualize blood vessels in angiograms, nonsteroidal anti-inflammatory drugs, and some antibiotic drugs. It is also common in people who suffer a devastating illness or overwhelming traumatic injury; in such cases it may be related to a more general organ failure known as *multiple organ dysfunction syndrome (MODS).*

Renal failure causes a multitude of problems. There is edema due to salt and water retention and metabolic acidosis due to an inability of the kidneys to excrete acidic substances. In the blood, urea builds up due to impaired renal excretion of metabolic waste products and potassium level rises, which can lead to cardiac arrest. Often, there is anemia because the kidneys no longer produce enough erythropoietin for adequate red blood cell production. Because the kidneys are no longer able to convert vitamin D to calcitriol, which is needed for adequate calcium absorption from the small intestine, osteomalacia also may occur.

Chronic renal failure (CRF) refers to a progressive and usually irreversible decline in glomerular filtration rate (GFR). CRF may result from chronic glomerulonephritis, pyelonephritis, polycystic kidney disease, or traumatic loss of kidney tissue. CRF develops in three stages. In the first stage, *diminished renal reserve,* nephrons are destroyed until about 75% of the functioning nephrons are lost. At this stage, a person may have no signs or symptoms because the remaining nephrons enlarge and take over the function of those that have been lost. Once 75% of the nephrons are lost, the person enters the second stage, called *renal insufficiency,* characterized by a decrease in

GFR and increased blood levels of nitrogen-containing wastes and creatinine. Also, the kidneys cannot effectively concentrate or dilute the urine. The final stage, called *end-stage renal failure,* occurs when about 90% of the nephrons have been lost. At this stage, GFR diminishes to 10–15% of normal, oliguria is present, and blood levels of nitrogen-containing wastes and creatinine increase further. People with end-stage renal failure need dialysis therapy and are possible candidates for a kidney transplant operation.

Polycystic Kidney Disease

Polycystic kidney disease (PKD) (pol'-ē-SIS-tik) is one of the most common inherited disorders. In PKD, the kidney tubules become riddled with hundreds or thousands of cysts (fluid-filled cavities). In addition, inappropriate apoptosis (programmed cell death) of cells in noncystic tubules leads to progressive impairment of renal function and eventually to end-stage renal failure.

People with PKD also may have cysts and apoptosis in the liver, pancreas, spleen, and gonads; increased risk of cerebral aneurysms; heart valve defects; and diverticula in the colon. Typically, symptoms are not noticed until adulthood, when patients may have back pain, urinary tract infections, blood in the urine, hypertension, and large abdominal masses. Using drugs to restore normal blood pressure, restricting protein and salt in the diet, and controlling urinary tract infections may slow progression to renal failure.

Urinary Bladder Cancer

Each year, nearly 12,000 Americans die from **urinary bladder cancer.** It generally strikes people over 50 years of age and is three times more likely to develop in males than females. The disease is typically painless as it develops, but in most cases blood in the urine is a primary sign of the disease. Less often, people experience painful and/or frequent urination.

As long as the disease is identified early and treated promptly, the prognosis is favorable. Fortunately, about 75% of the urinary bladder cancers are confined to the epithelium of the urinary bladder and are easily removed by surgery. The lesions tend to be low-grade, meaning that they have only a small potential for metastasis.

Urinary bladder cancer is frequently the result of a carcinogen. About half of all cases occur in people who smoke or have at some time smoked cigarettes. The cancer also tends to develop in people who are exposed to chemicals called aromatic amines. Workers in the leather, dye, rubber, and aluminum industries, as well as painters, are often exposed to these chemicals.

MEDICAL TERMINOLOGY

Azotemia (az-ō-TĒ-mē-a; *azot-* = nitrogen; *-emia* = condition of blood) Presence of urea or other nitrogen-containing substances in the blood.

Cystocele (SIS-tō-sēl; *cysto-* = bladder; *-cele* = hernia or rupture) Hernia of the urinary bladder.

Diabetic kidney disease A disorder caused by diabetes mellitus in which glomeruli are damaged. The result is the leakage of proteins into the urine and a reduction in the ability of the kidney to remove water and waste.

Dysuria (dis-Ū-rē-a; *dys-* = painful; *-uria* = urine) Painful urination.

Enuresis (en'-ū-RĒ-sis = to void urine) Involuntary voiding of urine after the age at which voluntary control has typically been attained.

Hydronephrosis (hī'-drō-ne-FRŌ-sis; *hydro-* = water; *-nephros-* = kidney; *-osis* = condition) Swelling of the kidney due to dilation of

the renal pelvis and calyces as a result of an obstruction to the flow of urine. It may be due to a congenital abnormality, a narrowing of the ureter, a kidney stone, or an enlarged prostate.

Intravenous pyelogram (in'-tra-VĒ-nus PĪ-e-lō-gram'; *intra-* = within; *-veno-* = vein; *pyelo-* = pelvis of kidney; *-gram* = record) (or **IVP**) Radiograph (x-ray) of the kidneys, ureters, and urinary bladder after venous injection of a radiopaque contrast medium.

Nephropathy (ne-FROP-a-thē; *nephro-* = kidney; *-pathos* = suffering) Any disease of the kidneys. Types include analgesic (from long-term and excessive use of drugs such as ibuprofen), lead (from ingestion of lead-based paint), and solvent (from carbon tetrachloride and other solvents).

Nocturnal enuresis (nok-TUR-nal en'-ū-RĒ-sis) Discharge of urine during sleep, resulting in bed-wetting; occurs in about 15% of

5-year-old children and generally resolves spontaneously, afflicting only about 1% of adults. It may have a genetic basis, as bed-wetting occurs more often in identical twins than in fraternal twins and more often in children whose parents or siblings were bed-wetters. Possible causes include smaller-than-normal bladder capacity, failure to awaken in response to a full bladder, and above-normal production of urine at night. Also referred to as **nocturia.**

Polyuria (pol′-ē-Ū-rē-a; *poly-* = too much) Excessive urine formation. It may occur in conditions such as diabetes mellitus and glomerulonephritis.

Stricture (STRIK-chur) Narrowing of the lumen of a canal or hollow organ, as may occur in the ureter, urethra, or any other tubular structure in the body.

Uremia (ū-RĒ-mē-a; *-emia* = condition of blood) Toxic levels of urea in the blood resulting from severe malfunction of the kidneys.

Urinary retention A failure to completely or normally void urine; may be due to an obstruction in the urethra or neck of the urinary bladder, to nervous contraction of the urethra, or to lack of urge to urinate. In men, an enlarged prostate may constrict the urethra and cause urinary retention. If urinary retention is prolonged, a catheter (slender rubber drainage tube) must be placed into the urethra to drain the urine.

CHAPTER REVIEW AND RESOURCE SUMMARY

Review

Resource

Introduction

1. The organs of the urinary system are the kidneys, ureters, urinary bladder, and urethra.
2. After the kidneys filter blood and return most water and many solutes to the bloodstream, the remaining water and solutes constitute urine.

Anatomy Overview - The Urinary System Overview

26.1 Overview of Kidney Functions

1. The kidneys regulate blood ionic composition, blood osmolarity, blood volume, blood pressure, and blood pH.
2. The kidneys also perform gluconeogenesis, release calcitriol and erythropoietin, and excrete wastes and foreign substances.

Anatomy Overview - Kidney Overview

26.2 Anatomy and Histology of the Kidneys

1. The kidneys are retroperitoneal organs attached to the posterior abdominal wall.
2. Three layers of tissue surround the kidneys: renal capsule, adipose capsule, and renal fascia.
3. Internally, the kidneys consist of a renal cortex, a renal medulla, renal pyramids, renal papillae, renal columns, major and minor calyces, and a renal pelvis.
4. Blood flows into the kidney through the renal artery and successively into segmental, interlobar, arcuate, and interlobular arteries; afferent arterioles; glomerular capillaries; efferent arterioles; peritubular capillaries and vasa recta; and interlobular, arcuate, and interlobar veins before flowing out of the kidney through the renal vein.
5. Vasomotor nerves from the sympathetic division of the autonomic nervous system supply kidney blood vessels; they help regulate the flow of blood through the kidney.
6. The nephron is the functional unit of the kidneys. A nephron consists of a renal corpuscle (glomerulus and glomerular or Bowman's capsule) and a renal tubule.
7. A renal tubule consists of a proximal convoluted tubule, a loop of Henle, and a distal convoluted tubule, which drains into a collecting duct (shared by several nephrons). The loop of Henle consists of a descending limb and an ascending limb.
8. A cortical nephron has a short loop that dips only into the superficial region of the renal medulla; a juxtamedullary nephron has a long loop of Henle that stretches through the renal medulla almost to the renal papilla.
9. The wall of the entire glomerular capsule, renal tubule, and ducts consists of a single layer of epithelial cells. The epithelium has distinctive histological features in different parts of the tubule. Table 26.1 summarizes the histological features of the renal tubule and collecting duct.
10. The juxtaglomerular apparatus (JGA) consists of the juxtaglomerular cells of an afferent arteriole and the macula densa of the final portion of the ascending limb of the loop of Henle.

Anatomy Overview - Blood Supply to the Kidney
Figure 26.3 - Internal Anatomy of the Kidneys
Figure 26.6 - Histology of a Renal Corpuscle
Exercise - Assemble the Urinary Tract
Exercise - Concentrate on Urinary System
Exercise - Paint the Nephron

26.3 Overview of Renal Physiology

1. Nephrons perform three basic tasks: glomerular filtration, tubular secretion, and tubular reabsorption.

Animation - Renal Filtration
Filtration Finale
Exercise - Magical Renal Ride
Exercise - Renal Regulation
Exercise - The Renal Trip

Review	**Resource**

26.4 Glomerular Filtration

1. Fluid that is filtered by glomeruli enters the capsular space and is called glomerular filtrate.
2. The filtration membrane consists of the glomerular endothelium, basal lamina, and filtration slits between pedicels of podocytes.
3. Most substances in blood plasma easily pass through the glomerular filter. However, blood cells and most proteins normally are not filtered.
4. Glomerular filtrate amounts to up to 180 liters of fluid per day. This large amount of fluid is filtered because the filter is porous and thin, the glomerular capillaries are long, and the capillary blood pressure is high.
5. Glomerular blood hydrostatic pressure (GBHP) promotes filtration; capsular hydrostatic pressure (CHP) and blood colloid osmotic pressure (BCOP) oppose filtration. Net filtration pressure (NFP) = GBHP – CHP – BCOP. NFP is about 10 mmHg.
6. Glomerular filtration rate (GFR) is the amount of filtrate formed in both kidneys per minute; it is normally 105–125 mL/min.
7. Glomerular filtration rate depends on renal autoregulation, neural regulation, and hormonal regulation. Table 26.2 summarizes regulation of GFR.

Animation - Renal Filtration
Animation - Hormonal Control of Blood Volume and Pressure
Animation - Water and Fluid Flow
Figure 26.9 - The Pressures That Drive Glomerular Filtration
Homeostatic Imbalances - The Case of the Thirsty Woman

26.5 Tubular Reabsorption and Tubular Secretion

1. Tubular reabsorption is a selective process that reclaims materials from tubular fluid and returns them to the bloodstream. Reabsorbed substances include water, glucose, amino acids, urea, and ions, such as sodium, chloride, potassium, bicarbonate, and phosphate (Table 26.3).
2. Some substances not needed by the body are removed from the blood and discharged into the urine via tubular secretion. Included are ions (K^+, H^+, and NH_4^+), urea, creatinine, and certain drugs.
3. Reabsorption routes include both paracellular (between tubule cells) and transcellular (across tubule cells) routes. The maximum amount of a substance that can be reabsorbed per unit time is called the transport maximum (T_m).
4. About 90% of water reabsorption is obligatory; it occurs via osmosis, together with reabsorption of solutes, and is not hormonally regulated. The remaining 10% is facultative water reabsorption, which varies according to body needs and is regulated by ADH.
5. Sodium ions are reabsorbed throughout the basolateral membrane via primary active transport.
6. In the proximal convoluted tubule, Na^+ ions are reabsorbed through the apical membranes via Na^+–glucose symporters and Na^+/H^+ antiporters; water is reabsorbed via osmosis; Cl^-, K^+, Ca^{2+}, Mg^{2+}, and urea are reabsorbed via passive diffusion; and NH_3 and NH_4^+ are secreted.
7. The loop of Henle reabsorbs 20–30% of the filtered Na^+, K^+, Ca^{2+}, and HCO_3^-; 35% of the filtered Cl^-; and 15% of the filtered water.
8. The distal convoluted tubule reabsorbs sodium and chloride ions via Na^+–Cl^- symporters.
9. In the collecting duct, principal cells reabsorb Na^+ and secrete K^+; intercalated cells reabsorb K^+ and HCO_3^- and secrete H^+.
10. Angiotensin II, aldosterone, antidiuretic hormone, atrial natriuretic peptide, and parathyroid hormone regulate solute and water reabsorption, as summarized in Table 26.4.

Animation - Renal Reabsorption and Secretion
Animation - Water Homeostasis
Animation - Hormonal Control of Blood Volume and Pressure

26.6 Production of Dilute and Concentrated Urine

1. In the absence of ADH, the kidneys produce dilute urine; renal tubules absorb more solutes than water.
2. In the presence of ADH, the kidneys produce concentrated urine; large amounts of water are reabsorbed from the tubular fluid into interstitial fluid, increasing solute concentration of the urine.
3. The countercurrent multiplier establishes an osmotic gradient in the interstitial fluid of the renal medulla that enables production of concentrated urine when ADH is present.

Animation - Water Homeostasis
Figure 26.19 - Mechanism of Urine Concentration

26.7 Evaluation of Kidney Function

1. A urinalysis is an analysis of the volume and physical, chemical, and microscopic properties of a urine sample. Table 26.5 summarizes the principal physical characteristics of normal urine.
2. Chemically, normal urine contains about 95% water and 5% solutes. The solutes normally include urea, creatinine, uric acid, urobilinogen, and various ions.
3. Table 26.6 lists several abnormal components that can be detected in a urinalysis, including albumin, glucose, red and white blood cells, ketone bodies, bilirubin, excessive urobilinogen, casts, and microbes.
4. Renal clearance refers to the ability of the kidneys to clear (remove) a specific substance from blood.

Homeostatic Imbalances - The Case of the Man with the Swollen Kidneys

Review	Resource

26.8 Urine Transportation, Storage, and Elimination

1. The ureters are retroperitoneal and consist of a mucosa, muscularis, and adventitia. They transport urine from the renal pelvis to the urinary bladder, primarily via peristalsis.
2. The urinary bladder is located in the pelvic cavity posterior to the pubic symphysis; its function is to store urine before micturition.
3. The urinary bladder consists of a mucosa with rugae, a muscularis (detrusor muscle), and an adventitia (serosa over the superior surface).
4. The micturition reflex discharges urine from the urinary bladder via parasympathetic impulses that cause contraction of the detrusor muscle and relaxation of the internal urethral sphincter muscle and via inhibition of impulses in somatic motor neurons to the external urethral sphincter.
5. The urethra is a tube leading from the floor of the urinary bladder to the exterior. Its anatomy and histology differ in females and males. In both sexes, the urethra functions to discharge urine from the body; in males, it discharges semen as well.

Anatomy Overview - Ureters, Urinary Bladder, and Urethra
Figure 26.22 - Comparison between Male and Female Urethras
Exercise - Pick the Urinary Process

26.9 Waste Management in Other Body Systems

1. Besides the kidneys, several other tissues, organs, and processes temporarily confine wastes, transport waste materials for disposal, recycle materials, and excrete excess or toxic substances.
2. Buffers bind excess H^+, the blood transports wastes, the liver converts toxic substances into less toxic ones, the lungs exhale CO_2, sweat glands help eliminate excess heat, and the gastrointestinal tract eliminates solid wastes.

Animation - Regulation of pH

26.10 Development of the Urinary System

1. The kidneys develop from intermediate mesoderm.
2. The kidneys develop in the following sequence: pronephros, mesonephros, and metanephros. Only the metanephros remains and develops into a functional kidney.

26.11 Aging and the Urinary System

1. With aging, the kidneys shrink in size, have a decreased blood flow, and filter less blood.
2. Common problems related to aging include urinary tract infections, increased frequency of urination, urinary retention or incontinence, and renal calculi.

Homeostatic Imbalances - The Case of the Consistently Full Bladder

SELF-QUIZ QUESTIONS

Fill in the blanks in the following statements.

1. The renal corpuscle consists of the _____ and _____.

2. Discharge of urine from the urinary bladder is called _____.

Indicate whether the following statements are true or false.

3. The most superficial region of the internal kidney is the renal medulla.

4. When dilute urine is being formed, the osmolarity of the fluid in the tubular lumen increases as it flows down the descending limb of the loop of Henle, decreases as it flows up the ascending limb, and continues to decrease as it flows through the rest of the nephron and collecting duct.

Choose the one best answer to the following questions.

5. Which of the following statements are *correct?* (1) Glomerular filtration rate (GFR) is directly related to the pressures that determine net filtration pressure. (2) Angiotensin II and atrial natriuretic peptide help regulate GFR. (3) Mechanisms that regulate GFR work by adjusting blood flow into and out of the glomerulus and by altering the glomerular capillary surface area available for filtration. (4) GFR increases when blood flow into glomerular capillaries decreases. (5) Normally, GFR increases very little when systemic blood pressure rises.
   (a) 1, 2, and 3   (b) 2, 3, and 4   (c) 3, 4, and 5
   (d) 1, 2, 3, and 5   (e) 2, 3, 4, and 5

6. Which of the following hormones affect Na^+, Cl^-, Ca^{2+}, and/or water reabsorption and/or K^+ secretion by the renal tubules? (1) angiotensin II, (2) aldosterone, (3) ADH, (4) atrial natriuretic peptide, (5) thyroid hormone, (6) parathyroid hormone.
   (a) 1, 3, and 5   (b) 2, 3, and 6   (c) 2, 4, and 5
   (d) 1, 2, 4, and 5   (e) 1, 2, 3, 4, and 6

7. Which of the following are features of the renal corpuscle that enhance its filtering capacity? (1) large glomerular capillary surface area, (2) thick, selectively permeable filtration membrane, (3) high capsular hydrostatic pressure, (4) high glomerular capillary pressure, (5) mesangial cells regulating the filtering surface area.
   (a) 1, 2, and 3   (b) 2, 4, and 5   (c) 1, 4, and 5
   (d) 2, 3, and 4   (e) 2, 3, and 5

8. Given the following values, calculate the net filtration pressure: (1) glomerular blood hydrostatic pressure = 40 mmHg, (2) capsular

hydrostatic pressure = 10 mmHg, (3) blood colloid osmotic pressure = 30 mmHg.
(a) −20 mmHg   (b) 0 mmHg   (c) 20 mmHg
(d) 60 mmHg   (e) 80 mmHg

9. The micturition reflex (1) is initiated by stretch receptors in the ureters, (2) relies on parasympathetic impulses from the micturition center in S2 and S3, (3) results in contraction of the detrusor muscle, (4) results in contraction of the internal urethral sphincter muscle, (5) inhibits motor neurons in the external urethral sphincter.
(a) 1, 2, 3, 4, and 5   (b) 1, 3, and 4   (c) 2, 3, 4, and 5
(d) 2 and 5   (e) 2, 3, and 5

10. Which of the following are mechanisms that control GFR? (1) renal autoregulation, (2) neural regulation, (3) hormonal regulation, (4) chemical regulation of ions, (5) presence or absence of a transporter.
(a) 1, 2, and 3   (b) 2, 3, and 4   (c) 3, 4, and 5
(d) 1, 3, and 5   (e) 1, 3, and 4

11. Place the route of blood flow through the kidney in the correct order:
(a) segmental arteries   (b) vasa recta
(c) arcuate arteries   (d) peritubular venules
(e) interlobular veins   (f) renal vein
(g) renal artery   (h) interlobar arteries
(i) peritubular capillaries   (j) efferent arterioles
(k) interlobar veins   (l) glomeruli
(m) arcuate veins   (n) afferent arterioles
(o) interlobular arteries

12. Place the route of filtrate flow in the correct order from its origin to the ureter:
(a) minor calyx
(b) ascending limb of loop of Henle
(c) papillary duct
(d) distal convoluted tubule
(e) major calyx
(f) descending limb of loop of Henle
(g) proximal convoluted tubule
(h) collecting duct
(i) renal pelvis

13. Match the following:
____ (a) cells in the last portion of the distal convoluted tubule and in the collecting ducts; regulated by ADH and aldosterone
____ (b) a capillary network lying in the glomerular capsule and functioning in filtration
____ (c) the functional unit of the kidney
____ (d) drains into a collecting duct
____ (e) combined glomerulus and glomerular capsule; where plasma is filtered
____ (f) the visceral layer of the glomerular capsule consisting of modified simple squamous epithelial cells
____ (g) cells of the final portion of the ascending limb of the loop of Henle that make contact with the afferent arteriole
____ (h) site of obligatory water reabsorption
____ (i) pores in the glomerular endothelial cells that allow filtration of blood solutes but not blood cells and platelets
____ (j) can secrete H^+ against a concentration gradient
____ (k) modified smooth muscle cells in the wall of the afferent arteriole

(1) podocytes
(2) glomerulus
(3) renal corpuscle
(4) proximal convoluted tubule
(5) distal convoluted tubule
(6) juxtaglomerular cells
(7) macula densa
(8) principal cells
(9) intercalated cells
(10) nephron
(11) fenestrations

14. Match the following:
____ (a) measure of blood nitrogen resulting from the catabolism and deamination of amino acids
____ (b) produced from the catabolism of creatine phosphate in skeletal muscle
____ (c) volume of blood that is cleared of a substance per unit of time
____ (d) can result from diabetes mellitus
____ (e) insoluble stones of crystallized salts
____ (f) usually indicates a pathological condition
____ (g) lack of voluntary control of micturition
____ (h) can be caused by damage to the filtration membranes

(1) incontinence
(2) renal calculi
(3) plasma creatinine
(4) BUN test
(5) albuminuria
(6) glucosuria
(7) renal plasma clearance
(8) hematuria

15. Match the following:

____ (a) membrane proteins that function as water channels

____ (b) a secondary active transport process that achieves Na$^+$ reabsorption, returns filtered HCO$_3^-$ and water to the peritubular capillaries, and secretes H$^+$

____ (c) stimulates principal cells to secrete more K$^+$ into tubular fluid and reabsorb more Na$^+$ and Cl$^-$ from tubular fluid

____ (d) enzyme secreted by juxtaglomerular cells

____ (e) reduces glomerular filtration rate; increases blood volume and pressure

____ (f) inhibits Na$^+$ and H$_2$O reabsorption in the proximal convoluted tubules and collecting ducts

____ (g) regulates facultative water reabsorption by increasing the water permeability of principal cells in the distal convoluted tubules and collecting ducts

____ (h) reabsorb Na$^+$ together with a variety of other solutes

____ (i) stimulates cells in the distal convoluted tubule to reabsorb more calcium into the blood

(1) angiotensin II
(2) atrial natriuretic peptide
(3) Na$^+$ symporters
(4) Na$^+$/H$^+$ antiporters
(5) aquaporins
(6) aldosterone
(7) ADH
(8) renin
(9) parathyroid hormone

CRITICAL THINKING QUESTIONS

1. Imagine the discovery of a new toxin that blocks renal tubule reabsorption but does not affect filtration. Predict the short-term effects of this toxin.

2. For each of the following urinalysis results, indicate whether you should be concerned or not and why: (a) dark yellow urine that is turbid; (b) ammonia-like odor of the urine; (c) presence of excessive albumin; (d) presence of epithelial cell casts; (e) pH of 5.5; (f) hematuria.

3. Bruce is experiencing sudden, rhythmic waves of pain in his groin area. He has noticed that, although he is consuming fluids, his urine output has decreased. From what condition is Bruce suffering? How is it treated? How can he prevent future episodes?

? ANSWERS TO FIGURE QUESTIONS

26.1 The kidneys, ureters, urinary bladder, and urethra are the components of the urinary system.

26.2 The kidneys are retroperitoneal because they are posterior to the peritoneum.

26.3 Blood vessels, lymphatic vessels, nerves, and a ureter pass through the renal hilum.

26.4 About 1200 mL of blood enters the renal arteries each minute.

26.5 Cortical nephrons have glomeruli in the superficial renal cortex, and their short loops of Henle penetrate only into the superficial renal medulla. Juxtamedullary nephrons have glomeruli deep in the renal cortex, and their long loops of Henle extend through the renal medulla nearly to the renal papilla.

26.6 This section must pass through the renal cortex because there are no renal corpuscles in the renal medulla.

26.7 Secreted penicillin is being removed from the bloodstream.

26.8 Endothelial fenestrations (pores) in glomerular capillaries are too small for red blood cells to pass through.

26.9 Obstruction of the right ureter would increase CHP and thus decrease NFP in the right kidney; the obstruction would have no effect on the left kidney.

26.10 *Auto-* means self; tubuloglomerular feedback is an example of autoregulation because it takes place entirely within the kidneys.

26.11 The tight junctions between tubule cells form a barrier that prevents diffusion of transporter, channel, and pump proteins between the apical and basolateral membranes.

26.12 Glucose enters a PCT cell via a Na$^+$–glucose symporter in the apical membrane and leaves via facilitated diffusion through the basolateral membrane.

26.13 The electrochemical gradient promotes movement of Na$^+$ into the tubule cell through the apical membrane antiporters.

26.14 Reabsorption of the solutes creates an osmotic gradient that promotes the reabsorption of water via osmosis.

26.15 This is considered secondary active transport because the symporter uses the energy stored in the concentration gradient of Na$^+$ between extracellular fluid and the cytosol. No water is reabsorbed here because the thick ascending limb of the loop of Henle is virtually impermeable to water.

26.16 In principal cells, aldosterone stimulates secretion of K$^+$ and reabsorption of Na$^+$ by increasing the activity of sodium–potassium pumps and number of leakage channels for Na$^+$ and K$^+$.

26.17 Aldosterone and atrial natriuretic peptide influence renal water reabsorption along with ADH.

26.18 Dilute urine is produced when the thick ascending limb of the loop of Henle, the distal convoluted tubule, and the collecting duct reabsorb more solutes than water.

26.19 The high osmolarity of interstitial fluid in the renal medulla is due mainly to Na$^+$, Cl$^-$, and urea.

26.20 Secretion occurs in the proximal convoluted tubule, the loop of Henle, and the collecting duct.

26.21 Lack of voluntary control over micturition is termed urinary incontinence.

26.22 The three subdivisions of the male urethra are the prostatic urethra, membranous urethra, and spongy urethra.

26.23 The kidneys start to form during the third week of development.

27

FLUID, ELECTROLYTE, AND ACID–BASE HOMEOSTASIS

FLUID, ELECTROLYTE, AND ACID–BASE HOMEOSTASIS *Regulating the volume and composition of body fluids, controlling their distribution throughout the body, and balancing the pH of body fluids are crucial to maintaining overall homeostasis and health.*

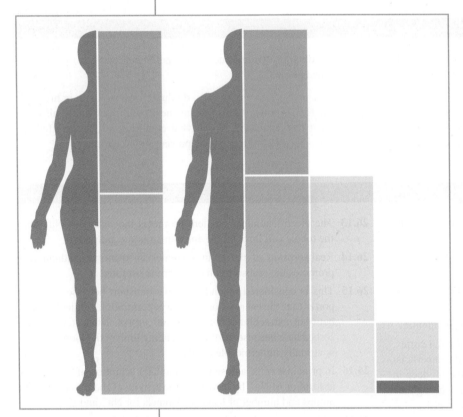

In Chapter 26 you learned how the kidneys form urine. One important function of the kidneys is to help maintain fluid balance in the body. The water and dissolved solutes throughout the body constitute the **body fluids.** Regulatory mechanisms involving the kidneys and other organs normally maintain homeostasis of the body fluids. Malfunction in any or all of them may seriously endanger the functioning of organs throughout the body. In this chapter, we will explore the mechanisms that regulate the volume and distribution of body fluids and examine the factors that determine the concentrations of solutes and the pH of body fluids.

?

Did you ever wonder how breathing can affect your body's pH?

27.1 FLUID COMPARTMENTS AND FLUID BALANCE

◉ **OBJECTIVES**

- Compare the locations of intracellular fluid (ICF) and extracellular fluid (ECF), and describe the various fluid compartments of the body.
- Describe the sources of water and solute gain and loss, and explain how each is regulated.
- Explain how fluids move between compartments.

In lean adults, body fluids constitute between 55% and 60% of total body mass in females and males, respectively (Figure 27.1). Body fluids are present in two main "compartments"—inside cells and outside cells. About two-thirds of body fluid is **intracellular fluid (ICF)** (*intra-* = within) or **cytosol,** the fluid within cells. The other third, called **extracellular fluid (ECF)** (*extra-* = outside), is outside cells and includes all other body fluids. About 80% of the ECF is **interstitial fluid** (*inter-* = between), which occupies the microscopic spaces between tissue cells, and 20% of the ECF is **plasma,** the liquid portion of the blood. Other extracellular fluids that are grouped with interstitial fluid include lymph in lymphatic vessels; cerebrospinal fluid in the nervous system; synovial fluid in joints; aqueous humor and vitreous body in the eyes; endolymph and perilymph in the ears; and pleural, pericardial, and peritoneal fluids between serous membranes.

Two general "barriers" separate intracellular fluid, interstitial fluid, and blood plasma.

1. The *plasma membrane* of individual cells separates intracellular fluid from the surrounding interstitial fluid. You learned in Chapter 3 that the plasma membrane is a selectively permeable barrier: It allows some substances to cross but blocks the movement of other substances. In addition, active transport pumps work continuously to maintain different concentrations of certain ions in the cytosol and interstitial fluid.

2. *Blood vessel walls* divide the interstitial fluid from blood plasma. Only in capillaries, the smallest blood vessels, are the walls thin enough and leaky enough to permit the exchange of water and solutes between blood plasma and interstitial fluid.

The body is in **fluid balance** when the required amounts of water and solutes are present and are correctly proportioned among the various compartments. **Water** is by far the largest single component of the body, making up 45–75% of total body mass, depending on age and gender.

Figure 27.1 Body fluid compartments.

🔑 The term body fluid refers to body water and its dissolved substances.

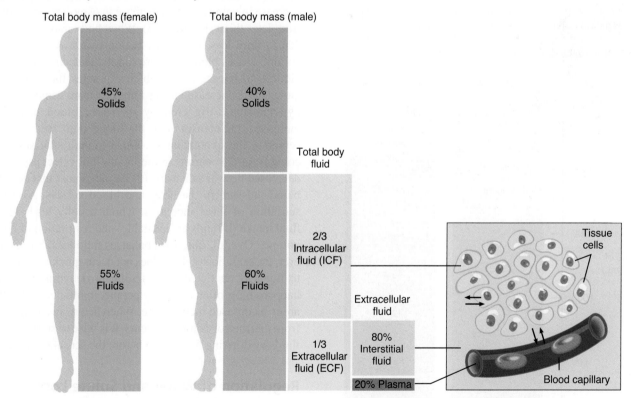

(a) Distribution of body solids and fluids in average lean, adult female and male

(b) Exchange of water among body fluid compartments

❓ What is the approximate volume of blood plasma in a lean 60-kg male? In a lean 60-kg female? (Note: One liter of body fluid has a mass of 1 kilogram.)

The processes of filtration, reabsorption, diffusion, and osmosis allow continual exchange of water and solutes among body fluid compartments (Figure 27.1b). Yet the volume of fluid in each compartment remains remarkably stable. The pressures that promote filtration of fluid from blood capillaries and reabsorption of fluid back into capillaries can be reviewed in Figure 21.7. Because osmosis is the primary means of water movement between intracellular fluid and interstitial fluid, the concentration of solutes in these fluids determines the *direction* of water movement. Because most solutes in body fluids are **electrolytes,** inorganic compounds that dissociate into ions, fluid balance is closely related to electrolyte balance. Because intake of water and electrolytes rarely occurs in exactly the same proportions as their presence in body fluids, the ability of the kidneys to excrete excess water by producing dilute urine, or to excrete excess electrolytes by producing concentrated urine, is of utmost importance in the maintenance of homeostasis.

Sources of Body Water Gain and Loss

The body can gain water by ingestion and by metabolic synthesis (Figure 27.2). The main sources of body water are ingested liquids (about 1600 mL) and moist foods (about 700 mL) absorbed from the gastrointestinal (GI) tract, which total about 2300 mL/day. The other source of water is **metabolic water** that is produced in the body mainly when electrons are accepted by oxygen during

Figure 27.2 Sources of daily water gain and loss under normal conditions. Numbers are average volumes for adults.

Normally, daily water loss equals daily water gain.

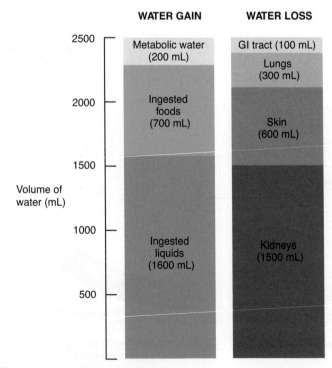

❓ **How does each of the following affect fluid balance: Hyperventilation? Vomiting? Fever? Diuretics?**

aerobic cellular respiration (see Figure 25.2) and to a smaller extent during dehydration synthesis reactions (see Figure 2.15). Metabolic water gain accounts for only 200 mL/day. Daily water gain from these two sources totals about 2500 mL.

Normally, body fluid volume remains constant because water loss equals water gain. Water loss occurs in four ways (Figure 27.2). Each day the kidneys excrete about 1500 mL in urine, the skin evaporates about 600 mL (400 mL through insensible perspiration—sweat that evaporates before it is perceived as moisture—and 200 mL as sweat), the lungs exhale about 300 mL as water vapor, and the gastrointestinal tract eliminates about 100 mL in feces. In women of reproductive age, additional water is lost in menstrual flow. On average, daily water loss totals about 2500 mL. The amount of water lost by a given route can vary considerably over time. For example, water may literally pour from the skin in the form of sweat during strenuous exertion. In other cases, water may be lost in diarrhea during a GI tract infection.

Regulation of Body Water Gain

The volume of metabolic water formed in the body depends entirely on the level of aerobic cellular respiration, which reflects the demand for ATP in body cells. When more ATP is produced, more water is formed. Body water gain is regulated mainly by the volume of water intake, or how much fluid you drink. An area in the hypothalamus known as the **thirst center** governs the urge to drink.

When water loss is greater than water gain, **dehydration**—a decrease in volume and an increase in osmolarity of body fluids—stimulates thirst (Figure 27.3). When body mass decreases by 2% due to fluid loss, mild dehydration exists. A decrease in blood volume causes blood pressure to fall. This change stimulates the kidneys to release renin, which promotes the formation of angiotensin II. Increased nerve impulses from osmoreceptors in the hypothalamus, triggered by increased blood osmolarity, and increased angiotensin II in the blood both stimulate the thirst center in the hypothalamus. Other signals that stimulate thirst come from (1) neurons in the mouth that detect dryness due to a decreased flow of saliva and (2) baroreceptors that detect lowered blood pressure in the heart and blood vessels. As a result, the sensation of thirst increases, which usually leads to increased fluid intake (if fluids are available) and restoration of normal fluid volume. Overall, fluid gain balances fluid loss. Sometimes, however, the sensation of thirst does not occur quickly enough or access to fluids is restricted, and significant dehydration ensues. This happens most often in elderly people, in infants, and in those who are in a confused mental state. When heavy sweating or fluid loss from diarrhea or vomiting occurs, it is wise to start replacing body fluids by drinking fluids even before the sensation of thirst occurs.

Regulation of Water and Solute Loss

Even though the loss of water and solutes through sweating and exhalation increases during exercise, elimination of *excess* body water or solutes occurs mainly by control of their loss in urine. The extent of *urinary salt (NaCl) loss* is the main factor that

Figure 27.3 Pathways through which dehydration stimulates thirst.

Dehydration occurs when water loss is greater than water gain.

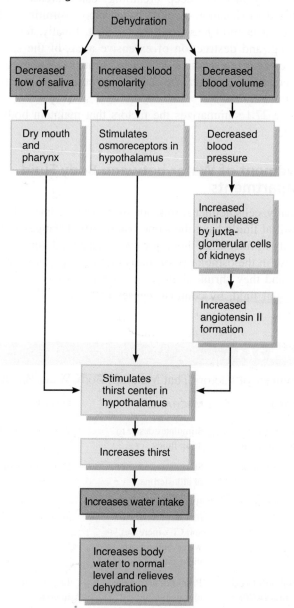

Does regulation of these pathways occur via negative or positive feedback? Why?

Figure 27.4 Hormonal regulation of renal Na$^+$ and Cl$^-$ reabsorption.

The three main hormones that regulate renal Na$^+$ and Cl$^-$ reabsorption (and thus the amount lost in the urine) are angiotensin II, aldosterone, and atrial natriuretic peptide.

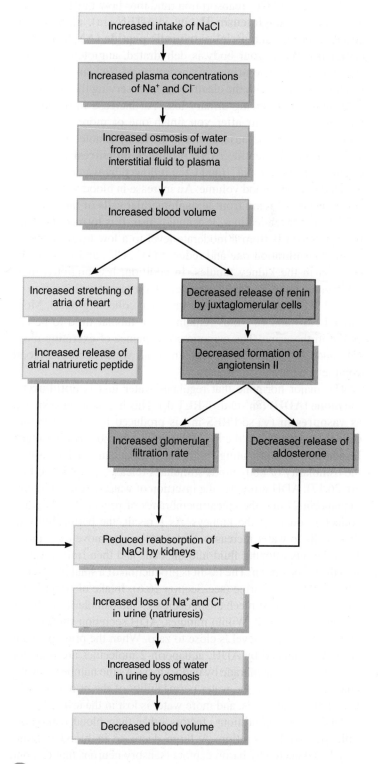

How does hyperaldosteronism (excessive aldosterone secretion) cause edema?

determines body fluid *volume*. The reason for this is that "water follows solutes" in osmosis, and the two main solutes in extracellular fluid (and in urine) are sodium ions (Na$^+$) and chloride ions (Cl$^-$). In a similar way, the main factor that determines body fluid *osmolarity* is the extent of *urinary water loss*.

Because our daily diet contains a highly variable amount of NaCl, urinary excretion of Na$^+$ and Cl$^-$ must also vary to maintain homeostasis. Hormonal changes regulate the urinary loss of these ions, which in turn affects blood volume. Figure 27.4 depicts the sequence of changes that occur after a salty meal. The increased intake of NaCl produces an increase in plasma levels of Na^+ and Cl^- (the major contributors to osmolarity of extracellular fluid). As a result, the osmolarity of interstitial fluid increases, which causes movement of water from intracellular fluid into interstitial fluid and then into plasma. Such water movement increases blood volume.

The three most important hormones that regulate the extent of renal Na^+ and Cl^- reabsorption (and thus how much is lost in the urine) are **angiotensin II** (an'-jē-ō-TEN-sin), **aldosterone** (al-DOS-ter-ōn), and **atrial natriuretic peptide (ANP)** (nā'-trē-ū-RET-ik). When your body is dehydrated, angiotensin II and aldosterone promote urinary reabsorption of Na^+ and Cl^- (and water by osmosis with the electrolytes), conserving the volume of body fluids by reducing urinary loss. An increase in blood volume, as might occur after you finish one or more supersized drinks, stretches the atria of the heart and promotes release of atrial natriuretic peptide. ANP promotes **natriuresis,** elevated urinary excretion of Na^+ (and Cl^-) followed by water excretion, which decreases blood volume. An increase in blood volume also slows release of renin from juxtaglomerular cells of the kidneys. When renin level declines, less angiotensin II is formed. Decline in angiotensin II from a moderate level to a low level increases glomerular filtration rate and reduces Na^+, Cl^-, and water reabsorption in the kidney tubules. In addition, less angiotensin II leads to lower levels of aldosterone, which causes reabsorption of filtered Na^+ and Cl^- to slow in the renal collecting ducts. More filtered Na^+ and Cl^- thus remain in the tubular fluid to be excreted in the urine. The osmotic consequence of excreting more Na^+ and Cl^- is loss of more water in urine, which decreases blood volume and blood pressure.

The major hormone that regulates water loss is **antidiuretic hormone (ADH)** (an'-tē-dī-ū-RET-ik). This hormone, also known as **vasopressin** (vā-sō-PRES-in), is produced by neurosecretory cells that extend from the hypothalamus to the posterior pituitary. In addition to stimulating the thirst mechanism, an increase in the osmolarity of body fluids stimulates release of ADH (see Figure 26.17). ADH promotes the insertion of water-channel proteins (aquaporin-2) into the apical membranes of principal cells in the collecting ducts of the kidneys. As a result, the permeability of these cells to water increases. Water molecules move by osmosis from the renal tubular fluid into the cells and then from the cells into the bloodstream. The result is production of a small volume of very concentrated urine (see Section 26.6). Intake of water in response to the thirst mechanism decreases the osmolarity of blood and interstitial fluid. Within minutes, ADH secretion shuts down, and soon its blood level is close to zero. When the principal cells are not stimulated by ADH, aquaporin-2 molecules are removed from the apical membrane by endocytosis. As the number of water channels decreases, the water permeability of the principal cells' apical membrane falls, and more water is lost in the urine.

Under some conditions, factors other than blood osmolarity influence ADH secretion. A large decrease in blood volume, which is detected by baroreceptors (sensory neurons that respond to stretching) in the left atrium and in blood vessel walls, also stimulates ADH release. In severe dehydration, glomerular filtra-tion rate decreases because blood pressure falls, so that less water is lost in the urine. Conversely, the intake of too much water increases blood pressure, causing the rate of glomerular filtration to rise and more water to be lost in the urine. Hyperventilation (abnormally fast and deep breathing) can increase fluid loss through the exhalation of more water vapor. Vomiting and diarrhea result in fluid loss from the GI tract. Finally, fever, heavy sweating, and destruction of extensive areas of the skin from burns can cause excessive water loss through the skin. In all of these conditions, an increase in ADH secretion will help conserve body fluids.

Table 27.1 summarizes the factors that maintain body water balance.

Movement of Water between Body Fluid Compartments

Normally, cells neither shrink nor swell because intracellular and interstitial fluids have the same osmolarity. Changes in the osmolarity of interstitial fluid, however, cause fluid imbalances. An increase in the osmolarity of interstitial fluid draws water out of cells, and they shrink slightly. A decrease in the osmolarity of interstitial fluid, by contrast, causes cells to swell. Changes in

---

TABLE 27.1

Summary of Factors That Maintain Body Water Balance

FACTOR	MECHANISM	EFFECT
Thirst center in hypothalamus	Stimulates desire to drink fluids.	Water gained if thirst is quenched.
Angiotensin II	Stimulates secretion of aldosterone.	Reduces loss of water in urine.
Aldosterone	By promoting urinary reabsorption of Na^+ and Cl^-, increases water reabsorption via osmosis.	Reduces loss of water in urine.
Atrial natriuretic peptide (ANP)	Promotes natriuresis, elevated urinary excretion of Na^+ (and Cl^-), accompanied by water.	Increases loss of water in urine.
Antidiuretic hormone (ADH), also known as vasopressin	Promotes insertion of water-channel proteins (aquaporin-2) into apical membranes of principal cells in collecting ducts of kidneys. As a result, water permeability of these cells increases and more water is reabsorbed.	Reduces loss of water in urine.

osmolarity most often result from changes in the concentration of Na^+.

A decrease in the osmolarity of interstitial fluid, as may occur after drinking a large volume of water, inhibits secretion of ADH. Normally, the kidneys then excrete a large volume of dilute urine, which restores the osmotic pressure of body fluids to normal. As a result, body cells swell only slightly, and only for a brief period. But when a person steadily consumes water faster than the kidneys can excrete it (the maximum urine flow rate is about 15 mL/min) or when renal function is poor, the result may be **water intoxication,** a state in which excessive body water causes cells to swell dangerously (Figure 27.5). If the body water and Na^+ lost during blood loss or excessive sweating, vomiting, or diarrhea is replaced by drinking plain water, then body fluids become more dilute. This dilution can cause the Na^+ concentration of plasma and then of interstitial fluid to fall below the normal range. When the Na^+ concentration of interstitial fluid decreases, its osmolarity also falls. The net result is osmosis of water from interstitial fluid into the cytosol. Water entering the cells causes them to swell, producing convulsions, coma, and possibly death. To prevent this dire sequence of events in cases of severe electrolyte and water loss, solutions given for intravenous or oral rehydration therapy (ORT) include a small amount of table salt (NaCl).

Figure 27.5 Series of events in water intoxication.

 Water intoxication is a state in which excessive body water causes cells to swell.

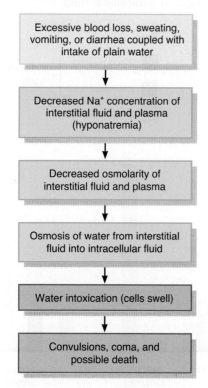

? **Why do solutions used for oral rehydration therapy contain a small amount of table salt (NaCl)?**

CLINICAL CONNECTION | *Enemas and Fluid Balance*

An **enema** (EN-e-ma) is the introduction of a solution into the rectum to draw water (and electrolytes) into the colon osmotically. The increased volume increases peristalsis, which evacuates feces. Enemas are used to treat constipation. Repeated enemas, especially in young children, increase the risk of fluid and electrolyte imbalances. •

✔ **CHECKPOINT**

1. What is the approximate volume of each of your body fluid compartments?
2. How are the routes of water gain and loss from the body regulated?
3. By what mechanism does thirst help regulate water intake?
4. How do angiotensin II, aldosterone, atrial natriuretic peptide, and antidiuretic hormone regulate the volume and osmolarity of body fluids?
5. What factors control the movement of water between interstitial fluid and intracellular fluid?

27.2 ELECTROLYTES IN BODY FLUIDS

◉ **OBJECTIVES**

• Compare the electrolyte composition of the three major fluid compartments: plasma, interstitial fluid, and intracellular fluid.
• Discuss the functions of sodium, chloride, potassium, bicarbonate, calcium, phosphate, and magnesium ions, and explain how their concentrations are regulated.

The ions formed when electrolytes dissolve and dissociate serve four general functions in the body. (1) Because they are largely confined to particular fluid compartments and are more numerous than nonelectrolytes, certain ions *control the osmosis of water between fluid compartments.* (2) Ions *help maintain the acid–base balance* required for normal cellular activities. (3) Ions *carry electrical current,* which allows production of action potentials and graded potentials. (4) Several ions *serve as cofactors* needed for optimal activity of enzymes.

Concentrations of Electrolytes in Body Fluids

To compare the charge carried by ions in different solutions, the concentration of ions is typically expressed in units of **milliequivalents per liter (mEq/liter)** (mil'-i-ē-KWIV-a-lents). These units give the concentration of cations or anions in a given volume of solution. One equivalent is the positive or negative charge equal to the amount of charge in one mole of H^+; a milliequivalent is one-thousandth of an equivalent. Recall that a mole of a substance is its molecular weight expressed in grams. For ions such as sodium (Na^+), potassium (K^+), and bicarbonate (HCO_3^-), which have a single positive or negative charge, the number of mEq/liter is equal to the number of mmol/liter. For ions such as calcium (Ca^{2+}) or phosphate (HPO_4^{2-}), which have two positive or negative charges, the number of mEq/liter is twice the number of mmol/liter.

Figure 27.6 compares the concentrations of the main electrolytes and protein anions in blood plasma, interstitial fluid, and intracellular fluid. The chief difference between the two extracellular fluids—blood plasma and interstitial fluid—is that blood plasma contains many protein anions, in contrast to interstitial fluid, which has very few. Because normal capillary membranes are virtually impermeable to proteins, only a few plasma proteins leak out of blood vessels into the interstitial fluid. This difference in protein concentration is largely responsible for the blood colloid osmotic pressure exerted by blood plasma. In other respects, the two fluids are similar.

The electrolyte content of intracellular fluid differs considerably from that of extracellular fluid. In extracellular fluid, the most abundant cation is Na^+, and the most abundant anion is Cl^-. In intracellular fluid, the most abundant cation is K^+, and the most abundant anions are proteins and phosphates (HPO_4^{2-}). By actively transporting Na^+ out of cells and K^+ into cells, sodium–potassium pumps (Na^+/K^+ ATPases) play a major role in maintaining the high intracellular concentration of K^+ and high extracellular concentration of Na^+.

Sodium

Sodium ions (Na^+) are the most abundant ions in extracellular fluid, accounting for 90% of the extracellular cations. The normal blood plasma Na^+ concentration is 136–148 mEq/liter. As we have already seen, Na^+ plays a pivotal role in fluid and electrolyte balance because it accounts for almost half of the osmolarity of extracellular fluid (142 of about 300 mOsm/liter). The flow of Na^+ through voltage-gated channels in the plasma membrane also is necessary for the generation and conduction of action potentials in neurons and muscle fibers. The typical daily intake of Na^+ in North America often far exceeds the body's normal daily requirements, due largely to excess dietary salt. The kidneys excrete excess Na^+, but they also can conserve it during periods of shortage.

The Na^+ level in the blood is controlled by aldosterone, antidiuretic hormone (ADH), and atrial natriuretic peptide (ANP). Aldosterone increases renal reabsorption of Na^+. When the blood plasma concentration of Na^+ drops below 135 mEq/liter, a condition called *hyponatremia*, ADH release ceases. The lack of ADH in turn permits greater excretion of water in urine and restoration of the normal Na^+ level in ECF. Atrial natriuretic peptide increases Na^+ excretion by the kidneys when the Na^+ level is above normal, a condition called *hypernatremia*.

Figure 27.6 Electrolyte and protein anion concentrations in plasma, interstitial fluid, and intracellular fluid. The height of each column represents milliequivalents per liter (mEq/liter).

The electrolytes present in extracellular fluids are different from those present in intracellular fluid.

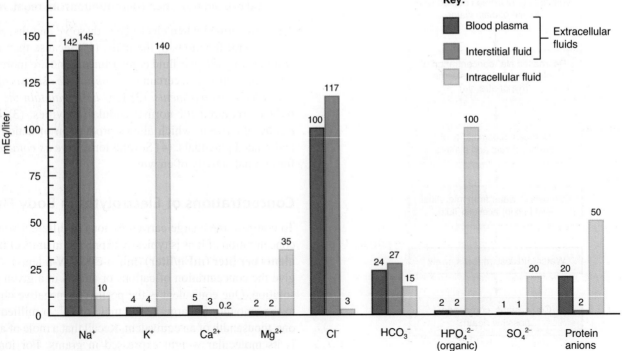

? **What cation and two anions are present in the highest concentrations in ECF and ICF?**

CLINICAL CONNECTION | *Indicators of Na⁺ Imbalance*

If excess sodium ions remain in the body because the kidneys fail to excrete enough of them, water is also osmotically retained. The result is increased blood volume, increased blood pressure, and **edema,** an abnormal accumulation of interstitial fluid. Renal failure and hyperaldosteronism (excessive aldosterone secretion) are two causes of Na⁺ retention. Excessive urinary loss of Na⁺, by contrast, causes excessive water loss, which results in **hypovolemia** (hī′-pō-vō-LĒ-mē-a), an abnormally low blood volume. Hypovolemia related to Na⁺ loss is most frequently due to the inadequate secretion of aldosterone associated with adrenal insufficiency or overly vigorous therapy with diuretic drugs. •

Chloride

Chloride ions (Cl^-) are the most prevalent anions in extracellular fluid. The normal blood plasma Cl^- concentration is 95–105 mEq/liter. Cl^- moves relatively easily between the extracellular and intracellular compartments because most plasma membranes contain many Cl^- leakage channels and antiporters. For this reason, Cl^- can help balance the level of anions in different fluid compartments. One example is the chloride shift that occurs between red blood cells and blood plasma as the blood level of carbon dioxide either increases or decreases (see Figure 23.23). In this case, the antiporter exchange of Cl^- for HCO_3^- maintains the correct balance of anions between ECF and ICF. Chloride ions also are part of the hydrochloric acid secreted into gastric juice. ADH helps regulate Cl^- balance in body fluids because it governs the extent of water loss in urine. Processes that increase or decrease renal reabsorption of sodium ions also affect reabsorption of chloride ions. (Recall that reabsorption of Na^+ and Cl^- occurs by means of Na^+–Cl^- symporters.)

Potassium

Potassium ions (K^+) are the most abundant cations in intracellular fluid (140 mEq/liter). K^+ plays a key role in establishing the resting membrane potential and in the repolarization phase of action potentials in neurons and muscle fibers; K^+ also helps maintain normal intracellular fluid volume. When K^+ moves into or out of cells, it often is exchanged for H^+ and thereby helps regulate the pH of body fluids.

The normal blood plasma K^+ concentration is 3.5–5.0 mEq/liter and is controlled mainly by aldosterone. When blood plasma K^+ concentration is high, more aldosterone is secreted into the blood. Aldosterone then stimulates principal cells of the renal collecting ducts to secrete more K^+ so excess K^+ is lost in the urine. Conversely, when blood plasma K^+ concentration is low, aldosterone secretion decreases and less K^+ is excreted in urine. Because K^+ is needed during the repolarization phase of action potentials, abnormal K^+ levels can be lethal. For instance, *hyperkalemia* (above-normal concentration of K^+ in blood) can cause death due to ventricular fibrillation.

Bicarbonate

Bicarbonate ions (HCO_3^-) are the second most prevalent extracellular anions. Normal blood plasma HCO_3^- concentration is 22–26 mEq/liter in systemic arterial blood and 23–27 mEq/liter in systemic venous blood. HCO_3^- concentration increases as blood flows through systemic capillaries because the carbon dioxide released by metabolically active cells combines with water to form carbonic acid; the carbonic acid then dissociates into H^+ and HCO_3^-. As blood flows through pulmonary capillaries, however, the concentration of HCO_3^- decreases again as carbon dioxide is exhaled. (Figure 23.23 shows these reactions.) Intracellular fluid also contains a small amount of HCO_3^-. As previously noted, the exchange of Cl^- for HCO_3^- helps maintain the correct balance of anions in extracellular fluid and intracellular fluid.

The kidneys are the main regulators of blood HCO_3^- concentration. The intercalated cells of the renal tubule can either form HCO_3^- and release it into the blood when the blood level is low (see Figure 27.8) or excrete excess HCO_3^- in the urine when the level in blood is too high. Changes in the blood level of HCO_3^- are considered later in this chapter in the section on acid–base balance.

Calcium

Because such a large amount of calcium is stored in bone, it is the most abundant mineral in the body. About 98% of the calcium in adults is located in the skeleton and teeth, where it is combined with phosphates to form a crystal lattice of mineral salts. In body fluids, calcium is mainly an extracellular cation (Ca^{2+}). The normal concentration of free or unattached Ca^{2+} in blood plasma is 4.5–5.5 mEq/liter. About the same amount of Ca^{2+} is attached to various plasma proteins. Besides contributing to the hardness of bones and teeth, Ca^{2+} plays important roles in blood clotting, neurotransmitter release, maintenance of muscle tone, and excitability of nervous and muscle tissue.

The most important regulator of Ca^{2+} concentration in blood plasma is parathyroid hormone (PTH) (see Figure 18.14). A low level of Ca^{2+} in blood plasma promotes release of more PTH, which stimulates osteoclasts in bone tissue to release calcium (and phosphate) from bone extracellular matrix. Thus, PTH increases bone *resorption*. Parathyroid hormone also enhances *reabsorption* of Ca^{2+} from glomerular filtrate through renal tubule cells and back into blood, and increases production of calcitriol (the form of vitamin D that acts as a hormone), which in turn increases Ca^{2+} *absorption* from food in the gastrointestinal tract. Recall that calcitonin (CT) produced by the thyroid gland inhibits the activity of osteoclasts, accelerates Ca^{2+} deposition into bones, and thus lowers blood Ca^{2+} levels.

Phosphate

About 85% of the phosphate in adults is present as calcium phosphate salts, which are structural components of bone and teeth. The remaining 15% is ionized. Three phosphate ions ($H_2PO_4^-$, HPO_4^{2-}, and PO_4^{3-}) are important intracellular anions. At the normal pH of body fluids, HPO_4^{2-} is the most prevalent form. Phosphates contribute about 100 mEq/liter of anions to intracellular fluid. HPO_4^{2-} is an important buffer of H^+, both in body fluids and in the urine. Although some are "free," most phosphate ions are covalently bound to organic molecules such as lipids (phospholipids), proteins, carbohydrates, nucleic acids (DNA and RNA), and adenosine triphosphate (ATP).

The normal blood plasma concentration of ionized phosphate is only 1.7–2.6 mEq/liter. The same two hormones that govern calcium homeostasis—parathyroid hormone (PTH) and calcitriol—also regulate the level of HPO_4^{2-} in blood plasma. PTH stimulates resorption of bone extracellular matrix by osteoclasts, which releases both phosphate and calcium ions into the bloodstream. In the kidneys, however, PTH inhibits reabsorption of phosphate ions while stimulating reabsorption of calcium ions by renal tubular cells. Thus, PTH increases urinary excretion of phosphate and lowers blood phosphate level. Calcitriol promotes absorption of both phosphates and calcium from the gastrointestinal tract. Fibroblast growth factor 23 (FGF 23) is a polypeptide paracrine (local hormone) that also helps regulate blood plasma levels of HPO_4^{2-}. This hormone decreases HPO_4^{2-} blood levels by increasing HPO_4^{2-} excretion by the kidneys and decreasing absorption of HPO_4^{2-} by the gastrointestinal tract.

Magnesium

In adults, about 54% of the total body magnesium is part of bone matrix as magnesium salts. The remaining 46% occurs as magnesium ions (Mg^{2+}) in intracellular fluid (45%) and extracellular fluid (1%). Mg^{2+} is the second most common intracellular cation (35 mEq/liter). Functionally, Mg^{2+} is a cofactor for certain enzymes needed for the metabolism of carbohydrates and proteins and for the sodium–potassium pump. Mg^{2+} is essential for normal neuromuscular activity, synaptic transmission, and myocardial functioning. In addition, secretion of parathyroid hormone (PTH) depends on Mg^{2+}.

Normal blood plasma Mg^{2+} concentration is low, only 1.3–2.1 mEq/liter. Several factors regulate the blood plasma level of Mg^{2+} by varying the rate at which it is excreted in the urine. The kidneys increase urinary excretion of Mg^{2+} in response to hypercalcemia, hypermagnesemia, increases in extracellular fluid volume, decreases in parathyroid hormone, and acidosis. The opposite conditions decrease renal excretion of Mg^{2+}.

Table 27.2 describes the imbalances that result from the deficiency or excess of several electrolytes.

People at risk for fluid and electrolyte imbalances include those who depend on others for fluid and food, such as infants, the elderly, and the hospitalized; individuals undergoing medical treatment that involves intravenous infusions, drainages or suctions, and urinary catheters; and people who receive diuretics, experience excessive fluid losses and require increased fluid intake, or experience fluid retention and have fluid restrictions. Finally, athletes and military personnel in extremely hot environments, postoperative individuals, severe burn or trauma cases, individuals with chronic diseases (congestive heart failure, diabetes,

chronic obstructive lung disease, and cancer), people in confinement, and individuals with altered levels of consciousness who may be unable to communicate needs or respond to thirst are also subject to fluid and electrolyte imbalances.

✔ **CHECKPOINT**

6. What are the functions of electrolytes in the body?
7. Name three important extracellular electrolytes and three important intracellular electrolytes and indicate how each is regulated.

27.3 ACID–BASE BALANCE

☉ **OBJECTIVES**

- Compare the roles of buffers, exhalation of carbon dioxide, and kidney excretion of H^+ in maintaining pH of body fluids.
- Define acid–base imbalances, describe their effects on the body, and explain how they are treated.

From our discussion thus far, it should be clear that various ions play different roles that help maintain homeostasis. A major homeostatic challenge is keeping the H^+ concentration (pH) of body fluids at an appropriate level. This task—the maintenance of acid–base balance—is of critical importance to normal cellular function. For example, the three-dimensional shape of all body proteins, which enables them to perform specific functions, is very sensitive to pH changes. When the diet contains a large amount of protein, as is typical in North America, cellular metabolism produces more acids than bases, which tends to acidify the blood. Before proceeding with this section of the chapter, you may wish to review the discussion of acids, bases, and pH in Section 2.4.

In a healthy person, several mechanisms help maintain the pH of systemic arterial blood between 7.35 and 7.45. (A pH of 7.4 corresponds to a H^+ concentration of 0.00004 mEq/liter = 40 nEq /liter.) Because metabolic reactions often produce a huge excess of H^+, the lack of any mechanism for the disposal of H^+ would cause H^+ in body fluids to rise quickly to a lethal level. Homeostasis of H^+ concentration within a narrow range is thus essential to survival. The removal of H^+ from body fluids and its subsequent elimination from the body depend on the following three major mechanisms:

1. *Buffer systems.* Buffers act quickly to temporarily bind H^+, removing the highly reactive, excess H^+ from solution. Buffers thus raise pH of body fluids but do not remove H^+ from the body.

2. *Exhalation of carbon dioxide.* By increasing the rate and depth of breathing, more carbon dioxide can be exhaled. Within minutes this reduces the level of carbonic acid in blood, which raises the blood pH (reduces blood H^+ level).

3. *Kidney excretion of H^+.* The slowest mechanism, but the only way to eliminate acids other than carbonic acid, is through their excretion in urine.

TABLE 27.2

Blood Electrolyte Imbalances

ELECTROLYTE*	DEFICIENCY		EXCESS	
	NAME AND CAUSES	SIGNS AND SYMPTOMS	NAME AND CAUSES	SIGNS AND SYMPTOMS
Sodium (Na$^+$) 136–148 mEq/liter	**Hyponatremia** (hī-pō-na-TRĒ-mē-a) may be due to decreased sodium intake; increased sodium loss through vomiting, diarrhea, aldosterone deficiency, or taking certain diuretics; and excessive water intake.	Muscular weakness; dizziness, headache, and hypotension; tachycardia and shock; mental confusion, stupor, and coma.	**Hypernatremia** may occur with dehydration, water deprivation, or excessive sodium in diet or intravenous fluids; causes hypertonicity of ECF, which pulls water out of body cells into ECF, causing cellular dehydration.	Intense thirst, hypertension, edema, agitation, and convulsions.
Chloride (Cl$^-$) 95–105 mEq/liter	**Hypochloremia** (hī-pō-klō-RĒ-mē-a) may be due to excessive vomiting, overhydration, aldosterone deficiency, congestive heart failure, and therapy with certain diuretics such as furosemide (Lasix®).	Muscle spasms, metabolic alkalosis, shallow respirations, hypotension, and tetany.	**Hyperchloremia** may result from dehydration due to water loss or water deprivation; excessive chloride intake; or severe renal failure, hyperaldosteronism, certain types of acidosis, and some drugs.	Lethargy, weakness, metabolic acidosis, and rapid, deep breathing.
Potassium (K$^+$) 3.5–5.0 mEq/liter	**Hypokalemia** (hī-pō-ka-LĒ-mē-a) may result from excessive loss due to vomiting or diarrhea, decreased potassium intake, hyperaldosteronism, kidney disease, and therapy with some diuretics.	Muscle fatigue, flaccid paralysis, mental confusion, increased urine output, shallow respirations, and changes in electrocardiogram, including flattening of T wave.	**Hyperkalemia** may be due to excessive potassium intake, renal failure, aldosterone deficiency, crushing injuries to body tissues, or transfusion of hemolyzed blood.	Irritability, nausea, vomiting, diarrhea, muscular weakness; can cause death by inducing ventricular fibrillation.
Calcium (Ca^{2+}) Total = 9.0–10.5 mg/dL; ionized = 4.5–5.5 mEq/liter	**Hypocalcemia** (hī-pō-kal-SĒ-mē-a) may be due to increased calcium loss, reduced calcium intake, elevated phosphate levels, or hypoparathyroidism.	Numbness and tingling of fingers; hyperactive reflexes, muscle cramps, tetany, and convulsions; bone fractures; spasms of laryngeal muscles that can cause death by asphyxiation.	**Hypercalcemia** may result from hyperparathyroidism, some cancers, excessive intake of vitamin D, and Paget's disease of bone.	Lethargy, weakness, anorexia, nausea, vomiting, polyuria, itching, bone pain, depression, confusion, paresthesia, stupor, and coma.
Phosphate (HPO$_4^{2-}$) 1.7–2.6 mEq/liter	**Hypophosphatemia** (hī-pō-fos'-fa-TĒ-mē-a) may occur through increased urinary losses, decreased intestinal absorption, or increased utilization.	Confusion, seizures, coma, chest and muscle pain, numbness and tingling of fingers, decreased coordination, memory loss, and lethargy.	**Hyperphosphatemia** occurs when kidneys fail to excrete excess phosphate, as in renal failure; can also result from increased intake of phosphates or destruction of body cells, which releases phosphates into blood.	Anorexia, nausea, vomiting, muscular weakness, hyperactive reflexes, tetany, and tachycardia.
Magnesium (Mg^{2+}) 1.3–2.1 mEq/liter	**Hypomagnesemia** (hī'-pō-mag'-ne-SĒ-mē-a) may be due to inadequate intake or excessive loss in urine or feces; also occurs in alcoholism, malnutrition, diabetes mellitus, and diuretic therapy.	Weakness, irritability, tetany, delirium, convulsions, confusion, anorexia, nausea, vomiting, paresthesia, and cardiac arrhythmias.	**Hypermagnesemia** occurs in renal failure or due to increased intake of Mg^{2+}, such as Mg^{2+}-containing antacids; also occurs in aldosterone deficiency and hypothyroidism.	Hypotension, muscular weakness or paralysis, nausea, vomiting, and altered mental functioning.

*Values are normal ranges of blood plasma levels in adults.

We will examine each of these mechanisms in more detail in the following sections.

The Actions of Buffer Systems

Most buffer systems in the body consist of a weak acid and the salt of that acid, which functions as a weak base. Buffers prevent rapid, drastic changes in the pH of body fluids by converting strong acids and bases into weak acids and weak bases within fractions of a second. Strong acids lower pH more than weak acids because strong acids release H^+ more readily and thus contribute more free hydrogen ions. Similarly, strong bases raise pH more than weak ones. The principal buffer systems of the body fluids are the protein buffer system, the carbonic acid–bicarbonate buffer system, and the phosphate buffer system.

Protein Buffer System

The **protein buffer system** is the most abundant buffer in intracellular fluid and blood plasma. For example, the protein hemoglobin is an especially good buffer within red blood cells, and albumin is the main protein buffer in blood plasma. Proteins are composed of amino acids, organic molecules that contain at least one carboxyl group (—COOH) and at least one amino group (—NH$_2$); these groups are the functional components of the protein buffer system. The free carboxyl group at one end of a protein acts like an acid by releasing H^+ when pH rises; it dissociates as follows:

$$
\underset{\substack{|\\H}}{\overset{R}{NH_2-C-COOH}} \longrightarrow \underset{\substack{|\\H}}{\overset{R}{NH_2-C-COO^-}} + H^+
$$

The H^+ is then able to react with any excess OH^- in the solution to form water. The free amino group at the other end of a protein can act as a base by combining with H^+ when pH falls, as follows:

$$
\underset{\substack{|\\H}}{\overset{R}{NH_2-C-COOH}} + H^+ \longrightarrow \underset{\substack{|\\H}}{\overset{R}{{}^+NH_3-C-COOH}}
$$

So proteins can buffer both acids and bases. In addition to the terminal carboxyl and amino groups, side chains that can buffer H^+ are present on 7 of the 20 amino acids.

As we have already noted, the protein hemoglobin is an important buffer of H^+ in red blood cells (see Figure 23.23). As blood flows through the systemic capillaries, carbon dioxide (CO_2) passes from tissue cells into red blood cells, where it combines with water (H_2O) to form carbonic acid (H_2CO_3). Once formed, H_2CO_3 dissociates into H^+ and HCO_3^-. At the same time that CO_2 is entering red blood cells, oxyhemoglobin (Hb—O$_2$) is giving up its oxygen to tissue cells. Reduced hemoglobin (deoxyhemoglobin) picks up most of the H^+. For this reason, reduced he-

moglobin usually is written as Hb—H. The following reactions summarize these relations:

$$
\underset{\substack{Water}}{H_2O} + \underset{\substack{Carbon\ dioxide\\(entering\ RBCs)}}{CO_2} \longrightarrow \underset{\substack{Carbonic\ acid}}{H_2CO_3}
$$

$$
\underset{\substack{Carbonic\ acid}}{H_2CO_3} \longrightarrow \underset{\substack{Hydrogen\ ion}}{H^+} + \underset{\substack{Bicarbonate\ ion}}{HCO_3^-}
$$

$$
\underset{\substack{Oxyhemoglobin\\(in\ RBCs)}}{Hb-O_2} + \underset{\substack{Hydrogen\ ion\\(from\ carbonic\\acid)}}{H^+} \longrightarrow \underset{\substack{Reduced\\hemoglobin}}{Hb-H} + \underset{\substack{Oxygen\\(released\ to\\tissue\ cells)}}{O_2}
$$

Carbonic Acid–Bicarbonate Buffer System

The **carbonic acid–bicarbonate buffer system** is based on the *bicarbonate ion* (HCO_3^-), which can act as a weak base, and *carbonic acid* (H_2CO_3), which can act as a weak acid. As you have already learned, HCO_3^- is a significant anion in both intracellular and extracellular fluids (see Figure 27.6). Because the kidneys also synthesize new HCO_3^- and reabsorb filtered HCO_3^-, this important buffer is not lost in the urine. If there is an excess of H^+, the HCO_3^- can function as a weak base and remove the excess H^+ as follows:

$$
\underset{\substack{Hydrogen\ ion}}{H^+} + \underset{\substack{Bicarbonate\ ion\\(weak\ base)}}{HCO_3^-} \longrightarrow \underset{\substack{Carbonic\ acid}}{H_2CO_3}
$$

Then, H_2CO_3 dissociates into water and carbon dioxide, and the CO_2 is exhaled from the lungs.

Conversely, if there is a shortage of H^+, the H_2CO_3 can function as a weak acid and provide H^+ as follows:

$$
\underset{\substack{Carbonic\ acid\\(weak\ acid)}}{H_2CO_3} \longrightarrow \underset{\substack{Hydrogen\ ion}}{H^+} + \underset{\substack{Bicarbonate\ ion}}{HCO_3^-}
$$

At a pH of 7.4, HCO_3^- concentration is about 24 mEq/liter and H_2CO_3 concentration is about 1.2 mmol/liter, so bicarbonate ions outnumber carbonic acid molecules by 20 to 1. Because CO_2 and H_2O combine to form H_2CO_3, this buffer system cannot protect against pH changes due to respiratory problems in which there is an excess or shortage of CO_2.

Phosphate Buffer System

The **phosphate buffer system** acts via a mechanism similar to the one for the carbonic acid–bicarbonate buffer system. The components of the phosphate buffer system are the ions *dihydrogen phosphate* ($H_2PO_4^-$) and *monohydrogen phosphate* (HPO_4^{2-}). Recall that phosphates are major anions in intracellular fluid and minor ones in extracellular fluids (see Figure 27.6). The dihydrogen phosphate ion acts as a weak acid and is capable of buffering strong bases such as OH^-, as follows:

$$
\underset{\substack{Hydroxide\ ion\\(strong\ base)}}{OH^-} + \underset{\substack{Dihydrogen\\phosphate\\(weak\ acid)}}{H_2PO_4^-} \longrightarrow \underset{\substack{Water}}{H_2O} + \underset{\substack{Monohydrogen\\phosphate\\(weak\ base)}}{HPO_4^{2-}}
$$

The monohydrogen phosphate ion is capable of buffering the H^+ released by a strong acid such as hydrochloric acid (HCl) by acting as a weak base:

$$H^+ + HPO_4^{2-} \longrightarrow H_2PO_4^-$$

Hydrogen ion (strong acid) — Monohydrogen phosphate (weak base) — Dihydrogen phosphate (weak acid)

Because the concentration of phosphates is highest in intracellular fluid, the phosphate buffer system is an important regulator of pH in the cytosol. It also acts to a smaller degree in extracellular fluids and buffers acids in urine. $H_2PO_4^-$ is formed when excess H^+ in the kidney tubule fluid combines with HPO_4^{2-} (see Figure 27.8). The H^+ that becomes part of the $H_2PO_4^-$ passes into the urine. This reaction is one way the kidneys help maintain blood pH by excreting H^+ in the urine.

Exhalation of Carbon Dioxide

The simple act of breathing also plays an important role in maintaining the pH of body fluids. An increase in the carbon dioxide (CO_2) concentration in body fluids increases H^+ concentration and thus lowers the pH (makes body fluids more acidic). Because H_2CO_3 can be eliminated by exhaling CO_2, it is called a **volatile acid.** Conversely, a decrease in the CO_2 concentration of body fluids raises the pH (makes body fluids more alkaline). This chemical interaction is illustrated by the following reversible reactions:

$$CO_2 + H_2O \rightleftharpoons H_2CO_3 \rightleftharpoons H^+ + HCO_3^-$$

Carbon dioxide — Water — Carbonic acid — Hydrogen ion — Bicarbonate ion

Changes in the rate and depth of breathing can alter the pH of body fluids within a couple of minutes. With increased ventilation, more CO_2 is exhaled. When CO_2 levels decrease, the reaction is driven to the left (blue arrows), H^+ concentration falls, and blood pH increases. Doubling the ventilation increases pH by about 0.23 units, from 7.4 to 7.63. If ventilation is slower than normal, less carbon dioxide is exhaled. When CO_2 levels increase, the reaction is driven to the right (red arrows), the H^+ concentration increases, and blood pH decreases. Reducing ventilation to one-quarter of normal lowers the pH by 0.4 units, from 7.4 to 7.0. These examples show the powerful effect of alterations in breathing on the pH of body fluids.

The pH of body fluids and the rate and depth of breathing interact via a negative feedback loop (Figure 27.7). When the blood acidity increases, the decrease in pH (increase in concentration of H^+) is detected by central chemoreceptors in the medulla oblongata and peripheral chemoreceptors in the aortic and carotid bodies, both of which stimulate the inspiratory area in the medulla oblongata. As a result, the diaphragm and other respiratory muscles contract more forcefully and frequently, so more CO_2 is exhaled. As less H_2CO_3 forms and fewer H^+ are present, blood pH increases. When the response brings blood pH (H^+ concentration) back to normal, there is a return to acid–base

Figure 27.7 Negative feedback regulation of blood pH by the respiratory system.

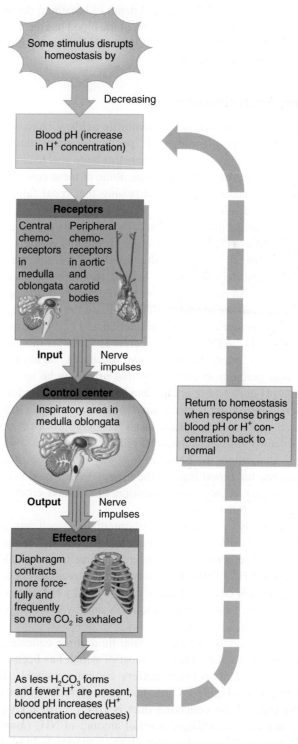

Exhalation of carbon dioxide lowers the H^+ concentration of blood.

? If you hold your breath for 30 seconds, what is likely to happen to your blood pH?

homeostasis. The same negative feedback loop operates if the blood level of CO_2 increases. Ventilation increases, which removes more CO_2, reducing the H^+ concentration and increasing the blood's pH.

By contrast, if the pH of the blood increases, the respiratory center is inhibited and the rate and depth of breathing decreases. A decrease in the CO_2 concentration of the blood has the same effect. When breathing decreases, CO_2 accumulates in the blood so its H^+ concentration increases.

Kidney Excretion of H^+

Metabolic reactions produce **nonvolatile acids** such as sulfuric acid at a rate of about 1 mEq of H^+ per day for every kilogram of body mass. The only way to eliminate this huge acid load is to excrete H^+ in the urine. Given the magnitude of these contributions to acid–base balance, it's not surprising that renal failure can quickly cause death.

As you learned in Chapter 26, cells in both the proximal convoluted tubules (PCT) and the collecting ducts of the kidneys secrete hydrogen ions into the tubular fluid. In the PCT, Na^+/H^+ antiporters secrete H^+ as they reabsorb Na^+ (see Figure 26.13). Even more important for regulation of pH of body fluids, however, are the intercalated cells of the collecting duct. The *apical* membranes of some intercalated cells include **proton pumps (H^+ ATPases)** that secrete H^+ into the tubular fluid (Figure 27.8). Intercalated cells can secrete H^+ against a concentration gradient so effectively that urine can be up to 1000 times (3 pH units) more acidic than blood. HCO_3^- produced by dissociation of H_2CO_3 inside intercalated cells crosses the basolateral membrane by means of Cl^-/HCO_3^- **antiporters** and then diffuses into peritubular capillaries (Figure 27.8a). The HCO_3^- that enters the blood in this way is *new* (not filtered). For this reason, blood leaving the kidney in the renal vein may have a higher HCO_3^- concentration than blood entering the kidney in the renal artery.

Interestingly, a second type of intercalated cell has proton pumps in its *basolateral* membrane and Cl^-/HCO_3^- antiporters in its apical membrane. These intercalated cells secrete HCO_3^- and reabsorb H^+. Thus, the two types of intercalated cells help maintain the pH of body fluids in two ways—by excreting excess H^+ when pH of body fluids is too low and by excreting excess HCO_3^- when pH is too high.

Some H^+ secreted into the tubular fluid of the collecting duct is buffered, but not by HCO_3^-, most of which has been filtered and reabsorbed. Two other buffers combine with H^+ in the collecting duct (Figure 27.8b). The most plentiful buffer in the tubular fluid of the collecting duct is HPO_4^{2-} (monohydrogen phosphate ion). In addition, a small amount of NH_3 (ammonia) also is present. H^+ combines with HPO_4^{2-} to form $H_2PO_4^-$ (dihydrogen phosphate ion) and with NH_3 to form NH_4^+ (ammonium ion). Because these ions cannot diffuse back into tubule cells, they are excreted in the urine.

Table 27.3 summarizes the mechanisms that maintain the pH of body fluids.

Figure 27.8 Secretion of H^+ by intercalated cells in the collecting duct. HCO_3^- = bicarbonate ion; CO_2 = carbon dioxide; H_2O = water; H_2CO_3 = carbonic acid; Cl^- = chloride ion; NH_3 = ammonia; NH_4^+ = ammonium ion; HPO_4^{2-} = monohydrogen phosphate ion; $H_2PO_4^-$ = dihydrogen phosphate ion.

Urine can be up to 1000 times more acidic than blood due to the operation of the proton pumps in the collecting ducts of the kidneys.

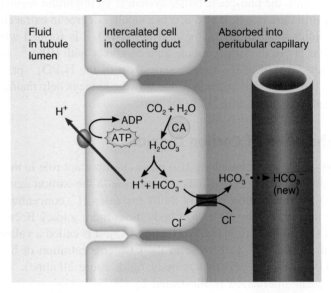

(a) Secretion of H^+

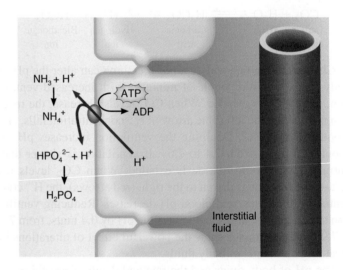

(b) Buffering of H^+ in urine

Key:

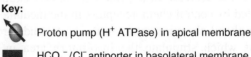 Proton pump (H^+ ATPase) in apical membrane

HCO_3^- /Cl^- antiporter in basolateral membrane

••▶ Diffusion

? **What would be the effects of a drug that blocks the activity of carbonic anhydrase?**

TABLE 27.3

Mechanisms That Maintain pH of Body Fluids

MECHANISM	COMMENTS
Buffer systems	Most consist of a weak acid and its salt, which functions as a weak base. They prevent drastic changes in body fluid pH.
Proteins	The most abundant buffers in body cells and blood. Hemoglobin inside red blood cells is a good buffer.
Carbonic acid–bicarbonate	Important regulator of blood pH. The most abundant buffers in extracellular fluid (ECF).
Phosphates	Important buffers in intracellular fluid and urine.
Exhalation of CO_2	With increased exhalation of CO_2, pH rises (fewer H^+). With decreased exhalation of CO_2, pH falls (more H^+).
Kidneys	Renal tubules secrete H^+ into urine and reabsorb HCO_3^- so it is not lost in urine.

Acid–Base Imbalances

The normal pH range of systemic arterial blood is between 7.35 (= 45 nEq of H^+/liter) and 7.45 (= 35 nEq of H^+/liter). **Acidosis** (or **acidemia**) is a condition in which blood pH is below 7.35; **alkalosis** (or **alkalemia**) is a condition in which blood pH is higher than 7.45.

The major physiological effect of acidosis is depression of the central nervous system through depression of synaptic transmission. If the systemic arterial blood pH falls below 7, depression of the nervous system is so severe that the individual becomes disoriented, then comatose, and may die. Patients with severe acidosis usually die while in a coma. A major physiological effect of alkalosis, by contrast, is overexcitability in both the central nervous system and peripheral nerves. Neurons conduct impulses repetitively, even when not stimulated by normal stimuli; the results are nervousness, muscle spasms, and even convulsions and death.

A change in blood pH that leads to acidosis or alkalosis may be countered by **compensation,** the physiological response to an acid–base imbalance that acts to normalize arterial blood pH. Compensation may be either *complete,* if pH indeed is brought within the normal range, or *partial,* if systemic arterial blood pH is still lower than 7.35 or higher than 7.45. If a person has altered blood pH due to metabolic causes, hyperventilation or hypoventilation can help bring blood pH back toward the normal range; this form of compensation, termed **respiratory compensation,** occurs within minutes and reaches its maximum within hours. If, however, a person has altered blood pH due to respiratory causes, then **renal compensation**—changes in secretion of H^+ and reabsorption of HCO_3^- by the kidney tubules—can help reverse the change. Renal compensation may begin in minutes, but it takes days to reach maximum effectiveness.

In the discussion that follows, note that both respiratory acidosis and respiratory alkalosis are disorders resulting from changes in the partial pressure of CO_2 (P_{CO_2}) in systemic arterial blood (normal range is 35–45 mmHg). By contrast, both metabolic acidosis and metabolic alkalosis are disorders resulting from changes in HCO_3^- concentration (normal range is 22–26 mEq/liter in systemic arterial blood).

Respiratory Acidosis

The hallmark of **respiratory acidosis** is an abnormally high P_{CO_2} in systemic arterial blood—above 45 mmHg. Inadequate exhalation of CO_2 causes the blood pH to drop. Any condition that decreases the movement of CO_2 from the blood to the alveoli of the lungs to the atmosphere causes a buildup of CO_2, H_2CO_3, and H^+. Such conditions include emphysema, pulmonary edema, injury to the respiratory center of the medulla oblongata, airway obstruction, or disorders of the muscles involved in breathing. If the respiratory problem is not too severe, the kidneys can help raise the blood pH into the normal range by increasing excretion of H^+ and reabsorption of HCO_3^- (renal compensation). The goal in treatment of respiratory acidosis is to increase the exhalation of CO_2, as, for instance, by providing ventilation therapy. In addition, intravenous administration of HCO_3^- may be helpful.

Respiratory Alkalosis

In **respiratory alkalosis,** systemic arterial blood P_{CO_2} falls below 35 mmHg. The cause of the drop in P_{CO_2} and the resulting increase in pH is hyperventilation, which occurs in conditions that stimulate the inspiratory area in the brain stem. Such conditions include oxygen deficiency due to high altitude or pulmonary disease, cerebrovascular accident (stroke), or severe anxiety. Again, renal compensation may bring blood pH into the normal range if the kidneys are able to decrease excretion of H^+ and reabsorption of HCO_3^-. Treatment of respiratory alkalosis is aimed at increasing the level of CO_2 in the body. One simple treatment is to have the person inhale and exhale into a paper bag for a short period; as a result, the person inhales air containing a higher-than-normal concentration of CO_2.

Metabolic Acidosis

In **metabolic acidosis,** the systemic arterial blood HCO_3^- level drops below 22 mEq/liter. Such a decline in this important buffer causes the blood pH to decrease. Three situations may lower the blood level of HCO_3^-: (1) actual loss of HCO_3^-, such as may occur with severe diarrhea or renal dysfunction; (2) accumulation of an acid other than carbonic acid, as may occur in ketosis (described in Clinical Connection: Ketosis in Chapter 25); or (3) failure of the kidneys to excrete H^+ from metabolism of dietary proteins. If the problem is not too severe, hyperventilation can help bring blood pH into the normal range (respiratory compensation). Treatment of metabolic acidosis consists of administering intravenous solutions of sodium bicarbonate and correcting the cause of the acidosis.

Metabolic Alkalosis

In **metabolic alkalosis,** the systemic arterial blood HCO_3^- concentration is above 26 mEq/liter. A nonrespiratory loss of acid or excessive intake of alkaline drugs causes the blood pH to increase above 7.45. Excessive vomiting of gastric contents, which results in a substantial loss of hydrochloric acid, is probably the most frequent cause of metabolic alkalosis. Other causes include gastric suctioning, use of certain diuretics, endocrine disorders, excessive intake of alkaline drugs (antacids), and severe dehydration. Respiratory compensation through hypoventilation may bring blood pH into the normal range. Treatment of metabolic alkalosis consists of giving fluid solutions to correct Cl^-, K^+, and other electrolyte deficiencies plus correcting the cause of alkalosis.

Table 27.4 summarizes respiratory and metabolic acidosis and alkalosis.

CLINICAL CONNECTION | *Diagnosis of Acid–Base Imbalances*

One can often pinpoint the cause of an acid–base imbalance by careful evaluation of three factors in a sample of systemic arterial blood: pH, concentration of HCO_3^-, and P_{CO_2}. These three blood chemistry values are examined in the following four-step sequence:

1. Note whether the pH is high (alkalosis) or low (acidosis).
2. Then decide which value—P_{CO_2} or HCO_3^-—is out of the normal range and could be the *cause* of the pH change. For example, *elevated pH* could be caused by *low* P_{CO_2} or high HCO_3^-.
3. If the cause is a *change in P_{CO_2}*, the problem is *respiratory;* if the cause is a *change in HCO_3^-*, the problem is *metabolic*.
4. Now look at the value that doesn't correspond with the observed pH change. If it is within its normal range, there is no compensation. If it is outside the normal range, compensation is occurring and partially correcting the pH imbalance. •

✔ CHECKPOINT

8. Explain how each of the following buffer systems helps to maintain the pH of body fluids: proteins, carbonic acid–bicarbonate buffers, and phosphates.
9. Define acidosis and alkalosis. Distinguish among respiratory and metabolic acidosis and alkalosis.
10. What are the principal physiological effects of acidosis and alkalosis?

27.4 AGING AND FLUID, ELECTROLYTE, AND ACID–BASE BALANCE

◉ OBJECTIVE

• Describe the changes in fluid, electrolyte, and acid–base balance that may occur with aging.

There are significant differences between adults and infants, especially premature infants, with respect to fluid distribution, regulation of fluid and electrolyte balance, and acid–base homeostasis. Accordingly, infants experience more problems than adults in these areas. The differences are related to the following conditions:

• *Proportion and distribution of water.* A newborn's total body mass is about 75% water (and can be as high as 90% in a premature infant); an adult's total body mass is about 55–60% water. (The "adult" percentage is achieved at about 2 years of age.) Adults have twice as much water in ICF as ECF, but the opposite is true in premature infants. Because ECF is subject to more changes than ICF, rapid losses or gains of body water are

---

TABLE 27.4

Summary of Acidosis and Alkalosis

CONDITION	DEFINITION	COMMON CAUSES	COMPENSATORY MECHANISM
Respiratory acidosis	Increased P_{CO_2} (above 45 mmHg) and decreased pH (below 7.35) if no compensation.	Hypoventilation due to emphysema, pulmonary edema, trauma to respiratory center, airway obstructions, or dysfunction of muscles of respiration.	Renal: increased excretion of H^+; increased reabsorption of HCO_3^-. If compensation is complete, pH will be within normal range but P_{CO_2} will be high.
Respiratory alkalosis	Decreased P_{CO_2} (below 35 mmHg) and increased pH (above 7.45) if no compensation.	Hyperventilation due to oxygen deficiency, pulmonary disease, cerebrovascular accident (CVA), or severe anxiety.	Renal: decreased excretion of H^+; decreased reabsorption of HCO_3^-. If compensation is complete, pH will be within normal range but P_{CO_2} will be low.
Metabolic acidosis	Decreased HCO_3^- (below 22 mEq/liter) and decreased pH (below 7.35) if no compensation.	Loss of bicarbonate ions due to diarrhea, accumulation of acid (ketosis), renal dysfunction.	Respiratory: hyperventilation, which increases loss of CO_2. If compensation is complete, pH will be within normal range but HCO_3^- will be low.
Metabolic alkalosis	Increased HCO_3^- (above 26 mEq/liter) and increased pH (above 7.45) if no compensation.	Loss of acid due to vomiting, gastric suctioning, or use of certain diuretics; excessive intake of alkaline drugs.	Respiratory: hypoventilation, which slows loss of CO_2. If compensation is complete, pH will be within normal range but HCO_3^- will be high.

much more critical in infants. Given that the rate of fluid intake and output is about seven times higher in infants than in adults, the slightest changes in fluid balance can result in severe abnormalities.

- *Metabolic rate.* The metabolic rate of infants is about double that of adults. This results in the production of more metabolic wastes and acids, which can lead to the development of acidosis in infants.

- *Functional development of the kidneys.* Infant kidneys are only about half as efficient in concentrating urine as those of adults. (Functional development is not complete until close to the end of the first month after birth.) As a result, the kidneys of newborns can neither concentrate urine nor rid the body of excess acids as effectively as those of adults.

- *Body surface area.* The ratio of body surface area to body volume of infants is about three times greater than that of adults. Water loss through the skin is significantly higher in infants than in adults.

- *Breathing rate.* The higher breathing rate of infants (about 30 to 80 times a minute) causes greater water loss from the lungs. Respiratory alkalosis may occur because greater ventilation eliminates more CO_2 and lowers the P_{CO_2}.

- *Ion concentrations.* Newborns have higher K^+ and Cl^- concentrations than adults. This creates a tendency toward metabolic acidosis.

By comparison with children and younger adults, older adults often have an impaired ability to maintain fluid, electrolyte, and acid–base balance. With increasing age, many people have a decreased volume of intracellular fluid and decreased total body K^+ due to declining skeletal muscle mass and increasing mass of adipose tissue (which contains very little water). Age-related decreases in respiratory and renal functioning may compromise acid–base balance by slowing the exhalation of CO_2 and the excretion of excess acids in urine. Other kidney changes, such as decreased blood flow, decreased glomerular filtration rate, and reduced sensitivity to antidiuretic hormone, have an adverse effect on the ability to maintain fluid and electrolyte balance. Due to a decrease in the number and efficiency of sweat glands, water loss from the skin declines with age. Because of these age-related changes, older adults are susceptible to several fluid and electrolyte disorders:

- *Dehydration* and *hypernatremia* often occur due to inadequate fluid intake or loss of more water than Na^+ in vomit, feces, or urine.

- *Hyponatremia* may occur due to inadequate intake of Na^+; elevated loss of Na^+ in urine, vomit, or diarrhea; or impaired ability of the kidneys to produce dilute urine.

- *Hypokalemia* often occurs in older adults who chronically use laxatives to relieve constipation or who take K^+-depleting diuretic drugs for treatment of hypertension or heart disease.

- *Acidosis* may occur due to impaired ability of the lungs and kidneys to compensate for acid–base imbalances. One cause of acidosis is decreased production of ammonia (NH_3) by renal tubule cells, which then is not available to combine with H^+ and be excreted in urine as NH_4^+; another cause is reduced exhalation of CO_2.

✔ CHECKPOINT

11. Why do infants experience greater problems with fluid, electrolyte, and acid–base balance than adults?

CHAPTER REVIEW AND RESOURCE SUMMARY

Review

Resource

27.1 Fluid Compartments and Fluid Balance

1. Body fluid includes water and dissolved solutes. About two-thirds of the body's fluid is located within cells and is called intracellular fluid (ICF). The other one-third, called extracellular fluid (ECF), includes interstitial fluid; blood plasma and lymph; cerebrospinal fluid; gastrointestinal tract fluids; synovial fluid; fluids of the eyes and ears; pleural, pericardial, and peritoneal fluids; and glomerular filtrate.

2. Fluid balance means that the required amounts of water and solutes are present and are correctly proportioned among the various compartments.

3. An inorganic substance that dissociates into ions in solution is called an electrolyte.

4. Water is the largest single constituent in the body. It makes up 45–75% of total body mass, depending on age, gender, and the amount of adipose tissue present.

5. Daily water gain and loss are each about 2500 mL. Sources of water gain are ingested liquids and foods, and water produced by cellular respiration and dehydration synthesis reactions (metabolic water). Water is lost from the body via urination, evaporation from the skin surface, exhalation of water vapor, and defecation. In women, menstrual flow is an additional route for loss of body water.

6. Body water gain is regulated by adjusting the volume of water intake, mainly by drinking more or less fluid. The thirst center in the hypothalamus governs the urge to drink. Although increased amounts of water and solutes are lost through sweating and exhalation during exercise, loss of excess body water or excess solutes depends mainly on regulating excretion in the urine. The extent of urinary NaCl loss is the main determinant of body fluid volume; the extent of urinary water loss is the main determinant

Overview of Fluids
Animation - Water Homeostasis
Animation - Water and Fluid Flow

of body fluid osmolarity. Table 27.1 summarizes the factors that regulate water gain and water loss in the body.

7. Angiotensin II and aldosterone reduce urinary loss of Na^+ and Cl^- and thereby increase the volume of body fluids. ANP promotes natriuresis, elevated excretion of Na^+ (and Cl^-), which decreases blood volume.

8. The major hormone that regulates water loss and thus body fluid osmolarity is antidiuretic hormone (ADH).

9. An increase in the osmolarity of interstitial fluid draws water out of cells, and they shrink slightly. A decrease in the osmolarity of interstitial fluid causes cells to swell. Most often a change in osmolarity is due to a change in the concentration of Na^+, the dominant solute in interstitial fluid.

10. When a person consumes water faster than the kidneys can excrete it or when renal function is poor, the result may be water intoxication, in which cells swell dangerously.

27.2 Electrolytes in Body Fluids

Concepts and Connections - Functions of Ions

1. Ions formed when electrolytes dissolve in body fluids control the osmosis of water between fluid compartments, help maintain acid–base balance, and carry electrical current.

2. The concentrations of cations and anions is expressed in units of milliequivalents/liter (mEq/liter). Blood plasma, interstitial fluid, and intracellular fluid contain varying types and amounts of ions.

3. Sodium ions (Na^+) are the most abundant extracellular ions. They are involved in impulse transmission, muscle contraction, and fluid and electrolyte balance. Na^+ level is controlled by aldosterone, antidiuretic hormone, and atrial natriuretic peptide.

4. Chloride ions (Cl^-) are the major extracellular anions. They play a role in regulating osmotic pressure and forming HCl in gastric juice. Cl^- level is controlled indirectly by antidiuretic hormone and by processes that increase or decrease renal reabsorption of Na^+.

5. Potassium ions (K^+) are the most abundant cations in intracellular fluid. They play a key role in the resting membrane potential and action potential of neurons and muscle fibers; help maintain intracellular fluid volume; and contribute to regulation of pH. K^+ level is controlled by aldosterone.

6. Bicarbonate ions (HCO_3^-) are the second most abundant anions in extracellular fluid. They are the most important buffer in blood plasma.

7. Calcium is the most abundant mineral in the body. Calcium salts are structural components of bones and teeth. Ca^{2+}, which are principally extracellular cations, function in blood clotting, neurotransmitter release, and contraction of muscle. Ca^{2+} level is controlled mainly by parathyroid hormone and calcitriol.

8. Phosphate ions ($H_2PO_4^-$, HPO_4^{2-}, and PO_4^{3-}) are principally intracellular anions, and their salts are structural components of bones and teeth. They are also required for the synthesis of nucleic acids and ATP and participate in buffer reactions. Their level is controlled by parathyroid hormone and calcitriol.

9. Magnesium ions (Mg^{2+}) are primarily intracellular cations. They act as cofactors in several enzyme systems.

10. Table 27.2 describes the imbalances that result from deficiency or excess of important body electrolytes.

27.3 Acid–Base Balance

Animation - Role of the Respiratory System in pH Regulation

Figure 27.7 - Negative Feedback Regulation of Blood pH

Figure 27.8 - Secretion of H^+ by Intercalated Cells

Exercise - Phix the pH

Concepts and Connections - Mechanisms of pH Balance

Concepts and Connections - Urinary System Regulation of pH

1. The overall acid–base balance of the body is maintained by controlling the H^+ concentration of body fluids, especially extracellular fluid.

2. The normal pH of systemic arterial blood is 7.35–7.45.

3. Homeostasis of pH is maintained by buffer systems, via exhalation of carbon dioxide, and via kidney excretion of H^+ and reabsorption of HCO_3^-. The important buffer systems include proteins, carbonic acid–bicarbonate buffers, and phosphates.

4. An increase in exhalation of carbon dioxide increases blood pH; a decrease in exhalation of CO_2 decreases blood pH.

5. In the proximal convoluted tubules of the kidneys, Na^+/H^+ antiporters secrete H^+ as they reabsorb Na^+. In the collecting ducts of the kidneys, some intercalated cells reabsorb K^+ and HCO_3^- and secrete H^+; other intercalated cells secrete HCO_3^-. In these ways, the kidneys can increase or decrease the pH of body fluids.

6. Table 27.3 summarizes the mechanisms that maintain pH of body fluids.

7. Acidosis is a systemic arterial blood pH below 7.35; its principal effect is depression of the central nervous system (CNS). Alkalosis is a systemic arterial blood pH above 7.45; its principal effect is overexcitability of the CNS.

Review   **Resource**

8. Respiratory acidosis and alkalosis are disorders due to changes in blood P_{CO_2}; metabolic acidosis and alkalosis are disorders associated with changes in blood HCO_3^- concentration.

9. Metabolic acidosis or alkalosis can be compensated by respiratory mechanisms (respiratory compensation); respiratory acidosis or alkalosis can be compensated by renal mechanisms (renal compensation). Table 27.4 summarizes the effects of respiratory and metabolic acidosis and alkalosis.

10. By examining systemic arterial blood pH, HCO_3^-, and P_{CO_2} values, it is possible to pinpoint the cause of an acid–base imbalance.

27.4 Aging and Fluid, Electrolyte, and Acid–Base Balance

1. With increasing age, there is decreased intracellular fluid volume and decreased K^+ due to declining skeletal muscle mass.

2. Decreased kidney function with aging adversely affects fluid and electrolyte balance.

SELF-QUIZ QUESTIONS

Fill in the blanks in the following statements.

1. The source of water that is derived from aerobic cellular respiration and dehydration synthesis reactions is _____ water.

2. In the carbonic acid–bicarbonate buffer system, the _____ acts as a weak base, and _____ acts as a weak acid.

Indicate whether the following statements are true or false.

3. The phosphate buffer system is an important regulator of pH in the cytosol.

4. The two compartments in which water can be found are plasma and cytosol.

Choose the one best answer to the following questions.

5. Which of the following statements are *true*? (1) Buffers prevent rapid, drastic changes in pH of a body fluid. (2) Buffers work slowly. (3) Strong acids lower pH more than weak acids because strong acids contribute fewer H^+. (4) Most buffers consist of a weak acid and the salt of that acid, which acts as a weak base. (5) Hemoglobin is an important buffer.
   (a) 1, 2, 3, and 5   (b) 1, 3, 4, and 5   (c) 1, 3, and 5
   (d) 1, 4, and 5   (e) 2, 3, and 5

6. Which of the following are *true* concerning ions in the body? (1) They control osmosis of water between fluid compartments. (2) They help maintain acid–base balance. (3) They carry electrical current. (4) They serve as cofactors for enzyme activity. (5) They serve as neurotransmitters under special circumstances.
   (a) 1, 3, and 5   (b) 2, 4, and 5   (c) 1, 4, and 5
   (d) 1, 2, and 4   (e) 1, 2, 3, and 4

7. Which of the following statements are *true*? (1) An increase in the carbon dioxide concentration in body fluids increases H^+ concentration and thus lowers pH. (2) Breath holding results in a decline in blood pH. (3) The respiratory buffer mechanism can eliminate a single volatile acid: carbonic acid. (4) The only way to eliminate nonvolatile acids is to excrete H^+ in the urine. (5) When the diet contains a large amount of protein, normal metabolism produces more acids than bases.
   (a) 1, 2, 3, 4, and 5 (b) 1, 3, 4, and 5   (c) 1, 2, 3, and 4
   (d) 1, 2, 4, and 5   (e) 1, 3, and 4

8. Concerning acid–base imbalances: (1) Acidosis can cause depression of the central nervous system through depression of synaptic transmission. (2) Renal compensation can resolve respiratory alkalosis or acidosis. (3) A major physiological effect of alkalosis is lack of excitability in the central nervous system and peripheral nerves. (4) Resolution of metabolic acidosis and alkalosis occurs through renal compensation. (5) In adjusting blood pH, renal compensation occurs quickly, but respiratory compensation takes days.
   (a) 1, 2, and 5   (b) 1 and 2   (c) 2, 3, and 4
   (d) 2, 3, and 5   (e) 1, 2, 3, and 5

9. Match the following:
   _____ (a) the most abundant cation in intracellular fluid; plays a key role in establishing the resting membrane potential
   _____ (b) the most abundant mineral in the body; plays important roles in blood clotting, neurotransmitter release, maintenance of muscle tone, and excitability of nervous and muscle tissue
   _____ (c) second most common intracellular cation; is a cofactor for enzymes involved in carbohydrate, protein, and Na^+/K^+ ATPase metabolism
   _____ (d) the most abundant extracellular cation; essential in fluid and electrolyte balance
   _____ (e) ions that are mostly combined with lipids, proteins, carbohydrates, nucleic acids, and ATP inside cells
   _____ (f) most prevalent extracellular anion; can help balance the level of anions in different fluid compartments
   _____ (g) second most prevalent extracellular anion; mainly regulated by the kidneys; important for acid–base balance
   _____ (h) substances that act to prevent rapid, drastic changes in the pH of a body fluid
   _____ (i) inorganic substances that dissociate into ions when in solution

   (1) sodium
   (2) chloride
   (3) electrolytes
   (4) bicarbonate
   (5) buffers
   (6) phosphate
   (7) magnesium
   (8) potassium
   (9) calcium

10. Match the following:
_____ (a) an abnormal increase in the volume of interstitial fluid
_____ (b) can occur during renal failure or destruction of body cells, which releases phosphates into the blood
_____ (c) the swelling of cells due to water moving from plasma into interstitial fluid and then into cells
_____ (d) occurs when water loss is greater than water gain
_____ (e) can be caused by excessive sodium in diet or with dehydration
_____ (f) condition that can occur as water moves out of plasma into interstitial fluid and blood volume decreases
_____ (g) can be caused by decreased potassium intake or kidney disease; results in muscle fatigue, increased urine output, changes in electrocardiogram
_____ (h) can occur from hypoparathyroidism
_____ (i) can be caused by emphysema, pulmonary edema, injury to the respiratory center of the medulla oblongata, airway destruction, or disorders of the muscles involved in breathing
_____ (j) can be caused by excessive water intake, excessive vomiting, or aldosterone deficiency
_____ (k) can be caused by actual loss of bicarbonate ions, ketosis, or failure of kidneys to excrete H^+
_____ (l) can be caused by excessive vomiting of gastric contents, gastric suctioning, use of certain diuretics, severe dehydration, or excessive intake of alkaline drugs
_____ (m) can be caused by oxygen deficiency at high altitude, stroke, or severe anxiety

(1) respiratory acidosis
(2) respiratory alkalosis
(3) metabolic acidosis
(4) metabolic alkalosis
(5) dehydration
(6) hypovolemia
(7) water intoxication
(8) edema
(9) hypokalemia
(10) hypernatremia
(11) hyponatremia
(12) hyperphosphatemia
(13) hypocalcemia

CRITICAL THINKING QUESTIONS

1. Robin is in the early stages of pregnancy and has been vomiting excessively for several days. She became weak, was confused, and was taken to the emergency room. What do you suspect has happened to Robin's acid–base balance? How would her body attempt to compensate? What electrolytes would be affected by her vomiting, and how do her symptoms reflect those imbalances?

2. Henry is in the intensive care unit because he suffered a severe myocardial infarction three days ago. The lab reports the following values from an arterial blood sample: pH 7.30, HCO_3^- = 20 mEq/liter, P_{CO_2} = 32 mmHg. Diagnose Henry's acid–base status and decide whether compensation is occurring.

3. This summer, Sam is training for a marathon by running 10 miles a day. Describe changes in his fluid balance as he trains.

ANSWERS TO FIGURE QUESTIONS

27.1 Plasma volume equals body mass × percent of body mass that is body fluid × proportion of body fluid that is ECF × proportion of ECF that is plasma × a conversion factor (1 liter/kg). For males, blood plasma volume = 60 kg × 0.60 × 1/3 × 0.20 × 1 liter/kg = 2.4 liters. Using similar calculations, female blood plasma volume is 2.2 liters.

27.2 Hyperventilation, vomiting, fever, and diuretics all increase fluid loss.

27.3 Negative feedback is in operation because the result (an increase in fluid intake) is opposite to the initiating stimulus (dehydration).

27.4 An elevated aldosterone level promotes abnormally high renal reabsorption of NaCl and water, which expands blood volume and increases blood pressure. Because of the increased blood pressure, more fluid filters out of capillaries and accumulates in the interstitial fluid, causing edema.

27.5 If a solution used for oral rehydration therapy contains a small amount of salt, both the salt and water are absorbed in the gastrointestinal tract, blood volume increases without a decrease in osmolarity, and water intoxication does not occur.

27.6 In ECF, the major cation is Na^+, and the major anions are Cl^- and HCO_3^-. In ICF, the major cation is K^+, and the major anions are proteins and organic phosphates (for example, ATP).

27.7 Holding your breath causes blood pH to decrease slightly as CO_2 and H^+ accumulate in the blood.

27.8 A carbonic anhydrase inhibitor reduces secretion of H^+ into the urine and reduces reabsorption of Na^+ and HCO_3^- into the blood. It has a diuretic effect and can cause acidosis (lowered pH of the blood) due to loss of HCO_3^- in the urine.

28 THE REPRODUCTIVE SYSTEMS

THE REPRODUCTIVE SYSTEMS AND HOMEOSTASIS *The male and female reproductive organs work together to produce offspring. In addition, the female reproductive organs contribute to sustaining the growth of embryos and fetuses.*

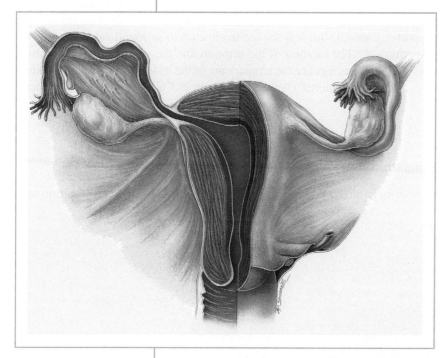

Sexual reproduction is the process by which organisms produce offspring by making germ cells called **gametes** (GAM-ēts = spouses). After the male gamete (sperm cell) unites with the female gamete (secondary oocyte)—an event called **fertilization** (fer′-til-i-ZĀ-shun)—the resulting cell contains one set of chromosomes from each parent. Males and females have anatomically distinct reproductive organs that are adapted for producing gametes, facilitating fertilization, and, in females, sustaining the growth of the embryo and fetus.

The male and female reproductive organs can be grouped by function. The **gonads**—testes in males and ovaries in females—produce gametes and secrete sex hormones. Various **ducts** then store and transport the gametes, and **accessory sex glands** produce substances that protect the gametes and facilitate their movement. Finally, **supporting structures,** such as the penis in males and the uterus in females, assist the delivery of gametes, and the uterus is also the site for the growth of the embryo and fetus during pregnancy.

Gynecology (gī-ne-KOL-ō-jē; *gyneco-* = woman; *-logy* = study of) is the specialized branch of medicine concerned with the diagnosis and treatment of diseases of the female reproductive system. As noted in Chapter 26, **urology** (ū-ROL-ō-jē) is the study of the urinary system. Urologists also diagnose and treat diseases and disorders of the male reproductive system. The branch of medicine that deals with male disorders, especially infertility and sexual dysfunction, is called **andrology** (an-DROL-ō-jē; *andro-* = masculine).

?

Did you ever wonder how breast augmentation and breast reduction are performed?

28.1 MALE REPRODUCTIVE SYSTEM

⦿ **OBJECTIVES**

- Describe the location, structure, and functions of the organs of the male reproductive system.
- Discuss the process of spermatogenesis in the testes.

The organs of the **male reproductive system** include the testes, a system of ducts (including the epididymis, ductus deferens, ejaculatory ducts, and urethra), accessory sex glands (seminal vesicles, prostate, and bulbourethral glands), and several supporting structures, including the scrotum and the penis (Figure 28.1). The testes (male gonads) produce sperm and secrete hormones. The duct system transports and stores sperm, assists in their maturation, and conveys them to the exterior. Semen contains sperm plus the secretions provided by the accessory sex glands. The supporting structures have various functions. The penis delivers sperm into the female reproductive tract and the scrotum supports the testes.

Scrotum

The **scrotum** (SKRŌ-tum = bag), the supporting structure for the testes, consists of loose skin and underlying subcutaneous layer that hangs from the root (attached portion) of the penis (Figure 28.1a). Externally, the scrotum looks like a single pouch of skin separated into lateral portions by a median ridge called the **raphe** (RĀ-fē = seam). Internally, the **scrotal septum** divides the scrotum into two sacs, each containing a single testis (Figure 28.2). The septum is made up of a subcutaneous layer and muscle tissue called the **dartos muscle** (DAR-tōs = skinned), which is composed of bundles of smooth muscle fibers. The dartos muscle is also found in the subcutaneous layer of the scrotum. Associated with each testis in the scrotum is the **cremaster muscle** (krē-MAS-ter = suspender), a series of small bands of skeletal muscle that descend as an extension of the internal oblique muscle through the spermatic cord to surround the testes.

The location of the scrotum and the contraction of its muscle fibers regulate the temperature of the testes. Normal sperm production requires a temperature about 2–3°C below core body tempera-

Figure 28.1 Male organs of reproduction and surrounding structures.

🔑 Reproductive organs are adapted for producing new individuals and passing on genetic material from one generation to the next.

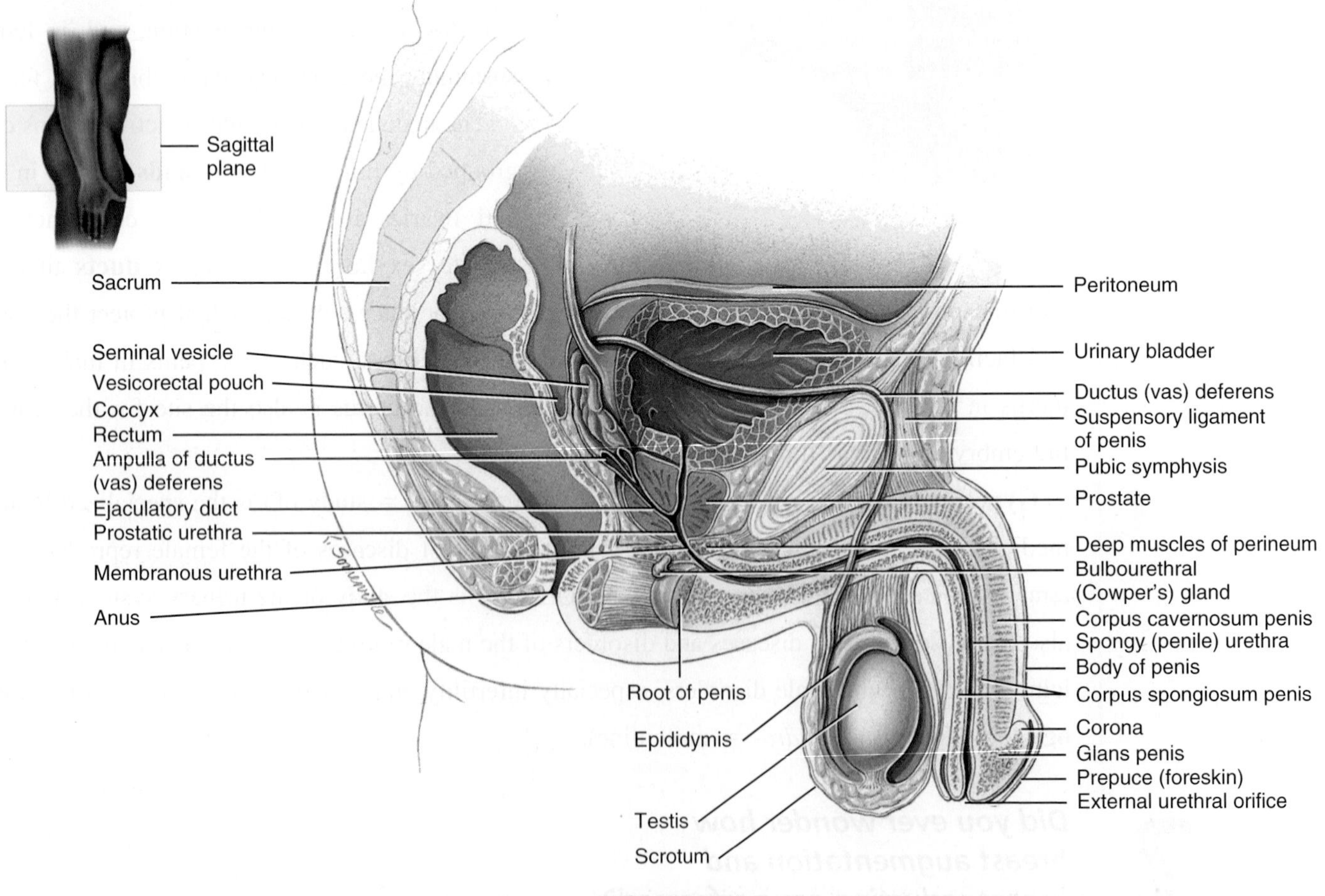

(a) Sagittal section

ture. This lowered temperature is maintained within the scrotum because it is outside the pelvic cavity. In response to cold temperatures, the cremaster and dartos muscles contract. Contraction of the cremaster muscles moves the testes closer to the body, where they can absorb body heat. Contraction of the dartos muscle causes the scrotum to become tight (wrinkled in appearance), which reduces heat loss. Exposure to warmth reverses these actions.

Testes

The **testes** (TES-tēz = witness), or **testicles,** are paired oval glands in the scrotum measuring about 5 cm (2 in.) long and 2.5 cm (1 in.) in diameter (Figure 28.3). Each testis (singular) has a mass of 10–15 grams. The testes develop near the kidneys, in the posterior portion of the abdomen, and they usually begin their descent into the scrotum through the inguinal canals (passageways in the anterior abdominal wall; see Figure 28.2) during the latter half of the seventh month of fetal development.

A serous membrane called the **tunica vaginalis** (TOO-nik-a vaj-i-NAL-is; *tunica* = sheath), which is derived from the peritoneum and forms during the descent of the testes, partially covers the testes. A collection of serous fluid in the tunica vaginalis is called a **hydrocele** (HĪ-drō-sēl; *hydro-* = water; *-kele* = hernia).

It may be caused by injury to the testes or inflammation of the epididymis. Usually, no treatment is required. Internal to the tunica vaginalis is a white fibrous capsule composed of dense irregular connective tissue, the **tunica albuginea** (al'-bū-JIN-ē-a; *albu-* = white); it extends inward, forming septa that divide the testis into a series of internal compartments called **lobules.** Each of the 200–300 lobules contains one to three tightly coiled tubules, the **seminiferous tubules** (sem'-i-NIF-er-us; *semin-* = seed; *-fer-* = to carry), where sperm are produced. The process by which the seminiferous tubules of the testes produce sperm is called spermatogenesis (sper'-ma-tō-JEN-e-sis; *genesis* = to be born).

The seminiferous tubules contain two types of cells: **spermatogenic cells** (sper'-ma-tō-JEN-ik), the sperm-forming cells, and **Sertoli cells** (ser-TŌ-lē), which have several functions in supporting spermatogenesis (Figure 28.4). Stem cells called **spermatogonia** (sper'-ma-tō-GŌ-nē-a; *-gonia* = offspring; singular is *spermatogonium*) develop from **primordial germ cells** (prī-MŌR-dē-al = primitive or early form) that arise from the yolk sac and enter the testes during the fifth week of development. In the embryonic testes, the primordial germ cells differentiate into spermatogonia, which remain dormant during childhood and actively begin producing sperm at puberty. Toward the lumen of the seminiferous tubule are layers of progressively more mature cells.

FUNCTIONS OF THE MALE REPRODUCTIVE SYSTEM

1. The testes produce sperm and the male sex hormone testosterone.
2. The ducts transport, store, and assist in maturation of sperm.
3. The accessory sex glands secrete most of the liquid portion of semen.
4. The penis contains the urethra, a passageway for ejaculation of semen and excretion of urine.

SUPERIOR

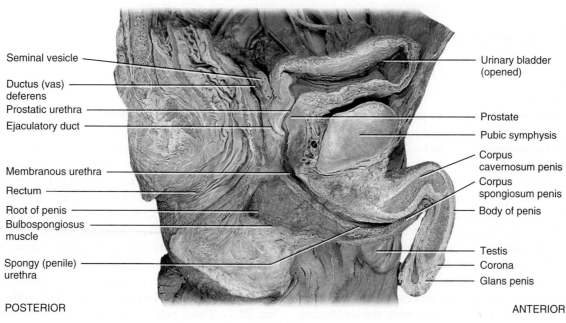

Seminal vesicle

Ductus (vas) deferens

Prostatic urethra

Ejaculatory duct

Membranous urethra

Rectum

Root of penis

Bulbospongiosus muscle

Spongy (penile) urethra

Urinary bladder (opened)

Prostate

Pubic symphysis

Corpus cavernosum penis

Corpus spongiosum penis

Body of penis

Testis

Corona

Glans penis

POSTERIOR

ANTERIOR

(b) Sagittal section

 What are the groups of reproductive organs in males, and what are the functions of each group?

Figure 28.2 The scrotum, the supporting structure for the testes.

The scrotum consists of loose skin and an underlying subcutaneous layer and supports the testes.

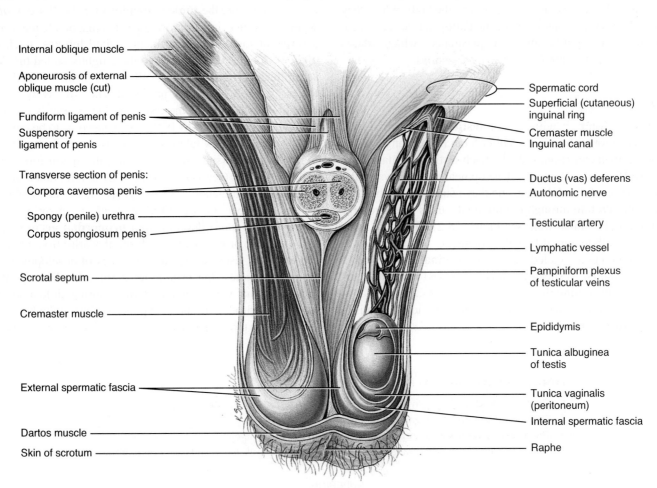

Internal oblique muscle

Aponeurosis of external oblique muscle (cut)

Fundiform ligament of penis

Suspensory ligament of penis

Transverse section of penis:

Corpora cavernosa penis

Spongy (penile) urethra

Corpus spongiosum penis

Scrotal septum

Cremaster muscle

External spermatic fascia

Dartos muscle

Skin of scrotum

Spermatic cord

Superficial (cutaneous) inguinal ring

Cremaster muscle

Inguinal canal

Ductus (vas) deferens

Autonomic nerve

Testicular artery

Lymphatic vessel

Pampiniform plexus of testicular veins

Epididymis

Tunica albuginea of testis

Tunica vaginalis (peritoneum)

Internal spermatic fascia

Raphe

Anterior view of scrotum and testes and transverse section of penis

Which muscles help regulate the temperature of the testes?

In order of advancing maturity, these are primary spermatocytes, secondary spermatocytes, spermatids, and sperm cells. After a **sperm cell,** or **spermatozoon** (sper′-ma-tō-ZŌ-on; *zoon* = life), has formed, it is released into the lumen of the seminiferous tubule. (The plural terms are *sperm* and *spermatozoa.*)

Embedded among the spermatogenic cells in the seminiferous tubules are large Sertoli cells or **sustentacular cells** (sus′-ten-TAK-ū-lar), which extend from the basement membrane to the lumen of the tubule. Internal to the basement membrane and spermatogonia, tight junctions join neighboring Sertoli cells to one another. These junctions form an obstruction known as the **blood–testis barrier** because substances must first pass through the Sertoli cells before they can reach the developing sperm. By isolating the developing gametes from the blood, the blood–testis barrier prevents an immune response against the spermatogenic cell's surface antigens, which are recognized as "foreign" by the immune system. The blood–testis barrier does not include spermatogonia.

Sertoli cells support and protect developing spermatogenic cells in several ways. They nourish spermatocytes, spermatids, and sperm; phagocytize excess spermatid cytoplasm as development proceeds; and control movements of spermatogenic cells and the release of sperm into the lumen of the seminiferous tubule. They also produce fluid for sperm transport, secrete the hormone inhibin, and regulate the effects of testosterone and FSH (follicle-stimulating hormone).

In the spaces between adjacent seminiferous tubules are clusters of cells called **Leydig (interstitial) cells** (LĪ-dig) (Figure 28.4). These cells secrete testosterone, the most prevalent androgen. An **androgen** is a hormone that promotes the development of masculine characteristics. Testosterone also promotes a man's *libido* (sexual drive).

Figure 28.3 Internal and external anatomy of a testis.

The testes are the male gonads, which produce haploid sperm.

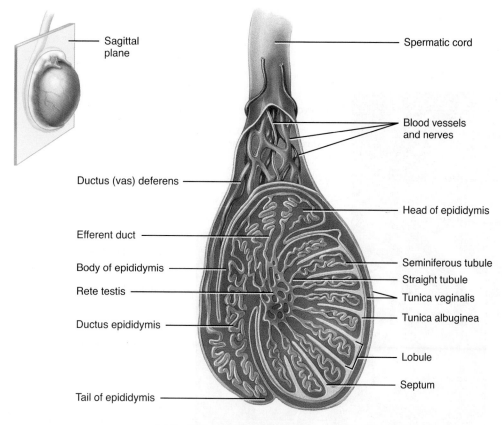

(a) Sagittal section of testis showing seminiferous tubules

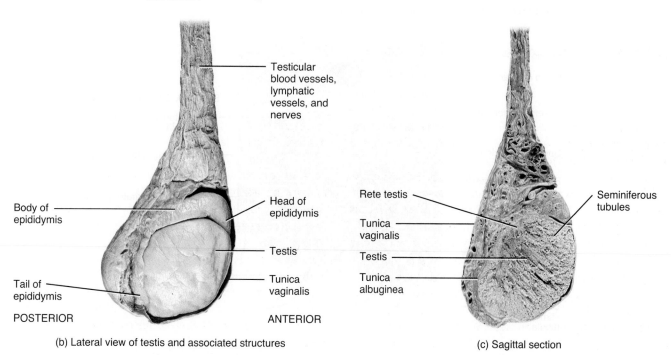

(b) Lateral view of testis and associated structures

(c) Sagittal section

? **What tissue layers cover and protect the testes?**

Figure 28.4 Microscopic anatomy of the seminiferous tubules and stages of sperm production (spermatogenesis). Arrows in (b) indicate the progression of spermatogenic cells from least mature to most mature. The (*n*) and (2*n*) refer to haploid and diploid numbers of chromosomes, respectively.

Spermatogenesis occurs in the seminiferous tubules of the testes.

Transverse plane

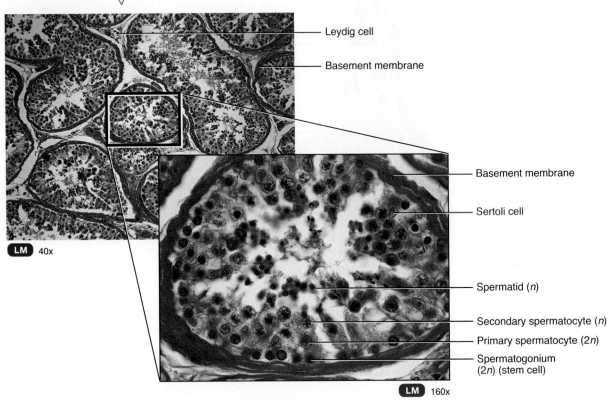

Leydig cell

Basement membrane

LM 40x

Basement membrane

Sertoli cell

Spermatid (*n*)

Secondary spermatocyte (*n*)

Primary spermatocyte (2*n*)

Spermatogonium (2*n*) (stem cell)

LM 160x

(a) Transverse section of several seminiferous tubules

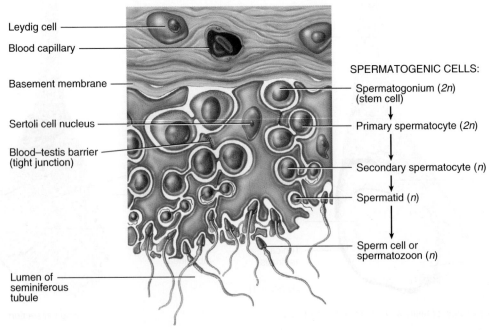

Leydig cell

Blood capillary

Basement membrane

Sertoli cell nucleus

Blood–testis barrier (tight junction)

Lumen of seminiferous tubule

SPERMATOGENIC CELLS:

Spermatogonium (2*n*) (stem cell)

↓

Primary spermatocyte (2*n*)

↓

Secondary spermatocyte (*n*)

↓

Spermatid (*n*)

↓

Sperm cell or spermatozoon (*n*)

? **Which cells secrete testosterone?**

(b) Transverse section of part of seminiferous tubule

CLINICAL CONNECTION | *Cryptorchidism*

The condition in which the testes do not descend into the scrotum is called **cryptorchidism** (krip-TOR-ki-dizm; *crypt-* = hidden; *-orchid* = testis); it occurs in about 3% of full-term infants and about 30% of premature infants. Untreated bilateral cryptorchidism results in sterility because the cells involved in the initial stages of spermatogenesis are destroyed by the higher temperature of the pelvic cavity. The chance of testicular cancer is 30–50 times greater in cryptorchid testes. The testes of about 80% of boys with cryptorchidism will descend spontaneously during the first year of life. When the testes remain undescended, the condition can be corrected surgically, ideally before 18 months of age. •

Spermatogenesis

Before you read this section, please review the topic of reproductive cell division in Chapter 3 in Section 3.7. Pay particular attention to Figures 3.33 and 3.34.

In humans, spermatogenesis takes 65–75 days. It begins with the spermatogonia, which contain the diploid ($2n$) number of chromosomes (Figure 28.5). Spermatogonia are types of *stem cells;* when they undergo mitosis, some spermatogonia remain near the basement membrane of the seminiferous tubule in an undifferentiated state to serve as a reservoir of cells for future cell division and subsequent sperm production. The rest of the spermatogonia lose contact with the basement membrane, squeeze through the tight junctions of the blood–testis barrier, undergo developmental changes, and differentiate into **primary spermatocytes** (SPER-ma-tō-sītz′). Primary spermatocytes, like spermatogonia, are diploid ($2n$); that is, they have 46 chromosomes.

Shortly after it forms, each primary spermatocyte replicates its DNA and then meiosis begins (Figure 28.5). In meiosis I, homologous pairs of chromosomes line up at the metaphase plate, and crossing-over occurs. Then, the meiotic spindle pulls one (duplicated) chromosome of each pair to an opposite pole of the dividing cell. The two cells formed by meiosis I are called **secondary spermatocytes.** Each secondary spermatocyte has 23 chromosomes, the haploid number (n). Each chromosome within a secondary spermatocyte, however, is made up of two chromatids (two copies of the DNA) still attached by a centromere. No replication of DNA occurs in the secondary spermatocytes.

In meiosis II, the chromosomes line up in single file along the metaphase plate, and the two chromatids of each chromosome separate. The four haploid cells resulting from meiosis II are called **spermatids** (SPER-ma-tids). A single primary spermatocyte therefore produces four spermatids via two rounds of cell division (meiosis I and meiosis II).

A unique process occurs during spermatogenesis. As spermatogenic cells proliferate, they fail to complete cytoplasmic separation (cytokinesis). The cells remain in contact via cytoplasmic bridges through their entire development (see Figures 28.4b and 28.5). This pattern of development most likely accounts for the synchronized production of sperm in any given area of the seminiferous tubule. It may also have survival value

in that half of the sperm contain an X chromosome and half contain a Y chromosome. The larger X chromosome may carry genes needed for spermatogenesis that are lacking on the smaller Y chromosome.

The final stage of spermatogenesis, **spermiogenesis** (sper′-mē-ō-JEN-e-sis), is the development of haploid spermatids into sperm. No cell division occurs in spermiogenesis; each spermatid becomes a single **sperm cell.** During this process, spherical spermatids transform into elongated, slender sperm. An acrosome (described shortly) forms atop the nucleus, which condenses and elongates, a flagellum develops, and mitochondria multiply. Sertoli cells dispose of the excess cytoplasm that sloughs off. Finally, sperm are released from their connections to Sertoli cells, an event known as **spermiation** (sper′-mē-Ā-shun). Sperm then enter the lumen of the seminiferous tubule. Fluid secreted by Sertoli cells pushes sperm along their way, toward the ducts of the testes. At this point, sperm are not yet able to swim.

Figure 28.5 Events in spermatogenesis. Diploid cells ($2n$) have 46 chromosomes; haploid cells (n) have 23 chromosomes.

Spermiogenesis involves the maturation of spermatids into sperm.

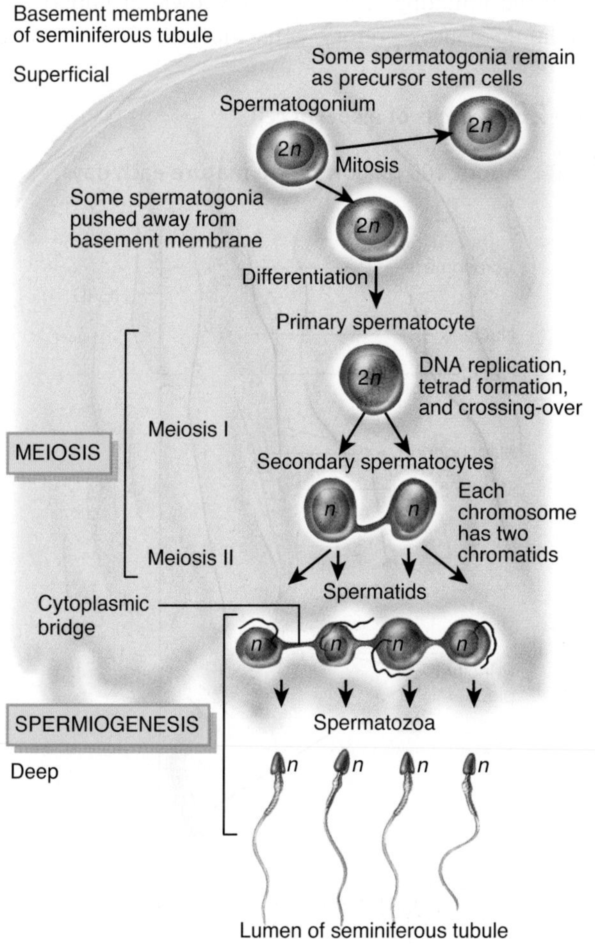

Basement membrane of seminiferous tubule

Superficial

Some spermatogonia remain as precursor stem cells

Spermatogonium

$2n$ — Mitosis — $2n$

Some spermatogonia pushed away from basement membrane — $2n$

Differentiation

Primary spermatocyte

$2n$ — DNA replication, tetrad formation, and crossing-over

MEIOSIS

Meiosis I

Secondary spermatocytes

n — n — Each chromosome has two chromatids

Meiosis II

Spermatids

Cytoplasmic bridge — n — n — n — n

SPERMIOGENESIS

Spermatozoa

Deep — n n n n

Lumen of seminiferous tubule

? **What is the outcome of meiosis I?**

Sperm

Each day about 300 million sperm complete the process of spermatogenesis. A sperm is about 60 μm long and contains several structures that are highly adapted for reaching and penetrating a secondary oocyte (Figure 28.6). The major parts of a sperm are the head and the tail. The flattened, pointed **head** of the sperm is about 4–5 μm long. It contains a **nucleus** with 23 highly condensed chromosomes. Covering the anterior two-thirds of the nucleus is the **acrosome** (AK-rō-sōm; *acro-* = atop; *-some* = body), a caplike vesicle filled with enzymes that help a sperm to penetrate a secondary oocyte to bring about fertilization. Among the enzymes are hyaluronidase and proteases. The **tail** of a sperm is subdivided into four parts: neck, middle piece, principal piece, and end piece. The **neck** is the constricted region just behind the head that contains centrioles. The centrioles form the microtubules that comprise the remainder of the tail. The **middle piece** contains mitochondria arranged in a spiral, which provide the energy (ATP) for locomotion of sperm to the site of fertilization and for sperm metabolism. The **principal piece** is the longest portion of the tail, and the **end piece** is the terminal, tapering portion of the tail. Once ejaculated, most sperm do not survive more than 48 hours within the female reproductive tract.

Hormonal Control of the Testes

Although the initiating factors are unknown, at puberty certain hypothalamic neurosecretory cells increase their secretion of

Figure 28.6 Parts of a sperm cell.

About 300 million sperm mature each day.

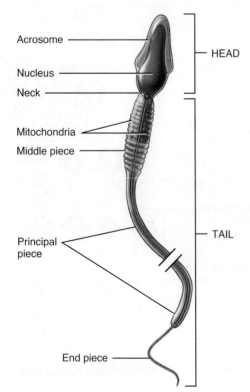

- Acrosome
- Nucleus
- Neck
- Mitochondria
- Middle piece
- Principal piece
- End piece
- HEAD
- TAIL

? **What are the functions of each part of a sperm cell?**

gonadotropin-releasing hormone (GnRH) (gō'-nad-ō-TRŌ-pin). This hormone in turn stimulates gonadotrophs in the anterior pituitary to increase their secretion of the two gonadotropins, **luteinizing hormone (LH)** (LOO-tē-in'-īz-ing) and **follicle-stimulating hormone (FSH).** Figure 28.7 shows the hormones and negative feedback loops that control secretion of testosterone and spermatogenesis.

LH stimulates Leydig cells, which are located between seminiferous tubules, to secrete the hormone **testosterone** (tes-TOS-te-rōn). This steroid hormone is synthesized from cholesterol in the testes and is the principal androgen. It is lipid-soluble and readily diffuses out of Leydig cells into the interstitial fluid and then into blood. Via negative feedback, testosterone suppresses secretion of LH by anterior pituitary gonadotrophs and suppresses secretion of GnRH by hypothalamic neurosecretory cells. In some target cells, such as those in the external genitals and prostate, the enzyme 5 alpha-reductase converts testosterone to another androgen called **dihydrotestosterone (DHT)** (dī-hī'-drō-tes-TOS-ter-ōn).

FSH acts indirectly to stimulate spermatogenesis (Figure 28.7). FSH and testosterone act synergistically on the Sertoli cells to stimulate secretion of **androgen-binding protein (ABP)** into the lumen of the seminiferous tubules and into the interstitial fluid around the spermatogenic cells. ABP binds to testosterone, keeping its concentration high. Testosterone stimulates the final steps of spermatogenesis in the seminiferous tubules. Once the degree of spermatogenesis required for male reproductive functions has been achieved, Sertoli cells release **inhibin,** a protein hormone named for its role in inhibiting FSH secretion by the anterior pituitary (Figure 28.7). If spermatogenesis is proceeding too slowly, less inhibin is released, which permits more FSH secretion and an increased rate of spermatogenesis.

Testosterone and dihydrotestosterone both bind to the same androgen receptors, which are found within the nuclei of target cells. The hormone–receptor complex regulates gene expression, turning some genes on and others off. Because of these changes, the androgens produce several effects:

- *Prenatal development.* Before birth, testosterone stimulates the male pattern of development of reproductive system ducts and the descent of the testes. Dihydrotestosterone stimulates development of the external genitals (described in Section 28.5). Testosterone also is converted in the brain to estrogens (feminizing hormones), which may play a role in the development of certain regions of the brain in males.

- *Development of male sexual characteristics.* At puberty, testosterone and dihydrotestosterone bring about development and enlargement of the male sex organs and the development of masculine secondary sexual characteristics. **Secondary sex characteristics** are traits that distinguish males and females but do not have a direct role in reproduction. These include muscular and skeletal growth that results in wide shoulders and narrow hips; facial and chest hair (within hereditary limits) and more hair on other parts of the body; thickening of the skin; increased sebaceous (oil) gland secretion; and enlargement of the larynx and consequent deepening of the voice.

- *Development of sexual function.* Androgens contribute to male sexual behavior and spermatogenesis and to sex drive (libido) in both males and females. Recall that the adrenal cortex is the main source of androgens in females.
- *Stimulation of anabolism.* Androgens are anabolic hormones; that is, they stimulate protein synthesis. This effect is obvious in the heavier muscle and bone mass of most men as compared to women.

A negative feedback system regulates testosterone production (Figure 28.8). When testosterone concentration in the blood increases to a certain level, it inhibits the release of GnRH by cells in the hypothalamus. As a result, there is less GnRH in the portal blood that flows from the hypothalamus to the anterior pituitary. Gonadotrophs in the anterior pituitary then release less LH, so

Figure 28.7 Hormonal control of spermatogenesis and actions of testosterone and dihydrotestosterone (DHT). In response to stimulation by FSH and testosterone, Sertoli cells secrete androgen-binding protein (ABP). Dashed red lines indicate negative feedback inhibition.

🔑 Release of FSH is stimulated by GnRH and inhibited by inhibin; release of LH is stimulated by GnRH and inhibited by testosterone.

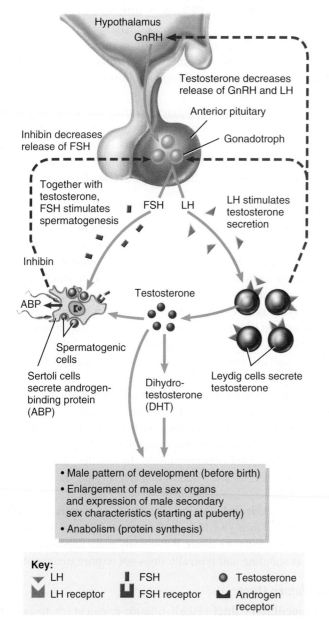

Key:
- ▼ LH
- ▮ FSH
- ● Testosterone
- ⩗ LH receptor
- ⊔ FSH receptor
- ⩗ Androgen receptor

❓ **Which cells secrete inhibin?**

Figure 28.8 Negative feedback control of blood level of testosterone.

🔑 Gonadotrophs of the anterior pituitary produce luteinizing hormone (LH).

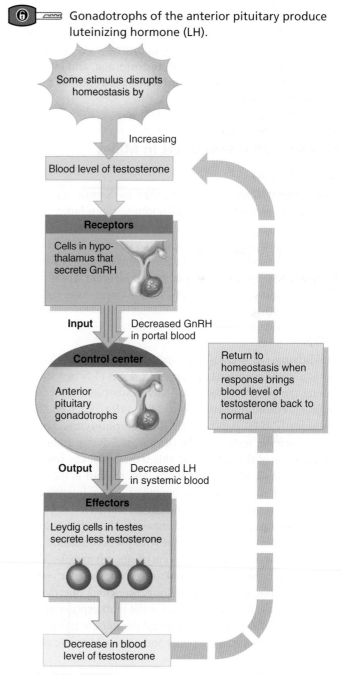

❓ **Which hormones inhibit secretion of FSH and LH by the anterior pituitary?**

the concentration of LH in systemic blood falls. With less stimulation by LH, the Leydig cells in the testes secrete less testosterone, and there is a return to homeostasis. If the testosterone concentration in the blood falls too low, however, GnRH is again released by the hypothalamus and stimulates secretion of LH by the anterior pituitary. LH in turn stimulates testosterone production by the testes.

✔ CHECKPOINT

1. Describe the function of the scrotum in protecting the testes from temperature fluctuations.
2. Describe the internal structure of a testis. Where are sperm cells produced? What are the functions of Sertoli cells and Leydig cells?
3. Describe the principal events of spermatogenesis.
4. Which part of a sperm cell contains enzymes that help the sperm cell fertilize a secondary oocyte?
5. What are the roles of FSH, LH, testosterone, and inhibin in the male reproductive system? How is secretion of these hormones controlled?

Reproductive System Ducts in Males

Ducts of the Testis

Pressure generated by the fluid secreted by Sertoli cells pushes sperm and fluid along the lumen of seminiferous tubules and then into a series of very short ducts called **straight tubules** (see Figure 28.3a). The straight tubules lead to a network of ducts in the testis called the **rete testis** (RĒ-tē = network). From the rete testis, sperm move into a series of coiled **efferent ducts** (EF-erent) in the epididymis that empty into a single tube called the **ductus epididymis.**

Epididymis

The **epididymis** (ep'-i-DID-i-mis; *epi-* = above or over; *-didymis* = testis) is a comma-shaped organ about 4 cm (1.5 in.) long that lies along the posterior border of each testis (see Figure 28.3a). The plural is **epididymides** (ep'-i-di-DIM-i-dēz). Each epididymis consists mostly of the tightly coiled **ductus epididymis.** The efferent ducts from the testis join the ductus epididymis at the larger, superior portion of the epididymis called the **head.** The **body** is the narrow midportion of the epididymis, and the **tail** is the smaller, inferior portion. At its distal end, the tail of the epididymis continues as the ductus (vas) deferens (discussed shortly).

The ductus epididymis would measure about 6 m (20 ft) in length if it were uncoiled. It is lined with pseudostratified columnar epithelium and encircled by layers of smooth muscle. The free surfaces of the columnar cells contain **stereocilia** (ster'-ē-ō-SIL-ē-a), which despite their name are long, branching microvilli (not cilia) that increase the surface area for the reabsorption of degenerated sperm. Connective tissue around the muscle layer attaches the loops of the ductus epididymis and carries blood vessels and nerves.

Functionally, the epididymis is the site of **sperm maturation,** the process by which sperm acquire motility and the ability to fertilize an ovum. This occurs over a period of about 14 days. The epididymis also helps propel sperm into the ductus (vas) deferens during sexual arousal by peristaltic contraction of its smooth muscle. In addition, the epididymis stores sperm, which remain viable here for up to several months. Any stored sperm that are not ejaculated by that time are eventually reabsorbed.

Ductus Deferens

Within the tail of the epididymis, the ductus epididymis becomes less convoluted, and its diameter increases. Beyond this point, the duct is known as the **ductus deferens** or **vas deferens** (DEF-er-enz) (see Figure 28.3a). The ductus deferens, which is about 45 cm (18 in.) long, ascends along the posterior border of the epididymis through the spermatic cord and then enters the pelvic cavity. There it loops over the ureter and passes over the side and down the posterior surface of the urinary bladder (see Figure 28.1a). The dilated terminal portion of the ductus deferens is the **ampulla** (am-PUL-la = little jar; Figure 28.9). The mucosa of the ductus deferens consists of pseudostratified columnar epithelium and lamina propria (areolar connective tissue). The muscularis is composed of three layers of smooth muscle; the inner and outer layers are longitudinal, and the middle layer is circular.

Functionally, the ductus deferens conveys sperm during sexual arousal from the epididymis toward the urethra by peristaltic contractions of its muscular coat. Like the epididymis, the ductus deferens also can store sperm for several months. Any stored sperm that are not ejaculated by that time are eventually reabsorbed.

Spermatic Cord

The **spermatic cord** is a supporting structure of the male reproductive system that ascends out of the scrotum (see Figure 28.2). It consists of the ductus (vas) deferens as it ascends through the scrotum, the testicular artery, veins that drain the testes and carry testosterone into circulation (the pampiniform plexus), autonomic nerves, lymphatic vessels, and the cremaster muscle. The spermatic cord and ilioinguinal nerve pass through the **inguinal canal** (ING-gwi-nal = groin), an oblique passageway in the anterior abdominal wall just superior and parallel to the medial half of the inguinal ligament. The canal, which is about 4–5 cm (about 2 in.) long, originates at the **deep (abdominal) inguinal ring,** a slitlike opening in the aponeurosis of the transversus abdominis muscle; the canal ends at the **superficial (subcutaneous) inguinal ring** (see Figure 28.2), a somewhat triangular opening in the aponeurosis of the external oblique muscle. In females, the round ligament of the uterus and ilioinguinal nerve pass through the inguinal canal.

The term **varicocele** (VAR-i-kō-sēl; *varico-* = varicose; *-kele* = hernia) refers to a swelling in the scrotum due to a dilation of the veins that drain the testes. It is usually more apparent when the person is standing and typically does not require treatment.

Ejaculatory Ducts

Each **ejaculatory duct** (ē-JAK-ū-la-tōr-ē; *ejacul-* = to expel) is about 2 cm (1 in.) long and is formed by the union of the duct from the seminal vesicle and the ampulla of the ductus (vas) deferens (Figure 28.9). The short ejaculatory ducts form just superior to the base (superior portion) of the prostate and pass inferiorly and anteriorly through the prostate. They terminate in the prostatic urethra, where they eject sperm and seminal vesicle secretions just before the release of semen from the urethra to the exterior.

Urethra

In males, the **urethra** (ū-RĒ-thra) is the shared terminal duct of the reproductive and urinary systems; it serves as a passageway for both semen and urine. About 20 cm (8 in.) long, it passes through the prostate, the deep muscles of the perineum, and the penis, and is subdivided into three parts (see Figures 28.1 and 26.22). The

Figure 28.9 Locations of several accessory reproductive organs in males. The prostate, urethra, and penis have been sectioned to show internal details.

The male urethra has three subdivisions: the prostatic, membranous, and spongy (penile) urethra.

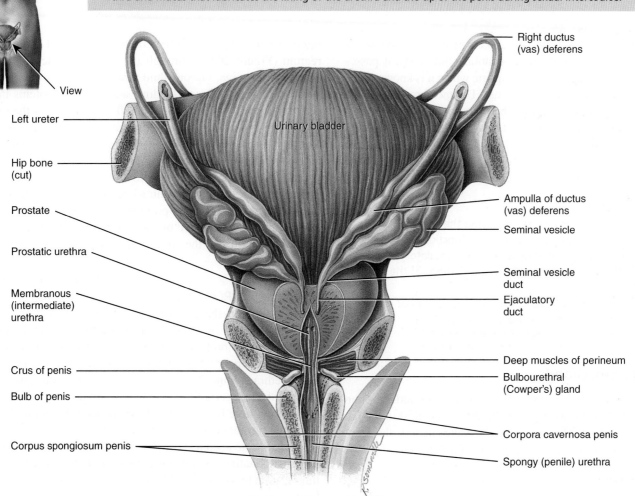

FUNCTIONS OF ACCESSORY SEX GLAND SECRETIONS

1. The seminal vesicles secrete an alkaline, viscous fluid that helps neutralize acid in the female reproductive tract, provides fructose for ATP production by sperm, contributes to sperm motility and viability, and helps semen coagulate after ejaculation.

2. The prostate secretes a milky, slightly acidic fluid that helps semen coagulate after ejaculation and subsequently breaks down the clot.

3. The bulbourethral (Cowper's) glands secrete an alkaline fluid that neutralizes the acidic environment of the urethra and mucus that lubricates the lining of the urethra and the tip of the penis during sexual intercourse.

View

Left ureter

Hip bone (cut)

Prostate

Prostatic urethra

Membranous (intermediate) urethra

Crus of penis

Bulb of penis

Corpus spongiosum penis

Urinary bladder

Right ductus (vas) deferens

Ampulla of ductus (vas) deferens

Seminal vesicle

Seminal vesicle duct

Ejaculatory duct

Deep muscles of perineum

Bulbourethral (Cowper's) gland

Corpora cavernosa penis

Spongy (penile) urethra

(a) Posterior view of male accessory organs of reproduction

FIGURE 28.9 CONTINUES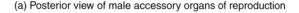

⬛ FIGURE 28.9 CONTINUED ▶

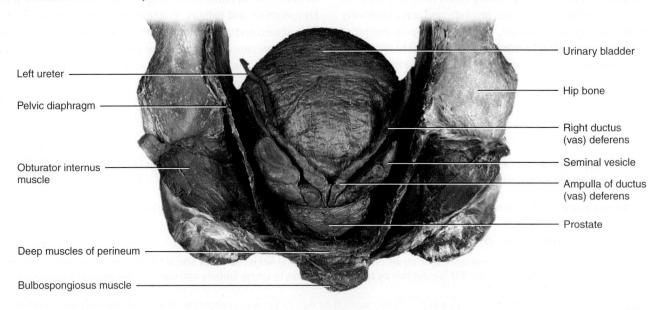

Left ureter

Pelvic diaphragm

Obturator internus
muscle

Deep muscles of perineum

Bulbospongiosus muscle

Urinary bladder

Hip bone

Right ductus
(vas) deferens

Seminal vesicle

Ampulla of ductus
(vas) deferens

Prostate

(b) Posterior view of male accessory organs of reproduction

❓ **What accessory sex gland contributes the majority of the seminal fluid?**

prostatic urethra (pros-TAT-ik) is 2–3 cm (1 in.) long and passes through the prostate. As this duct continues inferiorly, it passes through the deep muscles of the perineum, where it is known as the **membranous urethra** (MEM-bra-nus). The membranous urethra is about 1 cm (0.5 in.) in length. As this duct passes through the corpus spongiosum of the penis, it is known as the **spongy (penile) urethra,** which is about 15–20 cm (6–8 in.) long. The spongy urethra ends at the **external urethral orifice.** The histology of the male urethra may be reviewed in Section 26.8.

✔**CHECKPOINT**

6. Which ducts transport sperm within the testes?
7. Describe the location, structure, and functions of the ductus epididymis, ductus (vas) deferens, and ejaculatory duct.
8. Give the locations of the three subdivisions of the male urethra.
9. Trace the course of sperm through the system of ducts from the seminiferous tubules to the urethra.
10. List the structures within the spermatic cord.

Accessory Sex Glands

The ducts of the male reproductive system store and transport sperm cells, but the **accessory sex glands** secrete most of the liquid portion of semen. The accessory sex glands include the seminal vesicles, the prostate, and the bulbourethral glands.

Seminal Vesicles

The paired **seminal vesicles** (VES-i-kuls) or **seminal glands** are convoluted pouchlike structures, about 5 cm (2 in.) in length, ly-

ing posterior to the base of the urinary bladder and anterior to the rectum (Figure 28.9). Through the seminal vesicle ducts they secrete an alkaline, viscous fluid that contains fructose (a monosaccharide sugar), prostaglandins, and clotting proteins that are different from those in blood. The alkaline nature of the seminal fluid helps to neutralize the acidic environment of the male urethra and female reproductive tract that otherwise would inactivate and kill sperm. The fructose is used for ATP production by sperm. Prostaglandins contribute to sperm motility and viability and may stimulate smooth muscle contractions within the female reproductive tract. The clotting proteins help semen coagulate after ejaculation. Fluid secreted by the seminal vesicles normally constitutes about 60% of the volume of semen.

Prostate

The **prostate** (PROS-tāt; *prostata* = one who stands before) is a single, doughnut-shaped gland about the size of a golf ball. It measures about 4 cm (1.6 in.) from side to side, about 3 cm (1.2 in.) from top to bottom, and about 2 cm (0.8 in.) from front to back. It is inferior to the urinary bladder and surrounds the prostatic urethra (Figure 28.9). The prostate slowly increases in size from birth to puberty. It then expands rapidly until about age 30, after which time its size typically remains stable until about age 45, when further enlargement may occur.

The prostate secretes a milky, slightly acidic fluid (pH about 6.5) that contains several substances. (1) *Citric acid* in prostatic fluid is used by sperm for ATP production via the Krebs cycle. (2) Several *proteolytic enzymes,* such as *prostate-specific antigen (PSA),* pepsinogen, lysozyme, amylase, and hyaluronidase, eventually break down the clotting proteins from the seminal vesicles. (3) The function of the *acid phosphatase* secreted by the prostate is unknown. (4) *Seminalplasmin* in prostatic fluid is an antibiotic that can destroy bacteria. Seminalplasmin may help decrease the number of naturally occurring bacteria in semen and in the lower female reproductive tract. Secretions of the prostate enter the prostatic urethra through many prostatic ducts. Prostatic secretions make up about 25% of the volume of semen and contribute to sperm motility and viability.

Bulbourethral Glands

The paired **bulbourethral glands** (bul′-bō-ū-RĒ-thral), or **Cowper's glands** (KOW-pers), are about the size of peas. They are located inferior to the prostate on either side of the membranous urethra within the deep muscles of the perineum, and their ducts open into the spongy urethra (Figure 28.9). During sexual arousal, the bulbourethral glands secrete an alkaline fluid into the urethra that protects the passing sperm by neutralizing acids from urine in the urethra. They also secrete mucus that lubricates the end of the penis and the lining of the urethra, decreasing the number of sperm damaged during ejaculation. Some males release a drop or two of this mucus upon sexual arousal and erection. The fluid does not contain sperm cells.

Semen

Semen (= seed) is a mixture of sperm and **seminal fluid,** a liquid that consists of the secretions of the seminiferous tubules, seminal vesicles, prostate, and bulbourethral glands. The volume of semen in a typical ejaculation is 2.5–5 milliliters (mL), with 50–150 million sperm per mL. When the number falls below 20 million/mL, the male is likely to be infertile. A very large number of sperm is required for successful fertilization because only a tiny fraction ever reaches the secondary oocyte.

Despite the slight acidity of prostatic fluid, semen has a slightly alkaline pH of 7.2–7.7 due to the higher pH and larger volume of fluid from the seminal vesicles. The prostatic secretion gives semen a milky appearance, and fluids from the seminal vesicles and bulbourethral glands give it a sticky consistency. Seminal fluid provides sperm with a transportation medium, nutrients, and protection from the hostile acidic environment of the male's urethra and the female's vagina.

Once ejaculated, liquid semen coagulates within 5 minutes due to the presence of clotting proteins from the seminal vesicles. The functional role of semen coagulation is not known, but the proteins involved are different from those that cause blood coagulation. After about 10 to 20 minutes, semen reliquefies because prostate-specific antigen (PSA) and other proteolytic enzymes produced by the prostate break down the clot. Abnormal or delayed liquefaction of clotted semen may cause complete or partial immobilization of sperm, thereby inhibiting their movement through the cervix of the uterus. After passing through the uterus and uterine tube, the sperm are affected by secretions of the uterine tube in a process called **capacitation** (see Section 28.2). The presence of blood in semen is called **hemospermia** (hē-mō-SPER-mē-a; *hemo-* = blood; *-sperma* = seed). In most cases, it is caused by inflammation of the blood vessels lining the seminal vesicles; it is usually treated with antibiotics.

Penis

The **penis** (= tail) contains the urethra and is a passageway for the ejaculation of semen and the excretion of urine (Figure 28.10). It is cylindrical in shape and consists of a body, glans penis, and a root. The **body of the penis** is composed of three cylindrical masses of tissue, each surrounded by fibrous tissue called the **tunica albuginea** (al′-bū-JIN-ē-a) (Figure 28.10). The two dorsolateral masses are called the **corpora cavernosa penis** (*corpora* = main bodies; *cavernosa* = hollow). The smaller midventral mass, the **corpus spongiosum penis,** contains the spongy urethra and keeps it open during ejaculation. Skin and a subcutaneous layer enclose all three masses, which consist of erectile tissue. *Erectile tissue* is composed of numerous blood sinuses (vascular spaces) lined by endothelial cells and surrounded by smooth muscle and elastic connective tissue.

The distal end of the corpus spongiosum penis is a slightly enlarged, acorn-shaped region called the **glans penis;** its margin is the **corona** (kō-RŌ-na). The distal urethra enlarges within the glans penis and forms a terminal slitlike opening, the **external urethral orifice.** Covering the glans in an uncircumcised penis is the loosely fitting **prepuce** (PRĒ-poos), or **foreskin.**

The **root of the penis** is the attached portion (proximal portion). It consists of the **bulb of the penis,** the expanded posterior continuation of the base of the corpus spongiosum penis, and the **crura of the penis** (KROO-ra; singular is *crus* = resembling a leg), the two separated and tapered portions of the corpora cavernosa penis. The bulb of the penis is attached to the inferior surface of the deep muscles of the perineum and is enclosed by the bulbospongiosus muscle, a muscle that aids ejaculation. Each crus of the penis bends laterally away from the bulb of the penis to attach to the ischial and inferior pubic rami and is surrounded by the ischiocavernosus muscle (see Figure 11.13). The weight of the penis is supported by two ligaments that are continuous with the fascia of the penis. (1) The **fundiform ligament** (FUN-di-form) arises from the inferior part of the linea alba. (2) The **suspensory ligament of the penis** arises from the pubic symphysis.

CLINICAL CONNECTION | Circumcision

Circumcision (= to cut around) is a surgical procedure in which part of or the entire prepuce is removed. It is usually performed several days after birth, and is done for social, cultural, religious, and (more rarely) medical reasons. Although most health-care professionals find no medical justification for circumcision, some feel that it has benefits, such as a lower risk of urinary tract infections, protection against penile cancer, and possibly a lower risk for sexually transmitted diseases. Indeed, studies in several African villages have found lower rates of HIV infection among circumcised men. •

Figure 28.10 Internal structure of the penis. The inset in (b) shows details of the skin and fasciae.

The penis contains the urethra, a common pathway for semen and urine.

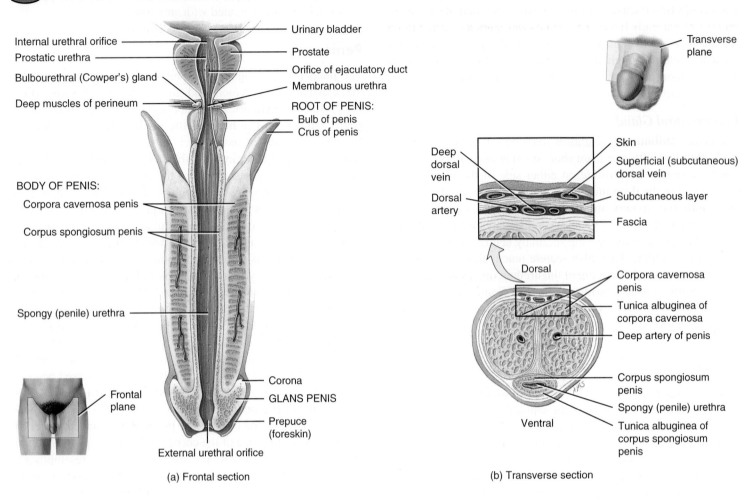

(a) Frontal section

(b) Transverse section

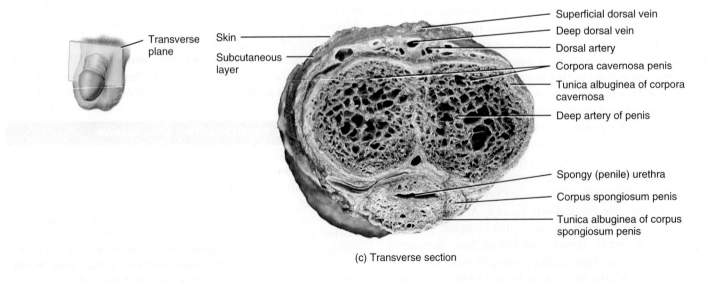

(c) Transverse section

Which tissue masses form the erectile tissue in the penis, and why do they become rigid during sexual arousal?

Upon sexual stimulation (visual, tactile, auditory, olfactory, or imagined), parasympathetic fibers from the sacral portion of the spinal cord initiate and maintain an **erection,** the enlargement and stiffening of the penis. The parasympathetic fibers produce and release nitric oxide (NO). The NO causes smooth muscle in the walls of arterioles supplying erectile tissue to relax, which allows these blood vessels to dilate. This in turn causes large amounts of blood to enter the erectile tissue of the penis. NO also causes the smooth muscle within the erectile tissue to relax, resulting in widening of the blood sinuses. The combination of increased blood flow and widening of the blood sinuses results in an erection. Expansion of the blood sinuses also compresses the veins that drain the penis; the slowing of blood outflow helps to maintain the erection.

The term **priapism** (PRĪ-a-pizm) refers to a persistent and usually painful erection of the penis that does not involve sexual desire or excitement. The condition may last up to several hours and is accompanied by pain and tenderness. It results from abnormalities of blood vessels and nerves, usually in response to medication used to produce erections in males who otherwise cannot attain them. Other causes include a spinal cord disorder, leukemia, sickle-cell disease, or a pelvic tumor.

Ejaculation (ē-jak-ū-LĀ-shun; *ejectus-* = to throw out), the powerful release of semen from the urethra to the exterior, is a sympathetic reflex coordinated by the lumbar portion of the spinal cord. As part of the reflex, the smooth muscle sphincter at the base of the urinary bladder closes, preventing urine from being expelled during ejaculation, and semen from entering the urinary bladder. Even before ejaculation occurs, peristaltic contractions in the epididymis, ductus (vas) deferens, seminal vesicles, ejaculatory ducts, and prostate propel semen into the penile portion of the urethra (spongy urethra). Typically, this leads to **emission** (ē-MISH-un), the discharge of a small volume of semen before ejaculation. Emission may also occur during sleep (nocturnal emission). The musculature of the penis (bulbospongiosus, ischiocavernosus, and superficial transverse perineus muscles), which is supplied by the pudendal nerve, also contracts at ejaculation (see Figure 11.13).

Once sexual stimulation of the penis has ended, the arterioles supplying the erectile tissue of the penis constrict and the smooth muscle within erectile tissue contracts, making the blood sinuses smaller. This relieves pressure on the veins supplying the penis and allows the blood to drain through them. Consequently, the penis returns to its flaccid (relaxed) state.

CLINICAL CONNECTION | *Premature Ejaculation*

A **premature ejaculation** is ejaculation that occurs too early, for example, during foreplay or on or shortly after penetration. It is usually caused by anxiety, other psychological causes, or an unusually sensitive foreskin or glans penis. For most males, premature ejaculation can be overcome by various techniques (such as squeezing the penis between the glans penis and shaft as ejaculation approaches), behavioral therapy, or medication. •

✔ CHECKPOINT

11. Briefly explain the locations and functions of the seminal vesicles, the prostate, and the bulbourethral (Cowper's) glands.
12. What is semen? What is its function?
13. Explain the physiological processes involved in erection and ejaculation.

28.2 FEMALE REPRODUCTIVE SYSTEM

◉ OBJECTIVES

• Describe the location, structure, and functions of the organs of the female reproductive system.
• Discuss the process of oogenesis in the ovaries.

The organs of the **female reproductive system** (Figure 28.11) include the ovaries (female gonads); the uterine (fallopian) tubes, or oviducts; the uterus; the vagina; and external organs, which are collectively called the vulva, or pudendum. The mammary glands are considered part of both the integumentary system and the female reproductive system.

Ovaries

The **ovaries** (= egg receptacles), which are the female gonads, are paired glands that resemble unshelled almonds in size and shape; they are homologous to the testes. (Here *homologous* means that two organs have the same embryonic origin.) The ovaries produce (1) gametes, secondary oocytes that develop into mature ova (eggs) after fertilization, and (2) hormones, including progesterone and estrogens (the female sex hormones), inhibin, and relaxin.

The ovaries, one on either side of the uterus, descend to the brim of the superior portion of the pelvic cavity during the third month of development. A series of ligaments holds them in position (Figure 28.12). The **broad ligament** of the uterus, which is a fold of the parietal peritoneum, attaches to the ovaries by a double-layered fold of peritoneum called the **mesovarium** (mez'-ō-VĀ-rē-um). The **ovarian ligament** anchors the ovaries to the uterus, and the **suspensory ligament** attaches them to the pelvic wall. Each ovary contains a **hilum** (HĪ-lum) or **hilus,** the point of entrance and exit for blood vessels and nerves along which the mesovarium is attached.

Histology of the Ovary

Each ovary consists of the following parts (Figure 28.13):

• The **germinal epithelium** (*germen* = sprout or bud) is a layer of simple epithelium (low cuboidal or squamous) that covers the surface of the ovary. We now know that the term germinal epithelium in humans is not accurate because it does not give rise to ova; the name came about because, at one time, people believed that it did. We have since learned that the cells that produce ova arise from the yolk sac and migrate to the ovaries during embryonic development.

Figure 28.11 **Female organs of reproduction and surrounding structures.**

The organs of reproduction in females include the ovaries, uterine (fallopian) tubes, uterus, vagina, vulva, and mammary glands.

FUNCTIONS OF THE FEMALE REPRODUCTIVE SYSTEM
1. The ovaries produce secondary oocytes and hormones, including progesterone and estrogens (female sex hormones), inhibin, and relaxin.
2. The uterine tubes transport a secondary oocyte to the uterus and normally are the sites where fertilization occurs.
3. The uterus is the site of implantation of a fertilized ovum, development of the fetus during pregnancy, and labor.
4. The vagina receives the penis during sexual intercourse and is a passageway for childbirth.
5. The mammary glands synthesize, secrete, and eject milk for nourishment of the newborn.

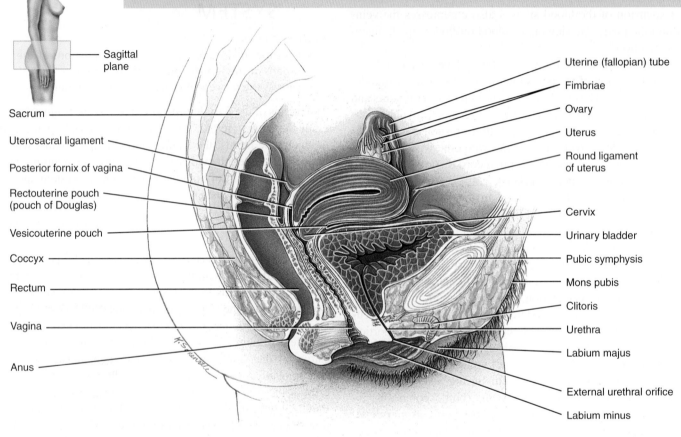

(a) Sagittal section

- The **tunica albuginea** is a whitish capsule of dense irregular connective tissue located immediately deep to the germinal epithelium.

- The **ovarian cortex** is a region just deep to the tunica albuginea. It consists of ovarian follicles (described shortly) surrounded by dense irregular connective tissue that contains collagen fibers and fibroblast-like cells called *stromal cells.*

- The **ovarian medulla** is deep to the ovarian cortex. The border between the cortex and medulla is indistinct, but the medulla consists of more loosely arranged connective tissue and contains blood vessels, lymphatic vessels, and nerves.

- **Ovarian follicles** (*folliculus* = little bag) are in the cortex and consist of **oocytes** (Ō-ō-sīts) in various stages of development,

plus the cells surrounding them. When the surrounding cells form a single layer, they are called **follicular cells** (fo-LIK-ū-lar); later in development, when they form several layers, they are referred to as **granulosa cells** (gran′-u-LŌ-sa). The surrounding cells nourish the developing oocyte and begin to secrete estrogens as the follicle grows larger.

- A **mature (graafian) follicle** (GRA-fē-an) is a large, fluid-filled follicle that is ready to rupture and expel its secondary oocyte, a process known as **ovulation.**

- A **corpus luteum** (= yellow body) contains the remnants of a mature follicle after ovulation. The corpus luteum produces progesterone, estrogens, relaxin, and inhibin until it degenerates into fibrous scar tissue called the **corpus albicans** (AL-bi-kanz = white body).

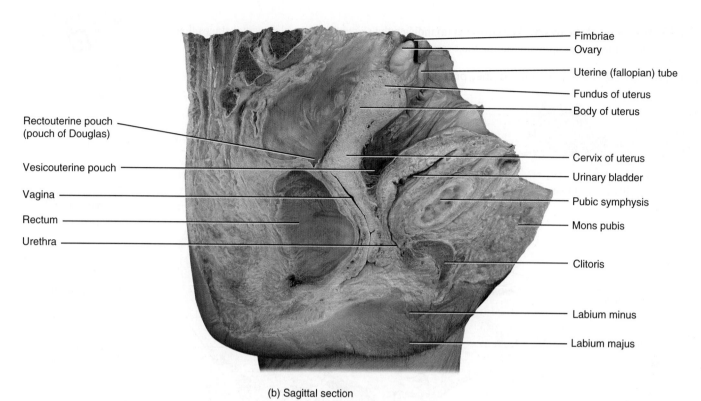

Fimbriae
Ovary
Uterine (fallopian) tube
Fundus of uterus
Body of uterus

Rectouterine pouch
(pouch of Douglas)

Vesicouterine pouch

Vagina

Rectum

Urethra

Cervix of uterus
Urinary bladder
Pubic symphysis
Mons pubis

Clitoris

Labium minus

Labium majus

(b) Sagittal section

? **Which structures in males are homologous to the ovaries, the clitoris, the paraurethral glands, and the greater vestibular glands?**

Figure 28.12 **Relative positions of the ovaries, the uterus, and the ligaments that support them.**

🔑 Ligaments holding the ovaries in position are the mesovarium, the ovarian ligament, and the suspensory ligament.

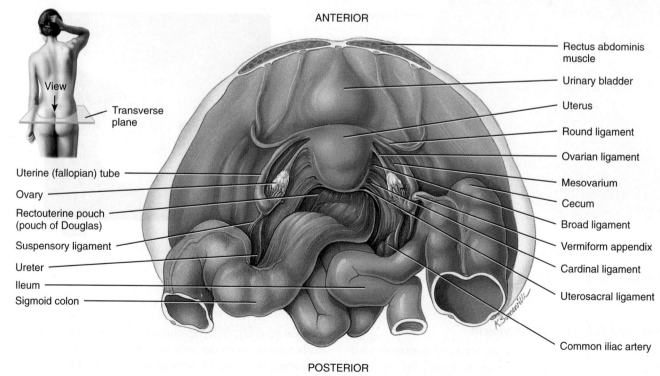

ANTERIOR

View

Transverse
plane

Rectus abdominis
muscle

Urinary bladder

Uterus

Round ligament

Ovarian ligament

Mesovarium

Cecum

Broad ligament

Vermiform appendix

Cardinal ligament

Uterosacral ligament

Common iliac artery

Uterine (fallopian) tube

Ovary

Rectouterine pouch
(pouch of Douglas)

Suspensory ligament

Ureter

Ileum

Sigmoid colon

POSTERIOR

Superior view of transverse section

? **To which structures do the mesovarium, ovarian ligament, and suspensory ligament anchor the ovary?**

Figure 28.13 Histology of the ovary. The arrows in (a) indicate the sequence of developmental stages that occur as part of the maturation of an ovum during the ovarian cycle.

🔑 The ovaries are the female gonads; they produce haploid oocytes.

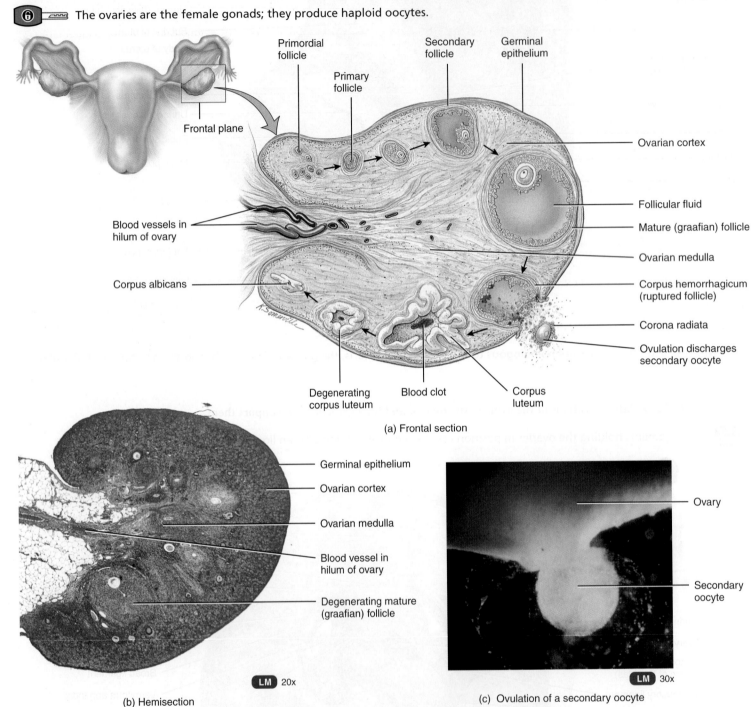

(a) Frontal section

(b) Hemisection   LM 20x

(c) Ovulation of a secondary oocyte   LM 30x

❓ **What structures in the ovary contain endocrine tissue, and what hormones do they secrete?**

Oogenesis and Follicular Development

The formation of gametes in the ovaries is termed **oogenesis** (ō-ō-JEN-e-sis; *oo-* = egg). In contrast to spermatogenesis, which begins in males at puberty, oogenesis begins in females before they are even born. Oogenesis occurs in essentially the same manner as spermatogenesis; meiosis (see Chapter 3) takes place and the resulting germ cells undergo maturation.

During early fetal development, primordial (primitive) germ cells migrate from the yolk sac to the ovaries. There, germ cells differentiate within the ovaries into **oogonia** (ō-ō-GŌ-nē-a; singular is **oogonium**). Oogonia are diploid (2*n*) stem cells that divide mitotically to produce millions of germ cells. Even before birth, most of these germ cells degenerate in a process known as **atresia** (a-TRĒ-zē-a). A few, however, develop into larger cells called **primary oocytes** that enter prophase of meiosis I during fetal development but do not complete that phase until after puberty. During this arrested stage of development, each primary oocyte is surrounded by a single layer of flat follicular cells, and the entire structure is called a **primordial follicle** (Figure 28.14a).

The ovarian cortex surrounding the primordial follicles consists of collagen fibers and fibroblast-like **stromal cells.** At birth, approximately 200,000 to 2,000,000 primary oocytes remain in each ovary. Of these, about 40,000 are still present at puberty, and around 400 will mature and ovulate during a woman's

Figure 28.14 Ovarian follicles.

As an ovarian follicle enlarges, follicular fluid accumulates in a cavity called the antrum.

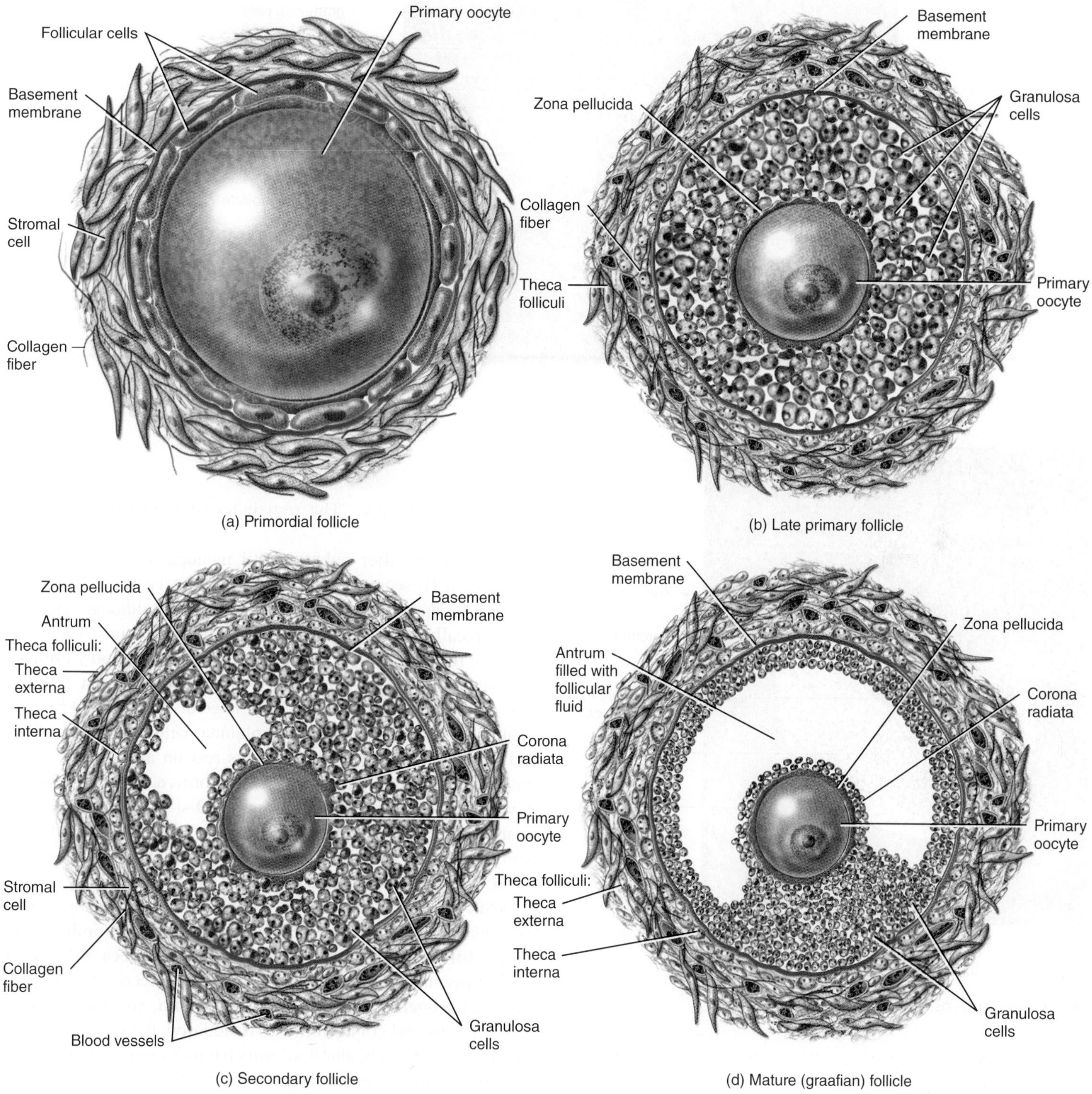

(a) Primordial follicle

(b) Late primary follicle

(c) Secondary follicle

(d) Mature (graafian) follicle

FIGURE 28.14 CONTINUES ▶

FIGURE 28.14 CONTINUED ▶

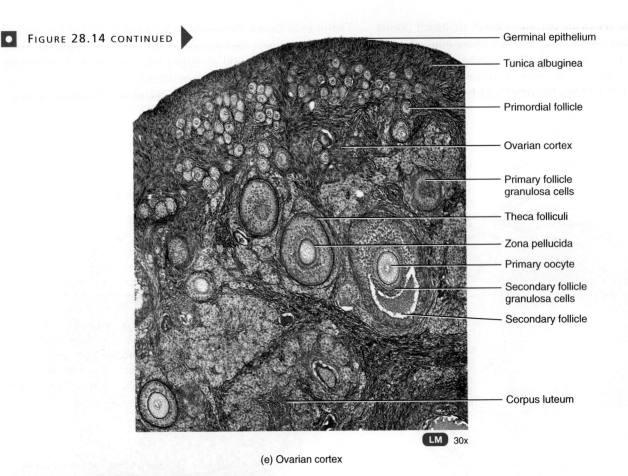

— Germinal epithelium
— Tunica albuginea
— Primordial follicle
— Ovarian cortex
— Primary follicle granulosa cells
— Theca folliculi
— Zona pellucida
— Primary oocyte
— Secondary follicle granulosa cells
— Secondary follicle
— Corpus luteum

LM 30x

(e) Ovarian cortex

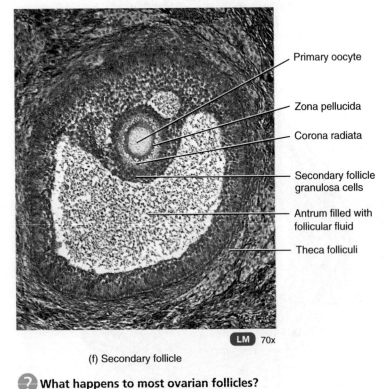

Primary oocyte
Zona pellucida
Corona radiata
Secondary follicle granulosa cells
Antrum filled with follicular fluid
Theca folliculi

LM 70x

(f) Secondary follicle

? **What happens to most ovarian follicles?**

reproductive lifetime. The remainder of the primary oocytes undergo atresia.

Each month after puberty until menopause, gonadotropins (FSH and LH) secreted by the anterior pituitary further stimulate the development of several primordial follicles, although only one will typically reach the maturity needed for ovulation. A few primordial follicles start to grow, developing into **primary follicles** (Figure 28.14b). Each primary follicle consists of a primary oocyte that is surrounded in a later stage of development by several layers of cuboidal and low-columnar cells called **granulosa cells.** The outermost granulosa cells rest on a basement membrane. As the primary follicle grows, it forms a clear glycoprotein layer called the **zona pellucida** (pe-LOO-si-da) between the primary oocyte and the granulosa cells. In addition, stromal cells surrounding the basement membrane begin to form an organized layer called the **theca folliculi** (THĒ-ka fo-LIK-ū-lī).

With continuing maturation, a primary follicle develops into a secondary follicle (Figure 28.14c). In a **secondary follicle,** the theca differentiates into two layers: (1) the **theca interna,** a highly vascularized internal layer of cuboidal secretory cells that secrete estrogens, and (2) the **theca externa,** an outer layer of stromal cells and collagen fibers. In addition, the granulosa cells begin to secrete follicular fluid, which builds up in a cavity called the **antrum** in the center of the secondary follicle. The innermost layer of granulosa cells becomes firmly attached to the zona pellucida and is now called the **corona radiata** (*corona* = crown; *radiata* = radiation) (Figure 28.14c).

The secondary follicle eventually becomes larger, turning into a **mature (graafian) follicle** (Figure 28.14d). While in this follicle, and just before ovulation, the diploid primary oocyte completes meiosis I, producing two haploid (*n*) cells of unequal size—each with 23 chromosomes (Figure 28.15). The smaller cell produced by meiosis I, called the **first polar body,** is essentially a packet of discarded nuclear material. The larger cell, known as the **secondary oocyte,** receives most of the cytoplasm. Once a secondary oocyte is formed, it begins meiosis II but then stops in metaphase. The mature (graafian) follicle soon ruptures and releases its secondary oocyte, a process known as **ovulation.**

At ovulation, the secondary oocyte is expelled into the pelvic cavity together with the first polar body and corona radiata. Normally these cells are swept into the uterine tube. If fertilization does not occur, the cells degenerate. If sperm are present in the uterine tube and one penetrates the secondary oocyte, however,

meiosis II resumes. The secondary oocyte splits into two haploid cells, again of unequal size. The larger cell is the **ovum,** or mature egg; the smaller one is the **second polar body.** The nuclei of the sperm cell and the ovum then unite, forming a diploid **zygote.** If the first polar body undergoes another division to produce two polar bodies, then the primary oocyte ultimately gives rise to three haploid polar bodies, which all degenerate, and a single haploid ovum. Thus, one primary oocyte gives rise to a single gamete (an ovum). By contrast, recall that in males one primary spermatocyte produces four gametes (sperm).

Table 28.1 summarizes the events of oogenesis and follicular development.

⚕ CLINICAL CONNECTION | *Ovarian Cysts*

An **ovarian cyst** is a fluid-filled sac in or on an ovary. Such cysts are relatively common, are usually noncancerous, and frequently disappear on their own. Cancerous cysts are more likely to occur in women over 40. Ovarian cysts may cause pain, pressure, a dull ache, or fullness in the abdomen; pain during sexual intercourse; delayed, painful, or irregular menstrual periods; abrupt onset of sharp pain in the lower abdomen; and/or vaginal bleeding. Most ovarian cysts require no treatment, but larger ones (more than 5 cm or 2 in.) may be removed surgically. •

✔ CHECKPOINT

14. How are the ovaries held in position in the pelvic cavity?
15. Describe the microscopic structure and functions of an ovary.
16. Describe the principal events of oogenesis.

Uterine Tubes

Females have two **uterine (fallopian) tubes,** or **oviducts,** that extend laterally from the uterus (Figure 28.16). The tubes, which measure about 10 cm (4 in.) long, lie within the folds of the broad ligaments of the uterus. They provide a route for sperm to reach an ovum and transport secondary oocytes and fertilized ova from the ovaries to the uterus. The funnel-shaped portion of each tube, called the **infundibulum** (in-fun-DIB-ū-lum), is close to the ovary but is open to the pelvic cavity. It ends in a fringe of fingerlike projections called **fimbriae** (FIM-brē-ē = fringe), one of which is attached to the lateral end of the ovary. From the infundibulum, the uterine tube extends medially and eventually inferiorly and attaches to the superior lateral angle of the uterus. The **ampulla** (am-PUL-la) of the uterine tube is the widest, longest portion, making up about the lateral two-thirds of its length. The **isthmus** (IS-mus) of the uterine tube is the more medial, short, narrow, thick-walled portion that joins the uterus.

Histologically, the uterine tubes are composed of three layers: mucosa, muscularis, and serosa. The mucosa consists of epithelium and lamina propria (areolar connective tissue). The epithelium contains ciliated simple columnar cells, which function as a "ciliary conveyor belt" to help move a fertilized ovum (or secondary oocyte) within the uterine tube toward the uterus, and nonciliated

Figure 28.15 Oogenesis. Diploid cells (2*n*) have 46 chromosomes; haploid cells (*n*) have 23 chromosomes.

🔑 In a secondary oocyte, meiosis II is completed only if fertilization occurs.

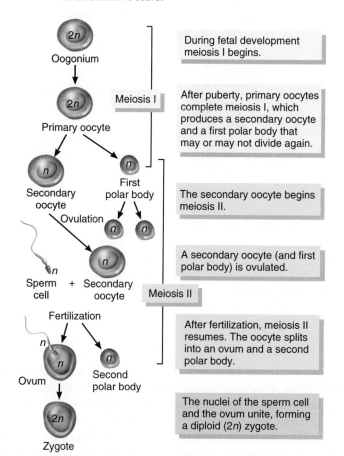

? How does the age of a primary oocyte in a female compare with the age of a primary spermatocyte in a male?

TABLE 28.1

Summary of Oogenesis and Follicular Development

AGE	OOGENESIS	FOLLICULAR DEVELOPMENT

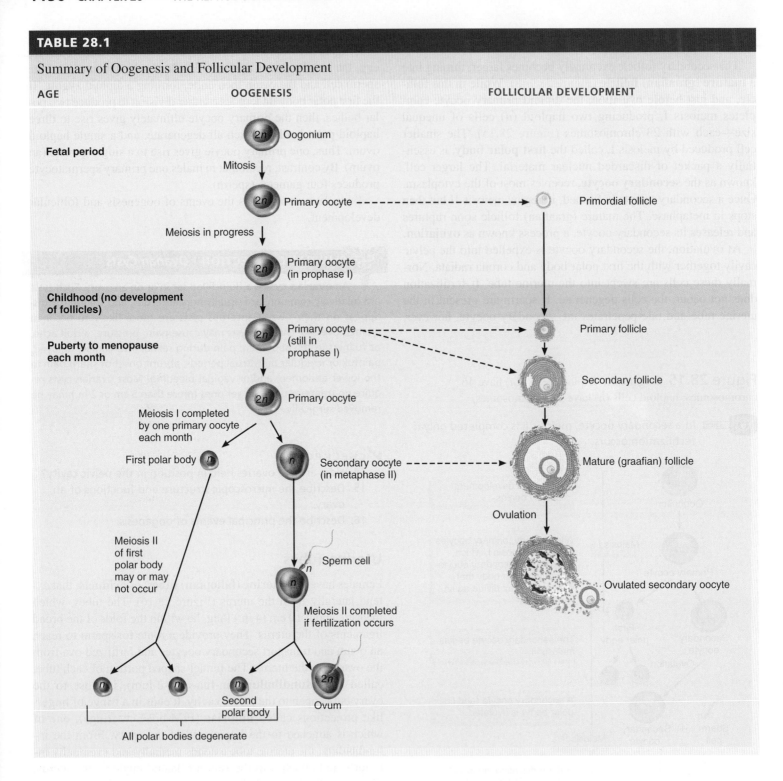

cells called **peg cells,** which have microvilli and secrete a fluid that provides nutrition for the ovum (Figure 28.17). The middle layer, the muscularis, is composed of an inner, thick, circular ring of smooth muscle and an outer, thin region of longitudinal smooth muscle. Peristaltic contractions of the muscularis and the ciliary action of the mucosa help move the oocyte or fertilized ovum toward the uterus. The outer layer of the uterine tubes is a serous membrane, the serosa.

After ovulation, local currents are produced by movements of the fimbriae, which surround the surface of the mature follicle just before ovulation occurs. These currents sweep the ovulated secondary oocyte from the peritoneal cavity into the uterine tube. A sperm cell usually encounters and fertilizes a secondary oocyte in the ampulla of the uterine tube, although fertilization in the peritoneal cavity is not uncommon. Fertilization can occur up to about 24 hours after ovulation. Some hours after fertilization, the nuclear materials

Figure 28.16 Relationship of the uterine (fallopian) tubes to the ovaries, uterus, and associated structures. In the left
side of the drawing the uterine tube and uterus have been sectioned to show internal structures.

 After ovulation, a secondary oocyte and its corona radiata move from the pelvic cavity into the infundibulum of the
uterine tube. The uterus is the site of menstruation, implantation of a fertilized ovum, development of the fetus, and
labor.

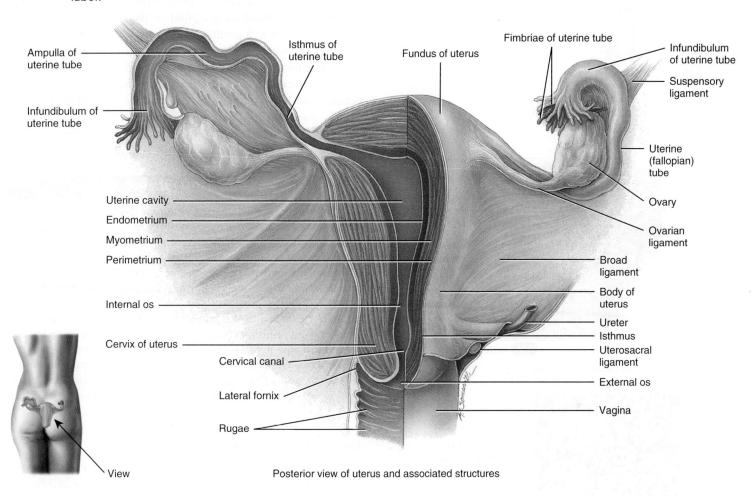

Posterior view of uterus and associated structures

View

? Where does fertilization usually occur?

of the haploid ovum and sperm unite. The diploid fertilized ovum is
now called a **zygote** and begins to undergo cell divisions while
moving toward the uterus. It arrives in the uterus 6 to 7 days after
ovulation. Unfertilized secondary oocytes disintegrate.

Uterus

The **uterus** (womb) serves as part of the pathway for sperm deposited in the vagina to reach the uterine tubes. It is also the site of
implantation of a fertilized ovum, development of the fetus during
pregnancy, and labor. During reproductive cycles when implantation does not occur, the uterus is the source of menstrual flow.

Anatomy of the Uterus

Situated between the urinary bladder and the rectum, the uterus is
the size and shape of an inverted pear (see Figure 28.16). In females

who have never been pregnant, it is about 7.5 cm (3 in.) long, 5 cm
(2 in.) wide, and 2.5 cm (1 in.) thick. The uterus is larger in females
who have recently been pregnant, and smaller (atrophied) when sex
hormone levels are low, as occurs after menopause.

Anatomical subdivisions of the uterus include (1) a dome-
shaped portion superior to the uterine tubes called the **fundus,**
(2) a tapering central portion called the **body,** and (3) an inferior
narrow portion called the **cervix** that opens into the vagina. Between the body of the uterus and the cervix is the **isthmus,** a constricted region about 1 cm (0.5 in.) long. The interior of the body
of the uterus is called the **uterine cavity,** and the interior of the
cervix is called the **cervical canal.** The cervical canal opens into
the uterine cavity at the **internal os** (*os* = mouthlike opening) and
into the vagina at the **external os.**

Normally, the body of the uterus projects anteriorly and superiorly over the urinary bladder in a position called **anteflexion**

Figure 28.17 Histology of the uterine (fallopian) tube.

Peristaltic contractions of the muscularis and ciliary action of the mucosa of the uterine tube help move the oocyte or fertilized ovum toward the uterus.

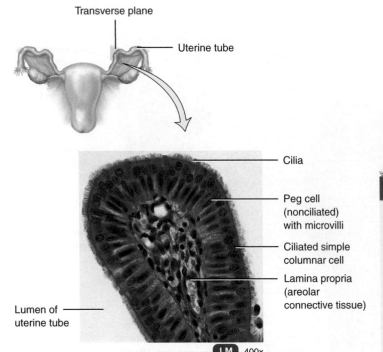

(a) Details of epithelium

LM 400x

Labels: Cilia; Peg cell (nonciliated) with microvilli; Ciliated simple columnar cell; Lamina propria (areolar connective tissue); Lumen of uterine tube; Transverse plane; Uterine tube

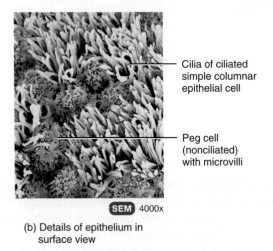

(b) Details of epithelium in surface view

SEM 4000x

Labels: Cilia of ciliated simple columnar epithelial cell; Peg cell (nonciliated) with microvilli

? What types of cells line the uterine tubes?

(an′-te-FLEK-shun; *ante-* = before). The cervix projects inferiorly and posteriorly and enters the anterior wall of the vagina at nearly a right angle (see Figure 28.11). Several ligaments that are either extensions of the parietal peritoneum or fibromuscular cords maintain the position of the uterus (see Figure 28.12). The paired **broad ligaments** are double folds of peritoneum attaching the uterus to either side of the pelvic cavity. The paired

uterosacral ligaments (ū′-ter-ō-SĀ-kral), also peritoneal extensions, lie on either side of the rectum and connect the uterus to the sacrum. The **cardinal (lateral cervical) ligaments** are located inferior to the bases of the broad ligaments and extend from the pelvic wall to the cervix and vagina. The **round ligaments** are bands of fibrous connective tissue between the layers of the broad ligament; they extend from a point on the uterus just inferior to the uterine tubes to a portion of the labia majora of the external genitalia. Although the ligaments normally maintain the anteflexed position of the uterus, they also allow the uterine body enough movement such that the uterus may become malpositioned. A posterior tilting of the uterus, called **retroflexion** (RET-rō-flek-shun; *retro-* = backward or behind), is a harmless variation of the normal position of the uterus. There is often no cause for the condition, but it may occur after childbirth.

> ⚕ **CLINICAL CONNECTION | *Uterine Prolapse***
>
> A condition called **uterine prolapse** (*prolapse* = falling down or downward displacement) may result from weakening of supporting ligaments and pelvic musculature associated with age or disease, traumatic vaginal delivery, chronic straining from coughing or difficult bowel movements, or pelvic tumors. The prolapse may be characterized as *first degree (mild)*, in which the cervix remains within the vagina; *second degree (marked)*, in which the cervix protrudes through the vagina to the exterior; and *third degree (complete)*, in which the entire uterus is outside the vagina. Depending on the degree of prolapse, treatment may involve pelvic exercises, dieting if a patient is overweight, a stool softener to minimize straining during defecation, pessary therapy (placement of a rubber device around the uterine cervix that helps prop up the uterus), or surgery. •

Histology of the Uterus

Histologically, the uterus consists of three layers of tissue: perimetrium, myometrium, and endometrium (Figure 28.18). The outer layer—the **perimetrium** (per′-i-MĒ-trē-um; *peri-* = around; *-metrium* = uterus) or serosa—is part of the visceral peritoneum; it is composed of simple squamous epithelium and areolar connective tissue. Laterally, it becomes the broad ligament. Anteriorly, it covers the urinary bladder and forms a shallow pouch, the **vesicouterine pouch** (ves′-i-kō-Ū-ter-in; *vesico-* = bladder; see Figure 28.11). Posteriorly, it covers the rectum and forms a deep pouch between the uterus and urinary bladder, the **rectouterine pouch** (rek-tō-Ū-ter-in; *recto-* = rectum) or **pouch of Douglas**—the most inferior point in the pelvic cavity.

The middle layer of the uterus, the **myometrium** (*myo-* = muscle), consists of three layers of smooth muscle fibers that are thickest in the fundus and thinnest in the cervix. The thicker middle layer is circular; the inner and outer layers are longitudinal or oblique. During labor and childbirth, coordinated contractions of the myometrium in response to oxytocin from the posterior pituitary help expel the fetus from the uterus.

The inner layer of the uterus, the **endometrium** (*endo-* = within), is highly vascularized and has three components: (1) An innermost layer composed of simple columnar epithelium (ciliated

Figure 28.18 Histology of the uterus.

The three layers of the uterus from superficial to deep are the perimetrium (serosa), the myometrium, and the endometrium.

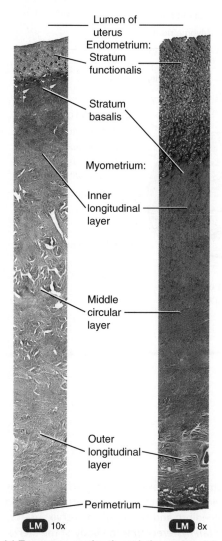

Lumen of uterus

Endometrium:
Stratum functionalis

Stratum basalis

Myometrium:

Inner longitudinal layer

Middle circular layer

Outer longitudinal layer

Perimetrium

LM 10x

LM 8x

(a) Transverse section through the uterine wall: second week of menstrual cycle (left) and third week of menstrual cycle (right)

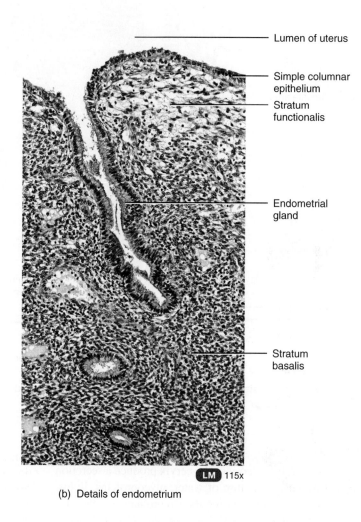

Lumen of uterus

Simple columnar epithelium

Stratum functionalis

Endometrial gland

Stratum basalis

LM 115x

(b) Details of endometrium

? **What structural features of the endometrium and myometrium contribute to their functions?**

and secretory cells) lines the lumen. (2) An underlying endometrial stroma is a very thick region of lamina propria (areolar connective tissue). (3) Endometrial (uterine) glands develop as invaginations of the luminal epithelium and extend almost to the myometrium. The endometrium is divided into two layers. The **stratum functionalis** *(functional layer)* lines the uterine cavity and sloughs off during menstruation. The deeper layer, the **stratum basalis** *(basal layer),* is permanent and gives rise to a new stratum functionalis after each menstruation.

Branches of the internal iliac artery called **uterine arteries** (Figure 28.19) supply blood to the uterus. Uterine arteries give off branches called **arcuate arteries** (AR-kū-āt = shaped like a bow)

that are arranged in a circular fashion in the myometrium. These arteries branch into **radial arteries** that penetrate deeply into the myometrium. Just before the branches enter the endometrium, they divide into two kinds of arterioles: **Straight arterioles** supply the stratum basalis with the materials needed to regenerate the stratum functionalis; **spiral arterioles** supply the stratum functionalis and change markedly during the menstrual cycle. Blood leaving the uterus is drained by the **uterine veins** into the internal iliac veins. The extensive blood supply of the uterus is essential to support regrowth of a new stratum functionalis after menstruation, implantation of a fertilized ovum, and development of the placenta.

Figure 28.19 Blood supply of the uterus. The inset shows histological details of the blood vessels of the endometrium.

6 Straight arterioles supply the materials needed for regeneration of the stratum functionalis.

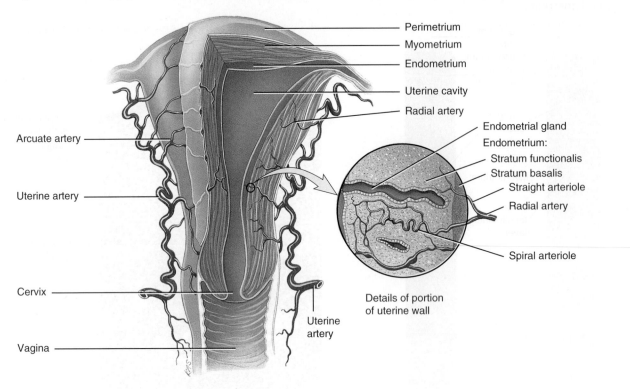

Anterior view with left side of uterus partially sectioned

? What is the functional significance of the stratum basalis of the endometrium?

Cervical Mucus

The secretory cells of the mucosa of the cervix produce a secretion called **cervical mucus,** a mixture of water, glycoproteins, lipids, enzymes, and inorganic salts. During their reproductive years, females secrete 20–60 mL of cervical mucus per day. Cervical mucus is more hospitable to sperm at or near the time of ovulation because it is then less viscous and more alkaline (pH 8.5). At other times, a more viscous mucus forms a cervical plug that physically impedes sperm penetration. Cervical mucus supplements the energy needs of sperm, and both the cervix and cervical mucus protect sperm from phagocytes and the hostile environment of the vagina and uterus. Cervical mucus may also play a role in **capacitation** (ka-pas'-i-TĀ-shun)—a series of functional changes that sperm undergo in the female reproductive tract before they are able to fertilize a secondary oocyte. Capacitation causes a sperm cell's tail to beat even more vigorously, and it prepares the sperm cell's plasma membrane to fuse with the oocyte's plasma membrane.

CLINICAL CONNECTION | *Hysterectomy*

Hysterectomy (hiss-ter-EK-tō-mē; *hyster-* = uterus), the surgical removal of the uterus, is the most common gynecological operation. It may be indicated in conditions such as fibroids, which are noncancerous tumors composed of muscular and fibrous tissue; endometriosis; pelvic inflammatory disease; recurrent ovarian cysts; excessive uterine bleeding; and cancer of the cervix, uterus, or ovaries. In a *partial (subtotal) hysterectomy,* the body of the uterus is removed but the cervix is left in place. A *complete hysterectomy* is the removal of both the body and cervix of the uterus. A *radical hysterectomy* includes removal of the body and cervix of the uterus, uterine tubes, possibly the ovaries, the superior portion of the vagina, pelvic lymph nodes, and supporting structures, such as ligaments. A hysterectomy can be performed either through an incision in the abdominal wall or through the vagina. •

✔CHECKPOINT

17. Where are the uterine tubes located, and what is their function?
18. What are the principal parts of the uterus? Where are they located in relation to one another?
19. Describe the arrangement of ligaments that hold the uterus in its normal position.
20. Describe the histology of the uterus.
21. Why is an abundant blood supply important to the uterus?

Vagina

The **vagina** (= sheath) is a tubular, 10-cm (4-in.) long fibromuscular canal lined with mucous membrane that extends from the exterior of the body to the uterine cervix (see Figures 28.11 and 28.16). It is the receptacle for the penis during sexual intercourse, the outlet for menstrual flow, and the passageway for childbirth. Situated between the urinary bladder and the rectum, the vagina is directed superiorly and posteriorly, where it attaches to the uterus. A recess called the **fornix** (= arch or vault) surrounds the vaginal attachment to the cervix. When properly inserted, a contraceptive diaphragm rests in the fornix, where it is held in place as it covers the cervix.

The **mucosa** of the vagina is continuous with that of the uterus. Histologically, it consists of nonkeratinized stratified squamous epithelium and areolar connective tissue that lies in a series of transverse folds called **rugae** (ROO-gē). Dendritic cells in the mucosa are antigen-presenting cells (described in Section 22.4). Unfortunately, they also participate in the transmission of viruses—for example, HIV (the virus that causes AIDS)—to a female during intercourse with an infected male. The mucosa of the vagina contains large stores of glycogen, the decomposition of which produces organic acids. The resulting acidic environment retards microbial growth, but it also is harmful to sperm. Alkaline components of semen, mainly from the seminal vesicles, raise the pH of fluid in the vagina and increase viability of the sperm.

The **muscularis** is composed of an outer circular layer and an inner longitudinal layer of smooth muscle that can stretch considerably to accommodate the penis during sexual intercourse and a child during birth.

The **adventitia,** the superficial layer of the vagina, consists of areolar connective tissue. It anchors the vagina to adjacent organs such as the urethra and urinary bladder anteriorly and the rectum and anal canal posteriorly.

A thin fold of vascularized mucous membrane, called the **hymen** (= membrane), forms a border around and partially closes the inferior end of the vaginal opening to the exterior, the **vaginal orifice** (see Figure 28.20). After its rupture, usually following the first sexual intercourse, only remnants of the hymen remain. Sometimes the hymen completely covers the orifice, a condition called **imperforate hymen** (im-PER-fō-rāt). Surgery may be needed to open the orifice and permit the discharge of menstrual flow.

Figure 28.20 Components of the vulva (pudendum).

The vulva refers to the external genitals of the female.

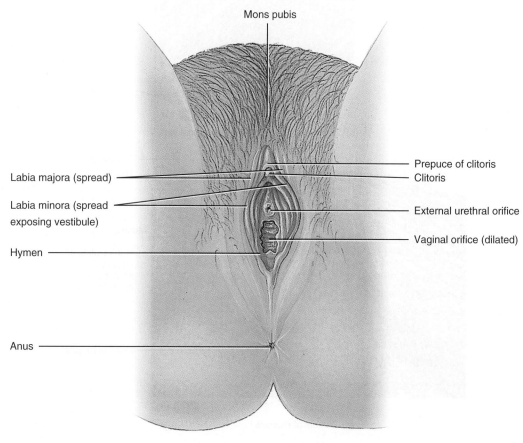

Mons pubis

Labia majora (spread)

Labia minora (spread exposing vestibule)

Hymen

Anus

Prepuce of clitoris

Clitoris

External urethral orifice

Vaginal orifice (dilated)

(a) Inferior view

FIGURE 28.20 CONTINUES

FIGURE 28.20 CONTINUED ▶

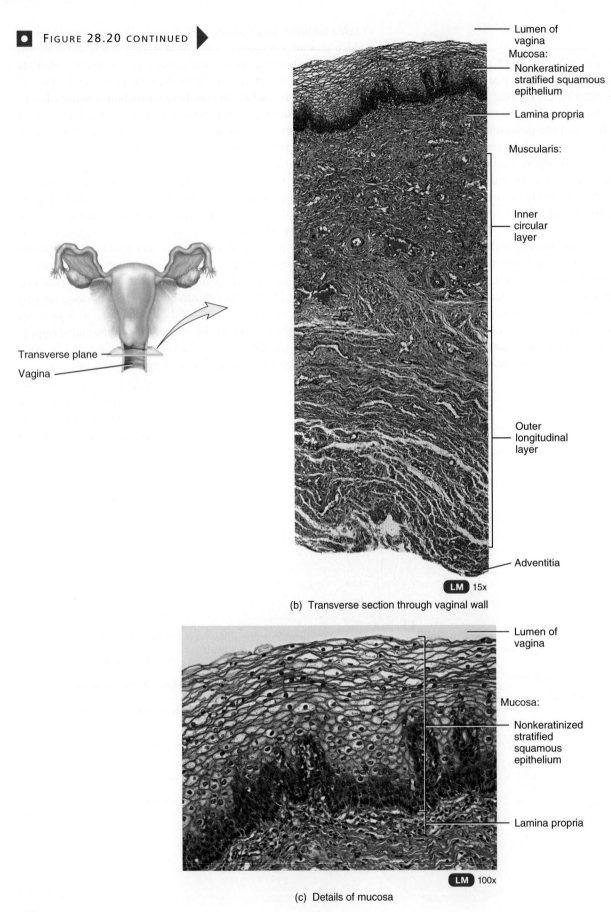

Lumen of vagina

Mucosa:
Nonkeratinized stratified squamous epithelium

Lamina propria

Muscularis:

Inner circular layer

Outer longitudinal layer

Adventitia

LM 15x

(b) Transverse section through vaginal wall

Transverse plane

Vagina

Lumen of vagina

Mucosa:
Nonkeratinized stratified squamous epithelium

Lamina propria

LM 100x

(c) Details of mucosa

? **What surface structures are anterior to the vaginal opening? Lateral to it?**

Vulva

The term **vulva** (VUL-va = to wrap around), or **pudendum** (pū-DEN-dum), refers to the external genitals of the female (Figure 28.20). The following components make up the vulva:

- Anterior to the vaginal and urethral openings is the **mons pubis** (MONZ PŪ-bis; *mons* = mountain), an elevation of adipose tissue covered by skin and coarse pubic hair that cushions the pubic symphysis.
- From the mons pubis, two longitudinal folds of skin, the **labia majora** (LĀ-bē-a ma-JŌ-ra; *labia* = lips; *majora* = larger), extend inferiorly and posteriorly. The singular term is *labium majus.* The labia majora are covered by pubic hair and contain an abundance of adipose tissue, sebaceous (oil) glands, and apocrine sudoriferous (sweat) glands. They are homologous to the scrotum.
- Medial to the labia majora are two smaller folds of skin called the **labia minora** (min-OR-a = smaller). The singular term is *labium minus.* Unlike the labia majora, the labia minora are devoid of pubic hair and fat and have few sudoriferous glands, but they do contain many sebaceous glands. The labia minora are homologous to the spongy (penile) urethra.
- The **clitoris** (KLI-to-ris) is a small cylindrical mass composed of two small erectile bodies, the *corpora cavernosa,* and nu-merous nerves and blood vessels. The clitoris is located at the anterior junction of the labia minora. A layer of skin called the **prepuce of the clitoris** (PRĒ-poos) is formed at the point where the labia minora unite and covers the body of the clitoris. The exposed portion of the clitoris is the **glans clitoris.** The clitoris is homologous to the glans penis in males. Like the male structure, the clitoris is capable of enlargement on tactile stimulation and has a role in sexual excitement in the female.

- The region between the labia minora is the **vestibule.** Within the vestibule are the hymen (if still present), the vaginal orifice, the external urethral orifice, and the openings of the ducts of several glands. The vestibule is homologous to the membranous urethra of males. The **vaginal orifice,** the opening of the vagina to the exterior, occupies the greater portion of the vestibule and is bordered by the hymen. Anterior to the vaginal orifice and posterior to the clitoris is the **external urethral orifice,** the opening of the urethra to the exterior. On either side of the external urethral orifice are the openings of the ducts of the **paraurethral glands** (par′-a-ū-RĒ-thral) or **Skene's glands** (SKĒNZ). These mucus-secreting glands are embedded in the wall of the urethra. The paraurethral glands are homologous to the prostate. On either side of the vaginal orifice itself are the **greater vestibular (Bartholin's) glands** (BAR-to-linz) (see Figure 28.21), which open by ducts into a groove between the hymen and labia minora. They produce a small quantity of

Figure 28.21 Perineum of a female. (Figure 11.13 shows the perineum of a male.)

 The perineum is a diamond-shaped area that includes the urogenital triangle and the anal triangle.

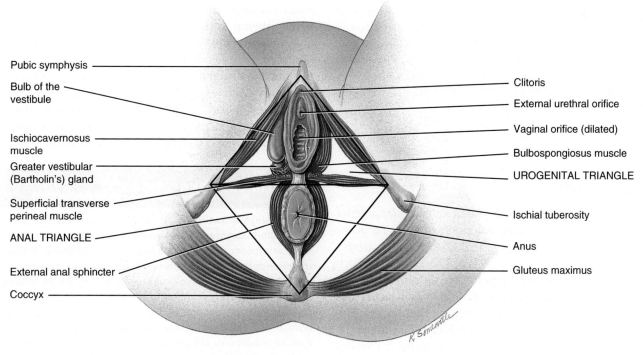

Inferior view

? **Why is the anterior portion of the perineum called the urogenital triangle?**

mucus during sexual arousal and intercourse that adds to cervical mucus and provides lubrication. The greater vestibular glands are homologous to the bulbourethral glands in males. Several **lesser vestibular glands** also open into the vestibule.

- The **bulb of the vestibule** (see Figure 28.21) consists of two elongated masses of erectile tissue just deep to the labia on either side of the vaginal orifice. The bulb of the vestibule becomes engorged with blood during sexual arousal, narrowing the vaginal orifice and placing pressure on the penis during intercourse. The bulb of the vestibule is homologous to the corpus spongiosum and bulb of the penis in males.

Table 28.2 summarizes the homologous structures of the female and male reproductive systems.

Perineum

The **perineum** (per′-i-NĒ-um) is the diamond-shaped area medial to the thighs and buttocks of both males and females (Figure 28.21). It contains the external genitals and anus. The perineum is bounded anteriorly by the pubic symphysis, laterally by the ischial tuberosities, and posteriorly by the coccyx. A transverse line drawn between the ischial tuberosities divides the perineum into an anterior **urogenital triangle** (ū′-rō-JEN-i-tal) that contains the external genitals and a posterior **anal triangle** that contains the anus.

CLINICAL CONNECTION | *Episiotomy*

In specific maternal-fetal situations that require a quick delivery, an **episiotomy** (e-piz-ē-OT-ō-mē; *episi-* = vulva or pubic region; *-otomy* = incision) may be performed. In the procedure a perineal cut made with surgical scissors. The cut may be made along the midline or at an angle of approximately 45 degrees to the midline. In effect, a straight, more easily sutured cut is substituted for the jagged tear that would otherwise be caused by passage of the fetus. The incision is closed in layers with sutures that are absorbed within a few weeks, so that the busy new mom does not have to worry about making time to have them removed. •

Mammary Glands

Each **breast** is a hemispheric projection of variable size anterior to the pectoralis major and serratus anterior muscles and attached to them by a layer of fascia composed of dense irregular connective tissue.

Each breast has one pigmented projection, the **nipple,** that has a series of closely spaced openings of ducts called **lactiferous ducts** (lak-TIF-e-rus), where milk emerges. The circular pigmented area of skin surrounding the nipple is called the **areola** (a-RĒ-ō-la = small space); it appears rough because it contains modified sebaceous (oil) glands. Strands of connective tissue called the **suspensory ligaments of the breast (Cooper's ligaments)** run between the skin and fascia and support the breast.

TABLE 28.2

Summary of Homologous Structures of the Female and Male Reproductive Systems

FEMALE STRUCTURES	MALE STRUCTURES
Ovaries	Testes
Ovum	Sperm cell
Labia majora	Scrotum
Labia minora	Spongy (penile) urethra
Vestibule	Membranous urethra
Bulb of vestibule	Corpus spongiosum penis and bulb of penis
Clitoris	Glans penis and corpora cavernosa
Paraurethral glands	Prostate
Greater vestibular glands	Bulbourethral (Cowper's) glands

These ligaments become looser with age or with the excessive strain that can occur in long-term jogging or high-impact aerobics. Wearing a supportive bra can slow this process and help maintain the strength of the suspensory ligaments.

Within each breast is a **mammary gland,** a modified sudoriferous (sweat) gland that produces milk (Figure 28.22). A mammary gland consists of 15 to 20 lobes, or compartments, separated by a variable amount of adipose tissue. In each lobe are several smaller compartments called **lobules,** composed of grapelike clusters of milk-secreting glands termed **alveoli** (al-VĒ-o-lī = small cavities) embedded in connective tissue. Contraction of **myoepithelial cells** (mī′-ō-ep′-i-THĒ-lē-al) surrounding the alveoli helps propel milk toward the nipples. When milk is being produced, it passes from the alveoli into a series of **secondary tubules** and then into the **mammary ducts.** Near the nipple, the mammary ducts expand to form sinuses called **lactiferous sinuses** (*lact-* = milk), where some milk may be stored before draining into a lactiferous duct. Each lactiferous duct typically carries milk from one of the lobes to the exterior.

CLINICAL CONNECTION | *Breast Augmentation and Reduction*

Breast augmentation (awg-men-TĀ-shun = enlargement), technically called **augmentation mammaplasty** (mam-a-PLAS-tē), is a surgical procedure to increase breast size and shape. It may be done to enhance breast size for females who feel that their breasts are too small, to restore breast volume due to weight loss or following pregnancy, to improve the shape of breasts that are sagging, and to improve breast appearance following surgery, trauma, or congenital abnormalities. The most commonly used implants are filled with either a saline solution or silicone gel. The incision for the implant is made under the breast, around the areola, in the armpit, or in the navel. Then a pocket is made to place the implant either directly behind the breast tissue or beneath the pectoralis major muscle.

Figure 28.22 **Mammary glands within the breasts.**

The mammary glands function in the synthesis, secretion, and ejection of milk (lactation).

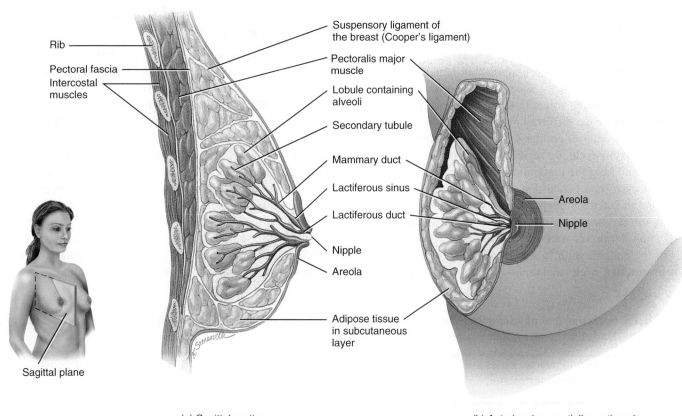

Rib

Pectoral fascia

Intercostal muscles

Suspensory ligament of the breast (Cooper's ligament)

Pectoralis major muscle

Lobule containing alveoli

Secondary tubule

Mammary duct

Lactiferous sinus

Lactiferous duct

Nipple

Areola

Adipose tissue in subcutaneous layer

Areola

Nipple

Sagittal plane

(a) Sagittal section

(b) Anterior view, partially sectioned

? **What hormones regulate the synthesis and ejection of milk?**

Breast reduction or **reduction mammaplasty** is a surgical procedure that involves decreasing breast size by removing fat, skin, and glandular tissue. This procedure is done because of chronic back, neck, and shoulder pain; poor posture; circulation or breathing problems; a skin rash under the breasts; restricted levels of activity; self-esteem problems; deep grooves in the shoulders from bra strap pressure; and difficulty wearing or fitting into certain bras and clothing. The most common procedure involves an incision around the areola, down the breast toward the crease between the breast and abdomen, and then along the crease. The surgeon removes excess tissue through the incision. In most cases, the nipple and areola remain attached to the breast. However, if the breasts are extremely large, the nipple and areola may have to be reattached at a higher position. •

CLINICAL CONNECTION | *Fibrocystic Disease of the Breasts*

The breasts of females are highly susceptible to cysts and tumors. In **fibrocystic disease** (fī-brō-SIS-tik), the most common cause of breast lumps in females, one or more cysts (fluid-filled sacs) and thickenings of alveoli develop. The condition, which occurs mainly in females between the ages of 30 and 50, is probably due to a relative excess of estrogens or a deficiency of progesterone in the postovulatory (luteal) phase of the reproductive cycle (discussed shortly). Fibrocystic disease usually causes one or both breasts to become lumpy, swollen, and tender a week or so before menstruation begins. •

The functions of the mammary glands are the synthesis, secretion, and ejection of milk; these functions, called **lactation** (lak-TĀ-shun), are associated with pregnancy and childbirth. Milk production is stimulated largely by the hormone prolactin from the anterior pituitary, with contributions from progesterone and estrogens. The ejection of milk is stimulated by oxytocin, which is released from the posterior pituitary in response to the sucking of an infant on the mother's nipple (suckling).

✔CHECKPOINT

22. How does the histology of the vagina contribute to its function?
23. What are the structures and functions of each part of the vulva?
24. Describe the components of the mammary glands and the structures that support them.
25. Outline the route milk takes from the alveoli of the mammary gland to the nipple.

28.3 THE FEMALE REPRODUCTIVE CYCLE

◉ **OBJECTIVE**

• Compare the major events of the ovarian and uterine cycles.

During their reproductive years, nonpregnant females normally exhibit cyclical changes in the ovaries and uterus. Each cycle takes about a month and involves both oogenesis and preparation of the uterus to receive a fertilized ovum. Hormones secreted by the hypothalamus, anterior pituitary, and ovaries control the main events. The **ovarian cycle** is a series of events in the ovaries that occur during and after the maturation of an oocyte. The **uterine (menstrual) cycle** is a concurrent series of changes in the endometrium of the uterus to prepare it for the arrival of a fertilized ovum that will develop there until birth. If fertilization does not occur, ovarian hormones wane, which causes the stratum functionalis of the endometrium to slough off. The general term **female reproductive cycle** encompasses the ovarian and uterine

cycles, the hormonal changes that regulate them, and the related cyclical changes in the breasts and cervix.

Hormonal Regulation of the Female Reproductive Cycle

Gonadotropin-releasing hormone (GnRH) secreted by the hypothalamus controls the ovarian and uterine cycles (Figure 28.23). GnRH stimulates the release of **follicle-stimulating hormone (FSH)** and **luteinizing hormone (LH)** from the anterior pituitary. FSH initiates follicular growth, while LH stimulates further development of the ovarian follicles. In addition, both FSH and LH stimulate the ovarian follicles to secrete estrogens. LH stimulates the theca cells of a developing follicle to produce androgens. Under the influence of FSH, the androgens are taken up by the granulosa cells of the follicle and then converted into estrogens. At midcycle, LH triggers ovulation and then promotes formation of the corpus luteum, the reason for the name luteinizing hormone. Stimulated by LH, the corpus luteum produces and secretes estrogens, progesterone, relaxin, and inhibin.

Figure 28.23 Secretion and physiological effects of estrogens, progesterone, relaxin, and inhibin in the female reproductive cycle. Dashed red lines indicate negative feedback inhibition.

🔑 The uterine and ovarian cycles are controlled by gonadotropin-releasing hormone (GnRH) and ovarian hormones (estrogens and progesterone).

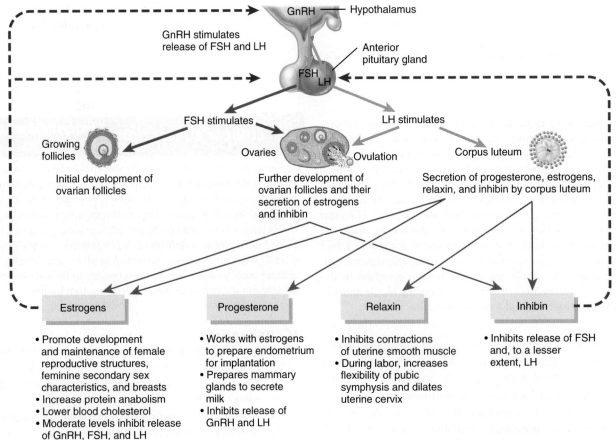

❓ **Of the several estrogens, which one exerts the major effect?**

At least six different estrogens have been isolated from the plasma of human females, but only three are present in significant quantities: *beta (β)-estradiol* (es-tra-DĪ-ol), *estrone,* and *estriol* (ES-trē-ol). In a nonpregnant woman, the most abundant estrogen is β-estradiol, which is synthesized from cholesterol in the ovaries.

Estrogens secreted by ovarian follicles have several important functions (Figure 28.23):

- Estrogens promote the development and maintenance of female reproductive structures, secondary sex characteristics, and the breasts. The secondary sex characteristics include distribution of adipose tissue in the breasts, abdomen, mons pubis, and hips; voice pitch; a broad pelvis; and pattern of hair growth on the head and body.

- Estrogens increase protein anabolism, including the building of strong bones. In this regard, estrogens are synergistic with human growth hormone (hGH).

- Estrogens lower blood cholesterol level, which is probably the reason that women under age 50 have a much lower risk of coronary artery disease than do men of comparable age.

- Moderate levels of estrogens in the blood inhibit both the release of GnRH by the hypothalamus and secretion of LH and FSH by the anterior pituitary.

Progesterone, secreted mainly by cells of the corpus luteum, cooperates with estrogens to prepare and maintain the endometrium for implantation of a fertilized ovum and to prepare the mammary glands for milk secretion. High levels of progesterone also inhibit secretion of GnRH and LH.

The small quantity of **relaxin** produced by the corpus luteum during each monthly cycle relaxes the uterus by inhibiting contractions of the myometrium. Presumably, implantation of a fertilized ovum occurs more readily in a "quiet" uterus. During pregnancy, the placenta produces much more relaxin, and it continues to relax uterine smooth muscle. At the end of pregnancy, relaxin also increases the flexibility of the pubic symphysis and may help dilate the uterine cervix, both of which ease delivery of the baby.

Inhibin is secreted by granulosa cells of growing follicles and by the corpus luteum after ovulation. It inhibits secretion of FSH and, to a lesser extent, LH.

Phases of the Female Reproductive Cycle

The duration of the female reproductive cycle typically ranges from 24 to 36 days. For this discussion, we assume a duration of 28 days and divide it into four phases: the menstrual phase, the preovulatory phase, ovulation, and the postovulatory phase (Figure 28.24).

Menstrual Phase

The **menstrual phase** (MEN-stroo-al), also called **menstruation** (men′-stroo-Ā-shun) or **menses** (MEN-sēz = month), lasts for roughly the first 5 days of the cycle. (By convention, the first day of menstruation is day 1 of a new cycle.)

EVENTS IN THE OVARIES Under the influence of FSH, several primordial follicles develop into primary follicles and then into secondary follicles. This developmental process may take several months to occur. Therefore, a follicle that begins to develop at the beginning of a particular menstrual cycle may not reach maturity and ovulate until several menstrual cycles later.

EVENTS IN THE UTERUS Menstrual flow from the uterus consists of 50–150 mL of blood, tissue fluid, mucus, and epithelial cells shed from the endometrium. This discharge occurs because the declining levels of progesterone and estrogens stimulate release of prostaglandins that cause the uterine spiral arterioles to constrict. As a result, the cells they supply become oxygen-deprived and start to die. Eventually, the entire stratum functionalis sloughs off. At this time the endometrium is very thin, about 2–5 mm, because only the stratum basalis remains. The menstrual flow passes from the uterine cavity through the cervix and vagina to the exterior.

Preovulatory Phase

The **preovulatory phase** (prē-OV-ū-la-tō-rē) is the time between the end of menstruation and ovulation. The preovulatory phase of the cycle is more variable in length than the other phases and accounts for most of the differences in length of the cycle. It lasts from days 6 to 13 in a 28-day cycle.

EVENTS IN THE OVARIES Some of the secondary follicles in the ovaries begin to secrete estrogens and inhibin. By about day 6, a single secondary follicle in one of the two ovaries has outgrown all the others to become the **dominant follicle.** Estrogens and inhibin secreted by the dominant follicle decrease the secretion of FSH, which causes other, less well-developed follicles to stop growing and undergo atresia. Fraternal (nonidentical) twins or triplets result when two or three secondary follicles become codominant and later are ovulated and fertilized at about the same time.

Normally, the one dominant secondary follicle becomes the **mature (graafian) follicle,** which continues to enlarge until it is more than 20 mm in diameter and ready for ovulation (see Figure 28.13). This follicle forms a blisterlike bulge due to the swelling antrum on the surface of the ovary. During the final maturation process, the mature follicle continues to increase its production of estrogens (Figure 28.24).

With reference to the ovarian cycle, the menstrual and preovulatory phases together are termed the **follicular phase** (fo-LIK-ū-lar) because ovarian follicles are growing and developing.

EVENTS IN THE UTERUS Estrogens liberated into the blood by growing ovarian follicles stimulate the repair of the endometrium; cells of the stratum basalis undergo mitosis and produce a new stratum functionalis. As the endometrium thickens, the short, straight endometrial glands develop, and the arterioles coil and lengthen as they penetrate the stratum functionalis. The thickness of the endometrium approximately doubles, to about 4–10 mm. With reference to the uterine cycle, the preovulatory phase is also termed the **proliferative phase** (prō-LIF-er-a-tiv) because the endometrium is proliferating.

Figure 28.24 **The female reproductive cycle.** The length of the female reproductive cycle typically is 24 to 36 days; the preovulatory phase is more variable in length than the other phases. (a) Events in the ovarian and uterine cycles and the release of anterior pituitary hormones are correlated with the sequence of the cycle's four phases. In the cycle shown, fertilization and implantation have not occurred. (b) Relative concentrations of anterior pituitary hormones (FSH and LH) and ovarian hormones (estrogens and progesterone) during the phases of a normal female reproductive cycle.

Estrogens are the primary ovarian hormones before ovulation; after ovulation, both progesterone and estrogens are secreted by the corpus luteum.

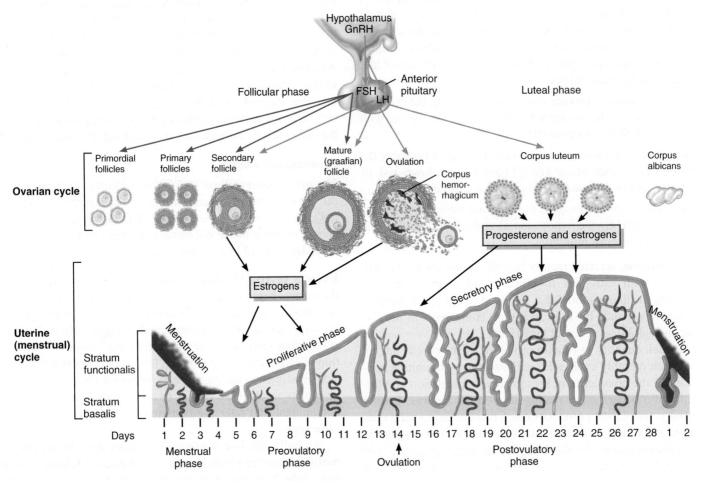

(a) Hormonal regulation of changes in the ovary and uterus

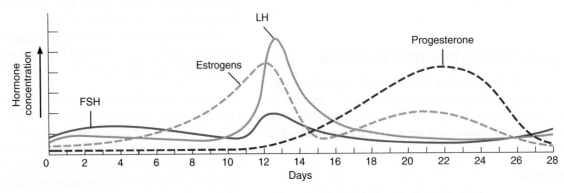

(b) Changes in concentration of anterior pituitary and ovarian hormones

Which hormones are responsible for the proliferative phase of endometrial growth, for ovulation, for growth of the corpus luteum, and for the surge of LH at midcycle?

Ovulation

Ovulation (ov′-ū-LĀ-shun), the rupture of the mature (graafian) follicle and the release of the secondary oocyte into the pelvic cavity, usually occurs on day 14 in a 28-day cycle. During ovulation, the secondary oocyte remains surrounded by its zona pellucida and corona radiata.

The *high levels of estrogens* during the last part of the preovulatory phase exert a *positive* feedback effect on the cells that secrete LH and gonadotropin-releasing hormone (GnRH) and cause ovulation, as follows (Figure 28.25):

1 A high concentration of estrogens stimulates more frequent release of GnRH from the hypothalamus. It also directly stimulates gonadotrophs in the anterior pituitary to secrete LH.

2 GnRH promotes the release of FSH and additional LH by the anterior pituitary.

3 LH causes rupture of the mature (graafian) follicle and expulsion of a secondary oocyte about 9 hours after the peak of the LH surge. The ovulated oocyte and its corona radiata cells are usually swept into the uterine tube.

From time to time, an oocyte is lost into the pelvic cavity, where it later disintegrates. The small amount of blood that sometimes leaks into the pelvic cavity from the ruptured follicle can cause pain, known as **mittelschmerz** (MIT-el-shmārts = pain in the middle), at the time of ovulation.

An over-the-counter home test that detects a rising level of LH can be used to predict ovulation a day in advance.

Postovulatory Phase

The **postovulatory phase** of the female reproductive cycle is the time between ovulation and onset of the next menses. In duration, it is the most constant part of the female reproductive cycle. It lasts for 14 days in a 28-day cycle, from day 15 to day 28 (see Figure 28.24).

EVENTS IN ONE OVARY After ovulation, the mature follicle collapses, and the basement membrane between the granulosa cells and theca interna breaks down. Once a blood clot forms from minor bleeding of the ruptured follicle, the follicle becomes the **corpus hemorrhagicum** (hem′-o-RAJ-i-kum; *hemo-* = blood; *rrhagic-* = bursting forth) (see Figure 28.13). Theca interna cells mix with the granulosa cells as they all become transformed into **corpus luteum cells** under the influence of LH. Stimulated by LH, the corpus luteum secretes progesterone, estrogen, relaxin, and inhibin. The luteal cells also absorb the blood clot. With reference to the ovarian cycle, this phase is also called the **luteal phase** (LOO-tē-al).

Later events in an ovary that has ovulated an oocyte depend on whether the oocyte is fertilized. If the oocyte *is not fertilized,* the corpus luteum has a lifespan of only 2 weeks. Then, its secretory activity declines, and it degenerates into a corpus albicans (see Figure 28.13). As the levels of progesterone, estrogens, and inhibin decrease, release of GnRH, FSH, and LH rises due to loss of negative feedback suppression by the ovarian hormones. Follicular growth resumes and a new ovarian cycle begins.

If the secondary oocyte *is fertilized* and begins to divide, the corpus luteum persists past its normal 2-week lifespan. It is "res-

Figure 28.25 High levels of estrogens exert a positive feedback effect (green arrows) on the hypothalamus and anterior pituitary, thereby increasing secretion of GnRH and LH.

At midcycle, a surge of LH triggers ovulation.

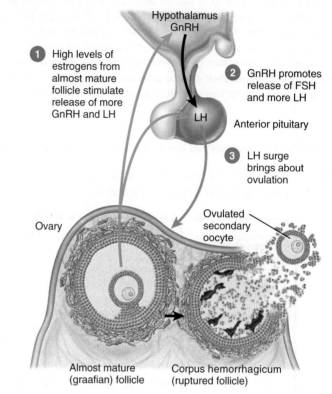

1 High levels of estrogens from almost mature follicle stimulate release of more GnRH and LH

Hypothalamus GnRH

2 GnRH promotes release of FSH and more LH

LH

Anterior pituitary

3 LH surge brings about ovulation

Ovary

Ovulated secondary oocyte

Almost mature (graafian) follicle

Corpus hemorrhagicum (ruptured follicle)

What is the effect of rising but still moderate levels of estrogens on the secretion of GnRH, LH, and FSH?

cued" from degeneration by **human chorionic gonadotropin (hCG)** (kō-rē-ON-ik). This hormone is produced by the chorion of the embryo beginning about 8 days after fertilization. Like LH, hCG stimulates the secretory activity of the corpus luteum. The presence of hCG in maternal blood or urine is an indicator of pregnancy and is the hormone detected by home pregnancy tests.

EVENTS IN THE UTERUS Progesterone and estrogens produced by the corpus luteum promote growth and coiling of the endometrial glands, vascularization of the superficial endometrium, and thickening of the endometrium to 12–18 mm (0.48–0.72 in.). Because of the secretory activity of the endometrial glands, which begin to secrete glycogen, this period is called the **secretory phase** of the uterine cycle. These preparatory changes peak about 1 week after ovulation, at the time a fertilized ovum might arrive in the uterus. If fertilization does not occur, the levels of progesterone and estrogens decline due to degeneration of the corpus luteum. Withdrawal of progesterone and estrogens causes menstruation.

Figure 28.26 summarizes the hormonal interactions and cyclical changes in the ovaries and uterus during the ovarian and uterine cycles.

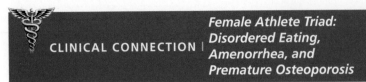

CLINICAL CONNECTION | *Female Athlete Triad: Disordered Eating, Amenorrhea, and Premature Osteoporosis*

The female reproductive cycle can be disrupted by many factors, including weight loss, low body weight, disordered eating, and vigorous physical activity. The observation that three conditions—disordered eating, amenorrhea, and osteoporosis—often occur together in female athletes led researchers to coin the term **female athlete triad.**

Many athletes experience intense pressure from coaches, parents, peers, and themselves to lose weight to improve performance. Hence,

they may develop disordered eating behaviors and engage in other harmful weight-loss practices in a struggle to maintain a very low body weight. **Amenorrhea** (a-men-ō-RĒ-a; *a-* = without; *-men-* = month; *-rrhea* = a flow) is the absence of menstruation. The most common causes of amenorrhea are pregnancy and menopause. In female athletes, amenorrhea results from reduced secretion of gonadotropin-releasing hormone, which decreases the release of LH and FSH. As a result, ovarian follicles fail to develop, ovulation does not occur, synthesis of estrogens and progesterone wanes, and monthly menstrual bleeding ceases. Most cases of the female athlete triad occur in young women with very low amounts of body fat. Low levels of the hormone leptin, secreted by adipose cells, may be a contributing factor.

Figure 28.26 Summary of hormonal interactions in the ovarian and uterine cycles.

Hormones from the anterior pituitary regulate ovarian function, and hormones from the ovaries regulate the changes in the endometrial lining of the uterus.

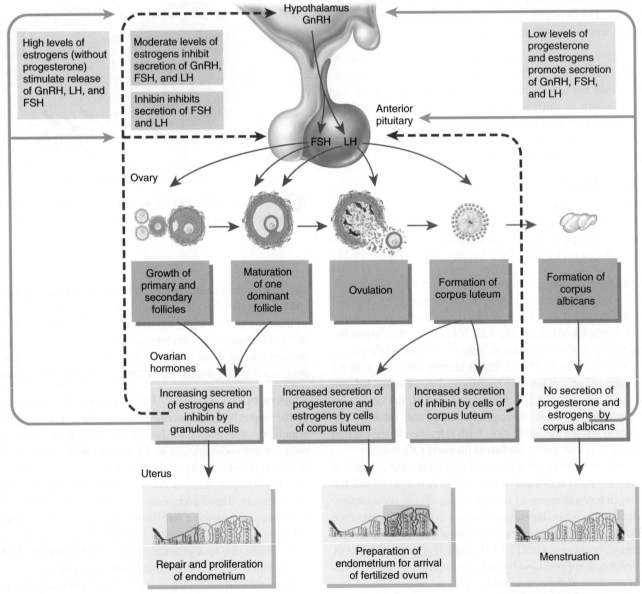

When declining levels of estrogens and progesterone stimulate secretion of GnRH, is this a positive or a negative feedback effect? Why?

Because estrogens help bones retain calcium and other minerals, chronically low levels of estrogens are associated with loss of bone mineral density. The female athlete triad causes "old bones in young women." In one study, amenorrheic runners in their twenties had low bone mineral densities, similar to those of postmenopausal women 50 to 70 years old! Short periods of amenorrhea in young athletes may cause no lasting harm. However, long-term cessation of the reproductive cycle may be accompanied by a loss of bone mass, and adolescent athletes may fail to achieve an adequate bone mass; both of these situations can lead to premature osteoporosis and irreversible bone damage. •

✔CHECKPOINT

26. Describe the function of each of the following hormones in the uterine and ovarian cycles: GnRH, FSH, LH, estrogens, progesterone, and inhibin.
27. Briefly outline the major events of each phase of the uterine cycle, and correlate them with the events of the ovarian cycle.
28. Prepare a labeled diagram of the major hormonal changes that occur during the uterine and ovarian cycles.

28.4 BIRTH CONTROL METHODS AND ABORTION

◉ OBJECTIVE

• Explain the differences among the various types of birth control methods and compare their effectiveness.

Birth control refers to restricting the number of children by various methods designed to control fertility and prevent conception. No single, ideal method of birth control exists. The only method of preventing pregnancy that is 100% reliable is **complete abstinence,** the avoidance of sexual intercourse. Several other methods are available; each has its advantages and disadvantages. These include surgical sterilization, hormonal methods, intrauterine devices, spermicides, barrier methods, and periodic abstinence. Table 28.3 provides the failure rates for various methods of birth control. Although it is not a form of birth control, in this section we will also discuss abortion, the premature expulsion of the products of conception from the uterus.

Birth Control Methods

Surgical Sterilization

Sterilization is a procedure that renders an individual incapable of further reproduction. The principal method for sterilization of males is a **vasectomy** (va-SEK-tō-mē; -ectomy = cut out), in which a portion of each ductus deferens is removed. In order to gain access to the ductus deferens, an incision is made with a scalpel (conventional procedure) or a puncture is made with special forceps (non-scalpel vasectomy). Next the ducts are located and cut, each is tied (ligated) in two places with stitches, and the portion between the ties is removed. Although sperm production continues in the testes, sperm can no longer reach the exterior. The

TABLE 28.3

Failure Rates for Several Birth Control Methods

METHOD	FAILURE RATES* (%)	
	PERFECT USE[†]	TYPICAL USE
Complete abstinence	0	0
Surgical sterilization		
Vasectomy	0.10	0.15
Tubal ligation	0.5	0.5
Non-incisional sterilization (Essure®)	0.2	0.2
Hormonal methods		
Oral contraceptives		
Combined pill	0.3	1–2
Seasonale®	0.3	1–2
Minipill	0.5	2
Non-oral contraceptives		
Contraceptive skin patch	0.1	1–2
Vaginal contraceptive ring	0.1	1–2
Emergency contraception	25	25
Hormone injections	0.3	1–2
Intrauterine devices (Copper T 380A®)	0.6	0.8
Spermicides (alone)	15	29
Barrier methods		
Male condom	2	15
Vaginal pouch	5	21
Diaphragm (with spermicide)	6	16
Cervical cap (with spermicide)	9	16
Periodic abstinence		
Rhythm	9	25
Sympto-thermal	2	20
No method	85	85

*Defined as percentage of women having an unintended pregnancy during the first year of use.
[†]Failure rate when the method is used correctly and consistently.

sperm degenerate and are destroyed by phagocytosis. Because the blood vessels are not cut, testosterone levels in the blood remain normal, so vasectomy has no effect on sexual desire or performance. If done correctly, it is close to 100% effective. The procedure can be reversed, but the chance of regaining fertility is only 30–40%. Sterilization in females most often is achieved by performing a **tubal ligation** (lī-GĀ-shun), in which both uterine tubes are tied closed and then cut. This can be achieved in a few different ways. "Clips" or "clamps" can be placed on the uterine tubes, the tubes can be tied and/or cut, and sometimes they are cauterized. In any case, the result is that the secondary oocyte cannot pass through the uterine tubes, and sperm cannot reach the oocyte.

Non-incisional Sterilization

Essure® is a non-incisional irreversible procedure that is an alternative to tubal ligation. In the Essure procedure, a soft micro-insert coil made of polyester fibers and metals (nickel–titanium and stainless steel) is inserted with a catheter into the vagina, through the uterus, and into each uterine tube. Over a three-month period, the insert stimulates tissue growth (scar tissue) in and around itself, blocking the uterine tubes. As with tubal ligation, the secondary oocyte cannot pass through the uterine tubes, and sperm cannot reach the oocyte. Unlike tubal ligation, Essure® does not require general anesthesia.

Hormonal Methods

Aside from complete abstinence or surgical sterilization, hormonal methods are the most effective means of birth control. **Oral contraceptives** (the pill) contain hormones designed to prevent pregnancy. Some, called *combined oral contraceptives (COCs),* contain both progestin (hormone with actions similar to progesterone) and estrogens. The primary action of COCs is to inhibit ovulation by suppressing the gonadotropins FSH and LH. The low levels of FSH and LH usually prevent the development of a dominant follicle in the ovary. As a result, levels of estrogens do not rise, the midcycle LH surge does not occur, and ovulation does not take place. Even if ovulation does occur, as it does in some cases, COCs may also block implantation in the uterus and inhibit the transport of ova and sperm in the uterine tubes.

Progestins thicken cervical mucus and make it more difficult for sperm to enter the uterus. *Progestin-only pills* thicken cervical mucus and may block implantation in the uterus, but they do not consistently inhibit ovulation.

Among the noncontraceptive benefits of oral contraceptives are regulation of the length of menstrual cycle and decreased menstrual flow (and therefore decreased risk of anemia). The pill also provides protection against endometrial and ovarian cancers and reduces the risk of endometriosis. However, oral contraceptives may not be advised for women with a history of blood clotting disorders, cerebral blood vessel damage, migraine headaches, hypertension, liver malfunction, or heart disease. Women who take the pill and smoke face far higher odds of having a heart attack or stroke than do nonsmoking pill users. Smokers should quit smoking or use an alternative method of birth control.

Following are several variations of *oral* hormonal methods of contraception:

- **Combined pill.** Contains both progestin and estrogens and is typically taken once a day for 3 weeks to prevent pregnancy and regulate the menstrual cycle. The pills taken during the fourth week are inactive (do not contain hormones) and permit menstruation to occur.
- **Seasonale®.** Contains both progestin and estrogens and is taken once a day in 3-month cycles of 12 weeks of hormone-containing pills followed by 1 week of inactive pills. Menstruation occurs during the thirteenth week.
- **Minipill.** Contains progestin only and is taken every day of the month.

Non-oral hormonal methods of contraception are also available. Among these are the following:

- **Contraceptive skin patch (Ortho Evra®).** Contains both progestin and estrogens delivered in a skin patch placed on the skin (upper outer arm, back, lower abdomen, or buttocks) once a week for 3 weeks. After 1 week, the patch is removed from one location and then a new one is placed elsewhere. During the fourth week no patch is used.
- **Vaginal contraceptive ring (NuvaRing®).** A flexible dough-nut-shaped ring about 5 cm (2 in.) in diameter that contains estrogens and progesterone and is inserted by the female herself into the vagina. It is left in the vagina for 3 weeks to prevent conception and then removed for one week to permit menstruation.
- **Emergency contraception (EC) (morning after pill).** Consists of progestin and estrogens or progestin alone to prevent pregnancy following unprotected sexual intercourse. The relatively high levels of progestin and estrogens in EC pills provide inhibition of FSH and LH secretion. Loss of the stimulating effects of these gonadotropic hormones causes the ovaries to cease secretion of their own estrogens and progesterone. In turn, declining levels of estrogens and progesterone induce shedding of the uterine lining, thereby blocking implantation. One pill is taken as soon as possible but within 72 hours of unprotected sexual intercourse. The second pill must be taken 12 hours after the first. The pills work in the same way as regular birth control pills.
- **Hormone injections (Depo-provera®).** An injectable progestin given intramuscularly by a health-care practitioner once every 3 months.

Intrauterine Devices

An **intrauterine device (IUD)** is a small object made of plastic, copper, or stainless steel that is inserted by a health-care professional into the cavity of the uterus. IUDs prevent fertilization from taking place by blocking sperm from entering the uterine tubes. The IUD most commonly used in the United States today is the Copper T 380A®, which is approved for up to 10 years of use and has long-term effectiveness comparable to that of tubal ligation. Some women cannot use IUDs because of expulsion, bleeding, or discomfort.

Spermicides

Various foams, creams, jellies, suppositories, and douches that contain sperm-killing agents, or **spermicides** (SPER-mi-sīds), make the vagina and cervix unfavorable for sperm survival and are available without prescription. They are placed in the vagina before sexual intercourse. The most widely used spermicide is *nonoxynol-9,* which kills sperm by disrupting their plasma membranes. A spermicide is more effective when used with a barrier method such as a male condom, vaginal pouch, diaphragm, or cervical cap.

Barrier Methods

Barrier methods use a physical barrier and are designed to prevent sperm from gaining access to the uterine cavity and uterine tubes. In addition to preventing pregnancy, certain barrier methods (male condom and vaginal pouch) may also provide some protection against sexually transmitted diseases (STDs) such as

AIDS. In contrast, oral contraceptives and IUDs confer no such protection. Among the barrier methods are the male condom, vaginal pouch, diaphragm, and cervical cap.

A **male condom** is a nonporous, latex covering placed over the penis that prevents deposition of sperm in the female reproductive tract. A **vaginal pouch,** sometimes called a **female condom,** is designed to prevent sperm from entering the uterus. It is made of two flexible rings connected by a polyurethane sheath. One ring lies inside the sheath and is inserted to fit over the cervix; the other ring remains outside the vagina and covers the female external genitals. A **diaphragm** is a rubber, dome-shaped structure that fits over the cervix and is used in conjunction with a spermicide. It can be inserted by the female up to 6 hours before intercourse. The diaphragm stops most sperm from passing into the cervix and the spermicide kills most sperm that do get by. Although diaphragm use does decrease the risk of some STDs, it does not fully protect against HIV infection because the vagina is still exposed. A **cervical cap** resembles a diaphragm but is smaller and more rigid. It fits snugly over the cervix and must be fitted by a health-care professional. Spermicides should be used with the cervical cap.

Periodic Abstinence

A couple can use their knowledge of the physiological changes that occur during the female reproductive cycle to decide either to abstain from intercourse on those days when pregnancy is a likely result, or to plan intercourse on those days if they wish to conceive a child. In females with normal and regular menstrual cycles, these physiological events help to predict the day on which ovulation is likely to occur.

The first physiologically based method, developed in the 1930s, is known as the **rhythm method.** It involves abstaining from sexual activity on the days that ovulation is likely to occur in each reproductive cycle. During this time (3 days before ovulation, the day of ovulation, and 3 days after ovulation) the couple abstains from intercourse. The effectiveness of the rhythm method for birth control is poor in many women due to the irregularity of the female reproductive cycle.

Another system is the **sympto-thermal method,** in which couples are instructed to know and understand certain signs of fertility. The signs of ovulation include increased basal body temperature; the production of abundant clear, stretchy cervical mucus; and pain associated with ovulation (*mittelschmerz*). If a couple abstains from sexual intercourse when the signs of ovulation are present and for 3 days afterward, the chance of pregnancy is decreased. A big problem with this method is that fertilization is very likely if intercourse occurs 1 or 2 days *before* ovulation.

Abortion

Abortion refers to the premature expulsion of the products of conception from the uterus, usually before the twentieth week of pregnancy. An abortion may be **spontaneous** (naturally occurring; also called a *miscarriage*) or **induced** (intentionally performed).

There are several types of induced abortions. One involves **mifepristone** (MIF-pris-tōn), called **miniprex** in the United States and **RU 486** in Europe. It is a hormone approved only for pregnancies 9 weeks or less when taken with misoprostol (a prostaglandin). Mifepristone is an antiprogestin; it blocks the action of progesterone by binding to and blocking progesterone receptors. Progesterone prepares the uterine endometrium for implantation and then maintains the uterine lining after implantation. If the level of progesterone falls during pregnancy or if the action of the hormone is blocked, menstruation occurs, and the embryo sloughs off along with the uterine lining. Within 12 hours after taking mifepristone, the endometrium starts to degenerate, and within 72 hours it begins to slough off. Misoprostol stimulates uterine contractions and is given after mifepristone to aid in expulsion of the endometrium.

Another type of induced abortion is called **vacuum aspiration** (suction) and can be performed up to the sixteenth week of pregnancy. A small, flexible tube attached to a vacuum source is inserted into the uterus through the vagina. The embryo or fetus, placenta, and lining of the uterus are then removed by suction. For pregnancies between 13 and 16 weeks, a technique called **dilation and evacuation** is commonly used. After the cervix is dilated, suction and forceps are used to remove the fetus, placenta, and uterine lining. From the sixteenth to twenty-fourth week, a **late-stage abortion** may be employed using surgical methods similar to dilation and evacuation or through nonsurgical methods using a saline solution or medications to induce abortion. Labor may be induced by using vaginal suppositories, intravenous infusion, or injections into the amniotic fluid through the uterus.

✔ CHECKPOINT

29. How do oral contraceptives reduce the likelihood of pregnancy?
30. How do some methods of birth control protect against sexually transmitted diseases?
31. What is the problem with developing an oral contraceptive pill for males?

28.5 DEVELOPMENT OF THE REPRODUCTIVE SYSTEMS

◉ OBJECTIVE

• Describe the development of the male and female reproductive systems.

The *gonads* develop from **gonadal ridges** that arise from growth of **intermediate mesoderm.** During the fifth week of development, the gonadal ridges appear as bulges just medial to the mesonephros (intermediate kidney) (Figure 28.27). Adjacent to the genital ridges are the **mesonephric ducts** (mez′-o-NEF-rik) or **Wolffian ducts** (WULF-ē-an), which eventually develop into structures of the reproductive system in males. A second pair of ducts, the **paramesonephric ducts** (par′-a-mes′-o-NEF-rik) or

Figure 28.27 Development of the internal reproductive systems.

The gonads develop from intermediate mesoderm.

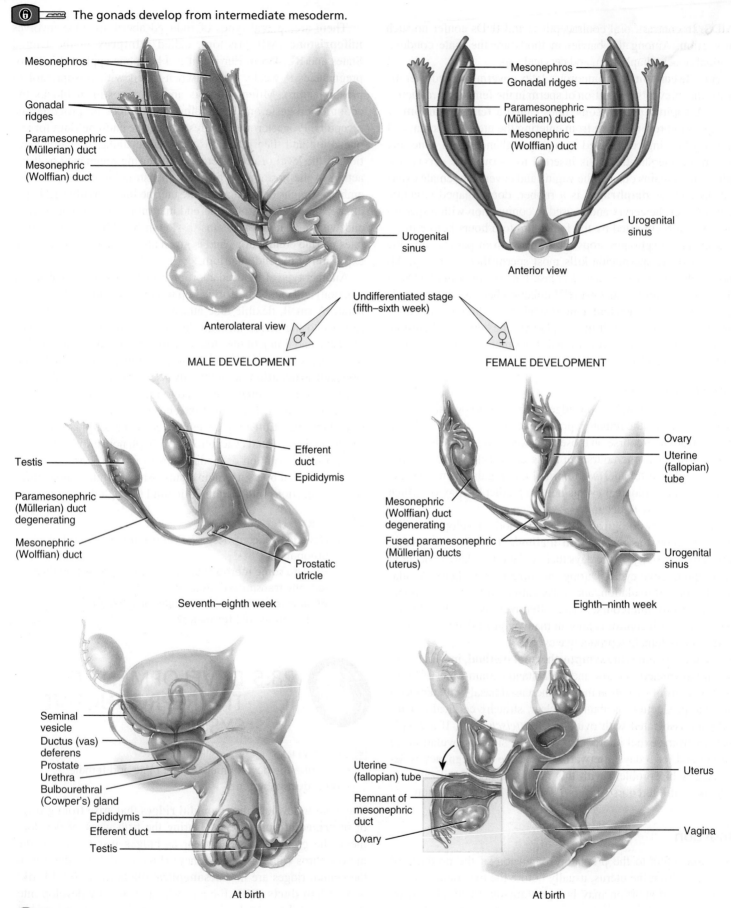

Undifferentiated stage (fifth–sixth week)

Anterolateral view ♂

Anterior view

MALE DEVELOPMENT

FEMALE DEVELOPMENT

Mesonephros
Gonadal ridges
Paramesonephric (Müllerian) duct
Mesonephric (Wolffian) duct
Urogenital sinus

Testis
Efferent duct
Epididymis
Paramesonephric (Müllerian) duct degenerating
Mesonephric (Wolffian) duct
Prostatic utricle

Seventh–eighth week

Ovary
Uterine (fallopian) tube
Mesonephric (Wolffian) duct degenerating
Fused paramesonephric (Müllerian) ducts (uterus)
Urogenital sinus

Eighth–ninth week

Seminal vesicle
Ductus (vas) deferens
Prostate
Urethra
Bulbourethral (Cowper's) gland
Epididymis
Efferent duct
Testis

At birth

Uterine (fallopian) tube
Remnant of mesonephric duct
Ovary
Uterus
Vagina

At birth

Which gene is responsible for the development of the gonads into testes?

1168

Müllerian ducts (mil-E-rē-an), develop lateral to the mesonephric ducts and eventually form structures of the reproductive system in females. Both sets of ducts empty into the urogenital sinus. An early embryo has the potential to follow either the male or the female pattern of development because it contains both sets of ducts and genital ridges that can differentiate into either testes or ovaries.

Cells of a male embryo have one X chromosome and one Y chromosome. The male pattern of development is initiated by a "master switch" gene on the Y chromosome named *SRY,* which stands for *Sex-determining Region of the Y* chromosome. When the *SRY* gene is expressed during development, its protein product causes the primitive Sertoli cells to begin to differentiate in the *testes* during the seventh week. The developing Sertoli cells secrete a hormone called **Müllerian-inhibiting substance (MIS),** which causes apoptosis of cells within the paramesonephric (Müllerian) ducts. As a result, those cells do not contribute any functional structures to the male reproductive system. Stimulated by human chorionic gonadotropin (hCG), primitive Leydig cells in the testes begin to secrete the androgen **testosterone** during the eighth week. Testosterone then stimulates development of the mesonephric duct on each side into the *epididymis, ductus (vas) deferens, ejaculatory duct,* and *seminal vesicle.* The *testes* connect to the mesonephric duct through a series of tubules that eventually become the *seminiferous tubules.* The *prostate* and *bulbourethral glands* are **endodermal** outgrowths of the urethra.

Cells of a female embryo have two X chromosomes and no Y chromosome. Because *SRY* is absent, the genital ridges develop into *ovaries,* and because MIS is not produced, the paramesonephric ducts flourish. The distal ends of the paramesonephric ducts fuse to form the *uterus* and *vagina;* the unfused proximal portions of the ducts become the *uterine (fallopian) tubes.* The mesonephric ducts degenerate without contributing any functional structures to the female reproductive system because of the absence of testosterone. The *greater* and *lesser vestibular glands* develop from endodermal outgrowths of the vestibule.

The *external genitals* of both male and female embryos (penis and scrotum in males and clitoris, labia, and vaginal orifice in females) also remain undifferentiated until about the eighth week. Before differentiation, all embryos have the following external structures (Figure 28.28):

1. **Urethral (urogenital) folds.** These paired structures develop from mesoderm in the cloacal region (see Figure 26.23).

2. **Urethral groove.** An indentation between the urethral folds, which is the opening into the urogenital sinus.

Figure 28.28 **Development of the external genitals.**

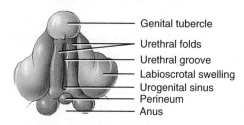
The external genitals of male and female embryos remain undifferentiated until about the eighth week.

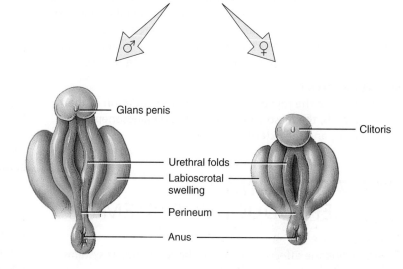

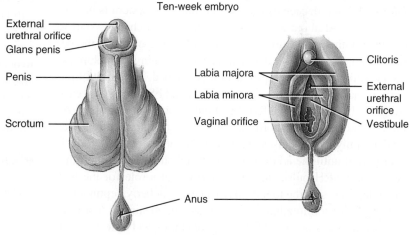

Which hormone is responsible for the differentiation of the external genitals?

3. **Genital tubercle.** A rounded elevation just anterior to the urethral folds.

4. **Labioscrotal swelling** (lā-bē-ō-SKRŌ-tal). Paired, elevated structures lateral to the urethral folds.

In male embryos, some testosterone is converted to a second androgen called **dihydrotestosterone (DHT).** DHT stimulates development of the urethra, prostate, and external genitals (scrotum and penis). Part of the genital tubercle elongates and develops into a penis. Fusion of the urethral folds forms the *spongy (penile) urethra* and leaves an opening to the exterior only at the distal end of the penis, the *external urethral orifice.* The labioscrotal swellings develop into the *scrotum.* In the absence of DHT, the genital tubercle gives rise to the *clitoris* in female embryos. The urethral folds remain open as the *labia minora,* and the labioscrotal swellings become the *labia majora.* The urethral groove becomes the *vestibule.* After birth, androgen levels decline because hCG is no longer present to stimulate secretion of testosterone.

✔ CHECKPOINT

32. Describe the role of hormones in differentiation of the gonads, the mesonephric ducts, the paramesonephric ducts, and the external genitals.

28.6 AGING AND THE REPRODUCTIVE SYSTEMS

◉ OBJECTIVE

• Describe the effects of aging on the reproductive systems.

During the first decade of life, the reproductive system is in a juvenile state. At about age 10, hormone-directed changes start to occur in both sexes. **Puberty** (PŪ-ber-tē = a ripe age) is the period when secondary sexual characteristics begin to develop and the potential for sexual reproduction is reached. The onset of puberty is marked by pulses or bursts of LH and FSH secretion, each triggered by a pulse of GnRH. Most pulses occur during sleep. As puberty advances, the hormone pulses occur during the day as well as at night. The pulses increase in frequency during a 3- to 4-year period until the adult pattern is established. The stimuli that cause the GnRH pulses are still unclear, but a role for the hormone leptin is starting to unfold. Just before puberty, leptin levels rise in proportion to adipose tissue mass. Interestingly, leptin receptors are present in both the hypothalamus and anterior pituitary. Mice that lack a functional leptin gene from birth are sterile and remain in a prepubertal state. Giving leptin to such mice elicits secretion of gonadotropins, and they become fertile. Leptin may signal the hypothalamus that long-term energy stores (triglycerides in adipose tissue) are adequate for reproductive functions to begin.

In females, the reproductive cycle normally occurs once each month from **menarche** (me-NAR-kē), the first menses, to **menopause,** the permanent cessation of menses. Thus, the female reproductive system has a time-limited span of fertility between menarche and menopause. For the first 1 to 2 years after menarche, ovulation only occurs in about 10% of the cycles and the luteal phase is short. Gradually, the percentage of ovulatory cycles increases, and the luteal phase reaches its normal duration of 14 days. With age, fertility declines. Between the ages of 40 and 50 the pool of remaining ovarian follicles becomes exhausted. As a result, the ovaries become less responsive to hormonal stimulation. The production of estrogens declines, despite copious secretion of FSH and LH by the anterior pituitary. Many women experience hot flashes and heavy sweating, which coincide with bursts of GnRH release. Other symptoms of menopause are headache, hair loss, muscular pains, vaginal dryness, insomnia, depression, weight gain, and mood swings. Some atrophy of the ovaries, uterine tubes, uterus, vagina, external genitalia, and breasts occurs in postmenopausal women. Due to loss of estrogens, most women experience a decline in bone mineral density after menopause. Sexual desire (libido) does not show a parallel decline; it may be maintained by adrenal sex steroids. The risk of having uterine cancer peaks at about 65 years of age, but cervical cancer is more common in younger women.

In males, declining reproductive function is much more subtle than in females. Healthy men often retain reproductive capacity into their eighties or nineties. At about age 55 a decline in testosterone synthesis leads to reduced muscle strength, fewer viable sperm, and decreased sexual desire. Although sperm production decreases 50–70% between ages 60 and 80, abundant sperm may still be present even in old age.

Enlargement of the prostate to two to four times its normal size occurs in most males over age 60. This condition, called **benign prostatic hyperplasia (BPH)** (hī-per-PLĀ-zē-a), decreases the size of the prostatic urethra and is characterized by frequent urination, nocturia (bed-wetting), hesitancy in urination, decreased force of urinary stream, postvoiding dribbling, and a sensation of incomplete emptying.

✔ CHECKPOINT

33. What changes occur in males and females at puberty?
34. What do the terms menarche and menopause mean?

• • •

To appreciate the many ways that the reproductive systems contribute to homeostasis of other body systems, examine *Focus on Homeostasis: The Reproductive Systems.* Next, in Chapter 29, you will explore the major events that occur during pregnancy and you will discover how genetics (inheritance) plays a role in the development of a child.

| --- | --- |
For all body systems 	The male and female reproductive systems produce gametes (oocytes and sperm) that unite to form embryos and fetuses, which contain cells that divide and differentiate to form all of the organ systems of the body.
Integumentary systems	Androgens promote the growth of body hair. Estrogens stimulate the deposition of fat in the breasts, abdomen, and hips. Mammary glands produce milk. Skin stretches during pregnancy as the fetus enlarges.
Skeletal system	Androgens and estrogens stimulate the growth and maintenance of bones of the skeletal system.
Muscular system	Androgens stimulate the growth of skeletal muscles.
Nervous system	Androgens influence libido (sex drive). Estrogens may play a role in the development of certain regions of the brain in males.
Endocrine system	Testosterone and estrogens exert feedback effects on the hypothalamus and anterior pituitary gland.
Cardiovascular system	Estrogens lower blood cholesterol level and may reduce the risk of coronary artery disease in women under age 50.
Lymphatic systems and immunity	The presence of an antibiotic-like chemical in semen and the acidic pH of vaginal fluid provide innate immunity against microbes in the reproductive tract.
Respiratory system	Sexual arousal increases the rate and depth of breathing.
Digestive system	The presence of the fetus during pregnancy crowds the digestive organs, which leads to heartburn and constipation.
Urinary system	In males, the portion of the urethra that extends through the prostate and penis is a passageway for urine as well as semen.

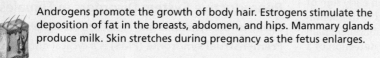

THE REPRODUCTIVE SYSTEMS

1171

DISORDERS: HOMEOSTATIC IMBALANCES

Reproductive System Disorders in Males

Testicular Cancer

Testicular cancer is the most common cancer in males between the ages of 20 and 35. More than 95% of testicular cancers arise from spermatogenic cells within the seminiferous tubules. An early sign of testicular cancer is a mass in the testis, often associated with a sensation of testicular heaviness or a dull ache in the lower abdomen; pain usually does not occur. To increase the chance for early detection of a testicular cancer, all males should perform regular self-examinations of the testes. The examination should be done starting in the teen years and once each month thereafter. After a warm bath or shower (when the scrotal skin is loose and relaxed) each testis should be examined as follows. The testis is grasped and gently rolled between the index finger and thumb, feeling for lumps, swellings, hardness, or other changes. If a lump or other change is detected, a physician should be consulted as soon as possible.

Prostate Disorders

Because the prostate surrounds part of the urethra, any infection, enlargement, or tumor can obstruct the flow of urine. Acute and chronic infections of the prostate are common in postpubescent males, often in association with inflammation of the urethra. Symptoms may include fever, chills, urinary frequency, frequent urination at night, difficulty in urinating, burning or painful urination, low back pain, joint and muscle pain, blood in the urine, or painful ejaculation. However, often there are no symptoms. Antibiotics are used to treat most cases that result from a bacterial infection. In **acute prostatitis,** the prostate becomes swollen and tender. **Chronic prostatitis** is one of the most common chronic infections in men of the middle and later years. On examination, the prostate feels enlarged, soft, and very tender, and its surface outline is irregular.

Prostate cancer is the leading cause of death from cancer in men in the United States, having surpassed lung cancer in 1991. Each year it is diagnosed in almost 200,000 U.S. men and causes nearly 40,000 deaths. The amount of PSA (prostate-specific antigen), which is produced only by prostate epithelial cells, increases with enlargement of the prostate and may indicate infection, benign enlargement, or prostate cancer. A blood test can measure the level of PSA in the blood. Males over the age of 40 should have an annual examination of the prostate gland. In a **digital rectal exam,** a physician palpates the gland through the rectum with the fingers (digits). Many physicians also recommend an annual PSA test for males over age 50. Treatment for prostate cancer may involve surgery, cryotherapy, radiation, hormonal therapy, and chemotherapy. Because many prostate cancers grow very slowly, some urologists recommend "watchful waiting" before treating small tumors in men over age 70.

Erectile Dysfunction

Erectile dysfunction (ED), previously termed **impotence,** is the consistent inability of an adult male to ejaculate or to attain or hold an erection long enough for sexual intercourse. Many cases of impotence are caused by insufficient release of nitric oxide (NO), which relaxes the smooth muscle of the penile arterioles and erectile tissue. The drug *Viagra*® (sildenafil) enhances smooth muscle relaxation by nitric oxide in the penis. Other causes of erectile dysfunction include diabetes mellitus, physical abnormalities of the penis, systemic disorders such as syphilis, vascular disturbances (arterial or venous obstructions), neurological disorders, surgery, testosterone deficiency, and drugs (alcohol, antidepressants, antihistamines, antihypertensives, narcotics, nicotine, and tranquilizers). Psychological factors such as anxiety or depression, fear of causing pregnancy, fear of sexually transmitted diseases, religious inhibitions, and emotional immaturity may also cause ED.

Reproductive System Disorders in Females

Premenstrual Syndrome and Premenstrual Dysphoric Disorder

Premenstrual syndrome (PMS) is a cyclical disorder of severe physical and emotional distress. It appears during the postovulatory (luteal) phase of the female reproductive cycle and dramatically disappears when menstruation begins. The signs and symptoms are highly variable from one woman to another. They may include edema, weight gain, breast swelling and tenderness, abdominal distension, backache, joint pain, constipation, skin eruptions, fatigue and lethargy, greater need for sleep, depression or anxiety, irritability, mood swings, headache, poor coordination and clumsiness, and cravings for sweet or salty foods. The cause of PMS is unknown. For some women, getting regular exercise; avoiding caffeine, salt, and alcohol; and eating a diet that is high in complex carbohydrates and lean proteins can bring considerable relief.

Premenstrual dysphoric disorder (PMDD) is a more severe syndrome in which PMS-like signs and symptoms do not resolve after the onset of menstruation. Clinical research studies have found that suppression of the reproductive cycle by a drug that interferes with GnRH (leuprolide) decreases symptoms significantly. Because symptoms reappear when estradiol or progesterone is given together with leuprolide, researchers propose that PMDD is caused by abnormal responses to normal levels of these ovarian hormones. *SSRIs* (selective serotonin reuptake inhibitors) have shown promise in treating both PMS and PMDD.

Endometriosis

Endometriosis (en'-dō-MĒ-trē-o'-sis; *endo-* = within; *metri-* = uterus; *-osis* = condition) is characterized by the growth of endometrial tissue outside the uterus. The tissue enters the pelvic cavity via the open uterine tubes and may be found in any of several sites—on the ovaries, the rectouterine pouch, the outer surface of the uterus, the sigmoid colon, pelvic and abdominal lymph nodes, the cervix, the abdominal wall, the kidneys, and the urinary bladder. Endometrial tissue responds to hormonal fluctuations, whether it is inside or outside the uterus. With each reproductive cycle, the tissue proliferates and then breaks down and bleeds. When this occurs outside the uterus, it can cause inflammation, pain, scarring, and infertility. Symptoms include premenstrual pain or unusually severe menstrual pain.

Breast Cancer

One in eight women in the United States faces the prospect of **breast cancer.** After lung cancer, it is the second-leading cause of death from cancer in U.S. women. Breast cancer can occur in males but is rare. In females, breast cancer is seldom seen before age 30; its incidence rises rapidly after menopause. An estimated 5% of the 180,000 cases diagnosed each year in the United States, particularly those that arise in younger women, stem from inherited genetic mutations (changes in the DNA). Researchers have now identified two genes that increase susceptibility to breast cancer: *BRCA*1 (*breast cancer 1*) and *BRCA*2. Mutation of *BRCA*1 also confers a high risk for ovarian cancer. In addition, mutations of the *p53* gene increase the risk of breast cancer in both males and females, and mutations of the androgen receptor gene are associated with the occurrence of breast cancer in some males. Because breast cancer generally is not painful until it becomes quite advanced, any lump, no matter how small, should be reported to a physician at once. Early detection—by breast self-examination and mammograms—is the best way to increase the chance of survival.

The most effective technique for detecting tumors less than 1 cm (0.4 in.) in diameter is **mammography** (mam-OG-ra-fē; -*graphy* = to record), a type of radiography using very sensitive x-ray film. The image of the breast, called a **mammogram** (see Table 1.3), is best obtained by compressing the breasts, one at a time, using flat plates. A supplementary procedure for evaluating breast abnormalities is **ultrasound.** Although ultrasound cannot detect tumors smaller than 1 cm in diameter (which mammography can detect), it can be used to determine whether a lump is a benign, fluid-filled cyst or a solid (and therefore possibly malignant) tumor.

Among the factors that increase the risk of developing breast cancer are (1) a family history of breast cancer, especially in a mother or sister; (2) nulliparity (never having borne a child) or having a first child after age 35; (3) previous cancer in one breast; (4) exposure to ionizing radiation, such as x-rays; (5) excessive alcohol intake; and (6) cigarette smoking.

The American Cancer Society recommends the following steps to help diagnose breast cancer as early as possible:

- All women over 20 should develop the habit of monthly breast self-examination.

- A physician should examine the breasts every 3 years when a woman is between the ages of 20 and 40, and every year after age 40.

- A mammogram should be taken in women between the ages of 35 and 39, to be used later for comparison (baseline mammogram).

- Women with no symptoms should have a mammogram every year after age 40.

- Women of any age with a history of breast cancer, a strong family history of the disease, or other risk factors should consult a physician to determine a schedule for mammography.

In November 2009, the United States Preventive Services Task Force (USPSTF) issued a series of recommendations relative to breast cancer screening for females at normal risk for breast cancer, that is, for females who have no signs or symptoms of breast cancer and who are not at a higher risk for breast cancer (for example, no family history). These recommendations are as follows:

- Women aged 50–74 should have a mammogram every 2 years.

- Women over 75 should not have mammograms.

- Breast self-examination is not required.

Treatment for breast cancer may involve hormone therapy, chemotherapy, radiation therapy, **lumpectomy** (lump-EK-tō-mē) (removal of the tumor and the immediate surrounding tissue), a modified or radical mastectomy, or a combination of these approaches. A **radical mastectomy** (mas-TEK-tō-mē; *mast-* = breast) involves removal of the affected breast along with the underlying pectoral muscles and the axillary lymph nodes. (Lymph nodes are removed because metastasis of cancerous cells usually occurs through lymphatic or blood vessels.) Radiation treatment and chemotherapy may follow the surgery to ensure the destruction of any stray cancer cells.

Several types of chemotherapeutic drugs are used to decrease the risk of relapse or disease progression. *Nolvadex®* (tamoxifen) is an antagonist to estrogens that binds to and blocks receptors for estrogens, thus decreasing the stimulating effect of estrogens on breast cancer cells. Tamoxifen has been used for 20 years and greatly reduces the risk of cancer recurrence. *Herceptin®*, a monoclonal antibody drug, targets an antigen on the surface of breast cancer cells. It is effective in causing regression of tumors and retarding progression of the disease. The early data from clinical trials of two new drugs, *Femara®* and *Amimidex®*, show relapse rates that are lower than those for tamoxifen. These drugs are inhibitors of aromatase, the enzyme needed for the final step in synthesis of estrogens. Finally, two drugs—tamoxifen and *Evista®* (raloxifene)—are being marketed for breast cancer *prevention.* Interestingly, raloxifene blocks estrogen receptors in the breasts and uterus but activates estrogen receptors in bone. Thus, it can be used to treat osteoporosis without increasing a woman's risk of breast or endometrial (uterine) cancer.

Ovarian Cancer

Even though **ovarian cancer** is the sixth most common form of cancer in females, it is the leading cause of death from all gynecological malignancies (excluding breast cancer) because it is difficult to detect before it metastasizes (spreads) beyond the ovaries. Risk factors associated with ovarian cancer include age (usually over age 50); race (whites are at highest risk); family history of ovarian cancer; more than 40 years of active ovulation; nulliparity or first pregnancy after age 30; a high-fat, low-fiber, vitamin A–deficient diet; and prolonged exposure to asbestos or talc. Early ovarian cancer has no symptoms or only mild ones associated with other common problems, such as abdominal discomfort, heartburn, nausea, loss of appetite, bloating, and flatulence. Later-stage signs and symptoms include an enlarged abdomen, abdominal and/or pelvic pain, persistent gastrointestinal disturbances, urinary complications, menstrual irregularities, and heavy menstrual bleeding.

Cervical Cancer

Cervical cancer, carcinoma of the cervix of the uterus, starts with **cervical dysplasia** (dis-PLĀ-sē-a), a change in the shape, growth, and number of cervical cells. The cells may either return to normal or progress to cancer. In most cases, cervical cancer may be detected in its earliest stages by a Pap test (see Clinical Connection: Papanicolaou Test in Chapter 4). Some evidence links cervical cancer to the virus that causes genital warts, human papillomavirus (HPV). Increased risk is associated with having a large number of sexual partners, having first intercourse at a young age, and smoking cigarettes.

Vulvovaginal Candidiasis

Candida albicans is a yeastlike fungus that commonly grows on mucous membranes of the gastrointestinal and genitourinary tracts. The organism is responsible for **vulvovaginal candidiasis** (vul-vō-VAJ-i-nal can-di-DĪ-a-sis), the most common form of **vaginitis** (vaj-i-NĪ-tis), inflammation of the vagina. Candidiasis is characterized by severe itching; a thick, yellow, cheesy discharge; a yeasty odor; and pain. The disorder, experienced at least once by about 75% of females, is usually a result of proliferation of the fungus following antibiotic therapy for another condition. Predisposing conditions include the use of oral contraceptives or cortisone-like medications, pregnancy, and diabetes.

Sexually Transmitted Diseases

A **sexually transmitted disease (STD)** is one that is spread by sexual contact. In most developed countries of the world, such as those of Western Europe, Japan, Australia, and New Zealand, the incidence of STDs has declined markedly during the past 25 years. In the United

States, by contrast, STDs have been rising to near-epidemic proportions; they currently affect more than 65 million people. AIDS and hepatitis B, which are sexually transmitted diseases that also may be contracted in other ways, are discussed in Chapters 22 and 24, respectively.

Chlamydia

Chlamydia (kla-MID-ē-a) is a sexually transmitted disease caused by the bacterium *Chlamydia trachomatis* (*chlamy-* = cloak). This unusual bacterium cannot reproduce outside body cells; it "cloaks" itself inside cells, where it divides. At present, chlamydia is the most prevalent sexually transmitted disease in the United States. In most cases, the initial infection is asymptomatic and thus difficult to recognize clinically. In males, urethritis is the principal result, causing a clear discharge, burning on urination, frequent urination, and painful urination. Without treatment, the epididymides may also become inflamed, leading to sterility. In 70% of females with chlamydia, symptoms are absent, but chlamydia is the leading cause of pelvic inflammatory disease. The uterine tubes may also become inflamed, which increases the risk of ectopic pregnancy (implantation of a fertilized ovum outside the uterus) and infertility due to the formation of scar tissue in the tubes.

Gonorrhea

Gonorrhea (gon-ō-RĒ-a) or **"the clap"** is caused by the bacterium *Neisseria gonorrhoeae*. In the United States, 1 million to 2 million new cases of gonorrhea appear each year, most among individuals aged 15–29 years. Discharges from infected mucous membranes are the source of transmission of the bacteria either during sexual contact or during the passage of a newborn through the birth canal. The infection site can be in the mouth and throat after oral–genital contact, in the vagina and penis after genital intercourse, or in the rectum after recto–genital contact.

Males usually experience urethritis with profuse pus drainage and painful urination. The prostate and epididymis may also become infected. In females, infection typically occurs in the vagina, often with a discharge of pus. Both infected males and females may harbor the disease without any symptoms, however, until it has progressed to a more advanced stage; about 5–10% of males and 50% of females are asymptomatic. In females, the infection and consequent inflammation can proceed from the vagina into the uterus, uterine tubes, and pelvic cavity. An estimated 50,000 to 80,000 women in the United States are made infertile by gonorrhea every year as a result of scar tissue formation that closes the uterine tubes. If bacteria in the birth canal are transmitted to the eyes of a newborn, blindness can result. Administration of a 1% silver nitrate solution in the infant's eyes prevents infection.

Syphilis

Syphilis, caused by the bacterium *Treponema pallidum* (trep-o-NĒ-ma PAL-i-dum), is transmitted through sexual contact or exchange of blood, or through the placenta to a fetus. The disease progresses through several stages. During the *primary stage*, the chief sign is a painless open sore, called a **chancre** (SHANG-ker), at the point of contact. The chancre heals within 1 to 5 weeks. From 6 to 24 weeks later, signs and symptoms such as a skin rash, fever, and aches in the joints and muscles usher in the *secondary stage*, which is systemic—the infection spreads to all major body systems. When signs of organ degeneration appear, the disease is said to be in the *tertiary stage*. If the nervous system is involved, the tertiary stage is called **neurosyphilis.** As motor areas become damaged extensively, victims may be unable to control urine and bowel movements. Eventually they may become bedridden and unable even to feed themselves. In addition, damage to the cerebral cortex produces memory loss and personality changes that range from irritability to hallucinations.

Genital Herpes

Genital herpes is an incurable STD. Type II herpes simplex virus (HSV-2) causes genital infections, producing painful blisters on the prepuce, glans penis, and penile shaft in males and on the vulva or sometimes high up in the vagina in females. The blisters disappear and reappear in most patients, but the virus itself remains in the body. A related virus, type I herpes simplex virus (HSV-1), causes cold sores on the mouth and lips. Infected individuals typically experience recurrences of symptoms several times a year.

Genital Warts

Warts are an infectious disease caused by viruses. *Human papillomavirus (HPV)* causes **genital warts,** which can be transmitted sexually. Nearly 1 million people a year develop genital warts in the United States. Patients with a history of genital warts may be at increased risk for cancers of the cervix, vagina, anus, vulva, and penis. There is no cure for genital warts. A vaccine (Gardasil®) against certain types of HPV that cause cervical cancer and genital warts is available and recommended for 11- and 12-year-old girls.

MEDICAL TERMINOLOGY

Castration (kas-TRĀ-shun = to prune) Removal, inactivation, or destruction of the gonads; commonly used in reference to removal of the testes only.

Colposcopy (kol-POS-kō-pē; *colpo-* = vagina; *-scopy* = to view) Visual inspection of the vagina and cervix of the uterus using a culposcope, an instrument that has a magnifying lens (between 5× and 50×) and a light. The procedure generally takes place after an unusual Pap smear.

Culdoscopy (kul-DOS-kō-pē; *-cul-* = cul-de-sac; *-scopy* = to examine) A procedure in which a culdoscope (endoscope) is inserted through the posterior wall of the vagina to view the rectouterine pouch in the pelvic cavity.

Dysmenorrhea (dis-men-ōr-Ē-a; dys- = difficult or painful) Pain associated with menstruation; the term is usually reserved to describe menstrual symptoms that are severe enough to prevent a woman from functioning normally for one or more days each month. Some cases are caused by uterine tumors, ovarian cysts, pelvic inflammatory disease, or intrauterine devices.

Dyspareunia (dis-pa-ROO-nē-a; *dys-* = difficult; *-para-* = beside; *-enue* = bed) Pain during sexual intercourse. It may occur in the genital area or in the pelvic cavity, and may be due to inadequate lubrication, inflammation, infection, an improperly fitting diaphragm or cervical cap, endometriosis, pelvic inflammatory disease, pelvic tumors, or weakened uterine ligaments.

Endocervical curettage (kū-re-TAHZH; *curette* = scraper) A procedure in which the cervix is dilated and the endometrium of the uterus is scraped with a spoon-shaped instrument called a curette; commonly called a *D and C* (dilation and curettage).

Fibroids (FĪ-broyds; *fibro-* = fiber; *-eidos* = resemblance) Noncancerous tumors in the myometrium of the uterus composed of muscular and fibrous tissue. Their growth appears to be related to high levels of estrogens. They do not occur before puberty and usually stop growing after menopause. Symptoms include abnormal menstrual bleeding and pain or pressure in the pelvic area.

Hermaphroditism (her-MAF-rō-dīt-izm) The presence of both ovarian and testicular tissue in one individual.

Hypospadias (hī'-pō-SPĀ-dē-as; *hypo-* = below) A common congenital abnormality in which the urethral opening is displaced. In males, the displaced opening may be on the underside of the penis, at the penoscrotal junction, between the scrotal folds, or in the perineum; in females, the urethra opens into the vagina. The problem can be corrected surgically.

Leukorrhea (loo'-kō-RĒ-a; *leuko-* = white) A whitish (nonbloody) vaginal discharge containing mucus and pus cells that may occur at any age and affects most women at some time.

Menorrhagia (men-ō-RA-jē-a; *meno-* = menstruation; *-rhage* = to burst forth) Excessively prolonged or profuse menstrual period. May be due to a disturbance in hormonal regulation of the menstrual cycle, pelvic infection, medications (anticoagulants), fibroids (noncancerous uterine tumors composed of muscle and fibrous tissue), endometriosis, or intrauterine devices.

Oophorectomy (ō'-of-ō-REK-tō-mē; *oophor-* = bearing eggs) Removal of the ovaries.

Orchitis (or-KĪ-tis; *orchi-* = testes; *-itis* = inflammation) Inflammation of the testes, for example, as a result of the mumps virus or a bacterial infection.

Ovarian cyst The most common form of ovarian tumor, in which a fluid-filled follicle or corpus luteum persists and continues growing.

Pelvic inflammatory disease (PID) A collective term for any extensive bacterial infection of the pelvic organs, especially the uterus, uterine tubes, or ovaries, which is characterized by pelvic soreness, lower back pain, abdominal pain, and urethritis. Often the early symptoms of PID occur just after menstruation. As infection spreads, fever may develop, along with painful abscesses of the reproductive organs.

Salpingectomy (sal'-pin-JEK-tō-mē; *salpingo* = tube) Removal of a uterine (fallopian) tube.

Smegma (SMEG-ma) The secretion, consisting principally of desquamated epithelial cells, found chiefly around the external genitals and especially under the foreskin of the male.

CHAPTER REVIEW AND RESOURCE SUMMARY

Review

Resource

Introduction

Anatomy Overview - The Reproductive System Overview

28.1 Male Reproductive System

1. Reproduction is the process by which new individuals of a species are produced and the genetic material is passed from generation to generation.

2. The organs of reproduction are grouped as gonads (produce gametes), ducts (transport and store gametes), accessory sex glands (produce materials that support gametes), and supporting structures (have various roles in reproduction). The male structures of reproduction include the testes (2), epididymidis (2), ducti (vasa) deferentia (2), ejaculatory ducts (2), seminal vesicles (2), urethra (1), prostate (1), bulbourethral (Cowper's) glands (2), and penis (1). The scrotum is a sac that hangs from the root of the penis and consists of loose skin and underlying subcutaneous layer; it supports the testes. The temperature of the testes is regulated by the cremaster muscles, which either contract to elevate the testes and move them closer to the pelvic cavity or relax and move them farther from the pelvic cavity. The dartos muscle causes the scrotum to become tight and wrinkled.

3. The testes are paired oval glands (gonads) in the scrotum containing seminiferous tubules, in which sperm cells are made; Sertoli cells (sustentacular cells), which nourish sperm cells and secrete inhibin; and Leydig (interstitial) cells, which produce the male sex hormone testosterone. The testes descend into the scrotum through the inguinal canals during the seventh month of fetal development. Failure of the testes to descend is called cryptorchidism.

4. Secondary oocytes and sperm, both of which are called gametes, are produced in the gonads. Spermatogenesis, which occurs in the testes, is the process whereby immature spermatogonia develop into sperm. The spermatogenesis sequence, which includes meiosis I, meiosis II, and spermiogenesis, results in the formation of four haploid sperm (spermatozoa) from each primary spermatocyte. Mature sperm consist of a head and a tail. Their function is to fertilize a secondary oocyte.

5. At puberty, gonadotropin-releasing hormone (GnRH) stimulates anterior pituitary secretion of FSH and LH. LH stimulates production of testosterone; FSH and testosterone stimulate spermatogenesis. Sertoli cells secrete androgen-binding protein (ABP), which binds to testosterone and keeps its concentration high in the seminiferous tubule. Testosterone controls the growth, development, and maintenance of sex organs; stimulates bone growth, protein anabolism, and sperm maturation; and stimulates development of masculine secondary sex characteristics. Inhibin is produced by Sertoli cells; its inhibition of FSH helps regulate the rate of spermatogenesis.

6. The duct system of the testes includes the seminiferous tubules, straight tubules, and rete testis. Sperm flow out of the testes through the efferent ducts. The ductus epididymis is the site of sperm maturation and storage. The ductus (vas) deferens stores sperm and propels them toward the urethra during ejaculation.

Anatomy Overview - Male Reproductive Anatomy
Anatomy Overview - Histology of the Testes
Anatomy Overview - Testicular Hormones
Animation - Spermatogenesis
Animation - Hormonal Control of Male Reproduction
Figure 28.3 - Internal and External Anatomy of a Testis
Figure 28.9 - Locations of Several Accessory Reproductive Organs in Males
Exercise - Spermatogenesis Selections
Exercise - Match the Male Hormones
Concepts and Connections - Regulation of Male Reproduction

Review

7. Each ejaculatory duct, formed by the union of the duct from the seminal vesicle and ampulla of the ductus (vas) deferens, is the passageway for ejection of sperm and secretions of the seminal vesicles into the first portion of the urethra, the prostatic urethra.

8. The urethra in males is subdivided into three portions: the prostatic, membranous, and spongy (penile) urethra.

9. The seminal vesicles secrete an alkaline, viscous fluid that contains fructose (used by sperm for ATP production). Seminal fluid constitutes about 60% of the volume of semen and contributes to sperm viability. The prostate secretes a slightly acidic fluid that constitutes about 25% of the volume of semen and contributes to sperm motility. The bulbourethral (Cowper's) glands secrete mucus for lubrication and an alkaline substance that neutralizes acid. Semen is a mixture of sperm and seminal fluid; it provides the fluid in which sperm are transported, supplies nutrients, and neutralizes the acidity of the male urethra and the vagina.

10. The penis consists of a root, a body, and a glans penis. Engorgement of the penile blood sinuses under the influence of sexual excitation is called erection.

28.2 Female Reproductive System

1. The female organs of reproduction include the ovaries (gonads), uterine (fallopian) tubes or oviducts, uterus, vagina, and vulva. The mammary glands are part of the integumentary system and also are considered part of the reproductive system in females.

2. The ovaries, the female gonads, are located in the superior portion of the pelvic cavity, lateral to the uterus. Ovaries produce secondary oocytes, discharge secondary oocytes (the process of ovulation), and secrete estrogens, progesterone, relaxin, and inhibin.

3. Oogenesis (the production of haploid secondary oocytes) begins in the ovaries. The oogenesis sequence includes meiosis I and meiosis II, which goes to completion only after an ovulated secondary oocyte is fertilized by a sperm cell.

4. The uterine (fallopian) tubes transport secondary oocytes from the ovaries to the uterus and are the normal sites of fertilization. Ciliated cells and peristaltic contractions help move a secondary oocyte or fertilized ovum toward the uterus.

5. The uterus is an organ the size and shape of an inverted pear that functions in menstruation, implantation of a fertilized ovum, development of a fetus during pregnancy, and labor. It also is part of the pathway for sperm to reach the uterine tubes to fertilize a secondary oocyte. Normally, the uterus is held in position by a series of ligaments. Histologically, the layers of the uterus are an outer perimetrium (serosa), a middle myometrium, and an inner endometrium.

6. The vagina is a passageway for sperm and the menstrual flow, the receptacle of the penis during sexual intercourse, and the inferior portion of the birth canal. It is capable of considerable stretching.

7. The vulva, a collective term for the external genitals of the female, consists of the mons pubis, labia majora, labia minora, clitoris, vestibule, vaginal and urethral orifices, hymen, and bulb of the vestibule, as well as three sets of glands: the paraurethral (Skene's), greater vestibular (Bartholin's), and lesser vestibular glands.

8. The perineum is a diamond-shaped area at the inferior end of the trunk medial to the thighs and buttocks.

9. The mammary glands are modified sweat glands lying superficial to the pectoralis major muscles. Their function is to synthesize, secrete, and eject milk (lactation).

10. Mammary gland development depends on estrogens and progesterone. Milk production is stimulated by prolactin, estrogens, and progesterone; milk ejection is stimulated by oxytocin.

Anatomy Overview - Female Reproductive Anatomy
Anatomy Overview - Histology of the Female Reproductive System
Anatomy Overview - Histology of the Mammary Glands
Figure 28.13 - Histology of the Ovary
Figure 28.16 - Relationship of the Uterine Tubes to the Ovaries, Uterus, and Associated Structures
Figure 28.22 - Mammary Glands
Exercise - Concentrate on Reproductive Structures

28.3 The Female Reproductive Cycle

1. The function of the ovarian cycle is to develop a secondary oocyte; the function of the uterine (menstrual) cycle is to prepare the endometrium each month to receive a fertilized egg. The female reproductive cycle includes both the ovarian and uterine cycles.

2. The uterine and ovarian cycles are controlled by GnRH from the hypothalamus, which stimulates the release of FSH and LH by the anterior pituitary. FSH and LH stimulate development of follicles and secretion of estrogens by the follicles. LH also stimulates ovulation, formation of the corpus luteum, and the secretion of progesterone and estrogens by the corpus luteum.

3. Estrogens stimulate the growth, development, and maintenance of female reproductive structures; stimulate the development of secondary sex characteristics; and stimulate protein synthesis. Progesterone works with estrogens to prepare the endometrium for implantation and the mammary glands for milk synthesis.

Anatomy Overview - Ovarian Hormones
Animation - Hormonal Control of Female Reproduction
Anatomy Overview - Hypothalamic Reproductive Hormones
Animation - Oogenesis
Figure 28.24 - The Female Reproductive Cycle
Exercise - Assemble the Cycle
Exercise - Match the Female Hormones
Exercise - Organize Oogenesis
Exercise - Assemble the Cycle

Review	Resource

4. Relaxin relaxes the myometrium at the time of possible implantation. At the end of a pregnancy, relaxin increases the flexibility of the pubic symphysis and helps dilate the uterine cervix to facilitate delivery.

5. During the menstrual phase, the stratum functionalis of the endometrium is shed, discharging blood, tissue fluid, mucus, and epithelial cells.

6. During the preovulatory phase, a group of follicles in the ovaries begins to undergo final maturation. One follicle outgrows the others and becomes dominant while the others degenerate. At the same time, endometrial repair occurs in the uterus. Estrogens are the dominant ovarian hormones during the preovulatory phase.

7. Ovulation is the rupture of the mature (graafian) follicle and the release of a secondary oocyte into the pelvic cavity. It is brought about by a surge of LH. Signs and symptoms of ovulation include increased basal body temperature; clear, stretchy cervical mucus; changes in the uterine cervix; and abdominal pain.

8. During the postovulatory phase, both progesterone and estrogens are secreted in large quantity by the corpus luteum of the ovary, and the uterine endometrium thickens in readiness for implantation.

9. If fertilization and implantation do not occur, the corpus luteum degenerates, and the resulting low levels of progesterone and estrogens allow discharge of the endometrium followed by the initiation of another reproductive cycle.

10. If fertilization and implantation do occur, the corpus luteum is maintained by hCG. The corpus luteum and later the placenta secrete progesterone and estrogens to support pregnancy and breast development for lactation.

Resource: Concepts and Connections - Regulation of Female Reproduction

28.4 Birth Control Methods and Abortion

1. Methods include complete abstinence, surgical sterilization (vasectomy, tubal ligation), non-incisional sterilization, hormonal methods (combined pill, minipill, contraceptive skin patch, vaginal contraceptive ring, emergency contraception, hormonal injections), intrauterine devices, spermicides, barrier methods (male condom, vaginal pouch, diaphragm, cervical cap), and periodic abstinence (rhythm and sympto-thermal methods).

2. Contraceptive pills of the combination type contain progestin and estrogens in concentrations that decrease the secretion of FSH and LH and thereby inhibit development of ovarian follicles and ovulation, inhibit transport of ova and sperm in the uterine tubes, and block implantation in the uterus.

3. An abortion is the premature expulsion from the uterus of the products of conception; it may be spontaneous or induced.

28.5 Development of the Reproductive Systems

1. The gonads develop from gonadal ridges that arise from growth of intermediate mesoderm. In the presence of the *SRY* gene, the gonads begin to differentiate into testes during the seventh week. The gonads differentiate into ovaries when the *SRY* gene is absent.

2. In males, testosterone stimulates development of each mesonephric duct into an epididymis, ductus (vas) deferens, ejaculatory duct, and seminal vesicle, and Müllerian-inhibiting substance (MIS) causes the paramesonephric duct cells to die. In females, testosterone and MIS are absent; the paramesonephric ducts develop into the uterine tubes, uterus, and vagina and the mesonephric ducts degenerate.

3. The external genitals develop from the genital tubercle and are stimulated to develop into typical male structures by the hormone dihydrotestosterone (DHT). The external genitals develop into female structures when DHT is not produced, the normal situation in female embryos.

Resource: Anatomy Overview - Developmental Stages

28.6 Aging and the Reproductive Systems

1. Puberty is the period when secondary sex characteristics begin to develop and the potential for sexual reproduction is reached.

2. The onset of puberty is marked by pulses or bursts of LH and FSH secretion, each triggered by a pulse of GnRH. The hormone leptin, released by adipose tissue, may signal the hypothalamus that long-term energy stores (triglycerides in adipose tissue) are adequate for reproductive functions to begin.

3. In females, the reproductive cycle normally occurs once each month from menarche, the first menses, to menopause, the permanent cessation of menses.

4. Between the ages of 40 and 50, the pool of remaining ovarian follicles becomes exhausted and levels of progesterone and estrogens decline. Most women experience a decline in bone mineral density after menopause, together with some atrophy of the ovaries, uterine tubes, uterus, vagina, external genitalia, and breasts. Uterine and breast cancer increase in incidence with age.

Review	Resource

5. In older males, decreased levels of testosterone are associated with decreased muscle strength, waning sexual desire, and fewer viable sperm; prostate disorders are common.

SELF-QUIZ QUESTIONS

Fill in the blanks in the following statement.

1. The period of time when secondary sexual characteristics begin to develop and the potential for sexual reproduction is reached is called _____. The first menses is called _____, and the permanent cessation of menses is called _____.

Indicate whether the following statements are true or false.

2. Spermatogenesis does not occur at normal core body temperature.

3. The route of sperm from the production in the testes to the exterior of the body is seminiferous tubules, straight tubules, rete testes, epididymis, ductus (vas) deferens, ejaculatory duct, prostatic urethra, membranous urethra, spongy urethra, external urethral orifice.

Choose the one best answer to the following questions.

4. Which of the following are functions of Sertoli cells? (1) protection of developing spermatogenic cells, (2) nourishment of spermatocytes, spermatids, and sperm, (3) phagocytosis of excess sperm cytoplasm as development proceeds, (4) mediation of the effects of testosterone and FSH, (5) control of movements of spermatogenic cells and release of sperm into the lumen of seminiferous tubules.
(a) 1, 2, 4, and 5   (b) 1, 2, 3, and 5   (c) 2, 3, 4, and 5
(d) 1, 2, 3, and 4   (e) 1, 2, 3, 4, and 5

5. Which of the following are *true?* (1) An erection is a sympathetic response initiated by sexual stimulation. (2) Dilation of blood vessels supplying erectile tissue results in erection. (3) Nitric oxide causes smooth muscle within erectile tissue to relax, which results in widening of blood sinuses. (4) Ejaculation is a sympathetic reflex coordinated by the sacral region of the spinal cord. (5) The purpose of the corpus cavernosa penis is to keep the spongy urethra open during ejaculation.
(a) 1, 2, and 3   (b) 1, 2, 3, 4, and 5   (c) 2 and 3
(d) 2, 4, and 5   (e) 1, 2, 3, and 4

6. Which of the following are *true* concerning estrogens? (1) They promote development and maintenance of female reproductive structures and secondary sex characteristics. (2) They help control fluid and electrolyte balance. (3) They increase protein catabolism. (4) They lower blood cholesterol. (5) In moderate levels, they inhibit the release of GnRH and the secretion of LH and FSH.
(a) 1, 4, and 5   (b) 1, 3, 4, and 5   (c) 1, 2, 3, and 5
(d) 1, 2, 3, and 4   (e) 1, 2, 3, 4, and 5

7. Which of the following statements are *correct?* (1) A sperm head contains DNA and an acrosome. (2) An acrosome is a specialized lysosome that contains enzymes that enable sperm to produce the ATP needed to propel themselves out of the male reproductive tract. (3) Mitochondria in the midpiece of a sperm produce ATP for sperm motility. (4) A sperm's tail, a flagellum, propels it along its way. (5) Once ejaculated, sperm are viable and normally are able to fertilize a secondary oocyte for 5 days.
(a) 1, 2, 3, and 4   (b) 2, 3, 4, and 5   (c) 1, 3, and 4
(d) 2, 4, and 5   (e) 2, 3, and 4

8. Which of the following statements are *correct?* (1) Spermatogonia are stem cells because when they undergo mitosis, some of the daughter cells remain to serve as a reservoir of cells for future mitosis. (2) Meiosis I is a division of pairs of chromosomes resulting in daughter cells with only one member of each chromosome pair. (3) Meiosis II separates the chromatids of each chromosome. (4) Spermiogenesis involves the maturation of spermatids into sperm. (5) The process by which the seminiferous tubules produce haploid sperm is called spermatogenesis.
(a) 1, 2, 3, and 5   (b) 1, 2, 3, 4, and 5   (c) 1, 3, 4, and 5
(d) 1, 2, 3, and 4   (e) 1, 3, and 5

9. Which of the following statements are *correct?* (1) Cells from the yolk sac give rise to oogonia. (2) Ova arise from the germinal epithelium of the ovary. (3) Primary oocytes enter prophase of meiosis I during fetal development but do not complete it until after puberty. (4) Once a secondary oocyte is formed, it proceeds to metaphase of meiosis II and stops at this stage. (5) The secondary oocyte resumes meiosis II and forms the ovum and a polar body only if fertilization occurs. (6) A primary oocyte gives rise to an ovum and four polar bodies.
(a) 1, 3, 4, and 5   (b) 1, 3, 4, and 6   (c) 1, 2, 4, and 6
(d) 1, 2, 4, and 5   (e) 1, 2, 5, and 6

10. Which of the following statements are *correct?* (1) The female reproductive cycle consists of a menstrual phase, a preovulatory phase, ovulation, and a postovulatory phase. (2) During the menstrual phase, small secondary follicles in the ovary begin to enlarge while the uterus is shedding its lining. (3) During the preovulatory phase, a dominant follicle continues to grow and begins to secrete estrogens and inhibin while the uterine lining begins to rebuild. (4) Ovulation results in the release of an ovum and the shedding of the uterine lining to nourish and support the released ovum. (5) After ovulation, a corpus luteum forms from the ruptured follicle and begins to secrete progesterone and estrogens, which it will continue to do throughout pregnancy if the egg is fertilized. (6) If pregnancy does not occur, then the corpus luteum degenerates into a scar called the corpus albicans, and the uterine lining is prepared to be shed again.
(a) 1, 2, 4, and 5   (b) 2, 4, 5, and 6   (c) 1, 4, 5, and 6
(d) 1, 3, 4, and 6   (e) 1, 2, 3, and 6

11. Oral contraceptives work by (1) causing a thickening of the cervical mucus, (2) blocking the uterine tubes, (3) inhibiting the release of FSH and LH, (4) preventing ovulation, (5) disrupting the plasma membranes of sperm, (6) irritating the endometrial lining so that it is inhospitable for fetal development.
(a) 3 only   (b) 3 and 4   (c) 1, 2, and 5
(d) 1, 3, and 4   (e) 1, 2, 3, 4, and 5

Au/Ed: Questions reordered to save # - OK? Pls. correct answer section.

SELF-QUIZ QUESTIONS **1179**

12. Match the following:

____ (a) the process during meiosis when portions of homologous chromosomes may be exchanged with each other

____ (b) refers to cells containing one-half the chromosome number

____ (c) the cell produced by the union of an egg and a sperm

____ (d) the degeneration of oogonia before and after birth

____ (e) a packet of discarded nuclear material from the first or second meiotic division of the egg

____ (f) refers to cells containing the full chromosome number

(1) zygote
(2) haploid
(3) diploid
(4) crossing-over
(5) polar body
(6) atresia

13. Match the following:

____ (a) modified sudoriferous glands involved in lactation

____ (b) a small, cylindrical mass of erectile tissue and nerves in the female; homologue of the male glans penis

____ (c) produce mucus in the female during sexual arousal and intercourse; homologous to the male bulbourethral glands

____ (d) the group of cells that nourish the developing oocyte and begin to secrete estrogens

____ (e) a pathway for sperm to reach the uterine tubes; the site of menstruation; the site of implantation of a fertilized ovum; the womb

____ (f) produces progesterone, estrogens, relaxin, and inhibin

____ (g) draw the ovum into the uterine tube

____ (h) the opening between the uterus and vagina

____ (i) muscular layer of uterus; responsible for expulsion of fetus from uterus

____ (j) mucus-secreting glands in the female that are homologous to the prostate gland

____ (k) the female copulatory organ; the birth canal

____ (l) passageway for the ovum to the uterus; usual site of fertilization; site of tubal ligation

____ (m) refers to the external genitals of the female

____ (n) the layer of the uterine lining that is partially shed during each monthly cycle

(1) follicle
(2) corpus luteum
(3) uterine tube
(4) fimbriae
(5) uterus
(6) cervix
(7) endometrium
(8) vagina
(9) vulva
(10) clitoris
(11) paraurethral glands
(12) greater vestibular glands
(13) mammary glands
(14) myometrium

14. Match the following:

____ (a) site of sperm maturation

____ (b) the male copulatory organ; a passageway for ejaculation of sperm and excretion of urine

____ (c) sperm-forming cells

____ (d) produce an alkaline substance that protects sperm by neutralizing acids in the urethra

____ (e) ejects sperm into the urethra just before ejaculation

____ (f) the supporting structure for the testes

____ (g) carries the sperm from the scrotum into the abdominopelvic cavity for release by ejaculation; is cut and tied as a means of sterilization

____ (h) the shared terminal duct of the reproductive and urinary systems in the male

____ (i) surrounds the urethra at the base of the bladder; produces secretions that contribute to sperm motility and viability

____ (j) produce testosterone

____ (k) supporting structure that consists of the ductus deferens, testicular artery, autonomic nerves, veins that drain the testes, lymphatic vessels, and cremaster muscle

____ (l) support and protect developing spermatogenic cells; secrete inhibin; form the blood–testis barrier

____ (m) secrete an alkaline fluid to help neutralize acids in the female reproductive tract; secrete fructose for use in ATP production by sperm

____ (n) contraction and relaxation move testes near to or away from pelvic cavity

____ (o) site of spermatogenesis

(1) spermatogenic cells
(2) Sertoli cells
(3) Leydig cells
(4) penis
(5) scrotum
(6) epididymis
(7) ductus deferens
(8) ejaculatory duct
(9) seminiferous tubules
(10) seminal vesicles
(11) prostate gland
(12) bulbourethral glands
(13) urethra
(14) spermatic cord
(15) cremaster muscle

CRITICAL THINKING QUESTIONS

1. Twenty-three-year-old Monica and her husband Bill are ready to start a family. They are both avid bicyclists and weight-lifters who carefully watch what they eat and pride themselves on their "buff" bodies. However, Monica is having difficulty becoming pregnant. Monica hasn't had a menstrual period for some time but informs the doctor that is normal for her. After consulting with her physician, the doctor tells Monica that she needs to cut back on her exercise routine and "put on some weight" in order to get pregnant. Monica is outraged because she figures she will gain enough weight when she is pregnant! Explain to Monica what has happened to her and why weight gain could help her achieve her goal of pregnancy.

2. The term "progesterone" means "for gestation (or pregnancy)." Describe how progesterone helps prepare the female body for pregnancy and helps maintain pregnancy.

3. After having borne five children, Mark's wife, Isabella, insists that he have a vasectomy. Mark is afraid that he will "dry up" and won't be able to perform sexually. How can you reassure him that his reproductive organs will function fine?

? ANSWERS TO FIGURE QUESTIONS

28.1 The gonads (testes) produce gametes (sperm) and hormones; the ducts transport, store, and receive gametes; the accessory sex glands secrete materials that support gametes; and the penis assists in the delivery and joining of gametes.

28.2 The cremaster and dartos muscles help regulate the temperature of the testes.

28.3 The tunica vaginalis and tunica albuginea are tissue layers that cover and protect the testes.

28.4 The Leydig (interstitial) cells of the testes secrete testosterone.

28.5 As a result of meiosis I, the number of chromosomes in each cell is reduced by half.

28.6 The sperm head contains the nucleus with 23 highly condensed chromosomes and an acrosome that contains enzymes for penetration of a secondary oocyte; the neck contains centrioles that produce microtubules for the rest of the tail; the midpiece contains mitochondria for ATP production for locomotion and metabolism; the principal and end pieces of the tail provide motility.

28.7 Sertoli cells secrete inhibin.

28.8 Testosterone inhibits secretion of LH, and inhibin inhibits secretion of FSH.

28.9 The seminal vesicles are the accessory sex glands that contribute the largest volume to seminal fluid.

28.10 Two tissue masses called the corpora cavernosa penis and one corpus spongiosum penis contain blood sinuses that fill with blood that cannot flow out of the penis as quickly as it flows in. The trapped blood engorges and stiffens the tissue, producing an erection. The corpus spongiosum penis keeps the spongy urethra open so that ejaculation can occur.

28.11 The testes are homologous to the ovaries; the glans penis is homologous to the clitoris; the prostate is homologous to the paraurethral glands; and the bulbourethral glands are homologous to the greater vestibular glands.

28.12 The mesovarium anchors the ovary to the broad ligament of the uterus and the uterine tube; the ovarian ligament anchors it to the uterus; the suspensory ligament anchors it to the pelvic wall.

28.13 Ovarian follicles secrete estrogens; the corpus luteum secretes progesterone, estrogens, relaxin, and inhibin.

28.14 Most ovarian follicles undergo atresia (degeneration).

28.15 Primary oocytes are present in the ovary at birth, so they are as old as the woman. In males, primary spermatocytes are continu-

ally being formed from stem cells (spermatogonia) and thus are only a few days old.

28.16 Fertilization most often occurs in the ampulla of the uterine tube.

28.17 Ciliated columnar epithelial cells and nonciliated (peg) cells with microvilli line the uterine tubes.

28.18 The endometrium is a highly vascularized, secretory epithelium that provides the oxygen and nutrients needed to sustain a fertilized egg; the myometrium is a thick smooth muscle layer that supports the uterine wall during pregnancy and contracts to expel the fetus at birth.

28.19 The stratum basalis of the endometrium provides cells to replace those that are shed (the stratum functionalis) during each menstruation.

28.20 Anterior to the vaginal opening are the mons pubis, clitoris, prepuce, and external urethral orifice. Lateral to the vaginal opening are the labia minora and labia majora.

28.21 The anterior portion of the perineum is called the urogenital triangle because its borders form a triangle that encloses the urethral (uro-) and vaginal (-genital) orifices.

28.22 Prolactin, estrogens, and progesterone regulate the synthesis of milk. Oxytocin regulates the ejection of milk.

28.23 The principal estrogen is β-estradiol.

28.24 The hormones responsible for the proliferative phase of endometrial growth are estrogens; for ovulation, LH; for growth of the corpus luteum, LH; and for the midcycle surge of LH, estrogens.

28.25 The effect of rising but moderate levels of estrogens is negative feedback inhibition of the secretion of GnRH, LH, and FSH.

28.26 This is negative feedback, because the response is opposite to the stimulus. A reduced amount of negative feedback due to declining levels of estrogens and progesterone stimulates release of GnRH, which in turn increases the production and release of FSH and LH, ultimately stimulating the secretion of estrogens.

28.27 The *SRY* gene on the Y chromosome is responsible for the development of the gonads into testes.

28.28 The presence of dihydrotestosterone (DHT) stimulates differentiation of the external genitals in males; its absence allows differentiation of the external genitals in females.

29 DEVELOPMENT AND INHERITANCE

DEVELOPMENT, INHERITANCE, AND HOMEOSTASIS *Both the genetic material inherited from parents (heredity) and normal development in the uterus (environment) play important roles in determining the homeostasis of a developing embryo and fetus and the subsequent birth of a healthy child.*

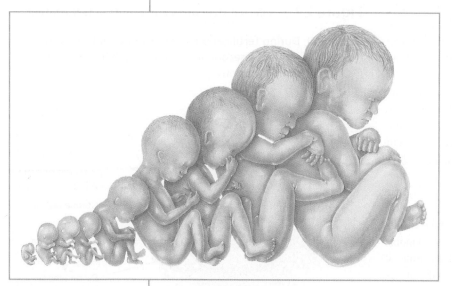

Developmental biology is the study of the sequence of events from the fertilization of a secondary oocyte by a sperm cell to the formation of an adult organism. **Pregnancy** is a sequence of events that begins with fertilization; proceeds to implantation, embryonic development, and fetal development; and ideally ends with birth about 38 weeks later, or 40 weeks after the last menstrual period.

Obstetrics (ob-STET-riks; *obstetrix* = midwife) deals with the management of pregnancy, labor, and the **neonatal period** (nē-ō-NĀ-tal), the first 28 days after birth. **Prenatal development** (prē-NĀ-tal; *pre-* = before; *-natal* = birth) is the time from fertilization to birth and is divided into three periods of three calendar months each, called **trimesters.**

1. The **first trimester** is the most critical stage of development, during which the rudiments of all the major organ systems appear, and also during which the developing organism is the most vulnerable to the effects of drugs, radiation, and microbes.

2. The **second trimester** is characterized by the nearly complete development of organ systems. By the end of this stage, the fetus assumes distinctively human features.

3. The **third trimester** represents a period of rapid fetal growth. During the early stages of this period, most of the organ systems are becoming fully functional.

In this chapter, we focus on the developmental sequence from fertilization through implantation, embryonic and fetal development, labor, birth, and the principles of inheritance (the passage of hereditary traits from one generation to another).

?

Did you ever wonder why the heart, blood vessels, and blood begin to form so early in the developmental process?

1181

29.1 EMBRYONIC PERIOD

⦿ **OBJECTIVE**

- Explain the major developmental events that occur during the embryonic period.

First Week of Development

The **embryonic period** extends from fertilization through the eighth week. The first week of development is characterized by several significant events including fertilization, cleavage of the zygote, blastocyst formation, and implantation.

Fertilization

During **fertilization** (fer'-ti-li-ZĀ-shun; *fertil-* = fruitful), the genetic material from a haploid sperm cell (spermatozoon) and a haploid secondary oocyte merges into a single diploid nucleus. Of the about 200 million sperm introduced into the vagina, fewer than 2 million (1%) reach the cervix of the uterus and only about 200 reach the secondary oocyte. Fertilization normally occurs in the uterine (fallopian) tube within 12 to 24 hours after ovulation. Sperm can remain viable for about 48 hours after deposition in the vagina, although a secondary oocyte is viable for only about 24 hours after ovulation. Thus, pregnancy is *most likely* to occur if intercourse takes place during a 3-day window—from 2 days before ovulation to 1 day after ovulation.

Sperm swim from the vagina into the cervical canal by the whiplike movements of their tails (flagella). The passage of sperm through the rest of the uterus and then into the uterine tube results mainly from contractions of the walls of these organs. Prostaglandins in semen are believed to stimulate uterine motility at the time of intercourse and to aid in the movement of sperm through the uterus and into the uterine tube. Sperm that reach the vicinity of the oocyte within minutes after ejaculation *are not capable* of fertilizing it until about seven hours later. During this time in the female reproductive tract, mostly in the uterine tube, sperm undergo **capacitation** (ka-pas-i-TĀ-shun; *capacit-* = capable of), a series of functional changes that cause the sperm's tail to beat even more vigorously and prepare its plasma membrane to fuse with the oocyte's plasma membrane. During capacitation, sperm are acted on by secretions in the female reproductive tract that result in the removal of cholesterol, glycoproteins, and proteins from the plasma membrane around the head of the sperm cell. Only capacitated sperm are capable of being attracted by and responding to chemical factors produced by the surrounding cells of the ovulated oocyte.

For fertilization to occur, a sperm cell first must penetrate two layers: the **corona radiata** (kō-RŌ-na = crown; rā-dē-A-ta = to shine), the granulosa cells that surround the secondary oocyte, and the **zona pellucida** (ZŌ-na = zone; pe-LOO-si-da = allowing passage of light), the clear glycoprotein layer between the corona radiata and the oocyte's plasma membrane (Figure 29.1a). The **acrosome** (AK-rō-sōm), a helmetlike structure that covers the head of a sperm (see Figure 28.6), contains several enzymes. Acrosomal enzymes and strong tail movements by the sperm help it penetrate the cells of the corona radiata and come in contact

with the zona pellucida. One of the glycoproteins in the zona pellucida, called ZP3, acts as a sperm receptor. Its binding to specific membrane proteins in the sperm head triggers the **acrosomal reaction,** the release of the contents of the acrosome. The acrosomal enzymes digest a path through the zona pellucida as the lashing sperm tail pushes the sperm cell onward. Although many sperm bind to ZP3 molecules and undergo acrosomal reactions, only the first sperm cell to penetrate the entire zona pellucida and reach the oocyte's plasma membrane fuses with the oocyte.

The fusion of a sperm cell with a secondary oocyte sets in motion events that block **polyspermy** (POL-ē-sper'-mē), fertilization by more than one sperm cell. Within a few seconds, the cell membrane of the oocyte depolarizes, which acts as a *fast block to polyspermy*—

Figure 29.1 Selected structures and events in fertilization.

🔓🔑 During fertilization, genetic material from a sperm cell and a secondary oocyte merge to form a single diploid nucleus.

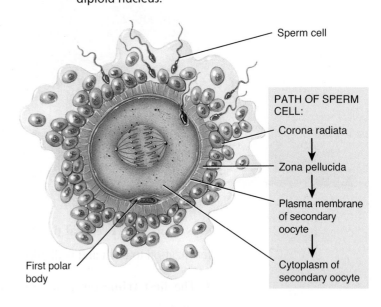

Sperm cell

PATH OF SPERM CELL:

Corona radiata

↓

Zona pellucida

↓

Plasma membrane of secondary oocyte

↓

Cytoplasm of secondary oocyte

First polar body

(a) Sperm cell penetrating secondary oocyte

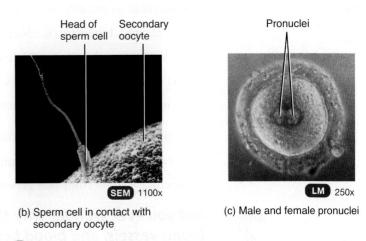

Head of sperm cell   Secondary oocyte

Pronuclei

SEM 1100x

LM 250x

(b) Sperm cell in contact with secondary oocyte

(c) Male and female pronuclei

? **What is capacitation?**

a depolarized oocyte cannot fuse with another sperm. Depolarization also triggers the intracellular release of calcium ions, which stimulate exocytosis of secretory vesicles from the oocyte. The molecules released by exocytosis inactivate ZP3 and harden the entire zona pellucida, events called the *slow block to polyspermy*.

Once a sperm cell enters a secondary oocyte, the oocyte first must complete meiosis II. It divides into a larger ovum (mature egg) and a smaller second polar body that fragments and disintegrates (see Figure 28.15). The nucleus in the head of the sperm develops into the **male pronucleus,** and the nucleus of the fertilized ovum develops into the **female pronucleus** (Figure 29.1c). After the male and female pronuclei form, they fuse, producing a single diploid nucleus, a process known as **syngamy** (SIN-ga-mē). Thus, the fusion of the haploid (*n*) pronuclei restores the diploid number (*2n*) of 46 chromosomes. The fertilized ovum now is called a **zygote** (*zygon* = yolk).

Dizygotic (fraternal) twins are produced from the independent release of two secondary oocytes and the subsequent fertilization of each by different sperm. They are the same age and in the uterus at the same time, but genetically they are as dissimilar as any other siblings. Dizygotic twins may or may not be the same sex. Because **monozygotic (identical) twins** develop from a single fertilized ovum, they contain exactly the same genetic material and are always the same sex. Monozygotic twins arise from separation of the developing cells into two embryos, which in 99% of the cases occurs before 8 days have passed. Separations that occur later than 8 days are likely to produce **conjoined twins,** a situation in which the twins are joined together and share some body structures.

Cleavage of the Zygote

After fertilization, rapid mitotic cell divisions of the zygote called **cleavage** (KLĒV-ij) take place (Figure 29.2). The first division of the zygote begins about 24 hours after fertilization and is completed about 6 hours later. Each succeeding division takes slightly less time. By the second day after fertilization, the second cleavage is completed and there are four cells (Figure 29.2b). By the end of the third day, there are 16 cells. The progressively smaller cells produced by cleavage are called **blastomeres** (BLAS-tō-mērz; *blasto-* = germ or sprout; *-meres* = parts). Successive cleavages eventually produce a solid sphere of cells called the **morula** (MOR-ū-la = mulberry). The morula is still surrounded by the zona pellucida and is about the same size as the original zygote (Figure 29.2c).

Blastocyst Formation

By the end of the fourth day, the number of cells in the morula increases as it continues to move through the uterine tube toward the uterine cavity. When the morula enters the uterine cavity on day 4 or 5, a glycogen-rich secretion from the glands of the endometrium of the uterus passes into the uterine cavity and enters the morula through the zona pellucida. This fluid, called **uterine milk,** along with nutrients stored in the cytoplasm of the blastomeres of the morula, provides nourishment for the developing morula. At the 32-cell stage, the fluid enters the morula, collects between the blastomeres, and reorganizes them around a large fluid-filled cavity called the **blastocyst cavity** (BLAS-tō-sist; *blasto-* = germ or

sprout; *-cyst* = bag), also called the **blastocoel** (BLAS-tō-sēl) (Figure 29.2e). Once the blastocyst cavity is formed, the developing mass is called the **blastocyst.** Though it now has hundreds of cells, the blastocyst is still about the same size as the original zygote.

During the formation of the blastocyst two distinct cell populations arise: the embryoblast and trophoblast (Figure 29.2e). The **embryoblast** (EM-brē-ō-blast′), or **inner cell mass,** is located internally and eventually develops into the embryo. The **trophoblast** (TRŌF-ō-blast; *tropho-* = develop or nourish) is the outer superficial layer of cells that forms the spherelike wall of the blastocyst. It will ultimately develop into the outer chorionic sac that surrounds

Figure 29.2 Cleavage and the formation of the morula and blastocyst.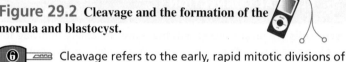

Cleavage refers to the early, rapid mitotic divisions of a zygote.

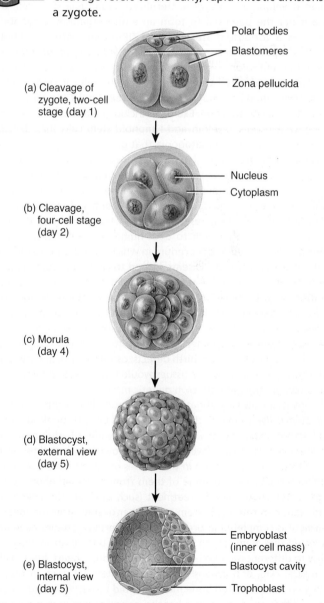

(a) Cleavage of zygote, two-cell stage (day 1)

Polar bodies
Blastomeres
Zona pellucida

(b) Cleavage, four-cell stage (day 2)

Nucleus
Cytoplasm

(c) Morula (day 4)

(d) Blastocyst, external view (day 5)

(e) Blastocyst, internal view (day 5)

Embryoblast (inner cell mass)
Blastocyst cavity
Trophoblast

❓ What is the histological difference between a morula and a blastocyst?

the fetus and the fetal portion of the placenta, the site of exchange of nutrients and wastes between the mother and fetus. On about the fifth day after fertilization, the blastocyst "hatches" from the zona pellucida by digesting a hole in it with an enzyme, and then squeezing through the hole. This shedding of the zona pellucida is necessary in order to permit the next step, implantation (attachment) into the vascular, glandular endometrial lining of the uterus.

Implantation

The blastocyst remains free within the uterine cavity for about 2 days before it attaches to the uterine wall. At this time the en-

dometrium is in its secretory phase. About 6 days after fertilization, the blastocyst loosely attaches to the endometrium in a process called **implantation** (im-plan-TĀ-shun) (Figure 29.3). As the blastocyst implants, usually in either the posterior portion

Figure 29.3 Relation of a blastocyst to the endometrium of the uterus at the time of implantation.

Implantation, the attachment of a blastocyst to the endometrium, occurs about 6 days after fertilization.

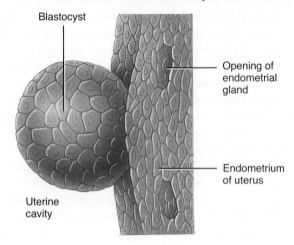

(a) External view of blastocyst, about 6 days after fertilization

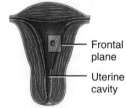

Frontal section through uterus

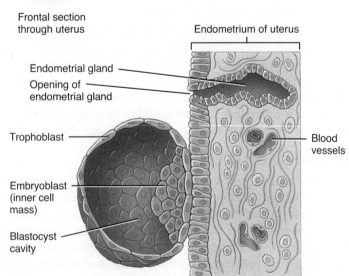

(b) Frontal section through endometrium of uterus and blastocyst, about 6 days after fertilization

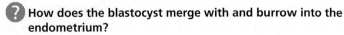

How does the blastocyst merge with and burrow into the endometrium?

CLINICAL CONNECTION | Stem Cell Research and Therapeutic Cloning

Stem cells are unspecialized cells that have the ability to divide for indefinite periods and give rise to specialized cells. In the context of human development, a zygote (fertilized ovum) is a stem cell. Because it has the potential to form an entire organism, a zygote is known as a *totipotent stem cell* (tō-TIP-ō-tent; *totus-* = whole; *-potentia* = power). Inner cell mass cells, called *pluripotent stem cells* (ploo-RIP-ō-tent; *plur-* = several), can give rise to many (but not all) different types of cells. Later, pluripotent stem cells can undergo further specialization into *multipotent stem cells* (mul-TIP-ō-tent), stem cells with a specific function. Examples include keratinocytes that produce new skin cells, myeloid and lymphoid stem cells that develop into blood cells, and spermatogonia that give rise to sperm. Pluripotent stem cells currently used in research are derived from (1) the embryoblast of embryos in the blastocyst stage that were destined to be used for infertility treatments but were not needed and from (2) nonliving fetuses terminated during the first 3 months of pregnancy.

On October 13, 2001, researchers reported cloning of the first human embryo to grow cells to treat human diseases. **Therapeutic cloning** is envisioned as a procedure in which the genetic material of a patient with a particular disease is used to create pluripotent stem cells to treat the disease. Using the principles of therapeutic cloning, scientists hope to make an embryo clone of a patient, remove the pluripotent stem cells from the embryo, and then use them to grow tissues to treat particular diseases and disorders, such as cancer, Parkinson disease, Alzheimer disease, spinal cord injury, diabetes, heart disease, stroke, burns, birth defects, osteoarthritis, and rheumatoid arthritis. Presumably, the tissues would not be rejected since they would contain the patient's own genetic material.

Scientists are also investigating the potential clinical applications of *adult stem cells*—stem cells that remain in the body throughout adulthood. Recent experiments suggest that the ovaries of adult mice contain stem cells that can develop into new ova (eggs). If these same types of stem cells are found in the ovaries of adult women, scientists could potentially harvest some of them from a woman about to undergo a sterilizing medical treatment (such as chemotherapy), store them, and then return the stem cells to her ovaries after the medical treatment is completed in order to restore fertility. Studies have also suggested that stem cells in human adult red bone marrow have the ability to differentiate into cells of the liver, kidney, heart, lung, skeletal muscle, skin, and organs of the gastrointestinal tract. In theory, adult stem cells from red bone marrow could be harvested from a patient and then used to repair other tissues and organs in that patient's body without having to use stem cells from embryos. •

of the fundus or the body of the uterus, it orients with the inner cell mass toward the endometrium (Figure 29.3b). About 7 days after fertilization, the blastocyst attaches to the endometrium more firmly, endometrial glands in the vicinity enlarge, and the endometrium becomes more vascularized (forms new blood vessels). The blastocyst eventually secretes enzymes and burrows into the endometrium and becomes surrounded by it.

Following implantation, the endometrium is known as the **decidua** (dē-SID-ū-a = falling off). The decidua separates from the endometrium after the fetus is delivered, much as it does in normal menstruation. Different regions of the decidua are named based on their positions relative to the site of the implanted blastocyst (Figure 29.4). The **decidua basalis** is the portion of the endometrium between the embryo and the stratum basalis of the uterus; it provides large amounts of glycogen and lipids for the developing embryo and fetus and later becomes the maternal part of the placenta. The **decidua capsularis** is the portion of the endometrium located between the embryo and the uterine cavity. The **decidua parietalis** (par-ri-e-TAL-is) is the remaining modified endometrium that lines the noninvolved areas of the rest of the uterus. As the embryo and later the fetus enlarges, the decidua capsularis bulges into the uterine cavity and fuses with the decidua parietalis, thereby obliterating the uterine cavity. By about 27 weeks, the decidua capsularis degenerates and disappears.

The major events associated with the first week of development are summarized in Figure 29.5.

Figure 29.4 Regions of the decidua.

The decidua is a modified portion of the endometrium that develops after implantation.

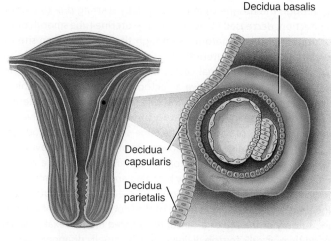

Frontal section of uterus · Details of decidua

Which part of the decidua helps form the maternal part of the placenta?

Figure 29.5 Summary of events associated with the first week of development.

Fertilization usually occurs in the uterine tube.

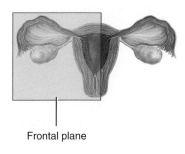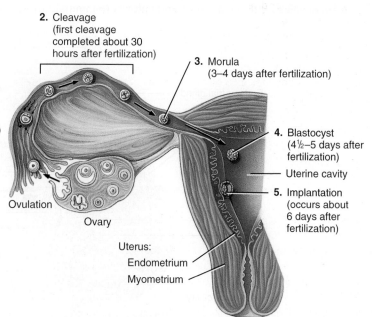

Frontal section through uterus, uterine tube, and ovary

In which phase of the uterine cycle does implantation occur?

CLINICAL CONNECTION | Ectopic Pregnancy

Ectopic pregnancy (ek-TOP-ik; *ec-* = out of; *-topic* = place) is the development of an embryo or fetus outside the uterine cavity. An ectopic pregnancy usually occurs when movement of the fertilized ovum through the uterine tube is impaired by scarring due to a prior tubal infection, decreased movement of the uterine tube smooth muscle, or abnormal tubal anatomy. Although the most common site of ectopic pregnancy is the uterine tube, ectopic pregnancies may also occur in the ovary, abdominal cavity, or uterine cervix. Women who smoke are twice as likely to have an ectopic pregnancy because nicotine in cigarette smoke paralyzes the cilia in the lining of the uterine tube (as it does those in the respiratory airways). Scars from pelvic inflammatory disease, previous uterine tube surgery, and previous ectopic pregnancy may also hinder movement of the fertilized ovum.

The signs and symptoms of ectopic pregnancy include one or two missed menstrual cycles followed by bleeding and acute abdominal and pelvic pain. Unless removed, the developing embryo can rupture the uterine tube, often resulting in death of the mother. Treatment options include surgery or the use of a cancer drug called methotrexate, which causes embryonic cells to stop dividing and eventually disappear. •

✔**CHECKPOINT**

1. Where does fertilization normally occur?
2. How is polyspermy prevented?
3. What is a morula, and how is it formed?
4. Describe the layers of a blastocyst and their eventual fates.
5. When, where, and how does implantation occur?

Second Week of Development

Development of the Trophoblast

About 8 days after fertilization, the trophoblast develops into two layers in the region of contact between the blastocyst and endometrium. These are a **syncytiotrophoblast** (sin-sīt´-ē-ō-TRŌF-ō-blast) that contains no distinct cell boundaries and a **cytotrophoblast** (sī-tō-TRŌF-ō-blast) between the embryoblast and syncytiotrophoblast that is composed of distinct cells (Figure 29.6a). The two layers of trophoblast become part of the chorion (one of the fetal membranes) as they undergo further growth (see Figure 29.11a inset). During implantation, the syncytiotrophoblast secretes enzymes that enable the blastocyst to penetrate the uterine lining by digesting and liquefying the endometrial cells. Eventually, the blastocyst becomes buried in the endometrium and inner one-third of the myometrium. Another secretion of the trophoblast is human chorionic gonadotropin (hCG), which has actions similar to LH. Human chorionic gonadotropin rescues the corpus luteum from degeneration and sustains its secretion of progesterone and estrogens. These hormones maintain the uterine lining in a secretory state, preventing menstruation. Peak secretion of hCG occurs about the ninth week of pregnancy, at which time the placenta is fully developed and produces the progesterone and estrogens that continue to sustain the pregnancy. The presence of hCG in maternal blood or urine is an indicator of pregnancy and is the hormone detected by home pregnancy tests.

Figure 29.6 Principal events of the second week of development.

About 8 days after fertilization, the trophoblast develops into a syncytiotrophoblast and a cytotrophoblast; the inner cell mass develops into a hypoblast and epiblast (bilaminar embryonic disc).

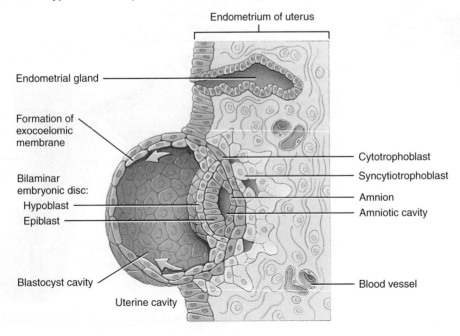

(a) Frontal section through endometrium of uterus showing blastocyst, about 8 days after fertilization

Development of the Bilaminar Embryonic Disc

Like those of the trophoblast, cells of the embryoblast also differentiate into two layers around 8 days after fertilization: a **hypoblast (primitive endoderm)** and **epiblast (primitive ectoderm)** (Figure 29.6a). Cells of the hypoblast and epiblast together form a flat disc referred to as the **bilaminar embryonic disc** (bī-LAM-in-ar = two-layered). Soon, a small cavity appears within the epiblast and eventually enlarges to form the **amniotic cavity** (am-nē-OT-ik; *amnio-* = lamb).

Development of the Amnion

As the amniotic cavity enlarges, a single layer of squamous cells forms a domelike roof above the epiblast cells called the **amnion** (AM-nē-on) (Figure 29.6a). Thus, the amnion forms the roof of the amniotic cavity, and the epiblast forms the floor. Initially, the amnion overlies only the bilaminar embryonic disc. However, as the embryonic disc increases in size and begins to fold, the amnion eventually surrounds the entire embryo (see Figure 29.11a inset), creating the amniotic cavity that becomes filled with **amniotic fluid.** Most amniotic fluid is initially derived from maternal blood. Later, the fetus contributes to the fluid by excreting urine into the amniotic cavity. Amniotic fluid serves as a shock absorber for the fetus, helps regulate fetal body temperature, helps prevent the fetus from drying out, and prevents adhesions between the skin of the fetus and surrounding tissues. The amnion usually ruptures just before birth; it and its fluid constitute the "bag of waters." Embryonic cells are normally sloughed off into amniotic fluid. They can be examined in a procedure called *amniocentesis,* which involves withdrawing some of the amniotic fluid that bathes the developing fetus and analyzing the fetal cells and dissolved substances (see Section 29.4).

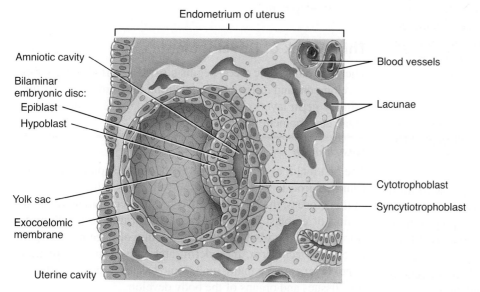

(b) Frontal section through endometrium of uterus showing blastocyst, about 9 days after fertilization

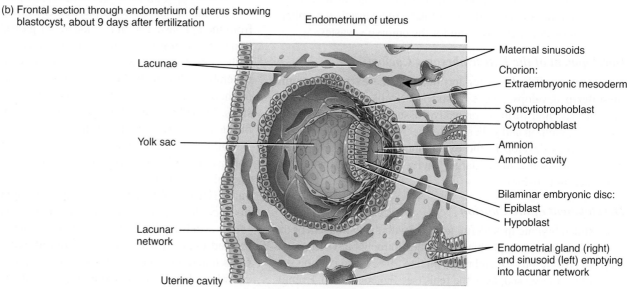

(c) Frontal section through endometrium of uterus showing blastocyst, about 12 days after fertilization

? **How is the bilaminar embryonic disc connected to the trophoblast?**

Development of the Yolk Sac

Also on the eighth day after fertilization, cells at the edge of the hypoblast migrate and cover the inner surface of the blastocyst wall (Figure 29.6a). The migrating columnar cells become squamous (flat) and then form a thin membrane referred to as the **exocoelomic membrane** (ek′-sō-sē-LŌ-mik; *exo-* = outside; *-koilos* = space). Together with the hypoblast, the exocoelomic membrane forms the wall of the **yolk sac,** the former blastocyst cavity during earlier development (Figure 29.6b). As a result, the bilaminar embryonic disc is now positioned between the amniotic cavity and yolk sac.

Since human embryos receive their nutrients from the endometrium, the yolk sac is relatively empty and small, and decreases in size as development progresses (see Figure 29.11a). Nevertheless, the yolk sac has several important functions in humans: supplies nutrients to the embryo during the second and third weeks of development; is the source of blood cells from the third through sixth weeks; contains the first cells (primordial germ cells) that will eventually migrate into the developing gonads, differentiate into the primitive germ cells, and form gametes; forms part of the gut (gastrointestinal tract); functions as a shock absorber; and helps prevent drying out of the embryo.

Development of Sinusoids

On the ninth day after fertilization, the blastocyst becomes completely embedded in the endometrium. As the syncytiotrophoblast expands, small spaces called **lacunae** (la-KOO-nē = little lakes) develop within it (Figure 29.6b).

By the twelfth day of development, the lacunae fuse to form larger, interconnecting spaces called **lacunar networks** (Figure 29.6c). Endometrial capillaries around the developing embryo become dilated and are referred to as **maternal sinusoids** (SĪ-nū-soyds). As the syncytiotrophoblast erodes some of the maternal sinusoids and endometrial glands, maternal blood and secretions from the glands enter the lacunar networks and flow through them. Maternal blood is both a rich source of materials for embryonic nutrition and a disposal site for the embryo's wastes.

Development of the Extraembryonic Coelom

About the twelfth day after fertilization, the **extraembryonic mesoderm** develops. These mesodermal cells are derived from the yolk sac and form a connective tissue layer (mesenchyme) around the amnion and yolk sac (Figure 29.6c). Soon a number of large cavities develop in the extraembryonic mesoderm, which then fuse to form a single, larger cavity called the **extraembryonic coelom** (SĒ-lom).

Development of the Chorion

The extraembryonic mesoderm, together with the two layers of the trophoblast (the cytotrophoblast and syncytiotrophoblast), forms the **chorion** (KŌ-rē-on = membrane) (Figure 29.6c). The chorion surrounds the embryo and, later, the fetus (see Figure 29.11a). Eventually it becomes the principal embryonic part of the placenta, the structure for exchange of materials between mother and fetus. The chorion also protects the embryo and fetus from the immune responses of the mother in two ways: (1) It secretes proteins that block antibody production by the mother. (2) It promotes the production of

T lymphocytes that suppress the normal immune response in the uterus. Finally, the chorion produces human chorionic gonadotropin (hCG), an important hormone of pregnancy (see Figure 29.16).

The inner layer of the chorion eventually fuses with the amnion. With the development of the chorion, the extraembryonic coelom is now referred to as the **chorionic cavity.** By the end of the second week of development, the bilaminar embryonic disc becomes connected to the trophoblast by a band of extraembryonic mesoderm called the **connecting (body) stalk** (see Figure 29.7). The connecting stalk is the future umbilical cord.

✔ **CHECKPOINT**

6. What are the functions of the trophoblast?
7. How is the bilaminar embryonic disc formed?
8. Describe the formation of the amnion, yolk sac, and chorion and explain their functions.
9. Why are sinusoids important during embryonic development?

Third Week of Development

The third embryonic week begins a 6-week period of very rapid development and differentiation. During the third week, the three primary germ layers are established and lay the groundwork for organ development in weeks 4 through 8.

Gastrulation

The first major event of the third week of development, **gastrulation** (gas-troo-LĀ-shun), occurs about 15 days after fertilization. In this process, the bilaminar (two-layered) embryonic disc, consisting of epiblast and hypoblast, transforms into a **trilaminar** (three-layered) **embryonic disc** consisting of three primary germ layers: the ectoderm, mesoderm, and endoderm. The **primary germ layers** are the major embryonic tissues from which the various tissues and organs of the body develop.

Gastrulation involves the rearrangement and migration of cells from the epiblast. The first evidence of gastrulation is the formation of the **primitive streak,** a faint groove on the dorsal surface of the epiblast that elongates from the posterior to the anterior part of the embryo (Figure 29.7a). The primitive streak clearly establishes the head and tail ends of the embryo, as well as its right and left sides. At the head end of the primitive streak a small group of epiblastic cells forms a rounded structure called the **primitive node.**

Following formation of the primitive streak, cells of the epiblast move inward below the primitive streak and detach from the epiblast (Figure 29.7b) in a process called **invagination** (in-vaj′-i-NĀ-shun). Once the cells have invaginated, some of them displace the hypoblast, forming the **endoderm** (*endo-* = inside; *-derm* = skin). Other cells remain between the epiblast and newly formed endoderm to form the **mesoderm** (*meso-* = middle). Cells remaining in the epiblast then form the **ectoderm** (*ecto-* = outside). The ectoderm and endoderm are epithelia composed of tightly packed cells; the mesoderm is a loosely organized connective tissue (mesenchyme). As the embryo develops, the endoderm ultimately becomes the epithelial lining of the gastrointestinal tract, respiratory tract, and several other organs. The mesoderm gives rise to muscles, bones, and other connective tissues, and the

Figure 29.7 Gastrulation.

Gastrulation involves the rearrangement and migration of cells from the epiblast.

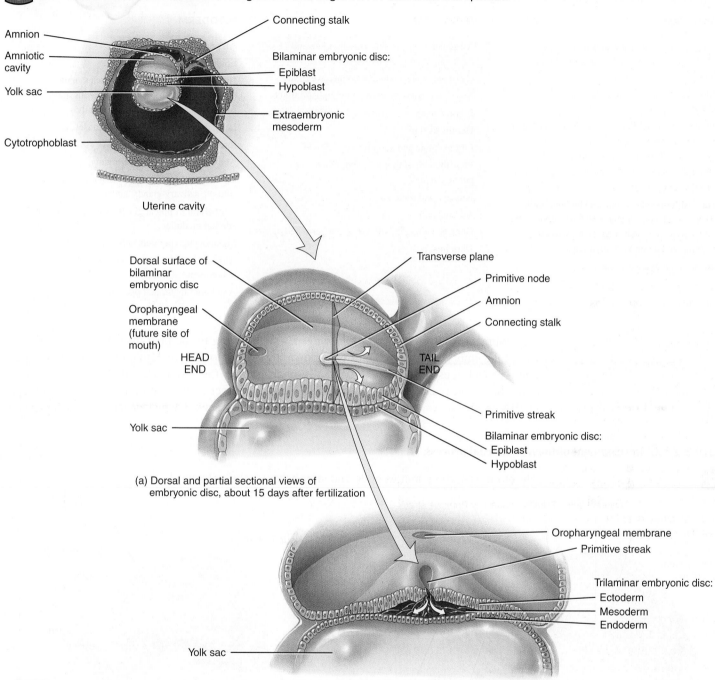

Amnion

Amniotic cavity

Yolk sac

Cytotrophoblast

Connecting stalk

Bilaminar embryonic disc:
Epiblast
Hypoblast

Extraembryonic mesoderm

Uterine cavity

Dorsal surface of bilaminar embryonic disc

Oropharyngeal membrane (future site of mouth)

HEAD END

Yolk sac

Transverse plane

Primitive node

Amnion

Connecting stalk

TAIL END

Primitive streak

Bilaminar embryonic disc:
Epiblast
Hypoblast

(a) Dorsal and partial sectional views of embryonic disc, about 15 days after fertilization

Oropharyngeal membrane

Primitive streak

Trilaminar embryonic disc:
Ectoderm
Mesoderm
Endoderm

Yolk sac

(b) Transverse section of trilaminar embryonic disc, about 16 days after fertilization

? **What is the significance of gastrulation?**

peritoneum. The ectoderm develops into the epidermis of the skin and the nervous system. Table 29.1 provides more details about the fates of these primary germ layers.

About 16 days after fertilization, mesodermal cells from the primitive node migrate toward the head end of the embryo and form a hollow tube of cells in the midline called the **notochordal process** (nō-tō-KOR-dal) (Figure 29.8). By days 22–24, the notochordal process becomes a solid cylinder of cells called the **notochord** (nō-

tō-KORD; *noto-* = back; *-chord* = cord). This structure plays an extremely important role in **induction** (in-DUK-shun), the process by which one tissue *(inducing tissue)* stimulates the development of an adjacent unspecialized tissue *(responding tissue)* into a specialized one. An inducing tissue usually produces a chemical substance that influences the responding tissue. The notochord induces certain mesodermal cells to develop into the vertebral bodies. It also forms the nucleus pulposus of the intervertebral discs (see Figure 7.24).

TABLE 29.1

Structures Produced by the Three Primary Germ Layers

ENDODERM	MESODERM	ECTODERM
Epithelial lining of gastrointestinal tract (except oral cavity and anal canal) and epithelium of its glands.	All skeletal and cardiac muscle tissue and most smooth muscle tissue.	All nervous tissue.
Epithelial lining of urinary bladder, gallbladder, and liver.	Cartilage, bone, and other connective tissues.	Epidermis of skin.
Epithelial lining of pharynx, auditory (eustachian) tubes, tonsils, tympanic (middle ear) cavity, larynx, trachea, bronchi, and lungs.	Blood, red bone marrow, and lymphatic tissue.	Hair follicles, arrector pili muscles, nails, epithelium of skin glands (sebaceous and sudoriferous), and mammary glands.
	Blood vessels and lymphatic vessels.	Lens, cornea, and internal eye muscles.
	Dermis of skin.	Internal and external ear.
Epithelium of thyroid gland, parathyroid glands, pancreas, and thymus.	Fibrous tunic and vascular tunic of eye.	Neuroepithelium of sense organs.
Epithelial lining of prostate and bulbourethral (Cowper's) glands, vagina, vestibule, urethra, and associated glands such as greater (Bartholin's) vestibular and lesser vestibular glands.	Mesothelium of thoracic, abdominal, and pelvic cavities.	Epithelium of oral cavity, nasal cavity, paranasal sinuses, salivary glands, and anal canal.
	Kidneys and ureters.	Epithelium of pineal gland, pituitary gland, and adrenal medullae.
	Adrenal cortex.	Melanocytes (pigment cells).
Gametes (sperm and oocytes).	Gonads and genital ducts (except germ cells).	Almost all skeletal and connective tissue components of head.
	Dura mater.	Arachnoid mater and pia mater.

Also during the third week of development, two faint depressions appear on the dorsal surface of the embryo where the ectoderm and endoderm make contact but lack mesoderm between them. The structure closer to the head end is called the **oropharyngeal membrane** (or-ō-fa-RIN-jē-al; *oro-* = mouth; *-pharyngeal* = pertaining to the pharynx) (Figure 29.8a, b). It breaks down during the fourth week to connect the mouth cavity to the pharynx and the remainder of the gastrointestinal tract. The structure closer to the tail end is called the **cloacal membrane** (klō-Ā-kul = sewer), which degenerates in the seventh

Figure 29.8 Development of the notochordal process.

The notochordal process develops from the primitive node and later becomes the notochord.

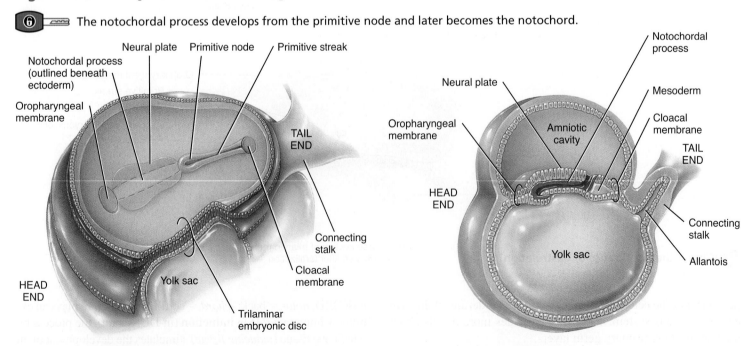

(a) Dorsal and partial sectional views of trilaminar embryonic disc, about 16 days after fertilization

(b) Sagittal section of trilaminar embryonic disc, about 16 days after fertilization

❓ **What is the significance of the notochord?**

week to form the openings of the anus and urinary and reproductive tracts.

When the cloacal membrane appears, the wall of the yolk sac forms a small vascularized outpouching called the **allantois** (a-LAN-tō-is; *allant-* = sausage) that extends into the connecting stalk (Figure 29.8b). In nonmammalian organisms enclosed in an amnion, the allantois is used for gas exchange and waste removal. Because of the role of the human placenta in these activities, the allantois is not a prominent structure in humans (see Fig-

ure 29.11a, inset). Nevertheless, it does function in the early formation of blood and blood vessels, and it is associated with the development of the urinary bladder.

Neurulation

In addition to inducing mesodermal cells to develop into vertebral bodies, the notochord also induces ectodermal cells over it to form the **neural plate** (Figure 29.9a). (Also see Figure 14.27.) By the end of the third week, the lateral edges of the neural plate

Figure 29.9 Neurulation and the development of somites.

Neurulation is the process by which the neural plate, neural folds, and neural tube form.

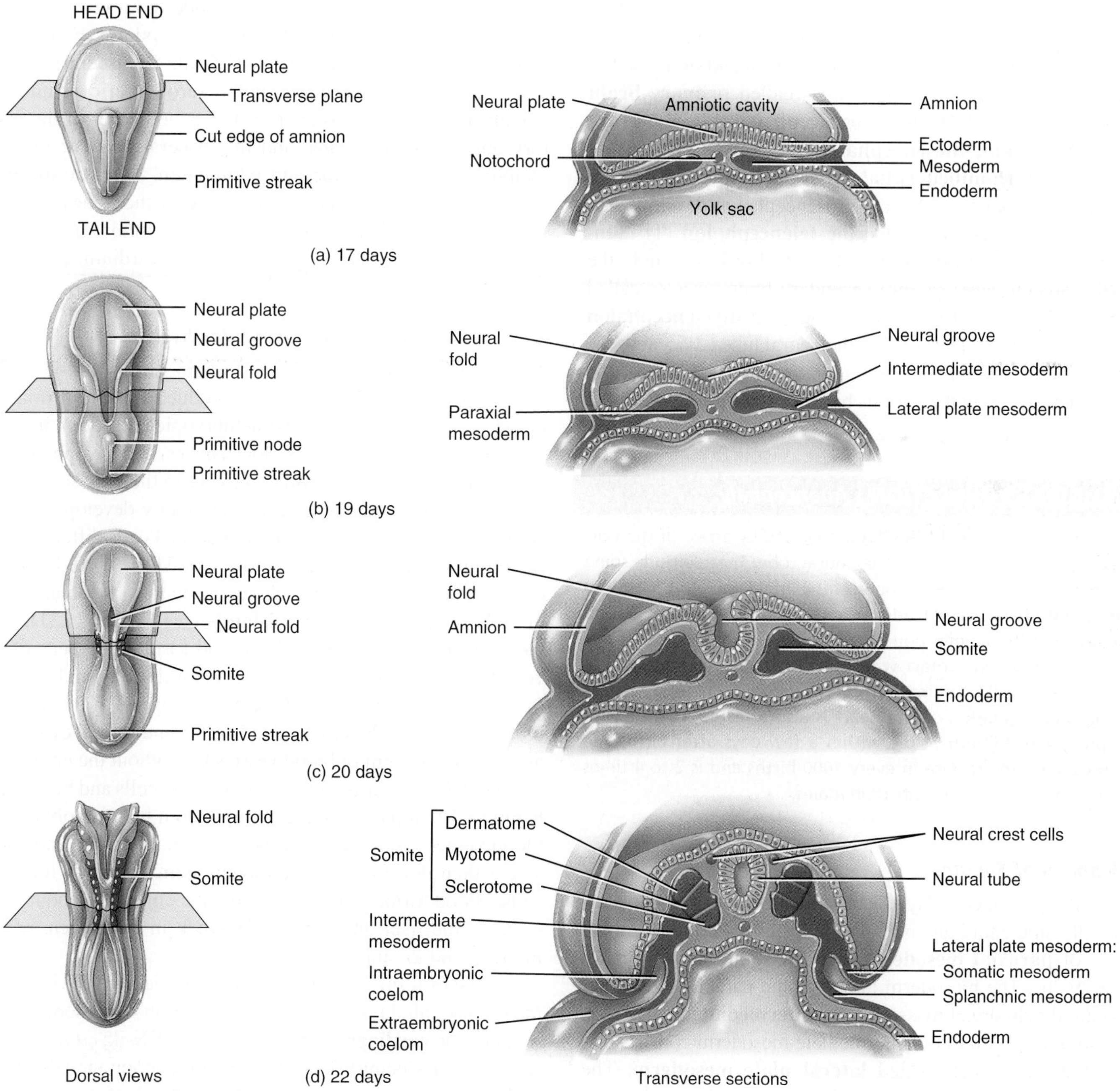

Which structures develop from the neural tube and somites?

become more elevated and form the **neural fold** (Figure 29.9b). The depressed midregion is called the **neural groove** (Figure 29.9c). Generally, the neural folds approach each other and fuse, thus converting the neural plate into a **neural tube** (Figure 29.9d). This occurs first near the middle of the embryo and then progresses toward the head and tail ends. Neural tube cells then develop into the brain and spinal cord. The process by which the neural plate, neural folds, and neural tube form is called **neurulation** (noor-oo-LĀ-shun).

As the neural tube forms, some of the ectodermal cells from the tube migrate to form several layers of cells called the **neural crest** (see Figure 14.27b). Neural crest cells give rise to all sensory neurons and postganglionic neurons of the peripheral nerves, the adrenal medullae, melanocytes (pigment cells) of the skin, arachnoid mater, and pia mater of the brain and spinal cord, and almost all of the skeletal and connective tissue components of the head.

At about 4 weeks after fertilization, the head end of the neural tube develops into three enlarged areas called **primary brain vesicles** (see Figure 14.28): the **prosencephalon** (PROS-en-sef′-a-lon) or **forebrain, mesencephalon** (mez-en-SEF-a-lon) or **midbrain,** and **rhombencephalon** (ROM-ben-sef′-a-lon) or **hindbrain.** At about 5 weeks, the prosencephalon develops into **secondary brain vesicles** called the **telencephalon** (TEL-en-sef′-a-lon) and **diencephalon** (dī-en-SEF-a-lon), and the rhombencephalon develops into secondary brain vesicles called the **metencephalon** (MET-en-sef′-a-lon) and **myelencephalon** (MĪ-el-en-sef′-a-lon). The areas of the neural tube adjacent to the myelencephalon develop into the spinal cord. The parts of the brain that develop from the various brain vesicles are described in Section 14.1.

✚ CLINICAL CONNECTION | Anencephaly

Neural tube defects (NTDs) are caused by arrest of the normal development and closure of the neural tube. These include spina bifida (discussed in Disorders: Homeostatic Imbalances in Chapter 7) and **anencephaly** (an-en-SEF-a-lē; *an-* = without; *-encephal* = brain). In anencephaly, the cranial bones fail to develop and certain parts of the brain remain in contact with amniotic fluid and degenerate. Usually, a part of the brain that controls vital functions such as breathing and regulation of the heart is also affected. Infants with anencephaly are stillborn or die within a few days after birth. The condition occurs about once in every 1000 births and is 2 to 4 times more common in female infants than males. •

Development of Somites

By about the seventeenth day after fertilization, the mesoderm adjacent to the notochord and neural tube forms paired longitudinal columns of **paraxial mesoderm** (par-AK-sē-al; *para-* = near) (Figure 29.9b). The mesoderm lateral to the paraxial mesoderm forms paired cylindrical masses called **intermediate mesoderm.** The mesoderm lateral to the intermediate mesoderm consists of a pair of flattened sheets called **lateral plate mesoderm.** The paraxial mesoderm soon segments into a series of paired, cube-shaped structures called **somites** (SŌ-mīts = little bodies). By the

end of the fifth week, 42–44 pairs of somites are present. The number of somites that develop over a given period can be correlated to the approximate age of the embryo.

Each somite differentiates into three regions: a **myotome** (MĪ-ō-tōm), a **dermatome,** and a **sclerotome** (SKLĒR-ō-tōm) (see Figure 10.17b). The myotomes develop into the skeletal muscles of the neck, trunk, and limbs; the dermatomes form connective tissue, including the dermis of the skin; and the sclerotomes give rise to the vertebrae and ribs.

Development of the Intraembryonic Coelom

In the third week of development, small spaces appear in the lateral plate mesoderm. These spaces soon merge to form a larger cavity called the **intraembryonic coelom** (SĒ-lom = cavity). This cavity splits the lateral plate mesoderm into two parts called the splanchnic mesoderm and somatic mesoderm (Figure 29.9d). **Splanchnic mesoderm** (SPLANGK-nik = visceral) forms the heart and the visceral layer of the serous pericardium, blood vessels, the smooth muscle and connective tissues of the respiratory and digestive organs, and the visceral layer of the serous membrane of the pleurae and peritoneum. **Somatic mesoderm** (sō-MAT-ik; *soma-* = body) gives rise to the bones, ligaments, blood vessels, and connective tissue of the limbs and the parietal layer of the serous membrane of the pericardium, pleurae, and peritoneum.

Development of the Cardiovascular System

At the beginning of the third week, **angiogenesis** (an′-jē-ō-JEN-e-sis; *angio-* = vessel; *-genesis* = production), the formation of blood vessels, begins in the extraembryonic mesoderm in the yolk sac, connecting stalk, and chorion. This early development is necessary because there is insufficient yolk in the yolk sac and ovum to provide adequate nutrition for the rapidly developing embryo. Angiogenesis is initiated when mesodermal cells differentiate into **hemangioblasts** (hē-MAN-jē-ō-blasts). These then develop into cells called **angioblasts,** which aggregate to form isolated masses of cells referred to as **blood islands** (see Figure 21.31). Spaces soon develop in the blood islands and form the lumens of blood vessels. Some angioblasts arrange themselves around each space to form the endothelium and the tunics (layers) of the developing blood vessels. As the blood islands grow and fuse, they soon form an extensive system of blood vessels throughout the embryo.

About 3 weeks after fertilization, blood cells and blood plasma begin to develop *outside* the embryo from hemangioblasts in the blood vessels in the walls of the yolk sac, allantois, and chorion. These then develop into pluripotent stem cells that form blood cells. Blood formation begins *within* the embryo at about the fifth week in the liver and the twelfth week in the spleen, red bone marrow, and thymus.

The heart forms from splanchnic mesoderm in the head end of the embryo on days 18 and 19. This region of mesodermal cells is called the **cardiogenic area** (kar-dē-ō-JEN-ik; *cardio-* = heart; *-genic* = producing). In response to induction signals from the underlying endoderm, these mesodermal cells form a pair of **endocardial tubes** (see Figure 20.19). The tubes then fuse to form a single **primitive heart tube.** By the end of the third week, the primitive heart tube bends on itself, becomes S-shaped, and begins to beat. It then joins blood vessels in other parts of the embryo, connecting stalk, chorion, and yolk sac to form a primitive cardiovascular system.

Development of the Chorionic Villi and Placenta

As the embryonic tissue invades the uterine wall, maternal uterine vessels are eroded and maternal blood fills spaces, called **lacunae** (la-KOO-nē), within the invading tissue. By the end of the second week of development, **chorionic villi** (kō-rē-ON-ik VIL-lī) begin to develop. These fingerlike projections consist of chorion (syncytiotrophoblast surrounded by cytotrophoblast) that projects into the endometrial wall of the uterus (Figure 29.10a). By the end of the third week, blood capillaries develop in the chorionic villi (Figure 29.10b). Blood vessels in the chorionic villi connect to the embryonic heart by way of the umbilical arteries and umbilical vein through the connecting (body) stalk, which will eventually become the umbilical cord (Figure 29.10c). The fetal blood capillaries within the chorionic villi project into the lacunae, which unite to form the **intervillous spaces** (in′-ter-VIL-us) that bathe the chorionic villi with maternal blood. As a result, maternal blood bathes the chorion-covered fetal blood vessels. Note, however, that maternal and fetal blood vessels do not join, and the blood they carry does not normally mix. Instead, oxygen and nutrients in the blood of the mother's intervillous spaces, which are the spaces between chorionic villi, diffuse across the cell membranes into the capillaries of the villi. Waste products such as carbon dioxide diffuse in the opposite direction.

Figure 29.10 Development of chorionic villi.

Blood vessels in chorionic villi connect to the embryonic heart via the umbilical arteries and umbilical vein.

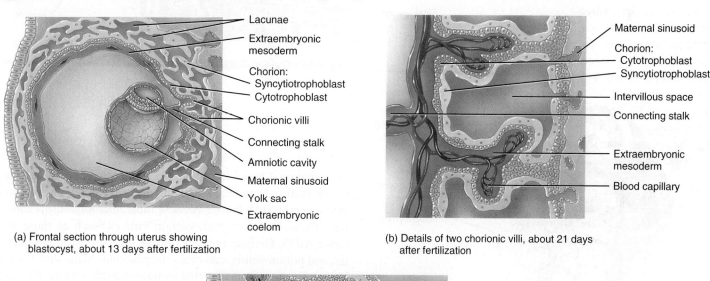

(a) Frontal section through uterus showing blastocyst, about 13 days after fertilization

Labels: Lacunae; Extraembryonic mesoderm; Chorion: Syncytiotrophoblast, Cytotrophoblast; Chorionic villi; Connecting stalk; Amniotic cavity; Maternal sinusoid; Yolk sac; Extraembryonic coelom

(b) Details of two chorionic villi, about 21 days after fertilization

Labels: Maternal sinusoid; Chorion: Cytotrophoblast, Syncytiotrophoblast; Intervillous space; Connecting stalk; Extraembryonic mesoderm; Blood capillary

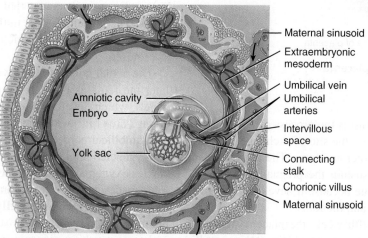

(c) Frontal section through uterus showing embryo and vascular supply, about 21 days after fertilization

Labels: Amniotic cavity; Embryo; Yolk sac; Maternal sinusoid; Extraembryonic mesoderm; Umbilical vein; Umbilical arteries; Intervillous space; Connecting stalk; Chorionic villus; Maternal sinusoid

? **Why is development of chorionic villi important?**

Figure 29.11 Placenta and umbilical cord.

The placenta is formed by the chorionic villi of the embryo and the decidua basalis of the endometrium of the mother.

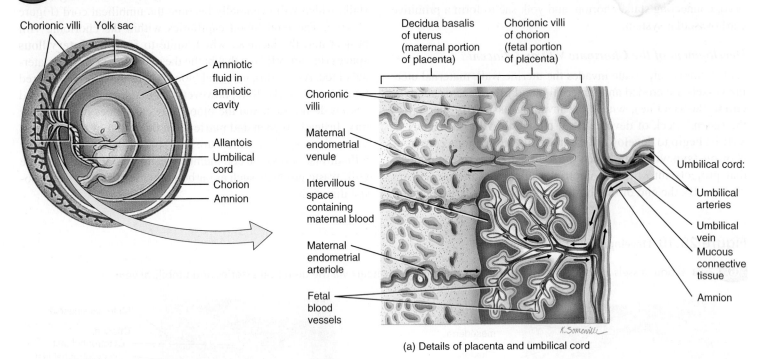

(a) Details of placenta and umbilical cord

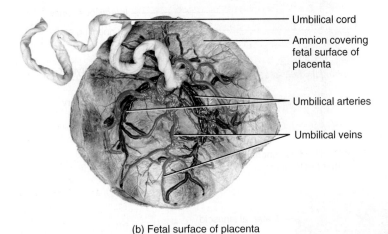

(b) Fetal surface of placenta

? **What is the function of the placenta?**

Placentation (plas-en-TĀ-shun) is the process of forming the **placenta** (pla-SEN-ta = flat cake), the site of exchange of nutrients and wastes between the mother and fetus. The placenta also produces hormones needed to sustain the pregnancy (see Figure 29.16). The placenta is unique because it develops from two separate individuals, the mother and the fetus.

By the beginning of the twelfth week, the placenta has two distinct parts: (1) the fetal portion formed by the chorionic villi of the chorion and (2) the maternal portion formed by the decidua basalis of the endometrium (Figure 29.11a). When fully developed, the placenta is shaped like a pancake (Figure 29.11b). Functionally, the placenta allows oxygen and nutrients to diffuse

from maternal blood into fetal blood while carbon dioxide and wastes diffuse from fetal blood into maternal blood. The placenta also is a protective barrier because most microorganisms cannot pass through it. However, certain viruses, such as those that cause AIDS, German measles, chickenpox, measles, encephalitis, and poliomyelitis, can cross the placenta. Many drugs, alcohol, and some substances that can cause birth defects also pass freely. The placenta stores nutrients such as carbohydrates, proteins, calcium, and iron, which are released into fetal circulation as required.

The actual connection between the placenta and embryo, and later the fetus, is through the **umbilical cord** (um-BIL-i-kul = navel), which develops from the connecting stalk and is usually about 2 cm (1 in.) wide and about 50–60 cm (20–24 in.) in length. The umbilical cord consists of two umbilical arteries that carry deoxygenated fetal blood to the placenta, one umbilical vein that carries oxygen and nutrients acquired from the mother's intervillous spaces into the fetus, and supporting mucous connective tissue called **Wharton's jelly** (WOR-tons) derived from the allantois. A layer of amnion surrounds the entire umbilical cord and gives it a shiny appearance (Figure 29.11). In some cases, the umbilical vein is used to transfuse blood into a fetus or to introduce drugs for various medical treatments.

In about 1 in 200 newborns, only one of the two umbilical arteries is present in the umbilical cord. It may be due to failure of the artery to develop or degeneration of the vessel early in development. Nearly 20% of infants with this condition develop cardiovascular defects.

After the birth of the baby, the placenta detaches from the uterus and is therefore termed the **afterbirth.** At this time, the umbilical cord is tied off and then severed. The small portion (about an inch) of the cord that remains attached to the infant begins to wither and falls off, usually within 12 to 15 days after birth. The area where the cord was attached becomes covered by a thin layer of skin, and scar tissue forms. The scar is the **umbilicus** (um-BIL-i-kus) or navel.

Pharmaceutical companies use human placentas as a source of hormones, drugs, and blood; portions of placentas are also used for burn coverage. The placental and umbilical cord veins can also be used in blood vessel grafts, and cord blood can be frozen to provide a future source of pluripotent stem cells, for example, to repopulate red bone marrow following radiotherapy for cancer.

CLINICAL CONNECTION | *Placenta Previa*

In some cases, the entire placenta or part of it may become implanted in the inferior portion of the uterus, near or covering the internal os of the cervix. This condition is called **placenta previa** (PRĒ-vē-a = before or in front of). Although placenta previa may lead to spontaneous abortion, it also occurs in approximately 1 in 250 live births. It is dangerous to the fetus because it may cause premature birth and intrauterine hypoxia due to maternal bleeding. Maternal mortality is increased due to hemorrhage and infection. The most important symptom is sudden, painless, bright-red vaginal bleeding in the third trimester. Cesarean section is the preferred method of delivery in placenta previa. •

✔ **CHECKPOINT**

10. When does gastrulation occur?
11. How do the three primary germ layers form? Why are they important?
12. What is meant by the term *induction?*
13. Describe how neurulation occurs. Why is it important?
14. What are the functions of somites?
15. How does the cardiovascular system develop?
16. How does the placenta form?

Fourth Week of Development

The fourth through eighth weeks of development are very significant in embryonic development because all major organs appear during this time. The term **organogenesis** (or'-ga-nō-JEN-e-sis) refers to the formation of body organs and systems. By the end of the eighth week, all the major body systems have begun to develop, although their functions for the most part are minimal. Organogenesis requires the presence of blood vessels to supply developing organs with oxygen and other nutrients. However, recent studies suggest that blood vessels play a significant role in organogenesis even before blood begins to flow within them. The endothelial cells of blood vessels apparently provide some type of developmental signal, either a secreted substance or a direct cell-to-cell interaction, that is necessary for organogenesis.

During the fourth week after fertilization, the embryo undergoes very dramatic changes in shape and size, nearly tripling its size. It is essentially converted from a flat, two-dimensional trilaminar embryonic disc to a three-dimensional cylinder, a process called **embryonic folding** (Figure 29.12a–d). The cylinder consists of endoderm in the center (gut), ectoderm on the outside (epidermis), and mesoderm in between. The main force responsible for embryonic folding is the different rates of growth of various parts of the embryo, especially the rapid longitudinal growth of the nervous system (neural tube). Folding in the median plane produces a **head fold** and a **tail fold;** folding in the horizontal plane results in the two **lateral folds.** Overall, due to the foldings, the embryo curves into a C-shape.

The head fold brings the developing heart and mouth into their eventual adult positions. The tail fold brings the developing anus into its eventual adult position. The lateral folds form as the lateral margins of the trilaminar embryonic disc bend ventrally. As they move toward the midline, the lateral folds incorporate the dorsal part of the yolk sac into the embryo as the **primitive gut,** the forerunner of the gastrointestinal tract (Figure 29.12b). The primitive gut differentiates into an anterior **foregut,** an intermediate **midgut,** and a posterior **hindgut** (Figure 29.12c). The fates of the foregut, midgut, and hindgut are described in Section 24.15. Recall that the oropharyngeal membrane is located in the head end of the embryo (see Figure 29.8). It separates the future pharyngeal (throat) region of the foregut from the **stomodeum** (stō-mō-DĒ-um; *stomo-* = mouth), the future oral cavity. Because of head folding, the oropharyngeal membrane moves downward and the foregut and stomodeum move closer to their final positions. When the oropharyngeal membrane ruptures during the fourth week, the pharyngeal region of the pharynx is brought into contact with the stomodeum.

In a developing embryo, the last part of the hindgut expands into a cavity called the **cloaca** (klo-Ā-ka) (see Figure 26.23). On the outside of the embryo is a small cavity in the tail region called the **proctodeum** (prok-tō-DĒ-um; *procto-* = anus) (Figure 29.12c). Separating the cloaca from the proctodeum is the **cloacal membrane** (see Figure 29.8). During embryonic development, the cloaca divides into a ventral urogenital sinus and a dorsal anorectal canal. As a result of tail folding, the cloacal membrane moves downward and the urogenital sinus, anorectal canal, and proctodeum move closer to their final positions. When the cloacal membrane ruptures during the seventh week of development, the urogenital and anal openings are created.

In addition to embryonic folding, development of somites, and development of the neural tube, five pairs of **pharyngeal** (fa-RIN-jē-al) or **branchial arches** (BRANG-kē-al; *branch* = gill) begin to develop on each side of the future head and neck regions (Figure 29.13) during the fourth week. These five paired structures begin their development on the twenty-second day after fertilization and form swellings on the surface of the embryo. Each

Figure 29.12 Embryonic folding.

Embryonic folding converts the two-dimensional trilaminar embryonic disc into a three-dimensional cylinder.

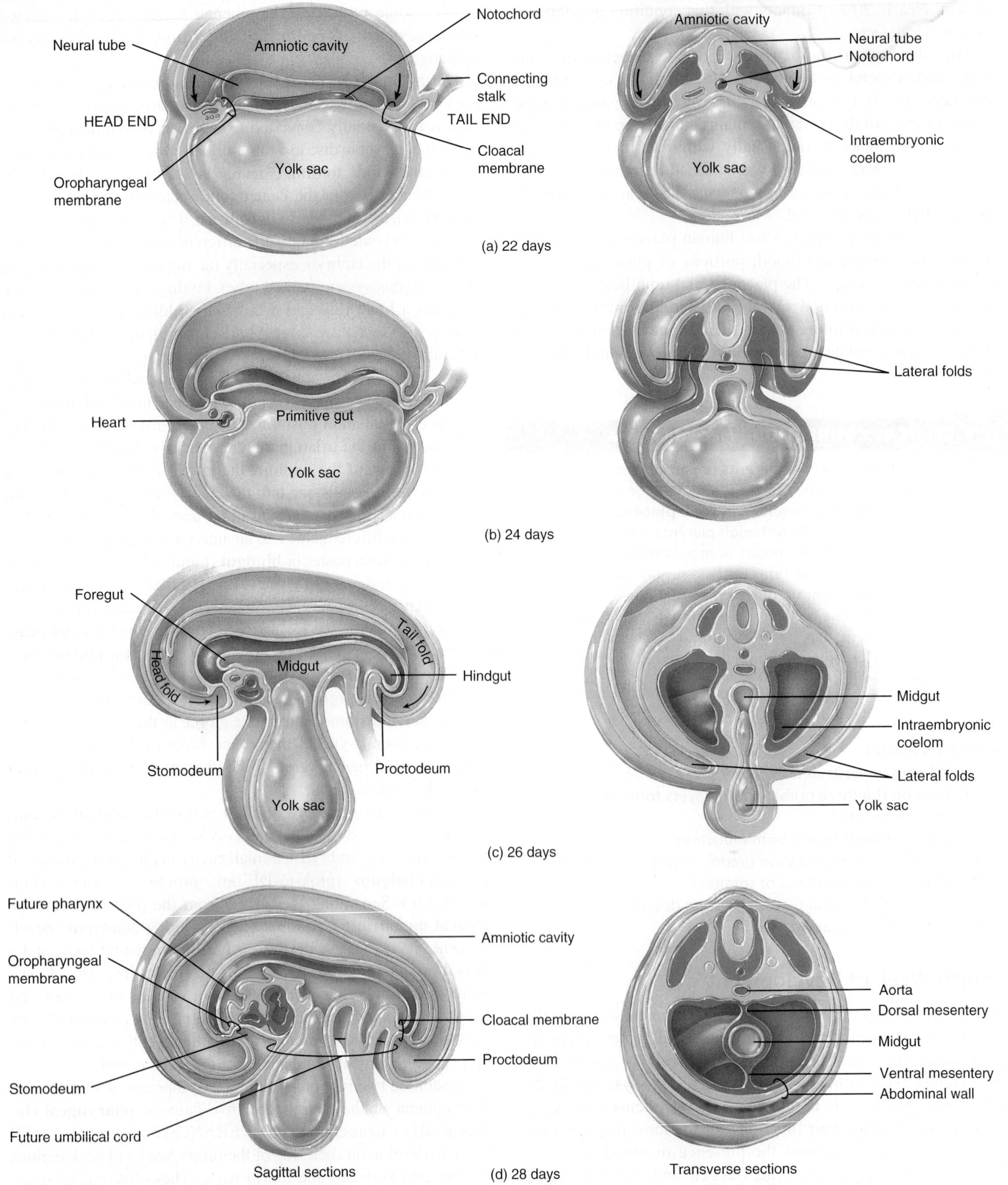

(a) 22 days

(b) 24 days

(c) 26 days

Sagittal sections   (d) 28 days   Transverse sections

? **What are the results of embryonic folding?**

pharyngeal arch consists of an outer covering of ectoderm and an inner covering of endoderm, with mesoderm in between. Within each pharyngeal arch there is an artery, a cranial nerve, skeletal cartilaginous rods that support the arch, and skeletal muscle tissue that attaches to and moves the cartilage rods. On the ectodermal surface of the pharyngeal region, each pharyngeal arch is separated by a groove called a **pharyngeal cleft** (Figure 29.13a). The

Figure 29.13 Development of pharyngeal arches, pharyngeal clefts, and pharyngeal pouches.

The five pairs of pharyngeal pouches consist of ectoderm, mesoderm, and endoderm and contain blood vessels, cranial nerves, cartilage, and muscle tissue.

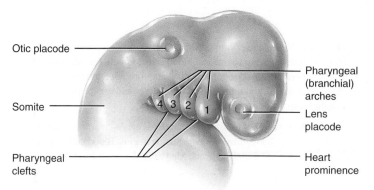

(a) External view, about 28-day embryo

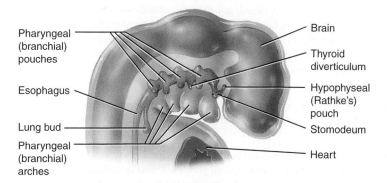

(b) Sagittal section, about 28-day embryo

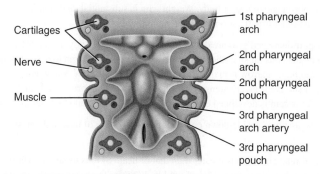

(c) Transverse section of the pharynx, about 28-day embryo

? Why are pharyngeal arches, clefts, and pouches important?

pharyngeal clefts meet corresponding balloonlike outgrowths of the endodermal pharyngeal lining called **pharyngeal pouches.** Where the pharyngeal cleft and pouch meet to separate the arches, the outer ectoderm of the cleft contacts the inner endoderm of the pouch and there is no mesoderm between (Figure 29.13b).

Just as the somite gives rise to specified structures in the body wall, each pharyngeal arch, cleft, and pouch gives rise to specified structures in the head and neck. Each pharyngeal arch is a developmental unit and includes a skeletal component, muscle, nerve, and blood vessels. In the human embryo, there are four obvious pharyngeal arches and two less distinct arches. Each of these arches develops into a specific and unique component of the head and neck region. For example, the first pharyngeal arch is often called the *mandibular arch* because it forms the jaws (the *mandible* is the lower jawbone).

The first sign of a developing ear is a thickened area of ectoderm, the **otic placode** (PLAK-ōd), or future internal ear, which can be distinguished about 22 days after fertilization. A thickened area of ectoderm called the **lens placode,** which will become the eye, also appears at this time (see Figure 29.13a).

By the middle of the fourth week, the upper limbs begin their development as outgrowths of mesoderm covered by ectoderm called **upper limb buds** (see Figure 8.18b). By the end of the fourth week, the **lower limb buds** develop. The heart also forms a distinct projection on the ventral surface of the embryo called the **heart prominence** (see Figure 8.18b). At the end of the fourth week the embryo has a distinctive **tail** (see Figure 8.18b).

Fifth through Eighth Weeks of Development

During the fifth week of development, there is a very rapid development of the brain, so growth of the head is considerable. By the end of the sixth week, the head grows even larger relative to the trunk, and the limbs show substantial development (see Figure 8.18c). In addition, the neck and trunk begin to straighten, and the heart is now four-chambered. By the seventh week, the various regions of the limbs become distinct and the beginnings of digits appear (see Figure 8.18d). At the start of the eighth week (the final week of the embryonic period), the digits of the hands are short and webbed, the tail is shorter but still visible, the eyes are open, and the auricles of the ears are visible (see Figure 8.18c). By the end of the eighth week, all regions of limbs are apparent; the digits are distinct and no longer webbed due to removal of cells via apoptosis. Also, the eyelids come together and may fuse, the tail disappears, and the external genitals begin to differentiate. The embryo now has clearly human characteristics.

✔ **CHECKPOINT**

17. How does embryonic folding occur?
18. How does the primitive gut form, and what is its significance?
19. What is the origin of the structures of the head and neck?
20. What are limb buds?
21. What changes occur in the limbs during the second half of the embryonic period?

29.2 FETAL PERIOD

◉ **OBJECTIVE**

• Describe the major events of the fetal period.

During the **fetal period** (from the ninth week until birth), tissues and organs that developed during the embryonic period grow and differentiate. Very few new structures appear during the fetal period, but the rate of body growth is remarkable, especially during the second half of intrauterine life. For example, during the last 2.5 months of intrauterine life, half of the full-term weight is added. At the beginning of the fetal period, the head is half the length of the body. By the end of the fetal period, the head size is only one-quarter the length of the body. During the same period, the limbs also increase in size from one-eighth to one-half the fetal length. The fetus is also less vulnerable to the damaging effects of drugs, radiation, and microbes than it was as an embryo.

A summary of the major developmental events of the embryonic and fetal periods is presented in Table 29.2 and illustrated in Figure 29.14.

TABLE 29.2

Summary of Changes during Embryonic and Fetal Development

TIME	APPROXIMATE SIZE AND WEIGHT	REPRESENTATIVE CHANGES
EMBRYONIC PERIOD		
1–4 weeks	0.6 cm (3/16 in.)	Primary germ layers and notochord develop. Neurulation occurs. Primary brain vesicles, somites, and intraembryonic coelom develop. Blood vessel formation begins and blood forms in yolk sac, allantois, and chorion. Heart forms and begins to beat. Chorionic villi develop and placental formation begins. The embryo folds. The primitive gut, pharyngeal arches, and limb buds develop. Eyes and ears begin to develop, tail forms, and body systems begin to form.
5–8 weeks	3 cm (1.25 in.) 1 g (1/30 oz)	Limbs become distinct and digits appear. Heart becomes four-chambered. Eyes are far apart and eyelids are fused. Nose develops and is flat. Face is more humanlike. Bone formation begins. Blood cells start to form in liver. External genitals begin to differentiate. Tail disappears. Major blood vessels form. Many internal organs continue to develop.
FETAL PERIOD		
9–12 weeks	7.5 cm (3 in.) 30 g (1 oz)	Head constitutes about half the length of fetal body, and fetal length nearly doubles. Brain continues to enlarge. Face is broad, with eyes fully developed, closed, and widely separated. Nose develops a bridge. External ears develop and are low set. Bone formation continues. Upper limbs almost reach final relative length but lower limbs are not quite as well developed. Heartbeat can be detected. Gender is distinguishable from external genitals. Urine secreted by fetus is added to amniotic fluid. Red bone marrow, thymus, and spleen participate in blood cell formation. Fetus begins to move, but its movements cannot be felt yet by the mother. Body systems continue to develop.
13–16 weeks	18 cm (6.5–7 in.) 100 g (4 oz)	Head is relatively smaller than rest of body. Eyes move medially to final positions, and ears move to final positions on sides of head. Lower limbs lengthen. Fetus appears even more humanlike. Rapid development of body systems occurs.
17–20 weeks	25–30 cm (10–12 in.) 200–450 g (0.5–1 lb)	Head is more proportionate to rest of body. Eyebrows and head hair are visible. Growth slows but lower limbs continue to lengthen. Vernix caseosa (fatty secretions of oil glands and dead epithelial cells) and lanugo (delicate fetal hair) cover fetus. Brown fat forms and is the site of heat production. Fetal movements are commonly felt by mother (quickening).
21–25 weeks	27–35 cm (11–14 in.) 550–800 g (1.25–1.5 lb)	Head becomes even more proportionate to rest of body. Weight gain is substantial, and skin is pink and wrinkled. Fetuses 24 weeks and older usually survive if born prematurely.
26–29 weeks	32–42 cm (13–17 in.) 1100–1350 g (2.5–3 lb)	Head and body are more proportionate and eyes are open. Toenails are visible. Body fat is 3.5% of total body mass and additional subcutaneous fat smoothes out some wrinkles. Testes begin to descend toward scrotum at 28 to 32 weeks. Red bone marrow is major site of blood cell production. Many fetuses born prematurely during this period survive if given intensive care because lungs can provide adequate ventilation and central nervous system is developed enough to control breathing and body temperature.
30–34 weeks	41–45 cm (16.5–18 in.) 2000–2300 g (4.5–5 lb)	Skin is pink and smooth. Fetus assumes upside-down position. Body fat is 8% of total body mass.
35–38 weeks	50 cm (20 in.) 3200–3400 g (7–7.5 lb)	By 38 weeks, circumference of fetal abdomen is greater than that of head. Skin is usually bluish-pink, and growth slows as birth approaches. Body fat is 16% of total body mass. Testes are usually in scrotum in full-term male infants. Even after birth, an infant is not completely developed; an additional year is required, especially for complete development of nervous system.

Throughout the text we have discussed the developmental anatomy of the various body systems in their respective chapters. The following list of these sections is presented here for your review.

- Integumentary System (Section 5.6)
- Skeletal System (Section 8.7)
- Muscular System (Section 10.11)
- Nervous System (Section 14.9)
- Endocrine System (Section 18.16)

- Heart (Section 20.8)
- Blood and Blood Vessels (Section 21.8)
- Lymphatic System and Immunity (Section 22.2)
- Respiratory System (Section 23.8)
- Digestive System (Section 24.15)
- Urinary System (Section 26.10)
- Reproductive Systems (Section 28.5)

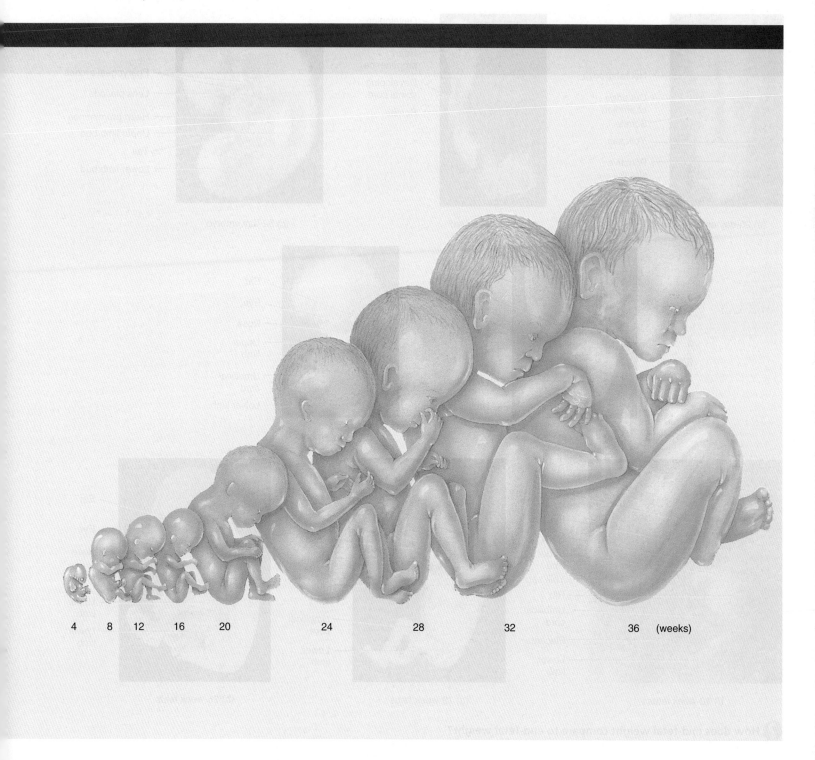

4   8   12   16   20   24   28   32   36   (weeks)

22. What are the general developmental trends during the fetal period?

23. Using Table 29.2 as a guide, select any one body structure in weeks 9 through 12 and trace its development through the remainder of the fetal period.

Figure 29.14 **Summary of representative developmental events of the embryonic and fetal periods.** The embryos and fetuses are not shown at their actual sizes.

🔒 Development during the fetal period is mostly concerned with the growth and differentiation of tissues and organs formed during the embryonic period.

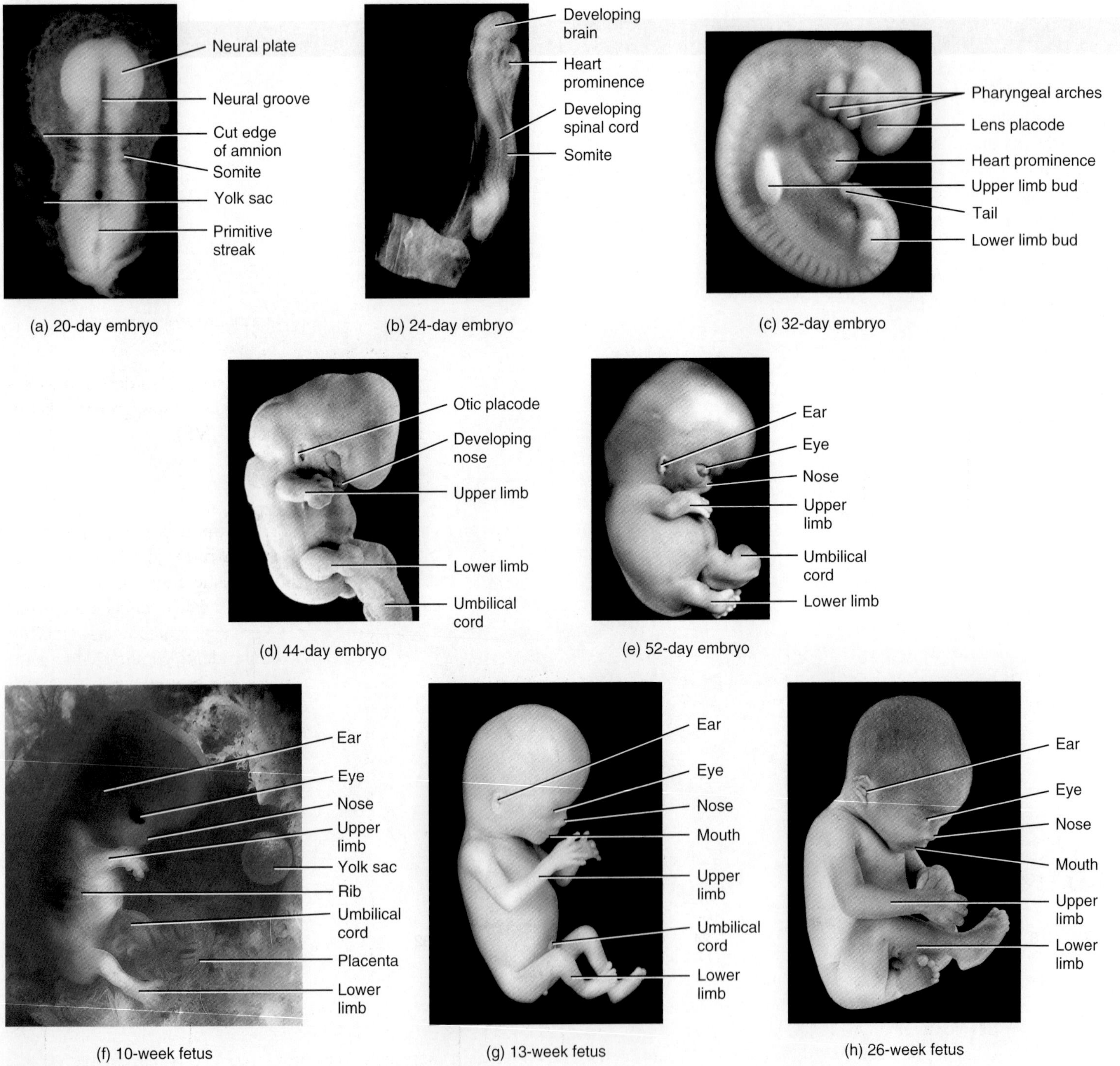

(a) 20-day embryo

Neural plate
Neural groove
Cut edge of amnion
Somite
Yolk sac
Primitive streak

(b) 24-day embryo

Developing brain
Heart prominence
Developing spinal cord
Somite

(c) 32-day embryo

Pharyngeal arches
Lens placode
Heart prominence
Upper limb bud
Tail
Lower limb bud

(d) 44-day embryo

Otic placode
Developing nose
Upper limb
Lower limb
Umbilical cord

(e) 52-day embryo

Ear
Eye
Nose
Upper limb
Umbilical cord
Lower limb

(f) 10-week fetus

Ear
Eye
Nose
Upper limb
Yolk sac
Rib
Umbilical cord
Placenta
Lower limb

(g) 13-week fetus

Ear
Eye
Nose
Mouth
Upper limb
Umbilical cord
Lower limb

(h) 26-week fetus

Ear
Eye
Nose
Mouth
Upper limb
Lower limb

❓ **How does mid-fetal weight compare to end-fetal weight?**

29.3 TERATOGENS

⦿ **OBJECTIVE**
- Define a teratogen and list several examples of teratogens.

Exposure of a developing embryo or fetus to certain environmental factors can damage the developing organism or even cause death. A **teratogen** (TER-a-tō-jen; *terato-* = monster; *-gen* = creating) is any agent or influence that causes developmental defects in the embryo. In the following sections we briefly discuss several examples.

Chemicals and Drugs

Because the placenta is not an absolute barrier between the maternal and fetal circulations, any drug or chemical that is dangerous to an infant should be considered potentially dangerous to the fetus when given to the mother. Alcohol is by far the number-one fetal teratogen. Intrauterine exposure to even a small amount of alcohol may result in **fetal alcohol syndrome (FAS),** one of the most common causes of mental retardation and the most common preventable cause of birth defects in the United States. The symptoms of FAS may include slow growth before and after birth, characteristic facial features (short palpebral fissures, a thin upper lip, and sunken nasal bridge), defective heart and other organs, malformed limbs, genital abnormalities, and central nervous system damage. Behavioral problems, such as hyperactivity, extreme nervousness, reduced ability to concentrate, and an inability to appreciate cause-and-effect relationships, are common.

Other teratogens include certain viruses (hepatitis B and C and certain papilloma viruses that cause sexually transmitted diseases); pesticides; defoliants (chemicals that cause plants to shed their leaves prematurely); industrial chemicals; some hormones; antibiotics; oral anticoagulants, anticonvulsants, antitumor agents, thyroid drugs, thalidomide, diethylstilbestrol (DES), and numerous other prescription drugs; LSD; and cocaine. A pregnant woman who uses cocaine, for example, subjects the fetus to higher risk of retarded growth, attention and orientation problems, hyperirritability, a tendency to stop breathing, malformed or missing organs, strokes, and seizures. The risks of spontaneous abortion, premature birth, and stillbirth also increase with fetal exposure to cocaine.

Cigarette Smoking

Strong evidence implicates cigarette smoking during pregnancy as a cause of low infant birth weight; there is also a strong association between smoking and a higher fetal and infant mortality rate. Women who smoke have a much higher risk of an ectopic pregnancy. Cigarette smoke may be teratogenic and may cause cardiac abnormalities as well as anencephaly (see Clinical Connection: Anencephaly). Maternal smoking also is a significant factor in the development of cleft lip and palate and has been linked with sudden infant death syndrome (SIDS). Infants nursing from smoking mothers have also been found to have an increased incidence of gastrointestinal disturbances. Even a mother's exposure to secondhand cigarette smoke (breathing air containing tobacco smoke) during pregnancy or while nursing predisposes her baby to increased incidence of respiratory problems, including bronchitis and pneumonia, during the first year of life.

Irradiation

Ionizing radiation of various kinds is a potent teratogen. Exposure of pregnant mothers to x-rays or radioactive isotopes during the embryo's susceptible period of development may cause microcephaly (small head size relative to the rest of the body), mental retardation, and skeletal malformations. Caution is advised, especially during the first trimester of pregnancy.

✔**CHECKPOINT**
24. What are some of the symptoms of fetal alcohol syndrome?
25. How does cigarette smoking affect embryonic and fetal development?

29.4 PRENATAL DIAGNOSTIC TESTS

⦿ **OBJECTIVE**
- Describe the procedures for fetal ultrasonography, amniocentesis, and chorionic villi sampling.

Several tests are available to detect genetic disorders and assess fetal well-being. Here we describe fetal ultrasonography, amniocentesis, and chorionic villi sampling (CVS).

Fetal Ultrasonography

If there is a question about the normal progress of a pregnancy, **fetal ultrasonography** (ul-tra-son-OG-ra-fē) may be performed. By far the most common use of diagnostic ultrasound is to determine a more accurate fetal age when the date of conception is unclear. It is also used to confirm pregnancy, evaluate fetal viability and growth, determine fetal position, identify multiple pregnancies, identify fetal–maternal abnormalities, and serve as an adjunct to special procedures such as amniocentesis. During fetal ultrasonography, a transducer, an instrument that emits high-frequency sound waves, is passed back and forth over the abdomen. The reflected sound waves from the developing fetus are picked up by the transducer and converted to an on-screen image called a **sonogram** (see Table 1.3). Because the urinary bladder serves as a landmark during the procedure, the patient needs to drink liquids before the procedure and not void urine to maintain a full bladder.

Amniocentesis

Amniocentesis (am′-nē-ō-sen-TĒ-sis; *amnio-* = amnion; *-centesis* = puncture to remove fluid) involves withdrawing some of the amniotic fluid that bathes the developing fetus and analyzing the fetal cells and dissolved substances. It is used to test for the presence of certain genetic disorders, such as Down syndrome (DS), hemophilia, Tay-Sachs disease, sickle-cell disease, and certain muscular dystrophies. It is also used to help determine survivability of the fetus. The test is usually done at 14–18 weeks of gestation. All gross chromosomal abnormalities and over 50 biochemical defects can be detected through amniocentesis. It can also reveal the baby's gender, which is important information for the diagnosis of sex-linked disorders, in which an abnormal gene carried by the mother affects her male offspring only (described in Section 29.10).

During amniocentesis, the position of the fetus and placenta is first identified using ultrasound and palpation. After the skin is prepared with an antiseptic and a local anesthetic is given, a hypodermic needle is inserted through the mother's abdominal wall and into the amniotic cavity within the uterus. Then, 10 to 30 mL of fluid and suspended cells are aspirated (Figure 29.15a) for microscopic examination and biochemical testing. Elevated levels of alpha-fetoprotein (AFP) and acetylcholinesterase may indicate failure of the nervous system to develop properly, as occurs in spina bifida or anencephaly (absence of the cerebrum), or may be due to other developmental or chromosomal problems. Chromosome studies, which require growing the cells for 2–4 weeks in a culture medium, may reveal rearranged, missing, or extra chromosomes. Amniocentesis is performed only when a risk for genetic defects is suspected, because there is about a 0.5% chance of spontaneous abortion after the procedure.

Chorionic Villi Sampling

In **chorionic villi sampling** or **CVS**, a catheter is guided through the vagina and cervix of the uterus and then advanced to the chorionic villi under ultrasound guidance (Figure 29.15b). About 30 milligrams of tissue is suctioned out and prepared for chromosomal analysis. Alternatively, the chorionic villi can be sampled by inserting a needle through the abdominal cavity, as performed in amniocentesis.

CVS can identify the same defects as amniocentesis because chorion cells and fetal cells contain the same genome. CVS offers several advantages over amniocentesis: It can be performed as early as 8 weeks of gestation, and test results are available in only a few days, permitting an earlier decision on whether to continue the pregnancy. However, CVS is slightly riskier than amniocentesis; after the procedure there is a 1–2% chance of spontaneous abortion.

Noninvasive Prenatal Tests

Currently, chorionic villi testing and amniocentesis are the only useful ways to obtain fetal tissue for prenatal testing of gene defects. While these invasive procedures pose relatively little risk when performed by experts, much work has been done to develop **noninvasive prenatal tests,** which do not require the penetration of any

Figure 29.15 Amniocentesis and chorionic villi sampling.

To detect genetic abnormalities, amniocentesis is performed at 14–16 weeks of gestation; chorionic villi sampling may be performed as early as 8 weeks of gestation.

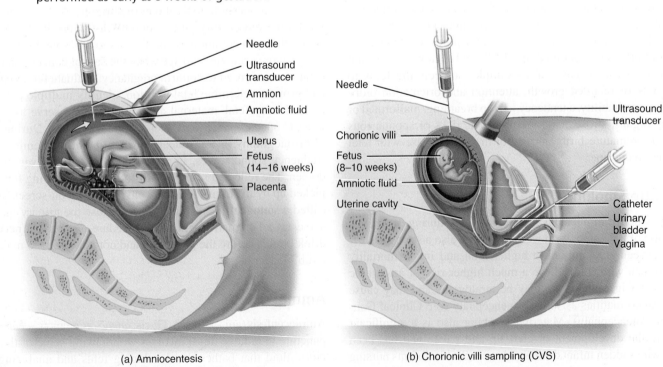

(a) Amniocentesis

(b) Chorionic villi sampling (CVS)

? **What information can be provided by amniocentesis?**

embryonic structure. The goal is to develop accurate, safe, more efficient, and less expensive tests for screening a large population.

The first such test developed was the **maternal alpha-fetoprotein (AFP) test** (AL-fa fē'-tō-PRŌ-tēn). In this test, the mother's blood is analyzed for the presence of AFP, a protein synthesized in the fetus that passes into the maternal circulation. The highest levels of AFP normally occur during weeks 12 through 15 of pregnancy. Later, AFP is not produced, and its concentration decreases to a very low level both in the fetus and in maternal blood. A high level of AFP after week 16 usually indicates that the fetus has a neural tube defect, such as spina bifida or anencephaly. Because the test is 95% accurate, it is now recommended that all pregnant women be tested for AFP. A newer test (Quad AFP Plus) probes maternal blood for AFP and three other molecules. The test permits prenatal screening for Down syndrome, trisomy 18, and neural tube defects; it also helps predict the delivery date and may reveal the presence of twins.

✔ CHECKPOINT

26. What conditions can be detected using fetal ultrasonography, amniocentesis, and chorionic villi sampling? What are the advantages of noninvasive prenatal tests?

29.5 MATERNAL CHANGES DURING PREGNANCY

◉ OBJECTIVES

• Describe the sources and functions of the hormones secreted during pregnancy.
• Describe the hormonal, anatomical, and physiological changes in the mother during pregnancy.

Hormones of Pregnancy

During the first 3 to 4 months of pregnancy, the corpus luteum in the ovary continues to secrete **progesterone** and **estrogens,** which maintain the lining of the uterus during pregnancy and prepare the mammary glands to secrete milk. The amounts secreted by the corpus luteum, however, are only slightly more than those produced after ovulation in a normal menstrual cycle. From the third month through the remainder of the pregnancy, the placenta itself provides the high levels of progesterone and estrogens required. As noted previously, the chorion secretes **human chorionic gonadotropin (hCG)** (kō-rē-ON-ik gō'-nad-ō-TRŌ-pin) into the blood. In turn, hCG stimulates the corpus luteum to continue production of progesterone and estrogens—an activity required to prevent menstruation and for the continued attachment of the embryo and fetus to the lining of the uterus (Figure 29.16a). By the eighth day after fertilization, hCG can be detected in the blood and urine of a pregnant woman. Peak secretion of hCG occurs at about the ninth week of pregnancy (Figure 29.16b). During the fourth and fifth months the hCG level decreases sharply and then levels off until childbirth.

The chorion begins to secrete estrogens after the first 3 or 4 weeks of pregnancy and progesterone by the sixth week. These hormones are secreted in increasing quantities until the time of birth (Figure 29.16b). By the fourth month, when the placenta is fully established, the secretion of hCG is greatly reduced, and the secretions of the corpus luteum are no longer essential. A high level of progesterone ensures that the uterine myometrium is relaxed and that the cervix is tightly closed. After delivery, estrogens and progesterone in the blood decrease to normal levels.

Relaxin, a hormone produced first by the corpus luteum of the ovary and later by the placenta, increases the flexibility of the pubic symphysis and ligaments of the sacroiliac and sacrococcygeal joints and helps dilate the uterine cervix during labor. Both of these actions ease delivery of the baby.

A third hormone produced by the chorion of the placenta is **human chorionic somatomammotropin (hCS)** (sō'-ma-tō-mam-ō-TRŌ-pin), also known as **human placental lactogen (hPL).** The rate of secretion of hCS increases in proportion to placental mass, reaching maximum levels after 32 weeks and remaining relatively constant after that. It is thought to help prepare the mammary glands for lactation, enhance maternal growth by increasing protein synthesis, and regulate certain aspects of metabolism in both mother and fetus. For example, hCS decreases the use of glucose by the mother and promotes the release of fatty acids from her adipose tissue, making more glucose available to the fetus.

The hormone most recently found to be produced by the placenta is **corticotropin-releasing hormone (CRH)** (kor'-ti-kō-TRŌ-pin), which in nonpregnant people is secreted only by neurosecretory cells in the hypothalamus. CRH is now thought to be part of the "clock" that establishes the timing of birth. Secretion of CRH by the placenta begins at about 12 weeks and increases enormously toward the end of pregnancy. Women who have higher levels of CRH earlier in pregnancy are more likely to deliver prematurely; those who have low levels are more likely to deliver after their due date. CRH from the placenta has a second important effect: It increases secretion of cortisol, which is needed for maturation of the fetal lungs and the production of surfactant (see "Alveoli" in Section 23.1).

CLINICAL CONNECTION | *Early Pregnancy Tests*

Early pregnancy tests detect the tiny amounts of human chorionic gonadotropin (hCG) in the urine that begin to be excreted about 8 days after fertilization. The test kits can detect pregnancy as early as the first day of a missed menstrual period—that is, at about 14 days after fertilization. Chemicals in the kits produce a color change if a reaction occurs between hCG in the urine and hCG antibodies included in the kit.

Several of the test kits available at pharmacies are as sensitive and accurate as test methods used in many hospitals. Still, false-negative and false-positive results can occur. A false-negative result (the test is negative, but the woman is pregnant) may be due to testing too soon or to an ectopic pregnancy. A false-positive result (the test is positive, but the woman is not pregnant) may be due to excess protein or blood in the urine or to hCG production due to a rare type of uterine cancer. Thiazide diuretics, hormones, steroids, and thyroid drugs may also affect the outcome of an early pregnancy test. •

Figure 29.16 Hormones during pregnancy.

🔒 ⊂▭ The corpus luteum produces progesterone and estrogens during the first 3–4 months of pregnancy, after which time the placenta assumes this function.

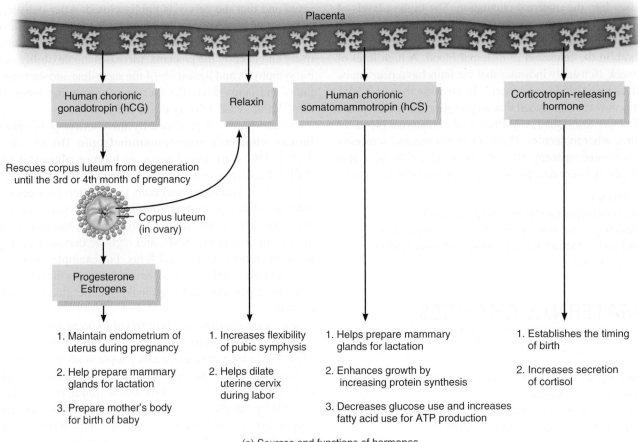

(a) Sources and functions of hormones

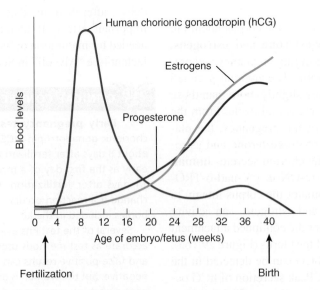

(b) Blood levels of hormones during pregnancy

❓ **Which hormone is detected by early pregnancy tests?**

Changes during Pregnancy

Near the end of the third month of pregnancy, the uterus occupies most of the pelvic cavity. As the fetus continues to grow, the uterus extends higher and higher into the abdominal cavity. Toward the end of a full-term pregnancy, the uterus fills nearly the entire abdominal cavity, reaching above the costal margin nearly to the xiphoid process of the sternum (Figure 29.17). It pushes the maternal intestines, liver, and stomach superiorly, elevates the diaphragm, and widens the thoracic cavity. Pressure on the stomach may force the stomach contents superiorly into the esophagus, resulting in heartburn. In the pelvic cavity, compression of the ureters and urinary bladder occurs.

Pregnancy-induced physiological changes also occur, including weight gain due to the fetus, amniotic fluid, the placenta, uterine enlargement, and increased total body water; increased storage of proteins, triglycerides, and minerals; marked breast enlargement in preparation for lactation; and lower back pain due to lordosis (hollow back).

Several changes occur in the maternal cardiovascular system. Stroke volume increases by about 30% and cardiac output rises by 20–30% due to increased maternal blood flow to the placenta and increased metabolism. Heart rate increases 10–15% and blood volume increases 30–50%, mostly during the second half of pregnancy. These increases are necessary to meet the additional demands of the fetus for nutrients and oxygen. When a pregnant woman is lying on her back, the enlarged uterus may compress the aorta, resulting in diminished blood flow to the uterus. Compression of the inferior vena cava also decreases venous return, which leads to edema in the lower limbs and may

Figure 29.17 Normal fetal location and position at the end of a full-term pregnancy.

The gestation period is the time interval (about 38 weeks) from fertilization to birth.

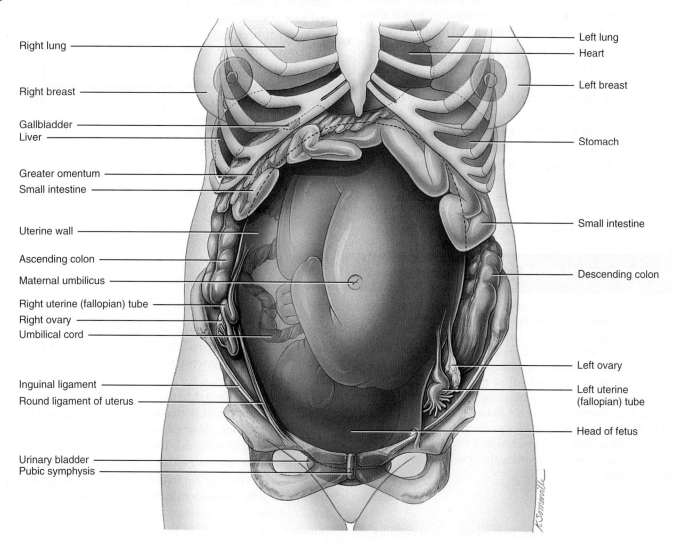

Anterior view of position of organs at end of full-term pregnancy

 What hormone increases the flexibility of the pubic symphysis and helps dilate the cervix of the uterus to ease delivery of the baby?

produce varicose veins. Compression of the renal artery can lead to renal hypertension.

Respiratory function is also altered during pregnancy to meet the added oxygen demands of the fetus. Tidal volume can increase by 30–40%, expiratory reserve volume can be reduced by up to 40%, functional residual capacity can decline by up to 25%, minute ventilation (the total volume of air inhaled and exhaled each minute) can increase by up to 40%, airway resistance in the bronchial tree can decline by 30–40%, and total body oxygen consumption can increase by about 10–20%. Dyspnea (difficult breathing) also occurs.

The digestive system also undergoes changes. Pregnant women experience an increase in appetite due to the added nutritional demands of the fetus. A general decrease in GI tract motility can cause constipation, delay gastric emptying time, and produce nausea, vomiting, and heartburn.

Pressure on the urinary bladder by the enlarging uterus can produce urinary symptoms, such as increased frequency and urgency of urination, and stress incontinence. An increase in renal plasma flow up to 35% and an increase in glomerular filtration rate up to 40% increase the renal filtering capacity, which allows faster elimination of the extra wastes produced by the fetus.

Changes in the skin during pregnancy are more apparent in some women than in others. Some women experience increased pigmentation around the eyes and cheekbones in a masklike pattern (chloasma), in the areolae of the breasts, and in the linea alba of the lower abdomen (linea nigra). Striae (stretch marks) over the abdomen can occur as the uterus enlarges, and hair loss increases.

Changes in the reproductive system include edema and increased vascularity of the vulva and increased pliability and vascularity of the vagina. The uterus increases from its nonpregnant mass of 60–80 g to 900–1200 g at term because of hyperplasia of muscle fibers in the myometrium in early pregnancy and hypertrophy of muscle fibers during the second and third trimesters.

CLINICAL CONNECTION | Pregnancy-Induced Hypertension

About 10–15% of all pregnant women in the United States experience **pregnancy-induced hypertension (PIH),** an elevated blood pressure that is associated with pregnancy. The major cause is **preeclampsia** (prē-ē-KLAMP-sē-a), an abnormal condition of pregnancy characterized by sudden hypertension, large amounts of protein in the urine, and generalized edema that typically appears after the twentieth week of pregnancy. Other signs and symptoms are generalized edema, blurred vision, and headaches. Preeclampsia might be related to an autoimmune or allergic reaction resulting from the presence of a fetus. Treatment involves bed rest and various drugs. When the condition is also associated with convulsions and coma, it is termed **eclampsia.** •

✔**CHECKPOINT**

27. List the hormones involved in pregnancy, and describe the functions of each.
28. What structural and functional changes occur in the mother during pregnancy?

29.6 EXERCISE AND PREGNANCY

◉ **OBJECTIVE**
• Explain the effects of pregnancy on exercise and of exercise on pregnancy.

Only a few changes in early pregnancy affect the ability to exercise. A pregnant woman may tire more easily than usual, or morning sickness may interfere with regular exercise. As the pregnancy progresses, weight is gained and posture changes, so more energy is needed to perform activities, and certain maneuvers (sudden stopping, changes in direction, rapid movements) are more difficult to execute. In addition, certain joints, especially the pubic symphysis, become less stable in response to the increased level of the hormone relaxin. As compensation, many mothers-to-be walk with widely spread legs and a shuffling motion.

Although blood shifts from viscera (including the uterus) to the muscles and skin during exercise, there is no evidence of inadequate blood flow to the placenta. The heat generated during exercise may cause dehydration and further increase body temperature. Especially during early pregnancy, excessive exercise and heat buildup should be avoided because elevated body temperature has been implicated in neural tube defects. Exercise has no known effect on lactation, provided a woman remains hydrated and wears a bra that provides good support. Overall, moderate physical activity does not endanger the fetus of a healthy woman who has a normal pregnancy. However, any physical activity that might endanger the fetus should be avoided.

Among the benefits of exercise to the mother during pregnancy are a greater sense of well-being and fewer physical complaints.

✔**CHECKPOINT**

29. Which changes in pregnancy have an effect on the ability to exercise?

29.7 LABOR

◉ **OBJECTIVE**
• Explain the events associated with the three stages of labor.

Labor is the process by which the fetus is expelled from the uterus through the vagina, also referred to as giving birth. A synonym for labor is **parturition** (par-toor-ISH-un; *parturit-* = childbirth).

The onset of labor is determined by complex interactions of several placental and fetal hormones. Because progesterone inhibits uterine contractions, labor cannot take place until the effects of progesterone are diminished. Toward the end of gestation, the levels of estrogens in the mother's blood rise sharply, producing changes that overcome the inhibiting effects of progesterone. The rise in estrogens results from increasing secretion by the placenta of corticotropin-releasing hormone, which stimulates the anterior pituitary gland of the fetus to secrete ACTH (adrenocorticotropic hormone). In turn, ACTH stimulates the fetal adrenal gland to secrete cortisol and dehydroepiandrosterone (DHEA) (dē-hī′-drō-ep′-ē-an-DROS-ter-ōn), the major adrenal androgen. The placenta then converts DHEA into an estrogen. High levels of estrogens cause the number of receptors for oxytocin on uterine muscle fibers to increase, and cause uterine muscle fibers to form gap junctions with one another. Oxytocin released by the posterior pituitary stimulates uterine contractions, and relaxin from the placenta assists by increasing the flexibility of the pubic symphysis and helping dilate the uterine cervix. Estrogen also stimulates the placenta to release prostaglandins, which induce production of enzymes that digest collagen fibers in the cervix, causing it to soften.

Control of labor contractions during parturition occurs via a positive feedback cycle (see Figure 1.4). Contractions of the uterine myometrium force the baby's head or body into the cervix, distending (stretching) the cervix. Stretch receptors in the cervix send nerve impulses to neurosecretory cells in the hypothalamus, causing them to release oxytocin into blood capillaries of the posterior pituitary gland. Oxytocin then is carried by the blood to the uterus, where it stimulates the myometrium to contract more forcefully. As the contractions intensify, the baby's body stretches the cervix still more, and the resulting nerve impulses stimulate the secretion of yet more oxytocin. With birth of the infant, the positive feedback cycle is broken because cervical distension suddenly lessens.

Uterine contractions occur in waves (quite similar to the peristaltic waves of the gastrointestinal tract) that start at the top of the uterus and move downward, eventually expelling the fetus. **True labor** begins when uterine contractions occur at regular intervals, usually producing pain. As the interval between contractions shortens, the contractions intensify. Another symptom of true labor in some women is localization of pain in the back that is intensified by walking. The most reliable indicator of true labor is dilation of the cervix and the "show," a discharge of a blood-containing mucus into the cervical canal. In **false labor,** pain is felt in the abdomen at irregular intervals, but it does not intensify and walking does not alter it significantly. There is no "show" and no cervical dilation.

True labor can be divided into three stages (Figure 29.18):

❶ Stage of dilation. The time from the onset of labor to the complete dilation of the cervix is the **stage of dilation.** This stage, which typically lasts 6–12 hours, features regular contractions of the uterus, usually a rupturing of the amniotic sac, and complete dilation (to 10 cm) of the cervix. If the amniotic sac does not rupture spontaneously, it is ruptured intentionally.

❷ Stage of expulsion. The time (10 minutes to several hours) from complete cervical dilation to delivery of the baby is the **stage of expulsion.**

❸ Placental stage. The time (5–30 minutes or more) after delivery until the placenta or "afterbirth" is expelled by powerful uterine contractions is the **placental stage.** These contractions also constrict blood vessels that were torn during delivery, reducing the likelihood of hemorrhage.

As a rule, labor lasts longer with first babies, typically about 14 hours. For women who have previously given birth, the average duration of labor is about 8 hours—although the time varies

Figure 29.18 Stages of true labor.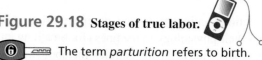

🔓 The term *parturition* refers to birth.

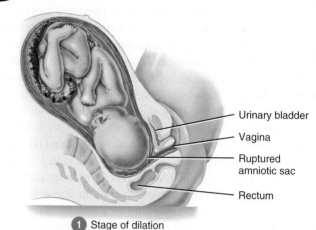

- Urinary bladder
- Vagina
- Ruptured amniotic sac
- Rectum

❶ Stage of dilation

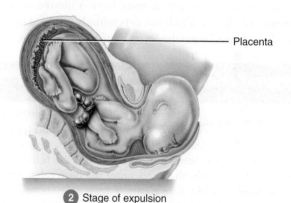

- Placenta

❷ Stage of expulsion

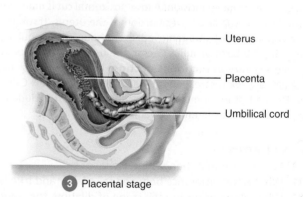

- Uterus
- Placenta
- Umbilical cord

❸ Placental stage

❓ What event marks the beginning of the stage of expulsion?

enormously among births. Because the fetus may be squeezed through the birth canal (cervix and vagina) for up to several hours, the fetus is stressed during childbirth: The fetal head is compressed, and the fetus undergoes some degree of intermittent hypoxia due to compression of the umbilical cord and the placenta during uterine contractions. In response to this stress, the fetal adrenal medullae secrete very high levels of epinephrine and norepinephrine, the "fight-or-flight" hormones. Much of the protection against the stresses of parturition, as well as preparation of the infant for surviving extrauterine life, is provided by these hormones.

Among other functions, epinephrine and norepinephrine clear the lungs and alter their physiology in readiness for breathing air, mobilize readily usable nutrients for cellular metabolism, and promote an increased flow of blood to the brain and heart.

About 7% of pregnant women do not deliver by 2 weeks after their due date. Such cases carry an increased risk of brain damage to the fetus, and even fetal death, due to inadequate supplies of oxygen and nutrients from an aging placenta. Post-term deliveries may be facilitated by inducing labor, initiated by administration of oxytocin (Pitocin®), or by surgical delivery (cesarean section).

Following the delivery of the baby and placenta is a 6-week period during which the maternal reproductive organs and physiology return to the prepregnancy state. This period is called the **puerperium** (pū-er-PER-ē-um). Through a process of tissue catabolism, the uterus undergoes a remarkable reduction in size, called **involution** (in-vō-LOO-shun), especially in lactating women. The cervix loses its elasticity and regains its prepregnancy firmness. For 2–4 weeks after delivery, women have a uterine discharge called **lochia** (LŌ-kē-a), which consists initially of blood and later of serous fluid derived from the former site of the placenta.

CLINICAL CONNECTION | Dystocia and Cesarean Section

Dystocia (dis-TŌ-sē-a; *dys-* = painful or difficult; *-toc-* = birth), or difficult labor, may result either from an abnormal position (presentation) of the fetus or a birth canal of inadequate size to permit vaginal delivery. In a **breech presentation,** for example, the fetal buttocks or lower limbs, rather than the head, enter the birth canal first; this occurs most often in premature births. If fetal or maternal distress prevents a vaginal birth, the baby may be delivered surgically through an abdominal incision. A low, horizontal cut is made through the abdominal wall and lower portion of the uterus, through which the baby and placenta are removed. Even though it is popularly associated with the birth of Julius Caesar, the true reason this procedure is termed a **cesarean section (C-section)** is because it was described in Roman law, *lex cesarea*, about 600 years before Julius Caesar was born. Even a history of multiple C-sections need not exclude a pregnant woman from attempting a vaginal delivery. •

✔ CHECKPOINT

30. What hormonal changes induce labor?
31. What is the difference between false labor and true labor?
32. What happens during the stage of dilation, the stage of expulsion, and the placental stage of true labor?

29.8 ADJUSTMENTS OF THE INFANT AT BIRTH

◉ OBJECTIVE
• Explain the respiratory and cardiovascular adjustments that occur in an infant at birth.

During pregnancy, the embryo (and later the fetus) is totally dependent on the mother for its existence. The mother supplies the fetus with oxygen and nutrients, eliminates its carbon dioxide and other wastes, protects it against shocks and temperature changes, and provides antibodies that confer protection against certain harmful microbes. At birth, a physiologically mature baby becomes much more self-supporting, and the newborn's body systems must make various adjustments. The most dramatic changes occur in the respiratory and cardiovascular systems.

Respiratory Adjustments

The reason that the fetus depends entirely on the mother for obtaining oxygen and eliminating carbon dioxide is that the fetal lungs are either collapsed or partially filled with amniotic fluid. The production of surfactant begins by the end of the sixth month of development. Because the respiratory system is fairly well developed at least 2 months before birth, premature babies delivered at 7 months are able to breathe and cry. After delivery, the baby's supply of oxygen from the mother ceases, and any amniotic fluid in the fetal lungs is absorbed. Because carbon dioxide is no longer being removed, it builds up in the blood. A rising CO_2 level stimulates the respiratory center in the medulla oblongata, causing the respiratory muscles to contract, and the baby to draw his or her first breath. Because the first inspiration is unusually deep, as the lungs contain no air, the baby also exhales vigorously and naturally cries. A full-term baby may breathe 45 times a minute for the first 2 weeks after birth. Breathing rate gradually declines until it approaches a normal rate of 12 breaths per minute.

Cardiovascular Adjustments

After the baby's first inspiration, the cardiovascular system must make several adjustments (see Figure 21.30). Closure of the foramen ovale between the atria of the fetal heart, which occurs at the moment of birth, diverts deoxygenated blood to the lungs for the first time. The foramen ovale is closed by two flaps of septal heart tissue that fold together and permanently fuse. The remnant of the foramen ovale is the fossa ovalis.

Once the lungs begin to function, the ductus arteriosus shuts off due to contractions of smooth muscle in its wall, and it becomes the ligamentum arteriosum. The muscle contraction is probably mediated by the polypeptide bradykinin, released from the lungs during their initial inflation. The ductus arteriosus generally does not close completely until about 3 months after birth. Prolonged incomplete closure results in a condition called **patent ductus arteriosus** (see Figure 20.23b).

After the umbilical cord is tied off and severed and blood no longer flows through the umbilical arteries, they fill with connective tissue, and their distal portions become the medial umbilical ligaments. The umbilical vein then becomes the ligamentum teres (round ligament) of the liver.

In the fetus, the ductus venosus connects the umbilical vein directly with the inferior vena cava, allowing blood from the placenta to bypass the fetal liver. When the umbilical cord is severed, the ductus venosus collapses, and venous blood from the viscera of the fetus flows into the hepatic portal vein to the liver and then via the hepatic vein to the inferior vena cava. The remnant of the ductus venosus becomes the ligamentum venosum.

At birth, an infant's pulse may range from 120 to 160 beats per minute and may go as high as 180 on excitation. After birth, oxygen use increases, which stimulates an increase in the rate of red blood cell and hemoglobin production. The white blood cell count at birth is very high—sometimes as much as 45,000 cells per microliter—but the count decreases rapidly by the seventh day. Recall that the white blood cell count of an adult is 5000–10,000 cells per microliter.

CLINICAL CONNECTION | *Premature Infants*

Delivery of a physiologically immature baby carries certain risks. A **premature infant** or "preemie" is generally considered a baby who weighs less than 2500 g (5.5 lb) at birth. Poor prenatal care, drug abuse, history of a previous premature delivery, and mother's age below 16 or above 35 increase the chance of premature delivery. The body of a premature infant is not yet ready to sustain some critical functions, and thus its survival is uncertain without medical intervention. The major problem after delivery of an infant under 36 weeks of gestation is respiratory distress syndrome (RDS) of the newborn due to insufficient surfactant. RDS can be eased by use of artificial surfactant and a ventilator that delivers oxygen until the lungs can operate on their own. •

✔CHECKPOINT

33. Why are respiratory and cardiovascular adjustments so important at birth?

29.9 THE PHYSIOLOGY OF LACTATION

◉ OBJECTIVE

• Discuss the physiology and hormonal control of lactation.

Lactation (lak-TĀ-shun) is the production and ejection of milk from the mammary glands. A principal hormone in promoting milk production is **prolactin (PRL),** which is secreted from the anterior pituitary gland. Even though prolactin levels increase as the pregnancy progresses, no milk production occurs because progesterone inhibits the effects of prolactin. After delivery, the levels of estrogens and progesterone in the mother's blood decrease, and the inhibition is removed. The principal stimulus in maintaining prolactin secretion during lactation is the sucking action of the infant. Suckling initiates nerve impulses from stretch receptors in the nipples to the hypothalamus; the impulses decrease hypothalamic release of prolactin-inhibiting hormone (PIH) and increase release of prolactin-releasing hormone (PRH), so more prolactin is released by the anterior pituitary.

Oxytocin causes release of milk into the mammary ducts via the **milk ejection reflex** (Figure 29.19). Milk formed by the glandular cells of the breasts is stored until the baby begins active suckling. Stimulation of touch receptors in the nipple initiates sensory nerve impulses that are relayed to the hypothalamus. In

response, secretion of oxytocin from the posterior pituitary increases. Carried by the bloodstream to the mammary glands, oxytocin stimulates contraction of myoepithelial (smooth muscle–like)

Figure 29.19 The milk ejection reflex, a positive feedback cycle.

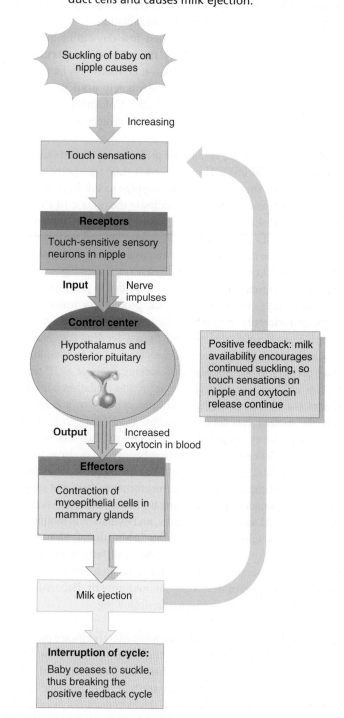

Oxytocin stimulates contraction of myoepithelial cells in the breasts, which squeezes the glandular and duct cells and causes milk ejection.

 What is another function of oxytocin?

cells surrounding the glandular cells and ducts. The resulting compression moves the milk from the alveoli of the mammary glands into the mammary ducts, where it can be suckled. This process is termed **milk ejection (let-down).** Even though the actual ejection of milk does not occur until 30–60 seconds after nursing begins (the latent period), some milk stored in lactiferous sinuses near the nipple is available during the latent period. Stimuli other than suckling, such as hearing a baby's cry or touching the mother's genitals, also can trigger oxytocin release and milk ejection. The suckling stimulation that produces the release of oxytocin also inhibits the release of PIH; this results in increased secretion of prolactin, which maintains lactation.

During late pregnancy and the first few days after birth, the mammary glands secrete a cloudy fluid called **colostrum.** Although it is not as nutritious as milk—it contains less lactose and virtually no fat—colostrum serves adequately until the appearance of true milk on about the fourth day. Colostrum and maternal milk contain important antibodies that protect the infant during the first few months of life.

Following birth of the infant, the prolactin level starts to return to the nonpregnant level. However, each time the mother nurses the infant, nerve impulses from the nipples to the hypothalamus increase the release of PRH (and decrease the release of PIH), resulting in a tenfold increase in prolactin secretion by the anterior pituitary that lasts about an hour. Prolactin acts on the mammary glands to provide milk for the next nursing period. If this surge of prolactin is blocked by injury or disease, or if nursing is discontinued, the mammary glands lose their ability to produce milk in only a few days. Even though milk production normally decreases considerably within 7–9 months after birth, it can continue for several years if nursing or **breast-feeding** continues.

Lactation often blocks ovarian cycles for the first few months following delivery, if the frequency of sucking is about 8–10 times a day. This effect is inconsistent, however, and ovulation commonly precedes the first menstrual period after delivery of a baby. As a result, the mother can never be certain she is not fertile. Breast-feeding is therefore an unreliable birth control measure. The suppression of ovulation during lactation is believed to occur as follows: During breast-feeding, neural input from the nipple reaches the hypothalamus and causes it to produce neurotransmitters that suppress the release of gonadotropin-releasing hormone (GnRH). As a result, production of LH and FSH decreases, and ovulation is inhibited.

A primary benefit of breast-feeding is nutritional: Human milk is a sterile solution that contains amounts of fatty acids, lactose, amino acids, minerals, vitamins, and water that are ideal for the baby's digestion, brain development, and growth. Breast-feeding also benefits infants by providing the following:

- *Beneficial cells.* Several types of white blood cells are present in breast milk. Neutrophils and macrophages serve as phagocytes, ingesting microbes in the baby's gastrointestinal tract. Macrophages also produce lysozyme and other immune system components. Plasma cells, which develop from B lymphocytes, produce antibodies against specific microbes, and T lymphocytes kill microbes directly or help mobilize other defenses.

- *Beneficial molecules.* Breast milk also contains an abundance of beneficial molecules. Maternal IgA antibodies in breast milk bind to microbes in the baby's gastrointestinal tract and prevent their migration into other body tissues. Because a mother produces antibodies to whatever disease-causing microbes are present in her environment, her breast milk affords protection against the specific infectious agents to which her baby is also exposed. Additionally, two milk proteins bind to nutrients that many bacteria need to grow and survive: B_{12}-binding protein ties up vitamin B_{12}, and lactoferrin ties up iron. Some fatty acids can kill certain viruses by disrupting their membranes, and lysozyme kills bacteria by disrupting their cell walls. Finally, interferons enhance the antimicrobial activity of immune cells.

- *Decreased incidence of diseases later in life.* Breast-feeding provides children with a slight reduction in risk of lymphoma, heart disease, allergies, respiratory and gastrointestinal infections, ear infections, diarrhea, diabetes mellitus, and meningitis.

- *Miscellaneous benefits.* Breast-feeding supports optimal infant growth, enhances intellectual and neurological development, and fosters mother–infant relations by establishing early and prolonged contact between them. Compared to cow's milk, the fats and iron in breast milk are more easily absorbed, the proteins in breast milk are more readily metabolized, and the lower sodium content of breast milk is more suited to an infant's needs. Premature infants benefit even more from breast-feeding because the milk produced by mothers of premature infants seems to be specially adapted to the infant's needs; it has a higher protein content than the milk of mothers of full-term infants. Finally, a baby is less likely to have an allergic reaction to its mother's milk than to milk from another source.

Years before oxytocin was discovered, it was common practice in midwifery to let a first-born twin nurse at the mother's breast to speed the birth of the second child. Now we know why this practice is helpful—it stimulates the release of oxytocin. Even after a single birth, nursing promotes expulsion of the placenta (afterbirth) and helps the uterus return to its normal size. Synthetic oxytocin (Pitocin) is often given to induce labor or to increase uterine tone and control hemorrhage just after parturition.

✔**CHECKPOINT**

34. Which hormones contribute to lactation? What is the function of each?

35. What are the benefits of breast-feeding over bottle-feeding?

29.10 INHERITANCE

◉ **OBJECTIVE**

- Define inheritance, and explain the inheritance of dominant, recessive, complex, and sex-linked traits.

As previously indicated, the genetic material of a father and a mother unite when a sperm cell fuses with a secondary oocyte to form a zygote. Children resemble their parents because they inherit traits passed down from both parents. We now examine some of the principles involved in that process, called inheritance.

Inheritance is the passage of hereditary traits from one generation to the next. It is the process by which you acquired your characteristics from your parents and may transmit some of your traits to your children. The branch of biology that deals with inheritance is called **genetics** (je-NET-iks). The area of health care that offers advice on genetic problems (or potential problems) is called **genetic counseling.**

Genotype and Phenotype

As you have already learned, the nuclei of all human cells except gametes contain 23 pairs of chromosomes—the diploid number (2*n*). One chromosome in each pair came from the mother, and the other came from the father. Each of these two homologues contains genes that control the same traits. If one chromosome of the pair contains a gene for body hair, for example, its homologue will contain a gene for body hair in the same position. Alternative forms of a gene that code for the same trait and are at the same location on homologous chromosomes are called **alleles** (a-LĒLZ). One allele of the previously mentioned body hair gene might code for coarse hair, and another might code for fine hair. A **mutation** (mū-TĀ-shun; *muta-* = change) is a permanent heritable change in an allele that produces a different variant of the same trait.

The relationship of genes to heredity is illustrated by examining the alleles involved in a disorder called **phenylketonuria (PKU)** (fen′-il-kē′-ton-OO-rē-a). People with PKU (see Clinical Connection: Phenylketonuria in Chapter 25) are unable to manufacture the enzyme phenylalanine hydroxylase. The allele that codes for phenylalanine hydroxylase is symbolized as *P;* the mutated allele that fails to produce a functional enzyme is represented by *p.* The chart in Figure 29.20, which shows the possible combinations of gametes from two parents who each have one *P* and one *p* allele, is called a **Punnett square.** In constructing a Punnett square, the possible paternal alleles in sperm are written at the left side and the possible maternal alleles in ova (or secondary oocytes) are written at the top. The four spaces on the chart show how the alleles can combine in zygotes formed by the union of these sperm and ova to produce the three different combinations of genes, or **genotypes** (JĒ-nō-tīps): *PP, Pp,* or *pp.* Notice from the Punnett square that 25% of the offspring will have the *PP* genotype, 50% will have the *Pp* genotype, and 25% will have the *pp* genotype. (These percentages are probabilities only; parents who have four children won't necessarily end up with one with PKU.) People who inherit *PP* or *Pp* genotypes do not have PKU; those with a *pp* genotype suffer from the disorder. Although people with a *Pp* genotype have one PKU allele (*p*), the allele that codes for the normal trait (*P*) masks the presence of the PKU allele. An allele that dominates or masks the presence of another allele and is fully expressed (*P* in this example) is said to be a **dominant allele,** and the trait expressed is called a dominant trait. The allele whose presence is completely masked (*p* in this example) is

Figure 29.20 Inheritance of phenylketonuria (PKU).

Genotype refers to genetic makeup; phenotype refers to the physical or outward expression of a gene.

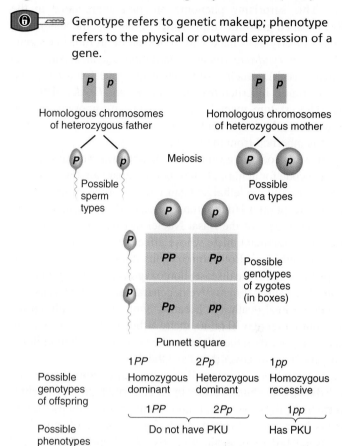

? If parents have the genotypes shown here, what is the chance that their first child will have PKU? What is the chance of PKU occurring in their second child?

said to be a **recessive allele,** and the trait it controls is called a recessive trait.

By tradition, the symbols for genes are written in italics, with dominant alleles written in capital letters and recessive alleles in lowercase letters. A person with the same alleles on homologous chromosomes (for example, *PP* or *pp*) is said to be **homozygous** (hō-mō-ZĪ-gus) for the trait. *PP* is homozygous dominant, and *pp* is homozygous recessive. An individual with different alleles on homologous chromosomes (for example, *Pp*) is said to be **heterozygous** (het′-er-ō-ZĪ-gus) for the trait.

Phenotype (FĒ-nō-tīp; *pheno-* = showing) refers to how the genetic makeup is expressed in the body; it is the physical or outward expression of a gene. A person with *Pp* (a heterozygote) has a different *genotype* from a person with *PP* (a homozygote), but both have the same *phenotype*—normal production of phenylalanine hydroxylase. Heterozygous individuals who carry a recessive gene but do not express it (*Pp*) can pass the gene on to their offspring. Such individuals are called **carriers** of the recessive gene.

Most genes give rise to the same phenotype whether they are inherited from the mother or the father. In a few cases, however, the phenotype is dramatically different, depending on the parental origin. This surprising phenomenon, first appreciated in the 1980s, is called **genomic imprinting.** In humans, the abnormalities most clearly associated with mutation of an imprinted gene are *Angelman syndrome* (mental retardation, ataxia, seizures, and minimal speech), which results when the gene for a particular abnormal trait is inherited from the mother, and *Prader-Willi syndrome* (short stature, mental retardation, obesity, poor responsiveness to external stimuli, and sexual immaturity), which results when it is inherited from the father.

Alleles that code for normal traits do not always dominate over those that code for abnormal ones, but dominant alleles for severe disorders usually are lethal and cause death of the embryo or fetus. One exception is Huntington disease (HD) (see Clinical Connection: Disorders of the Basal Nuclei in Chapter 16), which is caused by a dominant allele whose effects do not become manifest until adulthood. Both homozygous dominant and heterozygous people exhibit the disease; homozygous recessive people are normal. HD causes progressive degeneration of the nervous system and eventual death, but because symptoms typically do not appear until after age 30 or 40, many afflicted individuals will already have passed on the allele for the condition to their children by the time they discover they have the disease.

Occasionally an error in cell division, called **nondisjunction** (non'-dis-JUNK-shun), results in an abnormal number of chromosomes. In this situation, homologous chromosomes (during meiosis I) or sister chromatids (during anaphase of mitosis or meiosis II) fail to separate properly. See Figure 3.34. A cell from which one or more chromosomes have been added or deleted is called an **aneuploid** (AN-ū-ployd). A monosomic cell ($2n - 1$) is missing a chromosome; a trisomic cell ($2n + 1$) has an extra chromosome. Most cases of Down syndrome (see Disorders: Homeostatic Imbalances at the end of this chapter) are aneuploid disorders in which there is trisomy of chromosome 21. Nondisjunction usually occurs during gametogenesis (meiosis), but about 2% of Down syndrome cases result from nondisjunction during mitotic divisions in early embryonic development.

Another error in meiosis is a **translocation.** In this case, two chromosomes that are *not* homologous break and interchange portions of their chromosomes. The individual who has a translocation may be perfectly normal if no loss of genetic material took place when the rearrangement occurred. However, some of the person's gametes may not contain the correct amount and type of genetic material. About 3% of Down syndrome cases result from a translocation of part of chromosome 21 to another chromosome, usually chromosome 14 or 15. The individual who has this translocation is normal and does not even know that he or she is a "carrier." When such a carrier produces gametes, however, some gametes end up with a whole chromosome 21 plus another chromosome with the translocated fragment of chromosome 21. On fertilization, the zygote then has three, rather than two, copies of that part of chromosome 21.

Table 29.3 lists some dominant and recessive inherited structural and functional traits in humans.

TABLE 29.3

Selected Hereditary Traits in Humans

DOMINANT	RECESSIVE
Normal skin pigmentation	Albinism
Near- or farsightedness	Normal vision
PTC taster*	PTC nontaster
Polydactyly (extra digits)	Normal digits
Brachydactyly (short digits)	Normal digits
Syndactylism (webbed digits)	Normal digits
Diabetes insipidus	Normal urine excretion
Huntington disease	Normal nervous system
Widow's peak	Straight hairline
Curved (hyperextended) thumb	Straight thumb
Normal Cl⁻ transport	Cystic fibrosis
Hypercholesterolemia (familial)	Normal cholesterol level

*Ability to taste a chemical compound called phenylthiocarbamide (PTC).

Variations on Dominant–Recessive Inheritance

Most patterns of inheritance do not conform to the simple **dominant–recessive inheritance** we have just described, in which only dominant and recessive alleles interact. The phenotypic expression of a particular gene may be influenced not only by which alleles are present, but also by other genes and by the environment. Most inherited traits are influenced by more than one gene, and, to complicate matters, most genes can influence more than one trait. Variations on dominant–recessive inheritance include incomplete dominance, multiple-allele inheritance, and complex inheritance.

Incomplete Dominance

In **incomplete dominance,** neither member of a pair of alleles is dominant over the other, and the heterozygote has a phenotype intermediate between the homozygous dominant and the homozygous recessive phenotypes. An example of incomplete dominance in humans is the inheritance of **sickle-cell disease (SCD)** (Figure 29.21). People with the homozygous dominant genotype Hb^AHb^A form normal hemoglobin; those with the homozygous recessive genotype Hb^SHb^S have sickle-cell disease and severe anemia. Although they are usually healthy, those with the heterozygous genotype Hb^AHb^S have minor problems with anemia because half their hemoglobin is normal and half is not. Heterozygotes are carriers, and they are said to have *sickle-cell trait.*

Multiple-Allele Inheritance

Although a single individual inherits only two alleles for each gene, some genes may have more than two alternative forms; this is the basis for **multiple-allele inheritance.** One example of multiple-allele inheritance is the inheritance of the ABO blood group. The four blood types (phenotypes) of the ABO group—A, B, AB, and O—result from the inheritance of six combinations of

Figure 29.21 Inheritance of sickle-cell disease.

🔑 Sickle-cell disease is an example of incomplete dominance.

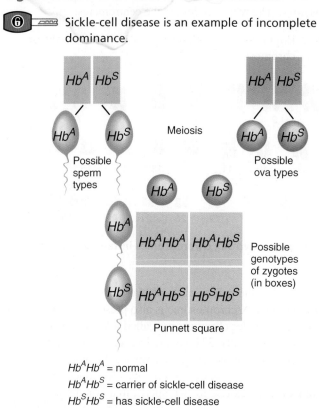

Hb^AHb^A = normal
Hb^AHb^S = carrier of sickle-cell disease
Hb^SHb^S = has sickle-cell disease

❓ **What are the distinguishing features of incomplete dominance?**

Figure 29.22 The 10 possible combinations of parental ABO blood types and the blood types their offspring could inherit. For each possible set of parents, the blue letters represent the blood types their offspring could inherit.

🔑 Inheritance of ABO blood types is an example of multiple-allele inheritance.

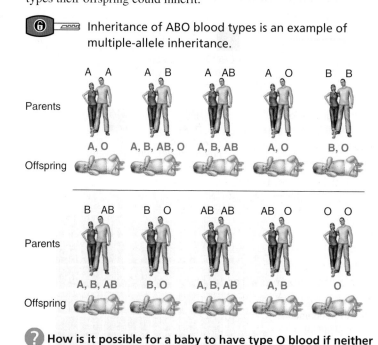

❓ **How is it possible for a baby to have type O blood if neither parent is type O?**

three different alleles of a single gene called the *I* gene: (1) allele I^A produces the A antigen, (2) allele I^B produces the B antigen, and (3) allele *i* produces neither A nor B antigen. Each person inherits two *I*-gene alleles, one from each parent, that give rise to the various phenotypes. The six possible genotypes produce four blood types, as follows:

Genotype	*Blood type (phenotype)*
I^AI^A or I^Ai	A
I^BI^B or I^Bi	B
I^AI^B	AB
ii	O

Notice that both I^A and I^B are inherited as dominant alleles, and *i* is inherited as a recessive allele. Because an individual with type AB blood has characteristics of both type A and type B red blood cells expressed in the phenotype, alleles I^A and I^B are said to be **codominant.** In other words, both genes are expressed equally in the heterozygote. Depending on the parental blood types, different offspring may have blood types different from each other. Figure 29.22 shows the blood types offspring could inherit, given the blood types of their parents.

Complex Inheritance

Most inherited traits are not controlled by one gene, but instead by the combined effects of two or more genes, a situation referred

to as **polygenic inheritance** (pol-ē-JĒN-ik; *poly-* = many), or the combined effects of many genes and environmental factors, a situation referred to as **complex inheritance.** Examples of complex traits include skin color, hair color, eye color, height, metabolism rate, and body build. In complex inheritance, one genotype can have many possible phenotypes, depending on the environment, or one phenotype can include many possible genotypes. For example, even though a person inherits several genes for tallness, full height potential may not be reached due to environmental factors, such as disease or malnutrition during the growth years. You have already learned that the risk of having a child with a neural tube defect is greater in pregnant women who lack adequate folic acid in their diet; this is also considered an environmental factor. Because neural tube defects are more prevalent in some families than in others, however, one or more genes may also contribute.

Often, a complex trait shows a continuous gradation of small differences between extremes among individuals. It is relatively easy to predict the risk of passing on an undesirable trait that is due to a single dominant or recessive gene, but it is very difficult to make this prediction when the trait is complex. Such traits are difficult to follow in a family because the range of variation is large, the number of different genes involved usually is not known, and the impact of environmental factors may be incompletely understood.

Skin color is a good example of a complex trait. It depends on environmental factors such as sun exposure and nutrition, as well as on several genes. Suppose that skin color is controlled by three separate genes, each having two alleles: *A, a*; *B, b*; and *C, c*

Figure 29.23 Complex inheritance of skin color.

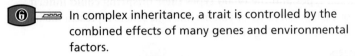 In complex inheritance, a trait is controlled by the combined effects of many genes and environmental factors.

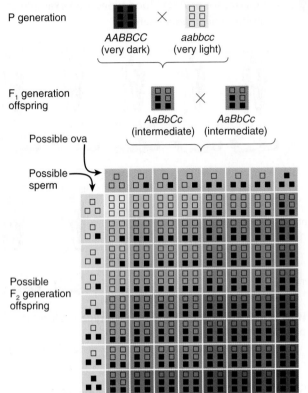

? **What other traits are transmitted by complex inheritance?**

(Figure 29.23). A person with the genotype *AABBCC* is very dark skinned, an individual with the genotype *aabbcc* is very light skinned, and a person with the genotype *AaBbCc* has an intermediate skin color. Parents having an intermediate skin color may have children with very light, very dark, or intermediate skin color. Note that the **P generation** (parental generation) is the starting generation, the **F₁ generation** (first filial generation) is produced from the P generation, and the **F₂ generation** (second filial generation) is produced from the F₁ generation.

Autosomes, Sex Chromosomes, and Sex Determination

When viewed under a microscope, the 46 human chromosomes in a normal somatic cell can be identified by their size, shape, and staining pattern to be members of 23 different pairs. An entire set of chromosomes arranged in decreasing order of size and according to the position of the centromere is called a **karyotype** (KAR-ē-ō-tīp; *karyo-* = nucleus; *-typos* = model) (Figure 29.24). In 22 of the pairs, the homologous chromosomes look alike and have the same appearance in both males and females; these 22 pairs are called **autosomes** (AW-tō-sōms). The two members of the 23rd pair are termed the **sex chromosomes;** they look different in males and fe-

Figure 29.24 Human karyotype showing autosomes and sex chromosomes. The white circles are the centromeres.

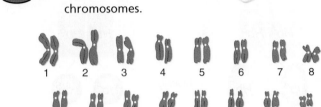 Human somatic cells contain 23 different pairs of chromosomes.

? **What are the two sex chromosomes in females and males?**

males. In females, the pair consists of two chromosomes called X chromosomes. One X chromosome is also present in males, but its mate is a much smaller chromosome called a Y chromosome. The Y chromosome has only 231 genes, less than 10% of the 2968 genes present on chromosome 1, the largest autosome.

When a spermatocyte undergoes meiosis to reduce its chromosome number, it gives rise to two sperm that contain an X chromosome and two sperm that contain a Y chromosome. Oocytes have no Y chromosomes and produce only X-containing gametes. If the secondary oocyte is fertilized by an X-bearing sperm, the offspring normally is female (XX). Fertilization by a Y-bearing sperm produces a male (XY). Thus, an individual's sex is determined by the father's chromosomes (Figure 29.25).

Both female and male embryos develop identically until about 7 weeks after fertilization. At that point, one or more genes set into motion a cascade of events that leads to the development of a male; in the absence of normal expression of the gene or genes, the female pattern of development occurs. It has been known since 1959 that the Y chromosome is needed to initiate male development. Experiments published in 1991 established that the prime male-determining gene is one called *SRY* (**sex-determining region of the Y chromosome**). When a small DNA fragment containing this gene was inserted into 11 female mouse embryos, three of them developed as males. (The researchers suspected that the gene failed to be integrated into the genetic material in the other eight.) *SRY* acts as a molecular switch to turn on the male pattern of development. Only if the *SRY* gene is present and functional in a fertilized ovum will the fetus develop testes and differentiate into a male; in the absence of *SRY*, the fetus will develop ovaries and differentiate into a female.

Case studies have confirmed the key role of *SRY* in directing the male pattern of development in humans. In some cases, phenotypic females with an XY genotype were found to have mutated *SRY* genes. These individuals failed to develop normally as males because their *SRY* gene was defective. In other cases, phenotypic males with an XX genotype were found to have a small piece of the Y chromosome, including the *SRY* gene, inserted into one of their X chromosomes.

Figure 29.25 Sex determination.

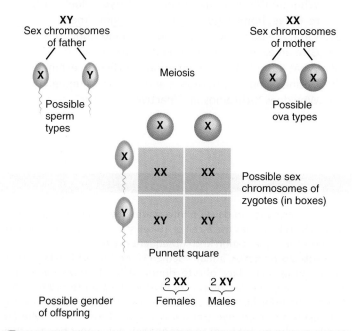

Sex is determined at the time of fertilization by the presence or absence of a Y chromosome in the sperm.

? **What are all chromosomes other than the sex chromosomes called?**

Sex-Linked Inheritance

In addition to determining the sex of the offspring, the sex chromosomes are responsible for the transmission of several nonsexual traits. Many of the genes for these traits are present on X chromosomes but are absent from Y chromosomes. This feature produces a pattern of heredity called **sex-linked inheritance** that is different from the patterns already described.

Red–Green Color Blindness

One example of sex-linked inheritance is **red–green color blindness,** the most common type of color blindness. This condition is characterized by a deficiency in either red- or green-sensitive cones, so red and green are seen as the same color (either red or green, depending on which cone is present). The gene for red–green color blindness is a recessive one designated c. Normal color vision, designated C, dominates. The C/c genes are located only on the X chromosome, so the ability to see colors depends entirely on the X chromosomes. The possible combinations are as follows:

Genotype	*Phenotype*
$X^C X^C$	Normal female
$X^C X^c$	Normal female (but carrier of recessive gene)
$X^c X^c$	Red–green color-blind female
$X^C Y$	Normal male
$X^c Y$	Red–green color-blind male

Only females who have two X^c genes are red–green color blind. This rare situation can result only from the mating of a color-blind male and a color-blind or carrier female. Because males do not have a second X chromosome that could mask the trait, all males with an X^c gene will be red–green color blind. Figure 29.26 illustrates the inheritance of red–green color blindness in the offspring of a normal male and a carrier female.

Traits inherited in the manner just described are called **sex-linked traits.** The most common type of **hemophilia** (hē-mō-FIL-ē-a)—a condition in which the blood fails to clot or clots very slowly after an injury—is also a sex-linked trait. Like the trait for red–green color blindness, hemophilia is caused by a recessive gene. Other sex-linked traits in humans are fragile X syndrome, nonfunctional sweat glands, certain forms of diabetes, some types of deafness, uncontrollable rolling of the eyeballs, absence of central incisors, night blindness, one form of cataract, juvenile glaucoma, and juvenile muscular dystrophy.

X-Chromosome Inactivation

Because they have two X chromosomes in every cell (except developing oocytes), females have a double set of all genes on the X chromosome. A mechanism termed **X-chromosome inactivation (lyonization)** in effect reduces the X-chromosome genes to a

Figure 29.26 An example of the inheritance of red–green color blindness.

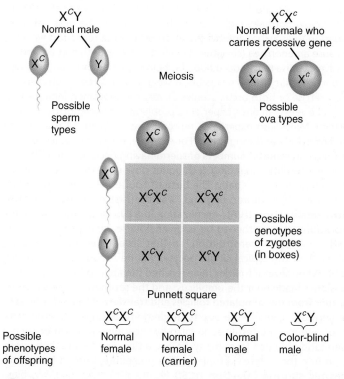

Red–green color blindness and hemophilia are examples of sex-linked traits.

? **What is the genotype of a red–green color-blind female?**

single set in females. In each cell of a female's body, one X chromosome is randomly and permanently inactivated early in development, and most of the genes of the inactivated X chromosome are not expressed (transcribed and translated). The nuclei of cells in female mammals contain a dark-staining body, called a **Barr body,** that is not present in the nuclei of cells in males. Geneticist Mary Lyon correctly predicted in 1961 that the Barr body is the inactivated X chromosome. During inactivation, chemical groups that prevent transcription into RNA are added to the X chromosome's DNA. As a result, an inactivated X chromosome reacts differently to histological stains and has a different appearance than the rest of the DNA. In nondividing (interphase) cells, it remains tightly coiled and can be seen as a dark-staining body within the nucleus. In a blood smear, the Barr body of neutrophils is termed a "drumstick" because it looks like a tiny drumstick-shaped projection of the nucleus.

✔ CHECKPOINT

36. What do the terms genotype, phenotype, dominant, recessive, homozygous, and heterozygous mean?
37. What are genomic imprinting and nondisjunction?
38. Give an example of incomplete dominance.
39. What is multiple-allele inheritance? Give an example.
40. Define complex inheritance and give an example.
41. Why does X-chromosome inactivation occur?

 DISORDERS: HOMEOSTATIC IMBALANCES

Infertility

Female infertility, or the inability to conceive, occurs in about 10% of all women of reproductive age in the United States. Female infertility may be caused by ovarian disease, obstruction of the uterine tubes, or conditions in which the uterus is not adequately prepared to receive a fertilized ovum. **Male infertility (sterility)** is an inability to fertilize a secondary oocyte; it does not imply erectile dysfunction (impotence). Male fertility requires production of adequate quantities of viable, normal sperm by the testes, unobstructed transport of sperm though the ducts, and satisfactory deposition in the vagina. The seminiferous tubules of the testes are sensitive to many factors—x-rays, infections, toxins, malnutrition, and higher-than-normal scrotal temperatures—that may cause degenerative changes and produce male sterility.

One cause of infertility in females is inadequate body fat. To begin and maintain a normal reproductive cycle, a female must have a minimum amount of body fat. Even a moderate deficiency of fat—10% to 15% below normal weight for height—may delay the onset of menstruation (menarche), inhibit ovulation during the reproductive cycle, or cause amenorrhea (cessation of menstruation). Both dieting and intensive exercise may reduce body fat below the minimum amount and lead to infertility that is reversible, if weight gain or reduction of intensive exercise or both occur. Studies of very obese women indicate that they, like very lean ones, experience problems with amenorrhea and infertility. Males also experience reproductive problems in response to undernutrition and weight loss. For example, they produce less prostatic fluid and reduced numbers of sperm having decreased motility.

Many fertility-expanding techniques now exist for assisting infertile couples to have a baby. The birth of Louise Joy Brown on July 12, 1978, near Manchester, England, was the first recorded case of **in vitro fertilization (IVF)**—fertilization in a laboratory dish. In the IVF procedure, the mother-to-be is given follicle-stimulating hormone (FSH) soon after menstruation, so that several secondary oocytes, rather than the typical single oocyte, will be produced (superovulation). When several follicles have reached the appropriate size, a small incision is made near the umbilicus, and the secondary oocytes are aspirated from the stimulated follicles and transferred to a solution containing sperm, where the oocytes undergo fertilization. Alternatively, an oocyte may be fertilized in vitro by suctioning a sperm or even a spermatid obtained from the testis into a tiny pipette and then injecting it into the oocyte's cytoplasm. This procedure, termed **intracytoplasmic sperma injection (ICSI)** (in′-tra-sī-tō-PLAZ-mik), has been used when infertility is due to impairments in sperm motility or to the failure of spermatids to develop into spermatozoa. When the zygote achieved by IVF reaches the 8-cell or 16-cell stage, it is introduced into the uterus for implantation and subsequent growth.

In **embryo transfer,** a man's semen is used to artificially inseminate a fertile secondary oocyte donor. After fertilization in the donor's uterine tube, the morula or blastocyst is transferred from the donor to the infertile woman, who then carries it (and subsequently the fetus) to term. Embryo transfer is indicated for women who are infertile or who do not want to pass on their own genes because they are carriers of a serious genetic disorder.

In **gamete intrafallopian transfer (GIFT)** the goal is to mimic the normal process of conception by uniting sperm and secondary oocyte in the prospective mother's uterine tubes. It is an attempt to bypass conditions in the female reproductive tract that might prevent fertilization, such as high acidity or inappropriate mucus. In this procedure, a woman is given FSH and LH to stimulate the production of several secondary oocytes, which are aspirated from the mature follicles, mixed outside the body with a solution containing sperm, and then immediately inserted into the uterine tubes.

Congenital Defects

An abnormality that is present at birth, and usually before, is called a **congenital defect.** Such defects occur during the formation of structures that develop during the period of organogenesis, the fourth through eighth weeks of development, when all major organs appear. During organogenesis stem cells are establishing the basic patterns of organ development, and it is during this time that developing structures are very susceptible to genetic and environmental influences.

Major structural defects occur in 2–3% of liveborn infants, and they are the leading cause of infant mortality, accounting for about 21% of infant deaths. Many congenital defects can be prevented by supplementation or avoidance of certain substances. For example, neural tube defects, such as spina bifida and anencephaly, can be prevented by having a pregnant female take folic acid. Iodine supplementation can prevent the mental retardation and bone deformation associated with cretinism. Avoidance of teratogens is also very important in preventing congenital defects.

Down Syndrome

Down syndrome (DS) is a disorder characterized by three, rather than two, copies of at least part of chromosome 21. Overall, one infant in 900 is born with Down syndrome. However, older women are more likely to have a DS baby. The chance of having a baby with this syndrome, which is less than 1 in 3000 for women under age 30, increases to 1 in 300 in the 35–39 age group and to 1 in 9 at age 48.

Down syndrome is characterized by mental retardation, retarded physical development (short stature and stubby fingers), distinctive facial structures (large tongue, flat profile, broad skull, slanting eyes, and round head), kidney defects, suppressed immune system, and malformations of the heart, ears, hands, and feet. Sexual maturity is rarely attained, and life expectancy is shorter.

MEDICAL TERMINOLOGY

Breech presentation A malpresentation in which the fetal buttocks or lower limbs present into the maternal pelvis; the most common cause is prematurity.

Conceptus (kon-SEP-tus) Includes all structures that develop from a zygote and includes an embryo plus the embryonic part of the placenta and associated membranes (chorion, amnion, yolk sac, and allantois).

Cryopreserved embryo (krī-ō-PRĒ-servd; *cryo-* = cold) An early embryo produced by in vitro fertilization (fertilization of a secondary oocyte in a laboratory dish) that is preserved for a long period by freezing it. After thawing, the early embryo is implanted into the uterine cavity. Also called a **frozen embryo.**

Deformation (dē-for-MĀ-shun; *de-* = without; *-forma* = form) A developmental abnormality due to mechanical forces that mold a part of the fetus over a prolonged period of time. Deformations usually involve the skeletal and/or muscular system and may be corrected after birth. An example is clubfeet.

Emesis gravidarum (EM-e-sis gra-VID-ar-um; *emeo* = to vomit; *gravida* = a pregnant woman) Episodes of nausea and possibly vomiting that are most likely to occur in the morning during the early weeks of pregnancy; also called **morning sickness.** Its cause is unknown, but the high levels of human chorionic gonadotropin (hCG) secreted by the placenta, and of progesterone secreted by the ovaries, have been implicated. If the severity of these symptoms requires hospitalization for intravenous feeding, the condition is known as **hyperemesis gravidarum.**

Epigenesis (ep-i-JEN-e-sis; *epi-* = upon; *-genesis* = creation) The development of an organism from an undifferentiated cell.

Fertilization age Two weeks less than the gestational age, since a secondary oocyte is not fertilized until about 2 weeks after the last normal menstrual period (LNMP).

Fetal alcohol syndrome (FAS) A specific pattern of fetal malformation due to intrauterine exposure to alcohol. FAS is one of the most common causes of mental retardation and the most common preventable cause of birth defects in the United States.

Fetal surgery A surgical procedure performed on a fetus; in some cases the uterus is opened and the fetus is operated on directly.

Fetal surgery has been used to repair diaphragmatic hernias and remove lesions in the lungs.

Gestational age (jes-TĀ-shun-al; *gestatus* = to bear) The age of an embryo or fetus calculated from the presumed first day of the last normal menstrual period (LNMP).

Karyotype (KAR-ē-ō-tīp; *karyo-* = nucleus) The chromosomal characteristics of an individual presented as a systematic arrangement of pairs of metaphase chromosomes arrayed in descending order of size and according to the position of the centromere (see Figure 29.24); useful in judging whether chromosomes are normal in number and structure.

Klinefelter's syndrome A sex chromosome aneuploidy, usually due to trisomy XXY, that occurs once in every 500 births. Such individuals are somewhat mentally disadvantaged, sterile males with undeveloped testes, scant body hair, and enlarged breasts.

Lethal gene (LĒ-thal JĒN; *lethum* = death) A gene that, when expressed, results in death either in the embryonic state or shortly after birth.

Metafemale syndrome A sex chromosome aneuploidy characterized by at least three X chromosomes (XXX) that occurs about once in every 700 births. These females have underdeveloped genital organs and limited fertility, and most are mentally retarded.

Primordium (prī-MOR-dē-um; *primus-* = first; *-ordior* = to begin) The beginning or first discernible indication of the development of an organ or structure.

Puerperal fever (pū-ER-per-al; *puer* = child) An infectious disease of childbirth, also called puerperal sepsis and childbed fever. The disease, which results from an infection originating in the birth canal, affects the mother's endometrium. It may spread to other pelvic structures and lead to septicemia.

Turner's syndrome A sex chromosome aneuploidy caused by the presence of a single X chromosome (designated XO); occurring about once in every 5000 births, it produces a sterile female with virtually no ovaries and limited development of secondary sex characteristics. Other features include short stature, webbed neck, underdeveloped breasts, and widely spaced nipples. Intelligence usually is normal.

CHAPTER REVIEW AND RESOURCE SUMMARY

WILEY PLUS

Review	Resource

29.1 Embryonic Period

1. Pregnancy is a sequence of events that begins with fertilization and proceeds to implantation, embryonic development, and fetal development. It normally ends in birth.
2. During fertilization a sperm cell penetrates a secondary oocyte and their pronuclei unite. Penetration of the zona pellucida is facilitated by enzymes in the sperm's acrosome. The resulting cell is a zygote. Normally, only one sperm cell fertilizes a secondary oocyte because of the fast and slow blocks to polyspermy.
3. Early rapid cell division of a zygote is called cleavage, and the cells produced by cleavage are called blastomeres. The solid sphere of cells produced by cleavage is a morula. The morula develops into a blastocyst, a hollow ball of cells differentiated into a trophoblast and an inner cell mass. The attachment

Resource

Animation - Fertilization
Anatomy Overview - Developmental Stages
Animation - Embryonic and Fetal Development
Figure 29.2 - Cleavage and the Formation of the Morula and Blastocyst

Review

of a blastocyst to the endometrium is termed implantation; it occurs as a result of enzymatic degradation of the endometrium. After implantation, the endometrium becomes modified and is known as the decidua. The trophoblast develops into the syncytiotrophoblast and cytotrophoblast, both of which become part of the chorion. The inner cell mass differentiates into hypoblast and epiblast, the bilaminar (two-layered) embryonic disc. The amnion is a thin protective membrane that develops from the cytotrophoblast.

Figure 29.5 - Summary of Events Associated with the First Week of Development

Figure 29.7 - Gastrulation

4. The exocoelomic membrane and hypoblast form the yolk sac, which transfers nutrients to the embryo, forms blood cells, produces primordial germ cells, and forms part of the gut. Erosion of sinusoids and endometrial glands provides blood and secretions, which enter lacunar networks to supply nutrition to and remove wastes from the embryo. The extraembryonic coelom forms within extraembryonic mesoderm. The extraembryonic mesoderm and trophoblast form the chorion, the principal embryonic part of the placenta.

5. The third week of development is characterized by gastrulation, the conversion of the bilaminar disc into a trilaminar (three-layered) embryo consisting of ectoderm, mesoderm, and endoderm. The first evidence of gastrulation is formation of the primitive streak, after which the primitive node, notochordal process, and notochord develop. The three primary germ layers form all tissues and organs of the developing organism. Table 29.1 summarizes the structures that develop from the primary germ layers. Also during the third week, the oropharyngeal and cloacal membranes form. The wall of the yolk sac forms a small vascularized outpouching called the allantois, which functions in blood formation and development of the urinary bladder.

6. The process by which the neural plate, neural folds, and neural tube form is called neurulation. The brain and spinal cord develop from the neural tube.

7. Paraxial mesoderm segments to form somites from which skeletal muscles of the neck, trunk, and limbs develop. Somites also form connective tissues and vertebrae.

8. Blood vessel formation, called angiogenesis, begins in mesodermal cells called angioblasts. The heart forms from mesodermal cells called the cardiogenic area. By the end of the third week, the primitive heart beats and circulates blood.

9. Chorionic villi, projections of the chorion, connect to the embryonic heart so that maternal and fetal blood vessels are brought into close proximity, allowing the exchange of nutrients and wastes between maternal and fetal blood. Placentation refers to formation of the placenta, the site of exchange of nutrients and wastes between the mother and fetus. The placenta also functions as a protective barrier, stores nutrients, and produces several hormones to maintain pregnancy. The actual connection between the placenta and embryo (and later the fetus) is the umbilical cord.

10. Organogenesis refers to the formation of body organs and systems and occurs during the fourth week of development. Conversion of the flat, two-dimensional trilaminar embryonic disc to a three-dimensional cylinder occurs by a process called embryonic folding. Embryonic folding brings various organs into their final adult positions and helps form the gastrointestinal tract. Pharyngeal arches, clefts, and pouches give rise to the structures of the head and neck. By the end of the fourth week, upper and lower limb buds develop, and by the end of the eighth week the embryo has clearly human features.

29.2 Fetal Period

1. The fetal period is primarily concerned with the growth and differentiation of tissues and organs that developed during the embryonic period.

2. The rate of body growth is remarkable, especially during the ninth and sixteenth weeks.

3. The principal changes associated with embryonic and fetal growth are summarized in Table 29.2.

Anatomy Overview - Developmental Stages
Animation - Embryonic and Fetal Development

29.3 Teratogens

1. Teratogens are agents that cause physical defects in developing embryos.

2. Among the more important teratogens are alcohol, pesticides, industrial chemicals, some prescription drugs, cocaine, LSD, nicotine, and ionizing radiation.

BBC Video: Teratogens

29.4 Prenatal Diagnostic Tests

1. Several prenatal diagnostic tests are used to detect genetic disorders and to assess fetal well-being. These include fetal ultrasonography, in which an image of a fetus is displayed on a screen; amniocentesis, the withdrawal and analysis of amniotic fluid and the fetal cells within it; and chorionic villi sampling (CVS), which involves withdrawal of chorionic villi tissue for chromosomal analysis.

2. CVS can be done earlier than amniocentesis, and the results are available more quickly, but it is slightly riskier than amniocentesis.

3. Noninvasive prenatal tests include the maternal alpha-fetoprotein (AFP) test to detect neural tube defects and the Quad AFP Plus test to detect Down syndrome, trisomy 18, and neural tube defects.

Review	Resource

29.5 Maternal Changes during Pregnancy

1. Pregnancy is maintained by human chorionic gonadotropin (hCG), estrogens, and progesterone.
2. Human chorionic somatomammotropin (hCS) contributes to breast development, protein anabolism, and catabolism of glucose and fatty acids.
3. Relaxin increases flexibility of the pubic symphysis and helps dilate the uterine cervix near the end of pregnancy.
4. Corticotropin-releasing hormone, produced by the placenta, is thought to establish the timing of birth, and stimulates the secretion of cortisol by the fetal adrenal gland.
5. During pregnancy, several anatomical and physiological changes occur in the mother.

Animation - Hormonal Regulation of Pregnancy
Figure 29.16 - Hormones during Pregnancy

29.6 Exercise and Pregnancy

1. During pregnancy, some joints become less stable, and certain physical activities are more difficult to execute.
2. Moderate physical activity does not endanger the fetus in a normal pregnancy.

29.7 Labor

1. Labor is the process by which the fetus is expelled from the uterus through the vagina to the outside. True labor involves dilation of the cervix, expulsion of the fetus, and delivery of the placenta.
2. Oxytocin stimulates uterine contractions via a positive feedback cycle.

Animation - Regulation of Labor and Birth
Animation - Positive Feedback Control of Labor
Figure 29.18 - Stages of True Labor
Exercise - Pregnancy, Birth, and Lactation

29.8 Adjustments of the Infant at Birth

1. The fetus depends on the mother for oxygen and nutrients, the removal of wastes, and protection.
2. Following birth, an infant's respiratory and cardiovascular systems undergo changes to enable them to become self-supporting during postnatal life.

29.9 The Physiology of Lactation

1. Lactation refers to the production and ejection of milk by the mammary glands.
2. Milk production is influenced by prolactin (PRL), estrogens, and progesterone.
3. Milk ejection is stimulated by oxytocin.
4. A few of the many benefits of breast-feeding include ideal nutrition for the infant, protection from disease, and decreased likelihood of developing allergies.

Animation - Regulation of Lactation

29.10 Inheritance

1. Inheritance is the passage of hereditary traits from one generation to the next.
2. The genetic makeup of an organism is called its genotype; the traits expressed are called its phenotype.
3. Dominant genes control a particular trait; expression of recessive genes is masked by dominant genes.
4. Many patterns of inheritance do not conform to the simple dominant–recessive patterns. In incomplete dominance, neither member of an allelic pair dominates; phenotypically, the heterozygote is intermediate between the homozygous dominant and the homozygous recessive. In multiple-allele inheritance, genes have more than two alternative forms. An example is the inheritance of ABO blood groups. In complex inheritance, a trait such as skin or eye color is controlled by the combined effects of two or more genes and may be influenced by environmental factors.
5. Each somatic cell has 46 chromosomes—22 pairs of autosomes and 1 pair of sex chromosomes.
6. In females, the sex chromosomes are two X chromosomes; in males, they are one X chromosome and a much smaller Y chromosome, which normally includes the prime male-determining gene, called *SRY*.
7. If the *SRY* gene is present and functional in a fertilized ovum, the fetus will develop testes and differentiate into a male. In the absence of *SRY*, the fetus will develop ovaries and differentiate into a female.
8. Red–green color blindness and hemophilia result from recessive genes located on the X chromosome. These sex-linked traits occur primarily in males because of the absence of any counterbalancing dominant genes on the Y chromosome.
9. A mechanism termed X-chromosome inactivation (lyonization) balances the difference in number of X chromosomes between males (one X) and females (two Xs). In each cell of a female's body, one X chromosome is randomly and permanently inactivated early in development and becomes a Barr body.
10. A given phenotype is the result of the interactions of genotype and the environment.

SELF-QUIZ QUESTIONS

Fill in the blanks in the following statements.

1. The three stages of true labor, in order of occurrence, are _____, _____, and _____.

2. Hormones produced by the _____ are responsible for maintaining the pregnancy during the first 3–4 months. The hormone responsible for preventing degeneration of the corpus luteum is _____ produced by the trophoblast.

3. Indicate the germ layers responsible for development of the following structures: (a) muscle, bone, and peritoneum: _____; (b) nervous system and epidermis: _____; (c) epithelial linings of respiratory and gastrointestinal tracts: _____ .

Indicate whether the following statement is true or false.

4. Labor is an example of a negative feedback cycle that ends with the birth of the infant.

Choose the one best answer to the following questions.

5. Which of the following are *true*? (1) During implantation the outer cell mass of the blastocyst orients toward the endometrium. (2) The decidua basalis provides glycogen and lipids for the developing fetus. (3) The decidua parietalis becomes the maternal part of the placenta. (4) During implantation, the syncytiotrophoblast secretes enzymes that allow the blastocyst to penetrate the uterine lining. (5) After fetal delivery, the decidua separates from the endometrium and is released from the uterus.
   (a) 2, 4, and 5   (b) 1, 2, and 3   (c) 2, 3, 4, and 5
   (d) 1, 2, 3, 4, and 5   (e) 1, 3, and 5

6. Which of the following are maternal changes that occur during pregnancy? (1) altered pulmonary function; (2) increased stroke volume, cardiac output, and heart rate, and decreased blood volume; (3) weight gain; (4) increased gastric motility, causing a delay in gastric emptying time; (5) edema and possible varicose veins.
   (a) 1, 2, 3, and 4   (b) 2, 3, 4, and 5   (c) 1, 3, 4, and 5
   (d) 1, 3, and 5   (e) 2, 4, and 5

7. Which of the following statements is *correct*? (a) Normal traits always dominate over abnormal traits. (b) Occasionally an error in meiosis called nondisjunction results in an abnormal number of chromosomes. (c) The mother always determines the sex of the child because she has either an X or Y gene in her oocytes. (d) Most patterns of inheritance are simple dominant–recessive inheritances. (e) Genes are expressed normally regardless of any outside influence such as chemicals or radiation.

8. Which of the following are *true* concerning fertilization? (1) The sperm first penetrate the zona pellucida and then the corona radiata. (2) The binding of specific membrane proteins in the sperm head to ZP3 causes the release of acrosomal contents. (3) Sperm are able to fertilize the oocyte within minutes after ejaculation. (4) Depolarization of the cell membrane of the secondary oocyte inhibits fertilization by more than one sperm. (5) The oocyte completes meiosis II after fertilization.
   (a) 1, 2, 4, and 5   (b) 1, 3, and 5   (c) 1, 2, 3, and 4
   (d) 1, 4, and 5   (e) 2, 4, and 5

9. Amniotic fluid (1) is derived entirely from a filtrate of maternal blood, (2) acts as a fetal shock absorber, (3) provides nutrients to the fetus, (4) helps regulate fetal body temperature, (5) prevents adhesions between the skin of the fetus and surrounding tissues.
   (a) 1, 2, 3, 4, and 5   (b) 2, 4, and 5   (c) 2, 3, 4, and 5
   (d) 1, 4, and 5   (e) 1, 2, 4, and 5

10. Which of the following structures develop during the fourth week after fertilization? (1) embryonic folding, (2) the neural tube, (3) otic placode (beginning of the ear), (4) beginning of the eyes, (5) upper and lower limb buds.
    (a) 1 and 2   (b) 1, 2, and 5   (c) 1, 2, 3, 4, and 5
    (d) 2, 3, and 5   (e) 1, 3, 4, and 5

11. Match the following:
    _____ (a) a fluid-filled sphere of cells
          that enters the uterine cavity
    _____ (b) cells produced by cleavage
    _____ (c) the developing individual from
          week nine of pregnancy
          until birth
    _____ (d) the outer covering of cells
          of the blastocyst
    _____ (e) membrane derived from
          trophoblast
    _____ (f) early divisions of the zygote
    _____ (g) a solid sphere of cells
          still surrounded by the
          zona pellucida
    _____ (h) event in which differentiation
          into the three primary germ
          layers occurs
    _____ (i) embryonic development of
          structures that will become
          the nervous system
    _____ (j) the formation of blood
          vessels to support the
          developing embryo
    _____ (k) result of the fusion of
          female and male pronuclei

    (1) cleavage
    (2) blastomeres
    (3) morula
    (4) angiogenesis
    (5) trophoblast
    (6) blastocyst
    (7) zygote
    (8) gastrulation
    (9) neurulation
    (10) chorion
    (11) fetus

12. Match the following:
_____ (a) stimulates the corpus luteum to continue production of progesterone and estrogens
_____ (b) increases the flexibility of the pubic symphysis and helps dilate the uterine cervix during labor
_____ (c) secreted by the placenta; helps establish the timing of birth and increases the secretion of cortisol for fetal lung maturation
_____ (d) helps prepare mammary glands for lactation; regulates certain aspects of maternal and fetal metabolism
_____ (e) stimulates uterine contractions; responsible for the milk ejection reflex
_____ (f) promotes milk synthesis and secretion; inhibited by progesterone during pregnancy

(1) oxytocin
(2) human chorionic somatomammotropin
(3) human chorionic gonadotropin
(4) prolactin
(5) corticotropin-releasing hormone
(6) relaxin

13. Match the following:
_____ (a) the penetration of a secondary oocyte by a single sperm cell
_____ (b) fertilization of a secondary oocyte by more than one sperm
_____ (c) the attachment of a blastocyst to the endometrium
_____ (d) the fusion of the genetic material from a haploid sperm and a haploid secondary oocyte into a single diploid nucleus
_____ (e) the induction by the female reproductive tract of functional changes in sperm that allow them to fertilize a secondary oocyte
_____ (f) the examination of embryonic or fetal cells sloughed off into the amniotic fluid
_____ (g) an abnormal condition of pregnancy characterized by sudden hypertension, large amounts of protein in urine, and generalized edema
_____ (h) noninvasive test that can detect fetal neural tube defects
_____ (i) the process of giving birth
_____ (j) the period of time (about 6 weeks) during which the maternal reproductive organs and physiology return to the prepregnancy state

(1) fertilization
(2) capacitation
(3) syngamy
(4) polyspermy
(5) implantation
(6) amniocentesis
(7) preeclampsia
(8) parturition
(9) puerperium
(10) maternal AFP test

14. Match the following:
_____ (a) the control of inherited traits by the combined effects of many genes
_____ (b) the two alternative forms of a gene that code for the same trait and are at the same location on homologous chromosomes
_____ (c) abnormal number of chromosomes due to failure of homologous chromosomes or chromatids to separate
_____ (d) inheritance based on genes that have more than two alternative forms; an example is the inheritance of blood type
_____ (e) a cell in which one or more chromosomes of a set is added or deleted
_____ (f) refers to an individual with different alleles on homologous chromosomes
_____ (g) traits linked to the X chromosome
_____ (h) permanent inheritable change in an allele that produces a different variant of the same trait
_____ (i) neither member of the allelic pair is dominant over the other, and the heterozygote has a phenotype intermediate between the homozygous dominant and the homozygous recessive
_____ (j) refers to how the genetic makeup is expressed in the body; the physical or outward expression of a gene
_____ (k) a homozygous dominant, homozygous recessive, or heterozygous genetic makeup; the actual gene arrangement
_____ (l) refers to a person with the same alleles on homologous chromosomes
_____ (m) inactivated X chromosome in females
_____ (n) heterozygous individuals who possess a recessive gene (but do not express it) and can pass the gene on to their offspring
_____ (o) interchange of portions of nonhomologous chromosomes
_____ (p) an allele that masks the presence of another allele and is fully expressed

(1) genotype
(2) phenotype
(3) alleles
(4) aneuploid
(5) incomplete dominance
(6) multiple-allele inheritance
(7) polygenic inheritance
(8) sex-linked inheritance
(9) homozygous
(10) heterozygous
(11) carriers
(12) dominant trait
(13) mutation
(14) nondisjunction
(15) translocation
(16) Barr body

15. Match the following:
_____ (a) the embryonic membrane that entirely surrounds the embryo
_____ (b) functions as an early site of blood formation; contains cells that migrate into the gonads and differentiate into the primitive germ cells
_____ (c) becomes the principal part of the embryonic placenta; produces human chorionic gonadotropin
_____ (d) modified endometrium after implantation has occurred; separates from the endometrium after the fetus is delivered
_____ (e) contains the vascular connections between mother and fetus
_____ (f) the fetal portion is formed by the chorionic villi and the maternal portion is formed by the decidua basalis of the endometrium; allows oxygen and nutrients to diffuse from maternal blood into fetal blood
_____ (g) serves as an early site of blood vessel formation
_____ (h) fingerlike projections of the chorion that bring maternal and fetal blood vessels into close proximity
_____ (i) plays an important role in induction whereby an inducing tissue stimulates the development of an unspecialized responding tissue into a specialized tissue

(1) decidua
(2) placenta
(3) amnion
(4) chorion
(5) allantois
(6) yolk sac
(7) notochord
(8) chorionic villi
(9) umbilical cord

CRITICAL THINKING QUESTIONS

1. Kathy is breast-feeding her infant and is experiencing what feels like early labor pains. What is causing these painful feelings? Is there a benefit to them?

2. Jack has hemophilia, which is a sex-linked blood clotting disorder. He blames his father for passing on the gene for hemophilia. Explain to Jack why his reasoning is wrong. How can Jack have hemophilia if his parents do not?

3. Alisa has asked her obstetrician to save and freeze her baby's cord blood after delivery in case the child needs a future bone marrow transplant. What is in the baby's cord blood that could be used to treat future disorders in the child?

ANSWERS TO FIGURE QUESTIONS

29.1 Capacitation is the group of functional changes in sperm that enable them to fertilize a secondary oocyte, which occur after the sperm have been deposited in the female reproductive tract.

29.2 A morula is a solid ball of cells; a blastocyst consists of a rim of cells (trophoblast) surrounding a cavity (blastocyst cavity) and an inner cell mass.

29.3 The blastocyst secretes digestive enzymes that eat away the endometrial lining at the site of implantation.

29.4 The decidua basalis helps form the maternal part of the placenta.

29.5 Implantation occurs during the secretory phase of the uterine cycle.

29.6 The bilaminar embryonic disc is attached to the trophoblast by the connecting stalk.

29.7 Gastrulation converts a bilaminar embryonic disc into a trilaminar embryonic disc.

29.8 The notochord induces mesodermal cells to develop into vertebral bodies and forms the nucleus pulposus of intervertebral discs.

29.9 The neural tube forms the brain and spinal cord; somites develop into skeletal muscles, connective tissue, and the vertebrae.

29.10 Chorionic villi help to bring the fetal and maternal blood vessels into close proximity.

29.11 The placenta participates in the exchange of materials between fetus and mother, serves as a protective barrier against many microbes, and stores nutrients.

29.12 As a result of embryonic folding, the embryo curves into a C-shape, various organs are brought into their eventual adult positions, and the primitive gut is formed.

29.13 Pharyngeal arches, clefts, and pouches give rise to structures of the head and neck.

29.14 Fetal weight doubles between the midfetal period and birth.

29.15 Amniocentesis is used primarily to detect genetic disorders, but it also provides information concerning the maturity (and survivability) of the fetus.

29.16 Early pregnancy tests detect elevated levels of human chorionic gonadotropin (hCG).

29.17 Relaxin increases the flexibility of the pubic symphysis and helps dilate the cervix of the uterus to ease delivery.

29.18 Complete dilation of the cervix marks the onset of the stage of expulsion.

29.19 Oxytocin also stimulates contraction of the uterus during delivery of a baby.

29.20 The odds that a child will have PKU are the same for each child—25%.

29.21 In incomplete dominance, neither member of an allelic pair is dominant; the heterozygote has a phenotype intermediate between the homozygous dominant and the homozygous recessive phenotypes.

29.22 A baby can have blood type O if each parent is heterozygous and has one i allele.

29.23 Hair color, height, and body build, among others, are traits passed on by complex inheritance.

29.24 The female sex chromosomes are XX, and the male sex chromosomes are XY.

29.25 The chromosomes that are not sex chromosomes are called autosomes.

29.26 A red–green color-blind female has an X^cX^c genotype.

MEASUREMENTS

U.S. Customary System

PARAMETER	UNIT	RELATION TO OTHER U.S. UNITS	SI (METRIC) EQUIVALENT
Length	inch	1/12 foot	2.54 centimeters
foot	12 inches	0.305 meter	
yard	36 inches	0.914 meter	
mile	5,280 feet	1.609 kilometers	
Mass	grain	1/1,000 pound	64.799 milligrams
dram	1/16 ounce	1.772 grams	
ounce	16 drams	28.350 grams	
pound	16 ounces	453.6 grams	
ton	2,000 pounds	907.18 kilograms	
Volume (Liquid)	ounce	1/16 pint	29.574 milliliters
pint	16 ounces	0.473 liter	
quart	2 pints	0.946 liter	
gallon	4 quarts	3.785 liters	
Volume (Dry)	pint	1/2 quart	0.551 liter
quart	2 pints	1.101 liters	
peck	8 quarts	8.810 liters	
bushel	4 pecks	35.239 liters	

International System (SI)

BASE UNITS			PREFIXES		
UNIT	QUANTITY	SYMBOL	PREFIX	MULTIPLIER	SYMBOL
meter	length	m	tera-	$10^{12} = 1,000,000,000,000$	T
kilogram	mass	kg	giga-	$10^{9} = 1,000,000,000$	G
second	time	s	mega-	$10^{6} = 1,000,000$	M
liter	volume	L	kilo-	$10^{3} = 1,000$	k
mole	amount of matter	mol	hecto-	$10^{2} = 100$	h
			deca-	$10^{1} = 10$	da
			deci-	$10^{-1} = 0.1$	d
			centi-	$10^{-2} = 0.01$	c
			milli-	$10^{-3} = 0.001$	m
			micro-	$10^{-6} = 0.000,001$	μ
			nano-	$10^{-9} = 0.000,000,001$	n
			pico-	$10^{-12} = 0.000,000,000,001$	p

Temperature Conversion

FAHRENHEIT (F) TO CELSIUS (C)

$$°C = (°F - 32) ÷ 1.8$$

CELSIUS (C) TO FAHRENHEIT (F)

$$°F = (°C × 1.8) + 32$$

U.S. to SI (Metric) Conversion

WHEN YOU KNOW	MULTIPLY BY	TO FIND
inches	2.54	centimeters
feet	30.48	centimeters
yards	0.91	meters
miles	1.61	kilometers
ounces	28.35	grams
pounds	0.45	kilograms
tons	0.91	metric tons
fluid ounces	29.57	milliliters
pints	0.47	liters
quarts	0.95	liters
gallons	3.79	liters

SI (Metric) to U.S. Conversion

WHEN YOU KNOW	MULTIPLY BY	TO FIND
millimeters	0.04	inches
centimeters	0.39	inches
meters	3.28	feet
kilometers	0.62	miles
liters	1.06	quarts
cubic meters	35.32	cubic feet
grams	0.035	ounces
kilograms	2.21	pounds

PERIODIC TABLE

The periodic table lists the known **chemical elements,** the basic units of matter. The elements in the table are arranged left-to-right in rows in order of their **atomic number,** the number of protons in the nucleus. Each horizontal row, numbered from 1 to 7, is a **period.** All elements in a given period have the same number of electron shells as their period number. For example, an atom of hydrogen or helium each has one electron shell, while an atom of potassium or calcium each has four electron shells. The elements in each column, or **group,** share chemical properties. For example, the elements in column IA are very chemically reactive, whereas the elements in column VIIIA have full electron shells and thus are chemically inert.

Scientists now recognize 117 different elements; 92 occur naturally on Earth, and the rest are produced from the natural elements using particle accelerators or nuclear reactors. Elements are designated by **chemical symbols,** which are the first one or two letters of the element's name in English, Latin, or another language.

Twenty-six of the 92 naturally occurring elements normally are present in your body. Of these, just four elements—oxygen (O), carbon (C), hydrogen (H), and nitrogen (N) (coded blue)—constitute about 96% of the body's mass. Eight others—calcium (Ca), phosphorus (P), potassium (K), sulfur (S), sodium (Na), chlorine (Cl), magnesium (Mg), and iron (Fe) (coded pink)—contribute 3.8% of the body's mass. An additional 14 elements, called **trace elements** because they are present in tiny amounts, account for the remaining 0.2% of the body's mass. The trace elements are aluminum, boron, chromium, cobalt, copper, fluorine, iodine, manganese, molybdenum, selenium, silicon, tin, vanadium, and zinc (coded yellow). Table 2.1 on page 30 provides information about the main chemical elements in the body.

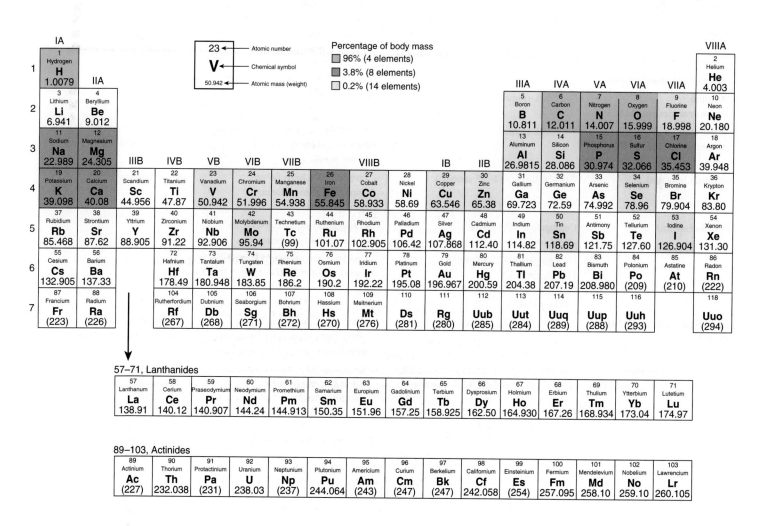

NORMAL VALUES FOR SELECTED BLOOD TESTS

The system of international (SI) units (Système Internationale d'Unités) is used in most countries and in many medical and scientific journals. Clinical laboratories in the United States, by contrast, usually report values for blood and urine tests in conventional units. The laboratory values in this Appendix give conventional units first, followed by SI equivalents in parentheses. Values listed for various blood tests should be viewed as reference values rather than absolute "normal" values for all well people. Values may vary due to age, gender, diet, and environment of the subject or the equipment, methods, and standards of the lab performing the measurement.

KEY TO SYMBOLS

g = gram
mg = milligram = 10^{-3} gram
μg = microgram = 10^{-6} gram
U = units
L = liter
dL = deciliter

mL = milliliter
μL = microliter
mEq/L = milliequivalents per liter
mmol/L = millimoles per liter
μmol/L = micromoles per liter
> = greater than; < = less than

Blood Tests

TEST (SPECIMEN)	U.S. REFERENCE VALUES (SI UNITS)	VALUES INCREASE IN	VALUES DECREASE IN
Aminotransferases (serum)			
Alanine aminotransferase (ALT)	0–35 U/L (same)	Liver disease or liver damage due to toxic drugs.	
Aspartate aminotransferase (AST)	0–35 U/L (same)	Myocardial infarction, liver disease, trauma to skeletal muscles, severe burns.	Beriberi, uncontrolled diabetes mellitus with acidosis, pregnancy.
Ammonia (plasma)	20–120 μg/dL (12–55 μmol/L)	Liver disease, heart failure, emphysema, pneumonia, hemolytic disease of newborn.	Hypertension.
Bilirubin (serum)	Conjugated: <0.5 mg/dL (<5.0 μmol/L)	Conjugated bilirubin: liver dysfunction or gallstones.	
	Unconjugated: 0.2–1.0 mg/dL (18–20 μmol/L) Newborn: 1.0–12.0 mg/dL (<200 μmol/L)	Unconjugated bilirubin: excessive hemolysis of red blood cells.	
Blood urea nitrogen (BUN) (serum)	8–26 mg/dL (2.9–9.3 mmol/L)	Kidney disease, urinary tract obstruction, shock, diabetes, burns, dehydration, myocardial infarction.	Liver failure, malnutrition, overhydration, pregnancy.
Carbon dioxide content (bicarbonate + dissolved CO_2) (whole blood)	Arterial: 19–24 mEq/L (19–24 mmol/L) Venous: 22–26 mEq/L (22–26 mmol/L)	Severe diarrhea, severe vomiting, starvation, emphysema, aldosteronism.	Renal failure, diabetic ketoacidosis, shock.

Blood Tests   Continued

TEST (SPECIMEN)	U.S. REFERENCE VALUES (SI UNITS)	VALUES INCREASE IN	VALUES DECREASE IN
Cholesterol, total (plasma) **HDL cholesterol** (plasma) **LDL cholesterol** (plasma)	<200 mg/dL (<5.2 mmol/L) is desirable >40 mg/dL (>1.0 mmol/L) is desirable <130 mg/dL (<3.2 mmol/L) is desirable	Hypercholesterolemia, uncontrolled diabetes mellitus, hypothyroidism, hypertension, atherosclerosis, nephrosis.	Liver disease, hyperthyroidism, fat malabsorption, pernicious or hemolytic anemia, severe infections.
Creatine (serum)	Males: 0.15–0.5 mg/dL (10–40 μmol/L) Females: 0.35–0.9 mg/dL (30–70 μmol/L)	Muscular dystrophy, damage to muscle tissue, electric shock, chronic alcoholism.	
Creatine kinase (CK), also known as **creatine phosphokinase (CPK)** (serum)	0–130 U/L (same)	Myocardial infarction, progressive muscular dystrophy, hypothyroidism, pulmonary edema.	
Creatinine (serum)	0.5–1.2 mg/dL (45–105 μmol/L)	Impaired renal function, urinary tract obstruction, giantism, acromegaly.	Decreased muscle mass, as occurs in muscular dystrophy or myasthenia gravis.
Electrolytes (plasma)	See Table 27.2 on page 1119		
Gamma-glutamyl transferase (GGT) (serum)	0–30 U/L (same)	Bile duct obstruction, cirrhosis, alcoholism, metastatic liver cancer, congestive heart failure.	
Glucose (plasma)	70–110 mg/dL (3.9–6.1 mmol/L)	Diabetes mellitus, acute stress, hyperthyroidism, chronic liver disease, Cushing's disease.	Addison's disease, hypothyroidism, hyperinsulinism.
Hemoglobin (whole blood)	Males: 14–18 g/100 mL (140–180 g/L) Females: 12–16 g/100 mL (120–160 g/L) Newborns: 14–20 g/100 mL (140–200 g/L)	Polycythemia, congestive heart failure, chronic obstructive pulmonary disease, living at high altitude.	Anemia, severe hemorrhage, cancer, hemolysis, Hodgkin's disease, nutritional deficiency of vitamin B_{12}, systemic lupus erythematosus, kidney disease.
Iron, total (serum)	Males: 80–180 μg/dL (14–32 μmol/L) Females: 60–160 μg/dL (11–29 μmol/L)	Liver disease, hemolytic anemia, iron poisoning.	Iron-deficiency anemia, chronic blood loss, pregnancy (late), chronic heavy menstruation.
Lactic dehydrogenase (LDH) (serum)	71–207 U/L (same)	Myocardial infarction, liver disease, skeletal muscle necrosis, extensive cancer.	

Lipids (serum)
   Total
   Triglycerides
	
	400–850 mg/dL (4.0–8.5 g/L)
	10–190 mg/dL (0.1–1.9 g/L)
		Hyperlipidemia, diabetes mellitus.	Fat malabsorption, hypothyroidism.

Platelet (thrombocyte) count (whole blood)	150,000–400,000/μL	Cancer, trauma, leukemia, cirrhosis.	Anemias, allergic conditions, hemorrhage.

Protein (serum)
   Total
   Albumin
   Globulin
	
	6–8 g/dL (60–80 g/L)
	4–6 g/dL (40–60 g/L)
	2.3–3.5 g/dL (23–35 g/L)
		Dehydration, shock, chronic infections.	Liver disease, poor protein intake, hemorrhage, diarrhea, malabsorption, chronic renal failure, severe burns.

Red blood cell (erythrocyte) count (whole blood)	Males: 4.5–6.5 million/μL Females: 3.9–5.6 million/μL	Polycythemia, dehydration, living at high altitude.	Hemorrhage, hemolysis, anemias, cancer, overhydration.
Uric acid (urate) (serum)	2.0–7.0 mg/dL (120–420) μmol/L	Impaired renal function, gout, metastatic cancer, shock, starvation.	
White blood cell (leukocyte) count, total (whole blood)	5,000–10,000/μL (See Table 19.3 on page 742 for relative percentages of different types of WBCs.)	Acute infections, trauma, malignant diseases, cardiovascular diseases. (See also Table 19.2 on page 741.)	Diabetes mellitus, anemia. (See also Table 19.2 on page 741.)

NORMAL VALUES FOR SELECTED URINE TESTS

Urine Tests

TEST (SPECIMEN)	U.S. REFERENCE VALUES (SI UNITS)	CLINICAL IMPLICATIONS
Amylase (2 hour)	35–260 Somogyi units/hr (6.5–48.1 units/hr)	Values increase in inflammation of the pancreas (pancreatitis) or salivary glands, obstruction of the pancreatic duct, and perforated peptic ulcer.
Bilirubin* (random)	Negative	Values increase in liver disease and obstructive biliary disease.
Blood* (random)	Negative	Values increase in renal disease, extensive burns, transfusion reactions, and hemolytic anemia.
Calcium (Ca²⁺) (random)	10 mg/dL (2.5 mmol/liter); up to 300 mg/24 hr (7.5 mmol/24 hr)	Amount depends on dietary intake; values increase in hyperparathyroidism, metastatic malignancies, and primary cancer of breasts and lungs; values decrease in hypoparathyroidism and vitamin D deficiency.
Casts (24 hour)		
Epithelial	Occasional	Values increase in nephrosis and heavy metal poisoning.
Granular	Occasional	Values increase in nephritis and pyelonephritis.
Hyaline	Occasional	Values increase in kidney infections.
Red blood cell	Occasional	Values increase in glomerular membrane damage and fever.
White blood cell	Occasional	Values increase in pyelonephritis, kidney stones, and cystitis.
Chloride (Cl⁻) (24 hour)	140–250 mEq/24 hr (140–250 mmol/24 hr)	Amount depends on dietary salt intake; values increase in Addison's disease, dehydration, and starvation; values decrease in pyloric obstruction, diarrhea, and emphysema.
Color (random)	Yellow, straw, amber	Varies with many disease states, hydration, and diet.
Creatinine (24 hour)	Males: 1.0–2.0 g/24 hr (9–18 mmol/24 hr) Females: 0.8–1.8 g/24 hr (7–16 mmol/24 hr)	Values increase in infections; values decrease in muscular atrophy, anemia, and kidney diseases.
Glucose*	Negative	Values increase in diabetes mellitus, brain injury, and myocardial infarction.
Hydroxycorticosteroids (17-hydroxysteroids) (24 hour)	Males: 5–15 mg/24 hr (13–41 μmol/24 hr) Females: 2–13 mg/24 hr (5–36 μmol/24 hr)	Values increase in Cushing's syndrome, burns, and infections; values decrease in Addison's disease.
Ketone bodies* (random)	Negative	Values increase in diabetic acidosis, fever, anorexia, fasting, and starvation.
17-ketosteroids (24 hour)	Males: 8–25 mg/24 hr (28–87 μmol/24 hr) Females: 5–15 mg/24 hr (17–53 μmol/24 hr)	Values decrease in surgery, burns, infections, adrenogenital syndrome, and Cushing's syndrome.

TEST (SPECIMEN)	U.S. REFERENCE VALUES (SI UNITS)	CLINICAL IMPLICATIONS
Odor (random)	Aromatic	Becomes acetonelike in diabetic ketosis.
Osmolality (24 hour)	500–1400 mOsm/kg water (500–1400 mmol/kg water)	Values increase in cirrhosis, congestive heart failure (CHF), and high-protein diets; values decrease in aldosteronism, diabetes insipidus, and hypokalemia.
pH* (random)	4.6–8.0	Values increase in urinary tract infections and severe alkalosis; values decrease in acidosis, emphysema, starvation, and dehydration.
Phenylpyruvic acid (random)	Negative	Values increase in phenylketonuria (PKU).
Potassium (K^+) (24 hour)	40–80 mEq/24 hr (40–80 mmol/24 hr)	Values increase in chronic renal failure, dehydration, starvation, and Cushing's syndrome; values decrease in diarrhea, malabsorption syndrome, and adrenal cortical insufficiency.
Protein* (albumin) (random)	Negative	Values increase in nephritis, fever, severe anemias, trauma, and hyperthyroidism.
Sodium (Na^+) (24 hour)	75–200 mg/24 hr (75–200 mmol/24 hr)	Amount depends on dietary salt intake; values increase in dehydration, starvation, and diabetic acidosis; values decrease in diarrhea, acute renal failure, emphysema, and Cushing's syndrome.
Specific gravity* (random)	1.001–1.035 (same)	Values increase in diabetes mellitus and excessive water loss; values decrease in absence of antidiuretic hormone (ADH) and severe renal damage.
Urea (random)	25–35 g/24 hr (420–580 mmol/24 hr)	Values increase in response to increased protein intake; values decrease in impaired renal function.
Uric acid (24 hour)	0.4–1.0 g/24 hr (1.5–4.0 mmol/24 hr)	Values increase in gout, leukemia, and liver disease; values decrease in kidney disease.
Urobilinogen* (2 hour)	0.3–1.0 Ehrlich units (1.7–6.0 μmol/24 hr)	Values increase in anemias, hepatitis A (infectious), biliary disease, and cirrhosis; values decrease in cholelithiasis and renal insufficiency.
Volume, total (24 hour)	1000–2000 mL/24 hr (1.0–2.0 liters/24 hr)	Varies with many factors.

* Test often performed using a **dipstick,** a plastic strip impregnated with chemicals that is dipped into a urine specimen to detect particular substances. Certain colors indicate the presence or absence of a substance and sometimes give a rough estimate of the amount(s) present.

ANSWERS

Answers to Self-Quiz Questions

Chapter 1   1. tissue 2. metabolism, anabolism, catabolism 3. intracellular fluid (ICF), extracellular fluid (ECF) 4. true 5. false 6. false 7. e 8. d 9. a 10. c 11. c 12. (a) 4, (b) 6, (c) 8, (d) 1, (e) 9, (f) 5, (g) 2, (h) 7, (i) 3, (j) 10 13. (a) 1, (b) 12, (c) 1, 6, (d) 6, (e)   , (f) 8, (g) 7 (h) 3, (i) 2, (j) 10 14. (a) 4, (b) 1, (c) 3, (d) 6, (e) 5, (f) 7, (g) 2 15. (a) 6, (b) 1, (c) 11, (d) 5, (e) 10, (f) 8, (g) 7, (h) 9, (i) 4, (j) 3, (k) 2

Chapter 2   1. 8 2. solid, liquid, gas 3. monosaccharides, amino acids 4. true 5. false 6. true 7. c 8. a 9. d 10. b 11. e 12. a 13. e 14. (a) 1, (b) 2, (c) 1, (d) 4, (e) 3 15. (a) 11, (b) 1, (c) 8, (d) 3, (e) 7, (f) 4, (g) 5, (h) 9, (i) 10, (j) 12, (k) 6, (l) 2

Chapter 3   1. plasma membrane, cytoplasm, nucleus 2. apoptosis, necrosis 3. Telomeres 4. UAG 5. false 6. true 7. true 8. e 9. c 10. c, g, i, b, d, k, f, j, a, e, h 11. a 12. c 13. (a) 2, (b) 3, (c) 5, (d) 7, (e) 6, (f) 8, (g) 1, (h) 4 14. (a) 2, (b) 9, (c) 3, (d) 5, (e) 11, (f) 8, (g) 1, (h) 6, (i) 10, (j) 7, (k) 13, (l) 4, (m) 12 15. (a) 3, (b) 9, (c) 1, (d) 5, (e) 11, (f) 4, (g) 8, (h) 7, (i) 2, (j) 10, (k) 6

Chapter 4   1. epithelial, connective, muscle, nervous 2. arrangement of cells in layers, cell shape 3. true 4. true 5. e 6. b 7. a 8. c 9. e 10. b 11. d 12. c 13. (a) 4, (b) 8, (c) 5, (d) 2, (e) 6, (f) 3, (g) 1, (h) 7 14. (a) C, (b) M, (c) N, (d) E, (e) C, (f) E, (g) M, (h) E, (i) C, (j) M, (k) N, (l) E, (m) C, (n) E, (o) M and N 15. (a) 3, (b) 5, (c) 8, (d) 13, (e) 9, (f) 7, (g) 11, (h) 6, (i) 2, (j) 4, (k) 10, (l) 12, (m) 1

Chapter 5   1. stratum lucidum 2. eccrine, ceruminous, apocrine 3. false 4. true 5. c 6. e 7. a 8. c 9. b 10. e 11. a 12. c 13. (a) 3, (b) 5, (c) 4, (d) 1 (e) 6, (f) 11, (g) 2, (h) 8, (i) 9, (j) 10, (k) 7 14. (a) 3, (b) 4, (c) 1, (d) 2 15. (a) 4, (b) 3, (c) 2, (d) 1, inflammatory, migratory, proliferative, maturation

Chapter 6   1. interstitial, appositional 2. hardness, tensile strength 3. true 4. true 5. true 6. d 7. a 8. e 9. c 10. a 11. (a) 1, (b) 4, (c) 3, (d) 2 12. (a) 2, (b) 6, (c) 4, (d) 5, (e) 7, (f) 3, (g) 1 13. (a) 3, (b) 9, (c) 8, (d) 1, (e) 5, (f) 4, (g) 6, (h) 7, (i) 12, (j) 2, (k) 11 (l) 10 14. (a) 3, (b) 7, (c) 6, (d) 1, (e) 4, (f) 2, (g) 5, (h) 9, (i) 8, (j) 10 15. (a) 12, (b) 4, (c) 8, (d) 6, (e) 3, (f) 9, (g) 13, (h) 10, (i) 7, (j) 5, (k) 2, (l) 11, (m) 1

Chapter 7   1. fontanels 2. pituitary gland 3. sacrum, coccyx 4. false 5. false 6. c 7. c 8. a 9. e 10. d 11. e 12. (a) 4, (b) 9, (c) 7, (d) 5, (e) 3, (f) 1, (g) 2, (h) 8, (i) 6 13. (a) 7, (b) 5, (c) 1, (d) 6, (e) 2, (f) 4, (g) 8, (h) 9, (i) 3, (j) 10, (k) 11, (l) 13, (m) 12 14. (a) 2, (b) 3, (c) 5, (d) 6, (e) 4, (f) 1, (g) 5, (h) 4, (i) 2, (j) 4, (k) 3 15. (a) 3, (b) 1, (c) 6, (d) 9, (e) 13, (f) 12, (g) 2, (h) 4, (i) 5, (j) 7, (k) 10, (l) 15, (m) 8, (n) 11, (o) 14

Chapter 8   1. metacarpals 2. ilium, ischium, pubis 3. true (lesser), false (greater) 4. false 5. true 6. b 7. c 8. e 9. c 10. a 11. b and e 12. a 13. (a) 2, (b) 6, (c) 9, (d) 7, (e) 4, (f) 5, (g) 8, (h) 10, (i) 1, (j) 3 14. (a) 3, (b) 8, (c) 4, (d) 11, (e) 9, (f) 13, (g) 5, (h) 6, (i) 10, (j) 14, (k) 2, (l) 1, (m) 7, (n) 12 15. (a) 4, (b) 3, (c) 3, (d) 6, (e) 7, (f) 1, (g) 3, (h) 2, (i) 5, (j) 9, (k) 8, (l) 2, (m) 4, (n) 6, (o) 7, (p) 9, (q) 6, (r) 3, (s) 4, (t) 9, (u) 4 and 5

Chapter 9   1. joint, articulation or arthrosis 2. arthroplasty 3. false 4. false 5. false 6. e 7. d 8. b 9. c 10. a 11. c 12. e 13. (a) 5, (b) 3, (c) 7, (d) 2, (e) 6, (f) 4, (g) 1 14. (a) 6, (b) 4, (c) 5, (d) 1, (e) 3, (f) 2 15. (a) 8, (b) 11, (c) 10, (d) 13, (e) 15, (f) 9, (g) 6, (h) 12, (i) 3, (j) 4, (k) 16, (l) 2, (m) 18, (n) 1, (o) 7, (p) 14, (q) 17, (r) 5

Chapter 10   1. motor unit 2. muscular atrophy, fibrosis 3. acetylcholine 4. true 5. true 6. e 7. a 8. c 9. e 10. d 11. b 12. (a) 5, (b) 6, (c) 9, (d) 7, (e) 2, (f) 4, (g) 10, (h) 3, (i) 1, (j) 8 13. (a) 7, (b) 10, (c) 9, (d) 12, (e) 8, (f) 11, (g) 6, (h) 1,

(i) 2, (j) 3, (k) 4, (l) 13, (m) 5 14. (a) 2, (b) 3, (c) 1, (d) 1 and 2, (e) 3, (f) 2, (g) 1, (h) 3, (i) 1 and 2, (j) 3, (k) 2 and 3, (l) 3 15. (a) 10, (b) 2, (c) 4, (d) 3, (e) 6, (f) 5, (g) 1, (h) 12, (i) 7, (j) 9, (k) 11, (l) 8

Chapter 11   1. buccinator 2. gastrocnemius, soleus, plantaris, calcaneus 3. true 4. true 5. b 6. c 7. c 8. a 9. e 10. e 11. (a) 6, (b) 2, (c) 8, (d) 5, (e) 3, (f) 1, (g) 7, (h) 4 12. (a) 13, (b) 9, (c) 8, (d) 6, (e) 3, (f) 11, (g) 10, (h) 1, (i) 2, (j) 7, (k) 12, (l) 4, (m) 5 13. (a) 6, (b) 3, (c) 7, (d) 4, (e) 2, (f) 9, (g) 5, (h) 1, (i) 8 14. (a) 10, (b) 1, (c) 9, (d) 8, (e) 12, (f) 17, (g) 2, (h) 6, (i) 8, (j) 14, (k) 5, (l) 4, (m) 2, (n) 15, (o) 1, (p) 11, (q) 13, (r) 12, (s) 7, (t) 16, (u) 11, (v) 17, (w) 16, (x) 15, (y) 3, (z) 10 15. (a) 3, (b) 1, (c) 2, (d) 1, (e) 2, (f) 3, (g) 3

Chapter 12   1. somatic, autonomic, enteric 2. sympathetic, parasympathetic 3. false 4. false 5. c 6. d 7. c 8. e 9. e 10. b 11. e 12. b 13. (a) 6, (b) 12, (c) 1, (d) 2, (e) 9, (f) 14, (g) 4, (h) 8, (i) 7, (j) 13, (k) 5, (l) 3, (m) 10, (n) 15, (o) 11 14. (a) 2, (b) 1, (c) 10, (d) 9, (e) 6, (f) 3, (g) 4, (h) 5, (i) 12, (j) 8, (k) 7, (l) 13, (m) 11 15. (a) 4, (b) 5, (c) 16, (d) 8, (e) 7, (f) 1, (g) 2, (h) 10, (i) 15, (j) 6, (k) 3, (l) 13, (m) 9, (n) 11, (o) 14, (p) 12

Chapter 13   1. mixed 2. sensory receptor, sensory neuron, integrating center, motor neuron, effector 3. true 4. false 5. c 6. c 7. a 8. c 9. d 10. e 11. a 12. d 13. (a) 1, (b) 8, (c) 4, (d) 2, (e) 11, (f) 1, (g) 6, (h) 5, (i) 3, (j) 9, (k) 1, (l) 12, (m) 7, (n) 2, (o) 10 14. (a) 14, (b) 12, (c) 13, (d) 1, (e) 2, (f) 5, (g) 11, (h) 8, (i) 10, (j) 9, (k) 15, (l) 4, (m) 7, (n) 3, (o) 6 15. (a) 2, (b) 1, (c) 3, (d) 4, (e) 1, (f) 5, (g) 3, (h) 2, (i) 4, (j) 1, (k) 2, (l) 4, (m) 3, (n) 5, (o) 1

Chapter 14   1. corpus callosum 2. frontal, temporal, parietal, occipital, insula 3. longitudinal fissure 4. false 5. true 6. d 7. c 8. d 9. e 10. d 11. e 12. (a) 3, (b) 5, (c) 6, (d) 8, (e) 11, (f) 10, (g) 7, (h) 9, (i) 1, (j) 4, (k) 2, (l) 12, (m) 1, (n) 8, (o) 5, (p) 7, (q) 12, (r) 10, (s) 9, (t) 1 and 2, (u) 3, 4, and 6, (v) 11 13. (a) 9, (b) 2, (c) 6, (d) 10, (e) 4, (f) 11, (g) 1, (h) 2, (i) 5, (j) 8, (k) 12, (l) 7, (m) 3, (n) 6 and 8, (o) 13, (p) 7, (q) 1 14. (a) 5, (b) 9, (c) 11, (d) 6, (e) 3, (f) 1, (g) 10, (h) 8, (i) 2, (j) 4, (k) 7 15. (a) 10, (b) 2, (c) 6, (d) 8, (e) 7, (f) 5, (g) 3, (h) 11, (i) 14, (j) 13, (k) 4, (l) 1, (m) 12, (n) 9, (o) 15

Chapter 15   1. acetylcholine, epinephrine or norepinephrine 2. thoracolumbar, craniosacral 3. true 4. true 5. d 6. e 7. b 8. c 9. e 10. a 11. a 12. c 13. e, b, g, f, d, a, c 14. (a) 3, (b) 2, (c) 1, (d) 1, (e) 2, (f) 3, (g) 3, (h) 1, (i) 4, (j) 2, (k) 5 15. (a) 2, (b) 1, (c) 1, (d) 2, (e) 1, (f) 1, (g) 2, (h) 2

Chapter 16   1. sensation, perception 2. decussation 3. false 4. true 5. c 6. 1 7. d 8. b 9. d 10. e 11. e 12. d 13. (a) 3, (b) 2, (c) 5, (d) 7, (e) 1, (f) 9, (g) 3, (h) 11, (i) 8, (j) 4, (k) 6, (l) 10 14. (a) 10, (b) 8, (c) 7, (d) 1, (e) 4, (f) 3, (g) 5, (h) 6, (i) 9, (j) 2 15. (a) 9, (b) 8, (c) 4, (d) 7, (e) 10, (f) 2, (g) 3, (h) 1, (i) 5, (j) 6, (k) 11, (l) 12

Chapter 17   1. sweet, sour, salty, bitter, umami 2. static, dynamic 3. true 4. false 5. d 6. a 7. d 8. b 9. b 10. c, j, k, d, h, l, e, b, f, i, a, m, g 11. c 12. a 13. (a) 1, (b) 5, (c) 7, (d) 6, (e) 8, (f) 2, (g) 4, (h) 3 14. (a) 3, (b) 5, (c) 8, (d) 2, (e) 7, (f) 4, (g) 1, (h) 6, (i) 9 15. (a) 2, (b) 11, (c) 14, (d) 13, (e) 3, (f) 10, (g) 6, (h) 12, (i) 4, (j) 5, (k) 9, (l) 1, (m) 7, (n) 8

Chapter 18   1. fight-or-flight response, resistance reaction, exhaustion 2. hypothalamus 3. less, more 4. false 5. true 6. b 7. e 8. d 9. a 10. e 11. c 12. a 13. (a) 8, (b) 2 (and 13, 18, 22), (c) 7, (d) 1, (e) 12, (f) 20, (g) 5, (h) 18, (i) 22, (j) 15 (and 11), (k) 3, (l) 17, (m) 21, (n) 6, (o) 13, (p) 11 (and 15), (q) 4, (r) 10, (s) 14, (t) 9, (u) 16, (v) 19 14. (a) 10, (b) 8, (c) 2, (d) 12, (e) 15, (f) 4, (g) 1, (h) 16, (i) 6, (j) 9, (k) 13, (l) 7, (m) 5, (n) 14, (o) 3, (p) 11 15. (a) 12, (b) 1, (c) 11, (d) 7, (e) 3, (f) 10, (g) 2, (h) 9, (i) 4, (j) 8, (k) 5, (l) 6

Chapter 19   1. serum 2. clot retraction 3. true 4. true 5. e 6. a 7. b 8. c 9. d 10. a 11. d 12. e 13. (a) 4, (b) 7, (c) 6, (d) 1, (e) 3, (f) 5, (g) 2 14. (a) 4, (b) 6, (c) 2, (d) 7, (e) 1, (f) 5, (g) 3 15. (a) 4, (b) 6, (c) 1, (d) 3, (e) 5, (f) 2

Chapter 20 1. left ventricle 2. systole, diastole 3. false 4. true 5. a 6. c 7. d 8. b 9. b 10. e 11. c 12. (a) 3, (b) 6, (c) 1, (d) 5, (e) 2, (f) 4 13. (a) 8, (b) 4, (c) 11, (d) 5, (e) 1, (f) 9, (g) 7, (h) 2, (i) 10, (j) 6, (k) 3 14. (a) 3, (b) 2, (c) 9, (d) 14, (e) 8, (f) 7, (g) 11, (h) 12, (i) 15, (j) 4, (k) 5, (l) 1, (m) 6, (n) 21, (o) 22, (p) 19, (q) 17, (r) 18, (s) 20, (t) 16, (u) 13, (v) 10 15. (a) 3, (b) 7, (c) 2, (d) 5, (e) 1, (f) 6, (g) 4 and 7

Chapter 21 1. carotid sinus, aortic 2. skeletal muscle pump, respiratory pump 3. true 4. true 5. b 6. a 7. c 8. e 9. a 10. d 11. (a) D, (b) C, (c) C, (d) D, (e) D, (f) C, (g) C, (h) C, (i) D, (j) D, (k) C 12. (a) 2, (b) 5, (c) 1, (d) 4, (e) 3 13. (a) 11, (b) 1, (c) 4, (d) 9, (e) 3, (f) 8, (g) 6, (h) 2, (i) 7, (j) 5, (k) 10, (l) 12, (m) 13 14. (a) 2, (b) 6, (c) 4, (d) 1, (e) 3, (f) 5 15. (a) 5, (b) 3, (c) 1, (d) 4, (e) 2, (f) 4, (g) 1, (h) 5, (i) 3

Chapter 22 1. skin, mucous membranes, antimicrobial proteins, natural killer cells, phagocytes 2. antigens 3. true 4. true 5. c 6. d 7. e 8. d 9. e 10. b 11. c 12. e, h, b, f, a, d, g, i, c 13. (a) 3, (b) 1, (c) 7, (d) 4, (e) 2, (f) 5, (g) 6 14. (a) 2, (b) 3, (c) 4, (d) 7, (e) 1, (f) 6, (g) 5

Chapter 23 1. oxyhemoglobin; dissolved CO_2, carbamino compounds (primarily carbaminohemoglobin), and bicarbonate ion 2. $CO_2 + H_2O \rightarrow H_2CO_3 \rightarrow H^+ + HCO_3^-$ 3. false 4. true 5. c 6. e 7. b 8. d 9. a 10. e 11. e, g, b, h, a, d, f, c 12. (a) 3, (b) 8, (c) 5, (d) 9, (e) 2, (f) 7, (g) 10, (h) 1, (i) 4, (j) 6 13. (a) 2, (b) 11, (c) 3, (d) 9, (e) 1, (f) 12, (g) 10, (h) 5, (i) 13, (j) 8, (k) 4, (l) 6, (m) 7 14. (a) 7, (b) 8, (c) 1, (d) 5, (e) 6, (f) 9, (g) 2, (h) 3, (i) 4 15. (a) 9, (b) 11, (c) 3, (d) 4, (e) 6, (f) 1, (g) 5, (h) 10, (i) 7, (j) 8, (k) 2

Chapter 24 1. monosaccharides; amino acids; monoglycerides, fatty acids; pentoses, phosphates, nitrogenous bases 2. diffusion, facilitated diffusion, osmosis, active transport 3. true 4. true 5. b 6. d 7. e 8. c 9. a 10. b 11. d 12. a 13. (a) 5, (b) 13, (c) 8, (d) 11, (e) 9, (f) 1, (g) 7, (h) 12, (i) 14, (j) 2, (k) 4, (l) 10, (m) 6, (n) 3 14. (a) 4, (b) 6, (c) 7, (d) 1, (e) 5, (f) 3, (g) 9, (h) 2, (i) 11, (j) 10, (k) 8

Chapter 25 1. hypothalamus 2. glucose 6-phosphate, pyruvic acid, acetyl coenzyme A 3. true 4. false 5. e 6. c 7. b 8. d 9. a 10. b 11. e 12. a 13. (a) 2 and 3, (b) 1, (c) 3 and 5, (d) 2, 4, 5, and 6, (e) 1, (f) 2, (g) 1, 4, and 6 14. (a) 9, (b) 12, (c) 11, (d) 10, (e) 4, (f) 13, (g) 3, (h) 5, (i) 8, (j) 1, (k) 6, (l) 2, (m) 7 15. (a) 17, (b) 15, (c) 8, (d) 19, (e) 1, (f) 7, (g) 4, (h) 10, (i) 16, (j) 14, (k) 11, (l) 2, (m) 13, (n) 6, (o) 20, (p) 9, (q) 5, (r) 18, (s) 3, (t) 12

Chapter 26 1. glomerulus, glomerular (Bowman's) capsule 2. micturition 3. false 4. true 5. d 6. e 7. c 8. b 9. e 10. a 11. g, a, h, c, o, n, l, j, i, b, d, e, m, k, f 12. g, f, b, d, h, c, a, e, i 13. (a) 8, (b) 2, (c) 10, (d) 5, (e) 3, (f) 1, (g) 7, (h) 4, (i) 11, (j) 9, (k) 6 14. (a) 4, (b) 3, (c) 7, (d) 6, (e) 2, (f) 8, (g) 1, (h) 5 15. (a) 5, (b) 4, (c) 6, (d) 8, (e) 1, (f) 2, (g) 7, (h) 3, (i) 9

Chapter 27 1. metabolic 2. bicarbonate ion, carbonic acid 3. truc 4. false 5. d 6. e 7. a 8. b 9. (a) 8, (b) 9, (c) 7, (d) 1, (e) 6, (f) 2, (g) 4, (h) 5, (i) 3 10. (a) 8, (b) 12, (c) 7, (d) 5, (e) 10, (f) 6, (g) 9, (h) 13, (i) 1, (j) 11, (k) 3, (l) 4, (m) 2

Chapter 28 1. puberty, menarche, menopause 2. true 3. true 4. e 5. c 6. a 7. c 8. b 9. a 10. e 11. d 12. (a) 4, (b) 2, (c) 1, (d) 6, (e) 5, (f) 3 13. (a) 13, (b) 10, (c) 12, (d) 1, (e) 5, (f) 2, (g) 4, (h) 6, (i) 14, (j) 11, (k) 8, (l) 3, (m) 9, (n) 7 14. (a) 6, (b) 4, (c) 1, (d) 12, (e) 8, (f) 5, (g) 7, (h) 13, (i) 11, (j) 3, (k) 14, (l) 2, (m) 10, (n) 15, (o) 9

Chapter 29 1. dilation, expulsion, placental 2. corpus luteum, human chorionic gonadotropin 3. mesoderm, ectoderm, endoderm 4. false 5. a 6. d 7. b 8. e 9. b 10. c 11. (a) 6, (b) 2, (c) 11, (d) 5, (e) 10, (f) 1, (g) 3, (h) 8, (i) 9, (j) 4, (k) 7 12. (a) 3, (b) 6, (c) 5, (d) 2, (e) 1, (f) 4 13. (a) 3, (b) 4, (c) 5, (d) 1, (e) 2, (f) 6, (g) 7, (h) 10, (i) 8, (j) 9 14. (a) 7, (b) 3, (c) 14, (d) 6, (e) 4, (f) 10, (g) 8, (h) 13, (i) 5, (j) 2, (k) 1, (l) 9, (m) 16, (n) 11, (o) 15, (p) 12 15. (a) 3, (b) 6, (c) 4, (d) 1, (e) 9, (f) 2, (g) 5, (h) 8, (i) 7

Answers to Critical Thinking Questions

Chapter 1

1. No. Computed tomography is used to look at differences in tissue density. To assess activity in an organ such as the brain, a positron emission tomography (PET) scan or a single-photo-emission computerized tomography (SPECT) scan would provide a colorized visual assessment of brain activity.

2. Stem cells are undifferentiated cells. Research using stem cells has shown that these undifferentiated cells may be prompted to differentiate into the specific cells needed to replace those which are damaged or malfunctioning.

3. Homeostasis is the relative constancy (or dynamic equilibrium) of the body's internal environment. Homeostasis is maintained as the body changes in response to shifting external and internal conditions, including those of temperature, pressure, fluid, electrolytes, and other chemicals.

Chapter 2

1. Neither butter nor margarine is a particularly good choice for frying eggs. Butter contains saturated fats that are associated with heart disease. However, many margarines contain hydrogenated or partially hydrogenated trans-fatty acids that also increase the risk of heart disease. An alternative would be frying the eggs in any of the mono- or polyunsaturated fats such as olive oil, peanut oil, or corn oil. Boiling or poaching eggs instead of frying them would reduce the fat content of his breakfast, as would eating only the egg whites (not the high-fat yolks).

2. High body temperatures can be life-threatening, especially in infants. The increased temperature can cause denaturing of structural proteins and vital enzymes. When this happens, the proteins become nonfunctional. If the denatured enzymes are required for reactions that are necessary for life, then the infant could die.

3. Simply adding water to the table sugar does not cause it to break apart into monosaccharides. The water acts as a solvent, dissolving the sucrose and forming a sugar-water solution. To complete the breakdown of table sugar to glucose and fructose would require the presence of the enzyme sucrase.

Chapter 3

1. Synthesis of mucin by ribosomes on rough endoplasmic reticulum, to transport vesicle, to entry face of Golgi complex, to transfer vesicle, to medial cisternae where protein is modified, to transfer vesicle, to exit face, to secretory vesicle, to plasma membrane where it undergoes exocytosis.

2. Since smooth ER inactivates or detoxifies drugs, and peroxisomes also destroy harmful substances such as alcohol, we would expect to see increased numbers of these organelles in Sebastian's liver cells.

3. In order to restore water balance to the cells, the runners need to consume hypotonic solutions. The water in the hypotonic solution will move from the blood, into the interstitial fluid, and then into the cells. Plain water works well; sports drinks contain water and some electrolytes (which may have been lost due to sweating) but will still be hypotonic in relation to the body cells.

Chapter 4

1. There are many possible adaptations, including: more adipose tissue for insulation; thicker bones for support; more red blood cells for oxygen transport; increased thickness of skin to prevent water loss; etc.

2. Infants tend to have a high proportion of brown fat, which contains many mitochondria and is highly vascularized. When broken down, brown fat produces heat that helps to maintain infants' body temperatures. This heat can also warm the blood, which then distributes the heat throughout the body.

3. Your bread-and-water diet is not providing you with the necessary nutrients to encourage tissue repair. You need proper amounts of many essential vitamins, especially vitamin C, which is required for repair of the matrix and blood vessels. Vitamin A is needed to help properly maintain epithelial tissue. Adequate protein is also needed in order to synthesize the structural proteins of the damaged tissue.

Chapter 5

1. The dust particles are primarily keratinocytes that are shed from the stratum corneum of the skin.

2. Tattoos are created by depositing ink into the dermis, which does not undergo shedding as the epidermis does. Although the tattoo will fade due to exposure to sunlight and the flushing away of ink particles by the lymphatic system, the tattoo is indeed permanent.

3. Chef Eduardo has damaged the nail matrix—the part of the nail that produces growth. Because the damaged area has not regrown properly, the nail matrix may be permanently damaged.

Chapter 6

1. Due to the strenuous, repetitive activity, Taryn has probably developed a stress fracture of her right tibia (lower leg bone). Stress fractures are due to repeated stress on a bone that causes microscopic breaks in the bone without any evidence of injury to other tissue. An x-ray would not reveal the stress fracture, but a bone scan would. Thus the bone scan would either confirm or negate the physician's diagnosis.

2. When Marcus broke his arm as a child, he injured his epiphyseal (growth) plate. Damage to the cartilage in the epiphyseal plate resulted in premature closure of the plate, which interfered with the lengthwise growth of the arm bone.

3. Exercise causes mechanical stress on bones, but because there is effectively zero gravity in space, the pull of gravity on bones is missing. The lack of stress from gravity results in bone demineralization and weakness.

Chapter 7

1. Inability to open mouth—damage to the mandible, probably at temporomandibular joint; black eye—trauma to the ridge over the supraorbital margin; broken nose—probably damage to the nasal septum (includes the vomer, septal cartilage, and perpendicular plate of the ethmoid) and possibly the nasal bones; broken cheek—fracture of zygomatic bone; broken upper jaw—fracture of maxilla; damaged eye socket—fracture of parts of the sphenoid, frontal, ethmoid, palatine, zygomatic, lacrimal and maxilla (all compose the eye socket); punctured lung—damage to the thoracic vertebrae, which have punctured the lung.

2. Due to the repeated and extensive tension on his bone surfaces, Bubba would experience deposition of new bone tissue. His arm bones would be thicker and with increased raised areas (projections) where the tendons attach his muscles to bone.

3. The "soft area" being referred to is the anterior fontanel, located between the parietal and frontal bones. This is one of several areas of fibrous connective tissue in the skull that has not ossified; it should complete its ossification at 18–24 months after birth. Fontanels allow flexibility of the skull for childbirth and for brain growth after birth. The connective tissue will not allow passage of water; thus no brain damage will occur through simply washing the baby's hair.

Chapter 8

1. There are several characteristics of the bony pelves that can be used to differentiate male from female: (1) The pelvis in the female is wider and more shallow than the male's; (2) the pelvic brim of the female is larger and more oval; (3) the pubic arch has an angle greater than 90°; (4) the pelvic outlet is wider than in a male; (5) the female's iliac crest is less curved and the ilium less vertical. Table 8.1 provides additional differences between female and male pelves. Age of the skeleton can be determined by the size of the bones, the presence or absence of epiphyseal plates, the degree of demineralization of the bones, and the general appearance of the "bumps" and ridges of bones.

2. Infants do have "flat feet" because their arches have not yet developed. As they begin to stand and walk, the arches should begin to develop in order to accommodate and support their body weight. The arches are usually fully developed by age 12 or 13, so Dad doesn't need to worry yet!

3. There are 14 phalanges in each hand: two bones in the thumb and three in each of the other fingers. Farmer White has lost five phalanges on his left hand (two in his thumb and three in his index finger), so he has nine remaining on his left and 14 remaining on his right for a total of 23.

Chapter 9

1. Katie's vertebral column, head, thighs, lower legs, lower arms, and fingers are flexed. Her forearms and shoulders are medially rotated. Her thighs and arms are adducted.

2. The knee joint is commonly injured, especially among athletes. The twisting of Jeremiah's leg could have resulted in a multitude of internal injuries to the knee joint but often football players suffer tearing of the anterior cruciate ligament and medial meniscus. The immediate swelling is due to blood from damaged blood vessels, damaged synovial membranes, and the torn meniscus. Continued swelling is a result of a buildup of synovial fluid, which can result in pain and decreased mobility. Jeremiah's doctor may aspirate some of the fluid ("draining the water off his knee") and might want to perform arthroscopy to check for the extent of the knee damage.

3. The condylar processes of the mandible passed anteriorly to the articular tubercles of the temporal bones, and this dislocated Antonio's mandible. It could be corrected by pressing the thumbs downward on the lower molar teeth and pushing the mandible backward.

Chapter 10

1. Muscle cells lose their ability to undergo cell division after birth. Therefore, the increase in size is not due to an increase in the number of muscle cells but rather is due to enlargement of the existing muscle fibers (hypertrophy). This enlargement can occur from forceful, repetitive muscular activity. It will cause the muscle fibers to increase their production of internal structures such as mitochondria and myofibrils and produce an increase in the muscle fiber diameter.

2. The "dark meat" of both chickens and ducks is composed primarily of slow oxidative (SO) muscle fibers. These fibers contain large amounts of myoglobin and capillaries, which accounts for their dark color. In addition, these fibers contain large numbers of mitochondria and generate ATP by aerobic respiration. SO fibers are resistant to fatigue and can produce sustained contractions for many hours. The legs of chickens and ducks are used for support, walking, and swimming (in ducks), all activities in which endurance is needed. In addition, migrating ducks require SO fibers in their breasts to enable them to have enough energy to fly for extremely long distances while migrating. There may be some fast oxidative–glycolytic (FOG) fibers in the dark meat. FOG fibers also contain large amounts of myoglobin and capillaries, contributing to the dark color. They can use aerobic or anaerobic cellular respiration to generate ATP and have high-to-moderate resistance to fatigue. These fibers would be good for the occasional "sprint" that ducks and chickens undergo to escape dangerous situations. In contrast, the white meat of a chicken breast is composed primarily of fast glycolytic (FG) fibers. FG fibers have lower amounts of myoglobin and capillaries that give the meat its white color. There are also few mitochondria in FG fibers, so these fibers generate ATP mainly by glycolysis. These fibers contract strongly and quickly and are adapted for intense anaerobic movements of short duration. Chickens occasionally use their breasts for flying extremely short distances, usually to escape prey or perceived danger, so FG fibers are appropriate for their breast muscle.

3. Destruction of the somatic motor neurons to skeletal muscle fibers will result in a loss of stimulation to the skeletal muscles. When not stimulated on a regular basis, a muscle begins to lose its muscle tone. Through lack of use, the muscle fibers will weaken, begin to decrease in size, and can be replaced by fibrous connective tissue, resulting in a type of denervation atrophy. A lack of stimulation of the breathing muscles (especially the diaphragm) from motor neurons can result in inability of the breathing muscles to contract, thus causing respiratory paralysis and possibly death of the individual from respiratory failure.

Chapter 11

1. All of the following could occur on the affected (right) side of the face: (1) drooping of eyelid—levator palpebrae superioris; (2) drooping of the mouth, drooling, keeping food in mouth—orbicularis oris, buccinator; (3) uneven smile—zygomaticus major, levator labii superioris, risorius; (4) inability to wrinkle forehead—occipitofrontalis; (5) trouble sucking through a straw—buccinator.

2. Bulbospongiosus, external urethral sphincter, and deep transverse perineal.

3. The rotator cuff is formed by a combination of the tendons of four deep muscles of the shoulder—subscapularis, supraspinatus, infraspinatus, and teres minor. These muscles add strength and stability to the shoulder joint. Although any of the muscles' tendons can be injured, the subscapularis is most often damaged. Dependent upon the injured muscle, Jose may have trouble medially rotating his arm (subscapularis), abducting his arm (supraspinatus), laterally rotating his arm (infraspinatus, teres minor), or extending his arm (teres minor).

Chapter 12

1. Smelling the coffee and hearing the alarm are somatic sensory, stretching and yawning are somatic motor, salivating is autonomic (parasympathetic) motor, stomach rumble is enteric motor.

2. Demyelinaton or destruction of the myelin sheath can lead to multiple problems, especially in infants and children whose myelin sheaths are still in the process of developing. The affected axons deteriorate, which will interfere with function in both the CNS and PNS. There will be lack of sensation and loss of motor control with less rapid and less coordinated body responses. Damage to the axons in the CNS can be permanent and Ming's brain development may be irreversibly affected.

3. Dr. Moro could develop a drug that: (1) is an agonist of substance P; (2) blocks the breakdown of substance P; (3) blocks the reuptake of substance P; (4) promotes the release of substance P; (5) suppresses the release of enkephalins.

Chapter 13

1. The needles will pierce the epidermis, dermis, and subcutaneous layer and then go between the vertebrae through the epidural space, the dura mater, the subdural space, the arachnoid mater, and into the CSF in the subarachnoid space. CSF is produced in the brain, and the spinal meninges are continuous with the cranial meninges.

2. The anterior gray horns contain cell bodies of somatic motor neurons and motor nuclei that are responsible for the nerve impulses for skeletal muscle contraction. Because the lower cervical region is affected (brachial plexus, C5–C8), you would expect that Sunil may have trouble with movement in his shoulder, arm, and hand on the affected side.

3. Allyson has damaged her posterior columns in the lower (lumbar) region of the spinal cord. The posterior columns are responsible for transmitting nerve impulses responsible for awareness of muscle position (proprioception) and touch—which are affected in Allyson—as well as other functions such as light pressure sensations and vibration sensations. Relating Allyson's symptoms to the distribution of dermatomes, it is likely that regions L4, L5, and S1 of her spinal cord were compressed.

Chapter 14

1. Movement of the right arm is controlled by the left hemisphere's primary motor area, located in the precentral gyrus. Speech is controlled by Broca's area in the left hemisphere's frontal lobe just superior to the lateral cerebral sulcus.

2. Nicky's right facial nerve has been affected; she is suffering from Bell's palsy due to the viral infection. The facial nerve controls contraction of skeletal muscles of the face, tear gland and salivary gland secretion, as well as conveying sensory impulses from many of the taste buds on the tongue.

3. You will need to design a drug that can get through the brain's blood–brain barrier (BBB). The drug should be lipid- or water-soluble. If the drug can open a gap between the tight junctions of the endothelial cells of the brain capillaries, it would be more likely to pass through the BBB. Targeting the drug to enter the brain in certain areas near the third ventricle (the circumventricular organs) might be an option as the BBB is entirely absent in those areas and the capillary endothelium is more permeable, allowing the blood-borne drug to more readily enter the brain tissue.

Chapter 15

1. Digestion and relaxation are controlled by increased stimulation of the parasympathetic division of the ANS. The salivary glands, pancreas, and liver will show increased secretion; the stomach and intestines will have increased activity; the gallbladder will have increased contractions; heart contractions will have decreased force and rate. Following is the nerve supply to each listed organ: salivary glands—facial nerves (cranial nerve VII) and glossopharyngeal nerves (cranial nerve IX); pancreas, liver, stomach, gallbladder, intestines and heart—vagus nerves (cranial nerve X).

2. Ciara experienced one of the "E situations" (emergency in her case), which has activated the fight-or-flight response. Some noticeable effects of increased sympathetic activity include an increase in heart rate, sweating on the palms, and contraction of the arrector pili muscles, which causes the goose flesh. Secretion of epinephrine and norepinphrine from the adrenal medullae will intensify and prolong the responses.

3. Mrs. Young needs to slow down the activity of her digestive system, which seems to be experiencing increased parasympathetic response. A parasympathetic blocking agent is needed. Because the stomach and intestines have muscarinic receptors, she needs to be provided with a muscarinic blocking agent (such as atropine), which will result in decreased motility in the stomach and intestines.

Chapter 16

1. Chemoreceptors in the nose detect odors. Proprioceptors detect body position and are involved in equilibrium. The chemoreceptors in the nose are rapidly adapting, whereas proprioceptors are slowly adapting. Thus the smell faded while the sensation of motion remained.

2. Thermal (heat) receptors in her left hand detect the stimulus. A nerve impulse is transmitted to the spinal cord through first-order neurons with cell bodies in posterior root ganglia. The impulses travel into the spinal cord, where the first-order neurons synapse with second-order neurons, whose cell bodies are located in the posterior gray horn of the spinal cord. The axons of the second-order neurons decussate to the right side in the spinal cord and then the impulses ascend through the lateral spinothalamic tract. The axons of the second-order neurons end in the ventral posterior nucleus of the right side of the thalamus, where they synapse with the third-order neurons. Axons of the third-order neurons transmit impulses to the specific primary somatosensory areas in the postcentral gyrus of the right parietal lobe.

3. When Marvin settled down for the night, he passed through Stages 1–3 of NREM sleep. Sleepwalking occurred when he was in Stage 4 (slow-wave sleep). Because this is the deepest stage of sleep, his mother was able to return him to his bed without awakening him. Marvin then cycled through REM and NREM sleep. His dreaming occurred during the REM phases of sleep. The noise of the alarm clock provided the sensory stimulus that stimulated the reticular activating system. Activation of this system sends numerous nerve impulses to widespread areas of the cerebral cortex, both directly and via the thalamus. The result is the state of wakefulness.

Chapter 17

1. Damage to the facial nerve would affect smell, taste, and hearing. Within the nasal epithelium and connective tissue, both the supporting cells and olfactory glands are innervated by branches of the facial nerve. Without input from the facial nerve, there will be a lack of mucus production required to dissolve odorants. The facial nerve also serves taste buds in the anterior two-thirds of the tongue, so damage can affect taste sensations. Hearing will be affected by a damaged facial nerve because the stapedius muscle, which is attached to the stapes, is innervated by the facial nerve. Contraction of the stapedius muscle helps to protect the inner ear from prolonged loud noises. Damage to the facial nerve will result in sounds that are excessively loud, resulting in more susceptibility to damage by prolonged loud noises.

2. With age, Gertrude has lost much of her sense of smell and taste due to a decline in olfactory and gustatory receptors. Since smell and taste are intimately linked, food no longer smells nor tastes as good to Gertrude. Gertrude has presbyopia, a loss of lens elasticity, which makes it difficult to read. She may also be experiencing age-related loss of sharpness of vision and depth perception. Gertrude's hearing difficulties could be a result of damage to hair cells in the organ of Corti or degeneration of the nerve pathway for hearing. The "buzzing" Gertrude hears may be tinnitus, which also occurs more frequently in the elderly.

3. Some of the eyedrops placed in the eye may pass through the nasolacrimal duct into the nasal cavity where olfactory receptors are stimulated. Because most "tastes" are actually smells, the child will "taste" the medicine from her eye.

Chapter 18

1. Yes, Amanda should visit the clinic, as these are serious signs and symptoms. She has an enlarged thyroid gland, or goiter. The goiter is probably due to hypothyroidism, which is causing the weight gain, fatigue, mental dullness, and other symptoms.

2. Amanda's problem is her pituitary gland, which is not secreting normal levels of TSH. Rising thyroxine (T_4) levels after the TSH injection indicates that her thyroid is functioning normally and able to respond to the increased TSH levels. If the thyroxine levels had not risen, then the problem would have been her thyroid gland.

3. Mr. Hernandez has diabetes insipidus caused by either insufficient production or release of ADH due to hypothalamus or posterior pituitary damage. He also could have defective ADH receptors in the kidneys. Diabetes insipidus is characterized by production of large volumes of urine, dehydration, and

increased thirst, but with no glucose or ketones present in the urine (which would be indicative of diabetes mellitus rather than diabetes insipidus).

Chapter 19

1. The broad spectrum antibiotics may have destroyed the bacteria that caused Shilpa's bladder infection but also destroyed the naturally occurring large intestine bacteria that produce vitamin K. Vitamin K is required for the synthesis of four clotting factors (II, VII, IX, and X). Without these clotting factors present in normal amounts, Shilpa will experience clotting problems until the intestinal bacteria reach normal levels and produce additional vitamin K.

2. Mrs. Brown's kidney failure is interfering with her ability to produce erythropoietin (EPO). Her physician can prescribe Epoetin alfa, a recombinant EPO, which is very effective in treating the decline in RBC production with renal failure.

3. A primary problem Thomas may experience is with clotting. Clotting time becomes longer because the liver is responsible for producing many of the clotting factors and clotting proteins such as fibrinogen. Thrombopoietin, which stimulates the formation of platelets, is also produced in the liver. In addition, the liver is responsible for eliminating bilirubin, produced from the breakdown of RBCs. With a malfunctioning liver, the bilirubin will accumulate, resulting in jaundice. In addition, there can be decreased concentrations of the plasma protein albumin, which can affect blood pressure.

Chapter 20

1. The dental procedures introduced bacteria into Gerald's blood. The bacteria colonized his endocardium and heart valves, resulting in bacterial endocarditis. Gerald may have had a previously undetected heart murmur, or the heart murmur may have resulted from his endocarditis. His physician will want to monitor his heart to watch for further damage to the valve.

2. Extremely rapid heart rates can result in a decreased stroke volume due to insufficient ventricular filling. As a result, the cardiac output will decline to the point where there may not be enough blood reaching the central nervous system. She initially may experience light-headedness but could lose consciousness if the cardiac output declines dramatically.

3. Mr. Perkins is suffering from angina pectoris and has several risk factors for coronary artery disease such as smoking, obesity, sedentary lifestyle, and male gender. Cardiac angiography involves the use of a cardiac catheter to inject a radiopaque medium into the heart and its vessels. The angiogram may reveal blockages such as atherosclerotic plaques in his coronary arteries.

Chapter 21

1. The hole in the heart was the foramen ovale, which is an opening between the right and left atria. In fetal circulation it allows blood to bypass the right ventricle, enter the left atrium, and join systemic circulation. The "hole" should close shortly after birth to become the fossa ovalis. Closure of the foramen ovale after birth will allow the deoxgenated blood from the right atrium to enter pulmonary circulation so that the blood can become oxygenated prior to entering systemic circulation. If closure doesn't occur, surgery may be required.

2. Michael is suffering from hypovolemic shock due to the loss of blood. The low blood pressure is a result of low blood volume and a subsequent decrease in cardiac output. His rapid, weak pulse is an attempt of the heart to compensate for the decrease in cardiac output through sympathetic stimulation of the heart and increased blood levels of epinephrine and norepinephrine. His pale, cool, and clammy skin is a result of sympathetic constriction of the blood vessels of the skin and sympathetic stimulation of sweat glands. The lack of urine production is due to increased secretion of aldosterone and ADH, both of which are produced to increase blood volume in order to compensate for Michael's hypotension. The fluid loss from his bleeding results in activation of the thirst center in the hypothalamus. His confusion and disorientation are caused by a reduced oxygen supply to the brain from the decreased cardiac output.

3. Maureen has varicose veins, a condition in which the venous valves become leaky. The leaking valves allow the backflow of blood and an increased pressure that distends the veins and allows fluid to leak into the surrounding tissue. Standing on hard surfaces for long periods of time can cause varicosities to develop. Maureen needs to elevate her legs when possible to counteract the effects of gravity on the blood flow in the lower legs. She could also utilize support hose, which add external support for the superficial veins, much like

skeletal muscle does for deeper veins. If the varices become severe, Maureen may require more extensive treatment such as sclerotherapy, radiofrequency endovenous occlusion, laser occlusion, or stripping.

Chapter 22

1. The influenza vaccination introduces a weakened or killed virus (which will not cause the disease) to the body. The immune system recognizes the antigen and mounts a primary immune response. Upon exposure to the same flu virus that was in the vaccine, the body will produce a secondary response, which will usually prevent a case of the flu. This is artificially acquired active immunity.

2. Mrs. Franco's lymph nodes were removed because metastasis of cancerous cells can occur through the lymph nodes and lymphatic vessels. Mrs. Franco's swelling is a lymphedema that is occurring due to the buildup of interstitial fluid from interference with drainage in the lymph vessels.

3. Tariq's physician would need to perform an antibody titer, which is a measure of the amount of antibody in the serum. If Tariq has previously been exposed to mumps (or been vaccinated for mumps), his blood should have elevated levels of IgG antibodies after this exposure from his sister. His immune system would be experiencing a secondary response. If he has not previously had mumps and has contracted mumps from his sister, his immune system would initiate a primary response. In that case, his blood would show an elevated titer of IgM antibodies, which are secreted by plasma cells after an initial exposure to the mumps antigen.

Chapter 23

1. Aretha's excess mucus production is causing blockage of the paranasal sinuses, which are used as hollow resonating chambers for singing and speech. In addition, her sore throat could be due to inflammation of the pharynx and larynx, which will affect their normal functions. Normally, the pharynx also acts as a resonating chamber and the true vocal cords, located in the larynx, vibrate for speech and singing. Inflammation of the true vocal cords (laryngitis) interferes with their ability to freely vibrate, which will affect both singing and speech.

2. In emphysema, there is destruction of the alveolar walls, producing abnormally large air spaces that remain filled with air during exhalation. The destruction of alveoli decreases the surface area for gas exchange across the respiratory membrane, resulting in a decreased blood O_2 level. Damage to the alveolar walls also causes a loss of elasticity, making exhalation more difficult. This can result in a buildup of CO_2. Cigarette smoke contains nicotine, carbon monoxide, and a variety of irritants, all of which affect the lungs. Nicotine constricts terminal bronchioles, decreasing the air flow into and out of the lungs; carbon monoxide binds to hemoglobin, reducing its ability to carry oxygen; irritants such as tar and fine particulate matter destroy cilia and increase mucus secretion, interfering with the ability of the respiratory passages to cleanse themselves.

3. The squirrel's nest blocked the passage of exhaust gas from the furnace, causing an accumulation of carbon monoxide (CO), a colorless, odorless gas, in the home. As they were sleeping, their blood was saturated with CO, which has a stronger affinity for hemoglobin than oxygen. As a result, the Robinsons became oxygen deficient. Without adequate oxygenation of the brain, the Robinsons died during their sleep.

Chapter 24

1. HCl has several important roles in digestion. HCl stimulates the secretion of hormones that promote the flow of bile and pancreatic juice. The presence of HCl destroys certain microbes that may have been ingested with food. HCl begins denaturing proteins in food, and provides the proper chemical environment for activating pepsinogen into pepsin, which breaks apart certain peptide bonds in proteins. It also helps in the action of gastric lipase, which splits triglycerides in fat molecules found in milk into fatty acids and monoglycerides.

2. Blockage of the pancreatic and bile ducts prevents pancreatic digestive enzymes and bile from reaching the duodenum. As a consequence, there will be problems digesting carbohydrates, proteins, nucleic acids, and lipids. Of particular concern is lipid digestion since the pancreatic juices contain the primary lipid-digesting enzyme. Fats will not be adequately digested, and Trey's feces will contain larger than normal amounts of lipids. In addition, the lack of bile salts will affect the body's ability to emulsify lipids and to form micelles required for absorption of fatty acids and monoglycerides (from lipid breakdown).

When lipids are not absorbed properly, then there will be malabsorption of the lipid-soluble vitamins (A, D, E, and K).

3. Antonio experienced gastroesophageal reflux. The stomach contents backed up (refluxed) into Antonio's esophagus due to a failure of the lower esophageal sphincter to fully close. The HCl from the stomach irritated the esophageal wall, which resulted in the burning sensation he felt; this is commonly known as "heartburn," even though it is not related to the heart. Antonio's recent meal worsened the problem. Alcohol and smoking both can cause the sphincter to relax, while certain foods such tomatoes, chocolate, and coffee can stimulate stomach acid secretion. In addition, lying down immediately after a meal can exacerbate the problem.

Chapter 25

1. Ingestion of cyanide affects cellular respiration. The cyanide binds to the cytochrome oxidase complex in the inner membrane of mitochondria. Blocking this complex interferes with the last step in electron transport in aerobic ATP production. Jane Doe's body would quickly run out of energy to perform vital functions, resulting in her death.

2. Glenn's total cholesterol and LDL levels are very high, while his HDL levels are low. Total cholesterol above 239 mg/dL and LDL above 159 mg/dL are considered high. The ratio of total cholesterol (TC) to HDL-cholesterol is a predictor of the risk of developing coronary artery disease. Glenn's TC to HDL is 15; a ratio above 4 is undesirable. His ratio places him at high risk of developing coronary artery disease. In addition, for every 50 mg/dL TC over 200 mg/dL, the risk of a heart attack doubles. Glenn needs to reduce his TC and LDL-cholesterol while raising his HDL-cholesterol levels. LDLs contribute to fatty plaque formation on coronary artery walls. On the other hand, HDLs help remove excess cholesterol from the blood, which helps decrease the risk of coronary artery disease. Glenn will need to reduce his dietary intake of total fat, saturated fats, and cholesterol, all of which contribute to raising LDL levels. Exercise will raise HDL levels. If those changes are not successful, drug therapy may be required.

3. The goal of weight loss programs is to reduce caloric intake so that the body utilizes stored lipids as an energy source. As part of that desired lipid metabolism, ketone bodies are produced. Some of these ketone bodies will be excreted in the urine. If no ketones are present, then Marissa's body is not breaking down lipids. Only through using fewer calories than needed will her body break down the stored fat and release ketones. Thus, she must be eating more calories than needed to support her daily activities—she is "cheating."

Chapter 26

1. Without reabsorption, initially 105–125 mL of filtrate would be lost per minute, assuming normal glomerular filtration rate. Fluid loss from the blood would cause a decrease in blood pressure, and therefore a decrease in GBHP. When GBHP dropped below 45 mmHg, filtration would stop (assuming normal CHP and BCOP) because NFP would be zero.

2. a. Although normally pale yellow, urine color can vary based upon concentration, diet, drugs, and disease. A dark yellow color would not necessarily indicate a problem, but further investigation may be needed. Turbidity or cloudiness can be caused by urine that has been standing for a period of time, from certain foods, or from bacterial infections. Further investigation is needed. b. Ammonia-like odor occurs when the urine sample is allowed to stand. c. Albumin should not be present in urine (or be present only in very small amounts) because it is too large to pass through the filtration membranes. The presence of high levels of albumin is cause for concern as it indicates damage to the filtration membranes. d. Casts are hardened masses of material that are flushed out in the urine. The presence of casts is not normal and indicates a pathology. e. The pH of normal urine ranges from 4.8 to 8.0. A pH of 5.5 is in normal range. f. Hematuria is the presence of red blood cells in the urine. It can occur with certain pathological conditions or from kidney trauma. Hematuria may occur if the urine sample was contaminated with menstrual blood.

3. Bruce has developed renal calculi (kidney stones), which are blocking his ureters and interfering with the flow of urine from the kidneys to the bladder. The rhythmic pains are a result of the peristaltic contractions of the ureters as they attempt to move the stones toward the bladder. Bruce can wait for the stones to pass, can have them surgically removed, or can use shock-wave lithotripsy to break apart the stones into smaller fragments that can be elimi-

nated with urine. To prevent future episodes, Bruce needs to watch his diet (limit calcium) and drink fluids, and may need drug intervention.

Chapter 27

1. The loss of stomach acids can result in metabolic alkalosis. Robin's HCO_3^- levels would be higher than normal. She would be hypoventilating in order to decrease her pH by slowing the loss of CO_2. Excessive vomiting can result in hyponatremia, hypokalemia, and hypochloremia. Both hyponatremia and hypokalemia can cause mental confusion.

2. (Step 1) pH = 7.30 indicates slight acidosis, which could be caused by elevated PCO_2 or lowered HCO_3^-. (Step 2) The HCO_3^- is lower than normal (20 mEq/liter), so (Step 3) the cause is metabolic. (Step 4) The PCO_2 is lower than normal (32 mmHg), so hyperventilation is providing some compensation. Diagnosis: Henry has partially compensated metabolic acidosis. A possible cause is kidney damage that resulted from interruption of blood flow during the heart attack.

3. Sam will experience increased fluid loss through increased evaporation from the skin and water vapor from the respiratory system through his increased respiratory rate. His insensible water loss will also increase (loss of water from mucous membranes of the mouth and respiratory system). Sam will have a decrease in urine formation.

Chapter 28

1. Monica's excessive training has resulted in an abnormally low amount of body fat. A certain amount of body fat is needed in order to produce the hormones required for the ovarian cycle. Several hormones are involved. Her amenorrhea is due to a lack of gonadotropin-releasing hormone, which in turn reduces the release of LH and FSH. Her follicles with their enclosed ova fail to develop and ovulation will not occur. In addition, synthesis of estrogens and progesterone declines from the lack of hormonal feedback. Usually a gain of weight will allow normal hormonal feedback mechanisms to return.

2. Along with estrogens, progesterone helps to prepare the endometrium for possible implantation of a zygote by promoting growth of the endometrium. The endometrial glands secrete glycogen, which will help sustain an embryo if one should implant. If implantation occurs, progesterone helps maintain the endometrium for the developing fetus. In addition, it helps prepare mammary glands to secrete milk. It inhibits the release of GnRH and LH, which stops a new ovarian cycle from occurring.

3. The ductus deferens is cut and tied in a vasectomy. This stops the release of sperm into the ejaculatory duct and urethra. Mark will still produce the secretions from his accessory glands (prostate, seminal vesicles, bulbourethral glands) in his ejaculate. In addition, a vasectomy does not affect sexual performance; he will be able to achieve erection and ejaculation, as those events are nervous system responses.

Chapter 29

1. As part of the feedback mechanism for lactation, oxytocin is released from the posterior pituitary. It is carried to the mammary glands where it causes milk to be released into the mammary ducts (milk ejection). The oxytocin is also transported in the blood to the uterus, which contains oxytocin receptors on the myometrium. The oxytocin causes contraction of the myometrium, resulting in the painful sensations that Kathy is experiencing. The uterine contractions can help return the uterus back to its prepregnancy size.

2. Sex-linked genetic traits, such as hemophilia, are present on the X chromosomes but not on the Y chromosomes. In males, the X chromosome is always inherited from the mother, and the Y chromosome from the father. Thus, Jack's hemophilia gene was inherited from his mother on his X chromosome. The gene for hemophilia is a recessive gene. His mother would need two recessive genes, one on each of her X chromosomes, to be hemophiliac. His father must carry the dominant (nonhemophiliac) gene on his X chromosome, so he also would not have hemophilia.

3. The cord blood is a source of pluripotent stem cells, which are unspecialized cells that have the potential to specialize into cells with specific functions. The hope is that stem cells can be used to generate cells and tissues to treat a variety of disorders. It is assumed that the tissues would not be rejected since they would contain the same genetic material as the patient—in this case Alisa's baby.

GLOSSARY

Pronunciation Key

1. The most strongly accented syllable appears in capital letters, for example, bilateral (bī-LAT-er-al) and diagnosis (dī-ag-NŌ-sis).

2. If there is a secondary accent, it is noted by a prime ('), for example, constitution (kon'-sti-TOO-shun) and physiology (fiz'-ē-OL-ō-jē). Any additional secondary accents are also noted by a prime, for example, decarboxylation (dē'-kar-bok'-si-LĀ-shun).

3. Vowels marked by a line above the letter are pronounced with the long sound, as in the following common words:

ā as in *māke*   ō as in *pōle*
ē as in *bē*   ū as in *cūte*
ī as in *īvy*

4. Vowels not marked by a line above the letter are pronounced with the short sound, as in the following words:

a as in *above* or *at*   o as in *not*
e as in *bet*   u as in *bud*
i as in *sip*

5. Other vowel sounds are indicated as follows:

oy as in *oil*
oo as in *root*

6. Consonant sounds are pronounced as in the following words:

b as in *bat*	m as in *mother*
ch as in *chair*	n as in *no*
d as in *dog*	p as in *pick*
f as in *father*	r as in *rib*
g as in *get*	s as in *so*
h as in *hat*	t as in *tea*
j as in *jump*	v as in *very*
k as in *can*	w as in *welcome*
ks as in *tax*	z as in *zero*
kw as in *quit*	zh as in *lesion*
l as in *let*	

A

Abdomen (ab-DŌ-men *or* AB-dō-men) The area between the diaphragm and pelvis.

Abdominal (ab-DOM-i-nal) **cavity** Superior portion of the abdominopelvic cavity that contains the stomach, spleen, liver, gallbladder, pancreas, kidneys, small intestine, and part of the large intestine.

Abdominal thrust manoeuvre A first-aid procedure for choking. Employs a quick, upward thrust against the diaphragm that forces air out of the lungs with sufficient force to eject any lodged material.

Abdominopelvic (ab-dom'-i-nō-PEL-vik) **cavity** Inferior to the diaphragm and subdivided into a superior abdominal cavity and an inferior pelvic cavity.

Abduction (ab-DUK-shun) Movement away from the midline of the body.

Abortion (a-BOR-shun) The premature loss **(spontaneous)** or removal **(induced)** of the embryo or nonviable fetus; miscarriage due to a failure in the normal process of developing or maturing.

Abscess (AB-ses) A localized collection of pus and liquefied tissue in a cavity.

Absorption (ab-SORP-shun) Intake of fluids or other substances by cells of the skin or mucous membranes; the passage of digested foods from the gastrointestinal tract into blood or lymph.

Accessory duct A duct of the pancreas that empties into the duodenum about 2.5 cm (1 in.) superior to the ampulla of Vater (hepatopancreatic ampulla). Also called the **duct of Santorini** (san'-tō-RĒ-nē).

Acetabulum (as'-e-TAB-ū-lum) The rounded cavity on the external surface of the hip bone that receives the head of the femur.

Acetylcholine (as'-ē-til-KŌ-lēn) **(ACh)** A neurotransmitter liberated by many peripheral nervous system neurons and some central nervous system neurons. It is excitatory at neuromuscular junctions but inhibitory at some other synapses (for example, it slows heart rate).

Achalasia (ak'-a-LĀ-zē-a) A condition, caused by malfunction of the myenteric plexus, in which the lower esophageal sphincter fails to relax normally as food approaches. A whole meal may become lodged in the oesophagus and enter the stomach very slowly. Distension of the oesophagus results in chest pain that is often confused with pain originating from the heart.

Achilles tendon *See* **Calcaneal tendon.**

Acini (AS-i-nē) Groups of cells in the pancreas that secrete digestive enzymes. *Singular* is **acinus** (AS-i-nus).

Acoustic (a-KOOS-tik) Pertaining to sound or the sense of hearing.

Acquired immunodeficiency syndrome (AIDS) A fatal disease caused by the human immunodeficiency virus (HIV). Characterized by a positive HIV-antibody test, low helper T cell count, and certain indicator diseases (for example, Kaposi's sarcoma, pneumocystis carinii pneumonia, tuberculosis, fungal diseases). Other symptoms include fever or night sweats, coughing, sore throat, fatigue, body aches, weight loss, and enlarged lymph nodes.

Acrosome (AK-rō-sōm) A lysosomelike organelle in the head of a sperm cell containing enzymes that facilitate the penetration of a sperm cell into a secondary oocyte.

Actin (AK-tin) A contractile protein that is part of thin filaments in muscle fibers.

Action potential (AP) An electrical signal that propagates along the membrane of a neuron or muscle fiber (cell); a rapid change in membrane potential that involves a depolarization followed by a repolarization. Also called a **nerve action potential** or **nerve impulse** as it relates to a neuron, and a **muscle action potential** as it relates to a muscle fiber.

Activation (ak'-ti-VĀ-shun) **energy** The minimum amount of energy required for a chemical reaction to occur.

Active transport The movement of substances across cell membranes against a concentration gradient, requiring the expenditure of cellular energy (ATP).

Acute (a-KŪT) Having rapid onset, severe symptoms, and a short course; not chronic.

Adaptation (ad'-ap-TĀ-shun) The adjustment of the pupil of the eye to changes in light intensity. The property by which a sensory neuron relays a decreased frequency of action potentials from a receptor, even though the strength of the stimulus remains constant; the decrease in perception of a sensation over time while the stimulus is still present.

Adduction (ad-DUK-shun) Movement toward the midline of the body.

Adenoids (AD-e-noyds) The pharyngeal tonsils.

Adenosine triphosphate (a-DEN-ō-sēn trī-FOS-fāt) **(ATP)** The main energy currency in living cells; used to transfer the chemical energy needed for metabolic reactions. ATP consists of the purine base *adenine* and the five-carbon sugar *ribose*, to which are added, in linear array, three *phosphate* groups.

Adhesion (ad-HĒ-zhun) Abnormal joining of parts to each other.

Adipocyte (AD-i-pō-sīt) Fat cell, derived from a fibroblast. Also called **fat cell** or **adipose cell.**

Adipose (AD-i-pōz) **tissue** Tissue composed of adipocytes specialized for triglyceride storage and present in the form of soft pads between various organs for support, protection, and insulation.

Adrenal cortex (a-DRĒ-nal KOR-teks) The outer portion of an adrenal gland, divided into three zones; the zona glomerulosa secretes mineralocorticoids, the zona fasciculata secretes glucocorticoids, and the zona reticularis secretes androgens.

Adrenal glands Two glands located superior to each kidney. Also called the **suprarenal** (soo′-pra-RĒ-nal) **glands.**

Adrenal medulla (me-DUL-a) The inner part of an adrenal gland, consisting of cells that secrete epinephrine, norepinephrine, and a small amount of dopamine in response to stimulation by sympathetic preganglionic neurons.

Adrenergic (ad′-ren-ER-jik) **neuron** A neuron that releases epinephrine (adrenaline) or norepinephrine (noradrenaline) as its neurotransmitter.

Adrenocorticotropic (ad-rē′-nō-kor-ti-kō-TRŌP-ik) **hormone (ACTH)** A hormone produced by the anterior pituitary that influences the production and secretion of certain hormones of the adrenal cortex. Also called **corticotropin** (kor′-ti-kō-TRŌ-pin).

Adventitia (ad-ven-TISH-a) The outermost covering of a structure or organ.

Aerobic (air-Ō-bik) Requiring molecular oxygen.

Aerobic (ār-Ō-bik) **cellular respiration** The production of ATP (36 molecules) from the complete oxidation of pyruvic acid in mitochondria. Carbon dioxide, water, and heat are also produced.

Afferent arteriole (AF-er-ent ar-TĒ-rē-ōl) A blood vessel of a kidney that divides into the capillary network called a glomerulus; there is one afferent arteriole for each glomerulus.

Agglutination (a-gloo-ti-NĀ-shun) Clumping of microorganisms or blood cells, typically due to an antigen–antibody reaction.

Aggregated lymphatic follicles Clusters of lymph nodules that are most numerous in the ileum. Also called **Peyer's** (PĪ-erz) **patches.**

Albinism (AL-bin-izm) Abnormal, nonpathological, partial, or total absence of pigment in skin, hair, and eyes.

Aldosterone (al-DOS-ter-ōn) A mineralocorticoid produced by the adrenal cortex that promotes sodium and water reabsorption by the kidneys and potassium excretion in urine.

All-or-none principle If a stimulus depolarizes a neuron to threshold, the neuron fires at its maximum voltage (all); if threshold is not reached, the neuron does not fire at all (none). Given above threshold, stronger stimuli do not produce stronger action potentials.

Allantois (a-LAN-tō-is) A small, vascularized outpouching of the yolk sac that serves as an early site for blood formation and development of the urinary bladder.

Alleles (a-LĒLZ) Alternate forms of a single gene that control the same inherited trait (such as type

A blood) and are located at the same position on homologous chromosomes.

Allergen (AL-er-jen) An antigen that evokes a hypersensitivity reaction.

Alopecia (al′-ō-PĒ-shē-a) The partial or complete lack of hair as a result of factors such as genetics, aging, endocrine disorders, chemotherapy, and skin diseases.

Alpha (AL-fa) **cell** A type of cell in the pancreatic islets (islets of Langerhans) that secretes the hormone glucagon. Also termed an **A cell.**

Alpha (α) receptor A type of receptor for norepinephrine and epinephrine; present on visceral effectors innervated by sympathetic postganglionic neurons.

Alveolar duct Branch of a respiratory bronchiole around which alveoli and alveolar sacs are arranged.

Alveolar macrophage (MAK-rō-fāj) Highly phagocytic cell found in the alveolar walls of the lungs. Also called a **dust cell.**

Alveolar sac A cluster of alveoli that share a common opening.

Alveolus (al-VĒ-ō-lus) A small hollow or cavity; an air sac in the lungs; milk-secreting portion of a mammary gland. *Plural* is **alveoli** (al-VĒ-ol-ī).

Alzheimer's (ALTZ-hī-mer) **disease (AD)** Disabling neurological disorder characterized by dysfunction and death of specific cerebral neurons, resulting in widespread intellectual impairment, personality changes, and fluctuations in alertness.

Amenorrhoea (ā-men-ō-RĒ-a) Absence of menstruation.

Amnesia (am-NĒ-zē-a) A lack or loss of memory.

Amnion (AM-nē-on) A thin, protective fetal membrane that develops from the epiblast; holds the fetus suspended in amniotic fluid. Also called the **"bag of waters."**

Amniotic (am′-nē-OT-ik) **fluid** Fluid in the amniotic cavity, the space between the developing embryo (or fetus) and amnion; the fluid is initially produced as a filtrate from maternal blood and later includes fetal urine. It functions as a shock absorber, helps regulate fetal body temperature, and helps prevent desiccation.

Amphiarthrosis (am′-fē-ar-THRŌ-sis) A slightly movable joint, in which the articulating bony surfaces are separated by fibrous connective tissue or fibrocartilage to which both are attached; types are syndesmosis and symphysis.

Ampulla (am-PUL-la) A saclike dilation of a canal or duct.

Ampulla of Vater *See* **Hepatopancreatic ampulla.**

Anabolism (a-NAB-ō-lizm) Synthetic, energy-requiring reactions whereby small molecules are built up into larger ones.

Anaemia (a-NĒ-mē-a) Condition of the blood in which the number of functional red blood cells or their haemoglobin content is below normal.

Anaerobic (an-ar-Ō-bik) Not requiring oxygen.

Anaesthesia (an′-es-THĒ-zē-a) A total or partial loss of feeling or sensation; may be general or local.

Anal (Ā-nal) **canal** The last 2 or 3 cm (1 in.) of the rectum; opens to the exterior through the anus.

Anal column A longitudinal fold in the mucous membrane of the anal canal that contains a network of arteries and veins.

Anal triangle The subdivision of the female or male perineum that contains the anus.

Analgesia (an-al-JĒ-zē-a) Pain relief; absence of the sensation of pain.

Anaphase (AN-a-fāz) The third stage of mitosis in which the chromatids that have separated at the centromeres move to opposite poles of the cell.

Anaphylaxis (an′-a-fi-LAK-sis) A hypersensitivity (allergic) reaction in which IgE antibodies attach to mast cells and basophils, causing them to produce mediators of anaphylaxis (histamine, leukotrienes, kinins, and prostaglandins) that bring about increased blood permeability, increased smooth muscle contraction, and increased mucus production. Examples are hay fever, hives, and anaphylactic shock.

Anastomosis (a-nas′-tō-MŌ-sis) An end-to-end union or joining of blood vessels, lymphatic vessels, or nerves.

Anatomic dead space Spaces of the nose, pharynx, larynx, trachea, bronchi, and bronchioles totaling about 150 mL of the 500 mL in a quiet breath (tidal volume); air in the anatomic dead space does not reach the alveoli to participate in gas exchange.

Anatomical (an′-a-TOM-i-kal) **position** A position of the body universally used in anatomical descriptions in which the body is erect, the head is level, the eyes face forward, the upper limbs are at the sides, the palms face forward, and the feet are flat on the floor.

Anatomy (a-NAT-ō-mē) The structure or study of the structure of the body and the relation of its parts to each other.

Androgens (AN-drō-jenz) Masculinizing sex hormones produced by the testes in males and the adrenal cortex in both sexes; responsible for libido (sexual desire); the two main androgens are testosterone and dihydrotestosterone.

Aneurysm (AN-ū-rizm) A saclike enlargement of a blood vessel caused by a weakening of its wall.

Angina pectoris (an-JĪ-na *or* AN-ji-na PEK-tō-ris) A pain in the chest related to reduced coronary circulation due to coronary artery disease (CAD) or spasms of vascular smooth muscle in coronary arteries.

Angiogenesis (an′-jē-ō-JEN-e-sis) The formation of blood vessels in the extraembryonic mesoderm of the yolk sac, connecting stalk, and chorion at the beginning of the third week of development.

Ankylosis (ang′-ki-LŌ-sis) Severe or complete loss of movement at a joint as the result of a disease process.

Antagonist (an-TAG-ō-nist) A muscle that has an action opposite that of the prime mover (agonist) and yields to the movement of the prime mover.

Antagonistic (an-tag-ō-NIST-ik) **effect** A hormonal interaction in which the effect of one hormone on a target cell is opposed by another hormone. For example, calcitonin (CT) lowers blood calcium level, whereas parathyroid hormone (PTH) raises it.

Anterior (an-TĒR-ē-or) Nearer to or at the front of the body. Equivalent to **ventral** in bipeds.

Anterior pituitary Anterior lobe of the pituitary gland. Also called the **adenohypophysis** (ad′-e-nō-hī-POF-i-sis).

Anterior root The structure composed of axons of motor (efferent) neurons that emerges from the anterior aspect of the spinal cord and extends laterally to join a posterior root, forming a spinal nerve. Also called a **ventral root.**

Anterolateral (an′-ter-ō-LAT-er-al) **pathway** Sensory pathway that conveys information related to pain, temperature, tickle, and itch. Also called **spinothalamic pathway.**

Antibody (AN-ti-bod′-ē) **(Ab)** A protein produced by plasma cells in response to a specific antigen; the antibody combines with that antigen to neutralize, inhibit, or destroy it. Also called an **immunoglobulin** (im-ū-nō-GLOB-ū-lin) or **Ig.**

Anticoagulant (an-tī-cō-AG-ū-lant) A substance that can delay, suppress, or prevent the clotting of blood.

Antidiuretic (an′-ti-dī-ū-RET-ik) Substance that inhibits urine formation.

Antidiuretic hormone (ADH) Hormone produced by neurosecretory cells in the paraventricular and supraoptic nuclei of the hypothalamus that stimulates water reabsorption from kidney tubule cells into the blood and vasoconstriction of arterioles. Also called **vasopressin** (vāz-ō-PRES-in).

Antigen (AN-ti-jen) **(Ag)** A substance that has immunogenicity (the ability to provoke an immune response) and reactivity (the ability to react with the antibodies or cells that result from the immune response); contraction of *anti*body *gen*erator. Also termed a **complete antigen.**

Antigen-presenting cell (APC) Special class of migratory cell that processes and presents antigens to T cells during an immune response; APCs include macrophages, B cells, and dendritic cells, which are present in the skin, mucous membranes, and lymph nodes.

Antioxidant A substance that inactivates oxygen-derived free radicals. Examples are selenium, zinc, beta carotene, and vitamins C and E.

Antrum (AN-trum) Any nearly closed cavity or chamber, especially one within a bone, such as a sinus.

Anuria (an-Ū-rē-a) Absence of urine formation or daily urine output of less than 50 mL.

Anus (Ā-nus) The distal end and outlet of the rectum.

Aorta (ā-OR-ta) The main systemic trunk of the arterial system of the body that emerges from the left ventricle.

Aortic (ā-OR-tik) **body** Cluster of chemoreceptors on or near the arch of the aorta that respond to changes in blood levels of oxygen, carbon dioxide, and hydrogen ions (H$^+$).

Aortic reflex A reflex that helps maintain normal systemic blood pressure; initiated by baroreceptors in the wall of the ascending aorta and arch of the aorta. Nerve impulses from aortic baroreceptors reach the cardiovascular center via sensory axons of the vagus (X) nerves.

Apex (Ā-peks) The pointed end of a conical structure, such as the apex of the heart.

Aphasia (a-FĀ-zē-a) Loss of ability to express oneself properly through speech or loss of verbal comprehension.

Apneustic (ap-NOO-stik) **area** A part of the respiratory center in the pons that sends stimulatory nerve impulses to the inspiratory area that activate and prolong inhalation and inhibit exhalation.

Apnoea (AP-nē-a) Temporary cessation of breathing.

Apocrine (AP-ō-krin) **gland** A type of gland in which the secretory products gather at the free end of the secreting cell and are pinched off, along with some of the cytoplasm, to become the secretion, as in mammary glands.

Aponeurosis (ap′-ō-noo-RŌ-sis) A sheetlike tendon joining one muscle with another or with bone.

Apoptosis (ap′-ōp-TŌ-sis *or* ap-ō-TŌ-sis) Programmed cell death; a normal type of cell death that removes unneeded cells during embryological development, regulates the number of cells in tissues, and eliminates many potentially dangerous cells such as cancer cells. During apoptosis, the DNA fragments, the nucleus condenses, mitochondria cease to function, and the cytoplasm shrinks, but the plasma membrane remains intact. Phagocytes engulf and digest the apoptotic cells, and an inflammatory response does not occur.

Appositional (a-pō-ZISH-o-nal) **growth** Growth due to surface deposition of material, as in the growth in diameter of cartilage and bone. Also called **exogenous** (eks-OJ-e-nus) **growth.**

Aqueous humour (A-kwē-us HŪ-mor) The watery fluid, similar in composition to cerebrospinal fluid, that fills the anterior cavity of the eye.

Arachnoid (a-RAK-noyd) **mater** The middle of the three meninges (coverings) of the brain and spinal cord. Also termed the **arachnoid.**

Arachnoid villus (VIL-us) Berrylike tuft of the arachnoid mater that protrudes into the superior sagittal sinus and through which cerebrospinal fluid is reabsorbed into the bloodstream.

Arbor vitae (AR-bor VĪ-tē) The white matter tracts of the cerebellum, which have a treelike appearance when seen in midsagittal section.

Arch of the aorta The most superior portion of the aorta, lying between the ascending and descending segments of the aorta.

Areola (a-RĒ-ō-la) Any tiny space in a tissue. The pigmented ring around the nipple of the breast.

Arm The part of the upper limb from the shoulder to the elbow.

Arousal (a-ROW-zal) Awakening from sleep, a response due to stimulation of the reticular activating system (RAS).

Arrector pili (a-REK-tor PĪ-lē) Smooth muscles attached to hairs; contraction pulls the hairs into a vertical position, resulting in "goose bumps."

Arrhythmia (a-RITH-mē-a) An irregular heart rhythm. Also called a **dysrhythmia.**

Arteriole (ar-TĒ-rē-ōl) A small, almost microscopic, artery that delivers blood to a capillary.

Arteriosclerosis (ar-tē-rē-ō-skle-RŌ-sis) Group of diseases characterized by thickening of the walls of arteries and loss of elasticity.

Artery (AR-ter-ē) A blood vessel that carries blood away from the heart.

Arthritis (ar-THRĪ-tis) Inflammation of a joint.

Arthrology (ar-THROL-ō-jē) The study or description of joints.

Arthroplasty (AR-thrō-plas′-tē) Surgical replacement of joints, for example, the hip and knee joints.

Arthroscopy (ar-THROS-kō-pē) A procedure for examining the interior of a joint, usually the knee, by inserting an arthroscope into a small incision; used to determine extent of damage, remove torn cartilage, repair cruciate ligaments, and obtain samples for analysis.

Arthrosis (ar-THRŌ-sis) A joint or articulation.

Articular (ar-TIK-ū-lar) **capsule** Sleevelike structure around a synovial joint composed of a fibrous capsule and a synovial membrane.

Articular cartilage (KAR-ti-lij) Hyaline cartilage attached to articular bone surfaces.

Articular disc Fibrocartilage pad between articular surfaces of bones of some synovial joints. Also called a **meniscus** (men-IS-kus).

Articulation (ar-tik-ū-LĀ-shun) A joint; a point of contact between bones, cartilage and bones, or teeth and bones.

Arytenoid (ar′-i-TĒ-noyd) **cartilages** A pair of small, pyramidal cartilages of the larynx that attach to the vocal folds and intrinsic pharyngeal muscles and can move the vocal folds.

Ascending colon (KŌ-lon) The part of the large intestine that passes superiorly from the caecum to the inferior border of the liver, where it bends at the right colic (hepatic) flexure to become the transverse colon.

Ascites (a-SĪ-tēz) Abnormal accumulation of serous fluid in the peritoneal cavity.

Association areas Large cortical regions on the lateral surfaces of the occipital, parietal, and temporal lobes and on the frontal lobes anterior to the motor areas connected by many motor and sensory axons to other parts of the cortex. The association areas are concerned with motor patterns, memory, concepts of word-hearing and word-seeing, reasoning, will, judgment, and personality traits.

Asthma (AZ-ma) Usually allergic reaction characterized by smooth muscle spasms in bronchi resulting in wheezing and difficult breathing. Also called **bronchial asthma.**

Astigmatism (a-STIG-ma-tizm) An irregularity of the lens or cornea of the eye causing the image to be out of focus and producing faulty vision.

Astrocyte (AS-trō-sīt) A neuroglial cell having a star shape that participates in brain development and the metabolism of neurotransmitters, helps form the blood–brain barrier, helps maintain the proper balance of K$^+$ for generation of nerve impulses, and provides a link between neurons and blood vessels.

Ataxia (a-TAK-sē-a) A lack of muscular coordination, lack of precision.

Atherosclerosis (ath-er-ō-skle-RŌ-sis) A progressive disease characterized by the formation in the walls of large and medium-sized arteries of lesions called atherosclerotic plaques.

Atherosclerotic (ath′-er-ō-skle-RO-tik) **plaque (PLAK)** A lesion that results from accumulated cholesterol and smooth muscle fibers (cells) of the tunica media of an artery; may become obstructive.

Atom Unit of matter that makes up a chemical element; consists of a nucleus (containing positively charged protons and uncharged neutrons) and negatively charged electrons that orbit the nucleus.

Atresia (a-TRĒ-zē-a) Degeneration and reabsorption of an ovarian follicle before it fully matures and ruptures; abnormal closure of a passage, or absence of a normal body opening.

Atrial fibrillation (Ā-trē-al fib-ri-LĀ-shun) **(AF)** Asynchronous contraction of cardiac muscle fibers in the atria that results in the cessation of atrial pumping.

Atrial natriuretic (nā′-trē-ū-RET-ik) **peptide (ANP)** Peptide hormone, produced by the atria of the heart in response to stretching, that inhibits aldosterone production and thus lowers blood pressure; causes natriuresis, increased urinary excretion of sodium.

Atrioventricular (AV) (ā′-trē-ō-ven-TRIK-ū-lar) **bundle** The part of the conduction system of the heart that begins at the atrioventricular (AV) node, passes through the cardiac skeleton separating the atria and the ventricles, then extends a short distance down the interventricular septum before splitting into right and left bundle branches. Also called the **bundle of His** (HISS).

Atrioventricular (AV) node The part of the conduction system of the heart made up of a compact mass of conducting cells located in the septum between the two atria.

Atrioventricular (AV) valve A heart valve made up of membranous flaps or cusps that allows blood to flow in one direction only, from an atrium into a ventricle.

Atrium (Ā-trē-um) A superior chamber of the heart.

Atrophy (AT-rō-fē) Wasting away or decrease in size of a part, due to a failure, abnormality of nutrition, or lack of use.

Auditory ossicle (AW-di-tō-rē OS-si-kul) One of the three small bones of the middle ear called the malleus, incus, and stapes.

Auditory tube The tube that connects the middle ear with the nose and nasopharynx region of the throat. Also called the **eustachian** (ū-STĀ-kē-an *or* ū-STĀ-shun) **tube** or **pharyngotympanic tube.**

Auscultation (aws-kul-TĀ-shun) Examination by listening to sounds in the body.

Autoimmunity An immunological response against a person's own tissues.

Autolysis (aw-TOL-i-sis) Self-destruction of cells by their own lysosomal digestive enzymes after death or in a pathological process.

Autonomic ganglion (aw′-tō-NOM-ik GANG-lē-on) A cluster of cell bodies of sympathetic or parasympathetic neurons located outside the central nervous system.

Autonomic nervous system (ANS) Visceral sensory (afferent) and visceral motor (efferent) neurons. Autonomic motor neurons, both sympathetic and parasympathetic, conduct nerve impulses from the central nervous system to smooth muscle, cardiac muscle, and glands. So named because this part of the nervous system was thought to be self-governing or spontaneous.

Autonomic plexus (PLEK-sus) A network of sympathetic and parasympathetic axons; examples are the cardiac, celiac, and pelvic plexuses, which are located in the thorax, abdomen, and pelvis, respectively.

Autophagy (aw-TOF-a-jē) Process by which worn-out organelles are digested within lysosomes.

Autopsy (AW-top-sē) The examination of the body after death.

Autorhythmic (aw′-tō-RITH-mik) **cells** Cardiac or smooth muscle fibers that are self-excitable (generate impulses without an external stimulus); act as the heart's pacemaker and conduct the pacing impulse through the conduction system of the heart; self-excitable neurons in the central nervous system, as in the inspiratory area of the brain stem.

Autosome (AW-tō-sōm) Any chromosome other than the X and Y chromosomes (sex chromosomes).

Axilla (ak-SIL-a) The small hollow beneath the arm where it joins the body at the shoulders. Also called the **armpit.**

Axon (AK-son) The usually single, long process of a nerve cell that propagates a nerve impulse toward the axon terminals.

Axon terminal Terminal branch of an axon where synaptic vesicles undergo exocytosis to release neurotransmitter molecules. Also called **telodendria** (tel′-o-DEN-drea).

B

B cell A lymphocyte that can develop into a clone of antibody-producing plasma cells or memory cells when properly stimulated by a specific antigen.

Babinski (ba-BIN-skē) **sign** Extension of the great toe, with or without fanning of the other toes, in response to stimulation of the outer margin of the sole; normal up to 18 months of age and indicative of damage to descending motor pathways such as the corticospinal tracts after that.

Back The posterior part of the body; the dorsum.

Ball-and-socket joint A synovial joint in which the rounded surface of one bone moves within a cup-shaped depression or socket of another bone, as in the shoulder or hip joint. Also called a **spheroid** (SFĒ-royd) **joint.**

Baroreceptor (bar′-ō-re-SEP-tor) Neuron capable of responding to changes in blood or **stretch receptor** air or fluid pressure. Also called a **pressoreceptor.**

Basal nuclei Paired clusters of grey matter deep in each cerebral hemisphere including the globus pallidus, putamen, and caudate nucleus. Together, the caudate nucleus and putamen are known as the corpus striatum. Nearby structures that are functionally linked to the basal nuclei are the substan-tia nigra of the midbrain and the subthalamic nuclei of the diencephalon.

Basement membrane Thin, extracellular layer between epithelium and connective tissue consisting of a basal lamina and a reticular lamina.

Basilar (BĀS-i-lar) **membrane** A membrane in the cochlea of the internal ear that separates the cochlear duct from the scala tympani and on which the spiral organ (organ of Corti) rests.

Basophil (BĀ-sō-fil) A type of white blood cell characterized by a pale nucleus and large granules that stain blue-purple with basic dyes.

Belly The abdomen. The gaster or prominent, fleshy part of a skeletal muscle.

Beta (BĀ-ta) **cell** A type of cell in the pancreatic islets (islets of Langerhans) in the pancreas that secretes the hormone insulin. Also called a **B cell.**

Beta (β) **receptor** A type of adrenergic receptor for epinephrine and norepinephrine; found on visceral effectors innervated by sympathetic postganglionic neurons.

Bicuspid (bī-KUS-pid) **valve** Atrioventricular (AV) valve on the left side of the heart. Also called the **mitral valve.**

Bilateral (bī-LAT-er-al) Pertaining to two sides of the body.

Bile (BĪL) A secretion of the liver consisting of water, bile salts, bile pigments, cholesterol, lecithin, and several ions; it emulsifies lipids prior to their digestion.

Bilirubin (bil-ē-ROO-bin) An orange pigment that is one of the end products of haemoglobin breakdown in the hepatocytes and is excreted as a waste material in bile.

Biopsy (BĪ-op-sē) The removal of a sample of living tissue to help diagnose a disorder, for example, cancer.

Blastocyst (BLAS-tō-sist) In the development of an embryo, a hollow ball of cells that consists of a blastocele (the internal cavity), trophoblast (outer cells), and embryoblast (inner cell mass).

Blastocyst (BLAS-tō-sist) **cavity** The fluid-filled cavity within the blastocyst. Also called the **blastocele.**

Blastomere (BLAS-tō-mēr) One of the cells resulting from the cleavage of a fertilized ovum.

Blastula (BLAS-tyū-la) An early stage in the development of a zygote.

Blind spot Area in the retina at the end of the optic (II) nerve in which there are no photoreceptors. Also called **optic disc.**

Blood The fluid that circulates through the heart, arteries, capillaries, and veins and that constitutes the chief means of transport within the body.

Blood–brain barrier (BBB) A barrier consisting of specialized brain capillaries and astrocytes that prevents the passage of materials from the blood to the cerebrospinal fluid and brain.

Blood clot A gel that consists of the formed elements of blood trapped in a network of insoluble protein fibers.

Blood island Isolated mass of mesoderm derived from angioblasts and from which blood vessels develop.

Blood pressure (BP) Force exerted by blood against the walls of blood vessels due to contraction of the heart and influenced by the elasticity of the vessel walls; clinically, a measure of the pressure in arteries during ventricular systole and ventricular diastole.

Blood reservoir (REZ-er-vwar) Systemic veins and venules that contain large amounts of blood that can be moved quickly to parts of the body requiring the blood.

Blood–testis barrier (BTB) A barrier formed by Sertoli cells that prevents an immune response against antigens produced by spermatogenic cells by isolating the cells from the blood.

Body cavity A space within the body that contains various internal organs.

Bolus (BŌ-lus) A soft, rounded mass, usually food, that is swallowed.

Bone remodelling Replacement of old bone by new bone tissue.

Bony labyrinth (LAB-i-rinth) A series of cavities within the petrous portion of the temporal bone forming the vestibule, cochlea, and semicircular canals of the inner ear.

Bowman's capsule *See* **Glomerular capsule.**

Brachial plexus (BRĀ-kē-al PLEK-sus) A network of nerve axons of the ventral rami of spinal nerves C5, C6, C7, C8, and T1. The nerves that emerge from the brachial plexus supply the upper limb.

Bradycardia (brād′-i-KAR-dē-a) A slow resting heart or pulse rate (under 50 beats per minute).

Brain The part of the central nervous system contained within the cranial cavity.

Brain stem The portion of the brain immediately superior to the spinal cord, made up of the medulla oblongata, pons, and midbrain.

Brain waves Electrical signals that can be recorded from the skin of the head due to electrical activity of brain neurons.

Broad ligament A double fold of parietal peritoneum attaching the uterus to the side of the pelvic cavity.

Broca's (BRŌ-kaz) **speech area** Motor area of the brain in the frontal lobe that translates thoughts into speech. Also called the **motor speech area.**

Bronchi (BRON-kī) Branches of the respiratory passageway including primary bronchi (the two divisions of the trachea), secondary or lobar bronchi (divisions of the primary bronchi that are distributed to the lobes of the lung), and tertiary or segmental bronchi (divisions of the secondary bronchi that are distributed to bronchopulmonary segments of the lung). *Singular* is **bronchus.**

Bronchial (BRON-kē-al) **tree** The trachea, bronchi, and their branching structures up to and including the terminal bronchioles.

Bronchiole (BRONG-kē-ōl) Branch of a tertiary bronchus further dividing into terminal bronchioles (distributed to lobules of the lung), which divide into respiratory bronchioles (distributed to alveolar sacs).

Bronchitis (brong-KĪ-tis) Inflammation of the mucous membrane of the bronchial tree; characterized by hypertrophy and hyperplasia of seromucous glands and goblet cells that line the bronchi which results in a productive cough.

Bronchopulmonary (brong′-kō-PUL-mō-ner-ē) **segment** One of the smaller divisions of a lobe of a lung supplied by its own branches of a bronchus.

Brunner's gland *See* **Duodenal gland.**

Buccal (BUK-al) Pertaining to the cheek or mouth.

Buffer system A weak acid and the salt of that acid (that functions as a weak base). Buffers prevent drastic changes in pH by converting strong acids and bases to weak acids and bases.

Bulb of penis Expanded portion of the base of the corpus spongiosum penis.

Bulbourethral (bul′-bō-ū-RĒ-thral) **gland** One of a pair of glands located inferior to the prostate on either side of the urethra that secretes an alkaline fluid into the cavernous urethra. Also called a **Cowper's** (KOW-perz) **gland.**

Bulimia (boo-LIM-ē-a *or* boo-LĒ-mē-a) A disorder characterized by overeating at least twice a week followed by purging by self-induced vomiting, strict dieting or fasting, vigorous exercise, or use of laxatives or diuretics. Also called **binge–purge syndrome.**

Bulk-phase endocytosis A process by which most body cells can ingest membrane-surrounded droplets of interstitial fluid.

Bundle branch One of the two branches of the atrioventricular (AV) bundle made up of specialized muscle fibers (cells) that transmit electrical impulses to the ventricles.

Bundle of His *See* **Atrioventricular (AV) bundle.**

Burn Tissue damage caused by excessive heat, electricity, radioactivity, or corrosive chemicals that denature (break down) proteins in the skin.

Bursa (BUR-sa) A sac or pouch of synovial fluid located at friction points, especially about joints.

Bursitis (bur-SĪ-tis) Inflammation of a bursa.

Buttocks (BUT-oks) The two fleshy masses on the posterior aspect of the inferior trunk, formed by the gluteal muscles.

C

Caecum (SĒ-kum) A blind pouch at the proximal end of the large intestine that attaches to the ileum.

Calcaneal (kal-KĀ-nē-al) **tendon** The tendon of the soleus, gastrocnemius, and plantaris muscles at the back of the heel. Also called the **Achilles** (a-KIL-ēz) **tendon.**

Calcification (kal′-si-fi-KĀ-shun) Deposition of mineral salts, primarily hydroxyapatite, in a framework formed by collagen fibers in which the tissue hardens. Also called **mineralization** (min′-e-ral-i-ZĀ-shun).

Calcitonin (kal-si-TŌ-nin) **(CT)** A hormone produced by the parafollicular cells of the thyroid gland that can lower the amount of blood calcium and phosphates by inhibiting bone resorption (breakdown of bone extracellular matrix) and by accelerating uptake of calcium and phosphates into bone matrix.

Calculus (KAL-kū-lus) A stone, or insoluble mass of crystallized salts or other material, formed within the body, as in the gallbladder, kidney, or urinary bladder.

Callous (KAL-lus) A growth of new bone tissue in and around a fractured area, ultimately replaced by mature bone. An acquired, localized thickening.

Calyx (KĀL-iks) Any cuplike division of the kidney pelvis. *Plural* is **calyces** (KĀ-li-sēz).

Canal (ka-NAL) A narrow tube, channel, or passageway.

Canaliculus (kan′-a-LIK-ū-lus) A small channel or canal, as in bones, where they connect lacunae. *Plural* is **canaliculi** (kan′-a-LIK-ū-lī).

Canal of Schlemm *See* **Scleral venous sinus.**

Cancer A group of diseases characterized by uncontrolled or abnormal cell division.

Capacitation (ka-pas′-i-TĀ-shun) The functional changes that sperm undergo in the female reproductive tract that allow them to fertilize a secondary oocyte.

Capillary (KAP-i-lar′-ē) A microscopic blood vessel located between an arteriole and venule through which materials are exchanged between blood and interstitial fluid.

Carbohydrate Organic compound consisting of carbon, hydrogen, and oxygen; the ratio of hydrogen to oxygen atoms is usually 2:1. Examples include sugars, glycogen, starches, and glucose.

Carcinogen (kar-SIN-ō-jen) A chemical substance or radiation that causes cancer.

Cardiac (KAR-dē-ak) **arrest** Cessation of an effective heartbeat in which the heart is completely stopped or in ventricular fibrillation.

Cardiac cycle A complete heartbeat consisting of systole (contraction) and diastole (relaxation) of both atria plus systole and diastole of both ventricles.

Cardiac muscle Striated muscle fibers (cells) that form the wall of the heart; stimulated by an intrinsic conduction system and autonomic motor neurons.

Cardiac notch An angular notch in the anterior border of the left lung into which part of the heart fits.

Cardiac output (CO) The volume of blood ejected from the left ventricle (or the right ventricle) into the aorta (or pulmonary trunk) each minute.

Cardinal ligament A ligament of the uterus, extending laterally from the cervix and vagina as a continuation of the broad ligament.

Cardiogenic area (kar-dē-ō-JEN-ik) A group of mesodermal cells in the head end of an embryo that gives rise to the heart.

Cardiology (kar-dē-OL-ō-jē) The study of the heart and diseases associated with it.

Cardiovascular (kar-dē-ō-VAS-kū-lar) **(CV) center** Groups of neurons scattered within the medulla oblongata that regulate heart rate, force of contraction, and blood vessel diameter.

Cardiovascular physiology Study of the functions of the heart and blood vessels.

Cardiovascular system System that consists of blood, the heart, and blood vessels.

Carotene (KAR-ō-tēn) Antioxidant precursor of vitamin A, which is needed for synthesis of photopigments; yellow-orange pigment present in the stratum corneum of the epidermis. Accounts for the yellowish coloration of skin. Also termed **beta carotene.**

Carotid (ka-ROT-id) **body** Cluster of chemoreceptors on or near the carotid sinus that respond to changes in blood levels of oxygen, carbon dioxide, and hydrogen ions.

Carotid sinus A dilated region of the internal carotid artery just superior to where it branches from the common carotid artery; it contains baroreceptors that monitor blood pressure.

Carpal bones The eight bones of the wrist. Also called **carpals.**

Carpus (KAR-pus) A collective term for the eight bones of the wrist.

Cartilage (KAR-ti-lij) A type of connective tissue consisting of chondrocytes in lacunae embedded in a dense network of collagen and elastic fibers and an extracellular matrix of chondroitin sulfate.

Cartilaginous (kar′-ti-LAJ-i-nus) **joint** A joint without a synovial (joint) cavity where the articulating bones are held tightly together by cartilage, allowing little or no movement.

Catabolism (ka-TAB-ō-lizm) Chemical reactions that break down complex organic compounds into simple ones, with the net release of energy.

Cataract (KAT-a-rakt) Loss of transparency of the lens of the eye or its capsule or both.

Cauda equina (KAW-da ē-KWĪ-na) A tail-like array of roots of spinal nerves at the inferior end of the spinal cord.

Caudal (KAW-dal) Pertaining to any tail-like structure; inferior in position.

Coeliac plexus (SĒ-lē-ak PLEK-sus) A large mass of autonomic ganglia and axons located at the level of the superior part of the first lumbar vertebra. Also called the **solar plexus.**

Cell The basic structural and functional unit of all organisms; the smallest structure capable of performing all the activities vital to life.

Cell biology The study of cellular structure and function. Also called **cytology.**

Cell cycle Growth and division of a single cell into two identical cells; consists of interphase and cell division.

Cell division Process by which a cell reproduces itself that consists of a nuclear division (mitosis) and a cytoplasmic division (cytokinesis); types include somatic and reproductive cell division.

Cell junction Point of contact between plasma membranes of tissue cells.

Cellular respiration The oxidation of glucose to produce ATP that involves glycolysis, acetyl coenzyme A formation, the Krebs cycle, and the electron transport chain.

Cementum (se-MEN-tum) Calcified tissue covering the root of a tooth.

Central canal A microscopic tube running the length of the spinal cord in the grey commissure. A circular channel running longitudinally in the center of an osteon (haversian system) of mature compact bone, containing blood and lymphatic vessels and nerves. Also called a **haversian** (ha-VER-shun) **canal.**

Central fovea (FŌ-vē-a) A depression in the center of the macula lutea of the retina, containing cones only and lacking blood vessels; the area of highest visual acuity (sharpness of vision).

Central nervous system (CNS) That portion of the nervous system that consists of the brain and spinal cord.

Centrioles (SEN-trē-ōlz) Paired, cylindrical structures of a centrosome, each consisting of a ring of microtubules and arranged at right angles to each other.

Centromere (SEN-trō-mēr) The constricted portion of a chromosome where the two chromatids are joined; serves as the point of attachment for the microtubules that pull chromatids during anaphase of cell division.

Centrosome (SEN-trō-sōm) A dense network of small protein fibers near the nucleus of a cell, containing a pair of centrioles and pericentriolar material.

Cephalic (se-FAL-ik) Pertaining to the head; superior in position.

Cerebellar peduncle (ser-e-BEL-ar pe-DUNG-kul) A bundle of nerve axons connecting the cerebellum with the brain stem.

Cerebellum (ser′-e-BEL-um) The part of the brain lying posterior to the medulla oblongata and pons; governs balance and coordinates skilled movements.

Cerebral aqueduct (SER-ē-bral AK-we-dukt) A channel through the midbrain connecting the third and fourth ventricles and containing cerebrospinal fluid. Also termed the **aqueduct of the midbrain.**

Cerebral arterial circle A ring of arteries forming an anastomosis at the base of the brain between the internal carotid and basilar arteries and arteries supplying the cerebral cortex. Also called the **circle of Willis.**

Cerebral cortex The surface of the cerebral hemispheres, 2–4 mm thick, consisting of grey matter; arranged in six layers of neuronal cell bodies in most areas.

Cerebral peduncle (pe-DUNG-kul or PĒ-dung-kul) One of a pair of nerve axon bundles located on the anterior surface of the midbrain, conducting nerve impulses between the pons and the cerebral hemispheres.

Cerebrospinal (se-rē′-brō-SPĪ-nal) **fluid (CSF)** A fluid produced by ependymal cells that cover choroid plexuses in the ventricles of the brain; the fluid circulates in the ventricles, the central canal, and the subarachnoid space around the brain and spinal cord.

Cerebrovascular (se-rē′-brō-VAS-kū-lar) **accident (CVA)** Destruction of brain tissue (infarction) resulting from obstruction or rupture of blood vessels that supply the brain. Also called a **stroke** or **brain attack.**

Cerebrum (se-RĒ-brum or SER-e-brum) The two hemispheres of the forebrain (derived from the telencephalon), making up the largest part of the brain.

Cerumen (se-ROO-men) Waxlike secretion produced by ceruminous glands in the external auditory meatus (ear canal). Also termed **earwax.**

Ceruminous (se-RŪ-mi-nus) **gland** A modified sudoriferous (sweat) gland in the external auditory meatus that secretes cerumen (ear wax).

Cervical ganglion (SER-vi-kul GANG-glē-on) A cluster of cell bodies of postganglionic sympathetic neurons located in the neck, near the vertebral column.

Cervical plexus (PLEK-sus) A network formed by nerve axons from the ventral rami of the first four cervical nerves and receiving gray rami communicantes from the superior cervical ganglion.

Cervix (SER-viks) Neck; any constricted portion of an organ, such as the inferior cylindrical part of the uterus.

Chemical reaction The formation of new chemical bonds or the breaking of old chemical bonds between atoms.

Chemistry (KEM-is-trē) The science of the structure and interactions of matter.

Chemoreceptor (kē′-mō-rē-SEP-tor) Sensory receptor that detects the presence of a specific chemical.

Chiasm (KĪ-azm) A crossing; especially the crossing of axons in the optic (II) nerve.

Chief cell The secreting cell of a gastric gland that produces pepsinogen, the precursor of the enzyme pepsin, and the enzyme gastric lipase. Also called a **zymogenic** (zī′-mō-JEN-ik) **cell.** Cell in the parathyroid glands that secretes parathyroid hormone (PTH). Also called a **principal cell.**

Cholecystectomy (kō′-lē-sis-TEK-tō-mē) Surgical removal of the gallbladder.

Cholecystitis (kō′-lē-sis-TĪ-tis) Inflammation of the gallbladder.

Cholesterol (kō-LES-te-rol) Classified as a lipid, the most abundant steroid in animal tissues; located in cell membranes and used for the synthesis of steroid hormones and bile salts.

Cholinergic (kō′-lin-ER-jik) **neuron** A neuron that liberates acetylcholine as its neurotransmitter.

Chondrocyte (KON-drō-sīt) Cell of mature cartilage.

Chondroitin (kon-DROY-tin) **sulfate** An amorphous extracellular matrix material found outside connective tissue cells.

Chordae tendineae (KOR-dē TEN-di-nē-ē) Tendonlike, fibrous cords that connect atrioventricular valves of the heart with papillary muscles.

Chorion (KŌ-rē-on) The most superficial fetal membrane that becomes the principal embryonic portion of the placenta; serves a protective and nutritive function.

Chorionic villi (kō-rē-ON-ik VIL-li) Fingerlike projections of the chorion that grow into the decidua basalis of the endometrium and contain fetal blood vessels.

Chorionic villi sampling (CVS) The removal of a sample of chorionic villus tissue by means of a catheter to analyze the tissue for prenatal genetic defects.

Choroid (KŌ-royd) One of the vascular coats of the eyeball.

Choroid plexus (PLEK-sus) A network of capillaries located in the roof of each of the four ventricles of the brain; ependymal cells around choroid plexuses produce cerebrospinal fluid.

Chromaffin (KRŌ-maf-in) **cell** Cell that has an affinity for chrome salts, due in part to the presence of the precursors of the neurotransmitter epinephrine; found, among other places, in the adrenal medulla.

Chromatid (KRŌ-ma-tid) One of a pair of identical connected nucleoprotein strands that are joined at the centromere and separate during cell division, each becoming a chromosome of one of the two daughter cells.

Chromatin (KRŌ-ma-tin) The threadlike mass of genetic material, consisting of DNA and histone proteins, that is present in the nucleus of a nondividing or interphase cell.

Chromatolysis (krō′-ma-TOL-i-sis) The breakdown of Nissl bodies into finely granular masses in the cell body of a neuron whose axon has been damaged.

Chromosome (KRŌ-mō-sōm) One of the small, threadlike structures in the nucleus of a cell, normally 46 in a human diploid cell, that bears the genetic material; composed of DNA and proteins (histones) that form a delicate chromatin thread during interphase; becomes packaged into compact rodlike structures that are visible under the light microscope during cell division.

Chronic (KRON-ik) Long term or frequently recurring; applied to a disease that is not acute.

Chronic obstructive pulmonary disease (COPD) A disease, such as bronchitis or emphysema, in which there is some degree of obstruction of airways and consequent increase in airway resistance.

Chyle (KĪL) The milky-appearing fluid found in the lacteals of the small intestine after absorption of lipids in food.

Chyme (KĪM) The semifluid mixture of partly digested food and digestive secretions found in the stomach and small intestine during digestion of a meal.

Ciliary (SIL-ē-ar′-ē) **body** One of the three parts of the vascular tunic of the eyeball, the others being the choroid and the iris; includes the ciliary muscle and the ciliary processes.

Ciliary ganglion (GANG-glē-on) A very small parasympathetic ganglion whose preganglionic axons come from the oculomotor (III) nerve and whose postganglionic axons carry nerve impulses to the ciliary muscle and the sphincter muscle of the iris.

Cilium (SIL-ē-um) A hair or hairlike process projecting from a cell that may be used to move the entire cell or to move substances along the surface of the cell. *Plural* is **cilia.**

Circadian (ser-KĀ-dē-an) **rhythm** The pattern of biological activity on a 24-hour cycle, such as the sleep–wake cycle.

Circle of Willis *See* **Cerebral arterial circle.**

Circular folds Permanent, deep, transverse folds in the mucosa and submucosa of the small intestine that increase the surface area for absorption. Also called **plicae circulares** (PLĪ-kē SER-kū-lar-ēs).

Circulation time The time required for a drop of blood to pass from the right atrium, through pulmonary circulation, back to the left atrium, through systemic circulation down to the foot, and back again to the right atrium.

Circumduction (ser-kum-DUK-shun) A movement at a synovial joint in which the distal end of a bone moves in a circle while the proximal end remains relatively stable.

Cirrhosis (si-RŌ-sis) A liver disorder in which the parenchymal cells are destroyed and replaced by connective tissue.

Cisterna chyli (sis-TER-na KĪ-lē) The origin of the thoracic duct.

Cleavage (KLĒV-ij) The rapid mitotic divisions following the fertilisation of a secondary oocyte, resulting in an increased number of progressively smaller cells, called blastomeres.

Clitoris (KLI-to-ris) An erectile organ of the female, located at the anterior junction of the labia minora, that is homologous to the male penis.

Clone (KLŌN) A population of identical cells.

Coarctation (kō′-ark-TĀ-shun) **of the aorta** A congenital heart defect in which a segment of the aorta is too narrow. As a result, the flow of oxygenated blood to the body is reduced, the left ventricle is forced to pump harder, and high blood pressure develops.

Coccyx (KOK-siks) The fused bones at the inferior end of the vertebral column.

Cochlea (KOK-lē-a) A winding, cone-shaped tube forming a portion of the inner ear and containing the spiral organ (organ of Corti).

Cochlear duct The membranous cochlea consisting of a spirally arranged tube enclosed in the bony cochlea and lying along its outer wall. Also called the **scala media** (SCA-la MĒ-dē-a).

Collagen (KOL-a-jen) A protein that is the main organic constituent of connective tissue.

Collateral circuit The alternate route taken by blood through an anastomosis.

Colliculus (ko-LIK-ū-lus) A small elevation.

Colon The portion of the large intestine consisting of ascending, transverse, descending, and sigmoid portions.

Colony-stimulating factor (CSF) One of a group of molecules that stimulates development of white blood cells. Examples are macrophage CSF and granulocyte CSF.

Colostrum (kō-LOS-trum) A thin, cloudy fluid secreted by the mammary glands a few days prior to or after delivery before true milk is produced.

Column (KOL-um) Group of white matter tracts in the spinal cord.

Common bile duct A tube formed by the union of the common hepatic duct and the cystic duct that empties bile into the duodenum at the hepatopancreatic ampulla (ampulla of Vater).

Compact (dense) bone tissue Bone tissue that contains few spaces between osteons (haversian systems); forms the external portion of all bones and the bulk of the diaphysis (shaft) of long bones; is found immediately deep to the periosteum and external to spongy bone.

Concha (KON-ka) A scroll-like bone found in the skull. *Plural* is **conchae** (KON-kē).

Concussion (kon-KUSH-un) Traumatic injury to the brain that produces no visible bruising but may result in abrupt, temporary loss of consciousness.

Conduction system A group of autorhythmic cardiac muscle fibers that generates and distributes electrical impulses to stimulate coordinated contraction of the heart chambers; includes the sinoatrial (SA) node, the atrioventricular (AV) node, the atrioven-

tricular (AV) bundle, the right and left bundle branches, and the Purkinje fibers.

Condyloid (KON-di-loyd) **joint** A synovial joint structured so that an oval-shaped condyle of one bone fits into an elliptical cavity of another bone, permitting side-to-side and back-and-forth movements, such as the joint at the wrist between the radius and carpals. Also called an **ellipsoidal** (ē-lip-SOYD-al) **joint.**

Cone (KŌN) The type of photoreceptor in the retina that is specialized for highly acute color vision in bright light.

Congenital (kon-JEN-i-tal) Present at the time of birth.

Conjunctiva (kon′-junk-TĪ-va) The delicate membrane covering the eyeball and lining the eyes.

Connective tissue One of the most abundant of the four basic tissue types in the body, performing the functions of binding and supporting; consists of relatively few cells in a generous extracellular matrix (the ground substance and fibers between the cells).

Consciousness (KON-shus-nes) A state of wakefulness in which an individual is fully alert, aware, and oriented, partly as a result of feedback between the cerebral cortex and reticular activating system.

Continuous conduction (kon-DUK-shun) Propagation of an action potential (nerve impulse) in a step-by-step depolarization of each adjacent area of an axon membrane.

Contraception (kon′-tra-SEP-shun) The prevention of fertilisation or impregnation without destroying fertility.

Contractility (kon′-trak-TIL-i-tē) The ability of cells or parts of cells to actively generate force to undergo shortening for movements. Muscle fibers (cells) exhibit a high degree of contractility.

Contralateral (KON-tra-lat-er-al) On the opposite side; affecting the opposite side of the body.

Conus medullaris (KŌ-nus med-ū-LAR-is) The tapered portion of the spinal cord inferior to the lumbar enlargement.

Convergence (con-VER-jens) A synaptic arrangement in which the synaptic end bulbs of several presynaptic neurons terminate on one postsynaptic neuron. The medial movement of the two eyeballs so that both are directed toward a near object being viewed in order to produce a single image.

Cornea (KOR-nē-a) The nonvascular, transparent fibrous coat through which the iris of the eye can be seen.

Corona (kō-RŌ-na) Margin of the glans penis.

Corona radiata The innermost layer of granulosa cells that is firmly attached to the zona pellucida around a secondary oocyte.

Coronary artery disease (CAD) A condition such as atherosclerosis that causes narrowing of coronary arteries so that blood flow to the heart is reduced. The result is **coronary heart disease (CHD),** in which the heart muscle receives inadequate blood flow due to an interruption of its blood supply.

Coronary circulation The pathway followed by the blood from the ascending aorta through the blood vessels supplying the heart and returning to the right atrium. Also called **cardiac circulation.**

Coronary sinus (SĪ-nus) A wide venous channel on the posterior surface of the heart that collects the blood from the coronary circulation and returns it to the right atrium.

Corpus albicans (KOR-pus AL-bi-kanz) A white fibrous patch in the ovary that forms after the corpus luteum regresses.

Corpus callosum (kal-LŌ-sum) The great commissure of the brain between the cerebral hemispheres.

Corpuscle of touch See **Meissner corpuscle.**

Corpus luteum (LOO-tē-um) A yellowish body in the ovary formed when a follicle has discharged its secondary oocyte; secretes oestrogens, progesterone, relaxin, and inhibin.

Corpus striatum (strī-Ā-tum) An area in the interior of each cerebral hemisphere composed of the caudate and putamen of the basal ganglia and white matter of the internal capsule, arranged in a striated manner.

Cortex (KOR-teks) An outer layer of an organ. The convoluted layer of grey matter covering each cerebral hemisphere.

Costal (KOS-tal) Pertaining to a rib.

Cramp A spasmodic, usually painful contraction of a muscle.

Cranial (KRĀ-nē-al) **cavity** A body cavity formed by the cranial bones and containing the brain.

Cranial nerve One of 12 pairs of nerves that leave the brain; pass through foramina in the skull; and supply sensory and motor neurons to the head, neck, part of the trunk, and viscera of the thorax and abdomen. Each is designated by a Roman numeral and a name.

Craniosacral (krā-nē-ō-SĀK-ral) **outflow** The axons of parasympathetic preganglionic neurons, which have their cell bodies located in nuclei in the brain stem and in the lateral grey matter of the sacral portion of the spinal cord.

Cranium (KRĀ-nē-um) The skeleton of the skull that protects the brain and the organs of sight, hearing, and balance; includes the frontal, parietal, temporal, occipital, sphenoid, and ethmoid bones.

Crista (KRIS-ta) A crest or ridged structure. A small elevation in the ampulla of each semicircular duct that contains receptors for dynamic equilibrium. *Plural* is **cristae.**

Crossing-over The exchange of a portion of one chromatid with another during meiosis. It permits an exchange of genes among chromatids and is one factor that results in genetic variation of progeny.

Crus (KRUS) **of penis** Separated, tapered portion of the corpora cavernosa penis. *Plural* is **crura** (KROO-ra).

Crypt of Lieberkühn *See* **Intestinal gland.**

Cryptorchidism (krip-TOR-ki-dizm) The condition of undescended testes.

Cuneate (KŪ-nē-āt) **nucleus** A group of neurons in the inferior part of the medulla oblongata in which axons of the cuneate fasciculus terminate.

Cupula (KU-pū-la) A mass of gelatinous material covering the hair cells of a crista; a sensory re-

ceptor in the ampulla of a semicircular canal stimulated when the head moves.

Cushing's syndrome Condition caused by a hypersecretion of glucocorticoids characterized by spindly legs, "moon face," "buffalo hump," pendulous abdomen, flushed facial skin, poor wound healing, hyperglycemia, osteoporosis, hypertension, and increased susceptibility to disease.

Cutaneous (kū-TĀ-nē-us) Pertaining to the skin.

Cyanosis (sī-a-NŌ-sis) A blue or dark purple discoloration, most easily seen in nail beds and mucous membranes, that results from an increased concentration of deoxygenated (reduced) haemoglobin (more than 5 g/dL).

Cyst (SIST) A sac with a distinct connective tissue wall, containing a fluid or other material.

Cystic (SIS-tik) **duct** The duct that carries bile from the gallbladder to the common bile duct.

Cystitis (sis-TĪ-tis) Inflammation of the urinary bladder.

Cytokinesis (sī′-tō-ki-NĒ-sis) Distribution of the cytoplasm into two separate cells during cell division; coordinated with nuclear division (mitosis).

Cytolysis (sī-TOL-i-sis) The rupture of living cells in which the contents leak out.

Cytoplasm (SĪ-tō-plasm) Cytosol plus all organelles except the nucleus.

Cytoskeleton Complex internal structure of cytoplasm consisting of microfilaments, microtubules, and intermediate filaments.

Cytosol (SĪ-tō-sol) Semifluid portion of cytoplasm in which organelles and inclusions are suspended and solutes are dissolved. Also called **intracellular fluid.**

D

Dartos (DAR-tōs) The contractile smooth muscular tissue deep to the skin of the scrotum.

Decidua (dē-SID-ū-a) That portion of the endometrium of the uterus (all but the deepest layer) that is modified during pregnancy and shed after childbirth.

Deciduous (dē-SID-ū-us) Falling off or being shed seasonally or at a particular stage of development. In the body, referring to the first set of teeth.

Decussation (dē′-ku-SĀ-shun) A crossing-over to the opposite (contralateral) side; an example is the crossing of 90% of the axons in the large motor tracts to opposite sides in the medullary pyramids.

Decussation (dē′-ku-SĀ-shun) **of pyramids** The crossing of most axons (90%) in the left pyramid of the medulla to the right side and the crossing of most axons (90%) in the right pyramid to the left side.

Deep Away from the surface of the body or an organ.

Deep abdominal inguinal (IN-gwi-nal) **ring** A slit-like opening in the aponeurosis of the transversus abdominis muscle that represents the origin of the inguinal canal.

Deep-vein thrombosis (usu.) The presence of a thrombus in a vein, usually a deep vein of the lower limbs.

Defaecation (def-e-KĀ-shun) The discharge of faeces from the rectum.

Deglutition (dē-gloo-TISH-un) The act of swallowing.

Dehydration (dē-hī-DRĀ-shun) Excessive loss of water from the body or its parts.

Delta cell A cell in the pancreatic islets (islets of Langerhans) that secretes somatostatin. Also termed a **D cell.**

Demineralisation (dē-min′-er-al-i-ZĀ-shun) Loss of calcium and phosphorus from bones.

Dendrite (DEN-drīt) A neuronal process that carries electrical signals, usually graded potentials, toward the cell body.

Dendritic (den-DRIT-ik) **cell** One type of antigen-presenting cell with long branchlike projections that commonly is present in mucosal linings such as the vagina, in the skin (Langerhans cells in the epidermis), and in lymph nodes (follicular dendritic cells).

Dental caries (KA-rēz) Gradual demineralisation of the enamel and dentin of a tooth that may invade the pulp and alveolar bone. Also called **tooth decay.**

Denticulate (den-TIK-ū-lāt) Finely toothed or serrated; characterized by a series of small, pointed projections.

Dentin (DEN-tin) The bony tissues of a tooth enclosing the pulp cavity.

Dentition (den-TI-shun) The eruption of teeth. The number, shape, and arrangement of teeth.

Deoxyribonucleic (dē-ok′-sē-rī-bō-nū-KLĒ-ik) **acid (DNA)** A nucleic acid constructed of nucleotides consisting of one of four bases (adenine, cytosine, guanine, or thymine), deoxyribose, and a phosphate group; encoded in the nucleotides is genetic information.

Depression (de-PRESH-un) Movement in which a part of the body moves inferiorly.

Dermal papilla (pa-PIL-a) Fingerlike projection of the papillary region of the dermis that may contain blood capillaries or corpuscles of touch (Meissner corpuscles).

Dermatology (der′-ma-TOL-ō-jē) The medical specialty dealing with diseases of the skin.

Dermatome (DER-ma-tōm) The cutaneous area developed from one embryonic spinal cord segment and receiving most of its sensory innervation from one spinal nerve. An instrument for incising the skin or cutting thin transplants of skin.

Dermis (DER-mis) A layer of dense irregular connective tissue lying deep to the epidermis.

Descending colon (KŌ-lon) The part of the large intestine descending from the left colic (splenic) flexure to the level of the left iliac crest.

Detrusor (de-TROO-ser) **muscle** Smooth muscle that forms the wall of the urinary bladder.

Developmental biology The study of development from the fertilized egg to the adult form.

Deviated nasal septum A nasal septum that does not run along the midline of the nasal cavity. It deviates (bends) to one side.

Diabetes mellitus (dī-a-BĒ-tēz MEL-i-tus) An endocrine disorder caused by an inability to produce or use insulin. It is characterized by the three "polys": polyuria (excessive urine production), polydipsia (excessive thirst), and polyphagia (excess eating).

Diagnosis (dī'-ag-NŌ-sis) Distinguishing one disease from another or determining the nature of a disease from signs and symptoms by inspection, palpation, laboratory tests, and other means.

Dialysis (dī-AL-i-sis) The removal of waste products from blood by diffusion through a selectively permeable membrane.

Diaphragm (DĪ-a-fram) Any partition that separates one area from another, especially the dome-shaped skeletal muscle between the thoracic and abdominal cavities. Also a dome-shaped device that is placed over the cervix, usually with a spermicide, to prevent conception.

Diaphysis (dī-AF-i-sis) The shaft of a long bone.

Diarrhoea (dī-a-RĒ-a) Frequent defecation of liquid faeces caused by increased motility of the intestines.

Diarthrosis (dī-ar-THRŌ-sis) A freely movable joint; types are gliding, hinge, pivot, condyloid, saddle, and ball-and-socket.

Diastole (dī-AS-tō-lē) In the cardiac cycle, the phase of relaxation or dilation of the heart muscle, especially of the ventricles.

Diastolic (dī-as-TOL-ik) **blood pressure (DBP)** The force exerted by blood on arterial walls during ventricular relaxation; the lowest blood pressure measured in the large arteries, normally about 80 mmHg in a young adult.

Diencephalon (DĪ-en-sef'-a-lon) A part of the brain consisting of the thalamus, hypothalamus, and epithalamus.

Differentiation (dif'-er-en-shē-Ā-shun) Development of a cell from an unspecialized to a specialized one.

Diffusion (di-FŪ-zhun) A passive process in which there is a net or greater movement of molecules or ions from a region of high concentration to a region of low concentration until equilibrium is reached.

Digestion (dī-JES-chun) The mechanical and chemical breakdown of food to simple molecules that can be absorbed and used by body cells.

Digestive system A system that consists of the gastrointestinal tract (mouth, pharynx, oesophagus, stomach, small intestine, and large intestine) and accessory digestive organs (teeth, tongue, salivary glands, liver, gallbladder, and pancreas). Its function is to break down foods into small molecules that can be used by body cells.

Dilate (DĪ-lāt) To expand or swell.

Diploid (DIP-loyd) **cell** Having the number of chromosomes characteristically found in the somatic cells of an organism; having two haploid sets of chromosomes, one each from the mother and father. Symbolized 2*n*.

Direct motor pathways Collections of upper motor neurons with cell bodies in the motor cortex that project axons into the spinal cord, where they synapse with lower motor neurons or interneurons in the anterior horns. Also called the **pyramidal pathways.**

Disease Any change from a state of health.

Disease Refers to an illness characterized by a recognizable set of signs and symptoms.

Dislocation (dis'-lō-KĀ-shun) Displacement of a bone from a joint with tearing of ligaments, tendons, and articular capsules. Also called **luxation** (luks-Ā-shun).

Dissect (di-SEKT) To separate tissues and parts of a cadaver or an organ for anatomical study.

Distal (DIS-tal) Farther from the attachment of a limb to the trunk; farther from the point of origin or attachment.

Diuretic (dī-ū-RET-ik) A chemical that increases urine volume by decreasing reabsorption of water, usually by inhibiting sodium reabsorption.

Divergence (dī-VER-jens) A synaptic arrangement in which the synaptic end bulbs of one presynaptic neuron terminate on several postsynaptic neurons.

Diverticulum (dī'-ver-TIK-ū-lum) A sac or pouch in the wall of a canal or organ, especially in the colon.

Dorsal ramus (RĀ-mus) A branch of a spinal nerve containing motor and sensory axons supplying the muscles, skin, and bones of the posterior part of the head, neck, and trunk.

Dorsiflexion (dor-si-FLEK-shun) Bending the foot in the direction of the dorsum (upper surface).

Down-regulation Phenomenon in which there is a decrease in the number of receptors in response to an excess of a hormone or neurotransmitter.

Dual innervation The concept by which most organs of the body receive impulses from sympathetic and parasympathetic neurons.

Duct of Santorini *See* **Accessory duct.**

Duct of Wirsung *See* **Pancreatic duct.**

Ductus arteriosus (DUK-tus ar-tē-rē-O-sus) A small vessel connecting the pulmonary trunk with the aorta; found only in the fetus.

Ductus (vas) deferens (DEF-er-ens) The duct that carries sperm from the epididymis to the ejaculatory duct. Also called the **seminal duct.**

Ductus epididymis (ep'-i-DID-i-mis) A tightly coiled tube inside the epididymis, distinguished into a head, body, and tail, in which sperm undergo maturation.

Ductus venosus (ve-NŌ-sus) A small vessel in the fetus that helps the circulation bypass the liver.

Duodenal (doo-ō-DĒ-nal) **gland** Gland in the submucosa of the duodenum that secretes an alkaline mucus to protect the lining of the small intestine from the action of enzymes and to help neutralize the acid in chyme. Also called a **Brunner's** (BRUN-erz) **gland.**

Duodenal papilla (pa-PIL-a) An elevation on the duodenal mucosa that receives the hepatopancreatic ampulla (ampulla of Vater).

Duodenum (doo'-ō-DĒ-num *or* doo-OD-e-num) The first 25 cm (10 in.) of the small intestine, which connects the stomach and the ileum.

Dura mater (DOO-ra MĀ-ter) The outermost of the three meninges (coverings) of the brain and spinal cord.

Dynamic equilibrium (ē-kwi-LIB-rē-um) The maintenance of body position, mainly the head, in response to sudden movements such as rotation.

Dysmenorrhoea (dis'-men-ō-RĒ-a) Painful menstruation.

Dysplasia (dis-PLĀ-zē-a) Change in the size, shape, and organization of cells due to chronic irritation or inflammation; may either revert to normal if stress is removed or progress to neoplasia.

Dyspnoea (DISP-nē-a) Shortness of breath; painful or laboured breathing.

E

Ectoderm The primary germ layer that gives rise to the nervous system and the epidermis of skin and its derivatives.

Ectopic (ek-TOP-ik) Out of the normal location, as in ectopic pregnancy.

Effector (e-FEK-tor) An organ of the body, either a muscle or a gland, that is innervated by somatic or autonomic motor neurons.

Efferent arteriole (EF-er-ent ar-TĒ-rē-ōl) A vessel of the renal vascular system that carries blood from a glomerulus to a peritubular capillary.

Efferent (EF-er-ent) **ducts** A series of coiled tubes that transport sperm from the rete testis to the epididymis.

Ejaculation (ē-jak-ū-LĀ-shun) The reflex ejection or expulsion of semen from the penis.

Ejaculatory (ē-JAK-ū-la-tō-rē) **duct** A tube that transports sperm from the ductus (vas) deferens to the prostatic urethra.

Elasticity (e-las-TIS-i-tē) The ability of tissue to return to its original shape after contraction or extension.

Electrocardiogram (e-lek'-trō-KAR-dē-ō-gram) **(ECG or EKG)** A recording of the electrical changes that accompany the cardiac cycle that can be detected at the surface of the body; may be resting, stress, or ambulatory.

Elevation (el-e-VĀ-shun) Movement in which a part of the body moves superiorly.

Embolus (EM-bō-lus) A blood clot, bubble of air or fat from broken bones, mass of bacteria, or other debris or foreign material transported by the blood.

Embryo (EM-brē-ō) The young of any organism in an early stage of development; in humans, the developing organism from fertilisation to the end of the eighth week of development.

Embryoblast (EM-brē-ō-blast) A region of cells of a blastocyst that differentiates into the three primary germ layers—ectoderm, mesoderm, and endoderm—from which all tissues and organs develop; also called an **inner cell mass.**

Embryology (em'-brē-OL-ō-jē) The study of development from the fertilized egg to the end of the eighth week of development.

Emesis (EM-e-sis) Vomiting.

Emigration (em'-i-GRĀ-shun) Process whereby white blood cells (WBCs) leave the bloodstream by rolling along the endothelium, sticking to it, and squeezing between the endothelial cells. Adhesion molecules help WBCs stick to the endothelium. Also known as **migration** or **extravasation.**

Emission (ē-MISH-un) Propulsion of sperm into the urethra due to peristaltic contractions of the ducts of the testes, epididymides, and ductus (vas) deferens as a result of sympathetic stimulation.

Emphysema (em-fi-SĒ-ma) A lung disorder in which alveolar walls disintegrate, producing abnormally large air spaces and loss of elasticity in the lungs; typically caused by exposure to cigarette smoke.

Emulsification (e-mul-si-fi-KĀ-shun) The dispersion of large lipid globules into smaller, uniformly distributed particles in the presence of bile.

Enamel (e-NAM-el) The hard, white substance covering the crown of a tooth.

Endocardium (en-dō-KAR-dē-um) The layer of the heart wall, composed of endothelium and smooth muscle, that lines the inside of the heart and covers the valves and tendons that hold the valves open.

Endochondral (en′-dō-KON-dral) **ossification** Bone formation within hyaline cartilage that develops from mesenchyme.

Endocrine (EN-dō-krin) **gland** A gland that secretes hormones into interstitial fluid and then the blood; a ductless gland.

Endocrine system (EN-dō-krin) All endocrine glands and hormone-secreting cells.

Endocrinology (en′-dō-kri-NOL-ō-jē) The science concerned with the structure and functions of endocrine glands and the diagnosis and treatment of disorders of the endocrine system.

Endocytosis (en′-dō-sī-TŌ-sis) The uptake into a cell of large molecules and particles in which a segment of plasma membrane surrounds the substance, encloses it, and brings it in; includes phagocytosis, pinocytosis, and receptor-mediated endocytosis.

Endoderm (EN-dō-derm) A primary germ layer of the developing embryo; gives rise to the gastrointestinal tract, urinary bladder, urethra, and respiratory tract.

Endodontics (en′-dō-DON-tiks) The branch of dentistry concerned with the prevention, diagnosis, and treatment of diseases that affect the pulp, root, periodontal ligament, and alveolar bone.

Endolymph (EN-dō-limf′) The fluid within the membranous labyrinth of the internal ear.

Endometriosis (en′-dō-me′-trē-Ō-sis) The growth of endometrial tissue outside the uterus.

Endometrium (en′-dō-MĒ-trē-um) The mucous membrane lining the uterus.

Endomysium (en′-dō-MĪZ-ē-um) Invagination of the perimysium separating each individual muscle fiber (cell).

Endoneurium (en′-dō-NOO-rē-um) Connective tissue wrapping around individual nerve axons.

Endoplasmic reticulum (en′-dō-PLAS-mik re-TIK-ū-lum) **(ER)** A network of channels running through the cytoplasm of a cell that serves in intracellular transportation, support, storage, synthesis, and packaging of molecules. Portions of ER where ribosomes are attached to the outer surface are called **rough ER;** portions that have no ribosomes are called **smooth ER.**

End organ of Ruffini *See* **Type II cutaneous mechanoreceptor.**

Endosteum (end-OS-tē-um) The membrane that lines the medullary (marrow) cavity of bones, consisting of osteogenic cells and scattered osteoclasts.

Endothelium (en′-dō-THĒ-lē-um) The layer of simple squamous epithelium that lines the cavities of the heart, blood vessels, and lymphatic vessels.

Enteric (en-TER-ik) **nervous system** A portion of the autonomic nervous system within the wall of the gastrointestinal tract, pancreas, and gallbladder. Its sensory neurons monitor tension in the intestinal wall and assess the composition of intestinal contents; its motor neurons exert control over the motility and secretions of the gastrointestinal tract.

Enteroendocrine (en-ter-ō-EN-dō-krin) **cell** A cell of the mucosa of the gastrointestinal tract that secretes a hormone that governs function of the GI tract; hormones secreted include gastrin, cholecystokinin, glucose-dependent insulinotropic peptide (GIP), and secretin.

Enzyme (EN-zīm) A substance that accelerates chemical reactions; an organic catalyst, usually a protein.

Eosinophil (ē-ō-SIN-ō-fil) A type of white blood cell characterized by granules that stain red or pink with acid dyes.

Ependymal (ep-EN-de-mal) **cells** Neuroglial cells that cover choroid plexuses and produce cerebrospinal fluid (CSF); they also line the ventricles of the brain and probably assist in the circulation of CSF.

Epicardium (ep′-i-KAR-dē-um) The thin outer layer of the heart wall, composed of serous tissue and mesothelium. Also called the **visceral pericardium.**

Epidemiology (ep′-i-dē-mē-OL-ō-jē) Study of the occurrence and transmission of diseases and disorders in human populations.

Epidermis (ep′-i-DERM-is) The superficial, thinner layer of skin, composed of keratinized stratified squamous epithelium.

Epididymis (ep′-i-DID-i-mis) A comma-shaped organ that lies along the posterior border of the testis and contains the ductus epididymis, in which sperm undergo maturation. *Plural is* **epididymides** (ep′-i-di-DIM-i-dēz).

Epidural (ep′-i-DOO-ral) **space** A space between the spinal dura mater and the vertebral canal, containing areolar connective tissue and a plexus of veins.

Epiglottis (ep′-i-GLOT-is) A large, leaf-shaped piece of cartilage lying on top of the larynx, attached to the thyroid cartilage; its unattached portion is free to move up and down to cover the glottis (vocal folds and rima glottidis) during swallowing.

Epimysium (ep-i-MĪZ-ē-um) Fibrous connective tissue around muscles.

Epinephrine (ep-ē-NEF-rin) Hormone secreted by the adrenal medulla that produces actions similar to those that result from sympathetic stimulation. Also called **adrenaline** (a-DREN-a-lin).

Epineurium (ep′-i-NOO-rē-um) The superficial connective tissue covering around an entire nerve.

Epiphyseal (ep′-i-FIZ-ē-al) **line** The remnant of the epiphyseal plate in the metaphysis of a long bone.

Epiphyseal plate The hyaline cartilage plate in the metaphysis of a long bone; site of lengthwise growth of long bones.

Epiphysis (e-PIF-i-sis) The end of a long bone, usually larger in diameter than the shaft (diaphysis).

Epiphysis cerebri (se-RĒ-brē) Pineal gland.

Episiotomy (e-piz′-ē-OT-ō-mē) A cut made with surgical scissors to avoid tearing of the perineum at the end of the second stage of labour.

Epistaxis (ep′-i-STAK-sis) Loss of blood from the nose due to trauma, infection, allergy, neoplasm, and bleeding disorders. Also called **nosebleed.**

Epithalamus (ep′-i-THAL-a-mus) Part of the diencephalon superior and posterior to the thalamus, comprising the pineal gland and associated structures.

Epithelial (ep-i-THĒ-lē-al) **tissue** The tissue that forms the innermost and outermost surfaces of body structures and forms glands.

Eponychium (ep′-o-NIK-ē-um) Narrow band of stratum corneum at the proximal border of a nail that extends from the margin of the nail wall. Also called the **cuticle.**

Equilibrium (ē-kwi-LIB-rē-um) The state of being balanced.

Erectile dysfunction (ED) Failure to maintain an erection long enough for sexual intercourse. Previously known as **impotence** (IM-pō-tens).

Erection (ē-REK-shun) The enlarged and stiff state of the penis or clitoris resulting from the engorgement of the spongy erectile tissue with blood.

Eructation (e-ruk′-TĀ-shun) The forceful expulsion of gas from the stomach. Also called **belching.**

Erythema (er-e-THĒ-ma) Skin redness usually caused by dilation of the capillaries.

Erythrocyte (e-RITH-rō-sīt) A mature red blood cell.

Erythropoietin (e-rith′-rō-POY-e-tin) **(EPO)** A hormone released by the juxtaglomerular cells of the kidneys that stimulates red blood cell production.

Eupnea (ŪP-nē-a) Normal quiet breathing.

Eustachian tube *See* **Auditory tube.**

Eversion (ē-VER-zhun) The movement of the sole laterally at the ankle joint or of an atrioventricular valve into an atrium during ventricular contraction.

Excitability (ek-sīt′-a-BIL-i-tē) The ability of muscle fibers to receive and respond to stimuli; the ability of neurons to respond to stimuli and generate nerve impulses.

Excretion (eks-KRĒ-shun) The process of eliminating waste products from the body; also the products excreted.

Exercise physiology Study of the changes in cell and organ function due to muscular activity.

Exhalation (eks-ha-LĀ-shun) Breathing out; expelling air from the lungs into the atmosphere. Also called **expiration.**

Exocrine (EK-sō-krin) **gland** A gland that secretes its products into ducts that carry the secretions into body cavities, into the lumen of an organ, or to the outer surface of the body.

Exocytosis (ek-sō-sī-TŌ-sis) A process in which membrane-enclosed secretory vesicles form inside the cell, fuse with the plasma membrane, and release their contents into the interstitial fluid; achieves secretion of materials from a cell.

Extensibility (ek-sten′-si-BIL-i-tē) The ability of muscle tissue to stretch when it is pulled.

Extension (eks-TEN-shun) An increase in the angle between two bones; restoring a body part to its anatomical position after flexion.

External Located on or near the surface.

External auditory (AW-di-tōr-ē) **canal** or **meatus** (mē-Ā-tus) A curved tube in the temporal bone that leads to the middle ear.

External ear The **outer ear,** consisting of the pinna, external auditory canal, and tympanic membrane (eardrum).

External nares (NĀ-rez) The openings into the nasal cavity on the exterior of the body. Also called the **nostrils.**

External respiration The exchange of respiratory gases between the lungs and blood. Also called **pulmonary respiration** or **pulmonary gas exchange.**

Exteroceptor (EKS-ter-ō-sep′-tor) A sensory receptor adapted for the reception of stimuli from outside the body.

Extracellular fluid (ECF) Fluid outside body cells, such as interstitial fluid and plasma.

Extracellular matrix (MĀ-triks) The ground substance and fibers between cells in a connective tissue.

Eyebrow The hairy ridge superior to the eye.

F

F cell A cell in the pancreatic islets (islets of Langerhans) that secretes pancreatic polypeptide.

Face The anterior aspect of the head.

Faeces (FĒ-sēz) Material discharged from the rectum and made up of bacteria, excretions, and food residue. Also called **stool.**

Falciform ligament (FAL-si-form LIG-a-ment) A sheet of parietal peritoneum between the two principal lobes of the liver. The ligamentum teres, or remnant of the umbilical vein, lies within its fold.

Falx cerebelli (FALKS′ ser-e-BEL-lī) A small triangular process of the dura mater attached to the occipital bone in the posterior cranial fossa and projecting inward between the two cerebellar hemispheres.

Falx cerebri (FALKS SER-e-brē) A fold of the dura mater extending deep into the longitudinal fissure between the two cerebral hemispheres.

Fascia (FASH-ē-a) Large connective tissue sheet that wraps around groups of muscles.

Fascicle (FAS-i-kul) A small bundle or cluster, especially of nerve or muscle fibers (cells). Also called a **fasciculus** (fa-SIK-ū-lus). *Plural is* **fasciculi** (fa-SIK-yoo-lī).

Fasciculation (fa-sik-ū-LĀ-shun) Abnormal, spontaneous twitch of all skeletal muscle fibers in one motor unit that is visible at the skin surface; not associated with movement of the affected muscle; present in progressive diseases of motor neurons, for example, poliomyelitis.

Fat A triglyceride that is a solid at room temperature.

Fatty acid A simple lipid that consists of a carboxyl group and a hydrocarbon chain; used to synthesize triglyceride and phospholipids.

Fauces (FAW-sēs) The opening from the mouth into the pharynx.

Feedback system (loop) A cycle of events in which the status of a body condition is monitored, evaluated, changed, remonitored, reevaluated, and so on.

Female reproductive cycle General term for the ovarian and uterine cycles, the hormonal changes that accompany them, and cyclic changes in the breasts and cervix; includes changes in the endometrium of a nonpregnant female that prepares the lining of the uterus to receive a fertilized ovum. Less correctly termed the **menstrual cycle.**

Fertilisation (fer-til-i-ZĀ-shun) Penetration of a secondary oocyte by a sperm cell, meiotic division of a secondary oocyte to form an ovum, and subsequent union of the nuclei of the gametes.

Fetal circulation The cardiovascular system of the fetus, including the placenta and special blood vessels involved in the exchange of materials between fetus and mother.

Fetus (usu., also foetus, informal) In humans, the developing organism *in utero* from the beginning of the third month to birth.

Fever An elevation in body temperature above the normal temperature of 37°C (98.6°F) due to a resetting of the hypothalamic thermostat.

Fibroblast (FĪ-brō-blast) A large, flat cell that secretes most of the extracellular matrix of areolar and dense connective tissues.

Fibrosis The process by which fibroblasts synthesize collagen fibers and other extracellular matrix materials that aggregate to form scar tissue.

Fibrous (FĪ-brus) **joint** A joint that allows little or no movement, such as a suture or a syndesmosis.

Fibrous tunic (TOO-nik) The superficial coat of the eyeball, made up of the posterior sclera and the anterior cornea.

Fight-or-flight response The effects produced on stimulation of the sympathetic division of the autonomic nervous system.

Filiform papilla (FIL-i-form pa-PIL-a) One of the conical projections that are distributed in parallel rows over the anterior two-thirds of the tongue and lack taste buds.

Filtration (fil-TRĀ-shun) The flow of a liquid through a filter (or membrane that acts like a filter) due to a hydrostatic pressure; occurs in capillaries due to blood pressure.

Filum terminale (FĪ-lum ter-mi-NAL-ē) Non-nervous fibrous tissue of the spinal cord that extends inferiorly from the conus medullaris to the coccyx.

Fimbriae (FIM-brē-ē) Fingerlike structures, especially the lateral ends of the uterine (fallopian) tubes.

Fissure (FISH-ur) A groove, fold, or slit that may be normal or abnormal.

Fixator A muscle that stabilizes the origin of the prime mover so that the prime mover can act more efficiently.

Fixed macrophage (MAK-rō-fāj) Stationary phagocytic cell found in the liver, lungs, brain, spleen, lymph nodes, subcutaneous tissue, and red bone marrow. Also called a **tissue macrophage** or **histiocyte** (HIS-tē-ō-sīt).

Flaccid (FLASS-id) Relaxed, flabby, or soft; lacking muscle tone.

Flagellum (fla-JEL-um) A hairlike, motile process on the extremity of a bacterium, protozoan, or sperm cell. *Plural is* **flagella** (fla-JEL-a).

Flatus (FLĀ-tus) Gas in the stomach or intestines; commonly used to denote expulsion of gas through the anus.

Flexion (FLEK-shun) Movement in which there is a decrease in the angle between two bones.

Follicle (FOL-i-kul) A small secretory sac or cavity; the group of cells that contains a developing oocyte in the ovaries.

Follicle-stimulating hormone (FSH) Hormone secreted by the anterior pituitary; it initiates development of ova and stimulates the ovaries to secrete oestrogens in females, and initiates sperm production in males.

Fontanel (fon-ta-NEL) A mesenchyme-filled space where bone formation is not yet complete, especially between the cranial bones of an infant's skull.

Foot The terminal part of the lower limb, from the ankle to the toes.

Foramen (fō-RĀ-men) A passage or opening; a communication between two cavities of an organ, or a hole in a bone for passage of vessels or nerves. *Plural is* **foramina** (fō-RAM-i-na).

Foramen ovale (fō-RĀ-men ō-VAL-ē) An opening in the fetal heart in the septum between the right and left atria. A hole in the greater wing of the sphenoid bone that transmits the mandibular branch of the trigeminal (V) nerve.

Forearm (FOR-arm) The part of the upper limb between the elbow and the wrist.

Fornix (FOR-niks) An arch or fold; a tract in the brain made up of association fibers, connecting the hippocampus with the mammillary bodies; a recess around the cervix of the uterus where it protrudes into the vagina.

Fossa (FOS-a) A furrow or shallow depression.

Fourth ventricle (VEN-tri-kul) A cavity filled with cerebrospinal fluid within the brain lying between the cerebellum and the medulla oblongata and pons.

Fracture (FRAK-choor) Any break in a bone.

Free radical An atom or group of atoms with an unpaired electron in the outermost shell. It is unstable, highly reactive, and destroys nearby molecules.

Frontal plane A plane at a right angle to a midsagittal plane that divides the body or organs into anterior and posterior portions. Also called a **coronal** (kō-RŌ-nal) **plane.**

Fundus (FUN-dus) The part of a hollow organ farthest from the opening.

Fungiform papilla (FUN-ji-form pa-PIL-a) A mushroomlike elevation on the upper surface of the tongue appearing as a red dot; most contain taste buds.

Furuncle (FŪ-rung-kul) A boil; painful nodule caused by bacterial infection and inflammation of a hair follicle or sebaceous (oil) gland.

G

Gallbladder A small pouch, located inferior to the liver, that stores bile and empties by means of the cystic duct.

Gallstone A solid mass, usually containing cholesterol, in the gallbladder or a bile-containing duct; formed anywhere between bile canaliculi in the liver and the hepatopancreatic ampulla (ampulla of Vater), where bile enters the duodenum. Also called a **biliary calculus.**

Gamete (GAM-ēt) A male or female reproductive cell; a sperm cell or secondary oocyte.

Ganglion (GANG-glē-on) Usually, a group of neuronal cell bodies lying outside the central nervous system (CNS). *Plural* is **ganglia** (GANG-glē-a).

Gastric (GAS-trik) **glands** Glands in the mucosa of the stomach composed of cells that empty their secretions into narrow channels called gastric pits. Types of cells are chief cells (secrete pepsinogen), parietal cells (secrete hydrochloric acid and intrinsic factor), surface mucous and mucous neck cells (secrete mucus), and G cells (secrete gastrin).

Gastroenterology (gas'-trō-en'-ter-OL-ō-jē) The medical specialty that deals with the structure, function, diagnosis, and treatment of diseases of the stomach and intestines.

Gastrointestinal (gas-trō-in-TES-ti-nal) **(GI) tract** A continuous tube running through the ventral body cavity extending from the mouth to the anus. Also called the **alimentary** (al'-i-MEN-tar-ē) **canal.**

Gastrulation (gas-troo-LĀ-shun) The migration of groups of cells from the epiblast that transform a bilaminar embryonic disc into a trilaminar embryonic disc with three primary germ layers; transformation of the blastula into the gastrula.

Gene (JĒN) Biological unit of heredity; a segment of DNA located in a definite position on a particular chromosome; a sequence of DNA that codes for a particular mRNA, rRNA, or tRNA.

Genetic engineering The manufacture and manipulation of genetic material.

Genetics The study of genes and heredity.

Genome (JĒ-nōm) The complete set of genes of an organism.

Genotype (JĒ-nō-tīp) The genetic makeup of an individual; the combination of alleles present at one or more chromosomal locations, as distinguished from the appearance, or phenotype, that results from those alleles.

Geriatrics (jer'-ē-AT-riks) The branch of medicine devoted to the medical problems and care of elderly persons.

Gestation (jes-TĀ-shun) The period of development from fertilisation to birth.

Gingivae (jin-JI-vē) Gums. They cover the alveolar processes of the mandible and maxilla and extend slightly into each socket.

Gland Specialized epithelial cell or cells that secrete substances; may be exocrine or endocrine.

Glans penis (glanz PĒ-nis) The slightly enlarged region at the distal end of the penis.

Glaucoma (glaw-KŌ-ma) An eye disorder in which there is increased intraocular pressure due to an excess of aqueous humour.

Glomerular (glō-MER-ū-lar) **capsule** A double-walled globe at the proximal end of a nephron that encloses the glomerular capillaries. Also called **Bowman's** (BŌ-manz) **capsule.**

Glomerular filtrate (glō-MER-ū-lar FIL-trāt) The fluid produced when blood is filtered by the filtration membrane in the glomeruli of the kidneys.

Glomerular filtration The first step in urine formation in which substances in blood pass through the filtration membrane and the filtrate enters the proximal convoluted tubule of a nephron.

Glomerular filtration rate The amount of filtrate formed in all renal corpuscles per minute. It averages 125 mL/min in males and 105 mL/min in females.

Glomerulus (glō-MER-ū-lus) A rounded mass of nerves or blood vessels, especially the microscopic tuft of capillaries that is surrounded by the glomerular (Bowman's) capsule of each kidney tubule. *Plural* is **glomeruli** (glō-MER-ū-li).

Glottis (GLOT-is) The vocal folds (true vocal cords) in the larynx plus the space between them (rima glottidis).

Glucagon (GLOO-ka-gon) A hormone produced by the alpha cells of the pancreatic islets (islets of Langerhans) that increases blood glucose level.

Glucocorticoids (gloo'-kō-KOR-ti-koyds) Hormones secreted by the cortex of the adrenal gland, especially cortisol, that influence glucose metabolism.

Glucose (GLOO-kōs) A hexose (six-carbon sugar), $C_6H_{12}O_6$, that is a major energy source for the production of ATP by body cells.

Glucosuria (gloo'-kō-SOO-rē-a) The presence of glucose in the urine; may be temporary or pathological. Also called **glycosuria.**

Glycogen (GLĪ-kō-jen) A highly branched polymer of glucose containing thousands of subunits; functions as a compact store of glucose molecules in liver and muscle fibers (cells).

Goblet cell A goblet-shaped unicellular gland that secretes mucus; present in epithelium of the airways and intestines.

Goiter (GOY-ter) An enlarged thyroid gland.

Golgi (GOL-jē) **complex** An organelle in the cytoplasm of cells consisting of four to six flattened sacs (cisternae), stacked on one another, with expanded areas at their ends; functions in processing, sorting, packaging, and delivering proteins and lipids to the plasma membrane, lysosomes, and secretory vesicles.

Golgi tendon organ *See* **Tendon organ.**

Gomphosis (gom-FŌ-sis) A fibrous joint in which a cone-shaped peg fits into a socket.

Gonad (GŌ-nad) A gland that produces gametes and hormones; the ovary in the female and the testis in the male.

Gonadotropic hormone Anterior pituitary hormone that affects the gonads.

Gout (GOWT) Hereditary condition associated with excessive uric acid in the blood; the acid crystallizes and deposits in joints, kidneys, and soft tissue.

Graafian follicle *See* **Mature follicle.**

Gracile (GRAS-īl) **nucleus** A group of nerve cells in the inferior part of the medulla oblongata in which axons of the gracile fasciculus terminate.

Greater oementum (ō-MEN-tum) A large fold in the serosa of the stomach that hangs down like an apron anterior to the intestines.

Greater vestibular (ves-TIB-ū-lar) **glands** A pair of glands on either side of the vaginal orifice that open by a duct into the space between the hymen and the labia minora. Also called **Bartholin's** (BAR-to-linz) **glands.**

Grey commisure (KOM-mi-shur) A narrow strip of grey matter connecting the two lateral gray masses within the spinal cord.

Grey matter Areas in the central nervous system and ganglia containing neuronal cell bodies, dendrites, unmyelinated axons, axon terminals, and neuroglia; Nissl bodies impart a gray color and there is little or no myelin in grey matter.

Grey ramus communicans (RĀ-mus kō-MŪ-ni-kans) A short nerve containing axons of sympathetic postganglionic neurons; the cell bodies of the neurons are in a sympathetic chain ganglion, and the unmyelinated axons extend via the gray ramus to a spinal nerve and then to the periphery to supply smooth muscle in blood vessels, arrector pili muscles, and sweat glands. *Plural* is **rami communicantes** (RĀ-mē kō-mū-ni-KAN-tēz).

Groin (GROYN) The depression between the thigh and the trunk; the inguinal region.

Gross anatomy The branch of anatomy that deals with structures that can be studied without using a microscope. Also called **macroscopic anatomy.**

Growth An increase in size due to an increase in (1) the number of cells, (2) the size of existing cells as internal components increase in size, or (3) the size of intercellular substances.

Gustation (gus-TĀ-shun). The sense of taste.

Gustatory (GUS-ta-tō'-rē) Pertaining to taste.

Gynaecology (gī'-ne-KOL-ō-jē) The branch of medicine dealing with the study and treatment of disorders of the female reproductive system.

Gynaecomastia (gīn'-e-kō-MAS-tē-a) Excessive growth (benign) of the male mammary glands due to secretion of oestrogens by an adrenal gland tumor (feminizing adenoma).

Gyrus (JI-rus) One of the folds of the cerebral cortex of the brain. *Plural* is **gyri** (JĪ-rī). Also called a **convolution.**

H

Haemangioblast (hē-MAN-jē-ō-blast) A precursor mesodermal cell that develops into blood and blood vessels.

Haematocrit (he-MAT-ō-krit) **(Hct)** The percentage of blood made up of red blood cells. Usually measured by centrifuging a blood sample in a graduated tube and then reading the volume of red blood cells and dividing it by the total volume of blood in the sample.

Haematology (hēm-a-TOL-ō-jē) The study of blood.

Haematoma (hē'-ma-TŌ-ma) A tumor or swelling filled with blood.

Haemodialysis (hē-mō-dī-AL-i-sis) Direct filtration of blood by removing wastes and excess electrolytes and fluid and then returning the cleansed blood.

Haemoglobin (hē′-mō-GLŌ-bin) **(Hb)** A substance in red blood cells consisting of the protein globin and the iron-containing red pigment heme that transports most of the oxygen and some carbon dioxide in blood.

Haemolysis (hē-MOL-i-sis) The escape of haemoglobin from the interior of a red blood cell into the surrounding medium; results from disruption of the cell membrane by toxins or drugs, freezing or thawing, or hypotonic solutions.

Haemolytic disease of the newborn (HDN) A hemolytic anaemia of a newborn child that results from the destruction of the infant's erythrocytes (red blood cells) by antibodies produced by the mother; usually the antibodies are due to an Rh blood type incompatibility. Also called **erythroblastosis fetalis** (e-rith′-rō-blas-TŌ-sis fe-TAL-is).

Haemophilia (hē′-mō-FIL-ē-a) A hereditary blood disorder where there is a deficient production of certain factors involved in blood clotting, resulting in excessive bleeding into joints, deep tissues, and elsewhere.

Haemopoiesis (hēm-ō-poy-Ē-sis) Blood cell production, which occurs in red bone marrow after birth. Also called **hematopoiesis** (hem′-a-tō-poy-Ē-sis).

Haemorrhage (HEM-o-rij) Bleeding; the escape of blood from blood vessels, especially when the loss is profuse.

Haemorrhoid (HEM-ō-royds) Dilated or varicosed blood vessels (usually veins) in the anal region. Also called **piles.**

Hair A threadlike structure produced by hair follicles that develops in the dermis. Also called a **pilus** (PĪ-lus).

Hair follicle (FOL-li-kul) Structure composed of epithelium and surrounding the root of a hair from which hair develops.

Hair root plexus (PLEK-sus) A network of dendrites arranged around the root of a hair as free or naked nerve endings that are stimulated when a hair shaft is moved.

Hand The terminal portion of an upper limb, including the carpals, metacarpals, and phalanges.

Haploid (HAP-loyd) **cell** Having half the number of chromosomes characteristically found in the somatic cells of an organism; characteristic of mature gametes. Symbolized *n*.

Hard palate (PAL-at) The anterior portion of the roof of the mouth, formed by the maxillae and palatine bones and lined by mucous membrane.

Haustra (HAWS-tra) A series of pouches that characterize the colon; caused by tonic contractions of the teniae coli. *Singular* is **haustrum.**

Haversian canal *See* **Central canal.**

Haversian system *See* **Osteon.**

Head The superior part of a human, cephalic to the neck. The superior or proximal part of a structure.

Hearing The ability to perceive sound.

Heart A hollow muscular organ lying slightly to the left of the midline of the chest that pumps the blood through the cardiovascular system.

Heart block An arrhythmia (dysrhythmia) of the heart in which the atria and ventricles contract independently because of a blocking of electrical impulses through the heart at some point in the conduction system.

Heart murmur (MER-mer) An abnormal sound that consists of a flow noise that is heard before, between, or after the normal heart sounds, or that may mask normal heart sounds.

Hemiplegia (hem-i-PLĒ-jē-a) Paralysis of the upper limb, trunk, and lower limb on one side of the body.

Hemodynamics (hē-mō-dī-NAM-iks) The forces involved in circulating blood throughout the body.

Hepatic (he-PAT-ik) Refers to the liver.

Hepatic duct A duct that receives bile from the bile capillaries. Small hepatic ducts merge to form the larger right and left hepatic ducts that unite to leave the liver as the common hepatic duct.

Hepatic portal circulation The flow of blood from the gastrointestinal organs to the liver before returning to the heart.

Hepatocyte (he-PAT-ō-cyte) A liver cell.

Hepatopancreatic (hep′-a-tō-pan′-krē-A-tik) **ampulla** A small, raised area in the duodenum where the combined common bile duct and main pancreatic duct empty into the duodenum. Also called the **ampulla of Vater** (FAH-ter).

Hernia (HER-nē-a) The protrusion or projection of an organ or part of an organ through a membrane or cavity wall, usually the abdominal cavity.

Herniated (HER-nē-ā′-ted) **disc** A rupture of an intervertebral disc so that the nucleus pulposus protrudes into the vertebral cavity. Also called a **slipped disc.**

Hiatus (hī-Ā-tus) An opening; a foramen.

Hilum (HĪ-lum) An area, depression, or pit where blood vessels and nerves enter or leave an organ. Also called a **hilus.**

Hinge joint A synovial joint in which a convex surface of one bone fits into a concave surface of another bone, such as the elbow, knee, ankle, and interphalangeal joints. Also called a **ginglymus** (JIN-gli-mus) **joint.**

Hirsutism (HER-soo-tizm) An excessive growth of hair in females and children, with a distribution similar to that in adult males, due to the conversion of vellus hairs into large terminal hairs in response to higher-than-normal levels of androgens.

Histamine (HISS-ta-mēn) Substance found in many cells, especially mast cells, basophils, and platelets, that is released when the cells are injured; results in vasodilation, increased permeability of blood vessels, and constriction of bronchioles.

Histology (his′-TOL-ō-jē) Microscopic study of the structure of tissues.

Holocrine (HŌ-lō-krin) **gland** A type of gland in which entire secretory cells, along with their accumulated secretions, make up the secretory product of the gland, as in the sebaceous (oil) glands.

Homoeostasis (hō′-mē-ō-STĀ-sis) The condition in which the body's internal environment remains relatively constant within physiological limits.

Homologous (hō-MOL-ō-gus) **chromosomes** Two chromosomes that belong to a pair. Also called **homologs.**

Hormone (HOR-mōn) A secretion of endocrine cells that alters the physiological activity of target cells of the body.

Horn An area of grey matter (anterior, lateral, or posterior) in the spinal cord.

Human chorionic gonadotropin (kō-rē-ON-ik gō-nad-ō-TRŌ-pin) **(hCG)** A hormone produced by the developing placenta that maintains the corpus luteum.

Human chorionic somatomammotropin (sō-mat-ō-mam-ō-TRŌ-pin) **(hCS)** Hormone produced by the chorion of the placenta that stimulates breast tissue for lactation, enhances body growth, and regulates metabolism. Also called **human placental lactogen (hPL).**

Human growth hormone (hGH) Hormone secreted by the anterior pituitary that stimulates growth of body tissues, especially skeletal and muscular tissues. Also known as **somatotropin** (sō′-ma-tō-TRŌ-pin) and **somatotropic hormone (STH).**

Hyaluronic (hī′-a-loo-RON-ik) **acid** A viscous, amorphous extracellular material that binds cells together, lubricates joints, and maintains the shape of the eyeballs.

Hymen (HĪ-men) A thin fold of vascularized mucous membrane at the vaginal orifice.

Hyperextension (hī′-per-ek-STEN-shun) Continuation of extension beyond the anatomical position, as in bending the head backward.

Hyperplasia (hī-per-PLĀ-zē-a) An abnormal increase in the number of normal cells in a tissue or organ, increasing its size.

Hypersecretion (hī′-per-se-KRĒ-shun) Overactivity of glands resulting in excessive secretion.

Hypersensitivity (hī′-per-sen-si-TI-vi-tē) Overreaction to an allergen that results in pathological changes in tissues. Also called **allergy.**

Hypertension (hī′-per-TEN-shun) High blood pressure.

Hyperthermia (hī′-per-THERM-ē-a) An elevated body temperature.

Hypertonia (hī′-per-TŌ-nē-a) Increased muscle tone that is expressed as spasticity or rigidity.

Hypertonic (hī′-per-TON-ik) Solution that causes cells to shrink due to loss of water by osmosis.

Hypertrophy (hī-PER-trō-fē) An excessive enlargement or overgrowth of tissue without cell division.

Hyperventilation (hī′-per-ven-ti-LĀ-shun) A rate of inhalation and exhalation higher than that required to maintain a normal partial pressure of carbon dioxide in the blood.

Hyponychium (hī′-pō-NIK-ē-um) Free edge of the fingernail.

Hypophyseal fossa (hī′-pō-FIZ-ē-al FOS-a) A depression on the superior surface of the sphenoid bone that houses the pituitary gland.

Hypophyseal (hī′-pō-FIZ-ē-al) **pouch** An outgrowth of ectoderm from the roof of the mouth from which the anterior pituitary develops.

Hypophysis (hī-POF-i-sis) Pituitary gland.

Hyposecretion (hī′-pō-se-KRĒ-shun) Underactivity of glands resulting in diminished secretion.

Hypothalamohypophyseal (hī′-pō-thal′-a-mō-hī-pō-FIZ-ē-al) **tract** A bundle of axons containing secretory vesicles filled with oxytocin or antidi- uretic hormone that extend from the hypothala- mus to the posterior pituitary.

Hypothalamus (hī′-pō-THAL-a-mus) A portion of the diencephalon, lying beneath the thalamus and forming the floor and part of the wall of the third ventricle.

Hypothermia (hī′-pō-THER-mē-a) Lowering of body temperature below 35°C (95°F); in surgical procedures, it refers to deliberate cooling of the body to slow down metabolism and reduce oxy- gen needs of tissues.

Hypotonia (hī′-pō-TŌ-nē-a) Decreased or lost muscle tone in which muscles appear flaccid.

Hypotonic (hī′-pō-TON-ik) Solution that causes cells to swell and perhaps rupture due to gain of water by osmosis.

Hypoventilation (hī-pō-ven-ti-LĀ-shun) A rate of inhalation and exhalation lower than that required to maintain a normal partial pressure of carbon dioxide in plasma.

Hypoxia (hī-POKS-ē-a) Lack of adequate oxygen at the tissue level.

Hysterectomy (hiss-te-REK-tō-mē) The surgical re- moval of the uterus.

I

Ileocaecal (il-ē-ō-SĒ-kal) **sphincter** A fold of mu- cous membrane that guards the opening from the ileum into the large intestine. Also called the **ileocecal valve.**

Ileum (IL-ē-um) The terminal part of the small in- testine.

Immunity (i-MŪ-ni-tē) The state of being resistant to injury, particularly by poisons, foreign pro- teins, and invading pathogens.

Immunoglobulin (im-ū-nō-GLOB-ū-lin) **(Ig)** An an- tibody synthesized by plasma cells derived from B lymphocytes in response to the introduction of an antigen. Immunoglobulins are divided into five kinds (IgG, IgM, IgA, IgD, IgE).

Immunology (im′-ū-NOL-ō-jē) The study of the responses of the body when challenged by antigens.

Imperforate (im′-PER-fō-rāt) Abnormally closed.

Implantation (im′-plan-TĀ-shun) The insertion of a tissue or a part into the body. The attachment of the blastocyst to the stratum basalis of the endometrium about 6 days after fertilisation.

Incontinence (in-KON-ti-nens) Inability to retain urine, semen, or faeces through loss of sphincter control.

Indirect motor pathways Motor tracts that convey information from the brain down the spinal cord for automatic movements, coordination of body movements with visual stimuli, skeletal muscle tone and posture, and balance. Also known as **ex- trapyramidal pathways.**

Induction (in-DUK-shun) The process by which one tissue (inducting tissue) stimulates the devel- opment of an adjacent unspecialized tissue (re- sponding tissue) into a specialized one.

Infarction (in-FARK-shun) A localized area of necrotic tissue, produced by inadequate oxygena- tion of the tissue.

Infection (in-FEK-shun) Invasion and multiplication of microorganisms in body tissues, which may be inapparent or characterized by cellular injury.

Inferior (in-FĒR-ē-or) Away from the head or to- ward the lower part of a structure. Also called **caudal** (KAW-dal).

Inferior vena cava (VĒ-na KĀ-va) **(IVC)** Large vein that collects blood from parts of the body inferior to the heart and returns it to the right atrium.

Infertility Inability to conceive or to cause concep- tion. Also called **sterility** in males.

Inflammation (in′-fla-MĀ-shun) Localized, pro- tective response to tissue injury designed to de- stroy, dilute, or wall off the infecting agent or injured tissue; characterized by redness, pain, heat, swelling, and sometimes loss of function.

Infundibulum (in-fun-DIB-ū-lum) The stalklike structure that attaches the pituitary gland to the hypothalamus of the brain. The funnel-shaped, open, distal end of the uterine (fallopian) tube.

Ingestion (in-JES-chun) The taking in of food, liq- uids, or drugs, by mouth.

Inguinal (IN-gwin-al) Pertaining to the groin.

Inguinal canal An oblique passageway in the ante- rior abdominal wall just superior and parallel to the medial half of the inguinal ligament that transmits the spermatic cord and ilioinguinal nerve in the male and round ligament of the uterus and ilioinguinal nerve in the female.

Inhalation (in-ha-LĀ-shun) The act of drawing air into the lungs. Also termed **inspiration.**

Inheritance The acquisition of body traits by trans- mission of genetic information from parents to offspring.

Inhibin A hormone secreted by the gonads that in- hibits release of follicle-stimulating hormone (FSH) by the anterior pituitary.

Inhibiting hormone Hormone secreted by the hy- pothalamus that can suppress secretion of hor- mones by the anterior pituitary.

Insertion (in-SER-shun) The attachment of a mus- cle tendon to a movable bone or the end opposite the origin.

Insula (IN-soo-la) A triangular area of the cerebral cortex that lies deep within the lateral cerebral fis- sure, under the parietal, frontal, and temporal lobes.

Insulin (IN-soo-lin) A hormone produced by the beta cells of a pancreatic islet (islet of Langerhans) that decreases the blood glucose level.

Integrins (IN-te-grinz) A family of transmembrane glycoproteins in plasma membranes that function in cell adhesion; they are present in hemidesmo- somes, which anchor cells to a basement mem- brane, and they mediate adhesion of neutrophils to endothelial cells during emigration.

Integumentary (in-teg-ū-MEN-tar-ē) Relating to the skin.

Integumentary (in-teg-ū-MEN-tar-ē) **system** A sys- tem composed of organs such as the skin, hair, oil and sweat glands, nails, and sensory receptors.

Intercalated (in-TER-ka-lāt-ed) **disc** An irregular transverse thickening of sarcolemma that contains desmosomes, which hold cardiac muscle fibers

(cells) together, and gap junctions, which aid in conduction of muscle action potentials from one fiber to the next.

Intercostal (in′-ter-KOS-tal) **nerve** A nerve supply- ing a muscle located between the ribs. Also called **thoracic nerve.**

Intermediate (in′-ter-MĒ-dē-at) Between two struc- tures, one of which is medial and one of which is lateral.

Intermediate filament Protein filament, ranging from 8 to 12 nm in diameter, that may provide structural reinforcement, hold organelles in place, and give shape to a cell.

Internal Away from the surface of the body.

Internal capsule A large tract of projection fibers lateral to the thalamus that is the major connec- tion between the cerebral cortex and the brain stem and spinal cord; contains axons of sensory neurons carrying auditory, visual, and somatic sensory signals to the cerebral cortex plus axons of motor neurons descending from the cerebral cortex to the thalamus, subthalamus, brain stem, and spinal cord.

Internal ear The inner ear or labyrinth, lying inside the temporal bone, containing the organs of hear- ing and balance.

Internal nares (NĀ-rez) The two openings posterior to the nasal cavities opening into the nasopharynx. Also called the **choanae** (kō-Ā-nē).

Internal respiration The exchange of respiratory gases between blood and body cells. Also called **tissue respiration** or **systemic gas exchange.**

Interneurons (in′-ter-NOO-ronz) Neurons whose ax- ons extend only for a short distance and contact nearby neurons in the brain, spinal cord, or a gan- glion; they comprise the vast majority of neurons in the body. Also called **association neurons.**

Interoceptor (IN-ter-ō-sep′-tor) Sensory receptor located in blood vessels and viscera that provides information about the body's internal environ- ment. Also called **visceroceptor.**

Interphase (IN-ter-fāz) The period of the cell cycle between cell divisions, consisting of the G_0 phase; G_1 (gap or growth) phase, when the cell is engaged in growth, metabolism, and production of sub- stances required for division; S (synthesis) phase, during which chromosomes are replicated; and G_2 phase.

Interstitial cell of Leydig *See* **Interstitial endocrinocyte.**

Interstitial (in′-ter-STISH-al) **endocrinocyte** A cell that is located in the connective tissue between seminiferous tubules in a mature testis that se- cretes testosterone. Also called an **interstitial cell of Leydig** (LĪ-dig).

Interstitial (in′-ter-STISH-al) **fluid** The portion of extracellular fluid that fills the microscopic spaces between the cells of tissues; the internal environment of the body. Also called **intercellu- lar** or **tissue fluid.**

Interstitial growth Growth from within, as in the growth of cartilage. Also called **endogenous** (en- DOJ-e-nus) **growth.**

Interventricular (in′-ter-ven-TRIK-ū-lar) **foramen** A narrow, oval opening through which the lateral ventricles of the brain communicate with the third ventricle. Also called the **foramen of Monro.**

Intervertebral (in′-ter-VER-te-bral) **disc** A pad of fibrocartilage located between the bodies of two vertebrae.

Intestinal gland A gland that opens onto the surface of the intestinal mucosa and secretes digestive enzymes. Also called a **crypt of Lieberkühn** (LĒ-ber-kūn).

Intracellular (in′-tra-SEL-yū-lar) **fluid** (**ICF**) Fluid located within cells.

Intrafusal (in′-tra-FŪ-sal) **fibers** Three to ten specialized muscle fibers (cells), partially enclosed in a spindle-shaped connective tissue capsule, that make up a muscle spindle.

Intramembranous (in′-tra-MEM-bra-nus) **ossification** Bone formation within mesenchyme arranged in sheetlike layers that resemble membranes.

Intramuscular injection An injection that penetrates the skin and subcutaneous layer to enter a skeletal muscle. Common sites are the deltoid, gluteus medius, and vastus lateralis muscles.

Intraocular (in′-tra-OK-ū-lar) **pressure** (**IOP**) Pressure in the eyeball, produced mainly by aqueous humour.

Intrinsic factor (**IF**) A glycoprotein, synthesized and secreted by the parietal cells of the gastric mucosa, that facilitates vitamin B_{12} absorption in the small intestine.

Invagination (in-vaj′-i-NĀ-shun) The pushing of the wall of a cavity into the cavity itself.

Inversion (in-VER-zhun) The movement of the sole medially at the ankle joint.

In vitro (VĒ-trō) Literally, in glass; outside the living body and in an artificial environment such as a laboratory test tube.

Ipsilateral (ip-si-LAT-er-al) On the same side, affecting the same side of the body.

Iris The colored portion of the vascular tunic of the eyeball seen through the cornea that contains circular and radial smooth muscle; the hole in the center of the iris is the pupil.

Irritable bowel syndrome (**IBS**) Disease of the entire gastrointestinal tract in which a person reacts to stress by developing symptoms (such as cramping and abdominal pain) associated with alternating patterns of diarrhoea and constipation. Excessive amounts of mucus may appear in faeces, and other symptoms include flatulence, nausea, and loss of appetite. Also known as **irritable colon** or **spastic colitis.**

Ischaemia (is-KĒ-mē-a) A lack of sufficient blood to a body part due to obstruction or constriction of a blood vessel.

Islet of Langerhans *See* **Pancreatic islet.**

Isotonic (ī′-sō-TON-ik) Having equal tension or tone. A solution having the same concentration of impermeable solutes as cytosol.

Isthmus (IS-mus) A narrow strip of tissue or narrow passage connecting two larger parts.

J

Jaundice (JON-dis) A condition characterized by yellowness of the skin, the white of the eyes, mu-

cous membranes, and body fluids because of a buildup of bilirubin.

Jejunum (je-JOO-num) The middle part of the small intestine.

Joint A point of contact between two bones, between bone and cartilage, or between bone and teeth. Also called an **articulation** or **arthrosis.**

Joint kinesthetic (kin′-es-THET-ik) **receptor** A proprioceptive receptor located in a joint, stimulated by joint movement.

Juxtaglomerular (juks-ta-glō-MER-ū-lar) **apparatus** (**JGA**) Consists of the macula densa (cells of the distal convoluted tubule adjacent to the afferent and efferent arteriole) and juxtaglomerular cells (modified cells of the afferent and sometimes efferent arteriole); secretes renin when blood pressure starts to fall.

K

Keratin (KER-a-tin) An insoluble protein found in the hair, nails, and other keratinized tissues of the epidermis.

Keratinocyte (ker-a-TIN-ō-sīt) The most numerous of the epidermal cells; produces keratin.

Kidney (KID-nē) One of the paired reddish organs located in the lumbar region that regulates the composition, volume, and pressure of blood and produces urine.

Kidney stone A solid mass, usually consisting of calcium oxalate, uric acid, or calcium phosphate crystals, that may form in any portion of the urinary tract. Also called a **renal calculus.**

Kinaesthesia (kin′-es-THĒ-zē-a) The perception of the extent and direction of movement of body parts; this sense is possible due to nerve impulses generated by proprioceptors.

Kinesiology (ki-nē-sē′-OL-ō-jē) The study of the movement of body parts.

Kinetochore (ki-NET-ō-kor) Protein complex attached to the outside of a centromere to which kinetochore microtubules attach.

Kupffer's cell *See* **Stellate reticuloendothelial cell.**

Kyphosis (kī-FŌ-sis) An exaggeration of the thoracic curve of the vertebral column, resulting in a "round-shouldered" appearance. Also called **hunchback.**

L

Labial frenulum (LĀ-bē-al FREN-ū-lum) A medial fold of mucous membrane between the inner surface of the lip and the gums.

Labia majora (LĀ-bē-a ma-JŌ-ra) Two longitudinal folds of skin extending downward and backward from the mons pubis of the female.

Labia minora (min-OR-a) Two small folds of mucous membrane lying medial to the labia majora of the female.

Labium (LĀ-bē-um) A lip. A liplike structure. *Plural* is **labia** (LA-bē-a).

Labour The process of giving birth in which a fetus is expelled from the uterus through the vagina.

Labyrinth (LAB-i-rinth) Intricate communicating passageway, especially in the internal ear.

Lacrimal canal A duct, one on each eyelid, beginning at the punctum at the medial margin of an eyelid and conveying tears medially into the nasolacrimal sac.

Lacrimal gland Secretory cells, located at the superior anterolateral portion of each orbit, that secrete tears into excretory ducts that open onto the surface of the conjunctiva.

Lacrimal sac The superior expanded portion of the nasolacrimal duct that receives the tears from a lacrimal canal.

Lactation (lak-TĀ-shun) The secretion and ejection of milk by the mammary glands.

Lacteal (LAK-tē-al) One of many lymphatic vessels in villi of the intestines that absorb triglycerides and other lipids from digested food.

Lacuna (la-KOO-na) A small, hollow space, such as that found in bones in which the osteocytes lie. *Plural* is **lacunae** (la-KOO-nē).

Lambdoid (LAM-doyd) **suture** The joint in the skull between the parietal bones and the occipital bone; sometimes contains sutural (Wormian) bones.

Lamellae (la-MEL-ē) Concentric rings of hard, calcified extracellular matrix found in compact bone.

Lamellated corpuscle *See* **Pacinian corpuscle.**

Lamina (LAM-i-na) A thin, flat layer or membrane, as the flattened part of either side of the arch of a vertebra. *Plural* is **laminae** (LAM-i-nē).

Lamina propria (PRŌ-prē-a) The connective tissue layer of a mucosa.

Langerhans (LANG-er-hans) **cell** Epidermal dendritic cell that functions as an antigen-presenting cell (APC) during an immune response.

Lanugo (la-NOO-gō) Fine downy hairs that cover the fetus.

Large intestine The portion of the gastrointestinal tract extending from the ileum of the small intestine to the anus, divided structurally into the caecum, colon, rectum, and anal canal.

Laryngopharynx (la-rin′-gō-FAIR-inks) The inferior portion of the pharynx, extending downward from the level of the hyoid bone that divides posteriorly into the oesophagus and anteriorly into the larynx. Also called the **hypopharynx.**

Larynx (LAIR-inks) The **voice box,** a short passageway that connects the pharynx with the trachea.

Lateral (LAT-er-al) Farther from the midline of the body or a structure.

Lateral ventricle (VEN-tri-kul) A cavity within a cerebral hemisphere that communicates with the lateral ventricle in the other cerebral hemisphere and with the third ventricle by way of the interventricular foramen.

Leg The part of the lower limb between the knee and the ankle.

Lens A transparent organ constructed of proteins (crystallins) lying posterior to the pupil and iris of the eyeball and anterior to the vitreous body.

Lesion (LĒ-zhun) Any localized, abnormal change in a body tissue.

Lesser oementum (ō-MEN-tum) A fold of the peritoneum that extends from the liver to the lesser curvature of the stomach and the first part of the duodenum.

Lesser vestibular (ves-TIB-ū-lar) **gland** One of the paired mucus-secreting glands with ducts that open on either side of the urethral orifice in the vestibule of the female.

Leukaemia (loo-KĒ-mē-a) A malignant disease of the blood-forming tissues characterized by either uncontrolled production and accumulation of immature leukocytes in which many cells fail to reach maturity (acute) or an accumulation of mature leukocytes in the blood because they do not die at the end of their normal life span (chronic).

Leukocyte (LOO-kō-sīt) A white blood cell.

Leydig (LĪ-dig) **cell** A type of cell that secretes testosterone; located in the connective tissue between seminiferous tubules in a mature testis. Also known as **interstitial cell of Leydig** or **interstitial endocrinocyte.**

Ligament (LIG-a-ment) Dense regular connective tissue that attaches bone to bone.

Ligand (LĪ-gand) A chemical substance that binds to a specific receptor.

Limbic system A part of the forebrain, sometimes termed the visceral brain, concerned with various aspects of emotion and behavior; includes the limbic lobe, dentate gyrus, amygdala, septal nuclei, mammillary bodies, anterior thalamic nucleus, olfactory bulbs, and bundles of myelinated axons.

Lingual frenulum (LIN-gwal FREN-ū-lum) A fold of mucous membrane that connects the tongue to the floor of the mouth.

Lipase An enzyme that splits fatty acids from triglycerides and phospholipids.

Lipid (LIP-id) An organic compound composed of carbon, hydrogen, and oxygen that is usually insoluble in water, but soluble in alcohol, ether, and chloroform; examples include triglycerides (fats and oils), phospholipids, steroids, and eicosanoids.

Lipid bilayer Arrangement of phospholipid, glycolipid, and cholesterol molecules in two parallel sheets in which the hydrophilic "heads" face outward and the hydrophobic "tails" face inward; found in cellular membranes.

Lipoprotein (lip′-ō-PRŌ-tēn) One of several types of particles containing lipids (cholesterol and triglycerides) and proteins that make it water soluble for transport in the blood; high levels of **low-density lipoproteins (LDLs)** are associated with increased risk of atherosclerosis, whereas high levels of **high-density lipoproteins (HDLs)** are associated with decreased risk of atherosclerosis.

Liver Large organ under the diaphragm that occupies most of the right hypochondriac region and part of the epigastric region. Functionally, it produces bile and synthesizes most plasma proteins; interconverts nutrients; detoxifies substances; stores glycogen, iron, and vitamins; carries on phagocytosis of worn-out blood cells and bacteria; and helps synthesize the active form of vitamin D.

Long-term potentiation (po-ten′-shē-Ā-shun) **(LTP)** Prolonged, enhanced synaptic transmission that occurs at certain synapses within the hippocampus of the brain; believed to underlie some aspects of memory.

Lordosis (lor-DŌ-sis) An exaggeration of the lumbar curve of the vertebral column. Also called **hollow back.**

Lower limb The appendage attached at the pelvic (hip) girdle, consisting of the thigh, knee, leg, ankle, foot, and toes. Also called the **lower extremity** or **lower appendage.**

Lumbar (LUM-bar) Region of the back and side between the ribs and pelvis; loin.

Lumbar plexus (PLEK-sus) A network formed by the anterior (ventral) branches of spinal nerves L1 through L4.

Lumen (LOO-men) The space within an artery, vein, intestine, renal tubule, or other tubular structure.

Lungs Main organs of respiration that lie on either side of the heart in the thoracic cavity.

Lunula (LOO-noo-la) The moon-shaped white area at the base of a nail.

Luteinising (LOO-tē-in′-īz-ing) **hormone (LH)** A hormone secreted by the anterior pituitary that stimulates ovulation, stimulates progesterone secretion by the corpus luteum, and readies the mammary glands for milk secretion in females; stimulates testosterone secretion by the testes in males.

Lymph (LIMF) Fluid confined in lymphatic vessels and flowing through the lymphatic system until it is returned to the blood.

Lymph node An oval or bean-shaped structure located along lymphatic vessels.

Lymphatic (lim-FAT-ik) **capillary** Closed-ended microscopic lymphatic vessel that begins in spaces between cells and converges with other lymphatic capillaries to form lymphatic vessels.

Lymphatic system (lim-FAT-ik) A system consisting of a fluid called lymph, vessels called lymphatics that transport lymph, a number of organs containing lymphatic tissue (lymphocytes within a filtering tissue), and red bone marrow.

Lymphatic tissue A specialized form of reticular tissue that contains large numbers of lymphocytes.

Lymphatic vessel A large vessel that collects lymph from lymphatic capillaries and converges with other lymphatic vessels to form the thoracic and right lymphatic ducts.

Lymphocyte (LIM-fō-sīt) A type of white blood cell that helps carry out cell-mediated and antibody-mediated immune responses; found in blood and in lymphatic tissues.

Lysosome (LĪ-sō-sōm) An organelle in the cytoplasm of a cell, enclosed by a single membrane and containing powerful digestive enzymes.

Lysosyme (LĪ-sō-zīm) A bactericidal enzyme found in tears, saliva, and perspiration.

M

Macrophage (MAK-rō-fāj) Phagocytic cell derived from a monocyte; may be fixed or wandering.

Macula (MAK-ū-la) A discolored spot or a colored area. A small, thickened region on the wall of the utricle and saccule that contains receptors for static equilibrium.

Macula lutea (LOO-tē-a) The yellow spot in the center of the retina.

Major histocompatibility (MHC) antigens Surface proteins on white blood cells and other nucleated cells that are unique for each person (except for identical siblings); used to type tissues and help prevent rejection of transplanted tissues. Also known as **human leukocyte antigens (HLA).**

Malignant (ma-LIG-nant) Referring to diseases that tend to become worse and cause death, especially the invasion and spreading of cancer.

Mammary (MAM-ar-ē) **gland** Modified sudoriferous (sweat) gland of the female that produces milk for the nourishment of the young.

Mammillary (MAM-i-ler-ē) **bodies** Two small rounded bodies on the inferior aspect of the hypothalamus that are involved in reflexes related to the sense of smell.

Marrow (MAR-ō) Soft, spongelike material in the cavities of bone. *Red bone marrow* produces blood cells; *yellow bone marrow* contains adipose tissue that stores triglycerides.

Mast cell A cell found in areolar connective tissue that releases histamine, a dilator of small blood vessels, during inflammation.

Mastication (mas′-ti-KĀ-shun) Chewing.

Mature follicle A large, fluid-filled follicle containing a secondary oocyte and surrounding granulosa cells that secrete oestrogens. Also called a **graafian** (GRAF-ē-an) **follicle.**

Meatus (mē-Ā-tus) A passage or opening, especially the external portion of a canal.

Mechanoreceptor (me-KAN-ō-rē-sep-tor) Sensory receptor that detects mechanical deformation of the receptor itself or adjacent cells; stimuli so detected include those related to touch, pressure, vibration, proprioception, hearing, equilibrium, and blood pressure.

Medial (MĒ-dē-al) Nearer the midline of the body or a structure.

Medial lemniscus (lem-NIS-kus) A white matter tract that originates in the gracile and cuneate nuclei of the medulla oblongata and extends to the thalamus on the same side; sensory axons in this tract conduct nerve impulses for the sensations of proprioception, fine touch, vibration, hearing, and equilibrium.

Median aperture (AP-er-choor) One of the three openings in the roof of the fourth ventricle through which cerebrospinal fluid enters the subarachnoid space of the brain and cord. Also called the **foramen of Magendie.**

Median plane A vertical plane dividing the body into right and left halves. Situated in the middle.

Mediastinum (mē′-dē-as-TĪ-num) The anatomical region on the thoracic cavity between the pleurae of the lungs that extends from the sternum to the vertebral column and from the first rib to the diaphragm.

Medulla (me-DOOL-la) An inner layer of an organ, such as the medulla of the kidneys.

Medulla oblongata (me-DOOL-la ob′-long-GA-ta) The most inferior part of the brain stem. Also termed the **medulla.**

Medullary (MED-ū-lar′-ē) **cavity** The space within the diaphysis of a bone that contains yellow bone marrow. Also called the **marrow cavity.**

Medullary rhythmicity (rith-MIS-i-tē) **area** The neurons of the respiratory center in the medulla oblongata that control the basic rhythm of respiration.

Meibomian gland *See* **Tarsal gland.**

Meiosis (mī-Ō-sis) A type of cell division that occurs during production of gametes, involving two successive nuclear divisions that result in cells with the haploid *(n)* number of chromosomes.

Meissner (MĪS-ner) **corpuscle** A sensory receptor for touch; found in dermal papillae, especially in the palms and soles. Also called a **corpuscle of touch.**

Melanin (MEL-a-nin) A dark black, brown, or yellow pigment found in some parts of the body such as the skin, hair, and pigmented layer of the retina.

Melanocyte (MEL-a-nō-sīt′) A pigmented cell, located between or beneath cells of the deepest layer of the epidermis, that synthesizes melanin.

Melanocyte-stimulating hormone (MSH) A hormone secreted by the anterior pituitary that stimulates the dispersion of melanin granules in melanocytes in amphibians; continued administration produces darkening of skin in humans.

Melatonin (mel-a-TŌN-in) A hormone secreted by the pineal gland that helps set the timing of the body's biological clock.

Membrane A thin, flexible sheet of tissue composed of an epithelial layer and an underlying connective tissue layer, as in an epithelial membrane, or of areolar connective tissue only, as in a synovial membrane.

Membranous labyrinth (mem-BRA-nus LAB-i-rinth) The part of the labyrinth of the internal ear that is located inside the bony labyrinth and separated from it by the perilymph; made up of the semicircular ducts, the saccule and utricle, and the cochlear duct.

Memory The ability to recall thoughts; commonly classifed as short-term (activated) and long-term.

Menarche (me-NAR-kē) The first menses (menstrual flow) and beginning of ovarian and uterine cycles.

Meninges (me-NIN-jēz) Three membranes covering the brain and spinal cord, called the dura mater, arachnoid mater, and pia mater. *Singular* is **meninx** (MEN-inks).

Menopause (MEN-ō-pawz) The termination of the menstrual cycles.

Menstruation (men′-stroo-Ā-shun) Periodic discharge of blood, tissue fluid, mucus, and epithelial cells that usually lasts for 5 days; caused by a sudden reduction in oestrogens and progesterone. Also called the **menstrual phase** or **menses.**

Merkel (MER-kel) **cell** Type of cell in the epidermis of hairless skin that makes contact with a Merkel (tactile) disc, which functions in touch.

Merkel disc Saucer-shaped free nerve endings that make contact with Merkel cells in the epidermis and function as touch receptors. Also called a **tactile disc.**

Merocrine (MER-ō-krin) **gland** Gland made up of secretory cells that remain intact throughout the process of formation and discharge of the secretory product, as in the salivary and pancreatic glands.

Mesenchyme (MEZ-en-kīm) An embryonic connective tissue from which all other connective tissues arise.

Mesentery (MEZ-en-ter′-ē) A fold of peritoneum attaching the small intestine to the posterior abdominal wall.

Mesocolon (mez′-ō-KŌ-lon) A fold of peritoneum attaching the colon to the posterior abdominal wall.

Mesoderm The middle primary germ layer that gives rise to connective tissues, blood and blood vessels, and muscles.

Mesothelium (mez′-ō-THĒ-lē-um) The layer of simple squamous epithelium that lines serous membranes.

Mesovarium (mez′-ō-VAR-ē-um) A short fold of peritoneum that attaches an ovary to the broad ligament of the uterus.

Metabolism (me-TAB-ō-lizm) All the biochemical reactions that occur within an organism, including the synthetic (anabolic) reactions and decomposition (catabolic) reactions.

Metacarpus (met′-a-KAR-pus) A collective term for the five bones that make up the palm.

Metaphase (MET-a-fāz) The second stage of mitosis, in which chromatid pairs line up on the metaphase plate of the cell.

Metaphysis (me-TAF-i-sis) Region of a long bone between the diaphysis and epiphysis that contains the epiphyseal plate in a growing bone.

Metarteriole (met′-ar-TĒ-rē-ōl) A blood vessel that emerges from an arteriole, traverses a capillary network, and empties into a venule.

Metastasis (me-TAS-ta-sis) The spread of cancer to surrounding tissues (local) or to other body sites (distant).

Metatarsus (met′-a-TAR-sus) A collective term for the five bones located in the foot between the tarsals and the phalanges.

Microfilament (mī-krō-FIL-a-ment) Rodlike protein filament about 6 nm in diameter; constitutes contractile units in muscle fibers (cells) and provides support, shape, and movement in nonmuscle cells.

Microglia (mī-KROG-lē-a) Neuroglial cells that carry on phagocytosis.

Microtubule (mī-krō-TOO-būl) Cylindrical protein filament, from 18 to 30 nm in diameter, consisting of the protein tubulin; provides support, structure, and transportation.

Microvilli (mī-krō-VIL-ī) Microscopic, fingerlike projections of the plasma membranes of cells that increase surface area for absorption, especially in the small intestine and proximal convoluted tubules of the kidneys.

Micturition (mik′-choo-RISH-un) The act of expelling urine from the urinary bladder. Also called **urination** (ū-ri-NĀ-shun).

Midbrain The part of the brain between the pons and the diencephalon. Also called the **mesencephalon** (mes′-en-SEF-a-lon).

Middle ear A small, epithelial-lined cavity hollowed out of the temporal bone, separated from the external ear by the eardrum and from the internal ear by a thin bony partition containing the oval and round windows; extending across the middle ear are the three auditory ossicles. Also called the **tympanic** (tim-PAN-ik) **cavity.**

Midline An imaginary vertical line that divides the body into equal left and right sides.

Midsagittal plane A vertical plane through the midline of the body that divides the body or organs into *equal* right and left sides. Also called a **median plane.**

Mineralocorticoids (min′-er-al-ō-KOR-ti-koyds) A group of hormones of the adrenal cortex that help regulate sodium and potassium balance.

Mitochondrion (mī-tō-KON-drē-on) A double-membraned organelle that plays a central role in the production of ATP; known as the "powerhouse" of the cell. *Plural* is **mitochondria.**

Mitosis (mī-TŌ-sis) The orderly division of the nucleus of a cell that ensures that each new nucleus has the same number and kind of chromosomes as the original nucleus. The process includes the replication of chromosomes and the distribution of the two sets of chromosomes into two separate and equal nuclei.

Mitotic spindle Collective term for a football-shaped assembly of microtubules (nonkinetochore, kinetochore, and aster) that is responsible for the movement of chromosomes during cell division.

Modality (mō-DAL-i-tē) Any of the specific sensory entities, such as vision, smell, taste, or touch.

Modiolus (mō-DĪ-ō′-lus) The central pillar or column of the cochlea.

Molecule (mol′-e-KŪL) A substance composed of two or more atoms chemically combined.

Monocyte (MON-ō-sīt′) The largest type of white blood cell, characterized by agranular cytoplasm.

Monounsaturated fat A fatty acid that contains one double covalent bond between its carbon atoms; it is not completely saturated with hydrogen atoms. Plentiful in triglycerides of olive and peanut oils.

Mons pubis (MONZ PŪ-bis) The rounded, fatty prominence over the pubic symphysis, covered by coarse pubic hair.

Morula (MOR-ū-la) A solid sphere of cells produced by successive cleavages of a fertilized ovum about 4 days after fertilisation.

Motor area The region of the cerebral cortex that governs muscular movement, particularly the precentral gyrus of the frontal lobe.

Motor end plate Region of the sarcolemma of a muscle fiber (cell) that includes acetylcholine (ACh) receptors, which bind ACh released by synaptic end bulbs of somatic motor neurons.

Motor neurons (NOO-ronz) Neurons that conduct impulses from the brain toward the spinal cord or out of the brain and spinal cord into cranial or spinal nerves to effectors that may be either muscles or glands. Also called **efferent neurons.**

Motor unit A motor neuron together with the muscle fibers (cells) it stimulates.

Mucosa-associated lymphatic tissue (MALT) Lymphatic nodules scattered throughout the lamina propria (connective tissue) of mucous membranes lining the gastrointestinal tract, respiratory airways, urinary tract, and reproductive tract.

Mucous (MŪ-kus) **cell** A unicellular gland that secretes mucus. Two types are mucous neck cells and surface mucous cells in the stomach.

Mucous membrane A membrane that lines a body cavity that opens to the exterior. Also called the **mucosa** (mū-KŌ-sa).

Mucus The thick fluid secretion of goblet cells, mucous cells, mucous glands, and mucous membranes.

Muscarinic (mus'-ka-RIN-ik) **receptor** Receptor for the neurotransmitter acetylcholine found on all effectors innervated by parasympathetic postganglionic axons and on sweat glands innervated by cholinergic sympathetic postganglionic axons; so named because muscarine activates these receptors but does not activate nicotinic receptors for acetylcholine.

Muscle An organ composed of one of three types of muscle tissue (skeletal, cardiac, or smooth), specialized for contraction to produce voluntary or involuntary movement of parts of the body.

Muscle action potential A stimulating impulse that propagates along the sarcolemma and transverse tubules; in skeletal muscle, it is generated by acetylcholine, which increases the permeability of the sarcolemma to cations, especially sodium ions (Na^+).

Muscle fatigue (fa-TĒG) Inability of a muscle to maintain its strength of contraction or tension; may be related to insufficient oxygen, depletion of glycogen, and/or lactic acid buildup.

Muscle spindle An encapsulated proprioceptor in a skeletal muscle, consisting of specialized intrafusal muscle fibers and nerve endings; stimulated by changes in length or tension of muscle fibers.

Muscle strain Tearing of skeletal muscle fibers or tendon. Also called a **muscle pull** or **muscle tear.**

Muscle tone A sustained, partial contraction of portions of a skeletal or smooth muscle in response to activation of stretch receptors or a baseline level of action potentials in the innervating motor neurons.

Muscular dystrophies (DIS-trō-fēz) Inherited muscle-destroying diseases, characterized by degeneration of muscle fibers (cells), which causes progressive atrophy of the skeletal muscle.

Muscularis (MUS-kū-la'-ris) A muscular layer (coat or tunic) of an organ.

Muscularis mucosae (mū-KŌ-sē) A thin layer of smooth muscle fibers that underlie the lamina propria of the mucosa of the gastrointestinal tract.

Muscular system Usually refers to the approximately 100 voluntary muscles of the body that are composed of skeletal muscle tissue.

Muscular system Usually refers to the voluntary muscles of the body that are composed of skeletal muscle tissue.

Muscular tissue A tissue specialized to produce motion in response to muscle action potentials by its qualities of contractility, extensibility, elasticity, and excitability; types include skeletal, cardiac, and smooth.

Musculoskeletal (mus'-kyū-lō-skel-ETAL) **system** An integrated body system consisting of bones, joints, and muscles.

Mutation (mū-TĀ-shun) Any change in the sequence of bases in a DNA molecule resulting in a permanent alteration in some inheritable trait.

Myasthenia (mī-as-THĒ-nē-a) **gravis** Weakness and fatigue of skeletal muscles caused by antibodies directed against acetylcholine receptors.

Myelin (MĪ-e-lin) **sheath** Multilayered lipid and protein covering, formed by Schwann cells and oligodendrocytes, around axons of many peripheral and central nervous system neurons.

Myenteric (mī-en-TER-ik) **plexus** A network of autonomic axons and postganglionic cell bodies located in the muscularis of the gastrointestinal tract. Also called the **plexus of Auerbach** (OW-er-bak).

Myocardial infarction (mī'-ō-KAR-dē-al in-FARK-shun) **(MI)** Gross necrosis of myocardial tissue due to interrupted blood supply. Also called a **heart attack.**

Myocardium (mī'-ō-KAR-dē-um) The middle layer of the heart wall, made up of cardiac muscle tissue, lying between the epicardium and the endocardium and constituting the bulk of the heart.

Myofibril (mī-ō-FĪ-bril) A threadlike structure, extending longitudinally through a muscle fiber (cell), consisting mainly of thick filaments (myosin) and thin filaments (actin, troponin, and tropomyosin).

Myoglobin (mī-ō-GLŌB-in) The oxygen-binding, iron-containing protein present in the sarcoplasm of muscle fibers (cells); contributes the red color to muscle.

Myogram (MĪ-ō-gram) The record or tracing produced by a myograph, an apparatus that measures and records the force of muscular contractions.

Myology (mī-OL-ō-jē) The study of muscles.

Myometrium (mī'-ō-MĒ-trē-um) The smooth muscle layer of the uterus.

Myopathy (mī-OP-a-thē) Any abnormal condition or disease of muscle tissue.

Myopia (mī-Ō-pē-a) Defect in vision in which objects can be seen distinctly only when very close to the eyes; nearsightedness.

Myosin (MĪ-ō-sin) The contractile protein that makes up the thick filaments of muscle fibers.

Myotome (MĪ-ō-tōm) A group of muscles innervated by the motor neurons of a single spinal segment. In an embryo, the portion of a somite that develops into some skeletal muscles.

N

Nail A hard plate, composed largely of keratin, that develops from the epidermis of the skin to form a protective covering on the dorsal surface of the distal phalanges of the fingers and toes.

Nail matrix (MĀ-triks) The part of the nail beneath the body and root from which the nail is produced.

Nasal (NĀ-zal) **cavity** A mucosa-lined cavity on either side of the nasal septum that opens onto the face at the external nares and into the nasopharynx at the internal nares.

Nasal septum (SEP-tum) A vertical partition composed of bone (perpendicular plate of ethmoid and vomer) and cartilage, covered with a mucous membrane, separating the nasal cavity into left and right sides.

Nasolacrimal (nā'-zō-LAK-ri-mal) **duct** A canal that transports the lacrimal secretion (tears) from the nasolacrimal sac into the nose.

Nasopharynx (nā'-zō-FAR-inks) The superior portion of the pharynx, lying posterior to the nose and extending inferiorly to the soft palate.

Neck The part of the body connecting the head and the trunk. A constricted portion of an organ, such as the neck of the femur or uterus.

Necrosis (ne-KRŌ-sis) A pathological type of cell death that results from disease, injury, or lack of blood supply in which many adjacent cells swell, burst, and spill their contents into the interstitial fluid, triggering an inflammatory response.

Negative feedback system A feedback cycle that reverses a change in a controlled condition.

Neonatal (nē-ō-NĀ-tal) Pertaining to the first 4 weeks after birth.

Neoplasm (NĒ-ō-plazm) A new growth that may be benign or malignant.

Nephron (NEF-ron) The functional unit of the kidney.

Nerve A cordlike bundle of neuronal axons and/or dendrites and associated connective tissue coursing together outside the central nervous system.

Nerve fiber General term for any process (axon or dendrite) projecting from the cell body of a neuron.

Nerve impulse A wave of depolarization and repolarization that self-propagates along the plasma membrane of a neuron; also called a **nerve action potential.**

Nervous system A network of billions of neurons and even more neuroglia that is organized into two main divisions. Central nervous system (brain and spinal cord) and peripheral nervous system (nerves, ganglia, enteric plexuses, and sensory receptors outside the central nervous system).

Nervous tissue Tissue containing neurons that initiate and conduct nerve impulses to coordinate homoeostasis, and neuroglia that provide support and nourishment to neurons.

Neuralgia (noo-RAL-jē-a) Attacks of pain along the entire course or branch of a peripheral sensory nerve.

Neural plate A thickening of ectoderm, induced by the notochord, that forms early in the third week of development and represents the beginning of the development of the nervous system.

Neural tube defect (NTD) A developmental abnormality in which the neural tube does not close properly. Examples are spina bifida and anencephaly.

Neuritis (noo-RĪ-tis) Inflammation of one or more nerves.

Neurofibral node *See* **Node of Ranvier.**

Neuroglia (noo-RŌG-lē-a) Cells of the nervous system that perform various supportive functions. The neuroglia of the central nervous system are the astrocytes, oligodendrocytes, microglia, and ependymal cells; neuroglia of the peripheral nervous system include Schwann cells and satellite cells. Also called **glial** (GLĒ-al) **cells.**

Neurohypophyseal (noo'-rō-hī'-pō-FIZ-ē-al) **bud** An outgrowth of ectoderm located on the floor of the hypothalamus that gives rise to the posterior pituitary.

Neurolemma (noo-rō-LEM-ma) The peripheral, nucleated cytoplasmic layer of the Schwann cell. Also called **sheath of Schwann** (SCHWON).

Neurology (noo-ROL-ō-jē) The study of the normal functioning and disorders of the nervous system.

Neuromuscular (noo-rō-MUS-kū-lar) **junction (NMJ)** A synapse between the axon terminals of a motor neuron and the sarcolemma of a muscle fiber (cell).

Neuron (NOO-ron) A nerve cell, consisting of a cell body, dendrites, and an axon.

Neurophysiology (NOOR-ō-fiz-ē-ol′-ō-jē) Study of the functional properties of nerves.

Neurosecretory (noo-rō-SĒK-re-tō-rē) **cell** A neuron that secretes a hypothalamic releasing hormone or inhibiting hormone into blood capillaries of the hypothalmus; a neuron that secretes oxytocin or antidiuretic hormone into blood capillaries of the posterior pituitary.

Neurotransmitter (noo′-rō-trans′-MIT-er) One of a variety of molecules within axon terminals that are released into the synaptic cleft in response to a nerve impulse and that change the membrane potential of the postsynaptic neuron.

Neurulation (noor-oo-LĀ-shun) The process by which the neural plate, neural folds, and neural tube develop.

Neutrophil (NOO-trō-fil) A type of white blood cell characterized by granules that stain pale lilac with a combination of acidic and basic dyes.

Nicotinic (nik′-ō-TIN-ik) **receptor** Receptor for the neurotransmitter acetylcholine found on both sympathetic and parasympathetic postganglionic neurons and on skeletal muscle in the motor end plate; so named because nicotine activates these receptors but does not activate muscarinic receptors for acetylcholine.

Nipple A pigmented, wrinkled projection on the surface of the breast that is the location of the openings of the lactiferous ducts for milk release.

Nociceptor (nō′-sē-SEP-tor) A free (naked) nerve ending that detects painful stimuli.

Node of Ranvier (RON-vē-ā) A space along a myelinated axon between the individual Schwann cells that form the myelin sheath and the neurolemma. Also called a **neurofibral node.**

Norepinephrine (nor′-ep-ē-NEF-rin) **(NE)** A hormone secreted by the adrenal medulla that produces actions similar to those that result from sympathetic stimulation. Also called **noradrenaline** (nor-a-DREN-a-lin).

Notochord (NŌ-tō-cord) A flexible rod of mesodermal tissue that lies where the future vertebral column will develop and plays a role in induction.

Nucleic (noo-KLĒ-ik) **acid** An organic compound that is a long polymer of nucleotides, with each nucleotide containing a pentose sugar, a phosphate group, and one of four possible nitrogenous bases (adenine, cytosine, guanine, and thymine or uracil).

Nucleolus (noo′-KLĒ-ō-lus) Spherical body within a cell nucleus composed of protein, DNA, and

RNA that is the site of the assembly of small and large ribosomal subunits. *Plural* is **nucleoli.**

Nucleosome (NOO-klē-ō-sōm) Structural subunit of a chromosome consisting of histones and DNA.

Nucleus (NOO-klē-us) A spherical or oval organelle of a cell that contains the hereditary factors of the cell, called genes. A cluster of unmyelinated nerve cell bodies in the central nervous system. The central part of an atom made up of protons and neutrons.

Nucleus pulposus (pul-PŌ-sus) A soft, pulpy, highly elastic substance in the center of an intervertebral disc; a remnant of the notochord.

Nutrient A chemical substance in food that provides energy, forms new body components, or assists in various body functions.

O

Obesity (ō-BĒS-i-tē) Body weight more than 20% above a desirable standard due to excessive accumulation of fat.

Oblique (ō-BLĒK) **plane** A plane that passes through the body or an organ at an angle between the transverse plane and either the midsagittal, parasagittal, or frontal plane.

Obstetrics (ob-STET-riks) The specialized branch of medicine that deals with pregnancy, labour, and the period of time immediately after delivery (about 6 weeks).

Oedema (usu., also edema) An abnormal accumulation of interstitial fluid.

Oesophagus (e-SOF-a-gus) The hollow muscular tube that connects the pharynx and the stomach.

Oestrogens (ES-tro-jenz) Feminizing sex hormones produced by the ovaries; govern development of oocytes, maintenance of female reproductive structures, and appearance of secondary sex characteristics; also affect fluid and electrolyte balance, and protein anabolism. Examples are β-estradiol, estrone, and estriol.

Olfaction (ōl-FAK-shun) The sense of smell.

Olfactory (ōl-FAK-tō-rē) Pertaining to smell.

Olfactory bulb A mass of grey matter containing cell bodies of neurons that form synapses with neurons of the olfactory (I) nerve, lying inferior to the frontal lobe of the cerebrum on either side of the crista galli of the ethmoid bone.

Olfactory receptor A bipolar neuron with its cell body lying between supporting cells located in the mucous membrane lining the superior portion of each nasal cavity; transduces odors into neural signals.

Olfactory tract A bundle of axons that extends from the olfactory bulb posteriorly to olfactory regions of the cerebral cortex.

Oligodendrocyte (OL-i-gō-den′-drō-sīt) A neuroglial cell that supports neurons and produces a myelin sheath around axons of neurons of the central nervous system.

Oliguria (ol′-i-GŪ-rē-a) Daily urinary output usually less than 250 mL.

Olive A prominent oval mass on each lateral surface of the superior part of the medulla oblongata.

Oncogene (ON-kō-jēn) Cancer-causing gene; it derives from a normal gene, termed a proto-

oncogene, that encodes proteins involved in cell growth or cell regulation but has the ability to transform a normal cell into a cancerous cell when it is mutated or inappropriately activated. One example is *p53*.

Oncology (on-KOL-ō-jē) The study of tumors.

Oogenesis (ō-ō-JEN-e-sis) Formation and development of female gametes (oocytes).

Oophorectomy (ō′-of-ō-REK-tō-me) Surgical removal of the ovaries.

Ophthalmic (of-THAL-mik) Pertaining to the eye.

Ophthalmologist (of′-thal-MOL-ō-jist) A physician who specializes in the diagnosis and treatment of eye disorders using drugs, surgery, and corrective lenses.

Ophthalmology (of-thal-MOL-ō-jē) The study of the structure, function, and diseases of the eye.

Optic (OP-tik) Refers to the eye, vision, or properties of light.

Optic chiasm (kī-AZM) A crossing point of the two branches of the optic (II) nerve, anterior to the pituitary gland. Also called **optic chiasma.**

Optic disc A small area of the retina containing openings through which the axons of the ganglion cells emerge as the optic (II) nerve. Also called the **blind spot.**

Optic tract A bundle of axons that carry nerve impulses from the retina of the eye between the optic chiasm and the thalamus.

Ora serrata (Ō-ra ser-RĀ-ta) The irregular margin of the retina lying internal and slightly posterior to the junction of the choroid and ciliary body.

Orbit (OR-bit) The bony, pyramidal-shaped cavity of the skull that holds the eyeball.

Organ A structure composed of two or more different kinds of tissues with a specific function and usually a recognizable shape.

Organelle (or-ga-NEL) A permanent structure within a cell with characteristic morphology that is specialized to serve a specific function in cellular activities.

Organism (OR-ga-nizm) A total living form; one individual.

Organogenesis (or′-ga-nō-JEN-e-sis) The formation of body organs and systems. By the end of the eighth week of development, all major body systems have begun to develop.

Orifice (OR-i-fis) Any aperture or opening.

Origin (OR-i-jin) The attachment of a muscle tendon to a stationary bone or the end opposite the insertion.

Oropharynx (or′-ō-FAR-inks) The intermediate portion of the pharynx, lying posterior to the mouth and extending from the soft palate to the hyoid bone.

Orthopaedics (or′-thō-PĒ-diks) The branch of medicine that deals with the preservation and restoration of the skeletal system, articulations, and associated structures.

Osmoreceptor (oz′-mō-rē-CEP-tor) Receptor in the hypothalamus that is sensitive to changes in blood osmolarity and, in response to high osmolarity (low water concentration), stimulates synthesis and release of antidiuretic hormone (ADH).

Osmosis (oz-MŌ-sis) The net movement of water molecules through a selectively permeable membrane from an area of higher water concentration to an area of lower water concentration until equilibrium is reached.

Osseous (OS-ē-us) Bony.

Ossicle (OS-si-kul) One of the small bones of the middle ear (malleus, incus, stapes).

Ossification (os′-i-fi-KĀ-shun) Formation of bone. Also called **osteogenesis.**

Ossification (os′-i-fi-KĀ-shun) **center** An area in the cartilage model of a future bone where the cartilage cells hypertrophy, secrete enzymes that calcify their extracellular matrix, and die, and the area they occupied is invaded by osteoblasts that then lay down bone.

Osteoblast (OS-tē-ō-blast′) Cell formed from an osteogenic cell that participates in bone formation by secreting some organic components and inorganic salts.

Osteoclast (OS-tē-ō-klast′) A large, multinuclear cell that resorbs (destroys) bone matrix.

Osteocyte (OS-tē-ō-sīt′) A mature bone cell that maintains the daily activities of bone tissue.

Osteogenic (os′-tē-ō-JEN-ik) **cell** Stem cell derived from mesenchyme that has mitotic potential and the ability to differentiate into an osteoblast.

Osteogenic layer The inner layer of the periosteum that contains cells responsible for forming new bone during growth and repair.

Osteology (os-tē-OL-ō-jē) The study of bones.

Osteon (OS-tē-on) The basic unit of structure in adult compact bone, consisting of a central (haversian) ca nal with its concentrically arranged lamellae, lacunae, osteocytes, and canaliculi. Also called a **haversian** (ha-VER-shun) **system.**

Osteoporosis (os′-tē-ō-pō-RŌ-sis) Age-related disorder characterized by decreased bone mass and increased susceptibility to fractures, often as a result of decreased levels of oestrogens.

Otic (Ō-tik) Pertaining to the ear.

Otolith (Ō-tō-lith) A particle of calcium carbonate embedded in the otolithic membrane that functions in maintaining static equilibrium.

Otolithic (ō-tō-LITH-ik) **membrane** Thick, gelatinous, glycoprotein layer located directly over hair cells of the macula in the saccule and utricle of the internal ear.

Otorhinolaryngology (ō-tō-rī′-nō-lar-in-GOL-ō-jē) The branch of medicine that deals with the diagnosis and treatment of diseases of the ears, nose, and throat.

Oval window A small, membrane-covered opening between the middle ear and inner ear into which the footplate of the stapes fits.

Ovarian (ō-VAR-ē-an) **cycle** A monthly series of events in the ovary associated with the maturation of a secondary oocyte.

Ovarian follicle (FOL-i-kul) A general name for oocytes (immature ova) in any stage of development, along with their surrounding epithelial cells.

Ovarian ligament (LIG-a-ment) A rounded cord of connective tissue that attaches the ovary to the uterus.

Ovary (Ō-var-ē) Female gonad that produces oocytes and the estrogen, progesterone, inhibin, and relaxin hormones.

Ovulation (ov′-ū-LĀ-shun) The rupture of a mature ovarian (graafian) follicle with discharge of a secondary oocyte into the pelvic cavity.

Ovum (Ō-vum) The female reproductive or germ cell; an egg cell; arises through completion of meiosis in a secondary oocyte after penetration by a sperm.

Oxyhemoglobin (ok′-sē-HĒ-mō-glō-bin) **(Hb—O₂)** Haemoglobin combined with oxygen.

Oxytocin (ok′-sē-TŌ-sin) **(OT)** A hormone secreted by neurosecretory cells in the paraventricular and supraoptic nuclei of the hypothalamus that stimulates contraction of smooth muscle in the pregnant uterus and myoepithelial cells around the ducts of mammary glands.

P

P wave The deflection wave of an electrocardiogram that signifies atrial depolarization.

Pacinian corpuscle (pa-SIN-ē-an) Oval-shaped pressure receptor located in the dermis or subcutaneous tissue and consisting of concentric layers of a connective tissue wrapped around the dendrites of a sensory neuron. Also called a **lamellated corpuscle.**

Palate (PAL-at) The horizontal structure separating the oral and the nasal cavities; the roof of the mouth.

Palpate (PAL-pāt) To examine by touch; to feel.

Pancreas (PAN-krē-as) A soft, oblong organ lying along the greater curvature of the stomach and connected by a duct to the duodenum. It is both an exocrine gland (secreting pancreatic juice) and an endocrine gland (secreting insulin, glucagon, somatostatin, and pancreatic polypeptide).

Pancreatic (pan′-krē-AT-ik) **duct** A single large tube that unites with the common bile duct from the liver and gallbladder and drains pancreatic juice into the duodenum at the hepatopancreatic ampulla (ampulla of Vater). Also called the **duct of Wirsung** (VĒR-sung).

Pancreatic islet (Ī-let) Cluster of endocrine gland cells in the pancreas that secretes insulin, glucagon, somatostatin, and pancreatic polypeptide. Also called an **islet of Langerhans** (LAHNG-er-hanz).

Papanicolaou (pa-pa-NI-kō-lō) **test** A cytological staining test for the detection and diagnosis of premalignant and malignant conditions of the female genital tract. Cells scraped from the epithelium of the cervix of the uterus are examined microscopically. Also called a **Pap test** or **Pap smear.**

Papilla (pa-PIL-a) A small nipple-shaped projection or elevation.

Paralysis (pa-RAL-a-sis) Loss or impairment of motor function due to a lesion of nervous or muscular origin.

Paranasal sinus (par′-a-NĀ-zal SĪ-nus) A mucuslined air cavity in a skull bone that communicates with the nasal cavity. Paranasal sinuses are located in the frontal, maxillary, ethmoid, and sphenoid bones.

Paraplegia (par-a-PLĒ-jē-a) Paralysis of both lower limbs.

Parasagittal plane (par-a-SAJ-i-tal) A vertical plane that does not pass through the midline and that divides the body or organs into *unequal* left and right portions.

Parasympathetic (par′-a-sim-pa-THET-ik) **division** One of the two subdivisions of the autonomic nervous system, having cell bodies of preganglionic neurons in nuclei in the brain stem and in the lateral gray horn of the sacral portion of the spinal cord; primarily concerned with activities that conserve and restore body energy.

Parathyroid (par′-a-THĪ-royd) **gland** One of usually four small endocrine glands embedded in the posterior surfaces of the lateral lobes of the thyroid gland.

Parathyroid hormone (PTH) A hormone secreted by the chief (principal) cells of the parathyroid glands that increases blood calcium level and decreases blood phosphate level. Also called **parathormone.**

Paraurethral (par′-a-ū-RĒ-thral) **gland** Gland embedded in the wall of the urethra whose duct opens on either side of the urethral orifice and secretes mucus. Also called **Skene's** (SKĒNZ) **gland.**

Parenchyma (pa-RENG-ki-ma) The functional parts of any organ, as opposed to tissue that forms its stroma or framework.

Parietal (pa-RĪ-e-tal) Pertaining to or forming the outer wall of a body cavity.

Parietal cell A type of secretory cell in gastric glands that produces hydrochloric acid and intrinsic factor. Also called an **oxyntic cell.**

Parietal pleura (PLOO-ra) The outer layer of the serous pleural membrane that encloses and protects the lungs; the layer that is attached to the wall of the pleural cavity.

Parkinson's disease (PD) Progressive degeneration of the basal nuclei and substantia nigra of the cerebrum resulting in decreased production of dopamine (DA) that leads to tremor, slowing of voluntary movements, and muscle weakness.

Parotid (pa-ROT-id) **gland** One of the paired salivary glands located inferior and anterior to the ears and connected to the oral cavity via a duct (Stensen's) that opens into the inside of the cheek opposite the maxillary (upper) second molar tooth.

Pars intermedia A small avascular zone between the anterior and posterior pituitary glands.

Parturition (par-toor-ISH-un) Act of giving birth to young; childbirth, delivery.

Patent (PĀ-tent) **ductus arteriosus (PDA)** A congenital heart defect in which the ductus arteriosus remains open. As a result, aortic blood flows into the lower-pressure pulmonary trunk, increasing pulmonary trunk pressure and overworking both ventricles.

Pathogen (PATH-ō-jen) A disease-producing microbe.

Pathological (path′-ō-LOJ-i-kal) **anatomy** The study of structural changes caused by disease.

Pathophysiology (PATH-ō-fez-ē-ol-ō-jē) Study of functional changes associated with disease and aging.

Pectinate (PEK-ti-nāt) **muscles** Projecting muscle bundles of the anterior atrial walls and the lining of the auricles.

Pectoral (PEK-tō-ral) Pertaining to the chest or breast.

Pedicel (PED-i-sel) Footlike structure, as on podocytes of a glomerulus.

Pelvic (PEL-vik) **cavity** Inferior portion of the abdominopelvic cavity that contains the urinary bladder, sigmoid colon, rectum, and internal female and male reproductive structures.

Pelvic splanchnic (PEL-vik SPLANGK-nik) **nerves** Consist of preganglionic parasympathetic axons from the levels of S2, S3, and S4 that supply the urinary bladder, reproductive organs, and the descending and sigmoid colon and rectum.

Pelvis The basinlike structure formed by the two hip bones, the sacrum, and the coccyx. The expanded, proximal portion of the ureter, lying within the kidney and into which the major calyces open.

Penis (PĒ-nis) The organ of urination and copulation in males; used to deposit semen into the female vagina.

Pepsin Protein-digesting enzyme secreted by chief cells of the stomach in the inactive form pepsinogen, which is converted to active pepsin by hydrochloric acid.

Peptic ulcer An ulcer that develops in areas of the gastrointestinal tract exposed to hydrochloric acid; classified as a gastric ulcer if in the lesser curvature of the stomach and as a duodenal ulcer if in the first part of the duodenum.

Percussion (pur-KUSH-un) The act of striking (percussing) an underlying part of the body with short, sharp taps as an aid in diagnosing the part by the quality of the sound produced.

Perforating canal A minute passageway by means of which blood vessels and nerves from the periosteum penetrate into compact bone. Also called **Volkmann's** (FŌLK-mans) **canal.**

Pericardial (per'-i-KAR-dē-al) **cavity** Small potential space between the visceral and parietal layers of the serous pericardium that contains pericardial fluid.

Pericardium (per'-i-KAR-dē-um) A loose-fitting membrane that encloses the heart, consisting of a superficial fibrous layer and a deep serous layer.

Perichondrium (per'-i-KON-drē-um) A covering of dense irregular connective tissue that surrounds the surface of most cartilage.

Perilymph (PER-i-limf) The fluid contained between the bony and membranous labyrinths of the inner ear.

Perimetrium (per'-i-MĒ-trē-um) The serosa of the uterus.

Perimysium (per-i-MĪZ-ē-um) Invagination of the epimysium that divides muscles into bundles.

Perineum (per'-i-NĒ-um) The pelvic floor; the space between the anus and the scrotum in the male and between the anus and the vulva in the female.

Perineurium (per'-i-NOO-rē-um) Connective tissue wrapping around fascicles in a nerve.

Periodontal (per'-ē-ō-DON-tal) **disease** A collective term for conditions characterized by degeneration of gingivae, alveolar bone, periodontal ligament, and cementum.

Periodontal ligament The periosteum lining the alveoli (sockets) for the teeth in the alveolar processes of the mandible and maxillae.

Periosteum (per'-ē-OS-tē-um) The covering of a bone that consists of connective tissue, osteogenic cells, and osteoblasts; is essential for bone growth, repair, and nutrition.

Peripheral (pe-RIF-er-al) Located on the outer part or a surface of the body.

Peripheral nervous system (PNS) The part of the nervous system that lies outside the central nervous system, consisting of nerves and ganglia.

Peristalsis (per'-i-STAL-sis) Successive muscular contractions along the wall of a hollow muscular structure.

Peritoneum (per'-i-tō-NĒ-um) The largest serous membrane of the body that lines the abdominal cavity and covers the viscera within it.

Peritonitis (per'-i-tō-NĪ-tis) Inflammation of the peritoneum.

Peroxisome (pe-ROKS-i-sōm) Organelle similar in structure to a lysosome that contains enzymes that use molecular oxygen to oxidize various organic compounds; such reactions produce hydrogen peroxide; abundant in liver cells.

Perspiration Sweat; produced by sudoriferous (sweat) glands and containing water, salts, urea, uric acid, amino acids, ammonia, sugar, lactic acid, and ascorbic acid. Helps maintain body temperature and eliminate wastes.

Peyer's patches *See* **Aggregated lymphatic follicles.**

pH A measure of the concentration of hydrogen ions (H^+) in a solution. The **pH scale** extends from 0 to 14, with a value of 7 expressing neutrality, values lower than 7 expressing increasing acidity, and values higher than 7 expressing increasing alkalinity.

Phagocytosis (fag'-ō-sī-TŌ-sis) The process by which phagocytes ingest and destroy microbes, cell debris, and other foreign matter.

Phalanx (FĀ-lanks) The bone of a finger or toe. *Plural* is **phalanges** (fa-LAN-jēz).

Pharmacology (far'-ma-KOL-ō-jē) The science of the effects and uses of drugs in the treatment of disease.

Pharynx (FAR-inks) The throat; a tube that starts at the internal nares and runs partway down the neck, where it opens into the oesophagus posteriorly and the larynx anteriorly.

Phenotype (FĒ-nō-tīp) The observable expression of genotype; physical characteristics of an organism determined by genetic makeup and influenced by interaction between genes and internal and external environmental factors.

Phlebitis (fle-BĪ-tis) Inflammation of a vein, usually in a lower limb.

Photopigment A substance that can absorb light and undergo structural changes that can lead to the development of a receptor potential. An example is rhodopsin. In the eye, also called **visual pigment.**

Photoreceptor Receptor that detects light shining on the retina of the eye.

Physiology (fiz'-ē-OL-ō-jē) Science that deals with the functions of an organism or its parts.

Pia mater (PĪ-a MĀ-ter *or* PĒ-a MA-ter) The innermost of the three meninges (coverings) of the brain and spinal cord.

Pineal (PĪN-ē-al) **gland** A cone-shaped gland located in the roof of the third ventricle that secretes melatonin. Also called the **epiphysis cerebri** (ē-PIF-i-sis se-RĒ-brē).

Pinealocyte (pin-ē-AL-ō-sīt) Secretory cell of the pineal gland that releases melatonin.

Pinna (PIN-na) The projecting part of the external ear composed of elastic cartilage and covered by skin and shaped like the flared end of a trumpet. Also called the **auricle** (OR-i-kul).

Pituicyte (pi-TOO-i-sīt) Supporting cell of the posterior pituitary.

Pituitary (pi-TOO-i-tār-ē) **gland** A small endocrine gland occupying the hypophyseal fossa of the sphenoid bone and attached to the hypothalamus by the infundibulum. Also called the **hypophysis** (hī-POF-i-sis).

Pivot joint A synovial joint in which a rounded, pointed, or conical surface of one bone articulates with a ring formed partly by another bone and partly by a ligament, as in the joint between the atlas and axis and between the proximal ends of the radius and ulna. Also called a **trochoid** (TRŌ-koyd) **joint.**

Placenta (pla-SEN-ta) The special structure through which the exchange of materials between fetal and maternal circulations occurs. Also called the **afterbirth.**

Plane joint A synovial joint having articulating surfaces that are usually flat, permitting only side-to-side and back-and-forth movements, as between carpal bones, tarsal bones, and the scapula and clavicle. Also called an **arthrodial** (ar-THRŌ-dē-al) **joint.**

Plantar flexion (PLAN-tar FLEK-shun) Bending the foot in the direction of the plantar surface (sole).

Plaque (PLAK) A layer of dense proteins on the inside of a plasma membrane in adherens junctions and desmosomes. A mass of bacterial cells, dextran (polysaccharide), and other debris that adheres to teeth (dental plaque). *See also* **Atherosclerotic plaque.**

Plasma (PLAZ-ma) The extracellular fluid found in blood vessels; blood minus the formed elements.

Plasma cell Cell that develops from a B cell (lymphocyte) and produces antibodies.

Plasma (cell) membrane Outer, limiting membrane that separates the cell's internal parts from extracellular fluid or the external environment.

Platelet (PLĀT-let) A fragment of cytoplasm enclosed in a cell membrane and lacking a nucleus; found in the circulating blood; plays a role in hemostasis. Also called a **thrombocyte** (THROM-bō-sīt).

Platelet plug Aggregation of platelets (thrombocytes) at a site where a blood vessel is damaged that helps stop or slow blood loss.

Pleura (PLOO-ra) The serous membrane that covers the lungs and lines the walls of the chest and the diaphragm.

Pleural cavity Small potential space between the visceral and parietal pleurae.

Plexus (PLEK-sus) A network of nerves, veins, or lymphatic vessels.

Plexus of Auerbach *See* **Myenteric plexus.**

Plexus of Meissner *See* **Submucosal plexus.**

Pluripotent (ploo-RI-pō-tent) **stem cell** Immature stem cell in red bone marrow that gives rise to precursors of all the different mature blood cells.

Pneumotaxic (noo-mō-TAK-sik) **area** A part of the respiratory center in the pons that continually sends inhibitory nerve impulses to the inspiratory area, limiting inhalation and facilitating exhalation.

Polycythaemia (pol′-ē-sī-THĒ-mē-a) Disorder characterized by an above-normal haematocrit (above 55%) in which hypertension, thrombosis, and haemorrhage can occur.

Polyunsaturated fat A fatty acid that contains more than one double covalent bond between its carbon atoms; abundant in triglycerides of corn oil, safflower oil, and cottonseed oil.

Polyuria (pol′-ē-Ū-rē-a) An excessive production of urine.

Pons (PONZ) The part of the brain stem that forms a "bridge" between the medulla oblongata and the midbrain, anterior to the cerebellum.

Portal system The circulation of blood from one capillary network into another through a vein.

Positive feedback system A feedback cycle that strengthens or reinforces a change in a controlled condition.

Postcentral gyrus Gyrus of cerebral cortex located immediately posterior to the central sulcus; contains the primary somatosensory area.

Posterior (pos-TĒR-ē-or) Nearer to or at the back of the body. Equivalent to **dorsal** in bipeds.

Posterior column–medial lemniscus pathway Sensory pathway that carries information related to proprioception, fine touch, two-point discrimination, pressure, and vibration. First-order neurons project from the spinal cord to the ipsilateral medulla in the posterior columns (gracile fasciculus and cuneate fasciculus). Second-order neurons project from the medulla to the contralateral thalamus in the medial lemniscus. Third-order neurons project from the thalamus to the somatosensory cortex (postcentral gyrus) on the same side.

Posterior pituitary Posterior lobe of the pituitary gland. Also called the **neurohypophysis** (noo-rō-hī-POF-i-sis).

Posterior root The structure composed of sensory axons lying between a spinal nerve and the dorsolateral aspect of the spinal cord. Also called the **dorsal (sensory) root.**

Posterior root ganglion (GANG-glē-on) A group of cell bodies of sensory neurons and their supporting cells located along the posterior root of a spinal nerve. Also called a **dorsal (sensory) root ganglion.**

Postganglionic neuron (pōst′-gang-lē-ON-ik NOO-ron) The second autonomic motor neuron in an autonomic pathway, having its cell body and dendrites located in an autonomic ganglion and its unmyelinated axon ending at cardiac muscle, smooth muscle, or a gland.

Postsynaptic (pōst-sin-AP-tik) **neuron** The nerve cell that is activated by the release of a neurotransmitter from another neuron and carries nerve impulses away from the synapse.

Pouch of Douglas *See* **Rectouterine pouch.**

Precapillary sphincter (SFINGK-ter) The distal-most muscle fiber (cell) at the metarteriole–capillary junction that regulates blood flow into capillaries.

Precentral gyrus Gyrus of cerebral cortex located immediately anterior to the central sulcus; contains the primary motor area.

Preganglionic (prē′-gang-lē-ON-ik) **neuron** The first autonomic motor neuron in an autonomic pathway, with its cell body and dendrites in the brain or spinal cord and its myelinated axon ending at an autonomic ganglion, where it synapses with a postganglionic neuron.

Pregnancy Sequence of events that normally includes fertilisation, implantation, embryonic growth, and fetal growth and terminates in birth.

Premenstrual syndrome (PMS) Severe physical and emotional stress ocurring late in the postovulatory phase of the menstrual cycle and sometimes overlapping with menstruation.

Prepuce (PRĒ-poos) The loose-fitting skin covering the glans of the penis and clitoris. Also called the **foreskin.**

Presbyopia (prez-bē-Ō-pē-a) A loss of elasticity of the lens of the eye due to advancing age with resulting inability to focus clearly on near objects.

Presynaptic (prē-sin-AP-tik) **neuron** A neuron that propagates nerve impulses toward a synapse.

Prevertebral ganglion (prē-VER-te-bral GANG-glē-on) A cluster of cell bodies of postganglionic sympathetic neurons anterior to the spinal column and close to large abdominal arteries. Also called a **collateral ganglion.**

Primary germ layer One of three layers of embryonic tissue, called ectoderm, mesoderm, and endoderm, that give rise to all tissues and organs of the body.

Primary motor area A region of the cerebral cortex in the precentral gyrus of the frontal lobe of the cerebrum that controls specific muscles or groups of muscles.

Primary somatosensory area A region of the cerebral cortex posterior to the central sulcus in the postcentral gyrus of the parietal lobe of the cerebrum that localizes exactly the points of the body where somatic sensations originate.

Prime mover The muscle directly responsible for producing a desired motion. Also called an **agonist** (AG-ō-nist).

Primitive gut Embryonic structure formed from the dorsal part of the yolk sac that gives rise to most of the gastrointestinal tract.

Primordial (prī-MŌR-dē-al) Existing first; especially primordial egg cells in the ovary.

Principal cell Cell type in the distal convoluted tubules and collecting ducts of the kidneys that is stimulated by aldosterone and antidiuretic hormone.

Proctology (prok-TOL-ō-jē) The branch of medicine concerned with the rectum and its disorders.

Progeny (PROJ-e-nē) Offspring or descendants.

Progesterone (prō-JES-te-rōn) A female sex hormone produced by the ovaries that helps prepare the endometrium of the uterus for implantation of a fertilized ovum and the mammary glands for milk secretion.

Prognosis (prog-NŌ-sis) A forecast of the probable results of a disorder; the outlook for recovery.

Prolactin (prō-LAK-tin) **(PRL)** A hormone secreted by the anterior pituitary that initiates and maintains milk secretion by the mammary glands.

Prolapse (PRŌ-laps) A dropping or falling down of an organ, especially the uterus or rectum.

Proliferation (prō-lif′-er-Ā-shun) Rapid and repeated reproduction of new parts, especially cells.

Pronation (prō-NĀ-shun) A movement of the forearm in which the palm is turned posteriorly.

Prophase (PRŌ-fāz) The first stage of mitosis during which chromatid pairs are formed and aggregate around the metaphase plate of the cell.

Proprioception (prō-prē-ō-SEP-shun) The perception of the position of body parts, especially the limbs, independent of vision; this sense is possible due to nerve impulses generated by proprioceptors.

Proprioceptor (PRŌ-prē-ō-sep′-tor) A receptor located in muscles, tendons, joints, or the internal ear (muscle spindles, tendon organs, joint kinesthetic receptors, and hair cells of the vestibular apparatus) that provides information about body position and movements.

Prostaglandin (pros′-ta-GLAN-din) **(PG)** A membrane-associated lipid; released in small quantities and acts as a local hormone.

Prostate (PROS-tāt) A doughnut-shaped gland inferior to the urinary bladder that surrounds the superior portion of the male urethra and secretes a slightly acidic solution that contributes to sperm motility and viability.

Proteasome (PRŌ-tē-a-sōm) Tiny cellular organelle in cytosol and nucleus containing proteases that destroy unneeded, damaged, or faulty proteins.

Protein An organic compound consisting of carbon, hydrogen, oxygen, nitrogen, and sometimes sulfur and phosphorus; synthesized on ribosomes and made up of amino acids linked by peptide bonds.

Prothrombin (prō-THROM-bin) An inactive blood-clotting factor synthesized by the liver, released into the blood, and converted to active thrombin in the process of blood clotting by the activated enzyme prothrombinase.

Proto-oncogene (prō′-tō-ON-kō-jēn) Gene responsible for some aspect of normal growth and development; it may transform into an oncogene, a gene capable of causing cancer.

Protraction (prō-TRAK-shun) The movement of the mandible or shoulder girdle forward on a plane parallel with the ground.

Proximal (PROK-si-mal) Nearer the attachment of a limb to the trunk; nearer to the point of origin or attachment.

Pseudopod (SOO-dō-pod) Temporary protrusion of the leading edge of a migrating cell; cellular projection that surrounds a particle undergoing phagocytosis.

Pterygopalatine ganglion (ter'-i-gō-PAL-a-tīn GANG-glē-on) A cluster of cell bodies of parasympathetic postganglionic neurons ending at the lacrimal and nasal glands.

Ptosis (TŌ-sis) Drooping, as of the eyelid or the kidney.

Puberty (PŪ-ber-tē) The time of life during which the secondary sex characteristics begin to appear and the capability for sexual reproduction is possible; usually occurs between the ages of 10 and 17.

Pubic symphysis A slightly movable cartilaginous joint between the anterior surfaces of the hip bones.

Puerperium (pū-er-PER-ē-um) The period immediately after childbirth, usually 4–6 weeks.

Pulmonary (PUL-mo-ner'-ē) Concerning or affected by the lungs.

Pulmonary circulation The flow of deoxygenated blood from the right ventricle to the lungs and the return of oxygenated blood from the lungs to the left atrium.

Pulmonary embolism (EM-bō-lizm) **(PE)** The presence of a blood clot or a foreign substance in a pulmonary arterial blood vessel that obstructs circulation to lung tissue.

Pulmonary oedema (usu., also edema) An abnormal accumulation of interstitial fluid in the tissue spaces and alveoli of the lungs due to increased pulmonary capillary permeability or increased pulmonary capillary pressure.

Pulmonary ventilation The inflow (inhalation) and outflow (exhalation) of air between the atmosphere and the lungs. Also called **breathing.**

Pulp cavity A cavity within the crown and neck of a tooth, which is filled with pulp, a connective tissue containing blood vessels, nerves, and lymphatic vessels.

Pulse (PULS) The rhythmic expansion and elastic recoil of a systemic artery after each contraction of the left ventricle.

Pupil The hole in the center of the iris, the area through which light enters the posterior cavity of the eyeball.

Purkinje (pur-KIN-jē) **fiber** Muscle fiber (cell) in the ventricular tissue of the heart specialized for conducting an action potential to the myocardium; part of the conduction system of the heart.

Pus The liquid product of inflammation containing leukocytes or their remains and debris of dead cells.

Pyloric (pī-LOR-ik) **sphincter** A thickened ring of smooth muscle through which the pylorus of the stomach communicates with the duodenum. Also called the **pyloric valve.**

Pyorrhoea (pī-ō-RĒ-a) A discharge or flow of pus, especially in the alveoli (sockets) and the tissues of the gums.

Pyramid (PIR-a-mid) A pointed or cone-shaped structure. One of two roughly triangular structures on the anterior aspect of the medulla oblongata composed of the largest motor tracts that run from the cerebral cortex to the spinal cord. A triangular structure in the renal medulla.

Pyramidal (pi-RAM-i-dal) **tracts (pathways)** *See* **Direct motor pathways.**

Q

QRS complex The deflection waves of an electrocardiogram that represent onset of ventricular depolarization.

Quadrant (KWOD-rant) One of four parts.

Quadriplegia (kwod'-ri-PLĒ-jē-a) Paralysis of four limbs: two upper and two lower.

R

Radiographic (rā'-dē-ō-GRAF-ik) **anatomy** Diagnostic branch of anatomy that includes the use of x rays.

Rami communicantes (RĀ-mē kō-mū-ni-KAN-tēz) Branches of a spinal nerve. *Singular* is **ramus communicans** (RĀ-mus kō-MŪ-ni-kans).

Rathke's pouch *See* **Hypophyseal pouch.**

Receptor A specialized cell or a distal portion of a neuron that responds to a specific sensory modality, such as touch, pressure, cold, light, or sound, and converts it to an electrical signal (generator or receptor potential). A specific molecule or cluster of molecules that recognizes and binds a particular ligand.

Receptor-mediated endocytosis A highly selective process whereby cells take up specific ligands, which usually are large molecules or particles, by enveloping them within a sac of plasma membrane. Ligands are eventually broken down by enzymes in lysosomes.

Recombinant DNA Synthetic DNA, formed by joining a fragment of DNA from one source to a portion of DNA from another.

Rectouterine (rek-tō-Ū-ter-in) **pouch** A pocket formed by the parietal peritoneum as it moves posteriorly from the surface of the uterus and is reflected onto the rectum; the most inferior point in the pelvic cavity. Also called the **pouch of Douglas.**

Rectum (REK-tum) The last 20 cm (8 in.) of the gastrointestinal tract, from the sigmoid colon to the anus.

Recumbent (re-KUM-bent) Lying down.

Red bone marrow A highly vascularized connective tissue located in microscopic spaces between trabeculae of spongy bone tissue.

Red nucleus A cluster of cell bodies in the midbrain, occupying a large part of the tectum from which axons extend into the rubroreticular and rubrospinal tracts.

Red pulp That portion of the spleen that consists of venous sinuses filled with blood and thin plates of splenic tissue called splenic (Billroth's) cords.

Referred pain Pain that is felt at a site remote from the place of origin.

Reflex Fast response to a change (stimulus) in the internal or external environment that attempts to restore homoeostasis.

Reflex arc The most basic conduction pathway through the nervous system, connecting a receptor and an effector and consisting of a receptor, a sensory neuron, an integrating center in the central nervous system, a motor neuron, and an effector. Also called **reflex circuit.**

Regional anatomy The division of anatomy dealing with a specific region of the body, such as the head, neck, chest, or abdomen.

Regurgitation (rē-gur'-ji-TĀ-shun) Return of solids or fluids to the mouth from the stomach; backward flow of blood through incompletely closed heart valves.

Relaxin (RLX) A female hormone produced by the ovaries and placenta that increases flexibility of the pubic symphysis and helps dilate the uterine cervix to ease delivery of a baby.

Releasing hormone Hormone secreted by the hypothalamus that can stimulate secretion of hormones of the anterior pituitary.

Renal (RĒ-nal) Pertaining to the kidneys.

Renal corpuscle (KOR-pus-l) A glomerular (Bowman's) capsule and its enclosed glomerulus.

Renal pelvis A cavity in the center of the kidney formed by the expanded, proximal portion of the ureter, lying within the kidney, and into which the major calyces open.

Renal physiology Study of the functions of the kidneys.

Renal pyramid A triangular structure in the renal medulla containing the straight segments of renal tubules and the vasa recta.

Reproduction (rē-prō-DUK-shun) The formation of new cells for growth, repair, or replacement; the production of a new individual.

Reproductive cell division Type of cell division in which gametes (sperm and oocytes) are produced; consists of meiosis and cytokinesis.

Respiration (res-pi-RĀ-shun) Overall exchange of gases between the atmosphere, blood, and body cells consisting of pulmonary ventilation, external respiration, and internal respiration.

Respiratory center Neurons in the pons and medulla oblongata of the brain stem that regulate the rate and depth of pulmonary ventilation.

Respiratory (res-PEER-a-tōr-ē) **physiology** Study of the functions of the air passageways and lungs.

Respiratory (res-PEER-a-tōr-ē) **system** System composed of the nose, pharynx, larynx, trachea, bronchi, and lungs that obtains oxygen for body cells and eliminates carbon dioxide from them.

Retention (rē-TEN-shun) A failure to void urine due to obstruction, nervous contraction of the urethra, or absence of sensation of desire to urinate.

Rete (RĒ-tē) **testis** The network of ducts in the testes.

Reticular (re-TIK-ū-lar) **activating system (RAS)** A portion of the reticular formation that has many ascending connections with the cerebral cortex; when this area of the brain stem is active, nerve impulses pass to the thalamus and widespread areas of the cerebral cortex, resulting in generalized alertness or arousal from sleep.

Reticular formation A network of small groups of neuronal cell bodies scattered among bundles of axons (mixed gray and white matter) beginning in the medulla oblongata and extending superiorly through the central part of the brain stem.

Reticulocyte (re-TIK-ū-lō-sīt) An immature red blood cell.

Reticulum (re-TIK-ū-lum) A network.

Retina (RET-i-na) The deep coat of the posterior portion of the eyeball consisting of nervous tissue (where the process of vision begins) and a pigmented layer of epithelial cells that contact the choroid.

Retinaculum (ret-i-NAK-ū-lum) A thickening of fascia that holds structures in place, for example, the superior and inferior retinacula of the ankle.

Retraction (rē-TRAK-shun) The movement of a protracted part of the body posteriorly on a plane parallel to the ground, as in pulling the lower jaw back in line with the upper jaw.

Retroperitoneal (re′-trō-per-i-tō-NĒ-al) External to the peritoneal lining of the abdominal cavity.

Rh factor An inherited antigen on the surface of red blood cells in Rh⁺ individuals; not present in Rh⁻ individuals.

Rhinology (rī-NOL-ō-jē) The study of the nose and its disorders.

Ribonucleic (rī-bō-noo-KLĒ-ik) **acid (RNA)** A single-stranded nucleic acid made up of nucleotides, each consisting of a nitrogenous base (adenine, cytosine, guanine, or uracil), ribose, and a phosphate group; major types are messenger RNA (mRNA), transfer RNA (tRNA), and ribosomal RNA (rRNA), each of which has a specific role during protein synthesis.

Ribosome (RĪ-bō-sōm) A cellular structure in the cytoplasm of cells, composed of a small subunit and a large subunit that contain ribosomal RNA and ribosomal proteins; the site of protein synthesis.

Right lymphatic (lim-FAT-ik) **duct** A vessel of the lymphatic system that drains lymph from the upper right side of the body and empties it into the right subclavian vein.

Rigidity (ri-JID-i-tē) Hypertonia characterized by increased muscle tone, but reflexes are not affected.

Rigor mortis State of partial contraction of muscles after death due to lack of ATP; myosin heads (cross-bridges) remain attached to actin, thus preventing relaxation.

Rod One of two types of photoreceptors in the retina of the eye; specialized for vision in dim light.

Root canal A narrow extension of the pulp cavity lying within the root of a tooth.

Root of penis Attached portion of penis that consists of the bulb and crura.

Rotation (rō-TĀ-shun) Moving a bone around its own axis, with no other movement.

Rotator cuff Refers to the tendons of four deep shoulder muscles (subscapularis, supraspinatus, infraspinatus, and teres minor) that form a complete circle around the shoulder; they strengthen and stabilize the shoulder joint.

Round ligament (LIG-a-ment) A band of fibrous connective tissue enclosed between the folds of the broad ligament of the uterus, emerging from the uterus just inferior to the uterine tube, extending laterally along the pelvic wall and through the deep inguinal ring to end in the labia majora.

Round window A small opening between the middle and internal ear, directly inferior to the oval window, covered by the secondary tympanic membrane.

Ruffini corpuscle A sensory receptor embedded deeply in the dermis and deeper tissues that detects the stretching of the skin.

Rugae (ROO-gē) Large folds in the mucosa of an empty hollow organ, such as the stomach or vagina.

S

Saccule (SAK-ūl) The inferior and smaller of the two chambers in the membranous labyrinth inside the vestibule of the internal ear containing a receptor organ for static equilibrium.

Sacral plexus (SĀ-kral PLEK-sus) A network formed by the ventral branches of spinal nerves L4 through S3.

Sacral promontory (PROM-on-tor′-ē) The superior surface of the body of the first sacral vertebra that projects anteriorly into the pelvic cavity; a line from the sacral promontory to the superior border of the pubic symphysis divides the abdominal and pelvic cavities.

Saddle joint A synovial joint in which the articular surface of one bone is saddle-shaped and the articular surface of the other bone is shaped like the legs of the rider sitting in the saddle, as in the joint between the trapezium and the metacarpal of the thumb.

Sagittal (SAJ-i-tal) **plane** A plane that divides the body or organs into left and right portions. Such a plane may be **midsagittal (median),** in which the divisions are equal, or **parasagittal,** in which the divisions are unequal.

Saliva (sa-LĪ-va) A clear, alkaline, somewhat viscous secretion produced mostly by the three pairs of salivary glands; contains various salts, mucin, lysosyme, salivary amylase, and lingual lipase (produced by glands in the tongue).

Salivary amylase (SAL-i-ver-ē AM-i-lās) An enzyme in saliva that initiates the chemical breakdown of starch.

Salivary gland One of three pairs of glands that lie external to the mouth and pour their secretory product (saliva) into ducts that empty into the oral cavity; the parotid, submandibular, and sublingual glands.

Sarcolemma (sar′-kō-LEM-ma) The cell membrane of a muscle fiber (cell), especially of a skeletal muscle fiber.

Sarcomere (SAR-kō-mēr) A contractile unit in a striated muscle fiber (cell) extending from one Z disc to the next Z disc.

Sarcoplasm (SAR-kō-plazm) The cytoplasm of a muscle fiber (cell).

Sarcoplasmic reticulum (sar′-kō-PLAZ-mik re-TIK-ū-lum) **(SR)** A network of saccules and tubes surrounding myofibrils of a muscle fiber (cell), comparable to endoplasmic reticulum; functions to reabsorb calcium ions during relaxation and to release them to cause contraction.

Satellite (SAT-i-līt) **cell** Flat neuroglial cells that surround cell bodies of peripheral nervous system ganglia to provide structural support and regulate the exchange of material between a neuronal cell body and interstitial fluid.

Saturated fat A fatty acid that contains only single bonds (no double bonds) between its carbon atoms; all carbon atoms are bonded to the maximum number of hydrogen atoms; prevalent in triglycerides of animal products such as meat, milk, milk products, and eggs.

Scala tympani (SKA-la TIM-pan-ē) The inferior spiral-shaped channel of the bony cochlea, filled with perilymph.

Scala vestibuli (ves-TIB-ū-lē) The superior spiral-shaped channel of the bony cochlea, filled with perilymph.

Schwann (SCHVON or SCHWON) **cell** A neuroglial cell of the peripheral nervous system that forms the myelin sheath and neurolemma around a nerve axon by wrapping around the axon in a jellyroll fashion.

Sciatica (sī-AT-i-ka) Inflammation and pain along the sciatic nerve; felt along the posterior aspect of the thigh extending down the inside of the leg.

Sclera (SKLE-ra) The white coat of fibrous tissue that forms the superficial protective covering over the eyeball except in the most anterior portion; the posterior portion of the fibrous tunic.

Scleral venous sinus A circular venous sinus located at the junction of the sclera and the cornea through which aqueous humour drains from the anterior chamber of the eyeball into the blood. Also called the **canal of Schlemm** (SHLEM).

Sclerosis (skle-RŌ-sis) A hardening with loss of elasticity of tissues.

Scoliosis (skō-lē-Ō-sis) An abnormal lateral curvature from the normal vertical line of the backbone.

Scrotum (SKRŌ-tum) A skin-covered pouch that contains the testes and their accessory structures.

Sebaceous (se-BĀ-shus) **gland** An exocrine gland in the dermis of the skin, almost always associated with a hair follicle, that secretes sebum. Also called an **oil gland.**

Sebum (SĒ-bum) Secretion of sebaceous (oil) glands.

Secondary sex characteristic A characteristic of the male or female body that develops at puberty under the influence of sex hormones but is not directly involved in sexual reproduction; examples are distribution of body hair, voice pitch, body shape, and muscle development.

Secretion (se-KRĒ-shun) Production and release from a cell or a gland of a physiologically active substance.

Selective permeability (per′-mē-a-BIL-i-tē) The property of a membrane by which it permits the passage of certain substances but restricts the passage of others.

Semen (SĒ-men) A fluid discharged at ejaculation by a male that consists of a mixture of sperm and the secretions of the seminiferous tubules, seminal vesicles, prostate, and bulbourethral (Cowper's) glands.

Semicircular canals Three bony channels (anterior, posterior, lateral), filled with perilymph, in which lie the membranous semicircular canals filled with endolymph. They contain receptors for equilibrium.

Semicircular ducts The membranous semicircular canals filled with endolymph and floating in the perilymph of the bony semicircular canals; they contain cristae that are concerned with dynamic equilibrium.

Semilunar (sem′-ē-LOO-nar) **(SL) valve** A valve between the aorta or the pulmonary trunk and a ventricle of the heart.

Seminal vesicle (SEM-i-nal VES-i-kul) One of a pair of convoluted, pouchlike structures, lying posterior and inferior to the urinary bladder and anterior to the rectum, that secrete a component of semen into the ejaculatory ducts. Also termed **seminal gland.**

Seminiferous tubule (sem′-i-NI-fer-us TOO-būl) A tightly coiled duct, located in the testis, where sperm are produced.

Sensation A state of awareness of external or internal conditions of the body.

Sensory area A region of the cerebral cortex concerned with the interpretation of sensory impulses.

Sensory neurons (NOO-ronz) Neurons that carry sensory information from cranial and spinal nerves into the brain and spinal cord or from a lower to a higher level in the spinal cord and brain. Also called **afferent neurons.**

Septal defect An opening in the atrial septum (atrial septal defect) because the foramen ovale fails to close, or the ventricular septum (ventricular septal defect) due to incomplete development of the ventricular septum.

Septum (SEP-tum) A wall dividing two cavities.

Serous (SĒR-us) **membrane** A membrane that lines a body cavity that does not open to the exterior. The external layer of an organ formed by a serous membrane. The membrane that lines the pleural, pericardial, and peritoneal cavities. Also called a **serosa** (se-RŌ-sa).

Sertoli (ser-TŌ-lē) **cell** A supporting cell in the seminiferous tubules that secretes fluid for supplying nutrients to sperm and the hormone inhibin, removes excess cytoplasm from spermatogenic cells, and mediates the effects of FSH and testosterone on spermatogenesis. Also called a **sustentacular** (sus′-ten-TAK-ū-lar) **cell.**

Serum Blood plasma minus its clotting proteins.

Sesamoid (SES-a-moyd) **bones** Small bones usually found in tendons.

Sex chromosomes The twenty-third pair of chromosomes, designated X and Y, which determines the genetic sex of an individual; in males, the pair is XY; in females, XX.

Sexual intercourse The insertion of the erect penis of a male into the vagina of a female. Also called **coitus** (KŌ-i-tus).

Sheath of Schwann *See* **Neurolemma.**

Shock Failure of the cardiovascular system to deliver adequate amounts of oxygen and nutrients to meet the metabolic needs of the body due to inadequate cardiac output. It is characterized by hypotension; clammy, cool, and pale skin; sweating; reduced urine formation; altered mental state; acidosis; tachycardia; weak, rapid pulse; and thirst. Types include hypovolemic, cardiogenic, vascular, and obstructive.

Shoulder joint A synovial joint where the humerus articulates with the scapula.

Sigmoid colon (SIG-moyd KŌ-lon) The S-shaped part of the large intestine that begins at the level of the left iliac crest, projects medially, and terminates at the rectum at about the level of the third sacral vertebra.

Sign Any objective evidence of disease that can be observed or measured, such as a lesion, swelling, or fever.

Sinoatrial (si-nō-Ā-trē-al) **(SA) node** A small mass of cardiac muscle fibers (cells) located in the right atrium inferior to the opening of the superior vena cava that spontaneously depolarize and generate a cardiac action potential about 100 times per minute. Also called the natural **pacemaker.**

Sinus (SĪ-nus) A hollow in a bone (paranasal sinus) or other tissue; a channel for blood (vascular sinus); any cavity having a narrow opening.

Sinusoid (SĪ-nū-soyd) A large, thin-walled, and leaky type of capillary, having large intercellular clefts that may allow proteins and blood cells to pass from a tissue into the bloodstream; present in the liver, spleen, anterior pituitary, parathyroid glands, and red bone marrow.

Skeletal muscle An organ specialized for contraction, composed of striated muscle fibers (cells), supported by connective tissue, attached to a bone by a tendon or an aponeurosis, and stimulated by somatic motor neurons.

Skeletal system Framework of bones and their associated cartilages, ligaments, and tendons.

Skene's gland *See* **Paraurethral gland.**

Skin The external covering of the body that consists of a superficial, thinner epidermis (epithelial tissue) and a deep, thicker dermis (connective tissue) that is anchored to the subcutaneous layer. Also called **cutaneous membrane.**

Skin graft The transfer of a patch of healthy skin taken from a donor site to cover a wound.

Skull The skeleton of the head consisting of the cranial and facial bones.

Sleep A state of partial unconsciousness from which a person can be aroused; associated with a low level of activity in the reticular activating system.

Sliding filament mechanism A model that describes muscle contraction in which thin filaments slide past thick ones so that the filaments overlap, causing shortening of a sarcomere, and thus shortening of muscle fibers and alternately shortening of the entire muscle.

Small intestine A long tube of the gastrointestinal tract that begins at the pyloric sphincter of the stomach, coils through the central and inferior part of the abdominal cavity, and ends at the large intestine; divided into three segments: duodenum, jejunum, and ileum.

Smooth muscle A tissue specialized for contraction, composed of smooth muscle fibers (cells), located in the walls of hollow internal organs, and innervated by autonomic motor neurons.

Sodium–potassium ATPase An active transport pump located in the plasma membrane that transports sodium ions out of the cell and potassium ions into the cell at the expense of cellular ATP. It functions to keep the ionic concentrations of these ions at physiological levels. Also called the **sodium–potassium pump.**

Soft palate (PAL-at) The posterior portion of the roof of the mouth, extending from the palatine bones to the uvula. It is a muscular partition lined with mucous membrane.

Somatic (sō-MAT-ik) **cell division** Type of cell division in which a single starting cell duplicates itself to produce two identical cells; consists of mitosis and cytokinesis.

Somatic motor pathway Pathway that carries information from the cerebral cortex, basal nuclei, and cerebellum that stimulates contraction of skeletal muscles.

Somatic nervous system (SNS) The portion of the peripheral nervous system consisting of somatic sensory (afferent) neurons and somatic motor (efferent) neurons.

Somatic sensory pathway Pathway that carries information from somatic sensory receptor to the primary somatosensory area in the cerebral cortex and cerebellum.

Somite (SŌ-mīt) Block of mesodermal cells in a developing embryo that is distinguished into a myotome (which forms most of the skeletal muscles), dermatome (which forms connective tissues), and sclerotome (which forms the vertebrae).

Spasm (SPAZM) A sudden, involuntary contraction of large groups of muscles.

Spasticity (spas-TIS-i-tē) Hypertonia characterized by increased muscle tone, increased tendon reflexes, and pathological reflexes (Babinski sign).

Spermatic (sper-MAT-ik) **cord** A supporting structure of the male reproductive system, extending from a testis to the deep inguinal ring, that includes the ductus (vas) deferens, arteries, veins, lymphatic vessels, nerves, cremaster muscle, and connective tissue.

Spermatogenesis (sper′-ma-tō-JEN-e-sis) The formation and development of sperm in the seminiferous tubules of the testes.

Sperm cell A mature male gamete. Also termed **spermatozoon** (sper′-ma-tō-ZŌ-on).

Spermiogenesis (sper′-mē-ō-JEN-e-sis) The maturation of spermatids into sperm.

Sphincter (SFINGK-ter) A circular muscle that constricts an opening.

Sphincter of Oddi *See* **Sphincter of the hepatopancreatic ampulla.**

Sphincter of the hepatopancreatic ampulla A circular muscle at the opening of the common bile and main pancreatic ducts in the duodenum. Also called the **sphincter of Oddi** (OD-ē).

Spinal (SPĪ-nal) **cord** A mass of nerve tissue located in the vertebral canal from which 31 pairs of spinal nerves originate.

Spinal nerve One of the 31 pairs of nerves that originate on the spinal cord from posterior and anterior roots.

Spinal shock A period from several days to several weeks following transection of the spinal cord that is characterized by the abolition of all reflex activity.

Spinothalamic (spī-nō-tha-LAM-ik) **tract** Sensory (ascending) tract that conveys information up the spinal cord to the thalamus for sensations of pain, temperature, itch, and tickle.

Spinous (SPĪ-nus) **process** A sharp or thornlike process or projection. Also called a **spine.** A sharp ridge running diagonally across the posterior surface of the scapula.

Spiral organ The organ of hearing, consisting of supporting cells and hair cells that rest on the basilar membrane and extend into the endolymph of the cochlear duct. Also called the **organ of Corti** (KOR-tē).

Splanchnic (SPLANK-nik) Pertaining to the viscera.

Spleen (SPLĒN) Large mass of lymphatic tissue between the fundus of the stomach and the diaphragm that functions in formation of blood cells during early fetal development, phagocytosis of ruptured blood cells, and proliferation of B cells during immune responses.

Spongy (cancellous) bone tissue Bone tissue that consists of an irregular latticework of thin plates of bone called trabeculae; spaces between trabeculae of some bones are filled with red bone marrow; found inside short, flat, and irregular bones and in the epiphyses (ends) of long bones.

Sprain Forcible wrenching or twisting of a joint with partial rupture or other injury to its attachments without dislocation.

Squamous (SKWĀ-mus) Flat or scalelike.

Starvation (star-VĀ-shun) The loss of energy stores in the form of glycogen, triglycerides, and proteins due to inadequate intake of nutrients or inability to digest, absorb, or metabolize ingested nutrients.

Static equilibrium (ē-kwi-LIB-rē-um) The maintenance of posture in response to changes in the orientation of the body, mainly the head, relative to the ground.

Stellate reticuloendothelial (STEL-āt re-tik′-ū-lō-en′-dō-THĒ-lē-al) **cell** Phagocytic cell bordering a sinusoid of the liver. Also called a **Kupffer** (KOOP-fer) **cell.**

Stem cell An unspecialized cell that has the ability to divide for indefinite periods and give rise to a specialized cell.

Stenosis (sten-Ō-sis) An abnormal narrowing or constriction of a duct or opening.

Stereocilia (ste′-rē-ō-SIL-ē-a) Groups of extremely long, slender, nonmotile microvilli projecting from epithelial cells lining the epididymis.

Sterile (STE-ril) Free from any living microorganisms. Unable to conceive or produce offspring.

Sterilisation (ster′-i-li-ZĀ-shun) Elimination of all living microorganisms. Any procedure that renders an individual incapable of reproduction (for example, castration, vasectomy, hysterectomy, or oophorectomy).

Stimulus Any stress that changes a controlled condition; any change in the internal or external environment that excites a sensory receptor, a neuron, or a muscle fiber.

Stomach The J-shaped enlargement of the gastrointestinal tract directly inferior to the diaphragm in the epigastric, umbilical, and left hypochondriac regions of the abdomen, between the oesophagus and small intestine.

Straight tubule (TOO-būl) A duct in a testis leading from a convoluted seminiferous tubule to the rete testis.

Stratum (STRĀ-tum) A layer.

Stratum basalis (ba-SAL-is) The layer of the endometrium next to the myometrium that is maintained during menstruation and gestation and produces a new stratum functionalis following menstruation or parturition.

Stratum functionalis (funk′-shun-AL-is) The layer of the endometrium next to the uterine cavity that is shed during menstruation and that forms the maternal portion of the placenta during gestation.

Stretch receptor Receptor in the walls of blood vessels, airways, or organs that monitors the amount of stretching. Also termed **baroreceptor.**

Striae (STRĪ-ē) Internal scarring due to overstretching of the skin in which collagen fibers and blood vessels in the dermis are damaged. Also called **stretch marks.**

Stroma (STRŌ-ma) The tissue that forms the ground substance, foundation, or framework of an organ, as opposed to its functional parts (parenchyma).

Subarachnoid (sub′-a-RAK-noyd) **space** A space between the arachnoid mater and the pia mater that surrounds the brain and spinal cord and through which cerebrospinal fluid circulates.

Subcutaneous (sub′-kū-TĀ-nē-us) Beneath the skin. Also called **hypodermic** (hī-pō-DER-mik).

Subcutaneous (subQ) layer A continuous sheet of areolar connective tissue and adipose tissue between the dermis of the skin and the deep fascia of the muscles. Also called the **hypodermis.**

Subdural (sub-DOO-ral) **space** A space between the dura mater and the arachnoid mater of the brain and spinal cord that contains a small amount of fluid.

Sublingual (sub-LING-gwal) **gland** One of a pair of salivary glands situated in the floor of the mouth deep to the mucous membrane and to the side of the lingual frenulum, with a duct (Rivinus') that opens into the floor of the mouth.

Submandibular (sub′-man-DIB-ū-lar) **gland** One of a pair of salivary glands found inferior to the base of the tongue deep to the mucous membrane in the posterior part of the floor of the mouth, posterior to the sublingual glands, with a duct (Wharton's) situated to the side of the lingual frenulum. Also called the **submaxillary** (sub′-MAK-si-ler-ē) **gland.**

Submucosa (sub-mū-KŌ-sa) A layer of connective tissue located deep to a mucous membrane, as in the gastrointestinal tract or the urinary bladder; the submucosa connects the mucosa to the muscularis layer.

Submucosal plexus A network of autonomic nerve fibers located in the superficial part of the submucous layer of the small intestine. Also called the **plexus of Meissner** (MĪZ-ner).

Substrate A molecule on which an enzyme acts.

Subthalamus (sub-THAL-a-mus) Part of the diencephalon inferior to the thalamus; the substantia nigra and red nucleus extend from the midbrain into the subthalamus.

Sudoriferous (soo′-dor-IF-er-us) **gland** An apocrine or eccrine exocrine gland in the dermis or subcutaneous layer that produces perspiration. Also called a **sweat gland.**

Sulcus (SUL-kus) A groove or depression between parts, especially between the convolutions of the brain. *Plural* is **sulci** (SUL-sī).

Superficial (soo′-per-FISH-al) Located on or near the surface of the body or an organ. Also called **external.**

Superficial subcutaneous inguinal (IN-gwi-nal) **ring** A triangular opening in the aponeurosis of the external oblique muscle that represents the termination of the inguinal canal.

Superior (soo-PĒR-ē-or) Toward the head or upper part of a structure. Also called **cephalic** or **cranial.**

Superior vena cava (VĒ-na KĀ-va) **(SVC)** Large vein that collects blood from parts of the body superior to the heart and returns it to the right atrium.

Supination (soo-pi-NĀ-shun) A movement of the forearm in which the palm is turned anteriorly.

Surface anatomy The study of the structures that can be identified from the outside of the body.

Surfactant (sur-FAK-tant) Complex mixture of phospholipids and lipoproteins, produced by type II alveolar (septal) cells in the lungs, that decreases surface tension.

Suspensory ligament (sus-PEN-so-rē LIG-a-ment) A fold of peritoneum extending laterally from the surface of the ovary to the pelvic wall.

Sutural (SOO-chur-al) **bone** A small bone located within a suture between certain cranial bones. Also called **Wormian** (WER-mē-an) **bone.**

Suture (SOO-chur) An immovable fibrous joint that joins skull bones.

Sympathetic (sim′-pa-THET-ik) **division** One of the two subdivisions of the autonomic nervous system, having cell bodies of preganglionic neurons in the lateral gray columns of the thoracic segment and the first two or three lumbar segments of the spinal cord; primarily concerned with processes involving the expenditure of energy.

Sympathetic trunk ganglion (GANG-glē-on) A cluster of cell bodies of sympathetic postganglionic neurons lateral to the vertebral column, close to the body of a vertebra. These ganglia extend inferiorly through the neck, thorax, and abdomen to the coccyx on both sides of the vertebral column and are connected to one another to form a chain on each side of the vertebral column. Also called **sympathetic chain, vertebral chain ganglia,** or **paravertebral ganglia.**

Symphysis (SIM-fi-sis) A line of union. A slightly movable cartilaginous joint such as the pubic symphysis.

Symptom (SIMP-tum) A subjective change in body function not apparent to an observer, such as pain or nausea, that indicates the presence of a disease or disorder of the body.

Synapse (SIN-aps) The functional junction between two neurons or between a neuron and an effector, such as a muscle or gland; may be electrical or chemical.

Synapsis (sin-AP-sis) The pairing of homologous chromosomes during prophase I of meiosis.

Synaptic (sin-AP-tik) **cleft** The narrow gap at a chemical synapse that separates the axon terminal of one neuron from another neuron or muscle fiber (cell) and across which a neurotransmitter diffuses to affect the postsynaptic cell.

Synaptic end bulb Expanded distal end of an axon terminal that contains synaptic vesicles. Also called a **synaptic knob.**

Synaptic vesicle Membrane-enclosed sac in a synaptic end bulb that stores neurotransmitters.

Synarthrosis (sin′-ar-THRŌ-sis) An immovable joint such as a suture, gomphosis, or synchondrosis.

Synchondrosis (sin′-kon-DRŌ-sis) A cartilaginous joint in which the connecting material is hyaline cartilage.

Syndesmosis (sin′-dez-MŌ-sis) A slightly movable joint in which articulating bones are united by fibrous connective tissue.

Synergist (SIN-er-jist) A muscle that assists the prime mover by reducing undesired action or unnecessary movement.

Synergistic (syn-er-JIS-tik) **effect** A hormonal interaction in which the effects of two or more hormones acting together is greater or more extensive than the sum of each hormone acting alone.

Synostosis (sin′-os-TŌ-sis) A joint in which the dense fibrous connective tissue that unites bones at a suture has been replaced by bone, resulting in a complete fusion across the suture line.

Synovial (si-NŌ-vē-al) **cavity** The space between the articulating bones of a synovial joint, filled with synovial fluid. Also called a **joint cavity.**

Synovial fluid Secretion of synovial membranes that lubricates joints and nourishes articular cartilage.

Synovial joint A fully movable or diarthrotic joint in which a synovial (joint) cavity is present between the two articulating bones.

Synovial membrane The deeper of the two layers of the articular capsule of a synovial joint, composed of areolar connective tissue that secretes synovial fluid into the synovial (joint) cavity.

System An association of organs that have a common function.

Systemic (sis-TEM-ik) Affecting the whole body; generalized.

Systemic anatomy The anatomical study of particular systems of the body, such as the skeletal, muscular, nervous, cardiovascular, or urinary systems.

Systemic circulation The routes through which oxygenated blood flows from the left ventricle through the aorta to all the organs of the body and deoxygenated blood returns to the right atrium.

Systole (SIS-tō-lē) In the cardiac cycle, the phase of contraction of the heart muscle, especially of the ventricles.

Systolic (sis-TOL-ik) **blood pressure (SBP)** The force exerted by blood on arterial walls during ventricular contraction; the highest pressure measured in the large arteries, about 120 mmHg under normal conditions for a young adult.

T

T cell A lymphocyte that becomes immunocompetent in the thymus and can differentiate into a helper T cell or a cytotoxic T cell, both of which function in cell-mediated immunity.

T wave The deflection wave of an electrocardiogram that represents ventricular repolarization.

Tachycardia (tak′-i-KAR-dē-a) An abnormally rapid resting heartbeat or pulse rate (over 100 beats per minute).

Tactile (TAK-tīl) Pertaining to the sense of touch.

Tactile disc *See* **Merkel disc.**

Target cell A cell whose activity is affected by a particular hormone.

Tarsal bones The seven bones of the ankle. Also called **tarsals.**

Tarsal gland Sebaceous (oil) gland that opens on the edge of each eyelid. Also called a **Meibomian** (mī-BŌ-mē-an) **gland.**

Tarsal plate A thin, elongated sheet of connective tissue, one in each eyelid, giving the eyelid form and support. The aponeurosis of the levator palpebrae superioris is attached to the tarsal plate of the superior eyelid.

Tarsus (TAR-sus) A collective term for the seven bones of the ankle.

Tectorial (tek-TŌ-rē-al) **membrane** A gelatinous membrane projecting over and in contact with the hair cells of the spiral organ (organ of Corti) in the cochlear duct.

Teeth (TĒTH) Accessory structures of digestion, composed of calcified connective tissue and embedded in bony sockets of the mandible and maxilla, that cut, shred, crush, and grind food. Also called **dentes** (DEN-tēz).

Telophase (TEL-ō-fāz) The final stage of mitosis.

Tendon (TEN-don) A white fibrous cord of dense regular connective tissue that attaches muscle to bone.

Tendon organ A proprioceptive receptor, sensitive to changes in muscle tension and force of contraction, found chiefly near the junctions of tendons and muscles. Also called a **Golgi** (GOL-jē) **tendon organ.**

Tendon reflex A polysynaptic, ipsilateral reflex that protects tendons and their associated muscles from damage that might be brought about by excessive tension. The receptors involved are called tendon organs (Golgi tendon organs).

Teniae coli (TĒ-nē-ē KŌ-lī) The three flat bands of thickened, longitudinal smooth muscle running the length of the large intestine, except in the rectum. *Singular* is **tenia coli.**

Tentorium cerebelli (ten-TŌ-rē-um ser′-e-BEL-ī) A transverse shelf of dura mater that forms a partition between the occipital lobe of the cerebral hemispheres and the cerebellum and that covers the cerebellum.

Teratogen (TER-a-tō-jen) Any agent or factor that causes physical defects in a developing embryo.

Terminal ganglion (TER-min-al GANG-glē-on) A cluster of cell bodies of parasympathetic postganglionic neurons either lying very close to the visceral effectors or located within the walls of the visceral effectors supplied by the postganglionic neurons. Also called **intramural ganglion.**

Testis (TES-tis) Male gonad that produces sperm and the hormones testosterone and inhibin. Also called a **testicle.**

Testosterone (tes-TOS-te-rōn) A male sex hormone (androgen) secreted by interstitial (Leydig) cells of a mature testis; needed for development of sperm; together with a second androgen termed **dihydrotestosterone** (dī-hī-drō-tes-TOS-ter-ōn) **(DHT),** controls the growth and development of male reproductive organs, secondary sex characteristics, and body growth.

Tetralogy of Fallot (tet-RAL-ō-jē of fal-Ō) A combination of four congenital heart defects: (1) constricted pulmonary semilunar valve, (2) interventricular septal opening, (3) emergence of the aorta from both ventricles instead of from the left only, and (4) enlarged right ventricle.

Thalamus (THAL-a-mus) A large, oval structure located bilaterally on either side of the third ventricle, consisting of two masses of grey matter organized into nuclei; main relay center for sensory impulses ascending to the cerebral cortex.

Thermoreceptor (THER-mō-rē-sep-tor) Sensory receptor that detects changes in temperature.

Thermoregulation Homeostatic regulation of body temperature through sweating and adjustment of blood flow in the dermis.

Thigh The portion of the lower limb between the hip and the knee.

Third ventricle (VEN-tri-kul) A slitlike cavity between the right and left halves of the thalamus and between the lateral ventricles of the brain.

Thoracic (thor-AS-ik) **cavity** Cavity superior to the diaphragm that contains two pleural cavities, the mediastinum, and the pericardial cavity.

Thoracic duct A lymphatic vessel that begins as a dilation called the cisterna chyli, receives lymph from the left side of the head, neck, and chest, left arm, and the entire body below the ribs, and empties into the junction between the internal jugular and left subclavian veins. Also called the **left lymphatic** (lim-FAT-ik) **duct.**

Thoracolumbar (thōr′-a-kō-LUM-bar) **outflow** The axons of sympathetic preganglionic neurons, which have their cell bodies in the lateral gray columns of the thoracic segments and first two or three lumbar segments of the spinal cord.

Thorax (THŌ-raks) The chest.

Thrombosis (THROM-BŌ-sis) The formation of a clot in an unbroken blood vessel, usually a vein.

Thrombus (THROM-bus) A stationary clot formed in an unbroken blood vessel, usually a vein.

Thymus (THĪ-mus) A bilobed organ, located in the superior mediastinum posterior to the sternum and between the lungs, in which T cells develop immunocompetence.

Thyroid cartilage (THĪ-royd KAR-ti-lij) The largest single cartilage of the larynx, consisting of two fused plates that form the anterior wall of the larynx. Also called **Adam's apple.**

Thyroid follicle (FOL-i-kul) Spherical sac that forms the parenchyma of the thyroid gland and consists of follicular cells that produce thyroxine (T_4) and triiodothyronine (T_3).

Thyroid gland An endocrine gland with right and left lateral lobes on either side of the trachea connected by an isthmus; located anterior to the trachea just inferior to the cricoid cartilage; secretes thyroxine (T_4), triiodothyronine (T_3), and calcitonin.

Thyroid-stimulating hormone (TSH) A hormone secreted by the anterior pituitary that stimulates the synthesis and secretion of thyroxine (T_4) and triiodothyronine (T_3). Also called **thyrotropin** (THĪ-rō-TRŌ-pin).

Thyroxine (thī-ROK-sēn) (**T_4**) A hormone secreted by the thyroid gland that regulates metabolism, growth and development, and the activity of the nervous system. Also called **tetraiodothyronine** (tet-ra-ī-ō-dō-THĪ-rō-nēn).

Tic Spasmodic, involuntary twitching of muscles that are normally under voluntary control.

Tissue A group of similar cells and their intercellular substance joined together to perform a specific function.

Tissue rejection Phenomenon by which the body recognizes the protein (HLA antigens) in transplanted tissues or organs as foreign and produces antibodies against them.

Tongue A large skeletal muscle covered by a mucous membrane located on the floor of the oral cavity.

Tonsil (TON-sil) An aggregation of large lymphatic nodules embedded in the mucous membrane of the throat.

Topical (TOP-i-kal) Applied to the surface rather than ingested or injected.

Torn cartilage A tearing of an articular disc (meniscus) in the knee.

Trabecula (tra-BEK-ū-la) Irregular latticework of thin plates of spongy bone tissue. Fibrous cord of connective tissue serving as supporting fiber by forming a septum extending into an organ from its wall or capsule. *Plural* is **trabeculae** (tra-BEK-ū-lē).

Trabeculae carneae (KAR-nē-ē) Ridges and folds of the myocardium in the ventricles.

Trachea (TRĀ-kē-a) Tubular air passageway extending from the larynx to the fifth thoracic vertebra. Also called the **windpipe.**

Tract A bundle of nerve axons in the central nervous system.

Transplantation (tranz-plan-TĀ-shun) The transfer of living cells, tissues, or organs from a donor to a recipient or from one part of the body to another in order to restore a lost function.

Transverse colon (trans-VERS KŌ-lon) The portion of the large intestine extending across the abdomen from the right colic (hepatic) flexure to the left colic (splenic) flexure.

Transverse fissure (FISH-er) The deep cleft that separates the cerebrum from the cerebellum.

Transverse plane A plane that divides the body or organs into superior and inferior portions. Also called a **cross-sectional** or **horizontal plane.**

Transverse tubules (TOO-būls) (**T tubules**) Small, cylindrical invaginations of the sarcolemma of striated muscle fibers (cells) that conduct muscle action potentials toward the center of the muscle fiber.

Tremor (TREM-or) Rhythmic, involuntary, purposeless contraction of opposing muscle groups.

Triad (TRĪ-ad) A complex of three units in a muscle fiber composed of a transverse tubule and the sarcoplasmic reticulum terminal cisterns on both sides of it.

Tricuspid (trī-KUS-pid) **valve** Atrioventricular (AV) valve on the right side of the heart.

Triglyceride A lipid that consists of one glycerol molecule and three fatty acid molecules.

Triglyceride (trī-GLI-ser-īd) A lipid formed from one molecule of glycerol and three molecules of fatty acids that may be either solid (fats) or liquid (oils) at room temperature; the body's most highly concentrated source of chemical potential energy. Found mainly within adipocytes. Also called a **neutral fat** or a **triacylglycerol.**

Trigone (TRĪ-gōn) A triangular region at the base of the urinary bladder.

Triiodothyronine (trī-ī-ō-dō-THĪ-rō-nēn) (**T_3**) A hormone produced by the thyroid gland that regulates metabolism, growth and development, and the activity of the nervous system.

Trophoblast (TRŌF-ō-blast) The superficial covering of cells of the blastocyst.

Tropic (TRŌ-pik) **hormone** A hormone whose target is another endocrine gland.

Trunk The part of the body to which the upper and lower limbs are attached.

Tubal ligation (lī-GĀ-shun) A sterilisation procedure in which the uterine (fallopian) tubes are tied and cut.

Tubular reabsorption The movement of filtrate from renal tubules back into blood in response to the body's specific needs.

Tubular secretion The movement of substances in blood into renal tubular fluid in response to the body's specific needs.

Tumor-suppressor gene A gene coding for a protein that normally inhibits cell division; loss or alteration of a tumor suppressor gene called *p53* is the most common genetic change in a wide variety of cancer cells.

Tunica albuginea (TOO-ni-ka al′-bū-JIN-ē-a) A dense white fibrous capsule covering a testis or deep to the surface of an ovary.

Tunica externa (eks-TER-na) The superficial coat of an artery or vein, composed mostly of elastic and collagen fibers. Also called the **adventitia.**

Tunica interna (in-TER-na) The deep coat of an artery or vein, consisting of a lining of endothelium, basement membrane, and internal elastic lamina. Also called the **tunica intima** (IN-ti-ma).

Tunica media (MĒ-dē-a) The intermediate coat of an artery or vein, composed of smooth muscle and elastic fibers.

Tympanic antrum (tim-PAN-ik AN-trum) An air space in the middle ear that leads into the mastoid air cells or sinus.

Tympanic (tim-PAN-ik) **membrane** A thin, semitransparent partition of fibrous connective tissue between the external auditory meatus and the middle ear. Also called the **eardrum.**

U

Umbilical (um-BIL-i-kul) **cord** The long, ropelike structure containing the umbilical arteries and vein that connect the fetus to the placenta.

Umbilicus (um-BIL-i-kus *or* um-bi-LĪ-kus) A small scar on the abdomen that marks the former attachment of the umbilical cord to the fetus. Also called the **navel.**

Upper limb The appendage attached at the shoulder girdle, consisting of the arm, forearm, wrist, hand, and fingers. Also called **upper extremity** or **upper appendage.**

Uraemia (ū-RĒ-mē-a) Accumulation of toxic levels of urea and other nitrogenous waste products in the blood, usually resulting from severe kidney malfunction.

Ureter (Ū-rē-ter) One of two tubes that connect the kidney with the urinary bladder.

Urethra (ū-RĒ-thra) The duct from the urinary bladder to the exterior of the body that conveys urine in females and urine and semen in males.

Urinalysis (ū-ri-NAL-i-sis) An analysis of the volume and physical, chemical, and microscopic properties of urine.

Urinary (Ū-ri-ner-ē) **bladder** A hollow, muscular organ situated in the pelvic cavity posterior to the pubic symphysis; receives urine via two ureters and stores urine until it is excreted through the urethra.

Urinary system A system that consists of the kidneys, ureters, urinary bladder, and urethra. The system regulates the ionic composition, pH, volume, pressure, and osmolarity of blood.

Urine The fluid produced by the kidneys that contains wastes and excess materials; excreted from the body through the urethra.

Urogenital (ū′-rō-JEN-i-tal) **triangle** The region of the pelvic floor inferior to the pubic symphysis, bounded by the pubic symphysis and the ischial tuberosities, and containing the external genitalia.

Urology (ū-ROL-ō-jē) The specialized branch of medicine that deals with the structure, function, and diseases of the male and female urinary systems and the male reproductive system.

Uterine cycle A series of changes in the endometrium of a nonpregnant female that prepares the lining of the uterus to receive a fertilized ovum. Also called **menstrual cycle.**

Uterine (Ū-ter-in) **tube** Duct that transports ova from the ovary to the uterus. Also called the **fallopian** (fal-LŌ-pē-an) **tube** or **oviduct.**

Uterosacral ligament (ū-ter-ō-SĀ-kral LIG-a-ment) A fibrous band of tissue extending from the cervix of the uterus laterally to the sacrum.

Uterus (Ū-te-rus) The hollow, muscular organ in females that is the site of menstruation, implantation, development of the fetus, and labour. Also called the **womb.**

Utricle (Ū-tri-kul) The larger of the two divisions of the membranous labyrinth located inside the vestibule of the inner ear, containing a receptor organ for static equilibrium.

Uvea (Ū-vē-a) The three structures that together make up the vascular tunic of the eye.

Uvula (Ū-vū-la) A soft, fleshy mass, especially the U-shaped pendant part, descending from the soft palate.

V

Vagina (va-JĪ-na) A muscular, tubular organ that leads from the uterus to the vestibule, situated between the urinary bladder and the rectum of the female.

Vallate papilla (VAL-āt pa-PIL-a) One of the circular projections that is arranged in an inverted V-shaped row at the back of the tongue; the largest of the elevations on the upper surface of the tongue containing taste buds. Also called **circumvallate papilla.**

Varicocele (VAR-i-kō-sēl) A twisted vein; especially, the accumulation of blood in the veins of the spermatic cord.

Varicose (VAR-i-kōs) Pertaining to an unnatural swelling, as in the case of a varicose vein.

Vas A vessel or duct.

Vasa recta (VĀ-sa REK-ta) Extensions of the efferent arteriole of a juxtamedullary nephron that run alongside the loop of the nephron (loop of Henle) in the medullary region of the kidney.

Vasa vasorum (va-SŌ-rum) Blood vessels that supply nutrients to the larger arteries and veins.

Vascular (VAS-kū-lar) Pertaining to or containing many blood vessels.

Vascular (venous) sinus A vein with a thin endothelial wall that lacks a tunica media and externa and is supported by surrounding tissue.

Vascular spasm Contraction of the smooth muscle in the wall of a damaged blood vessel to prevent blood loss.

Vascular tunic (TOO-nik) The middle layer of the eyeball, composed of the choroid, ciliary body, and iris. Also called the **uvea** (Ū-vē-a).

Vasectomy (va-SEK-tō-mē) A means of sterilisation of males in which a portion of each ductus (vas) deferens is removed.

Vasoconstriction (vāz-ō-kon-STRIK-shun) A decrease in the size of the lumen of a blood vessel caused by contraction of the smooth muscle in the wall of the vessel.

Vasodilation (vāz′-ō-dī-LĀ-shun) An increase in the size of the lumen of a blood vessel caused by relaxation of the smooth muscle in the wall of the vessel.

Vein A blood vessel that conveys blood from tissues back to the heart.

Vena cava (VĒ-na KĀ-va) One of two large veins that open into the right atrium, returning to the heart all of the deoxygenated blood from the systemic circulation except from the coronary circulation.

Ventral (VEN-tral) Pertaining to the anterior or front side of the body; opposite of dorsal.

Ventral ramus (RĀ-mus) The anterior branch of a spinal nerve, containing sensory and motor fibers to the muscles and skin of the anterior surface of the head, neck, trunk, and the limbs.

Ventricle (VEN-tri-kul) A cavity in the brain filled with cerebrospinal fluid. An inferior chamber of the heart.

Ventricular fibrillation (ven-TRIK-ū-lar fib-ri-LĀ-shun) (**VF** or **V-fib**) Asynchronous ventricular contractions; unless reversed by defibrillation, results in heart failure.

Venule (VEN-ūl) A small vein that collects blood from capillaries and delivers it to a vein.

Vermiform appendix (VER-mi-form a-PEN-diks) A twisted, coiled tube attached to the caecum.

Vermis (VER-mis) The central constricted area of the cerebellum that separates the two cerebellar hemispheres.

Vertebrae (VER-te-brē) Bones that make up the vertebral column.

Vertebral (VER-te-bral) **canal** A cavity within the vertebral column formed by the vertebral foramina of all the vertebrae and containing the spinal cord. Also called the **spinal canal.**

Vertebral column The 26 vertebrae of an adult and 33 vertebrae of a child; encloses and protects the spinal cord and serves as a point of attachment for the ribs and back muscles. Also called the **backbone, spine,** or **spinal column.**

Vesicle (VES-i-kul) A small bladder or sac containing liquid.

Vesicouterine (ves′-ik-ō-Ū-ter-in) **pouch** A shallow pouch formed by the reflection of the peritoneum from the anterior surface of the uterus, at the junction of the cervix and the body, to the posterior surface of the urinary bladder.

Vestibular (ves-TIB-ū-lar) **apparatus** Collective term for the organs of equilibrium, which include the saccule, utricle, and semicircular ducts.

Vestibular membrane The membrane that separates the cochlear duct from the scala vestibuli.

Vestibule (VES-ti-būl) A small space or cavity at the beginning of a canal, especially the inner ear, larynx, mouth, nose, and vagina.

Villus (VIL-lus) A projection of the intestinal mucosal cells containing connective tissue, blood vessels, and a lymphatic vessel; functions in the absorption of the end products of digestion. *Plural* is **villi** (VIL-ī).

Viscera (VIS-er-a) The organs inside the ventral body cavity. *Singular* is **viscus** (VIS-kus).

Visceral (VIS-er-al) Pertaining to the organs or to the covering of an organ.

Visceral effectors (e-FEK-torz) Organs of the ventral body cavity that respond to neural stimulation, including cardiac muscle, smooth muscle, and glands.

Vision The act of seeing.

Vitamin An organic molecule necessary in trace amounts that acts as a catalyst in normal metabolic processes in the body.

Vitreous (VIT-rē-us) **body** A soft, jellylike substance that fills the vitreous chamber of the eyeball, lying between the lens and the retina.

Vocal folds Pair of mucous membrane folds below the ventricular folds that function in voice production. Also called **true vocal cords.**

Volkmann's canal *See* **Perforating canal.**

Vulva (VUL-va) Collective designation for the external genitalia of the female. Also called the **pudendum** (poo-DEN-dum).

W

Wallerian (wal-LE-rē-an) **degeneration** Degeneration of the portion of the axon and myelin sheath of a neuron distal to the site of injury.

Wandering macrophage (MAK-rō-fāj) Phagocytic cell that develops from a monocyte, leaves the blood, and migrates to infected tissues.

White matter Aggregations or bundles of myelinated and unmyelinated axons located in the brain and spinal cord.

White pulp The regions of the spleen composed of lymphatic tissue, mostly B lymphocytes.

White ramus communicans (RĀ-mus kō-MŪ-ni-kans) The portion of a preganglionic sympathetic axon that branches from the anterior ramus of a spinal nerve to enter the nearest sympathetic trunk ganglion.

X

Xiphoid (ZĪ-foyd) Sword-shaped.

Xiphoid (ZĪ-foyd) **process** The inferior portion of the sternum (**breastbone).**

Y

Yolk sac An extraembryonic membrane composed of the exocoelomic membrane and hypoblast. It transfers nutrients to the embryo, is a source of blood cells, contains primordial germ cells that migrate into the gonads to form primitive germ cells, forms part of the gut, and helps prevent desiccation of the embryo.

Z

Zona fasciculata (ZŌ-na fa-sik′-ū-LA-ta) The middle zone of the adrenal cortex consisting of cells arranged in long, straight cords that secrete glucocorticoid hormones, mainly cortisol.

Zona glomerulosa (glo-mer′-ū-LŌ-sa) The outer zone of the adrenal cortex, directly under the connective tissue covering, consisting of cells arranged in arched loops or round balls that secrete mineralocorticoid hormones, mainly aldosterone.

Zona pellucida (pe-LOO-si-da) Clear glycoprotein layer between a secondary oocyte and the surrounding granulosa cells of the corona radiata.

Zona reticularis (ret-ik′-ū-LAR-is) The inner zone of the adrenal cortex, consisting of cords of branching cells that secrete sex hormones, chiefly androgens.

Zygote (ZĪ-gōt) The single cell resulting from the union of male and female gametes; the fertilized ovum.

CREDITS

Illustration Credits

Chapter 1 1.1, 1.11, 1.12: Kevin Somerville. 1.2–1.4, 1.8, 1.9: Imagineering. 1.5: Molly Borman. 1.6: Kevin Somerville/Imagineering. 1.7, Table 1.2: DNA Illustrations.

Chapter 2 2.1–2.25: Imagineering.

Chapter 3 3.1, 3.2, 3.16–3.20, 3.22–3.24, Table 3.2: Tomo Narashima. 3.3, 3.5–3.15, 3.21, 3.25–3.35: Imagineering.

Chapter 4 4.1, 4.9, Table 4.1–4.6: Kevin Somerville/Imagineering. 4.2, 4.4, 4.7, 4.8: Imagineering. 4.5, 4.6: Kevin Somerville.

Chapter 5 5.1a, 5.2–5.7: Kevin Somerville. 5.9: Kevin Somerville /Imagineering. 5.10: Imagineering.

Chapter 6 6.1, 6.4, 6.6a, Table 6.1: John Gibb. 6.2, 6.3, 6.5, 6.9: Kevin Somerville. 6.8, 6.10: Imagineering.

Chapter 7 7.1–7.24, Table 7.1, Table 7.4: John Gibb. 7.25: Imagineering.

Chapter 8 8.1–8.17, Table 8.1: John Gibb. 8.18: Kevin Somerville.

Chapter 9 9.1, 9.2, 9.3a, 9.10–9.16: John Gibb.

Chapter 10 10.1, 10.2, 10.17, 10.18: Kevin Somerville. 10.2d, 10.3, 10.4–10.8, 10.10–10.14: Imagineering. 10.9a-c: Kevin Somerville/Imagineering. 10.16: Imagineering.

Chapter 11 11.1, 11.2: John Gibb/Imagineering. 11.3–11.23: John Gibb.

Chapter 12 12.1a, 12.2a, 12.6–12.8: Kevin Somerville. 12.3–12.5, 12.9–12.28, Table 12.3: Imagineering.

Chapter 13 13.1–13.3, 13.5, 13.8b, 13.13: Kevin Somerville. 13.6, 13.10b: John Gibb. 13.7, 13.8a, 13.10a, 13.11: Steve Oh. 13.12 Imagineering. 13.14–13.18: Leonard Dank/Imagineering.

Chapter 14 14.1–14.3, 14.4b-c, 14.5–14.15, 14.27, 14.28, Table 14.3: Kevin Somerville. 14.4a, 14.4d, 14.16 Table 14.1: Imagineering. 14.17–14.26: Kevin Somerville/Richard Combs/Imagineering.

Chapter 15 15.1, 15.4, 15.6: Kevin Somerville/Imagineering. 15.2, 15.3, 15.7: Imagineering. 15.5: Kevin Somerville

Chapter 16 16.1, 16.3, 16.4, 16.8, 16.9, 16.12: Imagineering. 16.2, 16.13: Kevin Somerville. 16.5–16.7, 16.10–16.12: Kevin Somerville/ Imagineering.

Chapter 17 17.1, 17.7, 17.14, 17.21b-d, 17.22: Tomo Narashima. 17.2, 17.10, 17.17, 17.23, 17.26, Table 17.1, Table 17.2: Imagineering. 17.3: Molly Borman. 17.6: Sharon Ellis. 17.12, 17.13, 17.15, 17.16: Imagineering. 17.18–17.21a, 17.25a: Tomo Narashima/Imagineering. 17.24, 17.25b: Tomo Narashima/Sharon Ellis/Imagineering. 17.27, 17.28: Kevin Somerville/Imagineering.

Chapter 18 18.1, 18.21: Kevin Somerville. 18.5, 18.8, 18.10, 18.13a, 18.15, 18.18: Lynn O'Kelley/ Imagineering. 18.2–18.13c, 18.14, 18.16, 18.17, 18.19, 18.20: Imagineering.

Chapter 19 19.1, 19.3–19.6, 19.8, 19.9, 19.11–19.13, Table 19.2: Imagineering.

Chapter 20 20.1–20.5, 20.18–20.20, Table 20.1: Kevin Somerville. 20.7, 20.9, 20.14–20.16, 20.22, 20.23: Kevin Somerville/Imagineering. 20.10–20.13, 20.24: Imagineering

Chapter 21 21.1, 21.9, 21.12, 21.13, 21.17, 21.18a, 21.21, 21.23, 21.28, 21.29a, 21.30, 21.31: Kevin Somerville. 21.2, 20.5–20.8, 21.10, 21.11, 21.14–21.16 21.29b: Imagineering. 21.3, 21.4: Kevin Somerville/Imagineering. 21.18b, 21.19–21.22, 21.24–21.27: John Gibb. Exhibits: Imagineering/ John Gibb.

Chapter 22 22.1: Kevin Somerville/Imagineering/Richard Coombs. 22.2, 22.4, 22.9, 22.11–22.18, 22.20, 22.23: Imagineering. 22.3: John Gibb. 22.5, 22.7: Steve Oh. 22.6, 22.10: Kevin Somerville/ Imagineering. 22.8: Kevin Somerville. 22.19: Jared Schneidman Design.

Chapter 23 23.1, 23.2a, 23.10, 23.11: Kevin Somerville. 23.2b, 23.4a-c, 23.5, 23.6, 23.9: Molly Borman. 23.4d, 23.14–23.18, 23.23, 23.24: Imagineering. 23.13: John Gibb/Imagineering. 23.26: Kevin Somerville/Imagineering. 23.27: Imagineering.

Chapter 24 24.1, 24.2, 24.12, 24.16a-b, 24.18, 24.19, 24.24: Kevin Somerville. 24.4, 24.13, 24.16d-e, 24.21, 24.22, 24.25: Imagineering. 24.5, 24.7, 24.8, 24.10: Nadine Sokol. 24.6: DNA Illustrations. 24.11, 24.15: Steve Oh. 24.23: Molly Borman

Chapter 25 25.1–25.20: Imagineering.

Chapter 26 26.1, 26.2, 26.6, 26.8, 26.9, 26.23: Kevin Somerville. 26.3, 26.21: Steve Oh/Imagineering. 26.5, 26.10–26.20, Table 26.1: Imagineering. 26.22: Kevin Somerville/Imagineering.

Chapter 27 27.1–27.8: Imagineering

Chapter 28 28.1–28.4, 28.6, 28.9–28.14, 28.16, 28.19–28.22, 28.27, 28.28: Kevin Somerville. 28.5, 28.7, 28.8, 28.15, 28.23–28.26: Imagineering.

Chapter 29 29.1–29.13, 29.17, Table 29.2: Kevin Somerville. 29.15, 29.18: Kevin Somerville/ Imagineering. 29.16, 29.19–29.26: Imagineering.

Photo Credits

Chapter 1 Fig. 1.1 Rubberball Productions/ Getty Images. Fig. 1.8a Dissection Shawn Miller; Photograph Mark Nielsen. Fig. 1.8b Dissection Shawn Miller; Photograph Mark Nielsen. Fig. 1.8c Dissection Shawn Miller; Photograph Mark Nielsen. Fig. 1.12a Andy Washnik. Page 21 (left) Biophoto Associates/Photo Researchers. Page 21 (center) Breast Cancer Unit, Kings College Hospital, London/Photo Researchers, Inc. Page 21 (right) Zephyr/Photo Researchers, Inc. Page 22 (top left) Cardio-Thoracic Centre, Freeman Hospital, Newcastle-Upon-Tyne/ Photo Researchers, Inc. Page 22 (top center) CNRI/ Science Photo Library/Photo Researchers, Inc. Page 22 (top right) Science Photo Library/Photo Researchers, Inc. Page 22 (bottom left) Scott Camazine/Photo Researchers, Inc. Page 22 (bottom right) ISM/Phototake. Page 23 (center left) Simon Fraser/Photo Researchers, Inc. Page 23 (center right) ISM/Phototake. Page 23 (bottom) Courtesy Andrew Joseph Tortora and Damaris Soler. Page 24 (center left) Department of Nuclear Medicine, Charing Cross Hospital /Photo Researchers, Inc. Page 24 (center) SIU/Visuals Unlimited. Page 24 (center right) Dept. of Nuclear Medicine, Charing Cross Hospital/Photo Researchers, Inc. Page 24 (bottom right) ©Camal/Phototake.

Chapter 2 Fig. 2.23 Courtesy T.A. Steitz, Yale University.

Chapter 3 Fig. 3.4 Andy Washnik. Fig. 3.9 David Phillips/Photo Researchers, Inc. Fig. 3.13 ©Omikron/ Photo Researchers. Fig. 3.15a Albert Tousson/ Phototake. Fig. 3.15b Albert Tousson/Phototake. Fig. 3.15c Alexey Khodjakov/Photo Researchers, Inc. Fig. 3.16c David M. Phillips/Visuals Unlimited. Fig. 3.17b P. Motta/Photo Researchers, Inc. Fig. 3.17c David M. Phillips/Visuals Unlimited. Fig. 3.19b D. W. Fawcett/Photo Researchers, Inc. Fig. 3.20b Biophoto Associates/Photo Researchers, Inc. Fig. 3.22b David M. Phillips/Visuals Unlimited. Fig. 3.23b Don W. Fawcett/Visuals Unlimited. Fig. 3.24 D.W. Fawcett/Photo Researchers, Inc. Fig. 3.25b Andrew Syred/Photo Researchers, Inc. Fig. 3.32 Courtesy Michael Ross, University of Florida. Page 104 Steve Gschmeissner/Science Photo Library/Photo Researchers, Inc.

Chapter 4 All photos by Mark Nielsen except Fig 4.9a, which was taken by Michael Ross, University of Florida.

Chapter 5 Fig. 5.1b Courtesy Michael Ross, University of Florida. Fig. 5.1c David Becker/Photo Researchers, Inc. Fig. 5.1d Courtesy Andrew J. Kuntzman. Fig. 5.3b Mark Nielsen. Fig. 5.4b VVG/Science Photo Library/Photo Researchers, Inc. Fig. 5.8a Alain Dex/Photo Researchers, Inc. Fig. 5.8b

C1

Biophoto Associates/Photo Researchers, Inc. Fig. 5.9a Sheila Terry/Science Photo Library/Photo Researchers, Inc. Fig. 5.9b St. Stephen's Hospital/ Science Photo Library/Photo Researchers, Inc. Fig. 5.9c St. Stephen's Hospital/Science Photo Library/Photo Researchers, Inc. Fig. 5.11 Dr. P. Marazzi/Science Photo Library/Photo Researchers, Inc.

Chapter 6 Fig. 6.1b Mark Nielsen. Fig. 6.2 (left) CNRI/Photo Researchers, Inc. Fig. 6.2 (center) Dr. Richard Kessel and Randy Kardon/Tissues & Organs/Visuals Unlimited. Fig. 6.2 (right) Dr. Richard Kessel and Randy Kardon/Tissues & Organs/Visuals Unlimited. Fig. 6.6 Ralph Hutchings/Visuals Unlimited. Fig. 6.7a The Bergman Collection. Fig. 6.7b Mark Nielsen. Fig. 6.11a P. Motta/Photo Researchers, Inc. Fig. 6.11b P. Motta/Photo Researchers, Inc. Table 6.1a Courtesy Dr. Brent Layton. Table 6.1b Courtesy Per Amundson, M.D. Table 6.1c Courtesy Dr. Brent Layton. Table 6.1d Courtesy Dr. Brent Layton. Table 6.1e Courtesy Dr. Brent Layton. Table 6.1f Watney Collection/Phototake.

Chapter 7 Fig. 7.3b Mark Nielsen. Fig. 7.4c Mark Nielsen. Fig. 7.18b Mark Nielsen. Fig. 7.18c Mark Nielsen. Fig. 7.18d Mark Nielsen. Fig. 7.18e Mark Nielsen. Fig. 7.20b Mark Nielsen. Fig. 7.20c Mark Nielsen. Fig. 7.19b Mark Nielsen. Fig. 7.19c Mark Nielsen. Fig. 7.22b Mark Nielsen. Fig. 7.25a Princess Margaret Rose Orthopaedic Hospital/Photo Researchers, Inc. Fig. 7.25b Dr. P. Marazzi/Photo Researchers, Inc. Fig. 7.25c Custom Medical Stock Photo, Inc. Fig. 7.26 Center for Disease Control.

Chapter 9 Fig. 9.3b Mark Nielsen. Fig. 9.4 Mark Nielsen. Fig. 9.5 John Wilson White. Fig. 9.6 John Wilson White. Fig. 9.7 John Wilson White. Fig. 9.8 John Wilson White. Fig. 9.9 John Wilson White. Fig. 9.12d Dissection Shawn Miller, Photograph Mark Nielsen. Fig. 9.15d Dissection Shawn Miller, Photograph Mark Nielsen. Fig. 9.15g Dissection Shawn Miller, Photograph Mark Nielsen. Fig. 9.15h Dissection Shawn Miller, Photograph Mark Nielsen. Fig. 9.15i Dissection Shawn Miller, Photograph Mark Nielsen. Fig. 9.15j Dissection Shawn Miller, Photograph Mark Nielsen. Fig. 9.16b SIU/Visuals Unlimited. Fig. 9.16c ISM/Phototake. Fig. 9.16f Charles McRae/Visuals Unlimited.

Chapter 10 Fig. 10.5 Courtesy Hiroyouki Sasaki, Yale E. Goldman and Clara Franzini-Armstrong. Fig. 10.9d Don Fawcett/Photo Researchers, Inc. Fig. 10.15 John Wiley & Sons. Table 10.1 Courtesy Denah Appelt and Clara Franzini-Armstrong. Table 10.4 Biophoto Associates/Photo Researchers, Inc.

Chapter 11 Fig. 11.5c Dissection Shawn Miller, Photograph Mark Nielsen. Fig. 11.10d Dissection Nathan Mortensen and Shawn Miller; Photograph Mark Nielsen. Fig. 11.18g Andy Washnik. Fig. 11.21b Dissection Shawn Miller, Photograph Mark Nielsen.

Chapter 12 Fig. 12.2b Mark Nielsen. Fig. 12.8c Dennis Kunkel/Phototake. Fig. 12.8d Martin Rotker/Phototake.

Chapter 13 Fig. 13.1b Dissection Chris Roach; Photograph Mark Nielsen. Fig. 13.1c Dissection

Shawn Miller, Photograph Mark Nielsen. Fig. 13.3b Courtesy Michael Ross, University of Florida. Fig. 13.5b Dr. Richard Kessel and Randy Kardon/ Tissues & Organs/Visuals Unlimited. Fig. 13.6b Dissection Shawn Miller, Photograph Mark Nielsen. Table 13.1 Dissection Shawn Miller, Photograph Mark Nielsen. Table 13.1 Dissection Shawn Miller, Photograph Mark Nielsen. Table 13.1 Mark Nielsen and Shawn Miller. Table 13.1 Dissection Shawn Miller, Photograph Mark Nielsen. Page 529 (top) Bold Stock/ Age Fotostock America, Inc. Page 529 (just below top) Masterfile. Page 529 (center) Andersen Ross/ Getty Images, Inc. Page 529 (bottom) Masterfile.

Chapter 14 Fig. 14.1b Dissection Shawn Miller, Photograph Mark Nielsen. Fig. 14.4a Dissection Shawn Miller, Photograph Mark Nielsen. Fig. 14.8d Dissection Shawn Miller, Photograph Mark Nielsen. Fig. 14.11c Dissection Shawn Miller, Photograph Mark Nielsen. Fig. 14.12 N. Gluhbegovic and T.H. Williams, The Human Brain: A Photographic Guide, Harper and Row, Publishers, Inc. Hagerstown, MD, 1980. Table 14.2 Dissection Shawn Miller, Photograph Mark Nielsen. Table 14.2 Dissection Shawn Miller, Photograph Mark Nielsen. Table 14.2 Dissection Shawn Miller, Photograph Mark Nielsen. Table 14.2 Dissection Shawn Miller, Photograph Mark Nielsen. Table 14.2 Dissection Shawn Miller, Photograph Mark Nielsen. Table 14.2 Dissection Shawn Miller, Photograph Mark Nielsen. Fig. 14.17 Dissection Shawn Miller, Photograph Mark Nielsen. Fig. 14.18 Dissection Shawn Miller, Photograph Mark Nielsen. Fig. 14.19 Dissection Shawn Miller, Photograph Mark Nielsen. Fig. 14.19 Dissection Shawn Miller, Photograph Mark Nielsen. Fig. 14.19 Dissection Shawn Miller, Photograph Mark Nielsen. Fig. 14.20 Dissection Shawn Miller, Photograph Mark Nielsen. Fig. 14.21 Dissection Shawn Miller, Photograph Mark Nielsen. Fig. 14.22 Dissection Shawn Miller, Photograph Mark Nielsen. Fig. 14.23 Dissection Shawn Miller, Photograph Mark Nielsen. Fig. 14.24 Dissection Shawn Miller, Photograph Mark Nielsen. Fig. 14.25 Dissection Shawn Miller, Photograph Mark Nielsen. Fig. 14.26 Dissection Shawn Miller, Photograph Mark Nielsen.

Chapter 17 Fig. 17.1c Courtesy Michael Ross, University of Florida. Fig. 17.3d Mark Nielsen. Fig. 17.5 Geirge Diebold/Getty Images, Inc. Fig. 17.9 Paul Parker/Photo Researchers, Inc. Fig. 17.10c Mark Nielsen. Fig. 17.17 N. Gluhbegovic and T. H. Williams, The Human Brain: A Photographic Guide, Harper and Row, Publishers, Inc., Hagerstown, MD, 1980.

Chapter 18 Fig. 18.5 (top) Dissection Shawn Miller, Photograph Mark Nielsen. Fig. 18.5d Courtesy James Lowe, University of Nottingham, Nottingham, United Kingdom. Fig. 18.10b Mark Nielsen. Fig. 18.10c Dissection Shawn Miller, Photograph Mark Nielsen. Fig. 18.10d Dissection Shawn Miller, Photograph Mark Nielsen. Fig. 18.13b Mark Nielsen . Fig. 18.13d Dissection Shawn Miller, Photograph Mark Nielsen. Fig. 18.15c Dissection Shawn Miller, Photograph Mark Nielsen. Fig. 18.15d Mark Nielsen. Fig. 18.18 Mark Nielsen. Fig. 18.22a From New England Journal of Medicine, February 18, 1999,

vol. 340, No. 7, page 524. Photo provided courtesy of Robert Gagel, Department of Internal Medicine, University of Texas M.D. Anderson Cancer Center, Houston Texas Fig. 18.22b Lester Bergman/The Bergman Collection. Fig. 18.22c Lester Bergman/ The Bergman Collection. Fig. 18.22d Lester Bergman/ The Bergman Collection. Fig. 18.22e Biophoto Associates/Photo Researchers, Inc.

Chapter 19 Opener Juergen Berger/Photo Researchers, Inc. Fig. 19.2a Juergen Berger/Photo Researchers, Inc. Fig. 19.2b Mark Nielsen. Fig. 19.7a Courtesy Michael Ross, University of Florida. Fig. 19.7b Courtesy Michael Ross, University of Florida. Fig. 19.7c Courtesy Michael Ross, University of Florida. Fig. 19.7d Courtesy Michael Ross, University of Florida. Fig. 19.7e Courtesy Michael Ross, University of Florida. Fig. 19.10a Dennis Kunkel/ Visuals Unlimited. Fig. 19.10b Steve Gschmeissner/ Photo Researchers, Inc. Fig. 19.10c David Phillips/ Visuals Unlimited. Fig. 19.10d Dennis Kunkel/ Phototake. Fig. 19.14 Jean Claude Revy/Phototake. Fig. 19.15 Stanley Fleger/Visuals Unlimited.

Chapter 20 Fig. 20.1c Dissection Shawn Miller, Photograph Mark Nielsen. Fig. 20.3b Dissection Shawn Miller, Photograph Mark Nielsen. Fig. 20.4b Mark Nielsen. Fig. 20.4c Mark Nielsen. Fig. 20.6c Dissection Shawn Miller, Photograph Mark Nielsen. Fig. 20.6f Dissection Shawn Miller, Photograph Mark Nielsen. Fig. 20.6g Dissection Shawn Miller, Photograph Mark Nielsen. Fig. 20.8c Dissection Shawn Miller, Photograph Mark Nielsen. Fig. 20.17 Jim cummins/Taxi/Getty Images, Inc. Fig. 20.21 ©Vu/Cabisco/Visuals Unlimited. Fig. 20.21 W. Ober/Visuals Unlimited. Fig. 20.22d ©ISM/ Phototake.

Chapter 21 Fig. 21.1d Dennis Strete. Fig. 21.1e Courtesy Michael Ross, University of Florida. Fig. 21.5(top) Dissection Shawn Miller, Photograph Mark Nielsen. Fig. 21.5(bottom) Dissection Shawn Miller, Photograph Mark Nielsen. Exhibit 21.B Dissection Shawn Miller, Photograph Mark Nielsen. Fig. 21.19d Dissection Shawn Miller, Photograph Mark Nielsen. Fig. 21.21d Dissection Shawn Miller, Photograph Mark Nielsen. Fig. 21.25d Dissection Shawn Miller, Photograph Mark Nielsen.

Chapter 22 Fig. 22.5c Courtesy Michael Ross, University of Florida. Fig. 22.5b Courtesy Michael Ross, University of Florida. Fig. 22.6b Mark Nielsen. Fig. 22.6c Dissection Shawn Miller, Photograph Mark Nielsen. Fig. 22.7c Mark Nielsen. Fig. 22.9b Science Photo Library/Photo Researchers, Inc.

Chapter 23 Fig. 23.1b Dissection Shawn Miller, Photograph Mark Nielsen. Fig. 23.2c Dissection Shawn Miller, Photograph Mark Nielsen. Fig. 23.3 Courtesy Lynne Marie Barghesi. Fig. 23.5c CNRI/Photo Researchers, Inc. Fig. 23.6 Dissection Shawn Miller, Photograph Mark Nielsen. Fig. 23.8 Dissection Shawn Miller, Photograph Mark Nielsen. Fig. 23.10b Biophoto Associates/Photo Researchers, Inc. Fig. 23.11c Biophoto Associates/Photo Researchers, Inc.

Chapter 24 Fig. 24.1b Dissection Shawn Miller, Photograph Mark Nielsen. Fig. 24.4e Dissection

Shawn Miller, Photograph Mark Nielsen. Fig. 24.6b Mark Nielsen. Fig. 24.9 Mark Nielsen. Fig. 24.11b Dissection Shawn Miller, Photograph Mark Nielsen. Fig. 24.12 (inset) Hessler/Vu/Visuals Unlimited. Fig. 24.12c Mark Nielsen. Fig. 24.15d Dissection Shawn Miller, Photograph Mark Nielsen. Fig. 24.15e Dissection Shawn Miller, Photograph Mark Nielsen. Fig. 24.16c (top inset) Mark Nielsen. Fig. 24.16c Mark Nielsen. Fig. 24.18b Dissection Shawn Miller, Photograph Mark Nielsen. Fig. 24.19 Dissection Shawn Miller, Photograph Mark Nielsen. Fig. 24.20a Fred E. Hossler/Visuals Unlimited. Fig. 24.20b G. W. Willis/ Visuals Unlimited. Fig. 24.20c Courtesy Michael Ross, University of Florida. Fig. 24.20d Courtesy Michael Ross, University of Florida. Fig. 24.24c Courtesy Michael Ross, University of Florida. Fig. 24.24d Courtesy Michael Ross, University of Florida.

Chapter 26 Fig. 26.1b Dissection Shawn Miller, Photograph Mark Nielsen. Fig. 26.3b Mark Nielsen and Shawn Miller. Fig. 26.6b Dennis Strete. Fig. 26.8b Courtesy Michael Ross, University of Florida.

Chapter 28 Fig. 28.1b Dissection Shawn Miller, Photograph Mark Nielsen. Fig. 28.3b Dissection Shawn Miller, Photograph Mark Nielsen. Fig. 28.3c Dissection Shawn Miller, Photograph Mark Nielsen. Fig. 28.4a Mark Nielsen. Fig. 28.9b Dissection Shawn Miller, Photograph Mark Nielsen. Fig. 28.10c Dissection Shawn Miller, Photograph Mark Nielsen. Fig. 28.11b Dissection Shawn Miller, Photograph Mark Nielsen. Fig. 28.13b Courtesy Michael Ross, University of Florida. Fig. 28.13c Claude Edelmann/Photo Researchers, Inc. Fig. 28.14c Mark Nielsen. Fig. 28.14f Mark Nielsen. Fig. 28.17a Mark Nielsen. Fig. 28.17b P. Motta/Photo Researchers, Inc. Fig. 28.18a Mark Nielsen. Fig. 28.18b Courtesy Michael Ross, University of Florida. Fig. 28.20b Courtesy Michael Ross, University of Florida. Fig. 28.20c Mark Nielsen.

Chapter 29 Fig. 29.1b Don W. Fawcett/Photo Researchers, Inc. Fig. 29.1c Myriam Wharman/ Phototake. Fig. 29.11b Preparation and photography Mark Nielsen. Fig. 29.14a Photo provided courtesy of Kohei Shiota, Congenital Anomaly Research Center, Kyoto University, Graduate School of Medicine. Fig. 29.14b Courtesy National Museum of Health and Medicine, Armed Forces Institute of Pathology. Fig. 29.14c Courtesy National Museum of Health and Medicine, Armed Forces Institute of Pathology. Fig. 29.14d Courtesy National Museum of Health and Medicine, Armed Forces Institute of Pathology. Fig. 29.14e Courtesy National Museum of Health and Medicine, Armed Forces Institute of Pathology. Fig. 29.14f Photo by Lennart Nilsson/ Albert Bonniers Fîrlag AB, A Child is Born, Dell Publishing Company. Reproduced with permission. Fig. 29.14g Photo provided courtesy of Kohei Shiota, Congenital Anomaly Research Center, Kyoto University, Graduate School of Medicine. Fig. 29.14h Photo provided courtesy of Kohei Shiota, Congenital Anomaly Research Center, Kyoto University, Graduate School of Medicine.

INDEX

EPONYMS USED
IN THIS TEXT

In the life sciences, an eponym is the name of a structure, drug, or disease that is based on the name of a person. For example, you may be more familiar with the Achilles tendon than you are with its more anatomically descriptive term, the calcaneal tendon. Because eponyms remain in frequent use, this listing correlates common eponyms with their anatomical terms.

EPONYM	ANATOMICAL TERM	EPONYM	ANATOMICAL TERM
Achilles tendon	calcaneal tendon	Kupffer (KOOP-fer) cell	stellate reticuloendothelial cell
Adam's apple	thyroid cartilage	Leydig (LĪ-dig) cell	interstitial endocrinocyte
ampulla of Vater (VA-ter)	hepatopancreatic ampulla	loop of Henle (HEN-lē)	loop of the nephron
		Luschka's (LUSH-kaz) aperture	lateral aperture
Bartholin's (BAR-tō-linz) gland	greater vestibular gland		
Billroth's (BIL-rōtz) cord	splenic cord	Magendie's (ma-JEN-dēz) aperture	median aperture
Bowman's (BŌ-manz) capsule	glomerular capsule	Meibomian (mi-BŌ-mē-an) gland	tarsal gland
Bowman's (BŌ-manz) gland	olfactory gland	Meissner (MĪS-ner) corpuscle	corpuscle of touch
Broca's (BRŌ-kaz) area	motor speech area	Merkel (MER-kel) disc	tactile disc
Brunner's (BRUN-erz) gland	duodenal gland	Müllerian (mil-E rē-an) duct	paramesonephric duct
bundle of His (HISS)	artrioventricular (AV) bundle	organ of Corti (KOR-tē)	spiral organ
canal of Schlemm (SHLEM)	scleral venous sinus	Pacinian (pa-SIN-ē-an) corpuscle	lamellated corpuscle
circle of Willis (WIL-is)	cerebral arterial circle	Peyer's (PĪ-erz) patch	aggregated lymphatic follicle
Cooper's (KOO-perz) ligament	suspensory ligament of the breast	plexus of Auerbach (OW-er-bak)	myenteric plexus
		plexus of Meissner (MĪS-ner)	submucosal plexus
Cowper's (KOW-perz) gland	bulbourethral gland	pouch of Douglas	rectouterine pouch
crypt of Lieberkühn (LE-ber-kyūn)	intestinal gland	Purkinje (pur-KIN-jē) fiber	conduction myofiber
duct of Santorini (san'-tō-RĒ-nē)	accessory duct	Rathke's (rath-KĒZ) pouch	hypophyseal pouch
duct of Wirsung (VĒR-sung)	pancreatic duct	Ruffini (roo-FĒ-nē) corpuscle	type II cutaneous mechanoreceptor
Eustachian (yoo-STĀ-kē-an)	auditory tube		
		Sertoli (ser-TŌ-lē) cell	sustentacular cell
Fallopian (fal-LŌ-pē-an) tube	uterine tube	Skene's (SKĒNZ) gland	paraurethral gland
		sphincter of Oddi (OD-dē)	sphincter of the hepatopancreatic ampulla
gland of Littré (LĒ-tra)	urethral gland		
Golgi (GOL-jē) tendon organ	tendon organ	Volkmann's (FŌLK-manz) canal	perforating canal
Graafian (GRAF-ē-an) follicle	mature ovarian follicle		
		Wernicke's (VER-ni-kēz) area	auditory association area
Hassall's (HAS-alz) corpuscle	thymic corpuscle	Wharton's (HWAR-tunz) jelly	mucous connective tissue
Haversian (ha-VĒR-shun) canal	central canal	Wolffian duct	mesonephric duct
Haversian (ha-VĒR-shun) system	osteon	Wormian (WER-mē-an) bone	sutural bone
Heimlich (HĪM-lik) maneuver	abdomial thrust maneuver		
islet of Langerhans (LANG-er-hanz)	pancreatic islet		

COMBINING FORMS, WORD ROOTS, PREFIXES, AND SUFFIXES

Many of the terms used in anatomy and phsiology are compound words; that is, they are made up of word roots and one or more prefixes or suffixes. For example, *leukocyte* is formed from the word roots *leuk-* meaning "white", a connecting vowel (o), and *cyte* meaning "cell." Thus, a leukocyte is a white blood cell. The following list includes some of the most commonly used combining forms, word roots, prefixes, ad suffixes used in the study of anatomy and physiology. Each entry includes a usage example. Learning the meanings of these fundamental word parts will help you remember terms that, at first glance, may seem long or complicated.

COMBINING FORMS AND WORD ROOTS

Acous-, Acu- hearing Acoustics.
Acr- extremity Acromegaly.
Aden- gland Adenoma.
Alg-, Algia- pain Neuralgia.
Angi- vessel Angiocardiography.
Anthr- joint Arthropathy.
Aut-, Auto- self Autolysis.
Audit- hearing Auditory canal.

Bio- life, living Biopsy.
Blast- germ, bud Blastula.
Blephar- eyelid Blepharitis.
Brachi- arm Brachial plexus.
Bronch- trachea, windpipe Bronchoscopy.
Bucc- cheek Buccal.

Capit- head Decapitate.
Carcin- cancer Carcinogenic.
Cardi-, Cardia-, Cardio- heart Cardiogram.
Cephal- head Hydrocephalus.
Cerebro- brain Cerebrospinal fluid.
Chole- bile, gall Cholecystogram.
Chondr-, cartilage Chondrocyte.
Cor-, Coron- heart Coronary.
Cost- rib Costal.
Crani- skull Craniotomy.
Cut- skin Subcutaneous.
Cyst- sac, bladder Cystoscope.

Derma-, Dermato- skin Dermatosis.
Dura- hard Dura mater.

Enter- intestine Enteritis.

Erythr- red Erythrocyte.

Gastr- stomach Gastrointestinal.
Gloss- tongue Hypoglossal.
Glyco- sugar Glycogen.
Gyn-, Gynec- female, woman Gynecology.

Hem-, Hemat- blood Hematoma.
Hepar-, Hepat- liver Hepatitis.
Hist-, Histio- tissue Histology.
Hydr- water Dehydration.
Hyster- uterus Hysterectomy.

Ischi- hip, hip joint Ischium.

Kines- motion Kinesiology.

Labi- lip Labial.
Lacri- tears Lacrimal glands.
Laparo- loin, flank, abdomen Laparoscopy.
Leuko- white Leukocyte.
Lingu- tongue Sublingual glands.
Lip- fat Lipid.
Lumb- lower back, loin Lumbar.

Macul- spot, blotch Macula.
Malign- bad, harmful Malignant.
Mamm-, Mast- breast Mammography, Mastitis.
Meningo- membrane Meningitis.
Myel- marrow, spinal cord Myeloblast.
My-, Myo- muscle Myocardium.

Necro- corpse, dead Necrosis.
Nephro- kidney Nephron.
Neuro- nerve Neurotransmitter.

Ocul- eye Binocular.
Odont- tooth Orthodontic.
Onco- mass, tumor Oncology.
Oo- egg Oocyte.
Opthalm- eye Ophthalmology.
Or- mouth Oral.
Osm- odor, sense of small Anosmia.
Os-, Osseo-, Osteo- bone Osteocyte.
Ot- ear Otitus media.

Palpebr- eyelid Palpebra.
Patho- disease Pathogen.
Pelv- basin Renal pelvis.
Phag- to eat Phagocytosis.
Phleb- vein Phlebitis.
Phren- diaphragm Phrenic.
Pilo- hair Depilatory.
Pneumo- lung, air Pneumothorax.
Pod- foot Podocyte.
Procto- anus, rectum Proctology.
Pulmon- lung Pulmonary.

Ren- kidneys Renal artery.
Rhin- nose Rhinitis.

Scler-, Sclero- hard Atherosclerosis.
Sep-, Spetic- toxic condition due to micoorganisms Septicemia.
Soma-, Somato- body Somatotropin.
Sten- narrow Stenosis.
Stasis-, Stat- stand still Homeostasis.

Tegument- skin, covering Integumentary.
Therm- heat Thermogenesis.
Thromb- clot, lump Thrombus.

Vas- vessel, duct Vasoconstriction.

Zyg- joined, Zygote.

PREFIXES

A-, An- without, lack of, deficient Anesthesia.
Ab- away from, from Abnormal.
Ad-, Af- to, toward Adduction, Afferent neuron.
Alb- white Albino.
Alveol- cavity, socket Alveolus.
Andro- male, masculine Androgen.
Ante- before Antebrachial vein.
Anti- against Anticoagulant.

Bas- base, foundation Basal ganglia.
Bi- two, double Biceps.
Brady- slow Bradycardia.

Cata- down, lower, under Catabolism.
Circum- around Circumduction.
Cirrh- yellow Cirrhosis of the liver.
Co-, Con-, Com with, together Congenital.
Contra- against, opposite Contraception.
Crypt- hidden, concealed Cryptorchidism.
Cyano- blue Cyanosis.

De- down, from Deciduous.
Demi-, hemi- half Hemiplegia.
Di-, Diplo- two Diploid.
Dis- separation, apart, away from Dissection.
Dys- painful, difficult Dyspnea.

E-, Ec-, Ef- out from, out of Efferent neuron.
Ecto-, Exo- outside Ectopic pregnancy.
Em-, En- in, on Emmetropia.
End-, Endo- within, inside Endocardium.
Epi- upon, on, above Epidermis.
Eu- good, easy, normal Eupnea.
Ex-, Exo- outside, beyond Exocrine gland.
Extra- outside, beyond, in addition to Extracellular fluid.

Fore- before, in front of Forehead.

Gen- originate, produce, form Genitalia.
Gingiv- gum Gingivitis.

Hemi- half Hemiplegia.
Heter-, Hetero- other, different Heterozygous.
Homeo-, Homo- unchanging, the same, steady Homeostasis.
Hyper- over, above, excessive Hyperglycemia.
Hypo- under, beneath, deficient Hypothalamus.

In-, Im- in, inside, not Incontinent.
Infra- beneath Infraorbital.
Inter- among, between Intercostal.
Intra- within, inside Intracellular fluid.
Ipsi- same Ipsilateral.
Iso- equal, like Isotonic.

Juxta- near to Juxtaglomerular apparatus.

Later- side Lateral.

Macro- large, great Macrophage.
Mal- bad, abnormal Malnutrition.
Medi-, Meso- middle Medial.
Mega-, Megalo- great, large Magakaryocyte.
Melan- black Melanin.
Meta- after, beyond Metacarpus.
Micro- small Microfilament.
Mono- one Monounsaturated fat.

Neo- new Neonatal.